Encyclopedic Dictionary of Polymers

Jan W. Gooch (Ed.)

Encyclopedic Dictionary of Polymers

Volume 1
A-M

With 710 Figures and 38 Tables

Springer

Dr. Jan W. Gooch
2020 Howell Mill Road
C227
Atlanta, GA 30318
USA

ISBN 13: 978-0-387-31021-3
The electronic version of the whole set will be available under ISBN-13: 978-0-387-30160-0.
The print and electronic bundle of the whole set will be available under ISBN-13: 978-0-387-33502-5.

springer.com

Printed on acid-free paper SPIN: 11494034 2109 — 5 4 3 2 1 0

The editor would like to express his gratitude to
all individuals who made available their time
and resources in order to contribute to this book:
F. Joseph Schork; Lisa Deter-Hoskin,
Roger D. Hester, Paula T. Hammond and
by James W. Larsen.

Preface

The editor offers scientists, engineers, academia, and others interested in adhesives, coatings, elastomers, inks, plastics, and textiles a valuable communication tool within this book. In addition, the more recent innovations and biocompatible polymers and adhesives products have necessitated inclusion into any lexicon that addresses polymeric materials. Communication among scientific and engineering personnel has always been of critical importance, and as in any technical field, the terms and descriptions of materials and processes lag the availability of a manual or handbook that would benefit individuals working and studying in scientific and engineering disciplines. There is often a challenge when conveying an idea from one individual to another due to its complexity, and sometimes even the pronunciation of a word is different not only in different countries, but also in industries. Colloquialisms and trivial terms that find their way into technical language for materials and products tend to create a communications fog, thus unacceptable in today's global markets and technical communities.

The editor wishes to make a distinction between this book and traditional dictionaries, which provide a word and definition. The present book provides for each term a complete expression, chemical structures and mathematic expression where applicable, phonetic pronunciation, etymology, translations into German, French, and Spanish, and related figures if appropriate. This is a complete book of terminology never before attempted or published.

The information for each chemical entry is given as it is relevant to polymeric materials. Individual chemical species (e.g., ethanol) were taken from the *CRC Handbook of Chemistry and Physics* (2004), the Merck Index and other reference materials. The reader may refer to these references for additional physical properties and written chemical formulae. Extensive use was made of ChemDraw®, CambridgeSoft Corporation for naming and drawing chemical structures (conversion of structure to name and vice versa), which are included with each chemical entry where possible. Special attention was given to the IUPAC name that is often given with the common name for the convenience of the reader.

The editor assembled notes over a combined career in the chemical industries and academic institutions regarding technical communication among numerous colleagues and helpful acquaintances concerning expressions and associated anomalies. Presently, multiple methods of nomenclature are employed to describe identical chemical compounds by common and IUPAC names (e.g., acetone and 2-propanone) because the old systems (19th century European and trivial) methods of nomenclature exists with the modern International Union of Pure and Applied Chemistry, and the conflicts between them are not likely to relent in the near future including the weights and measures systems because some nations are reluctant to convert from English to metric and, and more recently, the International Systems of Units (SI). Conversion tables for converting other systems to the SI units are included in this book for this purpose. In addition, there are always differences in verbal pronunciation, but the reasons are not acceptable to prevent cogent communication between people sharing common interests.

In consideration of the many challenges confronting the reader, who must economize time investment, the structure of this book is optimized with regard the convenience of the reader as follows:

- Comprehensive table of contents
- Abbreviations and symbols
- Mathematics signs
- English, Greek, Latin, and Russian alphabets
- Pronunciation/phonetic symbols
- Main body of terms with entry term in English, French, German, and Italian
- Conversion factors
- Microbiology nomenclature and terminology
- References

The editor acknowledges the utilization of many international sources of information including journals, books, dictionaries, communications, and conversations with people experienced in materials, polymer science and engineering. A comprehensive reference section contains all of the sources of information used in this publication. Pronunciation, etymological, cross-reference, and related information is presented in the style of the 11th edn of the Merriam-Webster Dictionary, where known, for each term. The spelling for each term is presented in German, French, and Spanish where translation is possible. Each term in this book includes the following useful information:

- Spelling (in **bold** face) of each term and alternative spellings where more than one derivation is commonly used.
- Phonetic spelling \-\ using internationally published phonetic symbols, and this is the first book that includes phonetic pronunciation information missing in technical dictionaries that allows the reader to pronounce the term.
- Parts of speech in English following each phonetic spelling, e.g. *n.*, *adj.*
- Cross-references.
- Also called *example* in italics.
- Etymological information [-] for old and new terms that provides the reader the national origins of terms including root words, prefixes and suffixes; historical information is critical to the appreciation of a term and its true meaning.
- French, German, Italian, and Spanish spellings of the term {-}.
- A comprehensive explanation of the term.
- Mathematical expressions where applicable.
- Figures and tables where applicable.
- A comprehensive reference section is included for further research.

References are included for individual entries where a publication(s) is directly attributable to a definition or description. Not all of the references listed in the Reference section are directly attributable to entries, but they were reviewed for information and listed for the reader's information. Published dictionaries and glossaries of materials were very helpful for collecting information in the many diverse and smaller technologies of the huge field of polymers.

The editor is grateful that so much work has been done by other people interested in polymers.

The editor has attempted to utilize all relevant methods to convey the meaning of terms to the reader, because a term often requires more information than a standard entry in a textbook dictionary, so this book is dedicated to a complete expression. Terminology and correct pronunciation of technical terms is continuously evolving in scientific and industrial fields and too often undocumented or published, and therefore, not shared with others sometimes leading to misunderstandings. Engineering and scientific terms describe a material, procedure, test, theory or process, and communication between technical people must involve similar jargon or much will be lost in the translation as often has been the editor's experience. The editor has made an attempt to provide the reader who has an interested in the industries that have evolved from adhesives, coatings, inks, elastomers, plastics and textiles with the proper terminology to communicate with other parties whether or not directly involved in the industries. This publication is a single volume in the form of a desk-handbook that is hoped will be an invaluable tool for communicating in the spoken and written media.

Physics, electronic, and magnetic terms because they are related to materials and processes (e.g., *ampere*).

Biomolecular materials and processes have in the recent decade overlapped with polymer science and engineering. Advancements in polymeric materials research for biomolecular and medical applications are rapidly becoming commercialized, examples include biocompatible adhesives for sutureless tissue bonding, liquid dressings for wounds and many other materials used for *in vitro* and *in vivo* medical applications. To keep pace with these advancements, the editor has included useful terms in the main body that are commonly used in the material sciences for these new industries.

A microbiology section has been included to assist the reader in becoming familiar with the proper nomenclature of bacteria, fungi, mildew, and yeasts – organisms that affect materials and processes because they are ubiquitous in our environment. Corrosion of materials by micro-organisms is commonplace, and identification of a specific organism is critical to prevent its occurrence. Engineers and material scientists will appreciate the extensive sections on different types of micro-organisms together with a section dedicated to microbiology terminology that is useful for communicating in the jargon of biologists instead of referring to all organisms as "bugs."

New materials and processes, and therefore new terms, are constantly evolving with research, development and global commercialization. The editor will periodically update this publication for the convenience of the reader.

Statistics, numerical analysis other data processing, and experimental design terms are not addressed as individual terms, because they are not materials related, and the mathematical sciences possess an extensive, separate and distinct terminology while this book is primarily devoted to materials.

Table of Contents

Abbreviations and Symbols

Abbreviations	Symbols
An	Absorption (formerly extinction) ($=\log t_i^{-1}$)
A	Area
A	Surface
A	Helmholtz energy ($A = U - TS$)
A	Pre-exponential constant [in $k = A \exp(-E^{\ddagger}/RT)$]
A_2	Second virial coefficient
a	Exponent in the property/molecular weight relationship ($E^{\ddagger} = KM^{a}$); always with an index, e.g., a_{η}, a_s, etc.
a	Linear absorption coefficient, $a = l^{-1}$
Absolute	abs
Acre	Spell out
Acre-foot	acre-ft
Air horsepower	air hp
Alternating-current (as adjective)	a-c
A^m	molar Helmholtz energy
American Society for Testing and Materials	ASTM
Amount of a substance (mole)	n
Ampere	A or amp
Ampere-hour	amp-h
Amplitude, an elliptic function	am.
Angle	β
Angle, especially angle of rotation in optical activity	∝
Angstrom unit	Å
Antilogarithm	antilog
a_0	Constant in the Moffit-Yang equation
Area	*A*
Atactic	at
atomic weight	at. Wt.
Association	Assn.
Atmosphere	atm
Average	avg
Avogadro number	N_L
Avoirdupois	avdp
Azimuth	az or α
Barometer	bar
Barrel	bbl
Baumé	Bé
b_0	Constant in the Mofit-Yang equation
Board fee (feet board measure)	fbm
Boiler pressure	Spell out
Boiling point	bp
Boltzmann constant	k
Brake horsepower	bhp
Brake horsepower-hour	bhp-h
Brinell hardness number	Bhn
British Standards Institute	BSI
British thermal unit[1]	Btu or B
Bushel	bu
C	Heat capacity
c	Specific heat capacity (formerly; specific heat); c_p = specific isobaric heat capacity, c_v = specific isochore heat capacity
c	"Weight" concentration (=weight of solute divided by volume of solvent); IUPAC suggests the symbol ρ for this quantity, which could lead to confusion with the same IUPAC symbol for density

$\hat{c}$	Speed of light in a vacuum
$\hat{c}$	Speed of sound
Calorie	cal
Candle	c
Candle-hour	c-h
Candlepower	cp
Ceiling temperature of polymerization, °C	T_c
Cent	c or ¢
Center to center	c to c
Centigram	cg
Centiliter	cl
Centimeter or centimeter	cm
Centimeter–gram–second (system)	cgs
Centipoise	cP
Centistokes	cSt
Characteristic temperature	
Chemical	chem.
Chemical potential	μ
Chemical shift	δ
Chemically pure	cp
Circa, about, approximate	ca.
Circular	cir
Circular mils	cir mils
cis-tactic	ct
C^m	molar heat capacity
Coefficient	coef
Cologarithm	colog
Compare	cf.
Concentrate	conc
Conductivity	cond, λ
Constant	const
Continental horsepower	cont hp
Cord	cd
Cosecant	csc
Cosine	cos
Cosine of the amplitude, an elliptic function	cn
Cost, insurance, and freight	cif
Cotangent	cot
Coulomb	Spell out
Counter electromotive force	cemf
C_{tr}	Transfer constant ($C_{tr} = k_{tr}/k_p$)
Cubic	cu
Cubic centimeter (liquid, meaning milliliter. ml)	cu, cm, cm^3
Cubic centimeter	cm^3cubic expansion coefficient ∝
Cubic foot	cu ft
Cubic feet per minute	cfm
Cubic feet per second	cfs
Cubic inch	cu in.
Cubic meter	cu m or m^3
Cubic micron	cu μ or cu mu or $μ^3$
Cubic millimeter	cu mm or mm^3
Cubic yard	cu yd
Current density	Spell out
Cycles per second	Spell out or c
Cylinder	cyl
D	Diffusion coefficient
D_{rot}	Rotational diffusion coefficient
Day	Spell out
Decibel	db
Decigram	dg
Decomposition, °C	T_{dc}
Degree	deg or °
Degree Celsius	°C
Degree centigrade	C
Degree Fahrenheit	F or °
Degree Kelvin	K or none
Degree of crystallinity	∝
Degree of polymerization	X
Degree Réaumur	R
Delta amplitude, an elliptic function	dn
Depolymerization temperature	T_{dp}
Density	ρ
Diameter	diam
Dictionary of Architecture and construction	DAC
Diffusion coefficient	D
Dipole moment	p
Direct-current (as adjective)	d-c
Dollar	$

Dozen	doz
Dram	dr
Dynamic viscosity	ŋ
E	Energy (E_k = kinetic energy, E_p = potential energy, $E^{\ddagger}$ = energy of activation)
E	Electronegativity
E	Modulus of elasticity, Young's modulus ($E = \sigma_{ii}/\epsilon_{ii}$)
E	General property
E	Electrical field strength
e	Elementary charge
e	Parameter in the *Q-e* co-polymerization theory
e	Cohesive energy density (always with an index)
Edition	edn.
Editor, edited	ed
Efficiency	eff
Electric	elec
Electric polarizability of a molecule	∝
Electrical current strength	I
Electrical potential	V
Electrical resistance	R or X
Electromotive force	emf
Electronegativity	E
Elevation	el
Energy	E
Enthalph	H
Entropy	S
Equation	eq
Equivalent weight	equiv wt
Et alii (and others)	et al.
Et cetera	etc.
Excluded volume	u
excluded volume cluster integral	β
Exempli gratia (for example)	e.g.
Expansion coefficient	∝
External	ext
F	Force
f	Fraction (exclusing molar fraction, mass fraction, volume fraction)
f	Molecular coefficient of friction (e.g., f_s, f_D, f_{rot})
f	Functionality
Farad	Spell out or f
Federal	Fed.
Feet board measure (board feet)	fbm
Feet per minute	fpm
Feet per second	fps
Flash point	flp
Fluid	fl
Foot	ft
Foot-candle	ft-c
Foot-Lambert	ft-L
Foot-pound	ft-lb
Foot-pound-second (system)	fps
Foot-second (see cubic feet per second)	
Fraction	∫
Franc	fr
Free aboard ship	Spell out
Free alongside ship	Spell out
Free on board	fob
Freezing point	fp
Frequency	Spell out
Fusion point	fnp
G	Gibbs energy (formerly free energy or free enthalpy) ($G = H - TS$)
G	Shear modulus ($G = \sigma_{ij}$/*angle of shear)*
G	Statistical weight fraction ($G_l = g_l/\Sigma_l\, g_i$)
g	Gravitational acceleration
g	Statistical weight
g	*Gauche* conformation

g	Parameter for the dimensions of branched macromolecules
G^m	Molar Gibbs energy
Gallon	gal
Gallons per minute	gpm
Gallons per second	gps
Aauche conformation	g
Gibbs energy	G
Grain	Spell out
Gram	g
Gram-calorie	g-cal
Greatest common divisor	gcd
H	Enthalpy
H^m	Molar enthalpy
h	Height
h	Plank constant
Haversine	hav
Heat	Q
Heat capacity	C
Hectare	ha
Henry	H
High pressure (adjective)	h-p
Hogshead	hhd
Horsepower	hp
Horsepower-hour	hp-h
Hour	h or hr
Hundred	C
Hundredweight (112 lb)	cwt
Hydrogen ion concentration, negative logarithm of	pH
Hyperbolic cosine	cosh
Hyperbolic sine	sinh
Hyperbolic tangent	tanh
I	Electrical current strength
I	Radiation intensity of a system
i	Radiation intensity of a molecule
Ibidem (in the same place)	ibid.
Id est (that is)	i.e.

Inch	in.
Inch-pound	in-lb
Inches per second	ips
Indicated horsepower	ihp
Indicated horsepower-hour	ihp-h
Infrared	IR
Inside diameter	ID
Intermediate-pressure (adjective)	i-p
Internal	int
International Union of Pure and Applied Chemistry	IUPAC
Isotactic	it
J	Flow (of mass, volume, energy, etc.), always with a corresponding Index
Joule	J
K	General constant
K	Equilibrium constant
K	Compression modulus ($p = -K\ \Delta V/V_0$)
k	Boltzmann constant
k	Rate constant for chemical reactions (always with an index)
Kelvin	K (Not °K)
Kilocalorie	kcal
Kilocycles per second	kc
Kilogram	kg
Kilogram-calorie	kg-al
Kilogram-meter	kg-m
Kilograms per cubic meter	kg per cu m or kg/m^3
Kilograms per second	kgps
Kiloliter	kl
Kilometer or kilometer	km
Kilometer per second	kmps
Kilovolt	kV
Kilovolt-ampere	kVA
Kilowatt	kW

Kilowatthour	kWh
Knoop hardeness number	KHN
L	Chain end-to-end distance
L	Phenomenological coefficient
l	Length
Lambert	L
Latitude	lat or ϕ
Least common multiple	lcm
Length	l
Linear expansion coefficient	Y
Linear foot	lin ft
Liquid	liq
Lira	Spell out
Liter	l
Logarithm (common)	log
Logarithm (natural)	log. or ln
Kibgutyde	kibg. or λ
Loss angle	δ
Low-pressure (as adjuective)	l-p
Lumen	1*
Lumen-hour	1-h*
Luments per watt	lpW
M	"Molecular weight" (IUPAC molar mass)
m	Mass
Mass	spell out or m
Mass fraction	w
Mathematics (ical)	math
Maximum	max
Mean effective pressure	mep
Mean horizontal candlepower	mhcp
Meacycle	mHz
Megohm	MΩ
Melting point, -temperature	mp, T_m
Meter	m
Meter-kilogram	m-kg
Metre	m
Mho	spell out
Microampere	μA or mu A
Microfarad	μf
Microinch	μin.
Micrometer (formerly micron)	μm
Micromicrofarad	μμf
Micromicron	μμ
Micron	μ
Microvolt	μV
Microwatt	μW or mu W
Mile	Spell out
Miles per hour	mph
Miles per hour per second	mphps
Milli	m
Milliampere	mA
Milliequivalent	meq
Milligram	mg
Millihenry	mH
Millilambert	mL
Milliliter or milliliter	ml
Millimeter	mm
Millimeter or mercury (pressure)	mmHg
Millimicron	mμ or m mu
Million	Spell out
Million gallons per day	mgd
Millivolt	mV
Minimum	min
Minute	min
Minute (angular measure)	′
Minute (time) (in astronomical tables)	m
Mile	Spell out
Modal	m
Modulus of elasticity	E
Molar	M
Molar enthalpy	H^m
Molar Gibbs energy	G_m
Molar heat capacity	H_m
Mole	mol
Mole fraction	x
Molecular weight	Mol wt or M
Month	Spell out
N	Number of elementary particles (e.g., molecules, groups, atoms, electrons)

N_L	Avogadro number (Loschmidt's number)
n	Amount of a substance (mole)
n	Refractive index
Nanometer (formerly millimicron)	nm
National Association of Corrosion Engineers	NACE
National Electrical Code	NEC
Newton	N
Normal	N
Number of elementary particles	N
Occupational Safety and Health Administration	OSHA
Ohm	Ω
Ohm-centimeter	ohm-cm
Oil absorption	O.A.
Ounce	oz
Once-foot	oz-ft
Ounce-inch	oz-in.
Outside diameter	OD
Osomotic pressure	Π
P	Permeability of membranes
p	Probability
p	Dipole moment
$\mathbf{p}_i$	Induced dipolar moment
p	pressure
p	Extent of reaction
Paint Testing Manual	PTM
Parameter	Q
Partition function (system)	Q
Parts per billion	ppb
Parts per million	ppm
Pascal	Pa
Peck	pk
Penny (pency–new British)	p.
Pennyweight	dwt

Per	Diagonal line in expressions with unit symbols or (see Fundamental Rules)
Percent	%
Permeability of membranes	P
Peso	Spell out
Pint	pt.
Planck's constant ($E = h\nu$) (6.62517 $\pm$ 0.00023 $\times 10^{-27}$ erg s	h
Polymolecularity index	Q
Potential	Spell out
Potential difference	Spell out
pound	lb
Pound-foot	lb-ft
Pound-inch	lb-in.
Pound sterling	£
Pounds-force per square inch	psi
Pounds per brake horsepower-hour	lb per bhp-hr
Pounds per cubic foot	lb per cut ft
Pounds per square foot	psf
Pounds per square inch	psi
Pounds per square inch absolute	psia
Power factor	Spell out or pf
Pressure	p
Probability	p
Q	Quantity of electricity, charge
Q	Heat
Q	Partition function (system)
Q	Parameter in the *Q*–*e* co-polymerization equation
Q, *Q*	Polydispersity, polymolecularity index ($Q = \overline{M}_w/\overline{M}_n$)
q	Partition function (particles)

Quantity of electricity, charge	Q
Quart	qt
Quod vide (which see)	q.v.
R	Molar gas constant
R	Electrical resistance
R_G	Radius of gyration
R_n	Run number
R_ϑ	Rayleigh ratio
r	Radius
r_0	Initial molar ratio of reactive groups in polycondensations
Radian	Spell out
Radius	r
Radius of gyration	R_G
Rate constant	k
Rayleigh ratio	R_9
Rayleigh scattering	
Reactive kilovolt-ampere	kVA
Reactive volt-ampere	VA
Reference(s)	ref
Refractive index	n
Relaxation time	τ
Resistivity	ρ
Revolutions per minute	rpm
Revolutions per second	rps
Rod	Spell out
Root mean square	rms
S	Entropy
S^m	Molar entropy
S	Solubility coefficient
s	Sedimentation coefficient
s	Selectivity coefficient in osmotic measurements)
Saybolt Universal seconds	SUS
Secant	sec
Second	s or sec
Second (angular measure)	″
Second-foot (see cubic feet per second)	
Second (time) (in astronomical tables)	s
Second virial coefficient	A_2
Shaft horsepower	shp
Shilling	s
Sine	sin
Sine of the amplitude, an elliptic function	sn
Society	Soc.
Soluble	sol
Solubility coefficient	S
Solubility parameter	δ
Solution	soln
Specific gravity	sp gr
Specific heat	sp ht
Specific heat capacity (formerly: specific heat)	c
Specific optical rotation	[∝]
Specific volume	sp vol
Spherical candle power	scp
Square	sq
Square centimeter	sq cm or cm^2
Square foot	sq ft
Square inch	sq in.
Square kilometer	sq km or km^2
Square meter	sq m or m^2
Square micron	sq μ or μ^2
Square root of mean square	rms
Standard	std
Standard	Stnd.
Standard deviation	σ
Staudinger index	[η]
Stere	s
Syndiotactic	st
T	Temperature
t	Time
t	*trans* conformation
Tangent	tan
Temperature	T or temp
Tensile strength	ts
Threodiisotactic	tit

Thousand	M
Thousand foot-pounds	kip-ft
Thousand pound	kip
Ton	Spell out
Ton-mile	Spell out
Trans conformation	t
Trans-tactic	tt
U	Voltage
U	Internal energy
U^m	Molar internal energy
u	Excluded volume
Ultraviolet	UV
United States	US
V	Volume
V	Electrical potential
v	Rate, rate of reaction
v	Specific volume always with an index)
Vapor pressure	vp
Versed sine	vers
Versus	vs
Volt	v or V
Volt-ampere	VA
Volt-coulomb	Spell out
Voltage	U
Volume	V or vol.
Volume (of a publication)	Vol
W	Weight
W	Work
w	Mass function
Watt	w or W
Watthour	Wh
Watts per candle	wpc
Week	Spell out
Weight	W or w
Weight concentration*	c
Work	*y* yield
X	Degree of polymerization
X	Electrical resistance
x	Mole fractio *y* yield
Yard	yd
Year	yr
Young's	E
Z	Collision number
Z	*z* fraction
z	Ionic charge

z	Coordination number
z	Dissymmetry (light scattering)
z	Parameter in excluded volume theory
α	Angle, especially angle of rotation in optical activity
α	Cubic expansion coefficient $[\alpha = V^{-1} (\partial V/\partial T)_p]$
α	Expansion coefficient (as reduced length, e.g., α_L in the chain end-to-end distance or α_R for the radius of gyration)
α	Degree of crystallinity (always with an index)
α	Electric polarizability of a molecule
[α]	"Specific" optical rotation
β	angle
β	Coefficient of pressure
β	Excluded volume cluster integral
Γ	Preferential solvation
γ	Angle
γ	Surface tension
γ	Linear expansion coefficient
δ	Loss angle
δ	Solubility parameter
δ	Chemical shift
ϵ	Linear expansion $(\epsilon = \Delta l/l_0)$
ϵ	Expectation
ϵ_r	Relative permittivity (dielectric number)
η	Dynamic viscosity
$[\eta]$	Staudinger index (called J_0 in DIN 1342)
	Characteristic temperature, especially theta temperature

θ	Angle, especially angle of rotation
ϑ	Angle, especially valence angle
k	Isothermal compressibility [k = V^{-1} $(\partial V/\partial p)_T$]
k	Enthalpic interaction parameter in Solution theory
λ	Wavelength
λ	Heat conductivity
λ	Degree of coupling
μ	Chemical potential
μ	Moment
μ	Permanent dipole moment
ν	Moment, with respect to a reference value
ν	Frequency
ν	Kinetic chain length
φ	Shielding ratio in the theory of Random coils
Ξ	Partition function
Π	Osmotic pressure
ρ	Density
σ	Mechanical stress (σ_{ii} = normal stress, σ_{ij} = shear stress)
σ	Standard deviation
σ	Hindrance parameter
τ	Relaxation time
τ_i	Internal transmittance (transmission factor) (represents the ratio of transmitted to absorbed light)
ϕ	Volume fraction
$\varphi(r)$	Potential between two segments separated by a distance r
Φ	Constant in the viscosity-molecular-weight relationship
$[\Phi]$	"Molar" optical rotation
ξ	Interaction parameter in solution theory
ψ	Entropic interaction parameter in solution theory
ω	Angular frequency, angular velocity
Ω	angle
Ω	Probability
Ω	Skewness of a distribution

*Weight of solute divided by volume of solvent; IUPAC suggests the symbol ρ for this quantity, which could lead to confusion with the same IUPAC symbol for density.

Notations

The abbreviations for chemicals and polymer were taken from the "Manual of Symbols and Terminology for Physicochemical Quantities and Units," Pure Appl Chem 21*1) (1970), but some were added because of generally accepted use.

The ISO (International Standardization Organization) has suggested that all extensive quantities should be described by capital letters and all intensive quantities by lower-case letters. IUPAC does not follow this recommendation, however, uses lower-case letters for specific quantities.

The following symbols are used above or after a letter.

Symbols above letters

—: Signifies an average, e.g., $\overline{M}$ is the average molecular weight; more complicated averages are often indicated by $\langle\rangle$, e.g., $\langle R_G^2\rangle$ is another way of writing $\overline{(R_G^2)}_z$

~: Stands for a partial quantity, e.g., $\tilde{v}_A$ is the partial specific volume of the compound A; V_A is the volume of A, whereas $\tilde{V}_A^m$ xxx is the partial molar volume of A.

Superscripts

°: Pure substance or standard state

∞: Infinite dilution or infinitely high molecular weight *m:* Molar quantity (in cases where subscript letters are impractical)

(*q*): The *q* order of a moment (always in parentheses)

‡: Activated complex

Subscripts

Initial state	
1	Solvent
2	Solute
3	Additional components (e.g., precipitant, salt, etc.)
am	Amorphous
B	Brittleness
bd	Bond
cr	Crystalline
crit	Critical
cryst	Crystallization
e	Equilibrium
E	End group
G	Glassy state
i	Run number
i	Initiation
i	Isotactic diads
ii	Isotactic triads
Is	Heterotactic triads
j	Run number
k	Run number
m	Molar
M	Melting process
mon	Monomer
n	Number average
p	Polymerization, especially propagation
pol	polymer
r	General for average
s	Syndiotactic diads
ss	Syndiotactic triads
st	start reaction
t	Termination
tr	transfer
u	Monomeric unit
w	Weight average
z	*z* average

Prefixes

at	atactic
ct	*cis*-tactic
eit	erythrodiisotactic
it	isotactic
st	syndiotactic
tit	threodiisotactic
tt	*trans*-tactic

Square brackets around a letter signify molar concentrations. (IUPAC prescribes the symbol *c* for molar concentrations, but to date this has consistently been used for the mass/volume unit.)

Angles are always given by °.

Apart from some exceptions, the meter is not used as a unit of length; the units cm and mm derived from it are used. Use of the meter in macromolecular science leads to very impractical units.

Elemental symbols and atomic weights

Source: International Union of Pure and Applied Chemistry (IUPAC) 2001. Values from the 2001 table Pure Appl. Chem. 75: 1107–1122 (2003). The values of zinc, krypton, molybdenum and dysprosium have been modified. The *approved name* for element 110 is included, see Pure Appl. Chem. 75: 1613-1615 (2003). The *proposed name* for element 111 is also included.

A number in parentheses indicates the uncertainty in the last digit of the atomic weight.

1. Geological specimens are known in which the element has an isotopic composition outside the limits for normal material. The difference between the atomic weight of the element in such specimens and that given in the table may exceed the stated uncertainty.
2. Range in isotopic composition of normal terrestrial material prevents a more precise value being given; the tabulated value should be applicable to any normal material.
3. Modified isotopic compositions may be found in commercially available material because it has been subject to an undisclosed or inadvertent isotopic fractionation. Substantial deviations in atomic weight of the element from that given in the Table can occur.
4. Commercially available Li materials have atomic weights that range between 6.939 and 6.996; if a more accurate value is required, it must be

List of elements in name order

At. No.	Symbol	Name	Atomic wt.	Notes
89	Ac	Actinium	[227]	5
13	Al	Aluminium	26.981538(2)	
95	Am	Americium	[243]	5
51	Sb	Antimony	121.760(1)	1
18	Ar	Argon	39.948(1)	1, 2
33	As	Arsenic	74.92160(2)	
85	At	Astatine	[210]	5
56	Ba	Barium	137.327(7)	
97	Bk	Berkelium	[247]	5
4	Be	Beryllium	9.012182(3)	
83	Bi	Bismuth	208.98038(2)	
107	Bh	Bohrium	[264]	5, 6
5	B	Boron	10.811(7)	1, 2, 3
35	Br	Bromine	79.904(1)	
48	Cd	Cadmium	112.411(8)	1
55	Cs	Caesium	132.90545(2)	
20	Ca	Calcium	40.078(4)	1
98	Cf	Californium	[251]	5
6	C	Carbon	12.0107(8)	1, 2
58	Ce	Cerium	140.116(1)	1
17	Cl	Chlorine	35.453(2)	3
24	Cr	Chromium	51.9961(6)	
27	Co	Cobalt	58.933200(9)	
29	Cu	Copper	63.546(3)	2
96	Cm	Curium	[247]	5
110	Ds	Darmstadtium	[281]	5, 6
105	Db	Dubnium	[262]	5, 6
66	Dy	Dysprosium	162.500(1)	1
99	Es	Einsteinium	[252]	5
68	Er	Erbium	167.259(3)	1
63	Eu	Europium	151.964(1)	1
100	Fm	Fermium	[257]	5
9	F	Fluorine	18.9984032(5)	
87	Fr	Francium	[223]	5
64	Gd	Gadolinium	157.25(3)	1
31	Ga	Gallium	69.723(1)	
32	Ge	Germanium	72.64(1)	
79	Au	Gold	196.96655(2)	
72	Hf	Hafnium	178.49(2)	
108	Hs	Hassium	[277]	5, 6
2	He	Helium	4.002602(2)	1, 2
67	Ho	Holmium	164.93032(2)	
1	H	Hydrogen	1.00794(7)	1, 2, 3
49	In	Indium	114.818(3)	
53	I	Iodine	126.90447(3)	
77	Ir	Iridium	192.217(3)	
26	Fe	Iron	55.845(2)	

36	Kr	Krypton	83.798(2)	1, 3
57	La	Lanthanum	138.9055(2)	1
103	Lr	Lawrencium	[262]	5
82	Pb	Lead	207.2(1)	1, 2
3	Li	Lithium	[6.941(2)]	1, 2, 3, 4
71	Lu	Lutetium	174.967(1)	1
12	Mg	Magnesium	24.3050(6)	
25	Mn	Manganese	54.938049(9)	
109	Mt	Meitnerium	[268]	5, 6
101	Md	Mendelevium	[258]	5
80	Hg	Mercury	200.59(2)	
42	Mo	Molybdenum	95.94(2)	1
60	Nd	Neodymium	144.24(3)	1
10	Ne	Neon	20.1797(6)	1, 3
93	Np	Neptunium	[237]	5
28	Ni	Nickel	58.6934(2)	
41	Nb	Niobium	92.90638(2)	
7	N	Nitrogen	14.0067(2)	1, 2
102	No	Nobelium	[259]	5
76	Os	Osmium	190.23(3)	1
8	O	Oxygen	15.9994(3)	1, 2
46	Pd	Palladium	106.42(1)	1
15	P	Phosphorus	30.973761(2)	
78	Pt	Platinum	195.078(2)	
94	Pu	Plutonium	[244]	5
84	Po	Polonium	[209]	5
19	K	Potassium	39.0983(1)	1
59	Pr	Praseodymium	140.90765(2)	
61	Pm	Promethium	[145]	5
91	Pa	Protactinium	231.03588(2)	5
88	Ra	Radium	[226]	5
86	Rn	Radon	[222]	5
75	Re	Rhenium	186.207(1)	
45	Rh	Rhodium	102.90550(2)	
111	Rg	Roentgenium	[272]	5, 6
37	Rb	Rubidium	85.4678(3)	1
44	Ru	Ruthenium	101.07(2)	1
104	Rf	Rutherfordium	[261]	5, 6
62	Sm	Samarium	150.36(3)	1
21	Sc	Scandium	44.955910(8)	
106	Sg	Seaborgium	[266]	5, 6
34	Se	Selenium	78.96(3)	
14	Si	Silicon	28.0855(3)	2
47	Ag	Silver	107.8682(2)	1
11	Na	Sodium	22.989770(2)	
38	Sr	Strontium	87.62(1)	1, 2
16	S	Sulfur	32.065(5)	1, 2
73	Ta	Tantalum	180.9479(1)	
43	Tc	Technetium	[98]	5
52	Te	Tellurium	127.60(3)	1

65	Tb	Terbium	158.92534(2)	
81	Tl	Thallium	204.3833(2)	
90	Th	Thorium	232.0381(1)	1, 5
69	Tm	Thulium	168.93421(2)	
50	Sn	Tin	118.710(7)	1
22	Ti	Titanium	47.867(1)	
74	W	Tungsten	183.84(1)	
112	Uub	Ununbium	[285]	5, 6
116	Uuh	Ununhexium		see Note above
118	Uuo	Ununoctium		see Note above
114	Uuq	Ununquadium	[289]	5, 6
92	U	Uranium	238.02891(3)	1, 3, 5
23	V	Vanadium	50.9415(1)	
54	Xe	Xenon	131.293(6)	1, 3
70	Yb	Ytterbium	173.04(3)	1
39	Y	Yttrium	88.90585(2)	
30	Zn	Zinc	65.409(4)	
40	Zr	Zirconium	91.224(2)	1

determined for the specific material [range quoted for 1995 Tables 6.94 and 6.99].

5. Element has no stable nuclides. The value enclosed in brackets, e.g. [209], indicates the mass number of the longest-lived isotope of the element. However, three such elements (Th, Pa, and U) do have a characteristic terrestrial isotopic composition, and for these an atomic weight is tabulated.
6. The names and symbols for elements 112–118 are under review. The temporary system recommended by J Chatt (1979) Pure Appl. Chem. 51: 381–384 is used above. The names of elements 101–109 were agreed in 1997 [see Pure Appl. Chem. 69: 2471–2473 (1997)] and for element 110 in 2003 [see Pure Appl. Chem. 75: 1613–1615 (2003)]. The proposed name for element 111 is also included.

1. Geological specimens are known in which the element has an isotopic composition outside the limits for normal material. The difference between the atomic weight of the element in such specimens and that given in the table may exceed the stated uncertainty.
2. Range in isotopic composition of normal terrestrial material prevents a more precise value being given; the tabulated value should be applicable to any normal material.
3. Modified isotopic compositions may be found in commercially available material because it has been subject to an undisclosed or inadvertent isotopic fractionation. Substantial deviations in atomic weight of the element from that given in the table can occur.
4. Commercially available Li materials have atomic weights that range between 6.939 and 6.996; if a more accurate value is required, it must be determined for the specific material [range quoted for 1995 Tables 6.94 and 6.99].
5. Element has no stable nuclides. The value enclosed in brackets, e.g. [209], indicates the mass number of the longest-lived isotope of the element. However, three such elements (Th, Pa, and U) do have a characteristic terrestrial isotopic composition, and atomic weights are tabulated.
6. The names and symbols for elements 112–118 are under review. The temporary system recommended by J Chatt (1979) Pure Appl. Chem. 51: 381–384, is used above. The names of elements 101–109 were agreed in 1997 [see Pure Appl. Chem. 69: 2471–2473 (1997)] and for element 110 in 2003 [see Pure Appl. Chem. 75: 1613–1615 (2003)]. The proposed name for element 111 is also included.

Pronunciation symbols

ə	Banana, collide, abut
ˈə, ˌə	Humdrum, abut
ᵊ	Immediately preceding \l\, \n\, \m\, \ŋ \, as in batt**le**, mitt**en**, eat**en**, and sometimes op**en** \ˈō-pᵊm\, lock **and** key \-ᵊ ŋ-\; immediately following \l\, \m\, \r\, as often in French table, pris**me**, tit**re**
ər	furth**er**, m**er**g**er**, b**ir**d
ˈə-,ˈə-r	As in two different pronunciations of h**urr**y \ˈ hər-ē, \ˈ hə-rē\
a	m**a**t, m**a**p, m**a**d, g**a**g, sn**a**p, p**a**tch
ā	d**a**y, f**a**de, d**a**te, **a**orta, dr**a**pe, c**a**pe
ä	b**o**ther, c**o**t, and, with most American speakers, f**a**ther, c**a**rt
á	f**a**ther as pronounced by speakers who do not rhyme it with *bother*; French p**a**tte
aú	n**ow**, l**ou**d, **ou**t
b	ba**b**y, ri**b**
ch	**ch**in, nature \ˈnā-chər\
d	**d**i**d**, a**dd**er
e	b**e**t, b**e**d, p**e**ck
ˈē,ˌē	b**ea**t, nosebl**ee**d, **e**venly, **ea**sy
ē	eas**y**, meal**y**
f	**f**i**f**ty, cu**ff**
g	**g**o, bi**g**, **g**ift
h	**h**at, a**h**ead
hw	**wh**ale as pronounced by those who do not have the same pronunciation for both *whale* and *wail*
i	t**i**p, ban**i**sh, act**i**ve
ī	s**i**te, s**i**de, b**uy**, tr**i**pe
j	**j**ob, **g**em, e**dge**, **j**oin, **j**u**dge**
k	**k**in, coo**k**, a**ch**e
<u>k</u>	German i**ch**, Bu**ch**; one pronunciation of lo**ch**
l	**l**i**l**y, poo**l**
m	**m**ur**m**ur, di**m**, ny**m**ph
n	**n**o, ow**n**
ⁿ	Indicates that a preceding vowel or diphthong is pronounced with the nasal passages open, as in French *un bon vin blanc* \œⁿ-bōⁿvaⁿ-bläⁿ\
ŋ	sing \ˈ siŋ \, singer \ˈ siŋ-ər\, finger \ˈfiŋ-gər\, ink \ˈiŋk\
ō	b**o**ne, kn**ow**, b**eau**
ó	s**aw**, **a**ll, gn**aw**, c**au**ght
œ	French c**oe**uf, German H**ö**lle
<u>œ</u>	French f**eu**, German H**öh**le
ói	c**oi**n, destr**oy**
p	**p**e**pp**er, li**p**
r	**r**ed, ca**r**, **r**a**r**ity
s	**s**our**ce**, le**ss**
sh	as in **sh**y, mi**ss**ion, ma**ch**ine, spe**ci**al (actually, this is a single sound, not two); with a hyphen between, two sounds as in *grasshopper* \ˈgras-ˌhä-pər\
t	tie, attack, late, later, latter
th	as in thin, ether (actually, this is a single sound, not two); with a hyphen between, two sounds as in knighthood \ˈnīt-ˌh----d\
<u>th</u>	**th**en, ei**th**er, **th**is (actually, this is a single sound, not two)
ü	r**u**le, y**ou**th, **u**nion \ˈyün-yən\, few \ˈfyü\
ú	pull, wood, book, curable \ˈky ú r-ə-bəl\, fury \ˈfy----r-ē\
ue	German f**ü**llen, h**ü**bsch
<u>ue</u>	French r**ue**, German f**üh**len
v	**v**i**v**id, gi**ve**
w	**w**e, a**w**ay
y	**y**ard, **y**oung, cue \ˈkyü\, mute \ˈmyüt\, union \ˈyün-yən\
ʸ	indicates that during the articulation of the sound represented by the preceding character the front of the tongue has substantially the position it has for the articulation of the first sound of yard, as in French *digne* \dēnʸ\
Z	**z**one, rai**se**
zh	as in vi**s**ion, azure \ˈa-zhər\ (actually this is a single sound, not two).
\	reversed virgule used in pairs to mark the beginning and end of a transcription: \ˈpen\

ˈ	mark preceding a syllable with primary (strongest) stress: \ˈpen-mən-ˌship\
ˌ	mark preceding a syllable with secondary (medium) stress: \ˈpen-mən-ˌship\
-	mark of syllable division
( )	indicate that what is symbolized between is present in some utterances but not in others: *factory* \ˈfak-t(ə-)rē
÷	indicates that many regard as unacceptable the pronunciation variant immediately following: *cupola* \ˈkyü-pə-lə, ÷-ˌlō\

Explanatory notes and abbreviations

(date)	date that word was first recorded as having been used
[. . .]	etymology and origin(s) of word
{. . .}	usage and/or languages, including French, German, Italian and Spanish
adj	adjective
adv	adverb
B.C.	before Christ
Brit.	Britain, British
C	centigrade, Celsius
c	century
E	English
Eng.	England
F	French, Fahrenheit
Fr.	France
fr.	from
G	German
Gr.	Germany
L	Latin
ME	middle English
n	noun
neut.	neuter
NL	new Latin
OE	old English
OL	old Latin
pl	plural
prp.	present participle
R	Russian
sing.	singular
S	Spanish
U.K.	United Kingdom
v	verb

Languages

French, German and Spanish translations are enclosed in {--} and preceded by F, G, I and S, respectively; and gender is designated by f-feminine, m-masculine, n-neuter. For example: Polymer--{F polymere m} represents the french translation "polymere" of the English word polymer and it is in the masculine case. These translations were obtained from multi-language dictionaries including: Glenz W (ed) (2001) A glossary of plastics terminology in 5 languages, 5th edn. Hanser–Gardner Publications Inc., Cincinnati (with permission).

A

A

a \ā\ *n.* (1) SI abbreviation for prefix Atto-, (2) Symbol for acceleration.

"a" or "α" *n.* Redness–greenness coordinate in certain transformed color spaces, generally used as the Δ*a*, or difference in "*a*" between a specimen and a standard reference color. If "*a*" or Δ*a* is plus, then there is more redness than greenness; if "*a*" or Δ*a* is minus, then there is more greenness than redness. It is normally used with b or *β* as part of the chromaticity or chromaticity difference. McDonald and Roderick (1997) Colour physics for industry, 2nd edn. Society of Dyers and Colourists, West Yorkshire, England. Billmeyer FW and Saltzman M (1966) Principles of color technology. John Wiley and Sons Inc., New York.
See uniform chromaticity coordinates.

"a" Kubelka–Munk equation *n.* Mathematical constant characteristic of a color at complete hiding; dependent on the optical constants *K* and *S*: $a = ½(1/R\infty + R\infty) = 1 + K/S$. McDonald and Roderick (1997) Colour physics for industry, 2nd edn. Society of Dyers and Colourists, West Yorkshire, England.

au *n.* Abbreviation for atomic unit.

A *n.* Abbreviation for Ampere.

Å \ˈaŋ-strəm\ [Anders J. *Angstrom*] (1892) {*d* Angströmeinheit *f*, *f* unité f Angtröm, *s* unidad *f* Angström} *n.* A unit of length equal to 1×10^{-12}m. Abbreviation for deprecated Angstrom unit. Weast RC (ed) Handbook of chemistry and physics, 52nd edn. The Chemical Rubber Co., Boca Roton, FL.
See Angstrom unit.

A-Acid \ā-ˈa-səd\ [F. or L.; F. *acide*] (1626) *n.* $NH_2C_6H_4COOH$. Trade abbreviation for anthranilic acid, an intermediate used in the manufacture of the pigment, Lake Red D.

AATCC *n.* Abbreviation for the American Association of Textile Chemists and Colorists.

AB (=absolute). A prefix attached to the names of practical electrical units to indicate the corresponding unit in the old cgs system (emu), e.g., abampere and abvolt.

A-B-A model polymers *n.* Two phase block copolymers, predictable molecular weights, narrow molecular weight distribution, convenient end-capping, thermoplastic, anionically polymerized, i.e., Kraton-G[R] and Hytrel[R]. The B block is usually styrene that forms hard and amorphous domains.

A-B-A *n.* **Thermoplastic elastomers.** Three-block thermoplastic polymer elastomer, high-strength rubber, no vulcanization, completely soluble, two glass and two glass transition temperatures, i.e., styrene and butadiene.

Abbe' number \a-ˈbā, ˈa-ˌbā-\ [F, fr. LL *abbat-*, *abass*] (1530) *n.* The refractive index varies with the wavelength of incident light, and the abbe' number *ν* is given as a measure of this dispersion; and the capacity to separate the colors of white light increases as *ν* decreases.

Abbe' refractometer \-ˌrē-ˌfrak-ˈtä-mə-tər\ *n.* Common form of refractometer used for determining the refractive index of oils and other liquids, or of grease-like products, which are capable of liquefaction at moderate temperatures. Good accuracy is attainable in the range of 1.3–1.7, readings being given to the fourth decimal place. The prisms, which constitute the most important part of the instrument, and hence the liquid held between their faces, are capable of being maintained accurately at the temperature of the determination.

A

With the use of special liquids to form an optical seal to the prisms and a special technique of viewing, it is also used for determining the refractive index of solids such as plastics cast in sheets with polished surfaces and edges The refractometer measures the real part of the refractive index and thus helps to answer three different types of questions. First, and most simply, it is useful in the empirical identification of pure substances, it can act as a criterion of purity, and it serves in the quantitative analysis of solutions. These characterizations are made possible by the precision and accuracy of refractometers. Second, the evaluation of dipole moments of substances via measures of dielectric constant at a single temperature requires the knowledge of their refractive indexes. Third, refractive index measured as a function of wavelength, in concert with measurement of molar absorptivity characterize the optical properties of a given molecule. These measures in turn provide information on the electronic structures of molecules. As an example, refractometry can be useful in the determination of chain length and isomerism in organic molecules. The development of modern NMR and mass spectrometers has largely displaced the use of refractometry in such studies, giving less ambiguous answers regarding molecular structures, but at a great increase in instrumental complexity and cost. The modern Abbe refractometer invented at the Carl Zeiss Works was exclusively manufactured by Zeiss until the early 20th century. The explosive growth of laboratory work after World War I led a number of other companies to begin its manufacture as well, including Adam Hilgar and Stanley in Great Britain, and Spencer Lens Co., Bausch & Lomb, Gaertner, and Valentine in the USA Ernst Abbe constructed the first "Abbe" refractometer in 1869. Five years later, in 1874, he published a comprehensive booklet, and in it he discussed the theory of refractometer.

Abbozzo *adj.* Underpainting of an oil painting, either in monochrome or color. *Sometimes called bozzo or deadcoloring.*

Abcoulomb *n.* The abcouloumb, the emu of charge, is defined as the charge, which passes through a given surface in one second if a steady current of one abampere flows across the surface. Its dimensions are, therefore, $cm^{0.5}g^{0.5}$ which differ from the dimensions of the statcoulomb by a factor, which has the dimensions of speed. This relationship is connected with the fact that the ratio $2K_e/K_m$ must have the value of the square of the speed of light in any consistent system of units. It follows further that 1 abcouloumb $= 2.99793 \times 10^{10}$ statcoulomb, the speed of light in vacuo\being $(2.99793 \pm 0.000003) \times 10^{10}$ cm/s. Weast RC Handbook of chemistry and physics, 52nd edn. The Chemical Rubber Co., Boca Roton, FL.

Abegg's rule \ˈä-ˌbegz-\ [Abegg, Richard Wilhelm Heinrich; Danish chemist, major work on chemical valence] (1869–1910) *n.* A Chemistry: For a given chemical element (as sulfur) the sum of the absolute value of its negative valence of maximum value (as -2 for sulfur in H_2S) and its positive value (as $+6$ in H_2SO_4) is often equal to 8. For use in regard to a helical periodic system. This tendency is exhibited especially by the elements of the fourth–seventh groups and is known as Abegg's rule. General chemistry. Brookes/Cole, New York, 2003.

Aberration \ˌa-bə-ˈrā-shən\ [L. *aberrare*] (1594) *n.* In optical systems, the failure of light rays from one object point to converge to a single focal point.

See chromatic aberration and spherical aberration.

ABFA n. *See azobisformamide.*

Abherent \ab-ˈhir-ənt\ (adhesive) *n.* A coating or film applied to one surface to prevent or reduce its adhesion to another surface brought into intimate contact with it. Abherents applied to plastic films are often called *anti-blocking agents.* Those applied to molds, calendar rolls, etc., are sometimes called *release agents* or parting agents. Skeist I (1990, 1977, 1962) Handbook of adhesives. Van Nostrand Reinhold, New York.

Abhesive \-əb-ˈhē-siv, -ziv\ (1670) *n.* Material that resists adhesion; applied to surfaces to prevent sticking, heat-sealing, etc. Skeist I (1990, 1977, 1962) Handbook of adhesives. Van Nostrand Reinhold, New York.

Abietic acid \a-bē-ə-tek, a-səd\ *n.* $C_{19}H_{30}COOH$. A monocarboxylic acid derived from rosin. Plasticizers derived from it include hydroabietyl alcohol, hydrogenated methyl abietate, and methyl abietate.

Abietic acid, commercial grade *n.* $C_{20}H_{30}O_2$. Product consisting chiefly of rosin acids in substantially pure form, separated from rosin or tall oil commercially for specific purposes and in which abietic acid and its isomers are the principal components. Syn: Sylvic acid.

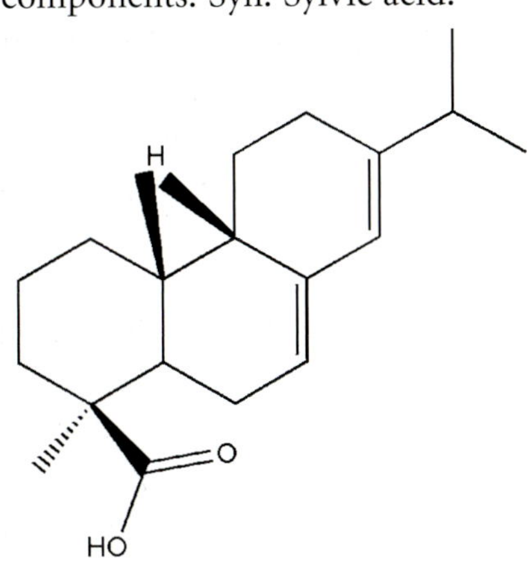

Abietates *n.* Esters or salts of abietic acid, a principal constituent of ordinary rosin from which the products of commerce are derived, no attempt being made to separate abietic acid from the other acids which rosin is likely to contain. Metallic abietates, as such, are rarely encountered under this name but generally as resonates. Esters of rosin, however, are commonly described as abietates and not as resonates. For example, methyl abietate (Trademark – Abalyn), a mixture of the methyl esters of the rosin acids. $C_{19}H_{29}COOCH_3$. The article of commerce is colorless to yellow, almost odorless, thick liquid. D_{20}^{20} 1.040bp 360–365°F with decompn. $^{n}D^{20}$ 1.530. Flash pt 180–218°C. Insoluble in water, miscible with usual organic solvents, also with aliphatic hydrocarbons. Dissolves ester gums, rosin, many synthetic resins as well as ethyl cellulose, rubber, etc., bp 360–365°F with decomposition; use as a solvent for ester gums, rosin, many synthetic resins, ethyl cellulose, rubber, etc.; in the manufacture of varnish resins; as ingredient in adhesives. Esters of rosin are described as abietates and include the methyl, ethyl, and benzyl derivatives, usually used as plasticizers. The ester abietates, which have enjoyed some popularity, are the methyl, ethyl, and benzyl derivatives. They are soft, resinous materials and are used chiefly as plasticizers. Langenheim JH (2003) Plant resins: chemistry, evolution ecology and ethnobotany. Timber Press, Portland OR. Wypych G (2003) Plasticizer's data base. Noyes Publication, New York. Paint: pigment, drying oils, polymers, resins, naval stores, cellulosics esters, and ink vehicles, vol 3. American Society for Testing and Material, 2001. Merck index, 13th edn. Merck and Company Inc., Whitehouse Station, NJ, 2001.

A

Ablation \a-ˈblā-shən\ (15c) *n.* Derived from the Latin *ablatio*, meaning "a carrying away", this term has been used by astrophysicists to describe the erosion and disintegration of meteors entering the atmosphere, and more recently by space scientists and engineers for the layer-by-layer decomposition of a plastic surface when heated quickly to a very high temperature. Usually, the decomposition is highly endothermic and the absorption of energy at the surface slows penetration of high temperature to the interior. In other words, it is the ability of a material such as a polymer to form a protective thermal layer when carbonized by extreme heat. {G ablative, F ablative, S ablative, I ablative}. Kidder RC (1994). Handbook of fire retardant coatings and fire testing services. CRC Press, Boca Raton, FL. Rosato DV (1992) Rosato's plastics encyclopedia and dictionary. Hanser–Gardner Publications, New York.

Ablative coatings *n.* Thick, mastic-like materials which absorb heat; they are designed to char and sacrifice themselves while protecting the metal substrate underneath. This type of coating is similar with Intumescent coatings that produce foam on exposure to high heat to protect the substrate, but do not char as ablative coatings. These coatings are used for missiles and re-entry rockets. Kidder RC (1994) Handbook of fire retardant coatings and fire testing services. CRC Press, Boca Raton, FL. Nelson G (1990) Fire and polymers: hazards identification and prevention. Oxford University Press, Oxford. *See also Ablative plastic.*

Ablative plastic *n.* Material which absorbs heat while part of it is being consumed by heat through a decomposition process (pyrolysis) which takes place near the surface exposed to the heat. Nelson G (1990) Fire and polymers: hazards identification and prevention. Oxford University Press, Oxford. Pittance JC (1990) Engineering plastics and composites. SAM International, Materials Park, OH.

ABL bottle *n.* A filament-wound test vessel about 46 cm in diameter and 61 cm long, subjected to rising internal hydrostatic pressure to determine the quality and strength of the composition from which it was made.

Abnormal crimp \(ˌ)ab-ˈnór-məl, əb- ˈkrimp\ *n.* A relative term for crimp that is either too low or too high in frequency and/or amplitude or that has been put into the fiber with improper angular characteristics.

ABR *n.* Copolymers from acrylic esters and butadiene.

Abraded yarn \ə-ˈbrādəd\ *n.* A filament yarn in which filaments have been cut or broken to create hairiness (fibrillation) to simulate the surface character of spun yarns. Abraded yarns are usually plied or twisted with other yarns before use. Kadolph SJ and Langford AL (2001) Textiles. Pearson Education, New York.

Abraser *n.* An instrument used for measuring resistance to abrasion using a sample on a turntable rotating under a pair of weighted abrading wheels that produce abrasion through side-slip. Koleske JV (ed) (1995). Paint and coating testing manual. American Society for Testing and Materials.

Abrasiometer *n.* One of the many devices used to test abrasion of a coating by using an air blast to drive an abrasive against the test film, or by rotating a film submerged in an abrasive, or by simply dropping a stream of abrasive onto the film. Koleske JV (ed) (1995). Paint and coating testing manual. American Society for Testing and Materials, www.gardco.com

Abrasion \ə-ˈbrā-zhən\ [ML *atrasion-, abrasion,* fr. L *abradere*] (1656) *n.* The wearing away of a surface in service by action such as rubbing, scraping, or erosion. {G Abrieb m, F abrasion f, S abrasión f, I abrasione f}.

Abrasion coefficient \-ˌkō-ə-ˈfi-shənt\ *n.* Method for reporting the result of an abrasion test using the falling sand abrasion tester, in which it is assumed that the abrasion resistance is proportional to the film thickness.

$$\text{Abrasion Coefficient} = \frac{W_1 - W_2}{T},$$

where W_1 is the grams of abrasive and holder before tests, W_2 the grams of abrasive and holder after test, and T is the thickness of coating in mils (0.001 in.) (0.025 mm). Koleske JV (ed) (1995). Paint and coating testing manual. American Society for Testing and Materials. Gardner–Sward handbook, MNL 17, 14th edn. ASTM, Conshohocken, PA.

Abrasion cycle *n.* The number of abrading motions or cycles to which a test specimen is subjected in a test of abrasion resistance. Paint and coating testing manual (Gardner–Sward handbook). MNL 17, 14th edn. ASTM, Conshohocken, PA, 1995.

Abrasion resistance *n.* (1) This test method (*see www.astm.org*) covers the determination of the resistance or organic coatings produced by an air blast of abrasive material on coatings applied a plane rigid substrate such as a glass or metal. (2) The ability of a coating to resist being worn away and to maintain its original appearance and structure as when subjected to rubbing, scraping, or erosion such as measured by the Taber Abraser The resistance to shearing of material from a surface, i.e., rubber has abrasion resistance from sand. The ability of a fiber or fabric to sustain wearing of its surface. (3) The ability of a material to withstand mechanical actions such as subbing, scraping, grinding, sanding, or erosion that tends progressively to remove material from its surface Gardner–Sward handbook. MNL 17, 14th edn. ASTM, Conshohocken, PA, 1995.
See abrasion.

Abrasion test *n.* Tests designed to determine the ability to withstand the effects of rubbing and scuffing.

Abrasive (1853) *n.* Any material which, by a process of grinding down, tends to make a surface smooth or rough.

Abrasive finishing *n.* (1) A method of removing flash, gate marks, and rough edges from plastics articles by means of grit-containing belts or wheels. The process is usually employed on large rigid or semirigid products with intricate surfaces that cannot be treated by tumbling or other more efficient methods of finishing. (2) To finish, dress, or decorate a surface using a material such as polishing grit.

Abrasive forming *n.* Formation of a part or shape using abrasives to chip away unwanted materials.

Abrasiveness \ə-ˈbrā-siv, zivˈnes\ (1875) *n.* The property of a substance that causes it to wear or scratch other surfaces with which it is in contact. Merriam-Webster's collegiate dictionary, 11th edn. Merriam-Webster Inc., Springfield, MA, 2004.

Abrasive wheels *n.* An abrasion material in the shape of a disk which is often turned on a power tool, i.e., abrasive polishing of granite.

Abraum A red ocher used to stain mahogany.

Abridged spectrophotometer \ə-ˈbrij ˌspek-trō-fə-ˈtä-mə-tər\ *n.* An instrument which

A

measures spectral transmittance or reflectance at a limited number of wavelengths, usually employing filters rather than a monochromator. Skoog DA, Holler FJ, Nieman TA (1997) Principles of instrumental analysis, Brooks/Coles, New York. Willard HH, Merritt LL, and Dean JA (1974) Instrumental methods of analysis. D. Van Nostrand Company, New York.

See filter spectrophotometer.

ABS \ˌā-(ˌ)bē-ˈes\ [*a*crylonitrile-*b*utadiene-*s*ytrene] (1966) *n.* Copolymer of acrylonitrile-butadiene-styrene segments. Abbreviation for acrylonitrile-butadiene-styrene.

See ABS resin.

Absolute \ˈab-sə ˌlüt\ [ME *absolut*, fr. L *absolutus*, fr. pp of *absolvere* to set free, absolve] (14c) *adj.* Adjective used to describe measurements in terms of fundamentally defined units. Merriam-Webster's collegiate dictionary, 10th edn. Merriam-Webster Inc., Springfield, MA, 2000.

Absolute alcohol *n.* Ethyl alcohol that has been refined by azeotropic distillation to 99.9% purity (200 proof). Other commercial ethanols contain about 5% water and may contain denaturants that make the alcohol undrinkable. Pure anhydrous ethyl alcohol (ethanol). The term is used to distinguish it from the several varieties of alcohol which are available, and which contain varying amounts of water and/or other impurities.

Absolute humidity (1867) *n.* The actual weight of water vapor contained in a unit weight of air.

See humidity, absolute.

Absolute pressure *See pressure.*

Absolute reflectance *n.* Reflectance measured relative to the perfect diffuser.

Absolute temperature *n.* Temperature measured from the absolute zero, at which all molecular motions cease; 0.0K (Kelvin) = −273.15°C. Whitten KW, Davis RE, Davis E, Peck LM., and Stanley GG (2003) General chemistry. Brookes/Cole, New York.

See Kelvin temperature scale.

Absolute units *n.* A system of units based on the smallest possible number of independent units. Specifically, one unit of force, work, energy and power not derived from or dependent on gravitation.

Absolute viscosity *n.* (1) Tangential force on unit area of either of two parallel planes at unit distance apart, when the space between the planes in filled with fluid (in question) and one of the planes moves with unit velocity in its own plane relative to the other. (2) Force required to move in opposite directions at a velocity of 1 m/s, two parallel plans of liquid, 1 m^2 in area and separated from each other by a distance of 1 m. The absolute viscosity is designated by the Greek letter η. Goodwin J. W., Goodwin J., and Hughes R. W. (2000) Rheology for chemists. Royal Society of Chemistry, Cambridge.

See poise, viscosity.

Absolute zero (1848) *n.* The temperature at which all particles in a substance are in their lowest energy states: 0 K or –273.15°C, the temperature at which all chemical activity ceases. It is equal to –273.15°C or –459.67°F. Absolute 0 K has never been achieved and does not exist in nature or perhaps not anywhere in the known universe. Serway RA, Faugh J S, Bennett CV (2005) College physics. Thomas, New York. Whitten KW, Davis RE, Davis E, Peck LM, Stanley GG (2003) General chemistry. Brookes/Cole, New York.

Abson *n.* ABS. Manufactured by Goodrich, USA.

Absorbance \əb-ˈsór-bən(t)s, -ˈzór-\ (1947) *n.* Logarithm of the reciprocal of spectral internal transmittance. The ability of a substance to transform radiant energy into a different of energy, usually with a resulting rise in temperature. Mathematically, absorbance is the negative logarithm to the base 10 of transmittance. Willard HH, Dean JA, Merritt LL (1995) Instrumental methods of analysis. Wadsworth, New York.

See Beer–Bouguer law and light absorbance.

Absorbency (1859) *n.* That property of a porous material, such as paper, which causes it to take up liquids or vapors (e.g., moisture) with which it is in contact. Merriam-Webster's collegiate dictionary, 11th edn. Merriam-Webster Inc., Springfield, MA, 2004.

Absorption [F & L; F, fr. L] (1741) *n.* (1) The penetration of a substance into the mass of another substance by chemical or physical action. (2) The process by which energy is dissipated within a specimen placed in a field of radiant energy. Since some part of the impinging energy may be transmitted through the specimen and another part be reflected, the energy absorbed will nearly always be less than that impinging. (3) The adhesion of a substance to the surface of a solid or liquid. Pollutants are extracted by adsorption on activated carbon or silica gel {G Absorption f, F absorption f, S absorción f, I assorbimento m}.

See, for example, water absorption.

Absorption coefficient *n.* Absorption of radiant energy for a unit concentration through a unit path-length for a specified wavelength and angle of incidence and viewing. Skoog DA, Holler FJ, Nieman TA (1997) Principles of instrumental analysis. Brooks/Coles, New York.

See absorption factor, Beer–Bouguer law, Kubelka–Munk theory, and Mie theory.

Absorption factor *n.* The ratio of the intensity loss by absorption to the total original intensity of radiation. If I_o represents the original intensity, I_r, the intensity of reflected radiation, I_t, the intensity of the transmitted radiation, then the absorption factor is given by the expression

$$\frac{I_o - (I_r + I_t)}{I_o}.$$

Also called coefficient of absorption. McDonald R (1997) Colour physics for industry, 2nd edn. Society of Dyers and Colouritst, West Yorkshire, England.

Absorption hygrometer *n.* Any one of several types of hygrometers containing a hygroscopic substance, the change in length, thickness, or mass of which is a measurable index of the humidity of the atmosphere.

Absorption, Lambert's law *n.* If I_o is the original intensity, I the intensity after passing through a thickness x of a material whose absorption coefficient is k, $I = I_o c^{-kx}$. The index of absorption k' is given by the relation $k = 4\pi k'n/\lambda'$, where n is the index of refraction and λ the wavelength in vacuo. The mass absorption is given by k/d when d is the density. The transmission factor is given by I/I_o. Barton AFM (1983) Handbook of solubility parameters and other cohesion parameters. Chemical Rubber Company Press, Boca Raton, FL.

Absorption, oil *n.* Oil absorption of a pigment or extender is recorded as the amount of vegetable drying oil required to convert a given mass or volume of the dry powder to a very stiff putty-like paste, which does not break or separate. It is more usually expressed as the pounds of refined linseed oil required for 100 pounds of pigment or g·100g.

Absorption spectrophotometry (spectrophotometry) *n.* An analytical technique utilizing the absorption of electromagnetic radiation by a specimen (or solution) as a

A

property related to the composition and quantity of a given material in the specimen. The radiation is usually is the ultraviolet, the visible, or the near-infrared portions of the electromagnetic spectrum. When the absorbing medium is in the gaseous state, the absorption spectrum consists of dark lines or bands, being the reverse of the mission spectrum of the absorbing gas. The spectrum of the transmitted light shows broad dark regions are that are not resolvable into lines and have no sharp or distinct edges when the absorbing medium is in the solid or liquid state. In quantitative spectrophotometry, the intensity of the radiation passing through a specimen or solution is compared with the intensity of the incident radiation and with radiation passing through a nonabsorbing solvent (*blank*). The percent absorbed by the solution is exponentially related to the solute concentration (Beer's law). Modern spectrophotometers are capable of generating nearly monochromatic radiation, so they can develop plots of percent absorption versus wavelength – *absorption spectra* – for the test compound. Skoog DA, Holler FJ, Nieman TA (1997) Principles of instrumental analysis. Brooks/Coles, New York. Willard HH., Dean JA., Merritt LL (1995) Instrumental methods of analysis. Wadsworth, New York.

See infrared spectrophotometry.

Absorption spectrum (1879) *n.* The spectrum obtained by the examination of light from a source, itself giving a continuous spectrum, after this light has passed through an absorbing medium in the gaseous state. The absorption spectrum will consist of dark lines or bands, being the reverse of the emission spectrum of the absorbing substance. The spectrum of the transmitted light shows broad dark regions that are not resolvable into lines and have no sharp or distinct edges when the absorbing medium is in the solid or liquid state. Skoog DA, Holler FJ, Nieman TA (1997) Principles of instrumental analysis. Brooks/Coles, New York. Willard HH, Merritt LL, Dean JA (1974) Instrumental methods of analysis. D. Van Nostrand Company, New York.

Absorption tinting strength *n. See tinting strength, absorption.*

Absorptive power or absorptivity *n.* For any body, the body is measured by the fraction of the radiant energy falling upon the body which is absorbed or transformed into heat. This ratio varies with the character of the surface and the wavelength of the incident energy. It is the ratio of the radiation absorbed by any substance to that absorbed under the same conditions by a black body. Fox AM (2001) Optical properties of solids. Oxford University Press, Oxford. Driggers RC, Edwards T, Co P (1998) Introduction to infrared and electro-optical systems. Artech House Inc., MA, USA.

ABS polymers *n.* Generic term for copolymers of polyblends from acrylonitrile, butadiene, and styrene.

ABS resin *n.* Any of a family of thermoplastics based on acrylonitrile, butadiene, and styrene combined by a variety of methods involving polymerization, graft copolymerization, physical mixtures, and combinations thereof. Hundreds of standard grades of ABS resins are available, and many special grades, alloyed or otherwise modified to yield unusual properties. The standard grades are rigid, hard and tough, and possess good impact strength. ABS compounds in pellet form can be extruded, blow molded, calendered, and injection molded. ABS powders are used as modifiers for other resins, for example, PVC.

Typical applications for ABS resins are household appliances, automotive parts, business-machines and telephone components, pipe and pipe fittings, packaging and shoe heels. Wickson EJ (1993) Handbook of polyvinyl chloride formulating. John Wiley and Sons Inc., New York. Harper CA (2002) Handbook of plastics, elastomers and composites, 4th edn. McGraw-Hill, New York.

Abut \ə-ˈbət\ [ME *abutten*, partly fr. OF *aboter* to border on, fr. *a-* (fr. L *ad-*) + *bout* blow, end, fr. *boter* to strike; partly fr. OF *abuter* to come to an end, fr. *a-* + *but* end] (15c) *v.* To adjoin at an end; to be contiguous.

Abvolt *n.* The cgs electromagnetic unit of potential difference and electromotive force. It is the potential difference that must exist between two points in order that one erg of work be done when one abcoulomb of charge is moved from one point to the other. One abvolt is 10^{-8}V. Weast RC Handbook of chemistry and physics, 52nd edn. The Chemical Rubber Co., Boca Roton, FL.

Acacia gum \ə-ˈkā-shə ˈgəm\ *n.* Water-soluble gum obtained from trees of the acacia species, as an exudation from incisions in the bark. It is water soluble and is used as an adhesive, thickening agent and for transparent paints. Whistler JN, BeMiller JN (eds) (1992) Industrial gums: polysaccharides and their derivatives. Elsevier Science and Technology Books, Amsterdam.

Also known as Gum Arabic.

Academy board \ə-ˈka-də-mē ˈbōrd\ *n.* A board which is given a surface in preparation for painting, primarily oil painting. It is made of paper containing chalk and size and has a face of pale gray or white ground, usually of a white lead, oil, and chalk mixture.

Accelerant \ik-ˈse-lə-rənt, ak-\ (1916) *n.* A chemical used to speed up chemical or other processes. For example, accelerants are used in dyeing triacetate and polyester fabrics. {G Beschleuniger m, F accélérateur m, S acelerador m, I acceleratore m}. Goldber DE (2003) Fundamentals of chemistry. McGraw-Hill Science/Engineering/Math, New York.

Accelerated aging *n.* Any set of conditions designed to produce in a short time the results obtained under normal conditions of aging. In accelerated aging test, the usual factors considered are heat, light, or oxygen either separately or combined. Koleske JV (ed) (1995) Paint and coating testing manual. American Society for Testing and Materials. Paint and coating testing manual (Gardner–Sward handbook), MNL 17, 14th edn. ASTM, Conshohocken, PA.

Accelerated life *See accelerated aging.*

Accelerated test *n.* A test procedure in which conditions such as temperature, humidity, and ultraviolet radiation are intensified to reduce the time required to obtain a deteriorating effect similar to one resulting from exposure to normal service conditions for much longer times.

Accelerated weathering *n.* Tests designed to simulate, but at the same time to intensify and accelerate, the destructive action of natural outdoor weathering on coatings films. The tests involve exposure to artificially produced components of natural weather, e.g., light, heat, cold, water vapor, rain, etc., which are arranged and repeated in a given cycle. There is no universally accepted test, and different investigators use different cycles. Paint and coating testing manual (Gardner–Sward

A

handbook), MNL 17, 14th edn., ASTM, Conshohocken, PA, 1995.

See artificial weathering.

Accelerated weathering machine *n.* Device intended to accelerate the deterioration of coatings by exposing them to controlled sources of radiant energy, heat, water, or other factors that may be introduced. Koleske JV (ed) (1995) Paint and coating testing manual. American Society for Testing and Materials. Paint and coating testing manual (Gardner–Sward handbook) MNL 17, 14th edn. ASTM, Conshohocken, PA, 1995.

See weatherometer and accelerated weathering.

Acceleration \ik-ˌse-lə-ˈrā-shən, (ˌ)ak-\ (1531) *n.* The time rate of change of velocity in either speed or direction. cgs unit – cm/s. Dimensions $[LT^{-2}]$.

See angular acceleration.

Acceleration due to gravity *n.* The acceleration of a body freely falling in a vacuum. The International Committee on Weights and Measures has adopted as a standard or accepted value, $980.665\,cm/s^2$ or $32.174\,ft/s^2$. Hartland S (ed) (2004) Surface and interfacial tension. CRC Press, Boca Raton, FL.

Acceleration due to gravity at any latitude and elevation *n.* If ϕ is the latitude and H the elevation in centimeters the acceleration in cgs units is, $g = 980.616 - 2.5928 \cos 2\phi + 0.0069 \cos^2 2\phi - 3.086 \times 10^{-6}\ H$ (Helmert's equation).

Accelerator (1611) *n.* (1) Any substance used in small proportion which increases the speed of a chemical reaction. In the paint industry, the term usually indicates materials that hasten the curing or cross-linking of a resin system. In the polyester resin field, it covers more specifically an additive which accelerates the action of the catalyst. (2) An organic or inorganic chemical which hastens the vulcanization of rubber, natural or synthetic, causing it to take place in a shorter time or at a lower temperature. Accelerators, particularly organic, are not mere catalysts of vulcanization, however, because they produce different and generally beneficial states of cure and different degrees of stability or resistance to chemical attack in the vulcanization. Odian GC (2004) Principles of polymerization. John Wiley and Sons Inc., New York. James F (ed) (1993) Whittington's dictionary of plastics. Technomic Publishing Co. Inc., Carley.

See catalyst and cross-linking agent.

Accommodation *n.* The adjustment of the eye to obtain maximum sharpness of the retinal imge for *n* object at which an observer is viewing. One of the important changes involves the shape of the eye lens.

Accra \ə-ˈkrä\ *n.* Natural copal resin of African origin. Langenheim JH (2003) Plant resins: chemistry, evolution ecology and ethnobotany. Timber Press, Portland, OR. Paint: pigment, drying oils, polymers, resins, naval stores, cellulosics esters, and ink vehicles, vol 3. American Society for Testing and Material, 2001.

Accroides *n.* Resinous accumulation which occurs on the leaf and stem of the *Xanthorrhoea* species. It is native to Australia and Tasmania. It appears on the market in red and yellow forms, both of which are soluble in industrial alcohol, and are used in spirit varnishes. The resin is also described as "Black Boy Gum", "Botany Bay Resin," "grass tree gum," "gum acaroid," "acaroid resin," "red gum," and "yacca" or "yacka" gum. Langenheim JH (2003) Plant resins: chemistry, evolution ecology and ethnobotany.

Timber Press, Portland, OR. Whistler JN, BeMiller JN (eds) (1992) Industrial gums: polysaccharides and their derivatives. Elsevier Science and Technology Books, Amsterdam. Langenheim JH (2003) Plant resins: chemistry, evolution ecology and ethnobotany. Timber Press, Portland, OR.

Accumulator \ə-ˈkyü-m(y)ə-ˌlā-tər\ (1748) *n.* Series of rolls which festoon strip metal on a continuous line both at the beginning and at the end. This allows the beginning or the end of the line to stop while the rest of the line is in operation. The accumulator actually accumulates a considerable length of strip, and gives a portion of its strip to whichever end is stopped.

Accumulator *n.* (1) In blow molding and injection molding, an auxiliary ram extruder providing fast parison delivery or fast mold filling. The accumulator cylinder is filled with plasticated melt from the main extruder between parison deliveries shots, and stores this melt until the plunger is called upon to deliver the next parison or shot. (2) A pressurized gas reservoir that stores energy in hydraulic systems. {G Akkumulator m, F accumulateur m, S acumulador m, I accumulatore m}. Strong AB (2000) Plastics materials and processing. Prentice-Hall, Columbus, OH.

Acenaphthene *n.* $C_{10}H_6(CH_2)_2$. Solid with an mp of 95°C, obtained from coal tar. It is a dyestuff intermediate.

Aceta *n.* Cellulose acetate, manufactured by Bayer, Germany.

Acetal \ˈa-sə-ˌtal\ *n* [G *Azetal*, fr. *azet-*, *acet-* + *Al*kohol] (1853). $CH_3CH(OC_2H_5)_2$. (1) A colorless, flammable liquid used in cosmetics and as a solvent. (2) Any of a class of compounds formed from aldehydes combined with alcohol. (3) A group of materials including polyoxymethylene (Delrin[R]).

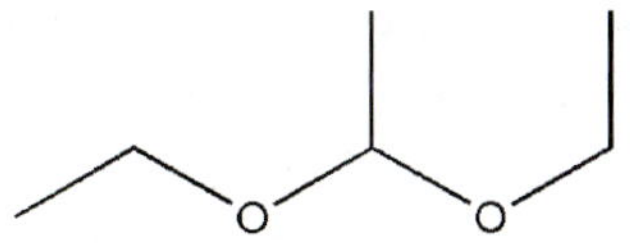

Acetaldehyde \ˌa-sə-ˈtal-də-hīd\ [ISV] (1877) (ethanal, ethyl aldehyde, acetic aldehyde) *n.* CH_3CHO. Low boiling liquid (21°C). A colorless, flammable liquid made by the hydration of acetylene, the oxidation or dehydrogenation of ethyl alcohol, or the oxidation of saturated hydrocarbons or ethylene (*See image*).

Acetaldehyde resin *n.* Product of auto-condensation of acetaldehyde.

Acetal formation, mechanism of *n.* The formation of the (–CHO–) repeat unit which after initiation (i.e., Lewis acids) forms –CHO+ and propagates to the polymer $(–CHO–)_n$.

Acetal resin (polyformaldehyde, polyoxymethylene, and polycarboxane) *n.* A thermoplastic produced by the addition polymerization of a aldehyde through the carbonyl function, yielding unbranched polyoxymethylene $(–O–CH_2–)n$ chains of great length. Examples are DuPont's "Delrin" and Hoechst–Celanese's "Celcon" (acetal copolymer based on trioxane). The acetal resins are among the strongest and stiffest of all thermoplastics, and are characterized by good fatigue life, resilience, low moisture sensitivity high resistance to solvents and chemicals, and good electrical properties. They may be processed by conventional injection molding and extrusion techniques, and fabricated by welding methods used for other thermoplastics. Their main area of application is industrial and mechanical products, e.g., gears,

A

rollers, and many automotive parts. Strong AB (2000) Plastics materials and processing. Prentice-Hall, Columbus, OH. James F (ed) (1993) Whittington's dictionary of plastics. Technomic Publishing Co. Inc., Carley.

Acetal resins *n.* High molecular weight, stable, linear polymers of formaldehyde; structurally, an oxygen atom joins the repeating units in an ether rather than ester-type link. These also include butyrals. Strong AB (2000) Plastics materials and processing. Prentice-Hall, Columbus, OH.

Acetamide \ə-ˈse-tə-ˌmīd, ˌa-sə-ˈta-ˌmīd\ [ISV] (1873) *n.* CH_3CONH_2. Amide of acetic acid with a melting point of 81°C and boiling point of 222°C. Generally used as a plasticizer for cellulose esters (*See image*).

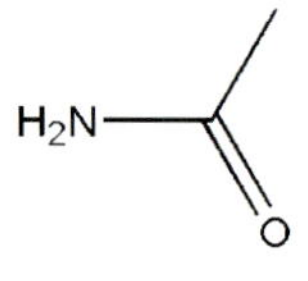

Acetate \ˈa-sə-ˌtāt\ (1827) *n.* (1) Generic name for fibers from cellulose-$2\frac{1}{2}$-acetate. (2) A salt or ester of acetic acid. (3) A generic name for cellulose acetate plastics, particularly for their fibers. Where at least 92% of the hydroxyl groups have been acetylated, the term *triacetate* may be used as the generic name of the fiber. (4) A compound containing the acetate group, $CH_3COO–$ (e.g., polyvinyl acetate).

Acetate chromes *n.* Lead chromate pigments prepared from lead acetate or basic lead acetate. Available as the lemon, primrose, medium, and orange shades. Kirk–Othmer encyclopedia of chemical technology: pigments–powders. John Wiley and Sons, New York, 1996. Solomon DH, Hawthorne DG (1991) Chemistry of pigments and fillers. Krieger Publishing Co., New York.

Acetate dope *n.* Term applied to cellulose acetate lacquers used for coating aircraft fabrics.

Acetate fiber *n.* A manufactured fiber in which the fiber-forming substance is cellulose acetate (FTC definition). Acetate is manufactured by treating purified cellulose refined from cotton linters and/or wood pulp with acetic anhydride in the presence of a catalyst. The resultant product, cellulose acetate flake, is precipitated, purified, dried, and dissolved in acetone to prepare the spinning solution. After filtration, the highly viscous solution is extruded through spinnerets into a column of warm air in which the acetone is evaporated, leaving solid continuous filaments of cellulose acetate. The evaporated acetone is recovered using a solvent recovery system to prepare additional spinning solution. The cellulose acetate fibers are intermingled and wound onto a bobbin or shippable métier cheese package, ready for use without further chemical processing. In the manufacture of staple fiber, the filaments from numerous spinnerets are combined into tow form, crimped, cut to the required length, and packaged in bales. Acetate fibers are environmentally friendly. Characteristics: Acetate fabrics are breathable, luxurious in appearance, fast-drying, wrinkle and shrinkage resistant, crisp, or soft in hand depending upon the end use. End uses: The end uses of acetate include women's and men's sportswear, evening wear, lingerie, dresses, blouses, robes, coats, other apparel, linings, draperies, bedspreads, upholstery, ribbons, formed fabrics, and filtration products. Complete textile glossary, Celanese Corporation, Three Park Avenue, New York, NY. Vincenti R (ed) (1994) Elsevier's textile dictionary. Elsevier Science and Technology Books. New York.

Acetate green *n.* Range of Brunswick or chrome greens derived from mixtures of acetate chrome and Prussian blue. These can be made wet or dry by mixing. The type made by wet mixing is preferred to paint making, as these greens have less tendency to partial separation in the film, a phenomenon known as "floating."

Acetate Rayon^{R} *n.* Cellulose acetate made from preswelling cellulose pulp with acetic acid followed by esterification with sulfuric acid–acetic anhydride mixture, the diacetate dissolved in acetone, the triacetate in methylene chloride and dry spun into fibers. Kadolph SJ, Langford AL (2001) Textiles. Pearson Education, New York.

Acetates *n.* (1) Metallic salts derived from acetic acid by interaction of the metallic oxide, hydroxide, carbonate with the acid, or by the esters derived by interaction of alcohols with acetic acid. Typical metallic salts are lead, cobalt, and manganese acetates. Common esters are ethyl, propyl, isopropyl, butyl, and amyl acetates. Acetate salts have the formula CH_3COOMe, where Me is a monovalent metal. Divalent metals like lead, etc., obviously combine with two acid radicals. The formula given above for the acetate salts apply also to esters, except that the Me becomes an alkyl radical. (2) It is also used colloquially for cellulose acetate plastics. Goldberg DE (2003) Fundamentals of chemistry. McGraw-Hill Science/Engineering/Math, New York. Vincenti R (ed) (1994) Elsevier's textile dictionary, 2003. Elsevier Science and Technology Books, Amsterdam.

Acetic acid \ə-ˈsē-tik, ˈa-səd\ (F *acétique* fr. L *acetum* vinegar] (1808) *n.* CH_3COOH (ethanoic acid, methanecarboxylic acid, and vinegar acid) CH_3COOH. A colorless liquid with the familiar taste and odor of vinegar, it is the chief constituent in dilute form. Acetic acid was originally derived by souring wine and beer, but is synthesized today by oxidation of acetaldehyde in the presence of a catalyst. Among the uses of acetic acid in the plastics industry is the manufacture of cellulosics plastics such as cellulose acetate (CA), CA butyrate and CA propionate, vinyl acetate, and acetate esters for plasticizing thermoplastics. It is a monobasic acid. Its mp is 16°C and bp, 118°C. In the paint industry, its chief applications are in the manufacture of metallic acetates used for the production of driers, and in the manufacture of acetate esters employed as solvents or plasticizers.(*See image*)

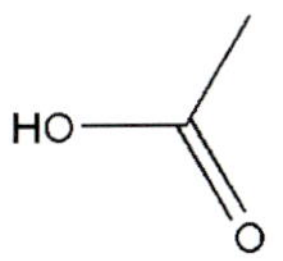

Goldberg DE (2003) Fundamentals of chemistry. McGraw-Hill Science/Engineering/Math, New York

Acetic aldehyde *See acetaldehyde.*

Acetic anhydride (1976) *n.* $(CH_3CO)_2O$. The pungent liquid may be thought of as the condensation product of two molecules of acetic acid by removal of one molecule of water, though in fact it is made by reaction of acetic acid with ketene, $CH_2{=}C{=}O$. It is a strong acetylating agent, used for many of the same purposes as its parent acid. It is an important reagent, which has wide application in the manufacture of many raw materials and intermediates for the paint trade, and also in analysis. It is used for the acetylation of hydroxyl groups as in the manufacture of acetyl ricinoleates, and in the manufacture of cellulose acetate. It has a bp of 137°C. Morrison RT, Boyd

RN (1992) Organic chemistry, 6th edn. Prentice-Hall, Englewood Cliffs, NJ. Paint/coatings dictionary, Blue Bell, PA: Federation of Societies for Coatings Technology, 1978.

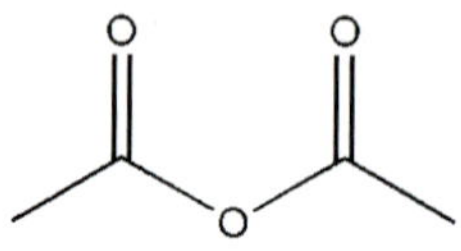

Acetic ester and acetic ether *n. See ethyl acetate.*

Acetic ether *See ethyl acetate.*

Acetocopal *n.* Product obtained when Congo Copal is reacted with acetic anhydride.

Acetone \ˈa-sə-ˌtōn\ [Gr *Azeton,* fr. L *acetum*] (1839) (dimethyl ketone, 2-propanone) *n.* CH_3COCH_3. The simplest and most important member of the ketone family of solvents. All the cellulosics plastics and polyvinyl chloride, polyvinyl acetate, polymethyl methacrylate, epoxies, and some thermosetting resins are soluble in acetone. It is also an intermediate in the production of bisphenol. It is a typical low-boiling ketone. It is a liquid which flashes at ordinary room temperature, has a bp of 57.5°C, fp of –15°C; and a Sp gr of 0.788 at 25°C. Ash M, Ash I (1996) Handbook of paint and coating raw materials: trade name products – Chemical products dictionary with trade name cross-references. Ashgate Publishing Ltd., New York: Weast RC Handbook of chemistry and physics. The Chemical Rubber Co., Boca Roton, FL.

Also known by its chemical names dimethyl ketone or propanone.

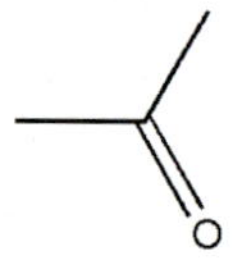

Acetone extraction *n.* In molded phenolic products, the amount of acetone-soluble material that can be extracted from the material is an indication of the degree of cure.

Acetone resin *n.* A synthetic resin produced by the reaction of acetone with materials such as phenol or formaldehyde.

Acetonyl acetone *n.* $CH_3COCH_2CH_2COCH_3$. Solvent containing two keto groups in each molecule. It has a boiling range of 188–193°C; a Sp gr of 0.973/15°C; a refractive index of 1.449; flp of 85°C (185°F); and vapor pressure of less than 2 mmHg at 30°C.

Acetophenone *n.* $CH_3COC_6H_5$. A solvent with a bp of 202°C; mp of 20°C; Sp gr of 1.023/25°C; flp of 83°C (180°F); and refractive index of 1.536.

Also known as phenyl methyl ketone or acetyl benzene (See image).

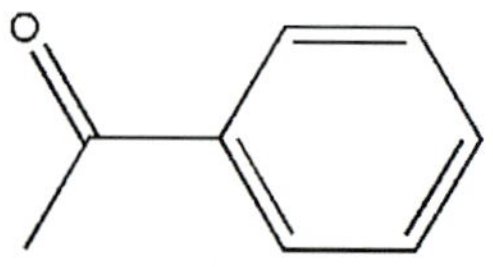

Acetyl \ə-ˈsē-tᵊl, ˈa-sə-; ˈa-sə-ˌtēl\ (1864) *n.* Monovalent radical CH_3CO–.

Acetylated congo *See acetocopal.*

Acetylated damar *n.* Product obtained when dammar is reacted with acetic anhydride.

Acetylation (1900) *vt.* Reaction wherein the hydrogen atom of a hydroxyl group is replaced by an acetyl radical (CH_3CO). It can really be regarded as an ester formation, except that it is a specific ester, the acetate, which is formed. When the acetyl value of castor or other oils containing free hydroxyl groups, or of the reaction mixture of monoglycerides, or alkyds, is determined, acetic anhydride is commonly used as the acetylating agent. This reacts with any free hydroxyl groups, whether they occur in the fatty acid chains of the vegetable oil or in unreacted polyhydric alcohols. This can also be written as a reaction with acetylene groups. Acetylene is an alkyne or HC≡CH. Morrison RT, Boyd RN (1992) Organic chemistry, 6th edn. Prentice-Hall, Englewood Cliffs, NJ.

Acetyl benzene \-ˈben-ˌzēn, ben-ˈ\.
See acetophenone.

Acetyl chloride *n.* CH_3OCl. Acetic acid in which the –OH group has been replaced by –Cl; a very active acetylating agent (*See image*).

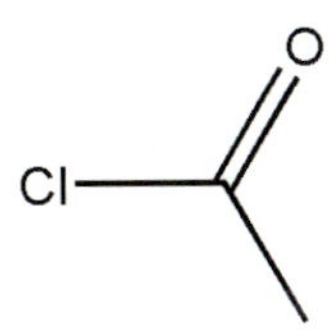

Acetyl coenzyme A (1952) *n.* A compound $C_{25}H_{38}N_7O_{17}P_3S$ formed as an intermediate in metabolism and active as a coenzyme in biological acetylations.

Acetyl cyclohexane sulfonyl peroxide *n.* A polymerization initiator, often used in conjunction with a dicarbonate such as di-sec-butyl peroxydicarbonate in vinyl chloride polymerization. The initiators have largely replaced enzoÿl and lauroÿl peroxides, the principal initiators in the early years of PVC production.
See acetyl cyclohexane sulfonyl peroxide.

Acetylene \ə-ˈse-t^{ə}l-ən, -t^{ə}l-ˌēn\ (1864) *n.* (ethyne) HC≡CH. A colorless (but not ordorless) gas obtained by reacting water with calcium carbide, CAC_2, or by cracking petroleum hydrocarbons. In the plastics industry, it is an important intermediate in the production of vinyl chloride, neoprene, acrylonitrile, and trichloroethylene.
See also polyacetylene.

Acetylene black *n.* Particularly pure form of carbon black pigment, made by the controlled combustion of acetylene in air under pressure. It is graphitic in nature and has high electrical conductivity. Donnet J-B, Wang M-J (1993) Carbon black. Marcel Dekker, New York.
See also carbon black.

Acetylene polymers *See polyacetylene.*

Acetyl number (value) *n.* Number of milligrams of potassium hydroxide required to neutralize the acetic acid set free from 1 g of an acetylated compound when the latter is subjected to hydrolysis.

Acetyl peroxide (diacetyl peroxide) *n.* $(CH_3CO)_2O_2$. A polymerization catalyst (*See image*).

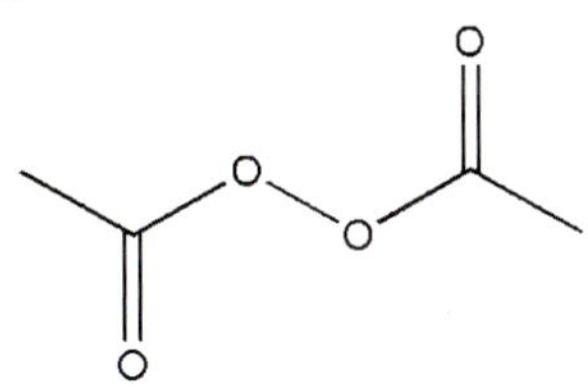

4-Acetyl resorcinol (2,4-dihydroxyacetophenone) *n.* $C_6H_3(OH)_2CO–CH_3$. A light stabilizer for plastics.

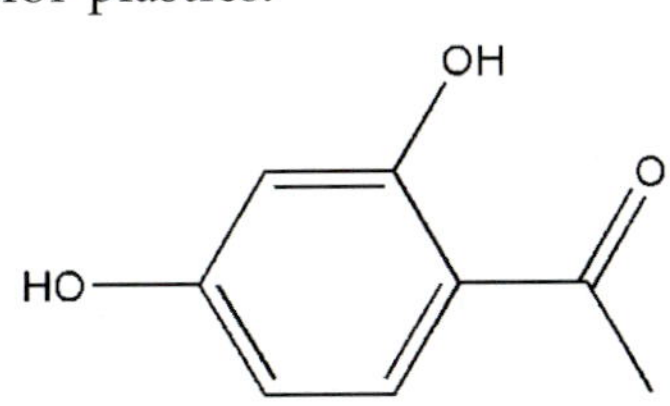

Acetyl ricinoleates *n.* Plasticizers, such as butyl and ethyl acetyl ricinoleates. They can be regarded as esters of acetylated ricinoleic acid. The hydroxy group in the ricinoleic chain has been acetylated or esterified, and the carboxylic group has also undergone esterification, but with an alcohol. Wypych G (2003) Plasticizer's data base. Noyes Publication, New York.

Acetyl triallyl citrate *n.* $CH_3COOC_3H_4(COOCH_2CH{=}CH_2)_3$. A cross-linking agent for polyesters and a polymerizable monomer. Easily polymerized with peroxide catalysts, it forms a clean, hard thermosetting resin.

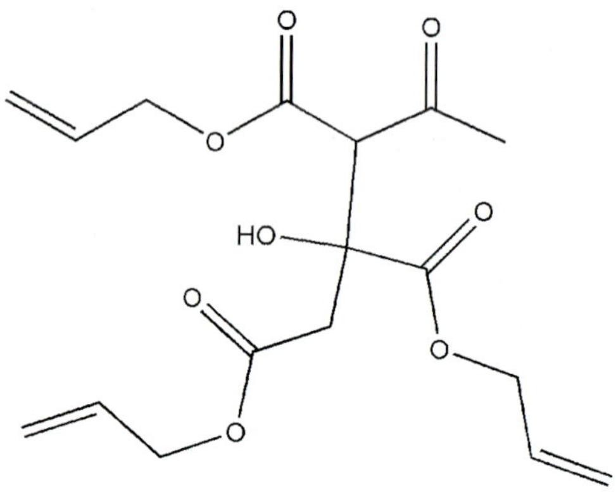

Acetyl triethyl citrate *n.* $CH_3COOC_3H_4(COOC_2H_5)_3$. A plasticizer produced by esterifying and acetylating citric acid, used in cellulose nitrate, cellulose acetate, and certain vinyls, e.g., polyvinyl acetate. It has been FDA-approved for food-contact use. Wypych G (ed) (2003) Plasticizer's data base. Noyes Publication, New York.

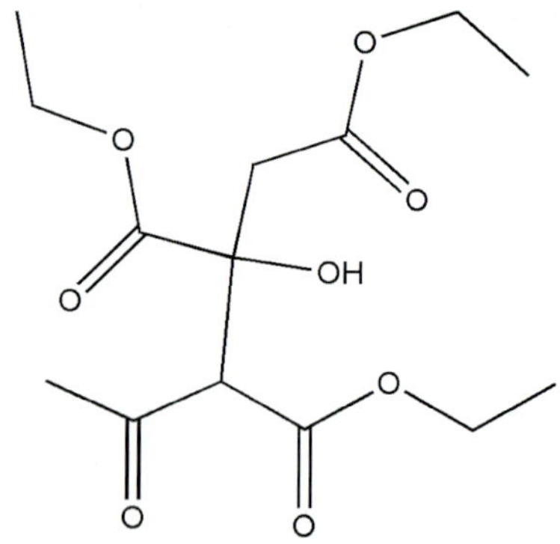

Acetyl tri-2-ethylhexyl citrate *n.* $CH_3COOC_3H_4(COOC_8H_{17})_3$. A plasticizer for vinyls, with limited compatibility for cellulose nitrate and ethyl cellulose.

Acetyl value *n.* The number of milligrams of potassium hydroxide (KOH) necessary to neutralize the acetic acid liberated by hydrolysis of 1 g of an acetylated compound. It can also be written as a measure of the degree of esterification or combination of acetyl radicals with cellulose in acetate or triacetate products.

Achromatic \ˌa-krə-ˈma-tik, (ˌ)ā-\ (1766) *adj.* Having no distinguishable hue; neutral. *See gray scale.*

Achromatic *n* **Objective.** An objective corrected spherically for one wavelength

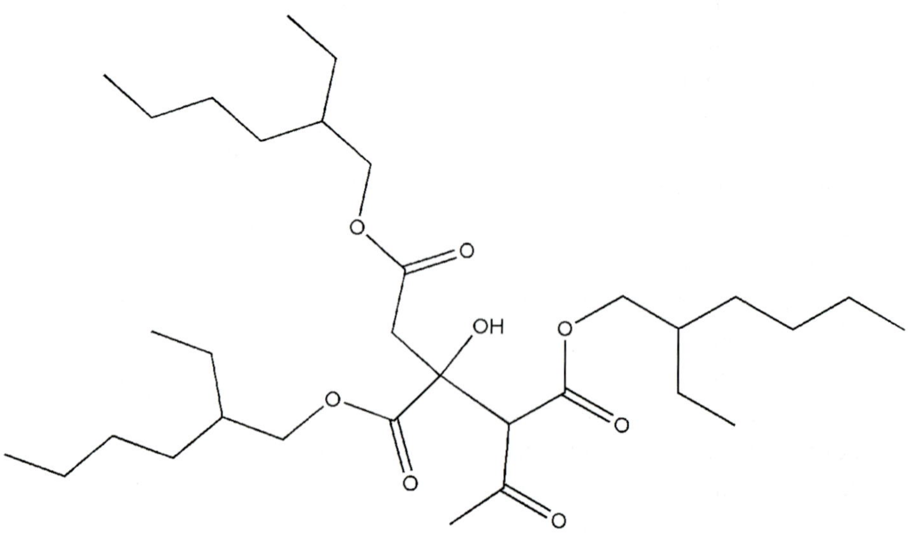

(usually green light) and chromatically for two wavelengths.

ACI *n.* Abbreviation for American Concrete Institute.

Acicular \ə'si-kyə-lər\ [L *acicula*] (1794) *adj.* Having a needle-like shape. It is a term applied chiefly to describe the shape of igment particles, which are long and slender. To be acicular, the length of the particle must be at least three times the width.

Acicular pigments *n.* Pigments whose particles are needle-shaped. Solomon DH, Hawthorne DG (1991) Chemistry of pigments and fillers. Krieger Publishing Co., New York.

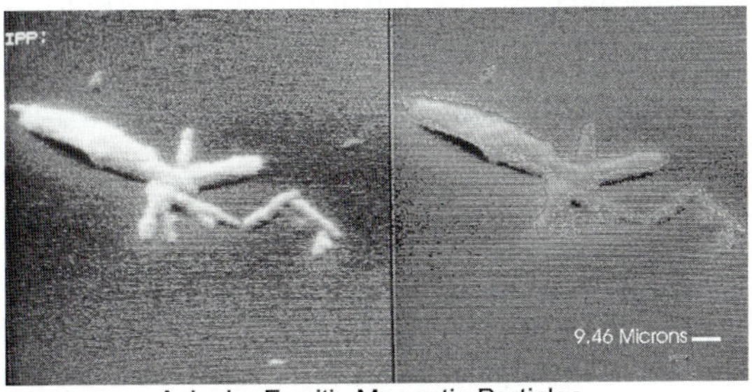

Acicular Ferritic Magnetic Particles

Acid \'a-səd\ [F *acide*, L *acidus* fr. *acēre*] (1626) *n.* (1) Inorganic compound characterized by an ionizable hydrogen atom. With organic acids, however, the definition must be extended to emphasize the ionizable hydrogen atom in question is directly attached, through an oxygen atom, to a carbon atom, which is also attached to another distinct oxygen atom. Thus, for example, acetic acid has the formula CH_3COOH, and it will be seen from this how the hydrogen atom is located. The inorganic acids, sometimes referred to as mineral acids, are written HCl, HNO_3, H_3PO_4, etc. All these hydrogen atoms, whether organic or inorganic, are capable of being substituted by a monovalent metallic radical, or by monovalent alkyl or aryl groupings. (2) Is a proton donor (Brønsted–Lowry). (3) Is an electron-pair acceptor (Lewis). (4) Increases the concentration of dissolved cations related to the solvent (solvent system) Goldberg DE (2003) Fundamentals of chemistry. McGraw-Hill Science/Engineering/Math, New York.

Acid acceptor *n.* A compound that acts as a stabilizer by chemically combining with acid that may be initially present in minute quantities in a plastic, or that may be formed during the decomposition in the resin. Odian GC (2004) Principles of polymerization. John Wiley and Sons Inc., New York.

See also stabilizer.

Acid catalysts *n.* Acids which may be either organic or inorganic, or salts from these acids which exhibit acidic characteristics. They are used to promote or accelerate chemical reactions, and find special applications in the manufacture and subsequent hardening of synthetic resins. Acid catalysts have been employed in the manufacture of polymerized drying oils, coumarone, urea, phenol- and melamine-formaldehyde resins, and in the cold-setting of compositions containing the last three named resins. Odian GC (2004) Principles of polymerization. John Wiley and Sons Inc., New York.

See catalyst and urea formaldehyde resins.

Acid curing (hardening) *n.* A process of curing or hardening resins through the use of acid catalysts. These are frequently employed with urea and melamine-formaldehyde resins. Odian GC (2004) Principles of polymerization. John Wiley and Sons Inc., New York.

Acid-Dyeable variants *n.* Polymers modified chemically to make them receptive to acid dyes. Vigo TL (1994) Textile processing, dyeing, finishing and performance. Elsevier Science, New York.

A

Acid dyes *n.* Term given to dyestuffs that possess acidic groupings, e.g., carboxy, sulfonic acid or both. They form salts with heavy metals like barium and calcium, and this reaction is used in lake formation. It is usually an azo, triarylmethane, or anthraquinone dye with acid substituents such as nitro, carboxy, or sulfonic acid. Acid dyes include such dyes as eosine, erioglaucine, fluorescein, naphthol yellow, ponceau, quinoline, tartrazine, etc. They precipitate with calcium, barium, or titanium glycerol chlorides or other metallic salts, to give suitable insoluble pigment dyestuffs, or they may be precipitated on bases to form lakes. Kadolph SJ, Langford AL (2001) Textiles. Pearson Education, New York. Wells K, Beal S, Woodburn C, Durant J, Brandimane J (1997) Fabric dyeing. Interweave Press Incorporated. Loveland, Colorado. Vigo TL (1994) Textile processing, dyeing, finishing and performance. Elsevier Science, New York.
See dyes.

Acid fading n. *See gas fading.*

Acid groups *n.* Functional groups having the properties of acids. In cellulose and its derivatives, these are usually carboxyl groups.

Acidic \ ə-ˈsi-dik, a-\ (1880) *adj.* A term describing a material having a pH of less than 7.0 in water. Whitten KW, Davis RE, Davis E, Peck LM, Stanley GG (2003) General chemistry. Brookes/Cole, New York.

Acidic solution *n.* An aqueous solution in which the concentration of hydrogen (hydronium) ions exceeds that of hydroxide ions. Whitten KW, Davis RE, Davis E, Peck LM, Stanley GG (2003) General chemistry. Brookes/Cole, New York.

Acidity \ə-ˈsi-də-tē, a-\ (1620) *n.* (1) Measure of the free acid present. (2) In oils, acidity denotes the presence of acid-type constituents whose concentrations are usually defined in terms of the neutralization number, called acid number. {G Säuregrad m, F taux d'acidité, taux m, S grado de acidez, grado m, I acidità f}. Dainth J (2004) Dictionary of chemistry. Oxford University Press, Oxford, UK. Whitten KW, Davis RE, Davis E, Peck LM, Stanley GG (2003) General chemistry. Brookes/Cole, New York.

Acid number or value *n* The number of milligrams of KOH required to neutralize the free acids in one gram of an oil, resin, varnish, or other substance; generally reported on the non-volatile.
See acid value.

Acidolysis *n.* A chemical reaction analogous to hydrolysis in which an acid plays a role similar to that of water.
See also ester interchange.

Acid recovery *n.* A reclamation process in chemical processing in which acid is extracted from a raw material, by-product, or waste product. In the manufacture of cellulose acetate, acetic acid is a major by-product. Acid recovery consists of combining all wash water containing appreciable acetic acid and concentrating it to obtain glacial acetic acid.

Acid refined linseed oil *n.* Linseed oil that has been treated with acid, usually sulfuric, to remove mucilaginous matter.

Acid resistance *n.* The ability of materials to withstand attack by acids, specifically strong mineral acids. The type of acid should be stated (i.e., organic or inorganic). Most plastics have excellent acid resistance. Tests for resistance of plastics to some acids are included in ASTM 543.

Acid salt *n.* A salt, the anion of which can serve as an acid by losing an H^+ (donating a proton). Goldberg DE (2003) Fundamentals of chemistry. McGraw-Hill Science/Engineering/Math, New York.

Acid sludge *n.* Residue which separates from mineral and related oils when they are refined with sulfuric acid.

Acid value *n.* The measure of the free-acid content of a substance. It is expressed as the number of milligrams of potassium hydroxide (KOH) neutralized by the free acid present in 1 g of the substance. This value, also called acid number, is sometimes used in connection with the end-group method of determining molecular weights of polyesters. It is also used in evaluating plasticizers, in which acid values should be as low as possible. Deligny P, Oldring PKT, Tuck N (2001) Resins for surface coatings, alkyds and polyester, vol 22. John Wiley and Sons, New York. Patton TC (1961) Alkyd resin technology. John Wiley and Sons Inc., New York. Martens CR (1961) Alkyd resins. Reinhold Publishing Co., New York. Wypych G (2003) Plasticizer's data base. Noyes Publication, New York.
See acid number.

Aclar *n.* Fluorinated polycarbonate film. Manufactured by Allied Chemical, USA.

ACM *n.* Copolymers from acrylic ester and 2-chlorovinyl ether.

Acoustical board \ə-ˈkü-stik or -sti-kəl-\ *n.* A low-density, sound-absorbing structural insulating board having a factory-applied finish and a fissured, felted-fiber, slotted or perforated surface pattern provided to reduce sound reflection. Harris CM (2005) Dictionary of architecture and construction. McGraw-Hill Co., New York.

Acoustical material *n.* Any material considered in terms of its acoustical properties. Commonly and especially, a material designed to absorb sound.

Acoustical paint *See anti-noise paint.*

Acoustical plaster *n.* A special low-density, sound-absorptive plaster, applied in the form of a finish-coat, to provide a continuous finished surface. Harris CM (2005) Dictionary of architecture and construction. McGraw-Hill Co., New York.

Acoustical tile *n.* An acoustical material in board form, usually having unit dimensions of 24 in. × 24 in. (approx. 61 cm × 61 cm) or less. Usually used on ceilings but also may be applied to sidewalls. Harris CM (2005) Dictionary of architecture and construction. McGraw-Hill Co., New York.

Acoustic coating *n.* Coating which absorbs or deadens sound. Harris CM (2005) Dictionary of architecture and construction. McGraw-Hill Co., New York.
See anti-noise paints.

Acoustic emission testing *n.* A non-destructive test for determining material or structural integrity by detecting and recording location, amplitude, and frequency of sound emissions as test loads are applied.

Acrilan \ˈa-krə-ˌlan, -lən\ *n.* Poly(acrylonitrile). Manufactured by Chemstrand Corp., Monsanto, USA. Tortora PG (2000) Fairchild's dictionary of textiles, 7th edn. Fairchild Publications, New York.

Acrolein \ə-ˈkrō-lē-ən\ [ISV *acr-* (fr. L *acr-*, *acer*) + L *olēre*] (1857) (propenal, acrylic or allyl aldehyde) *n.* $CH_2{=}CHCHO$. A liquid derived from the oxidation of allyl alcohol or propylene, used as an intermediate in the production of polyester resins and polyurethanes. It is an unsaturated liquid aldehyde with a bp of 52°C. It possesses a very pungent odor, and has strong lachrymatory properties.

Acrolein polymers and resins *n.* Homopolymers or copolymers of acrolein.

Acronal Dispersions based on uni- and copolymers of acrylic esters. Manufactured by BASF, Germany.

A

Acrylamide \ˌa-kr ᵊl- ˈa-ˌmīd\ [*acrylic* + *amide*] (1946) *n.* $CH_2{=}CHCONH_2$. A crystalline solid produced by hydrolysis of acrylonitrile; the monomer of polyacrylamide and a useful comonomer. Odian GC (2004) Principles of polymerization. John Wiley and Sons Inc., New York.

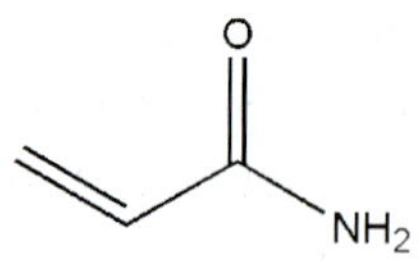

2-Acrylamido-2-methylpropanesulfonic acid (AMPS) *n.* A solid aliphatic sulfonic-acid monomer produced by Lubrizol Corp. Its homopolymers are water-soluble and hydrolytically stable. It can be incorporated into other polymers by crosslinking.

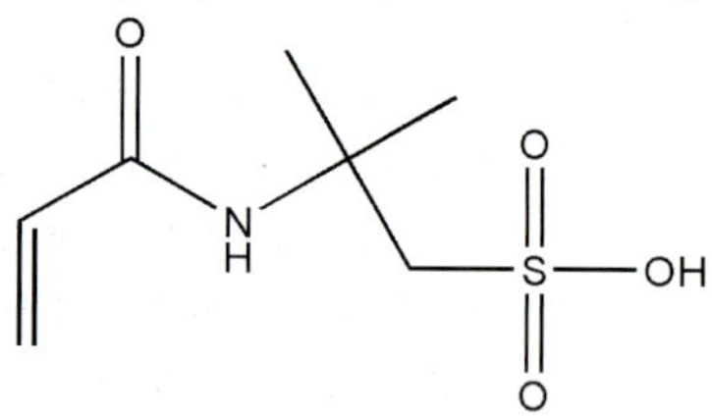

Acrylan rubber *n.* Butyl acrylate/5–10% acrylonitrile copolymer. Manufactured by Monomer Corp., USA.

Acrylate \ˈa-krə-ˌlāt\ (1873) *n.* Ester formed from acrylic acid. The term also applies to the metallic salts of this acid. Odian GC (2004) Principles of polymerization. John Wiley and Sons Inc., New York.

Acrylate elastomers *n.* Elastomeric formed from acrylate and elastomeric monomers such as butadiene. Harper CA (2002) Handbook of plastics, elastomers and composites, 4th edn. McGraw-Hill, New York.

Acrylate resin *See acrylic resin.*

Acrylic \ ə-ˈkri-lik\ [ISV *acro*kein + *-yl* + ¹*-ic*] (1855) *adj.* (1) The generic class of polymers and monomers (appox. 1942) derived from acrylic acid including polymethyl methacrylate. (2) Generic name for fibers from at least 85% poly(acrylonitrile). {D Acrylglas n, F verre acrylique, verre m, S vidrio crílico, vidrio m, I vetro acrilico, vetro m}. Odian GC (2004) Principles of polymerization. John Wiley and Sons Inc., New York. Morrison RT, Boyd, RN (1992) Organic chemistry, 6th edn. Prentice-Hall, Englewood Cliffs, NJ.

Acrylic acid (ca. 1855) (propenoic acid, vinylformic acid) *n.* $CH_2{=}CHCOOH$. A colorless, unsaturated acid, that polymerizes readily. The homopolymer is used as a thickener and textile-sizing agent and, cross-linked, as a cation-exchange resin. Acrylic-acid esters are widely used as monomers for acrylic resins. Mol wt., 72.06; mp, 14°C; bp, 141.0°C; Sp gr, 1.422.

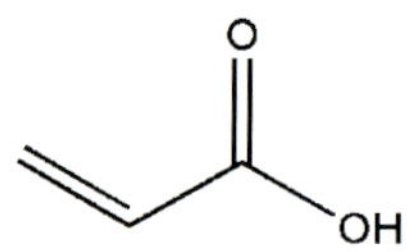

Odian GC (2004) Principles of polymerization. John Wiley and Sons Inc., New York.

Acrylic aldehyde *See acrolein.*

Acrylic coating polymers *n.* Acrylic resins or polymers such as polymethyl methacrylate used for formulating coating.

Acrylic ester (acryl ester) *n.* An ester of acrylic or methacrylic acid or of structural derivatives thereof. Polymers derived from these monomers range from soft, elastic, film-forming materials to hard plastics. They are readily polymerized as homopolymers or copolymers with many other monomers, contributing improved resistance to heat, light, and weathering. Some members of the acrylic-ester family (e.g., butylenes dimethacrylate and trimethylolpropane trimethacrylate) function as reactive plasticizers in PVC and elastomers. They serve as plasticizers during processing, then polymerize while curing to impart hardness to the finished article. Odian

GC (2004) Principles of polymerization. John Wiley and Sons Inc., New York. *See also acrylic resin.*

Acrylic esters *n.* Elastomeric polymers formed from acrylate and elastomeric monomers such as butadiene. Harper CA (2002) Handbook of plastics, elastomers and composites, 4th edn. McGraw-Hill, New York.

Acrylic fiber (1951) *n.* A manufactured fiber in which the fiber-forming substance is any long chain synthetic polymer composed of at least 85% by weight of acrylonitrile units [$-CH_2-CH(CN)-$] (FTC definition). Acrylic fibers are produced by two basic methods of spinning (extrusion), dry and wet. In the dry spinning method, material to be spun is dissolved is a solvent. After extrusion through the spinneret, the solvent is evaporated, producing continuous filaments, which later may be cut into staple, if desired. In wet spinning, the spinning solution is extruded into a liquid coagulating bath to form filaments, which are drawn, dried, and processed. Uses of acrylic fibers include floor coverings, blankets, and apparel uses such as suitings, pile fabrics, coats, collars, linings, dresses, and shirts. Vincenti R (1994) Elsevier's textile dictionary. Elsevier Science and Technology Books, New York. Kadolph SJ, Langford AL (2001) Textiles. Pearson Education, New York. Complete textile glossary. Celanese Corporation, Three Park Avenue, NY. Tortora PG, Merkel RS (2000) Fairchild's dictionary of textiles, 7th edn. Fairchild Publications, New York.

Acrylic foam *n.* A cellular polymer used for lining drapes and made by mixing an emulsified acrylic resin with compressed air in the ration of one part emulsion to four or five parts air, spreading the foam on a substrate, then drying in an oven. The emulsion may contain fillers and pigments to provide opacity, and a foaming aid such as ammonium stearate. When the coated fabric must have abrasion resistance for washing and cleaning, the acrylic foam can be crushed between rollers to partly collapse the cell structure. James F (ed) (1993) Whittington's dictionary of plastics. Technomic Publishing Co. Inc., Carley.

Acrylic latex *n.* Aqueous dispersion, thermoplastic or thermosetting, of polymers or copolymers of acrylic acid, methacrylic acid, esters of these acids, or acrylonitrile.

Acrylic plastics *n.* Thermoplastic or thermosetting plastics of polymers including copolymers of acrylic acid, methacrylic acid, esters of these acids or acrylonitrile.

Acrylic resin (1936) *n.* A polymer composed of acrylic or methacrylic esters, sometimes modified with non-acrylic monomers such as the ABS group. The acrylates may be methyl, ethyl, butyl, or 2-ethylhexyl. Usual methacrylates are the methyl, ethyl, butyl, lauryl, and stearyl. The resins may be in the form of molding powders or casting syrups, and are noted for their exceptional clarity and optical properties. Acrylics are widely used in lighting fixtures because they are slow-burning or even, with additives, self-extinguishing, and do not produce harmful smoke or gases in the presence of flame. Deligny P, Oldring PKT, Tuck N (2001) Resins for surface coatings, alkyds, and polyester, vol 22. John Wiley and Sons, New York. Paint: pigment, drying oils, polymers, resins, naval stores, cellulosics esters, and ink vehicles, vol 3. American Society for Testing and Material, Conshohocken, PA, 2001.

Acrylic rubber (AR) *n.* A synthetic rubber made at least partly from acrylonitrile, or

A

from ethyl acrylate copolymerized with many of the monomers or block polymers of the synthetic-rubber family. Harper CA (ed) (2002) Handbook of plastics, elastomers and composites, 4th edn. McGraw-Hill, New York.

Acrylide maroon *n.* This group includes the azo pigments based on acrylides of beta hydroxy naphthoic acid (e.g., toluidine maroon); they are characterized by their excellent soap, acid, and alkali resistance and good bake resistance. Poor bleed resistance in aromatic and alcohol solvents, poor light fastness in other than masstone shades (including metallics), low hiding power and high cost discourage their use except where chemical resistance requirements demand; to this extent they may be considered as specialty pigments. Herbst W, Hunger K (2004) Industrial organic pigments. John Wiley and Sons Inc., New York.

Acrylonitrile \ˌa-krə-lō-ˈnī-trəl, -ˌtrēl\ (1893) (propenenitrile, vinyl cyanide) *n.* (1) A monomer with the structure $CH_2{=}CH{-}CN$. It is most useful in copolymers. Its copolymer with butadiene is nitrile rubber, and several copolymers with styrene exist that are tougher than polystyrene. It is also used as a synthetic fiber and as a chemical intermediate. (2) A raw material for the manufacture of synthetic resins and rubbers. It is a liquid at room temperatures, with a bp of 77°C and flp of 0°C. Kadolph SJ, Langford AL (2001) Textiles. Pearson Education, New York. Odian G (2004) Principles of polymerization. Joihn Wiley and Sons, New York.

Also known as vinyl cyanide.

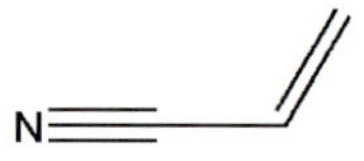

Acrylonitrile–butadiene copolymer (NBR) *n.* Any of a family of copolymers ranging from about 18 to 50% acrylonitrile, and sometimes including small amounts of a third monomer. The family includes the German materials perbunan and buna-N, and the nitrile rubbers. The outstanding property of this nitrile-rubber family is excellent resistance to oils, fats, and hydrocarbons such as motor fuels, making them useful for motor gaskets, abrasion linings, conveyor belts, and hoses for oils and fuels. Odain G (2004) Principles of polymerization. Joihn Wiley and Sons New York. Lenz RW (1967) Organic chemistry of synthetic high polymers. Interscience Publishers Inc., New York. James F (1993) Whittington's dictionary of plastics. Technomic Publishing Co. Inc., Carley.

Acrylonitrile-butadiene-styrene (ABS) *n.* Acrylonitrile and styrene liquids and butadiene gas polymerized together in a variety of ratios to produce the family of ABS resins.

Acrylonitrile-butadiene-styrene copolymers *n.* Terpolymer of three monomers, forming ABS.

Acrylonitrile-butadiene-styrene resin *n. See ABS resin and nitrile barrier resin.*

Acrylonitrile-chlorinated PE-styrene *n.* A terpolymer of three monomers by the same names.

Acrylonitrile-styrene copolymer *n.* Any of a group of copolymers that have the transparency of polystyrene, but with improved resistance to solvents and stress cracking.

ACS *n.* Abbreviation for the American Chemical Society, headquartered at 1155 16th St, NW, Washington, DC 20036. The Society's polymer chemistry division holds national meetings and publishes several journals.

A

Actinic degradation *n. See ultraviolet degradation.*

Actinic resistance *n. See ultraviolet resistance.*

Actinide series *n.* Elements of atomic numbers 89–103 analogous to the lanthanide series of the so-called rare earths.

Actinoid *n.* A member of the series of 14 elements following actinium in the periodic table. It is also called actinide.

Action \ˈak-shən\ *n.* Action is measured by the product of work by time. Cgs units of action are the erg-second and the joule-second. Dimensions (ML^2T^{-1}). Plank's quantum or constant of action is $6.62006876 \times 10^{-34}$ J.s. Giambattista A, Richardson R, Richardson RC, Richardson B (2003) College physics. McGraw Hill Science/Engineering/Math, New York.

Action stretch *n.* A term applied to fabrics and garments that give and recover in both the lengthwise and the widthwise directions. Action stretch is ideal for tight-fitting garments such as ski pants. Vincenti R (ed) (1994) Elsevier's textile dictionary. Elsevier Science and Technology Books, New York.

Activate \ˈak-tə-ˌvāt\ (1626) *v.* To put into a state of increased chemical activity.

Activated *adj.* Materials, specially treated to confer absorptive, adsorptive, or catalytic properties on them. Such substances include activated alumina, activated earths, and activated carbon.

Activated carbon (1921) *n.* (1) Any form of carbon characterized by high adsorptive capacity of gases, vapors, and colloidal solids. (2) A highly adsorbent powdered or granular carbon made usually by carbonization and chemical activation and used chiefly for purifying by adsorption. Whitten KW, Davis RE, Davis E, Peck LM, Stanley GG (2003) General chemistry. Brookes/Cole, New York.
Also known as activated charcoal.

Activated complex *n.* A short-lived combination formed by collision of reactant particles in an elementary process; also called transition state.

Activation (1626) *n.* The process of making more active: to make (as moleclules) reactive or more reactive; inducing radioactivity in a specimen by bombardment with neutron or other types of radiation; rendering a thermoplastic surface more receptive to printing inks, paints, and adhesives by chemical treatment such as carbon and alumina; corona discharge or flame treatment; energetic elevation of a molecule to a state in which it becomes ready to react with another molecule. {G Aktivierung f, F activation f, S activación f, I attivazione f}. Merriam-Webster's collegiate dictionary, 11th edn. Merriam-Webster Inc., Springfield, MA, USA, 2004. James F (ed) (1993) Whittington's dictionary of plastics. Technomic Publishing Co. Inc., Carley.

Activation energy (1940) (E, E_A) *n.* The energy required to facilitate reaction between two molecules or, in the context of the Eyring theory of low, the energy required to cause a molecule of liquid or chain segment of a polymer to "jump" from its present position to a nearby hold (i.e., an empty volume of molecular or chain-segment size) in the liquid. Activation energies are usually expressed per mole of substance (SI: J/mol) and are evaluated by fitting reaction-rate or flow data at several temperatures to an equation of the Arrhenius form, $k = Ae^{-Ea/RT}$. Phillip R Watson (1997) Physical chemistry. John Wiley and Sons Inc., New York.
See Arrhenius equation.

A

Activator (1626) *n.* (1) An agent added to the accelerator in natural or synthetic resins to enhance the action of the accelerator in the vulcanizing process. (2) A chemical additive used to initiate the chemical reaction in a specific mixture.
See accelerator and catalyst.

Activator, initiator *n.* It is usually necessary to add an activator (e.g., $FeSO_4 \cdot 7H_2$) to attain reasonable rates of polymerization when redox initiators (lauroÿl peroxide and fructose) are used at zero degree temperature or below.

Active mass *n.* The active mass of a substance is the number of gram molecular weights per liter in solution, or in gaseous form. Odian GC (2004) Principles of polymerization. John Wiley and Sons Inc. New York.

Active metal *n.* Metal in a condition of high chemical activity that is more susceptible to corrosion.

Active site *n.* A location on the surface of a heterogeneous catalyst or an enzyme at which reactant molecules can combine and react with low required activation energy. Smith MB, March J (2001) Advanced organic chemistry, 5th edn. John Wiley and Sons, New York. Odian GC (2004) Principles of polymerization. John Wiley and Sons Inc., New York.

Activity \ak-ˈti-və-tē\ (1530) *n.* A quantity which measures the parent or effective concentration (or, for a gas, partial pressure) of a species and which takes into account interparticle interactions which produce non-ideal behavior. At low concentrations (or pressures) activity is essentially equal to concentration (or pressure).

Activity coefficient *n.* A factor which, when multiplied by the molecular concentration yields the active mass. The activity coefficient is evaluated by thermodynamic calculations, usually from data on the emf of certain cells, or the lowering of the freezing point of certain solutions. It is a correction factor, which makes the thermodynamic calculations correct. Phillip R Watson (1997) Physical chemistry. John Wiley and Sons Inc., New York.

Acute bisectrix (Bx_a) The bisector of the acute optic axial angle for biaxial crystals. Rhodes G (1999) Crystallography made crystal clear: a guide for users of macromolecular models. Elsevier Science and Technology Books, New York.

ACV *n.* Abbreviation for Adams chromatic value.

Acyclic \(ˌ)ā-ˈsī-klik, -ˈsi-\ (1878) *n.* Open chain, not ring formation.

Acyl, acyl groups *n.* When the OH group is removed from a fatty acid molecule, the monovalent residue is described as an acyl radical. Examples of acyl radicals are CH_3CO–acetyl; C_2H_5CO=propionyl; and C_3H_7CO–butyrl. Radicals derived from carboxylic acids by removal of the hydroxyl group. Morrison RT, Boyd RN (1992) Organic chemistry, 6th edn. Prentice-Hall, Englewood Cliffs, NJ.

Acylation *vt* (1907). Formation or introduction of an acyl radical in or into an organic compound. An acyl group (RCO–) becomes attached to the aromatic ring, thus forming a ketone; the process is called acylation; includes Friedel-Craft acylation. Morrison RT, Boyd RN (1992) Organic chemistry, 6th edn. Prentice-Hall, Englewood Cliffs, NJ. Billmeyer FW, Saltzman M (1966) Principles of color technology. John Wiley and Sons Inc., New York.

Adams chromatic value color difference equation *n.* One of the transformations of CIE color space into another space which is perceived as visually more uniform. It is used to describe the difference between two similar colors: it is based on the Munsell value function.

$$Y_c/100 = 1.2219V_Y - 0.23111V_Y^2 +0.23951V_Y^3 - 0.021009V_Y^4+0.000840V_Y^5.$$

For determining the V_x, substitute $X_c/0.9804$ for Y_c, for determining V_z, substitute $Z_c/1.1810$ for Y_c in the above equation. The transformed space is then calculated according to the following equations:

$$a = V_X - V_Y,$$
$$b = 0.4\ (V_Z - V_Y) \text{ or } 0.4\ (V_Y - V_Z),$$
$$L = 0.23\ V_Y.$$

The total color difference, ΔE, is calculated as follows

$$\Delta E = \sqrt{(0.23\Delta V_Y + [\Delta(V_x - V_Y)]^2 + [0.4\Delta(V_z - V_Y)]^2}.$$

Note that if b is calculated as 0.4 $(V_z - V_Y)$, a plus number indicates blueness; if the b is calculated as 0.4 $(V_Y - V_z)$, a plus number indicates yellowness. The equation is generally used with a normalizing constant placed outside the brackets. If the components L, a, and b are used individually they are also multiplied by the normalizing constant, f. The normalizing constant or factor, f, is used to convert the numbers obtained to the same magnitude as those obtained with the NBD equation. The factor used is generally 40, although it may vary between 40 and 50 and must therefore be specified. Unless specified otherwise, the value 40 is assumed to have been used. Billmeyer FW, Saltzman M (1966) Principles of color technology. John Wiley and Sons Inc., New York. McDonald R (1997) Colour physics for industry, 2nd edn. Society of Dyers and Colourists, West Yorkshire, England.

Adaptation \ˌa-ˌdap-ˈtā-shən, –dəp-\ (1610) *n.* Changes in the sensitivity of the eye resulting from changes in the viewing light sources, thus enabling the eye to meet the needs of vision in a very wide range of conditions.

Adapter *n* (1801) (die adapter). In an extrusion setup, the portion of the die assembly that attaches the die to the extruder and provides, inside a flow channel for the molten plastic between the extruder and the die.

Adapter plate *n.* In injection molding, the plate holding the mold to the press frame or platen.

Adapter ring An annular retaining part of extrusion and injection apparatus.

ADC *See allyl diglycol carbonate.*

Addition polymer *n.* Polymer made by addition polymerization (IUPAC).

Addition polymerization *n.* A reaction in which unsaturated monomer molecules join together to provide a polymer in which the molecular formula of the repeating unit is identical with that of the monomer. The molecular weight (i.e., M_w, M_n) of the polymer so formed is thus the total of the molecular weights of all of the combined monomer units. Example: $n\mathrm{CH_2} = \mathrm{CH_2} \rightarrow (-\mathrm{CH_2CH_2})_n$, with molecular weight $= n \times 28.03$, and where $n = 100$, then 2803g/mol of polymer. Polymers formed from monomers without the lose of a small molecule unlike condensation polymerization, the composition of polymer is the same as the monomer, e.g., polyethylene. Odian GC (2004) Principles of polymerization. John Wiley and Sons Inc., New York. Lenz RW (1967) Organic chemistry of synthetic high polymers. Interscience, Publishers Inc., New York.

A

Additive \ˈa-də-tiv\j (1699) *adj.* A supplementary material combined with a base material to provide special properties. For example, pigments are used as dope additives to give color in mass dyeing. Examples are slip additives, pigments, stabilizers, and flame retardants. Also sometimes called Modifier {G Additiv n, F additif m, S aditivo m, I additivo m}. Bart J (2005) Additives in polymers: industrial analysis and applications. John Wiley and Sons, New York.

Additive color mixture *n.* Color which results when the same area of the retina of the eye is illuminated by lights of different spectral distribution, such as by two or more colored lights. Pigments such as paints obey substractive color mixtures whereas light (sources of illumination) obeys additive color mixture may result from addition of lights from two or more sources, by visual averaging of s colored dots (dot matrix) as on colored television screens or paper, or by additive color mixture rules. For example, yellow paint mixed with blue paint yeidls green paint, and yellow light mixed with blue light yields white light. McDonald Roderick (1997) Colour physics for industry, 2nd edn. Society of Deyes and Colourists, West Yorkshire, England. Billmeyer FW, Saltzman M (1966) Principles of color technology. John Wiley and Sons Inc., New York.

Additive reaction *n.* Chemical reaction in which two components join together to form a single reaction product. In a pure additive reaction, neither of the reactants undergoes molecular fission or splitting, but attaches itself to the other reactant intact. In other additive reactions, one of the reactants may split into two separate parts, each of which attaches itself to the the appropriate places of the other intact reactant. There is still, however, a single reaction product. Odian GC (2004) Principles of polymerization. John Wiley and Sons Inc., New York.

Adduct ə-ˈdəkt, a-\ [L *adductus,* pp of *adducere*] (ca. 1839) *vt.* (1) The cyclic product of an addition reaction between one unsaturated compound, such as a diene and another. (2) A crystalline mixture, not a true compound, in which molecules of one of the components are contained within the crystal-lattice framework of the other component. Such complexes are stable at room temperature but the entrapped component can escape when the mixture is melted or dissolved. Odian GC (2004) Principles of polymerization. John Wiley and Sons Inc., New York.

Adduct curing agent *See cross-linking agent.*

Adeps lanae, anhydrous *n.* Pharmaceutical name for lanolin or purified wood grease.

Adhere \ad-ˈhir, əd\ [MF *adhérer,* L *adhaerēre* fr. *ad-* + *haerēre*] (1536) *v.* To cause two surfaces to be held together by adhesion.

Adherend *n.* A body which is held to another body by an adhesive.
See also substrate.

Adherend preparation n. *See surface preparation.*

Adherometer An instrument that measures the force required to strip a coating from a metal surface. Paint/Coatings Dictionary. Federation of Societies for Coatings Technology, Philadelphia, 1978.

Adhesion \əd-ˈhē-zhən, əd-\ [F or L; F *adhésion,* fr. L *adhaesion-*, *adhaesio,* fr. *adhaerēre*] (1624) *n.* The state in which two surfaces are held together by interfacial forces, which may consist of valence forces or interlocking action, or both. One method for testing the strength of adhesive bonds. Skeist I (ed) (1990) Handbook of adhesives. Reinhold Publishing Co., New York.

See also adhesion, mechanical, and adhesion, specific {G Adhäsion f, F adhésion f, S adhesión f, I adesione f}.

Adhesion, mechanical *n.* Adhesion between surfaces in which the adhesive holds the parts together by interlocking action such as "contact cement."
See adhesion, specific.

Adhesion promoter *n.* A chemical coating that is applied to a substrate before it is coated with a plastic, to improve the adhesion of the plastic to the substrate. Adhesion promoters include materials such as silanes and silicones with hydrolysable groups on one end of their molecules that react with moisture to yield silanol groups, which in turn react with or adsorb to inorganic surfaces to enable strong bonds to be made. At the other ends of the molecules are reactive, but nonhydrolyzable groups that are compatible with resins or elastomers in adhesive formulations. Adhesion promoters are added to the adhesive as water or ethanol solutions. Skeist I (ed) (1990) Handbook of adhesives. Van Nostrand Reinhold, New York.

Adhesion, specific *n.* Adhesion between surfaces that are held together by valence forces of the same type as those that give rise to cohesion.
See adhesion, mechanical.

Adhesive (1670). *n, adj, adv. n.* Adhesives are used in to bond two or more surfaces. *Adj.* An adhesive material binds two or more surfaces. *Adv.* Two or surfaces or bonded adhesively. The above usages of the different forms of the term were derived from The following information was derived from. Adhesives used in all of these applications can be classified into five types. A monomeric cement contains a monomer of at least one of the polymers to be joined and is catalyzed so that a bond is produced by polymerization. A solvent cement is a product that dissolves the plastics being joined, forming strong intermolecular bonds, then evaporates. *Bonded adhesives* are solvent solutions of resins, sometimes containing plasticizers that dry at room temperature. *Elastomeric adhesives* contain natural or synthetic rubbers either dissolved in solvents or suspended in water or other liquid, and are dried at room or elevated temperatures. *Reactive adhesives* are those containing partly polymerized resins, e.g., epoxies, polyesters, or phenolics that cure with the aid of catalytic hardeners to form a bond (usually thermoset). Further, various descriptive adjectives are applied to the term, adhesive, to indicate certain characteristics as follows: *Physical form*—liquid adhesive, tape adhesive; *Chemical type*—silicate adhesive, resin adhesive; *Material bonded*—paper adhesive, metal-plastic adhesive, can-label adhesive; and *Conditions of use*—hot-setting adhesive. In addition, contact-adhesives, moisture cure adhesives (cyanoacrylate and silicone) are among the many types of specialty adhesives in the ever-growing number of products from the developing adhesives industry. Skeist I (ed) (1990) Handbook of adhesives. Van Nostrand Reinhold, New York. Skeist I (ed) (1990) Handbook of adhesives. Van Nostrand Reinhold, New York.

Adhesive activated yarns *n.* Yarns treated by the fiber manufacturer to promote better adhesion to another material such as rubber and/or allowing easier processing.

Adhesive assembly *n.* An adhesive that can be used for bonding parts, such as in the manufacture of a boat, airplane, furniture, etc. *Note*—The term, "assembly adhesive," is commonly used in the wood industry to distinguish such adhesives (formerly called "joint glues"). It is applied to adhesives used in fabricating finished structures or goods, or subassemblies thereof, as differentiated from adhesives used in the production of sheet materials for sale as such, for example, plywood, or laminates.

Adhesive cellular *See adhesive, foamed.*

Adhesive, cold-setting *n.* An adhesive that sets at temperatures below 20°C (68°F). *See also adhesive, hot-setting; adhesive, inter-mediate temperature setting; and adhesive, room temperature setting.*

Adhesive, contact *n.* An adhesive that is apparently dry to the touch and which will adhere to itself instantaneously upon contact; also called contact bond adhesive or dry bond adhesive.

Adhesive dispersion *n.* A two-phase system in which one phase is suspended in a liquid.

Adhesive, edge jointing *n.* Adhesive used to bond strips of veneer together by their edges in the formation of larger sheets.

Adhesive film *n.* A thin film of dry resin, usually a thermoset, used as an interleaf in the production of laminates such as plywood. Heat and pressure applied in the laminating process cause the film to bond the layers together.

Adhesive, foamed *n.* An adhesive, the apparent density of which has been decreased substantially by the presence of numerous gaseous cells dispersed throughout its mass.

Adhesive, heat activated *n.* A dry adhesive film that is rendered tacky or fluid by application of heat or heat and pressure to the assembly.

Adhesive, hot melt *n.* An adhesive that is applied in a molten state and forms a bond on cooling to a solid state.

Adhesive, hot-setting *n.* An adhesive that requires a temperature at or above 100°C (212°F) to set the adhesive.

Adhesive, intermediate temperature *n.* An adhesive that sets in the temperature range 31–99°C (87–211°F).

Adhesive migration *n.* In non-wovens, the movement of adhesive together with its carrier solvent in a fabric during drying, giving it a non-uniform distribution within the web, usually increasing to the outer layers.

Adhesive, multiple layer *n.* A film adhesive, usually supported, with a different adhesive composition on each side; designed to bond dissimilar materials such as the core to face bond of a sandwich composite.

Adhesive, pressure-sensitive *n.* A viscoelastic material that in solvent-free form remains permanently tacky. The material will adhere instantaneously to most solid surfaces and can be removalable (e.g., adhesive labels for garments).

Adhesive, room temperature *n.* An adhesive that sets in the temperature ranges of 20–30°C.

Adhesives *n.* In textiles, materials which cause fibers, yarns, or fabrics to stick together or to other materials.

Adhesive, separate application *n.* A term used to describe an adhesive consisting of two parts, one part being applied to one adherend and the other part to the other adherend and the two brought together to form a joint.

Adhesive, solvent *n.* An adhesive having a volatile organic liquid as a vehicle. *Note—*This term excludes water-based adhesives.

Adhesive, solvent activated *n.* A dry adhesive film that is rendered tacky just prior to use by application of a solvent.

Adhesive tape test *See tape test.*

Adhesive, warm-setting *n.* A term that is sometimes used as a Syn: intermediate temperature setting adhesive. *See adhesive, intermediate temperature setting.*

Adiabatic \ˌa-dē-ə-ˈba-tik, ˌā-ˌdī-ə-\ [Gk *adiabatos* impassable, fr. *a-* + *diabatos* passable, fr. *diabainein* to go across, fr. *dia-* + *bainein* to go] (1890) *adj.* Denoting a process or system in heat is not added or removed, but while work may be delivered to or by the system. Merriam-Webster's collegiate dictionary, 11th edn. Merriam-Webster Inc., Springfield, MA, USA, 2004. Ready RG (1996) Thermodynamics. Pleum Publishing Company, New York.

Adiabatic change *n.* A change which takes place with no gain or loss of heat. Ready RG (1996) Thermodynamics. Pleum Publishing Company, New York.

Adiabatic extrusion *See autothermal extrusion.*

Adipate plasticizer *n.* For plasticizers derived from adipic acid.

See

Benzyloctyl–diisobutyl–	Di-n-hexyl–
Bis(2,2,4-trimethyl-1,3-pentanediol)	Diisodecyl–
Monoisobutyrate–	Diisooctyl–
Dibutoxyethoxy ethyl–	Dimethoxyethyl–
Dibuto-xyethyl–	Di(methyl-cylo-Hexyl) –
Dibutyl–	
Dicapryl–	Dinonyl–
Diethoxyethyl–	Ditetrahydro-Furfuryl–
Dienthyl–	
Di(2-ethylhexyl)–	n-octyl-n-decyl–

All of the above are polypropylene adipates. Wypych G (2002) Plasticizer's data base. Noyes Publication, New York. James F (ed) (1993) Whittington's dictionary of plastics. Technomic Publishing Co. Inc., Carley.

Adipates *n.* Esters of adipic acid.

Adipic acid [ISV] (1877). (hexanedioic acid, 1.4-butanedicarboxylic acid) *n.* A dicarboxylic acid used in the production of polyamides, alkyd resins, and urethane foams. Esters of adipic acid are used as plasticizers and lubricants. It is used in the polymerization reaction to form nylon 66 polymers and in the manufacture of polyurethane foams acid value, 767, molecular weight, 146.1; mp, 151°C; and bp, 216°C/15 mmHg.

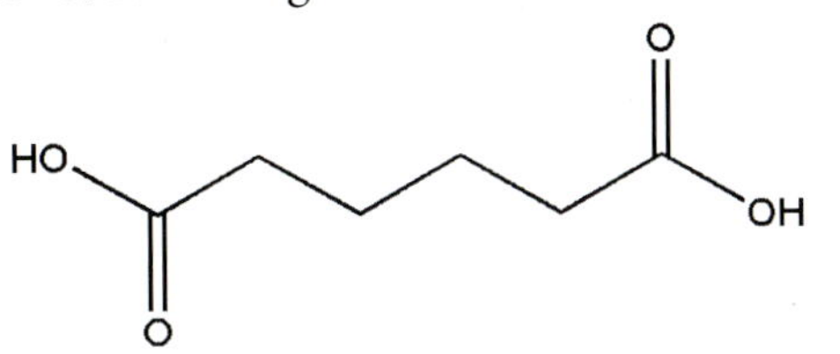

Adiponitrile carbonate (ADNC) *n. See* 5,5'-*tetramethylene di-*(1,3,4-*dioxazol–2–one*).

Adiprene *n.* Polyurethane elastomer. Manufactured by DuPont, USA.

ADNC Abbreviation for adiponitrile carbonate.

Adobe brick *n.* Large, roughly molded, sun dried clay brick of varying sizes.

Adronal *See cyclohexyl acetate.*

Adsorbed water *n.* Water which is held on the surface of a material by secondary bonding forces (e.g., hydrogen bonding), and its physical properties are substantially different from those of absorbed water or chemically combined water at the same temperature and pressure. This type of water is often called a hygroscopic film. Complete textile glossary. Celanese Acetate LLC, Three Park Avenue, New York, 2000. Goldberg DE (2003) Fundamentals of chemistry. McGraw-Hill Science/Engineering/Math, New York.

Adsorbent \- bənt\ *n* (1917). Substance offering a suitable active surface, upon which other substances may be adsorbed.

Adsorption \ad-ˈsórp shən, -ˈzórp\ [*ad* + *absorption*] (1882) *n.* (1) The concentration of molecules of a particular kind at the interface between two phases such as the pigment and vehicle in printing inks.

A

Adsorption can effectively remove a component such as the drier from an ink vehicle. (2) The attraction of gases, liquids, or solids to surface areas of textile fibers, yarns, fabrics, or any material. Leach RH, Pierce RJ, Hickman EP, Mackenzie MJ, Smith HG (1993) Printing ink manual, 5th edn. Blueprint, New York. Tortora PG (ed) (1997) Fairchild's dictionary of textiles. Fairchild Books, New York.

Adulteration \ə-ˌdəl-tə-ˈrā-shən\ (1506) *n.* Presence of inferior materials, which reduce the quality of the standard.

Advanced composite *n.* Polymer, resin, or other matrix-material system in which reinforcement is accomplished via high-strength, high-modulus materials in continuous filament form or is discontinuous form such as staple fibers, filberts, and in-situ dispersions. Harper CA (2002) Handbook of plastics, elastomers, and composites, 4th edn. McGraw-Hill, New York.

Advanced fiber *n.* Any reinforcing fibers characterized by either very high strength, modulus or high operating temperature, beyond those of more familiar fibers such as glass, nylon, and polyester. Examples are aramid, fibers of some metals, carbon, boron, silicon carbide and silicon nitride, and whiskers of metals and inorganics. Chung DD (1994) Carbon fiber compositers. Elsevier Science and Technology Books, New York. Pittance JC (ed) (1990) Engineering plastics and composites. SAM International, Materials Park, OH.

Advanced resin *n.* Any of a new multiclass of thermoplastics of carious chemical natures, distinguished from the established and more common plastic materials by higher strength, modulus, or serviceability at 200+ °C. These materials also command premium prices and some require special processing. Examples are polyimides, polyetheretherketone, liquid-crystal polymers, polytetrafluoroethylene, and polybenzimidazole. Odian GC (2004) Principles of polymerization. John Wiley and Sons Inc., New York. Modern plastics encyclopedia. McGraw-Hill/Modern Plastics, New York.

Advancing colors *n.* Colors that give the illusion of being closer to the eye than their complementary colors when both are adjacent and in the same plane; colors ranging from yellow-green to scarlet are examples.

Aeration cell *n.* Electrolytic cell in which the driving force to cause corrosion results from a difference in the amount of oxygen in solution at one point as compared to another. Corrosion is accelerated where the oxygen concentration is least or most.

Aerobic \ˌa(-ə)r-ˈō-bik, ˌe(-ə)r-\ (1884) *adj.* Term referring to processes that can occur only in the presence of oxygen.

Aerogel *n.* A porous, foam-like network with very small "cells," whose substance may be silica, a polymer or a carbonized polymer. Densities can range from half of solid values to as low as 3 mg/cm^3.

Aerograph *n.* A spray gun for coatings, using air.

Aerosol \ˈar-ə-ˌsäl, ˈer-, -ˌsȯl\ (1923) *n.* A suspension of liquid or solid particles in a gas (e.g., smoke, fog mist). In the packaging industry, the term means a self-contained sprayable product in which the propellant force is supplied by a compressed or liquidified gas (e.g., isopentane) for an "aerosol spray paint."

Aerosol coating *n.* A conveniently packaged spray coating in a sealed can. Pressure is supplied by compressed liquefied gas.

Aerugo *See verdigris.*

Aesthetics \es-ˈthe-tik, is-, *British usually* ēs-\ [Gr *ästhetisch*, NL *aestheticus*, Gk *aisthētikos* of sense of perception, fr. *aisthanesthai* to perceive] (1798) *adj.* In textiles, properties perceived by touch and sight, such as the hand, color, luster, drape, and texture of fabrics or garments.

Affine deformation *n.* A deformation in which each element in the volume distorts in the same way as does the volume as a whole.

Affinity \ə-ˈfi-nə -tē\ [ME *affinite*, MF *afinité*, L *affinitas*, fr. bordering on, related by marriage, fr. *ad* + *finis* end, border] (14c) *n.* With respect to an adhesive, affinity is attraction or polar similarity between the adhesive and an adherend.

African ochre *See pigment yellow* (42/77492).

After-bake *n.* A technique used with phenolic and amino resins to increase the output of a molding press by ejecting moldings before they are fully cured, subsequently completing the cure by baking them. After baking may also be done with fully cured parts to improve their electrical properties and heat resistance.

Afterburner \ˈbər-nər\ (1947) *n.* An air pollution abatement device that removes undesirable organic gases through incineration.

Aftercure *n.* A continuation of the process of curing or vulcanization after the cure has been carried to the desired degree and the source of heat removed, generally resulting in over-cure and a product less resistant to aging than properly cured products.

After-flame *n.* In an ignition test, persistence of flame after removal of the ignition source.

After-flame time *n.* The duration of after-flame.

Afterglow \ˈaf-tər ˌglō\ (1871) *n.* The flameless, glowing combustion of certain solid materials that occurs after the removal of an external source of ignition or after the cessation of combustion of the material.

After-tack *n.* Film defect in which the coated surface, having once reached a tack-free stage, subsequently develops a sticky condition. The effect may be due to syneresis (i.e., expulsion of liquid from a gel). Also applies to printing inks.

Aftertreatment *n.* Any treatment done after fabric production. In dyeing, it refers to treating dyed material in ways to improve properties; in nonwovens, it refers to finishing processes carried out after a web has been formed and bonded. Examples are embossing, creping, softening, printing, and dyeing.

Ag Chemical symbol for the element silver. Tortora PG (ed) (1889) Fairchild's dictionary of textiles. Fairchild Books, New York.

Agar (or agar-agar) \ˈa-gər\ [Malay *agar-agar*] (1889) *n.* Gelling hydrocolloid from sea algae plants which are polymers of galactose.

Ager A steam chamber used for ageing printed or padded material.

Age resistance Resistance to deterioration with time.

Agglomerate \ə-ˈglä-mə-ˌrāt\ [L *agglomerates*, pp of *agglomerare* to heap up, join fr. *ad-* + *glomer-*, *glomus* ball] (1684) *vt.* (1) Cluster of individual; particles. (2) Cluster of dry pigment particles held together by surface forces. The spaces between the particles are filled with air; may also be present in liquid paints of the pigment has not been properly dispersed. *See aggregate.*

Agglomeration (1774) *n.* Condition in which particles become united into clusters of individual particles. May be loosely used to refer to undispersed material.

Aggregate \ˈa-gri-gət\ [ME *aggregate*, fr. L *aggregatus*, pp of *aggregare* to add to, fr.

ad- + *greg-*, *grex* flock] (15c) *adj.* (1) A group of dry pigment particles held together by their surface forces; the spaces between the particles are filled with air. (2) An inert granular material, such natural sand, manufactured sand, gravel, crushed gravel, crushed stone, vermiculite, perlite, and air-cooled blast furnace slag, which when bound together into a conglomerate mass by a matrix, forms concrete or mortar. (3) In the reinforced-plastics industry, a mixture of a hard, fragmented material with an epoxy binder, used as a flooring or surfacing medium, or in epoxy tooling {*aggregation* G Aggregatzustand m, F état d'aggrégation, état m, S estado de agregación, estado m, I stato di aggregazione, stato m}.

See agglomerate.

Aggressive tack *See tack, dry.*

Aging (ageing) *n.* (1) The deterioration of textile or other materials caused by gradual oxidation during storage and/or exposure to light. (2) The oxidation stage of alkali-cellulose in the manufacture of viscose rayon from bleached wood pulp. (3) Originally, a process in which printed fabric was exposed to a hot, moist atmosphere. Presently, the term is applied to the treatment of printed fabric in moist steam in the absence of air. Aging is also used for the development of certain colors in dyeing, e.g., aniline black. (4) The process, or the results of, exposure of plastics to natural or artificial environmental conditions for a prolonged period of time. Storage of paints, varnishes, etc., under defined conditions of temperature, relative humidity, etc. in suitable containers, or of dry films of these materials, for the purpose of subsequent tests. Complete textile glossary. Celanese Acetate LLC, Three Park Avenue, New York, 2000. Glenz W (ed) (2001) A glossary of plastics terminology in five languages, 5th edn. Hanser–Gardner Publications Inc., Cinicinnati.

See also artificial aging and artificial weathering. See also maturing {G Alterung f, F vieillissement m, S envejecimiento m, I invecchiamento m}.

Aging, rubbers *n.* The process of oxidation and other degradations of rubber or elastomeric materials.

Aging time n. *See time, joint conditioning.*

Agitate \ˈa-jə-ˌtāt\ [L *agitatus*, pp of *agitare*, frequentative of *agere* to drive] (15c) *v.* To stir or to mix, as in the case of a dyebath or solution.

Agitation *n.* Process of mixing or stirring to achieve homogeneity, but not necessarily dispersion.

Agitator *n.* Mechanical device used for mixing or stirring.

AI *n.* Abbreviation of amide-imide (polymer). *See polyamide-imide resin.*

AIA *n.* Abbreviation for American institute of architects.

AICE Abbreviation for American institute of chemical engineers.

AIMA *n.* Abbreviation for acoustical and insulating materials association.

Air assist thermoforming *n.* Method of thermoforming in which air is employed to perform the sheet immediately prior to the final pull down onto the mold, using vacuum.

Air-assist vacuum forming *n.* A modification of the process of sheet thermoforming in which partial performing of the sheet is effected by air flow or air pressure before vacuum pull-down.

Air bag *n.* An automatically inflating bag in front of riders in an automobile to protect them from pitching forward in an accident. End use for manufactured textile fibers.

Air blowing *n.* Process employed for the production of low temperature blown oils or of high temperature isomerized oils. A large stream of air, sent through vegetable oils under suitable conditions, produces chemical changes in them. These changes are recognized by solubility characteristics, increased viscosity, and improved drying and polymerizing properties; but the changes which occur with a particular oil are governed by its chemical composition. Shahidi F, Bailey AE (eds) (2000) Bailey's industrial oil and fat products. John Wiley and Sons, New York. Paint: pigment, drying oils, polymers, resins, naval stores, cellulosics esters, and ink vehicles, vol 3. American Society for Testing and Material, 2001.

See blown bitumen, blown castor oil, blown oil, and blown stand oil.

Air brush *n.* Very small spray gun, not much larger than a fountain pen, designed as an artist's tool.

Air brushing *n.* Blowing color on a fabric or paper with a mechanized pneumatic brush.

Air bubble *n.* Dry bubble in coating film caused by entrapped air.

Air-bubble viscometer *n.* An instrument used to measure the viscosities of oils, varnishes and resin solutions by matching the rate of rise of an air bubble in the sample liquid with the rate of rise in one of a series of standard liquids, whose viscosities are known. The Gardner–Holt bubble viscometer is such an instrument (Paul N. Gardner, Company Inc., 316 N. E. Fifth Street, Pompano Beach, Fl, www.gardco.com).

Air cap *n.* Perforated housing for atomizing air at head of spray gun.

Air columns Frequency of vibration in (*see organ pipes*).

Air conditioning *n.* (1) A chemical process for sealing short, fuzzy fibers into a yarn. Fabrics made from air-conditioned yarns are porous. Because they allow more air circulation, these fabrics are also cooler. (2) Control of temperature and/or humidity in work or living space.

Air contamination *n.* Foreign substances introduced into the air which make the air impure.

Air cure *n.* Vulcanization which takes place at room temperature with the use of ultrafast or fast-acting accelerator.

Air drying *See drying.*

Air ducts *n.* Pipes that carry warn air or cold air to rooms and back to furnace or air conditioning system.

Air entangled yarns *See compacted yarns.*

Air entraining agents *n.* Natural wood resins, fats, inorganic materials, sulfonated compounds, and oils for air entrapment in concrete.

Air entrapment *n.* Inclusion of air bubbles in coating film or in other solids such as concrete.

See microvoids.

Air flotation *n.* Process used to separate light from heavy pigment particles by a strong current of air. The air stream is so arranged that the pigment particles are carried vertically from the grinding mill. By this means only the finest particles are carried away, the larger and heavier ones falling back to be re-ground.

Air forming *n.* A process in which air is used to separate and move fibers to fashion a web such as the Kroyer® process for short fibers, usually of wood pulp; or the Rando-Webber® process for staple-length fibers.

Air gap *n.* (1) In extrusion of film, sheet or a coating, the distance from the die opening to the nip formed by the pressure roll and the chill roll. (2) In the radio-frequency heating of plastics and corona treatment of films, the space between the electrode and the surface of the material.

Air jet A type of sandblasting gun in which the abrasive is conveyed to the gun by partial vacuum.

Air jet spinning *n.* A spinning system in which yarn is made by wrapping fibers around a core stream of fibers with compressed air. In this process, the fibers are drafted to appropriate sliver size, then fed to the air jet chambers where they are twisted, first in one direction, then in the reverse direction in a second chamber. They are stabilized after each twisting operation.

Air jet texturing *See texturing.*

Air-knife coating *n.* A coating technique especially suitable for thin coatings such as adhesives, wherein a high-pressure jet of air is forced through orifices in a knife to meter and control the thickness of the coating.

See also spread coating.

Air-laid nonwovens *n.* Fabrics made by an air-forming process. The fibers are distributed by air currents to give a random orientation within the web and a fabric with isotropic properties.

Airless blast deflashing *n.* The process of removing flash from molded parts by bombarding them with tiny nonabrading pellets that break off flash by impact.

See also blast finishing.

Airless spray *n.* A system of applying paint under high pressure in which the paint is broken up into droplets when it enters to lower-pressure region outside of the gun tip. Very little air is used as compared to conventional air spray.

Airless spraying *n.* Process of atomization of paint by forcing it through an orifice at high pressure. This effect is often aided by the vaporization of the solvents, especially if the paint has been previously heated. The term is not generally applied to those electrostatic spraying processes, which do not use air for atomization.

See also hydraulic spraying.

Air-lock *n.* Surface depression on a molded part, caused by trapped air between the mold surface and the plastic material.

Air loss *n.* Loss in mass by a plastic or coating on exposure to air at room temperature.

Air permeability *n.* The porosity or the ease with which air passes through material. Air permeability determines such factors as the wind resistance of sailcloth, the air resistance of parachute cloth, and the efficacy of various types of air filters. It also influences the warmth or coolness of a fabric.

Airplane fabric *n.* A plain, tightly woven, water-repellent fabric traditionally made of mercerized cotton. During World War I, the fabric was treated with a cellulose acetate dope and used to cover the wings, tail, and fuselage of airplanes. Today, similar fabrics made from nylon or polyester/cotton blends are used in rainwear and sportswear.

Air pollutants, hazardous *n.* Materials discharged into the atmosphere that have a proven relationship to increase human death rates.

Air pollution The presence of contaminants in the air in concentrations that interfere directly or indirectly with man's health, safety, or comfort.

Air quality control regions *n.* Geographical units of the country, as required by USA. Law, reflecting common air pollution problems, for purposes of reaching national standards.

Air-quality regulations *n.* Federal, state and/or local regulations constructed for the purpose of protecting air quality, e.g., low volatile organic compounds regulations.

Air quality standards *n.* The prescribed level of pollutants in air that cannot be exceeded during a specified time in a specified geographical area.

Air ring *n* (air-cooling ring). In the process of blowing tubular film, a circular manifold with one or more annular openings concentric with and just above the die lip that blows a uniform stream of air on or along the plastic tube. The air may be refrigerated.

Air sampling *n.* Determining quantities and types of atmospheric contaminants by measuring and evaluating a representative sample of air. The most numerous environmental hazards are chemical, and can be conveniently divided into (a) the particulates and (b) the gases or vapors. Particulates are mixtures or dispersions of solid or liquid particles in air and included dust, smoke, mist, and similar materials.

Air separation *See air flotation.*

Air shot (air purge) *n.* In injection molding, a shot made with the nozzle withdrawn from the mold, so that the expelled melt may fall freely and be caught for measurement of cup temperature or for weighing.

Airslip forming (airslip vacuum forming) *n.* A variation of snap-back forming in which a male mold is enclosed in a box so that, as the mold advances toward the softened sheet, air is trapped between mold and sheet. The ballooning sheet is thus kept from touching the mold during most of the latter's advance. At the end of the stroke, vacuum is applied, destroying the air cushion, and the sheet is sucked against the plug. *See also sheet thermoforming.*

Air-supported roof *n.* A fabric-based roofing system that is supported and held in place by air pressure.

Air vent *n.* A passageway, typically a fine groove or scratch, between a mold cavity and the outside edge of the mold face, that allows air to escape as melt is injected into the cavity.
See burn mark.

Al Chemical symbol for the element aluminum.

Alabaster \ˈa-lə-ˌbas-tər\ [ME *alabaster,* fr. MF, fr. L *alabaster* vase of alabaster, fr. Gk *alabastros*] (14c) *n.* Fine-grained, translucent variety of very pure gypsum, generally white or delicately shaded.

Alathon Ethylene/vinyl acetate copolymer. Manufactured by DuPont, USA.

Albatross \ˈal-bə-ˌtrós, -ˌträs\ [prob. alter. of obs. *alcatrace* frigate bird, fr. Sp or Pg *alcatraz* pelican, fr. Arabic *al-ghattās*, a kind of sea eagle] (1672) *n.* A soft, lightweight wool or wool blend fabric in a plain weave with a napped, fleecy surface that resembles in texture, the breast of the albatross. It is usually light-colored and is used in negligees, infants' wear, etc.

Albedo \al-ˈbē-(ˌ)dō\ [fr. L *albus*] (ca. 1859) *n.* The fraction of incident light or other electromagnetic radiation that is reflected by a surface.

Albertol, alberlat *n.* Modified phenolic resins. Manufactured by Chem. Werke Albert, Germany.

Albino bitumen *n.* Pale bitumen of petroleum origin, similar in general physical characteristics to the black types, but distinguished by its deep golden brown color.

Albite \ˈal-ˌbīt\ [Sw *albit,* fr. L *albus*] (ca. 1843) *n.* $NaAlSi_3O_8$. A widely distributed white feldspar. One of the common rock-forming plagioclase groups.

Albumin \al-ˈbyü-mən; ˈal-ˌbyü-, -byə-\ [ISV *albumen* + *-in*] (1869) *n.* Water-soluble protein derived from egg whites or animal blood, used in used in coatings and adhesives.

Alburnum Softer part of the wood between the inner part and the heart wood. Syn: sapwood.

A

Alchemy \ˈal-kə-mē\ [ME *alkamie, alquemie,* fr. MF or ML; MF *alquemie,* fr. ML *alchymia,* fr. Arabic *al-kīmiy ā',* fr. *al* the + *kīmiyā'* alchemy, from LGk *chēmeia*] (14c) *n.* The chemical period from about 300 BC to AD 1500 during which a central goal was the transmutation of base metals into gold.

Alcohol \ˈal-kə-ˌhól\ [Fr. L, fr. ML, powdered antimony, fr. OSp, from Arabic *al-kuÖul* the powdered antimony, from *kuÖl* kohl] (1672) *n.* (1) A generic term for organic compounds having the general structure ROH. In aliphatic alcohols, R has the formula C_nH_{2n+1}, as in methyl alcohol, CH_3OH, or *n*-butanol, C_4H_9OH. In more complex alcohols R may be other alkyl, acyclic or alkaryl groups. Alcohols are classified according to the number of –OH groups they contain–monohydric, dihydric, trihydric, or polyhydric. Dihydric alcohols are called glycols; trihydric alcohols are also known as *glycerols*; and the term polyol is used for any polyhydric alcohol. Alcohols have many important applications in the plastics industry. They are used directly as solvent applications in the plastics industry. They are used directly as solvent and diluents. Many esters of alcohols with organic acids are plasticizers. As intermediates, alcohols are used in the production of resins such as acrylics, alkyds, aminos, polyurethanes, and epoxies. (2) Specifically, ethanol, C_2H_5OH. Morrison RT, Boyd RN (1992) Organic chemistry, 6th edn. Prentice-Hall, Englewood Cliffs, NJ.

Alcohol, denatured *n.* Ethyl alcohol that has been adulterated with a toxic material such as acetaldehyde or benzene so as to render it unfit for human consumption but still useful as an industrial solvent or, occasionally, as a reactant.

Alcohol resistance *n.* Ability of a dried or cured coating to withstand the damaging effects of (ethyl) alcohol.

Alcohols *n.* A family of organic solvents containing the grouping C–OH, used in flexographic and gravure inks. The most common members of this group are: methyl (wood) alcohol, ethyl (grain) alcohol, Propyl and isopropyl alcohol.

Alcoholysis *n.* By analogy to *hydrolysis,* any chemical reaction in which an alcohol acts in a way similar to that of water. Its general chemical reaction involves an ester exchange. The cleavage of a C–C bond by the addition of an alcohol.
See also ester interchange.

Alcove \ˈal-ˌkōv\ [F *acôve,* fr. Sp *alcoba,* fr. Arabic *al-qubbah* the arch] (1676) *n.* A small recessed space, opening directly into a larger room.

Aldehyde \ˈal-də-ˌhīd\ [Gr *Aldehyde,* fr. NL *al. dehyde.,* abb. *alcohol dehydrogenatum* dehydrogenated alcohol] (ca. 1846) *n.* (1) A generic term for organic compounds containing a double-bonded oxygen and hydrogen bonded to the same terminal carbon atom of the molecule, i.e., the –CHO group. Thus they may be represented by the general formula RCHO. The simplest one is formaldehyde, HCHO, in which R is hydrogen. For all other, R represents a hydrocarbon radical. (2) Any of a class of highly reactive organic chemical compounds obtained by oxidation of primary alcohols, characterized by the common group –CHO and used in the manufacture of resins, dyes, and organic acids.

Aldehyde resin *n.* Synthetic resin made by treating various aldehydes with condensation agents. Phenol, urea, aniline, and melamine react readily with aldehydes, such as formaldehyde, which are also called *Aldehyde Resins* {G Aldehydharz n, F résine aldéhyde, résine f, S resina aldehídica, resina f I resina aldeidica, resina f}.

Aleurites *n.* Botanical name for a species of shrub or tree which provides vegetable oils used in varnish manufacture. The various *aleurites* include *A. fordii, A. montana, A. cordate, A. trisperma,* and *A. moluccana.* The first three yield tung oils, the fourth, bagillumbang oil, and the last, candlenut oil.

Aleuritic acid *n.* $C_{15}H_{28}(OH)_3COOH$, DL-erythro-9,10,6-trihydroxy-hexadecanoic acid. Acid obtained from the major constituent of the hard lac portion of shellac. It is present as an interester acid, of monocarboxylic type, which on hydrolysis with caustic soda yields a mixture of two sodium salts, sodium aleuritate and sodium lacollate.

Alfin catalyst *n.* A catalyst obtained from alkali alcoholates derived from a secondary alcohol, used for polymerizing olefins.

Alfin polymerization *n.* The "Alfin" catalyst (Hans-Georg Elias (1977) Macromolecules, Plenum Press, New York), consists of a suspension in an inert solvent like pentane of a mixture of an alklenylsodium compound (such as allyl sodium), an alkoxide of a secondary alcohol (such as isoproxide), and an alkali halide (such as sodium chloride); the catalyst is highly specific for the polymerization of dienes into the 1,4 forms.

Alfrey-price equation *n.* This copolymer equation enables you to predict the composition of the polymer as a function of the current ratio of monomers and the relative reactivities for monomers reacting with themselves or with the comonomer. *Also called the copolymerization equation*:

$$\frac{d[M_1]}{d[M_2]} = \frac{[M_1]}{[M_2]} \cdot \frac{r_1[M_1] + [M_2]}{[M_1] + r_2[M_2]}.$$

The ratio of rate constant, r, for homo-addition, k_{11} or k_{22}, over cross-addition, k_{12} or k_{21}, has been termed the reactivity ratio, r_1 or r_2, for the monomer. Odian GC (2004) Principles of polymerization. John Wiley and Sons Inc., New York.

Algae \ˈal-jē\ [L, seaweed] (1551) *n.* Unicellular or polycellular plants (*thallophytae*), a class of cryptograms, which live in fresh or saltwater and are distinguished from fungi by the presence of chlorophyll and response to photosynthesis (e.g., seaweeds, kelps). Black JG (2002) Microbiology, 5th edn. John Wiley and Sons Inc., New York. *See anti-fouling composition.*

Algicide (**algaecide**) \ˈal-jə-ˌsīd\ (1904) *n.* Chemical agent used to destroy algae.

Algin \ˈal-jən\ (1883) *n.* Any of various colloidal substances (as an alginate or alginic acid) derived from marine brown algae and used esp. as emulsifiers or thickeners. Soluble salts (sodium, ammonium, and

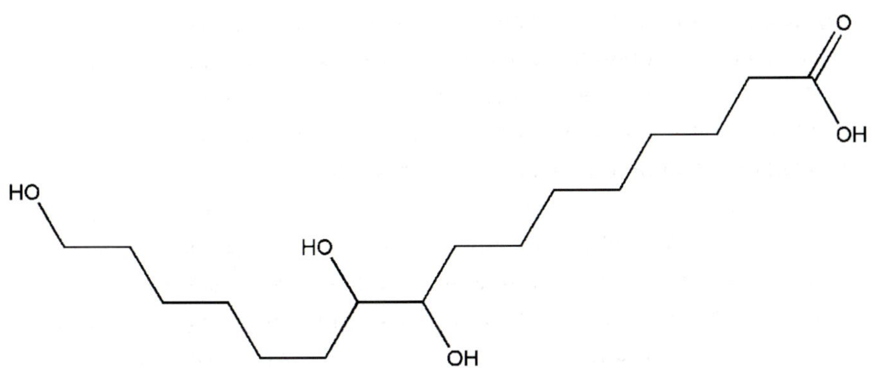

A

potassium) of a polyuronic acid (mannuronic acid), a polymer obtained from brown sea plants.

Alginate \ˈal-jə-ˌnāt\ (ca. 1909) *n.* Any derivative of alginic acid; alginates are used as emulsifying agents, as thickeners, and in films.

Alginate fiber *n.* Fiber formed from a metallic salt (normally calcium) of alginic acid, which is a natural polymer occurring in seaweed. Alginate fiber is soluble in water.

Alginates *See algin.*

Alginic acid *n.* Polysaccharide composed of beta D mannuronic acid residues found in certain seaweeds.

Algoflon Poly(tetrafluoroethylene). Manufactured by Montedison, Italy.

Aliphatic \ˌa-lə-ˈfat-ik\ [ISV, fr. Gk *aleiphat-*, aleiphar oil, fr. *aleiphein* to smear, perhaps akin to Gk *lipos* fat] (1889) *adj.* Designating a large class of organic compounds (and their radicals) having open-chain structures and consisting of the paraffin, olefin, and acetylene hydrocarbons and their derivatives. Examples are butane, isopropyl alcohol, many fats and oils, adipic acid, amyl acetate, ethylamines. The name also applies to petroleum products, which are straight-chain hydrocarbon derived from a paraffin-base crude oil {G aliphatisch, F aliphatique, S alifático, I alifatico}.

Aliphatic amino group *n.* $-NH_2$ radical when attached to a chain.

Aliphatic compounds *n.* A class of organic compounds which are composed of open chains of carbon atoms. These include paraffins, olefins, etc.

Aliphatic isocyanates *n.* Open chained structures containing the isocyanate group, $-N{=}C{=}O$. Aliphatic isocyanates are used extensively in paints and coatings, providing excellent UV stability, chemical resistance and hardness with flexibility to a variety of coatings and elastomers.

Aliphatic solvent *n.* Hydrocarbon solvents comprised primarily of paraffinic and cycloparaffinic (naphthenic) hydrocarbon compounds. Aromatic hydrocarbon content may range from less than 1% to about 35%.

Aliphatic solvents *n.* Organic liquids having an open chain hydrocarbon structure and *KB* values below 40. They are relatively poor solvents for printing ink resins. Examples are VM&P naphtha, textile spirits and mineral oils (*cf. aromatic solvents*).

Alizarin \ə-ˈli-zə-rən\ [prob. fr. F *alizarine*] (ca. 1835) *n.* 1,2-dihydroxyanthraquinone, the raw material in making many pigments. It is obtained from madder roots, or made synthetically, and is also used in the manufacture of lake pigments (*See image*).

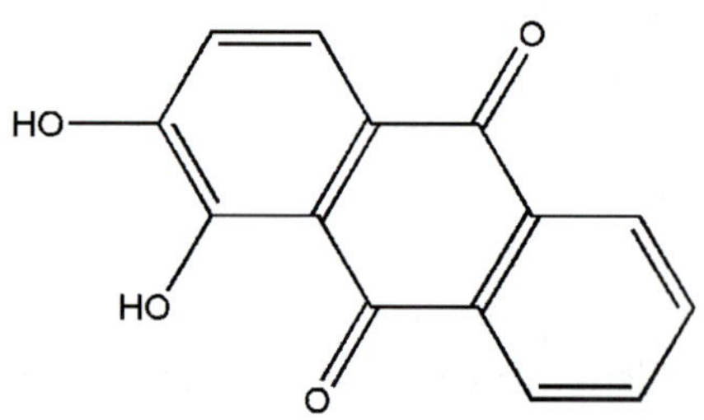

Alizarin madder lake *See alizarin red.*

Alizarin red *n.* Pigment Red 83 (58000). Light maroon pigment produced by treating alizarin with calcium and aluminum salts in the presence of alumina. It is identical with madder lake. Syn: alizarin madder lake, alizarin red B, and dihydroxy anthraquinone lake.

Alkali \ˈal-kə-ˌlī\ [ME, fr. ML, from Arabic *al-qili*] (14c) *n.* Any of the hydroxides and carbonates of the alkali metals (lithium, sodium, and potassium), and the radical ammonium. The term is also used more generally for any strong base in aqueous solution capable of forming salts.

See base.

Alkali blue *n.* Complex organic blue toner prepared by the phenylation of para rosaniline or fuchsine.

Alkali cellulose *n.* Also called regenerated cellulose, pure cellulose obtained from wood pulp, etc., reacted with strong alkali solutions at low temperatures, and is a further step in the reaction of mercerization; as a fiber is called rayon, viscose.

See regenerated cellulose.

Alkali metal (ca. 1885) *n.* A member of group AI in the periodic table.

Alkaline \ˈal-kə-lən, -ˌlīn\ (1677) *adj.* A term used to describe a material having a pH greater than 7.0 in water.

Alkaline catalysts *n.* Hydroxides of sodium, potassium, lithium, and ammonium, or salts derived from these metallic radicals, which exhibit alkaline characteristics. Gaseous ammonia can also be used, as well as a number of basic organic compounds. The most important reactions in which alkaline catalysts are involved are in the condensation of phenols with formaldehyde, the condensation of urea with formaldehyde, and in the isomerization of drying oils.

Alkaline-earth metal *n* (1903). Any of the bivalent strongly basic metals of group II of the periodic table comprising beryllium, magnesium, calcium, strontium, barium, and radium.

Known also as alkaline earth.

Alkali refined linseed oil *n.* Raw linseed oil treated with alkali to reduce the free acidity by formation of eater-soluble salts, which are subsequently removed by washing.

Alkali resistance *n.* (1) The ability of a plastic material to withstand the action of an alkali. ASTM D 543 lists several alkalis as reagents for testing the chemical resistance of plastics. (2) The degree to which a coating resists reaction with alkaline materials such as lime, cement, plaster, soap, etc.

Alkali-resistant paint *See alkali resistance.*

Alkali-resistant red (12315,12355) *n.* Azo pigments based on acrylides of beta hydroxy naphthoic acid and sometimes called naphthol reds; the distinction between them and the acrylide maroons is strictly tinctorial. For the most part, they have identical properties except the reds are somewhat easier grinding but poorer in bake resistance than the maroons. These reds find some application in latex paints, particularly for application over alkaline substrates; they are not sufficiently lightfast for exterior exposure.

Alkali-soluble resins *n.* These are generally lower molecular weight (than conventional lattices) polymers containing about 5–15% carboxyl groups which require amine and/or cosolvent to solubilizer them. These systems are generally dispersions of micelles rather than true solutions. Abbreviation for ASR.

Alkane \ˈal-ˌkān\ [*alkyl*+*-ane*] (1899) *n.* The generic term for any saturated, aliphatic hydrocarbon, i.e., a compound consisting of carbon and hydrogen only and containing no double or triple bonds. Linear alkanes are representable as C_nH_{2n+2} while cyclic alkanes have the general formula C_nH_2n. Examples are propane, C_3H_8, and cyclohexane, C_6H_{12}.

Alkane-imide resin *n.* A thermoplastic introduced by Raychem Corp under the trade name Polyimidal. The polymer retains high strength up to 200°C and melts at 302°C. It has good electrical properties, low water absorption, and high solvent resistance.

Alkathene Poly(ethylene) (high pressure) manufactured by ICI, Great Britain.

A

Synonyms: Ethylene polymer, Ambythene, Etherin, Hizex, Grez, Biocolene C, Epolene C and E.

Alkene \ˈal-ˌkēn\ [ISV *alkyl*+*ene*} (1899) *n.* An unsaturated hydrocarbon with the general formula C_nH_{2n}. Contains a C=C double bond. Same as olefin. Morrison RT, Boyd RN (1992) Organic Chemistry, 6th edn. Prentice-Hall, Englewood Cliffs, NJ.

Alkyd \ˈal-kəd\ [blend of *alkyl* and *acid*] (1929) *n.* Originally, alcohol and acid nouns formed the term "alkyd". Synthetic resins formed by the condensation of polyhydric alcohols with polybasic acids, including anhydrides. They may be regarded as complex esters. The most common polyhydric alcohol used is glycerol, and the most common polybasic acid is phthalic anhydride. Modified alkyds are those in which the polybasic acid is substituted in part by a monobasic acid, of which the vegetable oil fatty acids are typical. {G Alkydharz n, F résine alkyde, résine f, S resina alquídica, resina f, I resina alchidica, resina f}. Wicks ZW, Jones FN, Pappas SP (1999) Organic Coating. Wiley-Interscience, New York. Gooch JW (2002) Emulsification and Polymerization of Alkyd Resins. Kluwer Academic/Plenum Publishers, New York. Patton TC (1962) Alkyd resin technology. John Wiley and Sons Inc., New York.

Alkydal *n.* Trade name for a polyester resin, manufactured by Bayer, Germany.

Alkyd molding compound *n.* A compound based on an alkyd resin containing fillers, pigments, lubricants, and other additives. Alkyd molding compounds are chemically similar to polyesters, but the term alkyd is usually applied to those polyester formulations that use lesser quantities of monomers of the high-viscosity or dry types, resulting in free-flowing granular and nodular types. The compounds are used for applications requiring good electrical properties and long-term dimensional stability such as automotive distributor caps, rotors, and coil caps. Alkyds can be compression molded at low pressure, they cure rapidly, and they present no venting problems because no volatiles are liberated during cure.

Alkyd resin *n.* A polyester resin resulting from the condensation of a polyfunctional alcohol and acid, typically glycerine and phthalic anhydride. Today the term is mostly used for (1) resins modified with drying oils and used as vehicles for varnishes and paints; and (2) for cross-linking resins in alkyd molding compounds. The word *alkyd* is an acronym, from al- for alcohol, and –cid (changed to kyd) for acid. {G Alkydharz n, F résine alkyde, résine f, S resina alquídica, resina f, I resina alchidica, resina f}.
See glyptal.

Alkyl \ˈal-kəl\ (1882) *adj.* A general term for a monovalent aliphatic hydrocarbon radical, which may be represented as having been derived from an alkane by dropping one hydrogen from the formula, C_nH_{2n+2}. Examples of alkyl groups are C_2H_5– (ethyl) and $(CH_3)_2CH_2CH$– (isobutyl).

Alkyl aluminum compound *n.* Any of a family of organo-aluminum compounds widely used as catalysts in the Ziegler-process polymerization of olefins. Members include trialkyl compounds such as triethyl-, tripropyl-, and triisobutyl aluminums; alkyl aluminum hydrides such as diisobutyl aluminum hydride and diethyl aluminum hydride; and alkyl aluminum halides such as diethyl aluminum chloride.

Alkylaryl phosphate (octylphenyl phosphate) *n.* $(C_8H_{17}O)(C_6H_5O)PO$. A phosphate diester, a plasticizer for cellulose acetate butyrate, ethyl cellulose, polystyrene, and vinyl resins.

Alkylaryl phthalate *n.* Any of a family of diesters of phthalic acid containing two alkoxy groups, one aliphatic and one aromatic, used as plasticizers for cellulosic plastics, polymethyl methacrylate, polystyrene, and vinyl resins.

Alkylation \ˌal-kə-ˈlā-shən\ (1900) *n* The introduction of an alkyl radical into an organic molecule.

Alkyl group *n.* Monovalent aliphatic radicals derived from aliphatic hydrocarbons by removal of a hydrogen.

Alkyl phenolic resin *n.* Phenol–formaldehyde resin in which the phenol used has an alkyl group in the para position. In resins used in coatings, the most common are the tertiary butyl and tertiary amyl phenols.

Alkylthio cadmium *n.* A stabilizer for PVC.

Alkyne \ˈal-ˌkīn\ [*alkyl*+*-yne*, alter. of *–ine*] (ca. 1909) *n.* C_nH_{2n-2}. A hydrocarbon containing at least one pair of carbon atoms linked by a triple bond (–C≡C–). The simplest alkyne is acetylene, HC≡CH. *adj.* Signifying the presence in a compound of the triple bond.

Alligatoring \ˈa-lə-ˌgā -tər iŋ, ēŋ\ [S *el lagarto* the lizard, fr. *el* the (fr. L *ille* that + *lagarto* lizard, fr. (assumed) VL *lacartus*, fr. L *lacertus, lacerta*] (1579) *n.* A form of paint failure in which cracks form on the surface layer only. It is caused by the application of thick films where the underlying surface remains relatively soft. The effect is often caused during weather aging.

Allobar *n.* A form of an element differing in isotopic composition from the naturally occurring form.

Allomerism *n.* A similarity of crystalline form with a difference in chemical composition.

See polyallomer.

Alloprene *n.* Chlorinated rubber. Manufactured by ICI, Great Britain.

Allotonic Additive, which changes the surface tension of water.

Allotropes \ˈa-lə-ˌtrōp\ [ISV] (ca. 1889) *n.* Different forms of an uncombined element.

Allotropy \ə-ˈlä-trə-pē\ (1850) *n.* (1) The existence of a substance in two or more solid, liquid or gaseous forms due to differences in the arrangement of atoms of molecules. Examples are amorphous, graphite, and diamond forms of carbon; NO_2 and N_2O_4. (2) Property which an element or compound possesses, of existing in different forms which in themselves have different characteristics. These various forms are described as allotropic modifications. Carbon, for example, is found in an amorphous form as carbon black, and in the crystalline form as graphite, and as a diamond.

See monotropic and enantiotropic.

All-over pattern *n.* The typical effect produced by a wallcovering. A pattern in which the units of design are evenly distributed over a surface, without undue emphasis.

Allowable stress *n.* In engineering design, the maximum stress to which a structure or structural element may be subjected under the expected operating conditions. The allowable stress is normally less, by a sizeable factor of safety than the stress of the same type that would cause the member to fail under the same conditions.

Alloy \ˈa-ˌlói *also* ə-ˈlói\ [F *aloi*, fr. OF *alei*, fr. *aleir* to combine, fr. L *alligare* to bind] (1604) *n.* A blend of a polymer or copolymer with other polymers or elastomers. An important example is a blend of styrene-acrylonitrile copolymer with butadiene-acrylonitrile rubber. The term *polyblend* is sometimes used for such mixtures. Some

A

writers restrict the term allow to mixtures of polymers that form a single phase, reserving the term *blend* for nonhomogeneous mixture. The sale of plastics blends and alloys worldwide was 1.3 billion pounds (0.95 Tg) in 1987 and was predicted to have increased by more than 50% by 1992.

Allyd aldehyde *See acrolein.*

Allyl \ˈal-əl\ [ISV, fr. L *allium* garlic] (1854) *adj.* The group $CH_2{=}CH–$, e.g., allyl alcohol ($CH_2{=}CH–CH_2OH$). The unsaturated radical, $CH_2{=}CHCH_2–$, which upon liberation forms biallyl (1,5-hexadiene), a pungent, volatile liquid.

Allyl alcohol (propenyl alcohol, AA, 2-propene-1-ol) *n.* $CH_2{=}CH_CCH_2–OH$. A colorless liquid with a characteristic pungent odor synthesized by hydrolysis of allyl chloride (from propylene) with dilute caustic, or by the dehydration of propylene glycol. It is a basic material for all allyl resins, and its esters are used as plasticizers (*See image*).

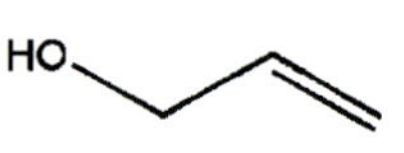

Allyl chloride (3-chloropropene, α-chloropropylene, AC) *n.* $CH_2{=}CH–CH_2Cl$. Used in the preparation of allyl alcohol and various thermosetting resins.

Allyl cyanide (3-butenoic acid nitrile, allyl carbylamine, vinylacetonitrile) *n.* $CH_2{=}CHCH_2CN$. Used as a crosslinking agent.

Allyl diglycol carbonate (ADC) *n.* A colorless, water-clear monomer that can be polymerized and cast into a variety of transparent, optical-grade products. It can be copolymerized with other unsaturated monomers such as vinyl acetate, maleic anhydride, and methyl methacrylate to produce polymers with a wide spectrum of properties.

Allyl diglycol carbonate resin *n.* A thermosetting-resin group with outstanding optical clarity, good mechanical properties, and the highest scratch resistance of all transparent plastics. The resins are made by polymerizing the monomer of the same name with catalysts such as benzoÿl peroxide or, preferable, diisopropyl peroxy dicarbonate.

Allyl esters *n.* Esters of allyl alcohol, used in the production of plasticizers and resins.

Allyl resins *n.* Resins formed by the addition polymerization of compounds containing the group $CH_2{=}CHCH_2$, such as esters of allyl alcohol with dibasic acids. They are commercially available as monomers, as partly polymerized prepolymers, and as molding compounds. The dominant compound in the family is diallyl phthalate (DAP). Others are diallyl isophthalate (DAIP), diallyl maleate (DAM), and diallyl chlorendate (DAC). The monomers and partial polymers may be cured with peroxide catalysts to thermosetting resins that are stable at high temperatures and have good solvent and chemical resistance. The molding compounds may be reinforced with glass fibers or other reinforcements, and are easily molded by compression- and transfer-molding methods. Carley JF (ed) (1993) Whittington's dictionary of plastics. Technomic Publishing Co. Inc.

Allyl starch *n.* Soft, gummy mass prepared by the reaction of starch with allyl chloride in the presence of strong alkali.

Almaciga *n.* Native Philippine name for manila copal.

Almond oil *n.* Vegetable non-drying oil, with an iodine value of about 99, and sp gr of 0.918/15°C.

Alpaca \al-ˈpa-kə\ [Sp, fr. Aymara *allpaqa*] (1811) *n.* (1) Long, fine hair from Alpaca sheep. (2) A fabric from alpaca fibers or blends (originally a cotton cloth with alpaca filling) that is used for dresses, coats, suits, and sweaters. It is also used as a pile lining for jackets and coats (the term has been incorrectly used to describe a rayon fabric).

Alpaca stitch *n.* A 1×1 purl-links stitch that is knit so that the courses run vertically instead of horizontally as the fabric comes off the knitting machine. A garment made with an alpaca stitch is not always 100% alpaca; it can be made of other natural or manufactured fibers.

Alpha- \ˈal-fə\ [ME, fr. L, fr. Gk, of Semitic origin; akin to Hebrew *āleph* aleph] (13c) *n.* A prefix, usually ignored in alphabetizing compound names, and usually abbreviated by the Greek letter α, signifying that the substitution is on the carbon atom immediately adjacent to the main functional group of the compound. An example is α-aminobutanol, $CH_3{-}CH_2CH(NH_2)CH_2OH$. Similarly, substituents on the second and third carbon atoms distant from the main functional group are designated β and γ, respectively.

Alpha cellulose *n.* One of three forms of cellulose. Alpha cellulose has the highest degree of polymerization and is the chief constituent of paper pulp and chemical dissolving-grade pulp. It is a colorless filler obtained by treating wood pulp with alkali, used in light-colored thermosetting resins such as urea formaldehyde and melamine formaldehyde. The material is sometimes treated with resinous agents to coat the individual particles and reduce water absorption of the finished articles.

Also see beta cellulose and gamma cellulose.

Alpha-hydroxypropionic acid *See lactic acid.*

Alpha-methylstyrene *n.* $C_6H_5C(CH_3){=}CH_2$. A colorless liquid, which can be polymerized by acids to resinous products and will copolymerize with other monomeric vinyl compounds. It has a bp of 165°C; Sp gr of 0.914 at 20°/4°.

Alpha olefins *n.* Olefins having 5–20 carbon atoms.

See olefin.

Alpha paper *n.* Paper made from purified wood cellulose, often beautifully preprinted when used as surfacing sheets of decorative laminates.

Alpha particle (1903) (alpha ray) *n.* A charged particle, essentially a helium nucleus, emitted during the radioactive decay of certain elements. Alpha particles have little penetrating power, typically dissipating their energy in passing through a few centimeters of air, but they can do harm if released within the human body. In simple form it is a radioactive emission consisting of two protons and two neutrons as in a ${}^4_2He$ nucleus.

Alpha-pinene *See pinene.*

Alpha-protein *n.* Protein obtained from Soya beans which is chiefly glycinin. The process of extraction of the protein from the beans involves dissolution of the protein and separation from cellulose, removal of carbohydrates, and hydrolysis of the protein to reduce its chemical complexity. Its uses are similar to those of casein, with which it has much in common, being employed in water paints, emulsions, and adhesives of all types.

Alsimag® *n.* Registered trademark of American Lava Corporation for ceramic materials. These materials are used in guides and discs on textile processing machines and fiber manufacturing equipment.

Alternating copolymer *n.* A polymer in which two different mer units alternate along the chain in a regular pattern of –A–B–A–B.–

See also graft and block polymer.

A

Alternating current (AC) *n.* Current in which the charge-flow periodically reverses, as opposed to direct current, and whose average value is zero. Alternating current usually implies a sinusoidal variation of current and voltage. This behavior is represented mathematically in various ways:

$$I = I_0 \cos(2\pi ft + \phi),$$
$$I = I_0 \angle\phi,$$
$$I = I_1 e^{jwt},$$

where f is the frequency; $w = 2\pi f$, the pulsatance, or radian frequency; ϕ the phase angle; I_o the amplitude; and I_1 the complex amplitude. In the complex rotation, it is understood that the actual current is the real part of I. For circuits involving also a capacitance C in farads and L in henrys, the impedance becomes,

$$\sqrt{R^2 + \left(2\pi fL - \frac{1}{2\pi fC}\right)^2}.$$

Also known as AC

Alternating polymers *n.* Copolymers composed of monomers in uniform alternating succession along the chain, which require highly specific copolymerization reactivity ratios.

Alternating strain amplitude *n.* Related through the complex modulus to the alternating stress amplitude.

Alternating stress *n.* A stress mode typical of fatigue tests in which the specimen is subjected to stress that varies sinusoidally between tension and compression, the two maximum stresses being equal in magnitude. In some tests, the stress cycles between zero and a tensile maximum, or other unsymmetrical limits. The term applies also to other modes of loading, such as bending and torsion.

Alternating-stress amplitude *n.* A test parameter of a dynamic fatigue test, others being frequency and environment. It is one half the algebraic difference between the highest and lowest stress in one full cycle.

Alternating twist *n.* A texturing procedure in which *S* and *Z* twist are alternately inserted in the yarn by means of a special heating arrangement.

Altitudes with the barometer *n.* If b_1 and b_2 denote the corrected barometer readings at two stations, t the mean of the temperatures, t_1 and t_2 of the air at the two stations, e_1 and e_2 the tension of water vapor at the two stations, h the mean height above sea level, Φ the latitude; then the difference in elevation in centimeters is $h = 1,843,000$ $(\log b_i - \log b_2)$ $(1 + 0.00367t)$ $(1 + 0.0026 \cos 2\phi + 0.00002h + 3/8k)$, where

$$k = 1/2(e_1/b_1 + c_2/b_2).$$

An approximate formula, sufficient for differences not over 1000 m is

$$H = 1,600,000\left(\frac{b_1 - b_2}{b_1 + b_2}\right)(1 + 0.004t).$$

Alumina \ə-ˈlü-mə-nə\ [NL, fr. L *alumin-. alumen* alum] (1801) (corundum) *n.* The oxide of aluminum, Al_2O_3, very refractory and next to diamond and boron nitride in hardness, obtained by the calcinations of bauxite. Alumina powder is used as a fire-retardant filler in plastics and, over the past two decades, alumina fibers have enjoyed increasing use as reinforcements for plastics, metals, and even ceramics. Its density is 3.965 g/cm^3.

Alumina-blanc fixe *n.* Composite color base used in the manufacture of certain types of lakes. It is made by first reacting aluminum sulfate with soda ash, and then adding barium chloride solution to form barium sulfate from the sodium sulfate derived from the first reaction.

Alumina fiber *n.* A class of reinforcing fibers available as whiskers or continuous filaments, with quite different properties. Whiskers are almost pure Al_2O_3 (corundum) and are grown by passing a stream of moist hydrogen over aluminum powder heated to 1300–1500°C. Their strength ranges from 4 to 24 GPa, modulus ranges from 400 to 1000 GPa, and they cost about US$15/g. Continuous filaments are lower in crystallinity and/or alumina content (densities range from 2.7 to 3.7 g/cm^3), tensile strengths range from 1.3 to 2.1 GPa, and moduli from 105 to 380 GPa, depending on the manufacturer.

Alumina hydrate *n.* A white inorganic pigment used as an extender in inks and noted for its transparency.
See aluminum oxide, hydrated.

Alumina trihydrate (aluminum hydroxide, aluminum hydrate, and hydrated aluminum oxide) *n.* $Al_2O_3{\cdot}3H_2O$ for $Al(OH)_3$. A white crystalline powder, alumina trihydrate accounts for about half of all flame retardants used in plastics. When heated above about 220°C, it releases water endothermically.

Aluminum (in UK, aluminium) \ə-ˈlü-mən-nəm\ [NL, fr. *alumina*] (1812) *n.*
See aluminum powder and aluminum paste.

Aluminum alkyl (aluminum trialkyl) *n.*
See alkyl aluminum compound.

Aluminum arsenite *n.* Admixture with alumina used as a base for lakes, its function being to improve the brilliance of the color.
See aluminum phosphate.

Aluminum chelate *n.* Chemically modified aluminum secondary butoxide, used as a curing agent for epoxy, phenolic, and alkyd resins.

Aluminum dihydroxy stearate *n.* $Al(OH)[OOC(CH_2)_{10}CHOH(CH_2)_5{-}CH_3]_2$. A white material used as a plastics lubricant.

Aluminum distearate $Al(OH)[OOC(CH_2)_{16}CH_3]_2$. A white powder used as a lubricant for plastics.

Aluminum hydrate *See aluminum oxide, hydrated.*

Aluminum ink *See silver ink.*

Aluminum isopropylate (aluminum isopropoxide). $Al[OCH(CH_3)_2]_3$. A white solid, crosslinking agent (*See image*).

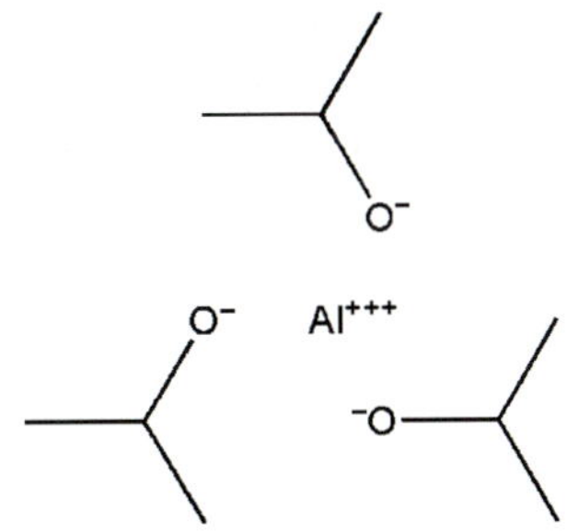

Aluminum mixing varnish *n.* Any of the vehicles for aluminum paste or powder which have the property of causing the metallic flakes of the pigment to "leaf" or float, and that retains this property for an indefinite period of time. They are characterized by low acid number, usually less than 15; but in some specifications, the acid number is required to be less than seven. Ready mixed aluminum vehicles are those that are: made without lead drier, used in factory prepared paints, and capable of retention of excellent leafing properties for very long periods of time. Ready-to-mix aluminum vehicles are mixed with aluminum paste or powder, and used within a few hours at the job site. Vehicles for non-leafing aluminum paints are rarely referred to as aluminum mixing varnishes.

Aluminum monostearate *n.* $Al(OH)_2[OOC(CH_2)_{16}CH_3]$. A white or yellowish-white powder, a stabilizer.

A

Aluminum naphthenate *n.* Pale brown, jelly-like product, containing approximately 4.0% of aluminum, although the precise metal content varies with the type of naphthenic acid used in its manufacture.

Aluminum oleate *n.* $Al[OOC(CH_2)_7CH{=}CH(CH_2)_7CH_3]_3$. A plastics lubricant.

Aluminum oxide *See aluminum oxide, hydrated.*

Aluminum oxide abrasive *n.* Produced in electric furnaces by purifying bauxite of a crystalline form and adding various amounts of titanium to impart extra toughness. The grain is a tough, durable abrasive characterized by the long life of its cutting edges. It is stable chemically and will not react with hydrochloric or nitric acids or with metals other than the alkaline metals.

See emery.

Aluminum oxide cloth *n.* Extremely hard, sharp, and enduring coated abrasive cloth particularly suited for finishing metals, capable of the hardest service.

Aluminum oxide, hydrated *n.* $Al(OH)_3$. Crystalline powder, balls, or granules. Used as a pigment or a base for organic lakes and as an extender for inks. Density, 2.4 g/cm^3 (19.8–20.2 lb/gal); O.A., 41–53 lb/100 lb. Particle size, 0.38–8.5 μm. Syn: alumina hydroxide, alumina trihydrate, alumina hydrate, hydrated alumina, gibb-site and transparent white.

Aluminum oxide paper *n.* Extremely hard, sharp and enduring coated abrasive paper, particularly suited for hardwoods, lacquer and metals.

Aluminum paint *n.* A coating consisting of a mixture of metallic aluminum pigment in powder or paste form, dispersed in a suitable vehicle.

Aluminum palmitate *n.* $Al_2(OH)_2[OOC(CH_2)_{14}CH_3]$. A plastics lubricant.

Aluminum paste *n.* Metallic aluminum flake pigment in paste form, consisting of aluminum, solvent, and various additives. The metallic aluminum pigment can be in the form of very small, coated leaves or amorphous powder, known under the respective designations of "leafing" and "non-leafing". *See leafing.*

Aluminum phosphate *n.* Admixture with alumina used as a base for lakes, its function being to improve the brilliance of the color. (*See image*).

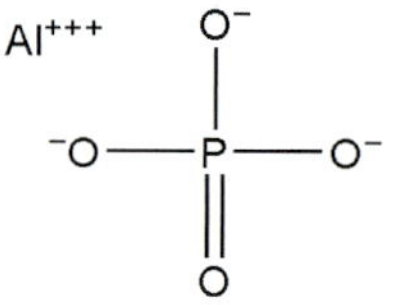

See aluminum arsenite

Aluminum pigment *See aluminum powder.*

Aluminum potassium silicate *n.* $3Al_2O_3 \cdot K_2O \cdot SiO_2 \cdot 2H_2O$. Pigment White 20 (77019). Generically known as mica, a complex hydrous aluminum silicate, based on several mineralogically related groups. The most common is Muscovite. Mica crystals have well-developed basic cleavage that permit slitting into thin flaky pigment particles. Widely used in paints, rubber and sealants. Density, 2.82 g/cm^3 (23.5 lb/gal); O.A., 56.74. Particle size 5–20 μm. Hardness (moh), 2.5. Syn: mica, ground muscovite, graphitic mica, and sericite.

Aluminum powder *n.* Minimum metal used as a pigment. Available in leafing or non-leafing forms. Density, 2.73 g/cm^3 (22.5 lb/gal).

Aluminum primer *n.* (1) Primer specifically formulated for aluminum metal. (2) Primer containing a proportion of aluminum pigment, but distinguished from aluminum paints in which the aluminum is designed to float to the top of the film giving metallic brilliance, a feature undesirable in a primer.

Aluminum primers are used on resinous timber or timber, which has been treated with oil-soluble wood preservatives.

Aluminum resinate *n.* Brown soft mass, insoluble in water; used for water proofing and as a drier in varnish.

Aluminum silicate *n.* Any of a large group of minerals with various proportions of Al_2O_3 and SiO_2, occurring naturally in clays. They are used as pigments and fillers in plastics.

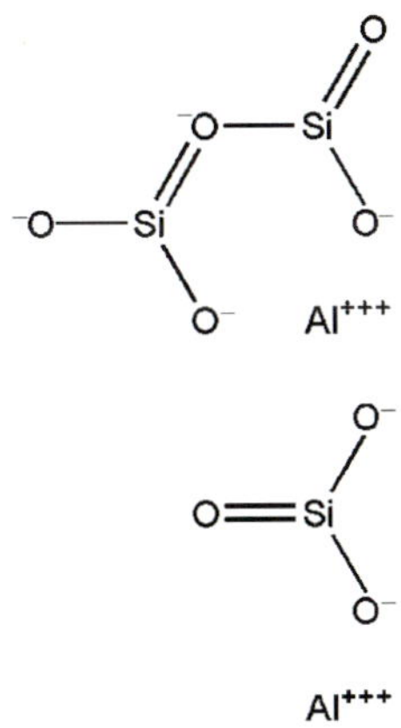

Aluminum silicate (clay) *n.* $Al_2O_3 \cdot 2SiO_2 \cdot 2H_2O$. Pigment white 19 (77005) white inert pigment of little color and opacity, obtained from certain natural deposits of china clay, kaolin, feldspar, and similar materials. Hydrated aluminum silicate, fine grain crystallized clay. Laminar structure with repeating alumina–silica configurations. Density, 2.58 g/cm^3 (21.5 lb/gal); O.A., 32.1–55 lb/100 gal. Particle size, 05.3.5 μm. Syn: Kaolin, hydrated aluminum silicate, china clay, pipe clay, and white bole.

Aluminum soaps *n.* Aluminum compounds of fatty acids and naphthenic acid, etc. These materials are soluble generally in paint vehicles and thinners, and are employed as thickening, flatting, and water proofing agents.

Aluminum stearate *n.* Complex salt or soap of aluminum and stearic acid. It is a white powder, which forms colloidal solutions or gels with drying oils and certain solvents. It is used as a flatting agent, anti-setting agent for pigments, and to help prevent penetration.

Amalgam \ə-ˈmal-gəm\ [ME *amalgame,* fr. MF, fr. ML *amalgama*] (15c) *n.* Solutions of certain metals in mercury, which is of considerable value in the preparation of unsupported films of surface coatings. If, for example, a coated tinned-iron sheet or tinfoil is rubbed with or immersed in mercury, an amalgam is formed between the tin coating and the mercury, The initially hard tin surface is replaced by a soft, semi-liquid coating of tin-amalgam, from which it is a relatively easy matter to detach dried coatings for further examination. The tin surface is said to be amalgamated.

Amber \ˈam-bər\ [ME *ambre,* fr. MF, fr. ML *ambra* fr. Arabic *anbar* ambergris] (14c) *n.* A natural fossil resin formed during the Oligocene age by exudation from a species of pine now extinct. Its empirical formula is $C_{10}H_{16}O$, it softens at about 150°C, can be fabricated and polished. It is a fossil resin found chiefly in the blue earth of East Prussia. It is primary an exudation of *Pinus* succinifera. It has been used in jewelry, cigarette holders, and pipe mouthpieces.

Amberlite Synthetic ion-exchange resins. Manufactured by Roehm & Haas, USA.

Ambient \ˈam-bē-ənt\ [L *ambient-, ambiens,* pp of *ambire* to go around, fr. *ambi-* + *ire* to go] (1596) *adj.* Completely surrounding; indicative of the surrounding environmental conditions such as temperature, pressure, atmosphere, etc. When no values are given, the temperature is presumed to be room temperature (18–23°C) and the atmosphere to be air at standard pressure (101.3 kPa).

A

Ambient air quality *n.* Average atmospheric purity, as distinguished from discharge measurement taken at the source of pollution. The general amount of pollution present in a broad area.

Ambient conditions *n.* A term used to denote the temperature, pressure, etc. of the surrounding air.
See atmospheric conditions.

Ambient cure *See self-curing.*

Ambient temperature *n.* (1) The temperature of the medium immersing an object. (2) The prevailing room temperature.

American gallon *See gallon, USA.*

American national standards institute *n.* (ANSI) address: 1430 Broadway, New York, NY 10018. Clearinghouse and national coordinator for voluntary standards of engineering, equipment, industrial processing, and safety. Formerly known as American Standards Association (ASA). In the plastics field, ANSI works closely with the society of the plastics industry (SPI) and the society of automotive engineers (SAE) to develop and publish standards for plastics materials, processing equipment, operations, and operating safety.

American process zinc oxide *n.* Zinc oxide pigment made directly from zinc ores. Sometimes called "direct" process.
See zinc oxide.

American society for testing and materials *n.* ASTM. A non-profit corporation formed for the development of standards on characteristics and performance of materials, products, systems, and services, and the promotion of related knowledge. In ASTM terminology, standards include test methods, definitions, recommended practices, classifications, and specifications.

American turpentine *n.* Light-colored, volatile, essential oil obtained from resinous exudates or resinous wood associated with living or dead coniferous trees, particularly of the genus, *Pinus*, or more commonly pines.
See turpentine.

American vermillion *n.* A pigment usually consisting of a lead molybdate or as a basic lead chromate (as chrome red).

Ameripol *n.* Poly(isoprene). Manufactured by Firestone, USA.

Ameripol SM *n.* *Cis*-1,4-poly(isoprene). Manufactured by Firestone, USA.

Amide \ˈa-ˌmīd, -məd\ [ISV, from New Latin *ammonia*] (ca. 1847) *n.* A compound containing the –$CONH_2$ group, formed by the reaction of an organic acid or an ester with ammonia. Except for formamide, all amides are crystalline solids at room temperature. Examples are acetamide, CH_3CONH_2, and urea, H_2NCONH_2.

Amide-imide resin *See amino resin.*

Amides \ˈa-ˌmīd, -məd\ [ISV, fr. L *ammonia*] (ca. 1847) *n.* Carboxylic acids in which the hydroxyl group of the acid is replaced by an amino or amido group (NH_2). Thus, acetamide (CH_3CONH_2) is obtained by substituting the hydroxyl (OH) of acetic acid with the NH_2 group. They may also be regarded as derivates of ammonia (NH_3), in which one of the hydrogen atoms is replaced by an acyl group.

Amido \ə-ˈmē-(ˌ)dō, ˈa-mə-ˌdō\ [ISV *amide*+-*o*-] (1877) *adj.* Terms "amido" and "amino" apply to the same grouping, NH_2. The former term is usually applied to the NH_2 group when it occurs in an acid amide.

Amine \ə-ˈmēn, ˈa-ˌmēn\ [ISR, fr. NL *ammonia*] (1863) *n.* Organic bases derived from the parent compound, ammonia (NH_3). The hydrogens of the ammonia may be substituted by alkyl groups, in which case the series of aliphatic bases is produced. Similarly, aromatic bases are formed when the hydrogens are substituted with aryl groups. Primary, secondary, tertiary and quaternary amines are formed as one, two, three, or four of the hydrogen atoms are substituted. Substitution of a fourth hydrogen atom is possible because it is considered available in the hypothetical compound, ammonium hydroxide (NH_4OH). A compound derived (in concept) from ammonia by substitution of one or more hydrogen atoms by a hydrocarbon radical. Amines in which one, two or all three of the ammonia hydrogens have been substituted are termed primary, secondary, and tertiary amines.

Amine end group *n.* The terminating ($-NH_2$) group of a nylon polymer chain. Amine end groups provide dye sites for polyamides.

Amine equivalent *n. See amine value.*

Amine equivalent weight *n.* Molecular weight of amine divided by the number of active hydrogens in the molecule.

Amine-furfural resin *See aniline-furfural resin.*

Amine nitrogen content *n.* Refer to federal test method 141a for test procedure.

Amine resin *n.* Synthetic resin derived from the reaction of urea, thiourea, melamine or allied compounds with aldehydes, particularly formaldehyde.
See amino resin.

Amines *n.* Organic bases derived from the parent compound, ammonia (NH_3) the hydrogens of the ammonia may be substituted by alkyl groups, in which case the series of aliphatic bases is produced. Similarly, aromatic bases are formed when the hydrogens are substituted with aryl groups. Primary, secondary, tertiary, and quaternary amines are formed as one, two, three or four of the hydrogen atoms are substituted. Substitution of a fourth hydrogen atom is possible because it is considered available in the hypothetical compound, ammonium hydroxide (NH_4OH).

Amine value *n.* The number of milligrams of potassium hydroxide equivalent to the fatty amine basicity in 1 g of sample. Syn: amine equivalent.

Amino- \ə-ˈmē-(ˌ)nō\ [ISV *amine*+*-o-*] (1904) *adj.* A prefix signifying the presence in a compound or resin of an $-NH_2$ or $=NH$ group.

Amino acid *n.* An organic acid containing an amino group attached to the carbon atom adjacent to the –COOH group, obtained by the hydrolysis of a protein or by synthesis. Examples are glycine, $CH_2(NH_2)-COOH$ (e.g., glycine $+NH_3-CH_2-COO^+$) and cysteine, $HSCH_2CH(NH_2)COOH$.

Amino acids (1898) *n.* Amino group and a carboxyl group, $H_2NCHRCOOH$, e.g., glycine $+NH_3-CH_2-COO^+$.

Amino coatings *n.* Made from amino resins which include resins from reacting urea, thiourea, melamine or allied compounds, usually with formaldehyde.

2-Amino-2-methyl-1-propanol *See amp.*

Amino plastics Plastics based on resins made by the condensation of amines, such as urea and melamine, with aldehydes.
See amino resin.

Aminoplasts *n.* Thermosetting resins made by the polycondensation of formaldehyde with a nitrogen compound and a higher aliphatic alcohol. The two general types are *urea-formaldehyde* and *triazine-formaldehyde.* Melamine is the triazine most often used. *See also amino resin.* {G Aminoplaste mpl, F aminoplastes mpl, S aminoplásticos mpl, I amminoplasti mpl}

γ-Aminopropyltriethoxy silane *n.* $NH_2(CH_2)_3Si(OC_2H_5)_3$. A silane coupling agent used in reinforced epoxy, phenolic, melamine and many thermoplastic resins.

A

Amino resin (polyalkene amide, aminoplast) *n.* A generic term for a group of nitrogen-rich polymers containing amino nitrogen or its derivatives. The starting amino-bearing material is usually reacted with formaldehyde to form a reactive monomer that is condensation-polymerized to a thermosetting resin. Included amino compounds are urea, melamine, copolymers of both with formaldehyde, and, of limited use, thiourea, aniline, dicyandiamide, toluenesulfonamide, benzoguanidine, ethylene urea, and acrylamide. Not included, because properties warrant separate classification, are polyamides of the nylon type, polyurethanes, polyacrylamide, and acrylamide copolymers. The most important members of the amino-resin family are melamine-formaldehyde and urea-formaldehyde resins. The basic resins are clear, water-white syrups or white powdered materials that can be dispersed in water to form colorless syrups. They cure to high temperatures with appropriate catalysts. Molding powders are made by adding fillers to the uncured syrups, forming a consistency suitable for compression and transfer molding. Amino resins are usually cured by baking and are blended with other resins (e.g., alkyds or epoxies). Amino resins are also cured by chemical means at normal air temperature, e.g., wood finishes.

AMMA *n.* Abbreviation for copolymers of acrylonitrile and methyl methacrylate. In Europe, written A/MMA.

Ammeter \ˈa-ˌmē-tər\ [*ampere*+-*meter*] (1882) *n.* Instrument for measuring the strength (amperage) of electric currents.

Ammine \ˈa-ˌmēn, a-ˈmēn\ [ISV *ammonia* +²-*ine*] (1897) *n.* The name given to ammonia, NH_3, when it serves as a ligand.

Ammonia cure *n.* A modification of a hot air pressure cure for rubber, often used for curing, in which ammonia gas is used to accelerate vulcanization and to prevent the deteriorating effect of air.

Ammonium caseinate *n.* Casein solubilized by ammonium hydroxide generally employing a hot water presoak.

Ammonium soaps *n.* Soaps formed by reaction of ammonia with the higher molecular weight fatty acids, such as ammonium oleate, linoleate, stearate, etc., used as wetting or emulsifying agents.

Amoora oil *n.* Semidrying vegetable oil with an iodine value of 135 and Sp gr of 0.939/15°C.

Amorphous \ə-ˈmór-fəs\ [Gk *amorphous*, fr. *a*-+*morphē* form] (ca. 1731) *adj.* Devoid of crystallinity or stratification. Most plastics are amorphous at processing temperatures, many retaining this state under all normal conditions. Lacking crystallinity.

Amorphous domain (and crystalline domains) *n.* Amorphous or noncrystalline portions of a solid polymer; conversely, crystalline a domains are nonamorphous. A single crystalline polymer chain can possess amorphous regions (domains), e.g., polyethylene.

Amorphous nylons *n.* When unsymmetrical monomers are use to synthesize polyamides (Nylons), the normal ability of the chains to crystallize can be disrupted and amorphous (often transparent) polymers are formed.

Amorphous silica *n.* SiO_2. A naturally occurring or synthetically produced pigment, characterized by the absence of pronounced crystalline structure, and which has no sharp peaks in its X-ray diffraction pattern. It may contain water of hydration or be an anhydrous type. It is used as an extender pigment, fatting agent, and as a desiccant in metal flake and metal powder coatings.

Amorphous solid *n.* A substance with the external appearance and characteristics of a solid, but with the irregular structure typical of a liquid; a highly supercooled liquid; a glass.

AMP *n.* Abbreviation for 2-amino-2-methyl-1-propanol. Used as a pigment dispersant or as a pH modifier.

Ampere \ˈam-ˌpir, -per\ [Andrè-Marie *Ampère*] (1881) *n.* (A) The primary electrical unit of the SI system, upon which all other electrical units are based. The ampere itself is defined as that current, which, if maintained in two long, parallel, fine wires located 1 m apart in a vacuum, will produce between these conductors a force of 2×10^{-7} N/m of length. Practically, an ampere is the current that flows between two points connected with an electric resistance of one ohm when their potential difference is 1 V.

Ampere's rule *n.* A positive charge moving horizontally is deflected by a force to the right if it is moving in a region where the magnetic field is vertically upward. This may be generalized to currents in wires by recalling that a current in a certain direction is equivalent to the motion of positive charges in that direction. The force felt by a negative charge is opposite to that felt by a positive charge.

Amphibole \ˈam(p)-fə-ˌbōl\ [F, fr. LL *amphibolus*, fr. Gr *amphibolos*] (ca. 1823) *n.* A group of asbestos minerals.

Amphoprotism *n.* The ability of a species either to gain or to lose a proton.

Amphoteric \ˌam(p)-fə-ˈter-ik\ [ISV, fr. Gk *amphoteros* each of two, fr. *amphō* both] (ca. 1849) *adj.* Designating an element or a compound that can behave either as an acid or a base, i.e., as an electron donor or an electron acceptor. Polymerization emulsifiers having both anionic and cationic groups are called amphoteric emulsifiers.

Amphoterism *n.* The ability of a substance to react either as an acid or as a base.

Amplitude \ˈam-plə-ˌtüd, -ˌtyüd\ (1542) *n.* The maximum value of the displacement in an oscillatory motion.

amu *n.* The atomic mass unit (amu), a unit of mass equal to 1/12 the mass of the carbon atom of mass number 12. On the atomic mass scale $^{12}C \equiv 12$.

$$
\begin{aligned}
1\,\text{amu} &= 931.4812(52)\,\text{MeV} \\
&= 1.660531(11) \times 10^{-27}\,\text{kg (SI units)} \\
&= 1.660531(11) \times 10^{-24}\,\text{g (cgs units)}.
\end{aligned}
$$

The numbers in parentheses are the standard deviation uncertainties in the last digits of the quoted value, computed on the basis of internal consistency.

Amyl \ˈa-məl\ [L *amy*lum + E *–yl*] (1850) *n.* The radical C_5H_{11}–, also known as *pentyl.* The amyl radical occurs in six isomeric forms, and the term amyl usually refers to any mixture of the isomers.

Amylaceous *adj.* Pertaining to, or of the nature of, starch; starchy.

Amyl acetate (ca. 1881) (banana oil, pear oil, and amylacetic ester) *n.* $CH_3COOC_5H_{11}$. A commercial solvent for several resins, including the cellulosics, vinyls, acrylics, polystyrene, and uncured alkyds and phenolics. It has a strong, fruity odor (hence its nicknames), and its main constituent is isoamyl acetate, but other isomers such as normal- and secondary-amyl acetates are present in amounts determined by the grade and origin. It is a medium boiling solvent. The commercial product has a boiling range of 120–145°C, a Sp gr of approximately 0.876, and a flp of 31°C (87°F). Syn: banana oil (*See image*).

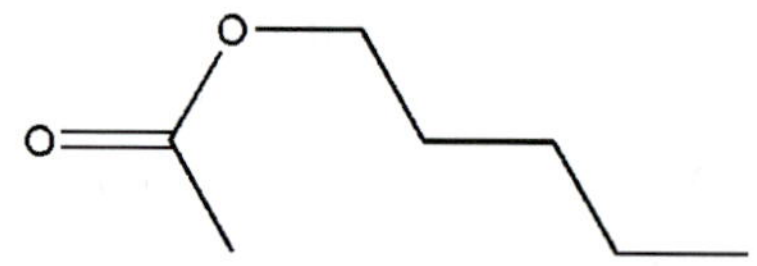

Amyl alcohol (1863) *n.* $C_5H_{11}OH$. Commercial alcohol having a boiling range of 128–132°C; Sp gr of 0.816/15°C; a flp of 43°C (110°F); and a vapor pressure of 10 mmHg at 20°C. It has some application in the manufacture of cellulose lacquers as a high boiling diluent (*See image*).

Amyl benzoate *n.* $C_6H_5COOC_5H_{11}$. High boiling solvent-plasticizer, with a bp of 261°C, and Sp gr of 0.994 (*See image*).

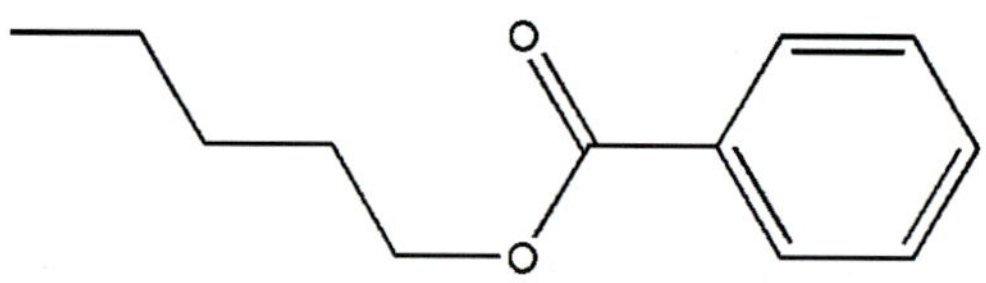

Amyl borate *n.* $(C_5H_{11})_3BO_3$. High boiling solvent-plasticizer. The commercial liquid has a boiling range of 250–260°C (*See image*).

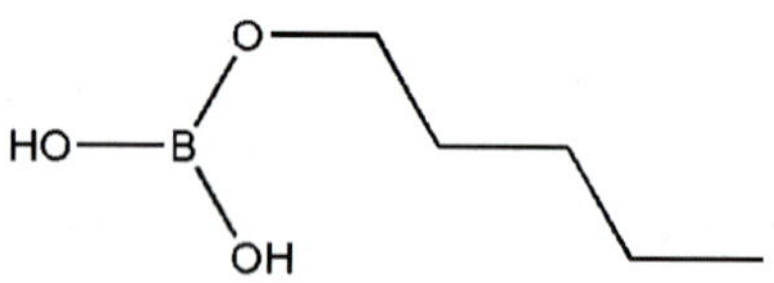

Amyl citrate *See citrates.*

Amyl ether *n.* $(C_5H_{11})_2O$. Amyl ether, or diamyl ether as it is correctly called, is a useful high boiling solvent of the ether series, with a boiling range of 170–190°C.

Amyl formate *n.* $HCOOC_5H_{11}$. A solvent for resins and cellulose derivatives. It is a medium boiling solvent with a bp of 13°C, and flp of 27°C.

Amyl lactate $CH_3CH(OH)COOC_5H_{11}$. Water white to pale yellow colored liquid used as an ester solvent. It plasticizes in lacquers and has a bp of 210°C.

Amyl nitrate (ca. 1881) *n.* $C_5H_{11}NO_2$. A pale yellow, pungent flammable liquid ester, of commercial amyl alcohol and nitrous acid.

Amyl oleate Solvent and plasticizer for cellulosic and vinyl resins.

Amyl propionate *n.* $CH_3CH_2COOC_5H_{11}$. Very strong solvent, similar in properties and uses to amyl acetate, but possessing a higher boiling range, 140–170°C. Its flp is about 41°C.

Amyl salicylate *See isoamyl salicylate.*

Amyl stearate *n.* $C_{17}H_{35}COOC_5H_{11}$. Pale yellow colored substance used as a high boiling plasticizer. It has a mp of 30°C and a bp of 360°C.

Amyl tartrate *n.* Plasticizer with strong solvent power and a bp of about 400°C.

Amyrin *n.* $C_{30}H_{50}O$ for both α and β modifications. Amyrins, isolated in α and β forms, are constituents of elemi resin. They are believed to be alcoholic in character.

Anacardic acid *n.* $C_{22}H_{32}O_3$. Ortho pentadecadienylsalicylic acid. Principal constituent of cashew nutshell liquid.

Anacardol *n.* $C_{21}H_{32}O$. Monohydroxy phenol, which closely resembles the anacardic acid from which it is readily obtained by decarboxylation when heated. The substituent side chain is the same for both compounds.

Anadonis green *See chromic oxide green.*

Anaerobic \ˌa-nə-ˈrō-bik; ˌan-ˌa(-ə), -ˌe(-ə)-\ (ca. 1881) *adj.* Free of oxygen and/or air. Used in connection with bacteria, developing without air. Important for anaerobic corrosion (tanks and ship bottoms).

Anaerobic adhesive *n.* An adhesive that cures only in the absence of air after being confined between assembled parts. An example is dimethacrylate adhesive, used for bonding assembly parts, locking screws and bolts, retaining gears and other shaft-mounted parts, and sealing threads and flanges.

Analyzer \ˈa-nᵊl-ˌīz\ [Prob. irreg. fr. analysis] (1587) *vt.* A second polarizing element inserted above a preparation. When its vibration direction is at right angles to the vibration direction of the polarizer, the field becomes black if no anisotropic specimen is on the stage.

Anamorphosis *n.* Distorted painting, which appears normal when viewed from the side. A form of "trick" painting fashionable in the 16th and 17th centuries.

Anatase \ˈa-nə-ˌtās, -ˌtāz\ [F, fr. Gk *anatasis* extension, fr. *anateinein* to extend, fr. *ana-* + *teinein* to stretch] (ca. 1828) (octahedrite) *n.* A crystalline ore of titanium dioxide.
See titanium dioxide, anatase.

Anchorage \ˈaŋ-k(ə-)rij\ (15c) *n.* Part of an insert that is molded inside of a plastic part and held fast by shrinkage of the plastic onto the insert's knurled surfaces. (1) Property or profile of metal or wood substrate to enhance the adhesion of a coating. (2) Adhesion of rubber to fiber, fabric, metal or other material to which the rubber compound is applied by calendering, welding, cement spreading, or other means.

Andrade creep *n.* A type of creep behavior in which the compliance of the sample is proportional to the cube root of the time under stress. A variety of polymers exhibit this behavior.

Anelasticity \ˌa-nᵊl-as-ˈti-sə-tē\ (1947) *n.* The dependence of elastic strain on both stress and time, resulting in a lag of strain behind stress. In materials subjected to cyclic stress, the anelastic effect causes damping.

Angel's hair *n.* Fibrous strands of material pulled away from thermoplastic films, particularly polypropylene, in heat-sealing and cutting operations that employ hot knives or wires. The angel's hair accumulates on the cutting mechanism, eventually affecting performance and requiring removal.

Angle \ˈaŋ-gəl\ [ME, fr. MF, fr. L *angulus*] (14c) *n.* The ratio between the arc and the radius of the arc. Units of angle, the radian, the angle subtended by an arc equal to the radius; the degree 1/360 part of the total angle about a point. Dimensions, –a numeric.

Angle head (offset head, and crosshead). An extruder head so designed that the principal direction of the extrudate makes an angle with the (extended) axis of the screw. *See also crosshead.*

Angle of contact *n.* Associated with the phenomenon of wetting. When a drop of liquid contacts a solid surface it can remain exactly spherical, in which case no wetting occurs; or it can spread out to a perfectly flat film, and in that latter case complete wetting occurs. The angle of contact is the angle between the tangent to the periphery of the drop at the point of contact with the solid, and the surface of the solid. When the drop spreads to a perfectly flat film, the angle of contact is zero. When it remains exactly spherical, the area of contact is a point only, and the angle of contact is 180°.

Angle of incidence *n.* Angle between the axis of an impinging light beam and the perpendicular to the object surface.

Angle of reflection *n.* Angle between the axis of a reflected light beam and the perpendicular to the object surface.

Angle of repose (angle of rest) *n.* The maximum angle that a conical pile of particles makes with the horizontal surface on which it rests. No ASTM test is listed for this important property of plastic powders and pellets. The smaller the angle of repose, the more easily does the material flow through hoppers and constrictions.

A

Angle of viewing Angle between the axis of a detected light beam and the perpendicular to the object surface.

Angle-ply laminate *n.* A laminate in which equal numbers of plies are oriented at equal plus and minus angles from the plies in the length direction, making the laminate orthotropic. The most commonly chosen angles are $+60°$, giving nearly equal strengths in all directions in the plane of the laminate.

See also cross laminate.

Angle press *n.* A hydraulic molding press equipped with horizontal and vertical rams, used in the production of complex moldings containing deep undercuts or side cavities.

Angocopalolic acid *n.* $C_{23}H_{36}O_3$. A constituent acid of Angola copal. It is an unsaturated monobasic acid and has a melting point of 85°C.

Angocopaloresenes *n.* Identified as both α and β forms, these are constituents of Angola copal. The former has a mp of about 64°C and the latter about 222°C; α angocopaloresene corresponds with $C_{30}H_{54}O_6$, and β angocopaloresene with $C_{25}H_{38}O_4$.

Angola copal *n.* Fossil copal of African origin.

Angora \aŋ-ˈgōr-ə\ (1852) *n.* (1) The hair of the Angora goat. The long, fine fibers are so smooth and soft that they must be combined with other fibers in weaving. (2) The hair of the Angora rabbit. The fine, lightweight hair is warm, and it is often blended with wool to decrease price and to obtain novelty effects in weaving. By law, the fiber must be described as Angora rabbit hair.

Angstrom unit (Å) \ˈaŋ-strəm\ [Anders J. *Angstrom*] (1892) {*G* Angströmeinheit *f*, *F* unité f Angtröm, *S* unidad *f* Angström} *n.* A unit of length equal to 1×10^{-10}m. A unit of linear measure named after A. J. Ångström, used especially in expressing the length of light waves, equal to one ten-thousandth of a micron, or one-hundred-millionth of a centimeter (1×10^{-8}cm). It has been replaced by the nanometer (nm). $1\,\text{Å} = 0.1\,\text{nm}$.

Angular acceleration *n.* The time rate of change of angular velocity either in angular speed or in direction of the axis of rotation (precession). cgs unit, one radian per second . Dimensions, $[T^{-2}]$. If the initial angular velocity is ω_t, the angular acceleration,

$$\alpha = \frac{\omega_t - \omega_o}{t}.$$

The angular velocity after time *t*,

$$\omega_t = \omega_o + a_t.$$

The angle swept out in time *t*,

$$\theta = \omega_o t + \frac{1}{2}\alpha t^2.$$

The angular velocity after movement through the arc θ,

$$\omega = \sqrt{w_o^2 + 2\alpha\,\theta}.$$

In the above equations, for angular displacement in radians, angular velocity will be in radians per second and angular acceleration in radians per squared seconds.

Angular aperture *n.* The angular aperture of an objective is the largest angular extent of wave surface, which it can transmit.

Angular aperture *n.* **AA** The largest angle between the image forming rays collected or transmitted by a lens system, e.g., objective or condenser of a microscope.

Angular harmonic motion or harmonic motion of rotation *n.* Periodic, oscillatory angular motion in which the restoring torque is proportional to the angular displacement. Torsional vibration.

Angular momentum or moment of momentum Quantity of angular motion measured by the product of the angular velocity and the moment of inertia. cgs unit, unnamed, its nature is expressed by gcm^2/s. Dimensions, $[ML^2T^{-1}]$. The angular momentum of a mass whose moment of inertia is I, rotating with angular velocity ω, is $I\omega$.

Angular velocity *n.* Time rate of angular motion about an axis. Cgs unit, one radian per second. Dimensions, $[T^{-1}]$. If the angle described in time t is θ, the angular velocity,

$$\omega = \frac{\theta}{t},$$

where θ in radians and t in seconds gives ω in radians per second.

Angular welding *See friction welding.*

Anhedral *adj.* Anhedral crystals are those whose growth has been impeded by adjacent crystals growing simultaneously, so that the development of plane faces is inhibited. Also crystals eroded, partly dissolved, or mechanically deformed to the point where nearly all traces of crystal faces have been removed. All anhedral crystals are irregularly shaped and do not have plane faces, cf., euhedral.

Anhydride \(ˌ)an-ˈhī-ˌdrīd\ (1863) *n.* (1) A compound from which water has been extracted. (2) An oxide of a metal (basic anhydride) or of a nonmetal (acidic anhydride) that forms a base or an acid, respectively, when united with water. (3) An organic compound made (conceptually) by the union of two acid molecules with the elimination of a molecule of water. In practice, organic anhydrides are usually produced by other reactions.

Anhydrite \-ˌdrīt\ [Gr *Anhydrit*, fr. Gk *anydros*] (ca. 1823) *n.* The mineralogical name for native anhydrous calcium sulfate which is often associated in nature with calcium sulfate dehydrate or gypsum. It occurs occasionally as an impurity in gypsum and plaster of Paris.

See calcium sulfate, anhydrous.

Anhydrous \-drəs\ [Gk *anydrods*, fr. *a-*+ *hydōr* water] (1819) *adj.* Perfectly dry; containing no water. It is generally accepted to mean free from water of any kind, whether as such in the free-state, or as water of crystallization or other tightly held water.

Anidex fiber *n.* A manufactured fiber in which the fiber-forming substance is any long chain synthetic polymer composed of at least 50% by weight of one or more esters of a monohydric alcohol and acrylic acid, ($CH_2{=}CH{-}COOH$) (FTC definition).

Anilides *n.* Acyl derivatives from aniline. Possibly the commonest anilide in the paint trade is acetanilide, $C_6H_5NHCOCH_3$.

Aniline \ˈa-nᵊl-ən\ [Gr *anilin*, fr. *Anil* indigo, fr. FP, fr. Arabic *an-nīl* the indigo plant. fr. Sanskrit fr. *nīlī* indigo, fr. feminine of *nīla* dark blue] (1850) (phenylamine, aminobenzene) *n.* $C_6H_5NH_2$. A colorless, oily liquid made by the reduction of nitrobenzene with iron chips and an acid catalyst. It is used in the production of aniline-formaldehyde resins and certain catalysts and antioxidants.

See dyes (See image).

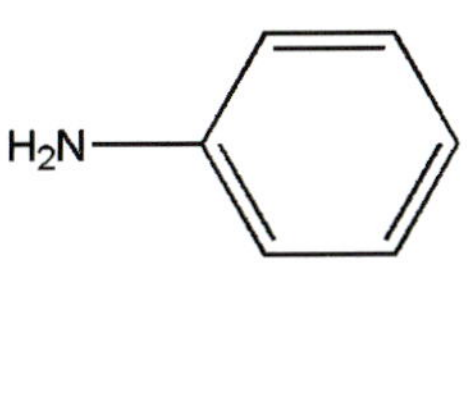

Aniline cloud point *See aniline point.*

A

Aniline dye *n.* Large class of synthetic dyes made from intermediates based upon or made from aniline.

Aniline-formaldehyde resin *n.* An aminoplastic that is made by condensing formaldehyde and aniline in an acid solution. The resins are thermoplastic and are used in making molded and laminated insulating materials with high dielectric strength and good chemical resistance.
See also amino resin.

Aniline-furfural resin *n.* Furfural yields resinous compounds not only with aniline but with aromatic amines generally. Rosin modified resins have also been produced. Amine-furfural resins are dark in color. Bleaching occurs on exposed to light. They possess the unique property of compatibility with cellulose acetate.

Aniline ink *n.* A fast-drying ink used for printing on cellophane, polyethylene, etc. Aniline inks were first made from solutions of coal-tar dyes in organic solvents, hence the name. Modern inks generally employ pigments rather than dyes.
See flexographic ink.

Aniline number *See aniline point.*

Aniline pigments *n.* Made from distillation products of coal tar, a by-product of coke and coal gas manufacture.
Also known as coal-tar colors.

Aniline point *n.* The minimum temperature at which a hydrocarbon solvent is completely soluble in an equal volume of freshly distilled aniline. Below this point, the mixture is cloudy and separates into two layers. It is used as a measure of solvent power of hydrocarbon solvents. Refer to ASTM D 1012. Syn: aniline cloud point and aniline number.

Aniline printing *See flexography.*

Anilox* roller Mechanically engraved intaglio from roller used in flexo presses to transfer a controlled film of ink from the contacting rubber-covered fountain roller to the rubber plate (or rubber-covered roller) which prints the web *Registered Trade Name, Inmont Corp.

Animal black (animal char, animal charcoal, boneblack) *n.* A form of charcoal derived from animal bones, used as a pigment. Three types are known: drop black; bone black; and ivory black. Drop black and bone black are produce by the calcinations of bones, but ivory black in its original cake form is virtually extinct.
See also carbon black, bone black, drop black, and ivory black.

Animal fats *n.* Include such products as lard, bone fat, tallow and butter fat. They are semisolid at ordinary air temperature, and consist chiefly of the glycerides of oleic and stearic acids.

Animal fibers *n.* Fibers of animal origin such as wool, alpaca, camel hair, and silk.

Animal oils *n.* Normally restricted to the animal foot oils and lard oil, and quite distinct from the fish oils. Neatsfoot oil is typical of the foot oils. The animal oils are characterized by the almost complete absence of acids more unsaturated than oleic acid, and by the presence of cholesterol. The presence of this latter compound enables them to be distinguished from the nondrying vegetable oils.

Animal waxes *n.* Obtained from a great variety of sources and have little in common, except absence of glycerides.

Animi *n.* Fossil copal, sometimes described as "goose-skin" animi because of the characteristic markings on its surface. It originated in Zanzibar and hence its alternative name, Zanzibar copal.
Also known as gum animi and mombassa gum.

Anion \ˈa-ˌnī-ən\ [Gk, neut. of *aniōn*, prp. of *anienai* to go up, fr. *ana-*+ *ienai* to go] (1834) *n.* An atom, molecule or radical that has gained an electron to become negatively charged. Anions in a liquid subjected to electric potential collect at the positive pole or anode.

Anion exchange resin *n.* Certain synthetic resins with the property of absorbing anions from aqueous salt solutions. *See ion-exchange resin.*

Anionic \ˌa-(ˌ)nī-ä-nik\ (ca. 1920) *adj.* Pertaining to a negatively (−) charged atom, radical, or molecule, or to any compound or mixture having negatively charged groups.

Anionic-cationic polymerization *n.* A method of preparing multi-block copolymers using a combination of anionic and cationic) (ion coupling) methods.

Anionic polymerization *n.* Polymerization using an anionic initiator (−charge), e.g., n-butyl lithium initiator to form polyisoprene. {G anionische Polymerisation f, F polymérisation anionique, polymérisation f, S polimerización aniónica, polimerización f, I polimerizzazione anionica, polimerizzazione f}. *See ionic polymerization.*

Anionic surfactant *n.* Surfactants, which give negatively charged ions in an aqueous solution.

Anisole *n.* $C_6H_5OCH_3$. Mixed aromatic–aliphatic ether. It has good solvent properties and a bp of 154°C.

Anisotropic \ˌa-ˌnī-sə-ˈträ-pik\ (1879) *adj.* (1) A transparent particle having different refractive indices depending on the vibration direction of light. (2) Said of materials whose properties, e.g., strength, refractive index, thermal conductivity, are unequal in different directions. Oriented thermoplastics and unidirectionally fiber-reinforced resins are typically anisotropic. Not isotropic. *See isotropic.*

Anisotropy *n.* The quality of being anisotropic; having directionally dependent properties. Exhibiting different optical properties when tested along axes in different directions. {G Anisotropie f, F anisotropie f, S.anisotropía f, I anisotropia f}. *See isotropic*

ANLAB color difference equation *n.* Abbreviation used primarily in Britain for Adams-Nickerson L, a, b Color difference equation. The normalizing factor is generally 40 and b is calculated as 0.4(V_Z−V_Y). *See Adams chromatic value color difference equation.*

ANM *n.* Copolymers from acrylic ester and acrylonitrile.

Annealing \ə-ˈnē(ə)l\ [ME *anelen* to set on fire, fr. OE *onîlan*, fr. *On*+ *îlan* to set on fire, burn, fr. *āl* fire; akin to OE *îled* fire, ON *eldr*] (1664) *v.* The process of relieving stresses in molded plastics, metal, or glass by heating to a predetermined temperature, maintaining this temperature for a set period of time, and slowly cooling the articles. Sometimes the articles are placed in jigs to prevent distortion as internal stresses are relieved during annealing.

Annual growth ring Growth layer of a tree put on in a single growth year. This includes springwood and summerwood.

Annular stop *n.* The opaque ring-shaped stop with a central opening placed usually in the objective back focal plane to give annular stop dispersion staining.

Anode \ˈa-ˌnōd\ [Gk *anodos* way up, fr. *ana-*+ *hodos* way] (1834) *n.* (1) The positive terminal of an electrical source to which electrons and negatively charged ions travel, negative for an electrolytic cell and positive for a galvanic cell. (2) The

A

electrode at which corrosion (oxidation) occurs, electron collection electrode.

Anprolene *See ethylene oxide.*

ANS *n.* Abbreviation for Adams–Nickerson–Stultz color difference equation.

ANSI *n.* Abbreviation for American National Standards Institute, which is the US. member of ISO.

Anthophyllite \ˌan(t)-thə-ˈfi-ˌlīt, (ˌ)an-ˈthä-fə-\ [Gr for *Anthophyllit*, fr. NL *anthophyllum*, fr. Gk *anthos*+*phyllon* leaf] (ca. 1828) *n.* A type of Asbestos, the major source which is in Finland. Anthophyllite is a natural magnesium iron silicate, formerly used as a filler in polypropylene to provide heat stability.

Anthracene \ˈan(t)-thrə-ˌsēn\ (1862) *n.* C_6H_4:$(CH)_2C_6H_4$. Yellow, crystalline solid obtained from the distillation of coal tar.

Anthracene oil *n.* Fraction obtained from the distillation of coal tar, with a boiling range in the region of 270°C, and containing a substantial proportion of anthracene and similar hydrocarbons.

Anthranilic acid \ˌan(t)-thrə-ˈni-lik-\ [ISV *anthr*acene+*anil*ine] (1881) *n.*
See A-acid.

Anthraquinone dyes \ˌan(t)-thrə-kwi-ˈnōn, -ˈkwi-ˌnōn\ [prob. fr. Fr, fr. *anthracene*+*quinone*] (1869) *n.* Dyes that have anthraquinone as their base and the carbonyl group (>C=O) as the chromophore. Anthraquinone-based dyes are found in most of the synthetic dye classes.
See dyes.

Antibacterial finish *n.* A treatment of a textile material to make it resistant to, or to retard growth of, bacteria.

Antibiotic \ˌan-ti-bī-ˈä-tik, -ˌtī-; ˌan-ti-bē-\ (1894) *adj.* A substance derived from a living organism capable of killing or incapacitating another organism.

Antiblocking agent (antiblock). An additive that is incorporated into resins and compounds to prevent surfaces of products (mainly films) from sticking to each other or to other surfaces. The term is not generally used for coatings, dusts, or sprays applied to surfaces for the same purpose, or as slip agents, after products have been formed. Antiblocking agents usually are finely divided, solid, infusible materials, such as silica, but be minerals or waxes. They function by forming minute, protruding asperities that maintain separating air spaces that interfere with adhesion.

Antibonding orbital *n.* The molecular orbital in which electrons have higher energies than in the unbonded atoms; an orbital characterized by a region of low electron probability density between the bonded atoms, producing a destabilizing effect on the molecule.

Antichlor *n.* A chemical, such as sodium thiosulfate, used to remove excess chlorine after bleaching.

Anticondensation paint *n.* Coating designed to minimize the effects of condensation of moisture under intermittently dry and humid conditions. Such a material normally has a matt textured finish and frequently contains heat-insulating material as a filler.

Anticorrosion paint or composition *n.* Coating used for preventing the corrosion of metals and, more particularly, specially formulated to prevent the rusting of iron, steel, and other metals.

Antifelting agents *n.* Products that prevent or minimize matting and compaction of textile materials.

Antifloating agent *n.* Additive used to prevent floating.

Antiflooding agent *n.* Additive used to prevent flooding.

Antifoaming agent *n.* An additive that reduces the surface tension of a solution or emulsion, thus inhibiting or modifying the formation of bubbles and foam. Commonly used are insoluble oils, dimethyl polysiloxanes and other silicones, certain alcohols, stearates, and glycols. In many polymerizations, these agents prevent foaming altogether. They are also used to delay foaming when producing cellular plastics. Irene Ash, Michael Ash (1993) Handbook of Industrial Surfactants, Ashgate Publishing, UK.

Antifogging agent *n.* An additive that prevents or reduces the condensation of fine droplets of water on a shiny surface. Such additives function as mild wetting agents that exude to the surface and lower the surface tension of water, thereby causing it to spread into a continuous film. Antifogging agents are much used in PVC wrapping films for meats and other moist foods. Examples are alkylphenol ethoxylates, complex polyol monoesters, polyoxyethylene esters of oleic acid, polyoxyethylene sorbitan esters of oleic acid, and sorbitan esters of fatty acids. Modern plastics encyclopedia. McGraw-Hill/Modern Plastics, New York.

Antifouling composition *n.* Paint-like composition used to prevent the growth of barnacles and other organisms on ships' bottoms. It usually contains substances, which are poisonous to such organisms in the early stages of growth. Such compositions are applied over protective paints, e.g., over anticorrosion paints in the case of steel ships.

Antigelling agent *n.* An additive that prevents a solution from forming a gel.

Antilivering agent *n.* Additive used to prevent the livering of a coating.
See livering.

Antimicrobial agent *See biocide.*

Antimony orange Sb_2S_3 Like antimony vermillion, which it resembles closely, it is a sulfide pigment, and can be prepared by precipitation from as aqueous antimony salt solution with hydrogen sulfide. Kirk-Othmer (1996) Encyclopedia of chemical technology: pigments-powders. John Wiley and Sons, New York. Solomon DH, Hawthorne DG (1991) Chemistry of pigments and fillers. Krieger Publishing Co., New York.

Antimony oxide *n.* Sb_2O_3. Pigment White 11 (77052). Antimony and oxygen combine to form trioxide, tetraoxide, and pentoxide. The most common antimony trioxide pigment is used as a flame retardant in paints, plastics and textiles. It contains, in addition, traces of the tetroxide (Sb_2O_4), and of the oxides of lead, arsenic, and iron. Density, 5.3–5.7 g/cm^3 (44.0–47.5 lb/gal); O.A., 11–14; particle size, 1 µm. Syn: antimony trioxide, antimony white, antimony sesquioxide, and flower of antimony. Kirk-Othmer (1996) Encyclopedia of chemical technology: pigments-powders. John Wiley and Sons, New York. Solomon DH, Hawthorne DG (1991) Chemistry of pigments and fillers. Krieger Publishing Co., New York.

Antimony trioxide (antimony white, flowers of antimony, and antimony oxide) *n.* Sb_2O_3. A very fine white powder made by vaporizing antimony metal in an oxidizing atmosphere, then cooling and collecting the oxide dust. Available in several ranges of particle size, it is used as a flame retardant and pigment in plastics, usually in synergistic combination with an organo-halogen compound. PVC is the biggest consumer. Kirk-Othmer (1996) Encyclopedia of chemical technology: pigments-powders. John Wiley and Sons, New York. Solomon DH,

Hawthorne DG (1991) Chemistry of pigments and fillers. Krieger Publishing Co., New York.

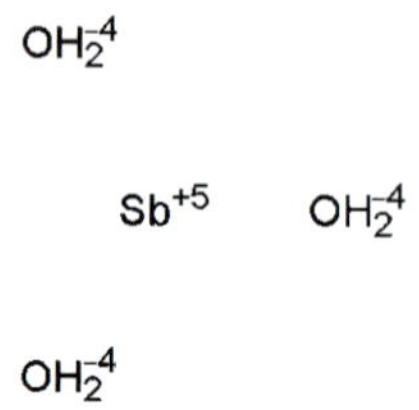

Antimony vermillion *n.* Sb_2S_3. Pigment varying on color from pale orange to deep crimson, with good opacity and tinting strength. It is prepared either by adding a solution of sodium thiosulfate to a solution of tartar emetic and tartaric acid, or to a solution of another suitable antimony salt. The pigment is not resistant to acids, alkalis, or heat, but, if properly made, is lightfast. It is sometimes referred to as Crimson Antimony. Kirk-Othmer (1996) Encyclopedia of chemical technology: pigments-powders. John Wiley and Sons, New York. Solomon DH, Hawthorne DG (1991) Chemistry of pigments and fillers. Krieger Publishing Co., New York.

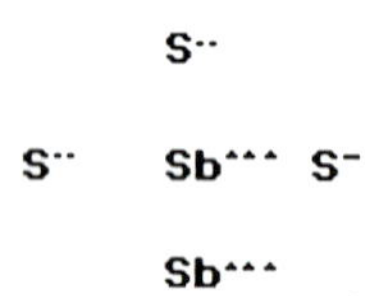

Antimony white n. *See antimony oxide.*

Antimony yellow *n.* $Pb_3(SbO_4)$. Essentially, lead antimonite, which may be considered to be chemically combined lead and antimony oxides. It varies in color from sulfur yellow to orange-yellow, depending upon the proportion of the two materials. Kirk-Othmer (1996) Encyclopedia of chemical technology: pigments-powders. John Wiley and Sons, New York.

Antinode \ˈan-ti-ˌnōd, ˈan-ˌtī-\ [ISV] (1882) *n.* A region or location of maximum disturbance in a standing wave. Giambattista A, Richardson R, Richardson RC, Richardson B (2003) College physics. McGraw Hill Science/Engineering/Math, New York.

Anti-noise paint *n.* Coating used for its sound deadening effect, on vibrating and naturally noisy machinery. It relies for its effectiveness on its very rough surface, produced by the incorporation of cork and similar granular, or fibrous materials. *Also called Acoustical Paint.*

Antioxidant \ˌan-tē-ˈäk-sə-dənt, ˌan-ˌtī-\ (1926) *n.* A substance that slows down the oxidation of oils, fats, etc., and thus helps to check deterioration: antioxidants are added commercially to foods, soaps, etc. Although the term technically applies to molecules reacting with oxygen, it is often applied to molecules that protect from any free radical molecule with unpaired electrons. Chemistry encyclopedia, www.ChemistryAbout.com. A substance incorporated in a material to inhibit oxidation at normal or elevated temperatures. Antioxidants are used mainly with natural and synthetic rubbers, petroleum-based resins, and other such polymers that oxidize readily due to structural unsaturation. However, some thermoplastics, namely polypropylene, ABS, rubber-modified polystyrene, acrylic and vinyl resins, also require protection by antioxidants for some uses. There are two main classes: (1) Those inhibit oxidation by reacting with chain-propagating radicals, such as hindered phenols that intercept free radicals. These are called primary antioxidants or free-radical scavengers. (2) Those decompose peroxide into non-radical and stable products; examples are phosphates and various sulfur compounds, e.g., esters of thiodipropionic acid. These are referred to as secondary antioxidants, or peroxide decomposers. {G Ozonschutzmittel n, F anti-ozone m, S antiozonante m, I antio-zonante m}. Merriam-Webster's collegiate dictionary, 10th edn. Springfield, MA, 2000. Glenz W (ed) (2001) A glossary of plastics terminology in five languages, 5th edn. Hanser–Gardner Publications Inc., Cinicinnati. James F (ed) (1993) Whittington's dictionary of plastics. Technomic Publishing Co. Inc., Carley.

Antioxidants *n.* Agents, which deter or retard autooxidation degration such as phenol and arylamines, agents react with the intermediate peroxy radicals. These agents retard the action of oxygen in drying oils and other substances subject to oxidation. {G Antioxidans n, F antioxydant m, S antioxidante m, I antiossidante m}. Glenz W (ed) (2001) A glossary of plastics terminology in five languages, 5th edn. Hanser–Gardner Publications Inc., Cinicinnati. Zaiko GE (ed) (1995) Degradation and stabilization of polymers. Nova Science Publishers Inc., New York.

Antiozonant \-ˈō-(ˌ)zō-nənt\ (1954) *n.* Substance that retards or prevents the action of ozone on elastomers when exposed under tension, either statically or dynamically, to air containing ozone.

Antiozonants Same antioxidants.

Anti-parallel spins Two spins, which are in opposite directions.

Antique finish *n.* Usually applied to nearly painted furniture or objects for the purpose of giving the appearance of mellowness of age.

Antiquing \(ˌ)an-ˈtēk\ (1923) *n.* (1) Treating or finishing of wood to make it look old. (2) Imparting a special coloring effect by the application of a base coat followed by a transparent or semitransparent colored glaze, which is either applied by brush, then partially wiped away or applied by wiping with a cloth.

Anti-sag agent *n.* Additive used to control sagging of a coating.
See thickening agent.

Antiseptic \ˌan-tə-ˈsep-tik\ [*anti-* + Gk *sēptikos*] (1751) *adj.* An agent used to destroy or restrain the growth of microorganisms.

Anti-septic wash *See fungicidal wash.*

Anti-settling agent *n.* Substance incorporated into pigmented paint to retard settling and to maintain uniform consistency during storage or, in dipping paints, during painting operations. These additives normally function by altering the rheological properties of the paint. Syn: suspending agent.

Anti-shatter composition *n.* Composition designed to prevent the fragmentation of glass and glass-like materials. The main use of these compositions is in the manufacture of spliterproof glass. They are applied as an adhesive, resilient sandwich between relatively thin sheets of glass. A violet blow may cause cracking of the glass, but the anti-shatter composition is designed to retain the various pieces in position. Properties include good permanent adhesion to the glass, a natural resilience such that it will absorb violet shocks without itself disintegrating, a permanent freedom from color, and absolute transparency.

Anti-silking agent *n.* Additive used to prevent silking, which is a special case of floating, resulting in the formation of parallel hairlike striations of differing colors, running through the length of the painted surface.

Anti-skinning agent *n.* Any material added to a coating or printing ink to prevent or retard the processes of oxidation or polymerization which results in the formation

A

of an insoluble skin on the surface of the coating or printing ink in a container.

Anti-slip paint *n.* Used for application to decks and other surfaces where the conditions conductive to slipping are present. It prevents slipping by providing a tough, rough surface. The roughness is usually achieved by the addition of sand or cork dust. *Also called non-skid paint.*

Antisoiling properties The properties of textile materials whereby they resist deposition of dirt and stains.

Antistaining properties The ability of a textile to resist the deposition of oil- or water-borne stains.

Anti-stat \-ˈstat *or* an-ti-stat-ic, -ˌsta-tik\ (1952) *adj.*

See antistatic agent.

Antistatic agent (antistat) *n.* A chemical (or other material) that imparts slight electrical conductivity to plastics compounds (or other polymeric formulation), preventing the accumulation of electrostatic charges on finished articles. The agent may be incorporated in the materials before molding (most preferable) or applied to their surfaces afterward. Antistatsic agents function either by being inherently conductive or by absorbing moisture from the air. Examples are long-chain aliphatic amines and amides, phosphate esters, quaternary ammonium salts, polyethylene glycols, polyethylene-glycol esters, and ethoxylated long-chain amines. Modern plastics encyclopedia. McGraw-Hill/Modern Plastics, New York, 1986-, 1990-, 1992-, 1993-editions.

Antistatic properties The ability of a textile material to disperse an electrostatic charge and to prevent the build up of static electricity.

Antitracking varnish *n.* When two points of high electrical potential difference are suitably located in contact with the surface of an organic insulating medium, the surface of the insulator may be degraded to carbon under the imposed electrical stresses, and a carbon track formed between the two points. Phenol-formaldehyde laminated boards are sometimes given a special coating of varnish at their exposed surfaces to prevent such carbonization. Varnishes used for such purposes are described as anti-tracking varnishes. American Society for Testing and Materials, www.astm.org, 100 Barr Harbor Drive, West Conshohocken, PA.

Antiwrinkling agent *n.* Material added to surface coating compositions to prevent the formation of wrinkles in films during drying.

Ant oil *See furfural.*

Antwerp blue *n.* Essentially, zinc ferrocyanide with a pale greenish-blue color. It is characterized by good opacity and almost complete absence of floating tendency.

AOCS Abbreviation for American Oil Chemists Society.

Apertured non-woven fabric *n.* A non-woven fabric having many small through-holes made by laying the fabric on a perforated plate or screen and applying fluid pressure.

APHA color scale *n.* Abbreviation for American Public Health Asscoation color scale developed for visually evaluating slight yellowness of solutions. It is based on dilutions of a platinum–cobalt standard solution of 500 APHA. This single-number color scale was originally developed by Hazen, and may also be referred to as the Hazen color scale.

API *n.* Abbreviation for American Petroleum Institute.

API gravity *n.* Measure of specific gravity of petroleum and petroleum products, defined by the following equation:

$$\text{API gravity, degrees} = \frac{141.5}{\text{specific gravity } 60/60°\text{F}} - 131.5.$$

Apochromat \ˌa-pə-krō-ˈmat\ [ISV] (1887) *adj.* A term applied to photographic and microscope objectives indicating the highest degree of color correction.

Apochromatic objective *n.* An objective corrected for spherical aberration at two wavelengths and for chromatic aberration at three wavelengths.

Apparent density The mass per unit volume of material including voids inherent in the material as tested, such as pellets, powders, foams, chopped film or fiber scrap. Usually expressed as lbs/cubic foot or g/cm^3. The term bulk density is synonymous for particulate materials. ASTM tests are D 1895 for pellets and powders, except PTFE powders, for which D 1457 applies; D 1622 for rigid foams.

See also density and bulk factor.

Apparent viscosity *n.* At any point in a fluid undergoing laminar shear, the nominal shear stress divided by apparent shear rate. In simple fluids, viscosity is a state property, depending only on composition, temperature, and pressure. In polymer melts and solutions, it is, nearly always, also dependent on the shear rate (or stress), hence the term *apparent viscosity.* The term is also applied to the quotient of the shear rate at the wall, which reduces to $\pi R^4 \cdot \Delta P/(8QL)$, where R and L are the radius and length of the tube, ΔP is the pressure drop through the tube, and Q is the volumetric flow rate. Elias HG (1977) Macromolecules, vols 1–2. Plenum Press, New York. Staudinger H, Heuer W (1930) A relationship between the viscosity and the molecular weight of polystyrene (German). Ber 63B:222-234.

Appearance \ə-ˈpir-ən(t)s\ (14c) *n.* Manifestations of the nature of objects and materials through visual attributes such as size, shape, color, texture, glossiness, transparency, opacity, etc.

Appliance finish *n.* Generally, thermoset coatings, which are characterized by their hardness, mar resistance, and good chemical resistance.

Application \ˌa-plə-ˈkā-shən\ [ME *applicacioun*, fr. L *application-*, application inclination, fr. *applicare*] (15c) *n.* Process by which surface coating compositions are transferred to a variety of surfaces, such as: brushing; spraying (cold and hot); dipping (simple immersion); dipping (assisted by application of vacuum and/or pressure); roller coating; flushing; and spreading.

Applicator \ˈa-plə-ˌkā-tər\ (1659) *n.* Device to deposit a film of a specified thickness, such as doctor blade, wire wound rod, drawdown bar, etc.

Applicator roll *n.* Roll in a roller coater, which applies the paint to a continuous strip.

Appliqué \ˌa-plə-ˈkā\ [Fr, pp of *appliquer* to put on, fr. L *applicare*] (1801) *n.* Pattern or picture, formed by laying various materials usually on a fabric ground. Especially suitable for dressmaking and needlework. A design or ornament applied to another surface. In wallpaper, cut-outs applied to a plain, textured, or figured background.

Appurtenance \ə-ˈpərt-nən(t)s, -ˈpər-tᵊn-ən(t)s\ (14c) *n.* Any built-in, nonstructural portion of a building, such as doors, windows, ventilators, electrical equipment, partitions, etc.

Apricot kernel mixed fatty acids *n.* These have a mp of 12–14°C; an acid value of 194; and an iodine value of 103.

Apricot kernel oil *n.* Non-drying oil with iodine value of 96.

A

Apron \ˈā-prən, -pərn\ {*often attributive*} [ME, alteration (resulting fr. false division of *a* napron) of *napron*, fr. MF *naperon*, dim. of *nape* cloth, mod. of L *mappa* napkin] (15c) *n.* A paved area, such as the juncture of a driveway with the street or with a garage entrance. Merriam-Webster's collegiate dictionary, 11th edn. Merriam-Webster Inc., Springfield, MA, 2004.

Apron mark *See decating mark.*

Aprotic solvent \(ˌ)ā-ˈprō-tik\ [*²a-* + *proto*n + *¹-ic*] (1931) *adj.* An organic solvent that neither donates protons to or accepts them from a substance dissolved in it. Benzene, C_6H_6, is such a solvent. Wypych G (2001) Handbook of solvents. Chemtec Publishing, New York.

Aprotic substance *n.* A substance that can act neither as an acid nor as a base. Goldberg DE (2003) Fundamentals of chemistry. McGraw-Hill Science/Engineering/Math, New York.

Aptitude, color *See color aptitude.*

A. Pullulans *See color aptitude.*

Aqua regia *n.* A powerful oxidizing and complexing solvent mixture composed of about three parts concentrated HCl to one part concentrated HNO_3.

Aquarelle \ˌa-kwə-ˈrel, ˌä-\ [F, fr. obs. I *acquarella* (now *acquerello*), fr. *acqua* water, fr. L *aqua*] (1869). (1) *n.* Drawing or print which has been colored with transparent water color washes. (2) Color that is made workable when mixed with water.

Aquatint \ˈa-kwə-ˌtint, ˈä-\ [I *acqua tinta* dyed water] (1782) *n.* Method of engraving which, like etching, involves the use of a mordant acid for biting a metal plate, but differing in that aquatint is used to render tonal instead of linear effects.

Aqueous \ˈa-kwē-əs, ˈa-\ [ML *acqueus*, fr. L *aqua*] (1646) *adj.* Water-containing or water-based.

Aqueous acrylic *See latex.*

Aqueous phase of emulsion The aqueous phase of an aqueous emulsion is water mixed with surfactant sometimes initiator and catalyst.

Aqueous thermoplastic emulsions *n.* Water based acrylic emulsions with noncrosslinked latex particles.

Aquo complex *n.* A complex in which water molecules serve as ligands.

AR *n.* Abbreviation for acrylic rubber.

Arabic (14c) n. *See acacia gum or gum Arabic.*

Arachis oil \ˈar-ə-kəs-\ [NL *arachis*, genus that includes the peanut, fr. Gk *arakis*, dim. of *arakos*, a legume] (ca. 1889) *n.* *See peanut oil.*

Arachne machine *n.* (1) A machine for producing loop-bonded nonwovens. The fabric is formed by knitting a series of warp yarns through a fiber web processed on a card. *Also see bonding.* (2) Stitch bonding.

Aragonite \ə-ˈra-gə-ˌnīt, ˈar-ə-g -\ [Gr *Aragonit*, fr. *Aragon*, Spain] (1803) *n.* *See calcium carbonate* (*See image*).

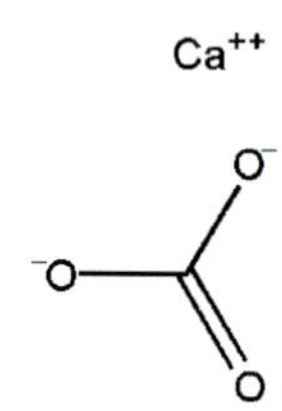

Aralac Albumin fiber. Manufactured by National Dairy Products, USA.

Araldite *n.* Epoxide resins. Manufactured by CBA, Switzerland.

Aramid \ˈar-ə-məd, -ˌmid\ [*aro*matic poly*amide*] (1972) *n.* Acronym for *ar*omatic poly*amide*, currently available in fiber foam only, having at least 85% of the amide groups bonded to two aromatic rings. DuPont's Kevlar® is poly(*p*-phenylene terephthalamide). Aramid fibers exhibit low flammability, high strength, and high modulus. Fabrics made from aramid fibers maintain their integrity at high temperatures. Such fabrics are used extensively in hot-air filters. Aramids are also found in protective clothing, ropes and cables, and tire cord.

Aramid fiber *n.* Any of a family of high-strength, high-modulus fibers made from aramid resin. DuPont's Kevlar® −49 and −29 are the best known. K-49's ultimate strength is 3.4 GPa (500 kpsi), modulus is 131 GPa (19 Mpsi), ultimate elongation is 2.4%. K-29 has about equal strength, but half the modulus and twice the elongation. Density of either is 1.44 g/cm^3. Strength/density is higher for either of these fibers than for any others except some whiskers.

Archimedes principle \ˌär-kə-ˈmē-dēz-\ *n.* A body wholly or partly immersed in a fluid is buoyed up by a force equal to the weight of the fluid displaced. A body of volume V (cm^3) immersed in a fluid of density ρ (g/cm^3) is buoyed up by a force in dynes

$$F = \rho g V,$$

where g is the acceleration due to gravity. A floating body displaces its own weight of liquid. Serway RA, Faugh JS, Bennett CV (2005) College physics. Thomas, New York.

Architectural coatings *n.* Coatings intended for on-site application to interior or exterior surfaces of residential, commercial, institutional, or industrial – as opposed to industrial coatings. They are protective and decorative finishes applied at ambient temperatures. Wicks ZN, Jones FN, Pappas SP (1999) Organic coatings science and technology, 2nd edn. Wiley-Interscience, New York.

Architecture \ˈär-kə-ˌtək-chər\ (1555) *n.* The art and science of designing and building structures, or large groups of structures, in keeping with aesthetic and functional criteria, and structures built in accordance with such principles.

Arcing \ˈär-kiŋ\ (1893) *v.* Swinging spray gun away from the perpendicular.

Arco microknife Instrument designed for testing scratch hardness and adhesion of coatings. A diamond point is weighted until it can penetrate a film to the metal substrate in two retracing steps. The weight necessary to achieve this cutting force for films of standard thickness is a measure of hardness.

Arc resistance *n.* The ability of a plastic material to maintain low conductivity along the path of exposure to a high-voltage electrical arc, usually stated in terms of the time required to render the material electrically conductive. Failure of the specimen may be caused by heating to incandescence, burning, tracking, or carbonization of the surface. Dissado LA, Fothergill CJ (eds) (1992) Electrical degradation and breakdown of polymers. Institution of Electrical Engineering (IEE), London. Emerson JA, Torkelson JM (eds) (1991) Optical and electrical properties of polymers: materials research society symposium proceedings, vol 24. Materials Research Society, Warrendale, PA. Ku CC, Liepins R (1987) Electrical properties of polymers. Hanser Publishers, New York.

Arc tracking *See tracking.*

Ardil *n.* Fiber from peanut protein. Manufactured by ICI, Great Britain.

Area, unit of The square centimeter. The area of a square whose sides are 1 cm in length. Other units of area are similarly derived. Dimensions, $[l^2]$.

Argillaceous \ˌär-jə-ˈlā-shəs\ (ca. 1731) *adj.* Composed primarily of clay or shale; clayey.

Argyle also ar-gyll \ˈär-ˌgī(ə)l, är-ˈ\ [*Argyle, Argyll,* branch of the Scottish clan of

A

Campbell, fr. whose tartan the design was adapted] (1899) *n.* A pattern consisting of diamond shapes of different colors knit in a fabric.

Arithmetic mean (arithmetic average, mean $\bar{x}$) *n.* (1) In statistics, the average of a set of measurements found by summing the measurements and dividing the sum by the number of measurements (number-average molecular weight is an arithmetic average). (2) The conceptual mean, μ, of the population from which a set of measurements was drawn, rarely known exactly. The sample mean, $\bar{x}$, is the most efficient estimator of the population mean, μ.

Arnaudon's green *See chromium oxide green.*

Arnite Poly(ethylene terephthalate) (as plastic). Manufactured by AKU, The Netherlands.

Aromatic compounds *n.* A class of organic compounds containing a resonant, unsaturated ring of carbon atoms. Included are benzene, naphthalene, anthracene, andtheir derivatives. The term *aromatic* stems from the fact that many of these compounds have an agreeable odor. Morrison RT, Boyd RN (1992) Organic chemistry, 6th edn. Prentice-Hall, Englewood Cliffs, NJ.

Aromatic content Percent aromatic hydrocarbons present in a solvent mixture or in a compound.

Aromatic electrophilic-substitution *n.* Typical reactions of the benzene ring since the ring is electron rich (a base) and reactants are electron poor; typical reactions include nitration, halaogenation, and sulonation. Morrison RT, Boyd RN (1992) Organic chemistry, 6th edn. Prentice-Hall, Englewood Cliffs, NJ.

Aromatic hydrocarbon *n.* A compound of carbon and hydrogen whose molecular structure contains one or more rings of six carbon atoms, with at least one of the rings containing alternating, resonant single and double bonds. Benzene, which is the simplest of the aromatic hydrocarbons, has the molecular formula C_6H_6. The family includes many solvents for plastics.

Aromatic petroleum residues *n. See diolefin resins.*

Aromatic polyamide *n. See aramid.*

Aromatic polyester *n.* A polyester that has aromatic rings in its chain, e.g., polyethylene terephthalate.

Aromatics (15c) *n.* Compounds containing at least one benzene ring. Benzene, toluol, Xylol or aromatic solvents.

Aromatic solvents *n.* Organic liquids having a cyclic or ring hydrocarbon structure and KB values over 40. They are good solvents for printing ink resins. Examples are toluol and xylol (cf. *Aliphatic solvents*). Use of aromatic solvents is severely restricted by Rule 66. Aromatic solvents containing less than 80% aromatic compounds are frequently designated as partial aromatic solvents.

Aromatization \ə-ˌrō-mə-tə-ˈzə-shən\ (15c) *vt.* Process of converting saturated or aliphatic hydrocarbons to aromatic hydrocarbons.

Arrhenius equation \ə-ˈrē-nē-əs, -ˈrā-\ *n.* (1) A classical equation describing how rates of chemical reactions increase with rising absolute temperature:

$$r = Ae^{-E/RT},$$

in which *r* is the reaction rate (in appropriate units), *A* the collision factor (in the same units as *r*), e the 2.71828..., *E* is the activation energy of the reaction (J/mol), *R* the universal molar-energy constant (8.3144J/(molK), and *T* is the absolute temperature (K). (2) An almost identical form, differing only in that the sign of the exponent is positive (viscosities *decrease* with rising temperature), has been used with good results to represent the temperature dependence of liquid viscosities, including those of polymer solutions and melts.

It, too, is often referred to as an Arrhenius equation. This form has also been successful in modeling the temperature dependence of creep failure and property retention during heat aging. For viscosity work, the logarithmic form of this equation is more convenient.

$$\ln(\mu/\mu_o) = E(T - T_o)/RTT_o.$$

here E, R, and T have the same meanings as above; μ_o represents the viscosity at a reference temperature, T_o, which is usually chosen to be within the temperature range over which the viscosities have been measured. Atkins PW, Atkins P, De Paula J (2001) Physical chemistry. W. H. Freeman Co., New York. Patton TC (1979) Paint flow and pigment dispersion: a rheological approach to coating and ink technology. John Wiley and Sons, New York. Miller ML (1966) Structure of polymers. The Reinhold Publishing Co., New York. Elias HG (1977) Macromolecules, vols 1–2. Plenum Press, New York.

Arrhenius plot *n.* A plot of the logarithm of the specific rate constant for a reaction against the reciprocal of the absolute temperature; useful for determining the activation energy of the reaction. Atkins PW, Atkins P, De Paula J (2001) Physical chemistry. W. H. Freeman Co., New York.

Arrhenius theory of electrolytic dissociation *n.* This theory states that the molecule of an electrolyte can give rise to two or more electrically charged atoms or ions. Atkins PW, Atkins P, De Paula J (2001) Physical chemistry. W. H. Freeman Co., New York.

Arsenic orange \ˈärs-nik ˈär-inj\ *n. See realgar.*

Artificial aging \ˌär-tə-ˈfi-shəl\ *n.* The accelerated testing of plastics to determine their changes in properties such as dimensional stability, water resistance, resistance to chemicals and solvents, light stability, and fatigue resistance.

Artificial daylight *n.* Term loosely applied to light sources, frequently equipped with filters, which are claimed to reproduce the color and spectral distribution of daylight. A more specific definition of the light source is to be preferred.

See correlated color temperature, color rendering index, and spectral power distribution curve.

Artificial stone *n.* Special concretes and tiles, artificially colored to simulate natural stone, obtained by mixing stone dust aggregate and chips with Portland cement.

Artificial turf *n.* A manufactured carpet having the appearance of grass. It is used to replace grass in sports arenas, yards, etc. *Also see recreational surfaces.*

Artificial ultramarine *n.* Synthetic ultramarine blue.

See ultramarine blue.

Artificial weathering *n.* A process of simulating weathering conditions such sunlight and rain using accelerated methods (e.g., Salt Fog Spray ASTM B 117). Artificial weathering is a method of utilizing exposure to cyclic laboratory conditions involving changes in temperature, relative humidity, and radiant energy, with or without direct water spray, in an attempt to produce changes in the material similar to those observed after long-term continuous outdoor exposure. American Society for Testing and Materials, www.astm.org, 100 Barr Harbor Drive, West Conshohocken,

A

PA. American Society for Testing and Materials, www.astm.org, 100 Barr Harbor Drive, West Conshohocken, PA.

Artists' colors *n.* Various paint media used by artists, such as oil paints, watercolors, gouache, tempera, encaustic, fresco, silicate esters, and latex.

Art linen *n.* A plain-weave, softly finished fabric used either bleached or unbleached as a base fabric for needlework.

Aryl \ˈar-əl\ [ISV *aromatic* + *-yl*] (1906) *adj.* Pertaining to monovalent aromatic groups, e.g., C_6H_5, phenyl.

ASA *See American national standards institute.*

Asbestine *See magnesium silicate, non-fibrous.*

Asbestos \as-ˈbes-təs, az-\ [ME *albestron* mineral supposed to be inextinguishable when set on fire, prob. fr. MF, fr. ML *asbeston,* alter. of L asbestos, fr. Gk, unslaked lime, fr. *asbestos* inextinguishable, fr. *a-* + *sbennynai* to quench] (1607) *n.* The commercial term for a family of fibrous mineral silicates comprising some 30 known varieties, of which six were, for many years, commercially important. They are of two general types, *serpentine* and *amphibole.* The serpentine type contains chrysotile that was most widely used as a reinforcement in thermosetting resins and laminates and, in finer form, as a filler in polyethylene, polypropylene nylons, and vinyls. Abestos used in flooring sheet and tiles. The amphiboles provided good chemical resistance and lower water absorption, but the outstanding properties provided all asbestos types were resistance to heat, fire retardance, and resistance to chemicals and for reinforcing plastics, and some brake linings for vehicles may still contain it. Extreme precautions must be taken in handling asbestos and asbestos-filled materials. The OSHA limit for such workers is 2fibers/m^3 of air, averaged over an 8-h shift. The use of asbestos products has declined in the United States due to health hazards associated with it. Hibbard MJ (2001) Mineralogy. McGraw-Hill Companies Inc., New York. James F (1993) Whittington's dictionary of plastics. Technomic Publishing Co. Inc., Carley.

Asbestos-cement *n.* A dense, rigid, board containing a high proportion of asbestos fibers bonded with Portland cement; resistant to fire, flame, and weathering; has low resistance to heat flow. Used as a building material in sheet form and corrugated sheeting. Harris CM (2005) Dictionary of architecture and construction. McGraw-Hill Co., New York.

Asbestos-cement board *n.* A dense, rigid, board containing a high proportion of asbestos fibers bonded with Portland cement; resistant to fire, flame, and weathering; has low resistance to heat flow. Used as a building material in sheet form and corrugated sheeting. Harris CM (2005) Dictionary of architecture and construction. McGraw-Hill Co., New York.

Asbestos plaster *n.* A fireproof insulating material generally composed of asbestos with Bentonite as the binder. Harris CM (2005) Dictionary of architecture and construction. McGraw-Hill Co., New York.

Asbestos sheeting *n.* A dense, rigid, board containing a high proportion of asbestos fibers bonded with Portland cement; resistant to fire, flame, and weathering; has low resistance to heat flow. Used as a building material in sheet form and corrugated sheeting. Harris CM (2005) Dictionary of architecture and construction. McGraw-Hill Co., New York.

Asbestos wallboard *n.* A dense, rigid, board containing a high proportion of asbestos fibers bonded with Portland cement; resistant to fire, flame, and weathering; has low resistance to heat flow. Used as a building material in sheet form and corrugated sheeting.

Ascaridole (1,4-peroxide-*p*-menthene-2). $C_{10}H_{16}O_2$. A naturally occurring peroxide with uses as a polymerization initiator.

Aseptic packaging *n.* A package that has been sterilized with gas, heat or radiation after it was sealed.

Ash \ˈash\ [ME *asshe,* fr. OE *æsc;* akin to OHGr *ask* ash, L *ornus* mountain ash] (before 12c) *n.* The mineral residue left after burning or decomposition of a sample of a substance.

Ash content The solid residue remaining after a substance has been incinerated or heated to a temperature sufficient to drive off all volatile or combustible substances.

Ashing *n.* A finishing process used to produce a satin-like finish on plastic articles, or to remove cold spots or teardrops from irregular surfaces, which cannot be reached by wet sanding. The part is applied to a loose muslin disk loaded with wet ground pumice, rotating at a lineal speed of about 20 m/s.

ASM international (American Society of Materials) *n.* A material science and engineering society headquartered in Materials Park, OH, www.asminternational.org.

Aspartic acid \ə-ˈspär-tik-\ [ISV, irreg. fr. L *asparagus*] (1863) *n* $COOHCH_2CH(NH_2)COOH$. Amino acid constituent of proteins of both animal and vegetable origin. In particular it is present in casein. It is a dibasic acid.

Aspect ratio (1907) *n.* (1) The ratio of length to diameter of a fiber or yarn bundle. (2) In tire production, the ratio of the height of the tire to its width. (3) In a rectangular structure, the ratio of the longer dimension to the shorter. Complete textile glossary. Celanese Acetate LLC, Three Park Avenue, New York, 2000.

Asphalt \ˈas-ˌfólt *also* ˈash-, *esp British* -ˌfalt\ [ME *aspalt,* fr. LL *asphaltus,* fr. Gk *asphaltos*] (14c) (bitumen) *n.* (1) A dark brown or black, bituminous, viscous material which gradually liquefies when heated. The predominating constituents are bitumens, all of which occur in the solid or semisolid form in nature or are obtained by refining petroleum, or which are combinations of the bitumens mentioned with each other or with petroleum or derivatives thereof. (2) A similar material obtained artificially in refining petroleum; used in built-up roofing systems as a water-proofing agent. (3) A mixture of such substances with an aggregate for use in paving. Usmani AM (1997) Asphalt science and technology. Marcel Dekker, New York.

Asphalt cut back *n.* Asphalt plus thinner; asphalt solution; asphalt coating formed by dissolving asphalt. Usmani AM (1997) Asphalt science and technology. Marcel Dekker, New York.

Asphalt emulsion *n.* A suspension or emulsion of ordinary asphalt in water. Such emulsions have attained wide use because, unlike straight asphalt, they do not have to be heated to be applied. The suspended asphalt is spread on in the usual way; and after the water has evaporated, the asphalt hardens into a continuous mass. Uses: highways; cement water-proofing; roofing compounds; and the like.
Also called emulsified Asphalt.

Asphaltenes *n.* Highly condensed hydrocarbon compounds present in bitumens and asphaltums. Little is known of their chemical structure, but they are usually characterized by their insolubility in low boiling aliphatic hydrocarbons. Their physical condition is solid or semisolid. Usmani

A

AM (1997) Asphalt science and technology. Marcel Dekker, New York.

Asphaltic bitumens *n.* May be either naturally occurring materials or otherwise obtained from the distillation of asphaltic base petroleum; they are readily soluble in aliphatic and aromatic hydrocarbons and have good film-forming properties. Usmani AM (1997) Asphalt science and technology. Marcel Dekker, New York.

Asphaltic pyrobitumens *n.* Include elaterite, impsonite, and wurtzlite. They are characterized by their hydrocarbon nature, infusibility, and insolubility in carbon disulfide. Oxygen-containing compounds are present only in small amounts. Usmani AM (1997) Asphalt science and technology. Marcel Dekker, New York.

Asphalt paint *n.* A liquid asphaltic product sometimes containing small amounts of other materials such as lampblack, aluminum flakes, and mineral pigments. Ash M, Ash I (1996) Handbook of paint and coating raw materials: trade name products – chemical products dictionary with trade name cross-references. Ashgate Publishing Ltd., New York.

Asphalt paper *n.* A paper sheet material that has been coated, saturated, or laminated with asphalt to increase its toughness and its resistance to water. Ash M, Ash I (1996) Handbook of paint and coating raw materials: trade name products – chemical products dictionary with trade name cross-references. Ashgate Publishing Ltd., New York.

Asphalt seal coat *n.* A bituminous coating, with or without aggregate, applied to the surface of a pavement to waterproof and preserve the surface and to improve the texture of a previously applied bituminous surface. Ash M, Ash I (1996) Handbook of paint and coating raw materials: trade name products – chemical products dictionary with trade name cross-references. Ashgate Publishing Ltd., New York.

Asphalt tile sealer *n.* Resin or plastic-type finish usually containing a water or alcohol carrier. Ash M, Ash I (1996) Handbook of paint and coating raw materials: trade name products – chemical products dictionary with trade name cross-references. Ashgate Publishing Ltd., New York.

Asphaltum (or asphalt) \as-ˈfól-təm, *esp British* -ˈfal-\ [ME *asphalt*, fr. LL *aspaltus*, fr. Gk *asphaltos*] (14c) *n.* US paint industry's term for asphalt. In Britain, the term is reserved for natural asphalt. Usmani AM (1997) Asphalt science and technology. Marcel Dekker, New York.

Asplit *n.* Phenoplast. Manufactured by Hoechst, Germany.

ASR *See alkali-soluble resins.*

Assembly \ə-ˈsem-blē\ [MD *assemblee*, fr. MF, fr. OF, fr. *assembler*] (14c) *n.* A group of materials or parts, including adhesive, which has been placed together for bonding or which has been bonded together.

Assembly adhesive *See adhesive, assembly.*

Assembly glue *See adhesive, assembly.*

Assembly of plastics *n.* Plastic parts may be joined to others by many methods. Self-tapping screws are made with special thread designs to suit specific resins. Threaded inserts to receive mounting screws may be molded in or installed by press-fitting or by means of self-tapping external threads. Press-fitting may be employed to join plastics to similar or dissimilar materials. Snap-fit joints are made by molding or machining an undercut in one part, and providing a lip to engage this undercut in the mating component.

A

Assembly time *n.* Elapsed time after the adhesive is spread and until the pressure becomes effective.

See time, assembly.

Associative thickener A thickener which obtains its efficiency presumably by association between thickener molecules or thickener and latex particles rather than through high molecular weight or chain stiffness of the thickener molecules themselves.

See also thickener.

A-stage *n.* An early stage in the preparation of certain thermosetting resins in which the material is still fusible and soluble in certain liquids. Sometimes referred to as a *resol.*

See also B-stage and C-stage.

A-stage thermosetting resins *n.* The first stage of novalac (phenolic) formation before crosslinking or curing.

Astigmatism \ə-ˈstig-mə-ˌti-zəm\ (1849) *n.* An error of spherical lenses peculiar to the formation of images by oblique pencils. The image of a point when astigmatism is present will consist of two focal lines at right angles to each other and separated by a measurable distance along the axis of the pencil. The error is not eliminated by reduction of aperture as is spherical aberration.

ASTM American society for testing materials. ASTM headquarters is located at 1916 Race St., Philadelphia, PA, may be the largest non-governmental, standards-writing body in the world, with 33,000 members. All information, standards, etc., for adhesives, coatings, inks, plastics and sealants are available on the official ASTM web site, www.astm.org.

Astrakhan cloth \ˈas-trə-kən, -ˌkan\ *adj.* A thick knit or woven fabric with loops or curls on the face. The base yarns are usually cotton or wool and the loops are made with fibers such as mohair, wool, and certain manufactured fibers. The face simulated the pelt of the astrakhan lamb. Vincenti R (ed) (1994) Elsevier's textile dictionary. Elsevier Science and Technology Books, New York.

Asymmetric \ˌa-sə-ˈme-tri-kəl\ [Gk *asymmetria* lack of proportion, fr. *asymmetros* ill-proportioned, fr. *a-* + *symmetrical*] (1690) *adj.* Of such a form that no point, line or plane exists about which opposite portions are congruent. The opposite of *symmetrical.* Merriam-Webster's Collegiate Dictionary, 11th edn. Merriam-Webster Inc., Springfield, MA, 2004.

Asymmetry *n.* In chemistry, a molecular arrangement in which a particular carbon atom is joined to four different groups.

Atactic \(ˌ)ā-ˈtak-tik\ [ISV ²*a-* + *-tactic*] (1957) *adj.* Pertaining to a polymer in which the pendant side groups, as $-CH_3$ in polypropylene, are randomly located around the main chain. It can also be defined as the term for polymer tacticity to indicate unsymmetrical and alternating substituent groups along a polymer chain {G ataktisch, F atactique, S atáctico, I atattico}.

Atactic block *n.* A block of chain units in a polymer of copolymer that has a random distribution of equal numbers of the possible configurational base units. Odian GC (2004) Principles of polymerization. John Wiley and Sons Inc., New York. Mark JE (ed) (1996) Physical properties of polymers handbook. Springer-Verlag, New York.

Atactic configuration *See atactic.*

Atactic polymer *n.* A polymer with molecules in which substituent groups or atoms are arranged at random around the backbone chain of atoms. The opposite of a *stereospecific polymer* and *isotactic polymer.*

A

Also see isotactic polymer, syndiotactic polymer, and tactic polymer.

ATE *n.* Abbreviation for Aluminum triethyl (triethyl-aluminum), a polymerization catalyst for olefins.

Atecticity *n.* The degree of random location that the side chains exhibit off the backbone chain of a polymer. Rosato DV (ed) (1992) Rosato's plastics encyclopedia and dictionary. Hanser–Gardner Publications, New York.

Athermal transformation *n.* A reaction that proceeds without thermal (not dependent on heat, enthalpy $\Delta H = 0$) activation as contrasted to isothermal transformation, which occurs at constant temperature. An *athermal* mixture of liquids or polymer in a solvent involves no enthalpy ($\Delta H = 0$) change in an ideal solution; and the Gibbs free energy ($-\Delta G$) is always negative for the solution (polymer dissolved in solvent) to occur while the entropy (ΔS) increases. Barton AFM (1983) Handbook of solubility parameters and other cohesion parameters. CRC Press, Boca Raton.

Atlac *n.* Polyester resin, manufactured by Atlas, USA.

Atmosphere (standard atmosphere) A unit of pressure; 1 atm $= 1.013 \times 10^5$ pascals (Pa) $=$ 760 mmHg.

Atmospheric conditions *n.* In general, the relative humidity, barometric pressure, and temperature existing at a given time.

Atmospheric fading *See gas fading.*

Atom \ˈa-təm\ [ME, fr. L *atomus,* fr. Gk *atomos,* fr. *atomos* indivisible, fr. *a-* + *temnein* to cut] (15c) *n.* The smallest particle of an element, which can enter into a chemical *combination.* All chemical compounds are formed of atoms, the difference between compounds being attributable to the nature, number, and arrangement of their constituent atoms. Goldberg DE (2003) Fundamentals of chemistry. McGraw-Hill Science/Engineering/Math, New York.

Atomic absorption *n.* An analytic method of measuring qualitatively/ quantitatively elements that is based on absorption of specific wave-lengths of radiation.

Atomic emission spectroscopy *n.* Same as atomic absorption, except radiation is emitted and measured as being indicative of specific atomic spectral characteristics. Willard HH, Merritt LL, Dean JA (1974) Instrumental methods of analysis. D. Van Nostrand Co., New York.

Atomic energy *n.* (1) The constitutive internal energy of the atom, which was absorbed when it was formed. (2) Energy derived from the mass converted into energy in nuclear transformations.

Atomic mass unit (1942) *n.* (1) The mass of a neutral atom of a nuclide. It is usually expressed in terms of the physical scale of atomic masses, that is, in atomic mass units (amu). (2) A unit of mass; one atomic mass unit (amu) is defined as $\frac{1}{12}$ the mass of one ${}^{12}_{6}C$ atom. Weast RC (ed) Handbook of chemistry and physics, 52nd edn. The Chemical Rubber Co., Boca Roton, FL. *See amu.*

Atomic number (1821) *n.* The number (Z) of protons within the atomic nucleus. The electrical charge of these protons determines the number and arrangement of the outer electrons of the atom and thereby the chemical and physical properties of the element.

Atomic structure *n.* According to the currently accepted view, the atom consists of a central part, called nucleus, and a number of *electrons* (called orbital or planetary electrons) circling about the latter, like planets about the sun.

The nucleus is of a high specific weight; it contains most of the mass of the entire atom (its mass is considered equal to the atomic mass) and is composed of positively charged particles, called *protons* (the number of which always equals the atomic number, and particles of 0 charge, called neutrons. The diameter of the nucleus is between 10^{-13} and 10^{-12}cm, and the relatively vast distance in which the orbital electrons circle about it is illustrated by the fact that this nuclear diameter is only 10^{-4} to 10^{-5} of the entire atomic diameter. While the nucleus carries an integral number of positive charges (an integral number of protons) each of $1{\cdot}6 \times 10^{-19}$ C, each electron carries one negative charge of $1{\cdot}6 \times 10^{-19}$C, and the number of orbital electrons is equal to the number of protons in the nucleus (i.e., to the atomic number, *Z*), so that the atom as a whole has a net charge of 0. The electrons are arranged in successive shells around the nucleus; the maximum number of electrons in each shell is determined by natural laws, and the extra nuclear electronic structure of the atom is characteristic of the element. The electrons in the inner shells are tightly bound to the nucleus; this inner structure can be altered by high-energy particles, γ-rays or radium, or X-rays. The electrons in the outer shells are responsible for the chemical properties of the element. Russell JB (1980) General chemistry. McGraw-Hill, New York. Whitten KW, Davis RE, Davis E, Peck LM, Stanley GG (2003) General chemistry. Brookes/Cole, New York.

Atomic theory (ca. 1847) *n.* All elementary forms of matter are composed of very small unit quantities called atoms. The atoms of a given element all have the same size and weight. The atoms of different elements have different sizes and weights. Atoms of the same or different elements unite with each other to form very small unit quantities of compound substances.

Atomic unit *n.* A mass equal to 1/12 the mass of an atom of carbon-12. The unit is deprecated by the SI system.

Atomic weight (1820) (atomic mass) *n.* The mass of an elemental isotope relative to that of the C-12 isotope of carbon, whose mass has been set at exactly 12.0000 atomic units (daltons). For most elements, the tabulated atomic weight is the average, weighted by natural mass abundance, over all the element's isotopes, so is never an integer. The actual mass-fraction of a gram-of one a u is the reciprocal of Avogadro's number. $1/(6.02283 \times 10^{23})$.

Attapulgite *n.* $(Mg, Al)_5Si_8O_{20}(OH)_2{\cdot}8H_2O$. A clay mineral, with the ideal formula, in which there is considerable replacement of magnesium by aluminum. Electron micrographs show single laths or needles and bundles of laths or needles oriented in random fashion, which structure is generally believed to account for the high degree of porosity and absorptivity characteristic of this mineral. The name is derived from the place name, Attapulgus, GA. *See hydrated magnesium aluminum silicate.*

Attenuation \ə-ˈten-yə-ˌwāt\ [L *attenuatus*, pp of *atfenuare* to make thin, fr. *ad-* + *tenuis*] (15c) *n.* (1) The process for making slim and slender, for example, the formation of fibers from molten glass. (2) The gradual diminution of intensity or amplitude of a damped vibration with distance or time. Merriam-Webster's collegiate dictionary, 10th edn. Springfield, MA, 2000. Giambattista A, Richardson R, Richardson

A

RC, Richardson B (2003) College physics. McGraw Hill Science/Engineering/Math, New York.

Atto- \ˈa-(ˌ)tō\ [ISV, fr. D or N *atten* eighteen, fr. ON *āttjān*; akin to OE *eahtatīene* eighteen]. The SI prefix meaning $\times 10^{-18}$.

Attrition mills *n.* Machines for reducing materials into smaller particles by grinding down by friction. Equipment used in shredding pulp prior to acetylation in the manufacture of acetate and triacetate fibers, equipment used in shredding pulp prior to acetylation.

AU *n.* Abbreviation for polyurethane elastomers with polyester segments.
See Angstrom unit.

Aufbau procedure *n.* The procedure of starting at the lowest of a set of energy levels and gradually adding electrons to progressively higher levels; "building-up" procedure.

Auger electron spectroscopy (AES) This technique is most powerful for providing analysis of the first few atom layers (10 Å or less) on the surface of the sample (AES explores the electronic energy levels in atoms. The term Auger-process has come to denote any electron de-excitation in which de-excitation energy is transferred to a second electron, the "Auger electron." An AES spectrum of alumina is shown. Thompson M, Baker MD, Tyson JF, Christie A (1985) Auger electron spectroscopy. John Wiley and Sons Inc., New York.

Aureobasidium Pullulans *n.* A common ubiquitous, omnivorous, and highly polymorphic species of fungus which is often the primary organism in defacement of paint films (black yeast). *A. Pullulans* is heterocaryotic since cells of most strains contain one–eight nuclei. It is now thought that *Pullularia Pullulans* and *A. Pullulans* are one and the same organism. Kirk PM, Cannon PF (2001) Fungi, 9th edn. CABI Publishing, Wallingford.

Aureolin *n.* Potassium cobaltinitrite yellow pigment. Syn: Cobalt yellow.

Austrian cinnabar *See chrome orange.*

Autoacceleration *n.* In some vinyl polmerizations, as the reaction approaches completion and the viscosity of the reaction medium rises, there is a rising rate of increase of molecular weight of the chains

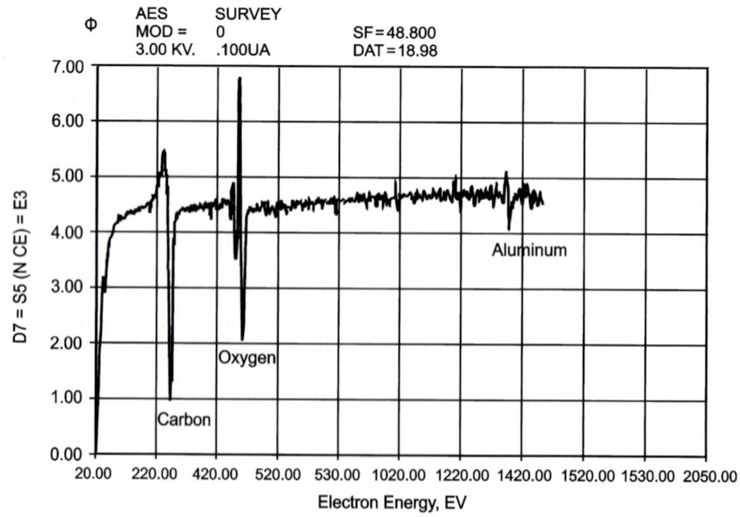

Auger Electron Spectroscopy Spectrum of Alumina, Al_2O_3.

that have not yet been terminated. This rising increase is called autoacceleration, or the *Trommsdorff effect*, or gel *effect*. In similar terms, the acceleration of a reaction, such polymerization, which continues to increase without external stimulus, i.e., gel effect and Trommsdorff effect.

Autoadhesion (tackiness) *n.* The ability of two continguous surfaces of the same material, when pressed together, to form a strong bond that prevents their separation at the place of contact.

Autocatalytic degradation *n.* A type of breakdown in which the initially generated products accelerate the rate at which later degradation proceeds.

Autoclave \ˈó-tō-ˌklāv\ [F, fr. *aut-* + L *clavis* key] (1876) *n.* A strong pressure vessel with a quick-opening door and means for heating and applying pressure to its contents. Autoclaves are widely used for bonding and curing reinforced-plastic laminates such as polyesters, epoxies, and phenolics. They are closed vessels for conducting a chemical reaction, sterilization or other operation under pressure and heat.

Autoclaveable *adj.* Capable of being sterilized in steam at two to three times standard atmospheric pressure with no change in properties.

Autoclave molding *See bag molding.*

Autodissociation (self-dissociation) *n.* The production of cations and anions by dissociation of solvent molecules without interaction with other species. Whitten KW, Davis RE, Davis E, Peck LM, Stanley GG (2003) General chemistry. Brookes/Cole, New York.

Auto-flex die *n.* Trade name for a type of sheet-extrusion die with a flexible lip in which each lip-adjusting bolt, which can either push against the lip or pull it, is paired with a nearby cartridge heater. When a signal from the beta-ray gauge that is constantly traversing the width of the sheet indicates that the sheet is too thick or too thin at a given point, the heater voltage at the relevant bolt is raised or lowered, expanding or contracting the bolt and decreasing or increasing the lip opening at that point. Shenoy AV (1996) Thermoplastics melt rheology and processing. Marcel Dekker, New York.James F (ed) (1993) Whittington's dictionary of plastics. Technomic Publishing Co. Inc., Carley.

Autogenous \ó-ˈtä-jə-nəs\ [Gk *autogenēs*, fr. *aut-* + *genēs* born, produced] (1846) *n.* A system processing independent of external influences, equivalent to autothermal extrusion. Shenoy AV (1996) Thermoplastics melt rheology and processing. Marcel Dekker, New York.

Autohesion Term which refers to the ability of two continguous surfaces of the same material to form a strong bond which prevents their separation at the place of contact. Also known as *Self-adhesion.* Skeist I (ed) (1990, 1977, 1962) Handbook of adhesives. Van Nostrand Reinhold, New York.

Autoignition temperature *n.* The temperature at which a combustible material will ignite and burn spontaneously under specified conditions. Troitzsch J (2004) Plastics flammability handbook: principle, regulations, testing and approval. Hanser–Gardner Publications, New York. Babrauskas V (2003) Ignition handbook. Fie Science Publishers, New York. Wypych G (ed) (2001) Handbook of solvents. Chemtec Publishing, New York. Tests for comparative flammability of liquids, UI 340. Laboratories Incorporated Underwriters, New York, 1997. Nelson G (1990) Fire and polymers: hazards identification and prevention. Oxford University Press, UK.

A

Automatic control *n.* In processing, control achieved by instruments that **Automatic mold.** A mold for compression, transfer, or injection molding that is equipped to perform all operations of the molding cycle, including ejection of the molded parts, in a completely automatic manner without human assistance.

Automatic press *n.* Hydraulic press for compression molding or an injection machine which operates continuously, being controlled mechanically, electrically, hydraulically, or by a combination of any of these methods.

Automatic profile control *n.* In film and sheet extrusion, a system for controlling the uniformity of thickness across the sheet. The main components are a traversing thickness sensor such as a beta-ray gauge, a computer and program that uses the sensor's signals to direct a mechanism that rotates the die-lip-adjusting bolts.

Automatic unscrewing mold *n.* A mold for making threaded products – bottle caps are typical – that incorporates a mechanism for unscrewing the product from the mold core (or vice versa) as the mold opens, thereby releasing the product. Strong AB (2000) Plastics materials and processing. Prentice-Hall, Columbus, OH.

Autoxidation or autooxidation \ó-ˌtäk-sə-ˈdā-shən\ (1883) *n.* Oxidation by direct combination with molecular oxygen (as in air) at ordinary temperatures as described in. The reaction of diatomic oxygen (O_2) with π-bonds to form intermediate peroxides, etc. and permanent chemical bonds that are responsible for drying or curing of oils in alkyd resins (e.g., the drying of linseed oil). A reaction in which one substance acts simultaneously as an oxidizing agent and a reducing agent; also called *disproportionation.* The area of interest to materials scientists is the polymerization of oils, oil based resins such as vegetable oil based alkyd resins and inks. Smith MB, March J (2001) Advanced organic chemistry, 5th edn. John Wiley and Sons, New York. Morrison RT, Boyd RN (1992) Organic chemistry, 6th edn. Prentice-Hall, Englewood Cliffs, NJ. Muizebelt WJ, Donkerbroek MWF, Nielsen JB, Hussem, Biedmond MEF (1998) Oxidative crosslinking of alkyd resins studied with mass spectroscopy and NMR using model compounds. J Coatings Tech 70 (876)83-92.

Autoxidation inhibitors *n.* Chemical agents, which inhibit the autoxidation (or oxidative polymerization) reactions, i.e., phenol and arylamines. Bart J (2005) Additives in polymers: industrial analysis and applications. John Wiley and Sons, New York. Zaiko GE (ed) (1995) Degradation and stabilization of polymers. Nova Science Publishers Inc., New York.

Autoxidative polymerization *n.* The reaction of oxygen with fatty acids, oils or other reactive materials to form higher molecular weight polymers (e.g., autoxidative polymerization of vegetable oils and emulsified vegetable oils). Gooch JW (2002) Emulsification and polymerization of alkyd resins. Kluwer Academic/Plenum Publishers, New York.

Autoxidation of polymers *n.* The reaction of oxygen with polymers to form peroxides, etc., which can result in crosslinking and/or degradation. Zaiko GE (ed) (1995) Degradation and stabilization of polymers. Nova Science Publishers Inc., New York.

Autothermal extrusion (adiabatic or autogenous extrusion) *n.* In screw extruders, a steady state of operation in which the increase in enthalpy of the plastic from feed throat to die entry is equal to the net

A

energy furnished by the drive to the screw. Strong AB (2000) Plastics materials and processing. Prentice-Hall, Columbus, OH.

Auxochrome *See chromophore.*

Average degree of polymerization *n.* $\overline{DP}$ Average degree of polymerization is equal to or twice for a chain growth polymerization, disproportionation or combination. Coleman MM, Strauss S (1998) Fundamentals of polymer science: an introductory text. CRC Press, Boca Raton, FL.

Average molecular weight (viscosity method) *n.* The molecular weight of polymeric materials determined by the viscosity of the polymer in solution at a specific temperature. This gives an average molecular weight of the molecular chains in the polymer independent of specific chain length. Slade PE (2001) Polymer molecular weights, vol 4. Marcel Dekker, New York. Billmeyer FW Jr (1984) Textbook of polymer science, 3rd edn. Wiley-Interscience, New York.Staudinger H, Heuer WA (1930) Relationship between the viscosity and the molecular weight of polystyrene (German). Ber 63B:222-234.

See number-average, viscosity-average, and weight-average molecular weight.

Average stiffness The ratio of change in stress to change in strain between two points on a stress–strain diagram, particularly the points of zero stress and breaking stress. Brown R (1999) Handbook of physical polymer testing, vol 50. Marcel Dekker, New York.

Also see modulus.

Average toughness *See toughness.*

Avogadro's law \ˌa-və-ˈgä-(ˌ)drō, ˌä-\ (1811) *np.* Equal volumes of different gases at the same pressure and temperature contain the same number of molecules.

Avogadro's number \ˌa-və-ˈgä-(ˌ)drōz-\ [Count Amedeo *Avogadro*] (1924) *np.* The number of atoms in exactly 12g of $^{12}_{6}C$; 6.02×10^{23}. The number of units in 1mol of units. Whitten KW, Davis RE, Davis E, Peck LM, Stanley GG (2003) General chemistry. Brookes/Cole, New York.

Avogadro's principle (or theory) *np.* The numbers of molecules present in equal volumes of gases at the same temperature and pressure are equal.

Avrami equation \ˈäv-rä-mē\ [M. Avrami, European, published 1939–1941] *n.* The time dependence of the overall crystallization of a polymer is described by the Avrami equation, where specific volume is related to crystalline and amorphous regions within the same polymer, then

$$(v_t - v_f)/(v_o - v_f) = \exp(-Kt^n),$$

where v_o and v_f are initial and limiting values of the specific volume, and v_t is the specific volume of at a time t, and K is a kinetic constant. This equation describes the rate of polymer crystallization at a temperature. The crystallinity is expressed as the volume fraction (ϕ) of the crystalline material. Elias HG (2003) An introduction to plastics. John Wiley and Sons, New York. Avrami M Kinetics of phase change I: general theory. J Chem Phys 7:1103; Kinetics of phase change II: transformation-time relations for random distribution of nuclei. 8:212; Kinetics of phase change III: granulation, phase change and microstructures. 9:177. Elias HG (2003) An introduction to plastics. John Wiley and Sons, New York.

Axial yarn *n.* A system of axis, crystallographic. One of several imaginary lines assumed in describing the positions of the planes by which a crystal is bounded, the positions of the atoms in the structure of the crystal and the directions associated with vectorial and tensorial physical properties. Longitudinal yarns in a

A

triaxial braid that are inserted between bias yarns. Vincenti R (ed) (1994) Elsevier's textile dictionary. Elsevier Science and Technology Books, New York.

Axminster carpet *n.* A machine-woven carpet in which successive weft-wise rows of pile are inserted during weaving according to a predetermined arrangement of colors. There are four main types of Axminster looms: Spool, Gripper, Gripper-Spool, and Chenille. Complete Textile Glossary. Celanese Corporation, Three Park Avenue, New York, NY.

Azelaic acid (nonanedioic acid, 1,7-heptanedicarboxylic acid) *n.* $HOOC(CH_2)_7COOH$. A yellowish-white crystalline powder, derived from a fatty acid such as oleic acid by oxidation with ozone. It is an intermediate used in the production of plasticizers, polyamides and alkyd resins. Acid value, 595.5; mol wt, 188.2; mp, 106°C; bp, 237°C/15 mmHg. For plasticizers derived from azelaic acid. Merck Index (2001) 13th edn. Merck and Company Inc., Whitehouse Station, NJ.

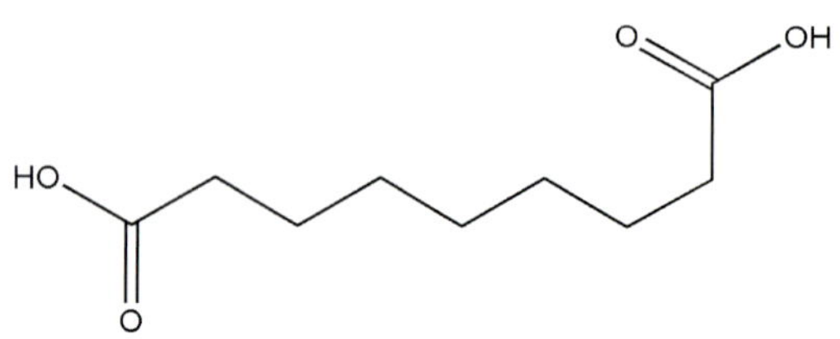

See dicyclohexyl-, di(2-ethylbutyl)-, di(2-ethylhexyl)-, di(2-ethylhexyl)-4-thio-, di-n-hexyl-, diisobutyl-, and diisooctyl azelates.

Azeotrope \ˈā-zē-ə-ˌtrōp\ [*a-* + *zeo-* (fr. Gk *zein* to boil) + *-trope*] (1938) *n.* A liquid mixture that is characterized by a constant minimum or maximum boiling point which is lower or higher than any of the components, and that distills without change in composition. Goldberg DE (2003) Fundamentals of chemistry. McGraw-Hill Science/Engineering/Math, New York. Weast RC (ed) Handbook of chemistry and physics, 52nd edn. The Chemical Rubber Co., Boca Raton, FL.

Azeotropic copolymer *n.* A copolymer in which the relative numbers of the different mer units are the same as in the mixture of monomers from which the copolymer was obtained. During a copolymerization reaction, the copolymer has the same composition as the monomer feed mixture, and a polymeric product of constant composition is formed throughout the copolymerization reaction. Kricheldorf HR, Swift G, Nuyken O, Huang SJ (2004) Handbook of polymer synthesis. CRC Press, Boca Raton, FL. Odian GC (2004) Principles of polymerization. John Wiley and Sons Inc., New York.

Azimuthal quantum number l \ˌa-zə-ˈmə-thəl-\ *n.* A quantum number hich specifies a subshell for an electron in an atom.

2(1-Aziridinyl)ethyl methacrylate *n.* A vinyl monomer that combines a reactive vinyl group with an aziridinyl functional group. It can be polymerized alone or with other vinyl monomers to yield polymers with pendant aziridinyl groups that promote adhesion of coatings to the polymer. Odian GC (2004) Principles of polymerization. John Wiley and Sons Inc., New York.

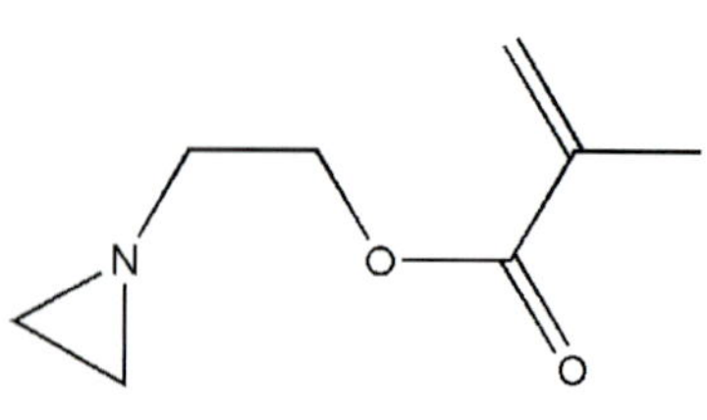

Azlon fiber *n.* A manufactured fiber in which the fiber-forming substance consists of any regenerated naturally occurring proteins

A

(FTC definition). Azlon is not currently produced in the United States.

Azobisformamide (ABFA, azodicarbonamide) *n.* $H_2NCON{=}NCONH_2$. An aliphatic azo compound widely used as a chemical blowing agent in PVC, polystyrene, polyolefins, many other plastics, and in natural and synthetic rubbers. It is nontoxic, odorless, nonstaining, and, unlike other organic blowing agents, it is self-extinguishing and does not support combustion. Since ABFA in the pure state decomposes at temperatures above 216°C, when used with heat-sensitive plastics such as PVC, an activator that lowers its decomposition temperature is added to the compound. Such activators are compounds of cadmium, zinc, and lead, which also act as heat stabilizers, either directly or synergistically with other stabilizers. Wickson EJ (ed) (1993) Handbook of polyvinyl chloride formulating. John Wiley and Sons Inc., New York. James F (ed) (1993) Whittington's dictionary of plastics. Technomic Publishing Co. Inc., Carley *(See image).*

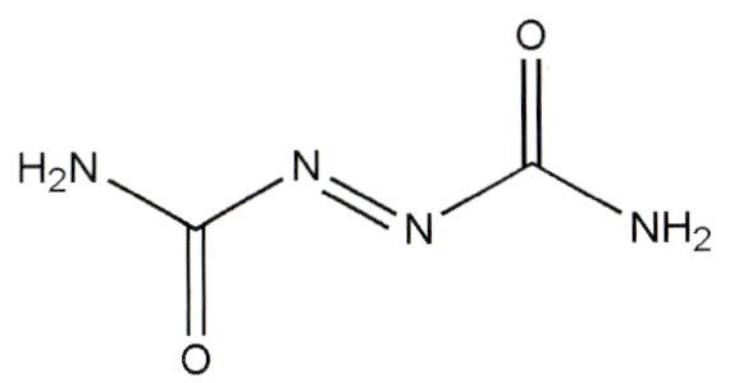

Azobis(isobutyronitrile) *n.* A blowing agent developed in Germany for use in rubber and PVC. It is non-staining and yields white PVC foam of fine uniform cell structure. However, its decomposition product, tetramethyl succinonitrile, is toxic and must be eliminated from the expanded product. For this reason the material is not used commercially in the USA as a blowing agent. It is used as a polymerization initiator. Odian GC (2004) Principles of polymerization. John Wiley and Sons Inc., New York. Elias HG (1977) Macromolecules, vols 1–2. Plenum Press, New York.

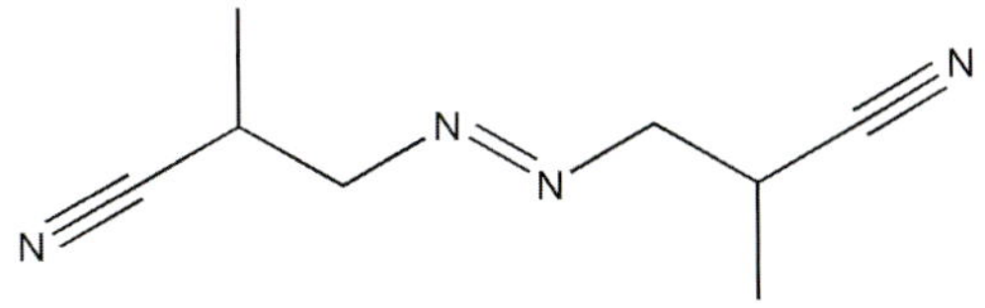

Azo-compound initiators *n.* Azo compounds (R–N=N–R) which decompose to form free radicals ($R+N_2+R$), which are capable of initiating polymerization.

Azodicarbonamide *See azobisformamide.*

Azo dye *n.* Any of an important family of dyes containing the –N=N– group, produced from amino compounds by the processes of diazotization and coupling. By varying the composition it is possible to produce acidic, basic, triazo, and tetrazo types, depending on the number of –N=N– groups in the molecule. Herbst W, Hunger K (2004) Industrial organic pigments. John Wiley and Sons Inc., New York.

Azo group *n.* The structural grouping, –N=N–.

Azo pigment n. *See benzidine yellows.*

Azure blue n. *See cobalt blue.*

Azurite \ˈa-zhə-ˌrīt\ [F, fr. azur] (ca. 1868) *n.* Natural blue pigment derived from the mineral, azurite, a basic copper carbonate. The mineral occurs in various parts of the world in secondary copper ore deposits where it is frequently associated with malachite, a green basic carbonate of copper. Perkins D (2001) Mineralogy. Prentice-Hall, New York. Kirk–Othmer (1996) Encyclopedia of chemical technology: pigments-powders. John Wiley and Sons, New York. Lewis PA (ed) (1985–1990) Pigment handbook, 2nd edn, vols 1–4. Wiley-Interscience, New York.

Also known as mountain blue.

B

b \ˈbē\ *n.* SI abbreviation for barn.

"b" or "β" *n.* Yellowness–blueness coordinate in color spaces, Δb, the difference in "b", between a specimen and a standard reference color, normally used with "a" or "α" as part of the chromaticity difference. If "b" is plus, there is more yellowness than blueness; if "b" is minus, there is more blueness than yellowness. The exception is Adams chromatic value. McDonald R (1997) Colour physics for industry, 2nd edn. Society of Dyers and Colourists, West Yorkshire, England.

See also uniform chromaticity scale diagram and color difference equations.

"b", Kubelka–Munk *n.* Mathematical constant characteristic of a color at complete opacity; dependent on the optical constants *K* and *S*. $b = [2(K/S) + (K/S)^2]^{1/2}$. McDonald R (1997) Colour physics for industry, 2nd edn. Society of Dyers and Colourists, West Yorkshire, England.

See "a", Kubelka–Munk equation.

B (1) Chemical symbol for the element boron. (2) Symbol for magnetic induction.

Ba *n.* Chemical symbol for the element barium.

Babo's law The addition of a non-volatile solid to a liquid in which it is soluble lowers the vapor pressure of the solvent in proportion to the amount of substance dissolved. Whitten KW, Davis RE, Davis E, Peck LM, Stanley GG (2003) General chemistry. Brookes/Cole, New York.

Back coating *n.* The application of latex or adhesive to the back of a carpet to anchor the tufts, usually followed immediately by addition of a secondary backing material such as woven jute or non-woven polypropylene. Complete textile glossary. Celanese Acetate LLC, New York, 2000.

Back draft (back taper, counterdraft) *n.* A slight undercut or tapered area in a mold tending to prevent removal of the molded part. Complete textile glossary. Celanese acetate LLC, New York, 2000.

See also undercut.

Backed cloth *n.* A material with an extra warp or filling added for weight and warmth. Satin-weave and twill-weave constructions are frequently used in the design of backed cloth because they are relatively resistant to the passage of air. Complete textile glossary. Celanese acetate LLC, New York, 2000.

Back filling *n.* A solution composed of varying amounts of cornstarch, China clay, talc, and tallow that is applied to the back side of low-grade, low-cost cloth to change its hand, improve its appearance, and increase its weight. Complete textile glossary. Celanese acetate LLC, New York, 2000.

Back focal plane *n.* A "plane" normal to the axis of a lens in which all back focal points lie. Moller KD (2003) Optics. Springer-Verlag, New York. Freir GD (1965) University physics. Appleton-Century-Crofts, New York.

Backing *n.* (1) A general term for any system of yarn, which interlaces on the back of a textile material. (2) A knit or woven fabric or plastic foam bonded to a face fabric. (3) A knot or woven fabric bonded to a vinyl or other plastic sheet material. (4) *See carpet backing.* Kadolph SJJ, Langford AL (2000) Textiles. Pearson Education, New York. Complete textile glossary. Celanese Acetate LLC, New York, 2000.

Backing away from fountain *n.* A condition in which an ink lacking in flow will not keep in contact with the fountain roller so

B

that the latter can transfer ink to the ductor roller. Eventually the prints become uneven, streaky, and weak. Leach RH, Pierce RJ, Hickman EP, Mackenzie MJ, Smith HG (eds) (1993) Printing ink manual, 5th edn. Blueprint, New York.

Backing coat *n.* In coil coating, the finish on the back side of continuous strip. Tracton AA (2005) Coatings technology handbook. Taylor and Francis Inc., New York.

Backing plate *n.* In injection molding, a plate used as a support for the cavity blocks, guide pins, bushing, etc. Sometimes called support plate.

Back paint *See back priming.*

Back pressure *n.* In extrusion, the head pressure. In screw-injection molding, the head pressure just before the valve opens to make the shot.

Back-pressure relief port *n.* A side channel in the head of an extruder, usually leading to a rupture disk through which the melt can escape if the pressure exceeds a safe limit.

Back priming *n.* The process of applying a coating to the back surface of construction materials prior to installation, usually for protection against the weather.

Back putty *See bed putty.*

Back sizing *See filler.*

Backstitch (1611) *n. See purl.*

Back taper *See back draft.*

Back warp *n.* The warp which, along with the back filling, actually forms the second face (back) of double, triple, or quadruple fabrics.

Back winding *n.* (1) Rewinding yarn or fiber from one type of package to another. (2) Winding yarn as it is deknit.

Bacteria \bak-ˈtir-ē-ə\ [plural of *bacterium*] (1884) *n.* Any of the numerous microscopic, spherical, rod-shaped, or spiral organisms of the class, *Schizomycetes.* Black JG (2002) Microbiology, 5th edn. John Wiley and Sons Inc., New York.

Bacterial corrosion *n.* A corrosion, which results from substances (e.g., ammonia or sulfuric acid) produced by the activity of certain bacteria. Baboian R (2002) Corrosion engineer's handbook, 3rd edn. NACE International – The Corrosion Society, Houston, TX.

Bactericidal \bak-ˌtir-ə-ˈsī-dᵊl\ (1878) *adj.* Capable of causing the death of bacteria. Black JG (2002) Microbiology, 5th edn. John Wiley and Sons Inc., New York.

Bactericidal fiber *n.* Fiber used for medical applications, socks, shoe liners, etc., in which bactericides are introduced directly into the fiber matrix as opposed to fiber simply having a bactericidal finish applied.

Bactericidal paint *n.* Coating, which discourages the multiplication of bacteria.

Bactericide (1878) *n.* An agent capable of destroying bacteria. Black JG (2002) Microbiology, 5th edn. John Wiley and Sons Inc., New York.
See also biocide.

Bacteriostat \-ˈtir-ē-ō-ˌstat\ (1920) *n.* An agent that, when incorporated in a plastics compound, will prevent the growth of bacteria on surfaces of articles made from the compound. Black JG (2002) Microbiology, 5th edn. John Wiley and Sons Inc., New York.

Bacteriostatic (1920) *n.* Inhibiting the growth of bacteria, but not bactericidal.

Bacteriostats (1920) *n.* Chemicals which inhibit the growth of bacteria cells, i.e., bacteriocidal chemicals.

Baddeleyite *See zirconium oxide.*

Baekeland, Dr. (1907) *n.* Inventor of the oldest family of phenolformaldehyde polymers. Odian GC (2004) Principles of polymerization. John Wiley and Sons Inc., New York.

Baffle \ˈba-fəl\ [prob. alter. of ME (Sc) *bawchillen* to denounce, discredit publicly] (ca. 1590) *vt.* A plug or other device inserted in a flow channel to restrict the flow or change its direction. A metal piece so placed in a rollercoater pan to direct the stream or flow of paint to the pick-up roller. Also used to direct air currents in an oven. Perry RH, Green DW (1997) Perry's chemical engineer's handbook, 7th edn. McGraw-Hill, New York.

Bagasse \be-ˈgas\ [Fr] (ca. 1826) (megass) *n.* A tough fiber derived from sugar cane, remaining after the sugar juice has been extracted. It is used as reinforcement in some laminates and molding powders. Vincenti, R (ed) (1994) Elsevier's textile dictionary. Elsevier Science and Technology Books, New York.

Bagging \ˈba-giŋ\ (1732) *n.* (1) A fabric woven in cylindrical or tubular form on an ordinary cam loom and used for grain bags, etc. (2) Fabric bulging caused by extension at the knees, elbows, etc., of a garment lacking dimensional stability.

Baggy cloth *n.* A fabric that does not lie flat, caused by sections of tight or loose yarns in either the warp or the filling.

Baggy selvage *See slack selvage.*

Bagillumbang oil *n.* Drying oil obtained from the species *Aleurites trisperma*, which grows in the Philippines.

Bag molding *n.* A method of forming and curing reinforced plastic laminates employing a flexible bag or mattress to apply pressure uniformly over one surface of the laminate. A perform comprising a fibrous sheet impregnated with an A- or B-state resin is placed over or in a rigid mold forming one surface of the article. The bag is applied to the upper surface, then pressure is applied by vacuum, in an autoclave, in a press, or by inflating the bag. Heat may be applied by steam in the autoclave, or through the rigid half of the mod. When an autoclave is used, the process is sometimes called autoclave molding.

Bail \ˈbā(ə)l\ [ME *beil, baile*, prob. of Scand origin; akin to Swedish *bygel* bow, hoop; akin to OE *būgan* to bend] (15c) *n.* Semicircular handle of a kettle, pail, or can.

Bailing machine *n.* Mechanical device designed to automatically attach bails to containers.

Bake hardness Increased hardness on heating due to elimination of retained solvent in thermoplastic coatings and thermal cross-linking in thermosetting coatings.

Bakelite \ˈbā-kə-ˌlīt, -ˌklīt\ n. (1) A trade name derived from the name of Leo H. Baekeland, a pioneering Belgian chemist who developed phenolic resins in the early 1900s. They are the oldest family of phenolformaldehyde polymers. The trade name was long used by the Bakelite Corporation, later absorbed by Union Carbide, who still uses the name for some of its resins. (2) Phenol-formaldehyde resins, manufactured by Bakelite Inc., USA.

Bake system *n.* Set of prime, intermediate, and/or top coats which require baking to effect a cured or dried film, and which together have been determined to yield the desired properties for a particular purpose.

Baking *n.* Process of drying or curing a coating by the application of heat in excess of 65°C (150°F). When below this temperature, the process is referred to as Forced Drying. Syn: stoving.

Baking finish *n.* Paint or varnish that requires baking at temperatures above 65°C (150°F), for the development of desired properties.

Baking schedule *n.* Set of related baking temperatures and baking times, which will

B

yield a cured or dried film from a coating materials such as a powder coating for maximum performance. The baking schedule is specific for each material because powder coatings are made from different materials and require different periods of flow time (fusion) and compete cure time (reaction) or program of time comprising temperature held at specific periods of time and minimum and maximum temperatures. Wicks ZN, Jones FN, Pappas SP (1999) Organic coatings science and technology, 2nd edn. Wiley-Interscience, New York.

Baking temperature *n.* Temperature above 65°C (150°F) at a specified time in a baking schedule, which has been determined to produce a cured or dried film having optimum desired properties for the coating.

Baking time *n.* Time at a specified temperature, above 65°C (150°F), in a baking schedule, which has been determined to produce a cured or dried film having optimum desired properties for the coating.

Baking vehicle (varnish) *n.* Vehicle specifically formulated for coatings, which are intended to be baked in order to affect a cured or dried film.

Balanced cloth *n.* A term describing a woven fabric with the same size yarn and the same number of threads per inch in both the warp and the filling direction.

Balanced design *n.* In reinforced plastics, a winding pattern so designed that the stresses in all filaments are equal. Pittance JC (ed) (1990) Engineering plastics and composites. SAM International, Materials Park, OH.

Balanced gating *n.* In multicavity injection molds, the objective is to fill all of the cavities simultaneously, to fill to the same final pressure, and then to have their gates freeze at the same time. Strong AB (2000) Plastics materials and processing. Prentice-Hall, Columbus, OH.

Balanced laminate *n.* A composite structure in which fiber layers laid at angles to the main axis occur in pairs, at equal ± angles, that may or may not be adjacent. Pittance JC (ed) (1990) Engineering plastics and composites. SAM International, Materials Park, OH.

Balanced reaction *n.* A reaction in which a state of equilibrium has been reached, and it can be made to proceed in one direction or another by adjusting the conditions. Conditions, which affect the direction of a reaction, are the concentrations of the reactants involved and the temperature and the pressure. Whitten KW, Davis RE, Davis E, Peck LM, Stanley GG (2003) General chemistry. Brookes/Cole, New York.

Balanced runners *n.* In a multicavity injection mold, the runners are balanced when the injected melt reaches all the cavity gates at the same instant after the start of injection. In practice, with identical cavities whose shape, size, number, and layout permit, all runner branches are given equal cross sections and corresponding branch lengths are made equal. Uniform metal temperature throughout is assumed.

Balanced solvents *n.* Combination of solvents designed to give a specified performance. Wypych G (ed) (2001) Handbook of solvents. Chemtec Publishing, New York.

Balanced twists *n.* In a plied yarn or cord, an arrangement of twist, which will not cause the yarn or cord to twist on itself of kink when held in an open loop.

Balata \bə-ˈlä-tə\ [Sp, fr. Carib] (1860) *n.* (1) Dried juice of a West Indian tree (*Mimusops globosa*). It has rubbery characteristics and is sometimes used as a substitute of gutta-percha. It resembles both this and ordinary rubber in that it consists of isoprene molecules (C_5H_8), but polymerized in its own particular way. Impure form of trans-polyisoprene. (2) *Trans*-1,4-poly(isoprene). Langenheim JH (2003) Plant resins: chemistry, evolution ecology and ethnobotany. Timber Press, Portland, OR.

Balata, natural *n.* A material identical in properties and composition to Gutta–Percha, obtained from trees in South America.

Balata, synthetic *n.* A stereospecific rubber, the trans isomer of polyisoprene, made by catalyzed addition polymerization of isoprene.

Bald spot Area or patch, usually in a wrinkle finish film, which has failed to wrinkle or give the desired optical effect.

Bale [ME, fr. MF, of Gr origin; akin to OHGr *balla* ball] (14c) *n.* Quantity of compressible articles or materials assembled into a shaped unit and bound with cord or metal ties.

Bale-and-ring test (ring-and-ball test) *n.* A method of determining the softening temperature of resins (*See www.astm.org*). A specimen is cast or molded in a metal ring of 16-mm inside diameter and 6.4-mm depth. This ring is placed upon a metal plate in a liquid bath heating at a controlled range, and a steel ball 9.5 mm in diameter weighting 3.5 g is placed in the center of the specimen. The softening point is considered to be the temperature of the liquid when the ball penetrates the specimen and touches the lower plate.

Baling machine *n.* Device used to form and fasten compressible materials into a shaped unit (bale).

Ball \ˈból\ [ME *bal*; akin to OE *bealluc* testis, OHGr *balla* ball, OE *bl* āwan to blow] (13c). When rubbing down paint, varnish films, or the like, with abrasive paper, the material removed by the abrasive action may be either in the form of a dry powder, or, if it has a tendency to softness or stickiness, will collect as relatively large lumps or balls. A film, which exhibits this latter phenomenon is said to ball. Paint/coatings dictionary. Federation of Societies for Coatings Technology, Philadelphia, Blue Bell, PA, 1978.

Ball and pebble mills *n.* A rotating mill for reducing size of materials (e.g., pigments and fillers) or pulverize them using hard metal or ceramic balls. Perry RH, Green DW (1997) Perry's chemical engineer's handbook, 7th edn. McGraw-Hill, New York.

Ball and ring method *n.* System of testing the melting and softening temperatures of asphalt, waxes, resins, and paraffin. A ring 15.875 mm in diameter is filled with the substance to be tested, and a steel ball 9.5 mm in diameter is placed upon the substance in the ring. The end point is the temperature at which the substance softens sufficiently to allow the ball to fall through the ring to the bottom of the beaker. Paint: pigment, drying oils, polymers, resins, naval stores, cellulosics esters, and ink vehicles, vol 3. American Society for Testing and Material, West Conshohocken, PA, 2001. Usmani AM (ed) (1997) Asphalt science and technology. Marcel Dekker, New York.
See also softening point.

Ball charge *n.* Volume of porcelain or steel balls loaded in a ball mill. It is generally one-third the total volume of the ball mill.

Ball mill (pebble mill) *n.* A cylindrical or conical shell rotating horizontally about its axis, partly filled with a grinding medium such as natural flint pebbles, ceramic pellets, or hard metal balls. The material to be ground is added to just fill, or slightly more than fill, the voids between the balls.

B

Water may or may not be added. The shell is rotated at a speed that causes the balls to cascade, thus reducing the particle sizes by repeated impacts. The operation may be batchwise or continuous. In the plastics industry, the term ball mill is reserved by some persons for mills containing metallic grinding media, and the term pebble mill for non-metallic media. For paint, steel balls are frequently used. Perry RH, Green DW (1997) Perry's chemical engineer's handbook, 7th edn. McGraw-Hill, New York. Weismantal GF (1981) Paint handbook. McGraw-Hill Corp. Inc., New York.

Ball milling *n.* A method of grinding and mixing material, with or without liquid, in a rotating cylinder or conical mill partially filled with grinding media such as balls or pebbles. Perry RH, Green DW (1997) Perry's chemical engineer's handbook, 7th edn. McGraw-Hill, New York.

Balloon \bə-ˈlün\. The curved paths of running yarns about the take-up package during spinning, down-twisting, plying, or winding, or while they are being withdrawn over-end from packages under appropriate yarn-winding conditions. Humphries M (2000) Fabric glossary. Prentice-Hall, Upper-Saddle River, NJ. Vincenti R (1994) Elsevier's textile dictionary. Elsevier science and technology books, New York.

Balloon fabric *n.* A plain-weave cloth having the same breaking strength in each direction. This fabric is made from fine (60–100s) combed yarn woven to constructions of 92 × 108–116 × 128. Vulcanized balloon fabric is used for air cells in planes and barrage balloons. Vincenti R (ed) (1994) Elsevier's textile dictionary. Elsevier Science and Technology Books, New York.

Ballotini *n.* Glass beads used in reflective paints. *See beaded* paint and traffic paint. Paint/coatings dictionary. Compiled by definitions Committee of the Federation of Societies for Coatings Technology, 1978.

Ball punch impact test *n.* Measure of the resistance of a coating to impact or shock. A ball punch is used for this test. Paint and coating testing manual (Gardner–Sward handbook) MNL 17, 14th edn. ASTM, Conshohocken, PA, 1995.

Ball rebound test *n.* A method for measuring the resilience of polymeric materials by dropping a steel ball on a specimen from a fixed height and observing the height of rebound. The difference between the two heights is proportional to the energy absorbed. By conducting tests over a range of temperature, results can indicate temperature of first- and second-order transitions, and effects of additives and plasticizers. Shah V (1998) Handbook of plastics testing technology. John Wiley and Sons, New York. Pittance JC (ed) (1990) Engineering plastics and composites. SAM International, Materials Park, OH. Harper CA (ed) (2002) Handbook of plastics. Elastomers and composites, 4th edn. McGraw-Hill, New York.

See relevant test methods from American Society for Testing and Materials, www.astm.org, West Conshohocken, PA.

Ball-up *n.* A term used in adhesive circles to describe the tendency of an adhesive to stick to itself.

Ball viscometer *See falling-ball viscometer.*

Ball wrap *n.* Parallel threads in the form of a twistless rope wound into a large ball. When wound mechanically with quick traverse a ball warp may be made in the form of a large cylindrical package.

Balmer series (of spectral lines) *n.* The wavelengths of a series of lines in the spectrum of hydrogen are given in angstroms by the equation

$$\lambda = 3646\frac{N^2}{N^2 - 4},$$

where N is an integer having values greater than 2. Weast RC (ed) (1971) Handbook of chemistry and physics, 59th edn, 1978–1979. The Chemical Rubber Co., Boca Raton, FL.

Balsa \ˈból-sə\ [Sp] (ca 1600) *n.* Wood from the tree *Ochroma lagopus*, Ecuador. Its density is only 0.12–0.2 g/cm^3, yet it has good strength, especially end-grain compressive strength, so it has found application as an interlayer in reinforced-plastics.
Sandwich structures. Hoadley RB (2000) Understanding wood. The Taunton Press, Newtown, CT.

Balsam \ˈból-səm\ [L *balsamum*, fr. Gk *balsamon*, prob. of Semitic origin; akin to Hebrew *bāshām* balsam] (before 12c) *n.* Oleoresinous exudations from plants, which are characterized by softness or a semi liquid consistency. They consist of mixtures of resin, essential oils, and other compounds. Typical balsams are Canada, gurjun, Peru, tolu, and storax balsams. Merriam-Webster's collegiate dictionary, 11th edn. Merriam-Webster Inc., Springfield, MA, 2004.
See oleoresin.

Baluster \ˈba-lə-stər\ [Fr *balustre*, fr. It *balaustro*, fr. *balaustra* wild pomegranatye flower, fr. L *balaustium*, fr. Gk *balaustion;* fr. its shape] (1602) *n.* (1) One of a number of short vertical members, often circular in section, used to support a stair handrail or a coping. (2) A post in a balustrade.
Also called banister.

Balustrade \ˈba-lə-strār\ [Fr, fr. It *balaustrata*, fr. *balaustro*] (1644) *n.* An entire railing system (as along the edge of a balcony) including a top rail and its balusters, and sometimes a bottom rail.

Banana liquid A solution of nitrocellulose in amyl acetate or similar solvent.

Banana oil *See amyl acetate.*

Banbury *n.* Compounding apparatus composed of a pair of contra rotating blades, which masticate the materials to form a homogeneous blend. An internal type heavy duty mixer.
See internal mixer.

Banbury mixer *n.* An intensive mixer originally used for rubber, and for many years used for mixing plastics such as cellulosics, vinyls, polyethylene, and others. It consists of two counter rotating, spiral-shaped blades encased in segments of cylindrical housing, the housing halves joined along internal ridges between the blades. Blades and housing may be cored for circulating heating or cooling liquids. A recent adaptation of the design, with connections to feed and discharge screws, permits continuous operation. Strong AB (2000) Plastics materials and processing. Prentice-Hall, Columbus, OH. Modern plastics encyclopedia. McGraw-Hill/Modern Plastics, New York, 1986-, 1990-, 1992-, 1993-edition.

Band heater *See heater band.*

Banding, heavy tow *n.* Non-uniform distribution of filaments across lowband width.

Bandle *n.* A coarse homespun linen made on narrow hand looms in Ireland.

Band spectrum *n.* Lines produced by molecular vibrations, which are so close together that they appear to be continuous.

Band wire *n.* A wire attached between dispensing and collecting vessels in order to equalize the electrical potential between them to dissipate electrostatic charge.

Banister \ˈba-nəs-tər\ [alter. of *bahuster*] (1667) *n. See baluster.*

Bank (1) In calendering and roll-milling, a cylindrical accumulation of working material in the nip of the rolls at the feed point. (2) Another name for a yarn creel.

B

Bank, rubber mill *n.* A relatively small amount of unvulcanized rubber compound rolling in the space between two mill rolls, while the major portion of the compound is bonded around the first roll. A pencil bank is a small, smooth rolling amount of compound between a calender roll and fabric surface being surface or friction coated. Vincenti R (ed) (1994) Elsevier's textile dictionary. Elsevier Science and Technology Books, New York.

Bar [Gr fr. Gk *baros*] (1910) *n.* A deprecated unit of pressure, long used in meteorological work as approximately one atmosphere, and actually equal to 0.987 standard atmosphere, i.e., 100 kPa.

Barathea \ˌbar-ə-ˈthē-ə\ [fr. *Barathea*] (1862) *n.* (1) A silk, rayon, or manufactured fiber necktie fabric with a broken rib weave and a characteristic pebbly appearance. (2) A fine, dress fabric with a silk warp and worsted filling, woven in a broken filling rib, which completely covers the warp. (3) A smooth-faced worsted uniform cloth with an indistinct twilled basket weave of fine two-ply yarns.

Barbender plastograph (PlastiCorder®) *n.* An instrument that continuously measures the torque exerted in shearing a polymer or compound specimen over a wider range of shear rates and temperatures, including those conditions anticipated in actual processing. The instrument records torque, time and temperature with a computer generated graphical representation referred to as a *plastigram*, from which one can infer processability. It shows the effect of additives and fillers, measures and records lubricity, plasticity, scorch, cure, shear and heat stability, and polymer consistency. The modern instrument is equipped with interchangeable heads for bench-scale studies of mixing and extrusion, and with computer control and data acquisition. Strong AB (2000) Plastics materials and processing. Prentice-Hall, Columbus, OH. Shah V (1998) Handbook of plastics testing technology. John Wiley and Sons, New York.

Barcol hardness *n.* The resistance of a material to penetration by a sharp steel point under a known load with an instrument called the Barcol Impressor. Direct readings are obtained on a scale from 0 to 100. The instrument has often been used as a way of judging the degree of cure of thermosetting resins and harder materials. The ASTM test, www.astm.org, is referred to as "Indentation Hardness of Rigid Plastics".
See indentation hardness.

Barefoot resin *See neat resin.*

Bare glass *n.* Glass (yarns, rovings, and fabrics) from which the sizing or finish has been removed: also, such glass before the application of sizing or finish.

Barex Copolymer from acrylonitrile and methyl methacrylate (3:1). Manufactured by Vistron, USA.

Bargeboard \ˈbärj-ˌbōrd, -ˌbórd\ (1833) *n.* An often ornamented and sloping board along a gable, covering the ends of roof timbers. Syn: vergeboard and gableboard. Merriam-Webster's collegiate dictionary, 11th edn. Merriam-Webster Inc., Springfield, MA, 2004.

Barite \ˈbar-ˌīt, ˌber-\ [Gr *barytēs* weight, fr. *barys*] (1868) *n. See barium sulfate.*

Barium \ˈbar-ē-əm, ˈber-\ [NL, fr. *bar-*] (1808) *n.* A silver-white malleable toxic bivalent metallic element of the alkaline-earth group that occurs only in combination. Whitten KW, Davis RE, Davis E, Peck LM, Stanley GG (2003) General chemistry. Brookes/Cole, New York.

Barium carbonate *n.* $BaCO_3$. A white compound, insoluble in water, occurring as a mineral or made by direct precipitation. It has a Sp gr of 4.275 and a mp of 1360°C. It is used in ceramics, fillers, and extenders. Syn: witherite (*See image*). Ash M, Ash I (1998) Handbook of fillers, extenders and dilutents. Synapse Information Resources Inc., New York.

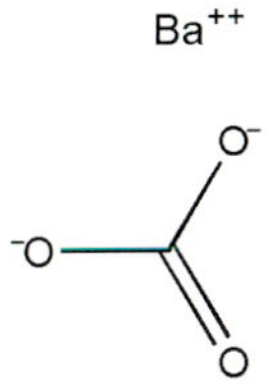

Barium chromate *See barium yellow.* Ash M, Ash I (1998) Handbook of fillers, extenders and dilutents. Synapse Information Resources Inc., New York.

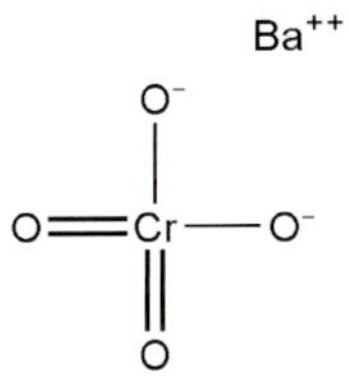

Barium chrome *See barium yellow.*

Barium-extended titanium dioxide *n.* Analogous to calcium-extended titanium dioxide, prepared by coprecipitating the $BaSO_4$ with TiO_2 or alternately made by a physical blending. In either case, the product contains 30% TiO_2 and 75% $BaSO_4$. It is no longer commercially available. Ash M, Ash I (1998) Handbook of fillers, extenders and dilutents. Synapse Information Resources Inc., New York.

Barium ferrite *See ferrite.*

Barium fillers *n.* Barium sulfate or other compound of barium, which serves as a filler for plastics compounding. Ash M, Ash I (1998) Handbook of fillers, extenders and dilutents. Synapse Information Resources Inc., New York.

Barium hydroxide monohydrate (barium monohydrate) $Ba(OH)_2–H_2O)$ *n.* A white powder used in the production of phenol-formaldehyde resins and barium soaps.

Barium lithol red *See lithol red.*

Barium lithol toner *n.* Barium salt of 2-amino-1-naphthalene sulfonic acid coupled to 2-naphthol.

Barium metaborate modified *n.* $BaB_2O_4 \cdot H_2O$. White crystalline pigment prepared by precipitation from aqueous solution; Sp gr of 3.3, density of 27.5 lb/gal refractive index of 1.55–1.60, O.A. of 30 lb/100 lb, and pH of 9.8–10.3. Ash M, Ash I (1998) Handbook of fillers, extenders and dilutents. Synapse Information Resources Inc., New York.

Barium naphthenate *n.* Metallic naphthenate used as a wetting agent for certain pigments, a hardener for some alkyds, and a thickener for oils.

Barium oleate *n.* Barium salt of oleic acid. Used for the prevention of chalking, the maintenance of color of titanium paints, and the dispersion of pigments in media during grinding.

Barium peroxide *n.* BaO_2 or $BaO_2 \cdot 8H_2O$. An oxidizing catalyst used in some polymerization reactions *(See image)*.

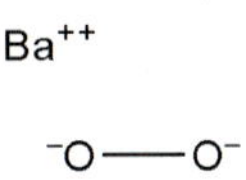

Barium plaster *n.* A special mill-mixed gypsum plaster containing barium salts; used to plaster walls of X-ray rooms.

Barium ricinoleate *n.* $Ba(OOCC_7H_4CH{=}CHCH_2CHOHC_5H_{10}CH_3)_2$. A heat stabilizer imparting good clarity, used most often in vinyl plastisols and organosols.

Barium stearate *n.* $Ba(C_{18}H_{35}O_2)_2$. White crystalline solid; insoluble in water or alcohol; mp of 160°C, and Sp gr of 1.145. Used

B

as a light and heat stabilizer in plastics. As a heat stabilizer, it is used particularly when sulfur staining is to be avoided. Also used as a lubricant where high temperatures are to be encountered.

Barium sulfate (1903) *n.* (1) $BaSO_4$. Natural pigment white 22 (77120). Mineral consisting essentially of barium sulfate. It may contain, in addition, small amounts of sulfates of calcium and strontium, common chalk, calcium fluoride, silica, and iron oxide. Used in paints as a filler or extender. Density, 4.5 g/cm^3 (37.5 lb/gal); O.A., 9; particle size 2–30 μm. Ash M, Ash I (1998) Handbook of fillers, extenders and dilutents. Synapse Information Resources Inc., New York.

Barium sulfates (barites, blanc fixe, heavy spar, permanent white, and terra ponderosa) $BaSO_4$. *n.* A white powder obtained from the mineral barite or synthesized chemically. One of the synthetic varieties, *blanc fixe*, is made by mixing aqueous solutions containing sulfate and barium ions. As a filler in plastics and rubbers, barium sulfate imparts opacity to X-rays but only a low order of optical opacity. Thus, it is useful as a filler when it is desired to increase specific gravity without adversely affecting the tinctorial power of pigments. Ash M, Ash I (1998) Handbook of fillers, extenders and dilutents. Synapse Information Resources Inc., New York.

Barium white *n. See barium sulfate, synthetic and natural.*

Barium yellow *n.* Pigment known also as barium chromate, barium chrome, or lemon yellow. It is essentially barium chromate made by adding a solution of sodium dichromate to barium chloride solution. Ash M, Ash I (1998) Handbook of fillers, extenders and dilutents. Synapse Information Resources Inc., New York.

Barking *n.* The removal of bark from wood prior to pulping.

Bar mold A mold in which the cavities are arranged in rows on separate bars that may be individually removed to facilitate stripping.

Barn (b) *n.* A miniscule area unit commensurate with the cross sections of atomic nuclei. One barn = 10^{-28} m^2. Nuclear cross sections range from about 0.01 to 1.5 b.

Barnacle \ˈbär-ni-kəl\ [ME *barnakille*, alter. of *bernake*, *bernekke*] (15c) *n.* Any of certain crustaceans of the group *cirripedia*, as the goose barnacles, the stalked species which cling to ship bottoms and floating timber, and the rock barnacles, the species which attach themselves to marine rocks.

Barre \ˈbär\ [Fr, fr. ML *barra*] (1936) *n.* A defect characterized by bars or streaks, fillingwise in woven fabrics or coursewise in weft-knit fabrics, caused by uneven tension in knitting, defective yarn, improper needle action, or other similar factors.

Barrel [ME *barel*, fr. MF *baril*] (14c) *n.* (1) The tubular main cylinder of an extruder, within which the screw rotates. (2) A container, agitated by rotation or vibration, used for tumbling moldings to remove flash and sharp edges. Also used for mixing of dry solids, e.g., pigments with resin pellets. (3) Standard unit of liquid volume in the petroleum industry. It is equal to 42 USA gallons, approximately 35 Imperial gallons; approximately 160 l.

Barrel finishing *See tumbling.*

Barrelling *See tumbling.*

Barrel mixing *See tumbling.*

Barrel polish *See tumbling.*

Barrier coat *n.* Coating used to isolate a paint system from the surface to which it is applied in order to prevent chemical or physical interaction between them, e.g., to prevent the paint solvent attacking the

underlying paint or to prevent bleeding from underlying paint or material. Wicks ZN, Jones FN, Pappas SP (1999) Organic coatings science and technology, 2nd edn. Wiley-Interscience, New York.
See tie coat and primer.

Barrier layer *n.* In multiplayer films, coextruded sheet, and blow-molded containers, a layer of polymer having very low permeability to the gases and/or vapors of interest for the application of the film, sheet, or container.

Barrier plastics *n.* Thermoplastics with low permeability to gases and/or vapors. Most important commercially are nitrile barrier resins. Several others, however, are based on various copolymers, some of which are more permeable than nitrile resins but are easier to process. A major application is bottles for carbonated beverages.

Barrier screw *See solids-draining screw.*

Barrier sheet *n.* An inner layer of a laminate, placed between the core and an outer layer.

Bar stock *n.* Standard lengths of plastics extrusion having simple cross-sectional shapes such as circular (rod stock), square, hexagonal, and rectangular of low aspect ratio, used in fabricating plastics parts by machining, welding, fastening, and adhesive bonding.

Barye *n.* cgs pressure unit $= 1\,\text{dyne/cm}^2$.

Barytes \ˈbar-ˌīt, ˈber-\ *See also blanc fixe.* Natural barium sulphate used as an ink pigment and a white extender. It is considerably more abrasive and gritty than precipitated barium sulphate (*See image*).

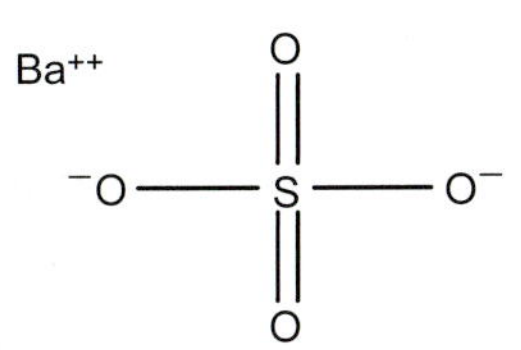

Base *n.* (1) The opposite of acid. Any molecule or ionic substance that can combine with a proton to produce a different substance (salts, soaps). (2) Reinforcing material (glass fiber, paper, cotton, asbestos, etc.) which is impregnated with resin in the forming of laminates and is used as an insulating support for an electrical printed pattern. (3) Metallic salt upon which coloring matter is precipitated to form the insoluble pigments called lakes. (4) A paste or liquid, which is to be thinned or tined. (5) In ink manufacture, a dispersion containing usually only one coloring matter, pigment or dye, properly dispersed in a vehicle. This is subsequently used for mixing to produce the desired end product. (6) *See baseboard.* (7) *See substrate and ground.* Whitten KW, Davis RE, Davis E, Peck LM, Stanley GG (2003) General chemistry. Brookes/Cole, New York. Leach RH, Pierce RJ, Hickman EP, Mackenzie MJ, Smith HG (eds) (1993) Printing ink manual, 5th edn. Blueprint, New York.

Baseboard (1853) *n.* (1) Board along the base of a wall usually finished with moldings. (2) A molding that conceals the joint between an interior wall and floor. Syn: for base, skirting board, base plate, mop board, scrub board, kick board, and washboard.

Basebox *n.* (1) In the metal-coating trade, a unit of area equal to $0.4861\,\text{m}^2$. (2) In can coating, the standard unit equal to 31,360 in.^{2} ($20.23\,\text{m}^2$).

Base coat *n.* (1) All plaster applied before the finish coat; may be a single coat or a scratch coat and a brown coat. (2) The first coat applied to a surface, as paint; a prime coat. (3) An initial coat applied to a wood surface before staining or otherwise finishing it.

Base fabric *n.* In coated fabrics, the underlying substrate.

Base plate *See baseboard.*

B

Base unit *n.* The smallest possible repeating unit of a polymer. Odian GC (2004) Principles of polymerization. John Wiley and Sons Inc., New York.

Basic *n.* Of an alkaline nature. Capable of uniting with an acid to form a salt.

Basic carbonate of lead *See carbonate white lead.*

Basic dye *n.* Having a slightly basic property due to the presence of aniline or a similar group. Industrial dye: chemistry, properties and applications. John Wiley and Sons, New York, 2003.

Basic fuchsin magenta *See fuchsin.*

Basic lead acetate *n.* Compound derived from lead oxide (litharge) by treatment with insufficient acetic acid for complete neutralization. It is used in the manufacture of acetate chromes. Gooch JW (1993) Lead based paint handbook. Plenum Press, New York.

Basic lead carbonate *n.* $2PbCO_3{\cdot}Pb(OH)_2$. A very effective heat stabilizer, used where toxicity is of no concern as in electrical-insulating compounds. Its use is limited because of its tendency to form blisters during processing and to cause spew when exposed to weather, also by rising concern about lead in the environment. Gooch JW (1993) Lead based paint handbook. Plenum Press, New York.
See carbonate white lead.

Basic lead chromate *n.* $PbCrO_4{\cdot}PbO$. Normal lead chromates are represented by the formula $PbCrO_4$, but if these are treated with alkalis, the color can be changed from the original bright yellow to an orange color, and basic chromates are produced. Gooch JW (1993) Lead based paint handbook. Plenum Press, New York.

Basic lead nitrate *n.* $Pb(NO_3)_2$. Compound prepared in a similar manner to basic lead acetate, but nitric acid is used in place of acetic acid to produce nitrate chromes. Gooch JW (1993) Lead based paint handbook. Plenum Press, New York.

Basic lead silico chromate *n.* Calcined basic lead chromate – basic lead silicate complex on a silica core, used as a corrosion-inhibiting pigment. Sp gr of 4.1; particle size of 7 μm; O.A. of 10–18 g/100 g; CrO_3 – 5.4 wt.%; SiO_2 – 47.6 wt.%; PbO – 47.0 wt.% (Gooch 1993).

Basic lead silicate *n.* $3PbO{\cdot}2SiO_2{\cdot}H_2O$. Pigment White 16 (77625). Pigment consisting of an adherent surface layer of basic lead silicate "cemented" to silica. Compositions and constants vary depending on commercial grade. Gooch JW (1993) Lead based paint handbook. Plenum Press, New York.

Basic lead sulfate *n.* $PbSO_4$, Pigment White 2 (77633). Pigment containing normal lead sulfate ($PbSO_4$) in "combination" with a proportion of basic lead, which is taken to be an oxide (PbO). It is a white pigment, which is used in considerable amounts in weather-resisting undercoats and finishes, usually in conjunction with carbonate white lead and/or zinc oxide. Its O.A. is 12–18 g/100 g and its Sp gr is 6.4.
Syn: sublimed white lead. Gooch JW (1993) Lead based paint handbook. Plenum Press, New York.

Basic pigment *n.* Any pigment which is capable of reacting with fatty acids to form soaps.

Basic solution *n.* An aqueous solution in, which the concentration of hydroxide ions exceeds that of hydrogen (hydronium) ions.

Basis weight *n.* The weight in pounds of a ream (500 sheets) of paper cut to a given standard size for that grade: 25 × 38 for book papers, 20 × 26 for cover papers, 22½ × 28½ or 22½ × 35 for bristols, 25½ × 30½ for index. For example, 500 sheets 25 × 38 of 80-lb coated will weigh 80 lb. (2) The weight of a unit area of fabric.

Examples are ounces per square yard and grams per square centimeter.

Basic zinc chromate *n.* $ZnCrO_4{\cdot}4Zn(OH)_2$. Yellow pigment used primarily for its corrosion-inhibiting properties. Syn: zinc tetroxy chromate.

Basket stitch (weave) *n.* In this knit construction, purl and plain loops are combined with a preponderance of purl loops in the pattern courses to give a basket-weave effect.

Bas-relief (basso-rilieva) *n.* Sculpture in low relief.

Basswood oil *n.* Semidrying oil. Sp gr of 0.938/15°C; iodine value of 111.0.

Bast fiber *n.* Any of a group of fibers taken from the inner barks of plants that run the length of the stem, is surrounded by enveloping tissue, and is cemented together by pectic gums. Included in the group are jute, flax, hemp, and ramie, some of which are used to reinforce plastics.

Batavia dammar or damar \bə-ˈtā-vē-ə ˈda-mər\ (1698) *n.* Gum or resin exported from Batavia (now called Jakarta), Indonesia. Batavia dammar has two subgrades: Padang or Sumatra.

Batch \ˈbach\ [ME *bache*; akin to OE *bacan* to bake] (15c) *n.* In Industry, a unit or quantity of production used in one complete operation. Merriam-Webster's collegiate dictionary, 11th edn. Merriam-Webster Inc., Springfield, MA, 2004).

Batch adjustment record *n.* A record of materials added to a paint batch to adjust viscosity, color, and such.

Batch polymerization *n.* The polymerization in a single and static, non-continuous volume of monomers. Odian GC (2004) Principles of polymerization. John Wiley and Sons Inc., New York.

Batch process *n.* Any process in which the charge is added intermittently in definite portions, or batches, the operations of he process and the removal of products being completed on each portion before the addition of the next; as opposed to continuous process.

Batch reactors *n.* A reaction vessel for batch reactions as opposed to continuous reactors. Smith JM (1981) Chemical engineering kinetics. McGraw-Hill Co., New York.

Batch record *n.* A record of all materials and proportions used to produce a batch of paint.

Bathrobe blanketing *n.* A double-faced fabric woven with a tightly twisted spun warp and two sets of soft spun filling yarns. The fabric is thick and warm and its filling yarns are frequently napped to produce a soft surface. Today's blankets are made of spun polyester, acrylic, or polyester/cotton blends. Kadolph SJJ, Langford AL (2001) Textiles. Pearson Education, New York.

Batik \bə-ˈtēk, ˈba-tik\ [Japanese *batik*] (1880) *n. See dyes.*

Batiste \bə-ˈtēst, ba-\ [F] (1697) *n.* (1) A sheer, woven, mercerized fabric of combed cotton or polyester/cotton resembling nainsook, only finer, with a lengthwise streak. (2) A rayon fabric decorated with dobby woven striped and jacquard florals. (3) A smooth, fine, woven fabric, lighter that challis and very similar to nun's veiling. Kadolph SJJ, Langford AL (2001) Textiles. Pearson Education, New York.

Batt \ˈbat\ (1871) *n.* Insulation in the form of a blanket, rather than loose filling.

Batten \ˈba-tən\ [alter. of ME *batent, bataunt* finished board, fr. MF *batant*, fr. pre part of *battre*] (1658) *n.* Small thin strips covering joints between wider boards on exterior building surfaces. Harris CM (2005) Dictionary of architecture and construction, McGraw-Hill Co., New York.

B

Battery \ˈba-t(ə-)-rē\ [MF *batterie*, fr. OF, fr. *battre* to beat, fr. L *battuere*] (1531) *n.* (1) A set of galvanic cells connected in series. (2) In common usage, a single commercial galvanic cell. Whitten KW, Davis RE, Davis E, Peck LM, Stanley GG (2003) General chemistry. Brookes/Cole, New York. Merriam-Webster's collegiate dictionary, 11th edn. Merriam-Webster Inc., Springfield, MA, 2004.

Batting \ˈba-tiŋ\ (1773) *n.* A soft, bulky assembly of fibers, usually carded. Battings are sold in sheets or rolls and used for warm interlinings, comforter stuffings, and other thermal or resiliency applications.

Batu *n.* East Indian semifossil resin, resembling the dammars. Soluble in aromatic or mixed aromatic–aliphatic hydrocarbons. Dissolves readily in warm vegetable oil. It is almost neutral. Acid value averages bout 30.

Baudouin test \bō-ˈdwa^n-\. Specific test for sesame oil, which gives a strong carmine coloration when a dilute alcoholic solution of furfural is added to a mixture of the oil and hydrochloric acid, and the whole violently shaken.

Baumé (Bé) \bō-ˈmā\ (1877) *adj.* A floating hydrometer method of measuring density of a liquid. The scale for floating hydrometer methods used to measure the specific gravity of a liquid (suspension, etc.). The depth of immersion is a linear function of the inverse of the density, www.photonics.com/dictionary. The Baume' scale is linear in inverse density. A dual transformation of specific gravity (*S*) for liquids devised by the French chemist Antoine Baumé for the graduation of hydrometers for the purpose of measuring percent salt (brine) in water; and, like Twaddell and API, becoming obsolete because they are empirical and have no scientific basis. Letting *S* equal the ratio of the density of the subject liquid at 15.6°C to that of water at the same temperature, the Baumé transformations are:
For liquids more dense than water, Bé $= 145\ [1-(1/S)]$, and for liquids less dense than water, Bé $= (140/S)-130$.
The hydrometer is used for determining the specific gravities of liquids, engraved, not directly in units of specific gravity, but in a scale consisting of Baumé degrees. These Baumé degrees, often written (Bé), can be converted to true specific gravities by reference to conversion tables.
(*See www.monashscientific.com and Baume'*).

Bauxite \ˈbók-ˌsīt, ˈbäk-\ [F *bauxite*, from Les *Baux*] (1861) *n.* $Al_2O_3.2H_2O$. Aluminum mineral, consisting chiefly of a hydrated aluminum oxide. The color of native bauxite may be white, brownish red, or gray. Used as the source of many aluminum compounds. Crude powdered bauxite has been used to some extent as filler for paints.

Bay *n.* Space between columns or supports of a building. A bay window is placed between such supports and usually projects outward. If curved or semicircular, it is called a bow window.

Bayardere *n.* A very broad term for stripes that run crosswise in a knit or woven fabric.

BBP *See butyl benzyl phthalate.*

BCF yarns *n.* Bulked continuous filament yarns for carpet trade, usually nylon, polypropylene, or polyester.

Be *n.* Chemical symbol for the element beryllium.

Bead \ˈbēd\ [ME *bede* prayer, prayer bead, fr. OE *bed*, *gebed* prayer, akin to OE *biddan* to entreat, pray] (before 12c) *n.* (1) Heavy accumulation of a coating, which occurs at the lower edge of a panel or other vertical surface as the result of excessive

flowing. (2) Solidified droplet of an oil–resin mix, withdrawn from a varnish or resin mix, for testing purposes. Instructions frequently include a statement that the oil–resin mix shall be heated until a hard or tough bead is obtained. (3) Glass spheres.
See ballotini.

Bead and butt *n.* Framed work in which the panel is flush with the framing and has a bead run on two edges in the direction of the grain; the ends are left plain.
Also known as bead, butt, and bead butt work.

Bead, butt and square *n.* Similar to bead and butt, but having the panels flush on the beaded face only, and showing square reveals on the other.

Beaded paint *n.* Traffic or marking paint with reflecting beads (ballotini) applied after striping or manufactured with beads and applied.

Beaded velvet *n.* Velvet with a cut-out pattern or a velvet pile effect, made on a Jacquard loom. This fabric is used primarily for evening wear.

Beader *n.* A device for rolling beads on the edges of thermoplastic sheets or cylinders.

Bead polymer *n.* A polymer in the form of nearly spherical particles about 1 mm in diameter.

Bead polymerization *n.* A type of olymerization identical to suspension polymerization, except that the monomer is dispersed as relatively large droplets in water or other suitable inert diluents by vigorous agitation.

Bead polymerization, suspension *n.* Through agitating or stirring with the aid of a dispersion agent, water insoluble monomers can be dispersed in water as fine droplets, oil soluble free radical initiators start polymerization in the droplets, the droplets grow and convert into beads or pearls.

Beads, reflective *See beaded paint and ballotini.*

Beam \ˈbēm\ [ME *beem,* fr. OE *bēam* tree, beam; akin to OHGr *boum* tree] (before 12c) *n.* (1) Timber or girder supporting a floor, roof, or ceiling, usually supported at each end by a pier, wall, post, or girder. A structural member whose prime function is to carry transverse loads, as a joist, girder, rafter, or purlin. (2) A cylinder of wood or metal, usually with a circular flange on each end, on which warp yarns are wound for slashing, weaving, and warp knitting.

Beam ceiling *n.* (1) A ceiling, usually of wood, made in imitation of exposed floor beams with the flooring showing between. (2) The underside of a floor, showing the actual beams, and finished to form a ceiling.

Beam dyeing machine *n.* A machine for dyeing warp yarns or fabrics that have been wound onto a special beam, the barrel of which is evenly perforated with holes. The dye liquor is forced through the yarn or fabric from inside to outside and vice versa. Kadolph SJJ, Langford AL (2001) Textiles. Pearson Education, New York. Industrial dye: chemistry, properties and applications, John Wiley and Sons, New York, 2003.

Beaming The operation of winding warp yarns onto a beam usually in preparation for slashing, weaving, or warp knitting.
Also called warping.

Beamroll *See beam.*

Bearding *n.* Fuzz on loop pile carpets usually resulting from poor anchorage or fiber snagging.

Bearing strength *n.* The ability of plastics sheets to sustain edgewise loads that are applied by pins, rods, or rivets used to assemble the sheets to other articles.

Beat frequencies *n.* The beat of two different frequencies of signals on a non-linear

B

circuit when they combine or beat together. It has a frequency equal to the difference of the two applied frequencies.

Beater *n.* (1) The machine which does most of the opening and cleaning work on a fiber picker and opener. Revolving at high speed, it beats against the fringe of fiber as the latter is fed into the machine. (2) A machine used in the paper industry for opening pulp and combining additives.

Beating-up *n.* The last operation of the loom in weaving, in which the last pick inserted in the fabric is "beat" into position against the preceding picks.

Beat(s) *n.* Two vibrations of slightly different frequencies f_1 and f_2 when added together, produce in a detector sensitive to both these frequencies, a regularly varying response which rises and falls at the "beat" frequency $f_b = f_1 - f_2$. It is important to note that a resonator, which is sharply tuned to f_b alone will not resound at all in the presence of these two beating frequencies.
See combination frequencies.

Beaumontage *n.* A resin, beeswax, and shellac mixture used for filling small holes or cracks in wood or metal. Paint: pigment, drying oils, polymers, resins, naval stores, cellulosics esters, and ink vehicles, vol 3. American Society for Testing and Material, West Conshohocken, PA, 2001.

Beaver cloth *n.* Made of high-quality wool, this heavy but soft fabric has a deep nap. Beaver cloth is frequently used in overcoats.

Becchi–Millian test *n.* Sensitive test for cottonseed oil. If cottonseed fatty acids are dissolved in 90% ethyl alcohol and a little dilute silver nitrate solution added, they separate, colored brownish or black, as the alcohol is boiled away on heating. It is claimed that as little as 1% of cottonseed oil is detected in a mixture by this test.

Beck *n.* A vessel for dyeing fabric in rope form, consisting primarily of a tank and a reel to advance the fabric.

Becke line *n.* The bright halo near the boundary of a transparent particle that moves with respect to that boundary as the microscope is focused through best focus.

Becke test *n.* The method for determining refractive index of a transparent particle by noting the direction in which the Becke line moves. The halo (Becke lines) will always move to the higher refractive index medium as the focus is raised. The halo crosses the boundary into the lower refractive index medium when the microscope is focused down. The particle must be illuminated with a narrow cone of axial light obtained by closing the aperture diaphragm of the condenser to a small aperture. Nesse WD (2003) Introduction to optical mineralogy. Oxford University Press, New York.

Beckacite *n.* Phenoplast, manufactured by Reichhold, USA.

Becquerel (Bq) \be-ˈkrel, ˌbe-kə-ˈrel\. The SI unit for rate of disintegration of a radioactive element, equal to 1 dis/s.

Bedford cord *n.* A rib-weave fabric with raised lengthwise cords produced by using stuffing threads in the warp. Since the fabric is strong and wears well, it is used for upholstery, suits, riding habits, and work clothes. Tortora PG, Merkel RS (2000) Fairchild's dictionary of, 7th edn. Fairchild Publications, New York.

Bed putty *n.* The glazier's putty under glass, on, which the glass is bedded. Syn: black putty.
See also face putty.

Beechnut oil *n.* A non-drying oil. Its iodine value averages 110, and its Sp gr, 0.992/15°C.

Beer–Bouguer law (Beer–Lambert law) *n.* Combined laws expression absorption of radiant energy of a single wavelength as a function of concentration and optical path length of absorbing material:

$$-\log T_i/cl = K,$$

where T_i is the internal transmittance, c the concentration, l the path length, and K is the unit absorption coefficient. The value $-\log T_i$ is sometimes called the optical density, OD, or simply the density, d; it is sometimes called the absorption, the absorbency, or the extinction. The unit absorption coefficient, K, is sometimes designated A, a, α, E, or ε. The units of concentration and path length must be specified. *See absorption coefficient.* Note – In Europe, it is called the law of Lambert–Beer. DeLevie R (1996) Principles of quanitative analysis. McGraw-Hill Higher Education, New York. Willard HH, Merritt LL, Dean JA (1974) Instrumental methods of analysis. D. Van Nostrand Co., New York.

Beer's law *np.* If two solutions of the same colored compound are made in the same solvent, one of which is, say, twice the concentration of the other, the absorption due to a given thickness of the first solution should be equal to that of twice the thickness of the second. Mathematically this may be expressed as $l_1c_1 = l_2c_2$ when the intensity of light passing through the two solutions is a constant and if the intensity and wavelength of light intensity and wavelength of light incident upon each solution are the same. DeLevie R (1996) Principles of quanitative analysis. McGraw-Hill Higher Education, New York. Willard, Merritt and Dean, 1974.

Beeswax \ˈbēz-ˌwaks\ (1676) *n.* Mixture of crude cerotic acid and myricin separated from honey. It is a soft natural wax, with a mp of 63°C, an acid value usually below 20, a saponification value of 90–95, and a very low iodine value of 6–13.

Beetling \ˈbē-təl-iŋ\ (ca. 1919) *adj.* A process in which round-thread linen or cotton fabric is pounded to give a flat effect. Beetled linen damask has an increased luster and a leather-like texture. Beetling is also used to give a thread-like or linen-like appearance to cotton.

Behenic acid $C_{22}H_{44}O_2$. Long chain, aliphatic acid. It has an mp of 81°C, a bp of 306°C/60 mmHg and an acid value of 164.8 (*See image*).

Belly benzoin \-ˈben-zə-wən\ *n.* Benzoin obtained as an exudation from the tree in the later years of its life. Belly benzoin is usually regarded as an inferior grade, obtained during the nine years following the tenth year of the tree's life.
See also benzoin.

Benard (vortex) cell *n.* Hexagon-shaped cell that is produced by the vortex action of solvent evaporation in thin films. During drying, all solvent paints and many varnishes exhibit Benard cell formation, i.e., roughly hexagonal cells, generally with a well-marked center. The whole cell is in movement with currents streaming up the center and flowing down the walls.

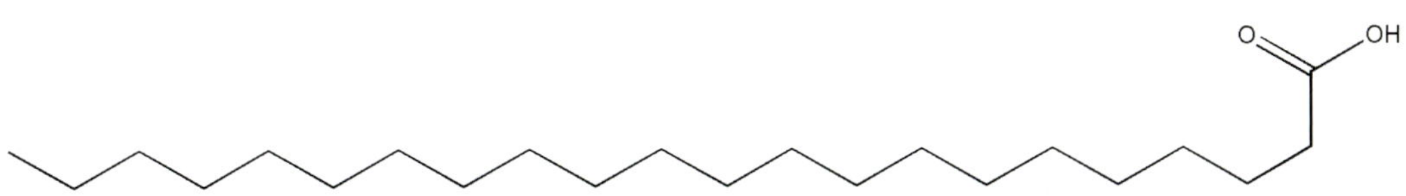

B

These current, because they affect pigments differently, lead to segregation and deposition of different components in different parts of the film. This phenomenon manifests itself in color and surface irregularities of the film. Weldon DG (2001) Failure analysis of paints and coatings. John Wiley and Sons, New York.

Bending length *n.* A measure of fabric stiffness based on how the fabric bends in one plane under the force of gravity. Paint/coatings dictionary. Federation of Societies for Coatings Technology, Philadelphia, Blue Bell, PA, 1978.

Bending modulus *n.* Maximum stress per unit area that a specimen can withstand without breaking when bent. For fibers, the stress per unit of linear fiber weight required to produce a specified deflection of a fiber.

Bending moment *n.* The resultant moment about the neutral axis of a beam or column, at any point along its span, of the system of forces that produce bending.

Bending rigidity *See flexural rigidity.*

Bending strength *See flexural strength.*

Bend test *See flexibility test.*

Benedict's test *n.* Special test designed to detect the pressure of reducing sugars, and it may also be applied to the products of hydrolysis of certain gums, starch, etc. Benedict's reagent consists of a solution of sodium citrate and sodium carbonate in water, to which a small quantity of copper sulfate is subsequently added. An orange precipitate is obtained when a few ml of the reagent is added to the solution of the suspected compound, if a reducing sugar is present. Whistler JN, BeMiller JN (eds) (1992) Industrial gums: polysaccharides and their derivatives. Elsevier Science and Technology Books, New York.

Bengaline \ˈbeŋ-gə-ˌlēn\ [F, from *Bengal*] *n.* A fabric similar to faille, only heavier, with a fine weave and widthwise cords. Originally, bengalines were made of a silk, wool, or rayon warp with a worsted or cotton filling and used for dresses, coats, trimmings, and draperies. Modern bengalines are made with filament acetate or polyester warps. Also, some bengalines have fine spun warps with 2- and 3-ply heavier spun yarns for filling cord effects. Vincenti R (ed) (1994) Elsevier's textile dictionary. Elsevier Science and Technology Books, New York.

Bengucopalic acid *n.* $C_{19}H_{30}O_2$. Monobasic acid which has been isolated from Benguela copal. It constitutes about 44% of the copal. The acid is said to resemble the congocopalic acid of Congo copal.

Bengucopalolic *n.* $C_{21}H_{32}O_3$. Major constituent acid of Benguela copal, being present to the extent of about 25% of the whole. It has a mp of 115°C.

Bengucopalresene *n.* Constituent of Benguela copal. Present in two forms, the α and β varieties. The former is present to the extent of about 5% and the latter to the extent of 15%.

Benguela copal *n.* Fairly hard natural copal obtained from West Africa. Sometimes known as Lisbon copal and, like other similar copals from this part of West Africa, gives satisfactory oleoresinous varnishes after running.

Beni \ˈbā-nē\ [Japanese]. A delicate pink or red pigment of vegetable origin.

Beni-ye [Japanese]. A print in which beni is the chief color used. Generally employed to describe all those two-color prints which immediately preceded the invention of polychrome printing.

Benne oil \ˈbe-nē\ [of African origin; akin to Malinke *bène* sesame] (1769) *n.*
See sesame oil.

Ben oil *n.* Non-drying vegetable oil. The physical properties recorded vary widely from different samples; for example, iodine values have been reported from 72 to 112.

Bentonite \ˈben-tᵊn-ˌīt\ (1898) *n.* Very fine-grained clay (a mixture and not a definite mineral type) derived from volcanic ash and consisting largely of montmorillonite mineral. Two classes of Bentonite are recognized: (1) sodium bentonite, which is a swelling type in water and (2) calcium bentonite or subbentonite, which exhibits little swelling in water.

See aluminum silicate. Also known as wilkinite.

Bentonite clay *See bentonite.*

Benzaldehyde \ben-ˈzal-də-ˌhīd\ [ISV] (1866) (benzoic aldehyde, oil of bitter almonds, benzoÿl hydride, benzene carbonal) *n.* C_6H_5CHO. A solvent, particularly for polyester and cellulosic plastics (*See image*).

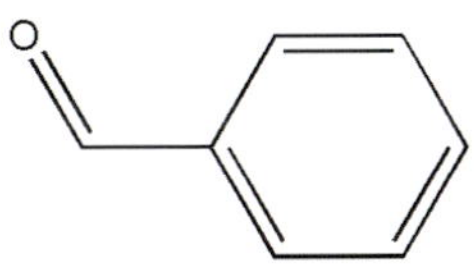

Benzene \ˈben-ˌzēn, ben-ˈ\ [ISV *benz-* + *-ene*] (ca. 1872) (benzol, phene) *n.* C_6H_6. The compound and building block of all aromatic organic chemistry. It took almost a century of investigation after its discovery by Faraday in 1823 to establish the structure of this extraordinarily stable ring: its system of resonant, alternating single and double bonds. Benzene is a solvent and intermediate in the production of phenolics, epoxies, styrene, and nylon. Hydrogenation of benzene yields cyclohexane, a solvent and raw material for preparing adipic acid, from which nylon is derived. As a solvent, benzene will dissolve ethyl cellulose, polyvinyl acetate, polymethyl methacrylate, polystyrene, coumarone-indene resins, and certain alkyds. Benzene is toxic and has been declared to be a carcinogen, so it requires very careful handling. Pure benzene has a bp of 80°C and a mp of 5.5°C. Its flp is below normal air temperature and its vapor pressure is about 118 mmHg at 30°C. Wypych G (ed) (2001) Handbook of solvents. Chemtec Publishing, New York. Merck index, 13th edn. Merck and Company Inc., Whitehouse Station, NJ, 2001.

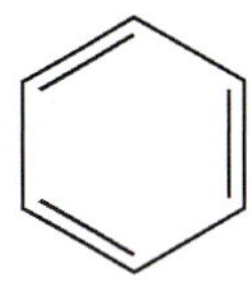

Benzene ring (phenyl ring) *n.* The six carbon atoms, diagramed as a hexagon, joined by alternating single and double bonds, each carbon with an attached hydrogen in the case of benzene itself, or with one or more hydrogens replaced by other atoms or radicals. The alternating double bonds, in either of two possible arrangements, may or may not be shown, depending on the expected audience's knowledge of organic chemistry. The Greek letter, φ, is also used as a symbol for *phenyl.* Morrison RT, Boyd RN (1992) Organic chemistry, 6th edn. Prentice-Hall, Englewood Cliffs, NJ.

Benzenesulfonylbutylamide *n.* $C_6H_5SO_2NHC_4H_9$. A plasticizer for cellulosics and polyvinyl acetate.

Benzenesulfonylhydrazide [4,4′-oxybis (benzenesulfonylhydrazide), OBSH] *n.* A blowing agent, a white crystalline solid that melts and begins to decompose near 105°C. It produces a white unicellular foam when incorporated in PVC plastisol but has a strong residual odor that does not evolve when it is used in rubbers. It is also used in epoxy and phenolic foams and serves as a cross-linking agent in rubber compositions and a rubber/resin blends.

Benzenoid \ˈben-ˌzēn, ben-ˈ\ [ISV *benz-* + *-ene*] (ca. 1872) *n.* Benzenoid compound is one, which has a structural resemblance to benzene.

Benzidine orange *n.* Metal-free diazo pigments based on dichlorobenzidine. They are highly transparent, bright in color, and low in cost due to their high tinctorial strength, but tend to bleed and fade upon exposure to light.

Benzidine yellow *n.* Benzidine or diarylide yellows are disazo pigment dyestuffs. They are approximately twice as strong, much more bleed and heat resistant and markedly inferior in lightfastness versus Hansa yellows.

Benzillic acid *n.* $(C_6H_5)_2C(OH)COOH$. It has a mol wt of 228.17 and a mp of 150°C. Known also as diphenyl glycolic acid (*See image*).

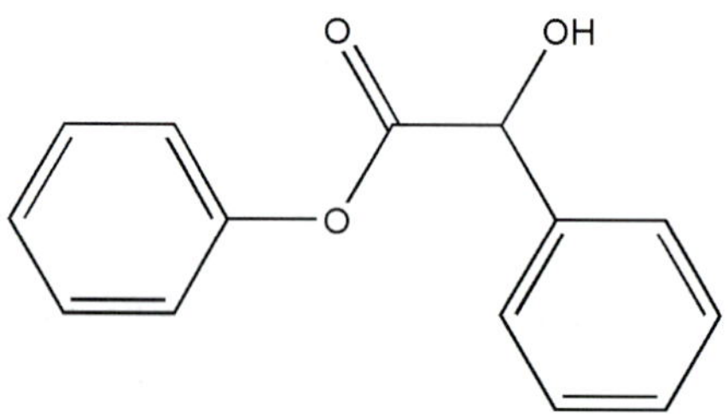

Benzine \ˈben-ˌzēn, ben-ˈ\ [Ger *Benzin*, fr. *benz-*] (1835) (deprecated) *n.* (1) This term is outdated and misleading, should not be used. (2) European term for gasoline.
See ligroin.

Benzoate fiber \ˈben-zə-ˌwāt\. Fiber with a silk-like hand made from a condensation polymer of *p*-(bhydroxyethoxy) benzoic acid.

Benzofuran \ˌben-zō-ˈfyur-ˌan, -fyu-ˈran\ (1946) *n.* A compound C_8H_60 found in coal tar and polymerized with indene to form thermoplastic resins used especially in adhesives and printing inks.
See coumarone.

Benzofuran resin *See coumarone-indene resin.*

Benzoguanamine *n.* (2,4′-diamino-6-phenyl-1,3,5-triazine) $C_6H_5N_3(NH_2)_2$. A crystalline compound that reacts with formaldehyde to give thermosetting resins with resistance to heat and alkalies, and gloss generally superior to hose of melamine-formaldehyde resins. Benzoguanamine resins are used for protective coatings, paper additives and finishes, laminating agents, textile finishes, and adhesives.

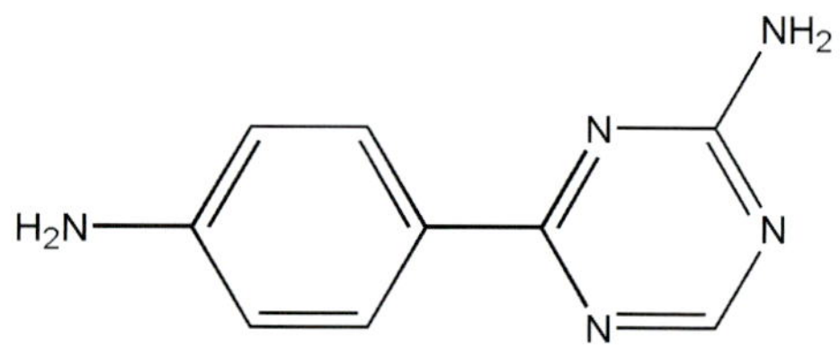

Benzoguanamine resins *n.* Resins based on benzoguanamine.

Benzoic acid \ben-ˈzō-ik-\ [ISV] (1791) (carboxybenzene, benzene carboxylic acid, phenylformic acid) *n.* C_6H_5COOH. A white, crystalline compound occurring naturally in benzoin gum and some berries, also synthesized from phthalic acid or toluene. It is used in making plasticizers such as 2-ethylhexyl-*p*-oxybenzoate, diethyleneglycol dibenzoate, dipropyleneglycol dibenzoate, ethyleneglycol dibenzoate, triethyleneglycol dibenzoate, polyethyleneglycol-(200)- and -(600)-dibenzoate, and benzophenone. It has a mp of 121°C, a bp of 249°C, and an acid value of 460. It is a useful peptizing agent for metallic soaps, either in their solid form or in hydrocarbon solutions. It is sometimes effective in reducing seeding, especially in spirit enamel containing small amounts of basic pigments.

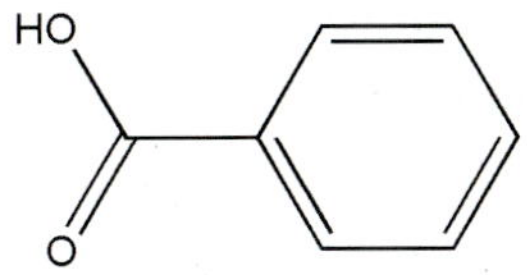

Benzoic ether \ben-ˈzō-ik-\ [ISV, fr. *benzoin*] (1791) *n. See ethyl benzoate.*

Benzol \ˈben-ˌzól, -ˌzōl\ [Gr, fr. *benz-* + *-ol*] (1838) *n.* Commercial solvent derived from the distillation of coal. It is a crude mixture of hydrocarbons, consisting chiefly of benzene, although it may contain in addition some toluene, xylene, carbon disulfide, thiophene, pyridine, acetonitrile, and paraffinic hydrocarbons. A 90%, or 90°, benzol is one from which 90% by volume can be distilled below 100°C.
Also can be spelled as benzole.

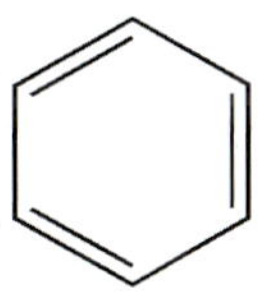

Benzoline *n.* (1) Gum benzoin, as it is more commonly known, is a spirit-soluble resin, which appears on the market as Siam, Palembang, Padang, Sumatra, and Penang benzoins. It has a very pleasant "vanilla" smell, and it is used partly on account of this, and partly as a plasticizing resin in spirit varnishes. Benzoin is obtained as an exudation from a tree as a result of deliberate incision. (2) (deprecated) Another name for petroleum benzine.
See also belly benzoin.

Benzoperoxide *See benzoyl peroxide.*

Benzophenone \ˌben-zō-fi-ˈnōn, -ˈfē-ˌnōn\ [ISV *benz-* + *phen-* + *-one*] (1885) *n.* (1) (diphenylketone) An involati le solvent and chemical intermediate. (2) Any of a family of UV stabilizers based on substituted 2-hydroxybenzophenone ("B"). Typical members are 4-methoxy-B, 4-octyloxy-B, 4-dodecyloxy-B, 2,2′-dihydroxy-4-methoxybenzophenone, and 2,2′-dihydroxy-4,4′-dimethoxybenzophenone. They function both as direct UV absorbers and, in the case of polyolefins, also as energy-transfer agents and radical scavengers. It is an aromatic ketone having excellent general compatibility. It is a solid at ordinary temperature, has a mp of 48°C, and a bp of 305°C. It is insoluble in water.
Known also as diphenyl ketone.

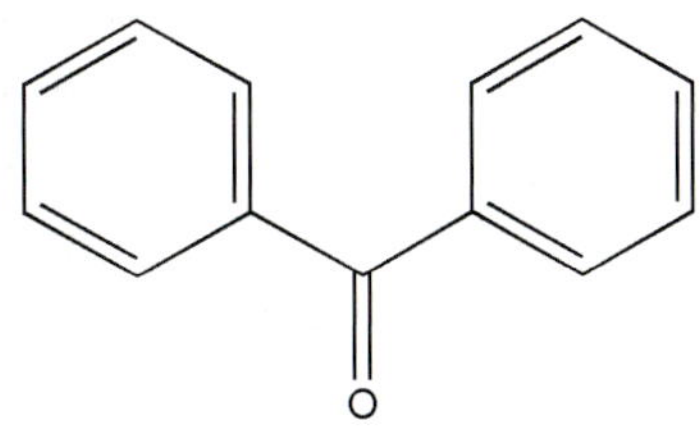

***p*-Benzoquinone** \-kwi-ˈnōn\ (1,4-benzoquinone, chinone). A yellow crystalline compound used, along with many of its derivatives, as an inhibitor in unsaturated polyester resins to prevent premature gelation during storage.

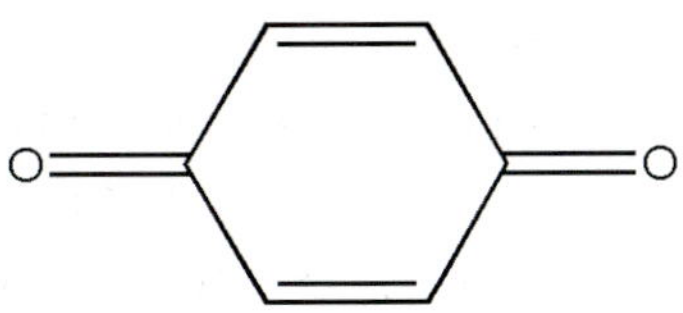

Benzotriazole *n.* (1) $C_6H_5N_3$. A double ring compound, parent to many derivatives. (2) Any of a family of UV stabilizers, derivatives of 2-(2′-hydroxyphenyl) benzotriazole, that function primarily as UV absorbers. Typical examples are 2-(2′-hydroxy-5′-methylphenyl)benzotriazole and

B

the corresponding 5′-*t*-octylphenyl analog. The benzotriazoles offer intense and broad UV absorption with a fairly sharp wavelength cutoff close to the visible region. The higher alkyl derivatives are less volatile and therefore more suitable for processing at higher temperatures.

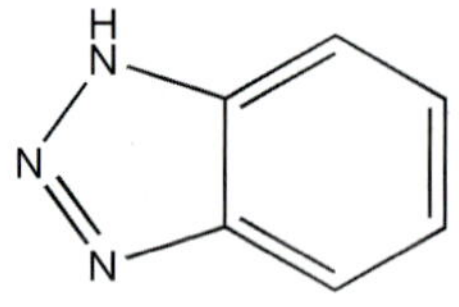

Benzoyl \ˈben-zə-ˌwil, -ˌzóil\ [Gr, from *Benzoësäure* benzoic acid + Gk *hylē* matter] (ca. 1855) *n.* C_6H_5CO–monovalent aryl radical.

Benzoyl peroxide (1924) (dibenzoÿl peroxide, DBP) *n.* A catalyst employed in the polymerization of polystyrene, styrene, vinyl, and acrylic resins. It is also a curing agent for polyester and silicone resins, usually used together with an accelerator such as dimethylaniline. It can be dispersed in diluents or plasticizers to diminish the explosion hazard associated with the dry product. Its natural state is colorless crystals of mp of 103°C. Odian GC (2004) Principles of polymerization. John Wiley and Sons Inc., New York.
Known also as benzoperoxide.

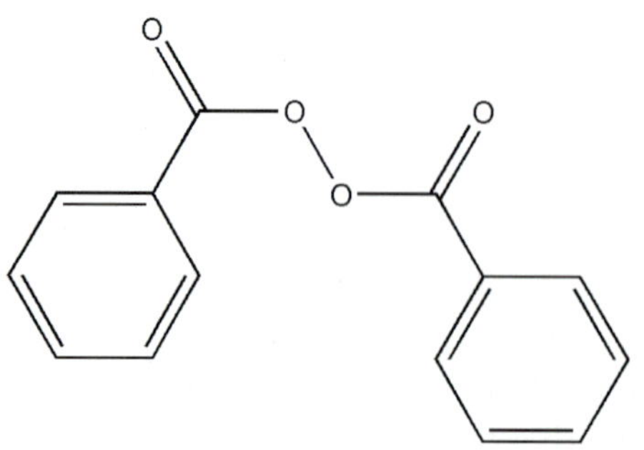

Benzyl \ˈben-ˌzēl, -zəl\ [ISV *benz-* + *-yl*] (1869) (α-tolyl) *n.* The radical $C_6H_5CH_2$–, which exists only in combination.

Benzyl abietate *n.* Very high boiling plasticizer derived from rosin. It is compatible with many other varnish and lacquer constituents. As a rosin derivative, it is susceptible to atmospheric oxidation and yellowing. (*See image*).
Sometimes known as benzyl resinate

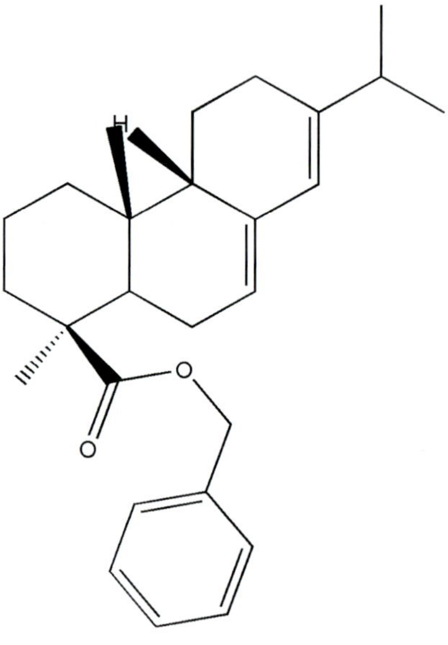

Benzyl acetate (phenylmethyl acetate) *n.* A colorless liquid with a pleasant aroma, a solvent for cellulosic resins.

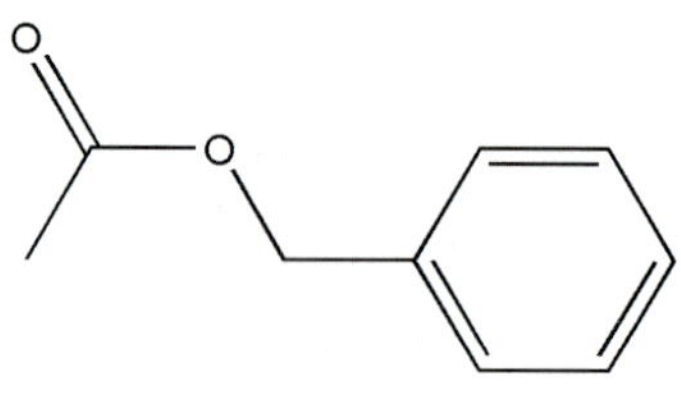

Benzyl alcohol (α-hydroxytoluene, phenylcarbinol) *n.* Has excellent solvent properties for a wide range of varnish and lacquer constituents. It has some application as a high boiling solvent, more particularly in baking finishes. It has a bp of 205°C, a flp of 300°C, and a vp of less than 1 mmHg at ordinary temperature.

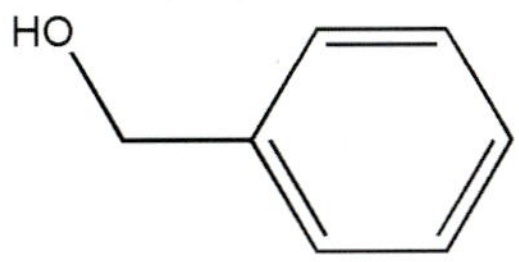

Benzyl benzoate *n.* A water-white liquid used as a plasticizer that freezes at room temperature. It has a mp of 19°C, and a bp of 323°C.

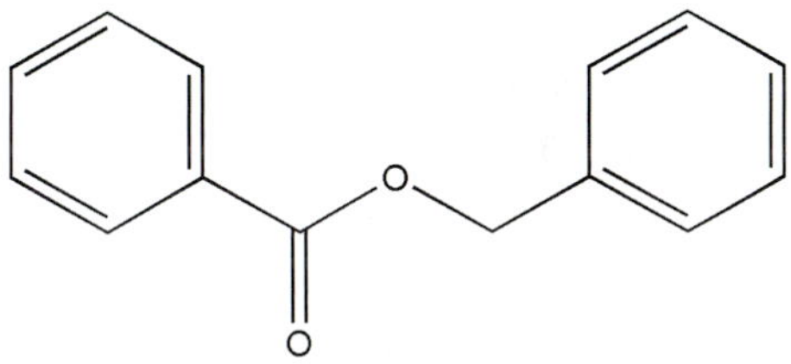

Benzyl butyl phthalate *n.* High melting solid, with a bp of 208–288°C/20 mmHg (*See image*).

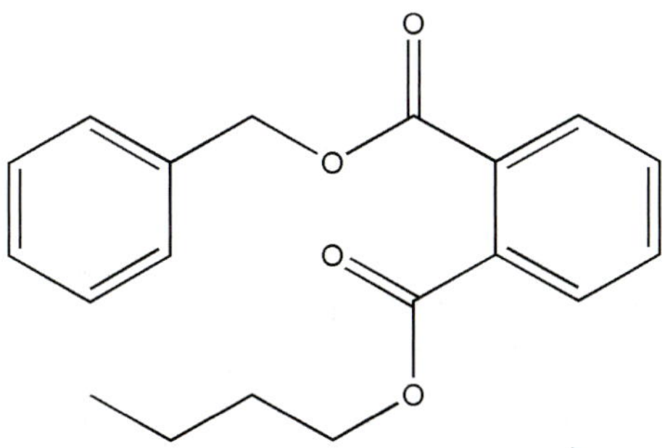

Benzyl butyrate *n.* A liquid with a heavy, fruity odor, a plasticizer.

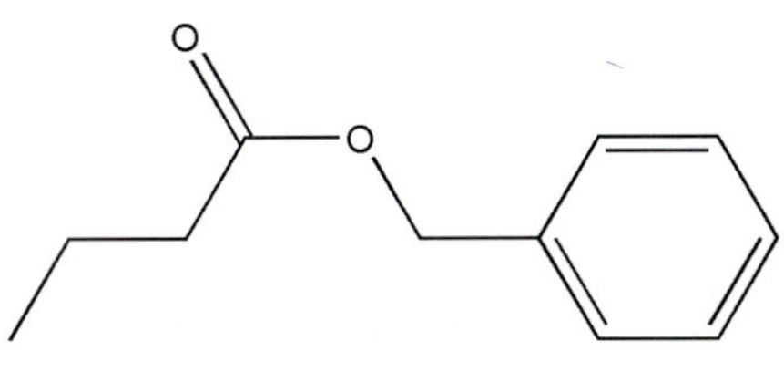

Benzyl cellosolve *n.* High boiling solvent with a bp of 256°C and Sp gr of 1.070/20°C.

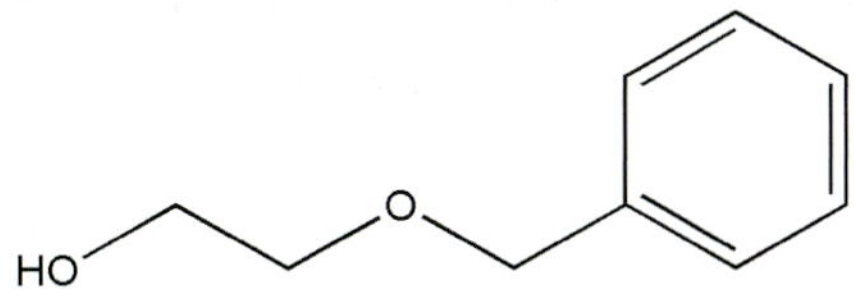

Benzyl cellulose *n.* A benzyl ether of cellulose, it is a cellulosic plastic used in lacquers. It also may be formulated for making films and compounds for molding and extrusion.

Benzyl formate *n.* A solvent for cellulosic resins.

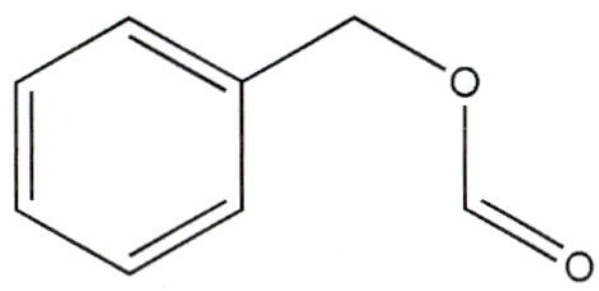

Benzyloctyl adipate (BOA) *n.* A plasticizer for polystyrene, vinyl, and cellulosic resins.

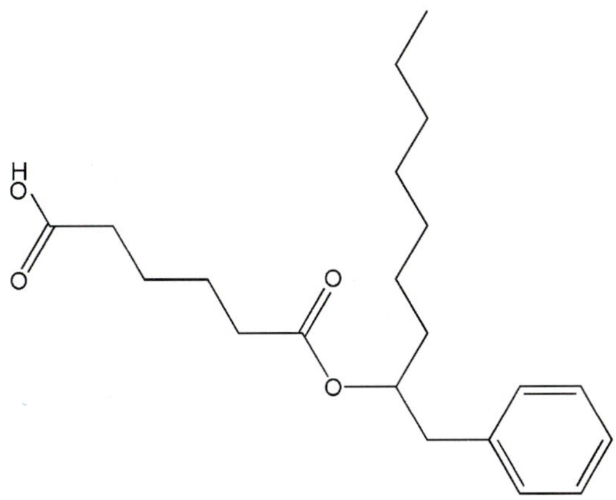

Benzyl phthalate *n.* A solid plasticizer: mp of 42°C, and bp of 277°C/15 mmHg. *Also known as dibenzyl phthalate.*

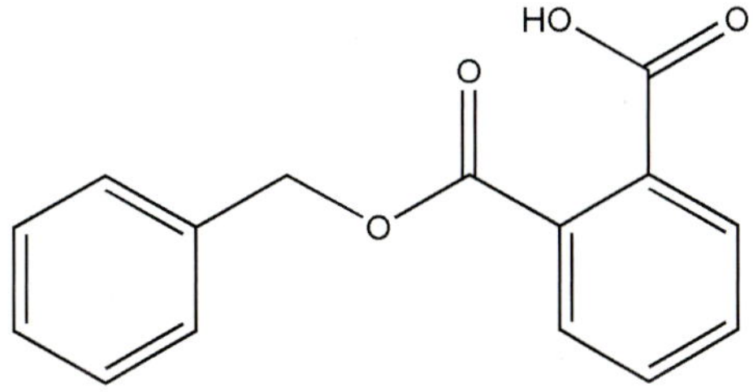

B

Benzyl resinate *See benzyl abietate.*

Benzyltrimethylammonium chloride *n.* A quaternary ammonium salt, a solvent for cellulosics, and a catalyst for phenolic resins.

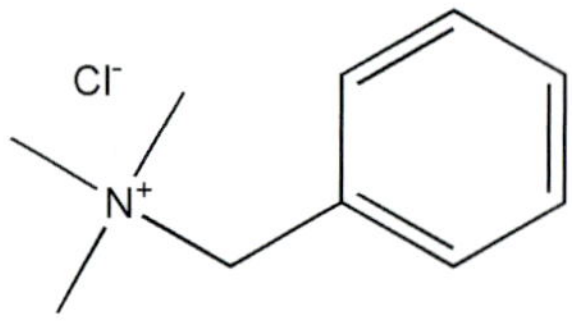

Ber *n.* One of the host trees to which the lac insect affixes itself.

Berlin blacks *n.* Low-cost form of air-drying blacks based on solutions of bitumens to which black pigments or fillers are added. They possess excellent opacity, but reduced gloss. Some Berlin blacks are semimatt.

Berlin blue *n.* A term used for any of the variety of iron-based blue pigments; Prussian blue.
See iron blue.

Berlin red A pigment consisting essentially of red iron oxide.

Bernoulli's theorem \bər-ˈnü-lēz-\ [Daniel *Bernoulli* † 1782 Swiss physicist]. At any point in a tube through which a liquid is flowing the sum of the pressure energy, potential energy, and kinetic energy is constant. If *p* is pressure; *h*, height above the reference plane; *d*, density of the liquid, and *v*, velocity of flow,

$$p + hdg + \frac{1}{2}dv^2 = \text{a constant.}$$

Serway RA, Faugh JS, Bennett CV (2005) College physics. Thomas, New York.

Bertholet principle of maximum work *n.* Of all possible chemical processes, which can proceed without the aid of external energy, that process which always takes place is accompanied by the greatest evolution of heat. This law holds good for low temperatures only and does not account for endothermic reactions. Russell JB (1980) General chemistry. McGraw-Hill, New York.

Bertrand lens *n.* An auxiliary, low-power lens which may be inserted in the bodytube of the microscope between the eyepiece and the objective for observing the back focal plane of the objective; useful for conoscopic observations in crystal optics and for checking microscope illumination quality.

Beryllium \bə-ˈri-lē-əm\ [NL, from Gk *bēryllion*] (ca. 1847) *n.* A steel-gray light strong brittle toxic bivalent metallic element used chiefly ads a hardening agent in alloys.

Beryllium copper *n.* Copper containing about 2.7% beryllium and 0.5% cobalt, used for blow molds and insertable injection-mold cavities. The small percentages of Be and Co greatly increase the strength and hardness of the copper whole preserving its high thermal conductivity and corrosion resistance. Beryllium copper is easily pressure cast and hobbed into mold cavities.

BESA Abbreviation for British engineering standards association.

BET Abbreviation for Brunauer, Emmett, and Teller, applied to an equation and method for determining the surface area of an absorbent, such as carbon.

Beta *adj.* A prefix, usually abbreviated as the Greek letter β and usually ignored in alphabetizing compound names, signifying that the so-labeled substitution is on the second carbon atom away from the main functional group of the molecule.
See alpha.

Beta cellulose *n.* One of the three forms of cellulose. It has a lower degree of polymerization than the alpha form. With gamma cellulose it is known as hemicellulose.
Also see alpha cellulose and gamma cellulose.

Beta gage (beta-ray gage) *n.* Gage consisting of two facing elements, a β-ray-emitting source, and a β-ray detector. When a sheet material (e.g., plastic) is passed between the elements, some of the β-rays are absorbed, according to the area density or the thickness of the sheet. Signals from the detecting element can be used to control equipment that automatically regulates the thickness. The most usual sources are krypton 85 and strontium 90. Also used for particular applications are cesium 137, promethium 147, and ruthenium 106.
See also thickness gauging.

Beta-minus (β^-) particle *n.* A radioactive emission consisting of a high-energy electron, ${}^{0}_{-1}e$.
Also known, simply, as a beta particle.

Beta particle (1904) *n.* A subatomic particle created at the instant of emission from a decaying radioactive atomic nucleus, having a mass (at rest) of 9.1095×10^{-28} g. A negatively charged beta particle is identical to an ordinary electron, and a positively charged one is identical to a positron. A stream of beta particles is called a *beta ray.* Such rays are use in equipment for measuring and controlling the thickness of plastics films, sheets, and other extrudates. Serway RA, Faugh JS, Bennett CV (2005) College physics. Thomas, New York.

Beta (b)-particle (beta ray) *n* (1902). One of the particles, which can be emitted by a radioactive atomic nucleus. It has a mass about $\frac{1}{1837}$ that of the proton. The negatively charged beta particle is identical with the ordinary electron, while the positively charged type (positron) differs from the electron in having equal but opposite electrical properties. The emission of an electron entails the change of a neutron into a proton inside the nucleus. The emission of a positron is similarly associated with the change of a proton into a neutron. Beta particles have no independent existence inside the nucleus, but are created at the instant of emissions.
See neutrino.

Beta-pinene *See pinene.*

Beta-plus (β^+) particle *n.* A radioactive emission (*particle*) consisting of a positron ${}^{0}_{1}e$.

Beta-ray gauge (beta gauge) *n.* A device for measuring the thickness of plastics films, sheets, or extruded shapes, consisting of a source of beta rays and a detecting element. When material is passed between the source and the detector, some of the rays are absorbed, the percent absorbed being a measure of the thickness of the material. Signals from the detecting element can be used to control equipment that automatically regulates the thickness. The most usual sources are krypton 85 and strontium 90. Also used for particular applications are cesium 137, promethium 147, and ruthenium 106.
See also thickness gauging.

Betatron \\ˈbā-tər-ˌträn\\ [ISV] (1941) *n.* An accelerator that uses an electrostatic field to impart high velocities to electrons. Energies of 5–6 MeV will produce X-rays equivalent in energy to gamma radiation of 12–20 g of radium.

Beton *n.* A kind of concrete; a mixture of lime, sand, and gravel.

Bevatron A six or more billion electron voltaccelerator of protons and other atomic particles. Energies of 5–6 MeV will produce X-rays equivalent in energy to gamma radiation of 12–20 g of radium. Makes use of a Cockcroft–Walton transformer cascade accelerator and a linear (q.v.) as well as an electromagnetic field in the build-up.

Bevel siding *See clapboard.*

BHT *n.* Abbreviation for butylated hydroxytoluene.
See di-tert-butyl-p-cresol.

B

Bias fabric *n.* A two-dimensional fabric that when oriented in the XY plane contains fibers that are aligned in a different direction, i.e., 45° to the *x*-axis fibers.

Bias filling *n.* A fabric defect in which the filling yarn does not run at a right angle to the warp. The principal cause is improper processing on the tenter frame.
Also see bow.

Bias ply *n.* A layer of reinforcing fiber, cloth, or sheet oriented at an angle, less than 90° and typically 45°, to the fiber direction in the main reinforcing layers.

Biaxial braid *n.* Braided structure with two yarn systems one running in one direction and the other in the opposite direction.

Biaxial crystals *n.* Anisotropic crystals in the orthorhombic, monoclinic and triclinic systems. They have three principal refractive indices, α, β, and γ.

Biaxial laminate *See bidirectional laminate.*

Biaxially oriented film *n.* Polymeric film (i.e., polyethylene) which as been strained or stretched in one direction which produces optical and physical birefringence or most importantly, high strength one the axial or long direction.

Biaxial orientation *n.* (1) The process of stretching hot plastic film or other article in two perpendicular directions, resulting in molecular alignment. (2) The state of the material that has been subjected to such stretching.
See also orientation and tentering.

Biaxial winding *n.* A method of filament winding in which the helical bands are laid in sequence, side-by-side, with no crossover of fibers.

Bibasic lead phthalate *n.* A heat and light stabilizer for vinyl insulation, opaque film and sheeting, and foam.

Bicarbureted hydrogen *See ethylene.*

Bicomponent fibers *See composite fibers.*

Bicomponent yarns *n.* Spun or filament yarns of two generic fibers or two variants of the same generic fiber.

Biconstituent fiber *n.* A fiber extruded from a homogeneous mixture of two different polymers. Such fibers combine the characteristics of the two polymers into a single fiber.

Bicyclic *n.* Consisting of two rings.

Biddiblack *n.* A particular type of mineral black, mined near Bideford, Devonshire, England.
See mineral black.

Bidirectional *n.* In two directions. (1) The term may be applied to illumination angles in three-dimensional space. Thus, the incident angle may be 45° relative to the material surface, but this designation does not describe the portion of the circumferential 45° angle possible. The term "bidirectional illumination" generally implies two incident beams separated circumferentially by 180°. (2) The term may also be applied to goniophotometric measurements, to the illumination and to the viewing angle. In either case, the angular distance is expressed as the degrees (or radians) from the perpendicular (normal) to the surface as 0°. Thus, the grazing angle approaches 90°.

Bidirectional fabric *n.* A fabric having reinforcing fibers in two directions, i.e., in the warp (machine) direction and filling (cross-machine) direction.

Bidirectional laminate *n.* A fiber-reinforced material in which the fibers are laid in two different directions, typically in the length and width directions. In particular such a laminate in which equal volumes of reinforcing fibers are laid in the two directions.

Bidirectional reflectance distribution function (BRDF) *n.* The ratio of radiance per unit irradiance, used for describing the

geometrical reflectance properties of the surface.
See bidirectional.

Bierbaum scratch hardness *See scratch hardness.*

Biff *See monkey.*

Bikerman boundary-layer theory *n.* The theory that adhesive bonding occurs through the formation of an adsorbed, strong boundary layer on the surface of the adhered.

Bilateral fibers *n.* Two generic fibers or variants of the same generic fiber extruded in a side-by-side relationship.

Billow forming *n.* A variant of thermoforming, in which the hot plastic sheet is clamped in a frame and expanded upward with mild air pressure against a male plug or female die as the plug or die descends into the frame. The process is suitable for thin-walled containers with high draw ratios.

Bimetallic cylinder *n.* In most modern extruders and injection machines, the barrel is lined, by centrifugal casting from the melt, with any of several white irons containing chromium and boron carbides and having hardnesses near Rockwell C65. Aster finish-grinding and polishing, the liner, about 1 mm thick, provides excellent resistance to wear or corrosion or both, depending on the formation. The best known trade name is XALOY®.

Bimetal plate *n.* In lithography, a plate in which the image area is copper or brass, and the non-image area is aluminum, stainless steel, or chromium.

Bimolecular process *n.* An elementary process in which the activated complex is formed as a result of the collision of two particles.

Bin activator *n.* A device that promotes the steady flow of granular or powdered plastics from storage bins or hoppers. Among the many types of equipment are vibrators or mallets acting upon the outside of the container, prodding devices or air jets acting directly on the material, inverted-cone baffles with vibrating means located at the bottom of the hopper, and other "live bottom" devices such as scrapers, rolls, and chains.

B

Bin cure *n.* Rubber term for partial or complete vulcanization of a mixed compound while stored in a bin or pile waiting for molding or further processing.
Also called pile burning or premature vulcanization.

Bin dischargers *n.* Restores flowability to materials, which are prone to packing during storage in bins, silos, or surge hoppers, and promotes free flow of materials, which tend to bridge or hang up.

Binder *n.* (1) Non-volatile portion of the liquid vehicle of a coating. It binds or cements the pigment particles together and the paint film as a whole to the material to which is it applied. (2) Component of an adhesive composition, which is primarily responsible for the adhesive forces, which hold two bodies together. (3) Resin or cementing constituent of a plastic compound which holds the other components together. The agent applied to glass mat or performs to bond the fibers prior to laminating or molding. (4) The components in an ink film which holds the pigment to the printed surface. Wicks ZN, Jones F, Pappas SP (1999) Organic coatings science and technology, 2nd edn. Wiley-Interscience, New York.
See also vehicle.

Binder content *n.* The weight of adhesive used to bond the fibers of a web together. Binder content is usually expressed as percent of fabric weight.

B

Binder demand *n.* That amount of binder needed to completely wet a pigment by displacing the air voids. This is determined primarily by: the particle size, shape, chemical composition, and density of the pigment; and the particle size, degree of polymerization and wetting properties of the binder. (Whereas the binder demand refers to a particular pigment-vehicle system, oil absorption is a numerical value assigned to a pigment based on a specific test method.)

Binder fibers *n.* Fibers that can act as an adhesive in a web because their softening point is relatively low compared with that of the other fibers in the material.

Binder ratio *See pigment/binder ratio.*

Binding varnish *n.* A term used by printers or lithographers to describe a viscous varnish in the ink that is used to toughen the dried film.

Bingham body *n.* Material displaying plastic flow.

Bingham liquid *n.* Liquid exhibiting plastic flow.
See plastic flow.

Bingham plastic *n.* A model for flow behavior in which no flow occurs until the shear stress exceeds a critical level called the yield value or yield stress above which shear rate is proportional to the stress (after the yield stress has been surpassed). When a Bingham plastic flows through a circular tube, there is a critical radius r_c at which the shear stress, $\Delta P r_c/(2L)$, equals the yield stress. All actual flow occurs between that radius and the wall radius, *R*, while from the center to r_c there is a solid plug carried along by the stream. Parfitt CD, Sing KSW (1976) Characterization of powder surfaces. Academic Press, London. Patton TC (1979) Paint flow and pigment dispersion: a rheological approach to coating and ink technology. John Wiley and Sons, New York.

Binodal *n.* Having or relating to two modes; especially, having or occurring with two statistical modes.

Biochemical oxygen demand (BOD) *n.* A standard test for estimating the degree of contamination of water supplies. It is expressed as the quantity of dissolved oxygen (in mg/l) required during stabilization of the decomposable organic matter by aerobic biochemical action.

Biocide \ˈbī-ə-ˌsīd\ (1947) *n.* An agent incorporated in or applied to the surfaces of plastics to destroy bacteria, fungi, marine organisms, etc. Some plastics, e.g., acetals, acrylics, epoxies, phenoxies, ABS, nylons, polycarbonate, polyesters, fluorocarbons, and polystyrene, are normally resistant to attack by bacteria or fungi. Others, e.g., alkyds, phenolics, low-density polyethylene, urethanes, and flexible vinyls can under some circumstances be affected by growth of these organisms on their surfaces. Even though the resins themselves might be resistant, additives such as plasticizers, stabilizers, fillers, and lubricants can serve as food for fungi and bacteria. Examples of biocides are organotins, brominated salicylanilides, mercaptans, quaternary ammonium compounds, and compounds of mercury, copper, and arsenic.

Biocompatibility \-kəm-ˌpa-tə-ˈbi-lə-tē\ (1971) *n.* Materials, which are compatible with living blood and tissue, which is important for artificial blood vessels, hearts, etc.

Biodegradable \-di-ˈgrā-də-bəl\ (1961) *adj.* The process of rapid decomposition as a result of the action of micro-organisms.

Biodegradable surfactant *n.* Surfactant which may be decomposed by biological action.

Biodegradation *n.* The gradual breakdown of plastics and matter by living organisms such as bacteria, fungi, and yeasts. Most of the commonly used plastics are essentially not biodegradable, exhibiting limited susceptibility to assimilation by microorganisms. An exception is polycaprolactam. However, the growing emphasis on environmental aspects of discarded plastics has stimulated research in ways of attaining biodegradability after a predetermined time period. One method is to add a UV-light sensitizer that causes photodegradation after a period of exposure to light, followed by breakup after prolonged exposure to the elements, after which bacteria will finish the job. A third method is the deliberate incorporation of weal links in the polymer chain, temporarily protected by a degradable stabilizer. Zaiko GE (ed) (1995) Degradation and stabilization of polymers. Nova Science Publishers Inc., New York.
See also photodegradation.

Biologically active polymers *n.* Polymers capable of specifically and reversibly binding to analytes, including molecules and cells. The biologically active polymers are also capable of releasing substances upon electrical stimulation.

Biomedical polymers *See biocompatibility.*

Biopolymer \ˌbī-ō-ˈpä-lə-mər\ (1961) *n.* (1) A copolymer in which there is irregularity with regard to the placement (relative locations within the polymer chain, in a linear copolymer) of two or three or more chemically different types of units. These units may be mers, in a product of addition polymerization, or residues of the condensing small molecules in a polycondensate. (2) A polymer produced by living organisms, such as cellulose, natural rubber, silk, rosin, and leather.

Biot number *n.* A dimensionless group, ht/k, important in convective heating and cooling of sheets. h is the heat-transfer coefficient at the sheet's surface, t the sheet thickness of half-thickness, and k is the thermal conductivity of the sheet material.

Biphenol *n.* **A** (4,4′-isopropylidenediphenol). An intermediate used in the production of epoxy, polycarbonate, and phenolic resins. The name was coined after the condensation reaction by which it may be formed – two (bis) molecules of phenol with one of acetone (A).

Biphenyl \(ˌ)bī-ˈfe-nᵊl, -ˈfē-\ [ISV] (1923) *n.* (diphenyl, phenyl benzene) $(C_6H_5)_2$. A stable, high-boiling (256°C) liquid long used as a heat-transfer medium.

Bipolymer *n.* A polymer derived from two species of monomer (IUPAC). The more commonly used Syn: *copolymer.*

Bird applicator *n.* Device used for the laboratory application of coating of a prescribed thickness to test panels. It is a machined steel bar with a fixed clearance having a beveled undercut to guide the coating material.
See applicator.

Bird's-eye *n.* (1) Small localized areas in wood with the fibers indented and otherwise contorted to form few to many small circular or elliptical figures remotely resembling birds' eyes on the tangential surface. Common in sugar maple and used for decorative purposes; rare in other hardwood species. (2) A generic term describing a cloth woven on a dobby loom, with a geometric pattern having a center dot resembling a bird's eye. Originally birds eye was made of cotton and used as a diaper cloth because of its absorbent qualities, but now the weave is made from a variety of fibers or fiber blends for many different end uses. (3) A speckled effect on the back of a

B

knit fabric resulting from the use of different colors on the face design. Vincenti R (ed) (1994) Elsevier's textile dictionary. Elsevier Science and Technology Books, New York.

Birefringence \ˌbī-ri-ˈfrin-jən(t)s\ [ISV] (1898) *n.* (1) (double refraction) The difference between any two refractive indices in a single material. When the refractive indices measured along three mutually perpendicular axes are identical, the material is said to be optically isotropic. Orientation of a polymer by drawing may alter the refractive index in the direction of draw so that it is no longer equal to that in the perpendicular directions, in which event the material is said to display birefringence. Crystalline polymers, normally birefringent, may become optically isotropic at their melting points. Studies of birefringence provide useful information regarding the shapes of molecules, degrees of orientation and crystal habits. (2) Property of anisotropic materials, those which possess different refractive indices according to the direction of vibration of light passing through a crystal; usually observed with the aid of a polarizing microscope. The degree of birefringence is expressed numerically as the difference between the highest and lowest indices of refraction, even though there may be more than two different indices of refraction. Fox AM (2001) Optical properties of solids. Oxford University Press, UK.

Bisacki *n.* Grade of *See lac.*

Bis(4-*t*-butylcyclohexyl)peroxy dicarbonate *n.* A catalyst of the organic-peroxide family, used in reinforced plastics and vinyl polymerization. Unlike other percarbonates, it does not require refrigeration for storage or handling.

Biscuit *See preform.*

Bis(ethoxyethoxyethyl) phthalate \ˈbis-e-ˈthäk-sē-e-ˈthäk-sē-ˈe-thəl-ˈtha-ˌlāt\ *n.* $C_6H_4(COOC_2H_4\ OC_2H_4OC_2H_5)$. A good primary plasticizer for polyvinyl acetate, nitrocellulose, cellulose acetate, and many other polymers (*See image*).

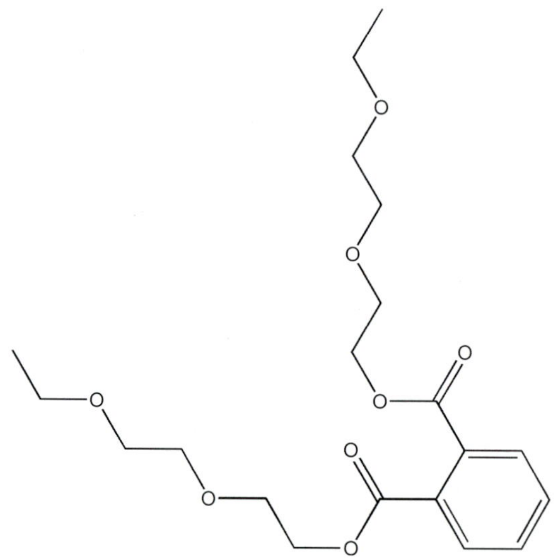

Bis(β-hydroxyethyl-γ-aminopropyltriethoxy silane *n.* A silane coupling agent used in reinforced epoxy resins, also in many reinforced thermoplastics such as PVC, polycarbonates, nylon, polypropylene, and polysulfones.

Bismaleimide resins *n.* Bismaleimide resins are low molecular substances (dry powders) containing imide structures already in the monomer form.

Bismarck brown *n.* Brown-colored dyestuff, which chemically is triaminoazobenzene. It dissolves in alcohol and water, to yield reddish-brown solutions. It has some application in the manufacture of mahogany stains.

Bisphenol A \ˈbis-ˈfē-ˌnōl, -ˌnól. fi-ˈ\ *n.* OH–Ar–$(CH_3)_2$ –Ar–OH, a diol which reacts with epichlorohydrin to form bisphenol epoxy resins. (4,4′-isopropylidenediphenol) An intermediate used in the production of epoxy, polycarbonate, and phenolic resins. The name was coined after the condensation reaction by which it may be formed – two (bis) molecules of phenol with one of acetone (A) (*See image*).

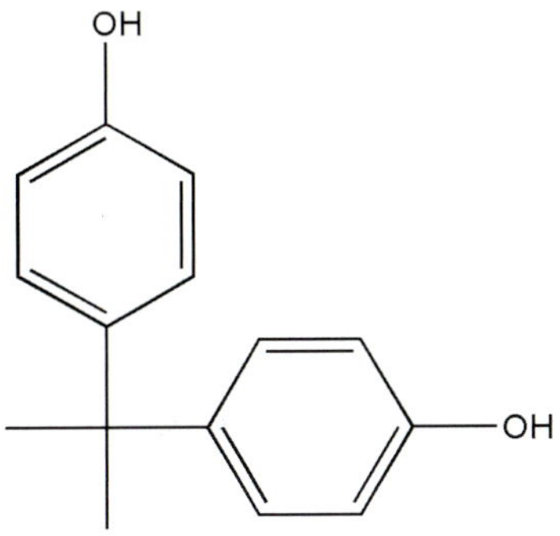

Bisphenol A and F *n.* Dihydroxydiphenyldimethylmethane, mol wt 224.1. Insoluble in water. Used in the manufacture of phenolic and epoxy resins. Condensation product formed by reaction of two (bis) molecules of phenol with acetone. This polyhydric phenol is a standard resin intermediate, along with epichlorohydrin, in the production of epoxy resins (*See image*).

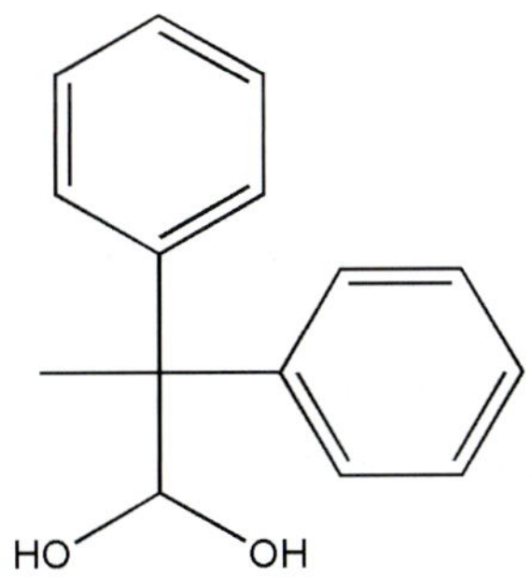

Bisphenol epoxy resins *n.* Resins based on bisphenol A.

BISRA *n.* Abbreviation for British iron and steel research association.

Bistre \ˈbis-tər\ [Fr *bistre*] (ca. 1751) *n.* Brown water color pigment, which is derived from the tarry soot of burned, resinous wood and beechwood. It is similar to asphaltum, in color and composition. The color varies from saffron yellow to brown-black, depending upon the source and treatment of the raw material.

Bis(tri-*n*-butyltin) oxide *n.* A liquid derived by the hydrolysis of tributyltin chloride, used to control the growth of most fungi, bacteria and marine organisms on plastics used in boat construction and in urethane foams.

Bis(tri-*n*-butyltin) sulfosalicylate *n.* An antimicrobial agent used in flexible PVC film and urethanes.

Bis(2,2,4-trimethyl-1,3-pentanediol) monoisobutyrate adipate *n.* A plasticizer for cellulosic resins and polystyrene.

Bite *n.* (1) The ability of an adhesive to penetrate surfaces and thereby produce an adhesive bond. (2) Syn: nip.

Bi-*tert*-butyl-*p*-cresol *n* (DBPC, butylated hydroxytoluene, and BHT) $[C(CH_3)_3]_2C_6CH_3H_2OH$. A white crystalline solid used as an antioxidant in polyethylene, vinyl monomers, and many other substances.

Bitter-almond oil, synthetic *See benzaldehyde.*

Bittiness *n.* Presence of material that hinders the appearance of a smooth and uniform coating film.

Bitty *n.* Said of coatings containing bits of skin or foreign matter of any type which project above the coating film and hinder the formation of a smooth and uniform film. Syn: peppery.
See also nibs.

Bitumen \bə-ˈtyu-mən, bī-, -ˈtu-, *esp British also* ˈbit-yə-\ [ME *bithumen* mineral pitch, fr. L *bitumen*-, bitumen] (15c) *n* Hydrocarbon material of natural or pyrogenous origin, or combinations of both, frequently accompanied by their non-metallic derivatives, which may be gaseous, liquid, semisolid, or solid and which is completely soluble in carbon disulfide.

See asphalt, brown bitumen, elaterite, gilsonite, and rafaelite.

Bitumens *n.* A naturally occurring, almost black materials that are also obtained in mineral-oil refining, it consists of a high molecular weight hydrocarbons dispersed in oil-like material.

Bituminous cement *n.* A black substance available in solid, semisolid, or liquid states at normal temperatures; composed of mixed indeterminate hydrocarbons; appreciably soluble only in carbon disulfide or other volatile liquid hydrocarbon; especially used in sealing built-up roofing and between joints and in cracks of concrete pavements.

Bituminous coating *n.* An asphalt or tar compound used to provide a protective finish for a surface.

Bituminous emulsion *n.* A suspension of minute globules of bituminous material in water or of minute globules of water in a liquid bituminous material; used as a protective coating against weather, especially where appearance is not important.

Bituminous paints *n.* (1) Originally, the class of paints consisting essentially of natural bitumens dissolved in organic solvents: they may or may not contain softening agents, pigments, and inorganic fillers. They are usually black or dark in color. Within recent years the term "bituminous" has, by common usage, come to include bitumen-like products such as petroleum asphalt. (2) A low-cost paint containing asphalt or coal tar, a thinner, and drying oils; used to waterproof concrete and to protect piping where bleeding of the asphalt is not a problem. Usmani AM (ed) (1997) Asphalt science and technology. Marcel Dekker, New York.

Bituminous varnish *See varnish.*

Biuret *n* (allophanamide, carbamylurea) $NH_2CONHCONH_2{\cdot}H_2O$. A white crystalline material derived from urea by heat or by reaction with an isocyanate. It is used primarily in analytical chemistry, but the biuret group is formed during some polymerization reactions, such as primary bonding in urethane elastomers. Odian GC (2004) Principles of polymerization. John Wiley and Sons Inc., New York.

Bivinyl *See butadiene.*

Black *n.* Ideally, the complete absorption of incident light; the absence of any reflection. In the practical sense, any color which is close to this ideal in a relative viewing situation, i.e., a color very low saturation and of low luminance.

Blackboard paint *n.* Rapid drying flat paint capable of permitting writing on it by chalk; usually black or green in color.

Black body *n.* The theory that atoms can exist for a duration solely in certain states, characterized by definite electronic orbits, i.e., by definite energy levels of their extranuclear electrons, and in these stationary states they do not emit radiation; the jump of an electron from an orbit to another of a smaller radius is accompanied by monochromatic radiation. A black body has an emissivity (e_B) of 1, indicating that a black body absorbs then emits all radiation at a wavelength. A black body is the opposite property of a perfect reflector of radiation at a wavelength that absorbs no radiation. Odian GC (2004) Principles of polymerization. John Wiley and Sons Inc., New York.

Black boy gum *See accroides.*

Black chalk A bluish black clay containing carbon.

Black dammar *See black east India.*

Black east India *n.* Dark colored resin collected in the East Indies. It resembles the dammars in several respects, notably in solubility and free acid content. Its alternative name is "black dammar".

Black iron oxide *n.* $FeO{\cdot}Fe_2O_3$ or Fe_3O_4. Pigment black 11 (77499). Two black oxides of iron are used, namely, the natural mineral, magnetite, and the artificially produced oxide. Both types are magnetic. Neither possesses outstanding opacity. Magnetite consists of about 95% Fe_3O_4. Artificial back oxide consists of a mixture of ferric oxide (Fe_2O_3) and ferrous oxide (FeO). Kirk–Othmer encyclopedia of chemical technology: pigments–powders. John Wiley and Sons, New York, 1996.

Black Japan *n.* Black varnish-like material in which the higher grades of asphalt or other resinous compositions are employed, blended with suitable drying oils to produce hard, glossy black coatings which become insoluble in mineral spirits. Weismantal GF (1981) Paint handbook. McGraw-Hill Corporation Inc., New York.

Black *n.* Another name for graphite.

Black light *n* (deprecated). Popular term for ultraviolet radiation without any visible radiation. It is an incorrect term because *light* is defined as *visible* radiation and ultraviolet radiation is not detected by the visual mechanism of the eye. Syn: ultraviolet radiation.

Black magnetic oxide *See black iron oxide.*

Black Orlon *n.* Orlon® (black), acrylic fiber.

Black out paint Opaque paint, usually black or dark in color, used during war time, on windows, skylights, etc., to prevent the passage of light to the outside of a building.

Black varnish *n.* Any varnish in which the resin component is substituted completely, or in part, by a petroleum bitumen, or natural asphaltum.

Blanc fixe \ˈblaŋk-ˈfiks\ [F] (1866) *n* A synthetic form of barium sulfate prepared by reacting aqueous solutions containing barium ions with others containing sulfate ions and precipitating the reaction product. It is used as a special-purpose filler to impart X-ray opacity and high specific gravity. Precipitated barium sulfate is used as a semi-transparent extender in printing inks.

Blanch \ˈblanch\ [ME *blaunchen*, fr. MF *blanchir*, fr. OF *blanche*, feminine of *blanc*] (15c) *v.* (1) To make white. (2) As distinct from blush or bloom, a pale milky cast on a coating film.

Blank *n.* (1) A piece punched or die-cut from a sheet and intended for further forming into its final shape. (2) In chemical analysis, a dummy sample or solution of the same matrix as the sample to be analyzed but containing none of the analyte sought.

Blanket *n.* An unquilted bedding fabric designed primarily to provide thermal insulation.

Blanket mark *See corrugation mark.*

Blanking (die cutting). The cutting of flat sheet stock to shape by striking it sharply with a punch while it is supported on a mating die. Punch presses are often used for the operation. An alternate method is to make the cut with a thin, sharp-edged, shaped steel blade called a steel-rule die. *See also die cutting.*

Blanking die *n.* A metal die used in the blanking process.

Blast cleaning *n.* Cleaning and roughening of a surface (particularly steel) by the use of natural or artificial grit or fine metal shot (usually steel), which is projected against a surface by compressed air. *Also known as power cleaning.*

Blast finishing *n.* The removal of flash from molded objects (and/or dulling their surfaces) by impinging media such as steel balls, crushed apricot pits, walnut shells or plastic pellets upon them with sufficient force to fracture the flash. When the material being deflashed is not sufficiently brittle at room temperature, the articles are first chilled to a temperature below their brittleness temperature. Typical blast-finishing machines consist of wheels rotating at high speeds, fed at their centers with the media, which are thrown out at high velocities against the objects.

Bleach *n.* (1) Refers to the method of measuring the tinctorial strength of an ink or toner, usually accomplished by mixing a small portion of the ink (or toner) with a large amount of white base and evaluating the tinctorial strength of the ink versus a control standard. (2) A whitening agent such as an aqueous solution of sodium hypochlorite commonly used for oxidizing or whitening clothes.

Bleached oil *n.* Oil, which has been refined by the acid, alkali, or mechanical process and in which the refining includes a treatment at about 88°C (190°F) with an adsorbent such as fuller's earth or activated charcoal, after which the oil is filtered.

Bleached shellac *n.* Substantially colorless product obtained by bleaching natural orange gum lac with chlorine or chlorine-containing agents, such as sodium or calcium hypochlorites.

Also known as bone dry bleached shellac.

Bleaching *n.* (1) Loss of color of a paint or varnish. This may be due to internal chemical or physical action in the paint itself, to influences from the surface on which it is applied or to weathering or contamination from the atmosphere. (2) Intentional lightening of the color of a material such as wood, vegetable oils, varnishes, etc. (3) Any of several processes to remove the natural and artificial impurities in fabrics to obtain clear whites for finished fabric or in preparation for dyeing and finishing.

See also fading and whitening in the grain.

Bleaching clay *n.* Clay that possesses decolorizing characteristics for use in refining of mineral, petroleum, vegetable, and animal oils from the spinneret hole involved.

Bleb \ˈbleb\ (1607) *n.* A blister or bubble on the face of a spinning jet, interrupting the extrusion of the filament.

Bleb rate *n.* The frequency of bleb formation in an extrusion operation.

Bleed (1) *n.* An escape passage at the parting line of a mold, similar to an air vent but deeper, serving to allow material to escape or bleed out. (2) The spreading or running of a pigment color by the action of a solvent.

See bleeding.

Bleed characteristics *See bleeding.*

Bleeding *n.* (1) The diffusion of colorants through a coating from a previously painted substrate due to the action of the vehicle or solvent or both. The action is dependent on the pigments; vehicles, and solvents of the systems. (2) Diffusion of a soluble colored substance from, into and through a coating from beneath, thus producing an undesirable staining or discoloration. Materials, which give rise to this effect are tannins or dyes in some types of wood; wood preservatives; bituminous coatings; pigment dyestuffs; and stains. (3) Diffusion of coloring matter from the substrate; also, the discoloration arising from such diffusion. In the case of printing ink, the spreading or running of a pigment color by the action of a solvent such as water or alcohol. (4) Migration to the rubber surface of an oil, wax, or plasticizer as a film or in drops, sometimes called sweating. Also a term applied to organic pigment colors if they migrate into an adjacent stock of a different color, or when they are removable at the surface by water or other solvents. (5) The spreading or migration of an ink component into an unwanted area. The terms *migration*, *crocking*, *blooming*, and *bronzing* are sometimes used loosely to describe the same phenomenon. (6) Loss of color by a fabric or yarn when immersed in water, a solvent, or a similar liquid medium, as a result of improper dyeing or the use of dyes of poor quality. Fabrics that bleed can cause staining of white or light shade fabrics in contact with them while wet. Martens CR (1968) Technology of paints, varnishes and lacquers. Reinhold Publishing Co., New York. Paint testing manual: physical and chemical examination of paints, varnishes, lacquers, and colors – stp 500. American Society for Testing and Materials, West Conshohocken, PA, 1973. Paint/coatings dictionary. Compiled by Definitions Committee of the Federation of Societies for Coatings Technology, 1978.

Bleeding pigment *n.* A pigment in a coating which is partially soluble in the solvent or vehicle portion of subsequent coats applied to it.

Bleedout In filament winding, the excess liquid resin that migrates to the surface of a winding.

Blemish *n.* Any surface imperfection of a coating or substrate.

Blend *n.* (1) A mixture, such as a mixture of solvents or a mixture of inks. (2) A yarn obtained when two or more staple fibers are combined in a textile process for producing spun yarns (e.g., at opening, carding, or drawing). (3) A fabric that contains a blended yarn (of the same fiber content) in the warp and filling.
See alloy.

Blender *n.* A round, softish brush of badger hair or similar material with a blunt tip, used for blending colors and removing brush marks left by coarser brushes. (2) A small laboratory mixer used to dispense pigment in a vehicle.
Also called softener.

Blenders *n.* Mixing of liquid/liquid and/or solids components by mixing in a vessel using paddles or other low shear rate means of mixing.

Blending *n.* (1) Any process in which two or more components or ingredients are physically intermingled without significant change of the physical states of the components. (2) The bringing together of two or more polymers, using whatever means may be needed, such that the final scale of segregation is microscopic or finer. (3) The combining of staple fibers of different physical characteristics to assure a uniform distribution of these fibers throughout the yarn.

Blending resin (extender resin) *n.* With respect to vinyl plastisols and organosols, a blending resin is one of larger particle size and lower cost than the dispersion resins normally used, a partial replacement for the primary resin. Blending resins are sometimes sued to achieve a better balance of properties other than cost.

Blinding *n.* Loss of luster of fibers after wet processing.

Blinding of lithographic plates *n.* Loss of ink-receptivity in the plates' image areas.

Blister *n.* (1) An imperfection on the surface of a plastic article caused by a pocket of air or gas beneath the surface. Note – A blister may be caused by insufficient adhesive; inadequate curing time, temperature or pressure; or trapped air, water, or solvent vapor. (2) A thermoformed canopy or pocket roughly hemispherical, for example an aircraft cockpit cover or a shape used in blister packaging.

Blister house *n.* House-like structure used as an accelerated test for blister resistance of coatings. Under controlled conditions, it attempts to simulate or accelerate blistering, which occurs on buildings.

Blistering *n.* Formation of dome-shaped projections in paints or varnish films resulting from local loss of adhesion and lifting of the film from the underlying surface.

Blistering of paints, determination of degree of *n.* Test method to evaluate the degree of blistering that may develop when paint systems are subjected to conditions which will cause blistering, using photographic reference standards.

Blistering resistance *n.* The ability of a coating to resist the formation of dome-shaped, liquid- or gas-filled projections in its film resulting from local loss of adhesion and lifting of the film from the underlying surface or coating.
See blistering.

Blister packaging *n.* The enclosing of articles in thermoformed, transparent "blisters" shaped to more or less fit the contours of the articles. The preformed blisters, usually slightly oversized to provide ample room, are made of thermoplastics such as vinyl, polystyrene, or cellulosic plastics. They are placed inverted in fixtures and loaded with the articles, and then cards coated with an adhesive are applied and sealed to the flanges between and around the blisters by means of heat and pressure.

Block *n.* A portion of a polymer molecule comprising many mer units that has at least one constitutional or configurational feature not present in the adjacent portions.

Block coat *See tie coat, transition primer, and barrier coat.*

Block copolymers *n.* (1) A copolymer with chains composed of shorter homopolymeric chains that are linked together. These "blocks" can alternate regularly or randomly. Such copolymers usually have some higher properties than either of the homopolymers or their physical blends. They have distinct blocks of polymer segments such as,

AAAABBBBAAAABBBAAA

sequencing of A and B monomers. When blocks are of different monomer species, the term *block copolymer* is used. (2) An essentially linear copolymer in which there are repeated segments of different chemical structure. Odian GC (2004) Principles of polymerization. John Wiley and Sons Inc., New York.

Block cutter *n.* A craftsman who, by hand, hammers into a roller the brass strips and felt from which the wallpaper will be printed.

Blocked curing agent *n.* A curing agent or hardener rendered unreactive, which can be reactivated as desired by physical or chemical means.

Block filler *n.* A pigmented coating of heavy consistency used to fill void spaces in concrete or cinder blocks, prior to the application of the top coat, in order to produce a smooth surface.

Block flooring *n.* End grain blocks glued down in decorative patterns.

Blocking *n.* (1) The undesirable sticking together of two painted surfaces when pressed together under normal conditions or under specified conditions of temperature, pressure, and relative humidity. (2) Undesired adhesion between touching layers of a material, such as occurs under moderate pressure and sometimes pressure and heat, during storage of fabrication. Such agents are called antiblocking agents.

Blocking point *n.* Lowest temperature at which two coated surfaces in mutual contact will tick together sufficiently to injure the surfaces permanently and/or prevent easy separation. Thus, blocking point is a major quality criterion for evaluating coating-grade petroleum waxes.

Blocking test *n.* Procedure for determining the tendency of painted surfaces to stick together (block) when stacked or placed in contact with each other under a weighted load. *See block resistance.*

Block press *n.* (1) A press used to agglomerate laminate squares under heat. The squares, which have been cut from laminated sheet, are crossed to combat the anisotropy that normally occurs during laminating. (2) A press used to mold very large blocks of polystyrene foam.

Block printing *n.* A process of printing with blocks on which the unit or design stands out in relief. Pear wood is traditionally used, and a different block is needed for each color. *Also called hard-blocking. See printing.*

Block resistance *n.* The resistance to blocking.

Blood compatibility *See biocompatibility.*

Blood red *n.* A pigment consisting essentially of red iron oxide.

Bloom \ˈblüm\ [ME *blome*, fr. ON *blōm*; akin to OE *blōwan* to blossom] (13c) *n.* A haziness which develops on high gloss surfaces resulting in scattering of the surface reflectance. One mechanism is by exudation of a component such as a plasticizer out of the paint film. (2) Undesirable deposit, which sometimes forms on a glossy coating, resulting in whitening or loss of gloss. (3) Similar to bleeding in that it is migration of liquids or solids to the surface of a rubber compound to cause a change of appearance in color or cloudiness at the surface. Waxes used in excess of their solubility point in rubber come to the surface as a wax bloom, as does sulfur that remains as an excess over the amount actually chemically combined with the rubber. (4) Undesirable exudation of pentachlorophenol on the surface of wood that has been treated with a preservative solution. (5) In the plastics industry, bloom is a visible exudation on the surface of a plastic, generally caused by lubricant, plasticizer, etc. (6) The fluorescence of mineral oil; the blue or purple cast apparent when the oil is spread in a thin film. (7) The appearance of brightness of a dyed fabric when the fabric is viewed across the top while held at eye level. Syn: FOG. Wicks ZN, Jones FN, Pappas SP (1999) Organic coatings science and technology, 2nd edn. Wiley-Interscience, New York. Paint/coatings dictionary. Compiled by Definitions Committee of the Federation of Societies for Coatings Technology, 1978. *See blush. See also oil and debloomed.*

Blooming *See opening.*

Bloom oil *n.* One of the liquid decomposition products obtained from the destructive distillation of rosin. It can be regarded as a rosin oil, and has a bp of about 270°C.

Blotch *See finishing spot.*

Blotching *See mottle.*

Blotch printing *See printing.*

Blowing *n.* Porosity or sponginess occurring during cure, either deliberately through use of a gas releasing material to form sponge or expanded rubber, or inadvertently due to entrapped moisture to cause undesirable porosity. The practice of injecting a gas into a hot and flowing resin to create a part or film.

Blowing *n* (of Plaster). Appearance on the surface of the plaster of conical hollows (pops or blows). They are due to the presence of particles of reactive material, which expand, after the plaster has set, with sufficient force to push out the plaster in front of the particles. They may occur in undercoats or finishing coats of plaster, or both.

Blowing agent (foaming agent) *n.* A substance that, alone or in combination with other substances, is capable of producing a cellular structure in a plastic or rubber mass. Thus, the term includes compressed gases that expand when pressure is released, soluble solids that leave pores when leached out, liquids that develop cells when they vaporize, and chemical agents that decompose or react under the influence of heat to form a gas. Liquid foaming agents include certain aliphatic and halogenated hydrocarbons, low-boiling alcohols, ethers, ketones, and aromatic hydrocarbons. The chemical blowing agents range from simple salts such as ammonium or sodium bicarbonate to complex nitrogen-releasing agents, of which azobisformamide (ABFA) is an important example.

Blown bitumen *n.* Generally regarded as the blown products of petroleum bitumens, which have been previously obtained as residues from the distillation of crude petroleum. The petroleum bitumens are subjected to high temperature blowing with air, which has profound effects on the properties of the material. Melting point is raised considerably, and the product becomes rubbery. For bitumens of the same melting point, the blown product has a higher penetration figure.
See bitumen and air blowing.

Blown castor oil *n.* Product of very high viscosity and reddish-brown color, obtained by blowing raw castor oil. One of its chief applications is as a plasticizer or softener for cellulosic compositions.

Blown film *See film blowing.*

Blown oils *n.* Vegetable or fish oil, which has been partially oxidized by blowing with a current of air while at an elevated temperature. The characteristics of the oil such as its increased viscosity and degree of oxidation can be controlled by time, temperature, and amount of air.

Blow-up ratio *n* (BUR). (1) In blow molding, the ratio of the largest diameter of a cavity into which a parison is to be blown to the outside diameter of the parison. (2) In blown-film extrusion, the ratio of the film diameter, before collapsing or gusseting, to the mean diameter of the die opening.

Blue asbestos (crocidolite) *n.* An iron-rich form of asbestos, fibers of which were long used in reinforced plastics when good chemical resistance was essential.

Blue basic lead sulfate (blue lead) *n.* A variant of white basic lead sulfate containing 78% monobasic lead sulfate. It is not actually a blue pigment, but has a dark gray or slate color.
See basic lead sulfate.

Blueing Neutralizing the yellow cast of certain white pigments or paints by adding a trace of blue, thereby increasing apparent whiteness.
See blue toner.

Blue stone *See copper sulfate.*

Blue toner *n.* Pigment or dye used in small quantities to neutralize the yellow cast of certain white pigments or coatings, increasing apparent whiteness; also used to make black coatings more jet black.

Blue verditer *n.* Basic copper carbonate. Syn: azurite.

Blue vitriol *See copper sulfate.*

Bluing *n.* A mold blemish in the form of a blue oxide film on the polished surface of a mold, caused by overheating.

Bluing off *v.* A term used by mold makers for the process of checking the accuracy of two mating surfaces by applying a thin coating of Prussian blue to one surface, pressing the coated surface against the other surface, and observing the areas of intimate contact where the blue color has transferred.

Blushing *n.* (1) Film defect, which appears as a milky opalescence as the film dries; can be a temporary or permanent condition. It is generally caused by rapid evaporation, moisture, or incompatibility. Syn: fog. (2) Milky opalescence which sometimes develops as a film of lacquer dries, and is due to the deposition of moisture from the air and/or precipitation of one or more of the solid constituents of the lacquer; usually confined to lacquers which dry solely by evaporation of solvent. Also applies to printing inks. (3) A white or grayish cast on high gloss paint; results from the precipitation of binder solids owing to incompatibility with water, oil, or solvent. *See also bloom.*

BMC *n.* Abbreviation for bulk molding compound.

BOA *See benzyl octyl adipate.*

Board *n.* (1) A heavy weight, thick sheet of paper or other fibrous substance, usually of a thickness greater than 6 mil (0.006 in.). (2) Lumber less than 2 in. (6 cm) thick and between 4 (10 cm) and 12 in. (30 cm) in width; a board less than 4 in. (10 cm) wide may be classified as a strip.

Board and batten *n.* A type of wall cladding for wood-frame houses: closely spaced, applied boards or sheets of plywood, the joints of which are covered by narrow wood strips.

Board and brace *n.* A type of carpentry work consisting of boards which are grooved along both edges and have thinner boards fitted between them.

Board foot A unit of cubic content, used in measuring lumber; equal in volume to a piece 1 ft^2 and 1 in. thick, or 2360 cm^3.

Boardy A term used to describe a fabric with a very stiff hand.

Bob-and-C \ˈbä-bən\ [origin unknown] (1530) *n.* A cylindrical or slightly tapered barrel, with or without flanges, for holding slubbings, rovings, or yarns.

BOBTEX® ICS yarn system *n.* A process for producing a simulated spun yarn by embedding individual fibers in a thermoplastic or adhesive coating on a filament yarn.

Bodanyl *n.* Poly(caprolactam). Manufactured by Feldmühle, Rorschach, Switzerland.

Bodied linseed oil *See bodied oil.*

Bodied oil *n.* In the strictest sense of the term, this is a drying oil which has been heat polymerized. The term is also used more broadly to describe drying oils which have viscosities greater than those of the oils in their natural condition, irrespective of how the increased viscosity is attained, e.g., blown oil or stand oil.

Body *n.* (1) A term used loosely in the paint and adhesives industries to denote overall consistency, i.e., a combination of viscosity, density, pastiness, tackiness, etc. (2) An aspect of fabric quality, akin to drape and hand. (3) A general term referring to viscosity, consistency, and flow of a vehicle or ink. (4) Used also to describe the increase in viscosity by polymerization of drying oils at high temperatures. (5) A practical term widely used to give a qualitative picture of consistency. For Newtonian liquids, both is the same as viscosity.

Body gum (#8 varnish) In printing inks, linseed oil that has been heat polymerized to a heavy, gummy state. It is commonly used as a bodying agent.

B

Body putty *n.* A paste-like mixture of plant resin, often a polyester, and a filler such as talc, used to smooth and repair metal surfaces such as auto bodies.

Bodying *n.* Increase in the apparent viscosity of a paint, varnish, resin, or lacquer, which occurs either deliberately during manufacture of adventitiously during storage.

Bodying agent *n.* A material added to an ink to increase its viscosity.
See thickener.

Boea *n.* One of the classes of Manila copal.

Boiled oil *n.* Linseed oil, and sometimes soya oil, which has been treated in any of the following ways: (a) Heated for a short time to slightly increase viscosity and shorten drying time; (b) Heated with compounds of lead and/or cobalt and/or manganese forming fatty acid salts of the metal(s), thus shortening the drying time; (c) By adding soluble driers with or without heating. Weismantal GF (1981) Paint handbook. McGraw-Hill Corporation Inc., New York.

Boilers Group name for nitrocellulose and lacquer solvents.

Boil-in-bag A type of food packaging foil, plastic, or laminate intend to be dropped into boiling water in order to cook the contents.

Boiling point Temperature at which the vapor pressure of a liquid is just slightly greater than the total pressure of the surroundings. The liquid as a consequence is rapidly converted from the liquid state into a vapor.

Boiling-point elevation *n.* A colligative property: the increase in boiling point of a solvent brought about by the presence of a solute. For each mole-weight of solute in solvent there is a boiling point elevation and the change in bp is represented by

$$\Delta T_{bp} = K_{bp} m,$$

where $K = 0.51$ for water, and m is the molality or mole-weight solute/1000g of solvent. Whitten KW, Davis RE, Davis E, Peck LM, Stanley GG (2003) General chemistry. Brookes/Cole, New York.

Blow molding The process of forming hollow articles by expanding a hot plastic element against the internal surfaces of a mold is blow molding. In its most common form, the process comprises extruding a tube (*parison*) downward between the opened halves of a metal mold, closing the mold to pinch off and seal the parison at top and bottom, injecting air through a needle inserted through the parison wall, cooling the mass by contact with a chilled mold, opening the mold and removing the formed article. Improvements on this process have increased the number of products and their properties. Strong AB (2000) Plastics materials and processing. Prentice-Hall, Columbus, OH. Ash M, Ash I (1982–1983) Encyclopedia of plastics polymers, and resins, vols 1–3. Chemical Publishing Co., New York. Pittance JC (ed) (1990) Engineering plastics and composites. SAM International, Materials Park, OH. Pittance JC (ed) (1990) Engineering plastics and composites. SAM International, Materials Park, OH. James F (ed) (1993) Whittington's dictionary of plastics. Technomic Publishing Co. Inc., Carley.

Boiling range Range between initial and final boiling temperatures of a solvent. Many solvents have no specific boiling point but distill over a definite range of temperature; e.g. VM&P Naphtha 250–300°F.

Bole *n.* Name frequently given in the arts to clay, either white or colored.

Bolster \ˈbōl-stər\ [ME, fr. OE; akin to OE *belg* bag] (before 12c) *n.* A spacer or filler in a mold.

Bolt *n.* A roll or piece of fabric of varying length.

Boltzmann constant *n.* The number that relates the average energy of a molecule to its absolute temperature, 1.380658×10^{-23} JK^{-1}, $1.3807 \times 10^{-16} ergK^{-1}$. Giambattista A, Richardson R, Richardson RC, Richardson B (2003) College physics. McGraw-Hill Science/Engineering/Math, New York.

Boltzmann superposition principle *n.* Strain is presumed to be a linear function of stress, so the total effect of applying several stresses is the sum of the effects of applying each one separately. Giambattista A, Richardson R, Richardson RC, Richardson B (2003) College physics. McGraw-Hill Science/Engineering/Math, New York.

Bolus alba \ˈbō-ləs ˌal-bə\. *See kaolin.*

BON *n.* Acronym for betaoxy-naphthoic acid, which can be coupled with a variety of amine compounds to form a soluble azo dye, which is then precipitated by any of several metals to form "Bon" reds or maroons.

Bon-arylamide red *n.* Any of any group of metal-free monazo pigments based on substituted 2-hydroxy-3-naphthoic acid.

Bond *n.* (1) *Chemical* – Force of attraction between atoms in a molecule. (2) *Adhesive* – Adhesion between to materials, such as between an adhesive and a given surface. (3) *Coatings* – Adhesion between a coating and a substrate to which it has been applied.
See also adhere.

Bond axis *n.* A line passing through the nuclei of two bonded atoms.

Bond coat *n.* Coating used to improve the adherence of succeeding coats.

Bond distance *n.* The distance between nuclei of two bonded atoms.
Also known as bond length.

BON pigment *n.* (1) Any of several brilliant reds and maroons widely used in plastics and rubbers, resistant to bleeding, migration, and crocking. The initials Bon stand for β-oxynaphthoic acid, the base raw material. (2) A related azo pigment made by coupling β-hydroxynaphthoic acid to an amine and forming the barium, calcium, strontium, or manganese salt thereof. The colors range from yellowish red to deep maroon.

Bond strength (adhesive) *n.* (1) Adhesive bond between joined substrates, the force required to break the bond divided by the bond area (Pa) (tensile-shear strength. The term adherence is frequently used in place of bond strength adhesion and bond). (2) Of fiber-reinforced laminates, the strength of the bond between fiber and matrix. (3) The degree of attraction between adjacent atoms within a molecule, usually expressed in J/mol. Skeist I (ed) (1990) Handbook of adhesives. Van Nostrand Reinhold, New York.
See peel adhesion.

Bonded adhesive *See adhesive.*

Bonded fabric *n.* (1) A wed of fibers held together by an adhesive medium that does not form a continuous film. (2) *See nonwoven fabric.*

Bonding *n.* (1) A process for adhesive laminating two or more fabrics or fabric and a layer of plastic foam. There are two methods: the flame method used for bonding foam and the adhesive method used for bonding face and backing fabrics. (2) One of several processes of binding fibers into thin sheets, webs, or battings by means of adhesives, plastics, or cohesion (self-bonding). Several types of bonding or bonding processes are listed below. Kadolph SJJ, Langford AL (2001) Textiles. Pearson Education, New York.

B

Point bonding *n.* The process of binding thermoplastic fibers into a onwoven fabric by applying heat and pressure so that a discrete pattern of fiber bonds is formed. Kadolph SJJ, Langford AL (2001) Textiles. Pearson Education, New York.

Print bonding *n.* A process of binding fibers into a non-woven fabric by applying an adhesive in a discrete pattern. Kadolph SJJ, Langford AL (2001) Textiles. Pearson Education, New York.

Saturation bonding *n.* A process of binding fibers into a non-woven fabric by soaking the web with an adhesive. Kadolph SJJ, Langford AL (2001) Textiles. Pearson Education, New York.

Spray bonding *n.* A process of binding fibers into a non-woven fabric involving the spray application of a fabric binder.

Stitch bonding *n.* A bonding technique for non-wovens in which the fibers are connected by stitches sewn or knitted through the web. Kadolph SJJ, Langford AL (2001) Textiles. Pearson Education, New York.
Also known as quilting.

Bonding action *n. See bond strength.*

Bonding equipment *n.* Equipment used for welding polymer parts such as ultrasonic welding; also cementing equipment.

Bonding orbital *n.* A molecular orbital in which electrons have lower energies than they would in the unbonded atoms; an orbital characterized by a region of high electron probability density between the bonded atoms, leading to a stabilizing effect on the molecule. Whitten KW, Davis RE, Davis E, Peck LM, Stanley GG (2003) General chemistry. Brookes/Cole, New York.

Bonding pair *n.* A pair of electrons shared between two atoms and constituting a covalent bond model). Whitten KW, Davis RE, Davis E, Peck LM, Stanley GG (2003) General chemistry. Brookes/Cole, New York.

Bonding resin *n.* Any resin used for bonding aggregates together, e.g., holt melt epoxy, such as foundry sands, grinding wheels, abrasive papers, asbestos brake linings, and concrete masses. Plywood adhesives are sometimes called bonding resins. Skeist I (ed) (1990) Handbook of adhesives. Van Nostrand Reinhold, New York.

Bonding with binder fibers *n.* Specially engineered low-melting point fibers are blended with other fibers in a web, so that a uniformly bonded structure can be generated at low temperature by fusion of the binder fiber with adjacent fibers. Skeist I (ed) (1990) Handbook of adhesives. Van Nostrand Reinhold, New York.

Bonds, chemical *n.* The electronic linkages between atoms such as –C–H, which can be covalent, ionic and others. Goldberg DE (2003) Fundamentals of chemistry. McGraw-Hill Science/Engineering/Math, New York.

Bone black *n.* Made by charring animal bones in closed retorts. Bone black is blue-black in color and is fairly smooth in texture. It contains carbon, calcium phosphate, and calcium carbonate. Kirk–Othmer encyclopedia of chemical technology: pigments–powders. John Wiley and Sons, New York, 1996.

Bone brown *n.* Pigment related to bone black. Both are obtained from bones by calcinations, but with bone brown the temperature of calcinations is much lower and less carbonization occurs. The pigment is not widely used.

Bone dry bleached shellac *n.* A bleached, light-colored orange gum shellac.
See bleached shellac.

Bone glue *n.* Impure gelatin prepared from bones. Skeist I (ed) (1990) Handbook of adhesives. Van Nostrand Reinhold, New York.

See gelatin and glue.

Bone oil *n.* Fatty oil obtained by the dry distillation of bones.

Bone white *n.* Pigment made by acclimating animal bones. Composed chiefly of tricalcium phosphate. Calcium carbonate and minor constituents make up the rest. Bone white is a grayish white and slightly gritty powder. Syn: bone ash. Kirk–Othmer encyclopedia of chemical technology: pigments–powders. John Wiley and Sons, New York, 1996.

Bookbinders' varnish *n.* Rapid drying varnish based on shellac, sandarac or gum mastic, or mixtures of these resins. Kirk–Othmer encyclopedia of chemical technology: pigments–powders. John Wiley and Sons, New York, 1996.

Booster ram *n.* A hydraulic ram used as an auxiliary to main ram of a molding press.

BOP Abbreviation for butyl octyl phthalate. *See butyl ethylhexyl phthalate.*

BOPP Abbreviation for biaxially oriented polypropylene.

Boralloy *n.* Boronitride, manufactured by Union Carbide, USA.

Borax \\ˈbōr-ˌaks, ˈbór-, -əks\\ [ME *boras*, fr. MF, fr. ML *borac-*, *borax*, fr. Arabic *būraq*, fr. Persian *būrah*] (14c) *n.* $Na_2B_4O_7{\cdot}10H_2O$. Natural hydrated sodium borate, found in salt lakes and alkali soils. Also, the commercial name for sodium borate. It is used in some fire-retardant paints, and to dissolve casein in distemper.

Border *n.* A narrow strip around an edge. A border wallcovering is used for trimming, or as a frieze, generally just under the ceiling. Harris CM (2005) Dictionary of architecture and construction. McGraw-Hill Co., New York.

See soffit.

Boric-acid ester *n.* One of a family of flame retardants for plastics, etc. and plasticizers. Examples are the trimethyl, tri-*n*-butyl, tricyclohexyl, and tri-*p*-cresyl borates. Kidder RC (1994) Handbook of fire retardant coatings and fire testing services. CRC Press, Boca Raton, FL.

Borneol \\ˈbór-nē-ˌól, -ˌōl\\ [ISV, fr. *Borneo*] (1876) *n.* $C_{10}H_{18}O$. Constituent of pine oil. A terpenic alcohol. It is a solid at ordinary temperature, with a mp of 203°C (397.4°F), bp of 212°C (413.6°F), Sp gr of 1.011. It is sometimes described as bornyl alcohol or borneo camphor. Solubility characteristics are good, and it has some application as a plasticizer.

Bornyl acetate *n.* A solvent and plasticizer for nitrocellulose.

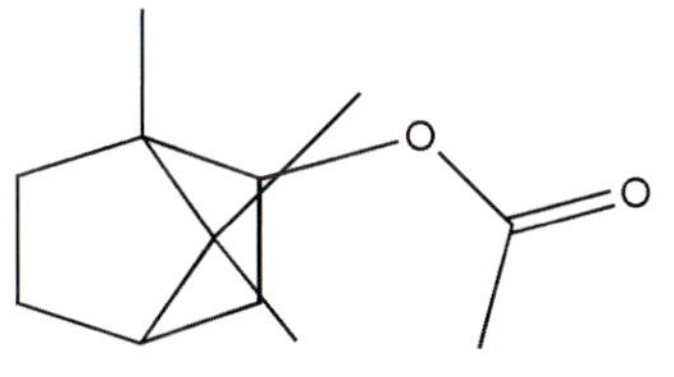

Boron-epoxy composite *n.* A composite in which boron fibers are embedded in an epoxy matrix. Modulus is about 200 GPa, tensile strength about 1.6 GPa, rising with fiber content. The combination of high specific modulus and high specific strength has made these composites attractive for aerospace vehicles in spite of their high cost. Harper CA (ed) (2002) Handbook of plastics, elastomers and composites, 4th edn. McGraw-Hill, New York.

Boron fiber *n.* An advanced reinforcing fiber produced by passing 10-µm, resistively heated tungsten wire through an atmosphere of boron trichloride and hydrogen. Hydrogen reduces the $BC1_3$ and the boron deposits on the wire to make a filament from 120 to 140 µm in diameter. Density

B

is low, 2.4–2.6 g/cm^3. Boron fiber is extremely strong and stiff with strength near 3.1 GPa (450 kpsi) and modulus near 400 GPa (58 Mpsi). The very high cost of these fibers (ca. $1/g) has limited their use. Harper CA (ed) (2002) Handbook of plastics, elastomers and composites, 4th edn. McGraw-Hill, New York.

Boron hydride *See diborane.*

Boron resins *n.* Esters derived from boric acids and polyhydric alcohols. Characterized by solubility in water.

Boron trifluoride etherate *n.* A metal halide $BF_3{\cdot}O(C_2H_5)_2$ which serves as a cationic initiator for the polymerization of monomers.

Boss *n.* (1) A protuberance provided on an article to add strength, facilitate alignment during assembly, or for attaching the article to another part. (2) That part of a drafting roll of largest diameter where the fibers are gripped. It may be an integral part of the roll, as in steel rolls, or it may have a covering of leather, cork, etc. In the former case, the boss is fluted.

Boston round *n.* A family of variously sized, blown bottles either glass or plastic, of circular-cylindrical shape with a short curved shoulder and length-to-diameter ratio in the body of about 1.7:1.

Boston stone, the *n.* Originally a paint mill imported around 1700 by a painter, Thomas Child. Later used as a landmark or central point from which distances from Boston were measured. It is the earliest known implement of the paint industry in America. The round stone was rolled back and forth in the trough of a larger flat stone underneath, in which oil and pigment were dispersed. The Boston Stone is the official insignia of the Federation of Societies for Coatings Technology. Paint/coatings dictionary. Federation of Societies for Coatings Technology, Philadelphia, Blue Bell, PA, 1978.

Botany bay resin *See accroides.*

Bottom *n.* (1) Ship's bottom. (2) Hard deposit in paint cans.

Bottom-drying *n.* Drying of a film from the bottom towards the top of the film, e.g., lead naphthenate is used as a bottom drier.

Bottoms *See tailings.*

Bouclé \bu-ˈklā\ [F *bouclé*] (1895) *n.* (1) A fabric woven or knit with bouclé yarns. (2) An uneven yarn of three piles one of which forms loops at intervals.

Bouclé yarn *n.* A novelty yarn with loops which give fabrics a rough appearance. Some bouclé yarns have cotton cores with other fibers wound around them. Bouclé yarns may be made from wool, cotton, silk, linen, manufactured fibers, or combinations of fibers. Vincenti R (ed) (1994) Elsevier's textile dictionary. Elsevier Science and Technology Books, New York.

Bouguer's (Lambert's) law *n.* Describes the proportionality between the absorption of radiant energy by a non-scattering material and its optical path length $-\log T_i/l = K$, where T_i is the internal transmittance, l the path length, and K is the unit absorption coefficient. Giambattista A, Richardson R, Richardson RC, Richardson B (2003) College physics. McGraw Hill Science/Engineering/Math, New York.

See Beer–Bougure law.

Bounce back *n.* A spray rebound.

Boundary surface *n.* A surface of constant electron probability density, ψ^2.

Bourrelet A double-knit fabric with raised loops running horizontally across the surface of the cloth giving a rippled or corded effect.

Bow *n.* (1) Distortion of a board in which the face is convex or concave longitudinally. (2) The greatest distance, measured parallel to the selvages, between a filling yarn and a straight line drawn between the points at which this yarn meets the selvages. Bow may be expressed directly in inches or as a percentage of the width of the fabric at that point.

Bow window *See bay.*

Boxed heart *n.* Term used when the pith falls entirely within the four faces anywhere in the length of a piece of wood.

Boxing *v.* Combining of two or more separate batches to one uniform batch.

Box loom *n.* A loom using two or more shuttles for weaving fabrics with filling yarns that differ in fiber type, color, twist, level, or yarn size. The box motion is automatic, changing from one shuttle to another. Examples of fabrics made on box looms are crepes and ginghams.

Box mark *n.* A fine line parallel to the filling caused by shuttle damage to a group of filling yarns.

Boyer-Beaman rule *n.* A rule of thumb stating that the ratio of polymer's glass-transition temperature to its melting temperature (T_g/T_m), both in Kelvin, usually lies between 0.5 and 0.7. For symmetrical polymers, such as polyethylene, it is close to 0.5; for unsymmetrical ones, such as polystyrene and polyisoprene, it is near 0.7.

Boyle's law \bóilz-\ [Robert *Boyle*] (ca. 1860) *np.* The part of the ideal-gas law, due to Robert Boyle, stating that the volume of a gas is inversely proportional to its pressure. This can also be stated as, at a constant temperature the volume of a given quantity of any gas varies inversely as the pressure to which the gas is subjected. For a perfect gas, changing from pressure p and volume v to pressure p' and volume v' without change of temperature, $pv = p'v'$. Atkins PW, De Paula J (2001) Physical chemistry, 7th edn. W. H. Freeman Co., New York.

Bozzo *See abbozzo.*

BPF *n.* Abbreviation for British plastics federation.

BR *n.* Poly(butadiene). Abbreviation for butadiene rubber (British standards institution).

See polybutadiene.

Br *n.* Chemical symbol for the element bromine.

Barbender plastograph *n.* A rheometer for measuring shear-strain with temperature relationships. Strong AB (2000) Plastics materials and processing. Prentice-Hall, Columbus, OH.

Braid *n.* (1) A narrow textile band, often used as trimming or binding, formed by plaiting several strands of yarn. The fabric is formed by interfacing the yarns diagonally to the production axis of the material. (2) In aerospace textiles, a system of three or more yarns which are interlaced in such a way that no two yarns are twisted around each other.

Braid angle *n.* The acute angle measured from the axis of a fabric or rope to a braiding yarn.

Braided fabric *n.* A narrow fabric made by crossing a number of strands diagonally so that each strand passed alternatively over or under one or more of the other strands. They are frequently used in shoelaces and suspenders.

Braiding *n.* The intertwining of three or more strands to make a cord. The strand form a regular diagonal pattern down the length of the cord.

Brake lining medium *n.* A medium for bonding together the chosen filling agent for the brake lining, e.g., asbestos.

Branched *n.* Refers to side chains attached to the main chain in molecular structure of polymers, as opposed to linear arrangement.

B

Branching *n.* (1) The growth of a new polymer chain from an active site on an established chain, in a direction different from that of the original chain. Branching occurs as a result of chain-transfer processes or from the polymerization of difunctional monomers, and is an important factor influencing polymer properties. (2) Two-dimensional polymers having relatively short side-chains branching from the main backbone. Odian GC (2004) Principles of polymerization. John Wiley and Sons Inc., New York.

Brashness *n.* Condition of wood characterized by low resistance to shock and by an abrupt failure across the grain without splintering.

Brayer *n.* A small hand roller for distributing ink.

Brazil wax *n.* A name sometimes applied to a carnauba wax.

Brazil-wood *n.* A natural red dye from the wood of *Caesalpinia braziliensis.*

Break The break in an oil is the flocculent material or foots which separate on long standing or upon heating.

Breakdown *n.* To soften or plasticize rubber by working it on a rubber mill or in an internal mixer. Same as to mill or masticate.

Breakdown voltage *n.* The voltage required, under specific conditions, to cause the failure of an insulating varnish or other insulating material.

See dielectric breakdown voltage and dielectric strength.

Breakfree oil *n.* Oil which has been refined mechanically by heating to about 88°C (190°F), adding an adsorbent such as fuller's earth or activated charcoal, agitating, and filtering.

Also known as non-break oil.

Break factor *n.* A measure of yarn strength calculated as: (1) the product of breaking strength times indirect yarn number, or (2) the product of breaking strength times the reciprocal of the direct yarn number.

Breaking *v.* Separation of an emulsion or latex into two phases.

Breaking extension *n.* Elongation necessary to cause rupture of a test specimen; the tensile strain at the moment of rupture. *See elongation.*

Breaking length *n.* A measure of the breaking strength of a yarn; the calculated length of a specimen whose weight is equal to its breaking load. The breaking length expressed in kilometers is numerically equal to the breaking tenacity expressed in grams-force per tex. Kadolph SJJ, Langford AL (2001) Textiles. Pearson Education, New York.

Breaking ratio *See break factor.*

Breaking strength (Load) *n.* (1) The maximum resultant internal force that resists rupture in a tension test. The expression "breaking strength" is not used for compression tests, bursting tests, or tear resistance tests in textiles. 2. The load (or force) required to break or rupture a specimen in a tensile test made according to a specified standard procedure.

Also see breaking load and ultimate strength.

Breaking tenacity *n.* The tensile stress at rupture of a specimen (fiber, filament, yarn, cord, or similar structure) expressed as Newton per tex, grams-force per tex, or gram-force per denier. The breaking tenacity is calculated from the breaking load and linear density of the unstrained specimen, or obtained directly from tensile testing machines, which can be suitably adjusted to indicate tenacity instead of breaking load for specimens of known linear density. Breaking tenacity expressed in grams-force per tex is numerically equal to breaking length expressed in kilometers. Shah V

(1998) Handbook of plastics testing technology. John Wiley and Sons, New York.

Break spinning *n.* A direct spinning process for converting manufactured fiber tows to spun yarn that incorporates prestretching and tow breaking with subsequent drafting and spinning in one operation.

Break-out *See smash.*

Break test(oils) *n.* Break in drying oils is a measure of the materials rendered insoluble in carbon tetrachloride under the conditions of test by a prescribed method.

Breathable film *n.* A film that is at least somewhat permeable to gases due to the presence of open cells throughout its mass or to minute perforations. Shah V (1998) Handbook of plastics testing technology. John Wiley and Sons, New York.

Brewster's law *n.* The tangent of the polarizing angle for a substance is equal to the index of refraction. The polarizing angle is that angle of incidence of which the reflected polarized ray is at right angles to the refracted ray. If n is the index of refraction and θ the polarizing angle, $n = \tan\theta$. Moller KD (2003) Optics. Springer-Verlag, New York.

Bridging *n.* (1) In the flow by gravity of powders or pellets in feed hoppers and the throats of extruders and injection molders, the stoppage of flow caused by the formation of an arch across the flow path.

Bridle *n.* Series of rolls that keeps the tension on the strip of a continuous line.

Bright *n.* The term applied to fibers whose luster has not been reduced by physical or chemical means; the opposite of dull or matte. Vincenti R (ed) (1994) Elsevier's textile dictionary. Elsevier Science and Technology Books, New York.

Brightener, optical *n.* fluorescent dye, or pigment, which absorbs UV radiation and re-emits light of a violet or bluish hue. Used to increase the luminance factor and to remove the yellowish of white or off-white materials. Bart J (2005) Additives in polymers: industrial analysis and applications. John Wiley and Sons, New York.

Brighteners *n.* Additives for providing optical brilliance to a plastic part. Bart J (2005) Additives in polymers: industrial analysis and applications. John Wiley and Sons, New York.

Brightening agents (optical brighteners, fluorescent bleaches, and optical whiteners) *n.* Brightening agents are chemicals used primarily in fibers but also to some extent in molded and extruded products, to overcome yellow casts and to enhance clarity or brightness. In contrast to bluing agents, which act by removing yellow light, the optical brighteners absorb ultraviolet rays and convert their energy into visible blue-violet light. Thus, they cannot be used ion compounds containing UV-absorbing agents. Optical brighteners are used in PVC sheet and film, fluorescent lighting fixtures, vinyl flooring, nylon fishing line, polyethylene bottles, etc. A few examples of optical brighteners are coumarins, naphthotriazolyl stilbenes, benzoxazolyl-, benzimidazolyl-, naphthylimide-, and diaminostilbene disulfonates. Bart J (2005) Additives in polymers: industrial analysis and applications. John Wiley and Sons, New York. *See brightener, optical.*

Brightfield illumination *n.* The usual form of microscope illumination with the image of the sample on a bright, evenly lighted field; or in vertical illumination, light reflected through the objective by means of prism or semireflecting plane glass. Loveland RP (1981) Photomicrograsphy. Krieger Publishing Co., New York.

Brightness *n.* (1) Brightness is measured by the flux emitted per unit emissive area as

B

projected on a plane normal to the line of sight. The unit of brightness is that of a perfectly diffusing surface giving out 1 lm/cm^2 of projected surface and is called the *Lambert*. The millilambert (0.001 L) is a more convenient unit. *Candle per square centimeter* is the brightness of a surface, which has, ion the direction considered, a luminous intensity of 1 c/cm^2. (2) (dprecated) The dimension of color that is referred to an achromatic scale, ranging from black to white, also called lightness or luminous reflectance or transmittance. Because of confusion with saturation, the use of this term should be discouraged. McDonald R (1997) Colour physics for industry, 2nd edn. Society of Dyers and Colourists, West Yorkshire, England.

Brightness flop *n.* Change in lightness with direction of observation.

Brilliance *n.* A subjective term referring to: (1) Clarity or freedom from visible suspended matter or opalescence in clear coatings. (2) Cleanness or lack of a muddy or dirty tone in pigmented coatings. (3) The combined effect of lightness and strength, or purity of tone of printing inks. Wicks ZN, Jones FN, Pappas SP (1999) Organic coatings science and technology, 2nd edn. Wiley-Interscience, New York.

Brilliant ultramarine Synthetic ultramarine blue.

See ultramarine blue.

Brine \ˈbrīn\ [ME, fr. OE *bryne*; akin to MD *brīne* brine] (before 12c) *n.* Concentrated aqueous salt solution. The term can be applied to solutions of any salt, but it is commonly applied to solutions of sodium chloride.

Brinell hardness \brə-ˈnel-\ [Johann A. *Brinell* † Swedish engineer] (1915) *n.* The hardness of a material as determined by pressing a hardened steel or carbide ball, 10 mm in diameter, into the specimen under a fixed load. The Brinell number is the load in kilograms divided by the area in mm^2 of the spherical impression formed by the ball. For non-ferrous materials, the prescribed load is 500 kg applied for 30 s. The Brinell test is used mostly for ductile metals, for which the Brinell number is related to yield strength in a simple way. Also known as Brinell hardness number. Shah V (1998) Handbook of plastics testing technology. John Wiley and Sons, New York.

Bring forward *n.* In repainting, to repair local defective areas with the appropriate paints so as to bring them into conformity with the surrounding areas before applying the finishing coats.

Brinkman number *n.* A dimensionless group relevant for heat transfer in flowing viscous liquids such as polymer melts. It is defined by $N_{Br} = \mu V^2/k\Delta T$, in which μ is the viscosity, V the velocity, k the thermal conductivity, and ΔT is the difference in temperature between the stream and the confining wall. The number represents the ratio of the rates of heat generation and heat conduction. Shenoy AV (1996) Thermoplastics melt rheology and processing. Marcel Dekker, New York.

Bristle \ˈbri-səl\ ME *bristil*, fr. *brust* bristle, fr. OE *byrst*; akin to OHGr *burst* bristle] (14c) *n.* (1) A generic term for a short, stiff, coarse fiber. It is frequently hog hair. (2) Any grade of nylon monofilament used in toothbrushes, hairbrushes, paintbrushes (*tapered bristle*), etc.

British gum *See dextrin.*

British standards institution A national organization (corresponding to the American National Standards Institute and the American Society for Testing and Materials), which establishes and publishes standard specifications and codes of practice.

British thermal unit (Btu) *n.* Before the introduction of the SI system, the Btu was variously defined as the quantity of heat required to raise the temperature of 1 lb (avoirdupois) of water 1°F, either at or near 39°F (the temperature of maximum density), at 50°F or 60°F, or averaged from 32 to 212°F ("mean Btu"). The Btu is defined as 1055.056 J.

Brittle \ˈbri-tᵊl\ [ME *britil*; akin to OE *brēotan* to break. ON *brjōta*] (14c) *adj.* Easily broken when bent rapidly or scratched as with a knife blade or the finger nail; the opposite of tough.

Brittle fracture An abrupt breaking of a material in which there is very little or no elongation or distortion of the part before failure. In a tensile test, the stress-strain graph is nearly linear to the breaking point. Grellman W, Seidler S (ed) (2001) Deformation and fracture behavior of polymers. Springer-Verlag, New York.

Brittleness *n.* The lack of flexibility and the tendency to crack when deformed or struck.

Brittleness temperature The temperature at which a plastic or elastomer breaks in cantilever-bending impact under specific conditions. The brittleness temperature is related to that of the glass transition, for those plastics with T_gs below room temperature, such as flexible PVC. Wickson EJ (ed) (1993) Handbook of polyvinyl chloride formulating. John Wiley and Sons Inc., New York.

Brittle point The temperature at which a polymer no longer exhibits viscoelastic properties. Shah V (1998) Handbook of plastics testing technology. John Wiley and Sons, New York.

Brittle point, brittle temperature Lowest temperature at which a material withstands an impact test under standardized conditions.

Brittle strength *n.* The strength of a polymer at the brittle point or temperature.

Broadcast \ˈbród-\ (1767) *adj.* To apply or sprinkle solid particles on an uncured coating surface.

Broadcloth \ -ˌklóth\ (15c) *n.* (1) Originally, a silk shirting fabric so named because it was woven in widths exceeding the usual 29 in. (2) A tightly woven, lustrous cotton or polyester/cotton blend fabric in a plain weave with a crosswise rib. It resembles poplin, but the rib is finer, and broadcloth always has more picks that poplin. The finest qualities are made with combed pima or Egyptian cotton. (3) A smooth, rich-looking, woolen fabric with a napped face and a twill back. Better grades have a glossy, velvety hand. Vincenti R (ed) (1994) Elsevier's textile dictionary. Elsevier Science and Technology Books, New York.

Broadgoods A fabrics-industry term for woven materials, including glass fabrics that are over 46 cm in width.

Broadloom \-ˌlüm\ (1925) *adj.* A term that refers to carpets woven in widths from 54 in. to 18 ft, as distinguished from narrow loom widths of 27–36 in. Vincenti R (ed) (1994) Elsevier's textile dictionary. Elsevier Science and Technology Books, New York.

Brocade \brō-ˈkād\ [S *brocado*, fr. Catalan *brocat*, fr. I *broccato*, fr. *broccare* to spur, brocade, fr. *brocco* small nail, fr. L *broccus* projecting] (1588) *n.* (1) A rich, Jacquard-woven fabric with an all over interwoven design of raised figures or flowers. The pattern is emphasized by contrasting surfaces or colors and often has gold or silver threads running through it. The background may be either a satin or a twill weave. (2) A term describing a cut-pile carpet having a surface texture created by mixing twisted and straight standing pile yarns. Vincenti R (ed) (1994) Elsevier's textile dictionary. Elsevier Science and Technology Books, New York.

Brocatelle \ˌbrä-kə-ˈtel\ [Fr, fr. It *broccatello*] (1669) *n.* A fabric similar to brocade with a satin or twill figure in high relief on a plain or satin background. Vincenti R (ed) (1994) Elsevier's textile dictionary. Elsevier Science and Technology Books, New York.

Broken color *n.* General multicolored effect brought about by the automatic merging of wet paints of various colors or by manipulation, which produce apparently accidental effects. Paint/coatings dictionary. Federation of Societies for Coatings Technology, Philadelphia, Blue Bell, PA, 1978.

Broken-color work *See antiquing.*

Broken end *n.* A broken, untied warp thread in a fabric. There are numerous causes, such as slubs, knots, improper shuttle alignment, shuttle hitting the warp shed, excessive warp tension, faulty sizing, and rough reeds, heddles, dropwires, and shuttles.

Broken pick *n.* A broken filling thread in a fabric. Usual caused include too much shuttle tension, weak yarn, or filling coming into contact with a sharp surface.

Broken selvage *See cut selvage.*

Broken white *n.* Off-white in which the initial pure whiteness of the principal pigment has been very slightly "broken" or modified in a yellow or brown direction.

Bronze \ˈbränz\ {*often attributive*} [F, fr. I *bronzo*] (1739) *n.* An appearance characteristic of some printed films in which the apparent color of the print depends upon the angles of viewing and illumination. Leach RH, Pierce RJ, Hickman, EP, Mackenzie MJ, Smith HG (eds) (1993) Printing ink manual, 5th edn. Blueprint, New York.

Bronze blue *n.* One of the names applied to the complex ferric ferrocyanide or iron blues. Leach RH, Pierce RJ, Hickman, EP, Mackenzie MJ, Smith HG (eds) (1993) Printing ink manual, 5th edn. Blueprint, New York.

Bronze gold powders *n.* Metallic powders made from alloys of copper. They range in color from bonze gold to green yellow. Solomon DH, Hawthorne DG (1991) Chemistry of pigments and fillers, Krieger Publishing Co., New York.

Bronze paint *n.* Paint incorporating bronze gold powder in a bronzing liquid. Solomon DH, Hawthorne DG (1991) Chemistry of pigments and fillers, Krieger Publishing Co., New York.

Bronze paste *n.* Paste consisting of fine metallic flakes in a volatile medium. The metal may be aluminum, copper, or alloys of copper. Solomon DH, Hawthorne DG (1991) Chemistry of pigments and fillers, Krieger Publishing Co., New York.

Bronze pigment *n.* Simulated bronze- or gold-colored pigments made by staining aluminum flakes with brown or yellow colorants. Solomon DH, Hawthorne DG (1991) Chemistry of pigments and fillers, Krieger Publishing Co., New York.

Bronzing *v.* (1) A subjective, descriptive, appearance term applied to the metal-like reflectance, which sometimes appears at the surface of non-metallic colored materials. It is perceived at the specular angle, by observing the image of a white light source, for example, and is characterized by a distinct hue of different dominant wavelength than the hue of the paint film itself. The origin of the selective specular reflectance observed is generally considered to be reflectance from very small particle size pigments partially separated from surrounding vehicle at or near the surface. (2) Characteristic metallic luster shown by certain highly colored pigments in full strength, e.g., certain iron and phthalocyanine blues. May also occur with bleeding type pigments in making finishes. (3) Application of imitation gold or other metals either in powder form or in leaves. (4) Printing with a sticky size and dusting same with finely powdered metal particles or flakes to give the appearance of metallic printing. *Also known as bronze dusting.* Leach RH, Pierce RJ, Hickman, EP, Mackenzie MJ, Smith HG (eds) (1993) Printing ink manual, 5th edn. Blueprint, New York.

Brookfield viscometer *n.* The Brookfield "Synchrolectric" viscometer is the most widely used instrument for measuring the viscosities of plastisols and other liquids, both Newtonian and non-Newtonian.

About a dozen models are available to accommodate subranges of viscosity in the overall range from 0.01 to 1000 Pas. It is portable and can be hand held. A synchronous motor provides four or eight spindle speeds by shifting gears. Near the tip- of the spindle, and concentric with it, is a horizontal disk whose drag torque in the liquid is detected by a torsion spring; a pointer indicates viscosity. By taking readings at different rotational speeds, one can estimate the pseudoplasticity of the liquid. For accurate work, the spindle guard should be removed and the diameter of the vessel containing the liquid should be at least five times that of the disk. Goodwin JW, Goodwin J, Hughes RW (2000) Rheology for chemists. Royal Society of Chemistry, UK. Pierce PE (1969) Rheology of coatings. J Paint Tech 41(533):383-395

Brookite \ˈbru-ˌkīt\ [*Henry J. Brooke* † 1859 English mineralogist] (1825) *n.* Rutile form of titanium dioxide characterized by its orthorhombic form. Its color varies from black to reddish brown. It is not used directly as a pigment.

Brown coat *n.* The coat of roughly finished plaster beneath the finish coat; in three-coat work, the second coat of plaster, applied over a scratch coat and covered by the finish coat; in two-coat work, the base-coat plaster applied over lath or masonry; may contain a greater proportion of aggregate than the scratch coat.

Also known as floating coat.

Browne heat test Heat polymerization test designed to determine the purity of tung and similar oils. The result is reported as the number of minutes required to cause gelation at the specific temperature (282°C). Paint: pigment, drying oils, polymers, resins, naval stores, cellulosics esters, and ink vehicles, vol 3. American Society for Testing and Material, West Conshohocken, PA, 2001.

Brown hard varnish *n.* Spirit varnish prepared from shellac dissolved in industrial alcohol. Cheaper grades may consist of shellac–rosin mixtures, spirit manila, spirit manila–rosin mixtures, or alcohol soluble rosin-modified phenolic resin.

Brownian motion \ˈbraú-nē-ən-\ [Robert *Brown* † 1858 Scottish botanist] (ca. 1889) *n.* Refers to the continual movement of extremely small particles suspended in a fluid. The accepted explanation is that movement of the particles is due to bombardment by molecules of the fluid. The average length of path L followed by a particle in any given direction at a given time t is directly proportional to the square root of the absolute temperature T and the time t; it is also inversely proportional to the square root of the viscosity η of the medium and to the diameter d of the particle (Avogadro's number N and the gas constant R enter in as constants):

$$L = \frac{2R}{N}\sqrt{\frac{Tt}{\eta d3\pi}}.$$

Whitten KW, Davis RE, Davis E, Peck LM, Stanley GG (2003) General chemistry. Brookes/Cole, New York.

Brownian movement *n.* A continuous agitation of particles in a colloidal solution caused by unbalanced impacts with molecules of the surrounding medium. The motion may be observed with a microscope when a strong beam of light is caused to traverse the solution across the line of sight. Whitten KW, Davis RE, Davis E, Peck LM, Stanley GG (2003) General chemistry. Brookes/Cole, New York.

Brown rot (1894) *n. See dry rot.*

Brunswick black \ˈbrənz-(ˌ)wik-\. Solution consisting of gilsonite, petroleum pitch, or similar material in spirit or aromatic hydrocarbons.

Brush \ˈbrəsh\ [ME *brusshe*, fr. MF *broisse*, fr. OF *broce*] (14c) *n.* A tool composed of bristles set into a handle; most often used to apply coatings. Bristles may be synthetic or natural. Weismantal GF (1981) Paint handbook. McGraw-Hill Corporation Inc., New York.

Brushability *n.* The ability or ease with which a coating can be brushed.

Brush (or roller) cleaner *n.* A combination of chemicals in which a brush or rollers may be soaked in order to permit cleaning with soap and water. Weismantal GF (1981) Paint handbook. McGraw-Hill Corporation Inc., New York.

Brush graining *n.* An imitation effect of wood grain; produced by drawing a clean dry brush through a dark liquid stain, applied over a dry, light base coat.

Brushing consistency *n.* Rheological conditions at which a coating can be properly applied to a substrate with a brush. Brushing is used on sweaters, scarves, knit underwear, wool broadcloths, etc.

Brush mark *n.* A small ridge or valley produced in a paint film by the combining action of the bristles of a brush.

Brushout The application of paint on a small surface for testing.

BS *n.* Abbreviation for British standard code of practice.

BSI *n.* Abbreviation for British standards institution.

Btu (BTU) *See British thermal unit.*

Bubble \ˈbə-bəl\ {*often attributive*} [ME *bobel*] (14c) *n.* (1) In blown-film extrusion, the expanding tube moving from the die to the collapsing rolls at the top of the tower. (2) A void within a molding or an extrusion.

Bubble coating *n.* Coating wherein the light scattering characteristics are derived partially or wholly from small bubbles or microvoids in a transparent binder. The air-filled or vapor-saturated voids having diameters in the range of the wavelength of light, are trapped in a water soluble medium such as soya protein. This type of coating has a low specific weight and high lightness. *See microvoids.*

Bubble forming *n.* A variant of sheet thermoforming, in which the plastic sheet is clamped in a frame suspended above a mold, heated, expanded into a blister shape with air pressure, then molded to its final shape by means of a descending plug applied to the blister and forcing it downward into the mold. Weismantal GF (1981) Paint handbook. McGraw-Hill Corporation Inc., New York.

Bubbler *n.* A device inserted into a mold force, cavity, or core that delivers.

Bubble tube viscometer *n.* One of the bubble viscometers which is used by comparing the speed of a rising air bubble in a liquid of unknown viscosity with that of a liquid of known viscosity, both liquids being in corked glass tubes of identical size, and at the same temperature. Syn: air bubble viscometer. Patton TC (1979) Paint flow and pigment dispersion: a rheological approach to coating and ink technology. John Wiley and Sons, New York. Paint and coating testing manual (Gardner-Sward Handbook) MNL 17, 14th edn. ASTM, Conshohocken, PA, 1995.

Bubbling *n.* Film defect, temporary or permanent, in which bubbles of air or solvent vapor, or both, are present in the applied film.

Bucket spinning *See pot spinning.*

Buckling (14c) *v.* (1) A crimping of the fibers in a composite material that may occur in glass-reinforced thermosets due to shrinkage of the resin during cure. (2) The principal mode of failure of axially loaded, slender structural members such as columns and panels.

Buckram \ˈbə-krəm\ [ME *buckeram*, fr. OF *boquerant*, from OP *bocaran*] (15c) *n.* A scrim fabric with a stiff finish, often used as interlining.

Buffer \ˈbə-fər\ {*often attributive*} [*buff*, v., to react like a soft body when struck] (1835) *n.* A solution which contains moderate or high concentrations of a Brønsted–Lowery conjugate acid–base pair; a solution whose pH does not change greatly in response to added acids or bases.

Buffer coat *n.* A finishing material applied over an old coating to protect it from solvent action of the new finish.

Buffing *n.* A surface-finishing method used with plastics and other materials in which the object is rubbed with cloths or cloth wheels that may contain fine, mild abrasives of waxes or both.
Also see tumbling.

Buffing compounds *n.* Polishing grit for developing a smooth surface.

Buhr mill *n.* Stone mill in which a viscous pigment-medium paste is dispersed between the flat surfaces of two large stones, one of which rotates.

Buhrstone \ˈbər-ˌstōn\ (1690) *n.* Type of porous flint, formed in France, used for millstones.

Build *n.* Real or apparent thickness, fullness, or depth of a dried film.

Building *n.* A more or less enclosed and permanent structure for housing, commerce, industry, etc., distinguished from mobile structures and those not intended for occupancy.

Building Block *n* (1846). A rectangular masonry unit, other than a brick, made of brunt clay, cement, concrete, glass, gypsum, or any other material suitable for use in building construction. Harris CM (2005) Dictionary of architecture and construction. McGraw-Hill Co., New York.

Building board *n.* Any sheet of building material, often faced with paper or vinyl; suitable for use as a finished surface on walls, ceilings, etc. Harris CM (2005) Dictionary of architecture and construction. McGraw-Hill Co., New York.

Building coats *n.* Coats used to build up a surface or surfaces before rubbing.

Building code *n.* A collection of rules and regulations adopted by authorities having appropriate jurisdiction to control the design and construction of buildings, alteration, repair, quality of materials, use and occupancy, and related factors of buildings within their jurisdiction; contains minimum architectural, structural, and mechanical standards for sanitation, public health, welfare, safety, and the provision of light and air.

B

Building paper *n.* A heavy, relatively cheap, durable paper, such as asphalt paper, used in building construction, especially in frame construction, to improve thermal insulation and weather protection and to act as a vapor barrier. Special types are: sheathing paper, used betweens sheathing and siding; floor lining paper, used between rough and finish floors. Harris CM (2005) Dictionary of architecture and construction. McGraw-Hill Co., New York.

Buildup *n.* A term applied to substantivity of dye for a textile material. It refers to the ability of a dye to produce deep shades.

Built-up roofing *n.* A continuous roof covering made up of laminations or plies of saturated or coated roofing felts, alternated with layers of asphalt or coal tar pitch and surfaced with a layer of gravel or slag in a heavy coat of asphalt or coal tar pitch or finished with a cap sheet; generally used on flat or low-pitched roofs. Harris CM (2005) Dictionary of architecture and construction. McGraw-Hill Co., New York.

Bulk density *n.* The density of a particulate material (granules, powder, flakes, chopped fiber, etc.) expressed as the ratio of weight to total volume, voids included.

Bulk development *n.* Any of various relaxation treatments to produce maximum bulk in textured or latent crimp yarns or in fabrics made there from. The essential conditions are heat, lubrication, movement, and the absence of tension. Bulk development may be accomplished during wet processing or may be a separate operation such as hot-air tumbling, steam-injection tumbling, or dry cleaning.

Bulk factor *n.* The ratio of the volume of a given mass of plastic particles to the volume of the same mass of material after molding or forming. The bulk factor is also equal to the ratio of the density after molding or forming to the apparent density (bulk density) of the material as received.

Bulking *n.* In the process of formulating coatings, the step wherein ingredient weights are converted to their volume equivalents.

Bulking value *n.* Reciprocal of apparent density. Solid volume of a unit weight of material, usually expressed as gallons per pound or l/kg. For practical purposes this is 0.120 divided by the specific gravity (ASTM). Syn: specific volume.
Also called specific volume.

Bulk material *n.* A material or product in large quantity such as a drum or sack.

Bulk modulus *n.* The modulus of volume elasticity, i.e., the resistance of a solid or liquid to change in volume with change in pressure, at constant temperature. The thermodynamic definition is:

$$B = -v(\partial P/\partial v)_T = -(\partial P/\partial \ln v)_T,$$

in which v is the specific volume, P the pressure, and T is the specific temperature. This may be approximated over an interval of pressure as $0.5{\cdot}(v_2+v_1){\cdot}(P_2-P_1)/(v_1-v_2)$. With plastics, B gradually diminishes as temperature rises and increase with rising pressure. Mark JE (ed) (1996) Physical properties of polymers handbook. Springer-Verlag, New York.

Bulk molding compound (BMC) *n. See premix.*

Bulk polymerization *n.* The polymerization of a monomer (mass polymerization) in the absence of any medium other than a catalyst or accelerator. The monomer is usually a liquid, but the term also applies to the polymerization of glass and solids in the absence of solvents or any other dispersing medium. Polystyrene, polymethyl methacrylate, low-density polyethylene, and styrene-acrylonitrile copolymers are examples of polymers most frequently produced by bulk polymerization. Acrylic monomers may be simultaneously polymerized and formed into products by conducting the polymerization in molds such as those for rods and sheets. Odian GC (2004) Principles of polymerization. John Wiley and Sons Inc., New York. Mark JE (ed) (1996) Physical properties of polymers handbook. Springer-Verlag, New York.

Bulk specific gravity *n.* The specific gravity of a porous solid when the volume of the solid used in the calculation includes both the permeable and impermeable voids. Compare specific gravity and bulk density.

Bulk yarn *n.* Yarn of glass (or other) fiber in bulk form, as opposed to roving, mat, or woven forms.

Bumping *n.* Sudden, explosive boiling following the superheating of a liquid.

Buna \ˈbyü-nə, ˈbü-\ *n.* Copolymer from butadiene and acrylonitrile, manufactured by Hüls, Germany.

See acrylonitrile-butadiene copolymer.

Buna rubber *n.* A brand of synthetic rubber made by polymerizing or copolymerizing butadiene with another material, as acrylonitrile, styrene, or sodium.

Buna-S or SS (GR-S) *n.* A synthetic elastomer produced by the copolymerization of butadiene and styrene. Manufactured by Hüls, Germany.

Also known as styrene-butadiene rubber.

Bunghold oil *See boiled oil.*

Bunghole \ˈbəŋ-ˌhōl\ (1571) *n.* Orifice, usually in a steel drum, a barrel, or a keg, which is stopped by a "bung" or large stopper.

Bunting \ˈbən-tiŋ\ [perhaps fr. E dialect bunt (to sift)] (1711) *n.* A soft, flimsy, loose-textured, plain weave cloth most frequently used in flags. Bunting was originally made from cotton or worsted yarns, but today's flags are made primarily from nylon or acrylic fibers.

BUR *n.* Abbreviation for blow-up ratio.

Burgundy pitch *See pitch, burgundy.*

Burl \ˈbər(-ə)l\ [ME *burle*, fr. (ass.) MF *bourle* tuft of wool, fr. (ass.) VL *burrula.* dimin. of LL *burra* shaggy cloth] (15c) *n.* Swirl or twist in the grain of wood that usually occurs near a know, but does not contain a knot.

Burlap \ˈbər-ˌlap\ [origin unknown] (ca. 1696) *n.* A coarse, heavy, plain weave fabric constructed from singles yarn of jute. Used for bags, upholstery lining, in curtains and draperies.

See jute.

Burling *v.* (1) The process of removing loose threads and knots from fabrics with a type of tweezers called a burling iron. (2) The process of correcting loose tufts and replacing missing tufts following carpet construction.

Burn mark *n.* Any visual sign of burning or charring at a particular spot on a part. In injection molding with poorly vented molds, burn marks can be caused by severe compression-heating of the air trapped in the mold cavity and consequent ignition of the molten plastic in contact with the hot air pocket (dieseling).

Burned *v.* Showing evidence of excessive heating during processing or use of a plastic, as evidenced by blistering, discoloration, charring, or distortion.

Burned finish *n.* Wood finish in which the hard portion of the grain stands out in relief, the effect being produced by using a blowtorch and a stiff-bristled brush.

Also known as fiery finish.

Burning-in *v.* Process of repairing scratches and other damaged places in a finish, by melting compounds (stick shellac) into the defect with a heated tool.

Burning off *v.* Removal of paint by a process in which the paint is softened by heat, e.g., from a flame, and then scraped off while still soft.

Burning rates *n.* (1) The oxidation in air of a polymer with time. (2) The speed at which a fabric burns. It can be expressed as the amount of fabric affected per unit time, in terms of distance or area traveled by the flame, afterglow, or char.
See flammability, oxygen-index flammability test, self-extinguishing.

Burnishing [ME, fr. MF *bruniss-*, stem of *brunir*, literally, to make brown, fr. *brun*] (14c) *adj, n.* Shiny or lustrous spots on a paint surface caused by rubbing or polishing.

Burnish resistance *n.* Resistance of a coating to an increase in gloss or sheen due to polishing or rubbing.

Burn-out printing *See printing.*

Burnt lime *See calcium oxide.*

Burnt sienna *See iron oxide, natural.*

Burnt umber *n.* Naturally occurring ferric oxide containing manganese dioxide and varying amounts of silica and alumina. Characteristic color is a rich, dark brown. Ferric oxide range, 45–65%; manganese dioxide range, 3–20%. Produced by calcining raw umber.
See umber.

Burr A device that assists in loop formation on circular-knitting machines equipped with spring needles.

Bursting strength *n.* Of rigid plastic tubing, the internal liquid pressure required to cause rupture of a test specimen. Tubes with internal diameters between 3.2 and 152 mm (1/8–6in.) may be tested and diameter and wall thickness must be reported. The term has much the same meaning for filament-wound pressure vessels. (2) The force required to rupture a fabric by distending it with a force applied at right angles to the plane of the fabric under specified condition. Bursting strength is a measure widely used for knit fabrics, non-woven fabrics, and felts where the constructions do not lend themselves to tensile tests. The two basic types of bursting tests are the inflated diaphragm method and the ball-bust method. Shah V (1998) Handbook of plastics testing technology. John Wiley and Sons, New York.

Bush kauri copal *n.* Special grade of the Kauri copal, which has fossilized on trees at the point of exudation and above ground level.

Butacite Poly(vinyl butyral). Manufactured by DuPont, USA.

Butcher's linen *n.* A plain weave, stiff fabric with thick and thin yarns in both the warp and the filling. The fabric was originally made of linen but is now duplicated in 100% polyester or a variety of blends such as polyester/rayon or polyester/cotton.

Butadiene \ˌbyü-tə-ˈdī-ˌēn, -ˌdī-ˈ\ [ISV *butane* + *di-* + *-ene*] (1900) *N.* Buta-1,3-diene, 1,3-butadiene, erythrene, vinylethylene, bivinyl, divinyl. $CH_2{=}CHCH{=}CH_2$. A gas, insoluble in water but soluble in alcohol and ether, obtained from cracking of petroleum, from coal-tar benzene, or from acetylene. It is widely used in the formation of copolymers with styrene, acrylonitrile, vinyl chloride and other monomers, imparting flexibility to the products made from them. Its homopolymer is a synthetic rubber. As noted it is a synthetic chemical compound, used principally in the manufacture of synthetic rubber, nylon, and latex paints (*See image*).

Butadiene-acrylonitrile copolymer (NBR) *See acrylonitrile-butadiene copolymer.*

Butadiene rubber (BR) *See polybutadiene.*

Butadiene styrene latex A synthetic latex similar to synthetic rubber; used for latex paints.

Also known as styrene butadiene.

Butadiene-styrene thermoplastics *See styrene-butadiene thermoplastic.*

Butaldehyde \ˈbət ˈal-də-ˌhīd\. *See butyraldehyde.*

Butanal *See butyraldehyde.*

1,4-Butanedicarboxylic acid *See adipic acid.*

1,3- or 1,4-Butanediol *See 1,3- or 1,4-butylene glycol.*

1,2,4-butanetriol *n.* A nearly colorless liquid, an intermediate for alkyd resins, and a plasticizer for cellulosics.

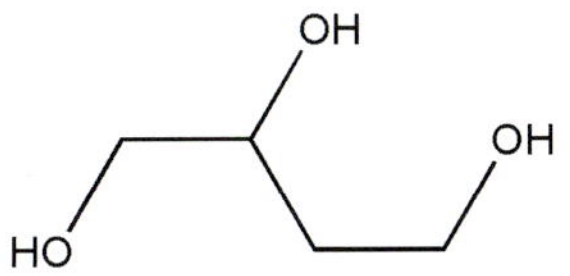

Butanoic acid *See butyric acid.*

1-Butanol \ˈbyü-tᵊn-ˌól\. *See n-butyl alcohol.*

Butene \ˈbyü-ˌtēn\ [ISV *butyl* + *-ene*] (1885) *n.* Any of the monounsaturated C_4 hydrocarbons listed below:

IUPAC name	Alternative names
1-butene	α-butylene
cis-2-butene	*cis*-β-butylene
trans-2-butene	*trans*-β-butylene
Methylpropene	Isobutylene, isobutene

The term butanes refers to the first three compounds above as a group. The butylenes are used as monomers for rubbery homopolymers and copolymers with styrene, acrylics other olefins, and vinyls. They are also used in adhesives for many plastics, and in making plasticizers. James F (ed) (1993) Whittington's dictionary of plastics. Technomic Publishing Co. Inc., Carley.

2-Butene-1,4-diol *n.* A nearly colorless, odorless liquid, an intermediate for alkyd resins, plasticizers, nylon, and a cross-linking agent for resins.

Buton *n.* Cross-linkable plastic produced at high polymerization temperatures from butadiene and styrene. Manufactured by Esso, Great Britain.

2-Butoxyethanol *See ethylene glycol monobutyl ether.*

2-Butoxyethyl pelargonic acid *n.* ($C_9H_{18}O_2$). A plasticizer for polystyrene, vinyl chloride polymers and copolymers, and cellulosics.

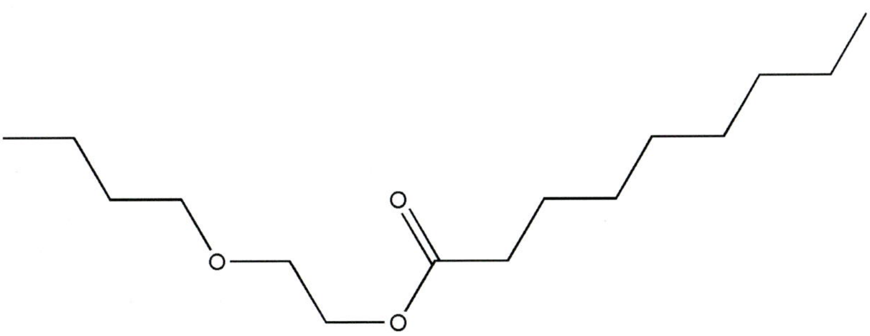

Butoxyethyl stearate *n.* Octadecanoic acid 2-butoxy ethylester. A high-boiling plasticizer for nitrocellulose, polystyrene, ethyl cellulose, and polyvinyl acetate.

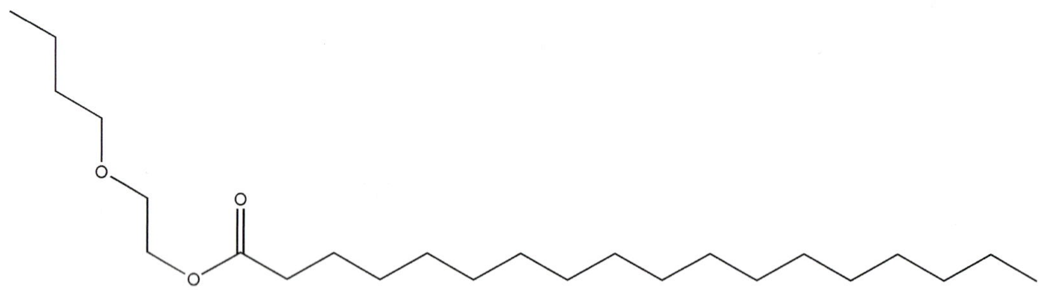

Butt *n.* To met without overlapping.

Butt fusion *n.* A method of joining pipe, sheet or other forms of a thermoplastic resin wherein the ends of the two pieces to be joined are heated into the lower end of the polymer's melting range, then rapidly pressed together and allowed to cool, forming a homogeneous bond. ASTM D 2657 describes a recommended practice for butt-joining polyolefin pipe by heat fusion.

Butt joint (1823) *n.* A joint made by fastening and/or bonding two surfaces that are perpendicular to the main surfaces of the parts being joined.

Button lac Lac refined in the form of buttons, which are usually about 3 in. (7.5 cm) in diameter and 1/4 in. (6 mm) in thickness.

Butt veneer *n.* Veneer having a strong curly figure, caused by roots coming into the trunk at all angles.

Butyl \ˈbyü-tᵊl\ [ISV *butyric* + *-yl*] (1869) *n.* (1) The radical C_4H_9-, occurring only in combination. (2) Abbreviation used by British standards institution for butyl rubber.

Butyl acetate *n.* $CH_3COOC_4H_9$. A pleasantly aromatic solvent of moderate strength for ethyl cellulose, cellulose nitrate, vinyls, polymethyl methacrylate, polystyrene, coumarone-indene resins, and certain alkyds and phenolics.

See n-butyl acetate.

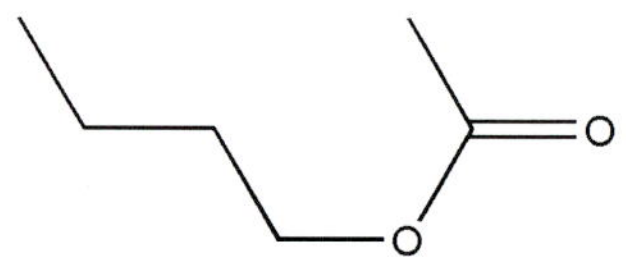

***n*-Butyl acetate** *n.* $CH_3COOC_4H_9$. Limpid, colorless liquid with fruity odor. Prepared by heating and distillation, after contact of *n*-butyl alcohol with acetic acid in the presence of a catalyst such as sulfuric acid. Solvent used in production of lacquers, natural gums and synthetic resins.

Butyl acetoxystearate *n.* $C_{24}H_{46}O_4$. A plasticizer similar to butyl acetyl ricinoleate,

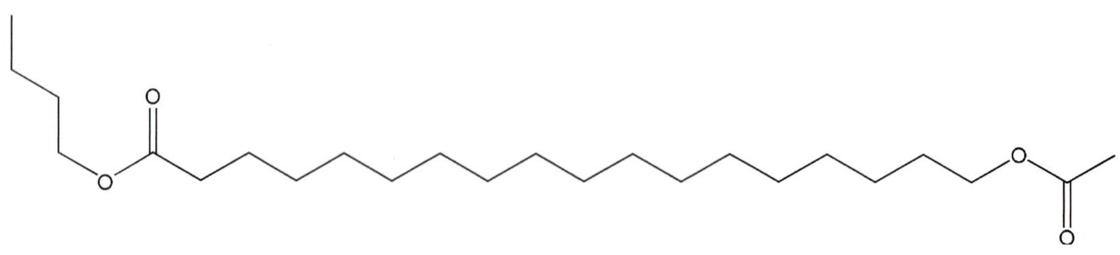

but with the double bond saturated. It is compatible with cellulosic and vinyl resins.

Butyl acetyl ricinoleate *n.* $C_{24}H_{44}O_4$. A yellow, oily liquid derived from castor oil, butyl alcohol, and acetic anhydride, used as a plasticizer, compatible with cellulosics and vinyls.

Butyl alcohol (ca. 1871) *n.* C_4H_9OH. Any of four flammable alcohols used in organic synthesis and as solvents.
See n-butyl alcohol.

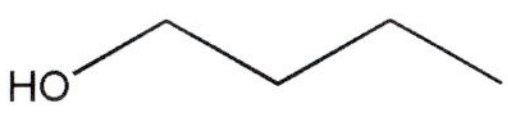

***n*-Butyl alcohol** *n.* Colorless liquid, with vinous, irritating odor which causes coughing. Solvent used for resins, coatings, shellac, and butylated melamine resins. Syn: 1-butanol and propyl carbinol.

***n*-Butyl acrylate** *n.* $CH_2{=}CHCOOC_4H_9$. Colorless liquid. A monomer used in the manufacture of synthetic resins and as a solvent for cellulose esters. Sp gr, 0.898; bp, 145°C.

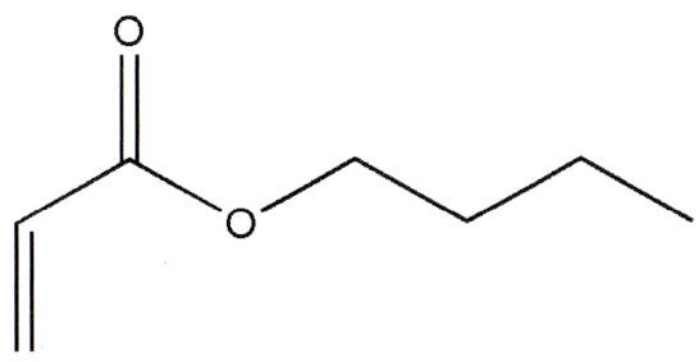

Butylated hydroxytoluene (di-*tert*-*p*-cresol, BHT) *n.* A white, crystalline solid, the most widely used antioxidant for plastics such as ABS and LDPE. It has been approved by the FDA for use in foods and food-packaging materials.

Butylated resin *n.* A resin containing the butyl radical, C_4H_9–.

Butyl benzenesulfonamide *n.* (*N*-*n*-butyl benzenesulfonamide) $C_6H_5SO_2$–NHC_4H_9. A plasticizer for some synthetic resins and an intermediate for resin manufacturer (*See image*).

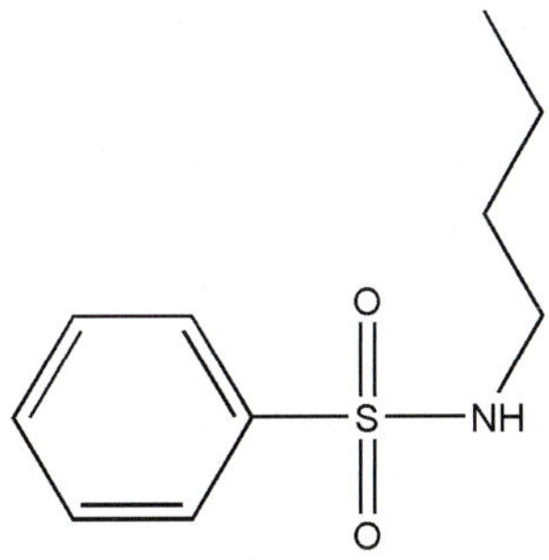

Butyl benzoate *n.* (*n*-butyl benzoate). A plasticizer and solvent for cellulosics.

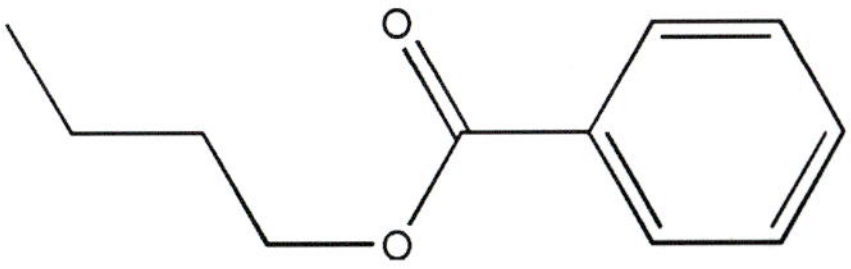

Butyl benzyl phthalate (BBP) *n.* $(C_4H_9OOC)C_6H_4(COOCH_2C_6H_5)$. A clear, oily liquid used as a plasticizer for cellulosic and vinyl resins. It imparts good stain resistance, low

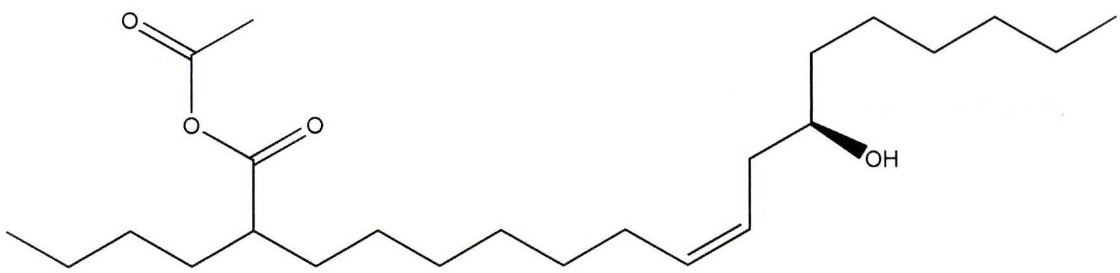

volatility at calendaring and extruding temperatures, low oil extraction and good heat and light stability.

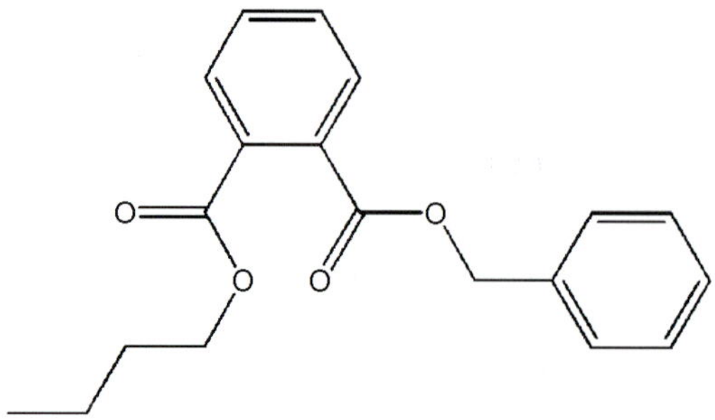

Butyl benzyl sebacate *n.* $C_4H_9OOC(CH_2)_8COOCH_2C_6H_5$. An ester-type plasticizer with a light straw color. It combines the desirable properties of dibenzyl sebacate and dibutyl sebacate.

Butyl borate *See tributyl borate.*

Butyl carbitol *n.* Diethylene glycol monobutyl ether, butyldiglycol, 2-[2-(1-butoxy) ethoxy] ethanol. Practically odorless oil, solvent, and water miscibile liquid used as a coalescent in latex paints; bp, 230.4°C; mp, −68.1°C; Sp gr, 0.9536.

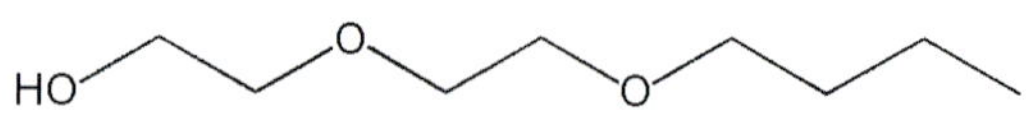

Butyl carbitol acetate *n.* $CH_3COO(CH_2CH_2O)_2C_4H_9$. Diethylene glycol monobutyl ether acetate.

Butyl cellosolve Ethylene glycol monobutyl ether, 2-butoxyethanol. Liquid soluble in 20 parts water; soluble in most organic solvents. It is used as a coalescent in latex paints; bp, 171–172°C; Sp gr, 0.901.

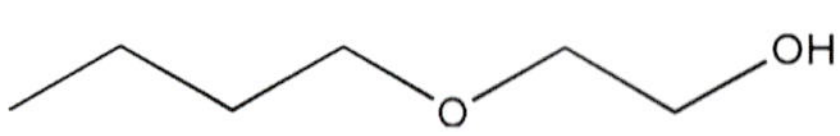

Butyl cellosolve acetate *n.* Ethylene glycol monobutyl ether acetate.

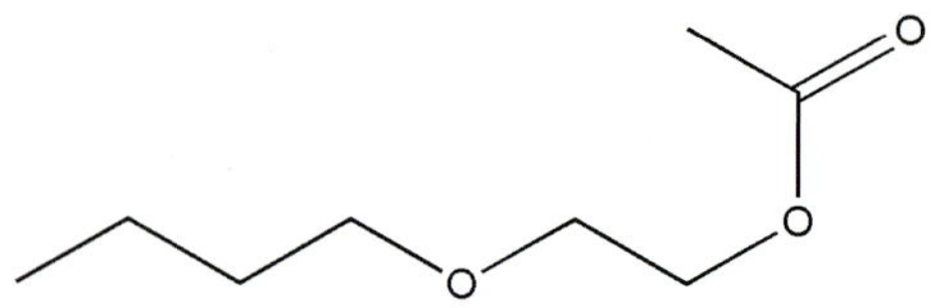

Butyl citrate *n.* Colorless or pale yellow, stable, odorless, non-volatile liquid. Plasticizer, anti-foam agent, solvent for cellulose nitrate.
Also known as tributyl citrate. See citrates (*See image*).

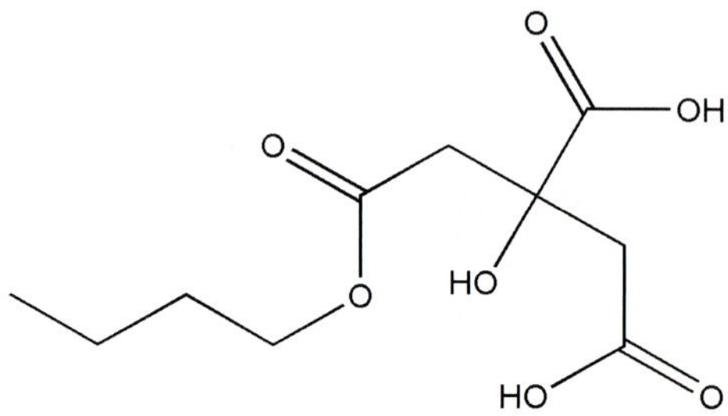

Butyl cyclohexyl phthalate *n.* A plasticizer for PVC, other vinyls, cellulosics, and polystyrene.

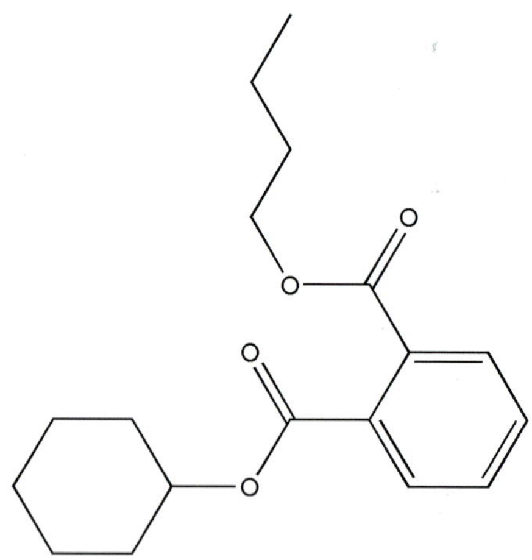

Butyl decyl phthalate *n.* $C_{22}H_{34}O_4$. A plasticizer for polystyrene, PVC, and vinyl chloride-acetate copolymers (*See image*).

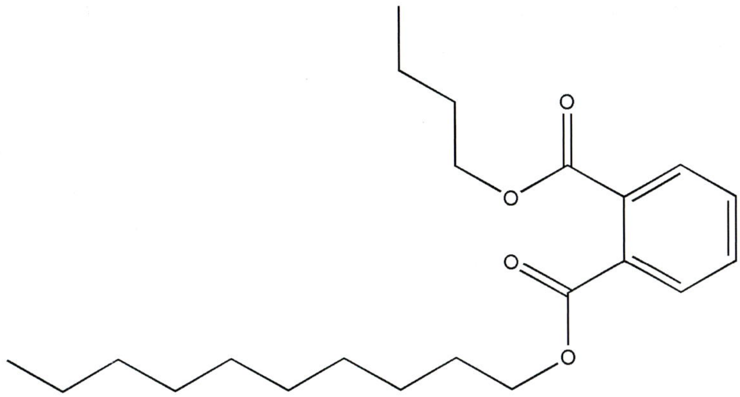

Butyl diglycol carbonate *n.* [diethylene-glycol bis(*n*-butylcarbonate)]. A colorless liquid of low volatility, used as a plasticizer with many resins (*See image*).

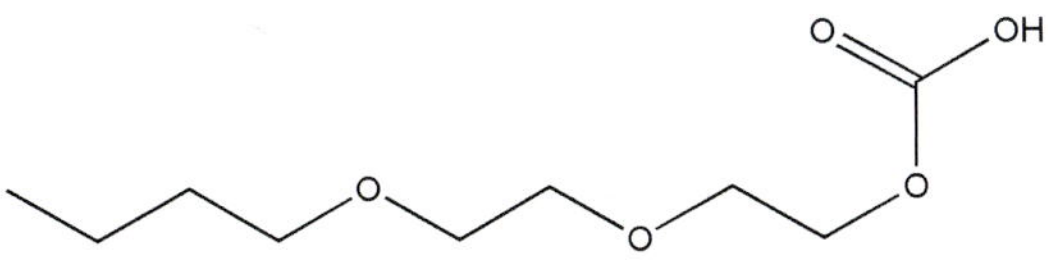

Butylene *See butene.*

1,3-Butylene dimethacrylate A polymerizable monomer used in PVC and rubber systems to obtain rigid or semi rigid products from materials that are normally flexible. The monomer acts as a plasicizer at room temperature and cross-links at processing temperatures (*See image*).

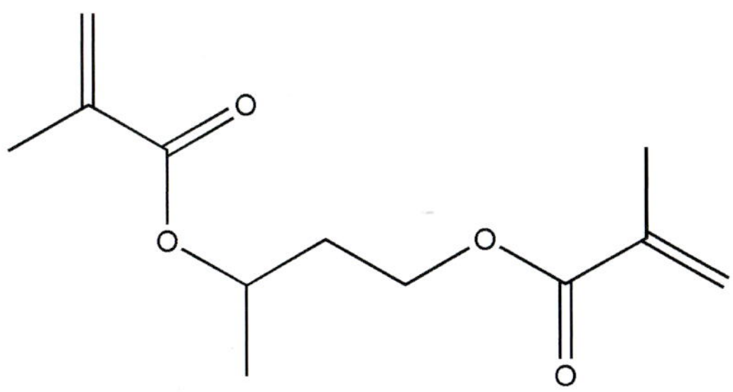

1,3-Butylene glycol *n.* (1,3-butanediol) $CH_3CHOHCH_2CH_2OH$. A colorless liquid made by catalytic hydrogenation of aldol (3-hydroxy-*n*-butyradehyde). Its most important use is as an intermediate in the manufacture of polyester plasticizers (*See image*).

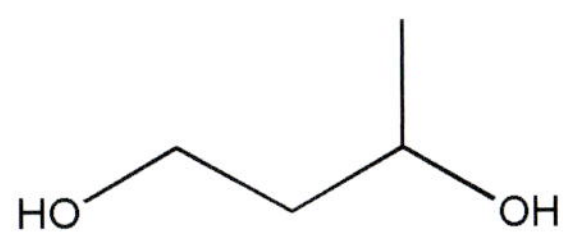

1,4-Butylene glycol *n.* (1,4-butanediol, tetramethylene glycol). A stable, hygroscopic, colorless liquid used in the production of polyesters by reaction with dibasic acids, and in the production of polyurethanes by reaction with diisocyanates (*See image*).

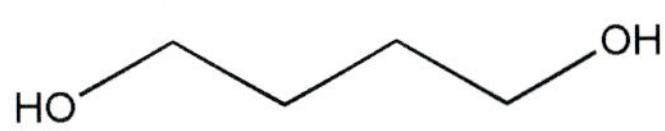

1,3-Butylene glycol adipate polyester *n* (Santicizer® 334F). A polymeric plasticizer for PVC.

Butyl epoxy stearate *n.* $C_{40}H_{78}O_5$. A plasticizer for PVC, imparting low-temperature flexibility (*See image*).

Butyl ethylhexyl phthalate *n.* (butyl octyl phthalate, BOP). A mixed ester of butanol and 2-ethylhexanol, widely used as a primary plasticizer for PVC compounds and plastisols, in which it performs like dioctyl

B

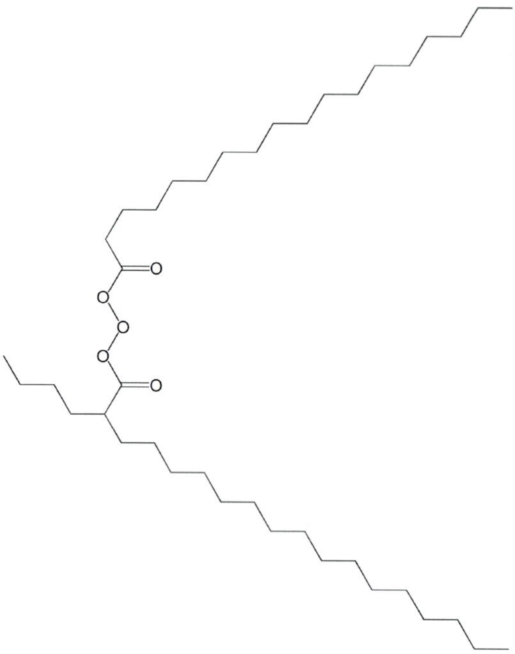

phthalate in most respects. It is also compatible with vinyl chloride-acetate copolymers, cellulose nitrate, ethyl cellulose, polystyrene, chlorinated rubber, and at lower concentrations, with polymethyl methacrylate (*See image*).

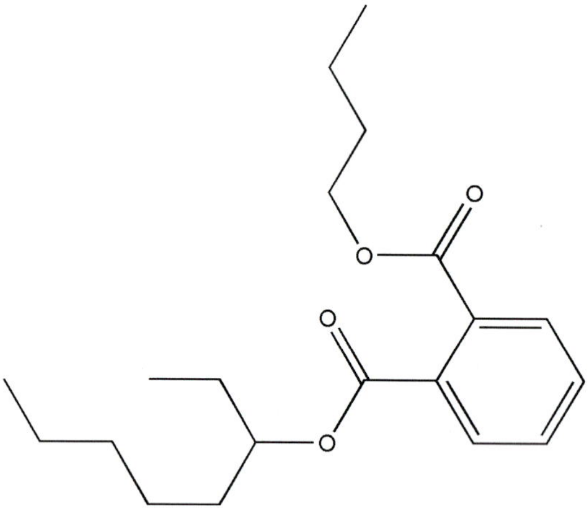

Butyl formate *n.* A solvent for several resins, including cellulose acetate (*See image*).

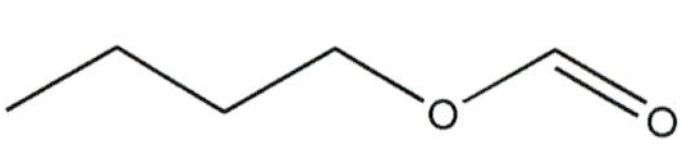

Butyl isodecyl phthalate *n.* (decyl butyl phthalate) $(C_4H_9OOC)C_6H_4\text{–}(COOC_{10}H_{21})$. A plasticizer for PVC and polystyrene.

Butyl isohexyl phthalate *n.* $(C_4H_9OOC)C_6H_4(COOC_6H_{13})$. A plasticizer for cellulosics, acrylic resins, polystyrene, PVC, and other vinyl resins.

Butyl lactate *n.* $CH_3CHOHCOOC_4H_9$. Slow evaporating liquid, which is miscible with most of the organic solvents. It is a solvent for nitrocellulose, ethyl cellulose, oils, dyes, natural gums, and synthetic resins.

Butyl octadecanoate *See butyl stearate.*

Butyl octyl phthalate *See butyl ethlyhexyl phthalate.*

Butyl oleate *n.* $CH_3(CH_2)_7CH{=}CH(CH_2)_7COOC_4H_9$. A solvent, plasticizer and lubricant, used mainly with neoprene and other synthetic rubbers, chlorinated rubber, and ethyl cellulose; also a mold lubricant (*See image*).

Butyl phthalate *See phthalate esters.*

Butyl phthalyl butyl glycollate *n.* A plasticizer with good light stability, used mainly with PVC and polystyrene, but compatible with most other thermoplastics. It has been approved by the FDA for contact with foods (*See image*).

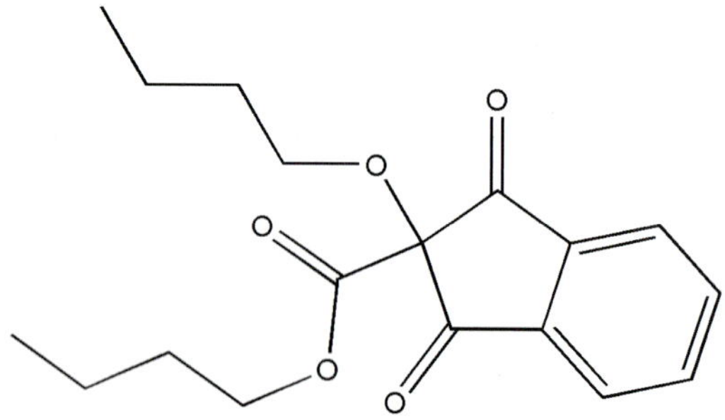

Butyl ricinoleate *n.* Yellow to colorless oleaginous liquid derived from castor oil fatty acid (ricinoleic acid) and butyl alcohol. Used as a plasticizer or lubricant (*See image*).

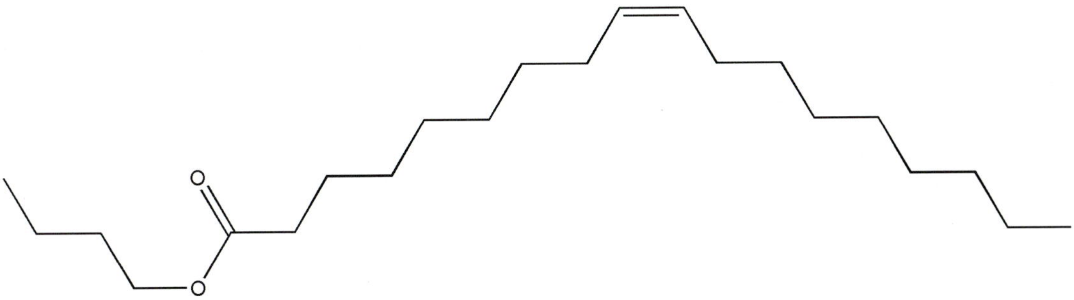

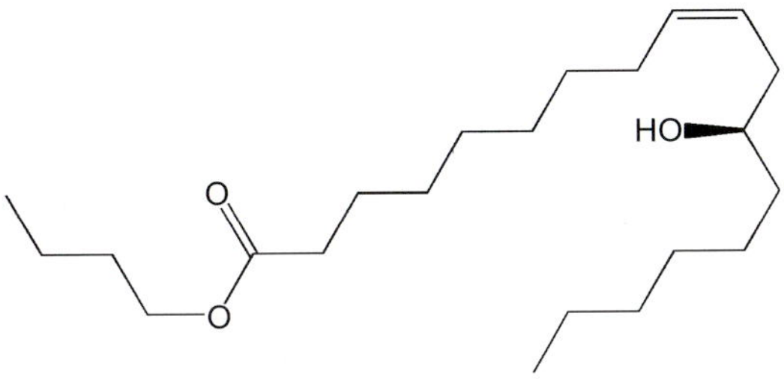

Butyl rubber *n* (1940). A synthetic elastomer produced by copolymerizing isobutylene with a small amount (ca 2%) of isoprene or butadiene. It has good resistance to hear, oxygen and ozone, and low gas permeability. Thus, it is widely used in inner tubes and to line tubeless tires. Butyl rubber is a vinyl polymer, and is very similar to polyethylene and polypropylene in structure, except that every other carbon is substituted with two methyl groups. It is made from the monomer isobutylene, by cationic vinyl polymerization. It can also go by the name of polyisobutylene. (1) Generic name for vulcanizable elastic copolymers of isobutylene and small amounts of diolefins. (2) Mixture of isobutylene, 98% and butadiene or isoprene, 2%. (3) Poly(isobutylene) with 2% isoprene. Manufactured by Bayer, Germany.

Butyl stearate (butyl octadecaonoate) *n.* $C_{17}H_{35}COOC_4H_9$. A mold lubricant and plasticizer, compatible with natural and synthetic rubbers, chlorinated rubber, and ethyl cellulose. It can be used in vinyls in very low concentrations as a non-toxic, secondary plasticizer and lubricant. In the production of polystyrene, butyl stearate is added to the emulsion polymerization to impart good flow properties to the resin (*See image*).

Butyl (tetra) titanate *n.* $Ti(OC_4H_9)_4$. Used for adhesion promoter, catalyst, and in heat-resistant paints.

Butyraldehyde \ˌbyü-tə-ˈral-də-ˌhīd\ [ISV] (ca. 1885) (butaldehyde, *n*-butanal, *n*-butyl aldehyde, butyric aldehyde) *n.* An aldehyde sometimes used in place of formaldehyde in the production of resins. Butyraldehyde reacts with polyvinyl alcohol to form polyvinyl butyral. It is used in the

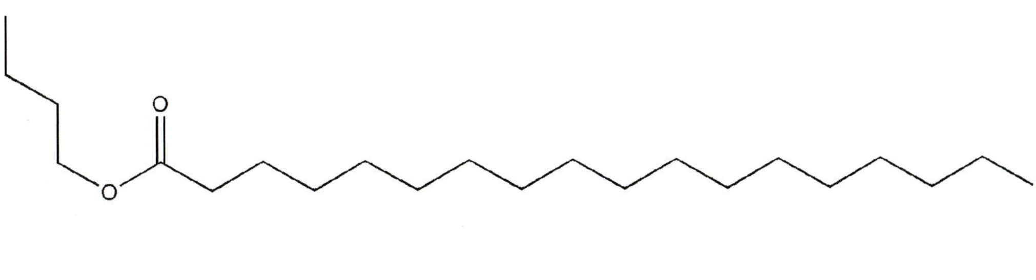

B

manufacture of rubber and synthetic resins. Sp gr, 0.817; bp, 75.7°C. Syn: butaldehyde, butyl aldehyde, and butyric aldehyde (*See image*).

Butyrate \ˈbyü-tə-ˌrāt\ (1873) *n.* (1) A salt or ester of butyric acid. (2) The industry nickname for cellulose acetate butyrate (CAB).

Butyrates *n.* Name applied to esters of butyric acid, such as cellulose butyrate, ethyl butyrate, etc.

Butyric acid (*n*-butyric acid, butanoic acid, ethylacetic acid, and propylformic acid). *n* $CH_3(CH_2)_2COOH$. A water soluble liquid with a strong butter-rancid odor, used in the production of cellulose acetate butyrate. Derivatives of butyric acid are used in the production of plasticizers for cellulosic plastics (*See image*).

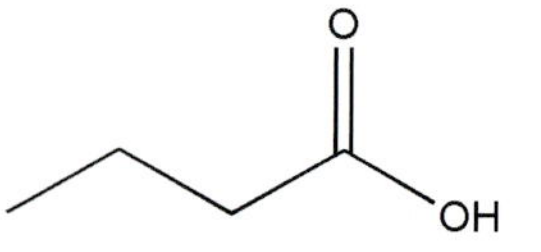

Butyric alcohol *See n-butyl alcohol.*

γ-Butyrolactone (butyrolactone) *n.* A hygroscopic, colorless liquid obtained by the dehydrogenation of 1,4-butanediol with the structure: It is a solvent for cellulosics, epoxy resins, and vinyl copolymers (*See image*).

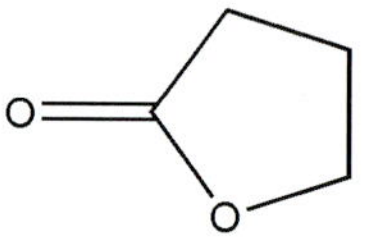

C

c *n.* \ˈsē\. (1) Abbreviation for SI prefix, centi-. (2) Abbreviation for Cubic.

C *n.* (1) Chemical symbol for the element carbon. (2) Abbreviation for Celsius or centigrade. (3) Symbol for electrical capacitance.

Ca *adj.* Circa; about, approximate.

Ca *n.* Chemical symbol for the element calcium.

CA *See cellulose acetate.*

C or Δc *n.* Abbreviations for chromaticity or chromaticity difference, receptivity. *See color difference equations.*

C 23 *n.* Ethylene/propylene copolymer. Manufactured by Montecatini, Italy.

CAB *n.* Cellulose acetobutyrate.

Cabinet *n.* A basic part of the manufactured-fiber spinning machine, where, in dry spinning, the filaments become solidified by solvent evaporation and, in melt spinning, the filaments are solidified by cooling.

Cabinet finish *n.* A varnished or polished hardwood interior finish as distinguished from a painted softwood finish.

Cabled yarn *n.* A yarn formed by twisting together two or more plied yarns.

Cable stitch *n.* A knit effect produced by crossing a group of stitches over a neighboring stitch group.

Cable twist *n.* A construction of thread, yarn, cord, or rope in which each successive twist is in the direction opposite the preceding twists; i.e., and S/Z/S or Z/S/Z construction.

Cable varnishes and lacquers *n.* Employed for impregnating the cloth or similar covering of cables. Normally, oil varnishes are used.

Cacahuananche oil *n.* This oil is obtained from the nuts of the tree, *Licania arborea.* So far as the usual laboratory tests are concerned, this oil and Brazilian oiticica oil are much alike. The raw oil becomes lard-like on aging but may be permanently liquefied by heat. The raw and lightly heat-treated oil wrinkles as it dries, similarly to oiticica and tung oils.

CAD/CAM *n.* Acronym for computer-aided design and computer-aided manufacturing.

Cadmium \ˈkad-mē-əm\ [NL, fr. L *cadmia* zinc oxide, fr. Gk *kadmeia*, fr. feminine of *kadmeios* Theban, from *Kadmos*] (1822) *n.* A bluish white malleable ductile toxic bivalent metallic element used in protective platings and in bearing metals.

Cadmium ethylhexanoate *n.* A metallic soap used as a stabilizer for vinyls, especially to avoid plate-out in calendaring compounds.

Cadmium green *n.* Pigment green 18. Composed of 93–94% hydrated chromium, oxide, the balance being cadmium yellow.

Cadmium lithopone *n.* Pigment red 108 (77202). Range of yellow, orange, and red colors consisting of either cadmium sulfide or cadmium red reduced on barium sulfate. They are usually produced by co-precipitation followed by calcinations. *See cadmium red.*

Cadmium/mercury lithopones *n.* Orange, red, and maroon pigments consisting of calcined co-precipitations of cadmium sulfide and mercury sulfide, with barium sulfate.

Cadmium/mercury sulfides *n.* CdSHgS. Pigment orange 23 (77201). A series of cadmium/mercury sulfides of mixed crystal composition. Colors vary according to CdS/HgS ratio from deep orange to maroon. Known for high heat stability, solvent insolubility, good chemical resistance, and excellent light fastness. CdS/HgS ratio, 89.1/10.9. Density 4.8–5.2 g/cm^3 (40–43 lb/gal); O.A., 33.2; particle size, 0.1–1.0 μm. Syn: Mercadium® orange, mercury cadmium, cadmium red/orange, and mercury/cadmium sulfide.

Cadmium orange *n.* Mixture or co-precipitate of cadmium sulfide and cadmium selenide used as a pigment.
See cadmium sulfide.

Cadmium plating *n.* Electrodeposition of cadmium on iron wire, steels articles, etc., to make them relatively rust-proof. This surface treatment makes the substrate more suitable for subsequent painting.

Cadmium pigment *n.* Any inorganic pigment based on cadmium sulfide or cadmium sulfoselenide, used widely in PVC, polystyrene, and polyolefins. Included are cadmium-maroon, -orange, -red, and-yellow. The cadmium pigments have good resistance to heat (up to 500°C) and to alkalis, and do not bleed. Light stability is good in solid colors, but may be poor when used for tints with white pigments. Resistance to acids is poor.

Cadmium red *n.* Pigment red 108 (77202). Pigments of calcined co-precipitations of compounds of cadmium, sulfur, and selenium; a brilliant red pigment, opaque, with good staining power, fast to light, unaffected by exposure to sulfur fumes; considerable resistance to heat. Density, 4.9 g/cm^3 (41 lb/gal); O.A., 20 lb/100 lb.

Cadmium ricinoleate *n.* $[CH_3(CH_2)_5-CHOHCH_2CH=CH(CH_2)_7COO]_2-Cd$. A white powder derived from castor oil, used as a heat stabilizer for vinyl chloride polymers and copolymers.

Cadmium selenide *See cadmium red and cadmium orange.*

Cadmium stearate *n.* $(C_{17}H_{35}COO)_2Cd$. A heat- and light-stabilizer, used when good clarity in transparent compositions is desired.

Cadmium sulfide (ca. 1893) (Cd/Zn)S or CdS Pigment yellow 37 (77199) *n.* Class of yellow, orange, and red pigments that include pure cadmium sulfides and blends of cadmium sulfides with ZnS and CdSe and used with $BaSO_4$ to form lithopone type pigments. They show excellent heat and good alkali resistance especially for baking finishes. Density, 4–64.7 g/cm^3 (38–39 lb/gal); O.A., 18–22 lb/100 lb. Syn: cadmium yellow and cadmium orange.

Cadmium yellow *See cadmium sulfide.*

Cage effect *n.* Free radicals that exist very close together in a solvent or monomer molecule called a "cage".

Caisson \ˈkā-ˌsän, -sən, British also kə-ˈsűn\ [Fr, fr. MF, fr. OP, fr. *caissa* chest, fr. L *capsa*] (ca. 1702) *n.* Recessed compartment in a vault or ceiling; recessed panel in a ceiling or soffit.

Cajeputene *See dipentine.*

Cake, press *n.* The thick cake of wet pigment that is withdrawn from the filter press.

Caking *n.* (1) Settling of pigment particles of a paint into a compact mass which is not easily redispersed by stirring. (2) In printing inks, caking is the collecting of pigment upon plates, rollers or blankets caused primarily by the inability of the vehicle to hold the pigment in suspension.
See piling.

Calcareous \kal-ˈkar-ē-əs, -ˈker-\ [L *calcarius* of lime, from *calc-*, *calx*] (1677) *adj.* Any material containing calcium or calcium compounds.

Calcicoater \ˈkal-sik ˈkotər\ *n.* (1) Vegetable oil product that has been reacted with lime to form a heavy gel-like soap, which has almost infinite tolerance for mineral spirits; used in flat wall paints, primers, and sealers. (2) Pigmented flat paints using such heavy gel-like soap vehicles.

Calcimine \ˈkal-sə-ˌmīn\ [alter. of *kalsomine*] (ca. 1859) *n.* Also spelled "kalsomine". Essentially, chalk and glue ready to mix with water. Used as a decoration for interior surfaces. It will not withstand washing. In Britain, it is referred to as powdered distemper.

Calcination \ˌkal-sə-ˈnā-shən\ *n.* (1) Process of heating or roasting a material to a high temperature, but below its fusing point, to cause it to lose moisture or other volatile material or to be oxidized or reduced. Originally, a heat process for the production of lime (CaO) from limestone. (2) The process of subjecting a sorptive mineral to prolonged heating at fairly high temperature, resulting in the removal of water, and an increase in the hardness, physical stability, and absorbent properties of the material.

Calcined clay *n.* China clay (kaolin) that has been heated until the combined water is removed and the plastic character of the clay is destroyed. This produces an air–solid interface within the particle, which increases hiding in he resulting coating. *See clay.*

Calcite \ˈkal-ˌsīt\ (1849) *n.* Naturally occurring form of calcium carbonate. It is an essential material of limestone, marble, and chalk.

Calcium acetate (vinegar salts, gray acetate, lime acetate, and brown acetate) *n.* $(CH_3COO)_2Ca{\cdot}H_2O$. A stabilizer.

Calcium carbide (ca. 1888) *n.* Usually a dark gray crystalline compound CaC_2 used for the generation of acetylene and for making calcium cyanamide.

Calcium carbonate (1873) (aragonite, calcite, chalk, limestone, lithographic stone, marble marl, travertine, and whiting) *n.* $CaCO_3$. Grades of calcium carbonate suitable as fillers for plastics are obtained from naturally occurring deposits as well as by chemical precipitation. The natural types are prepared by dry grinding, yielding particles usually over 20 μm, used in stiff products such as floor tiles; or by wet grinding, yielding particles under 16 μm, used in flexible products. The chemically precipitated types range from 0.05 to 11 μm in size, and are most often used in plastisols and highly flexible products. Both the wet-ground and precipitated types are available with coating such as resins, fatty acids, and calcium stearate. These coated grades have low oil absorption, of particular value in compounding plastisols. The calcium stearate coatings provide improved electrical properties, heat stability and lubricity during processing, which is beneficial in extrusion compounds.

Calcium carbonate fillers *n.* Fine powder of calcium carbonate (white) to fill spaces in a polymer or coating.

Calcium carbonate, natural *n.* $CaCO_3CaMg(CO_3)_2$. Pigment white 18 (77220). White extender pigment derived from natural chalk, limestone, or dolomite, consisting of calcium carbonate with up to about 44% magnesium carbonate. Density, 2.71 g/cm^3 (22.6 lb/gal); O.A., 6–15; particle size, 1.5–12 μm. Syn: calcite, limestone, whiting, marble flour, Paris white, chalk, ground oyster shells, iceland spar, and spanish white.

Calcium carbonate, synthetic *n.* $CaCO_3$. Pigment white 18 (77220). Calcium carbonate manufactured by a precipitation process in order to obtain a finer or more uniform particle size range. Four commercial processes are known. Density, 2.65 g/cm^3 (22.07 lb/gal); O.A., 28–58; particle size, 0.6–30 μm. Syn: precipitated calcium carbonate.

Calcium chloride (1885) *n.* $CaCl_2$. Salt used in the manufacture of some lakes and toners from acid dyestuffs, fireproof paints, sizing compounds, wood preservatives, snow melter, and anti-freeze.

Calcium glycerophosphate (calcium glycerinophosphate) *n.* $CaC_3H_7O_2–PO_4$. A white, crystalline powder, odorless and nearly tasteless, used as a stabilizer for plastics.

Calcium hydroxide (ca. 1889) *n.* $Ca(OH)_2$. A white crystalline strong alkali that is used in mortar and plaster.
Also known as slaked lime.

Calcium hypochlorite (ca. 1889) *n.* $CaCl_2O_2$. A white power used as a bleaching agent and disinfectant.

Calcium linoleate *n.* White amorphous powder soluble in alcohol and ether; insoluble in water. Used for waterproofing compounds, emulsifying agents, and as a stabilizer for oleoresinous paints.

Calcium lithol red *See lithol red.*

Calcium metasilicate *See calcium silicate, natural.*

Calcium naphthenate *n.* Calcium salt of naphthenic acids. Used as an auxiliary drier, dispersing and stabilizing aid.

Calcium octoate *n.* Calcium salt of 2-ethyl hexoic acid. Used as auxiliary drier, dispersing, and stabilizing aid.

Calcium oxide (ca. 1885) (calx, lime, quicklime, and burnt lime) *n.* CaO. A white powder with affinity for water, with which it combines to form calcium hydroxide. It has been used to remove traces of water in vinyl plastisols.
Also known as quicklime.

Calcium phosphate (1869) (calcium orthophosphate, tricalcium phosphate, tricalcic phosphate, and tertiary calcium phosphate) *n.* $Ca_3(PO_4)_2$. A stabilizer.

Calcium phosphate, dibasic (calcium biphosphate, acid calcium phosphate, and primary calcium phosphate) *n.* $CaH_4(PO_4)_2 \cdot H_2O$. A stabilizer.

Calcium phosphate, monobasic (dicalcium orthophosphate, bicalcic phosphate, and secondary calcium phosphate) $CaHPO_4$ or $CaHPO_4 \cdot H_2O$. A stabilizer.

Calcium resinate *n.* Calcium salt of rosin used as auxiliary drier and dispersing and stabilizing aid. Commonly known as Limed Rosin, which is really rosin, the acidity of which has been substantially neutralized.

Calcium ricinoleate *n.* $[CH_3(CH_2)_5CHOH\text{-}CH_2CH{=}CH(CH_2)_7COO]_2$. A white powder derived from castor oil, used as a non-toxic stabilizer for PVC.

Calcium silicate (1888) (wollastonite) *n.* $CaSiO_3$. A naturally occurring mineral found in metamorphic rocks, used as a reinforcing filler in low-density polyethylene, polyester, and other thermosetting molding compounds. It imparts smooth molded surfaces and low water absorption.

Calcium silicate, natural *n.* $CaSiO_3$. It has white color and acicular particles. It is characterized by high flatting action, combined with low oil absorption, and is used as an extender in paints, ceramics and plastics. Density, 2.9 g/cm^3 (24.2 lb/gal); O.A., 25–30; particle size, 7 μm (fine particle grade); hardness (moh), 4.5. Syn: wollastonite and calcium metasilicate.

Calcium silicate, synthetic *n.* $CaSiO_3 \cdot nH_2O$. Extender pigment with some dry hiding opacity. Density, 2.26 g/cm^3 (18.8 lb/gal); O.A., 280; particle size, 10–12 μm.

Calcium stearate *n.* $(C_{17}H_{35}COO)_2$. (1) A non-toxic stabilizer and lubricant. It is not often used alone because of its early color development, but is used in combination with zinc and magnesium derivatives and epoxides in manufacturing other non-toxic stabilizers. (2) Calcium salt of stearic acid. A metallic soap used in paints as wetting aid for pigments, flatting agents, and additive for improving sanding sealers.

Calcium sulfate (ca. 1885) (anhydrite) *n.* $CaSO_4{\cdot}2H_2O$. (1) A filler and white pigment. Notwithstanding its slight solubility in water, it is used to some extent in distempers and water paints, it forms the base of certain lakes and other pigments. The hydrated forms are known as *gypsum*, *terra alba*, and Plaster of Paris.

Calcium sulfate, anhydrous *n.* $CaSO_4$. Manufactured by calcining natural gypsum ($CaSO_4{\cdot}2H_2O$) or as byproducts from other precipitation processes. Notwithstanding its slight water solubility, it is used in distempers and water paints. It also forms the base of certain lakes and other pigments. Plaster of Paris is partially calcined gypsum. Density, 2.96 g/cm^3 (24.7 lb/gal); O.A., 20; particle size, 1.8–2.2 μm. Syn: anhydrite, dead burned calcium sulfate, and dehydrated gypsum.

Calcium sulfate hemihydrate *n.* CaSO4·½H_2O. *See gypsum (calcined) and plaster of Paris.*

Calcium tallate *n.* Calcium salt of tall oil acids used as an auxiliary drier and dispersing and stabilizing aid.

Calcium thiocyanate (calcium sulfocyanate, calcium rhodanate) *n.* Ca–$(SCN)_2{\cdot}3H_2O$. In water solution, a solubilizer for acrylic and cellulosic resins.

Calcium–zinc stabilizer *n.* Any of a family of stabilizers based on compounds and mixtures of compounds of calcium and zinc. Their effectiveness is limited, but they are among the few that have been approved by the FDA for materials to be contacted by foods.

Calendar *n.* (1) To produce or process sheets of material by pressing between a series of revolving heated rolls. (2) The machine performing the process of calendering.

Calender coating *n.* A roller coating on a film or other material. The process of applying plastics to substrates such as paper or fabric by passing both the substrate and a plastic film through calender rolls. *See coating.*

Calendered cloth *n.* Basis of varnished cloths for electrical insulation. The process of calendering cloths involves the passing of moist cloth between heavy rollers, resulting in a smooth or glazed appearance.

Calendered film and sheet *n.* Rolled film and sheet.

Calendered papers Wallpapers with hard coatings.

Calendering *n.* The process of forming thermoplastics sheeting and films by passing the material through a series of rigid, heated rolls and rolling to press out voids. Four rolls are typical. The gap between the last pair of heated rolls determines the thickness of the sheet. Subsequent chilled rolls cool the sheet. The plastic compound is usually premixed and plasticated on separate equipment, then fed continuously into the nip of the first pair of calender rolls.

Calendering rolls *n.* (1) The main cylinders on a calender. (2) Smooth or fluted rolls used on carious fiber-processing machines such as pickers and cards to compress the lap or sliver as it passes between them.

Calender with rubber *n.* A machine for impregnating fabric.

Calibrate \ˈka-lə-ˌbrāt\ (ca. 1864) *vt.* (1). To determine the exact relationship between the indicated or recorded readings of a measuring instrument or method and the true values of the quantities measured, over the instrument's range. (2) (mainly European usage). To bring the dimensions of an extruded tube or pipe within their specified ranges.

Calibration *n.* (1) The process of ascertaining the errors of a measuring technique or instrument by comparing its readings with the corresponding values of known standards or the readings of a more accurate technique or instrument. Calibration standards for the whole gamut of measurements are available from the USA National Institute of Standards and Technology (NIST), formerly the National Bureau of Standards, Gaithersbury, MD. "They can also furnish standard materials, some of them polymeric. *See NIST special publication 260*, "Standard Reference Materials Catalog". NIST also calibrates instruments on request. *See special publication 250*, "NIST Calibration Services Users' Guide". Both publications are available from the USA Government Printing Office in Washington DC. (2) *See calibrate* (2).

Calico \ˈka-li-ˌkō\ [Calicut, India] (1578) *n.* A plain, closely woven, inexpensive cloth, usually cotton or a cotton/manufactured fiber blend, characteristically having figured patterns on a white or contrasting background. Calico is typically used for aprons, dresses, and quilts.

Caliper \ˈka-lə-pər\ [alteration of *caliber*] (1588) *n.* (1) The thickness of film or sheet, typically stated in thousandths of an inch (mils). (2) Any of several types of precise instruments used to measure thickness, as micrometer caliper.

Calk \ˈkók, ˈkó-kər\ [prob. alter. of *calkin*, fr. ME *kakun*, fr. MD or ONF *calcain* heel, fr. L *calcaneum*, fr. *calc-*, *calx* heel] (1587) *n, vt. See caulk.*

Calking *See caulking compound.*

Calorie \ˈka-lə-rē, ˈkal-rē\ [F *calorie*, fr. L *calor* heat, fr. *calēre*] (1866) (small calorie) *n.* A deprecated, small unit of heat energy: the amount of heat required to raise the temperature of 1 g of water, at or near 4°C, 1°C. The ASTM "Standard for Metric Practice" (E 380) lists five slightly different calories, all nearly equal to 4.185 J. In future, most of these will fade away, leaving only the "calorie (International Table)", equal to 4.186800 J. The term calorie is also used loosely, especially in the nutrition field, to mean 1000 calories, the kilocalorie, or "large calorie".

Calorimeter *n.* An instrument for measuring the heat liberated (or absorbed) during chemical reactions or physical changes of state.

Calorimetry *n.* The process of measuring quantities of absorbed or evolved heat, often used to determine specific heat.

Calutron \ˈkal-yə-ˌträn\ [California University cyclotron] (1954) *n.* An apparatus operating on the principle of the mass spectrograph and used for separating U^{235} from U^{238}.

Calvary twill *n.* A pronounced, raised cord on a 63° twill weave characterizes this rugged cloth usually made from wool or wool blend yarns.

Cam *n.* A rotating or sliding piece or projection used to impart timed or periodic motion to other parts of a machine. It is used chiefly as a controlling or timing element in machines rather than as part of a power transmission mechanism. Cams are particularly important in both knitting and weaving machinery. Acronym for computer-aided manufacturing.

Camber \ˈkam-bər\ [F *cambrer*, fr. MF *camber*, fr. L *camur*] (1627) *v.* Strip under tension, as on a coating line that has a tendency to deviate from the horizontal or to slope to one side.

Cambogia *See gamboge.*

Cambric \ˈkām-brik\ [obs. Flemish *Kameryk*] (1530) *n.* A soft, white, closely woven, cotton or cotton blend fabric that has been calendered on the right side to give it a slight gloss. Cambric is used extensively for handkerchiefs.

Camel hair *n.* A term inaccurately applied to a fine haired brush, the hair of which is obtained from squirrels native to Russia and Siberia.

Camphol *See oxanilide.*

Camphor \ˈkam(p)-fər\ [ME *caumfre*, fr. AF, fr. ML *camphora*, fr. Arabic *kāfūr*, fr. Malay *kapur*] (*d*-2-camphanone, 2-keto-1,7,7-trimethylnorcamphane) *n.* C_{10}–H_{16}O. A bridged-ring, naturally occurring ketone with a mp of 175°C and a bp of 204°C with the structural formula: It is a colorless, aromatic, crystalline material originally derived from camphor oil but now mostly synthesized from pinene. It is used as a plasticizer for celluloid, cellulose nitrate, and lacquers. In 1870, the Hyatt brothers were awarded a USA patent for a horn-like compound of camphor and cellulose nitrate, which they called celluloid. The event marked the start of the USA plastics industry.

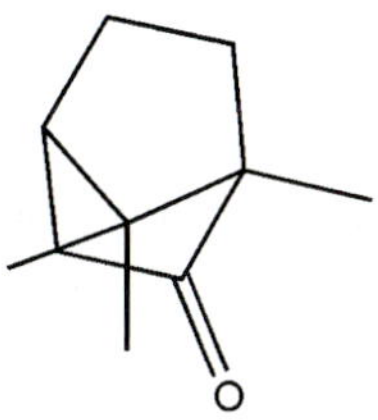

Camphoric acid *n.* $C_8H_{14}(COOH)_2$. A dibasic-acid plasticizer for cellulose nitrate, derived from Camphor by oxidizing the =C=O and adjacent =CH2 groups and opening the right-side ring.

Can *n.* (1) A cylindrical container, about 3 feet high and 10–12 in. in diameter, that is used to collect sliver delivered by a card, drawing frame, etc. (2) *See drying cylinders.*

CAN *n.* Abbreviation for cellulose acetate nitrate.

Canada balsam *n.* Oleoresin, which exudes naturally from *Pinus balsamea*, the Canadian balsam pine. It is essentially a resin dissolved in an essential oil. When freshly exuded it is a viscous liquid, but it hardens on exposure. Its chief use is for cementing lenses and other glass objects, because its refractive index, 1.53, is near that for glass. It is also used in the manufacture of fine lacquers.

Canada turpentine *See Canada balsam.*

Candela \kan-ˈdē-lə, -ˈde-, -ˈdā-; ˈkan-də-lə\ [L, candle] (1949) (cd) *n.* One of the six basic units of the SI system, the unit of luminous intensity, defined as the luminous intensity normal to the surface of 1/60 of a square centimeter of a black body at the temperature of freezing platinum (1772°C) under a pressure of 101.325 kPa. In older literature, the same abbreviation has been used for both candlepower and candle, which have different values.

Candelilla wax *n.* Natural wax obtained from the stems of a plant, which grows in Mexico and Texas. It varies in color from pale yellow-to-yellow or greenish-brown. It is one of the harder waxes, although not so hard as carnauba wax. Used in varnishes, electric insulation compositions, waterproofing and insect proofing, and paint removers.

Candle (new unit) *n.* 1/60 of the intensity of 1 cm^2 of a blackbody radiator at the temperature of solidification of platinum (2042 K).

Candle (unit of luminous intensity) *n.* Candle power is a measure of intensity of a source of light as compared with a standard candle.

See foot-candle.

Candlenut oil *n.* Oil derived from the nuts of the candlenut tree (*Aleurites moluccana*) of tropical Asia and Polynesia.

C

Candlewick fabric *n.* An unbleached muslin base fabric used to produce a chenille-like fabric by applying candlewick (heavy-plied yarn) loops and cutting the loops to give a fuzzy effect.

Cangle filter *n.* A small filter interposed between the spinning pump and spinning jet to effect final filtration of the spinning solution prior to extrusion.

Cangle water temperature *n.* The temperature of the water surrounding the candle filter or within the heating jacket during fiber extrusion.

Cannon–Fenske viscometer *n.* Capillary viscometer (calibrated) used for measuring relative flow time for liquids, useful for determining relative, specific and inherent viscosities of polymer solutions and extrapolation of intrinsic viscosities. It is convenient to clean and calibrate, but lacks correction due to different volume levels leading to weight due to gravity in the viscometer (*See image*).

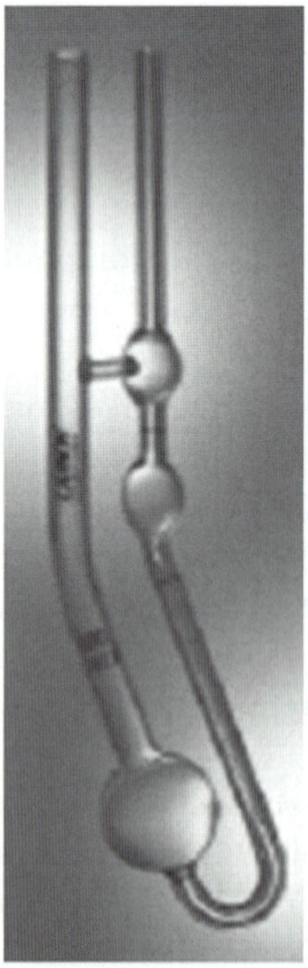

Canopy ceiling *n.* A decoration composed of a ceiling paper, or a sidewall paper such as a strip, in which the strips are cut in triangles and hung so that the apex will terminate in the center of the ceiling to produce a striking domed effect.

Can stability *See shelf life and storage stability.*

Cantilever-beam stiffness *n.* A method of determining stiffness of plastics by measuring the force and angle of bend of a cantilever beam made of the specimen material. The ASTM test is *D747*.
See also flexural modulus.

Canton flannel *n.* A heavy cotton or cotton blend material with a twilled face and a napped back. The fabric's strength, warmth, and absorbance make it ideal for interlinings and sleeping garments.

Canvas \ˈkan-vəs\ [ME *canevas*, fr. ONF. fr. (assumed VL *cannabaceus*, fr. L *cannabis*] (13c) *n, v, vt.* A coarse cloth made from cotton, hemp, or flax. It may be used for artist's canvas or a picture painted on canvas. *See duck.*

Canvas board *n.* A paper board with primed canvas fastened to one face.

Caoutchouc \ˈkaú-ˌchúk, -ˌchúk, -ˌchű\ [F, fr. obs. Sp *cauchuc* (now *caucho*), fr. Quechua *kawchu*] (1775) *n.* An early name for pure natural (raw) rubbers, still in use in the French literature.

CAP *See cellulose acetate propionate.*

Capacitance \kə-ˈpa-sə-tən(t)s\ [capacity] (1893) (C, electric capacity) *n.* The property of a system of two conducting surfaces (plates or foils, typically) separated by a non-conductor (*dielectric*) that permits storage of electric charge in proportion to the voltage difference between the conductors. The SI unit is the farad (F). A capacitor storing one coulomb of charge at a potential difference of 1°V has a capacitance of 1°F. Plastics are much used as capacitor dielectrics. Capacitance is measured by the charge, which must be communicated to a body to raise its potential one unit. Electrostatic unit capacitance is that which requires one electrostatic unit of charge to raise the potential one electrostatic unit. The farad = 0 × 1011 electrostatic units. A capacitance of one farad requires one coulomb of electricity to raise its potential 1°V. Dimensions $[\varepsilon\,\mathrm{L}]$, $[\mu^{-1}\mathrm{L}^{-1}\mathrm{T}^{2}]$. A conductor charged with a quantity Q to a potential V has a capacitance,

$$C = \frac{Q}{V}.$$

Capacitance of a spherical conductor of radius, r, is $C = Kr$. Capacitance of two concentric spheres of radii r and r' is

$$C = K\frac{rr'}{r - r'}.$$

Capacitance of a parallel plate condenser, the area of whose plates is A and the distance between them d,

$$C = \frac{KA}{4\pi d}.$$

Capacitances will be given in electrostatic units if the dimensions of condensers are substituted in cm. K is the dielectric constant of the medium.

Capillarity \\ˌka-pə-ˈlar-ə-tē\\ (1830) *n.* Force of attraction between like and dissimilar substances. It is exhibited, for example, by the rise of a liquid up a capillary tube, or by the wetting of solids by liquids.

Capillary action *n.* A phenomenon associated with surface tension and contact angle. Examples are the rise of liquids in capillary tubes and the action of blotting paper and wicks.

Capillary constant *n. Also known as specific cohesion,*

$$a^2 = \frac{2T}{(d_1 - d_2)q} = hr,$$

where T is the surface tension, d_1 and d_2, the densities of the two fluids, g the acceleration due to gravity, h is the height of the rise in a capillary tube of radius r.
See surface tension.

Capillary viscometers Seed for concentrated solutions or polymer melts and described just above, the other, described here, used

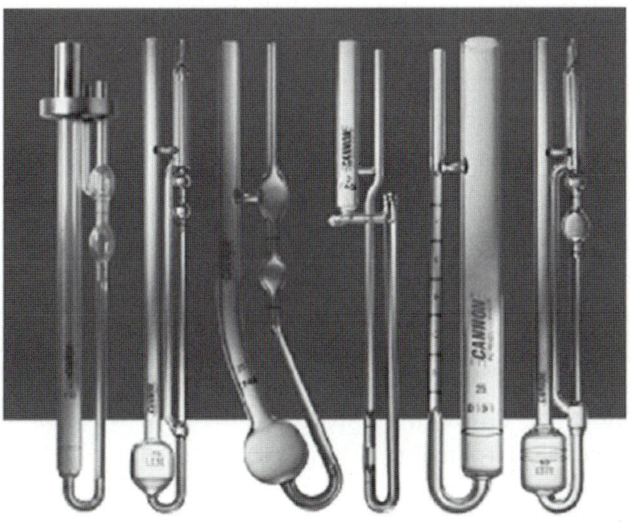

for measuring dilute-solution viscosities. The most widely used of the latter types employ a glass capillary tube and means for timing the flow of a measured volume of the solution (e.g., polymer in solvent) through the tube under the force of gravity. This time is then compared with the time taken for the same volume of pure solvent, or of another liquid of known viscosity, to flow through the same capillary. Relevant tests are from ASTM (www.astm.org).

See also dilute-solution viscosity and viscometer.

Capric acid \\ˈka-prik-\\ (decanoic acid, decylic acid, decoic acid) *n.* $CH_3(CH_2)_8COOH$. A plasticizer and an intermediate for resins (*See image*).

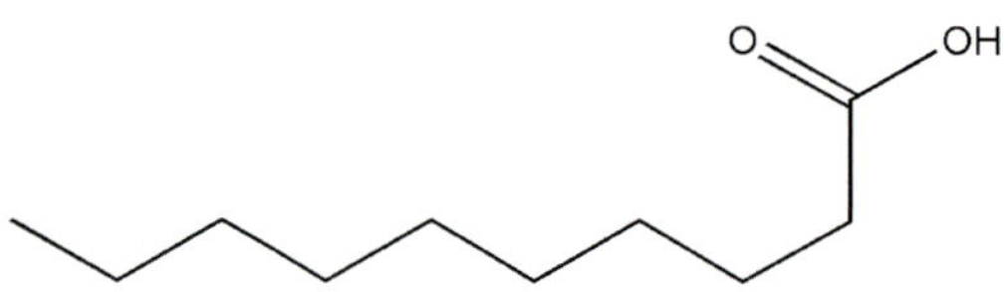

Caprolactam \\ˌka-prō-ˈlak-ˌtam\\ [*capro*ic acid + *lact*one + *am*ide] (1944) (ε-caprolactam, hexanoic acid-ε-amino lactam) *n.* (1) A cyclic amid having the structure shown below. When the ring is opened, caprolactam is polymerizable to a nylon resin known as nylon 6 or *polycaprolactam.* It is also used as a cross-linking agent for polyurethanes, and as a plasticizer. In the late 1960s it was found that caprolactam could be rotationally cast by heating the solid monomer to its melting point (ca 71°C) and introducing the molten monomer into a mold along with a catalyst, then heating and rotating the mold in the usual manner. The liquid gradually thickens and gels against the mold in the manner of a plastisol, and conversion to nylon 6 is accomplished in a few minutes. (2) A white, hygroscopic, crystalline solid or leaflets (*See image*).

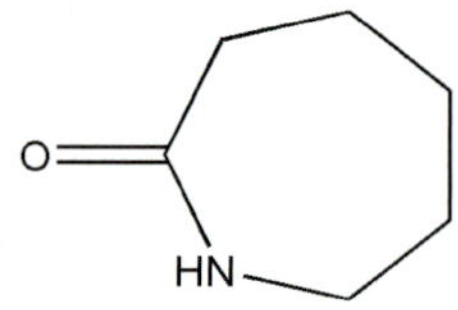

Cap spinning *n.* A system of spinning employing a stationary, highly polished metal cap just large enough to fit over the take-up bobbin, which revolves at a high rate of speed. The cap controls the build and imparts sufficient tension to the yarn for winding. The yarn is twisted and wound onto packages simultaneously.

Caprylic acid \\kə-ˈpri-lik\\ [ISV *capryl*] (1845) (octanoic acid, octoic acid, octylic acid, caprilic acid) *n.* $CH_3(CH_2)_6COOH$. A plasticizer and organic intermediate (*See image*).

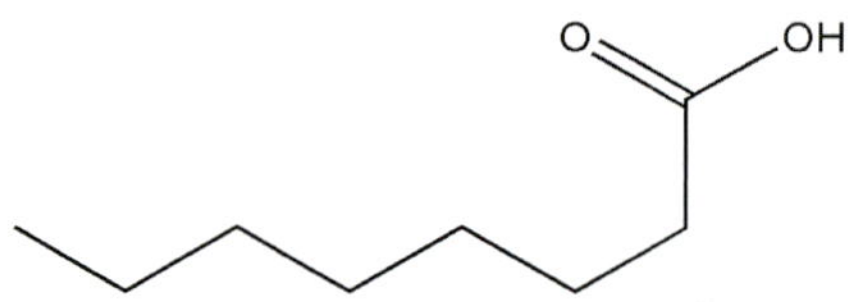

Caproic acid \\kə-ˈprō-ik-\\ *n.* $C_6H_{12}O_2$. Hexanoic acid. Natural unsaturated fatty acid found in oils or made synthetically and used in pharmaceuticals and flavors (*See image*).

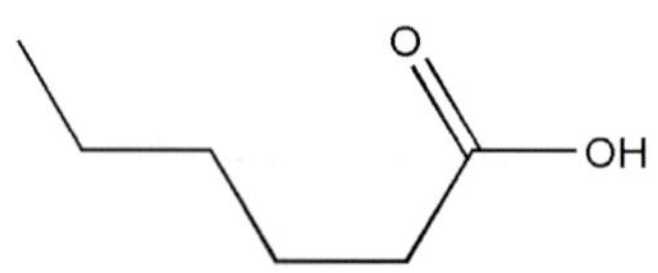

Capstan \\ˈkap-stən, -ˌstan\\ [ME, prob. fr. MF *cabestant*] (14c) *n.* In the plastics industry, a drum or pulley that controls the speed of a filament, wire, or web between production stages.

Captive production *n.* Production of materials or components by a manufacturer for its own use or for later incorporation in its products. Compare custom molder.

CAR Carbon fiber.

Carbamide \ˈkär-bə-ˌmīd, kär-ˈba-məd\ [ISV *carb-* + *amide*] (1865) *n. See urea.*

Carbamide phosphoric acid (urea phosphoric acid) *n.* $CO(NH_2)_2{\cdot}H_3PO_4$. A catalyst for acid-setting resins.

Carbamyl urea *See biuret.*

Carbanion \kär-ˈba-ˌnī-ən, -ˌnī-ˌän\ (1933) *n.* An organic ion carrying a negative charge (−) on a carbon atom.

Carbazole \ˈkär-bə-ˌzōl\ [ISV *carb-* + *az-* + *-ole*] (1887) (dibenzopyrrole, diphenylenimine) *n.* Pigment violet 23 (51319). Extracted from coal tar, unsaturated compound. (*See image*)

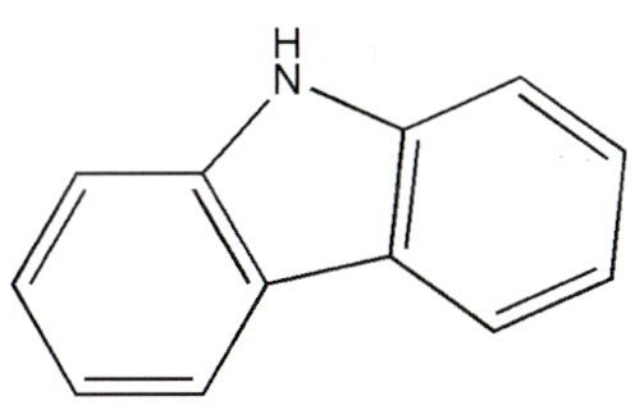

These pigments are manufactured in a variety of shades from red-shade to blue-shade violets. The latter appear more popular for toning (reddening) phthalocyanine blues. These pigments have excellent fastness properties even at high degrees of dilution but exhibit poorer color retention in metallized (aluminum flake) finishes than in light TiO_2 pastels. The are produced in both toner and lake (alumina hydrate) forms. They are used in the manufacture of poly(*N*-vinylcarbazole).

Carbenes *n.* Constituents of bituminous type products, which are insoluble in the chlorinated hydrocarbons, carbon tetrachloride and chloroform, but soluble in carbon disulfide.

Carbenium ions *n.* Stable carbenium ions, e.g., triphenyl-methyl and tropylium salts, and their merit is that can initiate the polymerization of certain olefins by direct addition.

Carbinol \ˈkär-bə-ˌnól, -ˌnōl\ [ISV, fr. obs. Gr *Karbin* methyl, fr. Gr *karb-*] (ca. 1885) *n.* (1) $-CH_2OH$. Monovalent primary alcohol radical. It may be part of an aliphatic or an aromatic alcohol. For example, ethyl alcohol (C_2H_5OH) can be described as methyl carbinol, or benzyl alcohol ($C_6H_5CH_2OH$) as phenyl carbinol. (2) It is sometimes used as a Syn: methanol.

Carbitol [2,(2-ethoxyethoxy)ethanol, ethyl cellosolve] $C_2H_5OCH_2{-}CH_2OCH_2CH_2OH$ and also written as $C_2H_5O(CH_2)_2O{-}(CH_2)_2OH$. This is a trade name of an active, water-soluble solvent, the monoethyl ether of diethylene glycol. Properties: bp, 195°C; sp gr, 0.990/20°C; flp, 96°C (205°F); vp, les than 0.1 mmHg/20°C. If the ethyl (C_2H_5) group is substituted by methyl (CH_3), butyl (C_4H_9) groups, etc., the corresponding methyl and butyl carbitols are obtained.

Carbodiimide polymers *n.* Self-addition of isocyanates with the aid of organic phosphine and arsine oxides as catalysts form yields of polymers referred to as *carbodiimides*; example, $nRN{=}C{=}NR$.

Carbohydrate \-ˌdrāt, -drət\ (ca. 1869) *n.* An aldehyde or ketone which is a polyol, or a polymer of these.

Carbohydrates *n.* Compounds containing carbon and the elements of water, and thus their only elements are carbon, hydrogen, and oxygen. They are all variations of the same basic formula: $C_m(H_2O)_n$. The carbohydrates include the mono-, di-, tri, tetra-, and poly-saccharides, as well as the conjugated saccharides.

Carbolic acid \kär-ˈbä-lik\ [ISV *carb-* + L *oleum* oil] (ca. 1859) *n.* A Syn: phenol.

Carbon \ˈkär-bən\ [F *carbone*, fr. L *carbon-*, *carbo* ember, charcoal] (1789) *n.* (1) A non-metallic tetravalent element, atomic no. 6, atomic wt. 12.011; the major bioelement. It has two natural isotopes, ^{12}C and ^{13}C (the former, set at 12.00000, being the standard for all molecular weights), and two artificial, radioactive isotopes of interest, ^{11}C and ^{14}C. The element occurs in three pure forms (diamond, graphite, and in the fullerines), in amorphous form (in charcoal, coke, and soot), and in the atmosphere as CO_2. Its compounds are found in all living tissues, and the study of its vast number of compounds constitutes most of organic chemistry. (2) A non-metallic chiefly tetravalent element found native (as in the diamond and graphite) or as a constituent of coal, petroleum, and asphalt, of limestone and other carbonates, atomic number 6, atomic weight 12.011 (longest living isotope). Smith MB, March J (2001) Advanced organic chemistry, 5th edn. John Wiley and Sons, New York.

Carbon 14 (radiocarbon) *n.* Radioactive carbon of mass number 14, naturally occurring bur usually made by irradiating calcium nitrate. It has been used as a β-ray source in gages for measuring the thickness of plastics films. Smith MB, March J (2001) Advanced organic chemistry, 5th edn. John Wiley and Sons, New York.

Carbon black (ca. 1889) *n.* Pigment black 7 (77266) finely divided carbon formed by any one of the following processes: (1) incomplete combustion of natural gas in burners under moving channel irons (channel carbon black). (2) Incomplete combustion of natural gas and petroleum in large, closed furnaces (furnace carbon black) (3) decomposition of gasses in large converters filled with hot refractory brick checker work. These carbon blacks vary in particle size and some of them may be surface treated. A generic term for the family of colloidal carbons. More specifically, carbon black is made by the partial combustion and/or thermal cracking of natural gas, oil. Or another hydrocarbon. Acetylene black is the carbon black derived from burning acetylene. Animal black is derived from bones of animals. Channel blacks are made by impinging gas flames against steel plates or channel irons (hence the name), from which the deposit is scraped at internals. Furnace black is the term sometimes applied to carbon blacks made in a refractory-lined furnace. *Lamp black*, the properties of which are markedly different from other carbon blacks, is made by burning heavy oils or other carbonaceous materials in closed systems equipped with settling chambers for collecting the soot. *Thermal black* is produced by passing natural gas through a heated brick checker work where it thermally cracks to form a relatively coarse carbon black. Carbon blacks are widely used as fillers and pigments in PVC, phenolics, polyolefins, and several other resins, also imparting resistance to ultraviolet rays. In polyethylene, carbon black acts as a cross-linking agent, in rubbers, as a reinforcement. Donnet J-B, Wang M-J (1993) Carbon black. Marcel Dekker, New York. Pierson HO (1994) Handbook of carbon, graphite, diamond and fullenes. Noyes Data Corporation/Noyes Corporation, New York.

Carbon blacks and graphite *n.* Carbon black is the amorphous form and graphite is the crystalline form of carbon. Pierson HO (1994) Handbook of carbon, graphite, diamond and fullenes. Noyes Data Corporation/Noyes Corporation, New York. Kirk–Othmer encyclopedia of chemical technology: pigments-powders. John Wiley and Sons, New York, 1996. Pierson HO (1994) Handbook of carbon, graphite, diamond and fullenes. Noyes Data Corporation/Noyes Corporation, New York.

Carbon dating (1951) *n.* Carbon dating is a variety of radioactive dating which is applicable only to matter that was one living and presumed to be in equilibrium with the atmosphere, taking carbon dioxide from the air for photosynthesis. Cosmic ray protons bombard nuclei n the upper atmosphere, producing neutrons that in turn bombard nitrogen, the radioactive isotope carbon-14 (^{14}C). The ^{14}C isotope combines with oxygen to form carbon dioxide and is incorporated into the cycle of living things. The ^{14}C isotope forms as at a rate that is constant, so that by measuring the radioactive emissions from once-living matter and comparing its activity with the equilibrium level of living things, a measurement of the time elapsed can be made (i.e., the activity of a sample can be directly compared to the equilibrium activity of living matter and the age calculated). The half-life of ^{14}C is about 5730 years by the emission of an electron of energy 0.016 MeV. This changes the number of the nucleus to seven, producing a nucleus of ^{14}N. The low activity limit of the ^{14}C limits age determinations to the order of 50,000 years by counting techniques, and extended to perhaps 1,000,000 years by accelerator techniques. Fossil fuels have no ^{14}C content, and atmospheric testing of nuclear weapons in the 1950s and 1960s increased the ^{14}C content of the atmosphere. Lowe JJ (1997) Radiocarbon dating. John Wiley and Sons, New York. Higham T, Ramsey B, Owen C (2004) Radiocarbon and archaeology. Oxford University School of Archaeology, UK. Levin HL (2005) The earth through time. John Wiley and Sons, New York.

Carbon dioxide (1873) *n.* CO_2. A gas commonly found in the atmosphere. In a solid state (below freezing temperature) it is called dry ice. Levin HL (2005) The earth through time. John Wiley and Sons, New York.

Also known as carbonic acid gas.

Carbon disulfide (1869) *n.* CS_2. A colorless flammable poisonous liquid used as a solvent for rubber and as an insect fumigant. *Also called carbon bisulfide.*

Carbon fiber (1960) *n.* (1) A high-tensile fiber or whisker made by heating rayon or polyacrylonitrile fibers or petroleum residues to appropriate temperatures. Fibers may be 7–8 μm in diameter and are more that 90% carbonized. (2) Any fiber consisting mainly of elemental carbon and increasingly used in reinforced-plastics products. They may be prepared by growing single crystals in a carbon electric arc under high-pressure inert gas; by growth from a vapor state via pyrolysis of a hydrocarbon gas; or by pyrolysis of organic fibers, the most widely used method. Polyacrylonitrile and rayon fibers are most commonly used as starting materials. The terms "carbon fibers" and "graphite fibers" are used somewhat interchangeably. However, PAN-based carbon fibers are 93–95% C by elemental analysis, whereas graphic fibers are usually 99 + % C. The difference is due mainly to the temperature of formation. 1315°C for fibers formed from PAN, while the high-modulus graphite fibers are graphitized at 3450°C. The higher the graphite content, the higher the elastic modulus but the lower the strength. Properties hold to very high temperatures in inert atmospheres (Properties transverse to the fiber length are much lower than along the length). In recent years, carbon fibers have become the leading reinforcement for high-performance composites. The less expensive carbon fibers produced from pitch have broadened the markets for these reinforcements. Strength and modulus range considerably depending on the supplier and grade, from 1.7 to 3.5 GPa (250–500 kpsi) for strength and from 230 to 830 GPa (34–120 Mpsi) for modulus. Chung DD (1994) Carbon fiber compositors. Elsevier Science and Technology Books, New York. Pierson HO (1994) Handbook of carbon, graphite, diamond and fullness. Noyes Data Corporation/ Noyes Corporation, New York.

Carbon 14 (1936) *n.* A heavy radioactive isotope of carbon of mass number 14 used in tracer studies and in dating archaeological and geological materials. Lowe JJ (1997) Radiocarbon dating. John Wiley and Sons, New York.

Carbon monoxide (1873) *n.* A colorless odorless very toxic gas CO that burns to carbon dioxide with a blue flame and is formed as a product of the incomplete combustion of carbon.

Carbon steel (1903) *n.* A strong hard steel that derives its physical properties from the presence of carbon.

Carbon tetrachloride (1866) (tetrachloromethane) *n.* CCl_4. A clear, dense pungent-smelling liquid similar to that of chloroform, a powerful solvent for many resins and miscible with most other organic solvents, and a starting compound in the synthesis of nylon 7. It has a sp gr of 1.629 and is non-flammable. Solvent action on oils and similar products and soft resins is good. It is frequently used to reduce fire hazards. Very toxic (no flash or fire point). Its use as a solvent has now been severely curtailed because of its high toxicity and known carcinogenicity. Syn: tetrachloromethane, perchloromethane.

Carbon 13 (1939) *n.* An isotope of carbon of mass number 13 that constitutes about 1/ 70 of natural carbon and is used as a tracer in spectroscopy utilizing nuclear magnetic resonance. Breitmaier E, Voelter W (1986) Carbon-13 NMR spectroscopy. John Wiley and Sons, New York.

Carbon 12 (1946) *n.* An isotope of carbon of mass number 12 that is the most abundant carbon isotope and is used as a standard for measurements of atomic weight. Goldberg DE (2003) Fundamentals of chemistry. McGraw-Hill Science/Engineering/Math, New York.

Carbonaceous \ˌkär-bə-ˈnā-shəs\ (1791) *adj.* Matter containing carbon.
See organic (1).

Carbon-arc lamp *n.* A type of fading lamp, which utilizes an arc between two carbon electrodes as the source of radiation.

Carbonate white lead *n.* $Pb_3O8C_2H_2$. Pigment white 1 (77597). Basic carbonate of lead, the composition of which approximates to the formula $Pb(OH)_2{\cdot}2PbCO_3$. It is the normal white lead of industry and is often given the above description to differentiate it from white lead sulfate, which is also used as a pigment. It is marketed in several grades designated according to the method of manufacture, e.g., stack, chamber, electrolytic or precipitated. The latter is further subdivided into high, medium, and low stain varieties. Syn: basic carbonate of lead, basic lead carbonate, cremnitz white, kremnitz white, and white lead. Gooch JW (1993) Lead based paint handbook. Plenum Press, New York.

Carbon–carbon composite *n.* A combination of carbon or graphite fibers in a carbon or graphite matrix, produced by impregnating a carbon- or graphite-fiber cloth or mat structure with a carbonizable binder such as pitch.

Carbon-fiber-reinforced plastics (CRP) *n.* Plastics, either thermosetting or thermoplastic – most commonly epoxies or high-performance resins – that contain carbon or graphite fibers. Chung DD (1994) Carbon fiber compositers. Elsevier Science and Technology Books, New York.

Carbonific *n.* Chemical compound which, upon decomposition, produces a mass of carbon which frequently occupies a volume much greater than the original unburned materials. Carbonifics function to produce the insulating, relatively incombustible properties of intumescent coatings.

Carbonium \kär-ˈbō-nē-əm-\ [*carb-* + *-onium*] (1942) *n.* A positively charged organic ion such as H_3C, having one less electron than a corresponding free radical and behaving chemically as if the positive charge were localized on the carbon atom.

Carbonium ions *n.* Positively charged organic molecules in which every atom has an octet of electrons which makes it more stable, e.g., $RCH{=}O^+R$.

Carbonium-ion polymerization *n.* Cationic $-C^+$ initiated polymerization reaction.

Carbonization *n.* Process of degrading organic matter to elemental carbon. The term is used chiefly in describing the decomposition of varnish films and the like, when subjected to elevated temperatures. Carbonization of films on stoving is always associated with pronounced darkening.

Carbonizing *n.* A chemical process for eliminating cellulosic material from wool or other animal fibers. The material is reacted with sulfuric acid or hydrogen chloride gas followed by heating. When the material is dry, the carbonized cellulose material is dust-like and can be removed.

Carbonyl (carbonyl group) \ˈkär-bə-ˌnil, -ˌnēl \ (1869) *n.* An organic functional group –$C{=}O$, found only in combination, as in aldehydes, ketones and organic acids.

Carbonyl addition–elimination *n.* The single most important type of reaction mechanism which has been applied to the preparation of step-growth polymers is the "addition–elimination reaction" of the carbonyl double bond of carboxylic acids and carboxylic acid derivatives; included in this general type of reaction are esterification amidation and anhydride formation from carboxylic acids, esters, amides, anhydrides and acid halides.

Carbonyl addition–substitution *n.* Step-growth polymerization schemes based on carbonyl addition-substitution reactions are almost entirely concerned with reaction of aldehydes; step-growth polymers prepared by this type of reaction include polyacetals, phenolformaldehyde polymers, ureaformaldehyde polymers, and melamineformaldehyde polymers.

Carbonyl group *n.* The bivalent radical, $-C{=}O$, especially in aldehydes or ketones.

Carbonyl value *n.* Keto group $-C{=}O$, is sometimes referred to as the carbonyl radical. Thus the amount of keto groups in a compound is its carbonyl value.

Carborane (dicarbadodecaborane) *n.* $C_2B_{10}H_{12}$. An icosahedral cage compound containing mostly boron, with two active hydrogens on the carbon atoms, and available in several isomers. It is polymerizable, but only the silicone copolymers have been manufactured.

Carborundum *See silicon carbide.*

Carbowax *n.* Poly(ethylene glycol), manufactured by Union Carbide, USA (now Dow Chemical).

Carboxy nitroso rubber (CNR) *n.* A fluorocarbon elastomer, synthesized as a terpolymer from tetrafluoroethylene trifluoronitrosomethane, and nitro-soperfluorobutyric acid. CNR has unique resistance to strong oxidizers and is nonflammable in pure oxygen, hence has found applications in the aerospace field. The gum can be processed on standard rubber-mixing equipment for molding, or dissolved for application by spraying, dipping or brushing.

Carboxyl end group *n.* The chain-terminating (–COOH) group found in polyamide and polyester polymers.

Carboxyl group *n.* The radical –COOH, characteristic of most organic acids.

Carboxylate ion *n.* The anion of a carboxylic acid.

Carboxylic *n.* Term for the –COOH group, the radical occurring in organic acids.

Carboxylic acid *n.* Organic acid which possesses one or more carboxyl groups (*See image*). The simplest member of the series is formic acid and has the general formula H·COOH.

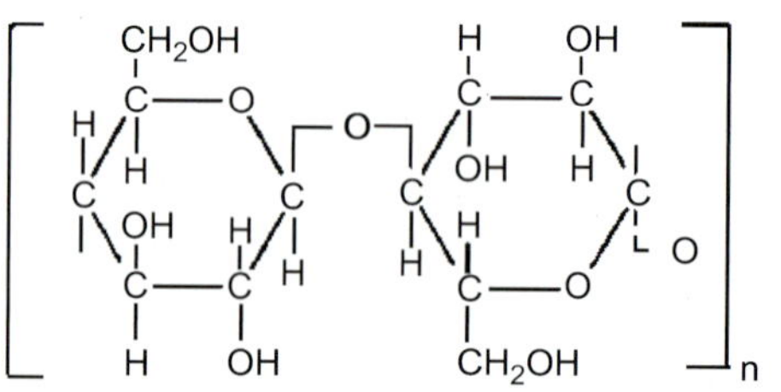

Carboxymethyl cellulose (CMC) \kär-ˌbäk-sē-ˌme-thəl, -ˈsel-yə-ˌlōs, -ˌlōz\ (1947) *n.* The common name for a cellulose ether of glycolic acid. It is an acid ether derivative of cellulose formed by the reaction of alkali cellulose with chloroacetic acid. It is usually marketed as a water-soluble sodium salt, more properly called sodium carboxymethyl cellulose. The sodium salt of this compound is commonly used as a stabilizer or an emulsifier. In the early literature, it is sometimes called cellulose glycolate or cellulose glycolic acid.

Carburizing *n.* A process for case-hardening steels in which the objects to be hardened are heated with carbonates and charcoal in the absence of air for bout 24 h at 840–950°C, then quenched in oil. The depth of hardening is about 1.7 mm and the surface hardness is from 50 to 55 on the Rockwell C scale.

Carcinogen \kär-ˈsi-nə-jən, ˈkär-sᵊn-ə-ˌjen\ (1853) *n.* Any material that has been tested and found to cause cancer in laboratory animals or that, through statistical studies, is correlated with the incidence of cancers in humans. An example from the plastics industry is vinyl chloride monomer, which is believed to have caused human liver cancer. PVC polymer, on the other hand, in non-carcinogenic. A list of known carcinogens is available from OSHA (USA Office of Health and Safety Administration). Merriam-Webster's collegiate dictionary, 11th edn. Merriam-Webster Inc., Springfield, MA, USA. Wickson EJ (ed) (1993) Handbook of polyvinyl chloride formulating. John Wiley and Sons Inc., New York.

Card choking *See cylinder loading.*

Card clothing *n.* The material used to cover the working surfaces of the card, i.e., cylinder and rolls or flats. The clothing consists of either wire teeth set in a foundation fabric or rubber, or narrow serrated metal flutes which are spirally arranged around the roll. The metallic wire has the appearance of band-saw blade.

Card conversion efficiency *n.* The efficiency of the carding process, expressed as a percentage obtained from ratio of sliver output to staple input.

Carded yarn *n.* A cotton yarn that has been carded but not combed. Carded yarns contain a wider range of fiber lengths and, as a result, are not as uniform or as strong as combed yarns. They are considerably cheaper and are used in medium and course counts.

Cardigan \ˈkär-di-gən\ [James Thomas Brudenell, 7th Earl of *Cardigan* † 1868 English soldier] (1868) *n.* (1) A modification of the rib-knitting stitch to allow tucking on one (half cardigan) or both (full cardigan) sets of needles. (2) A sweater that buttons down the front. Merriam-Webster's collegiate dictionary, 11th edn. Merriam-Webster Inc., Springfield, MA, USA.

Carding *n.* A process in the manufacture of spun yarns whereby the staple is opened, cleaned, aligned, and formed into a continuous, untwisted strand called a sliver.

Cardol *n.* One of the original constituents of cashew nutshell liquid, occurring to the extent of about 10%. It is a dihydroxy phenol, containing a side chain with two double bonds.

Care label *n.* The label that gives directions for cleaning, ironing, and otherwise maintaining a fabric of fiber product.

Carene *n.* Terpene hydrocarbon, which is a constituent of certain turpentines. Carene has been reported in German, Indian, Russian, and Finnish turpentines.

Cariflex *n.* Block copolymer of styrene/butadiene/styrene. Manufactured by Shell, The Netherlands.

Carlona *n.* Poly(ethylene), manufactured by Shell, The Netherlands.

Carlona pt *n.* Poly(propylene), manufactured by Shell, The Netherlands.

Carmine \ˈkär-mən, -ˌmīn\ [F *carmine*, fr. ML *carminium*, irreg. fr. Arabic *qirmiz* kermes + L *minium* cinnabar] (1712) *n.* Aluminum lake of a pigment from cochineal.

Carmine lake *n.* Natural Red 4 (75470). Barium lake of the dyestuff produced by coupling 7-amino-1-naphthalenesulfonic acid with R-sale (2-naphthol-3 : 6-disulfonic acid).

Carmine vermillion *See mercuric sulfide.*

Carnauba Wax \kär-ˈnó-bə, ˌkär-nə-ˈü-bə\ (1854) *n.* Extremely hard wax obtained from the leaves of a Brazilian tree. It appears on the market in several grades, such as bleached, yellow, fatty gray, and chalky gray. The wax has the following approximate constants: mp, 84°C, Sp gr, 0.998; acid value, 2; saponification value, 80; iodine value, 13. It is a constituent of wax polishes, and has some application in matt, overprinting, and baking varnishes. Its solubility is of a very low order in most varnish solvents and other constituents. Syn: Brazil wax.

Carnaubic acid *n.* $C_{23}H_{47}COOH$. One of the constituent acids for carnauba wax; monocarboxylic acid. Mp, 72°C.

Carnaubyl alcohol *n.* $C_{24}H_{49}OH$. Alcoholic constituent of carnauba wax. Mp, 69°C.

Carnot cycle *n.* A sequence of operations forming the working cycle of an ideal heat engine of maximum thermal efficiency. It consists of isothermal expansion, adiabatic expansion, isothermal compression, and adiabatic compression to the initial state.

Carotene (carotin) \ˈkar-ə-ˌtēn\ [ISV, fr. LL *carota* carrot] (1861) *n.* $C_{40}H_{56}$. Hydrocarbon found in linseed and other vegetable oils where it is believed to act as an anti-oxidant.

Carpet backing *n.* A primary backing through which the carpet tufts are inserted is always required for tufted carpets. The backing is usually made of woven jute or non-woven manufactured fiber fabrics. A secondary backing, again made of jute or manufactured fibers, is normally added at the latex back-coating stage. Carpet backings are an important end use for nonwoven fabrics.

Carpet underlay *n.* A separate fabric, which is used to provide cushioning for carpet. Carpet underlays are made of hair and jute, sponge rubber, bonded urethane, or foamed urethane.

Carpets *n.* Heavy functional and ornamental floor coverings consisting of pile yarns or fibers and a backing system. They may be tufted or woven.

Also see tufted carpet.

Carrageen \ˈkar-ə-ˌgēn\ [*Carragheen,* near Waterford, Irland] (1829) (sometimes spelled "Carrageenan" or "Carrageenin") *n.* Hydrocolloid obtained from a group of sea plants related to *Chondrus crispas,* colloquially called Irish moss or moss. It is a complex carbohydrate made up of galactose, dextrose, and levecose residue. Useful for mucilaginous and gelatizing qualities.

Also known as Irish moss, Irish gum, pearl moss, pig wrack, and rock salt moss.

Carrier *n.* (1) A product added to a dye bath to promote the dyeing of hydrophobic manufactured fibers and characterized by affinity for, and ability to swell, the fiber. (2) A moving holder for a package of yarn used on a braiding machine. (3) A term sometimes used to describe the tube or bobbin on which yarn is wound. Tortora PG (ed) (1997) Fairchild's dictionary of textiles. Fairchild Books, New York.

Carrierless dyeing variants *n.* Polymers that have been modified to increase their dyeability. Fibers and fabrics made from these polymers can be dyed at the boil without the use of carriers.

Carter lead *n.* White lead manufactured in the USA. by what is known as the Carter process. This is somewhat similar to the chamber process except that the metallic lead is used in a finely divided state instead of sheets. The process is therefore quicker, but the resulting product is very similar to chamber white lead.

Cartoon *n.* (1) Preliminary sketch or detailed drawing for a painting, mural, etc., either to scale or actual size. The design is transferred from the paper to the working surface. (2) Term also used to describe a comic drawing or an animated film based on a succession of comic, grotesque drawings.

Cartridge heater *n.* A rod-shaped electrical heating element, consisting of a metal outer shell, sealed within which is a Nichrome-wire coil embedded in a thermally stable, electrically insulating powder, such as magnesium oxide. Cartridge heaters come in a wide range of physical sizes and wattages and are used in heating and controlling the temperatures of dies and molds, injection nozzles, hot stampers, etc.

Carvone *n.* Ketone derived from the terpene dipentine.

Cascade coating *n.* A process for applying epoxy and other thermosets to objects such as electrical resistors and capacitors, in which finely powdered resin is poured over the preheated object. The article is usually rotated as the powder is applied.

Cascade control (piggy-back control) *n.* In automatic control, a system in which the output of one unit is the input of the next, the goal being to obtain closer control of an important, final variable by controlling more sensitive, linked variables. A cascade system may be of the open-loop or closed-loop type.

Case hardening *n.* Any of several processes by which the working surfaces of steel tools and molds are hardened after being machined in their softer, original states. It can

also be described as the surface hardening without thorough drying of the film.
See carburizing, flame hardening, and nitriding.

Casein *n.* (1) A protein usually obtained from milk. Used to make sizing, adhesive solutions, and coatings. Used as a binder in aqueous dispersions of pigments. (2) The protein substance occurring in milk and cheese. It can be obtained by treating skim milk with a dilute acid, but the type used mainly for plastics (rennet casein or paracasein) is made by treating warm skim milk with a rennet extract. Skeist I (ed) (1977) Handbook of adhesives. Reinhold Publishing Co., New York. Dainth J (2004) Dictionary of chemistry. Oxford University Press, New York.
See also casein plastic.

Casein finish *See casein paint and distemper.*

Casein paint *n.* Coatings in which casein replaces the ordinary drying oils, or is used as an emulsifying agent in emulsion paints. Both types may be thinned with water.

Casein plastic *n.* A family of thermosetting plastics derived from Casein, used widely in the early years of the plastics industry but less important now. Casein plastics have poor water resistance and dimensional stability, which limits their applications.

Caseinates *n.* Metallic salts derived from casein.

Casement A window sash that opens on hinges at the vertical edge.
See casement window.

Casement cloth *n.* A general term applied to lightweight, sheer fabrics used for curtains and for screening purposes and as a backing for heavy drapery fabrics of the decorative type. This type of fabric is sometimes made in small fancy weaves for dresswear.

Casement window *n.* Window hinged at the side to swing in or out.

Cashew net shell oil *n.* Natural oil from the shells of the nuts of the species *Anacardium occidentale*, containing a high proportion of cardol – a substituted phenol. By reaction of this oil with formaldehyde, a resin is produced which gives films with good chemical resistance.

Cashew nut shell liquid resin *n.* Resin derived from the liquid obtained from the shells of the nuts of the species, *Anacardium occidentale*, which grows chiefly in India and South America. The naturally occurring liquid is a mixture of a dihydroxy phenol cardol and anacardic acid. This acid readily decarboxylates on heating to yield a monohydroxy phenol with an unsaturated side chain, anacardol.

Cashmere *n.* The extremely soft hair of the Cashmere goat. Cashmere is often blended with sheep's wool in fabrics.

Casing *n.* A term coined by Bell Telephone Laboratories, an acronym for the process of Cross-linking by Activated Species of inert gases developed to impart printability and adhesive receptivity to polymers such as PTFE and polyethylene. In this process, articles are exposed to a flow activated inert gases in a glow-discharge tube, forming a shell of highly cross-linked molecules having high surface energy on the article surfaces.

Casing knife *n.* In paper handing, a knife used to trim wallpaper around casings, at moldings, baseboards, etc.

Cassel brown *See Vandyke brown.*

Cassel earth *See Vandyke brown.*

Cassel yellow *n.* $PbCl_2 \cdot 7PbO$. Lead oxychloride pigment.

Casson equation *n.* Rheology expression used to relate share rate, viscosity, and at infinite shear rate for dispesrsions (pigment coatings, etc.). Patton TC (1979) Paint flow and pigment dispersion: a rheological approach to coating and ink technology. John Wiley and Sons, New York.

C

Cassone painting *n.* Pictures used to decorate the sides of an Italian marriage chest (cassone). The most common subjects were scriptural, chivalrous, mythological, and heraldic.

Cast *n.* (1) A tinge; a subjective term used in combination with a directional indicator to describe a slight deviation from the norm in color or appearance. The term is most often applied to color, and refers to a small difference in hue. For example, four pure hues are generally recognized psychologically, red, green, blue, and yellow. Each is unique and contains no quality of the others. If a blue hue seems to contain some element of green that blue is said to have a green "cast". (2) The term may also be applied to appearance aspects other than color. For example, a material which is normally expected to be transparent (non-scattering) but which is slightly cloudy (due to a small amount of light scattering) is said to have a "milky cast" or a "cloudy cast". McDonald R (1997) Colour physics for industry, 2nd edn. Society of Dyers and Colourists, West Yorkshire, England. Skeist I (ed) (1977) Handbook of adhesives. Reinhold Publishing Co., New York.
See casting.

Cast coating *See coating.*

Cast embossing *n.* A process of casting films against an embossed temporary carrier. Vinyl plastisols, organosols, solutions, or lattices are used as film formers, which may be backed up with layers of foam or fabric. The temporary carrier is often paper, embossed with the desired pattern and treated so as to be easily stripped from the fused film or laminate.

Cast film *n.* Film produced by pouring or spreading a solution, hot melt, or dispersion of plastic material onto a temporary carrier – typically a polished metal roll, hardening the material by suitable means, and stripping the solidified film from the surface. Cellulosic, polystyrene, and vinyl films are often produced in this manner.

Cast film extruders *n.* Equipment used for preparing thin polymer films by use of one or two free-rolling casting and polishing rollers.

Cast film extrusion *See chill roll extrusion.*

Castanha de cotia kernel oil *n.* A drying oil from a tree indigenous to Brazil. It is believed to contain a substantial proportion of eleostearic acid, or acid of similar type. Its rate of polymerization is much slower than that of tung oil. Its iodine value is 154.

Cast-coated paper *n.* A paper or board, the coating of which is allowed to harden or set while in contact with a finished casting surface. Cast-coated papers have a high gloss.

Caster oil (dehydrated) *n.* Castor oil from which some chemically combined water has been removed to improve its drying properties.

Cast-in heater *n.* A type of heater used on cylindrical surfaces, such as extruded barrels, in which a rod-type heating element is bent to a semi-cylindrical shape and cast within an aluminum channel shape whose inner surface has the cylinder's radius. Pairs of such cast-in elements are then strapped tightly to the barrel with Monel bands and high-strength bolts, leaving the heater terminals exposed for electrical connections. Copper tubing may also be cast into the same elements, permitting circulation of water or, better, an non-volatile heat-transfer liquid, for barrel cooling.

Cast-in lining *See bimetallic cylinder.*

Casting *n.* Manufacturing process dating back to at least 4000 BC. for inexpensively producing large or complex parts. The process of forming solid or hollow articles from fluid plastic mixtures or resins by pouring or injecting the fluid into a mold or against a substrate with little or no pressure, followed by solidification and removal of the formed object. The finished product of a casting operation.

See also cast embossing, centrifugal-, film-, slush-, solid-, rotational-, and solvent casting; embedding, encapsulation, potting, and drawdown (2, *n*).

Casting plaster *See gypsum and calcined.*

Casting resin *n.* Resin which can be cast and hardened in a mold to form a shaped article.

Casting syrup (casting resin) *n.* Liquid monomers or partially polymerized polymers, usually containing catalysts or curing agents, capable of polymerizing to the solid state after they have been cast in molds. The materials most generally used as the acrylics, styrenes, polyesters, epoxies, silicones, and nylons.

Also called potting syrups when used for encapsulating articles such as electrical components or assemblies.

Castor oil (ricinus oil) *n.* A pale-yellowish oil derived from the seeds of the castor bean, *Ricinus communis*, and consisting essentially or ricinolein. Its principal characteristics are light color, relatively high specific gravity and viscosity, and its solubility is alcohol. It differs from other oils in that its composition is mostly hydroxy fatty acids. It is essentially a non-drying oil, but it may be converted to a drying oil by "chemical dehydration" by which a hydroxy group and an adjacent hydrogen atom are removed as water to form a drying oil fatty acid ester with two double bonds, one of them being conjugated. This dehydration yields what commonly is known as dehydrated castor oil. In its original undehydrated form, castor oil is well known for its use in resins and as a plasticizer for cellulose ester lacquers. It is an important starting material for plasticizers, certain nylons, and alkyd resins; and an ingredient in certain urethane forms. Merck Index, 13th edn. Merck and Company Inc., Whitehouse Station, NJ, 2001. Paint: pigment, drying oils, polymers, resins, naval stores, Cellulosics Esters, and ink vehicles, vol 3. American Society for Testing and Material, Philadelphia, PA, 2001.

CAT *n.* (1) Abbreviation for the ISCC color-matching amplitude test. This test was devised by the subcommittee for Problem 10 of the Inter-Society Color Council as a measure of discrimination ability for small differences in saturation. It is not a test for defective color vision. Results depend on experience as well as on inborn ability. It is sold by the Federation of Societies for Coatings Technology, Philadelphia, PA, USA. (2) Abbreviation for catalyst.

Cat eyes *n.* Undissolved globules of rubber in a cement generally made from only rubber and solvent. Expression also used for PVC sheet and paint films. etc.

Also called fish eyes. See pinhole.

Catalyst \ˈka-tᵊl-əst\ (1902) *n.* A substance that causes or accelerates a chemical reaction when added to the reactants in a minor amount, and that is not consumed in the reaction. A negative catalyst (inhibitor, retarder) decreases the rate of reaction or prevents it altogether.

See also hardener, inhibitor, acid catalysts, accelerator, curing agent, and initiator.

Catalysts and promoters *n.* Substances whose presence increases the rate of a chemical reaction. They are added in a

C

small quantity as compared to the amounts of primary reactants, and do not become a component part of the chain; they are referred to as an initiator. In some cases the catalyst functions by being consumed and regenerated; in other cases the catalyst seems not to enter the reaction and functions by virtue of surface characteristics of some kind. They are added to doped solvents to produce polymerization at room temperature or a temperature blow the softening point of the thermoplastic. A negative catalyst (inhibitor, retarder) slows down a chemical reaction.

Catalytic agent *n.* A substance, which by its mere presence alters the velocity of a reaction, and may be recovered unaltered in nature or amount at the end of the reaction.

Catalytic curing *n.* Mechanism by which a coating is cross-linked by the action of a catalyst as opposed to oxidation, etc. Examples of such systems are two-part (pot) and epoxies and polyurethanes.

Cataphoresis \ˌka-tə-fə-ˈrē-səs\ [NL] (1889) *n. See electrophoresis.*

Catch-up In lithography, the printing of non-image areas of a plate and is overcome generally by increasing the amount of fountain solution applied to it. It may occur as a run is started before the dampening adjustment is correctly set.

Caterpillar (caterpillar puller) *n.* A device used downstream of the extruder in extrusion of pipe and profiles, consisting of two driven and counter-rotating belts, having an elongated oval shape about 0.8 m long, with pads attached to the outsides of the belts. One of the belts is elevatable to adjust the clearance between them so as to firmly grip the extrudate being pulled away from the cooling tank, yet not so strongly as to deform it. Belt speed is adjustable over a wide range to accommodate different rates of extrusion.

Cathode \ˈka-ˌthōd\ [Gr *kathodos* way down, fr. *kata-* + *hodos* way] (1834) *n.* (1) In an electrolytic cell through which current is being forced by an external emf, the cathode is the negative electrode, giving up electrons to cations in the electrolyte. In a cell or battery *delivering current*, the cathode is the positive terminal. Also used in connection with cathodic protection. (2) In an electro-chemical cell, the electrode at which reduction occurs. (3) In a vacuum tube the cathode is the electrode from which electrons are emitted. The negatively charged electrode in a gas-discharge tube. *See anode.*

Cathode ray (1880) *n.* The stream of electrons emanating from the cathode in a discharge tube.

Cathode ray tube (1905) *n.* A vacuum tube in which a beam of electrons is projected on a fluorescent screen to produce a luminous spot at a point on the screen determined by the effect of the electron beam of a variable magnetic field within the tube.

Cathode sputtering *See vacuum metallizing.*

Cathodic corrosion *n.* Corrosion due to the development of alkalinity by a reaction at a cathode – sometimes experienced with aluminum and lead. Alkalinity from this source may also affect paints vulnerable to alkali attack.

See cell and electrolytic.

Cathodic polarization *n.* Portion of the reduction in the initial potential of a corrosion cell that occurs at the cathode.

Cathodic protection *n.* Reduction or elimination of corrosion of a metal achieved by making current flow to it from a solution by connecting it to the negative pole of some source of current. The source of the protective current may be a sacrificial metal, such as zinc, magnesium, or aluminum. The current may also be derived from a rectifier, generator, or battery applied through an appropriate anode which may be consumed by the applied current, as in the case of steel, or remain substantially unaffected by the current, as in the case of graphite or platinum. Uhlig HH (2000) Corrosion and corrosion control. John Wiley and Sons Inc., New York.

Cation \ˈkat-ˌī-ən\ [Gk *kation*, neut., prp. of *katienai* to go down cap, fr. *kata*-cata- + *ienai* to go] (1834) *n.* An atom, molecule or radical, usually in aqueous solution, that has lost an electron and has become positively charged. It can also be described as a positively charged atom or radical, which moves to the negative electrode or cathode during electrolysis. Metallic ions, such as iron and copper, etc., are cations.

Cationic \ˈkat-(ˌ)ī-ˈä-nik\ (ca. 1920) *adj.* Pertaining to any positively charged atom, radical, or molecule; or to any compound or mixture containing positively charged groups.

Cation-exchange resin *See ion-exchange resin.*

Cationic detergent *n.* A detergent that produces positively charged colloidal ions in solution.

Cationic dyeable variants *n.* Polymers modified chemically to make them receptive to cationic dyes.

Cationic dyes *See dyes, basic dyes.*

Cationic polymerization *n.* Process in which the active end of the growing polymer molecule is a positive ion. If the ion is a carbonium ion, it is referred to as carbonium ion polymerization. a polymerization reaction with a positive or cationic initiator, i.e., a cationic initiator.
See ionic polymerization.

Cationic surfactant *n.* Surfactant which gives a positively charged ion in aqueous solution.

Cauchy's dispersion formula *n*,

$$n = A + \frac{B}{\lambda^2} + \frac{C}{\lambda_4} + \cdots$$

An empirical expression giving an approximate relation between the refractive index n of a medium and the wavelength λ of the light; A, B, and C being constants for a given medium.

Caul *n.* A sheet of metal, wood, or other material used in laminating to apply and equalize pressure.
See platen.

Caulk (calk) *n.* To fill voids with plaster or semiplastic materials; to fill crevices in the adherend surface with adhesive materials; to provide a seal against moisture or solvent intrusion.

Caulking *See caulking compound.*

Caulking cartridge *n.* An expendable container made of plastic, fiberboard or metal, filled with caulking compound, for use in a caulking gun. A common type is 2 in. (5 cm) in diameter, approximately 8 in. (20 cm) long, and fitted with a plastic nozzle.

Caulking compound *n.* A soft, plastic, putty-like material, consisting of pigment and vehicle, used for sealing joints in buildings and other structures where normal structural movement may occur, or for preventing leakage. Caulking compound retains its plasticity for an extended period after application. It is usually available in two consistencies: "gun grade", for use with a caulking gun, and "knife grade", for application with a putty knife; extruded preformed shapes are also available.

Caulking gun *n.* A device for applying caulking compound by extrusion. In a hand gun, the required pressure is supplied mechanically by hand; in a pressure gun, the pressure required usually is greater and is supplied pneumatically.

C

Cause-and-effect diagram (fishbone diagram) *n.* A graphical way of analyzing a process, based on ideas and experiences of workers and engineers concerning the materials, machines, and methods of the process, in order to identify possible causes of product defects.

Caustic *n.* A strong chemical base.

Caustic potash (KOH) *n.* Archaic name for potassium hydroxide. A strong base.

Caustic soda (NaOH) *n.* Archaic name for sodium hydroxide. A strong base.

Cave painting *n.* The art produced in the form of paintings on the walls of caves from the beginning of the Old Stone Age to the end of the New Stone Age, around 3000 BC. The entire period covers a span of some 200,000 years. The Paleolithic cave paints at Altamira (Spain) and those at Lascaux (France) are believed to date between 40,000 and 10,000 BC.
See prehistoric art.

Cavity \ˈka-və-tē\ [MF *cavité*, fr. LL *cavitas*, fr. L *cavus*] (1541) *n.* A depression, or sometimes the set of matching or associated depressions, in a plastics molds that forms the outer surfaces of the cast or molded article(s). The cavity may surround a CORE, the portion of the mold that forms the inner surfaces of a hollow article.

Cavity side (British) *n.* The side of an injection mold that is adjacent to the nozzle.

Cavity-retainer plate *n.* A plate in a mold that holds the cavities and forces. Such plates are at the mold parting line and usually contain the guide pins and bushings.
Also called force-retainer plate.

Cavity-side part (USA) *n.* The stationary part of an injection mold.

Cavity-transfer mixer *n.* A two-piece device installed at the end of an extruder screw to accomplish both distributive and dispersive mixing. The *stator* is a barrel extension into whose inside surface is machined an array of many hemispherical cavities. The *rotor* is a screw extension whose exterior is similarly contoured. The lands of rotor and stator have the usual close clearance of screw and barrel. As the melt steam passes through, it is smeared between the lands and is repeatedly cut into small globs and recombined, passing from rotor to stator, stator to rotor, until it emerges.

CBA *n.* In the plastics-foam industry, abbreviation for chemical blowing agent.

cd *n.* SI abbreviation for Candela.

Cd *n.* Chemical symbol for the element cadmium.

CDP *n.* Abbreviation for cresyl diphenyl phosphate.

Cedar *n.* A durable softwood generally noted for decay resistance; includes Western red cedar, incense cedar, and eastern red cedar.

Cedar nut oil *n.* *Pinus cembra*, from the seeds of which this oil is derived, grows prolifically in several parts of the world. The main constituent acids are linoleic, linolenic, and oleic acids, and the oil possesses useful drying properties. Iodine values up to 160 have been reported.

Ceiling *n.* The overhead surface of a room, usually a covering or decorative treatment used to conceal the floor above or the roof.

Ceiling temperature *n.* The temperature above which polymerization will not occur, symbolized T_c.

Celanese acetate *See acetate fiber.*

Celcon *n.* Poly(formaldehyde) (from trioxane with some ethylene oxide). Manufactured by Celanese, USA.

Celestial blue pigment *n.* Iron blue precipitated on barytes. Usually only a few percent of blue is used.

Celestite *n.* Mineral, which is chiefly strontium sulfate. When ground, it has some application as a filler, being used as an alternative to barytes. The ground mineral is also known as *strontium white.*

Cell *n.* (1) A small etched depression in a gravure cylinder that carries the ink. (2) In the cellular-plastics industry, a single void produced by a blowing agent, by mechanically entrained gas, or by the evaporation of a volatile constituent. When the void is completely surrounded by polymer, the cell is said to e *closed.* A *completely open* cell has no wall membranes but is part of a three-dimensional network of connected fibers or rods.

Cell collapse *n.* A defeat in foamed plastics characterized by slumping and cratered surfaces, with the internal cells resembling a stack of leaflets when viewed in cross section under a microscope. The condition is caused by tearing of the cell walls, weakened by plasticization or other mechanism.

Cell, electroylic *n.* Source of electrical current that is responsible for corrosion, consisting of an anode and a cathode immersed in an electrolyte and electrically bonded together. The anode and cathode may be separate metals or dissimilar areas on the same metal. They will develop a difference in potential that causes current to flow and corrosion at the anode when the electrodes are in electrical contact with each other.

Cellidor *n.* Thermoplast based on Cellit. Manufactured by Bayer, Germany.

Cellit Cellulose acetate or acetobutyrate. Manufactured by Bayer, Germany.

Celloidin (celluidine, photoxylin) *n.* A form of cellulose nitrate made by precipitation from an ether-alcohol solution of collodion cotton.

See cellulose nitrate and collodion.

Cellon *n.* Cellulose acetate. Manufactured by Dynamit Nobel, Germany.

Cellophane *n.* (1) Regenerated cellulose film, chemically similar to rayon, made by mixing cellulose xanthate with dilute sodium hydroxide solution to form a viscose, then extruding the viscose into an acid bath for regeneration. Cellophane is coated on one or both sides to render it moisture proof and capable of being sealed with heat or solvent (The term *rayon* is used when the regenerated material is in fibrous form). Cellophane is widely used for packaging, most often with coatings of other polymers to overcome its tendency to absorb moisture and to improve the film's heat-sealability. Trade name for viscose film. (2) Hydrate cellulose from pump. Manufactured by Kalle, Germany.

Cellosolve *n.* $C_2H_5OCH_2CH_2OH$. Proprietary name for the monoethyl ether of ethylene glycol, ethoxyethanol. It is also the generic name for a comprehensive series of ethers or similar type such as methyl cellosolve, butyl cellosolve, etc. A relatively slow drying flexographic ink solvent frequently used as a retarder. Cellosolve is water miscible. Properties: bp, 135°C; Sp gr, 0.931/20°C; flp, 42°C (107°F); vp, 4 mmHg/20°C.

See ethylene glycol monoethyl ether.

Cellosolve® acetate *n.* $CH_3COOCH_2CH_2OCH_2CH_3$. Medium high boiling solvent (ethylene glycol monoethyl ether acetate). Properties: bp, 156°C (313°F); Sp gr, 0.975/20°C; flp, about 53°C (128°F); and vp, 1.5 mmHg/20°C.

Cellular mortar *See syntactic foam.*

Cellular plastic (expanded plastic, foamed plastic) *n.* A plastic with numerous cells of

gas distributed throughout its mass. The terms *cellular-*, *expanded-*, and *foamed plastic* are used synonymously. A cellular plastic may be produced by (1) incorporating a blowing agent that decomposes to liberate a gas; (2) mechanically whipping in a gas or vaporizable liquid; (3) by adding a water-soluble salt or a solvent-extractable agent to the mix prior to forming, then leaching out the agent after forming to leave voids; or (4) other techniques described under epoxy foam, phenolic foam, polystyrene foam, syntactic foam, urethane foam. Cellular plastics range in density from some slightly less than that of the parent resin to less than 0.01 g/cm^3. The cells may be open or closed, depending on the process and density.
See also structural foam.

Cellular striation *n.* In a cellular plastic, a layer of cells differing in size or nature from the majority of cells in the same mass.

Cellular vinyls *n.* Vinyls containing occulted gas in bubbles or cells. Cellular vinyls are used to form polymers with very low densities, down to below 0.1 g cm^{-3}, due to the high gas volume fraction, and hence can have exceptionally low thermal conductivities.

Cellulase *n.* Enzyme that attacks and breaks down cellulosic substrates such as wood, cellulosic thickeners in paint, etc.

Celluloid *n.* (1) An old trade name, now generic, for cellulose nitrate compounded with camphor and ethanol. The ethanol is removed after processing by heating, leaving behind the camphor, which toughens the compound. Originally the trade name and now the common name of a synthetic plastic made by mixing pyroxylin, or cellulose nitrate, with pigments and fillers in a solution of camphor in alcohol. (2) Cellulose nitrate, plasticized with camphor. Manufactured by Dynamit Nobel, Germany.

Celsius *See temperature, celsius.*

Cellulose \\ˈsel-yə-ˌlōs\\ [F, fr. *Cellule* living cell. Fr. NL *cellula*] (1848) *n.* $(C_6H_{10}O_5)_n$. (1) A polysaccharide $(C_6H_{10}O_5)_x$ of glucose units that constitutes he chief part of the cell walls of plants, occurs naturally in such fibrous products as cotton and kapok, and is the raw material of many manufactured good as paper and rayon. (2) A natural carbohydrate polymer of high molecular weight, having the structure shown below. Cellulose is a constituent of most higher plants (*spermatophyta*). Cotton is the purest natural form, containing about 90%; it occurs to a small extent in the animal kingdom. Chemically, cellulose is 1–4 glucan of high degree of polymerization. It is desirable to apply "cellulose" to this material only and to designate the predominantly cellulosic residue obtained by subjecting woody tissues to various pumping processes as "cellulosic residues", "cellulosic pumps", or the like. Cotton linters and wood pulp are the major sources of cellulose for cellulosic plastics (*See image*).

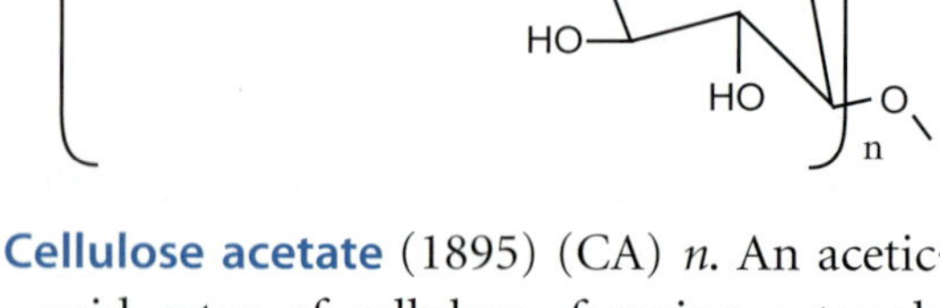

Cellulose acetate (1895) (CA) *n.* An acetic-acid ester of cellulose, forming a tough, transparent thermoplastic material when compounded with plasticizers. It is obtained by the action, under rigidly controlled conditions, of acetic acid and acetic anhydride on purified cellulose, usually obtained from cotton linters. All three available hydroxyl groups in each glucose unit of the cellulose can be acetylated, but in the material normally used for plastics it is usual to acetylate fully, then to lower the acetyl value by partial hydrolysis, leaving, on average, 2.4 acetate groups per C_6 unit. Cellulose acetate compounds are used when toughness, permanence, flame resistance, and transparency are required at moderate cost. However, they absorb up to 2.5% of atmospheric moisture, making them unsuitable for long-term outdoor exposure. (2) A clear thermoplastic material, usually in film form, made from cellulose and acetic acid. (3) Cellulose triacetate is derived from a reaction of acetic anhydride, acetic acid and little sulfuric acid; partial hydrolysis removes some of the acetate groups and degrades the chain to 200–300 repeat units and yields cellulose acetate (roughly a diacetate) which is the commercial product.

Cellulose acetate butyrate (CAB) *n.* A mixed ester produced by treating fibrous cellulose with butyric and acetic acids and anhydrides in the presence of sulfuric acid. CAB is generally supplied in the form of pellets prepared by mixing the molten ester with a plasticizer, then extruding and palletizing. It is one of the toughest of the cellulosic plastics, and has good transparency, colorability, weatherability, electrical properties and resistance to inorganic chemicals. It can be processed by extrusion, injection molding, blow molding, rotational molding, and thermoforming. Applications include pipe, tool handles, instrument housing, lighting, packaging film, and marine hardware.

Cellulose acetate propionate (CAP, cellulose propionate) *n.* A thermoplastic formed by treating fibrous cellulose with propionic and acetic acids and anhydrides in the presence of sulfuric acid. CAP is easily extruded and injection molded, forming tough, flexible products with shock resistance close to that of ethyl cellulose. In properties and applications it resembles the acetate butyrate rather than the straight acetate.

Cellulose acetobutyrate *See cellulose acetate butyrate.*

Cellulose esters *n.* Any derivative of cellulose in which the free hydroxyl groups attached to the cellulose chain have been replaced wholly or in part by acidic groups, e.g., nitrate, acetate, propionate, butyrate, or stearate groups. Esterification is effected by the use of a mixture of an acid with its anhydride in the presence of a catalyst such as sulfuric acid. Mixed esters of cellulose, e.g., cellulose acetate butyrate, are prepared by using mixed acids and mixed anhydrides.

Cellulose ethers *n.* A cellulose derivative based on the etherification products of cellulose, such as ethyl cellulose, methyl cellulose, and sodium carboxymethyl cellulose. *See cellulose thickeners.*

Cellulose ethers *n.* Derivatives of cellulose in which one or more of the hydroxyl hydrogens have been replaced by alkyl groups.

Cellulose fiber *n.* The fibrous material remaining after non-fibrous components of wood have been removed by the pulping and bleaching operations. Used in making paper, etc.

Cellulose glycolate *See carboxymethyl cellulose.*

Cellulose gum *See carboxymethyl cellulose.*

Cellulose lacquer *See lacquer.*

Cellulose nitrate (1880) (CN, nitrocellulose, NC, pyroxylin) *n.* Cellulose nitrate, dating back to the work of French chemist

C

Braconnet in 1833, is the oldest of the synthetic plastics. It is made by treating fibrous cellulose with a mixture of nitric and sulfuric acids, and was first used in the form of a lacquer (*See collodion*). In 1870, John Wesley Hyatt and his brother patented the use of plasticized cellulose nitrate as a solid, moldable material, the first commercial thermoplastic (celluloid). Camphor was the first (and is still the best) plasticizer for CN, although many camphor substitutes have been developed. Alcohol is normally used as a volatile solvent to assist in plasticization, after which it is removed. Molded products of CN are extremely tough, but highly flammable and subject to discoloration in sunlight. CN is amendable to many decorative variations. Its principal uses today are in knife handles, table-tennis balls, and eyeglass frames. A mixture of nitric and sulfuric acids converts cellulose into cellulose nitrate; pyroxylin is a less nitrated material and it has been useful for photographic film, collodion, and celluloid plastics.

Cellulose plastics *n.* Plastics based on derivatives of cellulose, such as esters (cellulose acetate) and ethers (ethyl cellulose).

Cellulose propionate *n.* Ester of cellulose and propionic acid.
See cellulose acetate propionate.

Cellulose thickeners *n.* Aqueous thickeners based on cellulose compounds such as carboxymethyl cellulose, hydroxyethyl cellulose, and methyl cellulose.

Cellulose triacetate *n.* A member of the cellulosic plastics family made by reacting purified cellulose with acetic anhydride in the presence of a catalyst in such a manner that at least 92% of the hydroxyl groups are replaced by acetyl groups. Because of its high softening point this material cannot be molded or extruded. Its major use is for casting films or spinning fibers from solutions, such as in a mixture of methylene chloride and methanol.

Cellulosic fiber *n.* A fiber composed of, or derived from, cellulose. Examples are cotton (cellulose), rayon (regenerated cellulose), acetate (cellulose acetate), and triacetate (cellulose triacetate).

Cellulosic plastic (cellulosic resin) *n.* Any of a family of thermoplastics made by substituting various chemical groups fore the hydroxy groups in the cellulose molecules of cotton and purified wood pulp.
See the following: cellulose acetate, cellulose acetate butyrate, cellulose acetate propionate, cellulose esters, cellulose nitrate, cellulose triacetate, ethyl cellulose, hydroxyethyl cellulose, and regenerated cellulose.

Cellulosics, resins *n.* Resins and polymers derived from cellulose $[C_6H_{10}O_5]_n$.

Celsius \ˈsel-sē-əs\ *adj.* One hundredth of the thermometer scale, divided into 100 degrees, in which 0°C is the freezing point of water, and 100°C is the boiling point at 1.013 bar. The preferred name according to the International System of Units (SI) is degree of Celsius (°C).

Cement *n.* A material or a mixture of materials (without aggregate) which, when in a plastic state, possesses adhesive and cohesive properties and hardens in place. Frequently, the term is used incorrectly for concrete, e.g., a "cement" block for concrete block (*n*) *See adhesive.* (*v*) *See bond.*

Cement-asbestos board *n.* A dense, rigid, non-combustible board containing a high proportion of asbestos fibers, which are bonded with Portland cement; highly resistant to weathering.
Also called asbestos-cement board.

Cement coating *See cement paint.*

Cement colorants *n.* Term used to describe colors with sufficient tinctorial strength and alkali resistance to be suitable for coloring Portland cement or concrete. The natural earth colors, synthetic iron oxide colors, chromium oxide, ultramarine, and some of the organic pigments are used for this purpose.

Cementing *n.* Joining plastics to themselves of dissimilar materials by means of solvents (dopes, or chemical cements. *Dope adhesives See solvent cementing*), comprise a solvent solution of a plastic similar to the plastic to be joined. *Chemical cements*, the only type suitable for thermosetting plastics, are based on monomers or semi-polymers or semi-polymers that polymerize in the joint to form a strong bond.
See adhesive.

Cementitious *n.* Having cementing properties.

Cement paint *n.* Paint supplied in dry powder form, based essentially on Portland cement, to which pigments are sometimes added for decorative purposes. This dry powder paint is mixed with water immediately before use.

Cement plaster *n.* (1) Plaster with Portland cement as the binder; sand and lime are added on job. Used for exterior work or in wet or high humidity areas. (2) In some regions, Gypsum Plaster.

Cement, Portland *See Portland cement.*

Cement, rubber *n.* An adhesive that is a dispersion or solution of raw or compounded rubber, or both, in a suitable liquid.

Cenobrium *See mercuric sulfide.*

Cenospheres *n.* Hollow microspheres in fly ash formed during combustion of coal in electric-power plants. They have had some use as a lower-cost substitute for glass microspheres in syntactic foams.

Centered cell *n.* A unit cell, which has entities (atoms, molecules, and ions) at locations in addition to the ell corners. A non-primitive cell.

Center-gated mold *n.* In injection molding, a mold in which each cavity is fed through an orifice at the center of the cavity. This type of gating is employed for items such as cups and bowls.

Centering mark *See clip mark.*

Center loop *See kink.*

Center of interest *n.* In a room, the principal focal point, architecturally speaking, such as, for example, a fireplace. In a wallcovering design, the dominant motif, usually hung at eye level in the central area of a room to establish a starting point for handing the rest of the wallcovering.

Centi-(c) the SI-approved prefix signifying multiplication by 10^{-2}.

Centigrade *See celsius.*

Centimeter \ˈsen-tə-ˌmē-tər\ (Brit. *centimetre*) *n.* A measure of length equal to a hundredth part of a meter (Brit. *metre*), or 0.3937 in.; abbreviated cm; an inch equals 2.54 cm.

Centipoise \-ˌpȯiz\ (cP) *n.* A deprecated, but still widely used viscosity unit, 0.01 poise. Water at 20°C has a viscosity of 1.002 cP. The SI equivalent is: 1 cP = 0.001 Pa s. Liquids of low viscosity are usually given in centipoise units.

Centistoke (cs) *n.* (1) A deprecated, but still used unit of kinematic viscosity, 0.01 Stoke, the approximate kinematic viscosity of water at 20°C. The SI equivalent is: $1\ \text{cs} = 10^{-6}\ \text{m}^2/\text{s}$. (2) One one-hundredth of a stoke, which is the unit of kinematic viscosity:

$$\nu = \frac{\eta}{\rho},$$

where η is the viscosity ρ is the density. Sine SI units $1\ \text{cSt} = 1\ \text{mm}^2/\text{s}$.

C

Central stop *n.* The opaque stop usually placed in the objective back focal plane to give central stop dispersion staining.

Centrifugal casting *n.* A method of forming plastic in which the dry or liquid plastic is placed in a rotatable container. It is heated to a molten condition by the transfer of hear through the walls of the container, and rotated such that the centrifugal force induced will force the molten plastic to conform to the configuration of the interior surface of the container.

Centrifugal clarifier *n.* Apparatus somewhat similar to a cream separator, used for clarifying clear and colored solutions by throwing out solid particles by means of centrifugal force. Clarifiers of this type are operated at high speed. The bulky and apparently undissolved particles are thrown aside into a collector.

Centrifugal coasting *n.* The process of forming tubes or other hollow cylindrical objects by introducing a measured amount of fluid resin dispersion into a rotatable container or mold, rotating the mold about the cylinder's axis at a speed high enough to force the fluid against all parts of the mold by centrifugal force, maintaining such rotation while solidifying the plastic by applicable means such as heating, then cooling if necessary, and removing the formed part. The fluid resin may be a dispersion such as a plastisol, or an A-stage thermoset with or without reinforcing strands.; This process should not be confused with rotational casting, which involves rotation at low speeds about one or more axes of rotation and gravity flow. Centrifugal casting is also used with metals, in particular, to manufacture bimetallic cylinders.

See also centrifugal molding.

Centrifugal impact mixer *n.* A device used for continuously mixing free-flowing dry blends, comprising a conical hopper in which are rotated at high speeds a rotor disk and a peripheral impactor. The material is fed to the center of the rotor, which throws it against the impactor blades, which in turn throws the material against fixed impactors at the extremities of the cone. From there, the material flows downward to a discharge orifice. Compare: high-intensity mixer.

Centrifugal molding *n.* A process similar to centrifugal casting except that the materials employed are dry, sinterable powders such as polyethylene. The powders are fused by heating the mold then solidified by cooling it.

Centrifugal pot *See pot spinning.*

Centrifugation *n.* A method for determining the distribution of molecular weights by spinning a solution of the specimen at a speed such that the molecules are not removed from the solvent but are held at a point where the centrifugal force tending to remove them is balanced by the dispersive forces caused by the thermal agitation.

Centrifuge *n.* Machine which exploits centrifugal force as a means of removing solid or semisolid particles from liquids.

Centripetal force *n.* The force required to keep a moving mass in a circular path. Centrifugal force is the name given to the reaction against centripetal force.

Cera alba *n.* Bleached or white beeswax.

Cera flava *n.* Yellow or unbleached beeswax.

Ceramic \sə-ˈra-mik, *esp Brit.* kə-\ [Gr *keramikos*, fr. *keramos* potter's clay, pottery] (1850) *adj.* Technology of producing fired clay and porcelain articles, their glazes, pigments, and modifiers.

Ceramic fiber *n.* A term embracing all reinforcing fibers made of refractory oxides such as Al_2O_3, BeO, MgO, $MgO{\cdot}Al_2O_3$, ThO_2, and ZrO_2. Although glasses are also ceramic materials, glass fibers are not generally included. Ceramic fibers are produced by chemical vapor deposition, melt drawing, spinning, and extrusion. Their main advantages are high strength and modulus, and resistance to high temperatures.

Ceramic fiber reinforcements *n.* Nonmetallic inorganic fibrous materials, available in a wide spectrum of forms, both continuous and discontinuous.

Ceramics *n.* A general term applied to the art or technique of producing products by a ceramic process.

Ceraplast *n.* Any reinforced thermoplastic, particularly polyethylene, containing ceramic or mineral particles that have been dispersed in the polymer melt to their ultimate size (no agglomerates) and completely enveloped in resin. Bonding of the envelope to the filler particles and the matrix polymer is aided by the addition of a small percentage of reactive monomer or resin precursor. It is believed that, in the extremely thin transition envelope, there is a smooth gradient of modulus from that of the particulate material to that of the polymer. The mechanical properties of ceraplasts are superior to those in which the same fillers have been conventionally incorporated.

Ceresine *n.* Non-crystalline wax made by refining ozokerite. Mp, 60–71°C (140–160°F).

Cerium naphthenate *n.* Rare earth drier for air drying and baking finishes; sometimes used to replace lead naphthenate.

Cermet \ˈsər-ˌmet\ [*cer*amic + *met*al] (1948) *n.* (1) Composite materials consisting of two components, one being either an oxide, carbide, boride or similar inorganic compound and the other a metallic binder. (2) Any refractory composition made by bonding grains of ceramics, metal carbides, nitrides, etc, with a metal. Co-deposition of cermets with nickel in the electroless-nickel process provides excellent wear resistance and chemical resistance to molds, dies, extruder screws and other tooling components used in the plastics industry.

Cerotic acid *n.* A common name for either heptacosanoic acid or hexacosanoic acid. The former has the formula: $C_{26}H_{53}$ COOH, and a mp of 82°C; the latter, the formula: $C_{25}H_{51}COOH$, and a mp of 88°C. Cerotic acid is a constituent of natural waxes, in which it occurs as the cerotate esters. Neocerotic acid, formula, $C_{24}H_{49}$ COOH, mp of 77.8°C, is known as a constituent acid of beeswax (*See image*).

Cerulean blue \sə-ˈrü-lē-ən\ *n.* Very complex pigment consisting essentially of a combination of cobalt and tin oxides. It is bluish-green in color and very stable.

Cerussite \sə-ˈrə-ˌsīt\ [Gr *Zerussit,* fr L *cerussa*] (1850) *n.* $PbCO_3$. Natural lead carbonate, found in the upper zone of lead deposits. Colorless, white, and gray. Sp gr of 6.55. Effervesces in nitric acid.

Ceryl alcohol *n.* $C_{26}H_{53}OH$. Alcohol obtained from Chinese wax (*See image*).

Cetoleic acid *n.* $CH_3(CH_2)_9CH{=}CH(CH_2)_9$ COOH. One of the constituent fatty acids of many fish oils, It has a single double bond.

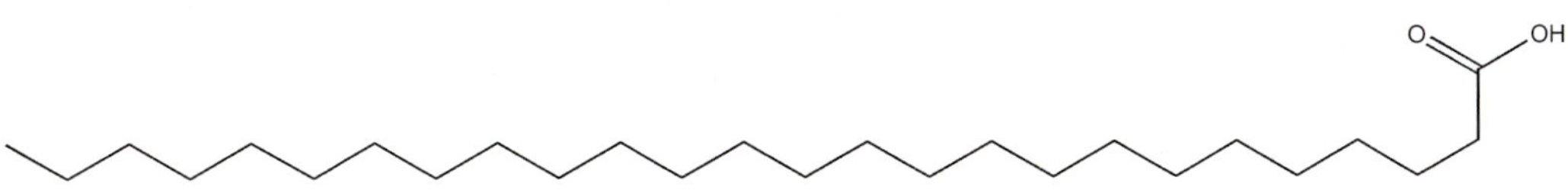

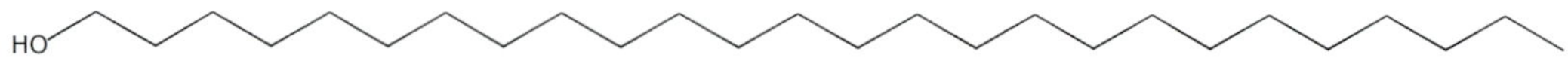

Cetraria *See Iceland moss.*

CF *n.* Cresol/formaldehyde resin.

CFK Man-made fiber-reinforced plastics.

CFRP *See carbon-fiber-reinforced plastics.*

C glass *n.* A type of glass fiber not quite as strong and stiff as e glass but having better chemical resistance. Its major constituents are SiO_2 65%, CaO 13%, Na_2O 8%, B_2O_3 5%, Al_2O_3 3%, and MgO 2%. Fiber density is 2.49 g/cm^3, tensile modulus is 71 GPa, and tensile strength is 3.2 GPa.

CGPM *n.* Abbreviation for Conférence Générate Des Poids Et Mesures, the international group that developed the system of weights and measures intended for worldwide use. The name Systéme International des Unités and the abbreviation SI were adopted by the 11th CGPM in 1960. For information on SI units *See SI* (2) *and the Appendix.*

cgs System Centimeter-gram-second system. Now replaced by SI units.

cgs-EM system of units System for measuring physical quantities in which the basic units are the centimeter, gram and second, and the numerical value of the magnetic constant, m, is unity. It is being replaced by the SI units.

Chafe mark *See abrasion mark.*

Chafed end *n.* A warp end that has been abraded during processing. It generally appears as a dull yarn often containing broken filaments.

Chaffer fabric *n.* A fabric, coated with unvulcanized rubber, that is wrapped around the bead section of the tire before vulcanization of the complete tire. The purpose of the chafer fabric is to maintain an abrasion-resistant later of rubber in contact with the wheel on which the tire is mounted.

Chai oil *n.* Mexican oil obtained from *Salvia hispanica.* It has excellent drying properties. A prominent characteristic of the oil is its high surface tension, which causes it to "crawl", cooking at 260°C (500°F) for a short time destroys this property. Its main constituent acids may make up more than 90% of the total acids. Properties: Sp gr, 0.9338/15°C; refractive index, 1.4855; iodine value, 196.3; saponification value 192.2.

Chain binders *n.* Yarns running in the warp direction on the back of a woven carpet, which hold construction yarns together.

Chain branching *n.* In a chain reaction, a step, which produces more chain carriers than it consumes.

Chain dyeing *See dyeing.*

Chain expansion factor *n.* The ratio of the length of a polymer chain in any given solvent to the length of the polymer chain in a theta solvent.

Chain extension *n.* Linear polymers (such as urethanes) of relatively low molecular weight may be extended by reaction of isocyanate endgroups with diamines to form interchain urea linkages, which, in turn add to other unreacted isocyanate endgroups to form biuret branching sites; reaction if isocyanate endgroups with diols to form new urethane linkages and extend the polymer chain, then, under more vigorous reaction conditions, by causing some still unreacted isocyanate endgroups to react with internal urethane linkages to form allophanate branching sites and cross-links; and by reaction of isocyanate endgroups with water to form urea linkages between linear chains, again followed by biuret cross-linking reactions.

Chain flexibility *n.* The ability of polymer molecules to assume a variety of configurations, arising from freedom of segments to rotate around C–C bonds. Polar side groups generally hinder rotation, making chains stiffer, while alkyl side chains tend to increase flexibility. Higher melting point materials are often characterized by relatively low chain flexibility (higher chains stiffness).

Chain, folded *n.* Chain folded is the conformation of a flexible polymer molecule when present in a crystal. The molecule exists and reenters the same crystal, frequently generating folds.

Chain isomerism *n.* Isomerism involving differences in skeletal chains of atoms in molecules.

Chain kinetics *n.* The study of rates and mechanisms of chemical reactions and of the factors on which they depend.

Chain length *n.* The number of monomeric or structural units in a linear polymer, or the main chain in a branched polymer. This can also be written as the total length of chain molecule, measured from atom to atom along the chain. Kinetic chain length *v.* gives the average number of monomer molecules that can be added to a free radical before the free radical is destroyed by a termination reaction. *Note*—This term should not be used for the direct distance between the ends of the molecule.

See also degree of polymerization.

Chain polymerization *n.* Polymerization processes are of two basic types: stepwise (or step reaction) and chain reaction. The kinetics of the two types of reaction are entirely different; the properties of the polymers they produce differ with respect to molecular weight distribution and usually, although not inevitably, differ in kind. The vast majority of chain reaction polymerize-tions are those generally known as addition polymerizations involving the conversion of vinyl-type monomers.

Chain reaction *n.* In general, any self-sustaining process, whether molecular or nuclear, the products of which are instrumental in, and directly contribute to the propagation of the process. Specifically, a *fission chain* reaction, where the energy liberated for particles produced (fission products) by the fusion of an atom cause the fusion of other atomic nuclei, which in turn propagate the fission reaction in the same manner. In other words, a reaction type characterized by the formation of products of a later step (chain carriers), which are reactants for an earlier step.

Chain scission *n.* A degradation mechanism (300–500°C) involving rupturing of bonds in the backboned of the polymer chain, while non-chain scission is concerned with all other bond-breaking reactions caused by pyrolysis.

Chain stiffness *See chain flexibility.*

Chain transfer *n.* Refers to the termination of a growing polymer chain and the start of a new one. The process is mediated by a chain transfer agent, which may be the monomer, initiator, solvent, polymer, or some species that has been added deliberately to affect chain transfer.

Chain-transfer agent *n.* A substance, used in polymerization that has the ability to stop the growth of a molecular chain by yielding an atom to the active radical at the end of the growing chain, but in turn being converted to another radical that *n* initiate the growth of a new chain. Examples are thiols and carbon tetrachloride. Such agents are useful for preventing the occurrence of too long chains and too high weight-average molecular weights.

C

Chair rail *n.* The topmost molding of a dado, placed on the wall at the height of a chair back as a protection.

Chalcogen \ˈkal-kə-jən\ [ISV *chalk-* bronze, ore (fr. Gk *chalkos*) + *-gen*; from the occurrence of oxygen and sulfur in many ores] (ca. 1961) *n.* Any of the elements oxytgen, sulfur, selenium and tellurium. A member of group VIA in the periodic table.

Chalk *n.* (1) A soft white mineral consisting essentially of calcium carbonate occurring as the remains of sea shells and minute marine organisms. (2) Soft naturally occurring form of calcium carbonate. The whiter varieties are powdered, washed, and dried. The product is known as whiting; the finer varieties as Paris white. (3) The product of chalking.
See chalking.

Chalkiness *n.* (1) A dull, whitened appearance sometimes associated with certain extra-dull colors. (2) A fillingwise fabric defect observed as bands varying luster or sheen.

Chalking *n.* (1) Formation of a friable powder on the surface of a paint film caused by the disintegration of the binding medium due to disruptive factors during weathering. The chalking of a paint film can be considerably affected by the choice and concentration of the pigment (BSI). It can also be affected by the choice of the binding medium. (2) A condition of a printing ink in which the pigment is not properly bound to the paper and can be easily rubbed off as a powder. (3) Formation of a powdery surface condition due to oxidation of the surface of rubber and release of pigments and fillers at the surface. Not to be confused with bloom, which looks similar.

Chalking resistance *n.* The ability of a coating to resist the formation of a friable powder on the surface of its film caused by the disintegration of the binding medium due to degradative weather factors. The chalking of a coating can be considerably affected by the choice and concentration of pigment and binding medium.
See chalking.

Chalk kauri *n.* Soft, white, powdery form of kauri copal.

Chalk masking *n.* The color change due to chalking and not due to a change in the colorant.

Chalk, precipitated *n.* Obtained as a by-product in water softening by sodium carbonate. It consists chiefly of calcium carbonate, but is often contaminated by appreciable quantities of calcium sulfate.

Challis \ˈsha-lē\ [prob. fr. the name *Challis*] (ca. 1837) *n.* A very soft, lightweight, plain-weave fabric, usually printed with a delicate floral pattern. The name is derived from the Anglo-Indian term "shalee" meaning soft.

Chamber white lead *n.* Particular type of white lead pigment, made by hanging strips or straps of lead in chambers, and subjecting them to the corrosive action of moist air, carbon dioxide, and acetic acid vapor. Chamber white lead differs from the stack form, usually being finer and of a better color.
See carter lead.

Chambray \ˈsham-ˌbrā, -brē\ [irreg. fr. *Cambrai*, France] (1814) *n.* (1) A plain woven-spun fabric, almost square (i.e., 80 × 76), with a colored warp and a white filling. Lightweight chambrays are used for shirts, dresses, and children's clothes. (2) A similar but heavier fabric of carded yam, used for work clothing.

Chameleon \kə-ˌmēl-yən\ [ME *camelion*, fr. MF, fr. L *chamaeleon*, fr. Gk *chamaileōn*, fr. *chamai* on the ground + *leōn* lion]

(14c) *n.* A variable multicolored effect achieved by using warp yarns of one color and two filling yarns of different colors in each shed. It is sometimes used in taffeta, faille, or poplin made from silk or manufactured filament yarns.

Change-can mixer (pony mixer) *n.* A type of planetary mixer comprising several paddle blades mounted on a vertical shaft rotating in one direction while the can or contained counter-rotates. The paddle shaft is usually mounted on a hinged structure so that it can be swung out of the can, permitting the can to b removed, emptied, and replaced easily. This type of mixer is employed for relatively small batches (12–480 L) of fluid dispersions and dry materials.

Change in filling *See mixed end or filling.*

Change in length on untwisting *n.* The increase or decrease in length measured when a specimen is untwisted. The change is expressed as the percentage extension or contraction of the nominal gauge length of the specimen, i.e., specimen length prior to untwisting.

Channel black *n.* Form of carbon black made from natural gas by the channel combustion process. The gas is burned with insufficient air in jets, and the flames are allowed to impinge on a cool, channeled metallic surface. The deposited carbon is then scraped from the channel after a certain period of burning. Because of air pollution control requirements, this type of black has been almost completely replaced by Furnace Black in the USA.

Channel depth (*h* or *H*) *n.* Of an extruder screw, at any point along its length, the radial distance between the flight-tip surface and the screw-root surface. In a screw section of constant depth, half the difference between the other (major) diameter of the screw and its root diameter.

Channel-depth ratio *n.* In an extruder screw, the ratio of the depth in the first turn of the screw at the feed end to the depth in the last turn at the delivery end. If the lead of the screw is constant, the channel–depth ratio is slightly larger than the channel–volume ratio (which follows).

Channel-volume ratio *n.* In an extruder screw, the ratio of the volume of the first turn of the screw at the feed end to the volume of the last turn at the delivery end. The term compression ratio is commonly used as a synonym in the extrusion industry.

Char (1, *n*) *n.* Animal or vegetable carbon black used as a decolorant in the process industries (2, *v*). To partly burn and blacken, especially the outside surface of a carbonaceous material.

Charcoal black *n.* Black pigment obtained from wood charcoal.

Charge (*n.*) The amount of material used to load a mold at one time or for one cycle. The amount may be expressed in either mass or volume units. In injection and transfer molding, the charge includes sprues and runners.

Charge-transfer *n.* Refers to the termination of a growing polymer chain and the start of a new one.

Char length *n.* In flammability testing, the distance from the edge of the sample exposed to the flame to the upper edge of the charred or void area.

Charles' law (Gay–Lussac's law) *n.* The temperature part of the ideal-gas law as follows: at constant pressure, the volume of any gas is directly proportional to its absolute temperature (*K*).

Charpy impact test *n.* A destructive test (ASTM D 256B) of impact resistance using a centrally notched test specimen 126 mm long and typically 12.7 mm square. The specimen is supported horizontally near its ends and struck on the side opposite the notch by a pendulum, having sufficient kinetic energy to break the specimen with one blow. The result is expressed as the quotient of the energy absorbed from the pendulum divided by the specimen width (J/cm or ft-lb/in).

Chase (shoe) *n.* An enclosure of any shape used to (1) Shrink-fit parts of a mold cavity in place. (2) Prevent spreading or distortion in hobbing, (3) Enclose an assembly of two or more parts of a split-cavity block and (4) A rectangular metal frame in which type and plates are locked up for letterpress printing.

Chaser *n.* Mill for dispensing pigments in binder to form very stiff pastes or putty. It consists of a circular pan in which one or two massive rollers run about central axes.

Chatki *n.* Stone mill for grinding shellac.

Chatter *n.* Transverse roll marks on roll coat-painted strip with varying film thicknesses, which are usually due to roll coater vibration or a non-concentric roller.

CHC *n.* Chlorohydrin copolymers (from epichlorohydrin and ethylene oxide).

Checking *n.* (1) That phenomenon manifested in paint films by slight breaks in the film that do not penetrate to the underlying surface. The break should be called a crack if the underlying surface is visible. Where precision is necessary in evaluating a paint film, checking may be described as visible (as seen with the naked eye) or as microscopic (as observed under a magnification of 10 diameters). (2) Equals crazing.

See cold-checking and checking resistance.

Checking resistance *n.* The ability of a coating to resist slight breaks in its film that do not penetrate to the underlying surface. The breaks should be called cracks if penetration extends through the underlying surface. Where precision is necessary in evaluating a coating film, checking may be described as visible (as seen with the naked eye) or as microscopic (as observed under minimum magnification of 10 diameters). *See cracking resistance. See checking.*

Checks *n.* (1) Rough surface due to fine cracks from weathering. (2) Roughness formed on calendered sheet when temperature of calender rolls is too low or when the sheet is chilled too suddenly.

Cheese *n.* (1) A supply of glass fiber wound into a cylindrical mass. (2) A cylindrical package of yarn wound on a flangeless tube.

Cheesecloth *n.* A low-count, plain weave, soft cotton, or cotton blend cloth also known as *gauze.*

Cheesy *adj.* Character of a paint or varnish film, which, although dry, is mechanically weak and rather soft.

Chelate \ˈkē-ˌlāt *also* ˈchē-\ (1826) *adj.* A compound comprised of metallic ions bound by a chelating agent.

Chelating agent *n.* (1) A term derived from the Greek word *chele*, meaning claw. Thus, a chelating agent is a substance whose molecules are capable of seizing and holding metallic ions in a claw like grip. (2) A sequestering or complexing agent that, in aqueous solution, renders a metallic ion inactive through the formation of an inner ring structure with the ion. (3) Organic compound that can remove many of the heavy metal cations from solution by forming soluble chelate compounds. Ethylenediaminetetraacetetic acid (EDTA) is typical. Syn: sequestering agent and chelate.

C

Chelation *n.* Reversible reaction of a metallic ion with a molecule or ion to form a complex molecule which does not have all or most of the characteristics of the original metallic ion. Syn: sequestration.

Chemical analysis *n.* Chemical analysis refers to the determination of chemical structure and chemically active species. It involves both direct measurements and use of specific compounds to achieve selective reactions of a component of the substance being analyzed; to produce a readily measurable species; or to determine a reactive end product.

Chemical change (reaction) *n.* A change in which one or more substances are transformed into one or more new substances.

Chemical composition *n.* Chemical composition is basically organic polymers that are very large molecules composed to chains of carbon atoms generally connected to hydrogen atoms (H), and often also to oxygen (O), nitrogen (N), chlorine (Cl), fluorine (F), and sulfur (S).

Chemical conversion coating *n.* A treatment, either chemical or electrochemical, of the metal surface to convert it to another chemical form which provides an insulating barrier of exceedingly low solubility between the metal and its environment, but which is an integral part of the metallic substrate. It provides greater corrosion resistance to the metal and increased adhesion of coatings applied to the metal. Examples are phosphate coatings on steel or zinc and chromate coatings on aluminum.

Chemical crimping *n.* A crinkled or puckered effect in fabric obtained by printing sodium hydroxide onto the goods in a planned design. When the material is washed, the part to which the paste has been applied will shrink and cause untreated areas to pucker. The same effect is obtained with a caustic resist print and a sodium hydroxide bath.

Chemical fiber *See manufactured fiber.*

Chemical finishing *n.* Processes in which additives are applied to change the aesthetic and functional properties of a material. Examples are the application of antioxidants, flame-retardant, wetting agents, and stain and water repellents.

Chemically foamed plastic *n.* A cellular plastic in which the cells are formed by thermal decomposition of a blowing agent or by the reaction of gas-liberating constituents. *See also cellular plastic.*

Chemical polymeric blends *n.* Mixtures of thermoplastic resins, etc.

Chemical pretreatment *See chemical conversion coating.*

Chemical properties *n.* Material characteristics that relate to the structure of a material and its formation from the elements.

Chemical property *n.* A property of a substance which can be described only by referring to a chemical reaction.

Chemical reactions *n.* Reactions between atoms and molecules to produce chemical compounds different from reactants. Chemical reactions may be elementary reactions or stepwise reactions.

Chemical resistance (reagent resistance) *n.* The ability of a plastic to maintain structural and esthetic integrity when exposed to acids, alkalis, solvents, and other chemicals. ASTM tests for chemical resistance of plastics include: C 581, chemical resistance of thermosetting resins used in glass–fiber-reinforced structures; D 543, resistance of plastics to chemical reagents; D 1239, resistance of plastic films to extraction by chemicals; D 1712, resistance of plastics to sulfide staining; D 21451, test

C

method for staining of polyvinyl chloride compositions by rubber-compounding ingredients; D 2299, determining relative stain resistance of plastics; D 3615, chemical resistance of thermoset molding compounds used in the manufacture of molded fittings; D 3681, Chemical Resistance of Reinforced-Thermosetting-Resin Pipe in a Deflected Condition; D 3753, Specification for Glass–Fiber-Reinforced Polyester Manholes; and D 4398, Determining the Chemical Resistance of Fiberglass-Reinforced Thermosetting Resins by One-Side Panel Exposure.

Chemical-resistant paint *See chemical resistance.*

Chemicals for electroplating *n.* Most plastics plated today are finished with a copper/nickel/chrome electroplate, but many other finishes are possible, such as bright brass antique brass, satin nickel, silver, black chrome, and gold. The actual composition of the electroplate is designed for the particular application of the electroplate.

Chemical shifts *n.* The variation of the resonance frequency of a nucleus in nuclear magnetic resonance (NMR) spectroscopy in consequence of its magnetic environment. through absorption arising from shielding and deshielding by electrons, value is ppm. The chemical shift of a nucleus, δ, is expressed in ppm by its frequency, ν_{cpd}, relative to a standard, ν_{ref}, and defined as $\delta = 10^6(\nu_{cpd} - \nu_{ref})/\nu_o$ where ν_o is the operating frequency of the spectrometer. For ^1H and ^{13}C NMR the reference signal is usually that of tetramethylsilane ($SiMe_4$).

Chemical stability *n.* Degree of resistance of a material to chemicals, such as acids, bases, solvents, oils, and oxidizing agents, and to chemical reactions, including those catalyzed by light.

Chemiluminescence \ˌke-mē-ˌlü-mə-ˈne-sᵊn(t)s, ˌkē-\ [ISV] (1889) *n.* Emission of light during a chemical reaction.

Chemisorption \ˈke-mi-ˌsórb, ˈke-, -ˌzórb\ [*chem.*- + *-sorb* (as in adsorb)] (1935) *vt.* Adsorption, particularly when irreversible, by chemical action rather than physical action.

Chemist \ˈke–mist\ [NL *chemista*, short for ML *alchimista*] (1562) *n.* (1) *obs.* Alchemist, one trained in chemistry (2) *Brit*: pharmacist. (3) A professional who possesses an earned bachelor's or higher degree with a major in a science from an accredited institution and who develops, applies, or communicates the principles of chemistry and exercises independent judgment and discretion in conceiving, planning, coordinating, or executing chemical projects, or who has experience in so doing. The chemical sciences deal with the composition, structure, and properties of substances and of the transformations they undergo.

Chemistry \ˈke-mə-strē\ *n.* The science of the compositions and structures of substances and of the changes, which these undergo. The study of molecules and the reactions they undergo.

Chenille \shə-ˈnē(ə)l\ [F, literally, caterpillar, fr. L *canicula*, dim. of *canis* dog; fr. its hairy appearance] (ca. 1739) *n.* (1). A yarn with a fuzzy pile protruding from all sides, cut from a woven chenille weft fabric. Chenille yarns are made from all fibers, and they are used as filling in fabrics and for embroidery, fringes, and tassels. (2) Fabric woven with chenille yarn.
Also see tufted fabric.

Cherry *n.* An even-textured, moderately high-density wood of the eastern USA, rich red-brown in color; takes a high luster; used for cabinetwork and paneling.

Cheviot \ˈshe-vē- ət, *esp British* ˈche-\ (1815) *n.* A rugged tweed made from uneven yarn, this fabric usually has a rather harsh hand.

Chevron \ˈshev-rən\ [ME, fr. MF, rafter, fr. (assumed) VL *caprion-*, *caprio* rafter, akin to L caper goat] (14c) *n.* A broad term applied to prints in zigzag stripes or to herringbone weaves.

Chicle \ˈchi-kəl, ˈchi-klē\ [AS, fr. Nahautl *tzictli*] (ca. 1889) *n.* Raw material for chewing gum (mixture of *trans*-1,4-poly (isoprene) + triterpenes. A natural product.

Chiffon *n.* A plain weave, lightweight, sheer, transparent fabric made from fine, highly twisted yarns. It is usually a square fabric, i.e., having approximately the same number of ends and picks and the same count in both warp and filling.

Chill-back *n.* A material capable of reducing the temperature of a batch of varnish or resin, either because of a considerable temperature difference between itself and the hot material, or because it absorbs heat in melting also.

Chillers *n.* A self-contained (individual or central) system comprised of a refrigeration unit and a coolant circulation mechanism consisting of a reservoir and a pump.

Chilling *n.* (1) In varnish or resin manufacture, the deliberate and rapid reduction of the temperature of the charge in a varnish or resin pot. (2) On a painted or varnished surface, a clouding of the surface or a reduction of luster as a result of the movement and cold air over the drying surface. *Also called quenching.*

Chill roll *n.* A metal roll – a shell within a shell with a relatively small clearance between the two – with water or other heat-transfer medium circulating through the so-formed annular space, used to cool an extruded or cast film or sheet prior to winding or cutting. The surface of the roll may be polished or textured to impart a finish to the plastic.

Chill-roll extrusion *n.* Film extrusion in which the molten film is drawn over cooled, polished rolls, imparting high gloss to the film. Syn: cast film extrusion.

Chimb *See chime.*

Chime *n.* The top ring of a coating container into which the friction top is pressed in order to effect an air-tight seal. Syn: chimb and chine.

Chimney *n.* An incombustible vertical structure containing one or more flues to provide draft for fireplaces, and to carry off gaseous products of combustion to the outside air from fireplaces, furnaces, or boilers.

China clay *n.* $Al_2O_3{\cdot}2SiO_2{\cdot}2H_2O$. Pigment white 19 (77005). A complex hydrated aluminum silicate produced by the breakdown of the mineral feldspar. The finest grades are used by the color and paint trades – often under proprietary names – and are utilized as a base for lakes and as an extender and anti-settling agent in paints. *See aluminum silicate. Also known as kaolin.*

China wood oil *See tung oil.*

Chinchilla cloth *n.* A heavy, twill weave, filling-pile fabric with a napped surface that is rolled into little tufts or nubs. The material is frequently double faced with a knitted or woven, plain or fancy back. Chinchilla cloth is used primarily in coats. The term is also used to refer to a knitted woolen fabric having a napped surface.

Chine *See chime.*

Chinese blue Term often applied to better grades of Prussian blue.
See milori blue and iron blue.

Chinese ink *See India ink.*

Chinese lac *n.* Resinous exudation of the wild Chinese tree, *Angia sinensis* or *Tssi-chau*, from which it is obtained by making deliberate incisions in the trunk. It is very susceptible to exposure, changing in color gradually, through red to black. The resinous exudation has very good solubility, blending satisfactorily with turpentine and alcohol.

Chinese red *See chrome orange.*

Chinese vegetable tallow Fatty substances from the kernels of the Chinese vegetable tallow tree, *Stillingia sebifera*. This is the same tree, which yields stillingia oil.

Chinese vermillion *See mercuric sulfide.*

Chinese wax An insect secretion, collected from certain evergreens, which are indigenous to China. Properties: mp, 81°C; Sp gr, 0.970; saponification value, 92; and a negligible iodine value. Known also as Insect Wax.

Chinese white (1) A paint using zinc oxide as the principal pigment. (2) Syn: zinc oxide.

Chino A cotton or cotton blend twill used by armies throughout the world for summer-weight uniforms. Chino is frequently dyed khaki.

Chinoiserie \shēn-ˈwäz-rē, -ˈwä-zə-; ˌshēn-ˌ-wäz-ˈrē, -ˌwä-zə-\ [F, fr. *chinois* Chinese, fr. *Chine* China] (1883) *n.* Originally, European designs "in the Chinese taste". Now loosely applied to almost any oriental form of decoration.

Chinon Graft copolymer of 70% acrylonitrile on 30% casein. Manufactured by Toyoba, Japan.

Chintz \ˈchin(t)s\ [earlier *chints*, plural of *chint*] (1614) *n.* A glazed fabric produced by friction calendering. Unglazed chintz is called cretonne.

Chip (1) The form of polymer feedstock used in fiber production. (2) The feedstock for a pulp digestor. (3) A defect in a non-woven fabric.

Also see flake.

Chip board *n.* Paperboard generally made from reclaimed paper stock and frequently used as a facing for partitions. The board is of moderately low density and should not be confused with particle board.

Chipping *n.* (1) Removal of paint, or rust and scale, by mechanical means. (2) Total or partial removal of a dried paint film in flakes by accidental damage or wear during service; in traffic paints, this failure is usually characterized by sharp edges and definite demarcation of the base area. (3) Spreading of white or colored chips or flakes during the application of seamless flooring. This procedure is used to impart a decorative effect to the flooring.

Chipping rating standards *n.* A set of photographic standards depicting the size and number of chips in each of several categories, as described in ASTM Method D-3170.

Chipping resistance *n.* The ability of a coating or layers of coatings to resist total or partial removal, usually in small pieces, resulting from impact by hard objects or from wear during service.

See chipping.

Chips *n.* (1) Term used to describe the size of resin particles. It represents a form intermediate between bold lumps and dust. (2) Dry combination of binder and pigment which in solvent forms paint, or which can be added to color plastics (e.g., vinyl chips and nitrocellulose chips).

Chirality \ˈkī-rəl\ [*chir-* + ^{1}*-al*] (1894) *adj.* The property of an organic molecule of not being identical with its mirror image; a compound whose molecules are chiral can exist as enantiomers, but non-chiral compounds cannot be enantiomers. All asymmetric molecules are chiral; however, not all chiral molecules are asymmetric since some having axes of rotational symmetry are chiral. Chiral and prochiral atoms are sites or potential sites, respectively, of stereoisomerism.

Chlorazol dyes *n.* Range of soluble dyestuffs, which are used in the manufacture of colored inks.

Chlorendic anhydride (1,4,5,6,7,7-hexachloro-5-norbornene-2,3-dicar-boxylic anhydride, HET anhydride, a difunctional acid anhydride) *n.* HET anhydride is a white, crystalline powder used as a hardening agent and flame retardant in epoxy, alkyd, and polyester resins (*See image*).

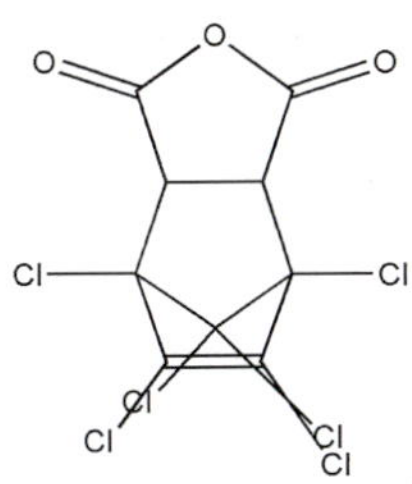

Chlorinated biphenyl (chlorinated diphenyl) *n.* Any of a group of plasticizers ranging from liquids to hard solids, used with polyvinylidene chloride and polystyrene. They are also used in conjunction with DOP as coplasticizers for PVC, and in conjunction with polyvinyl acetate, ethyl cellulose and other thermoplastics, as adhesives.

Chlorinated diphenyls (PCB) *n.* Range of chlorinated hydrocarbons varying in properties from very liquid plasticizers, through thick syrups to solid resins. They are characterized by excellent compatibility, stability, and absence of free acidity, high toxicity, and non-flammability.

Chlorinated hydrocarbon *n.* Any of a wide variety of liquids and solids resulting from the substitution or addition of chlorine in hydrocarbons such as methane, ethylene, and benzene. They are employed as solvents, plasticizers, and monomers for plastics manufacture.

Chlorinated hydrocarbons *n.* Powerful solvents, which include such members as chloroform, carbon tetrachloride, ethylene dichloride, methylene chloride, tetrachorethane, trichloroethylene, etc. Generally speaking, they are toxic. Their main applications include non-flammable paint removers, cleaning solutions, and special finishes where presence of residual solvent in the film is a disadvantage. Many are not prohibited in most countries.

Chlorinated paraffin (chlorocosane) *n.* Any of a family of yellow to light amber liquids produced by chlorinating a paraffin oil, with uses as secondary plasticizers for vinyls, polystyrene, polymethyl methacrylate, and coumarone-indene resins. Chlorinated paraffins also impart flame resistance to polyolefins, polystyrene, PVC, natural rubber, and unsaturated polyester resins.

Chlorinated paraffin wax *n.* Ordinary paraffin wax can be chlorinated, under certain conditions, to yield products which no longer resemble the parent wax, but which have assumed definite resinous characteristics. The amount of chlorine present determines the physical properties of the resultant waxes. These with smaller amounts of chlorine are liquids, whereas those with about 70% chlorine are brittle resins. Their applications are in fire-retarding compositions and as plasticizers.

Chlorinated para-nitraniline red *See chlorinated para reds.*

Chlorinated para reds *n.* Pigment red 4 (12085). There are two varieties of chlorinated para red: (1) Ortho-chlor-papa nitraniline and (2) Para-chlorotho nitraniline; both are diazotized and coupled to beta naphthol. Both are light yellow shade reds of the toluidine red orange.

C

Chlorinated polyether [poly-3,3-bis-(chloromethyl)oxacyclobutane, Penton®] *n.* A corrosion-resistant thermoplastic obtained by polymerization of chlorinated oxetane monomer, the oxetane being derived from pentaerythritol, to a high molecular weight (250,000–350,000). The polymer is obtained from pentaerythritol by preparing a chlorinated oxetane and polymerizing it to a polyether by means of opening the ring structure. The polymer is linear, crystalline, and extremely resistant to degradation at processing temperatures. It may be injection molded, extruded, or applied as a coating by the fluidized-bed method. The resin is widely used in valves, pumps, flow meters, etc., for chemical plants.

Chlorinated polyethylene (CPE) *n.* Any polyethylene modified by simple chemical substitution of chlorine on the linear backbone chain, CPEs range from rubbery amorphous elastomers at 35–40% Cl to hard, semicrystalline materials at 68–75% Cl. They are sometimes included with chlorinated natural and butyl rubbers under the term chlorinated rubbers. Certain CPEs are used as modifiers in PVC compounds to obtain better flexibility and toughness, particularly low-temperature toughness, greater latitude in compounding, and ease of processing.

Chlorinated polyvinyl chloride (CPVC) *n.* A PVC resin modified by post-chlorination. A series of such resins, known as "Hi-Temp Geon", is available from B F Goodrich Chemical Co. Compared to conventional rigid PVC, CPVC withstands service temperatures 20–30°C higher, is stronger, and has better chemical resistance. CPVC is mildly hygroscopic, so requires predrying before processing. With that proviso, it can be processed by all the methods used for rigid PVC with few modifications.

Chlorinated polyvinyl chloride resins *n.* CPVC is a plastic produced by the post-chlorination of PVC.

Chlorinated rubber (rubber chloride) *n.* (1) Natural rubber in which about two-thirds of the hydrogen atoms have been replaced by chlorine atoms. The resin is formed by the reaction of rubber with chlorine at about 100°C in an inert solvent or as a latex. Unlike rubber, the resulting product is readily soluble and yields solutions of low viscosity. It is sold as white powder, fibers, or as blocks. Commercial products generally contain about 65% chlorine. It has good chemical resistance properties, however, it tends to cobweb when sprayed. Now mostly chlorinated polymers are used, as 1-butene, polyethylene, etc. It has adhesive properties and, because of its good fire resistance, is used in paints. (2) Chlorinated natural rubber. Manufactured by Bayer, Germany.

Chlorine retention *n.* A characteristic of several resins and textile finishes whereby they retain some of the chlorine from bleach. On heating of the goods, the chlorine forms hydrochloric acid, causing tendering of the cloth. This is especially true of certain wrinkle resistant finishes for cotton and rayon.

Chlorobenzene \ˌklōr-ō-ˈben-ˌzēn, ˌklór-, -ben-ˈ\ [ISV] (ca. 1889) (chlorbenzene, chlorbenzol, chlorobenzol, phenylchloride) *n.* C_6H_5Cl. A solvent, and an intermediate in the production of phenol (*See image*).

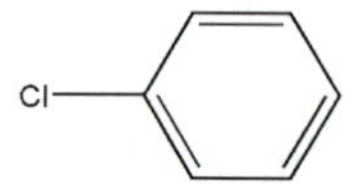

Chlorobutanol (chlorbutanol, 1,1,1-trichloro-2-methyl-2-propanol, acetone chloroform, trichloro-*tert*-butyl alcohol) *n.* $Cl_3CC(CH_3)_2OH$. A plasticizer for esters and ethers of cellulose (*See image*).

Chlorodiphenyl resin (chlorobiphenyl resin) *n.* Any resin made from chlorinated biphenyl, rosin or rosin ester, and the higher fatty acids. These resins are used as plasticizers and modifying resins in plastics, and in lacquers and varnishes.

Chloroethane *n.* Syn: ethyl chloride.

Chloroethene (chlorethylene) Syn: vinyl chloride.

Chlorofluorocarbon resin *n.* Any resin made by the polymerization of monomer(s) containing only carbon, chlorine, and fluorine. The principal member is polychlorotrifluoroethylene (pctfe).

Chlorofluorohydrocarbon resin *n.* A resin made by polymerization of monomer(s) containing only carbon, chlorine, fluorine, and hydrogen.

Chloroform \ˈklōr-ə-ˌfórm, ˈklór-\ [F *chloroforme*, fr. *chlor-* + *formyle* formyl; fr. its having been regarded as a trichloride of this group] (1838) (trichloromethane) *n.* $CHCl_3$. A pungent, toxic, dense liquid, useful as a solvent for epoxy resins and others. (2) Clean, colorless, volatile liquid. Sp gr of 1.485 (20/20°C); bp of 61.2°C; fp of 63.5°C; wt/gal of 12.29 lb (25°C); refractive index of 1.4422 (*See image*).

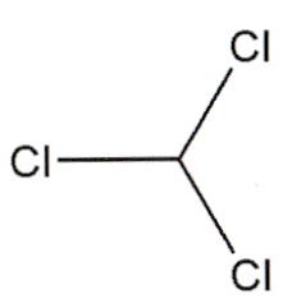

Chlorohydrin \ˌklōr-ə-ˌhī-drən, ˌklór-\ [ISV *chlor-* + *hydr-* + [1]*-in*] (ca. 1890) (α-chlorohydrin, 3-chloropropane-1,2-diol, glyceryl α-chlorohydrin) *n.* $ClCH_2CHOHCH_2OH$. A solvent, mainly for cellulosics (*See image*).

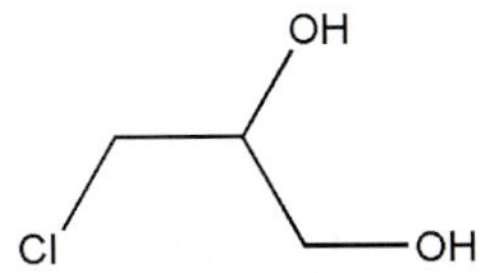

Chlorohydrin rubber *See epichlorohydrin rubber.*

α-Chloro-*meta*-nitroacetophenone *n.* $O_2NC_6H_4COOH_2Cl$. A bacteriostat and fungistat for plastics.

Chloronaphthalene oil *n.* Any of several nearly colorless oils derived by chlorinating naphthalene, used as plasticizers and flame retardants.

Chloroprene \-ˌprēn\ [*chlor-* + iso*prene*] (1931) *n.* $CH_2Cl–CHCl–CH=CH_2$. Polymerization of the monomer to poly(1-chloroprene) to form the synthetic chlorinated elastomer.

Chloroprene polymer *See neoprene.*

Chloropropylene oxide *n.* Syn: epichlorohydrin (*See image*).

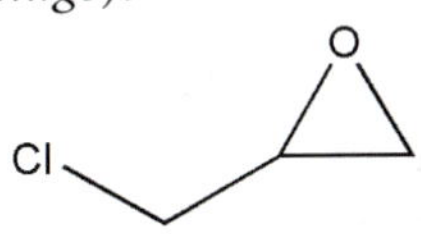

Chlorostyrenated polyester *n.* An unsaturated polyester resin made by reacting a fluid polyester with monochlorostyrene in place of styrene (*See polyester, unsaturated*). Monochlorostyrene is less volatile and more reactive than styrene, providing faster cure rates and increased flexural strength and modulus in glass–fiber laminates.

Chlorosulfonated polyethylene (CSM, CSPR) *n.* Polyethylene which has been reacted with a mixture of chlorine and sulfur dioxide under ultraviolet light irradiation. Polymer may be vulcanized to form a product with good ozone, heat, oxygen and weathering resistance. It is produced by simultaneous treatment, with sulfur dioxide and chlorine, of dissolved, radicalized polyethylene. A commercial product (Hypalon®, DuPont) contains 22–26% Cl and 1.3–1.7% S.

C

Chlorotrifluoroethylene *n.* $ClFC{=}CF_2$. A colorless gas. The monomer for polychlorotrifluoroethylene. It is obtained by either dehalogenation or dehydrohalogenation of saturated chlorofluorocarbons or chlorohydrocarbons, e.g., by reacting 1,1,2-trichlorotrifluoroethane with zinc. The monomer for the preparation of polychloro-trifluoroethylene by free radical polymerization in aqueous systems.

Choked coiler *n.* A condition in carding or drawing in which sliver is either puffy, badly condensed, or very uneven, leading to overloading of the coiler trumpets and causing work stoppage.

Choked flyers *n.* A situation in which roving will not pass through the flyer channels because of heavy or cockled conditions caused by such factors as uneven drafting, waste, over-cut fibers, and improper finish.

Choker bar (restrictor bar) *n.* A bendable metal bar incorporated in a sheet-extrusion die for controlling flow distribution and lateral sheet thickness, and for reducing stagnation in the melt. The shape of a flow passage between the choker bar and the lower die body is altered by turning bolts connected to the bar.

Cholesteric \ˌkō-lə-ˈster-ik, kə-ˈles-tə-rik\ [*cholesteric* relating to cholesterol, fr. F *cholesterique*] (1942) *adj. See liquid–crystal polymer.*

Chopped strand *n.* A type of glass–fiber reinforcement consisting of strands of individual glass fibers that have been chopped into lengths from 1 to 12 mm. The individual fibers are bonded together within the strands so that they remain in bundles after being cut.
See also roving.

Chopped-strand mat *n.* A mat formed from randomly oriented, chopped strands of glass and held together, just strongly enough for handling, by a binder.

Choppers *n.* Chopper guns, long cutters, roving cutters cut glass fibers into strands and shorter fibers to be used as reinforcements in plastic.

CHR Chlorohydrin elastomer [poly(epichlorohydrin)].

Christiansen effect *n.* When finely powdered substances, such as glass or quarts, are immersed in a liquid of the same index of refraction complete transparency can only be obtained for monochromatic light. If white light is employed the transmitted color corresponds to the particular wavelength for which the two substances, solid and liquid have exactly the same index of refraction. Due to differences in dispersion the indices of refraction will match for only a narrow band of the spectrum.

Chroma \ˈkrō-mə\ [Gk *chroma*] (ca. 1889) *n.* (1) Color intensity or purity of tone, being the degree of freedom from gray. (2) One of the three terms used in Munsell notations. It correlates approximately with the psychological dimension of saturation.

Chromatic aberration *n.* A lens receiving white light from an object will form the blue image closer to the lens than the red image, forming color fringes. This caused by refractive index variation with wavelength (dispersion), always blue greater than red.

Chromaticity *n.* The quality of color expressed as a function of wavelength and purity.

Chromaticity coordinates *n.* **CIE.** The ratios of each of the three tristimulus values *X*,

Y, and Z in relation to the sum of the three; designated as x, y, and z, respectively. They are sometimes referred to as the trichromatic coefficients. When written without subscripts they are assumed to have been calculated for Illuminant C and the 2° (1931) Standard Observer unless specified otherwise. If they have been obtained for other illuminants or observers, a subscript describing the observer of illuminant should be used. For example, x_{10} and y_{10} are chromaticity coordinates for the 10° observer and Illuminant *C*.

Chromaticity coordinates *n.* **General** The two dimensions of any color order system which exclude the lightness dimension and describe the chromaticity. Thus, x and y, or a and b, or u and v, or YB and RG, may be considered as chromaticity coordinates in various color spaces. Unless otherwise specified, the term chromaticity coordinates is assumed to refer to the CIE coordinates x, y, and z for Illuminant *C* and the 2° (1931) Standard Observer.

Also known as color coordinates.

Chromaticity diagram *n.* **CIE.** A two-dimensional graph of the chromaticity coordinates, x as the abscissa and y as the ordinate, which shows the spectrum locus (chromaticity coordinates of monochromatic light, 380–770 nm). It has many useful properties for comparing colors of both luminous and non-luminous materials.

Chromaticity diagram *n.* **General.** Plane diagram formed by plotting one of the chromaticity coordinates against the other.

Chromaticity difference diagrams *n.* Plane diagram formed by plotting the differences of one of the chromaticity coordinates against the differences of the other chromaticity coordinate, the differences being taken from the neutral point or, for a specific color, from the chromaticity coordinates for the standard of that color.

Chromaticness *n.* Combined hue and saturation of a color, with no reference to its lightness. Syn: chromaticity.

Chromatography (1937) *n.* In chemistry, analytical technique used for the chemical separation of mixtures and substances. A process in which a gas or liquid solution moves through a calibrated column containing a subdivided solid phase into which some components of the solution are absorbed, smaller molecules more quickly and thoroughly than larger ones. This is followed by pure carrier gas or solvent, the stream being monitored by a differential detector. Larger molecules emerge first, smaller ones later. The detector signal is proportional to the concentration of each species in the effluent. The process is mainly used for analysis of organic mixtures, but also for their separation. The name "chromatography" derives from the work of the Russian botanist N. Tswett, who first used the process to separate chloroplast pigments, obtaining colored bands on filter paper. Some variations of the process are *gas chromatography* (the gas mixture is passed through a porous bed, or through a capillary tube lined with an absorbent liquid or solid phase); paper chromatography (a drop of specimen is placed near one end of a porous paper); ion-exchange chromatography; thin-layer chromatography (the sample is placed on an absorbent cake spread on a smooth glass plate); and, important for plastics, size-exclusion chromatography.

Chrome green (brunswick green) *n.* $PbCrO_4xPbSO_4yFeNH_4Fe(CN)_6$. Any of a family of pigments ranging from light yellow-green through dark green, based on physical mixtures of chrome yellow and iron blue (a complex ammonium iron hexacyanoferrate). The amount of iron blue determines the shade, about 2% being used for light yellow-green, and up to 64% being used for dark greens. Variation can be achieved by pigment blends with extenders. Noted for good hiding and tinting strength, poor alkali resistance. Density, 4.1–5.4 g/cm^3 (34.2–45.0 lb/gal) O.A., 13–25; particle size, 0.2–1.2 μm. Syn: chromium green, lead chrome green, brunswick green and milori green.

Chromel and alumel *n.* Special high-nickel alloys that, when their wires are joined, develop high thermoelectric power, and are therefore useful for temperature measurement. "Type *K*" thermocouples are usually chosen for high-temperature work in oxidizing atmospheres.

Chrome ochre *See chromium oxide green.*

Chrome orange, light and deep *n.* $PbCrO_4 \cdot PbO$. Pigment orange 21 (77601). Chrome orange is chemically basic lead chromate. The hue ranges from light to deep orange, and the color is related to particle size. A pigment characterized by good opacity, fair lightfastness, and good gloss retention and durability. Disadvantages are poor alkali and acid resistance and poor high temperature stability. Density, 6.6–7.1 g/cm^3 (55.1–58.8 lb/gal); O.A., 9–12; particle size, 0.1–1.0 μm. Syn: orange chrome, Austrian cinnabar, Persian red, Chinese red, Derby red, American vermillion, and chrome red.

Chrome-orange pigment *n.* Any pigment based on basic lead chromate, $PbO \cdot PbCrO_4$, that is of a deep orange color.

Chrome oxide green *n.* A stable pigment based on anhydrous chromium oxide, Cr_2O_3. The form based on hydrated chromium oxide is called Guignet's Green. Syn: schnitzer's green. *See chromium oxide green.*

Chrome red *See basic lead chromate, chrome orange, and light and deep.*

Chrome vermillion *See molybdate orange.*

Chrome yellow *n.* A light resistant opaque yellow pigment composed essentially of lead chromate.

Chrome yellow, light, and primrose *n.* $xPbCrO_4 \cdot yPbSO_4$. Pigment yellow 34 (77603). Yellow pigments based on combinations of lead chromate and lead sulfate, The hue and tint strength being controlled by the ratio of chromate to sulfate during precipitation. In the light and lemon hues, the approximate ratio of x/y is 2.5/1 (2.5 $PbCrO_4 \cdot PbSO_4$); in the primrose hues, the x/y ratio is approximately 3.2/1. The pigments are characterized by good opacity, fair lightfastness, good glass retention and durability, and poor resistance to alkali, acid, and high temperature. Density, 5.44–6.09 g/cm^3 (17–39 lb/gal); O.A., 17–39; particle size, 0.1–0.8 μm. Syn: lemon chrome, golden chrome, middle chrome, Paris yellow, Leipsic yellow, and cologne yellow.

Chrome yellow, medium *n.* PbCrO4. Pigment yellow 34 (77600). Approaches chemically pure lead chromate. Characterized by good opacity, fair to good lightfastness, good gloss retention and durability, poor alkali resistance. Density, 5.58–6.04 g/cm^3 (46.6–50.3 lb/gal); O.A., 16–35; particle size, 0.15–1 μm.

Chrome yellow pigment (primrose chrome and permanent yellow) *n.* Any pigment based on normal lead chromate, $PbCrO_4$ that is characterized by a medium yellow color. Other shades ranging from light greenish-yellow to medium reddish-yellow are made by co-precipitating lead chromate with other insoluble compounds such as lead sulfate or lead phosphate.

Chromic acid (1800) *n.* H_2CrO_4. An acid analogous to sulfuric acid but known only in solution and in the form of its salts.

Chromic chloride (chromium chloride, chromium chloride hexahydrate) *n.* $CrCl_3$ or $[Cr(4H_2O)Cl_2]Cl{\cdot}2H_2O$. A catalyst for polymerizing olefins.

Chromic hydrate *See chromium hydroxide.*

Chromic oxide *See chromium oxide green.*

Chromium green *See chrome green.*

Chromium hydrate *See chromium hydroxide.*

Chromium hydroxide *n.* $Cr(OH)_3$. Green pigment, manufactured by adding a solution of ammonium hydroxide to the solution of a chromium salt. Syn: chromic hydrate and chromium hydrate.

Chromium oxide green *n.* Cr_2O_3. Pigment green 17 (77288). Most permanent green pigment, almost pure chromium sesquioxide. May be manufactured by reducing a chromate (e.g., sodium dichromate) with sulfur or carbonaceous materials. Outstanding lightfastness and resistance to acid, alkali and high temperatures often used as a colorant in cementitious products. Density, 5.1–5.4 g/cm^3 (42.5–45.0 lb/gal); O.A., 12–24; particle size 2–6 μm. Syn: arnaudon's green, chromium sesquioxide, chrome oxide green, vert emeraude (dull), anadonis green, chrome ochre, green cinnabar, green rouge, leaf green, oil green, dingler's green, and schnitzler's green.

Chromium oxide, hydrated *See hydrated chromium oxide.*

Chromium plating (chrome plating) *n.* An electrolytic process that deposits a hard, inert, smooth layer of chromium onto working surfaces of other metals for resistance to corrosion and wear. Extruder screws, chill rolls for sheet and film production, calendaring rolls, dies, and molds are commonly chromium plated.

Chromium sesquioxide *See chromium oxide green.*

Chromophore *n.* (1) A group such as –NO, –NO_2, or –N=N– that, when present in a molecule, enables the molecule to be transformed into a dye upon the introduction of an acid group. (2) Certain groups of atoms such as, –C=Cl, –C=N–, –N=N–, and –N=O, when present in an organic molecule, can give rise to colored compounds. Groups acting in this way are called chromophores. Although compounds containing chromophores are not necessarily colored. The hue is affected by the presence of other groups known as "auxochromes", such as –NH_2, –OH, –NO_2, –CH_3 and halogens –Cl, –Br, and –I.

Chrysotile \-ˌtīl\ [Gr *Chrysotil*, fr. *chrys-* + *-til* fiber, fr. Gk *tillein* to pluck] (1850) (serpentine) *n.* $Mg_3Si_2O_5(OH)_4$ for $3MgO{\cdot}2SiO_2{\cdot}2H_2O$. A hydrated magnesium orthosilicate, the chief constituent of *serpentine* asbestos. Chrysotile-bearing asbestos has been the most used type, once accounting for over 90% of the world production. Its fine and silky fibers, and mats and felts made therefrom, where widely used as fillers and reinforcements for plastics, providing excellent resistance to chemicals and fire. Many of its former uses are now prohibited because of the carcinogenicity of some types of asbestos. *See magnesium silicate and fibrous.*

Chute-feed system *n.* Pneumatic fiber transport system used in linking textile processing equipment or operations, especially opening, blending, and carding.

CI *n.* Abbreviation for color index.

C

Cibanoid *n.* Urea-formaldehyde resin. Manufactured by CIBA, Switzerland.

C.I.E *n.* Abbreviation for commission internationale d'éclair-AGE, the French name for the International Commission on Illumination. In older publications, the abbreviation ICI was used.

CIE chromaticity coordinates *See chromaticity coordinates, cie.*

CIE color difference equation *n.* May refer to the color difference equation provisionally adopted by the CIE in 1964, one of four then recommended for study. It incorporates a linear transformation into a more nearly uniform color space. The following are the equations used:

$$U^* = 13W^*(u - u_o),$$
$$V^* = 13W^*(v - v_o),$$
$$W^* = 25Y^{1/3} - 17 \quad (1 \le Y \le 100),$$
$$u = \frac{4X}{X + 15Y + 3Z} \text{ or } u = \frac{4x}{-2x + 12y + 3},$$
$$v = \frac{6X}{X + 15Y + 3Z} \text{ or } v = \frac{6y}{-2x + 12y + 3},$$

u_o and v_o refer to a nominally achromatic color. For the 1931 Standard Observer and Illuminant C, $u_o = 0.2009$ and $v_o = 0.3073$. The total color difference is calculated as

$$\Delta E = [(\Delta U^*)^2 = (\Delta V^*)^2 + (\Delta W^*)^2]^{1/2},$$

$\Delta U^* = U^*$ sample $-U^*$ standard, indicating redder if positive and greener if negative. $\Delta V^* = V^*$ sample $- V^*$ standard, indicating yellower if positive and bluer if negative. This equation is properly titled the 1964 CIE U^*, V^*, and W^* equation. Sward GG (ed) (1972) Paint testing manual: physical and chemical examination of paints, varnishes and lacquers, and colors, 13th edn. ASTM Special Technical Publication No, 500, American Society for Testing and Materials, Philadelphia, PA. Billmeyer FW, Saltzman M (1996) Principles of color technology. John Wiley and Sons Inc., New York.

CIE 1976 (*L* a* b) color difference equation** (from J Opt Soc Am 64:896) (1974) (Modified Adams–Nickerson).

$$\Delta E_{\text{CIE}}(L^*, a^*, b^*) = [(\Delta L^*)^2 + (\Delta a^*)^2 + (\Delta b^*)^2]^{1/2},$$

where

$$L^* = 25\left(\frac{100Y}{Y_o}\right)^{1/3} - 16 \quad (1 \le Y \le 100),$$
$$a^* = 500\left[\left(\frac{X}{X_o}\right)^{1/3} - \left(\frac{Y}{Y_o}\right)^{1/3}\right],$$
$$b^* = 200\left[\left(\frac{Y}{Y_o}\right)^{1/3} - \left(\frac{Z}{Z_o}\right)^{1/3}\right],$$

X, Y, and Z are the tristimulus values of the sample, X_o, Y_o, and Z_o define the color of the nominally white object color stimulus (the illuminant). $\Delta L^* = L^*$ for the sample – L^* for the standard; $\Delta a^* = a^*$ for the sample – a^* for the standard; $\Delta b = b^*$ for the sample – b^* for the standard.

CIE 1976 (*L*, u*, v) color difference equation** *n.* (from J Opt Soc Am 64:(1974).

$$\Delta E_{\text{CIE}}(L^*, u^*, v^*) = [(\Delta L^*)^2 + (\Delta u^*)^2 + (\Delta v^*)^2]^{1/2},$$

where

$$L^* = 25\left(\frac{100Y}{Y_o}\right)^{1/3} - 16 \quad (1 \le Y \le 100),$$
$$u^* = 13L^*(u' - u'_o),$$
$$v^* = 13L^*(v' - v'_o),$$

$$u' = \frac{4X}{X + 15Y + 3Z},$$
$$v' = \frac{9Y}{X + 15Y + 3Z},$$
$$u'_o = \frac{4X_o}{X_o + 15Y_o + 3Z_o},$$
$$v'_o = \frac{9Y_o}{X_o + 15Y_o + 3Z_o}.$$

X, *Y*, and *Z* are the tristimulus values of the sample. X_o, Y_o, and Z_o are the tristimulus values for the illuminant. $\Delta L = L^*$ for the sample – L^* for the standard; $\Delta u^* = u^*$ for the sample – u^* for the standard; $\Delta v^* - v^*$ for the sample – v^* for the standard.

CIE color notation system *n.* Colorimetric specification system based on stimulus-response characteristics adopted by the CIE in 1931. The current recommendations for the system may be obtained from the official publication, CIE Publication No. 15 (E-1.3.1) 1971, "Colorimetry Official Recommendation of the International Commission on Illumination", available from National Bureau of Standards and Technology, Washington, DC.

CIE luminosity curve *See luminosity curve.*

CIE standard observer *n.* The observer data adopted by the Commission Internationale d'Eclairage to represent the response of the average human eye, when light-adapted, to an equal-energy spectrum. Unless otherwise specified, the term applies to the data adopted in 1931 for a 2° field of vision. The data adopted in 1964, sometimes called the 1964 observer, were obtained for a 10°, annular field which excludes the 2° field of the 1931 observer functions.

CIE tristimulus values *n. See tristimulus values, cie.*

CIL flow test A capillary-rheometer test developed at Canadian Industries Ltd for characterizing the flow of thermoplastics. The reported flow unit is the amount of melt that is forced through a specified orifice per unit time when a suitably chosen force is applied. Similar to melt-flow index.

C

Cill *n.* Brit. term for sill.

Cim *n.* Acronym for computer-integrated manufacturing.

Cimene *See dipentene.*

Cinaper *n.* An obsolete form of cinnabar.

Cinder block *n.* A lightweight masonry unit made of cinder concrete; widely used or interior partitions.
See clinker block.

Cinnabar \ˈsi-nə-ˌbär\ [ME *cynabare*, fr. MF & L; MF *cenobre*, fr. L *cinnabaris*, from Gk *kinnabari*, of non-Indo-European origin; akin to Arabic *zinjafr* cinnabar] (14c) *n.* *See mercuric sulfide.*

Cinoper *n.* An obsolete form of cinnabar.

Cintz *n.* Wallpapers resembling the printed cotton materials from India once known as "chints", featuring brightly colored flowers.

Circuit *n.* In filament winding, one complete traverse of the fiber-feed mechanism of the winding machine; or a complete traverse of a winding band from one arbitrary point along the winding path to another point on a plane through the starting point and perpendicular to the axis.

Circular-knit fabric *n.* A tubular weft-knit fabric made of a circular-knitting machine.

Circular knitting *See knitting.*

Ciré \sə-ˈrā, sē-\ [F, fr. pp of *cirer* to wax, fr. *cire* wax, fr. L *cera*] (1921) *n.* A brilliant patent leather effect produced by application of wax, heat, and pressure.

Cis- A chemical prefix (Latin: "on this side"), usually ignored in alphabetizing lists, denoting an isomer in which certain atoms or groups are on the same side of a plane. Opposite of *trans-*.

C

Cis isomer *n.* Any isomer in which two identical atoms or groups are adjacent to each other or on the same side of a structure.

Cissing *n.* (1) A slight shrinkage of a glossy paint resulting in small cracks through which the undercoat may be seen; a mild form of crawling. British Syn: crawling. (2) A process for preparing a wood surface for graining by wetting with a sponge.
Also spelled "sissing".

Citrate plasticizer *n.* Any of a family of plasticizers derived from citric acid, HO_2C-$CH_2C(OH)(CO_2H)CH_2CO_2H$, noted for their low order of toxicity. Included citrates are: triethyl, tri(2-ethylhexyl), tricyclohexyl, tri-*n*-butyl, acetyl triethyl, acetyl tri-*n*-butyl, acetyl tri-*n*-octyl, *n*-decyl, and acetyl tri(2-ethylhexyl).

Citrates *n.* Esters derived from citric acid. The tri-ethyl, -butyl, and -amyl citrates form a series of plasticizers.

Citric acid *n.* $C_3H_4(OH)(COOH)_3$. Tricarboxylic, monohydroxy acid. One molecule of water is normally included in the formula. It has an mp of 154°C, but decomposes on further heating (*See image*).

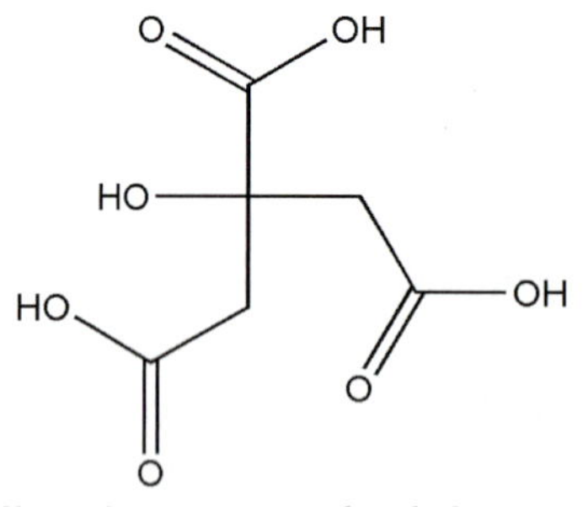

Citronella oil *n.* Essential oil, known also as oil of lemongrass. It possesses a strong lemon-like smell, and is used to a small extent as a deodorant.

CL *n.* Poly(vinyl chloride) fiber.

CLA *n.* Abbreviation for center-line average (BS 1134); indication of surface roughness.

Clamping capacity *n.* The largest rated projected area of cavities and runners that an injection or transfer press can safely hold closed at full molding pressure.

Clamping force *n.* In injection and transfer molding, the force applied to the mold to keep it closed, and opposing the pressure exerted by the injected plastic acting upon the projected area of cavities and runners. Per square centimeter of cavities and runners, at least 3.5 kN of clamping force is required.

Clamping plate *n.* A plate, fitted to a mold that secures the mold to the frame of the molding machine.

Clamping pressure *n.* In injection and transfer molding with a hydraulically operated mold, the hydraulic-fluid pressure applied to the mold ram to keep the mold closed during the molding cycle. Compare clamping force.

Clamps *n.* The parts of a testing machine that are used to hold a specimen while it is subjected to force.
Also called jaws.

Clamshell molding *n.* A term applied to the modern version of the oldest form of blow molding – preheating two sheets of plastic, placing them between halves of a split mold, closing the mold, drawing the sheets against their respective mold surfaces by means of vacuum, then completing the forming with air pressure between the sheets. The modern process, mechanized and conveyorized, is superior to blow molding from a parison for very large parts and for those in which uniformity of wall thickness is important.

Clapboard *n.* A wood siding commonly used as an exterior covering on a building of frame construction; applied horizontally and overlapped, with the grain running lengthwise; usually thicker along the lower edge than along the upper.
Also known as bevel siding and lap siding.

Claret red *n.* Azo pigment produced by coupling Tobias acid (2-naphthylamine-1-sulfonic acid) with 3-oxynaphthoic acid, followed by conversion to the calcium lake.

Clarifiers *n.* Additives that increase the transparency of a plastic material.

Clarifoil *n.* Cellulose acetate. Manufactured by British Celanese, Great Britain.

Clarity *n.* (1) In general, the optical property of being clear. (2) In acetate manufacture, a measure of the appearance of dope solutions, indicating the quality of the acetylation mixture. (3) In printing, the sharpness or definition of a print pattern.

Clarity of plastics *n.* The ability of a transparent material to transmit a clear image of an object when viewed through it, without any aberration.

Clash–Berg point *n.* The rising temperature at which the apparent modulus of rigidity of a specimen falls to 931 MPa, the end point of "flexibility" as defined by Clash and Berg in their studies of low-temperature flexibility. In a similar test described in ASTM D 1043, the deciding shear modulus is one-third the C–B value.

Class I, II, or III liquids *n.* Groupings of flammable and combustible liquids arbitrarily classified by boiling points and closed cup flash point methods, in order to specify particular procedures and types of electrical equipment to be used when handling them.

Classical mechanics *n.* A system of mechanics developed before the inception of quantum ideas; useful for describing the behavior of objects and particles much larger than atoms.

Clathrate \ˈkla-ˌthrāt\ [L *clathratus*, furnished with a lattice, fr. *clathrin* (plural) lattice, fr. Gk *kl*ēithron bar, fr. *kleiein* to close] (1906) *adj.* A "cage compound" in which atoms or molecules are trapped in a cage of covalently bonded atoms, but mot directly bonded to them.

Clay *n.* Any naturally occurring sediment rich in hydrated silicates of aluminum, predominating in particles of colloidal or near-colloidal size. There are many types of clays and clay-like minerals. Those of particular interest to the plastics industry are varieties refined by nature and man to a state of good color and particle-size distribution, such as kaolin (China clay). They are used as fillers in epoxy and polyester resins, PVC compounds, and urethane foams. Calcined clays are those that have been heated to a high temperature to drive off the chemically bound water, sometimes also surface-treated to improve their chemical inertness and moisture resistance. They are used primarily in vinyl insulation.

See aluminum silicate, kaolinite, and China clay.

Clean *n.* A complete lack of any visible non-uniformity sometimes referred to as seeds, when viewed in thin films by any macroscopic or microscopic use of visible light.

Cleaner *n.* (1) Detergent, alkali, acid or other cleaning material; usually water or steam borne. (2) Solvent for cleaning paint equipment.

Clear *n.* A complete lack of any visible non-uniformity when viewed in mass, in bottles or test tubes, by strong transmitted light.

Clear coating *n.* Transparent protective and/or decorative film.

Clearcole, clairecolle *n.* (1) Glue size in appropriate dilution with the addition of small quantity of whiting. It is used to reduce porosity of ceiling and other surfaces prior to the application of size-bound distemper. (2) A primer consisting of glue, water, and white lead or whiting. (3) A clear coating used in application of gold leaf.

C

Clearing *n.* The treatment of printed fabrics with a chemical solution to improve the appearance of the whites. In many cases the treatment also brightens the printed areas.
Also see reduction clearing.

Clear lacquer *See clear coating.*

Clear point *n.* With regard to vinyl plastisols, clear point is the rising temperature at which an unpigmented plastisol suddenly becomes transparent, signifying that the resin particles have completely dissolved in the warm plasticizer. This test is useful for determining the relative fusion temperatures of different plastisols.

Cleavage *n.* (1) Breaking of a laminate due to the separation of the strata. (2) Portion of the material adhering to the side of a container after the major portion has been drained. (3) The property of a crystalline substance of splitting along definite crystal planes.

Cleveland condensing humidity cabinet (QCT) *n.* An accelerated weathering apparatus which operates on a condensation type of water exposure at elevated temperature.

Cleveland open cup *n.* Device used in determining flash and fire points of petroleum products.

Cleve's acid *n.* Old name for α-naphthol-5-sulfonic acid. It was also known as *L acid* and is used in the manufacture of Helio Bordeaux BL.

Clicker die *n.* A cutting die for stamping out blanks from plastic sheet.

Clicker press *n.* A stamping press used with clicker dies to cut shapes from plastic sheet. Compare die cutting.

Clicking *n. See die cutting.*

Climate cabinet *n.* Any enclosure used to emulate selected climatic conditions.

Clinker block *n.* British term for cinder block.

Clipmark *n.* Visible deformation of selvage due to pressure from a tenter clip.

CLO *n.* A unit of thermal resistance. The insulation needed to keep an individual producing heat at the rate of 58 W/m^2 comfortable of 21°C air temperature with air movement of 0.1 m/s. One clo is roughly equal to the insulation value of typical indoor clothing.

Clockspring *n.* Slipping action in coil coating, usually at the center of a coil of the strip during the recoiling operation.

Cloisonné \ˌklói-zə-ˈnā, ˌklwä-\ [F, fr. pp of *cloisonner* to partition] (1863) *adj.* A surface decoration in which differently colored enamels or glazes are separated by fillets applied to the design outline. For porcelain enamel, the fillets are wire secured to the metal body; for tile and pottery, the fillets are made of ceramic paste, squeezed through a small-diameter orifice.

Cloqué fabric \klō-ˈkā, ˈklō-ˌ\ [F *cloqué*, fr. pp of *cloquer* to become blistered, fr. F dialect (Picard) *cloque* bell, bubble, fr. ML *clocca* bell] (1936) *n.* From the French term for blistered, it refers to any fabric whose surface exhibits an irregularly raised blister effect.

Closed assembly time *See time, assembly.*

Closed-cell foamed plastic (unicellular foam) *n.* A cellular plastic in, which interconnecting cells are too few to permit the bulk flow of fluids through the mass.

Closed-cell foams *n.* Individual cells are non-interconnecting. The ells are basically without access to the surrounding air of fluids; cells are not communicating.

Closed coat *n.* In coated abrasives, when the abrasive grains completely cover coat-side surface of the backings. Closed coats are designed for severe service and are used for most applications.
See also open coat.

Closed loop control *See feedback control.*

Close drying *n.* Material is said to be close drying when it does not show much fullness or body on the substrate. Often, the true fullness is not brought out until after rubbing. This is especially true with lacquer finishes.

Close grain *n.* Wood having narrow and inconspicuous annual growth rings. The term is sometimes used to designate wood having small and closely spaced pores, but in this sense the term "fine textured" is more often used.

Cloth *n.* A generic term embracing all textile fabrics and felts. Cloth may be formed of any textile fiber, wire, or other material, and it includes any pliant fabric woven, knit, felted, needled, sewn, or otherwise formed.

Cloud chamber *n.* An apparatus containing moist air or other gas which on sudden expansion condenses moisture to droplets on dust particles or other nuclei. Thus charged particles or ions in the space become nuclei and their numbers and behavior, when properly illuminated, may be studied.

Cloudiness *n.* The lack of clarity or transparency in a paint or varnish film.

Clouding *n.* Development in a clear varnish or lacquer film or liquid of an opalescence or cloudiness caused by the precipitation of insoluble matter or immiscibility of components.

Cloud point *n.* (1) In condensation polymerization, the temperature at which the first turbidity appears, caused by water separation when a reaction mixture is cooled. (2) In petroleum and other oils, the falling temperature at which the oil becomes cloudy, from precipitation, of wax or other solid. (3) Point at which a definite lack of clarity (cloudiness) appears when a liquid is subject to adulteration or when it is mixed with another substance, or the temperature at which a liquid becomes cloudy when it is cooled.

Cloudy web *n.* An uneven or irregular web from the doffer of a card.

Clumps In non-woven fabrics, an irregularly shaped grouping of fibers caused by insufficient fiber separation.

Clupanodonic acid *n.* Acid found in many fish oils. The acid possesses an unusually high iodine value – well in excess of 200. *Also called doscosapentanoic acid.*

CMC *n.* Abbreviation for carboxymethyl cellulose or for ceramic-metal composite.

CN *n.* Abbreviation for cellulose nitrate.

CNR *n.* Carboxynitroso rubber.

Co *n.* Chemical symbol for the element cobalt.

Coacervate *n.* \kō-ˈa-sər-ˌvāt\ [L *coacervatus*, pp of *coacervare* to heap up, fr. *co-* + *acervus* heap] (1929) *v.* An aggregate of colloidal droplets held together by electrostatic attractive forces.

Coacervation *n.* The separation of a polymer solution into two or more liquid phases, one of which is a polymer-rich liquid. The term was introduced to distinguish this phenomenon from the precipitation of a polymer solute in solid form. The process is used in microencapsulation by emulsifying or dispersing the material to be encapsulated with a solution of the polymer. By changing the temperature or concentration of the mixture, or by adding another polymer or solvent, a phase separation may be induced and the polymeric portion forms a thin coating on the external surfaces of the particles. After further treatment to solidify the polymeric wall, the capsules can be isolated in powder form by filtration. It is an intermediate stage between sol and gel formation.

C

Coagulant \kō-ˈa-gyə-lənt\ (1770) *n.* A substance that (1) initiates the formation of relatively large particles in a finely divided suspension or (2) assists in the formation of a gel, thus accelerating settling of the particles or their deposition on a substrate.

Coagulation *n.* (1) A physical or chemical action inducing transition from a fluid to a semi-solid or gel-like state, or the bringing together of small, individual particles into clumps. (2) Process whereby a fluid liquid is changed into a thickened, curdled or congealed mass. (3) Irreversible agglomeration of particles originally dispersed in a rubber latex.

Coagulation bath *n.* A liquid bath that serves to harden viscous polymer strands into solid fibers after extrusion through a spinneret. Used in wet spinning processes such as in rayon or acrylic fiber manufacture.

Coagulum *n.* An agglomerate of particles.

Coalesced filaments *n.* Filaments stuck together by design or accident during the extrusion process.

Coalescence *n.* The formation of a film of resinous or polymeric material when water evaporates from an emulsion or latex system, permitting contact and fusion of adjacent latex particles. Action of the joining of particles into a film as the volatile evaporates.

Coalescent (coalescing agent) *n.* Solvent with a high bp which, when added to a coating, aids in film formation via temporary plasticization (softening) of the vehicle.

Coal tar *n.* Coal tar which is also known as *aniline* is an oily, colorless, toxic liquid, which darkens upon exposure to air. It is soluble in water, alcohol, and ether, and forms a number of salts. It can also be described as a dark brown to black cementitious material produced by the destructive distillation of bituminous coal.

Coal tar colors *See aniline pigments.*

Coal tar epoxy *n.* Modified coal tar floor coating designed to bond to asphalt and concrete surfaces. High strength and resistant to most corrosive reagents.

Coal tar epoxy coating *n.* Coating in which binder or vehicle is a combination of coal tar with epoxy resin.

Coal tar hydrocarbons *n.* Aromatic hydrocarbons derived from coal tar, including benzene, toluene, xylene, naphtha, etc.

Coal tar pitch *n.* Distillation residue from coal tar. It varies considerably from a very soft to a very hard product. Fusion points vary from as low as 27°C (80°F) to as high as 232°C (450°F).

Coal tar pitch coatings *n.* Most plastics plated today are finished with a copper/nickel/chrome electroplate, but many other finishes are possible, such as bright brass, antique brass, satin nickel, silver, black chrome, and gold. The actual composition of the electroplate is designed for a particular application of the electroplate.

Coal-tar resin *See coumarone-indene resin.*

Coal tar-urethane coating *n.* Coating in which binder or vehicle is a combination of coal tar with a polyurethane resin.

Coarse end *See coarse thread.*

Coarse filling *See coarse thread.*

Coarse grain *n.* Wood with wide and conspicuous annual rings having considerable difference between springwood and summerwood. The term is sometimes used to designate wood and large pores, such as oak, ash, chestnut and walnut, but in this sense the term "coarse texture" is more often used.

Coarse pick *See coarse thread.*

Coarse thread *n.* A yarn larger in diameter than other yarns being used in the fabric.

Coat *n.* Paint, varnish or lacquer applied to a surface in a single application (one layer) to form a properly distributed film when dry. A coating system sexually consists of a number of coats separately applied in a predetermined order at suitable intervals to allow for drying or curing. It is possible with certain types of material to build up coating systems of adequate thickness and opacity by a more or less continuous process of application, e.g., wet-on-wet spraying. In this case, no part o the system can be defined as a separate coat in the above sense.

Coated abrasive *n.* A flexible-type backing upon which a film of adhesive holds and supports a coating of abrasive grains. The backing may be paper, cloth, vulcanized fiber or a combination of these materials. Various types of resin and hide glues are used as adhesives. The abrasives used are flint, emery, crocus, garnet, aluminum oxide and silicon carbide.

Coated fabric *n.* A cloth that has been impregnated and/or coated with a plastic material in the form of a solution, dispersion, hot melt, or powder. The term is sometimes used when a preformed film is applied to the fabric by calendaring, although such products are more properly termed laminates.

Coated paper *n.* A paper coated with clay, other white pigments and a suitable binder.

Coater Apparatus, which applies paint.

Coathanger die *n.* A sheet-, or film-extrusion die whose melt-distribution manifold has the obtuse-isoceles outline of a coathanger. This popular die design is said to yield uniform distribution of material across the full width of the extruded web, thus producing sheet of laterally more uniform thickness. Side-fed blow-molding dies and spiral-type dies for blown film may also be spoken as a coathanger dies.

Coating *n.* (1) Generic term for paints, lacquers, enamels, printing inks, etc. (2) A liquid, liquefiable or mastic composition, which is converted to a solid protective, decorative, or functional adherent film after application as a thin layer. (3) A composition, which when applied in thin layers, forms a non-tacky, adherent film that hides, protects, and/or decorates the substrate.

Coating, dip *n.* The process in, which a substrate is immersed in a solution (or dispersion) containing the coating material and withdrawn.

Coating methods *See:*

Air-knife coating	*Gravure-coating*
Calender coating	*Intumescent coating*
Cascade coating	*Kiss-roll coating*
Curtain coating	*Painting of plastics*
Decorating	*Plasma-spray coating*
Dip coating	*Printing on plastics*
Electrophoretic deposition	*Reverse-roll coating*
Electroplating on plastics	*Roller coating*
Electrostatic fluidized-bed coating	*Silver-spray process sinter coating*
Electrostatic spray coating	*Solution coating*
Extrusion coating	*Spray coating*
Flame-spray coating	*Spread coating*
Flocking	*Strippable coating*
Flow coating	*Transfer coating*
Fluidized-bed coating	*Urethane coatings*
Friction calendaring	*Vacuum metalizing*

Coating powders *n.* Finely divided, solid plastic materials, which are heat fusible and form relatively smooth, tough, electrical insulating coatings upon application to metal surfaces.

See powder coating.

Coating, spray *n.* The process in which a substrate is sprayed with the coating material.

Coating system *See coat.*

C

Cobalt \ˈkō-ˌbólt\ [Gr *Kobalt*, alter. of *Kobold*, literally, bobbin, fr. MHGr. *kobolt;* fr. its occurrence in silver ore, believed to be due to globins] (1683) *n.* A tough lustrous silver-white magnetic metallic element that is related to and occurs with iron and nickel and is used in alloys.

Cobalt aluminate *See cobalt blue.*

Cobalt blue (1835) *n.* $CoO{\cdot}Al_2O_3$. Pigment blue 28 (77346). Bright blue pigment, which is a complex product derived from oxides of cobalt and aluminum. It has excellent lightfastness and good chemical and heat resistance and with low opacity. Density, 4.2–4.3 g/cm^3 (35.0–35.8 lb/gal); O.A., 27–36; particle size, 0.5–1.0 μm. Syn: cobalt aluminate, king's blue, thenard's blue, azure blue, cobalt ultramarine, dumont's blue, enamel blue, gahn's ultramarine, Hungary blue, leitner's blue, leyden blue, Vienna blue, wentzel's blue, and zaffre.

Cobalt chloride (1885) *n.* A chloride of cobalt. The dichloride $CoCl_2$ that is blue when dehydrated, turns red in the presence of moisture, and is used to indicate humidity.

Cobalt drier *n.* One of the many organic cobalt salts (cobalt, naphthenate, cobalt octoate, etc.) which are soluble in paints and varnishes and used to speed the drying and hardening of the oil vehicle. They are also used to accelerate oxidation and polymerization of an ink film.

Also used as an initiator in polyesters.

Cobalt green *n.* Similar to cobalt blue, except that zinc oxide replaces wholly or partly the aluminum oxide in the latter. It does not contribute to good hiding power.

Also known as Rinmann's green or zinc green.

Cobalt naphthenate *See naphthenic acid.*

Cobalt 60 (1946) *n.* A heavy radioactive isotope of cobalt of the mass number 60 produced in nuclear reactors and used as a source of gamma rays (as for radiotherapy).

Cobalt ultramarine *n.* Red shade cobalt blue. *See cobalt blue.*

Cobalt yellow *See aureolin.*

Cobwebbing *n.* Production of fine filaments instead of the normal atomized particles when some coatings are sprayed. Although generally considered a defect in ordinary lacquers, use is made of this property to provide textured coating or a protective covering for equipment during storage. A cocoon is formed around the article by the pronounced cobwebbing action.

Cobwebbing (in gravure) *n.* A filmy, weblike build-up of dried ink or clear material on the doctor blade, ends of impression roll, or engraving.

Cobweb coating *See cobwebbing.*

Cocatalyst *n.* Chemicals which themselves are rather weak catalysts, but which greatly increase the activity of a given catalyst; also called promoters.

Cochineal \ˈkä-chə-ˌnēl, ˈkō-\ [MF & Sp; MF *cochenille*, fr. OSp *cochinilla* cochineal insect] (1582) *n.* Natural organic dyestuff made from the bodies of the female insect. *Coccus Cacti*, which lives on plants in Central and South America. The coloring principle is known as carminic acid and is generally laked. It is not fast to light.

Cochin oil *n.* Refined grade of coconut oil.

Cockled yarn *n.* Spun yarn in which some fibers do not lie parallel to the other fibers but instead are curled and kinked, forming a rough and uneven surface on the yarn. The general cause is fiber overcut to the extent that the drafting rolls catch and hold both ends of the fiber at the same time while attempting to draft, resulting in slippage or breakage (*Also see overcut*).

Cockling *n.* A crimpiness or pucker in yarn or fabric usually caused by lack of uniform quality in the raw material used, improper tension on yarn in weaving, or weaving together yarns of different numbers.

Coconut oil *n.* Expressed from the nut kernels of the coconut palms, the oil is composed of glycerides of lauric acid, capric acid, myristic, palmitic and oleic acids. Its main use is in non-yellow alkyds. Properties: Sp. gr 0.9190/25°C; refractive index, 1.4545/40°C; iodine value, 9; and saponification value, 255.

Co-cure *n.* To cure two or more different materials in one step.

Coefficient of cubical expansion *See coefficient of expansion.*

Coefficient of elasticity *n.* Reciprocal of Young's modulus in a tension test. A rarely seen Syn: modulus of elasticity.

Coefficient of expansion *n.* Ratio of increase in length, area, or volume of a substance for a 1°C rise in temperature.

Also known as Coefficient of Cubical Expansion.

Coefficient of friction *n.* Measure of the resistance to sliding of one surface in contact with another surface.

See friction.

Coefficient of plastic viscosity *See viscosity, plastic.*

Coefficient of scatter *See scatter, coefficient of.*

Coefficient of thermal conductivity *See thermal conductivity.*

Coefficient of thermal expansion *n.* The fractional change in length (or sometimes in volume, when specified) of a material for a unit change in temperature, as given by the equation:

$$CE = (1/L)\,dL/dT = d\ \ln L/dT \approx \Delta L/\Delta T.$$

See also volume coefficient of thermal expansion.

Coefficient of twist contraction *n.* The shortening of a yarn, due to twist, per unit length of untwisted yarn, usually expressed in percent.

Coefficient of viscosity (1866) *n. See viscosity* (2).

Coextrusion The process by which the outputs of two or more extruders are brought smoothly together in a feed block to form a single multiplayer stream that is fed to a die to produce a layered extrudate. The extruder streams may be split within the feed block to form dual layers, usually in a symmetrical arrangement about the center plane of the final sheet. Sheeting containing up to nine layers is commercially produced. Coextrusion is employed in film blowing, sheet and flat-film extrusion, blow molding, and extrusion coating. The advantage of coextrusion is that each ply imparts a desired characteristic property, such as stiffness, heat-sealability, impermeability, or resistance to some environment, all of which properties would be impossible to obtain with any single material. Layers of poorly compatible plastics can be coextruded by including a thin adhesive layer between them.

Coextrusion blow molding *n.* A variant of extrusion blow molding in, which the parison contains two or more layers of at least two materials.

See coextrusion.

Cogswell rheometer *See extensiometer.*

Cohesion *n.* Propensity of a single substance to adhere to itself; the internal attraction of molecular particles toward each other; the ability to resist partition from the mass;

internal adhesion; the force holding a single substance together.

Cohesive energy (density) *n.* For liquids the heat of vaporization per unit mass, divided by the specific volume, or the same quantity based on molar properties. Cohesive energy density is also equal to the square of the solubility parameter.

$$\mathrm{CED} = \Delta E_v / V_1.$$

For liquids, the heat of vaporization per unit mass, divided by the specific volume, or the same quantity based on molar properties. Cohesive energy density is also equal to the square of the solubility parameter.

Coil coating *n.* High-speed process, which applies paint and other coatings to a continuous flat coil of metal. A continuous coil of metal is unwound, cleaned, surface-treated, coated, heat-cured, cooled and rewound in one operation. The coated coil is subsequently unwound and formed into any number of products, such as house siding, Venetian blinds, and automotive and appliance parts.
Also called strip coating.

Coiling *n.* The depositing of sliver into cylindrical cans in helical loops. This arrangement permits easy removal for further processing.

Coiling soup *n.* In coating or spreading operations, a defect caused by the curling or turning motion of a bank of lacquer or other "soup" compound in front of a doctor knife, or similar application device, wherein streaks are formed on the surface of the coated fabric due to uneven application of the soup.

Coil yarn *See textured yarns.*

Coining *n.* A term borrowed from the metal-stamping industry for a process of forming integral hinges from plastics. In the case of a polypropylene article, the hinge is produced by molding a thin section between the two parts of the article to be hinged. Such a thin section cannot be molded easily in articles of nylon or acetal resin because of the difficulty of filling the half of the mold cavity opposite the gated half through the thin section. In the coining process, the area to be formed into a hinge is molded in a thickness suitable fore the molding process. Subsequently, the article is placed between bars in a press that quickly squeezes the plastic to the desired thickness. The material must be deformed beyond its compressive yield point but short of failure, so that is remains essentially stable with little recovery from the deformation. This rapid cold pressing produces a high degree of orientation that imparts high strength and flexibility to the integral hinge area.

Coinjection *n.* A process similar in its results to coextrusion but accomplished by modifications of the injection-molding process. By means of various nozzle and valving arrangement, two or more materials, can be injected either simultaneously or sequentially to form an article with an outer shell of one material with certain desired properties, the shell filled with another material to attain other desired properties such as reduced cost. Coinjection, like coextrusion, basically means that two or more different plastics are formed into a composite or laminated structure.

Coin marking *See metal marking.*

Coking *n.* In the running of copals, or sweating of asphaltums, and other materials, which naturally contain a proportion of infusible dirt or mineral matter, this extraneous matter is likely to escape the action of the stirrer. As a consequence, any organic matter associated with it is liable to be overheated, and carbonization occurs. This ultimately results in the formation of carbonaceous lumps, and it is then said to have coked.

Cold *n.* Dull, flat surface where a bright, lustrous one is desired.

Cold-bend test *n.* A test for measuring the flexibility of a plastic material at low temperatures. A specimen in a series is chilled to one of several specified low temperatures, then bent to a predetermined radius until the temperature at which half the specimens tested do not survive the bend has been identified.

Cold-checking *n.* Surface defect on lacquer, especially furniture finishes. It appears as a pattern of fine cracks developing on film surface, when the finished article is subjected to alternate hot and cold conditions. *See checking and temperature checking.*

Cold cracking *n.* Crazing and cracking of a coating subjected to low temperature or cold/ambient cycling.

Cold-curing *n.* Process of curing at normal atmospheric temperature (i.e., air dry).

Cold cut *n.* Dissolving a resin or other material in a suitable solvent by mechanical agitation without the application of heat.

Cold cycle *n.* Low temperature cycle during many accelerated tests.

Cold drawing (cold stretching) *n.* A stretching process performed at a temperature below a thermoplastic's melting range to orient the material and improve the tensile modulus and strength.

Cold flex *n.* The lowest temperature at which the test strip can be twisted through a 200° arc without breaking.

Cold flow *n.* Continuing distortion, deformation, or dimensional change which take place in materials under continuous load (constant stress) at ambient temperatures. Syn: compression set.
See creep.

Cold forming *n.* A group of processes by which sheets or billets of thermoplastics are formed into three-dimensional shapes at room temperature by processes used in the metal-working industry such as forging, brake-press bending, deep drawing, rolling, stamping, heading, and coining. The materials used, generally in relatively thick sections, include ABS, polycarbonate, polyolefins, and rigid PVC. When either the material or the forming dies are preheated, the preferred term is solid-phase forming.

Cold heading *n.* A process for forming short plastic rods into rivets by uniformly loading the projecting shaft end in compression while holding and containing the shaft trunk. All thermoplastics can be cold-heated but acetal and nylon are particularly suitable.

Cold molding *n.* A process similar to compression molding except that no heat is applied during the molding cycle. The formed part is subsequently cured by heating and cooling. A-stage phenolic resins and bituminous plastics are sometimes molded by his process.

Cold-parison blow molding *See blow molding.*

Cold pressing *n.* (1) Assembly method in which the bonded structures are held in place by pressure without the application of heat or drying air until the adhesive interface has solidified and reached proper shear proportions. (2) An extraction method for oils such as fish oil.

Cold-process roofing *n.* A bituminous roofing membrane, which consists of layers of coated felts that have been bonded with cold-applied asphalt roof cement and surfaced with an emulsified or cutback asphalt roof coating.

Cold-rolled steel *n.* Low-carbon, cold-reduced sheet steel.

Cold rubber *n.* A synthetic rubber made at a relatively low temperature (about 40°F or 4°C) and having greater strength and durability than that made at the usual temperature (about 120°F or 49°C.

Cold-runner injection molding (runnerless injection molding) *n.* Whereas in injection molding thermoplastics the runners are sometimes kept hot to reduce scrap (*See hot-runner mold*), with thermosets the runners are kept cooler than the cavities to prevent material from curing within the runner system. In cold-runner injection molding of thermosets the mold is divided into two sections: a heated section containing the cavities, and an insulated manifold section containing the injection sprue and runners. Material is fed from runners to cavities through very short gates or sub-sprues. The insulated manifold is maintained at a temperature high enough to soften the uncured material, generally in the vicinity of 90°C, but well below the curing temperature prevailing in the cavity section.

Cold-setting *n.* Resin or lacquer products, which, under the influence of suitable catalysts, are able to form suitable film properties without the application of heat.

Cold-setting adhesive *n.* Synthetic resin adhesive capable of hardening at normal room temperature in the presence of a hardener.

See adhesive, cold-setting.

Cold setting inks *n.* Solid inks, which must be melted and applied on a hot press. They solidify again on contact with unheated paper.

Cold slug *n.* The first material to enter an injection mold, so called because in passing through the sprue orifice it is cooled below the effective molding temperature. In some molds, a small well in the mold opposite the sprue catches the cold slug and thereby prevents it from entering the runner system.

Cold-slug well *n.* A small circular cavity, directly opposite the sprue opening in an injection mold that traps the cold slug.

Cold stretch *n.* Pulling operation, usually on extruded filaments, to improve tensile properties.

Cold test *n.* Refers to pour point. A low cold test means that the product has a low temperature pour point.

Cold water paint *n.* Paint in which the binder or vehicle portion is composed of soybean or other vegetable protein, glue, resin emulsion, or other similar material dispersed in water.

Collacral k *n.* Industrial poly(vinyl pyrrolidone). Manufactured by BASF, Germany.

Collage *n.* Literally, pasting. Employed to describe the practice of some 20th-century artists in introducing paper or other non-pigmented materials in painting. Basically, a pasting technique whereby pictorial images or patterns, and pieces of colored, textured material are superimposed on each other.

Collapse *n.* (1) Inadvertent densification of cellular material during manufacture resulting from breakdown of cell structure (ASTM D 883). (2) Inward contraction of the walls of a molded container, e.g., while cooling, resulting in permanent indentation.

Collet \ˈkä-lət\ [MF, dim. of *col* collar, fr. L *collum* neck] (1528) *n.* Rigid lateral contained for the mold-forming manual; a dam, a restriction box; the drive wheel that pulls glass fibers from the bushing [A forming tube is placed on the collet and a package of strand (forming cake) is wound up on the tube].

Colligative properties (Polymers) *n.* Vapor pressure lowering, freezing point depression, boiling point elevation and osmotic pressure, measurements or properties dependent on the number of molecules (i.e., pressure) and not the chemistry of the molecule.

Colligative property *n.* A property of a material or solution that depends mainly on the number, rather than the nature, of the molecules, atoms, or ions present. Examples are gas pressure and osmotic pressure of solutions.

Collimated roving *n.* Roving with strands that are more nearly parallel than those in standard roving, usually made by parallel winding.

Collision frequency factors *n.* The collision frequency factor p (number of collisions ($\sim 10^{11}$/s) is included in the "collision theory" to describe the temperature dependence of the rate constants of elementary reactions:

$$k_i = pZ\,\exp(-E/RT) = A\exp(-E/RT),$$

where Z is steric factor, E is Arrhenius activation energy ($-E/RT$) is the Boltzmann factor, and often p and Z are incorporated into one constant A; the second theory is the "transition state theory".Houston PL (2001) Chemical kinetics and reaction dynamics. McGraw-Hill Co., New York.

Collodion \kə-ˈlō-dē-ən\ [mod. of NL *collodium*, fr. Gk *kollōdēs* glutinous, fr. *kola* glue] (1851) *n.* A solution of cellulose nitrate in alcohol and ether used as a coating lacquer or to cast very thin films of cellulose nitrate on water ("microfilm").

Collodion cotton *See cellulose nitrate.*

Colloid \ˈkä-ˌlóid\ [ISV *coll-* + *-oid*] (ca. 1852) *n.* A dispersion of one phase in another in which the particles or units of the dispersed phase have at least one dimension which is larger than usual molecular dimensions, but two small to be observed visually. In other words, a phase dispersed to such a degree that the surface forces become an important factor in determining its properties. In general particles of colloidal dimensions are approximately 10 Å to 1 μm in size. Colloidal particles are often best distinguished from ordinary molecules due to the fact that colloidal particles cannot diffuse through membranes, which do allow ordinary molecules and ions to pass freely. Pertaining to colloid.

Colloidal clay *See bentonite.*

Colloidal dispersion *See colloid.*

Colloidal particle *n.* Large molecules or aggregates of smaller molecules in a size range of 1–500 nm (10^{-9} m).

Colloidal solution *n.* Any solution containing colloidal particles. Colloidal solutions are sometimes called sols.

Colloidal state *n.* Particular state in which any substance may exist under the proper conditions, determined by fineness of particle subdivision. The colloidal state is defined by a more or less well-marked ultramicroscopic zone in the scale of subdivision, the lower extreme of the zone approaching molecular dimensions, and the upper end gradually passing over into molecular aggregates (suspensions) visible under the ordinary microscope. Becher P (1989) Dictionary of colloid and surface science. Marcel Dekker, New York.

Colloid, colloidal state *n.* A state of a substance in the form of small particles dispersed in a medium; and colloid size particles that do not precipitate from a media; smoke and fog are colloidal dispersions.

Colloid, linear *n.* Any colloid with fibrous particles or macromolecules. The long, asymmetrical particles may be coiled up or, in some instances, they may be branched.

Colloid, micellar *n.* Colloid composed of micelles. A micellar colloid is an aggregation of molecules, usually arranged in a definite order, whose dimensions are less than 0.05 μm. Example: a colloidal solution of soap in water.

Colloid mill *n.* A device for preparing emulsions and reducing particle size, consisting of a high-speed rotor and a fixed or counter-rotating element in close proximity to the rotor. The liquid is conveyed continuously from a hopper to the space between the shearing elements, and then discharged into a receiver.
See also homogenizer.

Colloid, molecular *n.* Any colloid in which the colloidal particles are macromolecules.

Colloid, protective *See protective colloid.*

Colloid, reversible *n.* Colloid substance that can be converted from the gel to the solid form without the expenditure of chemical energy, that is, by simply warming the gel form or redissolving the dried substance in water. It is a colloid which forms a hydrophilic solid, such as gelatin, agar, gum arabi, starch; a protective colloid.

Colloid, sphere *n.* Colloid possessing more or less symmetrically shaped, corpuscular, and relatively compact particles.

Collotype \ˈkä-lə-ˌtīp\ [ISV] (1883) *n.* (1) A printing process utilizing a glass plate with a gelatin surface carrying the image to be reproduced. (2) A print made by this process.
Also called photogelatin process.

Colmonoy *See hard-facing alloy.*

Cologne earth, cologne brown *n.* A pigment, sometimes referred to erroneously as Vandyke brown, made from roasted American clays which contain ochre and bituminous matter.
See Vandyke brown.

Cologne spirits *See ethyl alcohol.*

Cologne yellow *See chrome yellow and light and primrose.*

Colonial spirit *n.* Another name for methyl alcohol.

Colophony, colophonium *n.* Old name for rosin; used in Europe.
See rosin.

Colorimeter, tristimulus *See tristimulus colorimeter.*

Color *n.* Color is a basic specification for plastic parts; one of its major properties is its ability to be integrally colored, almost without restriction; in viewing a colored object three factors observed: (1) The quality of light illuminating the object; (2) The ability of the object to absorb certain portions of the light and to reflect others; and (3) The sensitivity of the eye to the reflected illumination. Mcdonald R (ed) (1997) Colour physics for industry; colour physics for industry, 2nd edn. Society of Dyers and Colourists, West Yorkshire, England. Color, universal language and dictionary of names. NBS Special Publication 440, Stock No. 003-003-01705-1, Washington DC. Computer colorant formulation. Kuehni, Lexington Books, Farnsborough, England, 1975. Hardy CH (1936) Handbook of colorimetry. MIT Press, Cambridge, MA.

Color abrasion *n.* Color changes in localized areas of a garment resulting from differential wear.

Color amplitude test *n.* Test designed to examine any of the factors involved in color aptitude. Most widely used is the ISCC Color-matching aptitude text, or CAT, based on differences in chroma.

Colorant *n.* Any dye or pigment that can impart color to a plastic. The dyes are natural or synthetic compounds of submicroscopic or molecular size, soluble in most common solvents, yielding transparent colors. Their generally poor heat resistance and tendency to migrate limit their use as additives to a few families that are superior in heat resistance. However, dyes are sometimes used to post-color finished parts such as buttons and fibers. The pigments are organic and inorganic substances with larger particle sizes, rarely below 1 μm, and usually insoluble in the common solvents. Organic pigments produce translucent and nearly transparent colors, resist migration better than dyes, and are somewhat more, heat-resistant. Inorganic pigments are, with few exceptions, opaque and superior to organics in light-fastness, heat resistance and resistance to migration. Colorants are added to plastics by dry-coloring (simply tumbling the colorant with the base or compounded resin power or pellets); by extrusion coloring (extruding a dry-colored mixture and chopping it into pellets to be reprocessed); by masterbatching (*See color concentrate*); or by stirring colorants or dispersions thereof into liquid plastisols or resin systems. *See also*:

Bon pigment	*Organic pigment*
Fluorescent pigment	*Pearlescent pigment*
Flushed pigment	*Perylene pigment*
Glitter	*Phosphorescent pigment*
Inorganic pigment	*Phthalocyanine pigment*
Liquid colorant	*Quinacridone pigment*
Luminescent pigment	*Rhodamine*
Metallic-flake pigment	*Ultramarine-blue pigment*

Herbst W, Hunger K (2004) Industrial organic pigments. John Wiley and Sons Inc, New York. Kirk–Othmer encyclopedia of chemical technology: pigments-powders. John Wiley and Sons, New York, 1996. Paint/coatings dictionary. Compiled by Definitions Committee of the Federation of Societies for Coatings Technology, Philadelphia, PA, 1978.

Colorant dispenser *n.* Mechanical device used to disperse precisely measured volume amounts (usually) of colorants for the purpose of tinting or shading coatings bases.

Colorant match *n.* Color match made by using the same colorants in the match as were used in the standard.

Colorant mixture *n.* Mixture of colorants containing pigments, or dyestuffs, or both. The color of a mixture may be predicted by subtractive colorant mixture theories. *See subtractive colorant mixture, complex and simple.*

Colorants *n.* Colored pigments and dyes.

Color aptitude *n.* Ability to work with color, involving inborn factors as well as acquired abilities.

Color bleeding *See bleeding.*

Color blindness (deprecated) *n.* An incorrect term applied to defective color vision. An extremely small number of persons having complete lack of color response (are achromatopes). Most so-called color-blind persons are anomalous trichromats, seeing all three primaries but having responses, which are weaker than normal to one of the primaries. Persons who lack response to one primary are called dichromats, seeing only two primaries.

Color burn-out *n.* An objectionable change in the color of a printing ink which may occur either in bulk or on the printed sheet. In the former case it is associated primarily with tints, and is caused by a chemical reaction between certain components in the ink formulation. In the latter case it is generally caused by heat generated in a pile of printed material during the drying of an oxidizing type of ink.

Color center *n.* A kind of point defect in a crystal.

Color code *n.* A system of coloring piping and parts of equipment with various pre-coded colors for identification purposes.

Color compounding Formulation and preparation of colored materials for incorporation into molding of plastics.

Color concentrate *n.* A plastic compound that contains a high percentage of pigment, to be blended in precise amounts with the base resin or compound so that the correct final color will be achieved. The concentrate provides a clean and convenient method of obtaining accurate color shades in extruded and molded products. The term masterbatch is sometime used for color concentrate, as well as for concentrates of other additives.

See also multifunctional concentrate.

Color concentrates *n.* Concentrated color dyes and pigments.

Color constancy *n.* Relative independence of perceived object color to changes in color of the light source.

Color coordinates *See chromaticity coordinates, general.*

Color difference *n.* Magnitude and character of the difference between two colors under specified conditions.

Color difference equations *n.* Equations that transform CIE coordinates into a more uniform matrix such that a specified distance between two colors is more nearly proportional to the magnitude of an observed difference between them regardless of their hue. The total color difference is generally designated as ΔE, the lightness difference as ΔL, the total chromaticity difference as ΔC, the redness–greenness difference as Δa, $\Delta \alpha$, or ΔRG, and the yellowness–blueness difference as Δb, $\Delta \beta$, and ΔYB. The directional differences, such as ΔL, ΔRG, and ΔYB are frequently designated as the components of ΔE. Color difference equations may also be based on Munsell notations, although such equations are seldom used today. McDonald R (1997) Colour physics for industry, 2nd edn. Society of Dyers and Colourists, West Yorkshire, England. Billmeyer FW, Saltzman M (1996) Principles of color technology. John Wiley and Sons Inc., New York.

Colored cement *n.* A cement to which color pigment has been added.

Colorfast *n.* Fade-resistant.

Colorfastness *n.* Resistance to fading; i.e., the property of a dye to retain its color when the dyed (or printed) textile material is exposed to conditions or agents such as light, perspiration, atmospheric gases, or washing that can remove or destroy the color. A dye may be reasonably fast to one agent and only moderately fast to another. Degree of fastness of color is tested by standard procedures. Textile materials often must meet certain fastness specifications for a particular use.

See light resistance and lightfastness.

Color floating *See floating.*

Color gamut *See gamut, color.*

Color harmony manual *n.* Orderly array of colors spaced (approximately) according to the Ostwald color system, and made by the container corporation of America. Several editions were issued, each slightly different. It is no longer available.

See Ostwald color system.

Colorimeter \ˌkə-lə-ˈri-mə-tər\ [ISV] (ca. 1872) *n.* Instrument used to measure light reflected or transmitted by a specimen. Two general types of colorimeters are commonly used: (1) for measuring concentrations of colored materials for analytical purposes, and (2) for measuring quantities, which can be correlated with a psychophysical description of color. Generally, the second type should properly be referred to as a tristimulus colorimeter.

Colorimetric *n.* Adjective used to refer to measurements converted to psychophysical terms describing color or color relationships.

Colorimetric purity *See purity, colorimetric.*

Colorimetry (color-identification testing) *n.* Light measurements converted to a psychophysical description or notation, which can be correlated with visual evaluations of color and color differences.

Color index name *See color index name.*

Color index number *See color index number.*

Coloring aids *n.* Bisbenzoxazoles, triazinephenylcoumarins, and bis(styrl)-bisphenyls comprise the most widely used structural types of coloring additives. Used in conjunction with dyes and pigments, coloring aids absorb ultraviolet radiation and emit a blue-violet fluorescence, making polymer surfaces appear bright or cleaner.

Color in oil *See color index name.*

Color in varnish *See colorant.*

Colorist (1686) *n.* A person skilled in the art of color matching (colorant formulation) and knowledgeable concerning the behavior of colorants in a particular material; a tinter (in the American usage) or a shader. The word "colorist" is of European origin.

Coloristically *n.* A general term, which describes the nature of the comparison of two colorants in terms as used by a colorist. It may imply hue, saturation, or strength of any combination of these, as well as transparency, undertone, etc.

Coloristic properties *n.* A general term originating in Europe, which is derived from the characteristics, which a colorist sees and describes.

Color match Pair of colors exhibiting no perceptible difference when observed under specified conditions. The quality of an attempted match is described by the closeness of this ideal match.

Color matching *See matching, color.*

Color matching functions *n.* Relative amounts of three additive primaries required to match each wavelength of light. The term is generally used to refer to the CIE Standard Observer color matching functions designated $\bar{x} + \bar{y} + \bar{z}$.

See observer, standard.

Color measurement *n.* Color is a manufactured object is normally obtained by applying a colorant (dye or pigment) to al polymer substrate, such as textile, paper or paint medium. The appearance of such surface colors depends on (1) the nature of the prevailing illumination, (2) the interaction of the illuminating radiation with the colored species in the surface layers, and (3) the ability of the radiation that is transmitted, reflected and scattered from the colored surface to induce the sensation of color in the human eye/brain system. Color measurement consists of the physical measurement of light radiated, transmitted, or reflected by a specimen under specified conditions, and mathematically transformed into standardized colorimetric terms, which can be correlated with visual evaluations of colors relative to one another. McDonald R (ed) (1997) Colour physics for industry, 2nd edn. Society of Dyers and Colourists, West Yorkshire, England. Physical and chemical examination of paints, varnishes and lacquers and colors. Gardner Laboratory Inc., Bethesda, MD, 1946.

C

Color migration *n.* The movement of dyes or pigments through or out of a material.

Color mixture, additive *See additive color mixture.*

Color notion *n.* Orderly system of numbers, letters, or a combination of both, which serves to describe the relationship of colors in three-dimensional space. Thus, three dimensions must be included; for example, hue, value, and chroma of the Munsell System. Single dimensional notations, such as a yellowness sale, can be used only if the other two dimensions are fixed or described; two-dimensional notations can be used only if the third is fixed or described. *See color order systems.*

Color order systems *n.* Systems used to describe an orderly three-dimensional arrangement of colors. Three bases can be used for ordering colors: (1) an appearance basis, i.e., a psychological basis; in terms of hue, saturation, and lightness – an example is the Munsell System; (2) an orderly additive color mixture basis, i.e., a psychophysical basis – examples are the CIE System and the Ostwald System; and (3) an orderly subtractive color mixture basis – an example is the Plochere Color System, based on an orderly mixture of inks.

Color pigments *n.* Color materials for imparting color to concentrates and plastics parts.

Color proofs *n.* Prints, in color, either from the engraving or from the mounted plates.

Color rendering index *n.* Method for describing the effect that a particular light source has on the color appearance of objects, in comparison with their color appearance under a reference light source. The Illuminating Engineering Society of the USA and CIE Committee 1.3.2 have developed a method which is currently under study. A color rendering index of 100 indicates perfect agreement between the test source and the reference source.

Color retention *n.* The property that a material has when it is exposed to the elements and shows no signs of changing color.

Color space *n.* Three-dimensional solid enclosing all possible colors. The dimensions may be described in various geometries giving rise to various spacings with the solid. *See color order, systems and uniform color space.*

Color specification *n.* Term used loosely to describe either (1) the notation for a color standard using one of the color order systems, or (2) the allowable tolerance of samples from a standard in terms of color difference units. For (1), the best specification is a sample of the color desired.

Color stability *n.* The constancy of the characteristics of color in a plastic compound – hue, intensity, and saturation – in its products over their service lives in their design environments.
See light resistance.

Color standard *n.* An ink, wet sample, or printed proof to which another similar material is compared.

Color strength *n.* In printing ink, the effective concentration of coloring material per volume.
Also known as tinctorial strength.

Color stripper *n.* A chemical used to remove some or all of the dyestuffs from a fiber, yarn, or fabric so that a dyeing defect can be corrected, a shade lightened, or another color applied.

Color temperature *n.* Term used to describe the color of a Planckian (black-body) radiator. The color is expressed in chromaticity coordinates, x and y, determined from theoretical spectral power distribution curves, and the temperature is expressed in the absolute (Kelvin) scale. This term is sometimes used incorrectly to describe the color of "white" light sources.

Color testing instruments *n.* Colorimeters or color spectrophotometers for measuring wavelength and tristimulus values for characterizing color.

Color tolerance *n.* Limit of color difference from a standard, which is acceptable. It is generally expressed in terms of a particular color difference equation, which must be described, and may consist of a total color difference, ΔE, or directional color differences, or both.
See color difference equations.

Color vision, anomalous *n.* Term "anomalous" is used to imply the defective vision of an anomalous trichromat, an observer who sees three primaries but has weaker than normal response to one. The degree of the anomaly varies from slight to severe.
See color blindness.

Color vision, defective *n.* General term used to describe abnormal vision. The term is preferred to the common term "color blindness".
See color blindness and color vision, anomalous.

Color vision, normal *n.* Vision of observers requiring mixtures of three independent primary colors to color-match all colors, using quantities of the primaries sufficiently close to those required for the average observer. The observer with normal vision is referred to as a normal trichromat.

Color wash *n.* Each pigments, with or without whiting, lightly bound in glue size so as to facilitate ready removal since frequent removal is necessary, e.g., tinted lime wash.

Color *n.* British spelling of "colour"; used in English-speaking countries of the old British Empire, including Canada, Australia, New Zealand, etc.

Color index (C.I.) *n.* A publication of the Society of Dyers and Colourists (Great Britain) and the American Association of Textile Chemists and Colorists, which includes periodic additions and amendments. It provides C.I. generic names for classifying commercial dyes and pigments with respect to usage and C.I. numbers for classifying them with respect to chemical composition. It gives basic chemical and usage information, performance characteristics, and manufacturers of them.

Color index name *n.* Consists of the category, hue, and an identifying number. For example, one phthalocyanine green pigment is C.I. Pigment Green 7.
See preface for details.

Color index number *n.* A five-digit number which describes the chemical constitution. For example, phthalocyanine green, C.I. Pigment Green 7, is C.I. 74260 (all phthalocyanines are, or will be, numbered 74000–74999).

Colza oil \ˈkäl-zə, ˈkōl-\ [F, fr. D *koolzaad*, fr. MDu *coolsaet*, fr. *coole* cabbage + *saet* seed] (1712) *n.* Another name for rapeseed oil, usually reserved for refined grades.

Coma *n.* An aberration of spherical lenses, occurring in the case of oblique incidence, when the bundle of rays forming the image is unsymmetrical. The image of a point is comet shaped, hence the name.

Comb *n.* Thin spring-steel tool used in graining or combing.

Combed sliver *n.* A continuous band of untwisted fiber, relatively free of short fibers and trash, produced by combing card sliver.

Combed yarn *n.* A yarn produced from combed sliver.
Also see combing.

C

Combination *n.* The termination of a polymer chain by combining chains which doubles the molecular weight. Monomers of different chemical structure, which are initiated to polymerize and form copolymers.
See comonomers.

Combination fabric *n.* A fabric containing: (1) different fibers in the warp and filling (e.g., a cotton warp and a rayon filling), (2) ends of two or more fibers in the warp and/or filling, (3) combination yarns, (4) both filament yarn and spun yarn of the same or different fibers, or (5) filament yarns of two or more generic fiber types. Combination fabrics may be either knit or woven. They should not be confused with blend fabrics. Although blend fabrics also contain more that one fiber, the same intimately blended spun yarn is present in both warp and filling.

Combination frequencies *n.* Two vibrations of arbitrary frequencies f_1 and f_2 when applied simultaneously to a non-linear (distorting) device will excite it to a motion containing not only the original frequencies, but also members of a set of "combination" frequencies given by $f_c = mf_1 + nf_2$, where m and n are integers. A resonator sharply tuned to any one of these frequencies, which may be produced in the non-linear device will resound to it with an amplitude depending on the type of non-linearity. The superheterodyne radio receiver depends on this phenomenon.

Combination mold Syn: family mold.

Combination yarn *n.* A piled yarn containing two or more yarns that vary in fiber composition, content, and/or twist level; or plied yarn composed of both filament yarn and spun yarn.

Combined-oxide formula *n.* $3Na_2O \cdot 2CaO \cdot 3Al_2O_3 \cdot 2CO_2 \cdot 6SiO_2$, $Na_2O \cdot Al_2O_3 \cdot 4SiO_2 \cdot 2H_2O$. A formula, which represents the constituents as "oxides", with the metallic oxides preceding the acid anhydrides, each arranged in the order of increasing valence. Water of composition appears last. Examples: cancrinite and analcite.

Combined water *n.* The water chemically held, as water of crystallization, by the calcium sulfate dehydrate or hemihydrate crystal.

Combined yarn *See combination yarn.*

Combing *n.* (1) Act of partially removing a coat of wet paint with combs to imitate the grain of wood or other pattern. (2) A step subsequent to carding in cotton and worsted system processing which straightens the fibers and extracts neps, foreign matter, and short fibers. Combing produces a stronger, more even, more compact, finer, smoother yarn.

Combining volumes *n.* Under comparable conditions of pressure and temperature the volume ratios of gases involved in chemical reactions are simple whole numbers.

Combining weight *n.* The combining weight of an element or radical is its atomic weight divided by its valence. Syn: equivalent weight.

Combining weights, law of *n.* If the weights of elements which combine with each other be called their "combining weights", then elements always combine either in the ratio of their combining weights or of simple multiples of these weights.

Combustible liquid *n.* Any liquid having a flp at or about 37.8°C (100°F). Combustible liquids are divided into two classes: *Class II* and *Class III*. *Class II* liquids – Those with a flp at or above 37.8°C (100°F) and below 60°C (140°F), except a mixture having components with flp of 93.3°C (200°F) or higher, the volume of which make up 99% or more of the total volume of the mixture. *Class III* liquids – Those with flp at or above 60°C (140°F). *Class III* is subdivided into two classes: *Class IIIA* – Those with flp at or above 60°C (140°F) and below 93.3°C (200°F), except a mixture having components with flp of 93.3° (200°F) or higher, the volume of which make up 99% or more of the total volume of the mixture; *Class IIIB* – Those with flp at or above 93.3°C (200°F). When a combustible liquid is heated for use to within 16.7°C (30°F) of its flp, it is handled in accordance with the requirements for the next lower class of liquids. Wypych G (ed) (2001) Handbook of solvents. Chemtec Publishing, New York.

See flammable liquid.

Combustion analysis *n.* Any of several methods for quantitatively determining, by burning, the elemental composition of organic compounds, including plastics. First introduced in the 1830s by J. von Liebig, it was refined to permit accurate analysis of small samples (10–50 mg) by F. Pregel, who led the development of the microbalance. Modern combustion analysis is highly automated, but still relies on the microbalance.

Comfort *n.* Performance parameter of apparel referring to wearability. Encompasses such properties as wicking, stretch, hand, etc.

Comic inks *See news inks.*

Commercial allowance *n.* The commercial moisture regain plus a specific allowance for finish used in calculating the commercial or legal weight of a fiber shipment.

Commercial blast *See NACE no. 3.*

Commercial moisture regain *n.* An arbitrary value adopted as the moisture regain to be used in calculating the commercial or legal weight of a fiber shipment.

Commercial weight *n.* (1) In natural fibers, the dry weight of fibers or yarns plus the commercial moisture regain. (2) In manufactured fibers, the dry weight of staple spun yarns or filament yarns after scouring by prescribed methods, plush the commercial moisture regain.

Commercial xylene *See xylene.*

Commingled yarn \kə-ˈmiŋ-gəl, kä-\ (ca. 1626) *v.* In aerospace textiles, two or more continuous multifilament yarns, the filaments of which have been intermixed with each other without adding twist or otherwise disturbing parallel relationship of the combined filaments. Usually consists of a reinforcing yarn, such as graphite or glass, and a thermoplastic matrix yarn.

Comminute \ˈkä-mə-ˌnüt, -ˌnyüt\ [L *comminutus*, pp of *comminuere*, fr. *com-* + *minuere* to lessen] (1626) *v.* To pulverize or reduce to very small sizes, as by grinding.

Comminution *n.* Process by which aggregates are reduced to small size.

Common-ion effect *n.* The shifting of an ionic equilibrium due to the addition of an ion involved in the equilibrium; usually refers to the repression of dissociation of a weak electrolyte or the decrease in the solubility of an electrolyte brought about by the addition of an ion, which is a dissociation product.

Comoforming *n.* A fabrication process that combines vacuum-formed thermoplastic shapes with cold-molded fiberglass-reinforced resin to produce parts having excellent surface appearance and weatherability.

Comonomer \(ˌ)kō-ˈmä-nə-mər, -ˈmō-\ [*co-* + *monomer*] (1945) *n.* A monomer that is mixed with one or more other monomers for a polymerization reaction, to make a copolymer.

Compact *See powder compact.*

C

Compacted yarns *n.* Air-jet interlaced yarns. Since the entanglement serves only as a substitute for twist, the degree of interlace or tangle is not as great as in air-jet bulked yarns.

Compacting *v.* Compacting of a plastic material and the forcing of it through an orifice in more or less continuous fashion. *See extrusion.*

Compaction *See intermingling.*

Compactor *n.* A machine developed by Fabric Research Laboratories which is used to compact fabrics or to produce warp-stretch fabrics by means of forced crimp and/or shrinkage of the warp yarn.

Compact spinning process *n.* A term generally referring to a spinning process carried out using any one of the several small spinning machines of compact design offered by equipment vendors as "packaged" units in which spinning and subsequent processing (drawing, crimping, cutting, etc.) are linked.

Compatibility *n.* The ability of two or more substances to mix together or be joined without objectionable separation. In plastics technology the term is most often used in connection with plasticizers, but is also applied to resin pairs or to a resin and prospective compounding ingredients. ASTM tests for compatibility of plasticizers and PVC resins are D 2383 and D 3291. *See also loop test.*

Compatibility of plasticizers *n.* Miscibility and stability of one or more plasticizers to plastize a polymer or resin.

Compatibilizer *n.* A material that, added to a blend of ordinarily incompatible polymers, suppresses phase separation.

Compatible shrinkage *n.* A term used for bonded fabrics to indicate that the face fabric and lining have similar shrinkage. This is necessary to avoid puckering.

Compensating eyepiece *n.* An overcorrected eyepiece designed to compensate for certain undercorrections in fluorite and apochromatic objectives.

Compensator *n.* An anisotropic substance of known retardation superimposed on the field of view with its vibration directions 45° from the vibration directions of the polarizer and analyzer. When an anisotropic particle is positioned so its slow component is parallel to the slow component of the compensator, the retardations are added. When the particle and compensator slow components are perpendicular, the retardations are subtracted. Compensator retardations may be fixed or variable.

Compensator, full-wave plate (first-order red) *n.* A layer of quartz, selenite, calcite, or an oriented polymer film of the proper thickness to produce a retardation equivalent to about 530 nm, the first-order red.

Compensator, quarter-wave plate A thin mica (or other crystal) plate or an oriented polymer film of uniform thickness having a retardation of about 130 nm, first-order gray.

Compensator, quartz wedge *n.* A wedge, cut from quartz, having continuously variable retardation extending over several orders (usually 3–7) of interference colors. The retardation, which exactly compensates that of a crystal can be found by pushing in the wedge while counting orders until it reaches a position at which the interference color of the crystal appears black. Retardation can be compensated only when the flow component of the crystal and the wedge are perpendicular.

Complementary colors *n.* Two colors which, when mixed together in the proper proportions, result in a neutral color. Colored lights, which are complementary, when mixed additively, form white light and follow the laws of additive color mixture. Colorants, which are complementary, when mixed together, form black or gray and follow the laws of subtractive colorant mixture. With the exception of spectrophotometric complementaries, which exist only in theory, colors, which are complementary depend on the illuminant chromaticity considered as the neutral point. McDonald R (1997) Colour physics for industry, 2nd edn. Society of Dyers and Colourists, West Yorkshire, England.

Complete hiding *See hiding, complete.*

Complex *n.* An ion or, sometimes, molecule, which consists of a central atom or ion surrounded by some peripheral atoms, ions, or molecules (ligands) bonded to it.

Complexation *n.* The formation of a complex.

Complex dielectric constant *n.* Vectorial sum of the dielectric constant and the loss factor; analogous to complex shear modulus and to complex Young's modulus.

Complexes, catalyst- cocatalyst *n.* Stereospecific chemical complexes, usually derived from a transition metal halide and a metal hydride or a metal alkyl. An example is in stereospecific polymerization of propylene to crystalline polypropylene.

Complexing agent *See chelating agent sequestering agent.*

Complex modulus (complex dynamic modulus) *n.* A property of viscoelastic materials subjected to periodic variation or reversal of stress (stress cyling). The stress may be any of the three principal types; the material may be in the solid or liquid state (molten or concentrated solution). In such materials, the strain lags the applied stress in time and when the stress is periodic, the time lag is characterized by a phase angle, θ. The modulus – the ratio of stress to strain – is resolved into two parts, a "real" or in-phase part and an "imaginary" part lagging the real part by $\pi/2$ radians (90°). The resultant "envelope" that develops between the stress and lagging strain is call the hystereis of the stress–strain relationship. For example, in shear, the complex modulus is stated as $G = G' + iG''$, where $i = (-1)^{0.5}$. The vector sum of the two components is called the absolute (dynamic) modulus. The real part is often referred to as the "modulus" while the imaginary part is called the "dynamic viscosity". Both parts vary with frequency, both diminish with rising temperature, and, like static moduli, they are different for different modes of stress. Sepe MP (1998) Dynamic mechanical analysis. Plastics Design Library, Norwich, New York. Pittance JC (ed) (1990) Engineering plastics and composites. SAM International, Materials Park, OH.

Complex shear modulus *n.* Vectorial sum of the shear modulus and the loss modulus; analogous to complex dielectric constant.

Complex subtractive colorant mixture *See subtractive colorant mixture, complex.*

Complex Young's modulus Vectorial sum of Young's modulus and the loss modulus; analogous to the complex dielectric constant.

Compliance *n.* The degree to which a material deforms under stress; the reciprocal of the modulus. Thus, in each mode of stress, the material is characterized by three moduli and their reciprocals, three compliances. However, when the stress is varying, the "real" and "imaginary" parts of the complex compliances are *not* equal to the reciprocals of their counterparts in the complex modulus. Tensile compliance; the reciprocal of Young's modulus; shear compliance: the reciprocal of shear modulus.

Compliance, elastic *n.* Symbol *S*. An elastic constant, which is the ratio of a strain or strain component to a stress or stress component. For a perfectly elastic material it is the reciprocal of the elastic modulus. For a viscoelastic material the modulus and compliance are not reciprocally related due to their different time dependencies. Sepe MP (1998) Dynamic mechanical analysis. Plastics Design Library, Norwich, New York.

Component substances, law of *n.* Every material consists of one substance, or is a mixture of two or more substances, each of which exhibits a specific set of properties, independent of the other substances.

Composite \käm-ˈpä-zət, kəm-ˈ, *esp British* ˈkäm-pə-zit\ [L *compositus*, pp of *componere*] (1563) *adj.* (1) An article or substance made up of two or more distinct phases of different substances. In the plastics industry the term applies broadly to structures of reinforcing members (*dispersed phase*) incorporated in compatible resinous binders (*continuous phase*). Such composites are subdivided into classes on the basis of the reinforcing constituents: laminate, *particulate* (the dispersed phase consists of unlayered fibers); *flake* (flat flakes forming the dispersed phase); and *skeletal* (composed of a continuous skeletal matrix filled by a resin). (2) Hard or soft constructions in which the fibers themselves are consolidated to form structures rather than being formed into yarns. Rigidity of these constructions is controlled by the density, the modulus of the load-bearing fibers, and the fraction of fusible fibers. Strength is controlled by adhesion and shear-yield strength of the matrix unless fibers are bonded in a load-transferring matrix. (3) A structure made by laminating a non-woven fabric with another non-woven, with other materials, or by impregnating a non-woven fabric with resins. Pittance JC (ed) (1990) Engineering plastics and composites. SAM International, Materials Park, OH.

Composite fibers *n.* Fibers composed of two or more polymer types in a sheath-core or side-by-side (bilateral) relation.

Composite laminate *n.* A term sometimes applied to a laminated plastic bonded to a non-plastic material such as copper, vulcanized fiber, rubber, asbestos, lead, aluminum, etc. An example is the copper-clad laminated plastic used for printed-circuit boards.

Composite mold *n.* A mold in which different shapes are produced in one cycle from the several cavities.
See also family mold.

Composite molding *n.* The process of molding two or more materials in the same cavity in the same shot, but a combination of transfer and compression molding. For example, in making a ring gear, a loose nylon-fiber-filled material is loaded into an open mold around the tooth circle, the mold is closed, and then molten nylon is injected by transfer molding.

Composite pigment *n.* Pigment usually made by the mechanical operation of intermixing or blending two or more pigments.

Composites manufacturing association (CMA/SME) *n.* Formerly a subgroup of the Society of Manufacturing Engineers, CMA is now a separate entity with 3000 members to promote composites, publish books and tapes, and hold conferences. Its office is at 1 SME Dr, P.O. Box 390, Dearborn, MI.

Composite structures *n.* Any material made of more than one component. Composite structures are made from polymers, or from polymers along with other kinds of materials.

Composition *n.* (1) A synthetic material containing resins or/and elastomers, and perhaps other components, in specified percentages. (2) The list of constituents and their percentages in such a material.

Composition rollers *n.* Printing press rollers made primarily of glue and glycerin.

Composition roofing *See built-up roofing.*

Composition siding *See hardboard.*

Compound \ kəm-ˈ, ˈkäm-ˌ\ [ME *compounen*, MF *compondre*, fr. L *componere*, fr. *com-* + *ponere* to put] (14c) *n, v.* (1) A mixture of resin and the ingredients necessary to modify the resin to a form suitable for processing into finished articles having the desired performance properties. (2) In chemistry, a combination of atoms ionically or covalently bonded in fixed ratios to form a molecule. (3) To produce a plastic compound by blending ingredients with resins in intensive mixers, extruders, or dry-blenders. (4) A pure substance composed of atoms of different elements, whose components cannot be separated by physical means. Wickson EJ (ed) (1993) Handbook of polyvinyl chloride formulating. John Wiley and Sons Inc., New York.

Compound curve *n.* A surface having curvature in two principal directions. Simply curved surfaces, such as cylinders and cones, having only one direction of curvature, may be cut along an element and laid flat. Compound curves, such as spheres and hyperbolic paraboloids, cannot be laid flat without distortion no matter how they are cut. Structures having compound curvature (reinforced-plastic roofs, for example) have high stiffness for their mass.

Compounders *n.* Chemical combinations of materials which include all the materials necessary for the finished product. They include Bulk molding compounds (BMC), Sheet-molding compounds (SMC), and Thick molding compounds (TMC). Approximately one-third of all USA polymer production undergoes subsequent compounding.

Compounder's modulus *n.* Stiffness measurement extensively used by rubber technologists, expressed as "modulus at 300%" or "300% modulus" (any other percent elongation may be indicated, but 300% is commonly used). By this is meant the tensile stress at the indicated elongation. *See modulus of elasticity.*

Compounding *n.* Mixing basic resins with additives such as plasticizers, stabilizers, fillers, and pigments in a form suitable for processing into finished articles. In some areas of the industry the term includes fusion of the polymer, for example in the production of molding powders by extrusion and palletizing. In the plastisol industry, the compounding step ends with the preparation of the dispersion, fusion being part of the molding step.

Compounding extruders Extruders that make use of compounded polymer melt. The melt is produced through either distributive or dispersive mixing techniques.

Compounding, plastics *n.* Upgrading of polymers or polymer systems through melt mixing. A compounded plastic has hybrid properties, such as high gloss and good impact strength, or precision moldability and good stiffness.

Compounding screws *n.* A typical screw-extruder device, which make use of mixing enhancements such as barriers, flutes, waves, pins, or cavities to overcome the inherent shortcomings in distributive and dispersive mixing. Typically, the screw rotates and axially oscillates. Each turn in the screw's spiral is interrupted by three gaps. The resulting kneading flights are continuously wiped by stationary teeth in the machine's barrel. Uniform shear is introduced at low pressures; making the process particularly well-suited for heat- and shear-sensitive polymers.

Compounds *n.* Substances containing more than one constituent element and having properties, on the whole, different from those, which their constituents had as elementary substances. The composition of a given pure compound is perfectly definite, and is always the same no matter how that compound may have been formed.

Compreg *n.* A contraction of "**com**pressed im**preg**nated wood", usually referring to an assembly of veneer layers impregnated with a liquid resin and bonded under high pressure.

Compregnate *n.* To impregnate and simultaneously or subsequently compress, as in the production of compregs.

Compregnated wood *n.* Consolidation of the term "compressed-impregnated wood", referring usually to an assembly of layers of veneer impregnated with a liquid resin and bonded under very high pressures.
Also known as Compreg.

Compressed-air ejection *n.* The removal of a molding from its mold by means of a jet of compressed air.

Compressibility *n.* The relative change in volume per unit change in pressure; the reciprocal of bulk modulus.

Compression and transfer molding *n.* In TM the mold halves are brought together under pressure as in Compression molding. The charge of molding compound is then put into a pot and is driven from the pot through runners and gates into the mold cavities by means of a plunger.

Compression modeling *n.* A technique principally for thermoset plastic molding in which the molding compound is placed in the heated open mold cavity, mold is closed under pressure, causes the material to flow and compress.

Compression modulus *n.* The ratio of compressive stress to compressive strain below the proportional limit. While it is theoretically equal to Young's modulus determined from tensile testing, compressive modulus is usually somewhat greater in plastics.

Compression mold *n.* A mold used in the process of compression molding.

Compression molding *n.* A method of molding in which a thermosetting molding material, generally preheated, is placed in an open, heated mold cavity, the mold is closed with a top force or plug member, pressure is applied to force the material into intimate contact with all mold surfaces, and heat and pressure are maintained until the material has cured and solidified. The process most often employs thermosetting resins in a partly cured stage, either in the form of granules or putty-resins like masses, or sometimes in preformed shapes roughly conforming to the shape of the mold. Compression molding has also been used with thermoplastics, must notably phonograph records. In this process, the mold is cooled following the compression-flow stage. Compression molding is the oldest form of processing phenolic composite material. In its simplest form, it consists of a force and a cavity that make up a two-piece mold. The mold contains one or more cavities in the shape of the part to be molded. It is heated to 320–380°F, depending on the part's geometry and the closing speed of the press. Then the correct amount of composite material is added, and the two halves of the mold are brought together under 2,000–6,000 pounds of pressure/in.^{2} of molding area. The heat of the mold softens the phenolic material and, under pressure, it flows to form the shape of the mold. Strong AB (2000) Plastics materials and processing. Prentice-Hall, Columbus, OH. Harper CA (2000) Modern plastics encyclopedia. McGraw-Hill Professional, New York.

Compression-molding pressure *n.* (1) The force per unit of projected area applied to the molding material in a compression mold. The area is projected from all parts of the material under pressure during the complete closing of the mold. (2) The *hydraulic* pressure applied to the compression ram during molding.

Compression ratio *n.* In an extruder, the ratio of the volume of the first turn of the feed section to that of the last turn of the metering section. This ratio is a rough indication of the total compaction performed on the feedstock. More precisely called the *channel-volume ratio* or, for a screw of constant pitch, *channel-depth ratio.*

Compression set *n.* A permanent deformation remaining after release of a compressive stress. It is a property of interest in elastomers and cushioning materials, such as plastic foams. See, for example, ASTM tests D 395 and D 1565. Sometimes used to mean creep.
See cold flow.

Compression testing *n.* A method of determining behavior of a material subjected to a uniaxial compressive load.

Compression zone (compression section) *n.* The part of an extruder screw, connecting the feed and metering sections, in which the volume per turn is decreasing because of decreasing channel depth, or lead, or both. In some older designs, rarely made today, there were no distinct feed and metering sections and the volume per turn decreased over the entire screw length, so the compression zone was the entire screw. In two-stage screws used for vented operation, the rear metering zone is normally followed by a deep zone of *decompression*, then a second, short compression zone and second metering zone. Where decreasing channel depth is the means of volume reduction, it may be done with a conical screw-root profile or, more usual today, a helical root profile.

Compressive strength *n.* The load at which a test specimen fails in compression, divided by the original cross-sectional area perpendicular to the load. For rigid plastics, ASTM test D 695 is used; for rigid foams, D 1621. These tests also prescribe procedures for estimating compressive moduli. The actual mode of failure of a stiff material in a test of compressive strength is usually by diagonal shear.

Compressive stress *n.* The compressive load per unit area of perpendicular cross section carried by the specimen during a compression test.

Compton effect (Compton recoil effect) *n.* Elastic scattering of photons by electrons results in decrease infrequency and increase of wavelength of X-rays and γ-rays when scattered by free electrons.

Computer-aided design (CAD) *n.* Using a computer, and appropriate programs (software), in the engineering design – even to the production of finished working drawings – of parts, tools, molds, and assemblies.

C

Computer-aided manufacturing (CAM) *n.* Using a computer – usually dedicated and typically a minicomputer – with appropriate programs (software), to control parts or all of a manufacturing operation.

Computer numerical control (CNC, numerical control) *n.* The use of a dedicated small computer to implement the control of a process or, typically, a machining task.

Concavity factor *n.* The entire stress-strain curves of rubbers and elastomers that have no elastic limit are typically concave toward the stress axis (and convex to the strain axis). The concavity factor is the ratio (less than 1) between the energy beneath the extension curve to that beneath the straight line to the same final point.

Concentration *n.* (1) Amount of a substance expressed in relationship to the whole. (2) Act or process of increasing the amount of a given substance in relationship to the whole.

Concentration cell *n.* Electrolytic cell consisting of an electrolyte and two electrodes of the same metal or alloy that develop a difference in potential as a result of a difference in concentration of ions (most often metal ions) or oxygen at different points in a solution.

Concentricity *n.* The characteristic of circles or circular cylindrical surfaces of different radii having a common center. More loosely, the property, in *any* annular shape, of constant radial wall thickness. Concentricity is important in blown film, pipe and tubing, wire-coating, and many noncircular extrusions.

Conchoidal fracture *n.* Type of fracture seen when a mineral or other substance, such as glass, breaks to give irregularly curved, usually striated surface, and no cleavage along planes.

Concrete *n.* A composite material which consists essentially of a binding medium within which are embedded particles or fragments of aggregate; in Portland cement concrete, the binder is a mixture of Portland cement and water.
See cement.

Concrete block *n.* A hollow or solid concrete masonry unit consisting of Portland cement and suitable aggregates combined with water. Lime, fly ash, air-entraining agents, or other admixtures may be included. Sometimes incorrectly called cement block.

Concrete bond plaster *See bond plaster and concrete bond plaster.*

Concrete gun *n.* A spray gun used in applying freshly mixed concrete; compressed air forces the concrete along a flexible hose and through a nozzle.

Condensate *n.* Product obtained by cooling a vapor, such that it is converted either to a liquid or a solid.

Condensation *n.* (1) The process of reducing a gas or vapor to liquid or solid form. (2) A chemical reaction in which two or more molecules combine with the separation of water or some other simple molecule. If a polymer is formed the process is called *polycondensation.*

Condensation agent *n.* A chemical compound that acts as a catalyst and also furnishes a complement of material necessary for a polycondensation reaction to proceed.

Condensation polymer *n.* A polymer made by condensation polymerization.

Condensation polymerization *n.* A polymerization reaction in which water or some other simple molecule is eliminated from two or more monomer molecules as they combine to form the polymer or cross-links between polymer chains. Examples of resins so made are alkyds, phenol-formaldehyde and urea-formaldehyde, polyesters, polyamides, acetals, and polyphenylene oxide.

Condensation polymers *n.*, **by chain mechanisms** Step-growth polymers produced by a polymerization reaction in which the elimination of a small molecule, often water, has occurred, e.g., produced by a condensation polymerization. Important examples include the polyesters, polyamides and phenol-, urea- and melamine-formaldehyde polymers. Odian GC (2004) Principles of polymerization. John Wiley and Sons Inc., New York.

Condensation resin *n.* A resin formed by condensation polymerization.

Condenser *n.* The lens system mounted under the stage of the microscope to furnish a cone of light to the specimen. There are two basic types of condensers: bright-field and darkfield.

Condensers in parallel and series *n.* If c_1, c_2, c_2, etc. represent the capacitances of a series of condensers and C their combined capacitance,

when in parallel, $C = c_1 + c_2 + c_3 + \cdots$,

when in series, $\frac{1}{C} = \frac{1}{c_1} + \frac{1}{c_2} + \frac{1}{c_3} + \cdots$

Serway RA, Faugh JS, Bennett CV (2005) College physics. Thomas, New York.

Conditional match *n.* A pair of colors, which appear to match only under limited conditions, such as under a particular light source and a particular observer; a metameric match.

Conditioners *n.* A process of allowing textile materials (staple, tow, yarns, and fabrics) to reach hygroscopic equilibrium with the surrounding atmosphere. Materials may be conditioned in a standard atmosphere (65% RH, 70°F) for testing purposes or in arbitrary conditions existing in manufacturing or processing areas.

Conditioning *n.* Subjecting a test specimen to standard environmental and/or stress history prior to testing. Several "ambient" conditioning atmospheres, approved by ISO, are listed in ASTM specification E 171, and ASTM D 618 provides detailed information on conditioning. Plastics test specimens are usually conditioned at 23°C and 50% RH for several days or more. Test D 638 for tensile properties of plastics specifies ". . . . not less than 40 h. . . ." at these conditions.

Conditioning time *See time, joint conditioning.*

Conductance (1885) *n.* (1) The electrical term for the reciprocal of resistance, measured by the ratio of current flowing through a conductor to the difference of potential between its ends. The SI unit is the *siemens* (S), replacing the non-deprecated mho, to which it is exactly equal. ASTM D 257 prescribes tests for the resistance and conductance of electrical-insulating materials. (2) In heat transfer, the ratio of heat *flux* through a solid wall or a stagnant fluid film divided by the difference in temperature through the wall or film. The SI unit is $J/(m^2\,K)$. The reciprocal of resistance, is measured by the ratio of the current flowing through a conductor to the difference of potential between its ends. The practical unit of conductance, the mho, the conductance of a body through which 1 A of current flows when the potential difference is 1 V. The conductance of a body in mho is the reciprocal of the value of its resistance in ohms. Dimensions $[\varepsilon LT^{-1}]$, $[\mu^{-1} L^{-1} T]$. Weast RC (ed) (1978) CRC handbook of chemistry and physics, 59th edn. The Chemical Rubber Co., Boca Raton, FL.

C

Conductimetric analysis *n.* A method of analysis for certain ions in solution based on measurement of the solution's electrical conductivity.

Conducting polymer *n.* One of a class of polymers that, unlike most organic polymers, have high electrical conductivity. Polyacetylene is the best known member and at least one type has been prepared by doping that has conductivity almost equal to that of copper. While these polymers are expected to be useful in a wide range of devices, only a few commercial applications exist.

Also, see polyelectrolyte.

Conductive composite *n.* A composite material having a volume resistivity less than 500 Ω cm. Such composites may be created by adding metal or graphite powders to ordinary resins. Others are made from inherently conducting polymers. They are useful for static elimination, RF shielding, and in storage-battery components.

Conductive heat transfer *See heat transfer.*

Conductivity *n.* The current transferred across per unit potential gradient, $K = A/cm^2$ divided by V/cm. Reciprocal of resistivity.

Also known as specific conductance.

Conductivity (electrical) *n.* The reciprocal of volume resistivity; the conductance of a unit cube of material. The SI unit is siemens per meter (S/m). It is measured by the quantity of electricity transferred across unit area, per unit potential gradient per unit time. Reciprocal of resistivity. *Volume conductivity* or specific conductance, $k = 1/\rho$ where ρ is the volume resistivity. *Mass conductivity* $= k/d$ where d is density. *Equivalent conductivity* $\Lambda = k/c$ where c is the number of equivalents per unit volume of solution. *Molecular conductivity* $\mu = k/m$ where m is the number of moles per unit volume of solution. Dimensions: volume conductivity $[\varepsilon T^{-1}]$, $[\mu^{-1} L^{-2} T]$, mass conductivity $[\varepsilon M^{-1} L^3 T^{-1}]$, $[\mu^{-1} M^{-1} LT]$. $^{-1} M^{-1} LT]$. Ku CC, Liepins R (1987) Electrical properties of polymers. Hanser Publishers, New York.

Conductivity (thermal) *n.* Time rate of transfer of heat by conduction, through unit thickness, across unit area for unit difference of temperature. It is measured as cal/s/cm^2 for a thickness of 1 cm and a difference of temperature of 1°C. Dimensions $[MLT^{-3}\theta^{-1}]$. If the two opposite faces of a rectangular solid are maintained at temperatures, t_1 and t_2 the heat conducted across the solid of section a and thickness d in a time T will be,

$$Q = \frac{K(t_2 - t_1)at}{d},$$

where K is the constant depending on the nature of the substance, designated as the specific heat conductivity. K is usually given for Q in calories t_1 and t_2 in °C, a in cm^2, T in s, and d in cm. Ready RG (1996) Thermodynamics. Plenum Publishing Company, New York. Pethrick RA, Pethrick RA (eds)(1999) Modern techniques for polymer characterization. John Wiley and Sons, New York.

See heat conductivity.

Conductors *n.* A class of bodies, which are incapable of supporting electric strain. A charge given to a conductor spread to all parts of the body.

Cone *n.* A conical package of yarn, usually wound on a disposable paper core.

Cone mill *n.* Old type of mill used for dispersing pigments in media. They usually have conical hopes, and the grinding surfaces are also inclined like the surface of a cone.

Configuration *n.* The arrangement of atoms that characterizes a particular stereoisomer is called its configuration. Related chemical structures produced by the cleavage and reforming of covalent bonds; arrangement of polymers along a plastic molecule chain and the structural makeup of a chemical compound, especially with reference to the spatial relationship of the constituent atoms; *trans*- and *cis*-configurations of the carbon–carbon double bond are geometric isomers.

See conformation.

Configurational base unit *n.* A molecular repeating unit or mer whose configuration is defined at least at one site of stereoisomerism in the main chain of a polymer molecule. *Note*—In a regular polymer, a configurational base unit corresponds to the mer. For example, in regular polypropylene, the mer is $–CH(CH_3)CH_2–$ and the configurational base units are shown in the image, and these two configurational base units are enantiomeric to each other (mirror images).

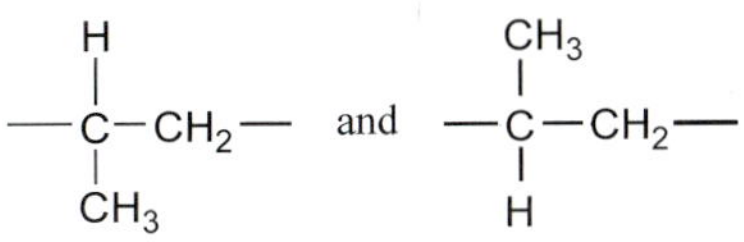

Configurational unit A molecular unit having one or more sites of defined stereoisomerism (IUPAC).

Configuration repeating unit *n.* The smallest set of one, two, or more successive configurational base units that prescribes configurational repetition at one of more sites of stereoisomerism in the main chain of a polymer molecule (IUPAC).

Conformal coating *See encapsulation.*

Conformation *n.* Different arrangements of atoms that can be converted into one another by rotation about single bonds are called conformations (e.g., *anti-*, *gauche-*conformations).

Congeal \kən-ˈjē(ə)l\ [ME *congelen*, fr. MF *congeler*, fr. L *congelare*, fr. *com-* + *gelare* to freeze] (14c) *v.* To change from a liquid or soft state to a solid or rigid state.

Congo copal *n.* Most important of the copals used for oil varnishes. It originates in the Belgian Congo (now Zaire). It becomes soluble in vegetable oils after running.

Congocopalic acid *See bengucopalic acid.*

Congo gum *n.* Fossilized gum resin obtained from the Congo region of Africa (now Zaire). Used mostly in manufacturing varnishes.

Congolene *n.* Bicyclic hydrocarbon, obtained from the oily layer resulting from the running of Congo copal.

Conical dry-blender *n.* A device consisting of two hollow cones joined at their bases by a short cylindrical section, mounted on a central shaft perpendicular to the conicylindrical axis. Material is charged and discharged at openings in the apexes of the cones. Mixing is accomplished by cascading, rolling, and tumbling of the charge as the chamber rotates bout the shaft.

Conical transition *n.* In a metering-type extruder screw, the root surface of the screw between the feed section and metering section having the shape of a cone whose diameter increases from that of the deeper feed section to that of the shallower metering section.

Coning The transfer of yarn from skeins or bobbins or other types of packages to cones.

Conjugate acid–base pair *n.* An acid and the base formed by removal of a proton from the acid, or a base and the acid formed by the addition of a proton to the base (Brønsted–Lowry).

Conjugated *n.* In organic chemistry, referring to the regular alternation of single and double bonds between carbon atoms. For example, in the conventional representation of the benzene molecule shown in the image where each single bond represents one pair of shared electrons, each double bond, two pairs.

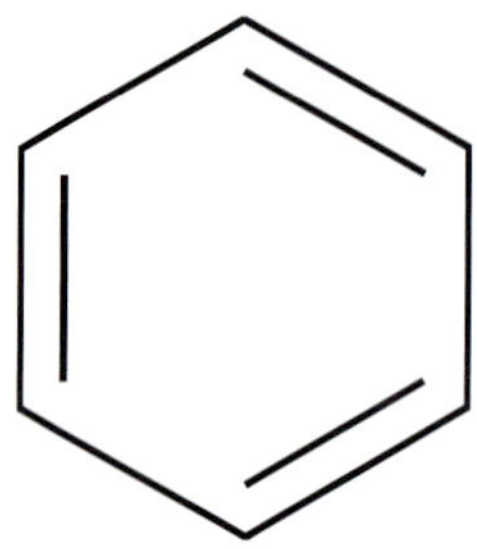

Conjugated diene polymerization *n.* Conjugated dienes often polymerize as bifunctional monomers with 1,4 addition or 1,2 addition. In this process, one double bond remains in the main chain for each monomer.

Conjugated double bonds *n.* A chemical term denoting double bonds separated from each other by a single bond. An example is the bonding in 1,3-butadiene, $CH_2{=}CH{-}CH{=}CH_2$.

Conjugated yarn *n.* A yarn made from conjugate filaments.

Conjugate fibers *n.* A two-component fiber with specific ability to crimp on hot or hot/wet treatment because of differential shrinkage.
Also see bilateral fibers.

Conjugate foci *n.* Under proper conditions light divergent from a point on or near the axis of a lens or spherical mirror is focused at another point. The point of convergence and the position of the source are interchangeable and are called conjugate foci.

Conjugate focil *n.* In an image-forming system, two fields are said to be conjugate with each other when one or more object fields are simultaneously in focus in a single plane, e.g., in Köhler illumination the field diaphram, specimen and ocular front focal plane.

Conjugation *n.* The location of the π orbital in such a way that it can overlap other orbitals within the molecule. Morrison RT, Boyd RN (1992) Organic chemistry, 6th edn. Prentice-Hall, Englewood Cliffs, NJ.

Conoscopic observation *n.* The study of the back focal plane of the objective by removing the eyepiece, by inserting a Bertrand lens, by examining the image at the eyepoint above the eyepiece with a magnifier or by using a phase telescope is called conoscopic because the observations are associated with the cone of light furnished by the condenser and viewed by the objective (cf., orthoscopic).

Consensus standard *n.* A standard developed according to a consensus agreement or general opinion among representatives of various interested or affected organizations and individuals.

Conservation of energy (Chem) *n.* In a chemical change there is no loss or gain but merely a transformation of energy from one form to another.

Consistency *n.* That property of a liquid adhesive by virtue of which it tends to resist deformation. *Note*—Consistency is not a fundamental property but is comprised of viscosity, plasticity, and other phenomena (*See also viscosity and viscosity coefficient*).

Consistency *n.* The density, firmness, viscosity, or resistance to flow of a substance, slurry, or aggregate.
See viscosity.

Consistency *n.* The property of a material or composition which is evidenced by its resistance to flow, represented by an undefined composite of properties, each measurable from the complete, force-rate flow curve as plastic viscosity, yield value and thixotropy. The term is applied to a variety of materials. For Newtonian liquids, consistency is simply viscosity. While "consistency" is an accepted rheological term, it has qualitative meaning only, and is used with qualifying adjectives as "buttery", "thin", "high", etc., in describing plastic flow. Usually measured in an empirical manner and in arbitrary units. Often used as a synonym for viscosity.

Consistometer *n.* An instrument for measuring the flow characteristics of a viscous or plastic material.
See viscometer and rheometer.

Consolidation *n.* Application of heat and pressure to form composite structures.

Constantan \ˈkän(t)-stən-ˌtan\ [fr. the fact that its resistance remains constant under change of temperature] (1903) *n.* An alloy containing about 55% copper and 45% nickel and having a low thermal coefficient of resistivity. Its main use in the plastics industry is in thermocouple wire with either iron or cooper as the mating element. Iron-constantan, Type J, and chromel–alumel. Type K, are widely used to sense temperatures in plastics-processing equipment.

Constant white *n.* Alternative name for blanc fixed.

Constitutional formula *n.* Device used to illustrate the composition of a chemical compound by displaying the individual atoms and radicals, joined together by valency linkages.

Constitutional repeating unit *n.* The smallest molecular unit whose repetition describes a regular polymer (IUPAC).

Constitutive equation *n.* In material science, an equation that relates stress in a material to strain or strain rate. Simple examples are (1) Hooke's law, which states that, in elastic solids, strain is directly proportional to stress, and (2) Newton's law of flow, which states that, in laminar shear flow, the shear rate is equal to the shear stress divided by the viscosity. Few plastic solids and liquids obey either of these laws.

Constitutive property *n.* A property, which depends on the constitution or structure of the molecule.

Consumer's risk *n.* In quality control and acceptance sampling, the risk of making a type II error, i.e., of accepting, under a given sampling plan, a lot that is of definitely unacceptable quality.

Contact adhesive *n.* A liquid adhesive that dries to a film that is not sticky to other materials but very sticky to itself. A typical contact adhesive is a neoprene elastomer mixed with either an organic-solvent vehicle or an aqueous dispersion medium. The adhesive is applied to both surfaces to be joined and dried at least partly. When pressed together with light to moderate pressure a bond of high initial strength results. Some definitions of "contact adhesive" stipulate that, for satisfactory bonding, the surfaces to be joined shall be no further apart than about 0.1 mm. Skeist I (ed) (1990) Handbook of adhesives. Van Nostrand Reinhold, New York.

Contact angle *n.* The angle between the edge of a liquid meniscus or drop and the solid surface with which it is in contact. A droplet placed on a horizontal solid surface may remain spherical or spread to a degree that is related to the surface energies of the two materials. The angle between the solid surface and the tangent to the droplet at the curve of contact with the surface is the contact angle, and an example of this measurement is shown. By virtue of measuring such droplet-contact angles for droplets of several liquids of different surface energies, the surface energy of the solid may be calculated as shown in the plot. Mittal KL (2003) Contact angle, wettability and adhesion, vols 1–3. VPS International Science Publishers, Zeist, The Netherlands. Oss CJ (1994) Interfacial forces in aqueous media.

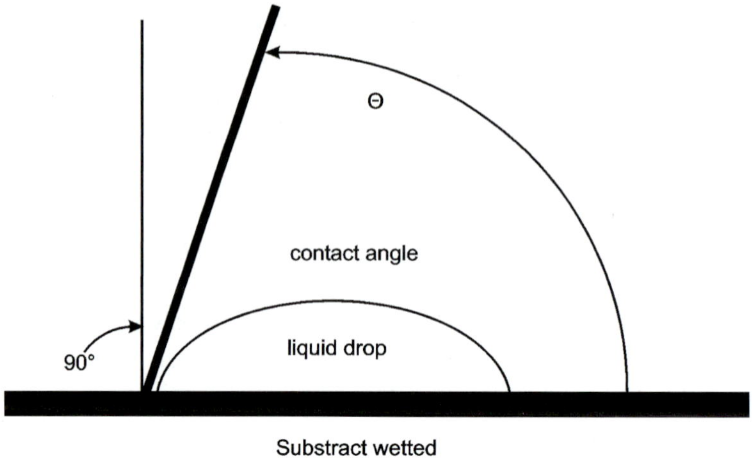

Marcel Dekker, New York. ASTM Test D 724–799, 2003.

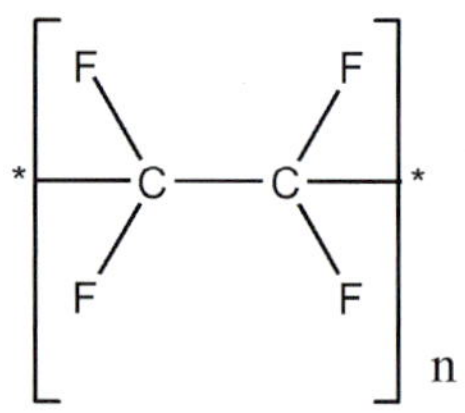

Contact laminating *See contact-pressure molding.*

Contact molding *n.* A process for molding reinforced plastics in which the reinforcement and plastic are placed on a mold.

Contact-pressure molding (contact molding) *n.* This term encompasses processes for forming shapes of reinforced plastics in which little or no pressure is applied during the forming and curing steps. It is usually employed in connection with the processes of sprayup and hand-layup molding when such processes do not include the application of pressure during curing.

Contact-pressure resin (contact resin, impression resin, low-pressure resin) *n.* A liquid resin that thickens or cross-links on heating and, components are an unsaturated monomer such as an allyl ester, or a mixture of styrene or other vinyl monomer with an unsaturated polyester or alkyd.

Contact resin *n.* A liquid resin that thickens or cross-links on heating and, when used for bonding laminates, requires little or no pressure. Typical components are an unsaturated monomer such as an allyl ester, or a mixture of styrene or other vinyl monomer with an unsaturated polyester or alkyd. *See contact-pressure resin.*

Contact resins *n.* Liquid resins, which thicken on heating and, when used for bonding laminates, require little or no pressure.

Container lining *n.* Protective coating applied to the inner walls of a container to prevent interaction between the contents and the material of construction of the container or to make the walls of the container impervious to the contents.

Continuous filament *n.* A single, flexible, small-diameter fiber of indefinite length. *See filament.*

Continuous-filament yarn *n.* A yarn formed by twisting together two or more – typically scores – of continuous filaments.

Continuous phase *n.* (1) In a suspension or emulsion the continuous phase refers to the liquid medium in which the solid or second-liquid particles are dispersed. The solid particles or droplets are called the *disperse phase.* (2) In a plastic filled or reinforced with solid particles, flakes or fibers, the binding resin is the continuous phase.

Continuous phase *n.* The medium or continuum in which the dispersed phase is contained.

Also called external phase.

Continuous polymerization n. In polymer manufacture, linkage of the various stages of polymerization so that materials flow without interruption from the addition of raw materials to delivery of the finished polymer from the system. Extrusion as film, chip or fiber may be linked to a continuous polymerization line. Because there is no break in the process while the transition from low molecular weight to high occurs, multiple stage reaction vessels may be required and accurate process control is critical.

Continuous roving *See roving.*

Continuous tone *n.* Tonal gradation without use of halftone dots.

Contour length *n.* The maximum value of the end-to-end distance of a polymer chain. If the chain consists of n links of each length l with valence angle θ, the chain length is $nl\sin(\theta/2)$.

Also known as displacement length.

Contraction *See take-up (twist) or take-up (yarn in fabric).*

Contraction allowance *See shrinkage allowance.*

Contrast ratio *n.* (1) Ratio of the reflectance of a dry paint film over a black substrate of 5% or less reflectance, to the reflectance of the same paint, equivalently applied and dried, over a substrate of 80% reflectance. (2) The ratio of the luminous reflectance, *Y* measured on a film over a black substrate, to the *Y* measure on the same film over a white substrate. The *Y* of the white and black must be specified. The contrast ratio will vary depending on the thickness of the film and on the concentrations of colorants. Sward GG (ed) (1972) Paint testing manual: physical and chemical examination of paints, varnishes and lacquers, and colors, 13th edn. ASTM Special Technical Publication No, 500, American Society for Testing and Materials, Philadelphia, PA.

Controlled-atmosphere packaging *n.* The packaging of a product in a gas other than air, typically an inert gas such as nitrogen.

Convalent bond *n.* A bond consisting of a pair of electrons shared between the bonded atoms.

Convection \kən-ˈvek-shən\ [LL *convection-*, *convectio*, fr. L *convehere* to bring together, fr. *com-* + *vehere* to carry] (ca. 1623) *n.* Any process by which heat energy or material exchange is effected by flow of the medium. Convection may be *natural*, driven by gravity and density or concentration differences at different points the medium; or *forced*, driven by pumps, blowers, or vibrating devices. Forced convection is almost always faster and more efficient because of the higher velocities and greater turbulence produced.

Conventional base unit of a polymer *n.* Base unit, defined without regard to steric isomerism (IUPAC).

Convergent die *n.* An extrusion die in which the internal channels of the die leading to the die orifice are decreasing in cross section in the direction of flow.

Conversation of energy, law of *n.* Energy can neither be created nor destroyed and therefore the total amount of energy in the universe remains constant.

C

Conversation of mass *n.* In all ordinary chemical changes, the total of the reactants is always equal to the total mass of the products.

Conversation of momentum, law of *n.* For any collision, the vector sum of the moment of the colliding bodies after collision equals the vector sum of their moment before collision. If two bodies of masses m_1 and m_2 have, before impact velocities v_1 and v_2 and after impact velocities u_1 and u_2

$$m_1 u_1 + m_2 u_2 = m_1 v_1 + m_2 v_2.$$

Conversion *n.* (1) In a chemical process, the molar percentage of any reactant, often the primary or most costly reactant that is changed into product. (2) In the packaging industry, the intermediate processing and fabrication of plastic film or sheeting into useful forms by slitting, die cutting, heat-sealing into bags, etc. for resale to packagers.

Conversion coating *See chemical conversion coating.*

Converted fabric *n.* A finished fabric as distinguished from greige fabric.

Converter An individual or organization, which buys greige fabrics and sells them as a finished product to cutters, wholesalers, retailers, and others. The converter arranges for the finishing of the fabric, namely bleaching, mercerizing, dyeing, printing, etc., to the buyers' specifications.

Conversion coating *See chemical conversion coating.*

Convertible coating *n.* Irreversible transformation of a coating after its film formation to a film insoluble in the solvent from which it was deposited. This can be effected by oxidation, thermal cross-linking or catalytic curing. This should not be confused with.

See non-convertible coating.

Convolution \ˌkän-və-ˈlü-shən\ (1545) *n.* (1) An irregular spiral or twisted condition characteristic of mature cotton fiber. It is visible under a microscopic. The finer fibers are generally more twisted than the coarser fibers. (2) Coil and curl in certain types of textured yarns which provide bulkiness to the yarn. Tortora PG (ed) (1997) Fairchild's dictionary of textiles. Fairchild Books, New York.

Cooling *n.* Processing highly radioactive materials to attain less radioactivity for subsequent use or handling.

Cooling channel *n.* A passageway provided in a mold, platen or die for circulating water or other cooling medium, in order to control the temperature of the metal surfaces in contact with the plastic being molded or extruded. Proper sizing and placement of cooling channels can do much to speed processing and optimize properties.

Cooling fixture (shrink fixture) *n.* A structure of wood or metal shaped to receive and restrain a part after its removal from a mold, so as to prevent distortion of the part while it is cooling.

Coordinate bond *n.* A covalent bond consisting of a pair of electrons donated by only one of the two atoms it joins.

Coordinate covalent bond A covalent bond in which both shared electrons appear to have been contributed by one atom.

Coordination catalyst *n.* A catalyst comprising a mixture of an organo-metallic compound, e.g., triethylaluminum, and a transition-metal compound, e.g., titanium tetrachloride. Often called Ziegler or Ziegler–Natta catalysts, they are used in polymerizing olefins and dienes.

Coordination catalysts *n.* Catalysts comprising a mixture of (a) an organo-metallic compound such as triethylaluminum or a transition-metal compound, such as titanium tetra-chloride. Known as Ziegler or Ziegler–Natta catalysts, they are used for the polymerization of olefins and dienes.

Coordination compound (Werner complex) *n.* A complex compound whose molecular structure contains a central atom bonded to other atoms by coordinate covalent bonds based on a shared pair of electrons, both of which are from a single atom or ion. A chelate is an special type of coordination compound.

Coordination number *n.* The number of atoms, ions, or molecules surrounding a central atom or ion in a complex, or sometimes, in a solid.

Coordination polymerization *n.* Polymerization of vinyl monomers using a catalyst comprising a transition metal salt and a metal alkyl; Ziegler–Natta catalysts for polymerization of ethylene to produce polyethylene.

Cop \ˈkäp\ [ME, fr. OE *copp*] (before 12c) *n.* (1) A headless tube upon which yarn or thread is wound. (2) Thread or yarn wound into the shape of a hollow cylinder with tapered ends. (3) Filling yarn wound upon a tapered tube (generally paper).

Copal \ˈkō-pəl\ [sp, fr. Nahuatl *copalli* resin] (1577) *n.* A fossil resin used in printing ink vehicles.

Copal esters *n.* These are normally regarded as the glyceride esters of run Congo copal, although the term can also be applied to esters of other copals.

Copal oil *n.* Another name for the condensed fumes obtained during the running of natural copals.

Copals *n.* Gum resins exuded from living plants and fossilized in the ground. Some of the opals are: Zanzibar, amber, kauri, Manila, and Congo.

Copal varnish *n.* A high-gloss varnish made with a drying oil, such as linseed oil, and copal.

Coping *n.* A protective cap, top, or cover of a wall, parapet, plaster, or chimney; often of stone, terra-cotta, concrete or wood.

Copolyamides *Also known as polyether block amide elastomers* – based on a block copolymer and a polyether. A wide range of grades and performance characteristics can be achieved by varying the polyamide and polyether blocks. These high-performance thermoplastic elastomers can withstand high heat and offer good heat-aging characteristics, long flex life, and excellent chemical resistance. Copolyamides are used in demanding medical applications such as catheters and in wire-and-cable jackets, automotive parts, and sporting goods.

Copolycondensation *n.* The copolymerization of two or more monomers by condensation polymerization.

Copolymer \(ˌ)kō-ˈpä-lə-mər\ (1936) *n.* This term usually, but not always, denotes a polymer two chemically distinct monomers. It is sometimes used for terpolymers, etc, containing more than two types of mer units. Three common types of copolymers are block copolymers, graft copolymers, and random copolymer. The IUPAC term for a polymer derived from two species of monomer, *bipolymer* eschews the foregoing ambiguity but is nevertheless rarely seen or heard.

See also bipolymer and terpolymer.

Copolymer equation *n.* $F_A/F_B = (r_A f_A/f_B + 1)/(r_B f_B/f_A + 1)$,

where f_A is the mole fraction of monomer A in feed, f_B the mole fraction of monomer B in feed, F_A the mole fraction of A in copolymer, F_B the mole fraction of B in copolymer, r_A the reactivity ratio of A, and r_B is the reactivity ratio of B.

C

Copolymerization *n.* The building up of linear or non-linear macromolecules (copolymers) in which many monomers, possessing molecules have one or more double bonds, have been located in every macromolecule of different size which constitutes the copolymerizate, following alternations which may be regular or not.
See polymerization.

Copolymers *n.* Polymers constructed from two different materials.

Copper acetoarsenite *Also known as Paris green, emerald green and Schweinfurt green.*

Copperas \ˈkä-p(ə-)rəs\ [ME *copperas*, fr. OF *couperose*, fr. ML *cuprosa*, prob. fr. *aqua cuprosa*, literally, copper water, fr. LL *cuprum*] (14c) *n.*
See ferrous sulfate.

Copper blue *n.* Another name for *cupri sulfide* (copper (II) sulfide). Syn: covellite.

Copper bronze *n.* Name given to a range of metallic powders ranging in color from "pale gold" to the deep "coin bronze". They consist generally of alloys of copper, but recently dyed aluminum powders of similar shades have been produced.
See also copper powder.

Copper carbonate $CuCO_3$ Material frequently used as a poison in antifouling paints.

Copper-clad laminate *n.* A laminated plastic surfaced with copper foil or plating, used for preparing printed circuits.

Copper driers (Deprecated) Copper salts of acids generally used for driers, such as naphthenic and 2-ethyl hexoic, which are used as fungicides and preservative additives in coatings, but do not function as driers. They are deep green and cannot be used in whites or light colors.

Copper powder *n.* Metallic copper and copper bronze powders have some application in anti-fouling paints, and in special primers. Copper bronzes of various colors are produced according to the amount of alloy metal present. In producing finishes from copper and copper bronze powders, great care in formulation is necessary, for these pigments are apt to cause inhibition of drying, discoloration of media, and gelation.

Copper resinate *n.* Green compound formed by dissolving copper acetate, verdigris or other copper salt in Venice turpentine, balsam, or similar resinous solutions.

Copper soap *n.* Combination of copper with a fatty acid.

Copper staining *n.* Usually caused by corrosion of copper screens, gutters or downspouts washing down on painted surfaces.

Copper sub-oxide Cu_2O. Another name for cuprous oxide [copper(I) oxide].

Copper sulfate *n.* $CuSO_4{\cdot}5H_2O$. Used occasionally in pigment manufacture, e.g., Para Brown.
Also known as blue vitriol and blue stone.

Copy *n.* Material, including art and text, submitted for reproduction. The term is also used to refer the final printed result.

Coral rubber *Cis*-1,4-poly(isoprene). Manufactured by Goodrich, USA.

Cord *n.* (1) The product formed by twisting together two or more plied yarns. (2) A rib on the surface of a fabric (e.g., corduroy and whipcord).

Corded selvage *See loopy selvage.*

Corduroy \ˈkór-də-ˌrói\ (ca. 1791) *n.* (1) In wallpaper, a very narrow strip imitating the fabric. The name is derived from the French "corde du roi" or king's cord. (2) A filling-pile fabric with ridges of pile (cords) running lengthwise parallel to the selvage.

Core *n.* (1) The central member of a laminate to which the faces of the sandwich are bonded. (2) A channel in a mold, extruder screw or cast-in heating element for circulation of heat-transfer media. (3) Part of a complex mold that forms undercut sections of a part, usually withdrawn to one side before the main members of the mod are opened. (4) The central member of a die for extruding pipe, tubing, wire-coating, or a parison to be blow-molded. (5) The central conductor in a coaxial cable. (6) All of an atom except for its valence shell of electrons; *also called the Kernel.*

Core and separator *See core* (4) above.

Core-bulked yarn *See textured yarns.*

Cored screw *n.* An extruder screw bored centrally from the rear to permit circulation of temperature-controlled liquid within all or part of the screw's length.

Core spinning *n.* The process of making a corespun yarn. It consists of feeding the core yarn (an elastomeric filament yarn, a regular filament yarn, a textured yarn, or a previously spun yarn) into the front delivery roll of the spinning frame and of covering the core yarn with a sheath of fibers during the spinning operation.

Core-spun yarn *n.* A yarn made by twisting fibers around a filament or a previously spun yarn, thus concealing the core. Core yarns are used in sewing thread, blankets, and socks and also to obtain novelty effects in fabrics.

Corfam® *n.* Permeable artificial leather from polyurethane/polyester/polyester fleece, manufactured by DuPont, USA. *See poromeric.*

Cork *n.* The outer bark of *Quercus suber*, a species of oak native to Mediterranean countries, having density in the range 0.22–0.26 g/cm^3. Cork is used as a core material [*See core* (1)] in sandwich structures and, in ground form, as a density-lowering filler in thermoplastic and thermosetting compounds for special applications such as flooring, ablative plastics, insulating compositions, and shoe inner soles.

Cork composite *n.* A compound consisting of ground cork, resins, and other additives and reinforcements, formed into rods, sheets, etc. Cork composites have relatively low density and are used in sporting goods, for thermal insulation, and ion ablative material.

Cork dust *n.* Very finely divided cork, which is used in anti-condensation paints.

Corkscrew twist *n.* A place in yarn or cord where uneven twist gives a corkscrew-like appearance.

Cornice *n.* (1) A horizontal molding or combination of moldings to finish the top of a wall. (2) Paper simulating the same.

Corn oil *n.* Oil obtained from the kernels of Indian corn, maize, *Zea mays.* It is semidrying, lying between cottonseed and soybean oils.

Corona \kə-ˈrō-nə\ [L, garland, crown, cornice] (1563) *n.* Cracks on the surface caused by ozone, usually spearing at right angles to a stress, checking, cutting or cracking.

Corona discharge *n.* (1) The flow of electrical energy from a conductor to or through the surrounding air or gas. The phenomenon occurs when the voltage difference is sufficient (>5000 V) to cause partial ionization of the gas. The discharge is characterized by a pale violet glow, a hissing noise, and the odor of ozone formed when the surrounding gas contains oxygen. Corona discharge occurs around high-voltage cables, thus making ozone resistance an important factor in compounding plastics for insulation of electrical wire and cables. (2) Curing method involving the bombardment of organic vapors with high-energy electrons at very low pressures in contact with the substrate to be coated.

See also casing, corona-discharge treatment and glow discharge.

Corona-discharge treatment *n.* A method of rendering inert plastics, primarily polyolefins, more receptive to inks, adhesives, and decorative coatings by subjecting their surfaces to a corona discharge. A typical method of treating film is to pas the film over a grounded metal cylinder above which is located a sharp-edged, high-voltage electrode spaced so as to leave a small air gap between the film and the electrode. The corona discharge oxidizes the film, forming polar groups on vulnerable sites, increasing the surface energy and making the film receptive to inks, etc.
See also flame treating.

Corona, internal *n.* In electrical cable, fault due to ionization of air between conductor and insulation.

Corona resistance *n.* Ability of a material to withstand the effects of corona discharge.

Correlated *n.* Different types of merchandise systematically related in color and design, as wallpaper with a fabric, or a group of papers designed to be used together.

Correlated color temperature Term used to describe the color of "white" light sources. Specifically, it is the temperature of the Planckian (black body) radiator, which produces the chromaticity most similar to that produced by the light source in question. The temperature is expressed in degrees on the absolute or Kelvin scale, or in mireds (micro-reciprocal degrees), 10^6/T.
See color temperature.

Correlation *n.* Relationship, degree of association, or index of prediction between two scores or sets of data. Measures the tendency of one score or set of data to vary concomitantly with the other (e.g., the tendency of students with high IQ to be above average in reading ability). The existence of a strong relationship – i.e., a high correlation – between two variables does not necessarily indicate that one has any causal influence on the other. Usually expressed as a decimal coefficient between −1.00 and +1.00 (Pearson v), where −1.00 indicates a perfect negative relationship, 0 indicates no relationship, and +1.00 indicates a perfect positive relationship. The Pearson v coefficient of correlation can assume any value on a continuum between −1.00 and +1.00.

Corrosion \kə-ˈrō-zhən\ [ME, fr. LL *corrosion-*, *corrosio* act of gnawing, fr. L *corrodere*] (14c) *n.* The deterioration of metal or of concrete by chemical or electrochemical reaction resulting from exposure to weathering, moisture, chemicals, or other agents in the environment in which it is placed. Corrosion. National Association of Corrosion Engineers, Houston, TX. Uhlig HH (1971) Corrosion and corrosion control. John Wiley and Sons Inc., New York. Uhlig HH (1948) Corrosion handbook. John Wiley and Sons Inc., New York.

Corrosion barriers *n.* A broad term applying to the ability of plastics to resist many environments, but in particular, attack by acids, bases, and oxidants.

Corrosion coating *See chemical conversion coating.*

Corrosion fatigue *n.* Reduction by corrosion of the ability of a metal to withstand cyclic or repeated stresses.

Corrosion-inhibiting paint *See anti-corrosion paint or composition.*

Corrosion-inhibitive pigment *n.* A pigment which when made into a paint has the property of minimizing corrosion of the substrate to which it is applied.

Corrosion inhibitor *n.* Any of a number of materials used to prevent the oxidation of metals; may be a coating applied to the surface, a paint undercoat, an additive or an element alloyed with the metal.

Corrosion potential *n.* Potential that a freely corroding metal or alloy exhibits in a particular solution.

Corrosion rate *n.* Speed at which a metal or alloy is wasted away because of corrosion.

Corrosion resistance *n.* A broad term applying to the ability of plastics to resist many environments, but in particular, attack by acids, bases, and oxidants.
See acid resistance, alkali resistance, artificial weathering, chemical resistance, deterioration, permanence, solvent resistance, stain resistance, sulfide straining, light resistance, volatile loss, and weathering.

Corrugation mark *n.* A fabric defect consisting of a crimped, rippled, wavy, pebbled, or cockled area in the fabric spoiling the uniformity of the texture.

Corundum \kə-ˈrən-dəm\ [Tamil *kuruntam*; akin to Sanskrit *kuruvinda* ruby] (1804) *n.* Natural aluminum oxide (including many gemstones), extremely hard, used as a filler in plastics to impart hardness, heat resistance, and abrasion resistance.

Corvic Poly(vinyl chloride). Manufactured by ICI, Great Britain.

Cosanic acids *n.* A series of fatty acids beginning with eicosanic acid ($C_{20}H_{40}O_2$) and ending with nanacosanic acid ($C_{29}H_{58}O_2$), each acid differing from its predecessor by one CH_2 group.

Cosmic rays *n.* Highly penetrating radiations which strike the earth, assumed to originate in interstellar space. They are classed as: primary, coming from the assumed source, and secondary, those induced in upper atmospheric nuclei by collision with primary cosmic rays.

Cosmotron A particle accelerator capable of given them energies to billions of electron volts.

Cosolvent *See coupling agent.*

COT *n.* The covering material used on various fiber-processing rolls, especially drawing rolls. Leather, cork, rubber, and synthetic materials are frequently employed.

Cottage steamer *n.* A chamber used for batch steaming of printed or dyed textiles. Cloth is looped on “poles” on a special cart, which fits into the steamer for processing.

Cotton *n.* Staple fibers, surrounding the seeds of various species of Gossypium.

Cotton count *n.* The yarn numbering system based on length and weight originally used for cotton yarns and now employed for most staple yarns spun on the cotton, or short-staple, system. It is based on a unit length of 840 yards, and the count of the yarn is equal to the number of 840-yard skeins required to weigh 1 pound. Under this system, the higher the number, the finer the yarn.
Also see yarn number.

Cotton fiber *n.* A unicellular, natural fiber composed of almost pure cellulose. As taken from plants, the fiber is found in lengths of 3/8 to 2 in. For marketing, the fibers are graded and classed for length, strength, and color.

Cotton-free *See dust-free.*

Cotton-free dry *See dust-free.*

Cotton linters *See linters.*

Cottonseed oil *n.* A semidrying oil obtained from the seeds of many types of plants of the genus, *Gossypium*. Its main constituent acids are linoleic (46%), palmitic (29%) and oleic (24%). As oil, it is rarely used in paint, but its fatty acids are used in the manufacture of alkyd resins.
See Becchi–Millian test.

Cottonseed pitch *See stearine pitch.*

Cotton system *n.* A process originally used for manufacturing cotton fiber into yarn,

C

and now also used extensively for producing spun yarns of manufactured fibers, including blends. Processing on the cotton system includes the general operations of opening, picking, carding, drawing, roving, and ring or mule spinning in the production of carded yarns. For combed yarns, three steps, culminating in combing, are included after the carding operation. There have been many modifications of this process, especially in recent years for the so-called "long draft", or "Casablancas", system. The cotton system is also proving to be the basis of many hybrid systems for handling wool yarns and for manufacturing other long-staple yarns.

Couette flow *n.* Shear flow in the annulus between two concentric cylinders, one of which is usually stationary while the other turns. By measuring the relative rotational velocity and the torque required to maintain steady flow, one can infer the viscosity of the liquid. Flow in the metering section of a single-screw extruder resembles Couette flow, modified by the presence of the flight and, normally, by the pressure rise along the screw.
See rotational viscometer.

Coulomb (C) \ˈkü-ˌläm\ [Charles A. de *Coulomb*] (1881) *n.* (1) A quantity of electricity defined in the SI system as equal to a current of 1 A flowing for 1 s, i.e., 1°C = 1 A/s. (2) Before SI, the quantity of electricity that must pass through a circuit to deposit 0.0011180 g of silver from a solution of silver nitrate. (3) The quantity of electricity on the positive plate of a 1 F capacitor when the potential difference between the plates is 1 V.

Coulomb's law *n.* (1854) The force of attraction or repulsion acting along a straight line between two electric charges is directly proportional to the product of the changes and inversely to the square of the distance between them.

Coumarin \ˈkü-mə-rən\ [F *coumarine*, fr. *coumarou* tonka bean tree, fr. S or P; S *cumarú*, fr. P, fr. Tupi *kumarú*] (1830) (cumarin) *n.* A dual-ring aromatic ketone, $C_9H_6O_2$, the sweet-smelling constituent of white clover, also produced synthetically. It is sometimes copolymerized with styrene to increase the deflection temperature above that of polystyrene (*See image*).

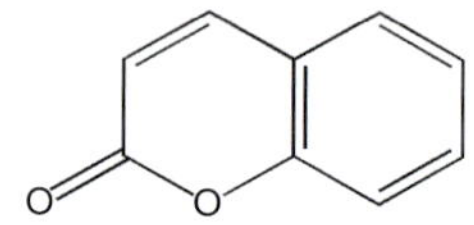

Coumarone (2,3-benzofuran, cumarone) *n.* C_8H_6O. Bicyclic ring compound. Parent substance for the coumarone resins. Properties: colorless liquid, aromatic odor; Sp gr of 1.078; mp of −18°C (−0.4°F); p of 177°C (351°F); insoluble in water; soluble in alcohol and ether; derived from the coal tar naphtha fraction boiling between 150 and 200°C (302–392°F), and having the structure as shown in the image. Syn: benzofuran and cumarone.
See coumarone-indene resins.

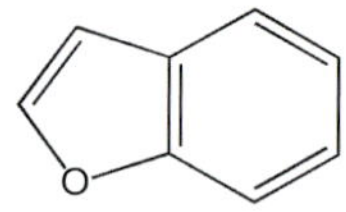

Coumarone-indene resin *n.* Any of a family of resins produced by polymerizing a coal-tar naphtha containing coumarone and indene. The naphtha is first washed with sulfuric acid to remove some impurities, then is polymerized in the presence of sulfuric acid or stannic chloride as a catalyst. Remaining impurities determine the quality of the resin, which can range from a clear, viscous liquid to a dark, brittle solid. Coumarone-indene resins have no commercial applications when used alone, but are used primarily as processing aids, extenders, and plasticizers with other resins and with rubbers.

Coumarone-indene resin (Cumar) *n.* Coal tar resins; indene resins; polycoumarone resins; polyindene resins. Resins obtained by heating mixtures of coumarone and indene (such as those that occur in the light-oil fraction from coal-tar refining) with sulfuric acid, so as to cause polymerization to thermoplastic materials with softening points of up to about 150°C (302°F). Properties: these vary from fairly viscous liquids to hard resins; color – pale yellow to nearly black; Sp gr of 1.05–1.10; soluble in hydrocarbon solvents, pyridine, acetone, carbon disulfide, and carbon tetrachloride; insoluble in water and alcohol. Used as components in aluminum paints, concrete curing compounds, pipe oils, rubber compounding, adhesives, chewing gum, printing inks, floor tile binding, and phonograph records.

Coumarone resins *n.* Any of the group of thermosetting resins derived by the polymerization of mixtures of coumarone and indene.

Count *n.* (1) A numerical designation of yarn size indicating the relationship of length to weight. (2) The number of warp yarns (ends) and filling yarns (picks) per inch in a woven fabric, or the number of wales and courses per inch in a knit fabric. For example, a fabric count of 68 × 52 indicates 68 ends/in. in the warp and 52 picks/in. in the filling.
Also see yarn number.

Counter-current *n.* Process used in many industries in which material to be dried or extracted is caused to flow against the stream of drying agent or extracting liquid. A typical example is the tunnel drying of pigments. Here the wet filter press cakes and the stream of hot air enters the tunnel from opposite ends. The driest air meets the practically dry pigment so that all moisture is removed. By the time the air has reached the other end, contact with the wet pigment insures that the maximum amount of moisture has been removed by the air stream.

Counterion \ˈkaun-tər-ˌī-ən, -ˌän\ (1940) *n.* An ion with a charge which is opposite in sign to that of some ion under consideration.

Counterions *n.* The oppositely charged component of *n* electron pair; each ion at Na^+ has a counter ion Cl^- when AnCl dissolves in water; the oppositely charged ion accompanying an ionic initiated polymerization reaction.

Country tar *See pine tar and kiln burned.*

Couple \ˈkə-pəl; "*couple of*" *if often* ˌkə-plə\ [ME, pair, bond, fr. OF *cople*, fr. L *copula* bond, fr. *co-* + *apere* to fasten] (13c) *n.* Two equal and oppositely directed parallel but not collinear forces acting upon a body to form a couple. The moment of the couple or torque is given by the product of one of the forces by the perpendicular distance between them. Dimension $[M L^2 T^{-2}]$.

Couple acting on a magnet *n.* Magnetic moment *ml* in a field of strength *H*. If the magnet is perpendicular to the direction of the field

$$C = Hml = HM$$

if the angle between the magnet and the field is θ

$$C = Hml \sin \theta.$$

The couple will be in dyne-cm for cgs electromagnetic units of *H*, *m*, and *l*.

Coupling *n.* The linking of a side effect to a principal effect. For composites an anisotropic laminate couples the shear to the normal components, while an unsymmetric

C

one couples curvature with extension. Poisson coupling links lateral contraction to axial extension. The joining together of two or more polymer molecules, which contain terminal chemically reactive groups, by reaction with a third, usually small, molecule capable of reaction with the polymer functional groups which are normally the same.

Coupling agent *n.* A chemical capable of reacting with both the reinforcement and the resin matrix of a composite material to form or promote a stronger bond at the interface. The agent may be applied from the gas phase or a solution to the reinforcing fiber, or added to the resin, or both. Harper CA (ed) (2002) Handbook of plastics, elastomers and composites, 4th edn. McGraw-Hill, New York.

Also called mutual solvent or cosolvent. See also silane coupling agent, titanate coupler, and adhesion promoter.

Coupon *n.* A representative specimen of a material or sheet product, cut from the product and set aside for testing.

Courlene *n.* Poly(ethylene) (fiber), manufactured by Courtaulds, Great Britain.

Courlene PY *n.* Poly(propylene) (fiber), manufactured by Courtaulds, Great Britain.

Course *n.* The row of loops or stitches running across a knit fabric, corresponding to the filling in woven fabrics.

Courtelle *n.* Poly(acrylonitrile). Manufactured by Courtaulds, Great Britain.

Covalent bond *n.* A bond that results from sharing electrons (e.g., H_2O).

Covalent solid *n.* A solid in which atoms are bonded covalently to form a giant extended network.

Cove ceiling *n.* A ceiling, which is rounded where it meets the wall.

Covellite \kō-ˈve-ˌlīt\ [F *covelline*, fr. Niccolò *Covelli* † 1829 Italian chemist] (1850) *n.* *See copper blue.*

Cover *n.* (1) The degree of evenness of thread spacing. (2) The degree to which underlying structure is concealed by the surface material, as in carpets, the degree to which pile covers backing. (3) The ability of a dye to conceal defects in fabric.

Coverage *n.* (1) Spreading rate generally expressed in ft^3/gal or m^2/l. In pigmented coatings, it is related to hiding power. In clear coatings, it refers to the area coated at a desired film thickness. (2) Description of the amount of paint per unit area, which must be applied to achieve a specified contrast ratio. (3) The surface area covered by a given quantity of ink or coating material. Paint/coatings dictionary. Compiled by Definitions Committee of the Federation of Societies for Coatings Technology, 1978.

See mileage and hiding power.

Cover factor *n.* The fraction of the surface area that is covered by yarns assuming round yarn shape.

Covering power *n.* (1) Term used occasionally in the paint industry as synonymous with hiding power but which actually has no precise meaning. (2) The ability of an ink to hide the material beneath (substrate) and to produce a uniform, opaque surface. *Also see opacity.*

Coverstock *n.* A lightweight non-woven material used to contain and conceal an underlying core material. Examples are the facing materials that cover the absorbent cores of diapers, sanitary napkins, and adult incontinence products.

Covert *n.* A medium weight to heavyweight wool or wool blend cloth woven with a steep twill from two or more shades of yarn-dyed fibers to produce a mottled or mélange effect.

Cowoven fabric *n.* In aerospace textiles, a fabric in which a reinforcing fiber and a matrix fiber are adjacent to each other as one end in the warp and/or filling direction.

CP *n.* (1) Abbreviation for cellulose propionate. (2) Abbreviation for chemically pure, a designation for laboratory chemicals now largely superseded by *analytical reagent.*
See cellulose acetate propionate.

cP *n.* Abbreviation for the deprecated but still widely used viscosity unit, centipoise.

CPE *n.* Abbreviation for chlorinated polyethylene.

CPET *n.* Abbreviation for crystalline polyethylene tere-phthalate.

CPVC *n.* (1) Abbreviations for chlorinated poly(vinyl chloride). (2) In the paint industry, abbreviations for critical pigment concentration, a source of confusion often encountered in the paint and color literature.

CR *n.* Abbreviation for chloroprene rubber (British Standards Institution).
See neoprene.

Cr *n.* Chemical symbol for the element chromium.

CR *n.* Poly(chloroprene).

CR-39 *n.* Abbreviation for "carbonate resin 39".
See diethylene glycol BIS-(allyl carbonate).

Crab *n.* A hand device used to stretch carpets in a small area.

Crabbing *n.* The process of heating wool or hair fabrics, under tension, in a hot or boiling liquid, then cooling under tension, to provide the fabric with dimensional stability for further wet processing.

Crack *n.* A defect in a woven fabric consisting of an open fillingwise streak extending partly or entirely across the fabric.

Cracking *n.* (1) Generally, the splitting of a dry paint or varnish film, usually as a result of aging. The following terms are used to denote the nature and extend of this defect: *hair-cracking.* Fine cracks which do not penetrate the top coat; they occur erratically and at random, *checking,* Fine cracks which do not penetrate the top coat and are distributed over the surface, giving the semibalance of a small pattern. *Cracking.* Specifically, a breakdown in which the cracks penetrate at least one coat and which may be expected to result ultimately in complete failure. *Crazing.* Resembles checking, but the cracks are deeper and broader. *Crocodilian* or *ligaturing.* A drastic type of crazing, producing a pattern resembling the hide of a crocodile. (2) The process of breaking down certain hydrocarbons into simpler ones of lower boiling points, by means of excess heat, distillation under pressure, etc., in order to give a greater yield of low boiling products than could be obtained by simple distillation. (3) Cracking is also the treatment of rubber, uncured and cured, by passing it through moving corrugated rolls, as in preparing tires and other vulcanized rubber for reclaiming. Hare CH (2001) Paint film degradation – mechanisms and control. Steel Structures Paint Council, Pittsburgh, PA. Koleske JV (1995) Paint and coating testing manual. American Society for Testing and Materials, Philadelphia, PA. Hess M (1965) Paint film defects. John Wiley and Sons Inc., New York.

Cracking resistance *n.* The ability of a coating to resist breaks of the film where the breaks extend through to the surface painted and the underlying surface is visible. The use of a minimum magnification of 10 diameters is recommended in cases where it is difficult to differentiate between cracking and checking.
See cracking and checking resistance.

Crackle finish *n.* Finish resulting from applying a top coat designed to shrink and crack and expose a more flexible undercoat, usually of a different color.

Crackle varnish Clear protective top coat applied over a crackle finish.

Crack mark *n.* A sharp break or crease in the surface of a coated or laminated fabric.

Crack stopper A method or material used or applied to delay the propagation of a potential or existing crack. Techniques include drilling of holes, installing a load-spreading doubler, or including in the design an interruption in part continuity.

Crammer-feeder (force feeder) *n.* A device fitted to the inlet port of an extruder that precompacts a low-density feedstock and propels it into the feed section of the extruder screw. As originally produced by Prodex Corp (now a division of HPM), the crammer-feeder consisted of a conical shell within which turned a decreasing-diameter screw driven independently of the extruder.

Crash *n.* A course fabric with a rough, irregular surface made from thick, uneven yarns.

Crater *n.* A small, shallow surface imperfection.

Cratering *n.* Formation of small bowl-shaped depressions in a paint or varnish film that may or may not expose the underlying surface.

Crawl Shrinkage of milled and calendered stock after removal from rolls.

Crawling *n.* (1) Defect in which a wet paint or varnish film recedes from small areas of the surface, leaving them apparently uncoated. British synonym is cissing. (2) The contraction of an ink film into drops after printing on a surface, which the ink does not wet completely.

Crawl space *n.* A shallow, unfinished space beneath the first floor of a house which has no basement, used for visual inspection and access to pipes and ducts. Also, a shallow space in the attic, immediately under the roof.

Crayon *n.* A small stick of pigment in oil or wax. Usually covered with paper for ease of handling. Certain crayons contain water-soluble dyes and are prepared in an aqueous medium.

Crazing *n.* An undesirable defect in plastics articles characterized by distinct surface cracks or minute frost-like internal cracks, resulting from stresses within the article that exceed the tensile strength of the plastic. Such stresses may result from molding shrinkage, or machining, flexing, impact shocks, temperature changes, or the action of chemicals and solvents. Hare CH (2001) Paint film degradation – mechanisms and control. Steel Structures Paint Council, Pittsburgh, PA. Koleske JV (1995) Paint and coating testing manual. American Society for Testing and Materials, Philadelphia, PA.
See also stress cracking.

Creaming of emulsion *n.* Separation of an emulsion into two layers of different concentration. The more concentrated top layer, i.e., the layer containing the greatest number of dispersed droplets per volume, has a creamy appearance. Gentle agitation of the two layers often effects uniformity of the emulsion.

Crease *n.* A break or line in a fabric generally caused by a sharp fold. Creases may be either desirable or undesirable, depending upon the situation. A crease may be intentionally pressed into a fabric by application of pressure and heat and sometimes moisture.

Crease recovery *See wrinkle recovery.*

Crease resistance *n.* Term used to indicate the capacity of a fabric to resist and/or to recover from, creases incidental to its usage.

Crease retention *n.* The ability of a fabric to maintain an inserted crease. Crease retention can be measured subjectively or by the relation of a crease in a subsequent state to the crease in the initial state. Crease retention may be strongly dependent on the conditions of use, e.g., normal wear, washing or tumble-drying.

Creel \ˈkrē(ə)l\ [ME *creille, crele*] (14c) *n.* The spool and its supporting structure on which continuous strands or rovings of reinforcing material are would for use in the filament-winding process.

Creeling *n.* The mounting of supply packages in a creel to feed fiber to a process, i.e., beaming or warping.

Creep \ˈkrēp\ [ME *crepe*n, fr. OE *crēopan*; akin to ON *krjūpa* to creep] (before 12c) *vt.* Due to its viscoelastic nature, a plastic subjected to a load for a period of time tends to deform more than it would from the same load released immediately after application. The degree of this deformation increases with the duration of the load and with rising temperature. Creep is the permanent deformation resulting from the prolonged application of a stress below the elastic limit. This deformation, after any time under stress is partly recoverable (*primary creep*) upon the release of the load and partly unrecoverable (*secondary creep*). Creep at room temperature is sometimes called *cold flow.* Elias HG (2003) An introduction to plastics. John Wiley and Sons, New York. Rosato DV (ed) (1992) Rosato's plastics encyclopedia and dictionary. Hanser-Gardner Publications, New York.

See also andrade creep, delayed deformation, drift, cold flow, compression set, tension set, and strain relaxation.

Creeping *n.* Spontaneous spreading of a liquid on a surface. In the case of an applied film of paint, varnish or lacquer, it refers to the spread of the wet film beyond the area to which it was applied.

Creep modulus *n.* The total deformation measured at constant load over a period of time divided into the applied stress. Creep modulus, if known, simplifies some plastic part designs by allowing the designer to use standard formulas for deformation in which the creep modulus, for the expected service life under load, replaces the conventional, short-time modulus.

Creep rubber *n.* Natural rubber of a pale- to dark amber color prepared by coagulating natural-rubber latex with acid, then milling this coagulum into sheets. The other basic form of solid natural rubber (i.e., ribbed sheet) is prepared by drying the latex on rolls in the presence of smoke.

Creep rupture *n.* The rupture of a plastic under a continuously applied stress that is less than the short-time strength. This phenomenon is caused by the viscoelastic nature of plastics. Creep-rupture tests are generally conducted over a series of loads ranging from those causing rupture within a few minutes to those requiring several years or more.

Creep strength *n.* The initial stress at which failure occurs after a measured time under load. Thus, creep strength (at any temperature) must be labeled with the time to failure. Like creep modulus, creep strength is useful to designers for applications in which plastic articles or members will carry sustained loads.

Cremnitz white *See carbonate white lead.*

Crenular cross section *See cross section.*

Crepe \ˈkrāp\ [F *crêpe*] (1797) *n.* A lightweight fabric characterized by a crinkling surface obtained by the use of:

(1) hard-twist filling yarns, (2) chemical treatment, (3) crepe weaves, and (4) embossing.

Crepe rubber *n.* A type of crude or sometimes synthetic rubber pressed into crinkled sheets.

Cresol \ˈkrē-ˌsól\ [ISV, irreg. fr. *cresote*] (ca. 1869) (hydroxytoluene, methylphenol) *n.* $H_3CC_6H_4OH$. An important family of coal-tar derivatives, occurring in ortho, meta, and para isomers, and used in the production of phenol-formaldehyde resins and tricresyl phosphate, an important plasticizer for PVC. Three cresols are possible, namely: (a) *o*-cresol, mp 30°C; bp, 191°C; (b) *m*-cresol, mp 4°C; bp, 205°C; (c) *p*-cresol, mp 36°C; bp, 201°C, and these are found together in the crude cresylic acid from coal tar. The main use of the cresols is in the manufacture of cresol-formaldehyde resins, and cresylic acid, rich in the meta isomer, is usually chosen for this purpose.

Known also as cresylic acid.

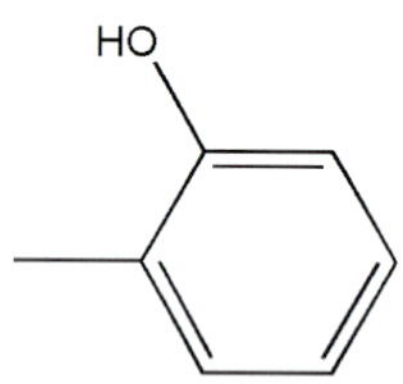

Cresol resin *n.* A phenolic-type resin obtained by condensing a cresol with an aldehyde.

Cresote *n.* Heavy, high boiling oil obtained from coal and wood tars. A major use for creosote is as a wood preservative, or as a base for the same. Syn: dead oil and pitch oil.

Cresyl diphenyl phosphate (CDP) *n.* (H_3C-$C_6H_4O)PO(C_6H_5O)_2$. A plasticizer for cellulosics, vinyl chloride polymers and copolymers, with a high degree of flame resistance and good low-temperature properties. It is also acceptable for use in food-packaging films. It is most often used, in low percentages of the total plasticizer, as a flame retardant.

Cresylic acid *n.* A term sometimes applied to mixture of *o*-, *m*-, and *p*-cresol, which are mildly acidic, but also including wider fractions of phenolic compounds derived from coal tar or petroleum that contain xylenols and other higher-boiling phenols in addition to the cresols. It is used in the production of phenolic resins and tricresyl phosphate.

See cresol (*See image*).

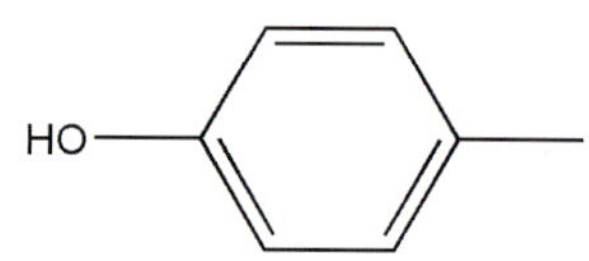

Cretonne \ˈkrē-ˌtän\ [F, fr. *Creton*, Normandy] (1870) *n. See chintz.*

Crevice corrosion *n.* Corrosion which occurs within an adjacent to a crevice formed by contact with another piece of the same or another metal or with a non-metallic material. When this occurs, the intensity of attack is usually more severe than on surrounding areas of the same surface.

Crimp *n.* The waviness of a fiber. It determines the capacity of fibers to cohere under light pressure. Crimp is measured by either the number of crimps or waves per unit length or by the decrease in length upon crimping directed by the uncrimped length, expressed as a percent. (2) The difference in distance between two points on an unstretched fiber and the same two points when the fiber is straightened under specified tension. Crimp is expressed as a percentage of the unstretched length. (3) The difference in distance between two points on a yarn as it lies in a fabric and the same two points when the yarn has been removed from the fabric and straightened under specified tension, expressed as a percentage of the distance between the two points as the yarn lies in the fabric.

Crimp amplitude *n.* The height of displacement of the fiber from its uncrimped condition.

Crimp deregistering *n.* The process of opening a tow band by causing the peaks and valleys of the crimp to lay randomly rather that uniformly.

Crimped yarn *See textured yarns* (4).

Crimp energy The amount of work required to uncrimp a fiber.

Crimp frequency The crimp level, or number of crimps per inch in yarn or tow.

Crimping The process of imparting crimp to tow or filament yarn.

Crimp setting *n.* An after treatment to set the crimp in yarn or fiber. Usually heat and steam are used, although the treatment may be chemical in nature.

Crimson antimony Syn: antimony vermillion.

Crimson lake *n.* Pigment derived from cochineal by precipitating the extract with aluminum and tin salts. It has a deep blue-red color. Alternative name is Florentine lake.

Crimson toner *n.* Azo pigment produced by coupling 4-aminotoluene-3-sulfonic acid with β-oxynaphthoic acid, followed by conversion to the calcium lake.

Crinkle *n.* (1) A wrinkled or puckered effect in fabric. It may be obtained either in the construction or in the finishing of the fabric. (2) The term is sometimes incorrectly used to describe the crimp of staple fiber.

Crinkle finish Syn: ripple finish.

Crinkle yarn *See textured yarns.*

Crinkling *See wrinkling.*

Crinoline \ˈkri-nᵊl-ən\ [F, fr. I *crinoline*, fr. *crino* horsehair (fr. L *crinis* hair) + *lino* flax, linen, fr. L *linum*] (1830) *n.* A stiff, heavily sized fabric used as an interlining or to support areas such as the edge of a hem.

Critical angle (C) (1873) *n.* The angle at which total reflection of a light ray passing from one medium to another occurs. The angle of incidence must be large (55–60°).

C

Critical conversion *n.* The degree of reaction at which the molecu*l*ar weight distribution curve first extends into the region of infinite molecular weight; gelation occurs after *critical conversion* and before the *upper limit of conversion.*

Critical damping *n.* In a damped vibrating system, damping so strong that the system, when displaced from rest, returns to rest in one-half cycle. Compare logarithmic decrement.

Critical length *See breaking length.*

Critical mass (1964) *n.* The minimum mass the fissile material must have in order to maintain a spontaneous fission chain reaction. For pure U^{235} it is computed to be about 20 pounds.

Critical micelle concentration (CMC) *n.* The point beyond which the concentration of single molecules of surfactants remains relatively constant. Much more surfactant may be dissolved to produce clear solutions, but the added increments form micelles in the solution instead of appearing as individual molecules.

Critical miscibility temperature *See flory temperature.*

Critical pigment volume concentration (CPVC) *n.* That level of pigmentation, PVC, value in the dry paint, where just sufficient binder is present to fill the voids between the pigment particles. At this level, a sharp break occurs in film properties such as scrub resistance, hiding, corrosion resistance, ease of stain removal, etc. Different requirements for each product would dictate different PVC or CPVC ratios. Ceiling paints, for instance, are not required to be very washable and can be formulated at or above CPVC, whereas gloss paints and many exterior formulations are designed well below their CPVC, where CPVC has no significance. CPVC has significance only in flat paints.

Critical point (ca. 1889) *n.* The temperature and pressure above which the liquid and gaseous states become indistinguishable.

Critical relative humidity *n.* The relative humidity at which the phase transition from the crystalline particle to the liquid droplet occurs.

Critical shear stress *n.* In extrusion, the shear stress in the melt at the die wall that signals the onset of melt fracture. The stress is on the order of 0.1–0.4 MPa.

Critical strain *n.* In a strength test, the strain at the yield point.

Critical surface tension *n.* The value of the surface tension of a liquid, γ_c, below which a drop of the liquid will wet and spread, forming a zero contact angle on a substrate whose surface energy is characterized by γ_c. This property is positively correlated with a polymer's solubility parameter.
See also surface tension and solubility parameter.

Critical temperature *n.* That temperature above which a gas cannot be liquefied by pressure alone. The pressure under which a substance may exist as a gas in equilibrium with the liquid at the critical temperature is the *critical pressure.*

Crocheting *n.* The interlocking of loops from a single thread with a hooked needle. Crocheting can be done either by hand or by machine.

Crocidolite \krō-ˈsi-dᵊl-ˌīt\ [Gr *Krokydolith,* fr. Gk *krokyd-*, *krokys* nap on cloth (akin to Gk *krekein* to weave) + Gr *–lith* –lite] (1835) *n.* Blue asbestos (Mineral Riebeckite).
See asbestos.

Crocking *n.* (1) *See bleeding.* (2) Physical transfer of color from one material to another. In ASTM D 1593, a test for resistance to crocking of flexible films subject the test film to rubbing with a 5-cm square of white cotton cloth.

Crocking *n.* (1) Removal of color on abrasion or rubbing. (2) Smudging, or rubbing off, of ink. (3) Color pigment in rubber which may not appear on the surface as a bloom but which will rub off and discolor an adjacent surface. (4) Staining of a white cloth by rubbing lightly over a colored surface.

Crocodiling *See cracking.*

Crocoisite, crocolite, crocoite Native lead chromate mineral of red color.

Crocus abrasive Either synthetic or natural iron oxide, crocus is the basis of the rouge used in many fine polishing and buffing operations. It is very soft, approximately 6 on the Mohs scale, bright red, and contains a small amount of silicon dioxide.

Crocus cloth *n.* An iron-oxide coated abrasive cloth, used as a polishing agent after most of the work has been done with emery or aluminum oxide, capable of giving a mirror-like finish to metal.

Crofon *n.* Optical fibers from polymethyl methacrylate and polyethylene, manufactured by DuPont, USA.

Crooked cloth *See baggy cloth.*

Cross coating *n.* Application of a coat of paint by a series of strokes or spray passes, each at right angles to the previous series.

Cross direction *n.* The width dimension, within the plane of the fabric, that is perpendicular to the direction in which the fabric is being produced by the machine.
See machine direction.

Cross dyeing *See dyeing.*

Cross-flow quency *n.* In cooling extruded polymer filaments, refers to cooling air directed from one side cross the path of the filaments. There may be some type of suction on the opposite side to remove the heated air.

Cross grain A pattern in which the fibers and other longitudinal elements deviate from a line parallel to the sides of the piece. Applies to either diagonal or spiral grain or a combination of the two.

Crosshead *n.* (1) A device that receives a molten stream of plastic emerging from an extruder, diverts the flow to a direction usually 45° or 90° from the axis of the extruder, and forms the extrudate to a shape such as a parison for blow molding or a jacket around a wire. An essential element of the crosshead is the mandrel, a tubular core with grooves of various shapes and held in place by a perforated plate, web, or spider legs. Material emerging from the space between the mandrel and the crosshead housing is given its final shape by means of a die mounted on the end of the crosshead. (2) The moving member of a testing machine.

Crosshead die *See crosshead.*

Crossing *n.* Method of obtaining even distribution of paint by means of a brush whereby the direction of brushing each series of strokes lies at right angles to that of the previous series.

Cross laminate *n.* A laminate in which the direction of greatest strength in some layers is perpendicular to that direction in other layers. A simple example is a laminate in which alternate layers contain unidirectional reinforcing fibers laid in perpendicular directions.

See laminated, cross.

Cross-linked polymerization *n.* A bond, atom, or group linking the chains of atoms in a polymer, protein, or other complex organic molecule.

Cross-linked polymers Polymers in which linear polymer chains are joined together by covalent chemical bonds.

Cross-linking *n.* Applied to polymer molecules, the setting up of chemical links among the molecular chains. When extensive, as in most cured thermosetting resins, cross-linking creates one infusible supermolecule from all the chains. Cross-linking can also occur between polymer molecules and other substances. For example, polyethylene can be cross-linked with carbon-black particles, which have sites to which polyethylene chains can link in the presence of a catalyst. The mixture of resin, filler, and catalyst can be molded as a thermoplastic, then transformed to a thermoset by cross-linking in the curing cycle. Cross-linking can be achieved by irradiation with high-energy electron beams, or by means of chemical cross-linking agents such as organic peroxides. The amount of cross-linking may be expressed in *x*-links per molecular weight or mass between *x*-links, weight between *x*-links sometimes referred to as cross-link density. Elias HG (2003) An introduction to plastics. John Wiley and Sons, New York. Elias HG (1977) Macromolecules, vols 1–2, Plenum Press, New York.

Cross-linking agent *n.* A substance that promotes or regulates intermolecular covalent bonding between polymer chains.

Cross-linking agents (curing agents) *n.* An additive used with a polymer in order to bring about cross-linking. In rubber technology, the cross-linking agents used are vulcanization ingredients (sulfur, peroxide, etc.). For thermosetting plastics, the cross-linking agents (hardeners) are amines or anhydrides for epoxy resins, or peroxides for unsaturated polyesters.

Cross-linking index *n.* The number of cross-linked units per primary polymer molecule, averaged over the whole specimen.

Cross-linking *n.* The stabilization of cellulosic or manufactured fibers through chemical reaction with certain compounds in such a way that the cellulose or manufactured polymer chains are bridged across or "cross-linked". Cross-linking improves such mechanical factors as wrinkle resistance. Random cross-linking in manufactured polymers is undesirable and leads to brittleness and loss of tensile strength.

Crossply *n.* A layer containing reinforcing fibers mainly running in one direction and perpendicular to the main fiber direction in adjacent layers.
See also cross laminate.

Cross section (nuclear cross section) *n.* A measure of the probability of a particular process. The nuclear cross section is expressed by a/bc, where a is the number of processes occurring, b the number of incident particles, and c the number of target nuclei/cm^3. There are nuclear cross sections for fusion, for slow neutron capture, for Compton collision, and for ionization by electron impact.

Cross stitch *See pinhole.*

Cross termination *n.* In free radical copolymerization, termination by reaction of two radicals terminated by monomer units of the opposite type, i.e., $\sim A^* + \sim B^* \rightarrow$ termination, by combination or disproportionation with rate constant k_{AB}. Cross-termination is often favored over termination by reaction between two like radicals due to polar effects.

Crotonaldehyde A colorless liquid synthesized by the aldol condensation of acetaldehyde, accompanied or followed by dehydration. It can be polymerized by triethylamine to a resin with film-forming properties, or copolymerized with many compounds. Other uses include solvent for PVC, short-stopped in the polymerization of vinyl chloride, and plasticizer synthesis.

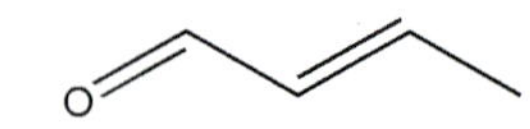

Crotonic acid (*trans*-2-butenoic acid, *trans*-*β*-methacrylic acid) *n.* CH_3–CH=CHCOOH. A white crystalline solid prepared by the oxidation of crotonaldehyde. It forms copolymers with vinyl acetate, used as hot-melt adhesives. Esters of crotonic acid are used as plasticizers for acrylic and cellulosic plastics (*See image*).

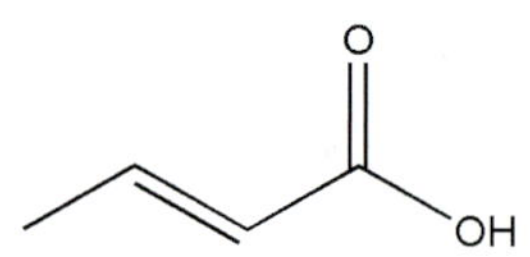

Crotonyl peroxide *n.* A catalyst for the polymerization of vinyl and vinylidene halides.

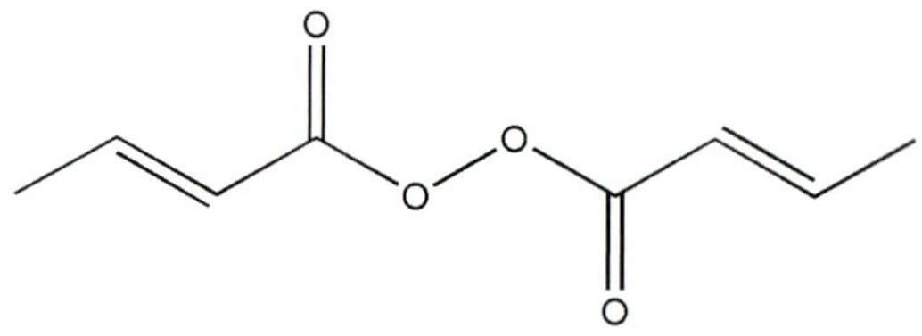

Crown *n.* (1) Of a calender roll, a gradual, small increase in the diameter of a roll toward the center to compensate for the slight deflection due to bending of the roll under pressure. (2) The much ore convex surface of a transmission-belt pulley designed to keep the belt running in the center of the pulley.

Crowsfeet *n.* A fabric defect consisting of breaks or wrinkles of varying degrees of intensity and size, resembling bird's footprints in shape, and occurring during wet processing of fabrics.

Crowsfooting *n.* (1) Type of film defect where small wrinkles occur in a pattern resembling that of a crow's foot. (2) Type of crystallization on the surface of a varnish or paint film after exposure to the gas test. *See wrinkling. See gas checking.*

Crude resinous liquid *See tall oil.*

Cryogenic \ˌkrī-ə-ˈje-nik\ (1896) *adj.* Pertaining to very low temperatures, usually temperatures below about −150°C (123 K). Evaluations of plastics at cryogenic temperatures are conducted for potential space applications.

Cryogenic finishing *n.* Process by which a material is cryogenically tempered (deep freezing below −300°F). Cryogenic finishing relieves stress in the substrate, thus creating a more uniform micro-structure which extends the life of the material.

Cryogenic grinding (freeze grinding) *n.* Thermoplastics are difficult to grind to small particle sizes at ambient temperatures because they soften, cohere in lumpy masses, and clog screens. When chilled by dry ice, liquid carbon dioxide, or liquid nitrogen, thermoplastics can be finely ground to powders suitable for electrostatic spraying and other powder processes.

Cryohydrate *n.* The solid, which separates when a saturated solution freezes. It contains the solvent and the solute in the same proportions as they were in the saturated solution.

Cryoscopy *n.* Measurement of polymer molecular weight based on freezing point depression.

Cryptometer *n.* An instrument for measuring the hiding power or opacity of pigmented composition. It functions on the principle that a film or wedge of the paint is obtained with a uniformly varying thickness. By adjusting the instrumentation, the thickness of film necessary to obliterate can be determined.

Crystal *n.* A homogeneous solid having an orderly and repetitive three-dimensional arrangement of its atoms. Crystalline polymers are never wholly crystalline but contain some amorphous material and many crystallites. The "ideal crystal" is a homogeneous portion of crystalline matter, whether bounded by faces or not. Crystalline matter is matter that possesses a triperiodic structure on the atomic scale. It is characterized by discontinuous vectorial properties that give rise to "crystal planes" (1) crystal growth (faces); (2) cohesion (cleavage planes); (3) twinning (twin planes); (4) gliding (gliding planes); and (5) X-ray electron, or neutron diffraction ("reflecting" planes); all of which are parallel to the lattice planes. Hibbard MJ (2001) Mineralogy. McGraw-Hill Companies Inc., New York.

Crystal lattice *n.* The regular, repeating arrangement of particles (atoms, ions, and molecules) in a crystal.

Crystalline \ˈkris-tə-lən\ [ME *cristallin*, fr. MF & L; MF, fr. L *crystallines*, fr. Gk *krystallinos*, fr. *krystallos*] (15c) *adj.* A substance (usually solid but can be liquid) in which the atoms or molecules are arranged in a definite pattern that is repeated regularly in three dimensions. Crystals tend to develop forms bounded by definitely oriented plane surfaces that are harmonious with their internal structure. They may belong to any of six crystal systems: cubic, hexagonal, tetragonal, orthorhombic, monoclinic, or triclinic.

Crystalline-amorphous structures *n.* The typical form of crystalline polymers. Crystalline polymers contain a large number of crystallites, as well as devoids of crystallinity. Most plastics are amorphous at processing temperatures, many retaining this state under all normal conditions.

Crystalline growth *n.* (1) The expansion and development of a crystal. The process involves diffusion of the crystallizing material to special sites on the surface of the crystal, incorporation of the molecules into the surface at these sites, and diffusion of heat away from the surface of the crystal. (2) The transformation of disoriented molecules, usually of the same substance, to a higher state of order. This process generally occurs rapidly for small molecules; however, the process is slow for polymer molecules and is arrested at temperatures below the glass transition temperature. Hibbard MJ (2001) Mineralogy. McGraw-Hill Companies Inc., New York.

Crystalline melting *n.* The heating of an ordered, crystalline structure polymer above its melts, it behaves like an amorphous polymer in that the molecular configurations become random.

Crystalline polymers *n.* Polymers containing both crystalline and amorphous material.

Crystalline silica *See silica, crystalline.*

Crystalline solid A true solid, one with a regular internal structure.

Crystallinity *n.* (1) A state of molecular structure in some resins attributed to the existence of solid crystals with a definite geometric form. Such structures are characterized by uniformity and compactness. (2) The percentage of a polymer sample that has formed crystals. Crystallinity in polymers occurs in two stages: nucleation and growth of nuclei. In most materials, crystallinity forms from growing nuclei to produce reoccurring/repeating structures that form lattices throughout the material; polymers can be crystalline, semicrystalline or non-crystalline (amorphous); crystallinity in a polymer exhibits a distinct endothermic energy transition or melting event (destruction of lattice structure) whereas a non-crystalline polymer shows a glass transition event or just softening.
See liquid-crystal polymer.

Crystallite \ˈkris-tə-ˌlīt\ [Gr Kristallit, fr. Gk krystallos] (1805) *n.* A perfect portion of an ordinary crystal; that is, a portion with its atoms and molecules arranged in a lattice free of defects. Ordinary crystals are composed of large numbers of crystallites, which may or may not be perfectly aligned with one another.

Crystallization (for ink) *n.* (1) A condition in which a dried ink film repels a second ink which must be printed on top of it. (2) Lacquers containing materials, which crystallize out from the medium as the solvent evaporates. (3) Materials in which advantage is taken of the tendency of certain drying oils, notably tung oil, to “frost” or “crystallize” when dried under certain conditions. Leach RH, Pierce RJ, Hickman EP, Mackenzie MJ, Smith HG (eds)(1993) Printing ink manual, 5th edn. Blueprint, New York.

Crystallized polyethylene terephthalate (CPET) *n.* PET resin to which a fractional percentage of nucleating agent has been added to encourage the development of crystallinity in extruded or molded products. The percent crystalline material is typically between 15 and 35, modulus and strength increase with crystalline content.
See also polyethylene terephthalate.

Crystal polystyrene *n.* Styrene homopolymer, which, through actually 100% amorphous, was so called because of its excellent clarity and the glitter of the early cube-cut pellets.
See polystyrene.

Crystal structures *n.* Homogeneous solids having an orderly and repetitive three-dimensional arrangement of its atoms.

CS *n.* Abbreviation for casein plastic.

CSMA Abbreviation for chemical specialties manufacturer association.

CSR Chlorosulfonated poly(ethylene).

C-stage *n.* The final stage in the reaction of certain thermosetting resins in which the material is relatively insoluble and infusible. Certain thermosetting resins in a fully cured adhesive layer are in this stage. Sometimes referred to as *Resite*.
See also A-stage and B-stage.

CTA *See cellulose triacetate.*

CTFE resin *See polychlorotrifluoroethylene.*

Cu Chemical symbol for the element copper (Latin: cuprum).

Cube root color difference equation *n.* Specific color difference equation. The following equations are those recommended by the CIE in 1967 for study:

$$L = 25.29G^{1/3} - 18.38,$$

where $G = 0.0010X + 1.05Y + 0.0004Z$,

$$a = K_a(R^{1/3} - G^{1/3}),$$

where $R = 1.1084X + 0.0852Y - 0.1454Z$ and $K_a = 105.0$ for $R < G$; $K_a = R > G$,

$$b = K_b(G^{1/3} - B^{1/3}),$$

where $B = 0.0062X + 0.0394Y + 0.8192Z$ and $K_b = 30.5$ for $B < G$; $K_b = 53.6$ for $B > A$,

$$\Delta E = [(\Delta L)^2 + (\Delta a)^2 + (\Delta b)^2]^{1/2}.$$

The differences (Δ's) are calculated as sample minus standard. McDonald R (1997) Colour physics for industry, 2nd edn. Society of Dyers and Colourists, West Yorkshire, England.
See color difference equations.

Cubic \ˈkyü-bik\ *adj.* (1) Having the appearance of a cube. (2) Of algebraic equations, containing the unknown to the power 3 and not higher. (3) Characterizing volume measure, as *cubic meter.* (4) The simplest of the six crystal systems, in which the three principal axes are mutually perpendicular and the atomic spacing is the same along all three.

Cull \ˈkəl\ [ME, fr. MF *cuillir,* fr. L *colligere* to bind together] (13c) *n.* (1) A rejected material or product. (2) In transfer molding, the material remaining in the transfer pot after the mold has been filled. A certain amount of cull is usually necessary for the operator to be confident that the cavity has been properly filled.

Culture *n.* (1) The process of securing the growth of fungi or other microorganisms upon artificial media. (2) The organisms resulting from the culturing process.

Cultured stone *n.* A term applied to decorative embedments of natural stones such as marble, granite, terrazzo, and slate in thermosetting resins. They are made by casting the resin, usually a polyester, in molds containing the stones. The embedments are used for counter tops, window sills, wall facings, flooring, giftware, etc.

Cumar gum *n.* A synthetic resin, used in varnishes to provide alkali-resistance properties.

Cumarone *See coumarone.*

Cumene (isopropylbenzene, isopropylbenzol, cumol) *n.* $C_6H_5CH(CH_3)_2$. A volatile liquid in the alkyl-aromatic family of hydrocarbons. It is used as a solvent and intermediate for the production of phenol, acetone, and α-methyl styrene; and as a catalyst for acrylic and polyester resins. Properties: bp, 153°C; sp gr, 0.862/20°C; refractive index, 1.506. *Known also as isopropyl benzene (See image).*

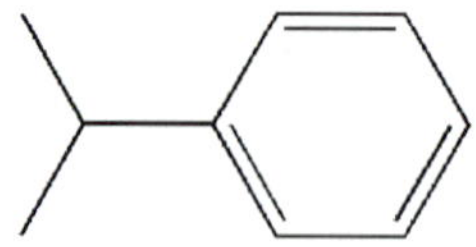

C

Cumene hydroperoxide *n.* $C_6H_5C(CH_3)_2$-OOH. A colorless liquid derived from an oxidize solution or emulsion of cumene, used as a polymerization catalyst.

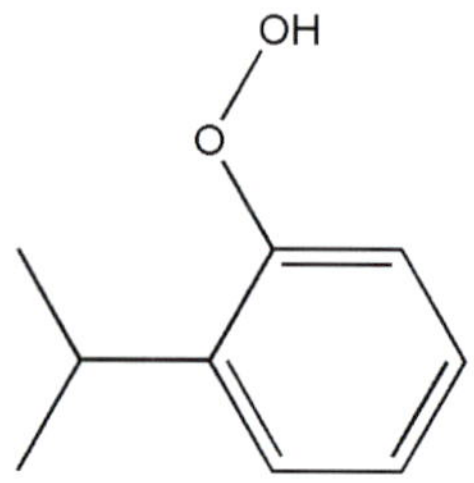

Cumylphenol derivative *n.* One of a group of polymer intermediates based on cumylphenol that offer higher performance at lower cost than nonylphenol competitors. The free phenol is an accelerator for amine hardeners of epoxy resins. Cumylphenyl acetate and the glycidyl ether are reactive in epoxy systems, giving enhanced strength. The benzoate of cumylphenol aids extrusion of PVC compounds.

Cup Efflux cup. *See ford cup.*

Cup-flow test *n.* A British standard test (B S 771) for measuring the flow properties of phenolic resins. A standard mold is charged with the specimen material and then closed under preset pressure. The time in seconds for the mold to close completely is the cup-flow index.

Cupioni A type of specialty or novelty yarn having slubs or enlarged sections of varying length.

Cuprammonium rayon *n.* A regenerated cellulose formed by dissolving cotton or wood-pulp linters in a solution of ammonia and copper oxide (from sulfate), then extruding the solution through spinnerets into warm water, where the filaments harden. The finest filaments (lowest denier) are made this way.

Cupric sulfide \ˈkyü-prink\ (1799) *adj. See copper blue.*

Cuprous (copper) oxide *n.* Cu_2O. Occurs in nature as the mineral, cuprite. Prepared commercially by furnace-reduction of a mixture of cupric oxide and copper. Used as a toxin in anti-fouling compositions.

Curing (1) In finishing fabrics, the process by which resins or plastics are set in or on textile materials, usually by heating. (2) In rubber processing, vulcanization. It is accomplished either by heat treatment or by treatment in cold sulfuryl chloride solution.

Cup temperature *n.* (1) In injection molding or extrusion, the measured temperature of a glob of melt collected at the injection nozzle or extrusion die in an insulated vessel and presumed to be equal to, or slightly less than, the average melt temperature leaving the machine. (2) (cup-mixing temperature, flow-average temperature In a flowing stream of fluid, the local product of velocity times temperature integrated over the stream cross section, said integral divided by the integral of the local velocity over the cross section. Strong AB (2000) Plastics materials and processing. Prentice-Hall, Columbus, OH.

Cup, weight per gallon *n.* Brass cup with a volume of exactly 83.3 ml used for quickly determining the wt/gal (or density) of a finished liquid product. Koleske JV (1995) Paint and coating testing manual. American Society for Testing and Materials, Philadelphia, PA. Wicks ZN, Jones FN, Pappas SP (1999) Organic coatings science and technology, 2nd edn. Wiley-Interscience, New York.

Curcas oil *n.* A vegetable oil obtained from the seeds of *Jatropha curcas*, grown in

Central America, and in the Cape Verde Isles, the Comores Isles and in some of the former Portuguese colonies, Siam and the East Indies. The percentage fatty acids as triglycerides are oleic, 62%; linoleic, 19%, and saturated fatty acid, the remainder.

Curdle \ˈkər-dᵊl\ [frequentative of ²*curd*] (1590) *v.* Syn: is coagulation. *See also coagulation and congeal.*

Cure *n.* (1) To change the properties of a polymeric system into a final, more stable, usable condition by the use of heat, radiation, or reaction with chemical additives (ASTM). (2) Synonymous with vulcanize. It includes time and temperature of vulcanization.

Cure, curing, also vulcanization *n.* To change the physical properties of a material by chemical reaction, which may be condensation, polymerization, or vulcanization.

Cure cycle *n.* The schedule of time periods at specified conditions to which a reacting thermosetting plastic or rubber composition is subjected to reach a specified property level.

Cure or curing *n.* Conversion of a wet coating or printing ink film to a solid film.

Cure stress *n.* Internal stress in cast or molded thermosetting parts, caused by unequal shrinkage in different sections of the parts. Depending on the directions of applied stress in service relative to the principal cure stress, parts may be considerably weaker, than their designers expected them to be.

Cure time *n.* The period of time that a reacting thermosetting plastic is exposed to specific conditions to reach a specified property level.

Curie \ˈkyúr-(ˌ)ē\ [Marie & Pierre *Curie*] (1910) *n.* Unit for measuring radioactivity. One Curie is the quantity of any radioactive isotope undergoing 3.7×10^{10} dis/s.

Curie's law The intensity of magnetization,

$$I = \frac{AH}{T},$$

where H is the magnetic field strength, T the absolute temperature, and A is the Currie's constant. Used for paramagnetic substances.

Curie point *n.* All ferro-magnetic substances have a definite temperature of transition at which the phenomena of ferro-magnetism disappear and the substances become merely paramagnetic. This temperature is called the "Curie Point" and is usually lower than the melting point. Serway RA, Faugh JS, Bennett CV (2005) College physics. Thomas, New York.

Curie–Weiss law *n.* the Curie law was modified by Weiss to state that the susceptibility of a paramagnetic substance above the Curie point varies inversely as the excess of the temperature above that point. This law is not valid at or below the Curie point. Serway RA, Faugh JS, Bennett CV (2005) College physics. Thomas, New York.

Curl *See kink.*

Curing agent (hardener) *n.* A substance or mixture of substances added to a plastic or rubber composition to promote or control the curing reaction. It is also an additive, which promotes the curing of a film. An agent that does not enter into the reaction is known as a *catalytic hardener* or *catalyst.* A *reactive curing agent* or *hardener* is generally used in much greater proportions than a catalyst, and is actually converted in the reaction. Syn: hardener.
See also accelerator, hardener, catalyst and cross-linking.

Curing agent blush *n.* A blushing, blooming or sweating caused by applying coatings such as amine cured epoxies under conditions of high humidity.

C

Curing agents for epoxy resins *n.* A catalytic or reactive agent that brings about polymerization causing cross-linking.

Curing temperature *n.* The temperature to which a thermosetting or elastomeric material is brought in order to commence and complete its final stage of cure. *See temperature, curing.*

Curing time (molding time) In the molding of thermosets, the time elapsing between the moment relative movement and between the mold parts creases, and the instant that pressure is released. Time necessary for curing.

Curl *n.* In paper, distortion of the unrestrained sheet due to differences in structure or coatings from one side to the other. The curl side is the concave side of the sheet.

Curling *n.* (1) *See coiling soup.* (2) Curling also refers to excessive warping of sheet goods, or distortion by uneven shrinkage. *See also crawl.*

Current \ˈkər-ənt\ [ME *curraunt*, fr. OF *currant*, pp of *courre* to run, fr. L *currere*] (Electric) *adj.* The rate of transfer of electricity. The transfer at the rate of 1 esu of electricity in 1 s is the electrostatic unit of current. The electromagnetic unit of current is a current of such strength that 1 cm of the wire in which it flows is pushed sideways with a force of 1 dyn when the wire is at right angles to a magnetic field of unit intensity. The practical unit of current is the ampere a transfer of 1 C/s, which is one-tenth the electromagnetic unit. The International ampere is the unvarying electric current which, when passed through a solution of silver nitrate in accordance with certain specifications, deposits silver at the rate of 0.00111800 g/s. The international ampere is equivalent to 0.999835 absolute ampere. The ampere-turn is the magnetic potential produced between the two faces of a coil of one turn carrying 1 A. Dimensions

$$\left[\varepsilon^{1/2}\,\mathrm{M}^{1/2}\,\mathrm{L}^{3/2}\,\mathrm{T}^{-2}\right],\quad \left[\mu^{-1/2}\,\mathrm{M}^{1/2}\,\mathrm{L}^{1/2}\,\mathrm{T}^{-1}\right].$$

Serway RA, Faugh JS, Bennett CV (2005) College physics. Thomas, New York. Lide DR (ed) (2004) CRC handbook of chemistry and physics. CRC Press, Boca Raton, FL.

Current density Magnitude of current per unit area of a metal surface, usually expressed in mA/ft^2 (mA/ft^2 or, A/m^2).

Current in a simple circuit *n.* The current in a circuit including an external resistance R and a cell of electromotive force E and internal resistance r,

$$I = \frac{E}{R + r}$$

if E is in volts and r and R in ohms the current will be in amperes. For two cells in parallel,

$$I = \frac{E}{R + r/2}.$$

For two cells in series,

$$I = \frac{2E}{R + 2r}.$$

Giambattista A, Richardson R, Richardson RC, Richardson B (2003) College physics. McGraw Hill Science/Engineering/Math, New York.

Curtain coating *n.* A method of applying paint to an object by moving it through a falling curtain of paint which may be used with low viscosity plastics or solutions, suspensions, or emulsions of plastics in which the substrate to be coated is passed through and perpendicular to a free flowing liquid "curtain", or "waterfall".

Curtaining Syn: sagging.

Curvature of field *n.* The image plane formed by a lens is naturally curved. While one part of the field will be in good focus, the rest will need refocusing to be sharp. While the eye may partially correct for this, a camera lens will not, and the final image as photographed will not be in perfect focus over the entire image plane.

Cushion back carpet *n.* (1) A unit of yarn number. The number of 100-yard lengths per pound avoirdupois of asbestos yarn or glass yarn, or the number of 300-yards lengths per pound avoirdupois of woolen yarn. (2) A length of woven cloth. (3) The number of needles per inch on a circular-knitting machine. A machine with 34 needles/in. is a 34-cut machine, and a fabric produced thereon is called a 34-cut fabric.

Cushion carpet A carpet with padding made as an integral part of the backing.

Custom color *n.* Special colors made by adding colorant to paint or by intermixing colors, which permits the retailer to match a color selected by the consumer.

Custom molder (Brit: trade molder) *n.* A firm specializing in the molding of items or components to the specifications o another firm that handles the sale and distribution of the item, or incorporates the custom-molded component into one of its own products.

Cut *n.* (1) An expression commonly used to designate a typographic printing plate. (2) To dilute with ink, lacquer or varnish with solvents or with clear base; to thin. (3) The proportion of shellac gum in alcohol. A shellac varnish is referred to as a 4 pound or five pound cut, or weight, which means 4 or 5 lbs of shellac gum to 1 gal of alcohol. (4) In the fiber industry, including glass and asbestos, the number of 100-yard lengths of fiber per pound. A now deprecated unit, 1 cut corresponds to a lineal density of 0.0049604 kg/m or 4960.4 texes.

Cut-layers As applied to laminated plastics, a condition of the surface of machined or ground rods and tubes and of sanded sheets in which cut edges of the surface layer or lower laminations are revealed (ASTM D 883).

Cut-off (flash groove, pinch-off) *n.* In compression molding, the line where the two halves of a mold come together, often along a sharp mating ridge and groove.

Cut-off saw (traveling cut-off) In extrusion of pipe, rod, and profiles, a circular saw that periodically swings forward, while moving downline at the same rate as the extruded product, to cut it into desired lengths. When the saw has completed its cut, it swings back at the same time quickly reversing travel to return to its starting point, poised for the next cut.

Cut pile A pile surface obtained by cutting the loops of yarn in a tufted or woven carpet.

Cut selvage *n.* A cut or break occurring only in the selvage. A cut selvage is caused by incorrect loom adjustment during weaving or improper edge construction. The term also refers to loose edges cut during shearing of the fabric.

Cut staple *v.* (1) An inferior cotton fiber that was accidentally cut because it was too damp during ginning. (2) A term sometimes used to denote staple of manufactured fibers.

Cut tape *See slit tape.*

Cutter (1) A mechanical device used to cut tow into staple. (2) A firm engaged in making up garments from finished fabrics. (3) A person employed in the wholesale garment industry whose specific work is to cut layers of fabric to be formed into garments. Cut velvet:
See beaded velvet.

C

Cutting in *v.* Painting of a surface adjacent to another surface which must not be painted, for example, painting the frames of a window and avoiding painting the glass, or the painting of an area on a previously painted surface.

Cut yarn A defective yarn, i.e., cut partially or completely through, resulting from malprocessing.

Cyanoacrylate adhesives *n.* alkyl -2-cyanocrylates polymerize rapidly via anionic initiation in the presence of weak bases (water, alcohol) at ambient temperatures; highly exothermic reaction yielding brittle polymers; very useful for quickly setting anaerobic adhesives; also useful for suture less topical tissue adhesives (e.g., Dermabond®); and industrial grades are commonly known as "super glue".

Cyanoguanidine *See dicyandiamide.*

Cyanuric acid \ˌsī-ə-ˈnúr-ik\ [*cyan-* + *urea*] (1838) (1,3,4-triazine-2,4,6-triol, tricyanic acid, tricarbimide) *n.* An acid evolved from the blowing agent, azodicarbonamide, when it decomposes. The acid is corrosive, and is the chief cause of plate-out on components of extruders in the structural-foam process when azodicarbonamide is used as the blowing agent (*See image*).

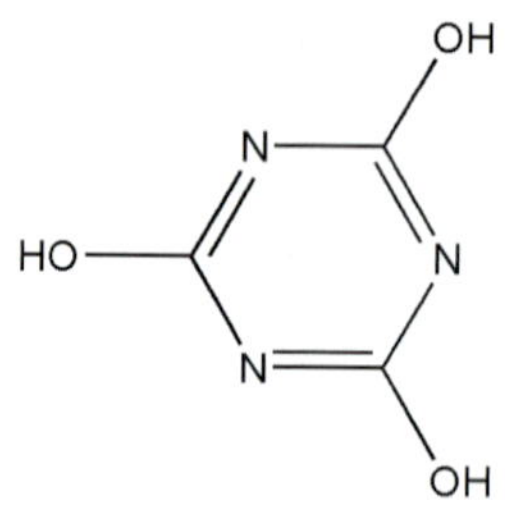

Cycle *n.* The series of sequential operations entering into a repeating batch process or part of the process. In a molding operation, cycle time is the average time elapsing, over several normal cycles, between a particular occurrence in one cycle and the same occurrence in the next cycle.

Cyclic *adj.* All cyclic compounds are recognized by their ring structure. Benzene is the simplest and most well-known cyclic compound, consisting of a single ring. Its monocyclic. When two or more rings are involved as with naphthalene, for example, the compound is bicyclic, tricycle, etc. Cyclic compounds may contain from three to six or more carbon atoms in the ring, e.g., Cyclopropane – three atoms; Cyclobutane – four atoms; Cyclopentane – five atoms; Cyclohexane – six atoms.

Cyclic diolefins resins *n.* Polymerized products obtained from the cracked distillates of petroleum. They are unsaturated products, with excellent solubility, acid, alkali, and water resistance. A special application is in metallic paint media.

Cyclic ketone resins *n.* Cyclohexanone is the primary constituent of these resins, and it is condensed either by the action of alkalis or acids. Mixed condensation products of cyclohexanone and aldehydes are also resinous. The cyclic ketone resins, which are available commercially are notable chiefly by reason of their extremely pale colors and characteristic smell. They are usually very readily soluble in vegetable oils.

Cyclic stress strain *n.* Repeated loading of a yarn on a tensile testing machine and the determination of the physical properties of the yarn during these cycles.

Cyclic trimmer *n.* Strictly, a polymer, in cyclic form, that contains three repeating groups. Cyclic trimmer is a by-product found in all commercial polyester and results in deposit buildup in package-dyeing equipment.

Cyclized rubber *n.* A thermoplastic resin produced by reacting natural rubber with stannic chloride or chlorostannic acid. This causes a reduction of the unsaturation and formation of condensed ring structures, typically with two or three rings being fused together. It is claimed (USA Patent 3,205,093) that a solution of cyclized rubber in toluene is one of the few lacquers known to adhere to polyolefins without their being pretreated or the lacquered surface being post-treated with radiation. Other uses include films and hot-melt coatings.

Cycloalkane A cyclic hydrocarbon with the general formula C_nH_{2n}.

Cyclohexane \ˌsī-klō-ˈhek-ˌsān\ [ISV] (ca. 1909) (hexamethylene, hexahydrobenzene) *n.* C_6H_{12}. A saturated hydrocarbon with a six-membered ring, cyclohexane is derived from the catalytic hydrogenation of benzene, and is used as a solvent for cellulosics and as an intermediate in the production of nylon (*See image*).

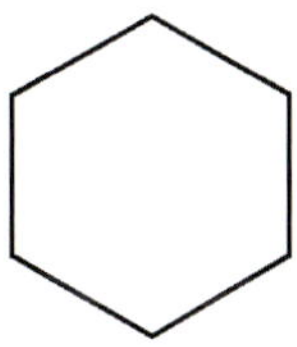

Cyclohexanol acetate *n.* $CH_3COOC_6H_{11}$. A non-flammable solvent for cellulosics and many other resins (*See image*).

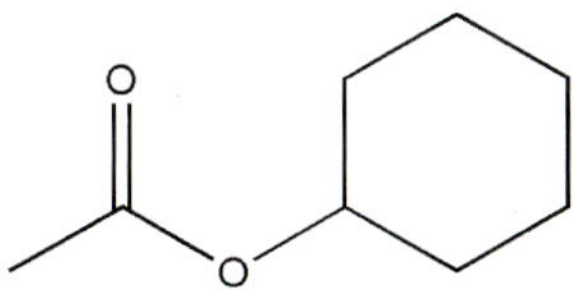

Cyclohexanone \-ˈhek-sə-ˌnōn\ (ca. 1909) (pimelic ketone, ketohexamethylene) *n.* $CH_2(CH_2)_4C{=}O$. A colorless liquid produced by the oxidation of cyclohexane or cyclohexanol. Its most important use is for the manufacturer of adipic acid for nylon 6/6, and caprolactam for nylon 6. It is also an excellent high-boiling, slowly evaporating solvent for many resins including cellulosics, acrylics and vinyls. It is one of the most powerful solvents for PVC, and is often used in lacquers to improve their adhesion to PVC (*See image*).

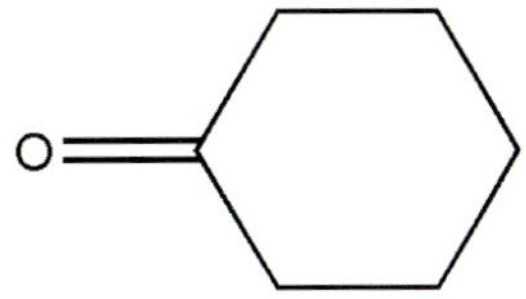

Cyclohexene oxide *n.* $C_6H_{10}O$. This epoxide is a highly reactive, colorless liquid that resembles ethylene oxide in most of its reactions. It is useful as an intermediate in the production of many organic chemicals used in plastics. Its epoxide structure is especially useful in applications where an HCI scavenger is required (*See image*).

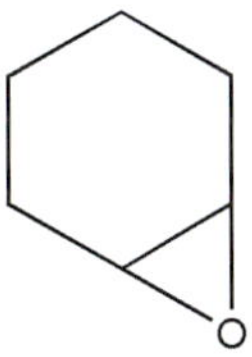

Cyclohexnol (hexahydrophenol) *n.* $CH_2(CH_2)_4CHOH$. A colorless, viscous liquid prepared by the oxidation of cyclohexane or by the hydrogenation of phenol. It is used as an intermediate in the production of nylon 6/6, it is a solvent for cellulosic resins, and it is an intermediate in the manufacture of phthalate-ester plasticizers.

Cyclohexyl methacrylate *n.* $H_2C{=}C(CH_3)COOC_6H_{11}$. A colorless monomer, polymerizable to resins for optical lenses and dental parts, and useful for potting electrical components (*See image*).

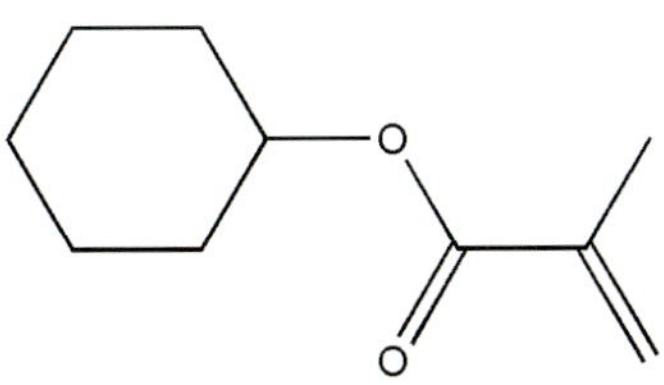

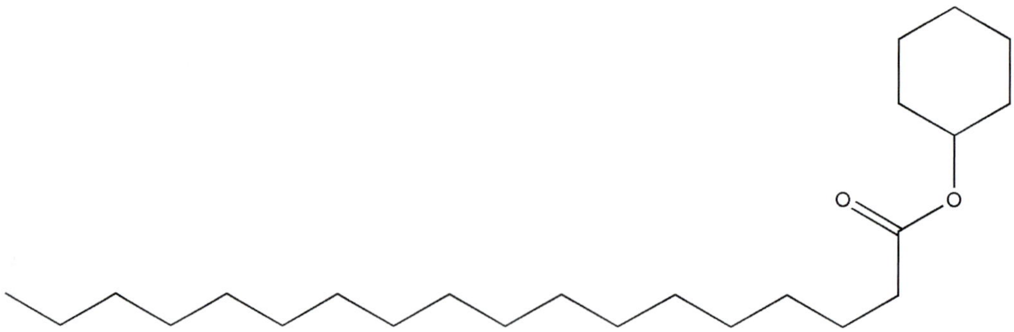

Cyclohexyl stearate *n.* $C_6H_{11}OOCC_{17}H_{35}$. A plasticizer for polystyrene, ethyl cellulose, and cellulose nitrate (*See image*).

Cycloparaffins \-ˈpar-ə-fən\ (1900) *n.* Ring compounds of saturated hydrocarbon type based on groupings of methylene radicals (CH_2). Typical cycloparaffins are cyclopropane, cyclobutane, cycloheptane. etc. The cycloparaffins have very good solvent properties, and are constituents of crude petroleum's.

Known also as naphthenes.

Cyclopentane (pentamethylene) *n.* C_5H_{10}. A solvent for cellulose ethers (*See image*).

Cyclopolymerization *n.* Polymerization in which ring structures are formed within the polymer chain.

Cyclotron \ˈsī-klə-ˌträn\ [*cycl-* + *-tron*; fr. the circular movement of the particles] (1935) *n.* The magnetic resonance accelerator for imparting very great velocities to heavier nuclear particles without the use of excessive voltages.

Cycolac *n.* ABS, manufactured by Marbon, USA.

Cycolon ABS, manufactured by Marbon, USA.

Cylinder *n.* (1) In carding, a large cast iron shell, with an outer diameter of 40–45 in., completely covered with card clothing on the surface. The shell is mounted rigidly on a shaft, which projects at each end to rest in bearings. The cylinder must be accurately balanced since it rotates at speeds of 160 rpm and higher. (2) The main roll, or pressure bowl, on roller printing machines. The engraved rolls that apply color are arranged around the cylinder (3) A slotted cylindrical housing for the needles in a circular-knitting machine. The number of slots per inch in the cylinder determines the cut of the machine. (4) *See drying cylinders.*

(*Also see printing, roller printing*).

Cylinder loading *n.* Fibers imbedded so deeply in the wire clothing on a card cylinder that they resist transfer to the doffer cylinder according to the normal fiber path through the card. Causes include improper finish, excess moisture, or static on the fiber. The fiber builds up to such an extent that the carding operation is adversely affected. In extreme cases, the card will be slowed or stopped.

Cylindrical weave *n.* A type of woven, knitted, or braided sleeve generated as reinforcement in tubular reinforced-plastic structures such as pipe. Typically, the reinforcement will constitute about half the volume of the finished product; but it can be much less, as in knit-reinforced garden hose.

Cymatic printing *n.* This proprietary process owned by KBC is a method in which the oscillations of a musical chord are "caught" on a quartz plate and the vibration patterns photographed. The patterns thus obtained are used in making unique print fabrics of unusual variety and originality. Vincenti R (ed) (1994) Elsevier's textile dictionary. Elsevier Science and Technology Books, New York.

Cynoper *See mercuric sulfide.*

D

d *n* \ˈdē\. (1) SI abbreviation for prefix deci-. (2) Abbreviation in use with SIU system for time interval of 1 day (=86,400 s).

D (1) Symbol for diameter. (2) Chemical symbol for the hydrogen isotope of atomic weight 2, deuterium.

D_{6500} or D_{65} *adj.* Refers to the daylight special power distribution curve with a correlated color temperature of 6500°K.
See daylight illuminates, cie, and correlated color temperature.

da *n.* SI abbreviation for prefix deca-.

Dabber \ˈdab-\ [ME *dabbe*] (14c) *n.* Dome-shaped brush of soft hair for applying spirit varnish or for polishing and finishing gilding.

DABCO (former trade name, now generic) *n.* Abbreviation for 1,4-diazabicyclo-2,2,2-octane.
See triethylenediamine.

DAC Abbreviation for diallyl chlorendate.

Dacron \ˈdā-ˌkrän\ {*trademark*}. Fiber from poly(ethylene terephthalate). Manufactured by DuPont, USA.

Dado \ˈdā-(ˌ)dō\ [It, die, plinth] (1664) *n.* (1) Lower part of an interior wall when paneled or decorated, customarily of chair rail height. (2) Paper that provides this architectural division.
See wainscot.

DAF Abbreviation for diallyl fumarate, a polymerizable monomer.

Dagincoleic acid *n.* $C_{22}H_{44}O_4$. Acid constituent of Borneo dammer. Mp, 170°C.

Dagingenoleic acid *n.* $C_{13}H_{26}O_3$. Acidic constituent of Borneo dammer. Mp 126°C.

DAIP *n.* Abbreviation for diallyl isophthalate resin.
See allyl resin.

Dalamar yellow *n.* Pigment yellow 74 and 65 (11741). An azo coupling of *m*-nitro-*o*-anisidine with acetoacet-*o*-anisidine; it is one of the growing group of close chemical relatives of Hansa yellow G (an azo coupling of *m*-nitro-*p*-toluidine with acetoacetanilide). The high strength toner is superior in tinting strength and fastness properties to Hansa yellow G.

Dalton's law of partial pressures *n.* The pressure exerted by a mixture of gases is equal to the sum of the separate pressures, which each gas would exert if it alone occupied the whole volume. This fact is expressed in the following formula:

$$PV = V(p_1 + p_2 + p_3, \text{ etc.}).$$

Giambattista A, Richardson R, Richardson RC, Richardson B (2003) College physics. McGraw-Hill Science/Engineering/Math, New York.

Dam *n.* A ridge around the perimeter of a mold that retains excess resin during pressing and curing.

DAM *n.* Abbreviation for diallyl maleate.
See allyl resin.

Damaged selvage *See cut selvage.*

Damask \ˈda-məsk\ [ME *damaske*, fr. ML *damascus*, fr *Damascus*] (14c) *n.* A firm, glossy, Jacquard-patterned fabric that may be made from linen, cotton, rayon, silk, or a combination of these with various manufactured fibers. Similar to brocade, but flatter and reversible, damask is used for napkins, tablecloths, draperies, and upholstery.

Dammar (Damar) \ˈda-mər\ [Malay] (1698) *n.* Natural resinous exudation from trees of the *Dipterocarpaceae* family, which grow chiefly in the East Indies (now Indonesia) and Malaya. Very pale in color (practically colorless to deep yellow). Soluble in hydrocarbons without the necessity of running. Average acid value of 30. A fossil resin used as an ingredient in printing ink varnishes.

D

Dammarolic acid *n.* $C_{56}H_{80}O_8$. Acidic constituent of dammar.

Dammar wax *See resene, β.*

Dammarylic acid *n.* $C_{38}H_{60}O_3$. Acidic constituent of dammar.

Dampener In lithography, cloth-covered, parchment paper, or rubber rollers that distribute the dampening solution to the press plate.

Dampending (in Tire Cord) *n.* The relative ability to absorb energy and deaden oscillation after excitation.

Dampening solution *See fountain solution.*

Damping *n.* Reduction of vibration amplitude due to viscous or frictional resistance within a material or to drag of its environment upon a structure. The vibration may be mechanical, sonic, or electronic. Energy dissipates per cycle is called hysteresis loss. *See also critical damping.*

Dancer roll *n.* A roller mounted on an axis, which is movable with respect to axes of other rollers in an apparatus, used to control or measure tension of a continuous web or strand as it passes through a series of rollers. Dancer rolls are used as tension-sensing devices in the extrusion coating of wire, and as tension-maintaining devices in film winding.

DAP Abbreviation for diallyl phthalate.

Darkfield illumination *n.* Incident or transmitted illumination of the specimen by indirect light whereby no direct light is admitted directly to the objective.

Dart impact test *See free-falling-dart test.*

Darvic *n.* Poly(vinyl chloride). Manufactured by ICI, Great Britain.

Dashpot \ˈdash-ˌpät\ (1861) *n.* (1) A device used for damping vibration and cushioning shock in hydraulic systems. Typically, a dashpot is a liquid- or gas-filled cylinder with a piston that is attached to a moving machine part. (2) A modeling concept useful in visualizing the mechanical behavior of viscoelastic materials, a purely viscous element that may operate alone or connected in series and/or parallel with springs and sliders.

Datiric acid \sə-ˈtir-ik\ *n.* $CH_3(CH_2)_{15}COOH$. Another name for margaric acid. A straight chain aliphatic acid. Mp, 59.5°C.

Daub \ˈdób, ˈdäb\ [ME, fr. MF *dauber*] (14c) *v.* To apply a coating by crude unskillful strokes.

Daylight \ˈdā-ˌlīt\ (13c) *n.* Natural illumination that is the result of various mixtures of direct sunlight covering a wide range of correlated color temperatures.

For precision in definitions, *See north light, artificial daylight, and daylight illuminants.*

Daylight fluorescent pigment *See fluorescent pigment.*

Daylight illuminants *n.* **CIE**. Series of illuminant spectral power distribution curves based on measurements of natural daylight and recommended by the CIE in 1965. Values are defined for the wavelength region 300–830 nm. They are described in terms of the correlated color temperature. The most important is D_{6500} because of the closeness of its correlated color temperature to that of Illuminant C, 6774°K. D_{7500} bluer than D_{6500} and D_{5500} yellower than D_{6500} are also used.

Daylight opening *n.* The clearance between two platens of a molding press when in the open position.

dB *n.* Abbreviation for decibel.

DBEP *n.* Abbreviation for dibutoxyethyl phthalate (*See image*).

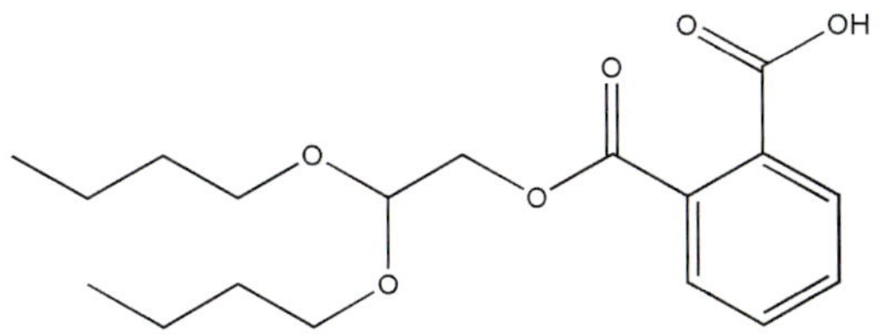

DBP *n.* Abbreviation for dibutyl phthalate (*See image*).

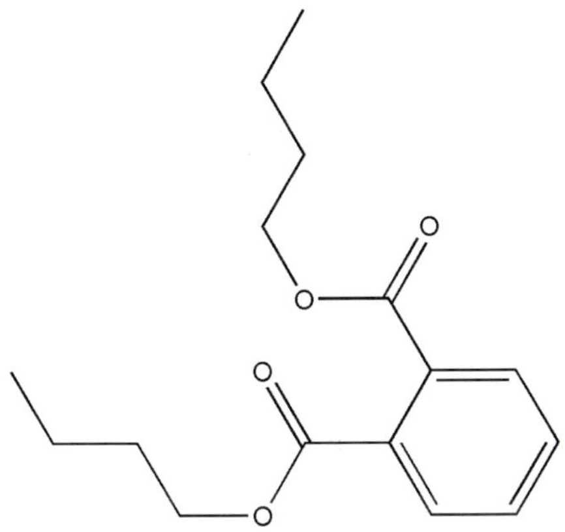

DBPC *n.* Abbreviation for di-*tert*-butyl-*p*-cresol.

DBS *n.* Abbreviation for dibutyl sebacate.

DBTDL *n.* Abbreviation for dibutyltin dilaurate.

DC drive (direct-current drive) *n.* A machine drive, particularly that of an extruder, powered by a direct-current motor. The availability of economical solid-state rectifiers, and the good torque-versus-speed characteristic and tight speed regulation of DC drives have made them the most popular choice today for variable-speed service.

DCHP *n.* Abbreviation for dicyclohexyl phthalate.

DCO *n.* Non-drying castor oil which has been converted into a drying oil by the catalytic removal of water from its principal fatty acid (80% ricinoleic acid), forming approximately equal quantities with 9–11 and 9–12 unsaturation. The term DCO rather than "dehydrated castor oil" has been suggested and is being used for this oil in order to reduce the tendency to confuse castor oil with DCO, which is entirely different in its properties from the oil from which it is derived.

DCP *n.* Abbreviation for dicapryl phthalate.

DDM *n.* Abbreviation for 4,4′-diaminodiphenyl methane.

DDP *n.* Abbreviation for didecyl phthalate.

DE *n.* Abbreviation for diatomaceous earth. *See diatomite.*

Dead burned calcium sulfate *See calcium sulfate, anhydrous.*

Deadcoloring *See abbozzo.*

Dead-end polymerization *n.* A free radical polymerization in which a very active initiator is used so that the rate of polymerization continuously decreases as initiator is consumed and a limiting conversion of monomer to polymer occurs.

Dead flat *n.* Of a finish, the quality of having no luster or gloss. Syn: lusterless.

Dead fold *n.* A fold that does not spontaneously unfold; a crease.

Dead oil *See creosote.*

Dead zone *n.* In control work, a range of the quantity sensed in which a small change in the quantity causes no change in the indication of it, nor any control action. Compare induction period.

Deaerate \ˌdē-ˈar-ˌāt\ (1791) *vt.* To remove air from a substance. Deaeration is an important step in the production of vinyl plastisols and most casting operations accomplished by subjecting the liquid to a moderate vacuum with or without agitation, to remove air that would cause objectionable bubbles or blisters in finished products.

Debloomed oil *See oil, debloomed.*

Debonding *n.* In a bonded joint, separation of the bonded surfaces. In a laminate, separation of layers or fibers from the matrix.

Deborah number \ˈde-b(ə-)rə ˈnəm-bər\ (De, N_{De}). The dimensionless ratio of the relaxation time of a viscoelastic fluid to a characteristic time for the flow process being considered. Relaxation time is often taken as the ratio of viscosity to modulus, while the characteristic time is a significant length divided by average velocity. The Deborah number has been useful in studying such phenomena as die-exit swell in extrusion. If De is high, elastic effects probably dominate; if De is low, the flow is essentially viscous.

Deburr \də-ˈbər\. To remove from a machined part rough edges or corners left by the machining operations. Compare deflashing.

Debye equation \də-ˈbī i-ˈkwā-zhən\. A relationship for the relative permittivity (ε) of a dielectric material as a function of the electronics and atomic polarizability (α_a) and the orientational polarizability terms.

$$(\varepsilon - 1)/(\varepsilon + 2) = (4\pi N_1/3)(\alpha_a + \mu^2/3kT),$$

where N_1 is the number of molecules present per cm^3, μ the dipole moment of the molecule concerned, k the Boltzmann's constant, and T is the temperature.

Deybe–Hűckle theory of strong electrolytes *n.* This theory assumes that strong electrolytes are completely dissociated in solution, but that the hydrated ions have enough residual attraction for each other that they are not truly independent of each other. The theory assumes that each ion has surrounding it on the average more ions of opposite charge, counterions, than of the same charge, and that the number of counterions near a given ion increases with increasing concentration. The presence of this diffuse ionic atmosphere relatively rich in counterions restricts the movement of each ion in the solution. The theory permits a quantitative prediction of how the conductivity should change with concentration, a prediction that agrees with experimental results. The theory also permits a calculation of the ionic-atmosphere effect on colligative properties. Russell JB (1980) General chemistry. McGraw-Hill Book Co, New York.

Deca- {*combining form*} [ME, fr. L. fr. Gk *deka-*, *dek-*, fr. *deka*] (da). SI-acceptable prefix meaning ×10.

Decabromodiphenyl oxide \-ˈbrō-(ˌ)mō-(ˌ)dī-ˌfe-nᵊl ˈäk-ˌsīd\ (deca-bromodiphenyl either) *n.* Diphenyl ether in, which all ten phenyl hydrogens have been replaced by bromine, containing 83% bromine. The commercial product, a free-flowing white powder, is much used as a flame retardant in high-impact polystyrene and structural foam, usually with synergistic antimony oxide. Typical would be 12% decabromodiphenyl oxide and 4% antimony oxide in the resin (*See image*).

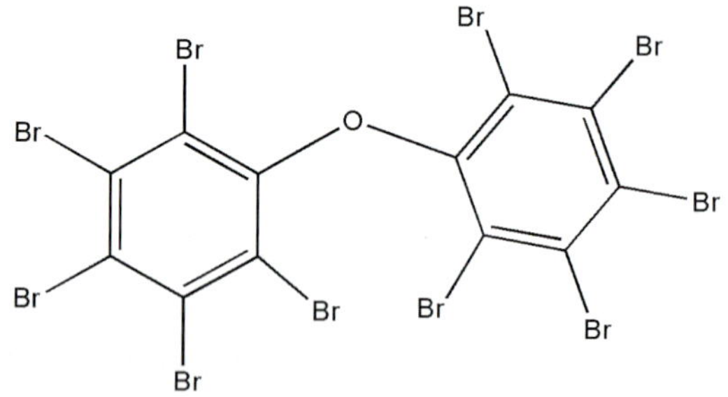

Decahydronaphthalene \-ˈhī-(ˌ)drō-ˈnaf-thə-ˌlēn\ (Decalin®) *n.* A saturated bicyclic hydrocarbon, $C_{10}H_{18}$, essentially two cyclohexane rings fused together, sharing two hydrogen atoms. The commercial product is a mixture of *cis-* and *trans-*isomers. It is a colorless liquid with an aromatic odor, derived by treating molten naphthalene with hydrogen in the presence of a catalyst.

It is a solvent for many resins. Hydrocarbon solvent with a boiling range for the commercial product (175–205°C), slightly higher than that for mineral spirit. Sp gr, 0.894/20°C; flp, 60°C (140°F); refractive index, 1.467; and vp, 2 mmHg/30°C. In Europe, Decaline.

Decalcomania \di-ˌkal-kə-ˈmā-nē-ə\ [F *décalcomanie*, fr. *décalquer* to copy by tracing (fr. *dé-* de- + *calquer* to trace, fr. It *calcare*, literally, to tread, fr. L) + *manie* mania, from LL *mania*] (1884) (decal) *n.* A printed design on a temporary carrier such as paper or film, and subsequently transferred to the item to be decorated. Deals are widely used for decorating many materials including plastics, paper, china, glass, furniture, etc. The imprint is adhered to plastic surfaces by means of a pressure-sensitive adhesive, solvent welding, or heat and pressure. Skeist I (ed) (1990) Handbook of adhesives. Van Nostrand Reinhold, New York.

DECALS *n.* An abbreviation for decalcomania. A printed design on a temporary carrier.

Decanedioic acid Syn: sebacic acid.

Decantation \ˌdē-ˌkan-ˈtā-shən\ [NL *decantare*, fr. L de- + ML *cantus* edge, fr. L, iron ring round a wheel] (1633) *n.* Or siphoning off the liquid from a precipitate or sediment or the upper layer of two immiscible liquids as a partial means of separating the phases.

Decarboxylate \-ˈbäk-sə-ˌlāt\ (1922) *vt.* To remove from an organic acid its carboxylic-acid groups.

Decarboxylation *n.* Removal of carboxyl groups from organic acids, usually as the result of heating.

Decating *See decatizing.*

Decating mark *n.* A crease mark or impression extending fillingwise across the fabric near the beginning or end of the piece.

Decatizing *n.* A finishing process in which fabric, wound tightly on a perforated roller, either has hot water circulated through it (wet decatizing), or has steam blown through it (dry decatizing). The process is aimed chiefly at improving the hand and removing wrinkles.

Decay \di-ˈka\ [ME, fr. ONF *decaïr*, fr. LL *decadere* to fall, sink, fr. L *de-* + *cadere* to fall] (15c) *n.* Diminution of a radioactive substance due to nuclear emission of alpha or beta particles, gamma rays or positrons. Giambattista A, Richardson R, Richardson RC, Richardson B (2003) College physics. McGraw-Hill Science/Engineering/Math, New York.

Deci- {*combining form*} [F *déci-*, fr. L *decimus* tenth, fr. *decem* ten] (d). The SI-approved prefix meaning × 0.1.

Decibel \ˈde-sə-ˌbel\ [ISV *deci-* + *bel*] (1928) (dB) *n.* In acoustics and electronic circuits, a change in sound intensity or circuit power by the factor $10^{0.1} = 1.259$. These changes correspond to changes in sound *pressure* or circuit *voltage* by the square root of the above factor, 1.122. In acoustics most measurements are referred to a sound-power level of 10^{-12} W, corresponding to a sound-pressure level of 0.00002 Pa, the presumed threshold of human audibility. One decibel is the least change in intensity detectable by normal ears. Thus, because of the exponential definition of the decibel, an 80-dB sound is ten times as intense as a 70-dB sound. OSHA regulations prescribe maximum sound levels permitted for employee exposure. In many plants, scrap grinders, which used to be major noise generators, now run much more quietly or are isolated by enclosures, or both. Giambattista A,

Richardson R, Richardson RC, Richardson B (2003) College physics. McGraw-Hill Science/Engineering/Math, New York.

Decitex *n.* One-tenth of a tex.

Deckle \ˈde-kəl\ [Gr *Deckel,* literally, cover, fr. *decken* to cover, fr. OHGr *decchen*] (1816) (deckle rod and cut-off plate) *n.* In extrusion of film or sheet, or extrusion coating, a small rod or plate attached to each end of the die that symmetrically shortens the length of the die opening, thus facilitating the production of a web of le than the maximum width.

D

Deck paint *n.* A paint having a high degree of resistance to mechanical wear; especially used on surfaces such as porch floors and ships' decks.

Declination \ˌde-klə-ˈnā-shən\ [ME *declinacioun,* fr. MF *declination,* fr. L *declination-, declinatio* angle of the heavens, turning aside] (14c) *n.* The angle between the vertical plane containing the direction of the earth's field at any point and a plane containing the geographic north and south meridian.

Decomposition \(ˌ)dē-ˌkäm-pə-ˈzi-shən\ [F *décomposer,* fr. *dé-* de + *composer* to compose] (ca. 1751) *n.* Breakdown of molecular structure by chemical or thermal action. With polymers, depending on the severity of conditions, decomposition products can range from subpolymers and oligomers down to monomers and even atoms and ions. In similar terms it is the chemical separation of a substance into two or more substances, which may differ from each other and from the original substances. Zaiko GE (ed) (1995) Degradation and stabilization of polymers. Nova Science Publishers Inc., New York.

See also degradation.

Decomposition temperature (T_{dc}) *n.* The temperature at which a polymer or other material (usually organic compound) decomposes by breaking chemical bonds, symbolized by (no symbol is available, but T_{dc} will be used for this purpose). This property should not be confused with the temperature of depolymerization, T_{dp}. {G Zersetzungstemperatur f, F température de décomposition, température f, S temperatura de descomposición, temperatura f, I temperatura di decomposizione, temperatura f}. Groenewoud WM (2001) Characterization of polymers by thermal analysis. Elsevier Science and Technology Books, New York. Lenz RW (1967) Organic chemistry of synthetic high polymers. Interscience Publishers Inc., New York.

Decomposition potential \-pə-ˈten(t)-shəl\ *n.* The minimum voltage, which must be applied across a pair of inert electrodes immersed in a medium in order to electrolyze the medium.

Decompression zone \ˌdē-kəm-ˈpre-shən ˈzōn\. In a vented extruder, the zone of deep flights immediately forward of the first metering zone and beneath the vent, where unwanted gases and vapors are released from the (typically) foaming melt. In some vented single-screw machines and most twin-screw machines, the two zones are separated by a dynamic value, whose resistance to flow helps to increase the superheating of the melt prior to decompression.

Décor \dā-ˈkór, di-ˈ; ˈde-ˌkór, ˈdā-ˌ\ [F *décor,* fr. *décorer* to decorate, fr. L *decorare*] (1897) *n.* The combination of materials, furnishing, and objects used in interior decorating to create an atmosphere or style.

Decorated \ˈde-kə-ˌrāt\ [L *decoratus,* pp of *decorare,* fr. *decor-,* decus ornament, honor] (1530) *vt.* Adorned, embellished, or made more attractive by means of color or surface detail.

Decorating *n.* The processes used for decorating plastics are defined under the following headings:

Airless spraying
Ashing
Decalcomania
Decorating
Double-shot molding
Electroless plating
Electrophoretic deposition
Electroplating on plastics
Electrostatic printing
Electrostatic spray coating
Embedment decorating
Embossing
Fill-and-wipe
Flexographic printing
Flocking
Flow coating
Gravure coating
Gravure printing
Hot stamping
In-mold decorating
Letterpress printing
Metallizing
Offset printing
Painting of plastics
Printing on plastics
Process
Roller coating
Rubber-plate printing
Screen printing
Second-surface
Spray coating
Spray-and-wipe painting
Thermographic-transfer
Vacuum metalizing
Valley printing

James F (ed) (1993) Whittington's dictionary of plastics. Technomic Publishing Co. Inc., Carley.

Decorative board (decorative laminate) *n.* A special term for laminates used in the furniture and cabinetry industries, which are defined by the Decorative Board Section of NEMA as "...a product resulting from the impregnation or coating of a decorative web of paper, cloth, or other carrying medium with a thermosetting type of resin and consolidation of one or more of these webs with a cellulosic substrate under heat and pressure of less than 500 psi" This includes all boards that were formerly called low-pressure melamine and polyester laminates, but does not include vinyls. "Cellulosic" here means impregnated paper, wood, or plywood.

Decorative painting *n.* Architectural or aesthetical painting.

Decorticating \(ˌ)dē-ˌkór-tə-ˈkā-iŋ\ [L *decortication-*, *decorticatio*, fr. *decorticare* to remove the bark from, fr. *de-* + *cortic-*, *cortex* bark] (ca. 1623) *n.* A mechanical process for separating the woody matter from the bast fiber of such plants as ramie and hemp.

Decoupage \ˌdā-(ˌ)kü-ˈpäzh\ [F *découpage*, literally, act of cutting out, fr MF, fr. *decouper* to cut out, fr. *de-* + *couper* to cut] (1946) *n.* Modern technique of "burying" a picture design, which has been affixed to a plain surface, in multiple coats of a clear, varnish-type coating. Many coats are applied until the surface no longer reveals the thickness of the design. This art-form is said to have developed from an old French peasant craft consisting of the decoration of a plain surface with cut-out designs of cloth or paper, and then coating it with shellac, varnish or clear lacquer. Another version is that the art of decoupage was originated by craftsmen of 18th century Europe in an effort to duplicate Chinese and Japanese lacquer-ware. Merriam-Webster's collegiate dictionary, 11th edn. Merriam-Webster Inc., Springfield, MA, USA, 2004.

Decyl butyl phthalate \-ˈbyü-tᵊl ˈtha-ˌlāt\. *See butyl isodecyl phthalate.*

Decyl-octyl methacrylate \-ˌme-ˈtha-krə-ˌlāt\. $H_2C{=}C(CH_3)COO\ (CH_2)_nCH_3$, with $n = 7–9$. A polymerizable mixed monomer for acrylic plastics (*See image*).

Decyl tridecyl phthalate *n.* $C_{10}H_{21}COO\ C_6H_4OOCC_{13}H_{27}$. A plasticizer for vinyls, cellulosics, and polystyrene (*See image*).

Deep \ˈdēp\ [ME *dep*, fr. OE dēop; akin to OHGr *tiof* deep, OE *dyppan* to dip] (before 12c) *adj.* Having depth (in color); often used improperly for dark.

Deep drawing *n.* The process of forming a thermoplastics sheet in a mold in which the ratio of depth to the shortest lateral opening is 1 : 1 or greater.

Deep-dyeing variants \-ˈver-ē-ənt\ *n.* Polymers that have been chemically modified to increase their dyeability. Fibers and fabrics made there from can be dyed to very heavy depth.

Defect \ˈdē-ˌfekt\ [ME, fr. MF, fr. L *defectus* lack, fr. *deficere* to desert, fail, fr.

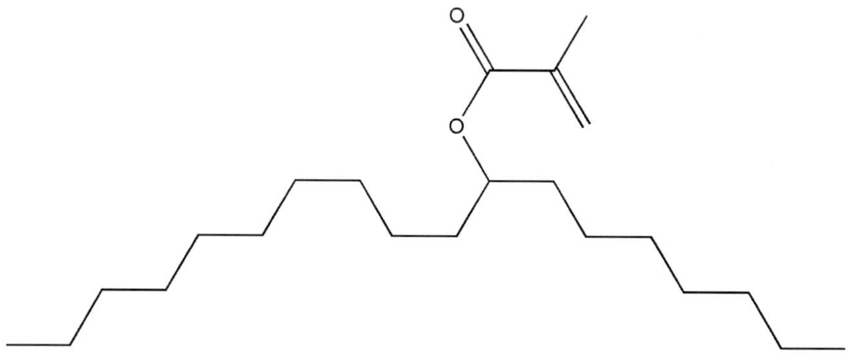

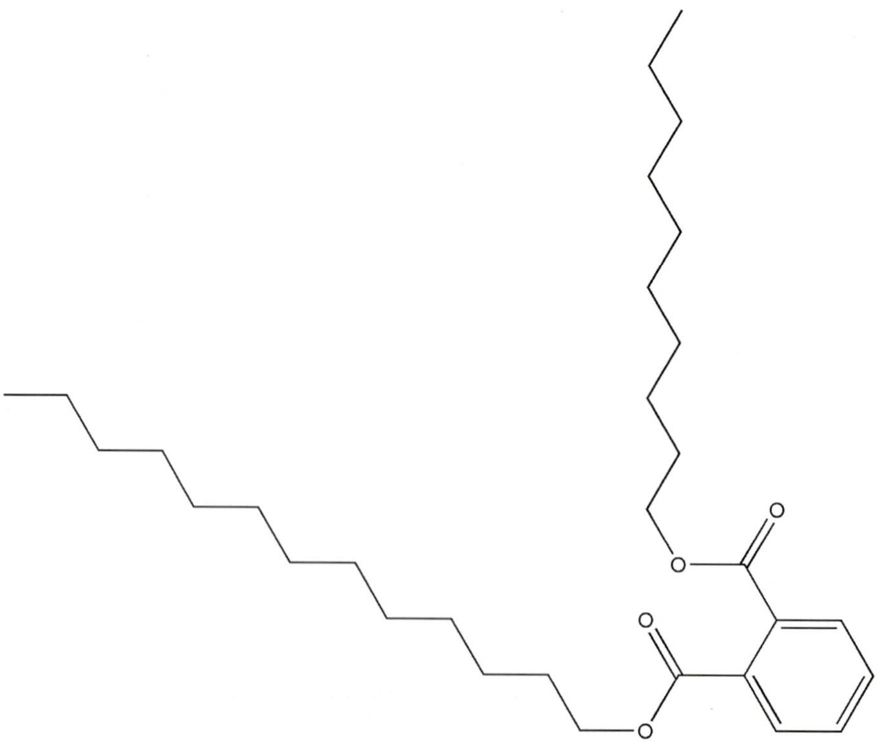

de- + *facere* to do] (15c) *n.* An internal irregularity or flaw in the structure of a crystal.

Defects *n.* A general term that refers to some flaw in a textile product that detracts from either performance or appearance properties.

Definite proportions, law of *n.* In every sample of each compound substance the proportions by weight of the constituent elements are always the same.

Deflashing \di-ˈflash-eŋ\ *n.* The removal of flash or rind left on molded plastic articles by spaces between the mold-cavity edges. Methods include tumbling, blast finishing, use of dry or wet abrasive belts, and hand methods using knives, scrapers, broaching tools, and files. Soft thermoplastic parts are sometimes deflashed by tumbling them while immersed in a severe coolant such as liquid nitrogen.

See also abrasive finishing and airless blast deflashing.

Deflashing equipment *n.* Equipment used to remove excess, unwanted material from a processed plastic.

Deflection temperature \di-ˈflek-shən ˈtem-pə(r)-ˌchúr\ *n.* According to ASTM D 648, the stress-labeled temperature at which a standard test bar, centrally loaded as a simple beam to develop a (theoretical) maximum stress of 455 or 1820 kPa (66 or 264 psi), and warmed at 2°C/min, deflects 0.25 mm. For many years this temperature was called the "heat-distortion point", a term now deprecated and fading from use {*Deflection* G Durchbiegung f, F déflexion f, S deflección f, I deflessione f}.

Deflocculant \de-ˈflä-kyə-ˌlāt\. An additive that prevents pigments in suspension from coalescing to form flocs.

See dispersant (dispersing agent). Also known as deflocculating agent.

Deflocculating agent *n.* A substance that breaks down agglomerates into primary particles or prevents the latter from combining into agglomerates (ISO).

Deflocculation *n.* State or condition of a dispersion of a solid in a liquid in which each solid particle remains independent and unassociated with adjacent particles. A deflocculated suspension shows zero or very low yield value.

Defoamer \de-ˈfōm- \ (defoaming agent) *n.* A substance that, when added in small percentages to a liquid containing gas bubbles, causes the small bubbles to coalesce into larger ones that rise to the surface and break. Additive used to reduce or eliminate foam formed in a coating or a coating constituent.

Deformation *n.* (1) Change in dimension or shape of a plastic product, particularly a test specimen. (2) In a tensile or compression test, the change in length of the specimen in the direction of applied force. Deformation divided by original length equals *elongation*, often expressed as a percent. (3) In a shear mode, the angle of shear in radians. {G Deformation f; Verformung f, F déformation f, S deformación f, I deformazione f}. Shah V (1998) Handbook of plastics testing technology. John Wiley and Sons, New York.

Degassing (breathing) *n.* In injection or transfer molding, the momentary opening and closing of a mold during the early stages of the cycle to permit the escape of air or gas from the heated compound.

See also deaerate.

Degating *n.* The removal of material left on a -plastic part formed by the passage between the runner and the cavity, i.e., the gate. The operation is sometimes performed automatically by a mold element. Otherwise, the gate may be removed by manual breaking or cutting, sometimes followed by sanding or burnishing.

Degating equipment *n.* Equipment used to separate the molded part, automatically or manually, from an injection molded solid runner system.

D

Degradable plastic *See biodegradation.*

Degradation \ˌde-grə-ˈda-shən\ (ca. 1535) *n.* A deleterious change in chemical structure, physical properties, or appearance of a plastic caused by exposure to heat (*thermal degradation*), light (*photodegradation*), oxygen (*oxidative degradation*), or weathering. The ability of plastics to withstand such degradation is called *stability.* Zaiko GE (ed) (1995) Degradation and stabilization of polymers. Nova Science Publishers Inc., New York. Dissado LA, Fothergill CJ (eds) (1992) Electrical degradation and breakdown of polymers. Institution of Electrical Engineering (IEE), London.
See also artificial weathering, autocatalytic degradation, biodegradation, corro-sion resistance, decomposition, deterioration, dew-cycle weathering test discoloration, pink staining, and xenon-arc aging.

Degras *n.* Natural dark wool grease from, which the various grades of lanolin are prepared.

Degreaser \(ˌ)dē-ˈgrē-sər\ *n.* Solvent or compounded material for removing oils, fats, or grease from a substrate. Also the apparatus in which this operation is carried out.

Degreasing *n.* Removal of grease, oil and other fatty matter (from metals, fabrics, etc.) by the use of solvents or chemical cleaners, electro, or heat processes.

Degree of cross-linking \di-ˈgrē əv ˈkrós-ˈlink-eŋ\ *n.* The fraction of mer units of a polymer that are cross-linked, equal to the quotient of the mer weight and the average molecular weight of segments between cross-links.

Degree of crystallinity \ˈkris-tə-ˈli-nə-tē\ *n.* The amount of crystalline material in a partially crystalline polymer, usually expressed as a percentage. For a particular polymer the value depends on the crystallization conditions, especially the rate of cooling and the crystallization temperature. Elias H-Georg (2003) An introduction to plastics. John Wiley and Sons, New York.

Degree of cure \ˈkyúr\ *n.* The extent to which curing or hardening of a thermosetting resin has progressed.
See also A-stage, B-stage, and C-stage.

Degree of efflorescence \ˌe-flə-ˈre-sᵊn(t)s\ *n.* (1) *Flatting and glossing* – A non-uniform decrease or increase in gloss noticed when the surface is illuminated and viewed at near grazing angles. There is no apparent change in color when viewed perpendicular to the surface. (2) *Fine efflorescence* – A barely discernible whitening of the surface when viewed perpendicularly. (3) *Medium efflorescence* – A readily noticeable whitening of the surface without a marked masking of the color. (4) *Heavy efflorescence* – A white deposit sufficient to mask the color.

Degree of esterification \-e-ˌster-ə-fə-ˈkā-shən\ *n.* The extent to which the acid groups of terephthalic and/or other acids have reacted with diols to form ester groups in polyester polymer production.

Degree of freedom \-ˈfrē-dəm\ (1867) *n.* The number of the variables determining the state of a system (usually pressure, temperature, and concentrations of the components) to which arbitrary values can be assigned.

Degree of metamerism \-mə-ˈta-mə-ˌri-zəm\ *n. See metamerism, degree of.*

Degree of orientation \-ˌōr-ē-ən-ˈtā-shən\ *n.* The extent of crystalline deformation after application of stress to a polymer.

Degree of polymerization \-pə-ˌli-mə-rə-ˈzā-shən\ (DP, chain length) *n.* The average number of monomer units per polymer molecule, $DP = M_n/M_{mer}$, where M_n is the polymer's number-average molecular weight and M_{mer} is the molecular weight of the mer (which may differ from monomer molecular weight). In a polymer having worthwhile mechanical properties, *DP* exceeds 500, and many commercial thermoplastics have *DP*s between 1000 and 10,000. Specifically, the average number of anhydroglucose units (or derivative units) per molecule of cellulose (or cellulose derivative). The type of average obtained depends upon the method used for the determination. Hence, the method must always be specified.
See also molecular weight.

Degumming \(ˌ)dē-ˈgəm-eŋ\ (1887) *vt.* The removal of gum from silk by boiling in a mildly alkaline solution. Usually accomplished on the knit or woven fabric.

DEHP (DOP) *n.* Abbreviation for di(2-ethylehexyl) phthalate.

Dehydrated castor oil \(ˌ)dē-ˈhī-ˌdrat-ˈed ˈkas-tər ˈói(ə)l\ *n.* A castor oil chemically treated to improve its drying qualities.
See DCO.

Dehydrated gypsum \ˈjip-səm\ *n.*
See calcium sulfate, anhydrous.

Dehydration *n.* The removal of water from a substance either by ordinary drying or heating, or by absorption, chemical action, condensation of water vapor, or by centrifugal force or filtration. The term is generally applied to the removal of combined water as, for instance, from hydrates.

Dehydroacetic acid \(ˌ)dē-ˌhī-drə-ə-ˈsē-tik\ *n.* Abbreviated is DHA. It is a rosin derivative present in hy-drogenated rosin. A heterocyclic compound used as a plasticizer, fungicide, and bactericide (*See image*).

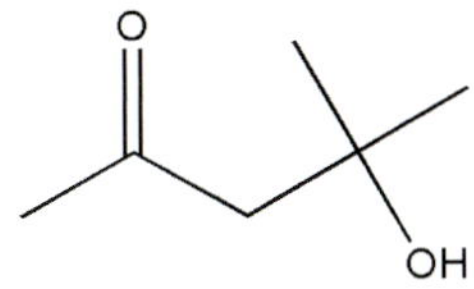

Dehydrogenated oils *n.* Semidrying oils which have been converted into drying oils by first treating them with a halogen (usually chlorine), then eliminating hydrogen halide. This results in conjugated double bonds in the fatty acids chain.

Dehydrogenation \ˌdē-(ˌ)hī-ˌdrä-jə-ˈnā-shən\ (1866) *n.* The removal of hydrogen from a compound by chemical means.
See also cracking (2).

Dehydrohalogenation \-ˌha-lə-jə-ˈnā-shən\ *n.* The splitting of hydrogen chloride or other hydrohalide from polymers such as PVC, by action of excessive heat or light.

Deionization \(ˌ)dē-ˈī-ə-nə-ˈsā-shən\ *n.* The removal of ions from a solution by ion exchange.

Deka- (da). A prefix permitted by SI and meaning × 10.

Delaminate \(ˌ)dē-ˈla-mə-ˌnāt\ *vt.* To separate existing layers.

Delaminated clay \ˈklā\ *n.* Pigment produced by mechanically separating the platelets of natural kaolin so that particles of approximately 1/4 μm thickness and up to 10 μm in diameter are produced. This pigment is used to increase the opacifying efficiency of titanium dioxide by keeping the titanium dioxide particles separated by the thickness of the calcined clay particle. This spacing allows each titanium dioxide particle to intercept a wavelength of visible light and thereby results in maximum light-scattering efficiency.

Delamination *n.* Separation of the wood surfaces at the bonded joints caused by a cohesive failure in the adhesive or a failure of the adhesive at the surface. When the separation occurs in the wood, even though

D

very close to the bonded joint, the separation is termed *wood failure* or *checking* rather than delamination. Magnification is often necessary to determine whether the failure is in the bond or in the wood. The term also applies to failure occurring between successive coatings.

Delated deformation \di-ˈlāt- ˌdē-ˌfór-ˈmā-shən\ *vt.* Deformation that is time-dependent and is exhibited by material subjected to a continuing load; creep. Delayed deformation may be recoverable following removal of the applied load.

Deleafing *n.* Loss of leafing of metallic pigments in paints, giving rise to reduced metallic luster.
See leafing.

Deliquescence \ˌde-liˈkwe-sənt\ [L *deliquescent-*, *deliquescens*, pp of *deliquscere*] (1791) *adj.* Property possessed by some materials of absorbing moisture from the air and forming a solution in the water so absorbed.

Deliquescent *n.* Said of some salts that are so strongly attractive of water that they can absorb enough from room air to liquefy.

Delocalized molecular orbital *n.* An orbital which extends over more than two atoms in a molecule, ion, or larger aggregate.

Delrin *n.* Poly(oxymethylene) (from formaldehyde). Manufactured by DuPont, USA.

Delta *E*, ΔE \ˈdel-tə\ *n.* Total color difference computed by use of a color difference equation. It is generally calculated as the square root of the sum of the squares of the chromaticity difference, ΔC, and the lightness difference, ΔL: $\Delta E = \left[(\Delta C)^2 + (\Delta L)^2\right]^{1/2}$.
See color difference equations.

Delustering *n.* Subduing or dulling the natural luster of a textile material by chemical or physical means. The term often refers to the use of titanium dioxide or other white pigments as delustrants in textile materials.

Delustrant *n.* A chemical agent used either before or after spinning to produce dull surfaces on synthetic fibers to obtain a more natural, silk like appearance.

Denaturation \(ˌ)dē-ˌnā-chə-ˈrā-shən\ (1685) *n.* Addition of unpleasant or toxic substances to a product to make it unfit for human consumption.

Denatured alcohol *n.* Ethyl alcohol to which denaturant (odorous and/or nauseating substances) has been added in small quantities to make it unpotable without diminishing its solvent power. Typical denaturants are benzene, acetaldehyde, and pyridine.

Dendrite \ˈden-ˌdrīt\ (1751) (dendritic) *n.* A branched, treelike crystal habit usually associated with rapid crystal growth.

Denier \di-ˈnī(-ə)r\ [ME *denere*, fr. MF *denier*, fr. L *denarius*, coin worth ten asses, fr. *denarius* containing ten, fr. *deni* ten each, fr. *decem* ten] (15c) *n.* A unit of weight expressing the size or coarseness but particularly the fineness of a relatively continuous fiber or yarn. The weight in grams of 9000 m of a fiber in the form of continuous filament. Although deprecated in SI, this is still the most widely used unit of lineal density in the textile industry to indicate fineness of natural or synthetic fibers. $1\ \text{denier} = 1.1111 \times 10^{-7}\ \text{kg/m}$. Joseph ML (1986) Textile science, 5th edn. CBS College Publishing, New York.
See also cut, grex number, and tex.

Denier per filament (DPF) \ˈfi-lə-mənt\ *n.* The denier of an individual continuous filament or an individual staple fiber if it were continuous. In filament yarns, it is the yarn denier divided by the number of filaments. Joespeh ML (1986) Textile science, 5th edn. CBS College Publishing, New York.

Denier variation *n.* Usually variation in diameter, or other cross-sectional dimension, along the length of a filament or bundle of filaments. It is caused by malfunction or lack of process control in fiber manufacturing and degrades resulting fabric appearance or performance. Joespeh ML (1986) Textile science, 5th edn. CBS College Publishing, New York.

Denim \ˈde-nəm\ [F (*serge*) *de Nîmes* serge of *Nîmes*, France] (1695) *n.* A firm 2 × 1 or 3 × 1 twill-weave fabric, often having a whitish tinge, obtained by using white filling yarns with colored warp yarns. Heavier weight denims, usually blue or brown, are used for dungarees, work clothes, and men's and women's sportswear. Lighter weight denims with softer finish are made in a variety of colors and patterns and are used for sportswear and draperies.

Densitometer \ˌden(t)-sə-ˈtä-mə-tər\ (1901) *n.* Instrument to measure the optical density of a transmitting material, or an instrument to measure the negative log of the reflectance of a reflecting material. Such instruments do not measure color. They are widely used in the graphic arts and photographic industries for process control. In spectrographic and dispersive X-ray analysis utilizing photography, they are adapted for scanning and plotting the density of the photographed transparency as a function of the wavelength.

Density \ˈden(t)-sə-tē\ (1603) (absolute density) *n.* Mass per unit volume of a substance. The SI unit is kg/m^3, but others in common use are g/cm^3, lb/ft^3, and lb/gallon. It is an important criterion in specifying some plastics, e.g., polyethylene, which can vary in density from 0.92 to 0.98 g/cm^3, with significant associated variation in properties such as modulus. Since density decreases with rising temperature, the temperature of measurement should be stated. *See www.astm.org* for current methods of determining density. *See also apparent density, bulk density, bulking value, gradient-tube density method, relative density, and specific gravity.*

Density, apparent *See apparent density.*

Density gradient \ˈgrā-dē-ənt\ *n.* A gradient of density such as in a gradient column for measuring density of plastic materials.

Density, optical Negative log of the transmittance.

See Beer–Bouguer law and absorption coefficient.

Densothene *n.* Poly(ethylene). Manufactured by Metal Box, Great Britain.

Dent *n.* On a loom, the space between the wires of a reed.

Deoxy- (desoxy-). A prefix denoting replacement of a hydroxyl group with hydrogen.

Deoxyribonucleic acid [*deoxyribose* + *nucleic acid*] (1938) *n.*

See DNA.

DEP Abbreviation for diethyl phthalate.

Depolymerization *n.* The reversion of a polymer to its monomer, or to a polymer of lower molecular weight. Such reversion occurs in most plastics when they are exposed to very high temperatures in the absence of air. Splitting of polymers into molecules or polymers of lower molecular weight. The temperature of depolymerization is symbolized as T_d. Odian GC (2004) Principles of polymerization. John Wiley and Sons Inc., New York. Lenz RW (1967) Organic chemistry of synthetic high polymers. Interscience Publishers Inc., New York.

Depolymerized rubber *n.* Rubber which has been extensively milled or otherwise treated so that it has sufficient solubility in aromatic hydrocarbons for the resulting solution to be used as a varnish or varnish constituent.

Deposit attack *n.* Excessive corrosion that may occur under solid substances that may be deposited on a metal surface and thus shield it from ready access to oxygen or ions in the solution. This is a form of concentration cell corrosion.

Deposition *n.* (1) Usually referred to in metal treatment as the amount of treatment by weight (mg/ft^2). (2) Process of applying a material to a base by means of vacuum, evaporation, sputtering, electrolysis, chemical reaction, vapor methods, etc.

Depropagation *n.* The sequential chain scission step during depolymerization responsible for the formation of monomer. Has a lower activation energy than propagation and hence is favored at high temperatures.

Depth of draw *n.* In thermoforming, the depth of the lowest point in the formed object relative to the clamped edge of the sheet. *Also see draw ratio* (2).

Depth of field (1911) *n.* The thickness of the preparation brought into good focus at a single setting of the focusing adjustment. It lies in the plane of the specimen and decreases with an increase in numerical aperture.

Depth of finish *n.* The appearance phenomenon of a coating having an "apparent" depth. *Also referred to as "seeing into the film".*

Depth of focus *n.* The thickness of the image of the preparation appearing in good focus at one setting of the focusing adjustment.

Depth of screw Syn: *flight depth.*

Depth of shade Concept which embodies the subjective or visual equality of concentration of colorants, dyes, or pigments. *See depth of shade, standard.*

Depth of shade, standard *n.* An arbitrarily chosen depth of shade for all hues, from which a uniform depth of shade may be determined for purposes of comparison. Depth of shade can then be described as a multiple or fraction of standard depth (developed primarily, and used most widely, in Europe).

Derby red \ˈdər-bē, *esp British* ˈdär-\ *n.* Basic lead chromate. *See chrome orange, light, and deep.*

Deregistering (Crimp) Process of disordering or disaligning the crimp in a tow band to produce bulk. *Also see threaded-roll process.*

Derived high polymer A polymer, which has been produced by chemical alteration of a primary high polymer or a natural high polymer. *See also primary high polymer.*

Dermatitis \ˌdər-mə-ˈtī-təs\ (1876) *n.* Inflammation or irritation of the skin. Industrial dermatitis is an occupational skin disease. There are two general types of skin reaction: primary irritation dermatitis and sensitization dermatitis.

Descaling *n.* Removal of mill sale or caked rust from steel by chemical and/or mechanical means, sometimes assisted by flame cleaning.

Desensitize \(ˌ)dē-ˈsen(t)-sə-ˌtīz\ (1898) *vt.* To treat non-image areas of a water lithographic plate to make them water-receptive and ink-repellant.

Desiccant \ˈde-se-kənt\ (1676) *n.* A substance capable of absorbing water vapor from air and other materials enclosed with it; a drying agent. Typically used to maintain low humidity in a storage or test vessel or package.

Desiccant dryers *n.* Materials, which absorb water from air (e.g., silica gel).

Desiccator \de-si-ˈkā-tər\ *n.* A plastic or heavy-glass bowl with a tight-fitting lid, containing a drying agent, in which a sample can be kept in a controlled, dry atmosphere, typically under a mild vacuum.

Designers color *n.* Opaque watercolor paints that dry flat and are not necessarily permanent.

Design life *n.* The period through which a part or product is expected to perform satisfactorily, remaining within preset performance tolerances.

Desmodur *n.* Isocyanate grades for polyurethane. Manufactured by Bayer, Germany.

Desmophen *n.* Polyester for polyurethanes. Manufactured by Bayer, Germany.

Desorption \-ˈsórp-shən\ (1924) *n.* The escape or loss of a substance previously absorbed into, or adsorbed onto, another.

Destaticization *n.* The treating of a plastic to minimize its tendency to accumulate static electric charge, or removal of charge.
See also antistatic agent, static eliminator, and soot-chamber test.

Destructive distillation \di-ˈstrek-tiv ˌdis-tə-ˈlā-shən\ (ca. 1831) *n.* Decomposition of an organic material by heating in the absence of air. Destructive distillation of wood, which produces charcoal and several small organic molecules, was once the major source of methanol.

Destructively distilled pine oil *See pine oil.*

Destructively distilled wood turpentine *See turpentine.*

Desulfurizing \(ˌ)dē-ˈsəl-fər-ˌīz-eŋ\ *vt.* An after treatment to remove sulfur from newly spun viscose rayon by passing the yarn through a sodium sulfide solution.

DETA *n.* Abbreviation for diethylenetriamine (*See image*).

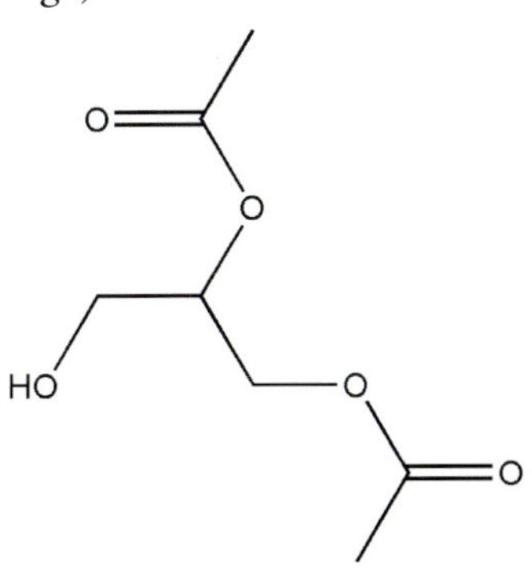

Detection limit *n.* Of chemical quantitative analyses, the least quantity of concentration of the substance being sought that can be detected with a stated level of confidence. This widely used term may be giving way to minimum detectable amount.

Detergent \-jənt\ (1616) *adj.* A surface-active agent that possesses the ability to clean soiled surfaces. *Anionic detergent* – A detergent that produces aggregates of negatively charged ions with colloidal properties. *Cationic detergent* – A detergent that produces aggregates of positively charged ions with colloidal properties. *Nonionic detergent* – A detergent that produces aggregates of electrically neutral molecules with colloidal properties.
See surfactants.

Deterioration \di-ˌtir-ē-ə-ˈrā-shən\ (ca. 1658) *n.* A usually gradual process of permanent impairment in the appearance or application of physical properties.
See degradation.

Deuterium \dü-ˈtir-ē-əm *also* dyü-\ [NL, fr. Gk *deuteros* second] (1933) *n.* Hydrogen 2; $^{2}_{1}H$; "heavy hydrogen." College physics. McGraw-Hill Science/Engineering/Math, New York, 2003.

Deuteron *n.* Nucleus of the deuterium atom or the ion of deuterium. Its structure – one neutron and one proton.

Developed dyes *n.* Dyes that are formed by the use of a developer. The substrate is first dyed in a neutral solution with a dye base, usually colorless. The dye is then diazotized with sodium nitrate and an acid and afterwards treated with a solution of β-naphthol, or a similar substance, which is the developer. Direct dyes are developed to produce a different shade or to improve washfastness or lightfastness.
See dyes.

Developing *n.* A stage in dyeing or printing in which leuco compounds, dyes, or dye intermediates are converted to the final, stable state or shade.

Developing ink *n.* A non-drying greasy ink composition specifically formulated for use in initial fixing or subsequent renewal of the image on a lithographic plate.

Devolatilization *n.* The removal from a resin of a substantial percentage of some unwanted volatile matter such as water, solvent, or monomer. Vented extruders have been widely used for this task.
See extraction extrusion.

Dew point (dew temperature) (ca. 1833) *n.* Temperature of the atmosphere at which the saturation vapor pressure equals the actual vapor pressure of the water vapor in the air and dew begins to form.

Dewaxing *n.* Process of removing waxes from natural resins (dammars, shellacs, etc.).

Dexel *n.* Cellulose acetate, manufactured by British Celanese, Great Britain.

Dexsil *n.* Poly(carborane siloxane), manufactured by Olin, USA.

Dextin Water-soluble adhesive or thickener prepared by the incomplete hydrolysis of starch by thermostatic (use of diastase), or acid methods. Syn: British gum.

Dextrorotatory \ˈdek-(ˌ)strō-ˈrō-t ə-ˌtōr-ē\ (1878) *adj.* To turn clockwise (sym. is D) or toward the right; *esp*: rotating the plane of polarization of light toward the right, <-crystals> – compare levorotatory (rotates light to the left). Morrison RT, Boyd RN (1992) Organic chemistry, 6th edn. Prentice-Hall, Englewood Cliffs, NJ.

DGEBA *n.* Abbreviation for diglycidyl ether of bisphenol.

D-Glass *n.* Glass with a high boron content, used for fibers in laminates that require a precisely controlled dielectric constant.

DHA *n.* Abbreviation for dehydroacetic acid (*See image*).

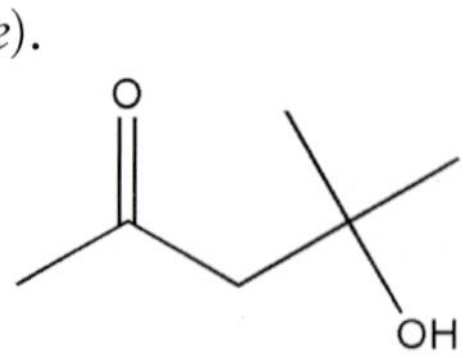

DHP *n.* Abbreviation for dihexyl phthalate. *See di(2-ethyl-butyl)phthalate.*

DHXP *n.* Same as DHP, preceding.

Di- A prefix meaning two or twice. The terms *bi-* and *bis-* are nearly equivalent, assigned with slight differences in meaning according to custom.

Diacetin *n.* (glyceryl diacetate. (1,3-diacetin), and 1,2-diacetin, a mixture of isomers. A water-soluble plasticizer, and a solvent for cellulose nitrate and cellulose acetate (*See image*).

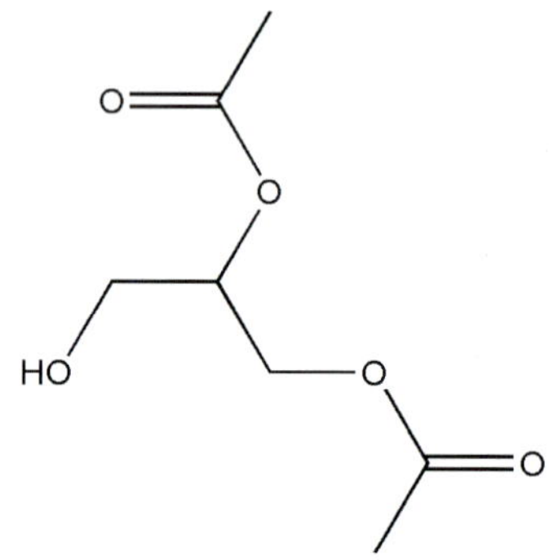

Diacetone alcohol (4-hydroxy-4-methyl-2-pentanone) *n.* A pleasant-smelling, colorless liquid, miscible with water and most organic liquids, used as a solvent for cellulosic, vinyl, and epoxy resins. Mol. wt., 116.16; Sp gr, 0.9406/20°C; bp, 169°C (760 mmHg); flp, 52°C (125°F). Syn: diacetone (*See image*).

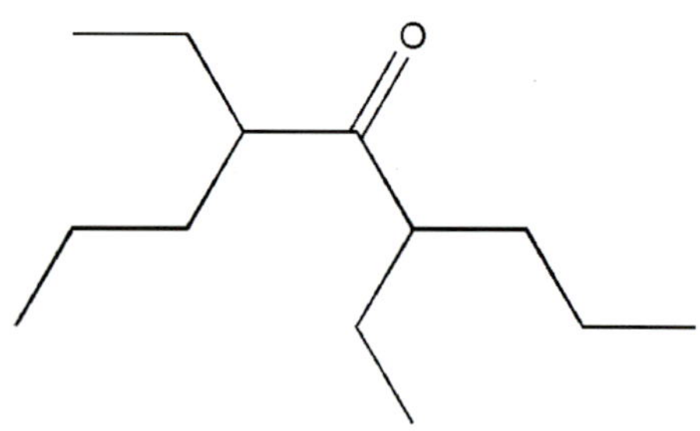

Diacetyl peroxide *See acetyl peroxide.*

Diafoam *n.* A term sometimes used for syntactic foam that contains gas bubbles in addition to microspheres.

Diagonal (45°C) flame test \dī-ˈa-gə-nəl\ *n.* *See flammability test.*

Diagonal grain *n.* Grain in which the longitudinal elements form an angle with the axis of the piece as a result of sawing at an angle with the bark of the tree or log; a form of cross grain.

Diagonal-grained wood *n.* Wood in which the fibers are at an angle with (i.e., diagonal to) the axis of a piece as a result of sawing at an angle with the axis of the tree.

Dial \ˈdī(-ə)l\ [ME *dyal*, fr. ML *dialis* clock wheel revolving daily mfr. L *dies* day] (15c) *n.* In a circular-knitting machine, a circular steel plate with radically arranged slots for needles. A knitting machine equipped with both a dial and a cylinder can produce double-knit fabrics. Tortora PG, Merkel RS (2000) Fairchild's dictionary of textiles, 7th edn. Fairchield Publications, New York.

Diallyl chlorendate (DAC) *n.* A reactive monomer used as a flame-resisting agent in diallyl phthalate, epoxy, and alkyd resins. It can be used in the monomeric form (a high-viscosity liquid); or in the polymeric form, alone or in conjunction with other flame retardants.

Diallyl esters *n.* Series of unsaturated esters, which polymerize rapidly to yield cross-linked resins in the presence of peroxides at relatively low temperatures (e.g., 40–100°C).

Diallyl isophthalate *n.* A polymerizable monomer, used in laminating and molding (*See image*).

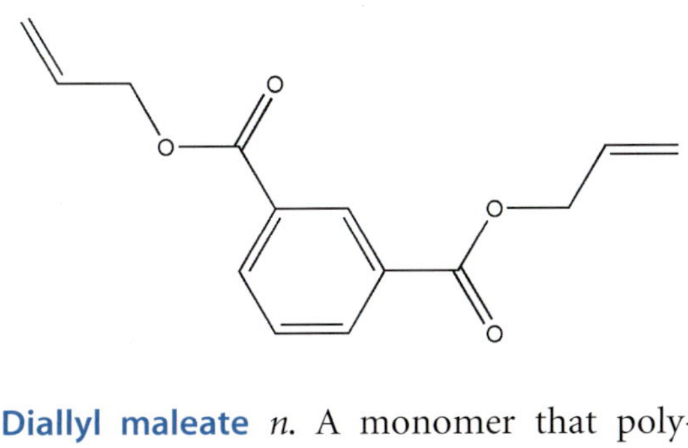

Diallyl maleate *n.* A monomer that polymerizes readily when exposed to light or temperatures above 50°C (*See image*).

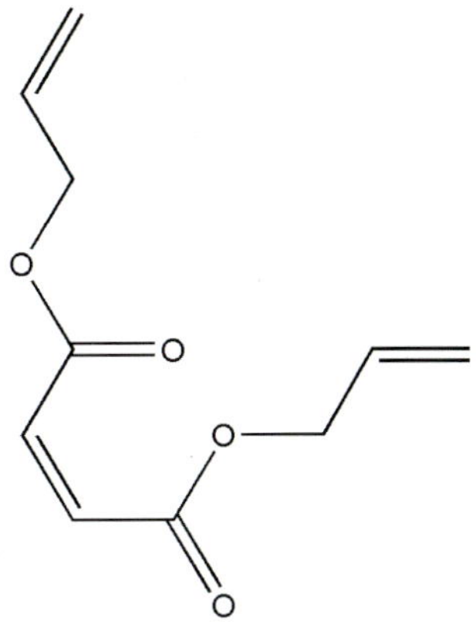

Diallyl phthalate (diallyl *o*-phthalate, DAP) *n.* In the monomeric form, DAP is a colorless liquid ester with a viscosity about equal to that of kerosene, widely used as a cross-linking monomer for unsaturated polyester resins, and as a polymerizable plasticizer for many resins. If polymerizes easily, either gradually or rapidly, increasing in viscosity until it finally becomes a clear, infusible solid. The name DAP is used for both the monomeric and polymeric forms. In the partly polymerized form, DAP is used in the production of thermosetting molding powders, casting resins, and laminates (*See image*).

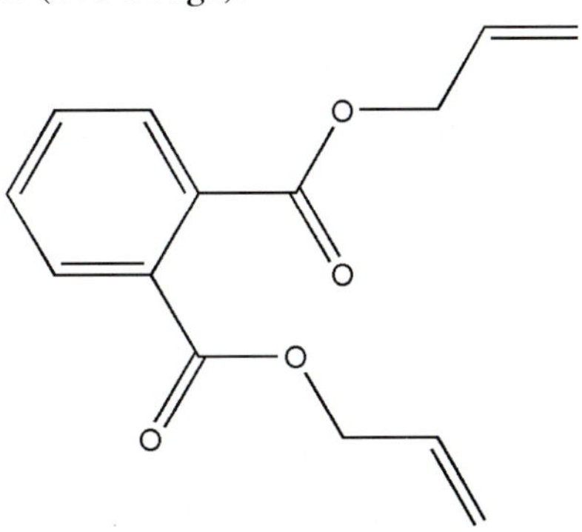

Diallyl phthalate resins *n.* Laminating resin prepared by free radical polymerization of diallyl phthalate. The polymer is highly cross-linked, with good thermal stability and retention of electrical properties under conditions of wet and dry heat.

Di-alphinyl phthalate *n.* A plasticizer derived by the esterification of primary aliphatic alcohols in the range C_7 to C_9. It is used in place of dioctyl phthalate in some applications.

Dialysis \dī-ˈa-lə-səs\ [NL, fr. Gk, separation, fr. *dialyein* to dissolve, fr. *dia-* + *lyein* to loosen] (1861) *n.* The separation of different-sized molecules by transport in solution through a semipermeable membrane, driven by the difference in chemical potential between the liquids separated by the membrane. The membranes may be any of a wide range of materials, e.g., parchment or microporous films of cellulosic polymers.

D

Diamagnetic materials \ˌdī-ə-mag-ˈne-tik\. Are those within which an externally applied magnetic field is slightly reduced because of an alteration of the atomic electron orbits produced by the field. Diamagnetism is an atomic-scale consequence of the Lenz law of induction. The permeability of diamagnetic material is slightly less than that of empty space.

Diamine \ˈdī-ə-ˌmēn\ [ISV] (1866) *n.* A compound with two amino groups. Hexamethylenediamine, one of the intermediates in the manufacture of nylon-66 salt, is an example of this chemical type.

4,4′-Diaminodiphenylmethane (DDM, methylenediamine, MD) *n.* A silvery, crystalline material obtained by heating formaldehyde aniline with aniline hydrochloride and aniline. It is used as a curing agent for epoxy resins and as an intermediate in making diisocyanates for urethane elastomers and foams by reaction with phosgene. Possible occupational hazards in the use of DDM are toxic hepatitis and liver damage (*See image*).

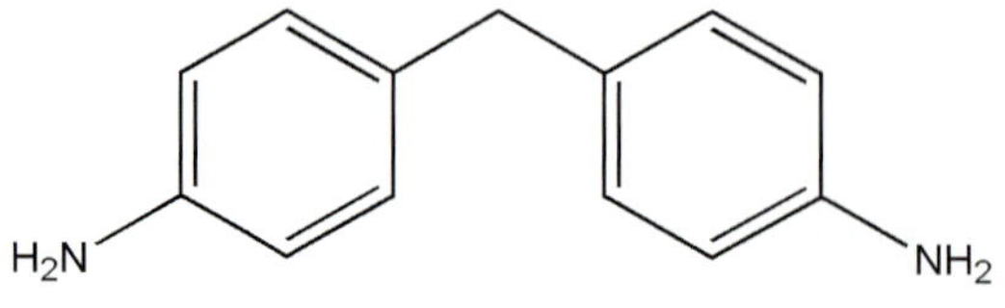

4,4′-Diaminodiphenylsulfone (DDS) *n.* An epoxy curing agent that gives the highest deflection temperatures among amine-cured epoxies.

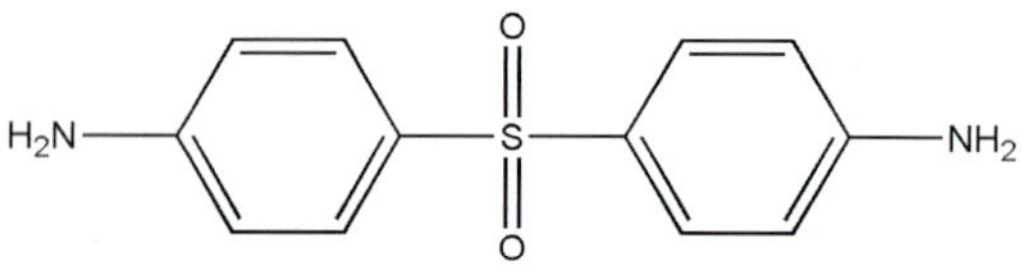

Diamyl phthalate *n.* $C_6H_4(COOC_5H_{11})_2$. A plasticizer derived by esterification of phthalic anhydride with amyl alcohol, compatible with most vinyls, polymethyl methacrylate, and cellulosics.

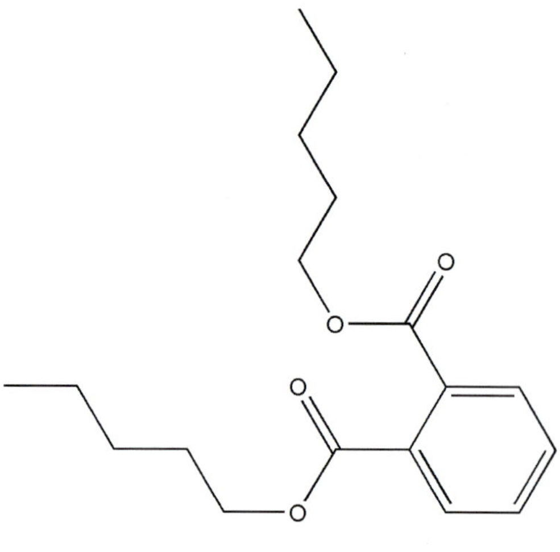

Diaphragm gate \ˈdī-ə-ˌfram\ (web gate) *n.* In an injection or transfer mold, a gate that forms a complete thin web across the opening of the part, used in the molding of annular and tubular objects.

Diarylide yellow A strong yellow toner used in many types of printing inks.
See benzidine yellow.

Diatomaceous calcite \ˌdī-ə-tə-ˈmā-shəs ˈkal-ˌsīt\ *n.* 2.4 $CaCO_3 \cdot SiO_2$ (approximate). Liquid blend of $CaCO_3$ and diatomaceous silica that is mined from a single large deposit located in Kansas. Used as an extruder. Sp gr, 2.54; density, 21.2 lb/gal; O.A., 25 lb/100 lb; particle size, 10 μm.

Diatomaceous earth (1883) *n. See diatomaceous silica.*

Diatomaceous silica *n.* $SiO_2·7H_2O$. Pigment White 27 (77811). A form of hydrous silica, processed from natural diatomate, a sedimentary rock of varying degrees of consolidation that is composed essentially of the fossilized siliceous skeletal remains of single-cell aquatic plant organisms called diatoms. Refractive index, 1.42; density, 2.00 g/cm^3 (16.7 lb/gal); particle size, 6–10 µm. Syn: diatomaceous earth, diatomite, and the obsolete names of kieselguhr, infusorial earth, tripoli.

Diatomite \di-ˈa-tə-ˌmīt\ (1887) (diatomaceous earth, DE, kieselguhr, infusorial earth, siliceous earth, and tripolite) *n.* The naturally occurring deposit of skeletons of small unicellular algae called *diatoms*, consisting of from 83 to 89% silica. Its many uses include fillers for plastics.

Diazo \dī-ˈa-(ˌ)zō\ [ISV *diaz-*, *diazo-*, fr. *di-* + az-] (1878) *adj.* –N=N– grouping. Diazo compounds are intermediates for manufacture of dyestuffs.

Diazo solution Solution resulting from treatment of an aromatic primary amine with sodium nitrite and mineral acid. Such solutions are stable only below 5°C and are employed as one of the ingredients in the manufacture of azo colors.

DIBA *n.* Abbreviation for diisobutyl adipate.

Dibasic \(ˌ)dī-ˈbā-sik\ (1868) *adj.* Pertaining to acids having two replaceable hydrogen atoms, e.g., sulfuric acid, H_2SO_4, or to acid salts in which two of the three hydrogens have been replaced by metal(s), e.g., K_2HPO_4.

Dibasic lead phosphite *n.* $2PbO·PbHPO_3·0.5H_2O$. A white, crystalline powder long used as a heat and light stabilizer for PVC and other chlorine-containing resins. It has good electrical properties, and acts as an antioxidant and screening agent for UV light. Less used today because of concern to keep lead out of the environment.

Dibasic lead stearate *n.* $2PbO·Pb(OOCC_{17}H_{35})_2$. A good heat stabilizer with lubricating properties.
See also lead stearate.

Dibenzyl ether $(C_6H_5CH_2)_2O$. A plasticizer for cellulose nitrate.

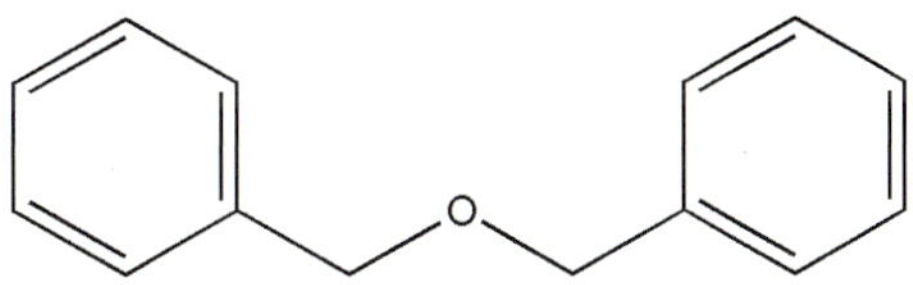

Dibenzyl phthalate *See benzyl phthalate.*

Dibenzyl sebacate *n.* $(C_6H_5CH_2OOC)_2(CH_2)_8$. A non-toxic plasticizer, often used in vinyl compounds for lining container closures.

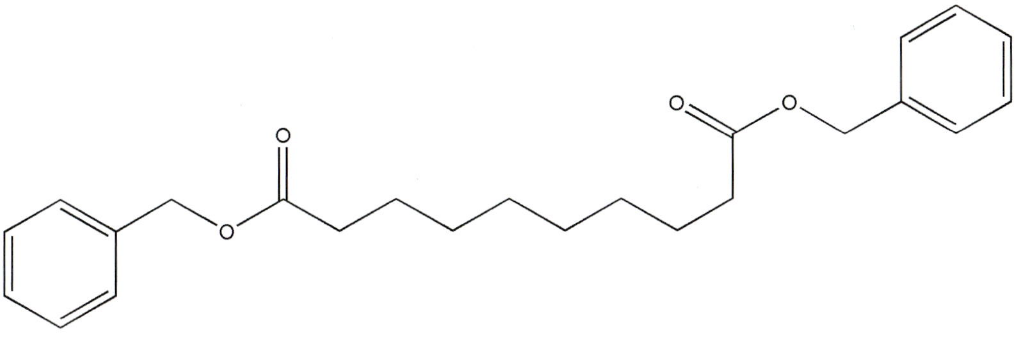

Diblock polymers *n.* Block copolymers consisting of two blocks, one of A repeating units and one of B repeating units. Thus its structure may be represented as follows:

AAA . . . AAAABBBB . . . BBB.

Diborane \-ˈbōr-ˌān\ *n.* B_2H_6 (boron hydride, boro-ethane). A colorless gas that has been used as a catalyst in the polymerization of ethylene.

DIBP Abbreviation for diisobutyl phthalate.

2,3-Dibromo-2-butene-1,4-diol *n.* (dibromobutenediol). $HOCH_2C(Br){=}C(BR)CH_2OH$. A low-molecular weight, chemically reactive, brominated primary glycol. It is used as a building block for condensation polymers that can be incorporated into a wide variety of polymers including esters, urethanes, and ethers. It is also used as a flame-retardant monomer for polyurethanes and thermoplastics, and as a substitute for methylene-bis-*o*-chloroniline (MOCA) in urethane foams.

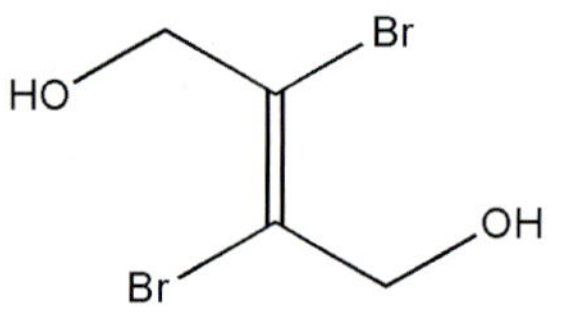

Dibromoneopentyl glycol *n.* A high-melting solid, available in powder or flake form, used as a flame retardant for polyester resins. A more convenient liquid material is made by using dibromoneopentyl glycol to form a polyester alkyd that is dissolved in styrene, the resulting liquid being more easily used in polyester reactors. The material is also adaptable to urethane foams, polymeric plasticizers, and coating resins.

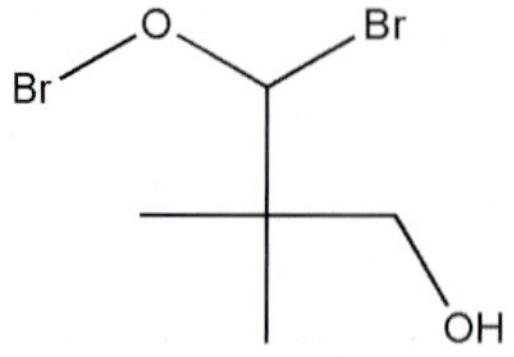

2,3-Dibromopropanol *n.* A brominated alcohol used as a component in making fire-retardant urethane foams.

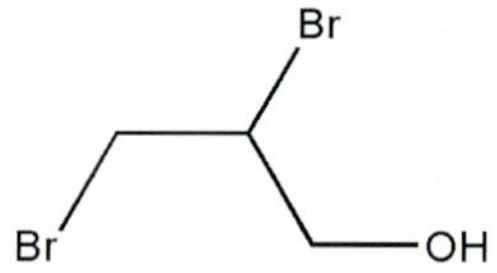

Dibutoxyethoxy ethyl adipate *n.* $[C_4H_9OC_2H_4OC_2H_4COO(CH_2)_2]_2$. A plasticizer for cellulose nitrate, ethyl cellulose, polyvinyl butyral, and polyvinyl acetate.

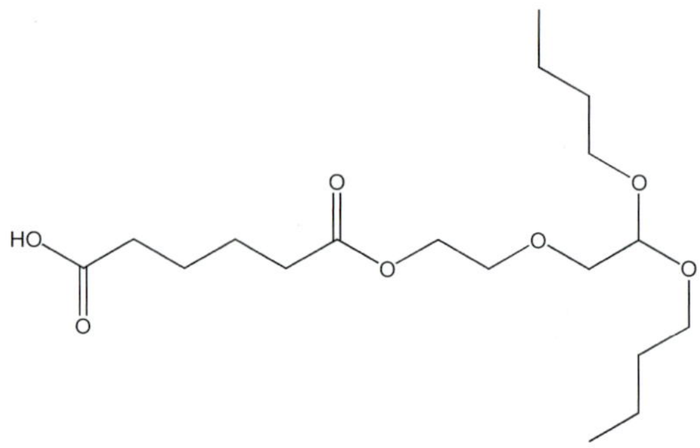

Dibutoxyethyl adipate *n.* $(C_2H_4COOC_2H_4OC_4H_9)_2$. A primary plasticizer for PVC and many other resins, imparting low-temperature flexibility and UV resistance. It is widely used as a plasticizer for polyvinyl butyral in the interlayer of safety glass.

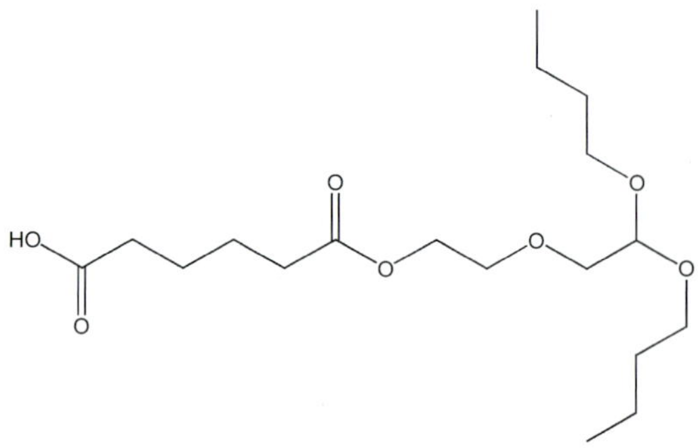

Dibutoxyethyl phthalate (DBEP) *n.* $C_6H_4(COOC_2H_4OC_4H_9)_2$. A primary plasticizer for vinyls, methacrylates, nitrocellulose, and ethyl cellulose, imparting low-temperature flexibility and UV resistance. Incorporation of up to 20% of DBEP into vinyl calendaring compounds eliminates defects such as streaks and blisters. In plastisols,

DBEP imparts low initial viscosity and low fusion temperatures.

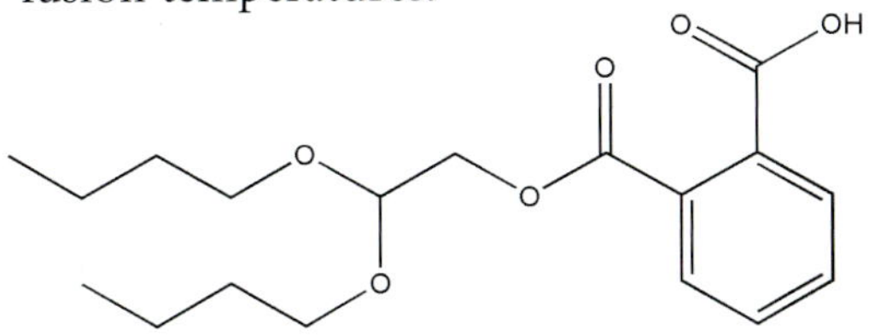

Dibutoxyethyl sebacate *n.* $(CH_2)_8(COOC_2H_4OC_4H_9)_2$. A primary plasticizer for PVC and PVAc, with low-temperature resistance.

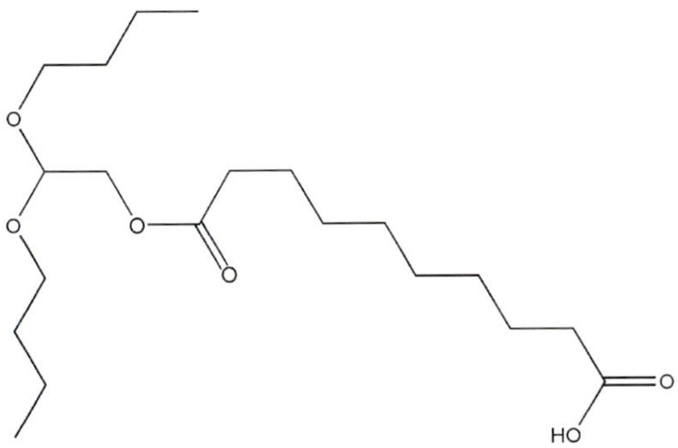

Dibutyl adipate \ˌdī-ˈbyü-təl-\ *n.* $(C_4H_9COO)_2(CH_2)_4$. A plasticizer for vinyl and cellulosic resins.

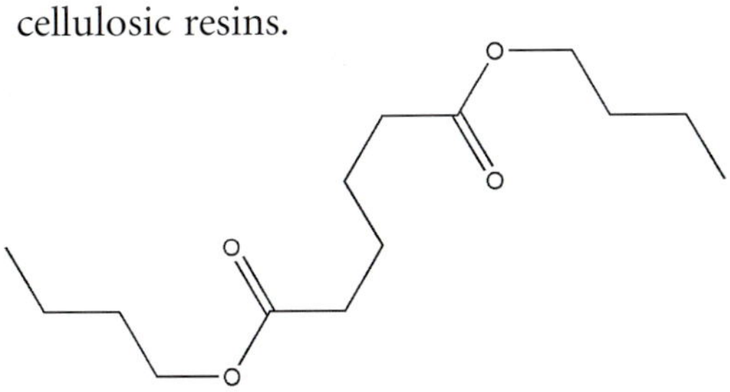

Dibutyl butyl phosphonate *n.* $C_4H_9P(O)(OC_4H_9)_2$. A plasticizer and antistatic agent.

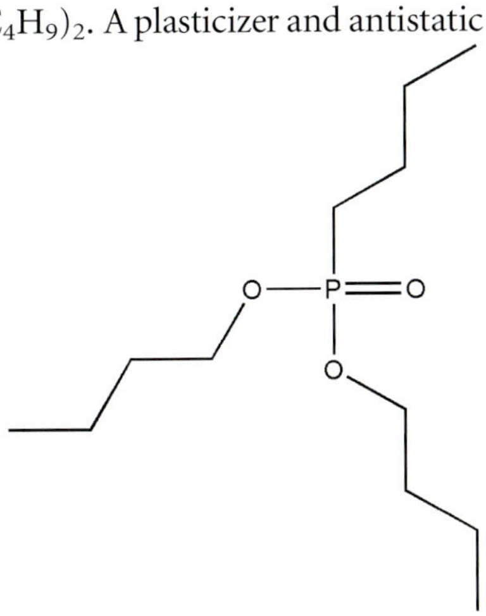

Dibutyl fumarate \fyú-ˌmar-ˈnāt\ *n.* $C_2H_2(COOC_4H_9)_2$. Derived from fumaric acid and used as a plasticizer for polyvinyl acetate, polyvinyl chloride, and PVC-AC copolymers.

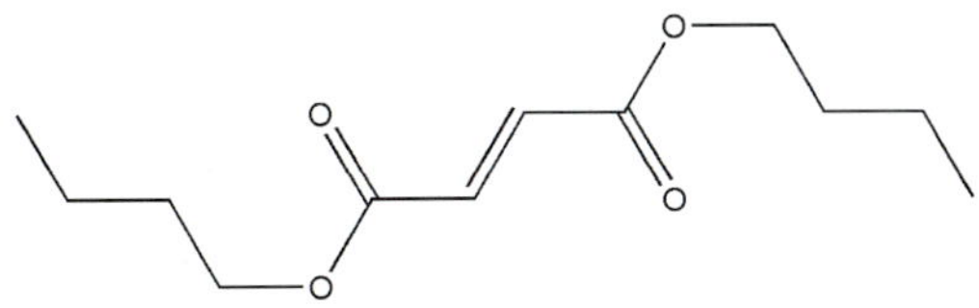

Dibutyl isosebacate *n.* $C_8H_{16}(COOC_4H_9)_2$. A plasticizer for vinyls and other thermoplastics.

Dibutyl phthalate [*phthal*ic acid + [1]-*ate*] (1925) (DBP) *n.* $C_6H_4(COOC_4H_9)_2$. One of the most widely used plasticizers for cellulose nitrate and other cellulose-ester and -ether lacquers and coatings. It is a primary plasticizer for many other resins, but its high volatility limits its use in vinyls. Sp gr. 1.0484 (20/20°C); flp, cleveland open cup, 171°C (340°F); bp, 340°C.

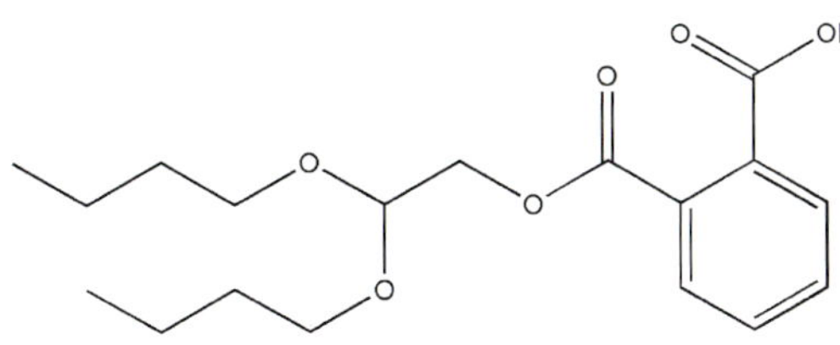

Dibutyl sebacate (DBS) *n.* $(CH_2)_8(COOC_4H_9)_2$. A plasticizer, one of the most effective of the sebacate family. It has good low-temperature properties, low volatility, and is compatible with vinyl chloride polymers and copolymers, polyvinyl butyral, and ethyl cellulose. It is non-toxic, suitable for uses in food wrappings.

Dibutyl succinate \ˈsək-sə-ˌnāt\ *n.* $(CH_2)_2(COOC_4H_9)_2$. A plasticizer for cellulosic resins.

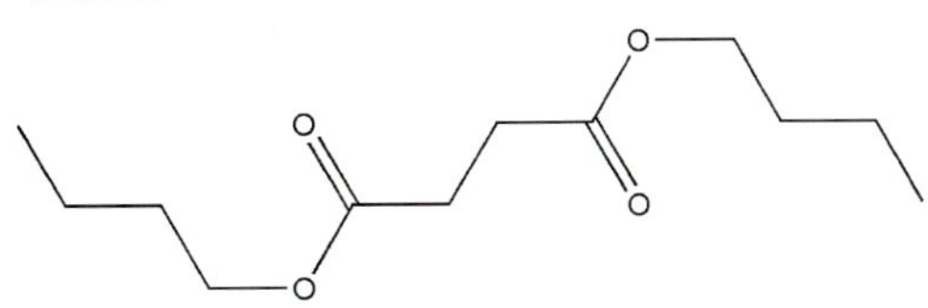

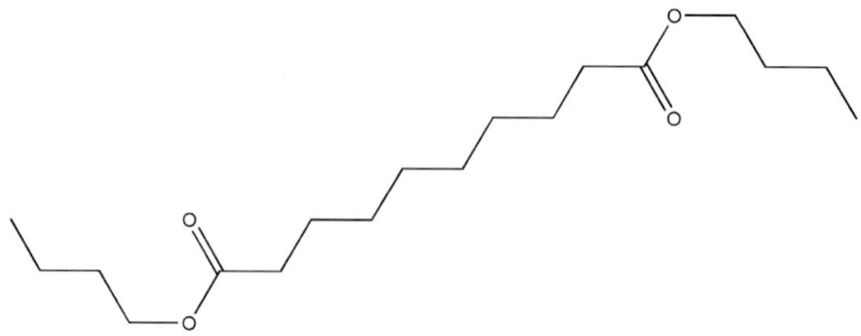

Dibutyl tartrate \\ˈtär-ˌtrāt\\ *n.* $(CHOH)_2(COOC_4H_9)_2$. A lubricant, plasticizer, and solvent for cellulosic resins.

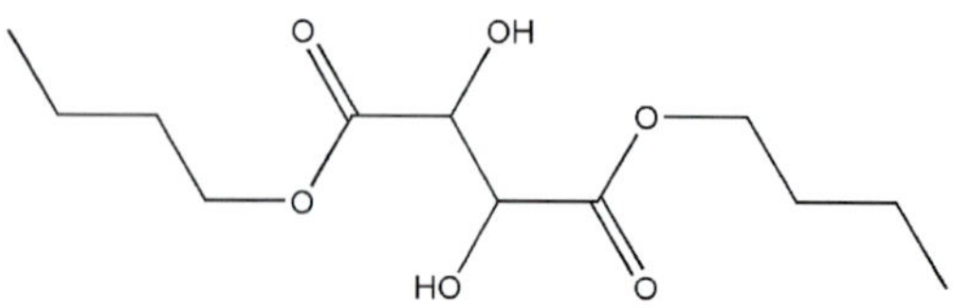

Dibutyltin bis(isooctylmercapto acetate) *n.* A stabilizer for rigid PVC, used primarily during the period from 1953 to 1970. Thereafter, improved butyltin and methyltin derivatives, and synergistic mixtures of them with other stabilizers, replaced this stabilizer for most uses.

Dibutyltin di-2-ethylhexoate *n.* $(C_4H_9)_2Sn(OOCC_7H_{18})_2$. A white, waxy solid made by reacting dibutyltin oxide with 2-ethylhexoic acid. Used as a catalyst for silicone curing and in polyether foams.

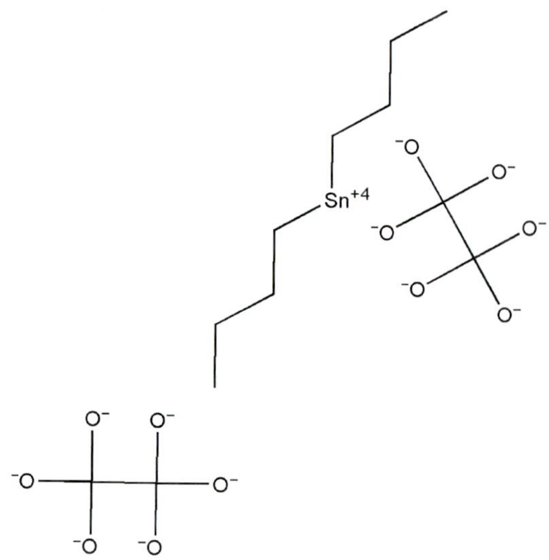

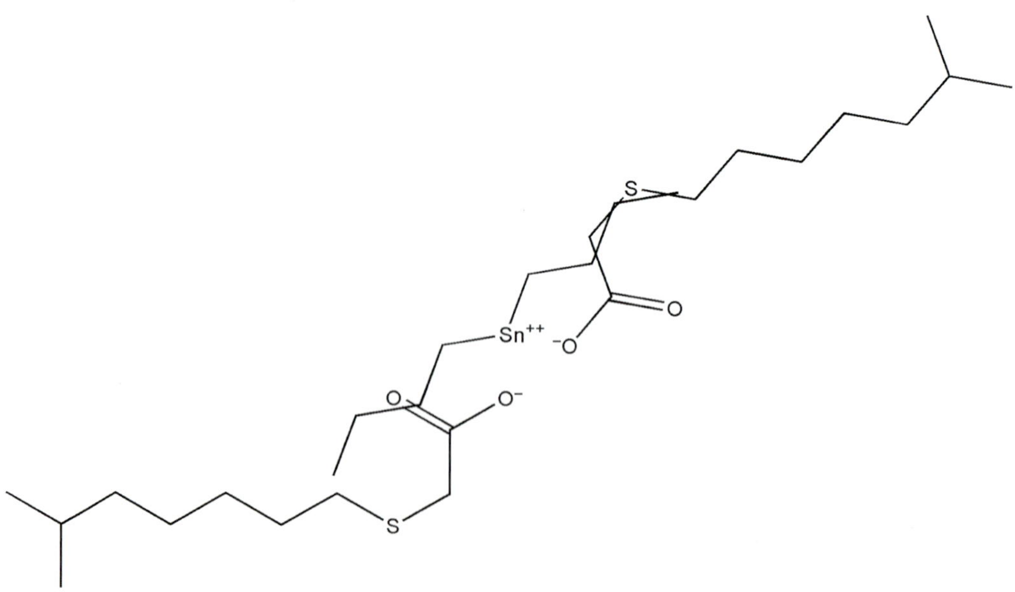

Dibutyltin dilaurate (DBTDL) *n.* $(C_4H_9)_2Sn(OOCC_{11}H_{23})_2$. A lubricating stabilizer for vinyl resins, a catalyst for urethane foams and for condensation polymerizations. It is used in vinyl compounds when good clarity is needed, and also imparts excellent light stability. However, it degrades at high processing temperatures.

Dibutyltin maleate *n.* $[(C_4H_9)_2Sn(OOCCH)_2]_\chi$. A white amorphous powder used as a condensation–polymerization catalyst and a stabilizer for PVC. The molecular weight of the material varies, and grades of lower molecular weight tend to be volatile and produce gases. The higher-weight polymers are very effective in rigid PVC.

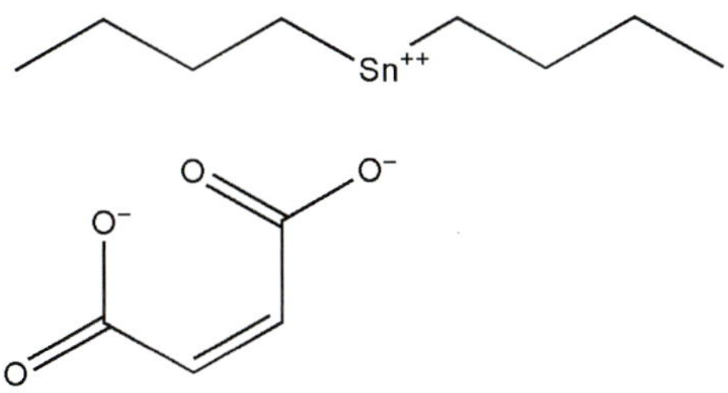

Dibutyltin sulfide *n.* $[(C_4H_9)_2SnS]_3$. An antioxidant and stabilizer for vinyl resins.

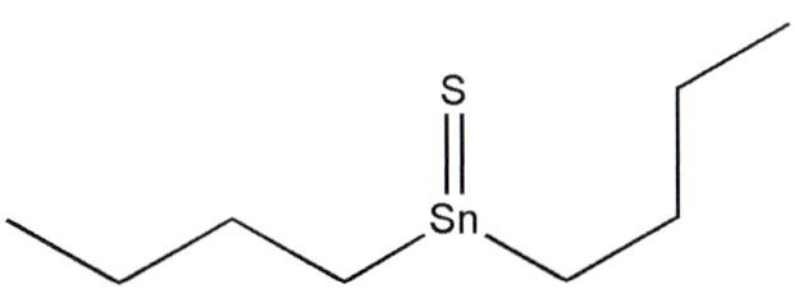

1,1-Dibutylurea *n.* (*N,N*-dibutylurea) $NH_2CON(C_4H_9)_2$. A polymerizable compound. When copolymerized with simple urea and formaldehyde, permanently thermoplastic resins are obtained.

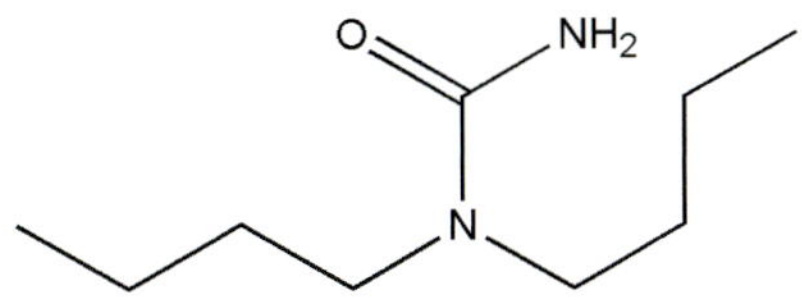

Dicapryl adipate *n.* $C_4H_8(COOC_8H_{17})_2$. A plasticizer for cellulosic and vinyl resins, yielding good low-temperature flexibility. Also compatible with polymethyl methacrylate and polystyrene.

Dicapryl phthalate *n.* [(DCP) di-(2-octyl) phthalate] $(C_8H_{17}COO)_2C_6\text{–}H_4$. A plasticizer for cellulosic and vinyl resins. It is similar to DOP and DIOP, but has low initial viscosity and is preferred in plastisols.

Dicapryl sebacate *n.* $(CH_2)_8(COOC_8H_{17})_2$. A plasticizer for vinyl resins and acrylonitrile rubbers, imparting good low-temperature flexibility.

Dicarboxylic acid \ˌdī-ˌkär-ˌbäk-ˈsi-lik\ *n.* Any of a large family of organic acids containing two carboxylic (–COOH) groups. Those of greatest importance in the plastics industry are the adipic, azelaic, glutaric,

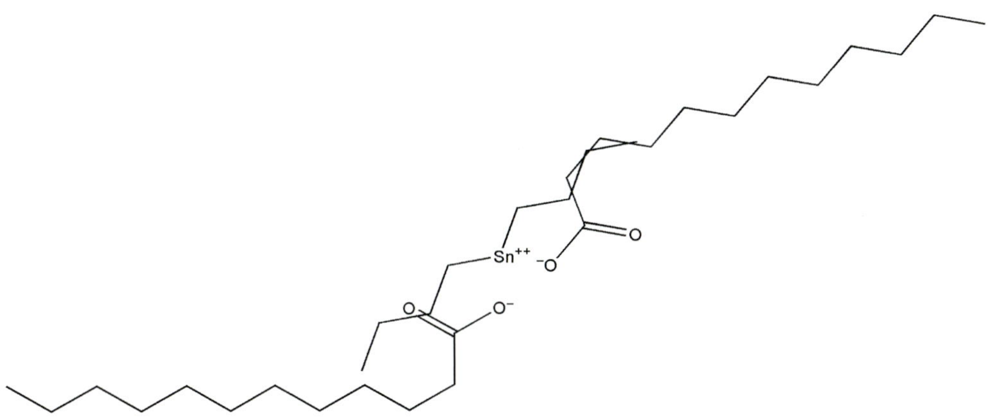

pimelic, sebacic, and succinic acids, esters of which are widely used as plasticizers that impart low-temperature flexibility. They are also used in the production of alkyd and polyester resins, polyurethanes, and nylons.

Dicers *n.* One of the two basic types of cold cutting systems. It consists of a die, a cooling area, a drying area (if water was used), and a cutting chamber.

Dicetyl ether \-ˈsē-təl ˈē-thər\ (dihexadecyl ether). A mold lubricant.

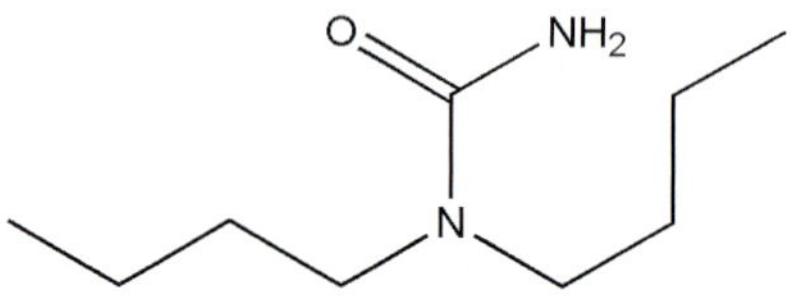

2,4-Dichlorobenzoyl peroxide *n.* A cross-linking agent for silicone elastomers. It is sold as a 40% active paste dispersed in silicone fluid (Candox®).

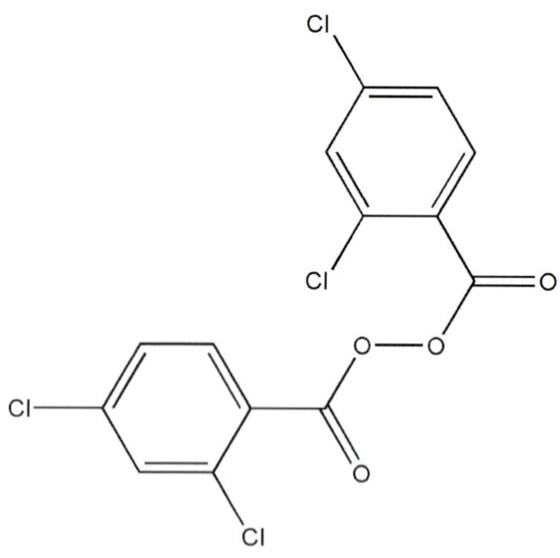

Dichlorodifluoromethane \-ˌdī-ˌflúr-ə-ˈme-ˌthān\ (1936) (Freon 12) *n.* CCl_2F_2. Long used as a refrigerant, aerosol propellant, and as a blowing agent for foamed plastics that was safer than inflammable hydrocarbons, this accused destroyer of stratospheric ozone is rapidly being phased out of all its former uses.

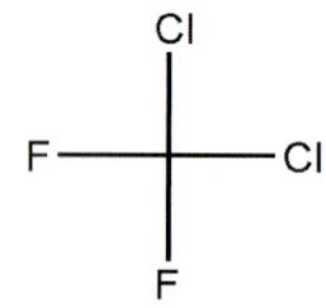

1,1-Dichloroethylene *n.* Syn: vinylidene chloride.

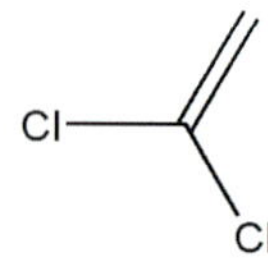

α-Dichlorohydrin *n.* $CH_2ClCHOHCH_2Cl$. A cellulosic-resin solvent.

Dichloromethane \(ˌ)dī-ˌklō-rō-ˈe-ˌthān\ (1936) *n.* Syn: methylene chloride.

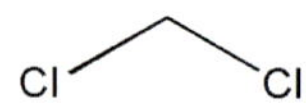

2,6-Dichlorostyrene *n.* $Cl_2C_6H_3CH{=}CH_2$. A monomer and comonomer used mainly in plastics research.

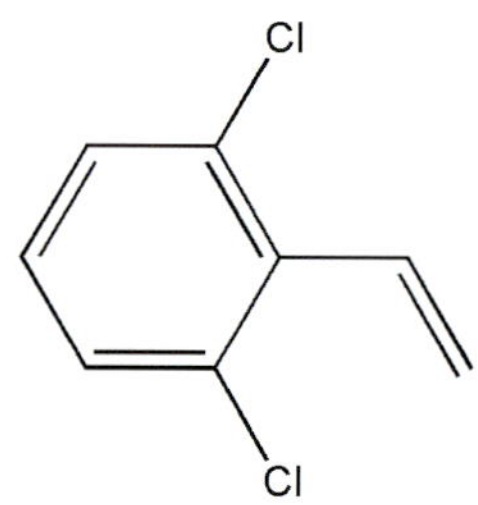

Dichlorotetrafluoroethane (Freon 114) *n.* ClF_2CCClF_2. A fluorocarbon blowing agent used when a low boiling point (3.6°C) is required.

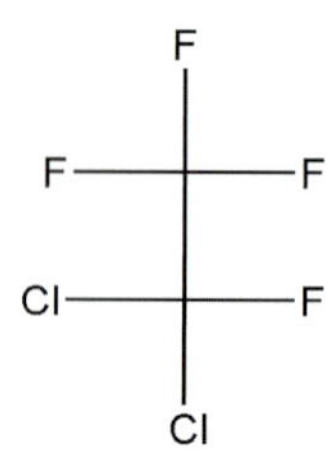

Dichroism \ˈdī-(ˌ)krō-ˌi-zəm\ (1819) *n.* (1). A property possessed by many doubly refracting crystals of exhibiting different colors when viewed from different directions. (2) Displaying different colors as a result of changes in concentration or thickness. Such changes do not result from changes in the absorption coefficients for the material, but rather on the change in perceived color resulting from more or less absorption across the visible spectrum as the concentration is changed. The term "dichroism" is frequently used (incorrectly) to describe this effect. Dichromism is predictable on the basis of the laws of subtractive colorant mixture.

Dicing *n.* The process of cutting thermoplastic sheets (or square strands) into cubical pellets for further processing.

Di-compounds *n. See under other headings, such as ethyl oxalate for diethyl oxalate, etc.*

Dicumyl peroxide *n.* $[C_6H_5C(CH_3)_2O]_2$. A vulcanizing agent for elastomers, also used to cross-link polyethylene.

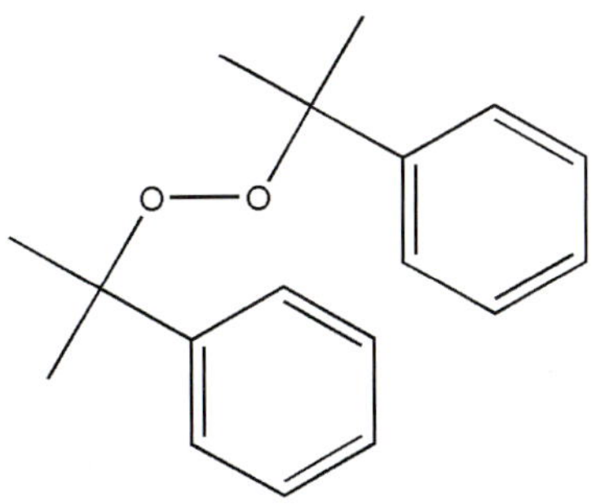

Dicyandiamide *n.* (cyanoguanidine) $H_2NC(=NH)NHCN$. The widely used, but incorrect name for the dimmer of cyanamide. Cyanoguanidine is used mainly in the production of melamine, but also as a stabilizer for vinyl resins and curing agent for epoxy resins.

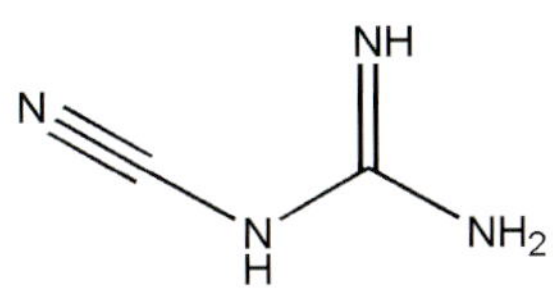

Dicyclohexyl azelate *n.* $C_6H_{11}OOCC_7H_{14}COOC_6H_{11}$. A plasticizer for PVC.

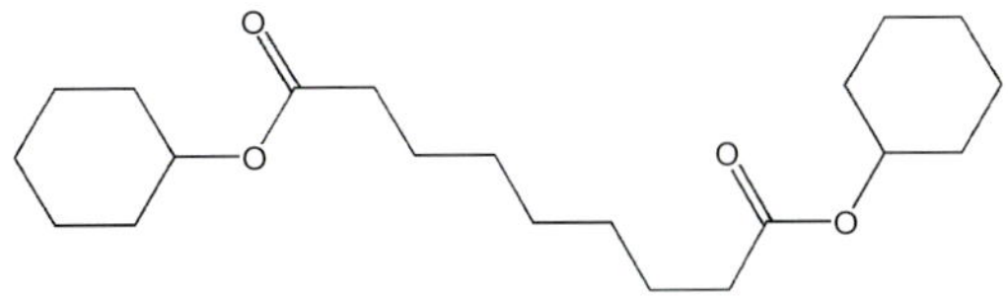

Dicyclohexyl phthalate (DCHP) *n.* $C_6H_4(COOC_6H_{11})_2$. A plasticizer for PVC and many other resins. It imparts good electrical properties, low volatility, low water and oil absorption, and resistance to extraction by hexane and gasoline. In vinyls DCHP is usually combined with other plasticizers. In cellulose nitrate, polystyrene, polymethyl methacrylate, and ethyl cellulose it serves as a primary plasticizer.

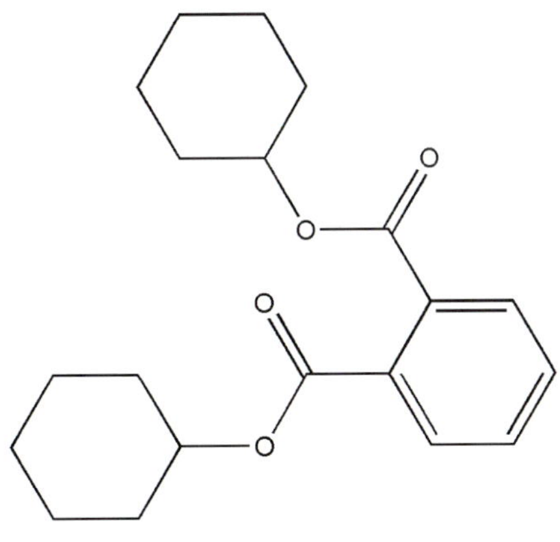

DIDA *n.* Abbreviation for diisodecyl adipate.

Didecyl adipate *n.* $C_4H_8(COOC_{10}H_{21})_2$. A plasticizer for PVC and cellulosics. Its most noteworthy properties are low-temperature flexibility, low volatility, and good electricals.

Didecyl ether *n.* $(C_{10}H_{21})_2O$. A processing and mold lubricant.

Didecyl phthalate (DDP) *n.* $C_6H_4(COOC_{10}H_{21})_2$. A primary plasticizer for vinyl resins, also compatible with polystyrene and cellulosics. It has the lowest specific gravity of the most common phthalate

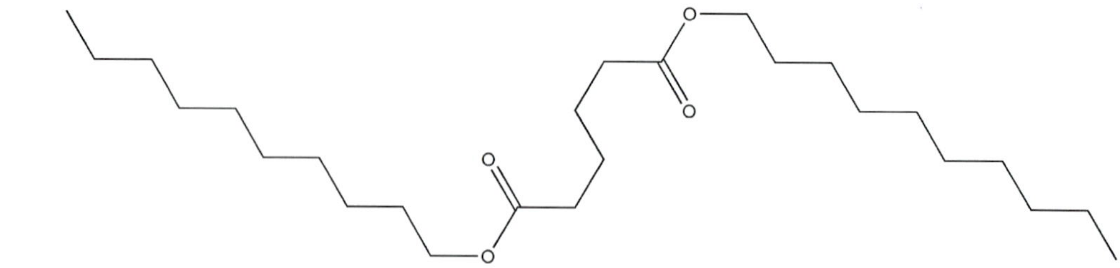

plasticizers, low volatility, and resistance to extraction by soapy water.

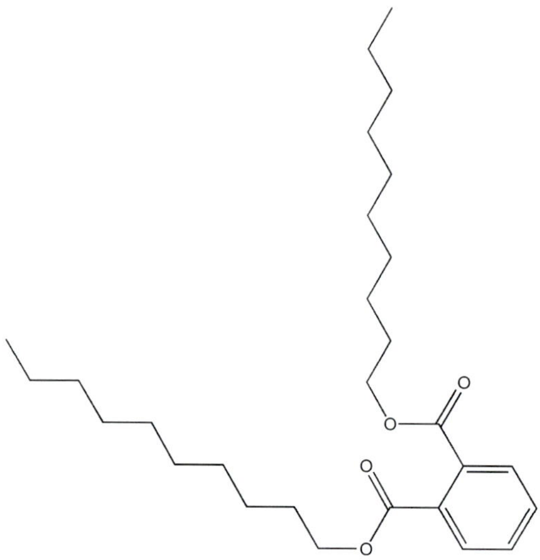

DIDG *n.* Abbreviation for diisodecyl glutarate.

DIDP *n.* Abbreviation for diisodecyl phthalate.

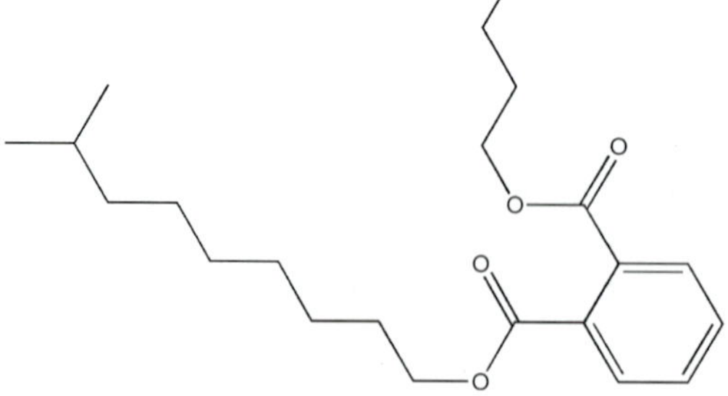

Die \ˈdī\ [ME *dī*, fr. MF *dé*] *n.* (1) Extrusion die. (2) The recessed block into which plastic material is injected or pressed, shaping the material to the desired form. The terms *mold cavity* or *cavity* are more often used. (3) Steel-rule die.

See die cutting.

Die adapter (extrusion) *n. See adapter.*

Die blade (die lip). In extrusion, a deformable member attached to a die body that determines the slot opening and that may be adjusted to produce uniform thickness across the film or sheet being made.

Die block *n.* That part of an extrusion die that holds the forming bushing and core.

Die body *n.* In the USA same as adapter. In Great Britain, the outer body or barrel of an extrusion die.

Die cart *n.* A sturdy, height-adjustable framework on casters designed to support a heavy extrusion die, such as a sheet die.

Die characteristic Of an extrusion die, the relationship between rate of flow of melt through the die, the pressure drop through the die, and the viscosity of the melt. It is defined by the flow equation

$$Q = K\,\Delta P/\eta,$$

in which Q is the volumetric flow rate, P the pressure drop, η the effective viscosity at the temperature and shear rate of the melt in the die, and K is the die characteristic. K is dependent on die dimensions and has the dimensions of volume. The die characteristic is linked to the screw characteristic by ΔP, which must be the same for both, whereas η usually will not be the same for both.

Dieckmann reaction *n.* Cyclization by way of a base catalyzed intramolecular casein type reaction of esters with active delta or epsilon ethylenes (*See image*).

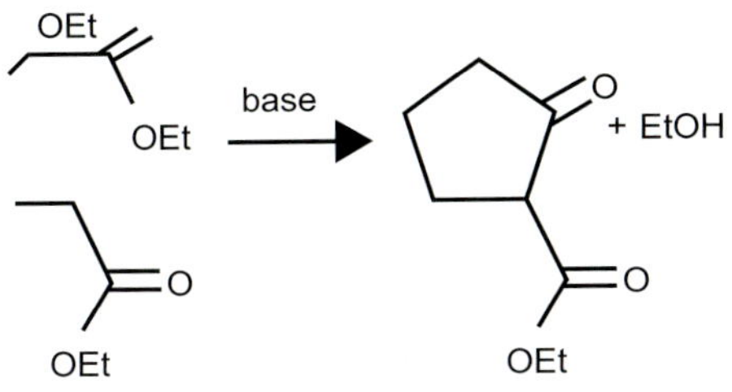

Die core *n.* The tapered element in an extrusion die for pipe or tubing that guides the material to the webs of the spider. *Sometimes called the torpedo* or *spreader.*

Die cutters *n.* Cuts shapes from sheet stock by sharply striking it with a shaped knife-edge, known as steel rule die.

Die cutting (blanking, clicking, and drinking) *n.* The process of cutting shapes from sheets of plastic by pressing a shaped knife-edge through one or several layers of sheeting. The dies are often called *steel-rule dies*, and pressure is applied smartly by hydraulic or mechanical means.

Die-entry angle *n.* At the entrance to the land of an extrusion die, the angle included between the adjacent, inside die surfaces. In many rheometer orifices and in some profile dies, the entry angle is 180°, the extreme of abruptness in actual use. Streamlined-entry dies stretch out the reduction and change of cross-sectional shape over a distance, so the included angle at the die lip may be 20–60°. Extreme gradualness is designed into high-speed wire-coating dies, where the approach to the lip is made in several states of decreasing taper, the final included angle being only 6–10°.

Die gap (die-lip opening) *n.* In film and sheeting dies, the perpendicular distance between the lips of the die land at any point along the width of the die, usually measured after the die has been brought up to temperature but before extrusion has begun. During extrusion, internal pressure inside the die causes the gap to enlarge slightly, more at the center than at the ends of a center-fed die, necessitating adjustments to attain uniform thickness across the film or sheet.

Die land *n.* In an extrusion die, the land is the portion of the die wherein the dimensions of the opening are constant from a point within the die to the discharge point, giving the extrudate its final shape.

Dielectric \ˌdī-ə-ˈlek-trik\ [*dia-* + *electric*] (1837) *n.* A material with electrical conductivity less than 10^{-6} S/cm (1 μS/cm), thus so weakly conductive that different parts of a sheet can hold different electrical charges. In radio-frequency heating, the term dielectric is used for the material being heated. The term is also used for the non-conductive material separating the conductive elements of a capacitor. Polymeric materials are widely employed as dielectrics. The two most important properties of a dielectric are its dielectric constant and dielectric strength, defined below. Values for some polymeric and other dielectrics at 60 Hz and room temperature are listed here.

Material	Dielectric constant	Dielectric strength (kV/mm)
Air	1.00054	0.8
Alumina*	3.5	1.6–6.3
Pyrex glass	4.5	13
Bond paper	3.5	14
Polyethylene	2.3	18–39

Polypropylene	2.1–2.7	18–26
Nylon 6/6	3.6–4.0	12–16
Glass-reinforced nylon 6/6	4.0–4.4	19
Crystal polystyrene	2.5–2.65	20–28
Acrylic resin	3.3–3.9	16
Phenoxy resin	4.1	16–20
Thermoplastic polyurethane	6–8	33–43
GP phenolic	5–10	12–17
Epoxy/glass	5.5	14

*At 1 MHz.

Harper CA (ed) (2002) Handbook of plastics, elastomers and composites, 4th edn. McGraw-Hill, New York.Ku CC, Liepins R (1987) Electrical properties of polymers. Hanser Publishers, New York.James F (ed) (1993) Whittington's dictionary of plastics. Technomic Publishing Co. Inc., Carley.

Dielectric absorption *n.* An accumulation of electrical charges within the body of an imperfect dielectric material when it is placed in an electric field. Ku CC, Liepins R (1987) Electrical properties of polymers. Hanser Publishers, New York.

Dielectric breakdown voltage (breakdown voltage and disruptive voltage) *n.* The voltage at which electrical breakdown of a specimen of electrical insulating material between two electrodes occurs under prescribed conditions.

Dielectric constant (1875) (permittivity constant) *n.* Between any two electrically charged bodies there is a force (attraction or repulsion) that varies according to the strength of the charges, q_1 and q_2, the distance between the bodies, r, and *a* characteristic of the medium separating the bodies (the dielectric) known as the dielectric constant, ε. The force is given by the

$$F = q_1 q_2/\varepsilon r^2.$$

For a vacuum, $\varepsilon = 1.0000$, values for some other materials are listed above. In practice the dielectric constant of a material is found by measuring the capacitance of a parallel-plate condenser using the material as the dielectric, then measuring the capacitance of the same condenser with a vacuum as the dielectric, and expressing the result as a ratio between the two capacitances.
When the dielectric is a polymeric material whose atoms or molecules may change their positions in an alternating electric field (a *polar* material), frictional energy is dissipated as heat and is characterized by the dissipation factor. Weast RC (ed) (1978) CRC handbook of chemistry and physics, 59th edn. The Chemical Rubber Co., Boca Raton, FL. James F (ed) (1993) Whittington's dictionary of plastics. Technomic Publishing Co. Inc., Carley. Ku CC, Liepins R (1987) Electrical properties of polymers. Hanser Publishers, New York.

Dielectric heat sealing *n.* A sealing process seed for fusing vinyl films and other thermoplastics films with sufficient dielectric loss whereby two layers of film are heated by dielectric heating and pressed together between two electrodes, an applicator and a platen, the films serving as the dielectric of the so-formed condenser. The applicator may be a pinpoint electrode as in "electronic sewing machines," a wheel, a moving belt, or a contoured blade. Frequencies employed range up to 200 MHz, but are usually 30 MHz or less to avoid interference problems. Harper CA (ed) (2002) Handbook of plastics, elastomers, and composites, 4th edn. McGraw-Hill, New York.
See also heat sealing.

Dielectric heating (1944) (electronic heating, RF heating, radio-frequency heating, high-frequency heating, and microwave heating) *n.* The process of heating poor (but polar) conductors of electricity (dielectrics) by means of high-frequency fields. At frequencies above 10 MHz sufficient heat for rapid sealing and welding of many plastics can be generated by low, safe voltages. The process of dielectric heating consists of placing the material to be heated between two shaped electrodes that are connected to a high-frequency current supply. These electrodes act as the plates of a capacitor and the material serves as the dielectric separating them. As the field changes polarity, charge-bearing atoms or groups of the dielectric undergo reorientation in an effort to keep their positive poles toward the electrode that is momentarily negative, thus generating molecular friction that is dissipated as heat. The theoretical rate of heating is given by the equation

$$P = 2\pi f C V^2 \tan \delta,$$

where P is the power input (W), f the electrical field frequency (Hz), C the capacitance (f), V the voltage difference between the electrodes (V), and $\tan \delta$ is the loss factor. The actual rate of heating will be somewhat less because of non-uniformity of the field, air gaps between electrodes and material, and losses to the surroundings. Dielectric heating is most effective for materials such as PVC and phenolics that have high loss factors because of their numerous polar groups. Non-polar plastics with low loss factors, such as polystyrene and polyethylene, are impractical to heat dielectrically. Ku CC, Liepins R (1987) Electrical properties of polymers. Hanser Publishers, New York. Harper CA (ed) (2002) Handbook of plastics, elastomers and composites, 4th edn. McGraw-Hill, New York.
See also dielectric heat sealing and microwave heating.

Dielectric loss *n.* Energy dissipated as heat within a polar material subjected to a rapidly alternating, strong electric field. Ku CC, Liepins R (1987) Electrical properties of polymers. Hanser Publishers, New York. *See dielectric heating.*

Dielectric phase angle *n.* The angular difference in phase between the alternating voltage (usually sinusoidal) applied to a dielectric and the resulting current. The angle is often symbolized by θ, the cosine of which is the power factor. Ku CC, Liepins R (1987) Electrical properties of polymers. Hanser Publishers, New York.

Dielectric properties of polymers *n.* This term usually means the lack of electrical conductivity or insulting value. Ku CC, Liepins R (1987) Electrical properties of polymers. Hanser Publishers, New York.

Dielectrics or insulators or non-conductors *n.* A class of bodies supporting an electric strain. A charge on one part of a non-conductor is not communicated to any other part.

Dielectric strength (electric strength) *n.* A measure of the voltage required to puncture an insulating material, expressed in volts per mil of thickness (SI: V/mm). The voltage is the root-mean-square voltage difference between the two electrodes in contact with opposite surfaces of the specimen at which electrical breakdown occurs under prescribed test conditions. Ku CC, Liepins R (1987) Electrical properties of polymers. Hanser Publishers, New York. Weast RC (ed) (1971) Handbook of chemistry and physics, 52nd edn. The Chemical Rubber Co., Boca Raton, FL.

Die lines *n.* In extrusion and blow molding, longitudinal marks caused y damaged die surfaces or by extrudate build-up on dies.

Die-lip build-up *n.* In extruding film and sheet, the gradual accumulation of a bead of resin on the face of the die, parallel to the slit. Similar build-up can occur on dies for shapes.

D

Diels–Alder polymers *n.* Polymers created as a result of a Diels–Alder reaction, by which a 1,3-diene reacts via addition to another unsaturated molecule (the dienophile) to form a cyclic adduct. Morrison RT, Boyd RN (1992) Organic chemistry, 6th edn. Prentice-Hall, Englewood Cliffs, NJ. Smith MB, March J (2001) Advanced organic chemistry, 5th edn. John Wiley and Sons, New York.

Die manifold *See manifold.*

Diene monomers \ˈdī-ˌēn ˈmä-nə-mər\ *n.* Any of a family of monomers based on unsaturated hydrocarbons having two double bonds.

Diene polymer *n.* Any of a family of polymers based on unsaturated hydrocarbons having two double bonds (*diolefins*). When the double bonds are separated by only one single bond, as in 1,3-butadiene ($CH_2{=}CH{-}CH{=}CH_2$), the diene and its double bonds are said to be *conjugated.* In an unconjugated diene the double bonds are separated by at least two single bonds and act more independently. The family includes polymers of butadiene, isoprene, cyclopentadiene, and copolymers with ethylene and propylene. Morrison RT, Boyd RN (1992) Organic chemistry, 6th edn. Prentice-Hall, Englewood Cliffs, NJ. Smith MB, March J (2001) Advanced organic chemistry, 5th edn. John Wiley and Sons, New York.

Diene polymers *n.* The family of polymers and copolymers based on unsaturated hydrocarbons or diolefins with two double bonds which includes ethylene, propylene, isoprene, butadiene, and cyclopentadiene.

Diene value or number *n.* Amount of maleic anhydride (expressed as equivalents of iodine), which will react with 100 parts of oil under specific conditions. It is a measure of the conjugated double bonds in the oil. *Also known as the maleic anhydride value or maleic value.*

Die plate *n.* (1) In injection molds, a member that is attached to the fixed or to the moving head of the press; *mold plate,* (2) In extrusion, especially of pellets or shapes, the die plate is that part of the die assembly that is bolted to the outlet of the die body and contains the orifices that form the melt into continuous strands or a particular cross-sectional shape.

Die pressure *n.* In extrusion, the pressure of the melt entering the die.

Dies, coextrusion *n.* Dies of nested configuration. Individual polymers flow through separate mandrel passages and are combined in the primary land area. Adjacent polymer layers must exhibit similar flows and adhesion characteristics to form the coextrusion.

Dieseling *See burn mark.*

Dies, engraving *See dies and hot stamping.*

Dies, extrusion *n.* Extrusion converts thermoplastics into a continuous stream of melt, which is then shaped by an extrusion die into uniform cross-sectional shape. The main classes of dies are for film, sheet, coating, pipe, and profiles.

Dies, hot stamping *n.* Method of depositing a decorative layer on the surface of a polymer, whereby a hot metal or rubber diem which bears the required design or lettering is placed in contact with the medium to be applied and the surface to be decorated. The application of heat melts the release coating and activates the adhesive. After a short contact, the die pressure is released and the film carrier separated from the plastic surface, leaving a decorative coating transferred in the area of the die.

Die spider *n.* In extrusion, the legs or webs supporting the die core within the head of an in-line pipe, tubing, or blown-film die, In many pipe dies, the spider legs are cored to permit application of air or water for cooling the mandrel.

Dies, pultrusion *n.* Polymer thermosetting process whereby fibers impregnated with a mixed solution of polymer and the necessary additives are pulled through a heated die. The fibers entering the die are generally saturated with the solution but are solid when exiting from the machine. The process can produce solid, open-sided, hollow shapes that can be cut to length and packaged for shipment.

Die stamping *See intaglio.*

Die swell *See extrudate swelling.*

Die-swell ratio (extrudate-swelling ratio) *n.* In extrusion, particularly in extruding parisons for blow molding, the ratio of the outer parison diameter or parison wall thickness to, respectively, the outer diameter of the parison die or the die gap. The ratio is affected by the polymer type, its temperature, the die geometry, and the extrusion rate. Some writers have defined the die-swell ratio as the ratio of the *cross section* of the extrudate shortly after emergence to that of the die opening.

2,2-Diethoxyacetophenone (DEAP) *n.* A photo-initiator used for curing acrylate coatings, either in an inert atmosphere or in air with UV light. It absorbs impinging light.

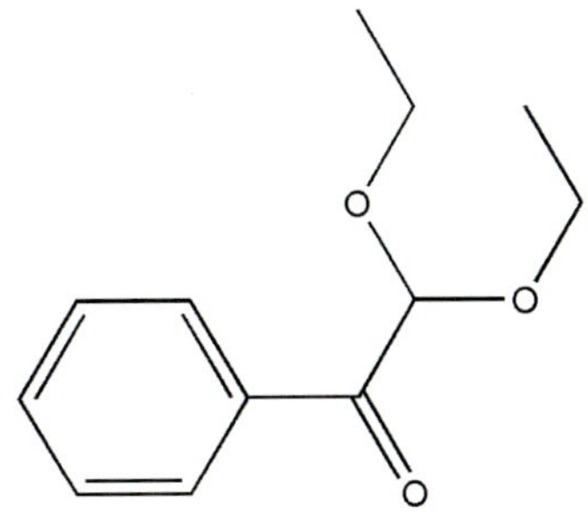

Diethoxyethyl adipate *n.* $(CH_2)_4(COOC_2H_4OC_2H_5)_2$. A plasticizer for cellulose resins.

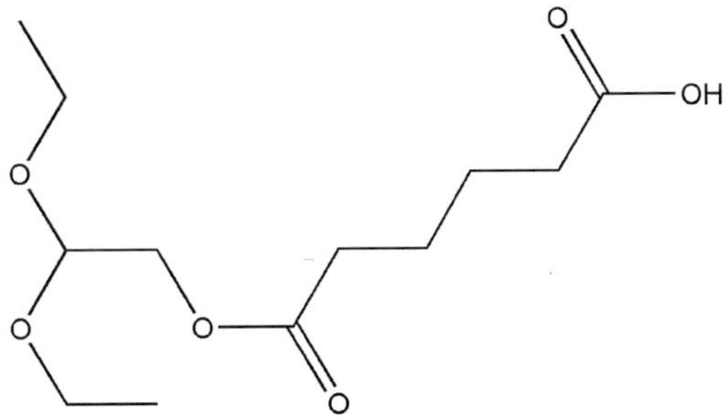

Diethoxyethyl phthalate *n.* $C_6H_4(COOC_2H_4OC_2H_5)_2$. A plasticizer for cellulosic and vinyl resins.

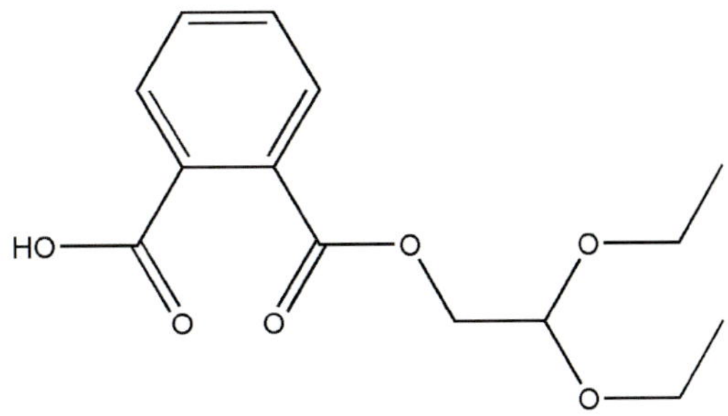

Diethyl adipate *n.* $(CH_2)_4(COOC_2H_5)_2$. A plasticizer for cellulosic resins.

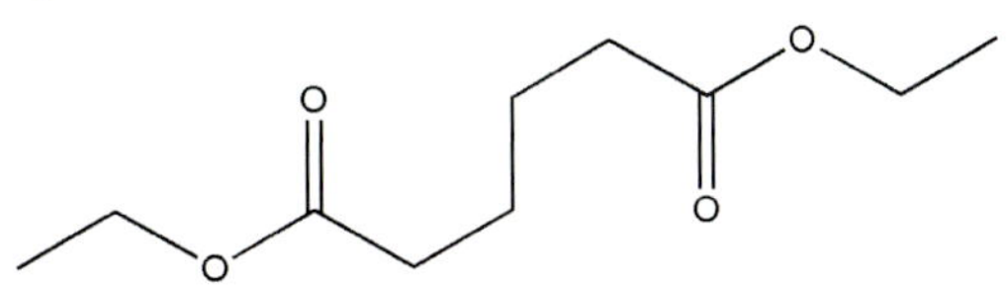

Diethylaluminum chloride *n.* A colorless liquid that bursts into flame instantly upon contact with air (pyrophoric) and reacts violently with water. It is used as a catalyst in olefin polymerization.

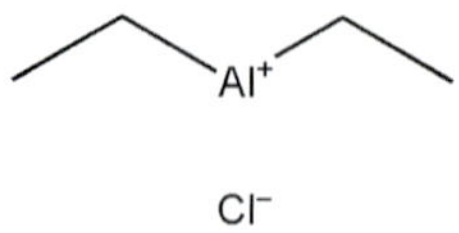

3-Diethylaminopropylamine *n.* An epoxy curing agent.

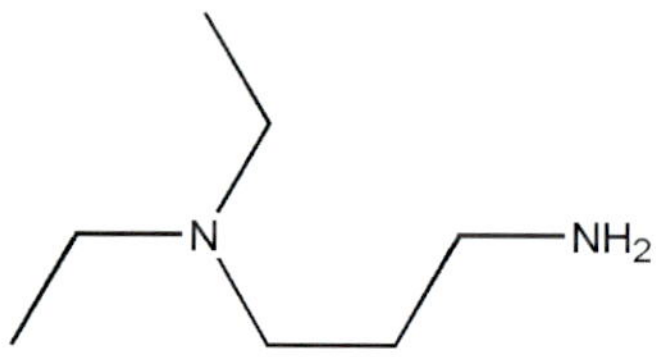

Di(2-ethylbutyl) azelate *n.* A plasticizer for PVC, its copolymers, and cellulose esters. It is very compatible and efficient in vinyls, finding its largest use in high-clarity film and sheeting.

Diethyl carbonate \ˈkär-bə-ˌnāt\ .tif) (ethyl carbonate) *n.* A solvent for cellulosic and many other resins.

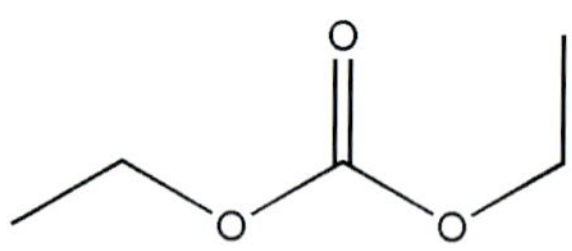

Diethylene glycol bis-(allyl carbonate) *n.* The monomer for some polycarbonate polymers.

Diethylene glycol dibenzoate *n.* plasticizer for cellulosic resins, polymethyl methacrylate, polystyrene, PVC, polyvinyl acetate, and other vinyls. It imparts good stain resistance.

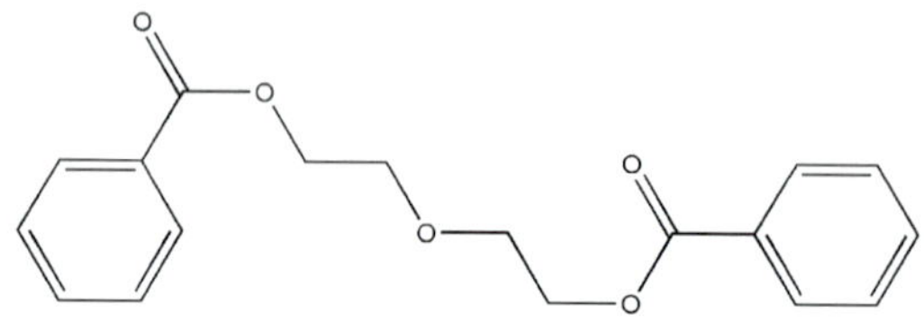

Diethylene glycol dipelargonate *n.* $[CH_3(CH_2)_7COOC_2H_4]_2O$. A simple diester of pelargonic acid used mainly as a secondary plasticizer for vinyl resins, but also as a plasticizer for cellulosics. Within the limits of its compatibility it provides economical low-temperature flexibility.

Diethylene glycol dipropionate *n.* A plasticizer for cellulosic plastics.

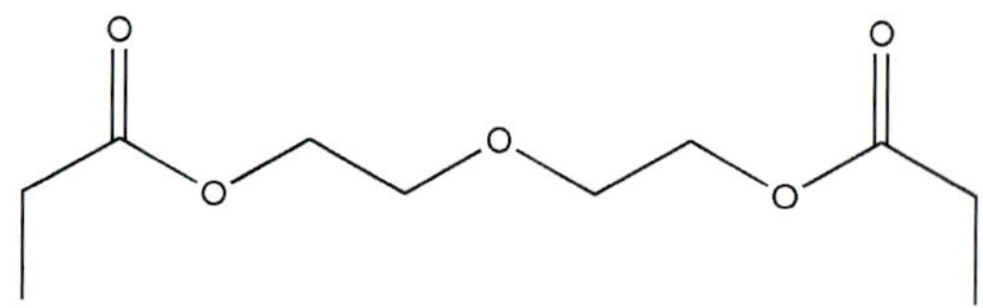

Diethylene glycol distearate *n.* $(C_{17}H_{35}COOC_2H_4)_2O$. A plasticizer for cellulose nitrate and ethyl cellulose.

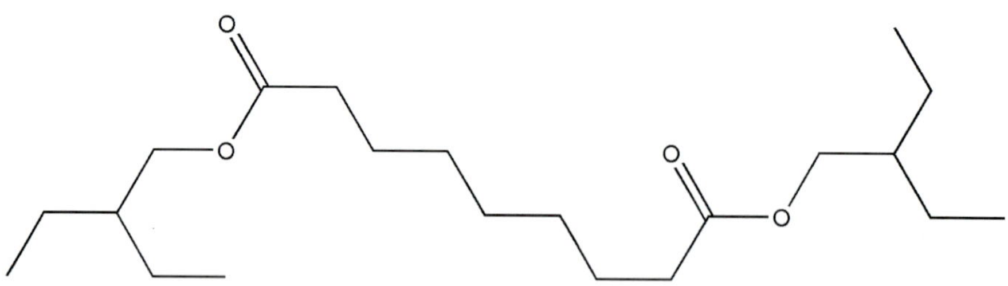

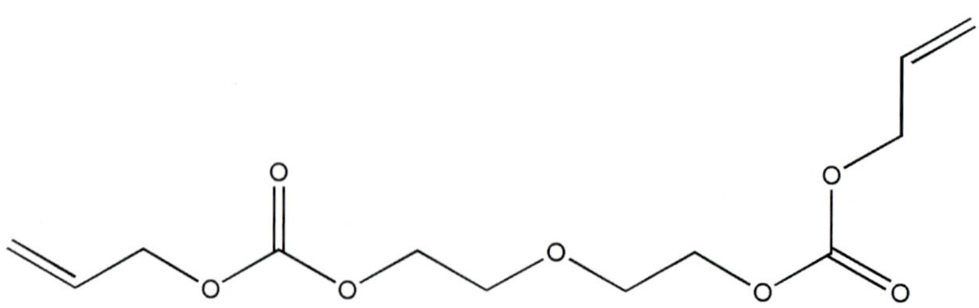

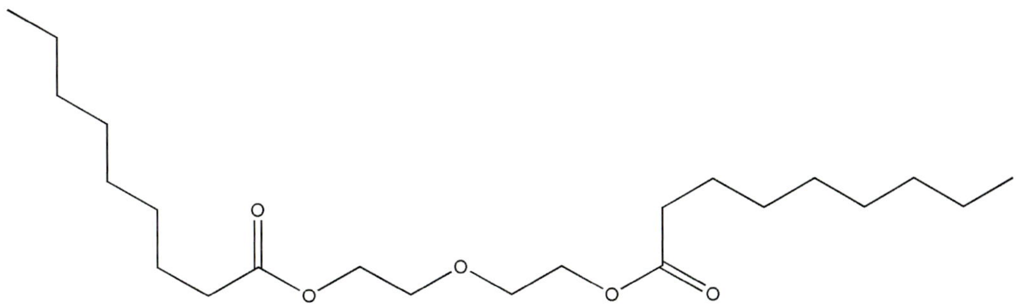

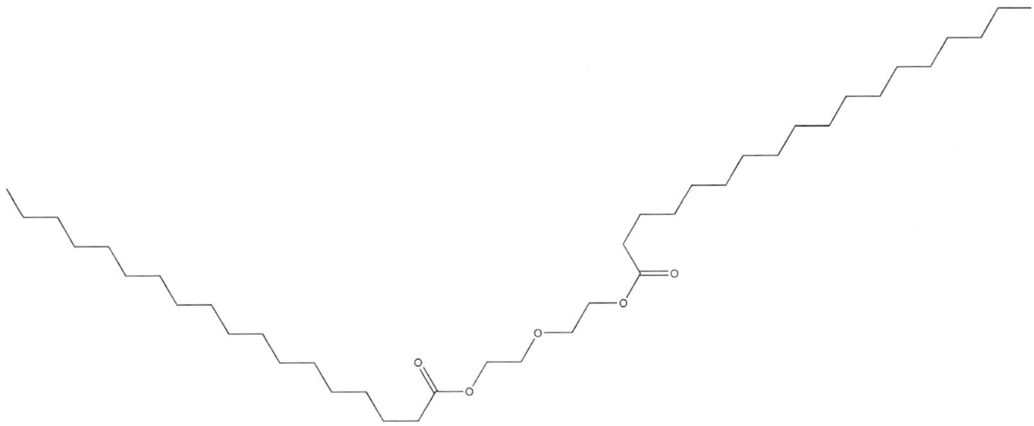

Diethylene glycol monoacetate *n.* HO $(CH_2)_2O(CH_2)_2OOCCH_3$. A solvent for cellulose nitrate and cellulose acetate.

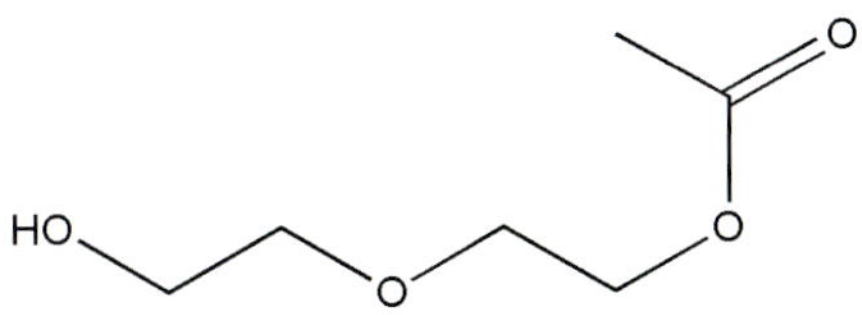

Diethylene glycol monobutyl ether *n.* $C_4H_9OCH_2CH_2OCH_2CH_2OH$. A solvent with a high boiling point, used in coatings when very slow drying rates are desired. It is also useful as a dispersant in vinyl organosols, and as an intermediate fore the production of plasticizers.
See butyl carbitol.

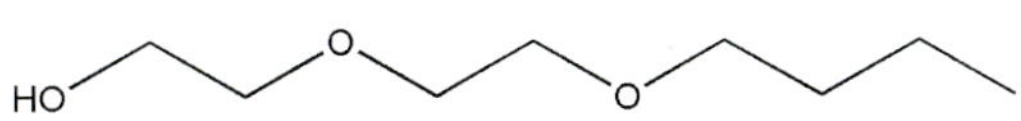

Diethylene glycol monobutyl ether acetate *See butyl carbitol acetate.*

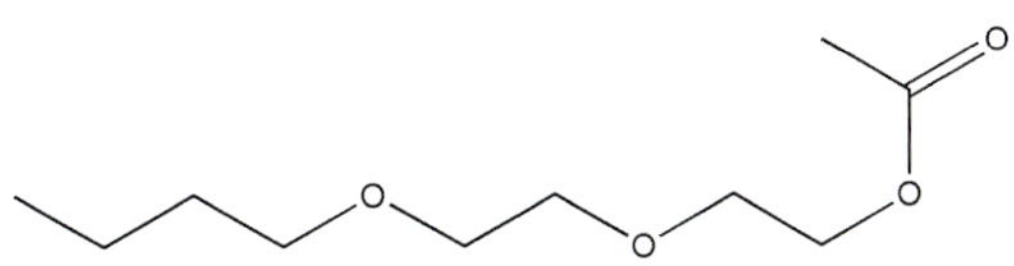

Diethylene glycol monolaurate *See diglycol laurate.*

Diethylene glycol monoricinoleate *See diglycol ricinoleate.*

Diethylenetriamine (DETA, 2,2′-diaminodiethylamine) *n.* A pungent liquid providing fast cures with epoxy resins, even at room temperature.

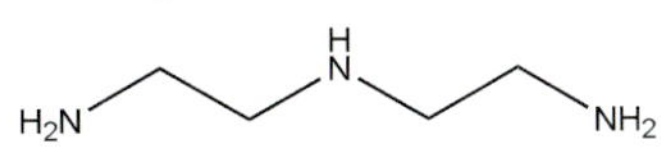

Diethyl ether \(ˌ)dī-ˈe-thəl-\ (ca. 1930) *n.* (1) A light volatile flammable liquid used as a sovlent and anesthetic. (2) Any of various organic compounds characterized by an oxygen atom attached to two carbon atoms.

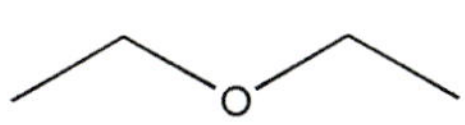

Di(2-ethylhexyl) adipate (dioctyl adipate, DOA) *n.* A primary plasticizer for vinyls, cellulose nitrate, polystyrene, and ethyl cellulose. In vinyls, DOA is often used in combination with phthalate and other plasticizers, imparting good resilience, low-temperature flexibility, and resistance to extraction by water. It is FDA-approved for use in vinyl food-packaging films.

Di(2-ethylhexyl) azelate *n.* (dioctyl azelate, DOZ) $(CH_2)_7[COOCH_2–CH(C_2H_5)C_4H_9]_2$. A plasticizer for vinyl chloride polymers and copolymers. It is one of the most compatible of the low-temperature, monomeric plasticizers, has low volatility, and imparts low extractability by water and soapy water. This plasticizer has been approved for food-contact use.

Di(2-ethylhexyl) 2-ethylhexyl phosphonate *n.* $C_8H_{17}PO(OC_8H_{17})_2$. A plasticizer and stabilizer.

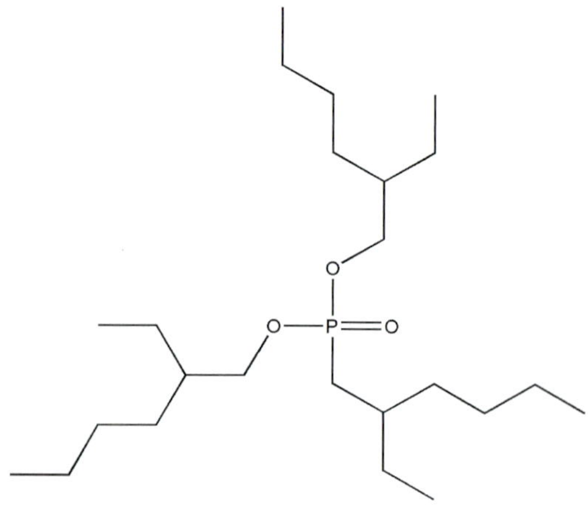

Di(2-ethylhexyl) hexahydrophthalate (dioctyl hexahydrophthalate) *n.* A light-colored liquid, a plasticizer for vinyls.

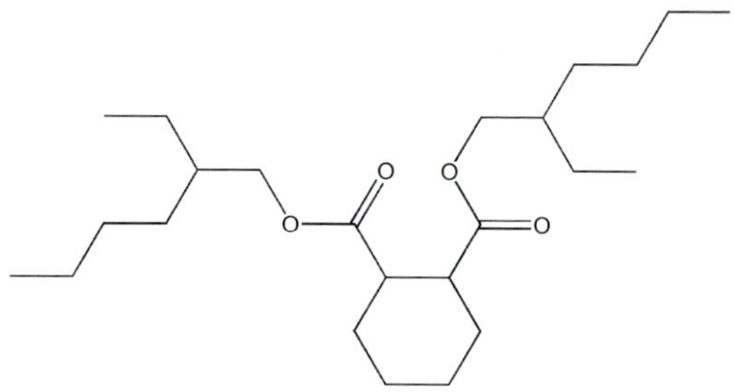

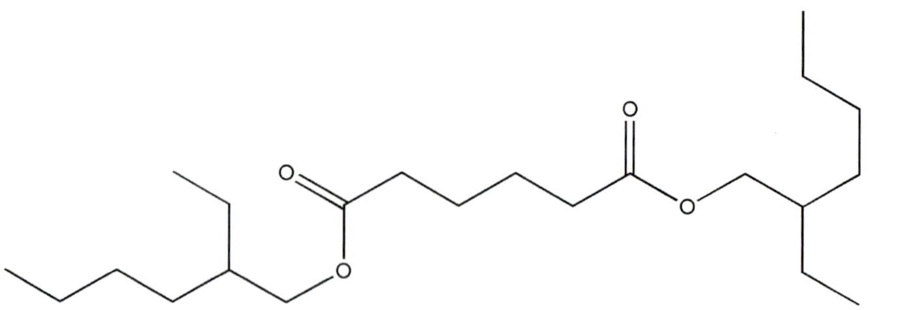

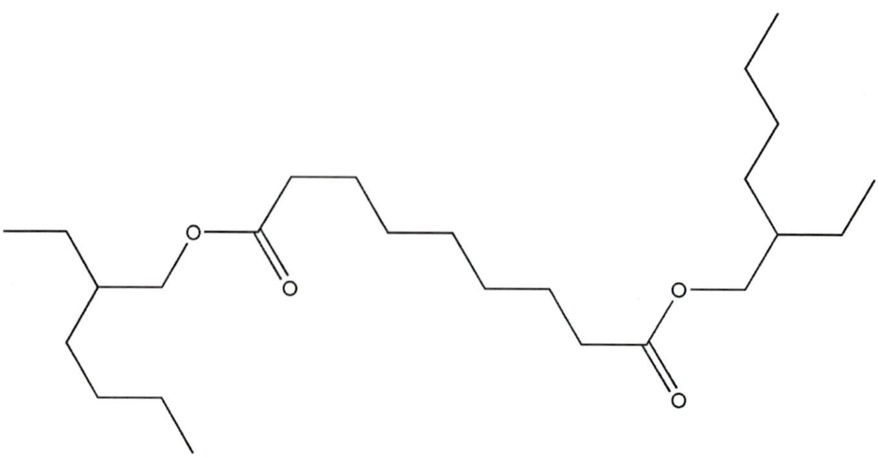

Di(2-ethylhexyl) isophthalate (dioctyl isophthalate, DIOP) *n.* A primary plasticizer for PVC, most notable for low volatility and its resistance to marring by nitrocellulose lacquers, in addition to good general-purpose properties. It is also compatible with polyvinyl butyral, vinyl chloride-acetate copolymers, cellulosic resins, polystyrene, and chlorinated rubber.

Di(2-ethylhexyl) phthalate (dioctyl phthalate, DOP) *n.* $C_6H_4[COO–CH_2CH(C_2H_5)C_4H_9]_2$. The most widely used plasticizer for PVC, also compatible with ethyl cellulose, cellulose nitrate, and polystyrene. It is generally recognized as imparting the best all-around good properties vinyls, and is often used as the standard against which other plasticizers are evaluated. DOP has been approved by the FDA for use in packaging films for greaseless foodstuffs of high water content.

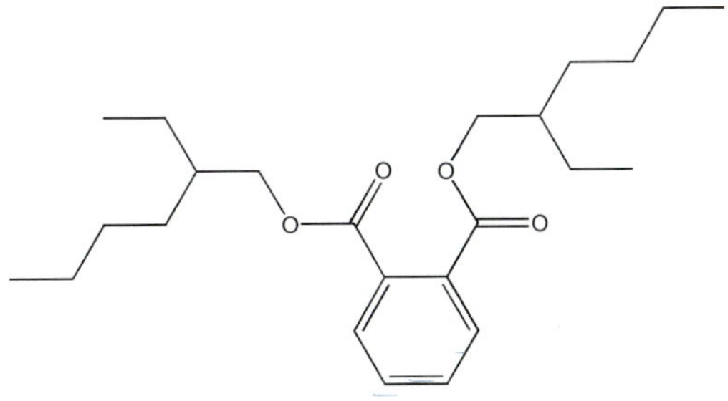

Di(2-ethylhexyl) sebacate (dioctyl sebacate) *n.* A plasticizer for vinyl chloride polymers and copolymers, cellulosic plastics, polystyrene, and polyethylene. It imparts good low-temperature properties.

Di(2-ethylhexyl) succinate (dioctyl succinate) *n.* A plasticizer.

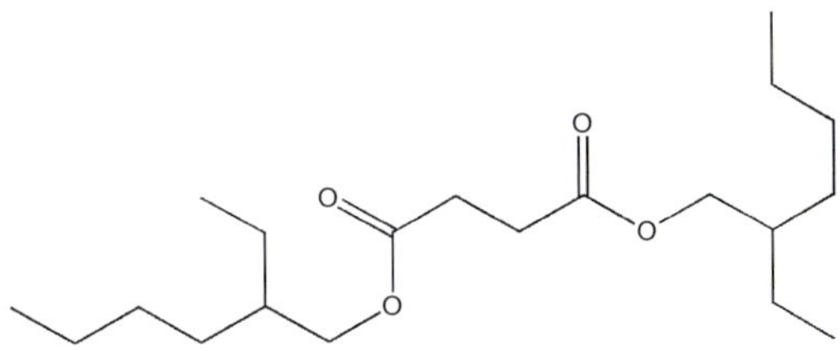

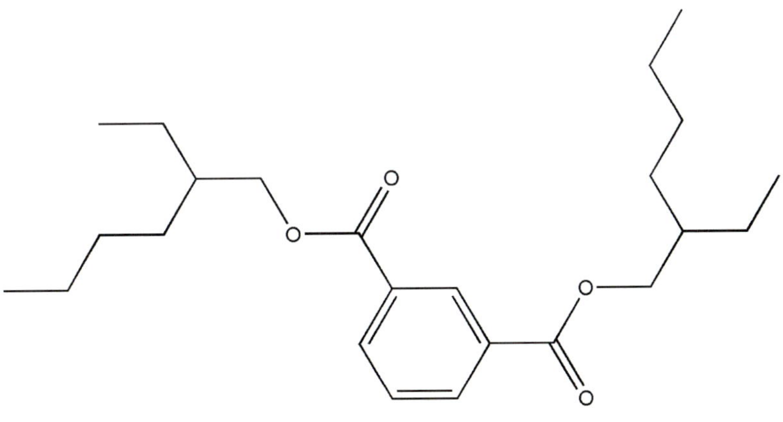

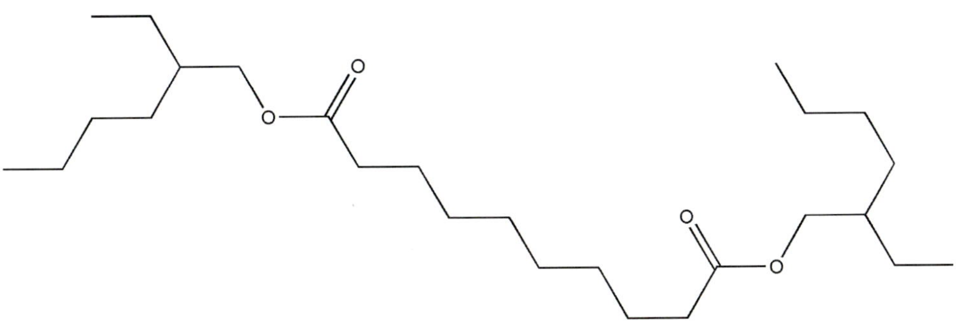

Di(2-ethylhexyl)-4-thioazelate *n.* $C_8H_{17}OOC(CH_2)_2CS(CH_2)_4COOC_8{-}H_{17}$. A plasticizer for cellulose nitrate, ethyl cellulose, polymethyl methacrylate, polystyrene, and vinyl resins.

Diethyl oxalate \ˈäk-sə-ˌlāt\. *See ethyl oxalate.*

Diethyl phthalate (ethyl phthalate) *n.* A plasticizer and solvent for nearly all thermoplastics and coumarone resins. It has been approved by FDA for use in food packaging.

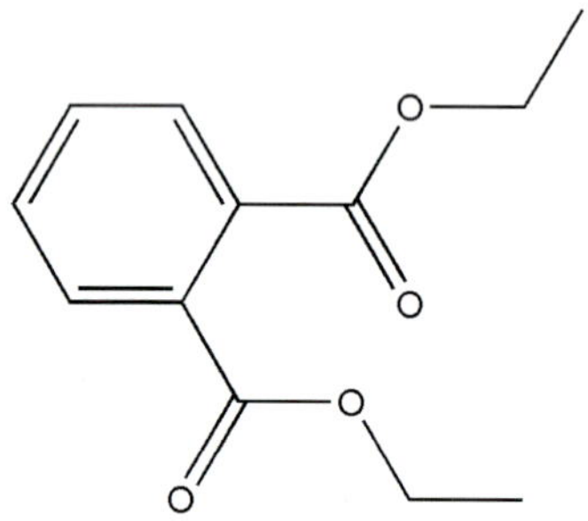

Diethyl sebacate *n.* A plasticizer with good low-temperature properties, compatible with PVC and many other thermoplastics.

Diethyl succinate *n.* Plasticizer for cellulosic resins.

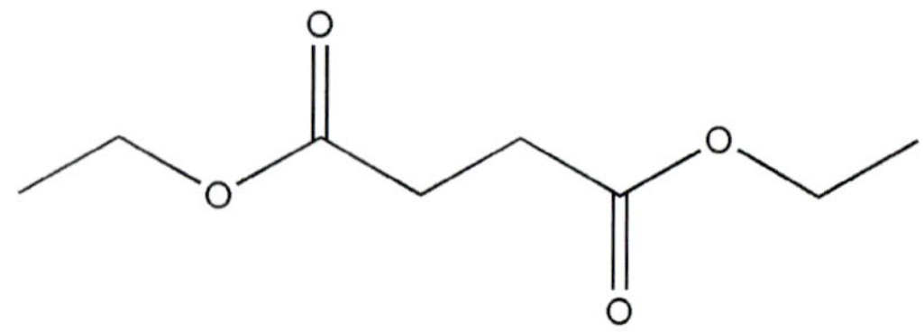

Diethyl tartrate *n.* A solvent and plasticize for cellulosic resins.

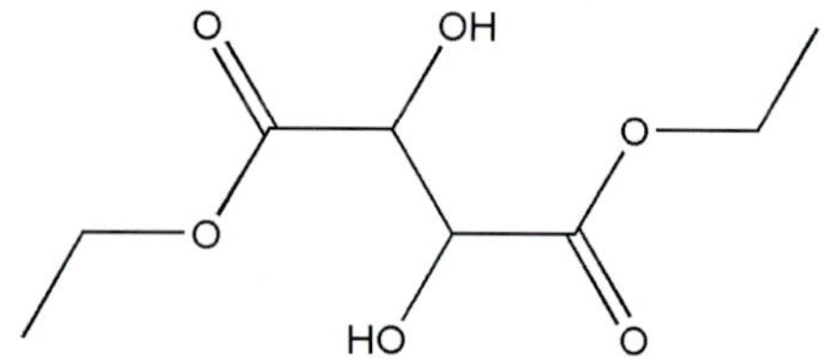

1,1-Diethylurea *n.* A white solid polymerizable with simple urea and formaldehyde to form permanently thermoplastic resins.

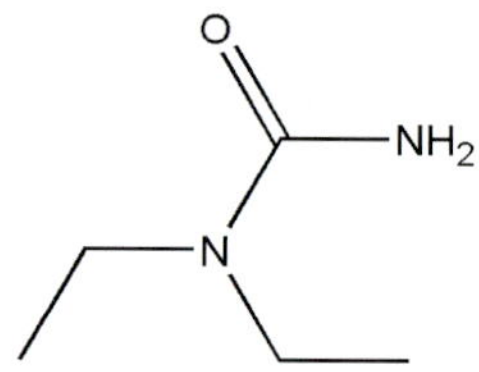

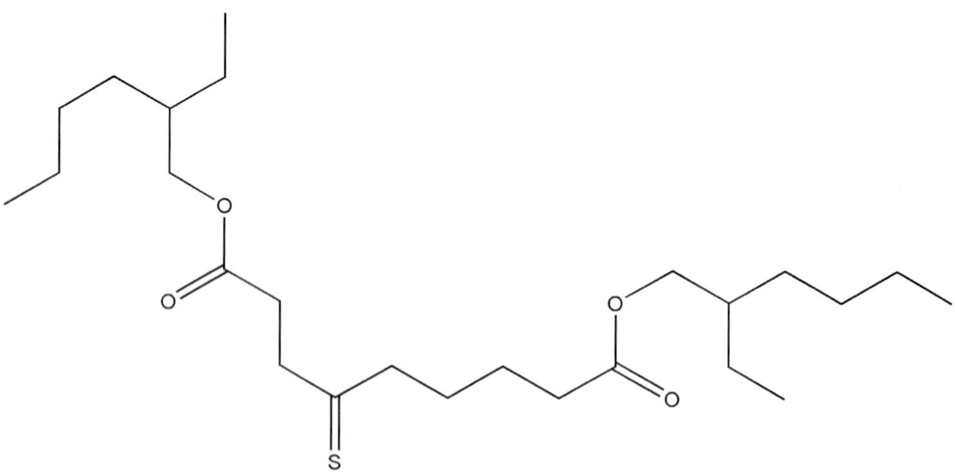

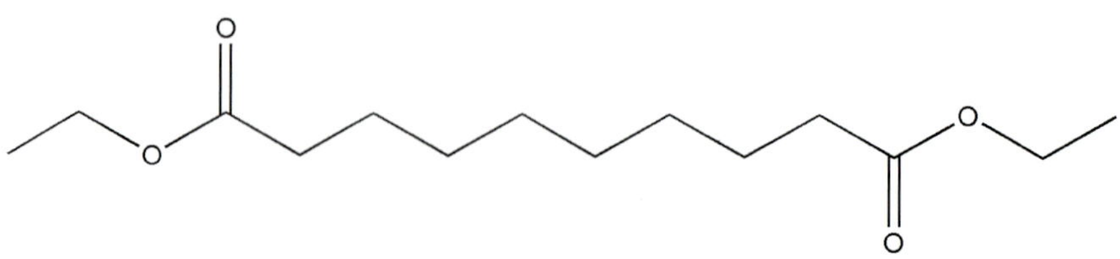

Diethyl zinc (1952) *n.* $C_4H_{10}Zn$. A volatile pyrophoric liquid organometallic compound used to catalyze polymerization reactions and to deacidify paper.

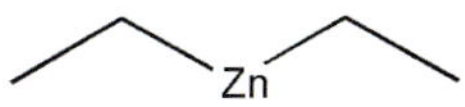

Differential refractometer *n.* An instrument used in connection with a chromatographic column in size-exclusion chromatography of polymers (in dilute solution). Two streams, one pure solvent, the other the eluting polymer solution, pass through the instrument, whose signal, proportional to the difference ion refractive indices of solution and solvent, is interpreted as polymer concentration. The instrument is also used to determine the rate of change of refractive index with concentration (d*n*/d*c*), an important parameter in the computation of weight-average molecular weights from light-scattering measurements in dilute polymer solutions.

Differential scanning calorimeter (DSC) *n.* An instrument that measures the rate of heat evolution of absorption of a specimen while it is undergoing a programmed temperature rise (Gooch, 1997). A recorder displays the data as a trace of increase in heat per increase in temperature (dq/dT) versus temperature. An example of a DSC thermogram of polypropylene is shown. The DSC has been used to study curing reactions and related properties of thermosetting resins, and heats of decomposition of resins (an example of a DSC instrument is the Perkin-Elmer Diamond DSC).

Differential thermal analysis (DTA) *n.* An analytical method in which the specimen material and an inert substance are heated concurrently in separate minipans and the difference in temperature between the two is recorded, along with the temperature of the inert substance (Gooch, 1997). DTA has been useful; in the study of phase transitions and curing and degradation reactions in polymers. An example of a DTA thermogram is shown. An example of a DTA instrument is the Perkin-Elmer Diamond TG/DTA (courtesy of Perkin-Elmer Inc, New York).

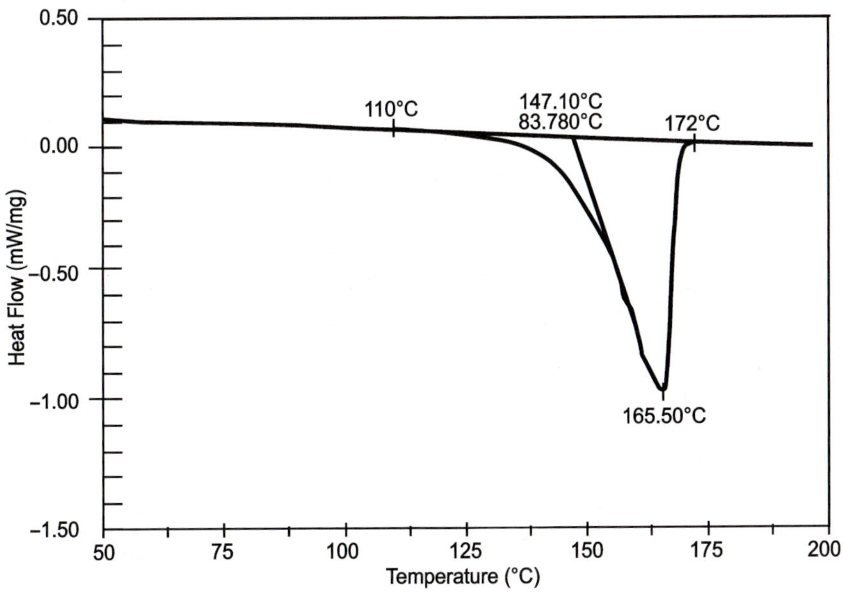

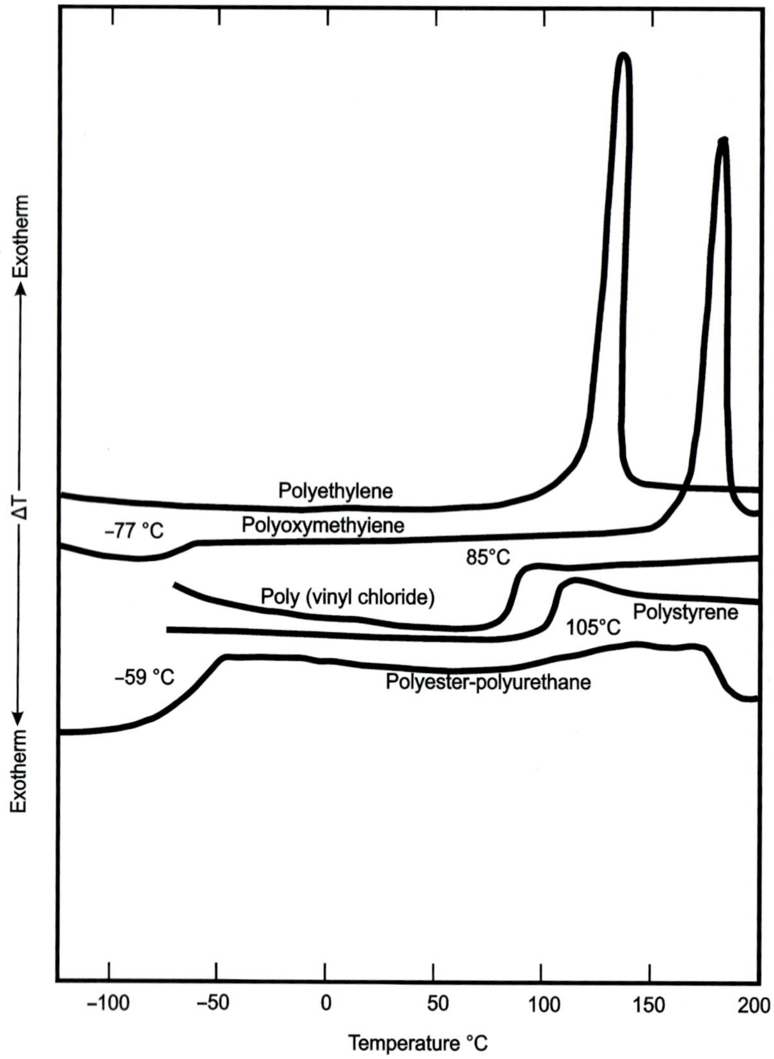

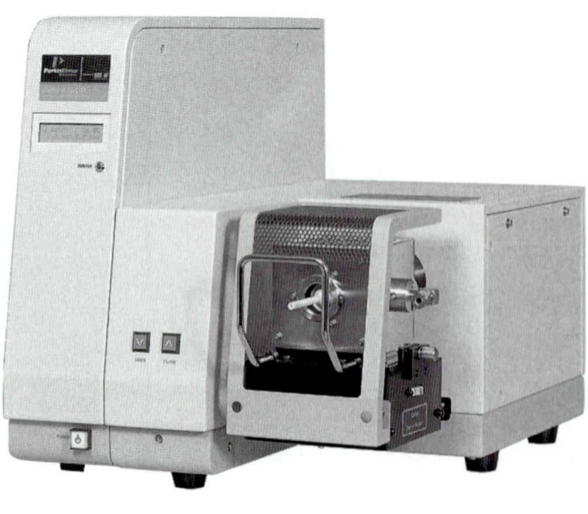

Differentiating solvent *n.* A solvent, which will discriminate among the strengths of acids or bases.

Diffraction \di-ˈfrak-shən\ [NL *diffraction-*, *diffractio*, fr. L *diffringere* to break apart, fr. *dis-* + *frangere* to break] (1671) *n.* Dispersion of light into a spectrum when it strikes an aperture or a slit of width in the same order of magnitude as the wavelength of the light. Diffraction depends on the wave nature of light and occurs in all wave motions. If monochromatic light strikes the edge of an opaque obstacle, bands of light and dark are observed near the edges of the beam. Diffraction is thus a special case of interference.

See dispersion, light, and diffraction grating.

Diffraction grating (1867) *n.* Device used to disperse a beam of electromagnetic radiation into its constituent wavelengths, i.e., for producing a spectrum. Grating may be prepared by ruling equidistant parallel lines onto a glass surface (transmission grating) or a metal surface (reflection grating). If s is the distance between the rulings, d the angle of diffraction, then the wavelength where the angle of incidence is 90° is (for the nth order spectrum),

$$\lambda = \frac{s \sin d}{n}.$$

If i is the angle of incidence, d the angle of diffraction, s the distance between the rulings, and n is the order of the spectrum, the wavelength is

$$\lambda = \frac{8}{n}(\sin i + \sin d).$$

Diffused light \di-ˈfyüz-\ [ME *diffused*, pp., fr. L *diffusus*, pp] *n.* Non-directional light.

Diffuse reflectance *n.* The ratio of the flux reradiated by diffuse reflection.

See reflectance, diffuse, and specular reflectance excluded.

Diffuse transmittance *See transmittance and diffuse.*

Diffusion *n.* The movement of a material in the body of a plastic. If the concentration (mass of solid per unit volume of solution) at on surface of a layer of liquid is d_1 and at the other surface d_2, the thickness of the layer h and the area under consideration A, then the mass of the substance which diffuses through the cross-section A in time t is,

$$m = \Delta A \frac{(d_2 - d_1)t}{h},$$

where Δ is the coefficient of diffusion.

Diffusion, autoacceleration *n.* In some vinyl polymerizations, as the reaction approaches completion and the viscosity of the reaction medium rises, there is a rising rate of increase of molecular weight of the chains that have not yet been terminated. This autoacceleration phenomena greatly affects the diffusivity of the polymer solution.

Diffusion bonding *n.* An experimental method of joining materials in which extremely flat and finely finished surfaces are clamped together under high vacuum for a period of hours or days, sometimes at a high temperature, allowing the atoms of each member to diffuse into the other. As of early 1992, the method had been used only with inorganic crystalline materials.

Diffusion constant *n* Symbol D. The constant of proportionality in Fick's laws of diffusion between the rate of diffusion and the concentration gradient (Fick's first law) and the variation of concentration with time and concentration gradient (Fick's second law).

Diffusion couple *n.* An assembly of two materials in such intimate contact that each diffuses into the other.

Diffusivity *n.* (1) (diffusion coefficient). The constant of proportionality in Fick's first law of diffusion, which states that the mass of molar rate of transport (flux) of one molecular species into another is equal to the diffusivity times the gradient of concentration. Several related units are in use, e.g., cm^2/s, ft^2/h. The SI unit is m^2/s, corresponding to flux in $mol/(s\,m^2)$ and gradient in $(mol/m^3)/m$. (2) Thermal diffusivity. Diffusivity or coefficient of diffusion is also given by Δ in the equation

$$\frac{dQ}{dt} = -\Delta\left(\frac{dc}{dx}\right)dy\,dz,$$

where dQ is the amount passing through an area dydz in the direction of x in a time dt where dc/dz is the rate of increase of volume concentration in the direction of x. Dimensions $[L^2T^{-1}]$.

See also Fick's law.

Diffusivity of heat *n.* It is given by Δ in the equation

$$\frac{dH}{dt} = -\Delta sd\frac{dT}{dx}dy\,dz,$$

where dH is the quantity of heat passing through the area dydz in the direction of x in a time dt. The rate of variation of temperature along x is given by dT/dx, s is specific heat and d, density. Dimensions $[L^2T^{-1}]$.

Digester \-ˈjes-tər\ (1614) *n.* Jacketed reaction vessel used to decompose, soften, or cook substances at high pressure and temperature.

Diglycidyl ether of bisphenol A (DGEBA) *n.* The main constituent of most commercial epoxy resins prior to curing, DGEBA is formed by the reaction of excess epichlorohydrin with bisphenol A in the presence of aqueous sodium hydroxide.

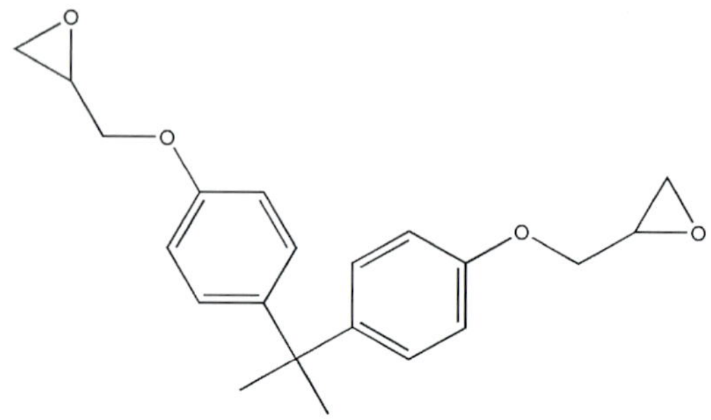

Diglycol laurate \-ˈglī-ˌkól ˈlór-ət\ (diethylene glycol monolaurate). A plasticizer for ethyl cellulose, cellulose nitrate, polyvinyl butyral, and vinyl chloride-acetate copolymers.

Diglycol ricinoleate *n.* (diethylene glycol monoricinoleate) $CH_3(CH_2)_5$–$CH(OH)CH_2CH{=}CH(CH_2)_7COOC_2H_4OC_2H_4OH$.

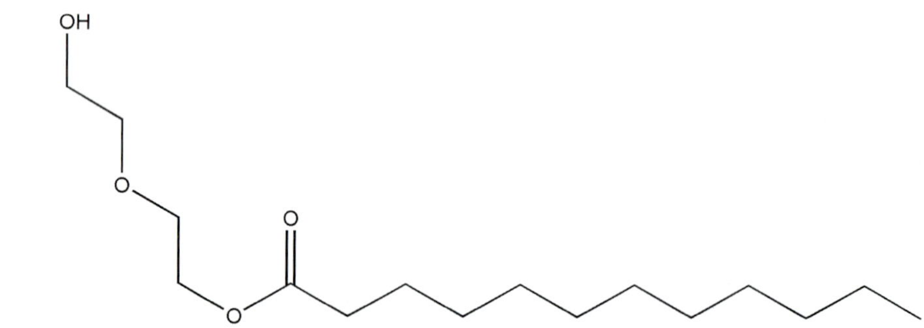

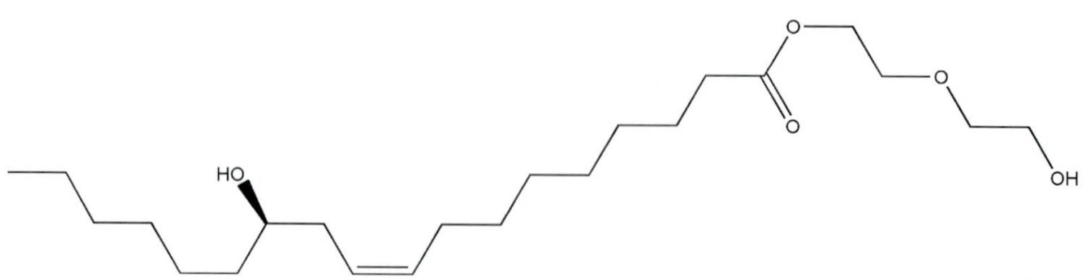

A plasticizer for ethyl cellulose and cellulose nitrate.

Dihalide *n.* A compound containing two halogen atoms (of valence = ~1) per molecule.

Dihexyl phthalate (DHXP) *See di(2-ethyl butyl) phthalate.*

Dihexyl sebacate *n.* A plasticizer for vinyls.

Dihydrodiethyl phthalate *n.* A plasticizer for PVC and cellulose nitrate, with partial compatibility with other thermoplastics.

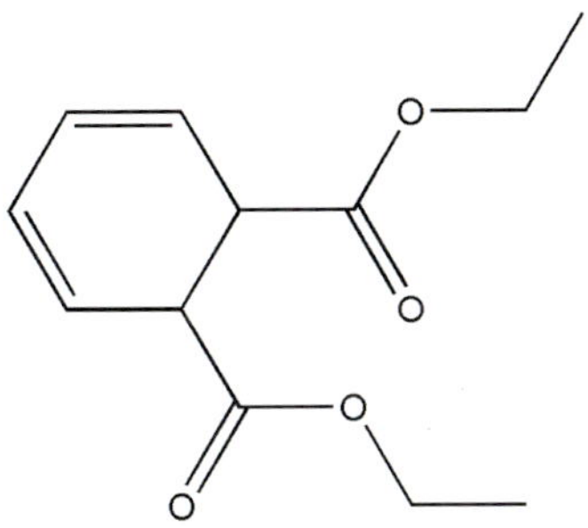

Dihydroxy anthraquinone lake \-ˌan(t)-thrə-kwi-ˈnōn-\ *n.*
See alizarin red.

Diiobutyl azelate *n.* $(CH_3)_2CHCH_2OOC(CH_2)_7$. A plasticizer for cellulosics, polymethyl methacrylate, PVC, and polyvinyl acetate.

Diisobutyl adipate (DIBA) *n.* A plasticizer for most synthetic resins, including cellulosics, PVC, and other vinyls, and FDA-approved for use in food-packaging films. In vinyls, it is a very active solvent and lowers processing temperatures to levels that permit elimination or lower percentages of stabilizers.

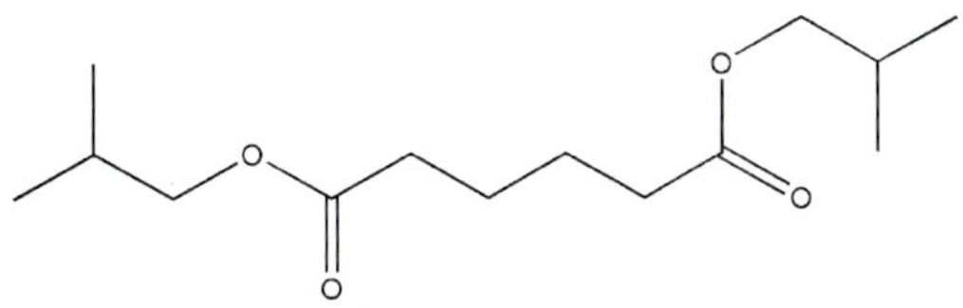

Diisobutyl aluminum chloride *n.* $[(CH_3)_2CHCH_2]AlCl$. A catalyst for polymerizing olefins.

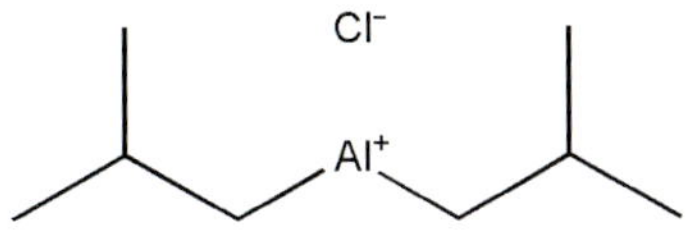

Diisobutyl ketone \-ˈkē-ˌtōn\ (2,6-dimethyl-4-heptanone) *n.* A high-boiling ketone with moderate solvent power for cellulose nitrate and vinyl copolymers. Having limited solvency for PVC, it is used as a viscosity modifier in organosols.

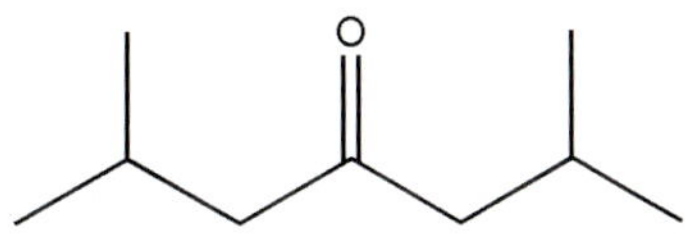

Diisobutyl phthalate (DIBP) *n.* $C_6H_4[COOCH_2CH(CH_3)_2]$. A plasticizer for vinyls, cellulosics, and polystyrene.

Diisocyanate *n.* Any compound containing two isocyanate (–NCO) groups, used in the production of polyurethanes. Many methods have been reported for synthesizing diisocyanates, but the one most widely used is reacting phosgene with an amine in a solvent. Toluene diisocyanate (TDI), the most commonly used, is an 80–20

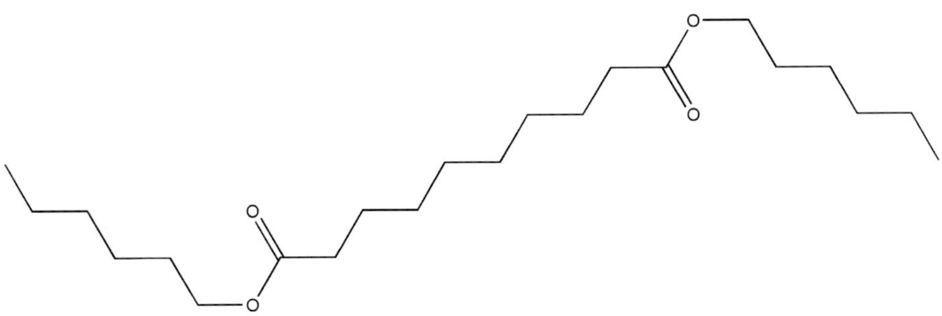

mixture of 2,4- and 2,6-toluene diisocyanate isomers. Also used are diphenylmethane-4,4′-diisocyanate (MDI), a modified toluene diisocyanate, and polymethylene polyphenyl isocyanate (PAPI). The diisocyanates are key ingredients in the production of urethane foams, fibers, coatings, and solid elastomers.

See also isocyanate and polyurethane.

Diisodecyl adipate (DIDA) *n.* $(-C_2H_4COOC_{10}H_{21})_2$. A plasticizer for PVC in lower concentrations, e.g. up to 30 phr, at which it imparts low-temperature flexibility and resistance to lacquer marring. It is often used in combination with phthalate and phosphate plasticizers. DIDA is completely compatible with vinyl chloride-acetate copolymers, cellulose acetate-butyrate with high butyral content, cellulose nitrate, ethyl cellulose, and chlorinated rubber. In polystyrene, it may be used up to 25 phr.

Diisodecyl-4,5-epoxy-tetrahydrophthalate *n.* A plasticizer for PVC that also acts as a stabilizer and fungistat. It is compatible with cellulose nitrate, ethyl cellulose, polymethyl methacrylate, polystyrene, and other vinyls polymers and copolymers.

Diisodecyl glutarate *n.* $(CH_2)_3(COOC_{10}H_{21})_2$. A plasticizer for PVC having low-temperature properties equal to those of dioctyl adipate but with lower volatility and greater resistance to soapy water.

Diisodecyl phthalate (DIDP) *n.* $C_6H_4(COOC_{10}H_{21})_2$. A general-purpose plasticizer for vinyl resins, imparting good water resistance and suitable for processing at high temperatures due to its low volatility. It is also compatible with most synthetic resins, e.g., cellulose nitrate, ethyl cellulose, and polystyrene.

Diisononyl phthalate (DINP) *n.* $C_6H_4(COOC_9H_{19})_2$. A plasticizer for PVC, cellulosics, and polystyrene. It has lower volatility than dioctyl phthalate with equivalent low-temperature performance and poorer efficiency.

Diisooctyl adipate (DIOA) *n.* $(-C_2H_4COOC_8H_{17})_2$. A primary plasticizer for vinyls, cellulose nitrate, polystyrene, polymethyl methacrylate, and ethyl cellulose. Its performance is similar to that of dioctyl phthalate.

Diisooctyl azelate (DIOZ) *n.* $C_7H_{14}(COOC_8H_{17})_2$. A plasticizer for cellulosics resins and polymers and copolymers of vinyl chloride. In vinyls it imparts good low-temperature properties and other characteristics similar to those obtained with dioctyl azelate.

Diisooctyl fumarate *n.* $(=CHCOOC_8H_{17})_2$. A plasticizer for PVC.

Diisooctyl isophthalate *n.* $C_6H_4(COOC_8H_{17})_2$. A plasticizer for PVC, also compatible with cellulose nitrate, ethyl cellulose, polystyrene, and other vinyl resins.

Diisooctyl monoisodecyl trimellitate *n.* A plasticizer for cellulosics and vinyl resins.

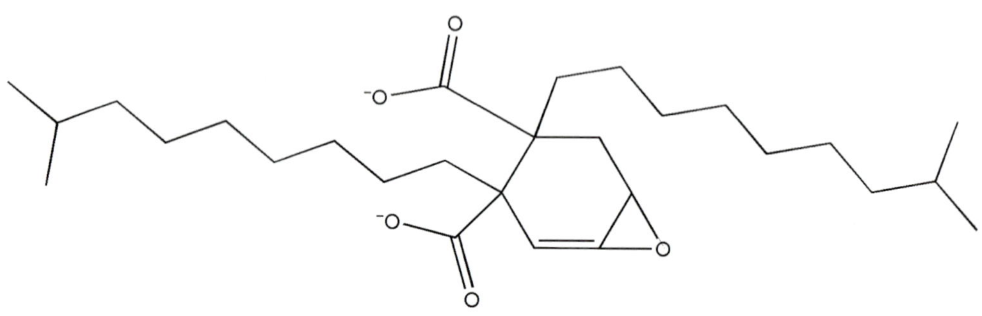

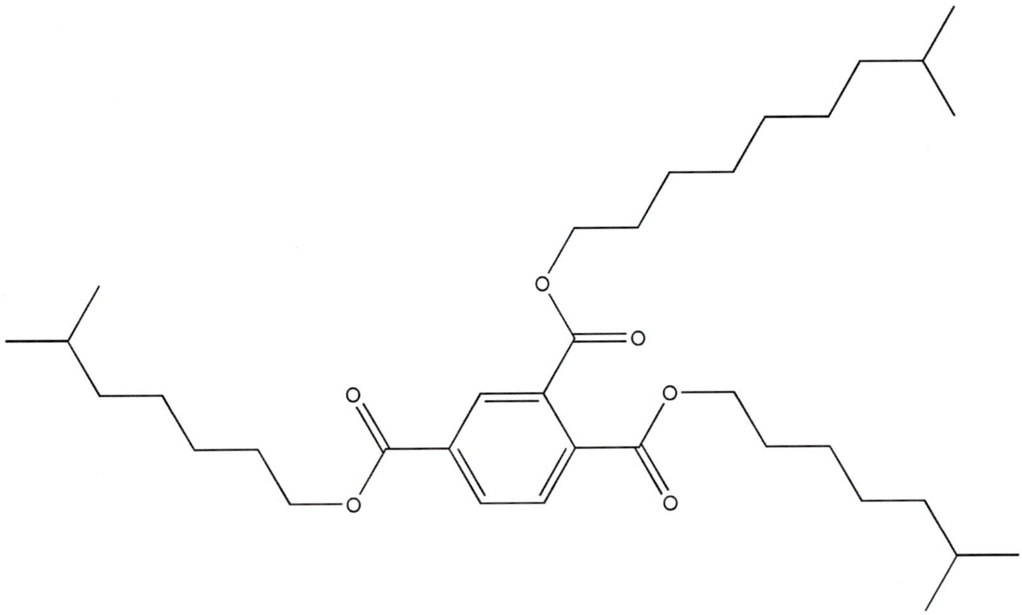

Diisooctyl phthalate (DIOP) *n.* $C_6H_4(COOC_8H_{17})_2$. A primary plasticizer for PVC, ethyl cellulose, cellulose nitrate, and polystyrene. In vinyls, its performance is similar to that of dioctyl phthalate except that it is slightly less volatile than DOP and produces better viscosity characteristics in plastisols. DIOP is FDA-approved for food-packaging materials and medical applications involving contact with water, but not with fats.

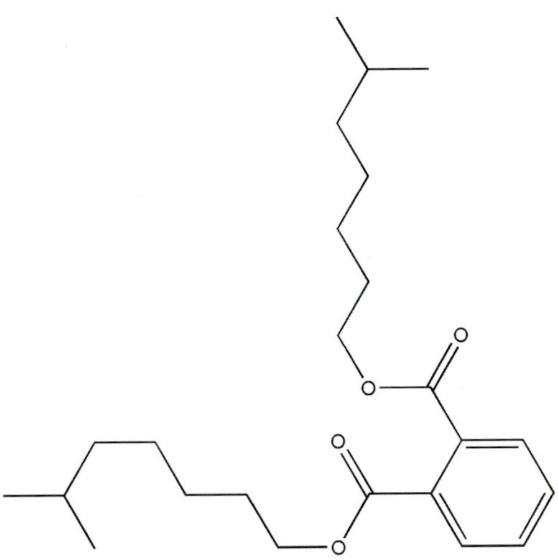

Diisooctyl Sebacate (DIOS) *n.* A plasticizer for vinyl and other resins.

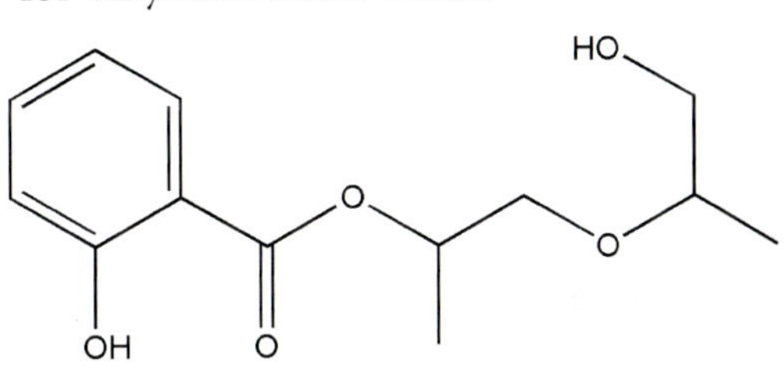

Diisopropylene glycol salicylate \sə-ˈli-sə-ˌlāt\ *n.* $HOCH_2CH(CH_3)CH(CH_3)CH_2O{-}OCC_6H_4OH$. An ultraviolet absorber in plastics.

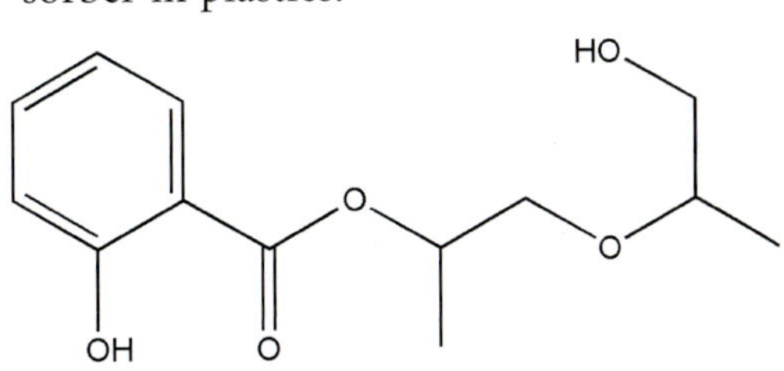

Dilatancy \dī-ˈlā-tᵊn(t)-sē\ (1885) *n.* Flow characterized by reversible, instantaneous increase in viscosity with increasing shear rate. The opposite of *pseudoplasticity.* Dilatancy in a pigment-vehicle system or pigment dispersions commonly results from high pigment loadings. The curve of the plot of shear stress versus shear rate is non-linear with shear stress increasing faster than the shear rate.

Also known as shear thickening.

Dilatometer \ˌdi-lə-ˈtä-mə-tər\ [ISV] (ca. 1883) *n.* An instrument for measuring the volume changes of a liquid or solid as the sample's temperature changes. A simple liquid dilatometer consists of a small bulb holding a known mass of sample and sealed to a graduated capillary of known diameter. As the sample is heated, its expansion is indicated by its rise up the capillary. From these measurements the volume coefficient of thermal expansion can be estimated (the inverse principle is used in liquid-in-glass thermometers).

Dilauryl ether (didodecyl ether) *n.* A lubricant for plastics processing.

Dilinoleic acid *n.* $C_{34}H_{62}(COOH)_2$. An unsaturated, dibasic acid used as a modifier in alkyd, nylon, and polyester resins.

Diluent \ˈdil-yə-wənt\ [L *diluent-*, *diluens*, pp of *diluere*] (ca. 1721) *n.* A substance that dilutes another substance. In an organosol, a diluent is a volatile liquid such as naphtha that has little or no solvating effect on the resin, but serves to lower the viscosity of the mix, and is evaporated during processing. In the paint industry, a diluent is any substance capable of thinning paints, varnishes, etc. The term is sometimes also used for a liquid added to a thermosetting resin to reduce its viscosity; and for an inert powdered substance added to an elastomers or resin merely to increase the volume (compare: filler).

See also extruder and thinner.

Dilute-solution viscosity (solution viscosity) *n.* (1) A catchall term that can mean any of the interrelated and quantitatively defined viscosity ratios of dilute polymer solutions or their absolute viscosities. (2) The kinematic viscosity of a solution as measured by timing the rate of efflux of a known volume of solution, by gravity flow, through a calibrated glass capillary that is immersed in a temperature-controlled bath. Two common types of viscometer are the Ostwald–Fenske and Ubbelohde. From the viscosities of the solution η and the solvent η_o, and the solution concentration c, five frequently mentioned "viscosities" (viscosity ratios, actually) can be derived, as follows:

Relative viscosity: $\eta_r = \eta/\eta_o$;
Specific viscosity: $\eta_{sp} = \eta_r - 1$;
Reduced viscosity: $\eta_{red} = \eta_{sp}/c$;
Inherent viscosity: $\eta_{inh} = (1n\eta_r)/c$;
Intrinsic viscosity: η = the limit as $c \to 0$ of η_{sp}/c, = the limit as $c \to 0$ of η_{inh}.

The intrinsic viscosity, because it is extrapolated to zero concentration from a series of measurements made at different concentrations, is independent of concentration. However, different solvents yield different intrinsic viscosities with the same polymer, so the solvent used must be identified. Some ASTM tests for viscosities of dilute polymer and plastics solutions are D 789, D 1243, D 1601, D 2857, and D 4603, all in Section 08. D 445, and D 446, describing the proper use of the viscometers mentioned above, are in Section 05.

Dilution ratio *n.* (1) As used in the surface-coatings industry, dilution ratio is the

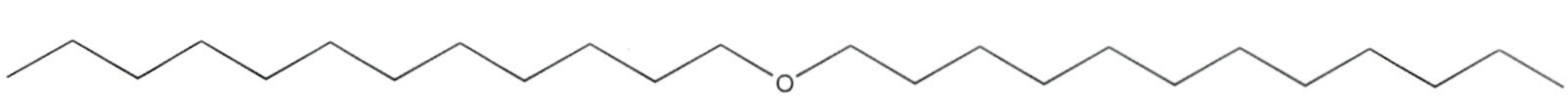

volume ratio of diluent to solvent in a blend of theses two constituents that just fails to completely dissolve 8.00 g of nitrocellulose in 100 ml of the blend. The procedure is described in ASTM D 1720, Section 06. It is used to determine the most economical, yet adequate amount of high-cost active solvent required in a nitrocellulose-lacquer system. (2) In most other contexts, dilution ration is the quotient of the concentration of the undiluted solute divided by that of the diluted solution, both concentrations in the same units.

Dimension \də-ˈmen(t)-shən *also* dī-\ [ME, fr. MF, fr. L *dimension-*, dimensio, fr. *dimetiri* to measure out, fr. *dis-* + *metiri* to measure] (14c) *n.* A geometric element in a design, such as length or angle, or the magnitude of such a quantity.

Dimensional formulae \ -ˈfór-myə-lə\ *n.* If mass, length, and time are considered fundamental quantities, the relation of other physical quantities and their units to these three may be expressed by a formula involving the symbols L, M, and T, respectively, with appropriate exponents. For example; the dimensional formula for volume would be expressed $[L^3]$; velocity $[LT^{-1}]$; force $[M\,LT^{-2}]$. Other fundamental quantities used in dimensional formulae may be indicated as follows: θ, temperature; ε, the dielectric constant of a vacuum; μ, the magnetic permeability of a vacuum.

Dimensional restorability *n.* The ability of a fabric to be returned to its original dimensions after laundering or dry cleaning, expressed in percent. For example, 2% dimensional restorability means that although a fabric may shrink more than this in washing, it can be restored to within 2% of its original dimensions by ordinary home pressing methods.

Dimensional stability *n.* The ability of a substance (wood, plastic, etc.) to retain its shape when subjected to varying degrees of temperature, moisture, pressure, or other stress.

Dimer \ˈdī-mər\ [ISV] (ca. 1926) *n.* (1) A molecule formed by union of two identical simpler molecules. For example, C_4H_8 is a dimmer of C_2H_4, as N_2O_4 is of NO_2. (2) A substance composed of dimmers.

Dimer acid *n.* A coined, generic term for high-molecular-weight, dibasic acids that combine and polymerize with alcohols and polyols to make plasticizers, etc. A trimer acid is analogous having three acid groups.

Dimer acids *n.* Liquid polycarboxylic acids produced by polymerization of unsaturated fatty acids. Pure dimmer acid is a C_{36} dicarboxylic acid. Commercial dimmer acids are composed of predominantly C_{36} dicarboxylic acid with minor amounts of C_{18} monocarboxylic acid and varying proportions of C_{54} tricarboxylic acid.

Dimerization *n.* State of polymerization when two similar molecules have united. The product is described as a dimmer.

Dimethoxyethyl adipate *n.* A plasticizer for cellulose-ester polymers.

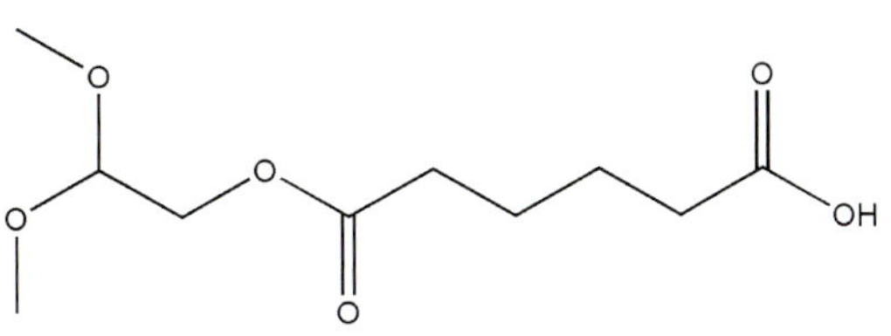

Di(2-methoxyethyl) phthalate *n.* A plasticizer, especially for cellulose acetate, but compatible, too, with other cellulosics, polystyrene, and vinyls.

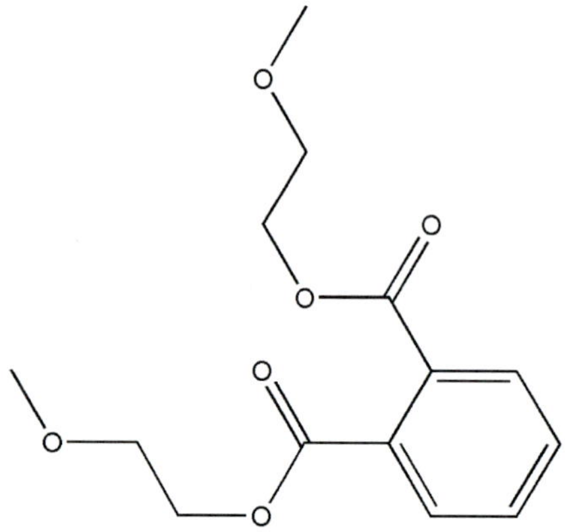

Dimethyl acetamide (DMAC) *n.* $CH_3CON(CH_3)_2$. A colorless liquid, a solvent for resins and plastics, a catalyst, and an intermediate.

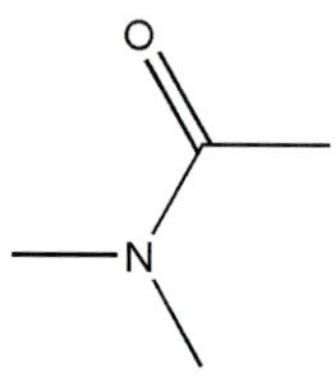

2-Dimethylamino ethanol (DMAE) *n.* A colorless liquid derived from ethylene oxide and dimethylamine, a catalyst for urethane foams. It has little odor and toxicity, and resists staining.

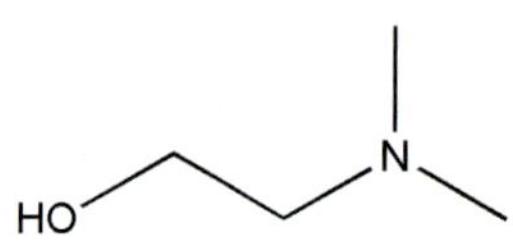

3-Dimethylaminopropylamine *n.* A colorless liquid used as a curing agent for epoxy resins.

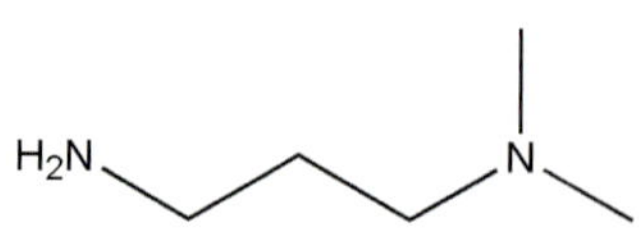

Dimethylaniline (DMA) *n.* The term usually means the tertiary amine, *N,N*-dimethylaniline, through ring-substituted isomers are known. This amine is useful as an accelerator in polyester molding and sprayup.

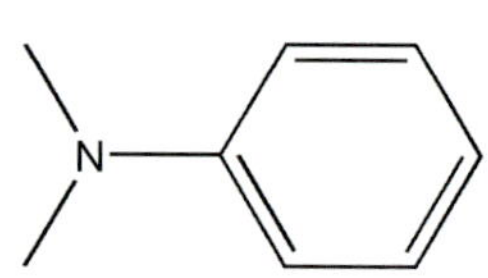

Di(methylcyclohexyl) adipate *n.* A plasticizer compatible with most thermoplastics.

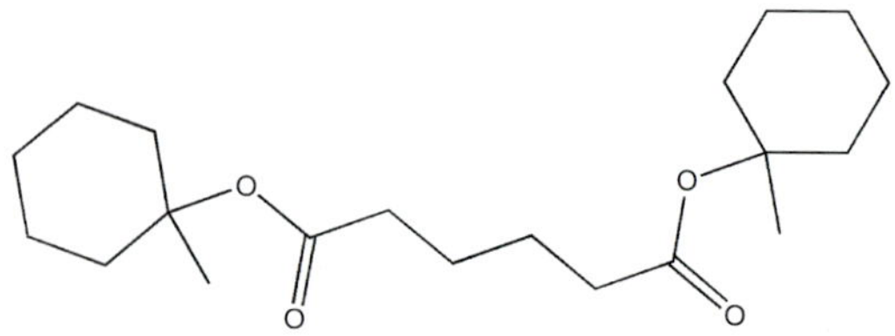

Di(methylcyclohexyl) phthalate *n.* A plasticizer for cellulosics, polystyrene, PVC, and other vinyl resins.

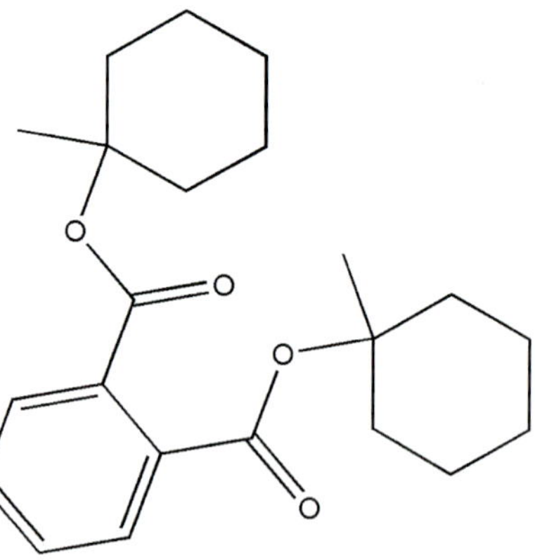

Dimethylformamide (DMF) *n.* A colorless and very active solvent for PVC, nylon, polyurethane, and many other resins and elastomers, with fairly low volatility. Its strong solvent power makes it useful as a solvent booster in coating, printing, and adhesive work, and in paint strippers. Because it is toxic and readily absorbed through the skin, DMF must be handled, and disposed of, with care.

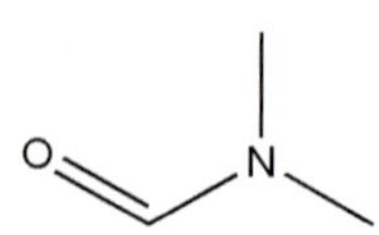

Dimethyl glutarate (DMG) *n.* A liquid chemical intermediate, a source of dicarboxylic acid, used in making plasticizers, polyester resins, synthetic fibers, films, adhesives, and solvents.

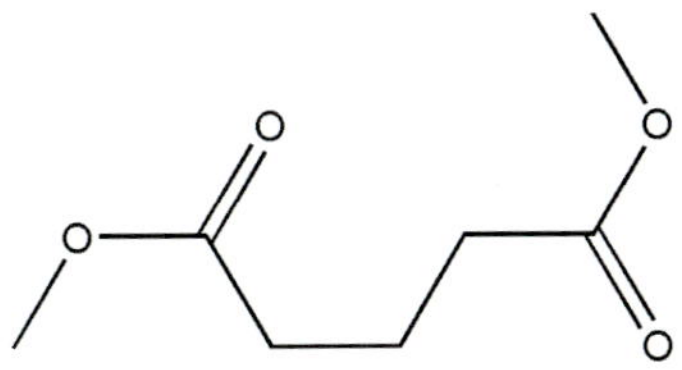

Dimethyl glycol phthalate *n.* $C_6H_4(COO CH_2CH_2OCH_3)_2$. A solvent and plasticizer for cellulosics resins.

Dimethylisobutylcarbinyl phthalate *n.* A plasticizer for most common thermoplastics.

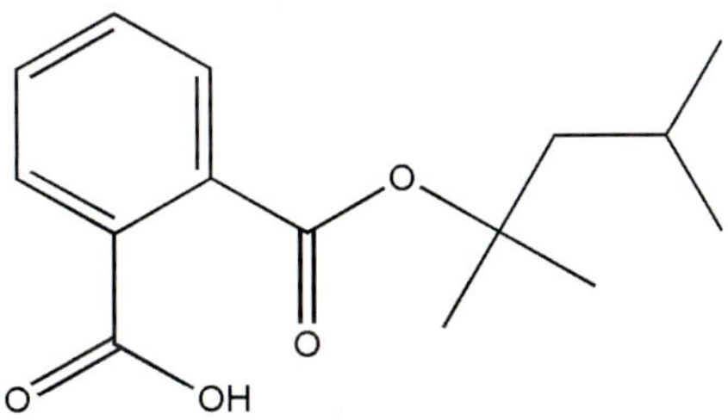

Dimethyl ketone Syn: acetone.

Dimethylol urea *n.* A colorless crystalline material resulting from the combination of urea and formaldehyde in the presence of salts or alkaline catalysts, the first or A-stage of urea-formaldehyde resin.

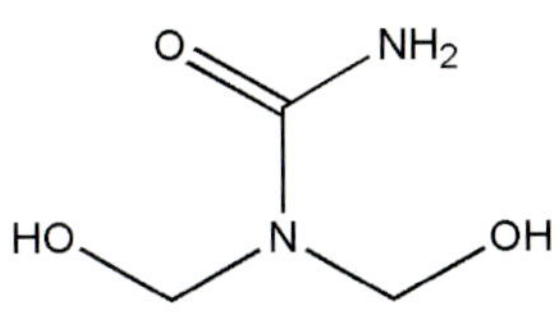

Dimethyl phthalate *n.* A non-toxic plasticizer for most common thermoplastics, but with limited compatibility with PVC. Plasticizer for nitrocellulose and cellulose acetate, resins, lacquers, and plastics. Colorless, odorless, lightfast, stable, and non-toxic liquid. Sp gr, 1.189; bp, 282°C; flp, 149°C (300°F).

Dimethyl polysiloxane *See polydimethylsiloxane.*

2,2-Dimethyl-1,3-propanediol *See neopentyl glycol.*

Dimethyl sebacate *n.* $[-(CH_2)_4COOCH_3]_2$. A solvent and plasticizer for cellulosics and vinyl resins, also compatible with most other thermoplastics.

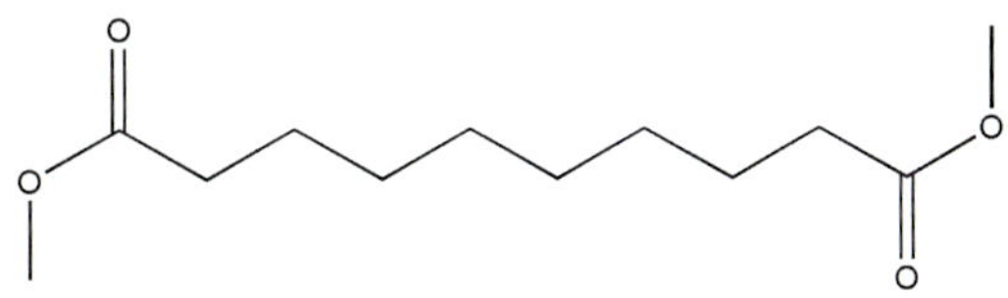

Dimethyl sulfoxide (DMSO) *n.* $(CH_3)_2SO$. An active polar solvent useful for dissolving such polar polymers as polyacrylonitrile and for certain polymerization reactions.

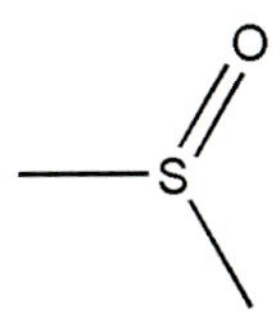

Dimethyl terephthalate (DMT) *n.* C_6H_4-1,4-$(COOCH_3)_2$. A white crystalline solid generally obtained by the oxidation of

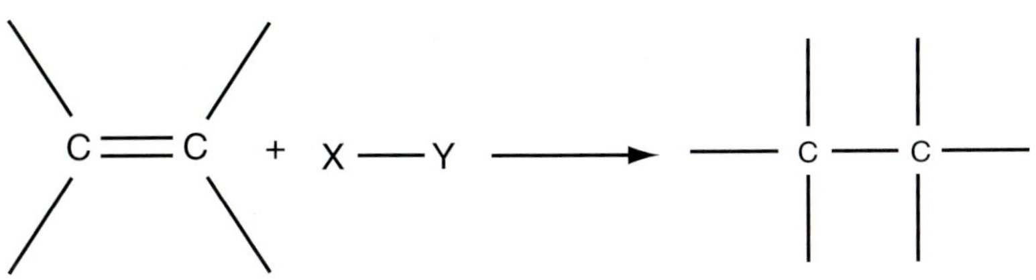

D

p-xylene. The carbomethoxy groups of DMT are typical of those attached to a benzene ring, and their ready participation in alcoholysis reactions is the basis for most of the uses of the material. DMT is used in making polyethylene terephthalate (PET) and polyester fibers therefrom.

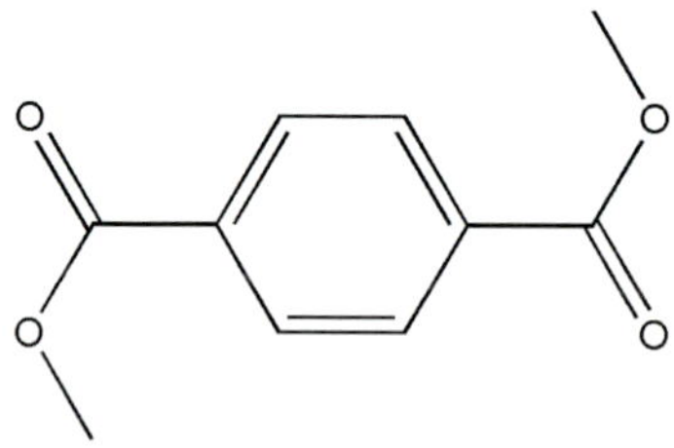

Dimethyl urea \yú-ˈrē-ə\. $(CH_3NH)_2CO$. Primary condensation product of urea and formaldehyde. When condensed in the presence of alcohols, it forms oil-soluble resins. With mildly acid salts, it may be used as an adhesive.

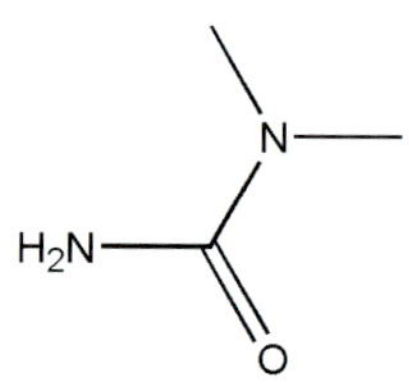

Diminution of pressure at the side of a moving stream *n.* If a fluid of density *d* moves with a velocity *v*, the diminution of pressure due to the motion is (neglecting viscosity),

$$p = \frac{1}{2} dv^2$$

Dimity \ˈdi-mə-tē\ [alter. of ME *demyt*, fr. ML *dimitum*, fr. MGk *dimitos* of double threat, fr. Gk *di-* + *mitos* warp thread] (1570) *n.* A sheer, thin, spun cloth that sometimes has cords or stripes woven in. It is used for aprons, pinafores, and many types of dress goods.

Dimorphous \(ˌ)dī-ˈmór-fəs\ [Gk *dimorphos* having two forms, fr. *di-* + *-morphos*-morphous] (1832) *adj.* Material which can exist in two distinct crystalline forms having different melting points.

Dimple \ˈdim-pəl\ [ME *dympull*; akin to OHGr *tumphilo* whirlpoor, OE *dyppan* to dip] (1602) *v.* A depression or indentation on a surface. Syn: sink mark.

DIN *n.* (1) Abbreviation for Deutsches industrie norm (German Industry Standard). (2) Also abbreviation for Deutsches institut für normung (formerly DNA).

DIN color difference equation *n.* Color difference equation based on the DIN color order system. It is described in English by Richter in J Opt Soc Am 45:223.

DIN color system *n.* Official German standard color system (DIN 6164). The color solid is described in terms of hue (DIN – Farbton), abbreviated F or T, saturation (DIN – Sättigung), abbreviated S, and relative lightness (DIN – Dunkelstufe), abbreviated D.

Dingler's green *See chromium oxide green.*

Di-*n*-hexyl adipate *n.* $(CH_2)_4(COOC_6H_{13})_2$. An important low-temperature plasticizer for synthetic rubbers and several plastics, including some cellulosics, PVC, and polyvinyl acetate.

Di-*n*-hexyl azelate (DNHZ) *n.* $(CH_2)_7(COOC_6H_{13})_2$. A plasticizer for cellulose and vinyl resins. It has been approved for food-contact use, has low volatility and good compatibility.

Dinitraniline *n.* Base which, when diazotized and coupled with *β*-napthol, forms permanent red 2G.

Dinitraniline orange *n.* Pigment orange 5 (12075). A monoazo pigment dyestuff prepared by coupling diazotized dinitro aniline to *β*-naphthol. It bleeds in organic solvents and has fair heat resistance. Its lightfastness in masstone is good. The good chemical resistance, freedom from toxic metals and high color intensity of dinitraniline orange encourage its use over the less expensive chrome and molybdate oranges in some specialized finishes not requiring good bake or bleed resistance.

Dinitrosopentamethylenetetramine (DNPT) *n.* A blowing agent widely used for foam rubber, but of limited use in the plastics industry due to its high decomposition temperature and unpleasant residual odor.

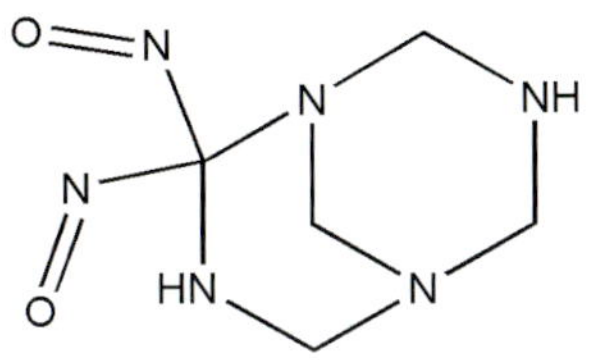

Dinitrosoterephthalamide (DNTA) *n.* A chemical blowing agent for vinyls, liquid polyamide resins, and silicone rubbers. It is especially noted for its low decomposition exotherm.

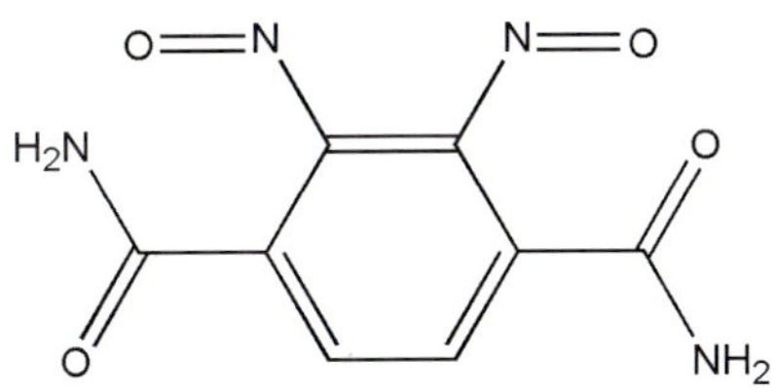

Dinking *See die cutting.*

Di-*n*-octyl, *n*-decyl phthalate (DNODP) *n.* A mixed plasticizer for PVC and several other thermoplastics.

Di-*n*-octyltin maleate polymer *n.* Like the preceding tin stabilizer, this one, too, has been FDA-approved for use in food-packaging compositions up to 3 phr.

Di-*n*-octyltin-*S,S*′-bis(isooctyl mercapto-acetate) *n.* A stabilizer for PVC that has been approved for use in food-grade bottles up to 3 phr when made to certain purity specifications.

Dinonyl adipate (DNA) *n.* A non-alcohol ester used as a plasticizer for cellulosic, acrylic, styrene, and vinyl polymers.

Dinonyl phthalate (DNP) *n.* A general-purpose plasticizer for vinyl resins, with low volatility and good electrical properties.

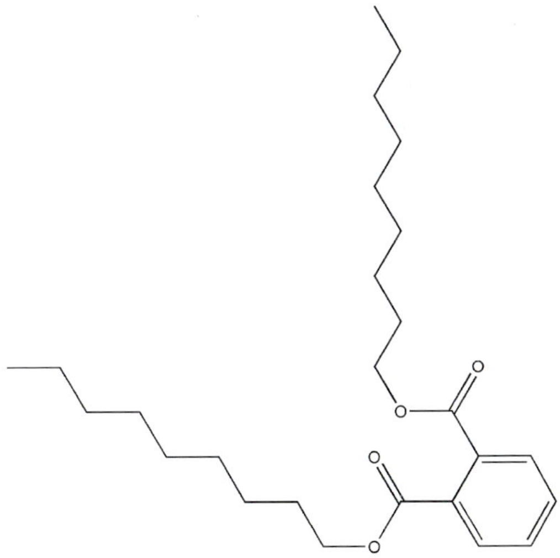

DINP *n.* Abbreviation for diisononyl phthalate.

DIOA *n.* Abbreviation for diisooctyl adipate.

Dioctyl *See di(2-ethylhexyl)* – for several compounds for which this shorter prefix is more commonly used.

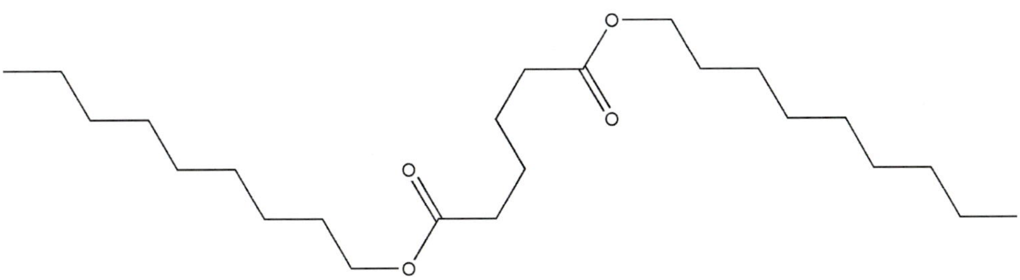

Dioctyl ether *n.* A mold and processing lubricant.

Dioctyl fumarate (DOF) *n.* An unsaturated plasticizer for vinyl resins.

Di(2-octyl) phthalate (DOP) *n.* *See dicapryl phthalate* from which, by addition polymerization, is prepared the transparent, water-white polymer CR-39, one of the allyl-resin family. Since it is difficult and tricky to handle, the resin has found little commercial use aside from optical applications. For these, its excellent optical properties and resistance to scratching, and its low density – half that of glass – have given it wide use in eyeglasses. Other uses include optical filters, instrument windows, welders' masks, and large windows in atomic-energy plants.

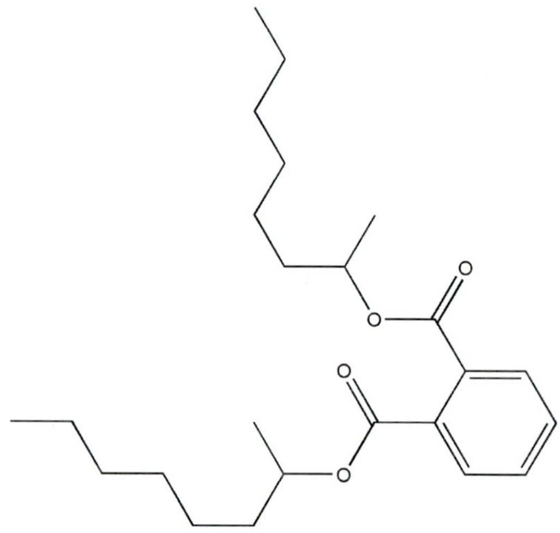

Dioctyl terephthalate (DOTP) *n.* Although the physical properties of DOTP are similar to those of DOP (the ortho isomer), DOTP is less volatile, imparts slightly better

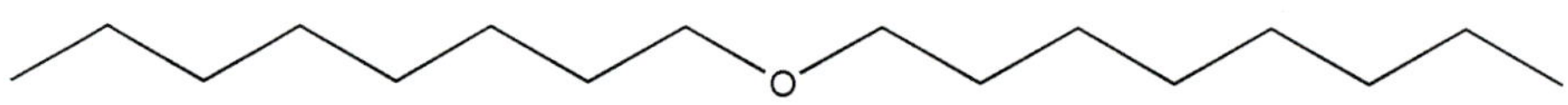

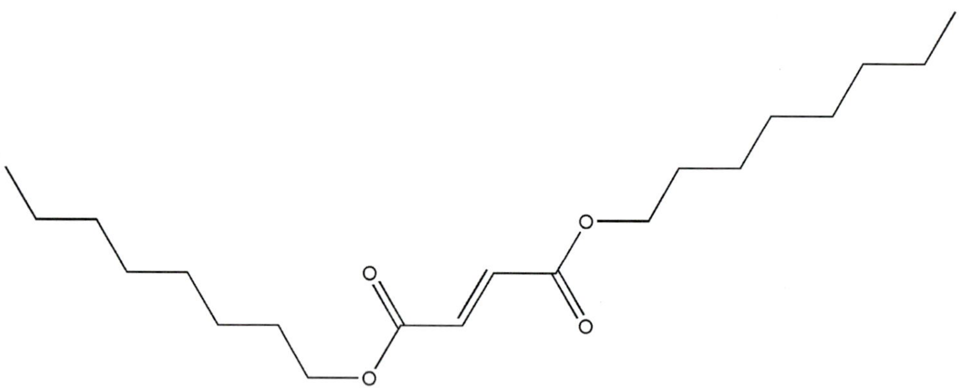

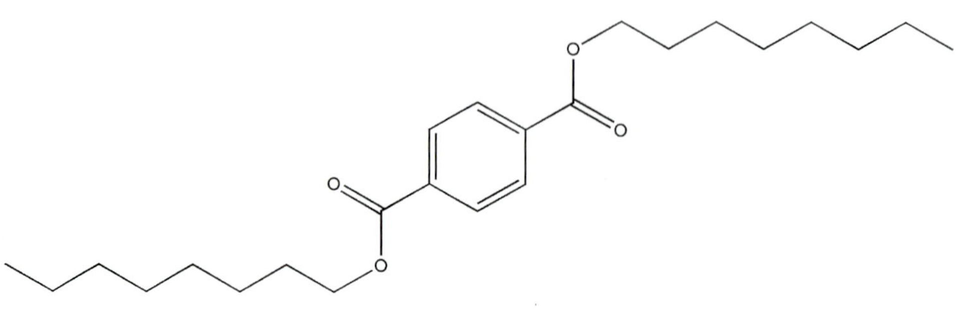

low-temperature flexibility, and is more resistant to lacquer marring.

Dioctyltin stabilizer *See organotin stabilizer.*

Diofan *n.* Dispersion of copolymers of vinylidene chloride. Manufactured by BASF, Germany.

Diol \ˈdī-ˌōl\ [ISV] (1923) *n.* An acronym for dihydric alcohol, i.e., an alcohol containing two hydroxyl (–OH) groups. Ethylene glycol, $HOCH_2CH_2OH$, and 1,5-pentanediol. $HO(CH_2)_5OH$, are examples.

Diolefin polymer \dī-ˈō-lə-fən\ *n. See diene polymer.*

Diolefin resins *n.* More generally known as aromatic petroleum residues. By-products from the cracking of petroleum.

Diolen *n.* Fiber from poly(ethylene terephthalate). Manufactured by Glanzstoff, Germany.

DIOP *n.* Abbreviation for diisooctyl phthalate.

DIOS *n.* Abbreviation for diisooctyl sebacate.

Dioxan (dioxane) \dī-ˈäk-ˌsān\ [ISV] (1912) *n.* Cyclic 1,4 diethylene dioxide. $CH_2CH_2OCH_2CH_2O$. Bp, 101°C; Sp gr, 1.035/20°C; fp, 11°C (52°F); refractive index, 1.423; vp, 26 mmHg/20°C. A water-miscible solvent for cellulose esters.

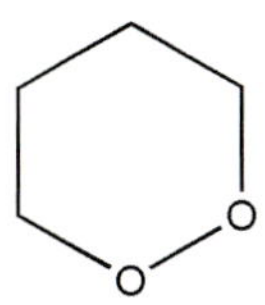

1,4-Dioxane (diethylene ether, diethylene dioxide, and dioxyethylene ether) *n.* $OCH_2CH_2OCH_2CH_2$. A solvent for cellulose esters and other plastics. Dioxane is rarely used today because of its suspected carcinogenicity.

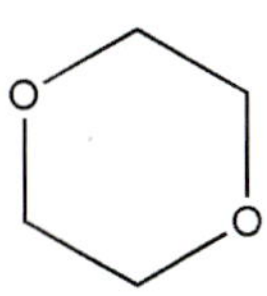

Di-*o*-xenyl phenyl phosphate *n.* $(C_6H_5C_6H_4O)_2(C_6H_5O)PO$. A plasticizer for cellulosics, polystyrene, and vinyls.

DIOZ *n.* Abbreviation for diisooctyl azelate.

Dip \ˈdip\ (1599) *n.* The angle measured in a vertical plane between the direction of the earth's magnetic field and the horizontal.

Dip [ME *dippen*, fr. OE *dyppan*; akin to OHGr *tupfen* to ash, Lithuanian *dubus* deep] (before 12c) *v.* (1) Immersion of a textile material in some processing liquid. The term is usually used in connection with padding or slashing process. (2) The rubber compound with which tire cords and other in-rubber textiles are treated to give improved adhesion to rubber.

Dipcoat *n.* A paint or plastic coating, which is applied by completely immersing an article in a tank of the coating.

Dip coating *n.* A coating process wherein the object to be coated, preheated or at room temperature, depending on the materials, is dipped into a tank of fluid resin, solution, or dispersion, withdrawn and subjected to further heat or drying to solidify the deposit. *See fluidized-bed coating* for a similar process employing powdered resin. *See coating, dip.*

Dip dyeing *See dyeing.*

Dipentene (commercial product) *n.* Optically inactive form of the monocyclic terpene hydrocarbon limonene. Commercial dipentenes contains a substantial portion of other monocyclic and bicyclic, as well as some oxygenated terpenes having closely related boiling ranges. They are generally obtained by fractional distillation from the crude oils recovered in the several commercial methods of utilizing pine wood, also by isomerization during the chemical processing of terpenes. The four kinds of commercial dipentene are: (1) *Steam-distilled dipentene* – From the crude oleoresinous extract used for the processing of related steam-distilled wood naval stores. (2) *Sulfate dipentene* – From the crude condensate of the vapors generated in the digestion of wood in the sulfate paper pulp process. (3) *Destructively distilled dipentene* – From the lighter portions of the oil, recovered during the destructive distillation of pine wood. (4) *Chemically processed dipentene* – Recovered as a by-product in connection with the chemical treatment and conversion of other terpenes. Used as a solvent, but also has anti-skinning properties. Sp gr, 0.850; flash point, 52°C (Cleveland Open Cup); boiling range, 175 (347°F) to 188°C (370°F). (5) Pure Compound, *di*-limonene, *p*-mentha-1,8-diene $C_{10}H_{16}$, acyclic diene, a colorless liquid, Sp gr, 0.84; bp, 176°C; molecular weight, 136.12, with an orange/lemony odor, used as a solvent for coumarone and alkyd resins, rubber, and natural resins.

Dip forming (dip molding) *n.* A process similar to dip coating, except that the fused, cured or dried deposit is stripped from the dipping mandrel. As most frequently used for making vinyl-plastisol articles, the process comprises dipping into the plastisol a preheated form shaped to the desired inside dimensions of the finished article allowing the plastisol to gel in a layer of the desired thickness against the form surface; withdrawing the coated form; heating the deposit to fuse the layer; cooling and stripping off the deposit. Some articles may be inverted after stripping so that the textured inside surface becomes the external surface of the finished article.

Dip-grained wood *n.* Wood, which has single waves or undulations of the fibers, such as occur around knots and pitch pockets.

Diphenyl \(ˌ)dī-ˈfe-nᵊl\ *n.* Liquid whose vapor is used as a heat transfer agent in chemical processes. At low pressures, the temperatures obtainable with diphenyl vapor are higher than those obtainable with steam. Syn: biphenyl.

Diphenylamine [ISV] (1872) (DPA) *n.* $(C_6H_5)_2NH$. A crystalline solid. Used as a stabilizer for several plastics. Also, a weak secondary base used as a dyestuff intermediate and as an indicator.

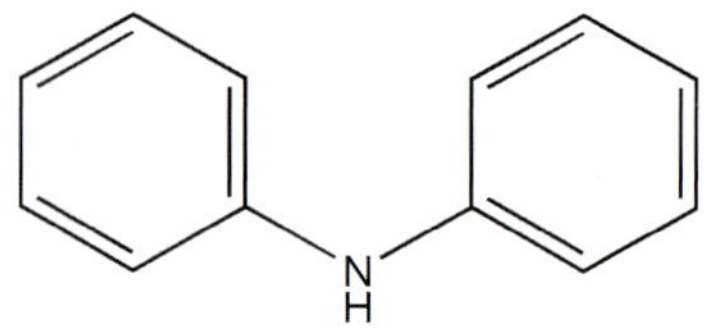

Diphenyl carbonate *n.* $(C_6H_5O)_2CO$. The monomer from which polycarbonates are produced.

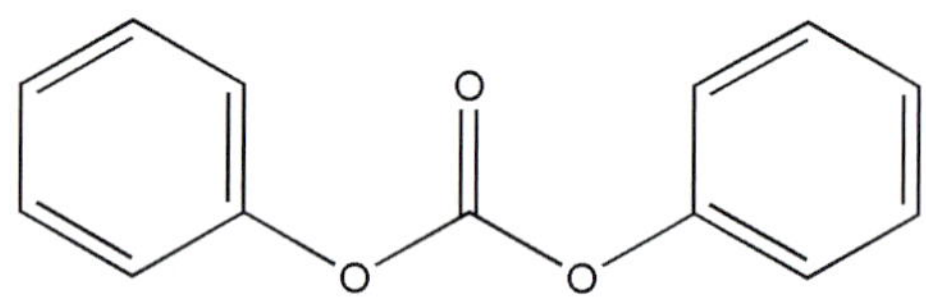

Diphenyl decyl phosphite *n.* A nearly colorless liquid used as a stabilizer for vinyl and polyolefin resins.

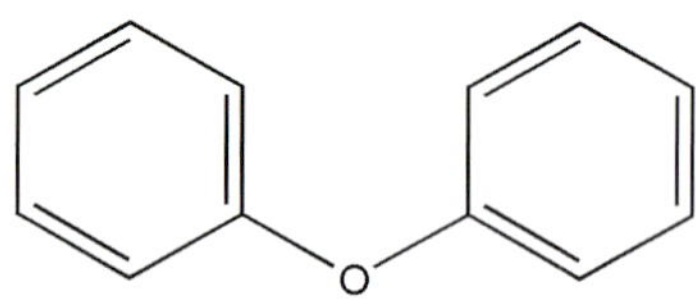

Diphenyl ether *n.* A plasticizer. Bp, 252°C; mp, 28°C.

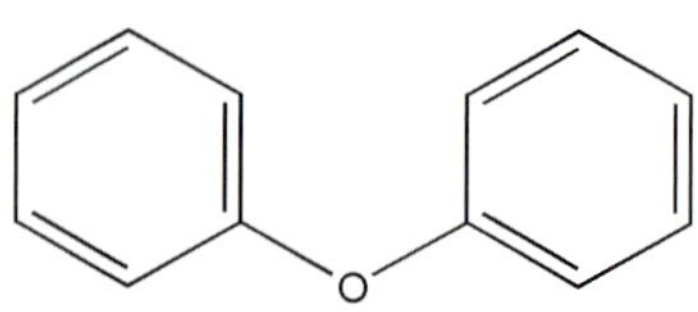

Diphenyl ketone *See benzophenone.*

Diphenyl mono-*o*-xenyl phosphate *n.* $(C_6H_5O)_2(C_6H_5C_6H_4O)PO$. A plasticizer for cellulosics plastics, polystyrene, and, with limited compatibility, vinyls.

Diphenyl octyl phosphate (DPOP) *n.* $(C_6H_5O)_2(C_8H_{17}O)PO$. A flame-retardant plasticizer for PVC and cellulosics resins. It has been approved by DDA for use in food packaging.

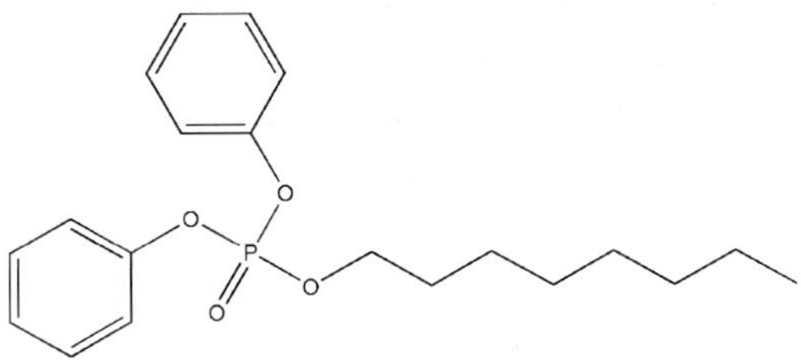

Diphenyl phosphate *n.* A permanent plasticizer. *Also called DPP.*

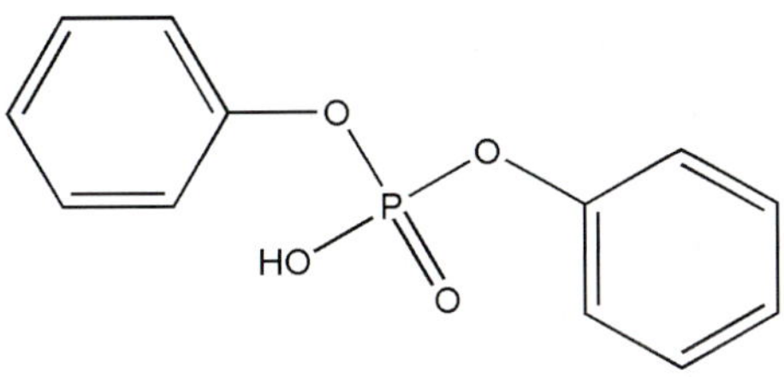

Diphenyl phthalate (DPP) *n.* A powder that melts at 75°C. It is used as a solid plasticizer for rigid PVC, cellulosics, and other resins.

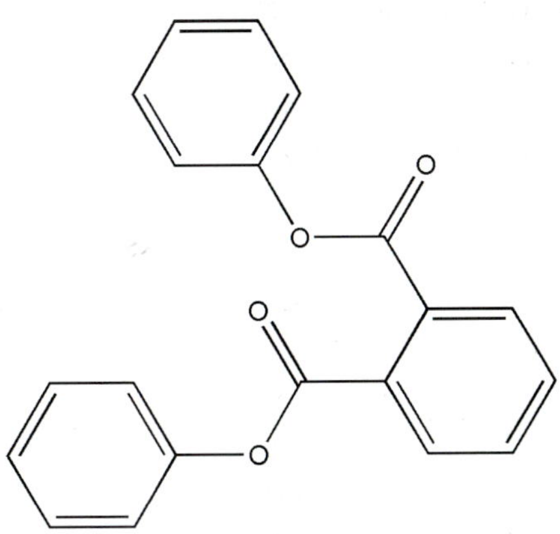

Diphenyl xylenyl phosphate *n.* A plasticizer for cellulose acetate-butyrate, cellulose nitrate, polystyrene, PVC, and vinyl chloride-acetate copolymers.

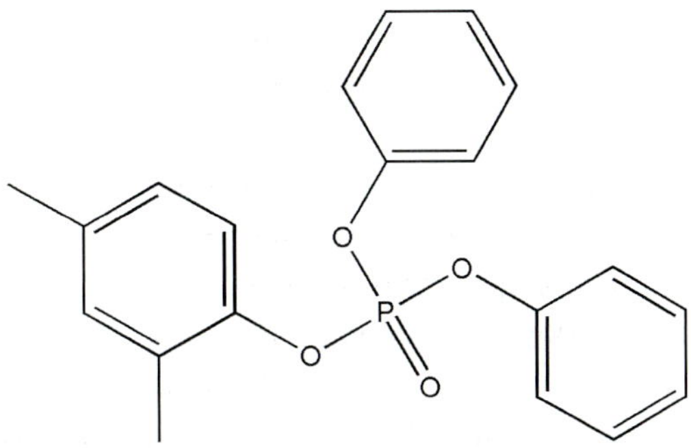

Diphosphoglyceric acid \(ˌ)dī-ˈfäs-fō-gli-ˌser-ik-\ (1959) *n.* A diphosphate of glyceric acid that is an important intermediate in photosynthesis and in glycolysis and fermentation.

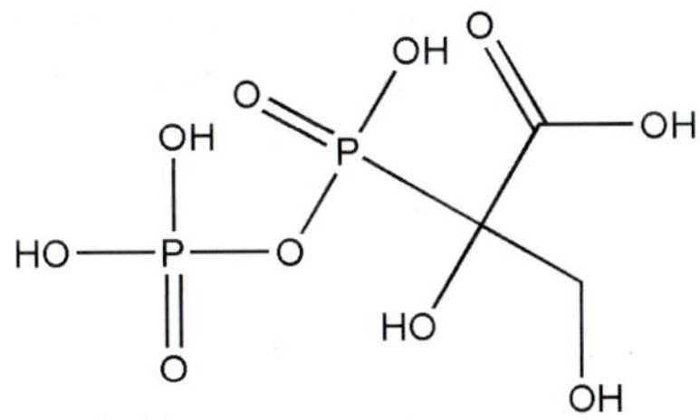

Dip molding *See dip forming.*

Dipole \ˈdī-ˌpōl\ [ISV] (1912) *n.* (1) A combination of two electrically or magnetically charged particles of opposite sign that are separated by a small distance. (2) Any system of charges, such as a circulating electric current, having the property (a) that no net force acts upon it in a uniform field; or, (b) a torque proportional to sin θ, where θ is the angle between the dipole axis and a uniform field, *does* act on the charges. In polymers, atoms such as Cl bearing negative charges induce opposite charge in neighboring atoms, thus creating dipoles. Tiny movements of these dipoles in reaction to the changing directions of rapidly alternating fields are the basis for dielectric heating, and (c) it produces a potential, which is proportional to the inverse square of the distance from it. Weast RC (ed) Handbook of chemistry and physics, 52nd edn. The Chemical Rubber Co., Boca Raton, FL.

D

Dipole bonds *n.* Chemical bonds formed when the intramolecular interactions are polar covalent (uneven sharing of electron pair). The strength of the dipole–dipole interaction increases with the polarity of the participating molecules.

Dipole–dipole forces *n.* Forces between polar molecules.

Dipole moment (1926) *n.* The product of the magnitude of the charge at one end of a dipole times the distance between the opposite charges. A mathematical entity; the product of one of the charges of a dipole unit by the distance separating the two dipolar charges. In terms of the definition of a dipole (2), the dipole moment p is related to the torque T, and the field strength E (or B) through the equation $T = p \times E$.

Dipole moment, molecular *n.* It is found from measurements of dielectric constant (i.e., by its temperature dependence, as in the *Debye equation for total polarization*) that certain molecules have permanent dipole moments. These moments are associated with transfer of charge within the molecule, and provide valuable information as to the molecular structure.

Dip penetration *n.* The degree of saturation through a tire cord after impregnation with an adhesive.

Dip pickup *n.* The amount of adhesive applied to a tire cord by dipping, expressed as a percentage of the weight of the cord before dipping.

Dipping *n.* Method of application in which the complete immersion of an article in the coating is followed by draining. Paints, lacquers, and stains designed for such use are usually designated "dipping". This process may be carried out either at ordinary or elevated temperatures.
See coating, dip.

Dipropylene glycol *n.* (2,2′-dihydroxydipropyl ether) $(CH_3CHOH–CH_2)_2O$. A high-boiling glycol ether with a low order of toxicity, widely used as a solvent and chemical intermediate. As a solvent it is used with cellulose acetate and nitrate, and is one of the few known solvents for polyethylene. Thus it is used in screening tests to identify polyethylene. As an intermediate, dipropylene glycol reacts with dibasic acids to form alkyd resins, polyester plasticizers, and urethane-foam intermediates.

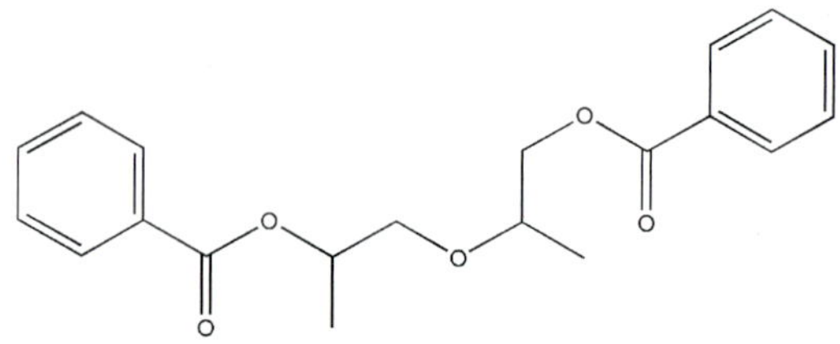

Dipropylene glycol dibenzoat A plasticizer for PVC, also compatible with most common thermoplastics, imparting good stain resistance.

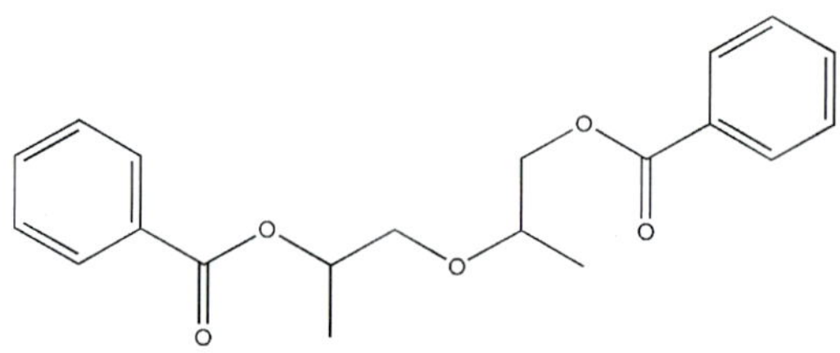

Dipropylene glycol monosalicylate (salicylic acid, dipropylene glycol monoester) *n.* A light-colored oil used in ultraviolet-screening agents and plasticizers.

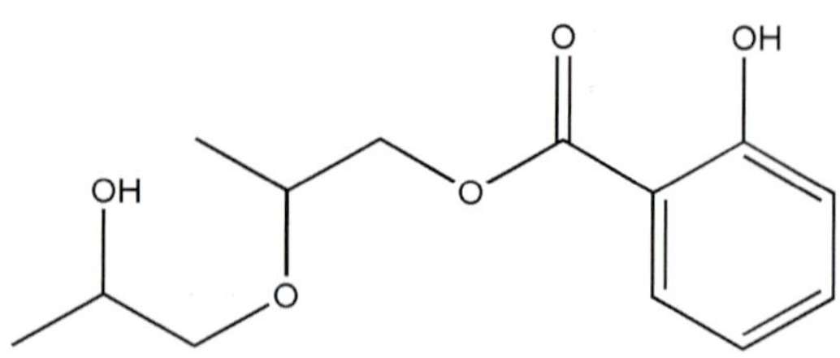

Dipropyl ketone *n.* A stable, colorless liquid, a solvent for many resins. Bp, 143°C; flp, 49°C (120°C); vp. 5 mmHg/20°C.

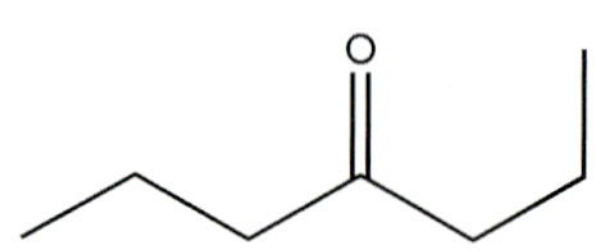

Dipropyl oxalate \-ˈäk-sə-ˌlāt\ *n.* An active, high-boiling (211°C) solvent.

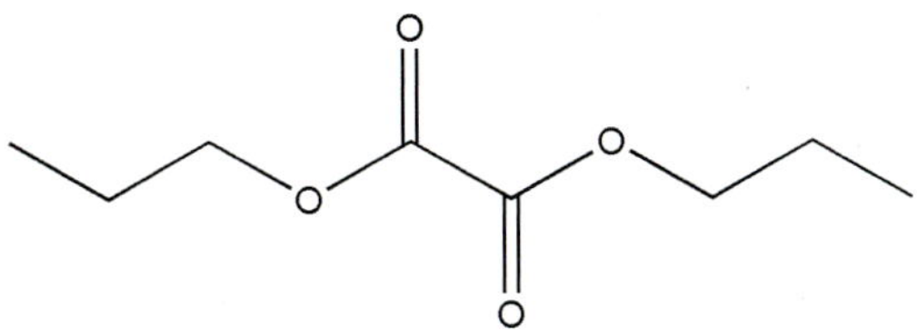

Dipropyl phthalate *n.* A plasticizer for cellulose acetate and cellulose acetate-butyrate.

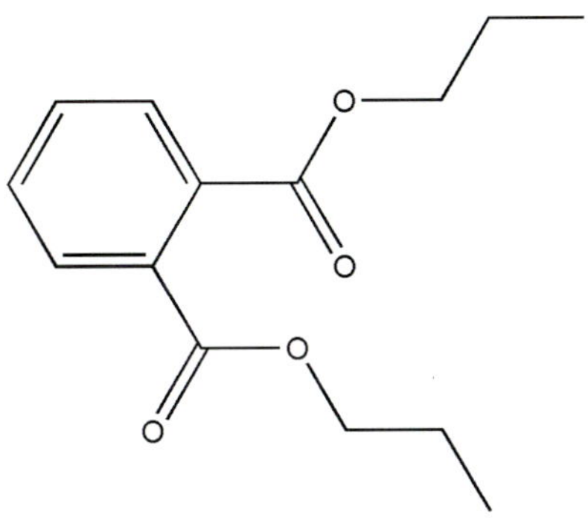

Dipropyl succinate *n.* Plasticizer. Bp, 246°C; Sp gr, 1.016/4°C.

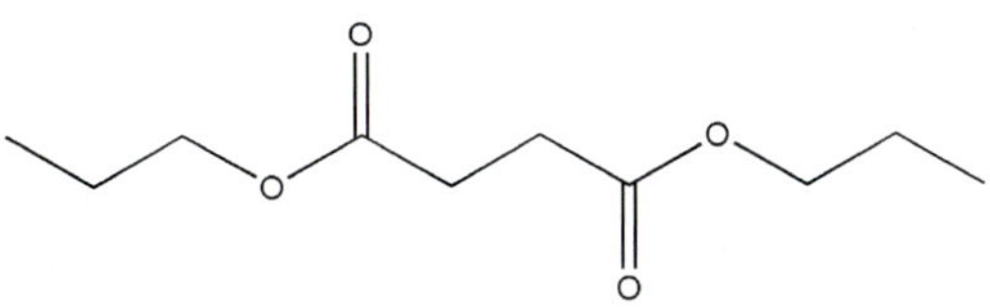

Diprotic acid *n.* An acid with two available H^+, or two denotable protons.

Dip treating *n.* The process of passing fiber, cord, or fabric through an adhesive bath, followed by drying and heat-treating of the adhesive-coated fiber to obtain better adhesion.

Di-*p*-xylylene (DPX, DPXN) *n.* $(-CH_2C_6H_4CH_2-)_2$. The stable dimmer of *p*-xylylene, a white powder. When heated to 600°C in an evacuated chamber, the monomer is regenerated and instantly polymerizes to form a tough, impervious film on any cold surface between the heating pot and the vacuum source.
See parylene.

Diradicals \-ˈra-di-kəls\ *n.* Molecular species having two unpaired electrons, in which at least two different electronic states with different multiplicities [electron-paired (singlet state) or electron-unpaired (triplet state)] can be identified. For example, $H_2C\cdot{-}CH_2C\cdot H_2$ propane-1,3-diyl (trimethylene).

Direct coating *n.* The simplest method of coating, this procedure involves spreading the coating with a knife. The moving fabric substrate is usually supported by a roller or a sleeve. The gap between the knife and the fabric determines coating thickness.

Direct dyes *n.* A class of dyestuffs that are applied directly to the substrate in a neutral or alkaline bath. They produce full shades on cotton and linen without mordanting and can also be applied to rayon, silk, and wool. Direct dyes give bright shades but exhibit poor washfastness. Various after treatments are used to improve the washfastness of direct dyes, and such dyes are referred to as "after treated direct colors".
See dyes.

Direct esterification *n.* In the production of polyethylene terephthalate, the process in which ethylene glycol is reacted with terephthalic acid to form bis-*p*-hydroxyethyl terephthalate monomer with the generation of water as a by-product.

Direct gate *n.* A gate that has the same cross section as the runner, i.e., the absence of a distinguishable gate.

Directionally oriented fabrics *n.* Rigid fabric constructions containing inlaid warp or fill yarns held in place by a warp-knit structure. Used in geotextiles, coated fabrics, composites, etc.

Directional reflectance *See reflectance, directional.*

Direction of twist *See twist, direction of.*

Direct lighting *n.* Lighting, which is so controlled by a reflector that the major portion of the light directly reaches the spot to be illuminated without redirection, diffusion or reflection.

D

Direct printing *See printing.*

Dirt \ˈdərt\ [ME *drit,* fr. ON; akin to OE *drītan,* to defeat] (13c) *n.* Disfiguring foreign material other than microorganisms on or embedded in a dried coating.
Also called soil.

Dirt collection *n.* The accumulation of dust, dirt, and other foreign matter on a paint surface.

Dirt resistance *n.* The ability of a coating to resist soiling by foreign material, other than micro-organisms, deposited on or embedded in the dried coating.

Disbond *v.* In an adhesive-bonded joint, to separate at the bond surface. *n* Such as a separation.

Discharge-inception voltage *n.* In a dielectric-strength test, the voltage at which discharges begin in the voids within the specimen.

Discharge printing *See printing.*

Discoloration \(ˌ)dis-ˌkə-lə-ˈrā-shən\ (1642) *n.* (1) Any change from an initial color possessed by a plastic. (2) A lack of uniformity in color where color should be uniform over the whole area of a plastic object. In the second sense, where they are applicable, one may use the more definite terms *mottle, segregation,* or *two-tone.* Discoloration can be caused by inadequate blending of ingredients, by overheating, exposure to light, irradiation, or chemical action.

Discolored pick *See mixed end or filling.*

Discontinuities *See holydays.*

Discontinuous phase *See dispersed phase.*

Dished *n.* Showing a symmetrical concave distortion of a flat or curved surface of a plastic object, so that, as normally viewed, it appears more concave than its design calls for. Opposite of domed.
See also warp and sink mark.

Dislocation \ˌdis-(ˌ)lō-ˈkā-shən\ *n.* A structural defect in which the lattice planes in a crystal is incomplete or warped.
Also known as a line defect.

Disorder, crystalline *n.* Defect in the lattice of a crystal, characterized by local faults in the atomic arrangement.

Dispersant \di-ˈspər-sənt\ (1941) *n.* A dispersing agent, often of a surface active chemical, that promotes formation of a dispersion or maintains a state of dispersion by preventing settling or aggregation.

Dispersant *n.* In an organosol, a liquid component that has a solvating or peptizing action on the resin, thus aiding in dispersing and suspending it. Additive that increases the stability of a suspension of powders (pigments) in a liquid medium.
Also known as dispersing agent.

Dispersed *n.* Finely divided or colloidal in nature.

Dispersed phase *n.* That phase in an emulsion or suspension which is broken down into droplets or discrete particles and dispersed throughout the other or continuous phase.
Also called discontinuous phase.

Disperse dyes *n.* A class of slightly water-soluble dyes originally introduced for dyeing acetate and usually applied from fine aqueous suspensions. Disperse dyes are widely used for dyeing most of the manufactured fibers.
See dyes.

Disperse phase *n.* In a suspension or emulsion, the "disperse phase" refers to the particles of one material individually dispersed, more or less stably, in the continuously connected domain of another material, usually a liquid, known as the *continuous phase.*

Disperse system *n.* Any system in which particles (of any size and state) are dispersed in a homogeneous medium. A colloidal solution represents a special kind of disperse system.

Dispersing agent *n.* A material added, usually in relatively small percentage, to a suspending medium to promote and maintain the separation of discrete, fine particles of solids or liquids. Dispersants are used, for example, in the wet grinding of pigments and for suspending water-insoluble dyes.

See dispersant.

Dispersion curve *n.* A graph of the refractive index (ordinate) as a function of wavelength of radiation (abscissa).

Dispersion, degree of *n.* Quantity varying reciprocally with aggregate size. The larger the aggregates, the lower the degree of dispersion.

Dispersion forces *n.* (1) Weak forces between atoms or molecules due to momentary fluctuations in their electronic charge-cloud distributions. (2) The force of attraction between molecules possessing no permanent dipole. The interaction energy is given by

$$U_{\mathrm{p}} = -\frac{3}{4}h\frac{V_{\mathrm{o}}\alpha^2}{r^6},$$

where h is the Planck's constant, V_{o} the characteristic frequency of the molecule, r the distance between the molecules, and α is the polarizability.

Also called London forces.

Dispersion, light *n.* (1) The separation of light into separate wavelengths by means of a dispersion device such as a prism or diffraction grating. Fundamentally, dispersion is related to different wavelengths of light in a given medium. (2) The variation of refractive index with color (or wavelength) of light. The spreading of white light into its component colors when passing through a glass prism is due to dispersion, which, in turn, is due to the fact that the refractive index of transparent substances is lower for long wavelengths than for short wavelengths. A measure of dispersion is ν defined as:

$$\nu = \frac{n_{\mathrm{D}} - 1}{n_{\mathrm{F}} - n_{\mathrm{C}}},$$

where n_{D} is the refractive index at 589 nm (yellow); $n_{\mathrm{F}} = 486$ nm (blue); $n_{\mathrm{C}} = 656$ nm (red).

See dispersion curve.

Dispersion, particles and liquids (14c) *n.* (1) A system of two or more phases comprising one or more finely divided materials distributed in another material. Types of dispersions are *emulsions* (liquids in liquids), *suspensions* (solids in liquids), *foams* (gases in liquids or solidified liquids), and *aerosols* (liquids in gases). In the plastics industry, the term dispersion usually denotes a finely divided solid dispersed in a liquid or in another solid. Examples are fillers and pigments in molding compounds, plastisols and organosols. This term is used loosely in many ways with special meanings, which are often implied but not defined. Since the term is decidedly generic and includes all types of colloidal systems, it should not be used on the assumption that it refers to a desirable feature of one type of dispersion. For example, in paint and printing ink technology a "good" dispersion usually means that a pigment is finely divided and deflocculated in the vehicle. Where aggregates are present in the dispersed phase, the material is considered without exception a "poor"

D

dispersion regardless of the state of flocculation. Where the pigment particles are finely divided but flocculated, the material has been called variously a "good" and a "poor" dispersion. (2) A mixing process in which the particles or globules of the discontinuous phase are reduced in size. This may be accomplished by breaking solid particles, as in ball-milling, or by shearing viscous regions, as in kneading or extrusion. (3) Two-phase system in which one phase, called the dispersed phase, is permanently distributed as small particles through the second phase called the continuous phase. Leach RH, Pierce RJ (1993) Printing ink handbook, national association of printing ink manufacturers Inc., 5th edn. Kluwer Academic Press, London, UK. Leach RH, Pierce RJ, Hickman EP, Mackenzie MJ, Smith HG (eds) (1993) Printing ink manual, 5th edn. Blueprint, New York. Paint/Coatings dictionary, Federation of societies for coatings technology, Philadelphia, Blue Bell, PA, 1978. Carley JF (ed) (1993) Whittington's dictionary of plastics. Technomic Publishing Co. Inc.
See also segregation.

Dispersion, pigment *n.* Suspension of pigment particles uniformly in a medium such as a paint vehicle, plastic matrix, etc. The process of dispersing the pigment involves the separation of individual pigment particles, and coating them with the medium.

Dispersion resin *n.* A special type of PVC resin with very small spherical particles, usually 1 μm or less in diameter, permitting them to be mixed with plasticizers by simple stirring techniques. They are used in compounding plastisols and organosols.
See organosol and plastisol.

Dispersion resins *n.* Used as the primary vinyl resin in the production of plastisols and organosols. These resins are fine, white powders, with an average particle size of 1 μm and a bulk density of approximately 15–20 pounds/ft^3. The fine particle size, combined with the dense nature of the particle allows these products to be easily dispersed in plasticizers and solvents. The specific selection of a dispersion resin is primarily based on the fusion, rheological and appearance requirements of the end product and the manufacturing process.

Dispersions *n.* Systems of two or more phases comprising one or more finely divided materials distributed in another material. Types of dispersion are emulsions (liquids in liquids), suspensions (solids in liquids), foams (gases in liquids or solidified liquids), and aerosols (liquids in gases). Patton TC (1965) Paint flow and pigment dispersion. Interscience Publishers Inc. Solomon DH, Hawthorne DG (1991) Chemistry of pigments and fillers. Krieger Publishing Co., New York. Ash M, Ash I (1998) Handbook of fillers, extenders and dilutents. Synapse Information Resources Inc., New York. Kirk-Othmer (1996) Encyclopedia of chemical technology: pigments–powders. John Wiley and Sons, New York. Parfitt GD (1969) Dispersion of powders in liquids. Elsevier Publishing Co., New York.

Dispersive mixing *n.* Any mixing process in which the principal action is reduction of the size of particles or interlayers. The progress of the process is judged by two criteria, intensity of segregation and scale of segregation.

Dispersive power *n.* If n_1 and n_2 are the indices of refraction for wavelengths λ_1 and λ_2 and n the mean index or that for sodium light, the dispersive power for the specified wavelength is

$$\bar{\omega} = \frac{n_2 - n_1}{n - 1}.$$

Dispersoid \-ˈspər-ˌsóid\ (1911) *n.* Particles of a dispersion.

Displacement \di-ˈsplā-smənt\ (1611) *n.* A reaction in which an elementary substance displaces and sets free a constituent element from a compound.

Displacement angle *n.* In filament winding, the angle whose tangent equals the quotient of the advancement distance of the winding ribbon on the equator after one complete turn, divided by (π × the equatorial diameter).

Displacement or elongation (at an instant) *n.* The distance of a vibrating or oscillating particle from its position of equilibrium.

Disproportionation \-ˌpōr-shə-ˈnā-shən\ (ca. 1929) *n.* A reaction in which one substance acts simultaneously as an oxidizing agent and a reducing agent; auto-oxidation. It can also be written as the termination by chain transfer between macroradicals, to produce a saturated and an unsaturated polymer molecule.

Dissipation \ˌdi-sə-ˈpā-shən\ *n.* The loss modulus in a plastic part when imparted with rapid, cyclic changes (or even reversals) of stress. The product of mechanical dissipation is heat, which can raise the temperature of the part and cause it to weaken, creep rapidly, or even fail prematurely. Dissipation can also apply to electrical systems, whereby a material with small dissipation will tend to better insulate heat. This is a desirable property in electrical insulations for high-frequency applications because it minimizes the waste of electrical energy as heat.

Dissipation factor (electrical) *n.* The ratio of the conductance of a capacitor in which the test material is the dielectric to its susceptance; or the ratio of its parallel reactance to its parallel resistance. Most plastics have a low dissipation factor, a desirable property in electrical insulations for high-frequency applications because it minimizes the waste of electrical energy as heat. On the other hand, for those plastics with high dissipation factors, e.g., PVC and phenolic, the property provides a fast and economical method of even heating using microwaves and is the basis of electronic preheating of thermosetting molding powders and of electronic sealing of flexible-vinyl films.
See dielectric heating.

Dissipation factor (mechanical) *n.* The ratio of the loss modulus to the modulus of elasticity in a plastic part undergoing rapid, cyclic changes (or even reversals) of stress. As in electrical dissipation, the product of mechanical dissipation is heat, heat that can raise the temperature of the part and cause it to weaken, creep rapidly, or even fail prematurely.

Dissociation \(ˌ)di-ˌsō-sē-ˈā-shən\ (1611) *n.* The splitting apart of a molecule to form two fragments; the reaction of an electrolyte with a solvent to form ions. *Sometimes called ionization.*

Dissociation constant *n*, K_{diss}. The equilibrium constant for dissociation equilibrium.

Dissymmetry \(ˌ)di(s)-ˈsi-mə-trē\ (1845) *n.* Present in a molecule that does not have mirror image symmetry but does have a translational or rotational axis of symmetry.

Distearyl ether (dioctadecyl ether) A mold lubricant.

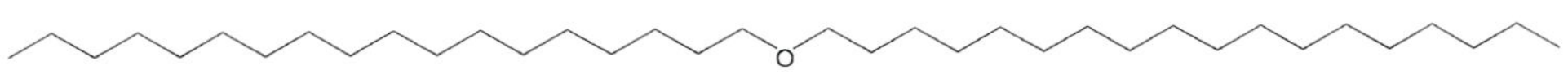

D

Distemper \dis-ˈtem-pər\ [obs. *distemper*, v., dilute, mix to produce distemper, fr. ME, fr. MF *destemprer*, fr. L *dis-* + temperare] (1632) *n.* Heavily pigmented, matt drying composition, capable of being thinned with water, in which the binding medium consists essentially of either glue or casein or similar sizing material.

Distemper blush *n.* A flat brush, 5–10 in. wide, well packed, with long bristle.

Distemper colors *n.* Colorants suitable for tinting distempers. The two chief requirements are lightfastness in reduced form and fastness of alkali.

Distemper, oil-bound *See oil-bound distemper.*

Distensibility \-ˈsten(t)-sə-bəl\ [*disten-* (fr. L *distensus*, pp of *distendere*) + *-ible*] (ca. 1828) *adj.* Ability to be stretched.
See elongation.

Distillation \ˌdis-tə-ˈlā-shən\ (14c) *n.* Process of separation consisting of vaporizing a liquid and collecting the vapor, which is then usually condensed to a liquid.

Distillation range *n.* Temperature range over which a mixture of liquids is distilled.

Distinctness-of-image gloss *n.* The sharpness with which image outlines are reflected by the surface of an object.

Distortion \di-ˈstór-shən\ (1581) *n.* (1) A change in the shape of a solid body, often associated with temperature differences or stress gradients within the body. (2) An apparent change of shape as perceived through an optically imperfect, transparent membrane or reflected from an imperfect mirror.

Distressing *n.* In antiquing, the furniture process designed to produce a more authentic look of age or craftsmanship. It attempts to simulate the small visible marks of abuse picked up by a piece of furniture during a long period of use. In simulating natural marring or aging of some kind, it may involve the use of gouges, hammers, chains, and blunt or pointed instruments such as common car keys, a belt buckle or sharp pencil point. Other effects may be achieved with a fine brush or a pointed wood instrument dipped into ink or glaze and placed on a surface. If no natural counterpart to an effect exists, it is not authentically aged and should be considered a novelty finish or accent piece.

Distributing roller *n.* A rubber covered roller, which conveys ink from the fountain to the ink drum of a rotating press.

Distribution law *n.* A substance distributes itself between two immiscible solvents so that the ratio of its concentrations in the two solvents is approximately a constant (and equal to the ratio of the solubilities of the substance in each solvent). The above statement requires modification is more than one molecular species is formed.

Distribution length *n.* In fibers, a graphic or tabular presentation of the proportion or percentage (by number or by weight) of fibers having different lengths.

Distributive mixing *n.* Any mixing operation in which the principal action is convection rather than dispersion, resulting in elements of the different phases being more intimately blended, but without much size reduction. Typical is blending of masterbatch pigmented pellets with virgin pellets prior to extrusion or molding, by tumbling.

Di-*tert*-butyl peroxide *n.* A member of the alkyl peroxide family, used as an initiator in vinyl chloride polymerization, polyester reactions, and as a crosslinking agent. A stable liquid used as a catalyst for polymerizations at high temperatures of a variety of olefin and vinyl monomers, e.g., ethylene, styrene, and styrenated alkyds.

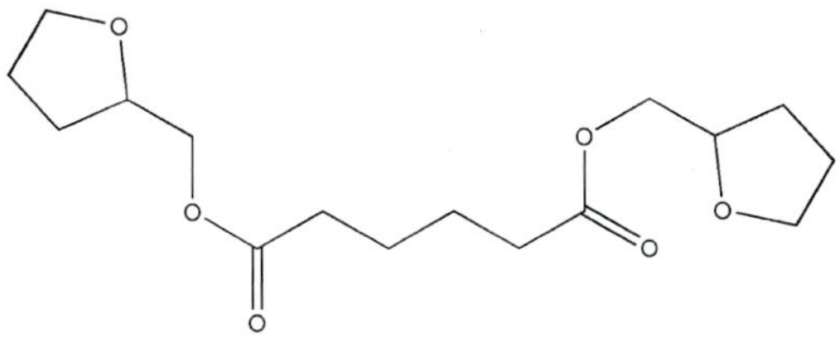

Ditetrahydrofurfuryl adipate *n.* A heterocyclic plasticizer for cellulose acetate-butyrate.

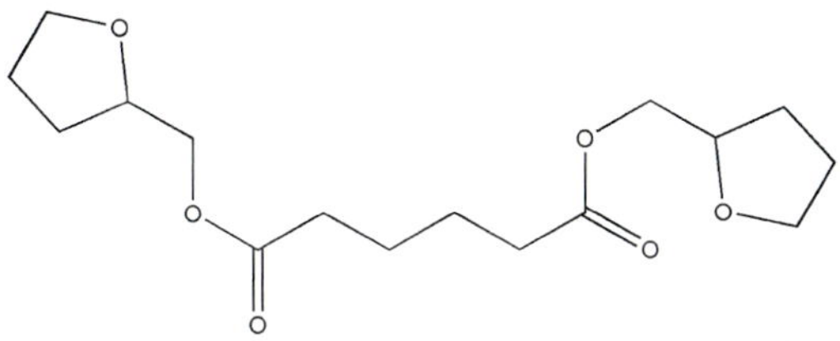

Ditridecyl phthalate (DTDP) *n.* $C_6H_4(COOC_{13}H_{27})_2$. A primary plasticizer for PVC, also compatible with cellulosics and polystyrene. In vinyls, it imparts resistance to high temperatures and to extraction by hot soapy water, excellent flexibility, and anti-fogging properties.

Ditridecyl thiodipropionate *n.* $(C_{13}H_{27}OOCCH_2CH_2)_2S$. A stabilizer, plasticizer, and softening agent.

Diundecyl phthalate (DUP) *n.* A plasticizer characterized by low volatility and good low-temperature properties compared to other phthalates.

Divergent die \də-ˈvər-jənt-\ *n.* A die for hollow articles in which the internal channels leading to the orifice increase in cross section toward the lip.

Divided threadline extrusion *n.* Spinning of two separate threadlines from one spinneret.

Divinyl B Syn: butadiene.

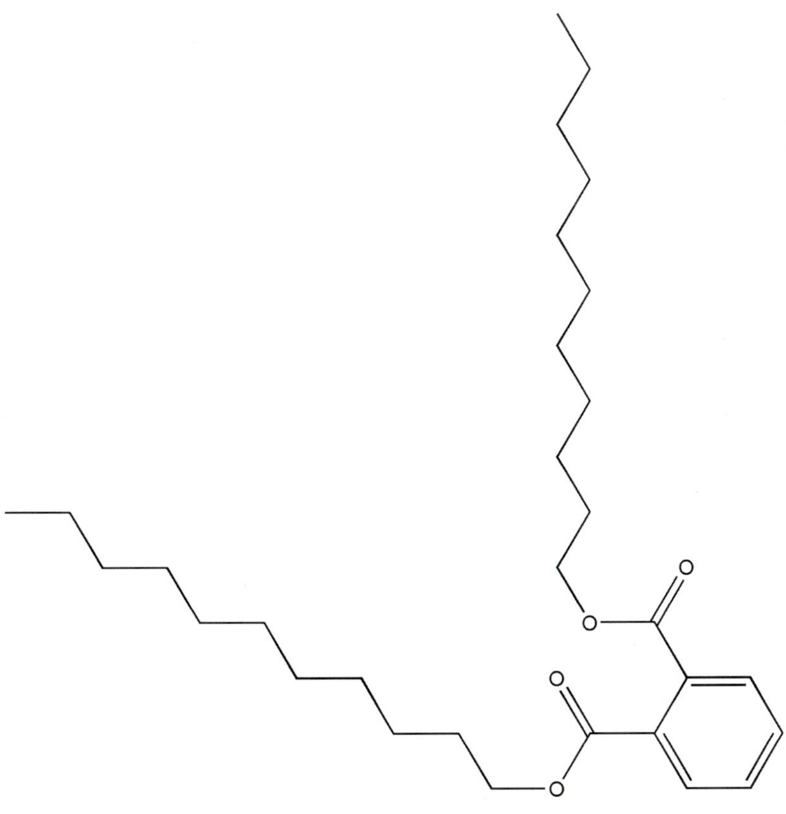

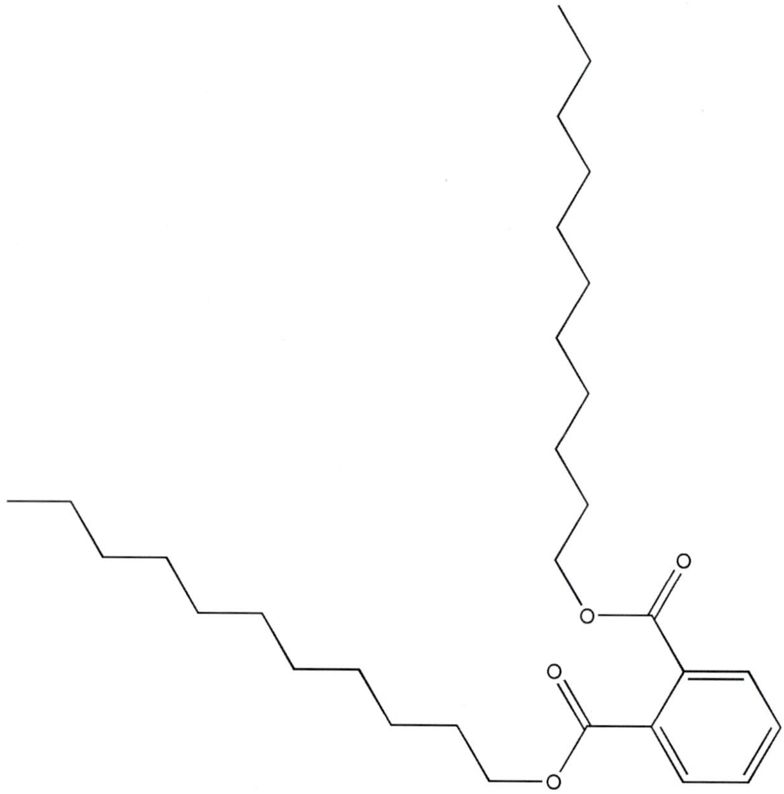

Divinylbenzene (DVB, vinylstyrene) *n.* A monomer derived from styrene, used in making ion-exchange resins, synthetic rubbers, and casting resins. The commercial product contains 55% mixed *m*- and *p*-isomers, the rest being ethylvinylbenzenes. It is often used along with styrene as a reactive monomer in the production of polyester resins, to which it imparts a higher degree of crosslinking and superior chemical resistance.

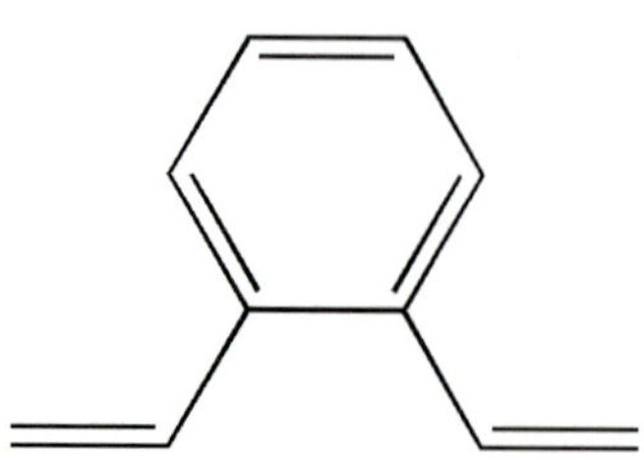

Divinyl polymerization, polymers *n.* Polymerization of a monomer, which contains two vinyl groups. When the two double bonds are conjugated (as in 1,3-dienes) the polymerization is called diene polymerization.

DIY *n.* Abbreviation for "do it yourself".

Dl-limonene *See dipentine.*

DLVO theory *n.* A dispersion stabilizing theory in which zeta potential is used to explain that as two particles approach one another their ionic atmospheres begin to overlap and a repulsion force is developed. The theory was prepared by Dergaquin and Landau in the Soviet Union and Verwey and Overbeck in the Netherlands, thus the DLVO.

DMA *n.* Abbreviation for dimethylaniline.

DMEP *n.* Abbreviation for di(2-methoxyethyl) phthalate.

DMF *n.* Abbreviation for dimethyl formamide.

DMG *n.* Abbreviation for dimethyl glutarate.

DMSO *n.* Abbreviation for dimethyl sulfoxide.

DMT *n.* Abbreviation for dimethyl terephthalate.

DNA *n.* Abbreviation for dinonyl adipate.

DNHZ *n.* Abbreviation for di-*n*-hexyl azelate.

DNODA *n.* Abbreviation for di-*n*-octyl-*n*-decyl adipate.
See n-octyl-n-decyl adipate.

DNODP *n.* Abbreviation for di(*n*-octyl-*n*-decyl) phthalate.

DNP *n.* Abbreviation for dimethyl phthalate.

DNP *n.* Abbreviation for dinonyl phthalate.

DNPT *n.* Abbreviation for dinitrosopentamethylenetetr-amine.

DNTA *n.* Abbreviation for dinitrosoterephthalamide.

DOA *n.* Abbreviation for dioctyl adipate or di (2-ethylhexyl) adipate.

Dobby \ˈdä-bē\ [perhaps fr. *Dobby*, nickname for *Robert*] (1878) *n.* (1) A mechanical attachment on a loom. A dobby controls the harnesses to permit the weaving of geometric figures. (2) A loom equipped with a dobby. (3) A fabric woven on a dobby loom.

Doctor \ˈdäk-tər\ [ME *doctour* teacher, doctor, fr. MF & ML; MF, fr. ML *doctor*, fr. L, teacher, fr. *docēre* to teach] *n.* (1) Device for spreading a thin film of even thickness on base material. In the paint industry, it is used to prepare paint and varnish films of even and predetermined thickness. (2) In gravure printing and coating, a blade that scrapes off the excess ink or lacquer from the surface of the etched cylindrical roll just prior to printing, leaving the "cells" filled with ink or lacquer {*Doctor knife* G Abstreifmesser n; Rakelmesser n, F racle f, S cuchilla de recubrimiento, cuchilla f, I raschiatore m; racla f}.

Doctor-bar blade *n.* A scraper mechanism that regulates the amount of adhesive on the spreader rolls or on the surface being coated.

Doctor mark or streak *n.* Streak or ridge in coated fabrics caused by a damaged doctor blade.
Also called knife mark.

Doctor roll *n.* Roller mechanism, which is revolving at a different surface speed, or in an opposite direction, resulting in a wiping action for regulating the adhesive supplied to the spreader roll.

Doctor streak *n.* A defect in printed fabrics consisting of a wavy white or colored streak in the warp direction. It is caused by a damaged or improperly set doctor blade on the printing machine.

Doctor test *n.* Method for detecting the presence of mercaptan sulfur in gasolines, naphthas, and kerosenes.

Dodecanoic acid *See lauric acid.*

4-Dodecyloxy-2-hydroxbenzophenone *n.* An ultraviolet inhibitor for polyethylene and polypropylene, also suggested as suitable for PVC, polystyrene, polyesters, and surface coatings such as those based on cellulosic and acrylic resins.

Doeskin finish \ˈdō-ˌskin\ *n.* A soft low nap that is brushed in one direction. Cloth with this type of finish is used on billiard tables and in men's wear.

DOF *n.* Abbreviation for dioctyl fumarate.

DOFF A set of full bobbins produced by one machine (a roving frame, a spinning frame, or a manufactured filament-yarn extrusion machine).

Doffer *n.* (1) The last or delivery cylinder of the card from which the sheet of fibers is removed by the doffer comb. (2) An operator who removes full bobbins, spools, containers, or other packages from a machine and replaces them with empty ones.

D

Doffer comb *n.* A reciprocating comb, the teeth of which oscillate close to the card clothing of the doffer to strip the web of fibers from the card.

Doffer loading *n.* Fibers imbedded so deeply into the doffer wire clothing that the doffer comb cannot dislodge them to form a traveling web.

Doffing *n.* The operation of removing full packages, bobbins, spools, roving cans, caps, etc., from a machine and replacing them with empty ones.

Dogbone *n.* A slang term for the dumbbell shape of tensile-test specimens having relatively wide ends for gripping and a narrower, gage-length center within which all plastic deformation and breakage occur. *See tensile bar.*

Dogskin *See orange peel* (coatings defect).

Doily \ˈdói-lē\ [*Doily* or *Doyley fl* 1711 London draper] (1711) *n.* In filament winding, the planar reinforcement that is applied to a local area between windings to provide extra strength in an area where a cut-out is to be made; for example, a port opening.

DOIP *n.* Abbreviation for dioctyl isophthalate. *See di(2-ethylhexyl) isophthalate.*

Dolomite \ˈdō-lə-ˌmīt\ [F, fr. *Déodat* de *Dolomieu* † 1801 F geoloist] (1794) *n* $CaCO_3{\cdot}MgCO_3$. A mineral having the empirical composition, 1 mol of calcium carbonate and 1 mol of magnesium carbonate. Any calcium carbonate rock containing 20% or more of magnesium carbonate. *See calcium carbonate, natural.*

Domain \dō-ˈmān\ [ME *domayne*, fr. MF *domaine, demaine*, fr. L *dominium*, fr. *dominus*] (15c) *n.* A region in a ferromagnetic substance within which the magnetic moments of all the atoms are aligned.

Domains *n.* A term used in non-crystalline systems in which the chemically different sections of the chain separate, generating two or more amorphous phases.

Domed *n.* Having a symmetrical convex protrusion in the surface. Opposite of dished.

Dominant wavelength *n.* **CIE.** Wavelength of spectrally pure light which, when additively mixed with the illuminant and adjusted for luminance, would match the color. It is expressed in wavelength units, nm (or μm), and is determined graphically from the chromaticity diagram or by suitable computation. The dominant wavelength of the non-spectral colors, the purples, is described in terms of the dominant wavelength of the green, to which they are the additive complements, followed by the letter c. Thus, a dominant wavelength designation of 520c describes a purple, which is additively complementary to a green having a dominant wavelength of 520 nm. Dominant wavelength describes the hue of a color, although the visual evaluation of constant hue does not necessarily coincide with constant dominant wavelength. Commission Internationale De L'Eclairage (CIE), CIE Central Bureau, Kegelgasse 27, A-1030, Wien, Austria. [English: International Commission on Illumination, Vienna, Austria], www.cie.co.at/headerbasic.html.

Domino \ˈdä-mə-ˌnō\ *n.* Early French wallpaper made in small sheets, originally imitating marble, later patterned.

Donegal \ˌdä-ni-ˈgól\ *n.* A tweed fabric with colorful slubs woven in, donegal is used for suits and coats.

Doorframe *n.* An assembly built into a wall consisting of two upright members (jambs) and a head (lintel) over the doorway; encloses the doorway and provides support on which to hang the door.

DOP *n.* Abbreviation for dioctyl phthalate. *See di(2-ethylhexyl) phthalate.*

Dope \\ˈdōp\\ [D *doop* sauce, fr. *dopen* to dip; akin to OE *dyppan* to dip] (1807) *n.* (1) A solution of a cellulosic plastic, historically cellulose nitrate, used for treating fabrics. (2) Cellulose ester lacquer for adhesive or coating purposes.

See spinning solution.

Dope cotton *n.* Designation of high viscosity nitrocellulose, i.e., 20–30, 40, and 70 s nitrocellulose.

See nitrocellulose.

Dope dyed *See dyeing, mass-colored.*

Doping *n.* The addition of controlled small amounts of a foreign substance to an otherwise pure substance.

Doppler effect \\ˈdä-plər\\ [Christian J. *Doppler*] (1905) (Light) *n.* The apparent change in the wavelength of light, produced by the motion in the line of sight of either the observer or the source of light.

Doppler effects *n.* Effects on the apparent frequency of a wave train produced (1) by motion of the source toward or away from the stationary observer, and (2) by motion of the observer toward or from the stationary source; the motion in each case being with reference to the (supposedly stationary) medium. For sound waves, the observed frequency f_o, in cycles/s, is given by

$$f_o = \frac{v + w - v_o}{v + w - v_s} f_s,$$

where v is the velocity of sound in the medium, v_o the velocity of the observer, v_s the velocity of the source, w the velocity of the wind in the direction of sound propagation, and f_s is the frequency of source. For optical waves

$$f_o = f_s \sqrt{\frac{c + v_r}{c - v_r}},$$

where v_r is the velocity of the source relative to the observer and c is the speed of light.

Dormer \\ˈdór-mər\\ [MF *dormeor*, fr. L *dormitorium*] (1592) *n.* Gable-like projection or a window set upright in a sloping roof, or the roofed projection in which this window is set.

DOS *n.* Abbreviation for dioctyl sebacate.

See di(2-ethylhexyl) sebacate.

Dosimeter \\dō-ˈsi-mə-tər\\ [LL *dosis* + ISV -*meter*] (1938) *n.* detector worn by workers to measure the amount of an environmental agent, usually radioactivity, but sometimes noise or noxious gases, to which the worker has been exposed during a working shift or longer period of time.

Dot *n.* The individual element of a halftone printing plate.

DOTP *n.* Abbreviation for dioctyl terephthalate.

Dotted Swiss (ca. 1924) *n.* A sheer cotton or cotton blend fabric with small dot motif, dotted Swiss is used for dress goods, curtains, baby clothes, etc.

Double back *n.* A secondary backing glued to the back of carpet, usually to increase dimensional stability.

Double bond *n.* A type of covalent bond (pi or π bond in C=C), common in organic chemistry, in which two pairs of electrons are shared between two elements. The double bond may be symbolized either by ":" or "=", as in ethylene, $CH_2{:}CH_2$, or dimethyl ketone $(CH_3)_2C{=}O$. Elements possess a definite number of valency linkages or bonds, and these are employed in uniting with other elements. When all these bonds are completely employed in such unions, saturated compounds result. If, however, a residual bond or linkage remains, this may be involved with another in forming a reactive or unsaturated double bond. If a hydrogen atom is removed from each of the carbon atoms, two spare linkages occur, and an unsaturated double bond results. Double bonds are centers of great reactivity, and are able to accept addition of suitable elements such as hydrogen, chlorine, etc., to yield fully saturated compounds. The reactivity of double bonds in drying oils is responsible for their drying (autoxidation) and polymerizing properties.

Double-bond addition *n.* Chemical reaction whereby a double-bonded molecule (typically an alkene) is reacted with a reagent to form a single product. When an alkene undergoes addition, two groups add to the carbon atoms of the double bond, and the carbons become saturated.

Double cloth construction *n.* Two fabrics are woven in the loom at the same time, one fabric on top of the other, with binder threads holding the two fabrics together. The weave on the two fabrics can be different.

Double decomposition *n.* Double decomposition consists of a simple exchange of the parts of two substances to form two new substances.

Double end *n.* Two ends woven as one in a fabric. A double end may be intentional for fabric styling, or accidental, in which case a fabric defect results.

Double helix (1954) *n.* Physical structure of DNA, whereby two complementary strands are joined by hydrogen bonds between the base pairs. This double strand coils into a helical arrangement.
See DNA.

Double-hung sash *n.* Window divided into two sections, one lowering from the top and the other rising from the bottom.

Double knit fabric (1895) *n.* A fabric produced on a circular-knitting machine equipped with two sets of latch needles situated at right angles to each other (dial and cylinder).

Double pick *See mispick.*

Doubler (13c) *v.* In filament winding, a local area with extra wound reinforcement, either wound integrally with the part or wound separately and bonded to the part.

Double-ram press *n.* A press for injection or transfer molding in which two distinct systems of the same kind (hydraulic or mechanical), or of different kinds, create respectively the injection or transfer force and the clamping force.

Double refraction (1831) *n.* The refraction of light in two slightly different directions to form two rays or vector components. Each ray is polarized, and their vibration directions are perpendicular to each other. Furthermore, each ray has a different velocity, and therefore a different refractive index.
See birefringence.

Double-screw extruder *See extruder, twin-screw.*

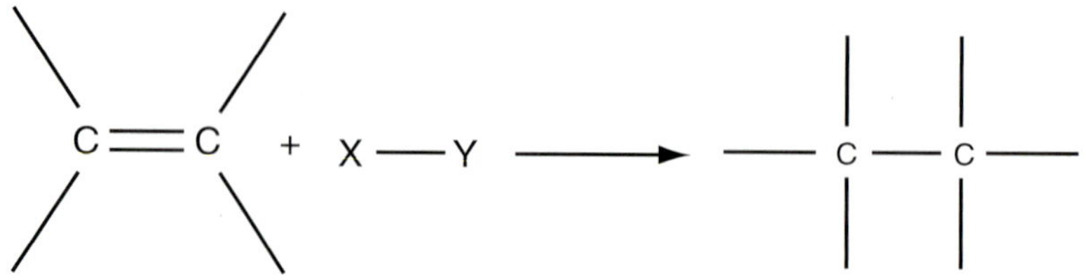

Double seam can *n.* Can formed by interlocking the edges of both the end and the body, double folding and rolling firmly together.

Double selvage *See rolled selvage.*

Double-shot molding (two-shot molding, insert molding, two-color molding, and over-molding) *n.* A process for making two-color or two-material parts by means of successive molding operations. The basic process includes the steps of injection molding one part, transferring this part to a second mold as an insert, and molding the second material against the first. Examples of parts made by double-shot molding are computer keys, pushbuttons, telephone-keypad buttons, and other such products in which indicia must resist heavy wear and remain permanently legible. In a modification of the process, cups and the like with differently colored insides and outsides are made automatically by means of a machine equipped with two injection molders and a swinging platen carrying two cup cores indexed around a central tie-bar, bringing the molds into position for each successive shot. This arrangement permits simultaneous molding of both shots. Some of these products are made today by coinjection.

Double-skinned sheet *n.* A plastic sheet consisting of a relatively thick center section bonded on each face to thinner layers that differ in color or composition from the center section. Today such sheets are usually made by coextrusion of the three plastic layers and, in many cases, two additional adhesive layers between the core and the skin.

Double spread *See spread.*

Doubletone ink *n.* A type of printing ink which produces the effect of two-color printing with a single impression. These inks contain a soluble toner, which bleeds out to produce a secondary color.

Double weave *n.* A fabric woven with two systems of warp or filling threads so combined that only one is visible on either side. Cutting the yarns that hold the two cloths together yields two separate cut pile fabrics.

D

Doubling *n.* (1) A process for combining several strands of sliver, roving, or yarn in yarn manufacturing. (2) The process of twisting together two or more singles or plied yarns, i.e., plying. (3) A British term for twisting. (4) The term doubling is sometimes used in a sense opposite to singling. This is unintentional plying. (5) A yarn, considerably heavier that normal, produced by a broken end becoming attached to and twisting into another end.

Dough \ˈdō\ [ME *dogh*, fr. OE *dāg*; akin to OHGr *teic* dough, *fingere* to shape, Gk *teichos* wall] (dough-molding compound) *n.* A term sometimes used for a reinforced-plastic mixture of dough-like consistency in an uncured or partly cured state. A typical dough consists of polyester resin, glass fiber, calcium carbonate, lubricants and catalysts. The compounds are formed into products either by hand layup or, more usually, by compression molding.

Douppioni *n.* A rough or irregular yarn made of silk reeled from double or triple cocoons. Fabrics of douppioni have an irregular appearance with long, thin slubs. Douppioni-like yarns are now being spun from polyester and/or rayon staple.

Dowel \ˈdaú(-ə)l\ [ME *dowle*; akin to OHGr *tubili* plug, LGk *typhos* wedge] (14c) (dowel pin) *n.* A hardened steel pin, usually having a slight taper, used to maintain alignment between two or more parts of a mold or machine.

Dowel bushing *n.* A hardened steel insert in the portion of a mold that receives the dowel pin.

Downdraft metier *n.* A dry-spinning machine in which the airflow within the drying cabinet is in the same direction as the yarn path (downward).

Downgrade *n.* In quality control, the lowering of the grade and/or value of a product due to the presence of defects.

Downtwister *n.* A cap, ring, or flyer twisting frame.

Downtwisting *n.* A process for inserting twist into yarn in which the yarn passes downward from the supply package (a bobbin, cheese, or cone) to the revolving spindle. The package or packages of yarn to be twisted are positioned on the creel, and the ends of yarn are led downward through individual guides and stop motions to the positively driven feed roll and from there to the revolving take-up package or bobbin, which inserts twist.

Dowtherm® *n.* The trade name (Dow Chemical Co., TX, USA) of a liquid heat-transfer medium consisting of biphenyl and phenyl ether in eutectic ratio. High-boiling and stable, it is less used today than formerly because of its suspected carcinogenicity. Bp, 258°C. Used for heat transfer.

DOZ *n.* Abbreviation for dioctyl azelate. *See di(2-ethylhexyl) azelate.*

DP *n.* Abbreviation for degree of polymerization.

DPA *n.* Abbreviation for diphenylamine.

DPCF (DPCP) *n.* Abbreviation for diphenyl cresyl phosphate, a plasticizer.

DPP *n.* Abbreviation for diphenyl phthalate.

Draft *n.* A slight taper in a mold wall, proceeding inward from the parting surface to the bottom of the cavity, designed to facilitate removal of the molded object from the mold. When the taper is reversed, tending to impede removal of the article, the term *back draft*, or *reverse draft* is employed. In weaving, a pattern or plan for drawing-in.

Draft angle *n.* In a mold, the angle in the profile plane with its vertex at the bottom of the mold or cavity, between the side of the mold and the vertical plane. In most cases, 1° is adequate for smooth ejection of parts from the mold.

Drafting *See drawing* (1).

Draft ratio *n.* The ratio between the weight or length of fiber fed into various machines and that delivered from the machines in spun yarn manufacture. It represents the reduction in bulk and weight of stock, one of the most important principles in the production of yarn from staple fibers.

Drag (brush drag) *n.* Resistance encountered when applying a coating by brush. The British term for this condition is "gummy".

Drag flow *n.* (1) In general, the laminar flow of a viscous liquid that is contained between two surfaces, one of which is moving relative to the other. (2) In the metering section of an extruder screw, the rate of drag flow is the component of total material flow in the down-channel direction caused by the relative motion between the screw and cylinder. If the screw were discharging freely, the output rate would be equal to the drag-flow rate. (3) In a wire-coating die, the flow generated by the relative motion of the wire through the stationary die. In both extruders and wire-coating dies, pure drag flows almost never occur, but are altered in complex ways by opposing or augmenting pressure fields and the non-Newtonia nature of the melts.

Dragged in filling *See pulled-in filing.*

Dragon's blood *n.* Dark red resinous exudation, obtained from the surface of the fruit of the *Calamus draco* and similar species. Used to some extent as a colorant in coatings.

Drainage fabrics *See geotextiles.*

Dralon *n.* Poly(acrylonitrile). Manufactured by Bayer, Germany.

Drape \ˈdrāp\ [prob. back-form. fr. *drapery*] (1847) *v.* (1) With reference to plastics films and coated fabrics, their ability to hang without creases and to form graceful folds when used as draperies, shower curtains, and the like. (2) In sheet thermoforming, the ability of the preheated sheet to conform to the mold under the influence of gravity.

Drape-assist frame *n.* In sheet thermoforming, a frame made from thin wires or thick bars shaped to the periphery of the depressed areas of the mold and suspended above the sheet to be formed. During forming, the assist frame drops down, drawing the softened sheet tightly into the mold and thereby preventing webbing between high areas of the mold and permitting closer spacing in arrays of multiple molds.

Drape forming (drape vacuum forming, drape thermoforming) *n.* Forming a thermoplastic sheet into three-dimensional articles by clamping the sheet in a movable frame, heating the sheet, then lowering it to drape over the high points of a male mold. Vacuum is then applied to complete the forming.

See also sheet thermoforming.

Draw back *n.* A crossed end; an end broken during warping that when repaired was not free or was tied in with an adjacent end or ends overlapping the broken end. The end draws or pulls back when unwound on the slasher.

Also see sticker (1).

Draw crimping *See draw-texturing.*

Drawdown *n.* (1) A film of ink deposited on paper by a smooth edged blade to evaluate the undertone and masstone of the ink. Syn: pulldown. (2) A thin film of even thickness on base material cast by means of a drawdown bar. Syn: casting. (3) In extrusion, the process of pulling the extrudate away from the die at a lineal speed greater than the average velocity of the melt in the die, thus reducing extrudate's cross-sectional dimensions. The term is also used by blow molders to denote the decrease in parison diameter and wall thickness due to gravity.

Draw down *n.* The amount by which manufactured filaments are stretched following extrusion.

Also see drawing (2).

Drawdown bar *n.* Rectangular metal bar designed to deposit a specified thickness of wet coating film on test panels or other substrates.

Drawdown ratio *n.* In extrusion of fiber spinning, the ratio of the cross-sectional area of the die opening to that of the finished product. In making sheet or cast film, where the sheet width is nearly equal to the width of the die opening, the ratio of the thickness of the die opening to that of the final sheet is sometimes spoken of as the drawdown ratio.

Draw frame blends *n.* Blends of fibers made at the draw frame by feeding in ends of appropriate card sliver. This method is used when blend uniformity is not a critical factor.

Drawinella *n.* Cellulose triacetate. Manufactured by Wacker, Germany.

Drawing *n.* The process of stretching a thermoplastic filament, sheet, or rod to reduce its cross-sectional area and/or to improve its physical properties by orientation.

D

Drawing *n.* (1) The process of attenuating or increasing the length per unit weight of laps, slivers, slubbings, or rovings. (2) The hot or cold stretching of continuous filament yarn or tow to align and arrange the crystalline structure of the molecules to achieve improved tensile properties.

Drawing in *n.* In weaving, the process of threading warp ends through the eyes of the heddles and the dents of the reed.

Drawn tow *n.* A zero-twist bundle of continuous filaments that has been stretched to achieve molecular orientation (tows for staple and spun yarn application are usually crimped).

Drawout *n.* A method of application in which the coating is applied, usually on paper, by spreading the pigment-vehicle mass evenly with a broad knife. Useful for examination of masstone, undertone, opacity, printing strength, and in some cases, fadeometer ratings of pigments.

Draw process *n.* Process for shaping a tubular or solid cylindrical specimen. The process involves passing the material through a die where it is then gripped by a suitable device, which can pull it forward on a mechanical or hydraulic bench. A reduction in the diameter (solid specimens) and wall thickness (tubular specimens) results. Multi-die drawing is necessary in wire production.

Draw ratio *n.* (1) A measure of the degree of stretching during the orientation of a fiber or filament, expressed as the ratio of the cross-sectional area of the undrawn material to that of the drawn material. (2) In monofilament manufacture, the filament is wrapped several times around a vertical, driven roll, passed through a warming oven, then wrapped again around a second roll running faster than the first one. In this way the filament is stretched and oriented and its cross section is reduced. The ratio of the surface speed of the second, faster roll to that of the first equals the draw ratio.

Draw resonance *n.* A phenomenon occurring in film and filament extrusion in which the extrudate is drawn into a quenching bath at a certain critical speed that creates a cyclic pulsation in the cross-sectional area of the extrudate. The pulsation increases with rising drawing speed until the filament or film eventually breaks at the bath surface. Draw resonance has been observed while extruding polypropylene, polyethylene, and polystyrene.

Draw sizing *n.* A system linking draw warping and sizing in a continuous process. A typical system includes the following elements: (1) creel, (2) eyelet board, (3) warp-draw machine, (4) intermingler, (5) tension compensator and break monitor, (6) sizing bath, (7) dryers, and (8) waxing and winding units.

Draw-texturing *n.* In the manufacture of thermoplastic fibers, the simultaneous process of drawing to increase molecular orientation and imparting crimp to increase bulk.

Draw-twisting *n.* The operation of stretching continuous filament yarn to align and order the molecular and crystalline structure in which the yarn is taken up by means of a ring-and-traveler device that inserts a small amount of twist (usually 0.25–0.5 turn/in.) into the drawn yarn.

Draw-warping *n.* A process in which a number of threadlines, usually 800–2000 ends of POY feedstock, are oriented under essentially equal mechanical and thermal conditions by a stretching stage using variable speed rolls, then directly wound onto the beam. This process gives uniform end-to-end properties.

Draw-winding *n.* The operation of stretching continuous filament yarn to align or order molecular and crystalline structure. The drawn yarn is taken up on a parallel tub or cheese, resulting in a zero-twist yarn.

Dressed lumber *n.* Lumber having one or more of its faces planed smooth.

Also called dressed stuff and surfaced lumber.

Dressing compound *n.* Bituminous liquid used hot or cold for dressing the exposed surface of roofing felt.

Dribbling or drooling *n.* A condition sometimes occurring in injection molding between shots in which melt drips from the withdrawn nozzle. Drooling was a common problem with 6/6 nylon in the 1950s because of depolymerization caused by moisture pickup in feed hoppers and overheating in the old, torpedo-type heating cylinders.

Drier (1528) *n.* A composition, which accelerates the drying of oil, paint, printing ink, or varnish. They consist mainly of metallic salts, which exert a catalytic effect on the oxidation and polymerization of the oil vehicles employed and are available in both solid and liquid forms.

Also called siccative.

Drier absorption *See drier dissipation.*

Drier dissipation *n.* A loss in catalytic power of a drier due to a physical absorption or a chemical reaction with certain pigments.

Driers drop-weight test (falling-weight test) *n.* Any test of impact resistance in which known weights are dropped once or repeatedly on the test specimen. Examples are ASTM D 4272 (plastic films), D 3029, D 4226, and D 4495 (rigid PVC sheet and parts), and F 736, Section 15 (polycarbonate sheet).

Also see free-falling-dart test.

Drift *See overspray and creep.*

Drill (1743) *n.* A strong denim-like material with a diagonal 2 × 1 weave running toward the left selvage. Drill is often called khaki when it is dyed that color.

Drop *n.* One vertical descent of a scaffold.

Drop black *n.* Another name for bone black. *See bone black.*

Drop ceiling *n.* A form of decoration in which the ceiling paper is brought down to a suitable depth on the walls of a room and divided from the sidewall by a border or molding. Gives the illusion of a lower ceiling.

Drop cloth *n.* Cover used to protect floors and furniture, etc., from paint spillage and droppings.

Dropped stitches *n.* A defect in knit cloth characterized by recurrent cuts in one or more wales of a length of cloth.

Drop siding *n.* An exterior wall cladding of wooden boards (or strips of other material, such as aluminum or vinyl), which are tongued and grooved or rabbeted and overlapped so that the lower edge of each board interlocks with a groove in the board immediately below it. *Also known as novelty siding and rustic siding.*

Drop stitch *n.* (1) An open design made in knitting by removing some of the needles at set intervals. (2) A defect in knit fabric.

Drop wires *n.* A stop-motion device utilizing metal wires suspended from warp or creeled yarns. When a yarn breaks, the wire drops, activation the switch that stops the machine.

Drum coloring *See dry coloring.*

Drum extruder *n.* A plasticating machine having a rotating cylindrical element inside a concentric or eccentric housing. Pellets or powder are fed into the gap between drum and housing at the top. Shear action melts the solids and the melt exists through a slot about 270° around from the feed point, usually passing into a pressure-developing device that can form an extruded product. A wiper bar above the die prevents melted material from recirculating.

Drum tumbler *n.* A device used to mix plastic pellets with color concentrates and/or regrind. The materials are charged into cylindrical drums that are tumbled end-over-end or rotated about an inclined axis for a time sufficient to thoroughly blend the ingredients.

D

Dry *vt.* To change the physical state of an adhesive on an adherend by the loss of solvent constituents by evaporation or absorption, or both. A film is considered dry when it feels firm to the finger, using moderate pressure. *See also cure and set.*

Dry back *n.* The change in color or finish of an ink film as it dries.

Dry blend *n.* A dry, free-flowing mixture of resin powder, typically PVC, with plasticizers and other additives, prepared by blending the components in a large, closed mixing bowl with a high-speed rotor at its bottom, stopping the action at temperatures comfortably below the fluxing point. Dry blends are generally more economical feedstocks for extrusion than molding powders and pellets made by plasticating and extrusion, but in some cases have been difficult to process.

Dry-blend *v.* To combine ingredients, typically in a high-speed mixer, to produce a dry blend.

Dry brush *v.* Technique in which paint or ink is applied sparingly with a semidry brush.

Dry bulk *See apparent density.*

Dry cleaning (1817) *v.* Removing dirt and stains from fabrics or garments by processing in organic solvents (chlorinated hydrocarbons or mineral spirits).

Dry color *v.* Term used loosely to describe dry colored pigments. A concentrated pigment or coloring matter suitable for grinding or dispersing and use in coatings.

Dry coloring *v.* The process of combining colorants to molding compounds and resin pellets by tumble-blending them with dyes, pigments, or color concentrates. This process enables custom molders and extruders to carry a large inventory of uncolored compound, preparing smaller batches of colored compounds to customers' specifications.

Dryer (1528) *n.* A mechanical device designed to accelerate the drying of inks. Equipment used to remove moisture or water from plastic material or substrates during processing.

Dry filling *n.* The application of finishing chemicals to dry fabric, usually by padding. Dry forming: The production of fiber webs by methods that do not use water or other liquids, i.e., air-laying or carding.

Dry-hard *n.* Film is considered dry-hard when any mark left by the thumb, exerting maximum downward pressure on a film, is completely removed by a light polishing with a soft cloth.

Dry-hiding *n.* Increase in the hiding power of paint, which occurs in the drying process. It is most significant in non-glossy paints, which, after drying, have pigments or extender pigments protruding from the surface of the paint vehicle, which scatter the incident light and add to the hiding.

Dry hydrate *n.* Calcium hydroxide; hydrated lime powder.

Dry ice (1925) *n.* Solidified carbon dioxide. *See carbon dioxide.*

Drying *n.* (1) Process of change of coatings from the liquid to the solid state, due to evaporation of the solvent, physico–chemical reactions of the blinding medium, or a combination of these causes. (2) Process of removing moisture from pigments. It is carried out either in a current of hot air or in a vacuum.

Drying cylinders *n.* Any of a number of heated revolving cylinders for drying fabric or yarn. They are arranged either vertically or horizontally in sets, with the number varying according to the material to be dried. They are often internally heated with steam and Teflon coated to prevent sticking.

Drying of ink *n.* The conversion of an ink film to a solid state. This can be accomplished by any of the following means, either singly or in combination: oxidation, evaporation, polymerization, penetration, gelation, and precipitation.

Drying oil *n.* Any of several plant-derived polyunsaturated oils, such as linseed, oiticica, and tung, that, when exposed to air, form dry, tough, durable films. Linseed oil to be used in paints and varnishes today is usually boiled with cobalt or manganese salts of linoleic or naphthenic acids to shorten its "drying" time.

Drying temperature *See temperature drying.*

Drying time *n.* (1) Time required for an applied film of a coating to reach the desired stage of cure, hardness, or non-tackiness (*See dry, dry-hard, dust-free time, set-to-touch time, sand-dry, through-dry, dry-to-handle time, touch-dry, tack-free, surface drying, dry-to-recoat time, and dry-to-sand*). (2) The time required for an ink to form a tack-free surface after being applied to the paper, or other printed surface.

Dry-laid *n.* Non-woven web made from dry fiber. Usually refers to fabrics from carded webs versus air-laid non-wovens, which are formed from random webs.

Dry laminate *n.* A laminate containing insufficient resin for complete bonding of the reinforcement.

Dry layup *n.* The construction of a laminate by layering preimpregnated, partly cured reinforcements in or on a mold, usually followed by bag molding or autoclave molding.

Dry offset *n.* An indirect letterpress process in which the ink is transferred from a relief plate to a blanket and then to the stock. *Also known as letterset.*

Dry purge *n.* In extrusion, preparatory to shutting down operation after the feed hopper has been emptied, running the extruder until no more plastic emerges from the head end.

Dry rot (1795) *n.* Special type of brown rot, causing underspread damage in buildings. In the USA, the causal organism is *Portia incrassate.*

Dry spinning *See spinning.*

Dry spot *n.* An imperfection in reinforced plastics, an area of incomplete surface film where the reinforcement has not been wetted with resin (ASTM D 883).

Dry spray *n.* Overspray or bounceback; sand finish due to spray particles being partially dried before reaching the surface.

Dry strength *n.* The strength of an adhesive joint determined immediately after drying or curing under specified conditions or after a period of conditioning in a standard laboratory atmosphere.
See also wet strength and strength, dry.

Dry strippable paper *See strippable coating.*

Dry tack *See tack, dry.*

Dry-through *n.* Film is considered dry-through when no loosening, detachment, wrinkling or other distortion of the film occurs when the thumb is borne downward while simultaneously turning the thumb through an angle of 90° in the plane of the film. The arm of the operator is kept in a vertical straight line from the wrist to the shoulder and maximum pressure is exerted by the arm.

Dry-to-handle time *n.* Time interval between application and ability to pick up without damage.

See drying time.

Dry-to-recoat time *n.* Time interval between the application of the coating and its ability to receive the next coat satisfactorily.

Dry-to-sand *n.* That stage of drying when a coating can be sanded without the sandpaper sticking or clogging.

Dry-to-touch time *n.* Interval between application and tack-free time.

See set-to-touch time.

Dry wall *n.* (1) An interior wall, constructed with a dry-wall finish material such as gypsum board or plywood. (2) In masonry construction, a self-supporting rubble or ashlar (squared building stone) wall built without mortar.

Dry winding *n.* Filament winding with preimpregnated roving, as distinguished from *wet winding* in which unimpregnated roving is pulled through a resin just prior to winding on a mandrel.

DSC Differential scanning calorimetry, measurement melting temperatures, glass transition temperature and heat of melting; lower temperature range than DTA.

See differential scanning calorimetry.

DSTDP *n.* Abbreviation for distearylthiodipropionate.

DTA *n.* Differential thermal analysis measurement of melting temperature; generates a plot of "heat versus temperature" and semiquantiative, but higher temperature range than DSC. DTA is capable of measuring events (e.g., melting, crystallization, and decomposition) but not heat of melting or crystallization whereas DSC is quantitative and appropriate for thermally characterizing polymers. Abbreviation for differential thermal analysis.

DTDP *n.* Abbreviation for ditridecyl phthalate.

Dual-sensor control *n.* An improved system for controlling cylinder and plastic temperatures in extruders. For each heating zone, there are two temperature sensors, one in a shallow well slightly beneath the heater, the other deep, just outside the lining layer and near the plastic. An average of the two signals is used to control the electrical heat input (Eurotherm/Welex).

Dubbing out *n.* (1) Filling in hollow and irregular surfaces and leveling walls with plaster before regular plasterwork. (2) Forming, very roughly, a plaster cornice, before the final plaster coat is applied.

Duck \ˈdək\ [Du *doek* cloth; akin to OHGr *tuoh* cloth] (1640) *n.* A compact, firm, heavy, plain weave fabric with a weigh of 6–50 oz/yard^2. Plied yarn duck has plied yarn in both warp and filling. Flat duck has a warp of two single yarns woven as one and a filling of either single or plied yarn.

Duck eye *See spinning.*

Ductile fracture \ˈdək-tᵊl\ [MF & L, MF, fr. *ductilis,* from *ducere*] (14c) (ductile rupture) *adj.* The breaking or tearing, most commonly in tension, of a test specimen or part after considerable unrecoverable stretching (plastic strain) has occurred. Since the mode of fracture depends on conditions as well as material, the distinction between ductile and brittle fracture, which latter occurs after relatively little, recoverable strain, is not always clear. Low temperatures, especially below the glass transition (T_g), and high rates of strain favor brittle behavior, while the opposites favor ductile behavior.

Ductility *n.* The amount of plastic strain that a material can undergo before rupture. A ductile material generally shows a yield point.

Ductor roller *n.* The roller which is in intermittent contact with the fountain roller and transfers ink to the distribution system of the press. On a lithographic press it is also the roller, which transfers the fountain solution to the dampening rollers.

Dull \ˈdəl\ [ME *dul*; akin to OE *dol* follish, OIr *dall* blind] (13c) *adj.* A term applied to manufactured fibers that have been chemically or physically modified to reduce their normal luster. Matte; opposite of bright; low in luster.

Dullness *n.* (1) Lack of luster or gloss. (2) Colors of low chroma or saturation.

Dull rubbing *n.* Rubbing a dried film of a coating to dull finish, with an abrasive paper, pumice, steel wool and oil or water.

Dulmadge mixing section *n.* Invented by F. E. Dulmadge, this mixing section, usually located near the end of an extruder or injection screw, is really three short sections separated by gaps with no flights. In each section, the regular screw flight is replaced by about 20 narrow, closely spaced flight segments with lead angles of about 70°. The melt approaching the first section is subdivided into 20 substreams that, on exiting, are swirled circumferentially before entering the second section, and again, on leaving that one, before entering the third. The main goal was to improve melt-temperature uniformity.

Dulong and Petit, law of *n.* The specific heats of the several elements are inversely proportional to their atomic weights. The atomic heats of solid elements are constant and approximately equal to 6.3. Certain elements of low atomic weight and high melting point have, however, much lower atomic heats at ordinary temperatures.

Dumas nitrogen *n.* A long used method for determining the percentage of nitrogen in organic materials, invented by A. Dumas in 1830. The sample was first oxidized with red-hot copper oxide, followed by reduction of NO_x–N_2 over hot copper. Following absorption of CO_2 in KOH solution and condensation and absorption of most of the water, the volume of N_2 was measured. The method has largely been supplanted by automated analyzers that directly determine C, H, and N spectroscopically and find O by difference.

Dumbbell *n.* A piece of rubber cut in the shape of a dumbbell used for physical testing.
See dogbone.

Dumbells *n.* A defect frequently seen in wet-formed non-woven fabrics; an unusually long fiber will become entangled with groups of regular-length fibers at each end, thus producing a dumbbell-shaped clump.

Dumont's blue *np. See cobalt blue.*

Dungaree \ˌdeŋ-gə-ˈrē\ [Hindi *dugrī*] (1673) *n.* A term describing a coarse denim-type fabric, usually dyed blue, that is used for work overalls.

Duplex paper *n.* Wallpaper, which consists of two separate papers pasted together used to create a highly embossed effect.

Duplex printing *See printing.*

Duplicate cavity plate *n.* A removable plate that retains cavities, used where two-plate operation is necessary for loading inserts, etc.

Durability *n.* Degree to which paints and paint materials withstand the destructive effect of the conditions to which they are subjected.

D

Durable press *n.* A term describing a garment that has been treated so that it retains its smooth appearance, shape, and creases or pleats in laundering. In such garments no ironing is required, particularly if the garment is tumble-dried. Durable press finishing is accomplished by several methods; two of the most common are the following: (1) A fabric that contains a thermoplastic fiber and cotton or rayon may be treated with a special resin that, when cured, imparts the permanent shape to the cotton or rayon component of the fabric. The resin-treated fabric may be precured (cured in finishing and subsequently pressed in garment form at a higher temperature to achieve the permanent shape) or postcured (not cured until the finished garment has been sewn and pressed into shape). In both cases, the thermoplastic fiber in the garment is set in the final heat treatment. This fiber, when heat-set, also contributes to the permanence of the garment shape, but the thermoplastic component of the blend is needed for strength since the cotton or rayon component is somewhat degraded by the durable-press treatment. (2) Garments of a fabric containing a sufficient amount of a thermoplastic fiber, such as polyester, nylon, or acrylic, may be pressed with sufficient pressure and time to achieve a permanent garment shape.

Also see ease-of-care, permanent finish, and wash-and-wear.

Duranit *n.* Butadiene/styrene copolymer. Manufactured by Hüls, Germany.

Durene *n.* (durol, 1,2,4,5-tetramethylbenzene) $C_6H_2(CH_3)_4$. A substance occurring in coal tar, but usually prepared from xylene and methyl chloride in the presence of $AlCl_3$. It has been patented (USA 4,000,120) as an additive to make packaging films of polyolefins and polystyrene photodegradable under direct action of sunlight.

Durethan *n.* Polyamides or polyurethanes. Manufactured by Bayer, Germany.

Durette *n.* Fiber of iosphthalic acid and *m*-phenylene diamine. Manufactured by Monsanto, USA.

Durometer \dú-ˈrä-mə-tər *also* dyu\ [L *durus* hard] (ca. 1879) *n.* An instrument used for measuring the hardness of rubber, plastic, or protective coatings.

See ASTM test D 2240, also indentation hardness.

Durometer hardness *See indentation hardness.*

Dust \ˈdəst\ [ME, fr. OE *dūst*; akin to OHGr *tunst* storm, and prob. to L *fumus* smoke] (before 12c) *n.* One of the size gradings of various gums, being composed of very small pieces, almost as fine as dust.

Dust free *n.* (1) Descriptive of the stage in the drying of a paint or varnish film at which dust will no longer stick to the surface. (2) That stage of drying when cotton fibers, that have been dropped onto the film from a height of one inch, can be removed by blowing lightly across the surface of the film.

Also referred to as cotton-free or cotton-free dry.

Dust-free time *n.* The time required for a freshly applied paint or compound to form a skin on its surface so that dust will not adhere to it.

Dusting bronzes *See bronzing.*

Dust-resistant *n.* A term applied to a fabric that has been tightly woven so that it resists dust penetration.

Dutch metal *n.* Leaves of bright brass which are used for overlaying in the same manner in which gold leaf is applied.

Dutch pink *n.* Tint of red. The name is also used for certain yellow lakes, prepared from quercitron, Persian berries, or similar natural yellow coloring matters.

Dutch process *n.* Old name for stack process for manufacture of white lead.

Dutral *n.* Ethylene/propylene copolymer, manufactured by Montecatini, Italy.

DVB *n.* Abbreviation for divinylbenzene.

Dwell \ˈdwel\ [ME, fr. OE *dwellan* to go astray, hinder, akin to OHGr *twellen* to tarry] (13c) *vt.* (1) A pause in the application of pressure to a mold, made just before the mold has completely closed, to allow the escape of gas from the molding material. (2) In filament winding, the time that the traverse mechanism is stationary while the mandrel continues to rotate to the appropriate point for a new traverse to begin. (3) In heat sealing, dwell time is the time during which pressure and heat (or microwave energy) are applied to the area to be sealed.

Dwell time *n.* The time during a process in which a particular substance remains in one location (e.g., the time during which molten polymer remains in a spinning pack).

Dye \ˈdī\ [ME *dehe*, fr. OE *dēah*, *dēag*] (before 12c) *n.* An intensely colored substance that imparts color to a substrate to which it is applied. Retention of the dye in the substrate may be by means of adsorption, solution, mechanical bonding, or by ionic or covalent chemical bonding. The dye substance is devoid of crystal structure. Dyes used for coloring plastics usually dissolve in the plastic melt, unlike pigments, which remain dispersed as undissolved particles. Merriam-Webster's collegiate dictionary, 11th edn. Merriam-Webster Inc., Springfield, MA, USA, 2004. Complete textile glossary. Celanese Acetate LLC, New York, 2000. Vincenti R (ed) (1994) Elsevier's textile dictionary. Elsevier Science and Technology Books, New York.

Dye, acid *See acid dyes.*

Dye, basic *See basic dye.*

Dye fleck *n.* (1) An imperfection in fabric caused by residual undissolved dye. (2) A defect caused by small sections of undrawn thermoplastic yarn that dye deeper that the drawn yarn.

Dyeing *n.* To add soluble colorants that either form a chemical bond with the substrate or become closely associated with it by a physical process in order to change the color. Below is a listing of different types of fabric dyeing and a definition of each. Vincenti R (ed) (1994) Elsevier's textile dictionary. Elsevier Science and Technology Books, New York.

Dyeing, chain *n.* A method of dyeing yarns and fabrics of low tensile strength of tying them end-to-end and running them through the dyebath in a continuous process.

Dyeing, cross *n.* A method of dyeing blend or combination fabrics to two or more shades by the use of dyes with different affinities for the different fibers.

Dyeing, high-temperature *n.* A dyeing operation in which the aqueous dyebaths are maintained at temperatures greater than 100°C by use of pressurized equipment. Used for many manufactured fibers.

Dyeing, ingrain *adj* (1766). Term used to describe yarn or stock that is dyed in two or more shades prior to knitting or weaving to create blended color effects in fabrics.

Dyeing, jet *n.* High-temperature piece dyeing in which the dye liquor is circulated via a Venturi jet thus providing the driving force to move the loop of fabric.

Dyeing, mass-colored *n.* A term to describe a manufactured fiber (yarn, staple, or tow) that has been colored by the introduction of pigments or insoluble dyes into the polymer melt or spinning solution prior to extrusion. Usually, the colors are fast to most destructive agents.

Dyeing, muff A *n.* form of yarn dyeing in which the cone has been removed.

D

Package dyeing *See dyeing, yarn dyeing.*

Dyeing, pad *n.* A form of dyeing whereby a dye solution is applied by means of a padder or mangle.

Dyeing, piece *n.* The dyeing of fabrics "in the piece", i.e., in fabric form after weaving or knitting as opposed to dyeing in the form of yarn or stock.

Dyeing, pressure *n.* Dyeing by means of forced circulation of dye through packages of fiber, yarn, or fabric under superatmospheric pressure.

Dyeing, reserve *n.* (1) A method of dyeing in which one component of a blend or combination fabric is left undyed. The objective is accomplished by the use of dyes that have affinity for the fiber to be colored but not for the fiber to be reserved. (2) A method of treating yarn or fabric so that in the subsequent dyeing operation the treated portion will not be dyed.

Dyeing, skein *n.* The dyeing of yarn in the form of skeins, or hanks.

Dyeing, solution *n. See dyeing, mass-colored.*

Dyeing, solvent *n.* A dyeing method based on solubility of a dye in some liquid other than water, although water may be present in the dyebath.

Dyeing, space *n.* A yarn-dyeing process in which each strand is dyed with more that one color at irregular intervals. Space dyeing produces an effect of unorganized design in subsequent fabric form. The two primary methods are knit-de-knit and warp printing.

Dyed, space *n. See dyeing, mass-colored.*

Dyeing, stock *n.* The dyeing of fibers in staple form.

Dyeing, thermal fixation *n.* A *process* for dyeing polyester whereby the color is diffused into the fiber by means of dry heat.

Dyeing, union *n.* A method of dyeing a fabric containing two or more fibers or yarns to the same shade so as to achieve the appearance of a solid colored fabric.

Dyeing, yarn *n.* The dyeing of yarn before the fabric is woven or knit. Yarn can be dyed in the form of skeins, muff, packages, cheeses, cakes, chain-wraps, and beams.

Dyes, basic *n.* A class of positive-ion-carrying dyes known for their brilliant hues. Basic dyes are composed of large-molecule, water-soluble salts that have a direct affinity for wool and silk and can be applied to cotton with a mordant. The fastness of basic dyes on these fibers is very poor. Basic dyes are also used on basic-dyeable acrylics, modacrylics, nylons, and polyesters, on which they exhibit reasonably good fastness.

Dyes, cationic *See dyes, basic dyes.*

Dyes, fiber-reactive dyes *n.* A type of water-soluble anionic dye having affinity for cellulose fibers. In the presence of alkali, they react with hydroxyl groups in the cellulose and thus are liked with the fiber. Fiber-reactive dyes are relatively new dyes and are used extensively on cellulosics when bright shades are desired.

Dyeing, gel *n.* Passing a wet-spun fiber that is in the gel state (not yet at full crystallinity or orientation) through a dyebath containing dye with affinity for the fiber. This process provides good accessibility of the dye sites.

Dyes, macromolecular *n.* A group of inherently colored polymers. They are useful both as polymers and as dyes with high color yield. The chromophores fit the recognized CI classes, i.e., azo, anthraquinone, etc., although not all CI classes are represented. Used for mass dyeing, hair dyes, writing inks, etc. Tortora PG (ed) (1997) Fairchild's dictionary of textiles. Fairchild Books, New York.

Dyes, metallized *n.* A class of dyes that have metals in their molecular structure. They are applied from an acid bath.

Dyes, naphthol *n.* A type of azo compound formed on the fiber by first treating the fiber with a phenolic compound. The fiber is then immersed in a second solution containing a diazonium salt that reacts with the phenolic compound to produce a colored azo compound. Since the phenolic compound is dissolved in caustic solution, these dyes are mainly used for cellulose fiber, although other fibers can be dyed by modifying the process.
Also see dyes, developed dyes.

Dyes, premetallized *n.* Acid dyes that are treated with coordinating metals such as chromium. This type of dye has much better wetfastness than regular acid dye. Premetallized dyes are used on nylon, silk, and wool.

Dyes, sulfur *n.* A class of water-insoluble dyes that are applied in a soluble, reduced form from a sodium sulfide solution and are then reoxidized to the insoluble form on the fiber. Sulfur dyes are mainly used on cotton for economical dark shades of moderate to good fastness to washing and light. They generally give very poor fastness to chlorine.

Dyes, vat *n.* A class of water-insoluble dyes which are applied to the fiber in a reduced, soluble form (leuco compound) and then reoxidized to the original insoluble form. Vat dyes are among the most resistant dyes to both washing and sunlight. They are widely used on cotton, linen rayon, and other cellulosic fibers. Tortora PG (ed) (1997) Fairchild's dictionary of textiles. Fairchild Books, New York.

Asphalt overlay fabrics *n.* Fabric systems installed between the old and new asphalt layers during pavement resurfacing. The fabric absorbs the tack coat sprayed on the old surface thus forming a permanent moisture barrier to protect the subgrade from strength loss due to water intrusion. The fabric system also helps retard reflective cracking by serving as a flexible layer to diffuse stress.

Drainage fabrics *n.* Fabrics used as filter media or separators in subsurface drainage systems. The fabric is installed around the drainage pipe, or coarse stone in the drain; it allows water to pass freely from the soil to the drain, but prevents soil particles from migrating into the drain system. These fabrics are also used as outer coverings in prefabricated drainage composites and serve the same function as a filtering medium.

Erosion control fabrics *n.* Fabrics used in the stabilization of embankments and the containment of silt run-off from erodible slopes. In embankment stabilization, the fabric functions as a filter medium behind stabilizing rip-rap revetments. In siltation control, the fabric acts as a filter to contain silt while allowing excess water to drain freely. In turf reinforcement, the mat is used to retain soil while allowing roots and stems to grow through. In fabric-forming systems for the construction of revetments, a double-layer, water-permeable fabric is positioned, then pumped full of structural grout. These systems are alternatives to rip-rap.

Reinforcement fabrics *n.* Fabric system used in the construction of steep slopes and retaining walls. By stabilizing the soil mass, they reduce the stress on the retaining wall with corresponding decrease in load-bearing requirements for the wall design. In slope reinforcement, the stabilization permits steeper slope construction than would be possible based on soil properties.

D

Subsurface stabilization fabrics *n.* Fabrics used in the construction on access roads, railroads, parking, and storage areas over soft, unstable soil. The fabric is placed between the subgrade and the stabilizing fill material where it provides separation of subsoil and fill, filtration of moisture at the subsoil/fill interface, and added tensile reinforcement of the compacted fill.

Dyeing auxiliaries *n.* Various substances that can be added to the dyebath to aid dyeing. They may necessary to transfer the dye from the bath to the fiber or they may provide improvements in leveling, penetration, etc. Also call dyeing assistants.

Dyeing plastics *n.* Process of adding color to plastics to make them attractive in appearance. Color is produced by introduction of either dyes or pigments, which produce color by selectively absorbing, transmitting, reflecting, and scattering specific areas of light energy from wave bands that constitute white light.

Dye migration *See migration* (1).

Dye, mordant *n.* Term given to a range of dyestuffs, the colors of which are developed when precipitated onto bases to form lakes. Often a salt-forming reaction takes place between the dyestuff and the base.

Dye pigments *n.* Dyes that by nature are insoluble in water and can be used directly as pigments without any chemical transformation.

Dye range *n.* A broad term referring to the collection of dye and chemical baths, drying equipment, etc., in a continuous-dyeing line.

Dyes *n.* Substances that add color to textiles. They are incorporated into the fiber by chemical reaction, absorption, or dispersion. Dyes differ in their resistance to sunlight, perspiration, washing, gas, alkalies, and other agents; their affinity for different fibers; their reaction to cleaning agents and methods; and their solubility and method of application. Various classes and types are listed below.
Also see color index (CI).

Dyes and dyestuffs *n.* Organic coloring substances. Pigment dyestuffs are virtually insoluble in water and in the usual paint and varnish vehicles. Some examples of pigment dyestuffs are toluidine red, phthalocyanine blue, and arylamide yellow. Some dyes are soluble in water and organic solvents and are classed according to their solubilities, for example, water-soluble dyes, spirit-soluble dyes, and oil-soluble dyes. Harper CA (2000) Modern plastics encyclopedia. McGraw Hill Professional, New York.

Dye sites *n.* Functional groups within a fiber that provide sites for chemical bonding with the dye molecule. Dye sites may be either in the polymer chain or in chemical additives included in the fiber. Kadolph SJJ, Langford AL (2001) Textiles. Pearson Education, New York.

Dye, spirit-soluble *See spirit-soluble dye.*

Dyestuff *n.* Colorant which can be fixed firmly to a material to be dyed so as to be more or less fast to light, water, soap, etc.

Dynamic adhesion *n.* The ability of a cord-to-rubber bond to resist degradation resulting from flexure.

Dynamic fatigue *n.* Usually the same as *fatigue.* The "dynamic" modifier is sometimes used by persons who think of creep and creep failure as "static fatigue".

See alternating stress amplitude. See entries at fatigue.

Dynamic mechanical analyzer *n.* An instrument that can test in an oscillating-flexural mode over a range of temperature and frequency to provide estimates of the "real", i.e., in-phase, and "imaginary", i.e., out-of-phase parts of the complex modulus. The real part is the elastic component, the imaginary part is the loss component. The square root of the sum of their squares is the complex modulus. With polymers, the components and the modulus are usually dependent on both temperature and frequency. ASTM D 4065 spells out the standard practice for reporting dynamic mechanical properties of plastics. An example of a DMA thermogram of different Perkin-Elmer Inc., manufactures the Diamond DMA instrument. Polymer films is shown. Sepe MP (1998) Dynamic mechanical analysis. Plastics Design Library, Norwich, New York.
See also mechanical spectrometer

Dynamic mechanical properties *n.* (1) The stress–strain properties of a material when subjected to an applied sinusoidally varying stress or strain. For a perfectly elastic material the strain response is immediate and the stress and strain are in phase. For a viscous fluid, stress and strain are 90° out of phase. (2) The mechanical properties of composites as deformed under periodic forces such as dynamic modulus, loss modulus and mechanical damping or internal friction. Sepe MP (1998) Dynamic mechanical analysis. Plastics Design Library, Norwich, New York.

Dynamic mechanical spectrum *n.* The information obtained from testing with a mechanical spectrometer. A plot or tabulation of complex modulus or its components versus frequency of oscillation or temperature or both. The mode of stress may be tensile/compressive, flexural, or torsional (shear). Because both abscissa (frequency) and ordinates can range widely, bilogarithmic plots are usual. Sepe MP (1998)

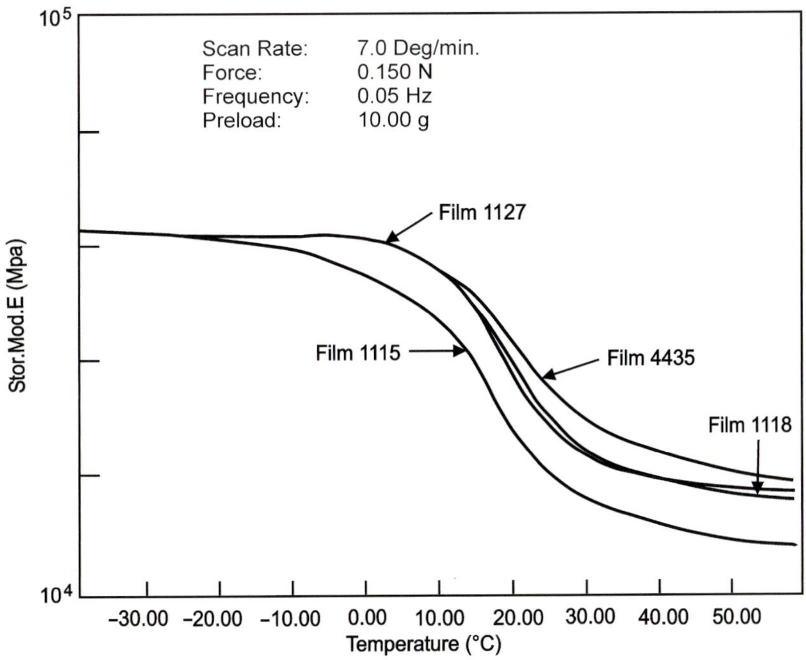

Dynamic mechanical analysis. Plastics Design Library, Norwich, New York.

Dynamic stress *n.* A stress whose magnitude and/or direction vary with time, typically cyclically and sinusoidally.

Dynamic stress relaxometer *n.* An instrument that measures the relaxation response of an elastomeric material to a prescribed shear deformation over a range of temperature. Basic elements of the instrument are a cone-shaped stator cavity and a conical rotor, both electrically heated. The sample is placed in the stator, which rises to a position to form a constant specimen thickness, forcing out the excess material. After heating to the desired temperature, the rotor is rotated quickly through a small angle, to a known shear deformation. The subsequent drop-off torque, which results from the relaxation of stress within the sample, is recorded over the time it takes for it to decay.

Dynamic valve *n.* A device sometimes incorporated in an extruder head to control flow by adjusting the clearance between the conical elements, one stationary, the other rotating with the screw. Dynamic valves of various designs have also been used between the stages of two-stage, vented extruders as an aid to balancing the flow rates in the stages and preventing extrusion out the vent.

Dynamic viscosity *n.* (1) Absolute viscosity as distinguished from kinematic viscosity. (2) In sinusoidally varying shear, the part of the stress in phase with the rate of strain, divided by the strain rate. Shah V (1998) Handbook of plastics testing technology. John Wiley and Sons, New York.

See also viscosity.

Dynapoint process *n.* A continuous computer-controlled process for manufacturing tufted carpets with intricate patterns from undyed yarn. The carpet is dyed as it is tufted and the colors and pattern are clearly visible through the primary backing of the carpet.

Dyne *n.* In the (now deprecated) cgs system of units, the force required to accelerate a mass of 1 g by 1 cm/s^2. The dyne $= 1 \times 10^{-5}$ N. Lide DR (ed) (2004) CRC handbook of chemistry and physics. CRC Press, Boca Raton, FL.

Dynel *n.* Vinyl chloride/acrylonitrile copolymer. Manufactured by Union Carbide, USA.

Dypnone *n.* (phenyl α-methyl styryl ketone, 1,3-diphenyl-2-butene-1- one) $C_6H_5COC\,HC(CH_3)C_6H_5$. A plasticizer and ultraviolet absorber.

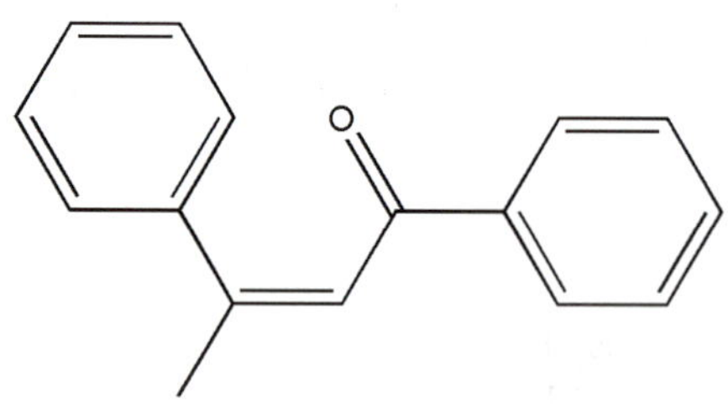

E

e \ˈē\ {often capitalized, often attributive} (before 12c) *n.* (1) The base of natural logarithms, 2.71828. (2) The charge on an electron, 1.60199×10^{-18} C.

ΔE, Δe *n.* The total color difference computed with a color difference equation. It is generally calculated as the square root of the sum of the squares of the chromaticity difference, ΔC, and the lightness difference, ΔL:

$$\Delta E = [(\Delta C)^2 + (\Delta L)^2]^{1/2}.$$

E *n.* (1) SI abbreviation for prefix exa-. (2) Symbol commonly used for modulus of elasticity in tension (*See modulus of elasticity*). For activation energy in the Arrhenius equation and for electric potential. Sepe MP (1998) Dynamic mechanical analysis. Plastics Design Library, Norwich, New York. Connors KA (1990) Chemial kinetics. John Wiley and Sons, New York.

EA *n.* Segmented polyurethane fiber.

EAA *n.* Abbreviation for ethylene–acrylic acid copolymer.

Earth pigment *n.* Class of pigments, which are usually mined directly from the earth, dried, generally Calcined and ground. Typical examples are red and yellow oxides of iron, yellow ochre, raw and burnt siennas, and raw and burnt umbers. Syn: mineral pigments and natural pigments.

Ease-of-care *n.* A term used to characterize fabrics that, after laundering, can be restored to their original appearance with a minimum of ironing or other treatment. An ease-of-care fabric generally wrinkles only slightly upon laundering. Vincenti R (ed) (1994) Elsevier's textile dictionary. Elsevier Science and Technology Books, New York.
Also see durable press and wash and wear.

Eave \ˈēvz] [ME *eves* (singular), fr. OE *efes*; akin to OHGr *obasa* portico, OE *ūp* up] (before 12c) *n.* Lower part of a roof projecting beyond the wall undernearth.
See soffit.

Ebonite \ˈe-be-ˌnit\ (1861) *n.* A hard material made by sulfur vulcanization of rubber.

Ebonize \-ˌnīz\ (ca. 1828) *vt.* To blacken with paint or stain to look like ebony.

Ebulliometry *n.* Method of measuring molecular weight of polymers under 20,000; based on vapor pressure lowering and boiling point elevation. Pethrick RA, Dawkins JW (eds) (1999) Modern techniques for polymer characterization. John Wiley and Sons, New York.

EC *n.* Abbreviation for ethyl cellulose.

EC *n.* Ethyl cellulose.

Ecology \iˈkä-lə-jē-, e-\ (Gr *Ökologie*, fr. *öko-* eco- + -logie –logy (1873) *n.* The interrelationships of living things to one another and to their environment.

Ecosystem \ˌsis-təm\ (1935) *n.* The interaction system of a biological community and its non-living environment.

ECTFE *See poly(ethylene-chlorotrifluoroethylene).*

EDC *n.* Abbreviation for ethylene dichloride.

Eddy current \ˈe-dē ˈkər-ənt\ (1887)(Foucault current) *n.* The current induced in a mass of conducting material by a varying magnetic field.
Also called Foucault current.

Edge bead *n.* In some cast-film and sheet-extrusion operations, the narrow border at the edge of the sheet, usually somewhat thicker, that must be trimmed off prior to winding or stacking the product.

Edge crimping *See texturing, edge crimping method.*

Edge dislocation *n.* A dislocation in which a layer of particles in the crystal is incomplete.

Edge grain *n.* Wood or veneer so sawed that the annual rings form an angle of 45–90° with the surface of the piece.
Also called vertical grain.

Edge roll *n.* The curl that develops on the edge of a single-knit fabric preventing it from lying flat.

E

Edge runner mill *n.* Mill consisting of a horizontal, or substantially horizontal, circular pan which may be of cast iron or stone-lined. One or more circular rollers, which may also be of cast iron or suitable stone, are made to rotate edgewise around the pan. The material to be ground, which may be dry pigment or other powder or pigment paste, is disintegrated between the surfaces of the rollers and the pan. Sometimes referred to as putty chaser.

Edgewise (1677) *adv.* Plane of the laminate perpendicular to the laminations.

Edistir *n.* Poly(styrene). Manufactured by Montedison, Italy.

EDM *n.* Abbreviation for electrical-discharge machining.

EEA *n.* Copolymer from ethylene and ethyl acrylate. Abbreviation for ethylene–ethyl acrylate copolymer.

Effective modulus *n.* Syn: creep modulus.

Effervescence \ˌe-fər-ˈves\ [L *effervescere*, fr. *ex-* + *fervescere* to begin to boil, inchoative of *fervēre* to boil] (1784) *vt.* The vigorous evolution of gas, which accompanies some chemical reactions, e.g., treatment of a carbonate with certain mineral acids.

Efficiency \i-ˈfi-shən-sē\ (1633) *n.* (1) The efficiency of a machine or process, the ratio of the energy delivered to that supplied. (2) The effectiveness of operation or production, particularly as related to some standard for same. (3) The reciprocal of the unit cost of items for sale (*cost efficiency*, *economy*). Merriam-Webster's collegiate dictionary, 11th edn. Merriam-Webster Inc., Springfield, MA, 2004.
See also efficiency of reinforcement and plasticizer efficiency.

Efficiency of reinforcement (fiber-efficiency factor) *n.* The percentage of fiber in a reinforced-plastic structure or part contributing to the property of concern. For example, in a unidirectionally reinforced bar, the theoretical efficiency for Young's modulus and fiber-direction tensile strength is 100%. For a sheet molded from chopped-strand mat with all fibers randomly oriented in the sheet plane, the efficiency for in-plane properties is 37%. With chopped strands randomly oriented in three dimensions, efficiency falls to 20%.

Efflorescence \-ˈre-sᵊn(t)s\ (1626) *n.* An encrustation of soluble salts, commonly white, deposited on the surface of coatings, stone, brick, plaster, or mortar; usually caused by salts or free alkalies leached from mortar or adjacent concrete as moisture moves through it, *also known as laitance.* {G Ausblühen n, F efflorescence f, S eflorescencia f, I efflorescenza f}. Weismantal GF (1981) Paint handbook. McGraw-Hill Corporation Inc., New York.

Effluent limitations *n.* Any restrictions, established by the government or by management, on quantities, rates and concentrations of chemical, physical, biological, and other constituents which are discharged from paint sources.

Efflux \ˈe-ˌfləks\ [ML *effluxus*, fr. *effluere*] (1647) *n.* The process of flowing out, in any given system.

Efflux viscometer *n.* A cup type viscometer containing an orifice. Whatever precision and accuracy an efflux instrument may have is essentially dependent upon the dimension of the orifice. The closer an orifice resembles a capillary, the more accurate is the instrument.

See viscometer, ford cup.

Effusion \i-ˈfyü-zhən\ (15c) *n.* The passage of a substance through a small orifice.

Eggshell \ˈeg-ˌshel\ (1835) *adj.* (1) Gloss lying between semigloss and flat. Generally thought to be between 20 and 35 as determined with a 60° glossmeter. (2) An off white color.

Eggshell finish *n.* (1) Low sheen (seminat) surface which exhibits surface reflectance (gloss) similar to that from an eggshell. Thus it is a gloss lying between flat and semigloss and is generally thought to be between 20 and 35, as determined with a 60° gloss meter. (2) Paint, which exhibits eggshell gloss.

E glass *n.* A low-alkali ($Na_2O + K_2O = 0.6\%$) borosilicate glass, like Pyrex®, with good electrical properties, the most widely used glass–fiber reinforcement for plastics. Major constituents are: SiO_2, 54%; CaO, 17%; Al_2O_3, 15%; B_2O_3, 8%; MgO, 4.7%. The fibers are vulnerable to surface abrasion so are always sized before stranding. Average fiber properties are: density = 2.54 g/cm^3; tensile modulus (axial) = 72 GPa; tensile strength = 3.5 GPa.

Egyptian asphaltum \i-ˈjip-shən ˈas-ˈfól-təm\ *n. See syrian asphaltum.*

Egyptian blue *n.* Blue pigment, of ancient origin, which is substantially a copper silicate, obtained by calcinations.

EHMWPE *n.* Abbreviation for extra-high-molecular-weight polyethylene, any of a subfamily of linear PE resins having molecular weights in the range 250,000–1,500,000.

See also polyethylene.

Eicosanic acid *See cosanic acids.*

Eilers equation *n.* A modification of the Einstein equation relating the viscosity η_f of a Newtonian liquid filled with spherical particles to the viscosity of the pure liquid η_o and extending to concentrations above 10% of filler. It is

$$\frac{\eta_n}{\eta_o} = \left[\frac{1 + 1.25f}{1 - Sf}\right]^2,$$

where f is the volume fraction of spheres and S is an empirical coefficient usually between 1.2 and 1.3. Sudduth RD (2003) J Appl Sci 48(1):25–36.

See also Mooney equation.

Einstein equation *n.* An equation relating the viscosity η_f of a sphere-filled, Newtonian liquid to that of the unfilled liquid η_o, for volume fractions f of spheres up to about 10%. It is

$$\eta_{rel} = \eta_f/\eta_o = 1 + k_E f,$$

where k_E is the Einstein coefficient = 2.5 for spheres. The Einstein equation has been extended to higher concentrations by adding terms in f^2, f^3, etc., and by correcting for the limiting volume fraction f_m that can be filled by uniform spheres. A versatile model of this type is

$$\eta_{rel} = \frac{1 + af + bf^2 + cf^3}{1 - (f/f_m)},$$

where a, b, and c are the empirical constants and $f_m \approx 0.7$. The Mooney equation is a form of this model for which b and $c = 0$. Kamide K, Dobashi T (2000) Physical chemistry of polymer solutions. Elsevier, New York.

See also the Eilers equation and Frankel–Acrivos equation.

Einstein theory for mass–energy equivalence *n.* The equivalence of a quantity of mass m and a quantity of energy E by the formula $E = mc^2$. The conversion factor c^2 is the square of the velocity of light. Serway RA, Faugh JS, Bennett CV (2005) College physics. Thomas, New York.

Ejection ram *n.* A small supplementary hydraulic ram fitted to a molding press to operate piece-ejection device.

Ejector pin (ejector sleeve, knockout pin, KO pin) *n.* A rod, pin, or sleeve that pushes a molding off a force or out of a cavity of a mold. Attached to an ejector bar or plate, it is actuated by the ejector rod(s) of the press or by auxiliary hydraulic or compressed-air cylinders.

Ejector plate *n.* A plate that backs up the ejector pins and holds the ejector assembly together.

Ejector-return pin (return pin, surface pin, safety pin, and position pushback) *n.* A projection, usually one of several that push back the ejector assembly as the mold closes.

Ejector rod *n.* A bar that actuates the ejector assembly when a mold is opened.

Elaidamide *n.* $CH_3(CH_2)_7CH{=}CH(CH_2)_7CONH_2$. The amide of *trans*-9-octadecenoic acid, a steroisomer of oleamide used in fractional percentages as a slip agent for polyethylene to be made into film.

Elaidic acid *n.* $CH_3(CH_2)_7CH{=}CH(CH_2)_7COOH$ (octadec-9-enoic acid, *trans*-9-octadeconic acid). Properties: mp, 51°C; bp, 225°C/10 mmHg; iodine value, 90.1.

Elastic \i-ˈlas-tik\ [NL *elasticus*, fr. LGk *elastos* ductile, beaten, fr. Gk *elaunein* to drive beat out; prob. akin to Gk *ēlythe* he went, OI *luid*] (1674) *adj.* Ability of a substance to return to its approximate original shape or volume after a distorting force on the substance has been removed. Billmeyer FW Jr (1984) Textbook of polymer science, 3rd edn. Wiley-Interscience, New York.
See also elasticity.

Elastic compliance *n.* The inverse of the Young's modulus.

Elastic constant *n.* Any of the several constants of a constitutive relationship between stress (of any mode) and strain in a material. For an isotropic material stressed in its elastic range, there are (at any temperature) four interrelated constants: tensile modulus *E*, shear modulus *G*, bulk modulus *B*, and Poisson's ratio μ. Two expressions of the relations are

$$G = E/2(1+\mu) \text{ and } B = E/3(1-2\mu).$$

More constants are needed to define the behavior of non-isotropic materials. Shah V (1998) Handbook of plastics testing technology. John Wiley and Sons, New York.

Elastic deformation *n.* A change in dimensions of an object under load that is fully recovered when the load is released. That part of the total strain in a stressed body that disappears upon removal of the stress.
See also plastic deformation.

Elastic design *n.* Engineering design for load-bearing members based on the assumption that stress and strain are proportional and will be kept well within the elastic range, with working stresses set at half or less of the yield stress. Elastic design based on short-time test measurements may be applicable to the design of plastic products that will be loaded intermittently and for short periods. However, the universal phenomenon of creep in plastics means that, under sustained loads, viscoelastic rather an elastic behavior is the

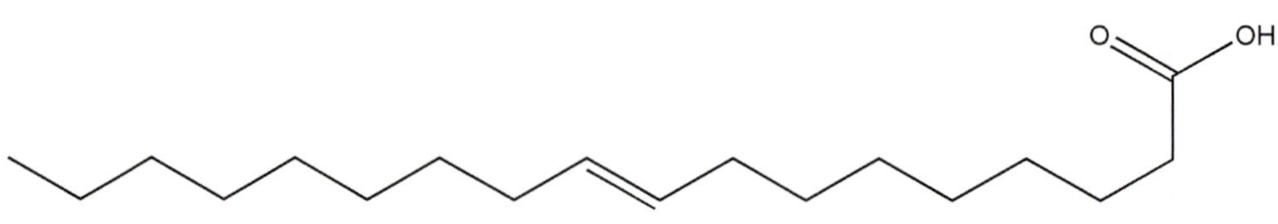

norm. Even so, elastic design methods can often be used by employing creep modulus and creep strength in place of short-time parameters. Shah V (1998) Handbook of plastics testing technology. John Wiley and Sons, New York.

Elasticity (1664) *n.* A property that defines the extent to which a material resist small deformations from which a material recovers completely when deforming force is removed. When the deformation is proportional to the applied load, the material is said to exhibit Hookean elasticity or ideal elasticity. Elasticity equals stress divided by strain. Shah V (1998) Handbook of plastics testing technology. John Wiley and Sons, New York. Elias HG (1977) Macromolecules, vols 1–2. Plenum Press, New York. Weast RC (ed) (1978) CRC handbook of chemistry and physics, 59th edn. The Chemical Rubber Co., Boca Raton, FL.

Elasticity, Hookean *n.* The simplest form of stretching elasticity is the Hookean law (e.g., a coiled steel spring). By taking the derivative of the free energy with respect to the relative area change one obtains from the Hookean law that the surface tension is proportional to the surface area (at fixed number of molecules). Shah V (1998) Handbook of plastics testing technology. John Wiley and Sons, New York. Weast RC (ed) (1971) Handbook of chemistry and physics, 52nd edn. The Chemical Rubber Co., Boca Raton, FL.

Elasticized fabric *n.* A fabric that contains elastic threads. Such fabrics are used for girdles, garters, and similar items. Vincenti R (ed) (1994) Elsevier's textile dictionary. Elsevier Science and Technology Books, New York.

Elasticizer *n.* A compounding additive that contributes elasticity to a resin such as chlorinated polyethylenes and chlorinated copolymers of ethylene and propylene are blended with PVC compositions to provide elasticity. Wickson EJ (ed) (1993) Handbook of polyvinyl chloride formulating. John Wiley and Sons Inc., New York. Nass LI (ed) (1976) Encyclopedia of PVC. Marcel Dekker Inc., New York.

Elastic limit *n.* The greatest stress that a material can experience which, when released, will result in no permanent deformation. With some polymers this limit can be well above the proportional limit. Billmeyer FS Jr (1962) Textbook of polymer science. John Wiley and Sons Inc., New York. Elias HG (1977) Macromolecules, vols 1–2. Plenum Press, New York. Miller ML (1966) Structure of polymers. The Reinhold Publishing Co., New York. Brown R (1999) Handbook of physical polymer testing, vol 50. Marcel Dekker, New York.

See also yield point.

Elastic-melt extruder *See extruder, elastic-melt.*

Elastic memory *See memory.*

Elastic moduli *n* (1904) (*Young's modulus by bending*). Bar supported at both ends. If a flexure s is produced by the weight of mass m, added midway between the supports separated by a distance l, for a rectangular bar with vertical dimensions of cross-section a, and horizontal dimension b, the modulus is,

$$M = \frac{mgl^3}{4sa^3b}.$$

For a cylindrical bar of radius r,

$$M = \frac{mgl^3}{12\pi r^4 s}.$$

For a bar supported at one end. In the case of a rectangular bar as described above,

$$M = \frac{4mgl^3}{sa^3b}.$$

For a round bar supported at one end,

$$M = \frac{4mgl^3}{3\pi r^4 s}.$$

Brown R (1999) Handbook of physical polymer testing, vol 50. Marcel Dekker, New York.

Elastic moduli (*Modulus of rigidity*) *n.* If a couple ($C = mgx$) produces a twist of θ radians in a bar of length l and radius r, the modulus is

$$M = \frac{2Cl}{\pi r^4 \theta}.$$

The substitution in the above formulae for the elastic coefficients of m in grams, g in cm/s^2, l, a, b, and r in cm, s in cm, and C in dyne-cm will give moduli in dyn/cm^2. The dimensions of elastic moduli are the same as of stress $[ML^{-1}T^{-2}]$. Weast RC (ed) (1971) Handbook of chemistry and physics, 52nd edn. The Chemical Rubber Co., Boca Raton, FL.

Elastic moduli (*Coefficient of restitution*) *n.* Two bodies moving in the same straight line, with velocities v_1 and v_2, respectively, collide and after impact move with velocities v_3 and v_4. The coefficient of restitution is

$$C = \frac{v_4 - v_3}{v_2 - v_1}.$$

Weast RC (ed) (1971) Handbook of chemistry and physics, 52nd edn. The Chemical Rubber Co., Boca Raton, FL.

Elastic modulus *See modulus of elasticity.*

Elastic nylon *See nylon* 6/10.

Elastic polyamides *n.* Elements that are combined by chemical processes into long-chain polymers that constitute the fiber-forming substance known as polyamides. Harper CA (ed) (2002) Handbook of plastics, elastomers, and composites, 4th edn. McGraw-Hill, New York.

Elastic recovery *n.* That fraction of a given deformation that behaves elastically. A perfectly elastic material has a recovery of 100% while a perfectly plastic material has no elastic recovery. Elastic recovery is an important property in films used for stretch packaging because it relates directly to the ability of a film to hold a load together. Retention of the elastic-recovery stress over a period of time is also important. Shah V (1998) Handbook of plastics testing technology. John Wiley and Sons, New York.

Elastic solid *n.* Solid in which, for all values of the shearing stress below the rupture (shear strength), the strain is fully determined by the stress regardless of whether the stress is increasing or decreasing.

Elastodynamic extruder *See extruder, elastic-melt.*

Elastomer \-tər-mər\ [*elastic* + *-o-* + *-mer*] (ca. 1939) *n.* Generally, a material that at room temperature can be stretched repeatedly to at least twice its original length and, immediately upon release of the stress, returns with force to its approximate original length. Stretchability and deformation with returnability is one criteria by which materials called plastics in commerce are distinguished from elastomers and rubbers. More explicitly, polymers that stretch and recover reproducibly without permanent change in dimensions (or deformation) are rubber or elastic materials. Elastomers are useful and divese substances tht easily form various rubbery shapes. Many industries rely on parts made from elastomers, especially automobiles, sports equipment, electronics, electrical equipment, and assembly line factories (www.dupontelastomers.com). Another criterion is that, unlike thermoplastics that can be repeatedly softened and hardened by heating and cooling without substantial change in properties, most elastomers are given their final properties by mastication with fillers, processing aids, antioxidants, curing agents, and others followed by vulcanization (curing) at elevated temperatures that cross-links the molecular chains and causes them to be thermosetting systems. However, a few elastomers are thermoplastic systems. Polymers usually considered to be elastomers, at least in some of their forms, are listed in the follow table. {G Elastomer n, F élastomère m, S elastómero m, I elastomero m}.

Chemical name	Abbreviations
Acrylonitrile–chloroprene copolymer	NCR
Acrylonitrile–isoprene copolymer	NIR
Butadiene–acrylonitrile copolymer	GR-N, NBR, and PBAN
Chlorinated polyethylene	CPE
Chlorosulfonated polyethylene	CSM, CSR, and CSPR
Ethylene ether polysulfide	EOT
Ethylene–ethyl acrylate copolymer	EEA, E/EA
Ethylene polysulfide	ET
Ethylene–propylene copolymer	EPM, EPR
Ethylene–propylene-diene terpolymer	EPD, EPDM, EPT, and EPTR
Fluoroelastomer (any)	FPM
Fluorosilicone	FVSI
Hexafluoropropylene–vinylidene fluoride copolymer	FPM
Isobutene–isoprene copolymer	Butyl, GR-I
Organopolysiloxane	SI
Acrylic ester–butadiene copolymer	ABR, AR
Polybutadiene	BP, BR, and CBR
Polychloroprene	CR
Polyepichlorohydrin	CO, CHR
Polyisobutene	PIB
Polyisoprene, natural	NR
Polyisoprene, synthetic	CI, IR, and PIP
Polyurethane (polyester)	AU, PUR
Polyurethane (polyether)	EU
Polyurethane (polyether and polyester)	TPU
Styrene–butadiene copolymer	SBR, GR-S
Styrene–chloroprene copolymer	SCR
Polyethylene–butyl graft copolymer	TPO
Styrene–butadiene–styrene triblock polymer	SBS

A beneficial property of elastomers is that they can be compounded or joined with other materials to strengthen certain characteristics. Other kinds of polymers may be installed next to various other materials, such as metal, hard plastic, or different kinds of rubber, with excellent adhesion. Harper CA (ed) (2002) Handbook of plastics, elastomers, and composites, 4th edn. McGraw-Hill, New York. Harper CA (ed) (2002) Handbook of plastics, elastomers, and composites, 4th edn. McGraw-Hill, New York. Elias HG et al. (1983) Abbreviations for thermoplastics, thermosets, fibers, elastomers, and additives. Polym News 9:101–110. James F (ed) (1993) Whittington's dictionary of plastics. Technomic Publishing Co. Inc., Carley. Harper CA (ed) (2002) Handbook of plastics, elastomers, and composites, 4th edn. McGraw-Hill, New York. Skeist I (ed) (1990) Handbook of adhesives. Van Nostrand Reinhold, New York.

Elastomeric *adj.* Rubber like, and relating to or having the properties of elastomers.

E

Elaterite \i-ˈla-tər-ˌrīt\ [Gr *Elaterit,* fr. Gk *elatēr*] (1826) *n.* Infusible asphaltic pyrobitumen of petroleum origin, sometime described as mineral rubber or mineral Caoutchouc.

Electret \i-ˈlek-trət\ [*electr*icity + magn*et*] (1885) *n.* A disk of polymeric material that has been electrically polarized so that one side has a positive charge and the other a negative charge, analogous to a permanent magnet. Electrets may be formed of poor conductors such as polymethyl methacrylate, polystyrene, nylon, and polypropylene, by heating and cooling them in the presence of a strong electric field. Bar-Cohen Y (ed) (2001) Electroactive polymer (EAP) actuators as artificial muscles. SPIE Press, Bellingham, Washington. Ku CC, Liepins R (1987) Electrical properties of polymers. Hanser Publishers, New York.

Electrical conductivity *n.* (1) (Symbol is λ expressed in units of $\Omega^{-1}\,m^{-1}$) A measure of the ease of transporting electric charge from one point to another in an electric field (Ku and Liepins, 1987). (2) The reciprocal of resistivity (resistivity symbol is ρ expressed in units of $\Omega\,m$). (3) $R=\rho l/A$, where R is the resistance, Ω; ρ the resistivity, $\Omega\,m$; l the length of material, m; and A is the area of cross-section of material, m^2. {G elektrische Leitfähigkeit f, F conductivité électrique, conductivité f, S conductividad eléctrica, conductividad f, I conduttività elettrica, conduttività f}. Serway RA, Faugh JS, Bennett CV (2005) College physics. Thomas, New York. Emerson JA, Torkelson JM (eds) (1991) Optical and electrical properties of polymers: materials research society symposium proceedings, vol 24. Materials Research Society, Warrendale, PA. Ku CC, Liepins R (1987) Electrical properties of polymers. Hanser Publishers, New York. Seanor DA (1982) Electrical conduction in polymers. Academic Press, New York.

Electrical-discharge machining (EDM, spark erosion) *n.* A method of machining molds and extrusion dies in which a conductive tool (often brass) has the inverse shape of the cavity or hole to be machined. A high-voltage DC difference is applied between the tool and the work piece. Capacitive discharge erodes steel from the work piece at about eight times the rate that the tool itself is eroded. Roughing and finishing tools are used. The process is accurate, produces good detail, can cut thin, deep slots, and can be used with hardened steels, thus averting the distortion sometimes caused by hardening after conventional machining. Metal removal is relatively slow. James F (ed) (1993) Whittington's dictionary of plastics. Technomic Publishing Co. Inc., Carley.

Electrical finish *n.* A finish designed to increase or maintain electrical resistivity of a textile material.

Electrical insulation *n.* Material with very low conductivity, which surrounds active electrical devices. Common electrical insulation chemicals are fluorine-containing polymers. Dissado LA, Fothergill CJ (eds) (1992) Electrical degradation and breakdown of polymers. Institution of Electrical Engineering (IEE), London. Ku CC, Liepins R (1987) Electrical properties of polymers. Hanser Publishers, New York.

Electrically conductive polymers *n.* Electrical properties of polymers are their responses when an electric field is applied to them. Business-machine housings, structural components, and static-control accessories often require plastics that have some degree of electrical conductivity. Additives and fillers imparting such conductivity are metal powders, carbon black, carbon

fibers, and metallized-glass fibers and spheres. Inherently conducting polymers are naturally conducting while filled polymers owe their conductivity to the electrical properties of the filler materials. Ku CC, Liepins R (1987) Electrical properties of polymers. Hanser Publishers, New York.
See also conducting polymer.

Electrical measurements *n.* Analog electronics deal with electrical signals that move through a continuous range of voltages. Digital electronics, on the other hand, deal with electrical signals that only assume fixed voltages. Giambattista A, Richardson R, Richardson RC, Richardson B (2003) College physics. McGraw-Hill Science/Engineering/Math, New York.

Electrical resistance *n.* (1) The electrical resistance is the ohmic resistance to the flow of electrical current. Related properties of plastics are insulation resistance, surface resistivity, and volume resistivity. (2) The ability of plastics to withstand various electrical stresses. Giambattista A, Richardson R, Richardson RC, Richardson B (2003) College physics. McGraw-Hill Science/Engineering/Math, New York.
See arc resistance, break-down voltage, corona resistance, and dielectric strength.

Electric field intensity *n.* The electric field intensity is measured by the force exerted on unit charge. Unit field intensity is the field, which exerts the force of 1 dyn on unit positive charge. Dimensions $[\varepsilon^{-1/2}\,\mathrm{M}^{1/2}\,\mathrm{L}^{-1/2}\,\mathrm{T}^{-1}]$, $[\mu^{-1/2}\,\mathrm{M}^{1/2}\,\mathrm{L}^{-1/2}\,\mathrm{T}^{-1}]$. The field intensity or force exerted on unit charge at a point distance r from a charge q in a vacuum

$$H = \frac{q}{r}.$$

If the dielectric in the above cases is not a vacuum the dielectric constant ε must be introduced. The formula becomes

$$H = \frac{q}{\varepsilon r^2}.$$

The value of ε is frequently considered unity for air. If the dielectric constant of a vacuum is considered unity the value for air at 0°C and 760 mm pressure is 1.000567. Weast RC (ed) (1971) Handbook of chemistry and physics, 52nd edn. The Chemical Rubber Co., Boca Raton, FL.

Electric strength *n.* Another term for dielectric strength.

Electrochemical cell *n.* Any device which converts electrical into chemical energy, or vice versa.

Electroactive polymers (EAP) *n.* Polymers that respond to electrical and magnetic stimulus with a significant change in shape and size. An example of an electroactive polymer is poly(vinylidene fluoride) (PVDF) that has been used for pressure sensitive (piezoelectricity) devices such as pressure sensors. Also, some polymers respond to photonic and thermal stimulus. Bar-Cohen Y (ed) (2001) Electroactive polymer (EAP) actuators as artificial muscles. SPIE Press, Bellingham, Washington.

Electrochemical equivalent *n.* In an electrolytic cell, the mass of a metal deposited with the passage of 1 C of electricity. In the SI system, the coulomb is defined as 1 A s. Goldberg DE (2003) Fundamentals of chemistry. McGraw-Hill Science/Engineering/Math, New York.

Electrocoating *n.* An organic finish, which is often applied as a prime coat on steel.
See electrodeposition.

Electrocuring *n.* Process, which uses an electron beam to cure organic coatings applied to commercial products on a continuous production line.

Electrode \i-ˈlek-ˌtrōd\ (1834) *n.* A terminal member in an electrical circuit designed to promote an electrical field between it and another electrode. In the plastics industry electrodes are used in microwave heat sealing and surface treating of films. One of the electrodes may be a press platen or a roll. Goldberg DE (2003) Fundamentals of chemistry. McGraw-Hill Science/Engineering/Math, New York. Weast RC (ed) (1971) Handbook of chemistry and physics, 52nd edn. The Chemical Rubber Co., Boca Raton, FL.

Electrodeposition (1882) *vt.* Method of paint application in which an article to be coated which is an electrical conductor is made one of the electrodes in a tank of water-thinned paint. The other electrode is generally a metal such as copper. The two electrodes are connected to a source of electric power, the polarity of the article to be coated being of the opposite sign to that on the particles in the liquid paint in the tank. The charged particles move towards the articles under the influence of the electric field, and when they give up their charge at the electrode (article) they are deposited and ultimately form a continuous film of paint. Weismantal GF (1981) Paint handbook. McGraw-Hill Corporation Inc., New York.

Electrodeposition coating The process of using a water-borne electrodeposition (E-Coat) paint process instead of a conventional organic solvent-based spray {G Elektrotauchlackierung f, F revêtement par électrodéposition, revêtement m, S revestimiento por electrodeposición, revestimiento m, I rivestimento per elettrodeposizione, rivestimento m}.

Electrode potential *n.* (1) The voltage associated with a half-reaction written, by convention, as a reduction; a reduction potential. (2) The difference in potential between an electrode and the immediately adjacent electrolyte referred to some standard electrode as zero potential.

Electroformed mold *n.* A mold made by electroplating a model, which is subsequently removed from the metal deposit. The deposit is sometimes reinforced with cast or sprayed metal backings to increase its strength and rigidity. Such molds are used in slush casting of vinyl plastisols and other forming processes done at low pressures.

Electroforming *n.* A method of making molds for plastics processes, usually those employing low or moderate pressures, in which a pattern made of preplated wax or flexible material is electroplated.

Electrokinetic potential *See zeta potential.*

Electroless plating *n.* The deposition of metals on a catalytic surface from solution without an external source of current. The process is used as a preliminary step in preparing plastic articles for conventional electroplating. After cleaning or etching, the plastic surface is immersed in solutions that react to precipitate a catalytic metal in situ, for example first in an acidic stannous chloride solution, then into a solution of palladium chloride. Palladium is reduced to its catalytic metallic state by the tin. Another way of producing a catalytic surface is to immerse the plastic article in a colloidal solution of palladium followed by immersion in an accelerator solution. The electroless plating bath is a solution of several components, including nickel or copper salts, chelating agents, stabilizers, and reducers. Metal reduced from the salt plates on the active sites on the palladium surface, the plated plastic being removed from the bath when the thickness of the electroless deposit is from 3 to 7 μm. The plastic article thus treated can now be plated with nickel or copper by the electroless method, forming a conductive surface that then can be plated with other metals by conventional electroplating.

Electroless plating equipment *n.* Equipment used in the deposition and formation of a continuous metallic film on a non-conductive plastic surface without the use of an electric current.

Electroluminescence \i-ˌlek-trō-ˌlü-mə-ˈne-sᵊn(t)s\ (ca. 1909) *n.* Generation of light by high-frequency electrical discharge through a gas or by applying an alternating current to a phosphor {G Elektrolumineszenz f, F électroluminescence f, S electroluminiscencia f, I elettroluminescenza f}.

Electrolysis \-ˌträ-lə-səs\ (1834) *n.* The passage of an electric current through a medium to produce a chemical change. If a current i flows for a time t and deposits a metal whose electrochemical equivalent is e, the mass deposited is $m = eit$. The value of e is usually given for mass in gram, i in ampere, and t in second.

Electrolyte (1834) *n.* A substance which produces ions when dissolved solution or when fused, thereby becoming electrically conducting {G Elektrolyt m, F électrolyte m, S electrólito m, I elettrolito m}.

Electrolytic cell *n.* An electrochemical cell in which electrical energy is used to produce chemical change; a cell in which electrolysis takes place.

See cell, electrolytic.

Electrolytic dissociation or ionization theory *n.* When an acid, base or salt is dissolved in water or any other dissociating solvent, a part or all of the molecules of the dissolved substance are broken up into parts called ions, some of which are charged with positive electricity and are called cations, and an equivalent number of which are charged with negative electricity and are called anions.

Electrolytic solution tension theory (Heimholtz double layer theory) *n.* When a metal, or any other substance capable of existing in solution as ions, is placed in water or any other dissociating solvent, a part of the metal or other substances passes into solution in the form of ions, thus leaving the remainder of the metal or substances charged with an equivalent amount of electricity of opposite sign from that carried by the ions. This establishes a difference in potential between the metal and the solvent in which it is immersed.

Electrolytic white lead *n.* Form of white lead, chemically similar to the stack and chamber types, which is made by the electrolysis of a solution of a lead salt followed by treatment with carbon dioxide gas.

Electromagnetic adhesive *n.* An intimate blend of a material that absorbs electromagnetic energy with a thermoplastic of the same composition as the sections to be bonded. The adhesive is applied in the form of a liquid, a ribbon, a wire, or a molded gasket to one of the surfaces to be joined. The two surfaces are brought into contact, then the adhesive is rapidly heated by eddy currents induced by a high-frequency induction coil placed close to the joint. This melts the adhesive which, after cooling, bonds the surfaces together.

Electromagnetic spectrum (ca. 1934) *n.* The entire range of wavelengths or frequencies of electromagnetic radiation from the shortest gamma rays to the longest radio waves and including visible light.

Electromagnetic welding *See induction welding.*

Electromagnetic wave (1908) *n.* One of the waves that are propagated by simultaneous periodic variations of electric and magnetic field intensity and that include radio waves, infrared, visible light, ultraviolet, X-rays, and gamma rays.

Electromotive force *n.* The difference in electric potential that causes current to flow in a circuit. The SI unit is the volt (V), defined as the difference of potential between two points of a conductor carrying a constant current of 1 A when the power dissipated between the two points equals 1 W. Thus, in SI, 1 V = 1 W/A.

E

Electromotive force *n.* This force is defined as that which causes a flow of current. The electromotive force of a cell is measured by the maximum difference of potential between its plates. The electromagnetic unit of potential difference is that against which 1 erg of work is done in the transfer of electromagnetic unit quantity. The volt is that potential difference against which 1 J of work is done in the transfer of 1 C. One volt is equivalent to 10^8 electromagnetic units of potential. The international volt is the electrical potential which when steadily applied to a conductor whose resistance is one international ohm will cause a current of one international ampere to flow. The international volt = 1.00033 absolute volts. The electromotive force of a Weston standard cell is 1.0183 int. volts at 20°C. Dimensions $[\varepsilon^{-1/2} M^{1/2} L^{-1/2} T^{-1}]$, $[\mu^{-1/2} M^{1/2} L^{3/2} T^{-1}]$.

Electromotive series *n.* A list of the metals arranged in the decreasing order of their tendencies to pass into ionic form by losing electrons.

Electron \i-ˈlek-ˌträn\ [*electr-* + ²*-on*] (1891) *n.* A (perhaps) fundamental subatomic particle with a very low mass and a unit negative electrical charge; found in the extranuclear region of an atom. The electron is a small particle having a unit negative electrical charge, a small mass, and a small diameter. Its charge is (4.80294 + 0.00008) × 10^{-10} absolute electrostatic units, it mass $\frac{1}{1837}$ that of the hydrogen nucleus, and its diameter about 10–12 cm. Every atom consists of one nucleus and one or more electrons. Cathode rays and beta rays are electrons.

Electron An elementary subnuclear particle having a unit negative electrical charge, 1.602191 × 10^{-19} C, its mass (at rest) is 1/1837 that of a hydrogen nucleus, or 9.10953 × 10^{-28} g, and its diameter is 5.6359 × 10^{-13} cm. Every atom consists of a nucleus of protons and neutrons, and as many orbiting electrons as there are protons in the nucleus. Cathode rays and beta rays are electrons.

Electron affinity *n.* The quantity of energy released when a gaseous, isolated, ground-state atom (or, sometimes, ion) gains an electron.

Electron beam *n.* A stream of electrons in an electron optical system.

Electron-beam machines *n.* The key to the technology is electrobeam (or E-beam) lithography. E-beam lithography is a technique for creating extremely fine patterns (much smaller than can be seen by the naked eye) required by the modern electronics industry for integrated circuits. Derived from the early scanning-electron microscopes, the technique consists of scanning a beam of electrons across a surface covered with a thin film, called a resist. The electrons produce a chemical change in this resist, which allows the surface to be patterned.

Electron-beam radiation *n.* Magnetically accelerated electrons focused by electric fields have been used for cross-linking polyethylene in special applications such as wire coating, to improve modulus and temperature resistance. Treatment levels must be carefully controlled since overexposure will cause degradation. Electron beams are also used to cure epoxy-resin coatings, eliminating the need for photoinitiators.

Electron capture *n.* A mode of radioactive decay in which an electron from the extranuclear region, usually the *K* shell, is captured by a nucleus.

Electron curing *n.* A method for curing polymer matrix composites.

Electron-deficient compound *n.* A compound in which insufficient electrons are available to bond all the atoms with conventional (two-center) covalent bonds.

Electron diffraction *n.* Used to identify crystalline substances based on the spacing of atomic planes within their structures.

Electronegativity *n.* The relative tendency of a bonded atom to attract electrons to itself.

Electron gas (ca. 1929) *n.* The delocalized electrons in a metal.

Electron lens (1931) *n.* A device for focusing a beam of electrons by means of an electric or a magnetic field.

Electronic heating *See dielectric heating.*

Electronic treating *See corona-discharge treatment.*

Electron microprobe *n.* An instrument that utilizes the bombardment of a small sample with a beam of high-energy electrons to determine the composition, elemental identification and quantification of a material (e.g., Al, O, etc.).

Electron microscopy (1932) *n.* Electron microscopy is applied to observe phase domain of a size of 50–1000 Å. This comes true with transmission electron microscopy by applying dyeing techniques such as oxidizing the unsaturated domain with OsO_4 and RuO_4.

Electron multiplier (1936) *n.* A device that utilizes secondary emission of electrons for amplifying a current of electrons.

Electron paramagnetic resonance *n.* A method to investigate the behavior of samples containing unpaired electrons (free radicals or compounds comprising an ion whose outer electronic shell is incomplete) in an applied magnetic field.

Electron probe (1962) *n.* A microprobe that uses an electron beam to induce X-ray emissions in a sample.

Electron spectroscopy *n.* A surface specific technique utilizing the emission of low energy electrons in the Auger process.

Electron tube (1922) *n.* An electronic device in which conduction by electrons takes place through a vacuum or a gaseous medium within a sealed glass or metal container and which has various uses based on the controlled flow of electrons.

Electron volt (1930) (eV) *n.* The kinetic energy acquired by any charged particle carrying unit electronic charge when it falls through a potential difference of 1 V. One eV is equal to 1.60219×10^{-19} J. Multiples of this unit in common use are the keV (10^3), MeV (10^6), and GeV (10^9). The GeV is also written BeV.

Electrophile \i-ˈlek-trə-ˌfīl\ (1943) *n.* An atom or group of atoms, which appears to seek electrons in its reactions.

Electrophoresis \-trə-fə-ˈrē-səs\ [NL] (1911) *n.* The movement of suspended particles through a fluid or gel under the action of an electromotive force applied to electrodes in contact with the suspension.

Electrophoretic deposition *n.* A direct-current process analogous to electroplating, used to coat electrically conductive articles with plastics, deposited from aqueous lattices or dispersions. The cathode may be a non-corrodible metal such as stainless steel, generally serving as the container in which the process is performed. The DC potential is usually under 100 V. The deposited coatings are baked to remove residual water. Among available polymer lattices suitable for the process are PVC, polyvinylidene chloride, acrylics, nylons, polyesters, polytetrafluoroethylene, and polyethylene.

Electroplating chemicals *n.* Copper, gold, silver, chromium, and nickel are generally used as the conductive metal for plating. Acrylonitrile–butadiene–styrene resins have been most widely used for electroplated articles. Others in commercial use for the process include cellulose acetate, some grades of polypropylene, polysulfones, polycarbonate, polyphenylene oxide, nylons, and rigid PVC.

Electroplating on plastics *n.* Articles of almost any of the common plastics can be plated by conventional processes used on metals after their surfaces have been rendered conductive by precipitation of silver or other conductor. A layer of copper is usually applied first, followed by a final plating of gold, silver, chromium, or nickel. Acrylonitrile–butadiene–styrene resins have been most widely used for electroplated articles. Others in commercial use for the process include cellulose acetate, some grades of polypropylene, polysulfones, polycarbonate, polyphenylene oxide, nylons, and rigid PVC.

(*See electroless plating*). *See also metallizing.*

Electroplating plastics *n.* Plastics can be plated by conventional processes used for metals, after their surfaces have been rendered conductive by precipitation of silver or other conductive substance.

Electrostatic coating *n.* A coating that creates, electrical charges, and disperses particles or droplets toward the target by a variety of methods {G elektrostatisches Pulversprühverfahren n, F revêtement de poudre électrostatique, revêtement m, S recubrimiento por polvo electrostático, recubrimiento m, I rivestimento con polvere elettrostatica, rivestimento m}.

Electrostatic detearing *n.* Process of removing blobs and the thick edges of paint from an article, which has been coated by dipping. The process consists of passing the dipped article, after a limited period of draining, over a grid at a high electrical potential. The blobs and thick edges of paint are removed from the article by attraction to the grid.

Electrostatic fluidized-bed coating *n.* A process combining elements of the fluidized-bed method of coating and electrostatic spraying. Pointed electrodes are inserted through the porous bottom of a fluidized-bed container. When the bed is aerated in the usual manner, a potential of about 100 kV is applied between the electrodes and ground. The associated charge repels the fluidized plastic particles into the space above the bed, from which they are attracted to a grounded article to be coated. The article may be at room temperature when inserted in the powder bed, the coating temporarily adhering by electrostatic charge. Subsequent heating fuses the coating.

Electrostatic forces *n.* The forces that exist between particles, which are electrically charged.

Electrostatic printing (electrostatography) *n.* Contactless printing by any of several methods based on electrostatic principles. Electrophotography depends on light or other electromagnetic energy and photo-semiconductors, which are non-conductors of electricity in the dark and conductors when exposed to electromagnetic radiation. Electrography involves the use of a dielectric image, stencil, or facsimile scanning for source to form the image.

Electrostatic printing *n.* A printing process employing electrostatic charge to transfer powdered ink from an electrically charged stencil to a plastic film or sheet. The film to be printed is interposed between a grounded metal plate and the stencil. Areas corresponding to those not to be printed are masked on the stencil as in conventional screen printing. The powdered ink is brushed on the back side of the screen, where it receives a charge propelling it toward the grounded plate as an image cloud until intercepted by the film. Post-heating is usually required to fuse the ink to the substrate.

Electrostatic spray coating *n.* A spraying process that employs electrical charges to direct the paths of atomized particles to the work surface. Dry plastic powders are charged with the static electricity as they emerge from a spray gun, the nozzle of which is attached to the negative terminal of a high-voltage DC power supply. The charged particles are attracted to the grounded object, which must be at least slightly electrically conductive. The powder coating is subsequently heated to obtain a smooth, homogeneous layer.

Electrostatic spraying *n.* (1) A process where paint spray is blown through an electrostatic field. (2) A system of applying paint in which the sprayed paint droplets are given an electrical charge that results in their attraction to the grounded work piece.

Electrostatic spraying *n.* Methods of application spraying in which an electrostatic potential is created between the article to be coated and the atomized paint particles. The charged particles of paint are attracted to the article being painted and are then deposited and discharged. The electrostatic potential is used in some processes to aid the atomization of the paint.

E

Element \ˈe-lə-mənt\ [ME, fr. OF & L; OF, fr. L *elementum*] (13c) *n.* A pure substance composed of atoms each having the same atomic number (number of protons). An element cannot be chemically decomposed. Substances, which cannot be decomposed by the ordinary types of chemical change, or made by chemical union.

Elemental analyses *n.* Method of substrate characterization, by which molecules are broken down to their individual elements.

Elementary process A single step of a reaction mechanism.

Elemi Gum \ˈe-lə-mē ˈgəm\ [NL *elimi*, prob. fr. Arabic al *lāmi* the elemi] (1543) *n.* Any of various natural oleoresins derived from certain tropical trees, especially *Canarium Iuzonicum* of the Philippines, and used in making varnishes and inks. It is soluble in most organic solvents, and used chiefly to impart elasticity and adhesion to lacquers and varnishes.

Eleomargaric acid *n.* Another name for eleostearic acid.

Eleostearic acid *n.* $CH_3(CH_2)_3(CH{=}CH)_3(CH_2)_7COOH$. Ocatadeca-9,11,13-trienoic acid. *cis, trans, trans*, 9,11,13-Octadecatrienoic acid. Principal constituent acid of tung oil, characterized by the presence of three conjugated double bonds. Two forms are known, namely α and β types, with mps of 49 and 71°C, respectively (*See image*).

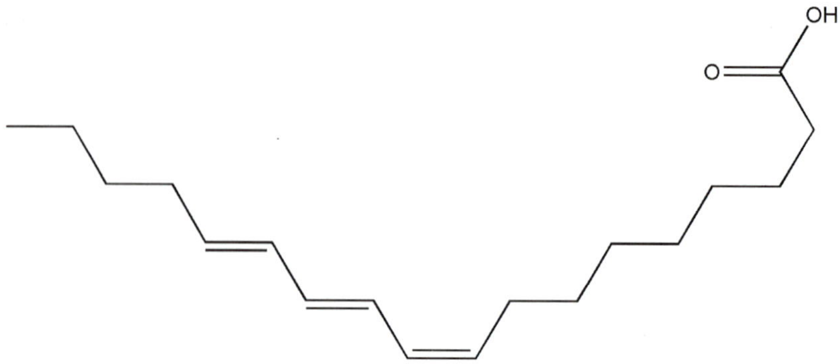

Eleostearic acid

Elimination reaction *n.* A reaction in which atoms or groups on adjacent atoms in a molecule are removed to leave a double or triple bond between the atoms.

Ellis model *n.* A three-constant model of pseudoplastic flow that merges Newton's law of flow, applicable at very low shear rates, with the power law at high rates and provides a smooth transition between the two. For one-dimensional flow the equation is

$$-\frac{dv_z}{dx} = \frac{\tau_{xz}}{\eta_o}\left[1 + \left(\frac{\tau_{xz}}{\tau_{1/2}}\right)^{\alpha-1}\right],$$

where v_z is the *z*-directed velocity perpendicular to coordinate *x*, τ_{xz} is the opposing shear stress, η_o is the zero-shear (Newtonian) viscosity, a is an exponent larger than 1, approximately equal to the reciprocal of the flow-behavior index and $\tau_{1/2}$ is the shear stress at which the viscosity (=shear stress/shear rate) is half η_o.

Elmendorf tear strength *n.* The Elmendorf tear tester, originally developed to test papers and fabrics, has been adapted for plastics films in ASTM D 1922. Acting by gravity, a calibrated pendulum swings through an arc, tearing the specimen from a precut slit. The energy absorbed is indicated by a pointer and scale. Some other modes of measuring tear resistance are spelled out in ASTM D 1004, and D 2582.

Elongation \(ˌ)ē-ˌlóŋ-ˈgā-shən\ (14c) *n.* In tensile testing, the fractional increase in length of a marked test length as the test specimen is stretched and stress rises. At any point during the test, percent nominal elongation = 100 × the increase in gage length/original length. Ultimate elongation is the elongation just prior to rupture of the sample and is the "elongation at break" reported in most properties tables.
See also true strain.

Elongational flow (extensional flow) *n.* Flow caused by stretching a material, usually a hot melt, as in fiber drawing, film blowing, parison draw down, and biaxial stretching of sheet. This flow is always accompanied by a reduction in cross section. The rate of elongation, at any moment during one-dimensional, elongational flow, is given by

$$de/dt = \left(\frac{1}{s}\right)\left(\frac{ds}{dt}\right) = \frac{d(\ln s)}{dt},$$

where *e* is the true elongation and *s* is the strand length at time *t*.

Elongational viscosity (extensional viscosity, Trouton viscosity) *n.* The viscosity that characterizes an element undergoing elongational flow (above). It is equal to the tensile stress divided by the rate of elongation and for polymers it depends on the rate, but may increase with rate, unlike the usual reduction of shear viscosity with rate. Tensile viscosities are apt to be many times larger than shear viscosities for the same resin, temperature, and deformation rate. Values in the range of 10^4–10^7 Pa s have been reported. For Newtonian liquids, the elongation viscosity is three times the shear viscosity (at the same temperature).

Elongation at break *n.* The increase in length when the last component of the specimen breaks.

Eluent \ˈel-yə-wənt\ [L *eluent-*, *eluens*, pp of *eluere*] (1941) *n.* In gas and liquid chromatography, the fluid that carries the solute out of the column.

Elution [L *elutus*, pp of *eluere* to wash out, fr. *e-* + lavere to wash] (1731) *n.* The removal, in chromatography, of the species adsorbed on the column matrix by a flowing liquid or gas.

Elutriate \ē-ˈlü-trē-ˌāt\ [L *elutri atus*, pp of *elutriare* to put in a vat, perhaps fr. *elutrum* vat, fr. Gk *elytron* reservoir, literally, covering] (ca. 1727) *vt.* To purify, separate, or remove by washing.

Elutriation *n.* The separation of less massive particles from a sample of distributed particle sizes or densities by upward flow of a liquid or gas.

Elvanol *n.* Poly(vinyl alcohol), manufactured by DuPont, USA.

EMA *n.* Abbreviation for ethylene–acrylic acid copolymer.

Embedding *n.* The process of encasing an article in a resinous mass, performed by placing the article in a simple mold, pouring a liquid resin into the mold to completely submerge the article, sometimes under vacuum so as to suck out hidden air bubbles, curing the resin, and removing the encased article from the mold. In the case of electrical components, the lead wires or terminals may protrude from the embedment. The main difference between embedding and potting is that in potting, the model is a container that remains fixed to the resinous mass. The liquid resin may contain microspheres to reduce the final mass of the embedment.
See also encapsulation, impregnation, and potting.

E

Embedment decorating *n.* A technique fur decorating reinforced-plastics articles in which a mat or web of fibrous material printed with a design is embedded in the surface of the article and covered with a transparent gel coat. The technique can be adapted for use in hand lay-up, continuous laminating, pultrusion, and matched-die molding.

Embossed paper *n.* Wallpaper run through rollers with raised areas, to provide a light relief effect.

Embossing *n.* Any of several techniques used to crease depressed patterns in plastics films or sheeting. In the case of cast film, embossing can be accomplished directly by casting on an inverse-textured belt or roll. Calendered films are frequently embossed by rolls following the calender rolls. Other films or coated fabrics can be embossed subsequent to manufacture by reheating them and passing them through embossing rolls, or compressing them between plates. Extruded sheets, up to 3 mm or thicker, are commonly embossed as the sheets emerge from the extruder with an embossing roll on the takeoff.

Embroidery \im-ˈbrói-d(ə)rē\ (14c) *n.* Ornamental designs worked on a fabric with threads. Embroidery may be done either by hand or by machine.

Emerald green \ˈem-rəld ˈgrēn\ (1646) *n.* Name applied to two distinctly different green pigments: complex copper acetoarsenite (*also known as Paris green and*

Schweinfurt green); and chromium hydroxide.

See hydrated chromium oxide.

Emerald oxide of chromium *See hydrated chromium oxide.*

Emery \ˈem-rē, ˈe-mə-\ [ME, fr. MF *emeri*, fr. OIt *smiriglio*, fr. ML *smiriglum*, fr. Gk *smyrid-*, *smyris*] (15c) *n.* A mixture of mostly corundum and some magnetite that, because of its great hardness, is widely used for grinding and polishing, including tumble-polishing of plastics moldings. Emery is available in many grades from coarse to extremely fine, as powder and bonded to paper and cloth.

Emery abrasive *n.* A natural composition of corundum and iron oxide, found in large deposits both in USA and the near East. It is magnetic and partly soluble in hydrochloric acid. The grains are blocky, cut slowly, and tend to polish the material being abraded.

Emery cloth *n.* Coated abrasive cloth used for light polishing of metal or for removing rust spots or similar light work; not recommended for large metal surfaces or tough metal, where aluminum oxide cloth is much more satisfactory.

EMI *n.* Abbreviation for 2-ethyl-4-methylimidazole.

Emission \e-ˈmi-shən\ (1607) *n.* (1) Discharges into the air by a pollution source as distinguished from effluents, which are discharged into water. (2) The emitting of radiation. Emission depends on the temperature to which a material is heated relative to the temperature of its surroundings, on the time, and on the nature of the surface. Basically, therefore, emission is a net rate at which the body emits radiation. As an adjective, it is used to describe the characteristic radiation emitted by elements in a spectroscope, e.g., emission spectrum.

Emission factor *n.* The average amount of pollutants emitted from a polluting source per unit of material produced.

Emissive power (or emissivity) *n.* Emissive power is measured by the energy radiated from unit area of a surface in unit time for unit difference of temperature between the surface in question and surrounding bodies. For the cgs system the emissive power is given in ergs per second per square centimeter with the radiating surface at 1° absolute and the surroundings at absolute zero.

See radiation formula.

Empirical formula \-i-kəl ˈfór-myə-lə\ (1885) *n.* A formula expressing the simplest whole-number ratio of atoms of each element in a compound without providing information on the grouping of the atoms. For example, the empirical formula of oleic acid is $C_{18}H_{34}O_2$. Also called simplest formula.

Emulgator *See emulsifier.*

Emulsification \i-ˌməl-sə-fə-ˈkā-shən\ (1859) *n.* (1) In lithography, a condition resulting from the distribution of fountain solution in the ink. Excessive emulsification will produce poor printing. (2) The process of dispersing one liquid in another (the liquids being mutually insoluble or sparingly soluble in each other). When water is one of the liquids, two types of emulsions are possible: oil-in-water (water is the continuous state), and water-in-oil. The term "oil" describes any organic liquid sparingly soluble in water.

Emulsified asphalt *See asphalt emulsion.*

Emulsifier (1888) *n.* Substance that intimately mixes, modifies the surface tension of colloidal droplets, and disperses dissimilar materials ordinarily immiscible, such as oil and water, to produce a stable emulsion. The emulsifier has the double task of promoting the emulsification and of stabilizing the finished products. {G Emulgator m, F émulsifiant m, S emulsionante m, I emulsionante m}.

Also known as emulsifying agent and emuglator

Emulsifiers *n.* A surface-active agent promoting the formation and stabilization of an emulsion.

Emulsifying agent *n.* A substance used to facilitate the formation of an emulsion from two or more immiscible liquids, and/or to promote the stability of the emulsion. As surface-active agents, emulsifiers act to reduce interfacial tensions between the several phases. They also act as protective colloids to promote stability.

See emulsifier.

Emulsifying agents *n.* Materials used to facilitate the preparation of emulsions and to improve their stability.

Emulsion \i-ˈməl-shən\ [NL *emulsion-*, *emulsio*, fr. L *emulgēre* to milk out, from *e-* + *mulgēre* to milk; akin to OE *melcan* to milk, Gk *amelgein*] (1612) *n.* Strictly, an emulsion is a two-phase, substantially permanent, intimate mixture of two incompletely miscible liquids, one of which is dispersed as finite globules in the other. However, in plastics and other industries the term is sometimes broadened to include colloidal suspensions of solids such as waxes and resins in liquids. In an emulsion, the liquid that forms globules is known as the dispersed, discontinuous, or internal phase. The surrounding liquid is called the continuous or external phase. The dispersed phase may be held in suspension by mechanical agitation or by the addition of small amounts of emulsifying agents {G Emulsion f, F émulsion f, S emulsión f, I emulsione f}.

Emulsion paint *n.* Paint, the vehicle of which is an emulsion of binder in water. The binder may be oil, oleoresinous varnish, resin or other emulsifiable binder. Not to be confused with a latex paint in which the vehicle is a latex.

Emulsion polymerization *n.* A polymerization process in which the monomer or mixture of monomers is emulsified in a low-viscosity aqueous medium by means of soaps or other surface-active, solubilizing and emulsifying agents. The emulsion does not require intensive stirring as in suspension polymerization, and produces polymers of higher molecular weight than those produced by bulk or suspension processes. The polymers remain in emulsion, and must be recovered from the latex by freezing or chemical precipitation. The polymerization medium usually also contains a water-soluble initiator, catalyst, or chain-transfer agent. Examples of polymers produced from emulsions are polyvinyl acetate, acrylo-nitrile–styrene terpolymer, PVC, polyethylene, acrylics, and polystyrene.

Emulsion spinning *n.* Process by which a polymer matrix suspension of very fine particles is subjected to a high temperature sintering and drawing process, during which the matrix polymer is burnt off and the dispersed particles are coalesced.

Emulsion stabilizer *See protective colloid.*

Emulsoid \i-ˈməl-ˌsóid\ (ca. 1909) *n.* Colloidal particle, which takes up water. Syn: hydrophile.

Enamel \i-ˈna-məl\ [ME, fr. MF *enamailler*, fr. *en-* + esmail enamel, of Germanic origin; akin to OH Gr *smelzan* to melt] (14c) *vt.* (1) A dispersion of pigment in a liquid that forms a solid adherent film, on the surface to which it is applied, by means of oxidation polymerization, or other chemical reaction. The liquid vehicle of an enamel usually contains a thermosetting resin and a solvent. An initial soft film is formed by evaporation of the solvent, then the film hardens or cures at room temperature or during baking. (2) A class of substances having similar composition to glass with the addition of stannic oxide, SnO_2, and other infusible substances to render the enamel opaque.

E

Enamel blue *See cobalt blue.*

Enantiomer \i-ˈnan-tē-ə-mər\ [Gk *enangtios* + E –*mer*] (ca. 1929) *n.* Either of a pair of chemical compounds whose molecular structures have a mirror-image relationship to each other. An asymmetric molecule that is the mirror image of its stereoisomer. The two isomers are given the prefixes *dextro-* and *levo-*, e.g., *d-* and *l*-lactic acid. The physical properties of pure enantiomers are equal within experimental error, yet mixtures of the two, called racemic mixtures, may have different properties. For example, 50–50 *di*-lactic acid melts 20°C lower than its pure enantiomers.

Enantomeric \-ˌnan-t ē-ə-ˈmer-ik\ *adj.*

Enantiomeric configurational unit *n.* Either of two stereoisomeric groups in a polymer that are mirror images at the plane containing the main-chain bonds.

Enantiomorph \i-ˈnan-tē-ə-ˌmórf\ [GK *enantios* opposite (fr. *enanti* facing, fr. *en* in + *anti* against) + ISR - morph] (1885) *n.* Enantiomer; also, either of a pair of crystals (as of quartz) that are structural mirror images.

Enantiotropic *n.* Compound or an element able to exist in two distinct crystalline forms, depending upon its maintenance above or below a certain transition temperature.

Encapsulating \in-ˈkap-sə-ˌlāt-iŋ\ (1876) *v.* The process of combining elements to create a new entity.

Encapsulation *n.* The process of applying a fairly thick coating that conforms to the shape of the coated object. The coating, of either thermoplastic or thermosetting resin, may be applied by brushing, dipping, spraying, or thermoforming. The process is much used for the protection and insulation of electrical components and assemblies.
See also embedding, impregnation, potting, and microencapsulation.

Encapsulization *n.* The enclosure of adhesive particles with a protective film that prevents them from coalescing until such time as proper pressure or solvation is applied.

Encaustic \in-ˈkó-stik\ [*encaustic*, adj, fr. L *encausticus*, fr. Gk *enkaustikos*, fr. *enkaiein* to burn in, fr. *en-* + *kaiein* to burn] (1601) *n.* (1) Painted with a mixture of a paint solution and wax, which, after application, is set by heat. (2) Colors, which have been applied to brick, glass, porcelain, and tile and set by the application of heat.

Encounter *n.* The period of time during which two reactant particles are trapped by a solvent-molecule cage in a liquid-solution reaction.

End *n.* (1) A strand of roving consisting of a given number of filaments gathered together. The group of filaments is considered to be an end or strand before twisting, and a yarn after the twist has been applied. (2) An individual warp yarn, thread, or fiber. (3) A short length or remnant of fabric.

End-capping *n.* Conversion, by chemical reaction, of the end groups of polymer chains to less reactive, more stable groups, thus preventing "unzipping" of chains and rendering the polymer itself more stable to processing. An example is the conversion of –OH end groups in polyoxymethylene to acetate groups.

End grain *n.* The surface of timber exposed when a tree is felled or when timber is cross-cut in any other way.

End group *n.* A chemical group or radical forming the end of a polymer chain. These are normally different from the repeating unit group or groups. Although end groups constitute a minute fraction of the polymer, they may vary considerably from the main-chain chemical structure and may exert effects on polymer properties that are stronger than one would expect from their numbers.

End-group analysis *n.* The quantitative determination, by chemical or spectrographic methods, of the number of end groups present in a sample of polymer. With linear, unbranched polymers, the mass of the sample divided by half the measured number of end groups equals the number-average molecular weight.

Endo- {*combining form*} [F, fr. Gk, fr. *endon* within; akin to Gk *en* in, OL *indu*, Hittite *andan* within]. (1) A chemical prefix denoting an inner position, for example in a ring rather than in a side chain, or attached as a bridge within a ring. (2) When prefixing "-thermic", denoting that the reaction so labeled takes heat from the surroundings when proceeding from left to right. In both sense, the opposite of Exo-.

Endothermic \ˌen-də-ˈthər-mik\ [ISV] (1884) *adj.* Pertaining to a chemical reaction or an operation that is accompanied by the absorption of heat. Opposite of exothermic.

Endothermic reaction *n.* A reaction which absorbs heat.

End out A void caused by a missing warp yarn.

End point (1899) *n.* (1) Maximum distillation temperature when a substance is distilled. (2) Stoichiometric point as shown by an indicator, potentiometer or other means, in a titration. (3) Required values of viscosity, acid values, etc., the attainment of which indicates the conclusion of a process in resin or varnish manufacturing.

End-to-end distance *n.* The square root of the average square of the distance from the one of the polymer chain to the other end of the polymer chain.

Endurance limit (fatigue limit) *n.* The stress level below which a specimen will withstand cyclic stress indefinitely without exhibiting fatigue failure. Rigid, elastic, low-damping materials such as thermosetting plastics and some crystalline thermoplastics do not exhibit endurance limits.

Endurance ratio *n.* The ratio of the endurance limit, under cyclic stress reversal, to the short-time, static strength of a material. If the mode of stress is not specified, tension/compression may be presumed.

Energy \ˈe-nər-jē\ [LL *energia* fr. Gk *energeia* activity, fr. *energos* active, fr. *en-* in + *ergon* work] (1599) *n.* The capability of doing work. Potential energy is energy due to position of one body with respect to another or to the relative parts of the same body. Kinetic energy is energy due to motion. cgs units, the erg, the energy expended when a force of 1 dyne acts through a distance of 1 cm; the joule is 1×10^7 ergs. The potential energy of a mass m, raised through a distance h, where g is the acceleration due to gravity is

$$E = mgh.$$

The kinetic energy of mass m, moving with a velocity v, is

$$E = \frac{1}{2}mv^2.$$

Energy will be given in ergs if m is in g, g in cm/s^2, h in cm and v in cm/s.

Energy absorption *n.* The energy required to break or elongate a fiber to a certain point.

Energy, *E* *n.* The ability to do work.

Energy dispersive X-ray analysis (EDXRA) This method of elemental analysis (microanalysis) is often used in conjunction with scanning electron microscopy (SEM). An

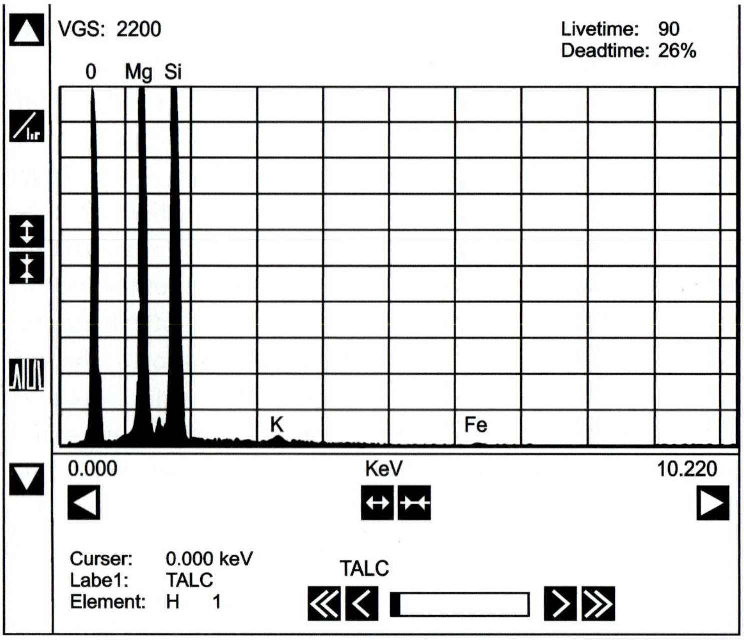

Energy dispersive X-ray analysis (EDXRA) spectrograph of talc mica

example of an EDXRA spectrograph is shown. Staniforth M, Goldstein J, Echlin P, L ifshin E, Newbury DA (2003) Scanning electron microscopy and X-ray microanalysis. Springer-Verlag, New York.

Energy (free energy) *n.* Usable power (as heat or electricity); also; the resources for producing such power.

Energy of a charge *n.* In ergs where Q is the change and V the potential in electrostatic units.

$$E = \frac{1}{2} QV.$$

Energy of rotation *n.* If a mass whose moment of inertia about an axis is I, rotates with angular velocity ω about this axis, the kinetic energy of rotation will be

$$E = \frac{1}{2} I\omega^2.$$

Energy will be given in ergs if I is in g cm^2 and ω in rad/s.

Energy of the electric field *n.* If H is the electric field intensity in electrostatic units and K the specific inductive capacity, the energy of the field in ergs/cm^3 is

$$E = \frac{KH^2}{8\pi}.$$

Energy-to-break *n.* The total energy required to rupture a yarn or cord.

Engineering plastic *n.* (1) A broad term covering those plastics, with or without fillers and reinforcements that have mechanical, chemical, electrical, and/or thermal properties suitable for industrial applications. R. B. Seymour, an outstanding authority, defined them as polymers thermoplastic or thermosetting, that maintain their dimensional stability and major mechanical properties in the temperature range 0–100°C. He listed the "big five" (among neat resins) as nylons, polycarbonate, acetals, polyphenylene ether, and thermoplastic polyesters. Among many others are acrylics, fluorocarbons, phenoxy, acrylonitrile–butadiene–styrene terpolymer, polyaryl either, polybutylene, chlorinated polyether, polyether, and many polymers reinforced with advanced fibers. At the high end of the spectrum of performance, usable temperature range, and price, unreinforced engineering plastics are dubbed advanced resins.

Engineering plastics *n.* Plastics which are modified by using modifiers, additives, fillers and reinforcements.

Engineering polymers *n.* Polymers referred to as "commodity thermoplastics", accounting for 85% of total plastics production. The major engineering polymers are ethylene, propylene, styrene, and vinyl chloride, which offer the optimum balance of easy processability and moderate properties at low cost.

English gallon *See imperial gallon.*

English vermillion *See mercuric sulfide.*

Engraved-roll coating *See gravure coating.*

Engraving (1601) *n.* Machine-printing (of wallpaper) with etched-out rollers to obtain subtle and fine-effects not possible with surface printing.

Enhancement ratio *n.* In a filled or reinforced plastic, the ratio of the modulus of the filled material to that of the neat resin. The ratio is likely to be higher at low strains than high strains because of slippage between filler and matrix.

Enjay-butyl *n.* Isobutylene/isoprene copolymer, manufactured by Enjay, USA.

Enkatherm *n.* Poly(terephthaloyl oxamidrazone), manufactured by AKZO, The Netherlands.

Entangled yarns *See compacted yarns.*

Entangling *n.* (1) A method of forming a fabric by wrapping and knotting fibers in a web about each other, by mechanical means, or by the use of jets of pressurized water, so as to bond the fibers (*also see hydroentangling and spunlaced fabric*). (2) *See intermingling.*

Entering *n.* The process of threading each warp yarn on a loom beam through a separate drop wire, heddle, and reed space in preparation for weaving. This process may be done by hand or by a semiautomatic machine.

Enthalpy \ˈen-ˌthal-pē\ *n* [*en-* + Gk *thalpein* to heat] (ca. 1924) (heat content) *n.* Thermodynamically, the enthalpy of a system is the sum of the internal energy and the pressure–volume product multiplied by the pressure $H = E + pv$ where H is the enthalpy or heat content, E the internal energy of the system, p the pressure, and v is the volume. We are usually concerned about changes in enthalpy rather than absolute values. Enthalpies of polymers are usually stated per unit mass, e.g., kJ/kg, and are ordinarily referred to room temperature, 20–25°C, at which temperature enthalpy is arbitrarily set to zero. The rate of change of enthalpy with temperature is heat capacity.

Enthalpy (*H*) *n.* A thermodynamic quantity, which is useful for describing heat exchanges taking place under constant-pressure conditions. The enthalpy of a system is defined as the sum of its energy, E, and its pressure–volume product: $H = E + pv$

Enthalpy of formulation (ΔH_f) *n.* The enthalpy change for a reaction in which compound is formed from its uncombined elements.

Entrance angle (entry angle) *n.* In an extrusion diem the total included angle, never more than 180°, of the main converging surfaces of the flow channel leading to the land area of the die.

Entropy, S \ˈen-tre-pe\ [ISV [2]*en-* + Gk *tropē* change, literally, turn, fr. *trepein* to turn]

(1875) *n.* (1) A measure of the unavailable energy in a thermodynamic system, commonly expressed in terms of its changes on an arbitrary scale, the entropy of water at 0°C being assigned the value of zero. (2) When a system (or sample of material) is heated or cooled from one temperature, T_1, to another T_2, the change in its entropy is given by the equation

$$\Delta S = \int_{T_1}^{T_2} \frac{dQ}{T}.$$

(3) Also, entropy is a thermodynamic quantity, which measures the degree of disorder or randomness in a system.

Entropy, of elasticity *n.* A measure of the unavailable energy in a closed thermodynamic system that is also usually considered to be a measure of the system's disorder and that is a property of the system's state and is related to it in such a manner that a reversible change in heat in the system produces a change in the measure which varies directly with the heat change and inversely with the absolute temperature at which the change takes place.

Environment \in-ˈvī-rə(n)-mənt\ (1827) *n.* The sum of all external conditions and influences affecting the life, development, and survival of an organism.

Environmental protection agency (EPA) *n.* Charged by the clean air amendments of 1970 to the air quality act of 1967 with establishing minimum air quality standards which must be incorporated in regional state standards. These standards are essentially similar to those first promulgated by the Los Angeles region of California in their Rule 66.

Environmental stress cracking *n.* The formation of internal or external cracks in a plastic caused by tensile stresses well below its short-time strength, and induced by exposure to heat, solvent vapor, or chemically active solutions. ASTM test F 1248 describes the measurement of the environmental stress-cracking resistance of polyethylene pipe in the presence of a surface-active agent. Other ASTM tests treating this subject are D 1693, D 2561, D 2951, and, in Section 15.03, F 484, F 791, and F 1164.

Enzyme \ˈen-ˌzīm\ [Gr *Enzym*, fr. MGk *enzymos* leavened, fr. Gk *en-* + *zymē* leaven] (1881) *n.* Any of numerous proteins or conjugated proteins produced by, and functioning as, biochemical catalysts in living organisms.

EOS Abbreviation for equation of state.

Eosine \ˈē-ə-sən\ [ISV, fr. Gk *ēōs* dawn] (1866) *n.* $C_{20}H_8Br_4O_5$. Red acid dye used in the manufacture of lake pigments. Sodium salt of brominated fluorescein.
Also see phloxine.

EP *n.* (1) Abbreviation for epoxy resin or epoxide. (2) (Usually E/P) Abbreviation sometimes used for copolymers of ethylene and propylene.

EPA *See environmental protection agency.*

EPDM *n.* Elastomer from ethylene, propylene, and a diene. Abbreviation for elastomeric terpolymer from ethylene, propylene, and a conjugated diene.
See ethylene–propylene rubber.

Epi- (1) A prefix signifying a chemical compound or group differing from a parent compound or group by having a bridge connection. (2) EPI:
See epichlorohydrin.

Epichlorohydrin \ˌe-pi-ˌklōr-ə-ˈhī-drən\ (ca. 1891) (chloropropylene oxide, EPI) *n.* An active solvent for cellulosic and other resins, and a key reactant for epoxy resins having the structure shown below. EPI is highly reactive with polyhydric phenols such as bisphenol A and forms glycidyls with many compounds containing active hydrogens. Epichlorohydrin (ECH) is used primarily in the manufacture of epoxy resins and synthetic glycerol. The main other uses include the production of epichlorohydrin elastomers, polyamide–epichlorohydrin resins, water treatment chemicals, and a variety of glycidyl derivatives.

Epichlorohydrin rubber (CO, CEO) *n.* Any of several elastomers comprising polymers and copolymers of epichlorohydrin, with good high-temperature resistance, low-temperature flexibility, resistance to fuels, oils, and ozone, and low gas permeability. The homopolymer (CO) is a saturated, aliphatic polyether with a chloromethyl side chain. The ECO type is an equimolar copolymer of epichlorohydrin and ethylene oxide.

Epikote *n.* Epoxide resin, manufactured by Shell, The Netherlands.

Episulfide *n.* The sulfur analog of epoxides in which the sulfur is part of a ring. The two most important members are ethylene sulfide and propylene sulfide. Episulfides are starting materials for polysulfide rubber.

Epitropic fibers *n.* Fibers with an altered surface property, e.g., electrically conducting, abrasive, etc.

EPM *n.* Elastomer from ethylene and propylene. Abbreviation for ethylene–propylene rubber.

Epon *n.* Epoxide resin, manufactured by Shell, The Netherlands.

Epoxidation \(ˌ)-e-ˌpäk-sə-ˈda-shən\ (1944) *n.* A chemical reaction in which an oxygen atom is joined to an olefinically unsaturated molecule to form a cyclic, three-membered either. The products of epoxidation are known as oiranes or epoxides.

Epoxide (1930) *n.* Any compound containing the oxirane structure, a three-membered ring containing two carbon atoms and one oxygen atom. The most important members are ethylene oxide and propylene oxide.

Epoxide equivalent *n.* The mass of resin in grams that contains 1 g-equivalent of epoxide.

Epoxidized soybean oil *See* epoxy plasticizer.

Epoxy [*epi-* + *oxy*] (1916) (epoxy group, oxirane group) *adj.* In textiles, a compound used in durable-press applications for white fabrics. It provides chlorine resistance but causes loss of tensile strength. A label denoting an oxygen atom joined in a ring with two carbon atoms, as shown below.

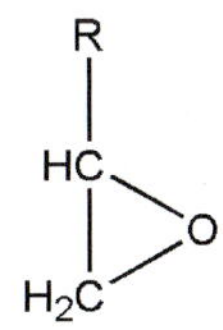

Epoxy adduct *n.* Resin having all the required amine incorporated but requiring additional epoxy resin for curing.

***β*-(3,4-Epoxycyclohexyl) ethyltrimethoxy silane** *n.* A coupling agent for reinforced polyester, epoxy, phenolic, melamine, and many thermoplastics.

Epoxy ester *n.* An epoxy resin partially esterified with fatty acids, rosin, etc.; single package epoxy.

Epoxy foam *n.* Two basic types of epoxy foams are in use, chemical foams and syntactic foams. Chemical-foam compositions contain the resin, curing agent, blowing agent, wetting agent, and a small percentage of an inert organic compound such as toluene to dissipate the exothermic heat of curing, and thus control the foaming action. Because foaming is rapid, the curing agent is withheld until all the other ingredients have been mixed, to be added just prior to casting. These systems may also contain amine-terminated polyamide resins to impart resiliency to the foam. In syntactic foams, the voids are provided by hollow phenolic microspheres and the resin does not foam but acts as a binder for the spheres. Epoxy foams are used in casting, in potting and encapsulating of electrical assemblies, in insulating coatings for chemical-storage tanks, and in cores of laminates for aircraft and boats.

Epoxy–novolac resin *n.* A two-step resin made by reacting epichlorohydrin with a phenol-formaldehyde condensate. Such resins are also known as thermoplastic, B-stage phenolic resins that are in a state of partial cure. Whereas bisphenol-based epoxy resins contain up to two epoxy groups per molecule, the epoxy novolacs may have seven or more such groups, producing more tightly crosslinked structures in the cured resins. Thus they are stronger and superior in other properties.

Epoxy number *n.* The number of gram-equivalents of epoxy groups per 100 g of polymer, equal to 1/100 of the reciprocal of the epoxides equivalent.

Epoxy paint *n.* Paint based on an epoxy resin.

Epoxy plasticizer (epoxides plasticizer) *n.* Any of a large family of plasticizers obtained by the epoxidation of vegetable oils or fatty acids. The two main types are (a) epoxidized unsaturated triglycerides, e.g., soybean oil and linseed oil; and (b) epoxidized esters of unsaturated fatty acids, e.g., oleic acid, or butyl-, octyl-, or decyl- esters. Most epoxy plasticizers have a heat-stabilizing effect and they are often used for stabilization in conjunction with other stabilizers. Epoxidized oils generally have good resistance to extrusion and migration and low volatility, but they cannot be used as sole plasticizers in unfilled vinyl compounds and hence are not considered to be primary plasticizers. Certain epoxidized soybean oils have been FDA-approved for food-contact use.

Epoxy plastics *n.* Plastics based on resins made by the reaction of epoxides or oxiranes with other materials such as amines, alcohols, phenols, carboxylic acids, acid anhydrides, and unsaturated compounds.

Epoxy resin (1950) *n.* Any of a family of thermosetting resins containing the oxirane group (*See epoxy for structure*). Originally made by condensing epichlorohydrin and bisphenol A, epoxy resins are not more generally formed from low-molecular-weight diglycidyl ethers of bisphenol A and modifications thereof; or, another type, by the oxidation of olefins with peracetic acid. Depending on molecular weight, the resins range from liquids to solids. The liquids, used for casting, potting, coating, and adhesives, are cured with amines, polyamides, anhydrides, or other catalysts. The solid resins are often modified with other resins and unsaturated fatty acids. Epoxy resins are widely used in reinforced plastics, have strong adhesion to glass fibers. Epoxies based on epoxidized heterocyclic hydantoin are useful in electrical composites because their thermal expansion coefficient can be matched to that of copper. Their low viscosities are effective in wetting the various reinforcing materials used with them. Fast-curing epoxies are based on the diglycidyl ether of 4-methylol resorcinol (DGEMR). The methylol group appears to effectively catalyze the curing reactions. This resin is curable with all types of conventional epoxy hardeners including aliphatic and aromatic amines, anhydrides, and amidoamines. DGEMR cures approximately thirty times as fast as a conventional bisphenol A epoxy and two to five times as fast as older fast-gelling epoxies, and at lower temperatures. DGEMR may be formulated with flexibilizers and fillers without prolonging gel time. These same properties make the resin well suited for adhesives, coatings, and low-temperature applications.

See also epoxy–novolac resin.

Epoxy resins *n.* Plastic or resinous materials used for strong, fast-setting adhesives, as heat resistant coatings and binders, etc. Cross-linking resins based on the reactivity of the epoxides group. One common type is the resin made from epichlorohydrin and bisphenol A. Aliphatic polyols such as glycerol may be used instead of the aromatic bisphenol A or bisphenol F.

Epoxy stabilizer (epoxides stabilizer) *n.* Most epoxy plasticizers also serve as stabilizers because of the ability of the epoxides group to accept HCl, or to serve as an intermediate, in the presence of metallic salts, to convert HCl to a metallic chloride. Epoxy stabilizers are most often used in conjunction with barium–cadmium and other stabilizers, with which they have a synergistic effect.

EPR *n.* Abbreviation for ethylene–propylene rubber.

EPS *n.* Abbreviation for expanded polystyrene. *See polystyrene foam.*

Epsom salts \ˈep-səm-\ [*Epsom*, England] (1876) *n.* $MgSO_4{\cdot}7H_2O$. Magnesium sulfate. Water-soluble magnesium salt for the preparation of precipitated soaps. Rhombic, colored or white; Sp gr, 1.68.

Equalizer rod A metal rod wound with a fine wire around its axis so that ink can be drawn down evenly and at a given thickness across a piece of paper.

Also called a Meyer rod or bar. See wire-wound rod.

Equant *n.* A shape having nearly equal dimensions.

Equation of state (EOS) *n.* For an ideal gas, if the pressure and temperature are constant, the volume of of the gas depends on the mass, or amount of gas. Then, a single property called the gas density (ratio of mass/volume). If the mass and temperature are held constant, the product of pressure and volume are observed to be nearly constant for a real gas. The product of pressure and volume is exactly for an ideal gas. This relationship between pressure and volume is called Boyle's Law. Finally, if the mass and pressure are held constant, the volume is directly proportional to the temperature for an ideal gas. This relationship is called Charles and Gay–Lussac's law. The gas laws of Boyle and Charles and Gay–Lussac can be combined into a single equation of state: $PV = nRT$, where P is pressure, V volume, T absolute temperature, n number of moles and R is the universal gas constant. Anerodynamicists us a different form of the equation of state that is specialized of air. Regarding polymers and monomers, equation of state is an equation giving the specific volume (v) of a polymer from the known temperature and pressure and, sometimes, from its morphological form. An early example is the modified Van der Waals form, successfully tested on amorphous and molten polymers. The equation is:

$$v = b + RT/M(P + \pi),$$

where b is the "unfree" specific volume occupied by the polymer molecules, roughly above 90% of V, R the universal molar-energy constant, T the absolute temperature, P is the pressure, M an empirical molecular weight that for several thermoplastics has been closely equal to the mer weight, and π is an empirical internal pressure much larger than the highest injection-molding pressures. A number of more complex models have since been introduced and tested against experimental data. Atkins PW, Atkins P, De Paula J (2001) Physical chemistry. W. H. Freeman Co., New York. Perry RH, Green DW (1997) Perry's chemical engineer's handbook, 7th edn. McGraw-Hill, New York. James F (ed) (1993) Whittington's dictionary of plastics. Technomic Publishing Co. Inc., Carley.

E

Equilibrium \ˌē -kwə-ˈli-brē-əm\ [L *aequilbrium*, fr. *aequilibris* being in equilibrium, fr. *aequi-* + *libra*, weight, balance] (1608) *n.* A state of balance between opposing forces or actions that is either static (as in a body acted on by forces whose resultant is zero) or dynamic (as in a reversible chemical reaction when the velocities in both directions are equal) Chemical: a state of affairs in which a chemical reaction and its reverse reaction are taking place at equal velocities, so that the concentrations of reacting substances remain constant.

Equilibrium condition *n.* The condition, that the mass-action expression equals the value of the equilibrium constant for a reaction, which is satisfied when the reaction system is at equilibrium.

Equilibrium constant (1929) *n.* The product of the concentrations (or activities) of the substances produced at equilibrium in a chemical reaction divided by the product of concentrations of the reacting substances, which concentration raised to that power which is the coefficient of the substance in the chemical equation.

Equivalence point \i-ˈkwiv-lən(t)s ˈpóint\ *n.* The state in a titration at which equal numbers of equivalents of reactants and products have been mixed.

Equivalent, acid–base *n.* The quantity of an acid (or base), which will furnish (or react with) 1 mol of H^+.

Equivalent conductance *n.* The equivalent conductance of an electrolyte is defined as the conductance of a volume of solution containing one equivalent weight of dissolved substance when placed between two parallel electrodes 1 cm apart, and large enough to contain between them all of the solution. Λ is never determined directly, but is calculated from a specific conductance. If *C* is the concentration per cubic centimeter is *C*/1000, and the volume containing one equivalent of the solute is, therefore, 1000/*C*. Since, L_s is the conductance of a centimeter cube of the solution, the conductance of 1000/*C* cc, and hence Λ will be

$$\Lambda = \frac{1000 L_s}{C}.$$

Equivalent, redox *n.* The quantity of an oxidizing agent (or a reducing agent) which will accept (or furnish) 1 mol of electrons.

Equivalent single yarn number *See yarn number, equivalent single.*

Equivalent weight (1904) (combining weight, equivalent mass) *n.* The atomic or formula weight of a given element or ion divided by its valence in a reaction under consideration. Elements entering into combination always do so in quantities proportional to their equivalent weights. In oxidation–reduction reactions the equivalent weights of the reacting entities are dependent upon the change in oxidation numbers of the particular substances.

Erioglaucine dye *See acid dyes.*

Erlangen blue *n.* Another name for Prussian blue.

Erode \i-ˈrōd\ [L *erodere* to eat away, fr. *e-* + *rodere* to gnaw] (1612) *v.*
See etch.

Erosion (1541) *n.* (1) Wearing away of the top coating of a painted surface, e.g., by chalking or by the abrasive action of windborne particles of grit, which may result in exposure of the underlying surface. (2) Phenomenon manifested in paint films by the wearing away of the finish to expose the substrate or undercoat. The degree of failure is dependent on the amount of substrate or undercoat visible. Erosion occurs as the result of chalking or by the abrasive action of windborne particles of grit.

Erosion breakdown *n.* In an electrical-conductor insulation, deterioration caused by chemical attack of corrosive chemicals such as ozone and nitric acid that are formed by corona discharge from a high-voltage cable. This breakdown can occur even the most chemically polymers, such as fluorocarbons, after long exposure to the condition.

Erosion control fabrics *See geotextiles.*

Erosion resistance *n.* The ability of a coating to withstand being gradually worn away by chalking or by the abrasive action of water or windborne particles of grit. The degree of resistance is dependent upon the amount of coating retained. *See erosion.*

Erucamide *n.* $CH_3(CH_2)_7CH{=}CH(CH_2)_{11}COONH_2$. The amide of *cis*-13-docosenoic acid, used in fractional percentages as a slip agent in polyethylene film resins.

Erucyl alcohol *n.* $CH_3(CH_2)_7CH{=}CH(CH_2)_{11}CH_2OH$. A monounsaturated, fatty alcohol used as a mold lubricant.

Erucic acid \i-ˈrü-sik-\ [NL *Eruca*, genus of herbs, fr. L, colewort] (1869) *n.* $CH_3(CH_2)_7CH{=}CH(CH_2)_{11}COOH$. Unsaturated fatty acid found in many vegetable oils. Properties: mp, 33°C; bp, 281°C/30 mmHg; Sp gr, 0.860/55°C; and iodine value, 75.2. *Also known as cis-13-docosenoic acid.*

Erythrene *n.* 1-3 butadiene. Old syn: butadiene.

Erythritol *n.* $H(CHOH)_4H$. Tetrahydric alcohol. Properties: mp, 112°C; bp, 330°C.

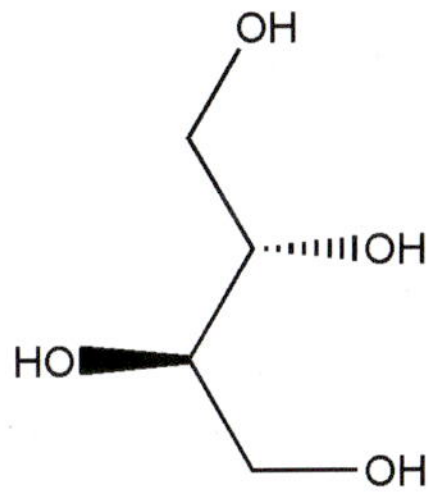

ESC *n.* Abbreviation for environmental stress cracking.

ESCA Energy spectroscopy chemical analysis. An example of an ESCA spectrum for

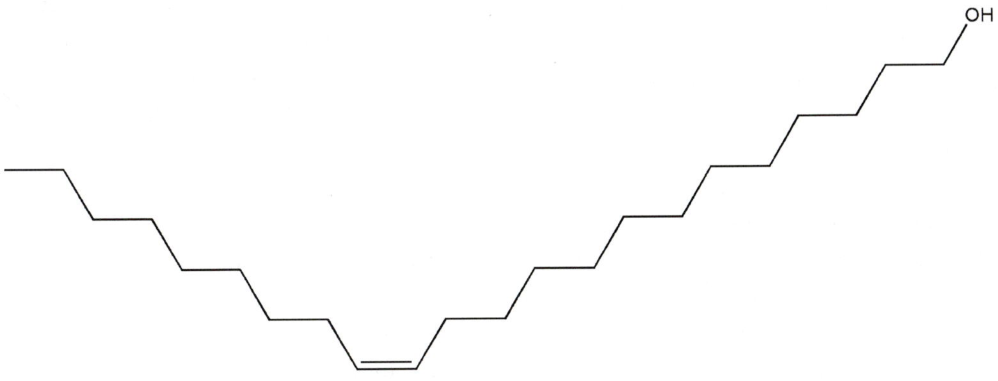

Erucic acid

chloride is shown. Siegbahn K (1967) ESCA, atomic, molecular and solid state structure studied by means of electron spectroscopy. Almquist & Siks, Uppsala, Sweden.

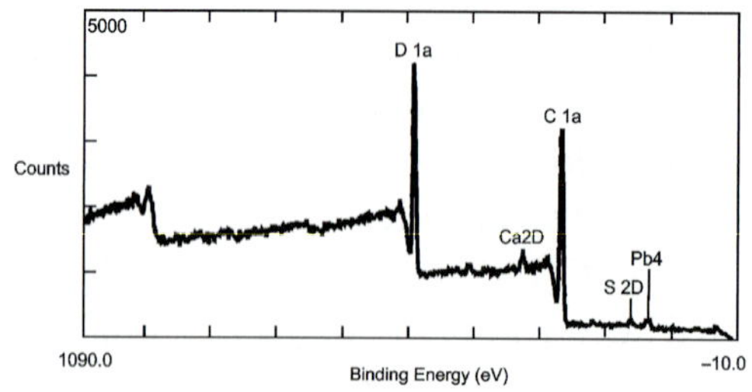

Escaping tendency *n.* The tendency shown by a substance to escape from its phase to another.

ESO *n.* Abbreviation for epoxidized soybean oil.
See epoxy plasticizers.

Esparto wax \is-ˈpär-(ˌ)tō-\. Hard vegetable wax obtained from esparto grass, indigenous to North Africa and parts of Spain. It is used to some extent as a substitute for carnauba, candellia and similar hard waxes. Properties: mp, approximately 71°C; acid value, 30, saponification value, 75.

ESR *n.* Abbreviation for electron-spin-resonance spectroscopy.

Essential oils *n.* Volatile oils or essences derived from vegetation and characterized by distinctive odors and a substantial measure of resistance to hydrolysis. Chemically, essential oils are often principally terpenes. Some essential oils are nearly pure single compounds. Some contain resins in solution and are then called oleoresins or balsams. Shahidi F, Bailey AE (eds) (2005) Bailey's industrial oil and fat products. John Wiley and Sons, New York. Langenheim JH (2003) Plant resins: chemistry, evolution ecology and ethnobotany. Timber Press, Portland, OR. Paint: pigment, drying oils, polymers, resins, naval stores, cellulosics esters, and ink vehicles, vol 3. American Society for Testing and Material, Philadelphia, PA, 2001.

Estane *n.* Polyurethane. Manufactured by Goodrich, USA.

Ester \ˈes-tər\ [Gr. fr. *Essigäther* ethyl acetate, fr. *Essig* vinegar + *Äther* ether] (ca. 1852) *n.* An organic compound with the general formula R–C–O–R′ analogous to an inorganic salt. Esters are formed by reacting an acid with an alcohol, or by the exchange of a replaceable hydrogen atom of an acid for an organic alkyl radical. Esters of many monofunctional alcohols and organic acids are oily, fruity-smelling liquids, forming important families of solvents and plasticizers. When the alcohol selected is polyfunctional, that is, contains two or more –OH groups, and the acid is di- or polybasic, long chains of repeating units can be formed by their reaction. These are polyesters.

Ester gum *n.* Resin made from rosin or rosin acids and a polyhydric alcohol, such as glycerol or pentaerythritol. Used as an ingredient in certain printing ink varnishes. Whistler JN, BeMiller JN (eds) (1992) Industrial gums: polysaccharides and their derivatives. Elsevier Science and Technology Books, New York.

Esterification \-ˌster-ə-tə-ˈkā-shən\ (ca. 1905) *n.* The chemical process of combining an acid and an alcohol to form an ester. Cellulose acetate is an ester formed by the reaction of acetic acid and the hydroxyl groups of cellulose. Polyethylene terephthalate, the most common fiber-forming polyester, is a product of esterification of terephthalic acid with ethylene glycol. Morrison RT, Boyd RN (1992) Organic chemistry, 6th edn. Prentice-Hall, Englewood Cliffs, NJ.

Ester interchange (ester exchange) *n.* A reaction between an ester and another compound in which occurs an exchange of alkoxy or acyl groups, resulting in the formation of a different ester. When an ester is reacted with an alcohol, the process is called alcoholysis; reaction between an ester and an acid is called acidolysis. Tester interchanges are used in producing plasticizers, polyvinyl alcohol, acrylics, polyesters, and polycarbonates. Morrison RT, Boyd RN (1992) Organic chemistry, 6th edn. Prentice-Hall, Englewood Cliffs, NJ.

See transesterification.

Ester number *n.* Difference between the saponification value and the acid number. Therefore, it is the number of mg of potassium hydroxide necessary to saponify the neutral esters in 1 g of fat, wax, or oil.

Esters *n.* A class of organic liquids, used as solvents, which are the products of the reactions of organic acids and alcohols. Examples are ethyl acetate, dioctyl phthalate.

Esthetics *See aesthetics.*

Eta (η) *n.* Greek letter; symbol for absolute viscosity.

Etch \ˈech\ [D *etsen,* fr. Gr. *ätzen* to etch, corrode, fr. OH Gr *azzen* to feed; akin to OH Gr *ezzan* to eat] (1634) *v.* Wear away or roughen a surface with, or as if with, an acid. Erode is sometimes used as a synonym.

Etching (1634) *v.* (1) In lithography, the use of acidic substances to produce a surface in the non-printing areas that is receptive to the fountain solution but not to the ink. (2) In engraving, a treatment with acid or by mechanical means to make certain areas considerably lower than the surface of the engraving.

See printing, burn-out printing.

Etching primer *See wash primer.*

ETFE *n.* Abbreviation for ethylene–tetrafluoroethylene co-polymer.

Ethanal Syn: acetaldehyde.

Ethanite *n.* Synthetic rubber made by the interaction of potassium polysulfide and ethylene dichloride.

Ethanol \ˈe-thə-ˌnól, -ˌnōl, *British also* ˈē-\ *n* (1900) *n.* Syn: ethyl alcohol.

Ethanolamine \ˌe-thə-ˈnä-lə-ˌmēn\ (1897) *n.* $NH_2(CH_2)_2OH$. Bp, 171°C/150 mmHg. Known also as monoethanolamine, colamine, 2-amino ethanol, and betahyroxyamine.

Ethanolurea *n.* $NH_2CONHCH_2CH_2OH$. A white compound melting at 71–74°C. It condenses with formaldehyde to form permanently thermoplastic, water-soluble resins. Simple urea can be incorporated in the condensation reaction to give modified resins with any desired degree of water solubility and flexibility, both of which properties increase with urea content.

Ethene \ˈe-ˌthēn\ (1873) *n.* IUPAC's name for ethylene.

Ethenoid plastics *n.* (1) Plastics made from monomers containing the polymerizable double-bond group C=C, for example ethylene. Thermosetting ethenoid resins are made from monomers for linear polymers capable of giving crosslinked structures as a result of double-bond polymerization. (2) A British generic term that includes acrylic, vinyl, and styrene plastics.

Ether \ˈē-thər\ [ME, fr. L *aether,* fr. Gk *aithēr,* fr. *aithein* to ignitem blaze; akin to OE *ād* pyre] (14c) *n.* (1) Any organic compound in which an oxygen atom is interposed between two carbon atoms or organic radicals in the molecular structure. Ethers are often derived from alcohols by elimination of one molecule of water from two molecules of alcohol. (2) Specifically, diethyl ether $(C_2H_5)_2O$.

Etherified urea resins *n.* Urea resins in which the methylol groups have been etherified with suitable alcohols. Alcohols commonly used include butyl and octyl alcohols. The etherified resins are characterized by improved hydrocarbon tolerance, some types permitting substantial additions of aliphatic hydrocarbons.

Ethocel *n.* Cellulose ether, manufactured by Dow, USA.

Ethyl \ˈe-thəl\ [Gr *Ethyl* (now *Äthyl*), fr. *Äther* ether + *-yl*] (1838) *n.* $–C_2H_5$. Monovalent alkyl radical.

Ethyl abietate *n.* $C_{19}H_{29}COOC_2H_5$. Amber colored viscous liquid, which hardens upon oxidation. Properties: bp, 350°C; Sp gr. 1.03; flp, 171°C (340°F).

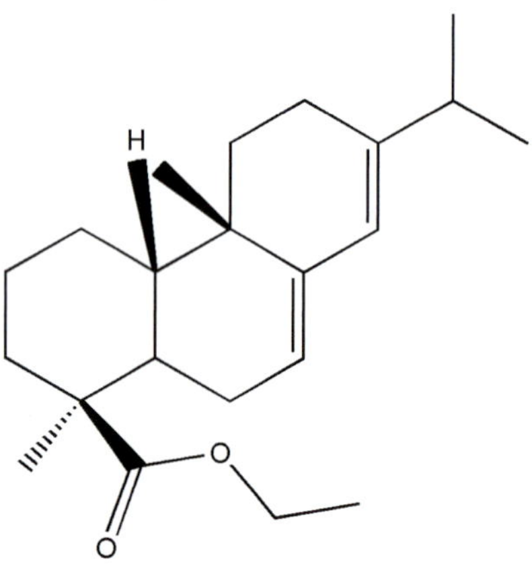

Ethyl acetanilide (ethyl phenylacetamide) *n.* A substitute for camphor in the manufacture of celluloid.

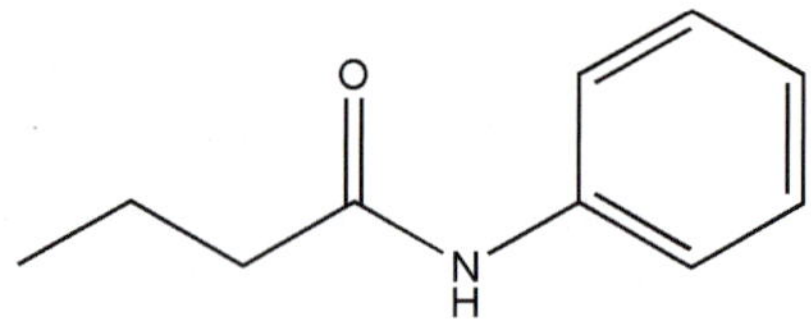

***n*-Ethyl acetanilide** *n.* $C_6H_5N(C_2H_5)CO CH_3$. Properties: bp, 258°C; flp, 124°C (255°F). Substitute for camphor in the nitrocellulose industries.

Ethyl acetate (1874) *n.* $CH_3COOC_2H_5$. A colorless liquid made by heating acetic acid and ethyl alcohol in the presence of sulfuric acid, then distilling. It is a powerful solvent for ethyl cellulose, polyvinyl acetate, cellulose acetate–butyrate, acrylics, polystyrene, and coumarone–indene resins. It is also used in flexographic and rotogravure inks. Although it is highly flammable, it is the least toxic of common industrial solvents. Properties: bp, 77°C; Sp gr, 0.901/20°C; flp. 0.56°C (31°F); refractive index, 1.373; vp, 77 mmHg/20°C. Also known as acetic ether and acetic ester.

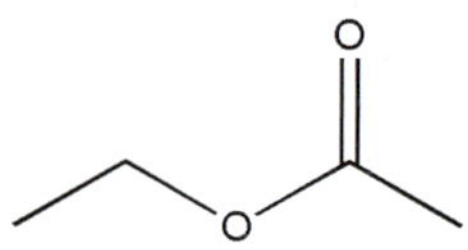

Ethylacetic acid *n.* Syn: butyric acid.

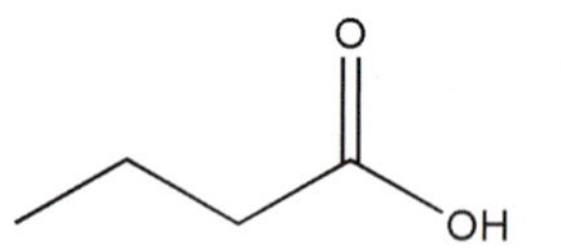

Ethyl acetyl ricinoleate *n.* $C_{17}H_{32}(OCOCH_3)COOC_2H_5$. Properties: bp, 400°C; Sp gr, 0.931.

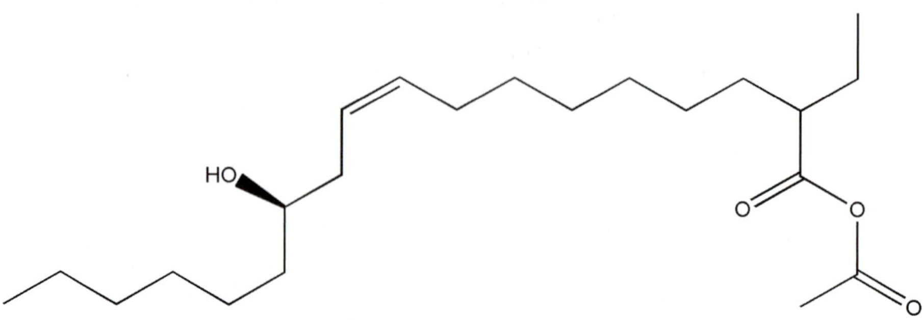

Ethyl acetyl ricinoleate

Ethyl acrylate *n.* $CH_2{=}CHCOOC_2H_5$. A colorless liquid, insoluble in water; miscible with most organic solvents. Used in the manufacture of synthetic resins. Sp gr, 0.9283; bp, 101°C.

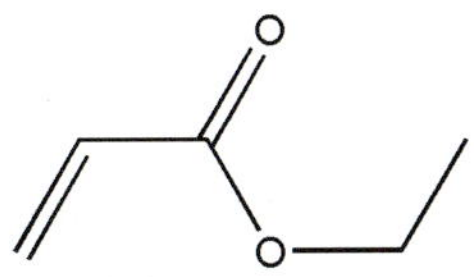

Ethyl alcohol (1869) (alcohol, ethanol, and grain alcohol) *n.* An alcohol used, in denatured form, as a solvent for ethyl cellulose, polyvinyl acetate, and polyvinyl butyrate. The industrial grade of undenatured alcohol usually contains 5 wt% water. Properties: bp, 78°C; Sp gr. 0.7938/15°C; refractive index, 1.367. The pure compound is called absolute alcohol.

Also known as alcohol, grain alcohol, ethanol, fermentation alcohol, spirit of wind, ethyl hydroxide, cologne sprits, and EtOH.

Ethyl aldehyde *n.* Syn: acetaldehyde.

Ethyl aluminum dichloride *n.* $C_2H_5AlCl_2$. A clear, yellow, flammable liquid, a catalyst for olefin polymerization.

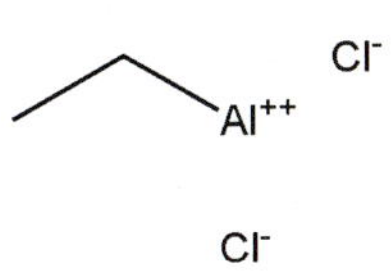

Ethyl aluminum sesquichloride *n.* $(C_2H_5)_3Al_2Cl_3$. A catalyst for olefin polymerization.

Ethylbenzene \ˌe-thil-ˈben-ˌzēn\ [ISV] (1873) *n.* Colorless liquid with an aromatic odor. Used as an intermediate in styrene production and as a solvent. Properties: molecular weight, 106.16; bp, 136°C; Sp gr, 0.8673.

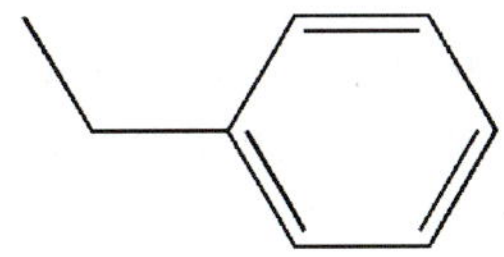

Ethyl benzoate (benzoic ether) *n.* A colorless liquid derived by heating ethyl alcohol and benzoic acid in the presence of sulfuric acid. It is a solvent for cellulosics.

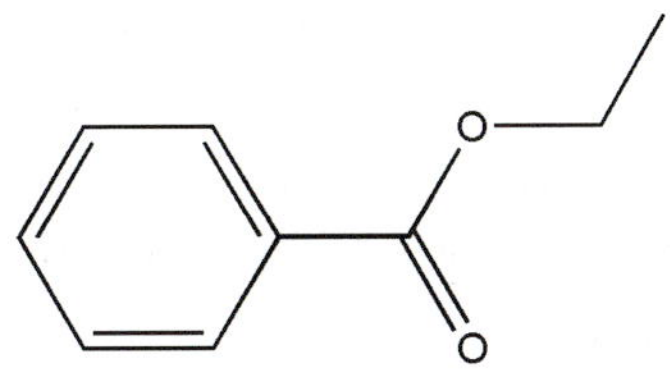

2-Ethylbutyl acetate *n.* A solvent for cellulose nitrate.

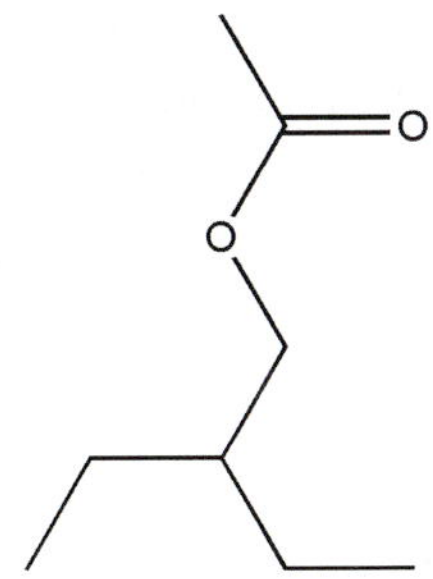

Ethyl butyrate (ethyl butanoate) *n.* A solvent for cellulosics.

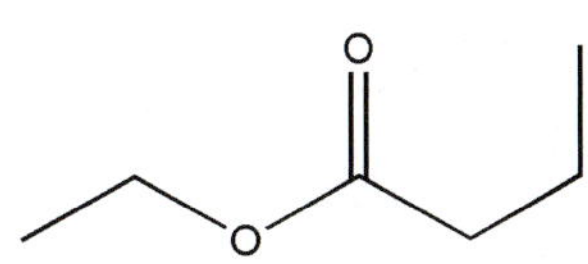

Ethyl carbamate *n.* Syn: urethane.

Ethyl carbonate (glycol carbonate, 1,3-dioxolan-2-one) *n.* A solvent for many polymers and resins. Syn: diethyl carbonate.

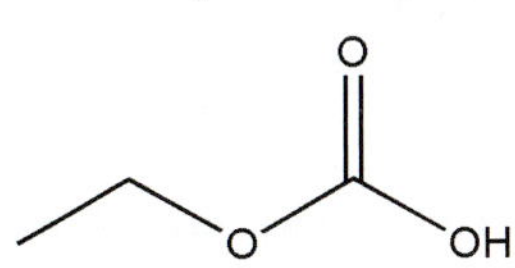

Ethyl cellulose (1936) (EC) *n.* An ethyl ether of cellulose formed by reacting cellulose steeped in alkali with ethyl chloride; it is a white granular thermoplastic resin. Since the repeating units are etheric, it is chemically different from other cellulosics, which are esters, and is therefore not compatible with them. EC resin can be injection molded, extruded, cast into film, or used as a coating material. It has the lowest density of all cellulosic plastics, good toughness and impact resistance, and is dimensionally stable over a wide temperature range. Syn: carbitol.

Ethyl chloride (ca. 1891) (chloroethane) *n.* A colorless gas at ambient conditions, used in the production of ethyl cellulose by reaction with sodium cellulose.

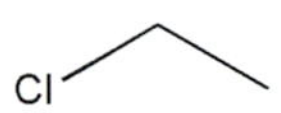

Ethyl citrate *See triethyl citrate.*

Ethylene (Ethene) (bicarburetted hydrogen, ethene) *n.* A colorless, flammable gas derived by cracking petroleum and by distillation from natural gas. In addition to serving as the monomer for polyethylene, it has many uses in the plastics industry including the synthesis of ethylene oxide, ethyl alcohol, ethylene glycol (used in making alkyd and polyester resins), ethyl chloride, and other ethyl esters. Properties: molecular weight, 28; bp, −1.025°C; Sp gr of liquid, 0.610/0°C.

Ethylene-acid copolymer resins *n.* Resins that are flexible, specialty thermoplastics created by high-pressure copolymerization of ethylene (E) and methacrylic acid (MAA) or acrylic acid (AA).

Ethylene–acrylic acid copolymer (EAA) *n.* Either of the block or random copolymers of ethylene and acrylic acid whose ionic character gives strong adhesion to metals and other surfaces. Their toughness has created uses in multilayer packaging films and golf-ball covers. Two trade names are DuPont's Surlyn® (block) and Dow's EAA (random).

***N,N′*-Ethylene bis-stearamide** (Acrawax C®) *n.* A lubricant used in acrylonitrile–butadiene–styrene resins, PVC, and polystyrenes.

Ethylene-bis tris-(2-cyanoethyl) phosphonium bromide (ECPB) *n.* A flame retardant for thermoplastics. In polymethyl methacrylate, 20% ECPB caused the resin to become opaque and reduced its burning rate to zero.

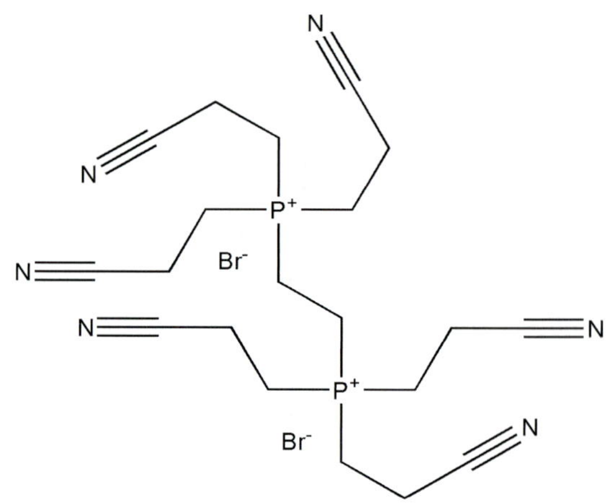

Ethylene carboxylic acid *n.* A little used Syn: acrylic acid.

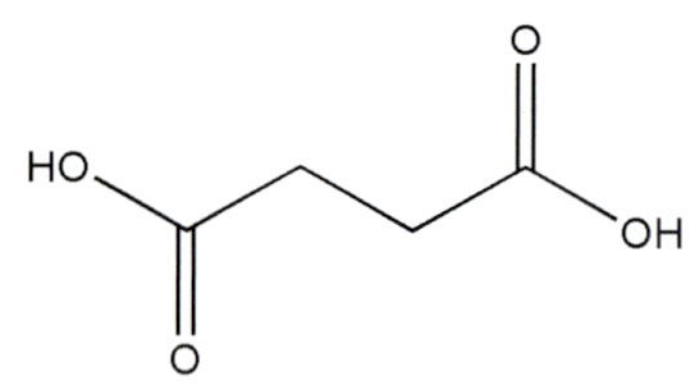

Ethylene chloride *n.* Syn: ethylene dichloride.

Ethylene chlorotrifluoroethylene copolymer (ECTFE, E/CTFE) *n.* A fluoroplastic with good mechanical, thermal, electrical, processing, and resistance properties.

Ethylene dichloride *n.* $ClCH_2CH_2Cl$. Colorless, oily liquid used for metal degreasing; paint and varnish remover. Properties: bp, 83°C; Sp gr, 1.252/20°C; vp, 65 mmHg/20°C; mol wt. 99.

Also known as 1,2-dichlorethane.

Ethylene–ethyl acrylate copolymer (EEA, E/EA) *n.* A family of elastomeric resins similar in appearance to polyethylene, but possessing properties like those of rubber and flexible vinyls.

Ethylene–ethyl acrylate resins ethylene glycol *n.* CH_2OHCH_2OH. A type of alcohol, completely miscible in water, used in latex and water-based paints to provide stability when frozen; used in heating and cooling systems as a fluid for transferring heat. Bp; 197°C; Sp gr, 1.12/4°C; flp, 111°C (232°F). Used as a diol in manufacturing resins.

Also known as ethylene alcohol.

Ethylene glycol (1901) (ethanediol, ethylene alcohol, and glycol) *n.* $HOCH_2CH_2$–OH. A clear, syrupy liquid used as a solvent for cellulosics, particularly cellophane, and in the production of alkyd resins and polyethylene terephthalate.

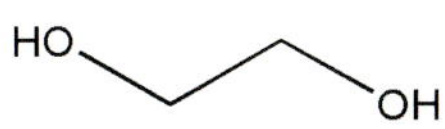

Ethylene glycol diacetate *n.* A very slowly evaporating solvent for cellulosic and acrylic resins, sometimes used as a fugitive plasticizer for vinyls and acrylics.

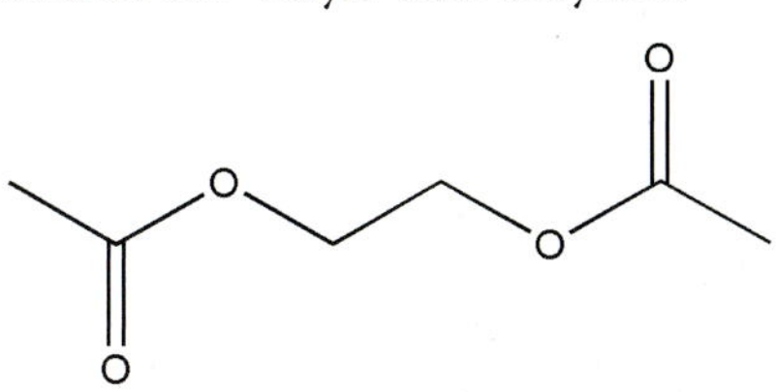

Ethylene glycol dibenzoate *n.* A plasticizer for cellulosic resins, having limited compatibility with some vinyl resins.

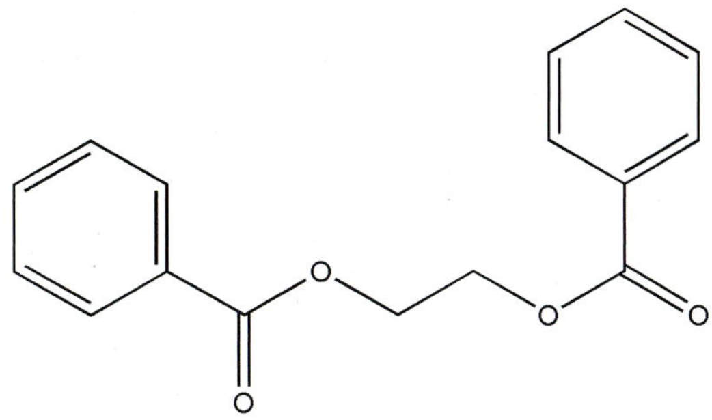

Ethylene glycol dibutyrate (glycol dibutyrate). A plasticizer for cellulosic plastics.

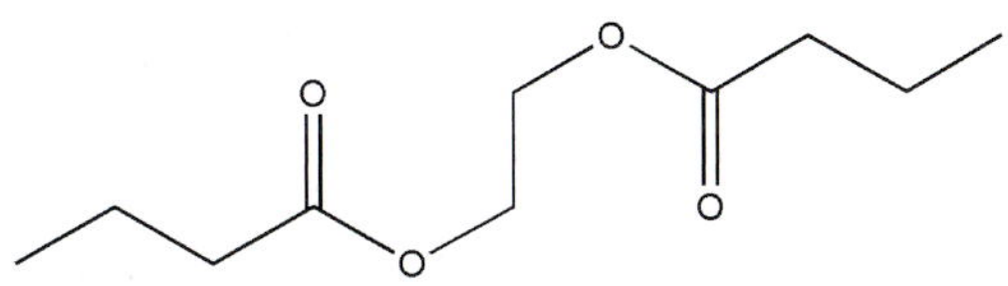

Ethylene glycol dipropionate (glycol propionate) *n.* A plasticizer for cellulosic resins.

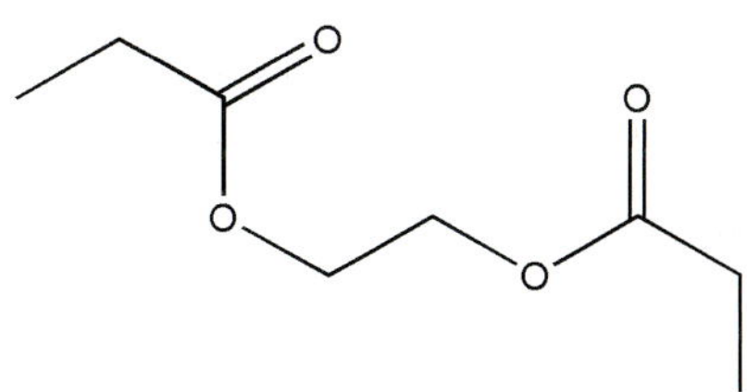

Ethylene glycol monoacetate (glycol monoacetate). A solvent for cellulose nitrate and cellulose acetate.

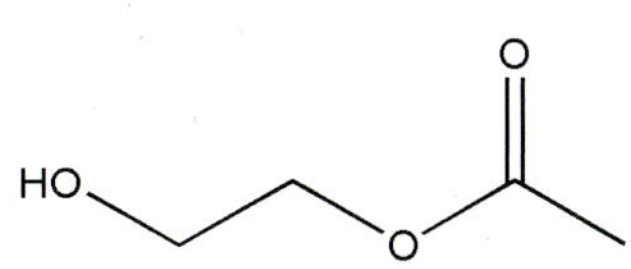

Ethylene glycol monobenzyl ether (benzyl cellosolve) *n.* A solvent for cellulose acetate.

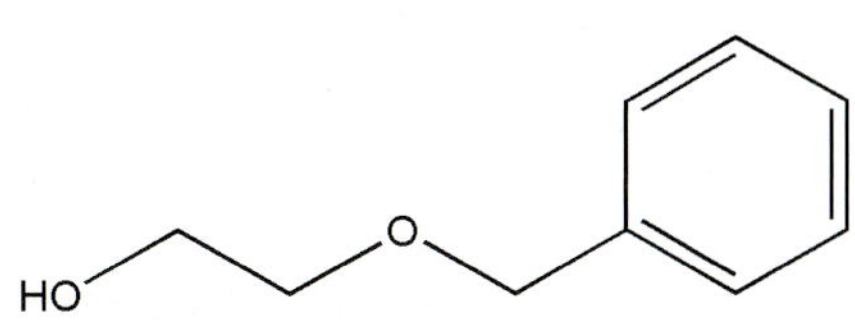

E

Ethylene glycol monobutyl ether (2-butoxyethanol, butyl cellosolve) *n.* A colorless liquid used as a solvent for cellulosic, phenolic, alkyd, and epoxy resins, especially in varnish and other coating formulations.

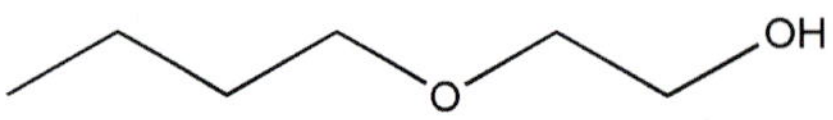

Ethylene glycol monobutyl ether acetate *n.* A colorless liquid with a fruity aroma, used as a high-boiling solvent for cellulose nitrate, epoxy resins, and as a film-coalescing aid for polyvinyl-acetate latex.

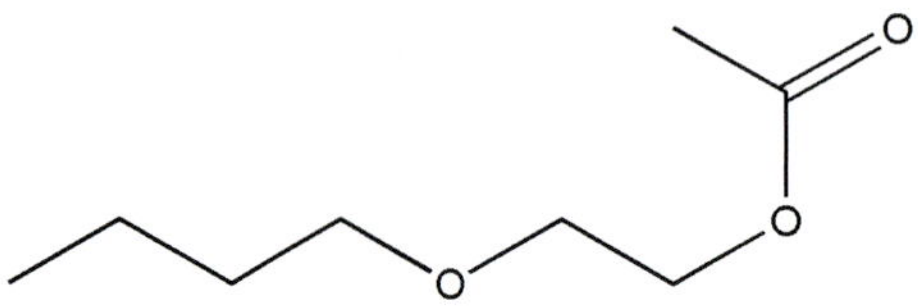

Ethylene glycol monobutyl ether laurate *n.* A plasticizer for cellulosics, polystyrene, and vinyls.

Ethylene glycol monobutyl ether oleate *n.* A plasticizer for cellulose nitrate, ethyl cellulose, and PVC.

Ethylene glycol monobutyl ether stearate *n.* A plasticizer for cellulose nitrate, ethyl cellulose, polystyrene, and polyvinyl butyral.

Ethylene glycol monoethyl ether (cellosolve, ethyl cellosolve) *n.* A solvent for cellulose nitrate, phenolic, alkyd, and epoxy resins. It is colorless, nearly odorless, has a low evaporation rate, and imparts good flow properties to coatings.

Ethylene glycol monoethyl ether acetate (cellosolve acetate) *n.* $C_2H_5OC_2H_4OOCCH_3$. A solvent for cellulose nitrate, ethyl cellulose, vinyl polymers, and copolymers, polymethyl methacrylate,

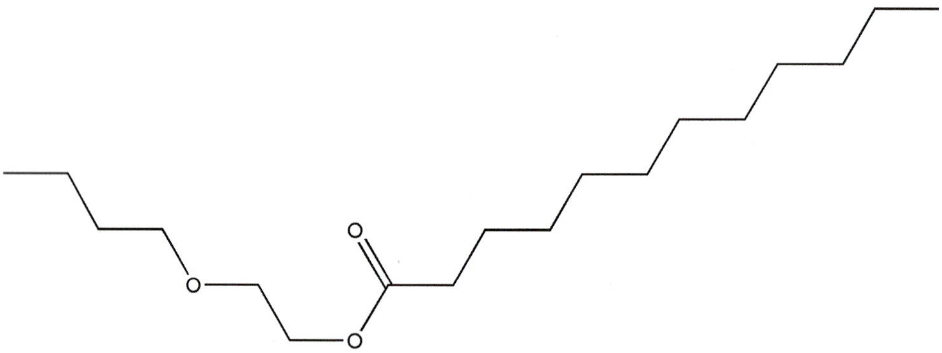

Ethylene glycol monobutyl ether laurate

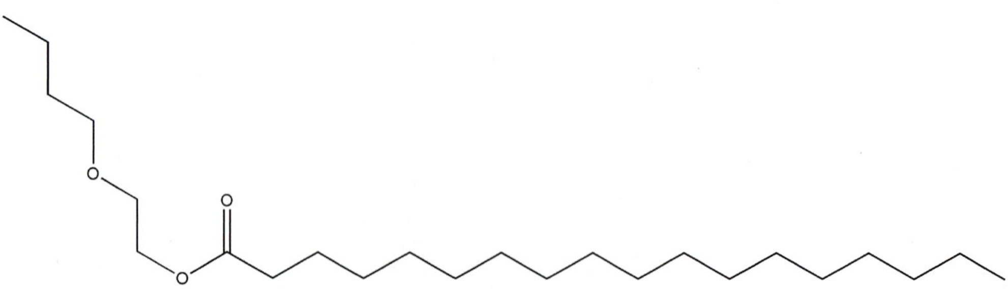

Ethylene glycol monobutyl ether stearate

polystyrene, epoxy, coumarone–indene and alkyd resins.

Ethylene glycol monoethyl ether laurate *n.* A plasticizer for cellulosic and vinyl resins, and polystyrene.

Ethylene glycol monoethyl ether ricinoleate *n.* A plasticizer.

Ethylene glycol monomethyl ester *n.* $CH_3OC_2H_4OOCC_{17}\text{–}H_{35}$. A plasticizer for cellulosics and polystyrene, having limited compatibility with other thermoplastics.

Ethylene glycol monomethyl ether (2-methoxyethanol, methyl cellosolve) *n.* A solvent for cellulose esters.

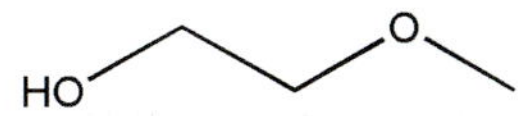

Ethylene glycol monomethyl ether myristate *n.* A plasticizer for cellulosic plastics, PVC, and polyvinyl butyral.

Ethylene glycol monomethyl ether oleate *n.* A plasticizer for cellulosic and vinyl resins.

Ethylene glycol monooctyl ether *n.* A solvent for cellulose esters, and a plasticizer.

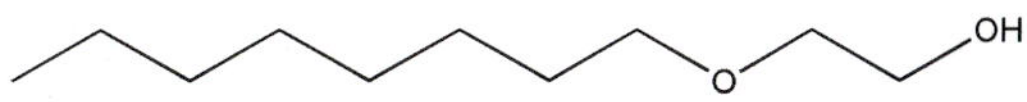

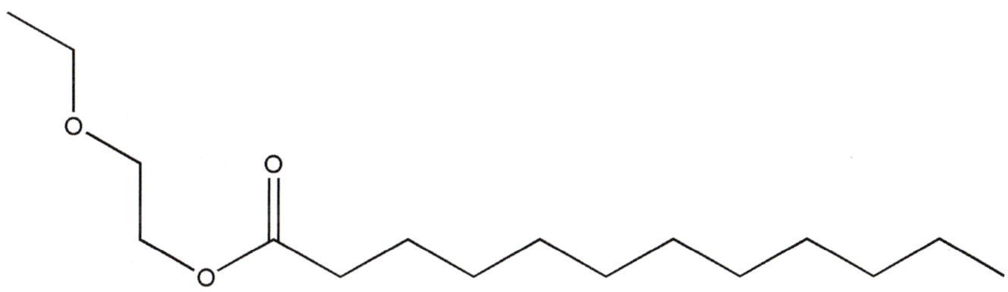

Ethylene glycol monoethyl ether laurate

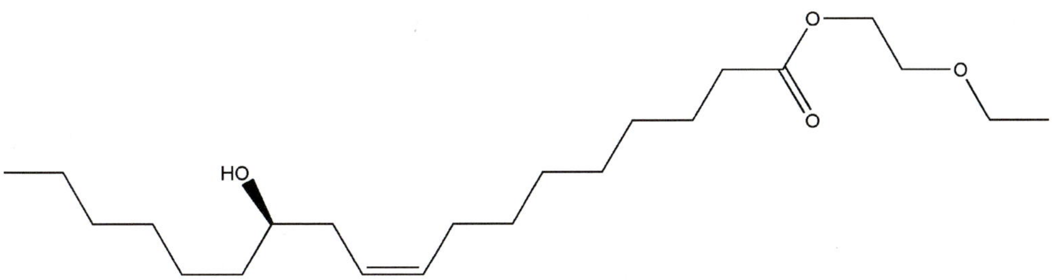

Ethylene glycol monoethyl ether ricinoleate

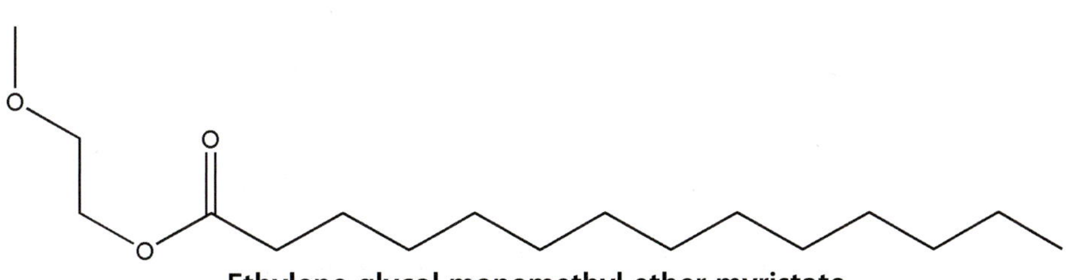

Ethylene glycol monomethyl ether myristate

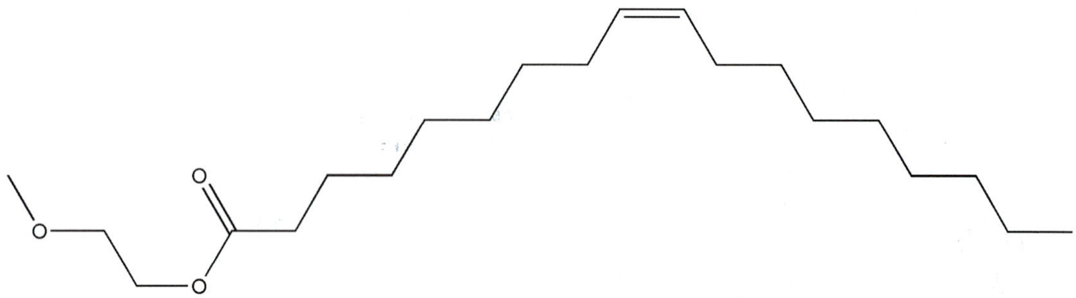

Ethylene glycol monomethyl ether oleate

Ethylene glycol monophenyl ether *n.* A solvent for cellulosics, vinyls, phenolics, and alkyd resins.

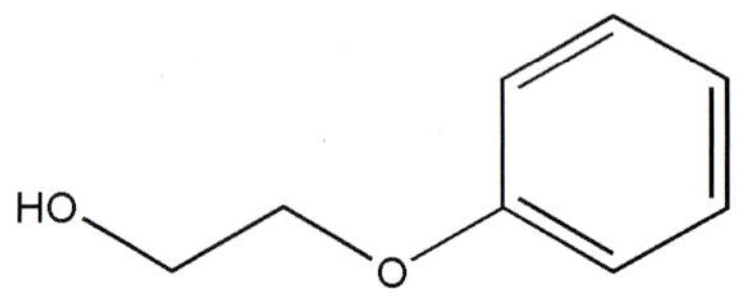

Ethylene glycol monoricinoleate (ethylene glycol ricinoleate) *n.* Plasticizer and an intermediate for urethane polymers.

Ethylene–methyl acrylate copolymer (EA, E/EA) *n.* An elastomer vulcanizable with peroxides or diamines. It resists attack by oils and temperatures to 175°C.

Ethylene oxide (1898) (epoxyethane) *n.* A three-membered ring compound with the formula H_2COCH_2, colorless, flammable gas at ordinary room temperature and pressure. It is used in organic synthesis, especially in the production of ethylene glycol and it is the starting material for the manufacture of acrylonitrile and nonionic surfactants. Bp, 10.7°C; Sp gr, 0.869. Syn: oxirane and anprolene.

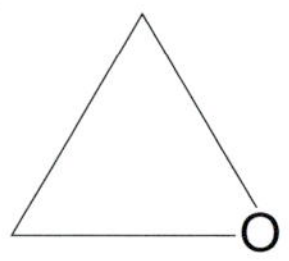

Ethylene plastic *See polyethylene.*

Ethylene, polymerization *n.* A chemical reaction in which two or more molecules combine to form larger molecules that contain repeating structural units.

Ethylene–propylene rubber (E/P, EPDM, SPM, and EPR) *n.* Any of a group of elastomers obtained by the stereospecific copolymerization of ethylene and propylene (EOM, or of these two monomers and a third monomer such as an unconjugated diene (EPDM). Their properties are similar to those of natural rubber in many respects, and they have been proposed as

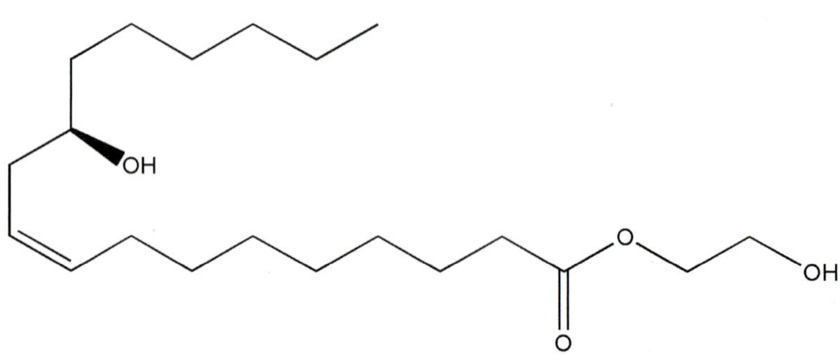

Ethylene glycol monoricinoleate

potential substitutes for natural rubber in tires.

Ethylene–tetrafluoroethylene copolymer *n.* A copolymer of ethylene and tetrafluoroethylene (DuPont Tefzel®), ETFE is readily processed by extrusion and injection molding. It has excellent resistance to heat, abrasion, chemicals, and impact, with good electrical properties.

Ethylene–urea resin *n.* A type of amino resin.

Ethylene–vinyl acetate copolymer (EVA, E/VAC) *n.* Any copolymer containing mainly ethylene with minor proportions of vinyl acetate. They retain many of the properties of polyethylene but have considerably increased flexibility, elongation and impact resistance. They resemble elastomers in many ways, but can be processed as thermoplastics.

Ethylene–vinyl acrylate resins *n.* Co-polymers of the polyolefins family derived from random co-polymerization of acetate and ethylene.

Ethylene–vinyl alcohol copolymer (EVAL, E/VAL, and EVOH) *n.* A family of copolymers made by hydrolyzing ethylene–vinyl acetate copolymers with high VA content. Those containing about 20–35% ethylene are useful as barriers to many vapors and gases, though not to water. Because of their water sensitivity, they are usually sandwiched between layers of other polymers.

Ethyl formate *n.* Solvent for cellulose acetate.

2-Ethyl-1,3-hexanediol *n.* A stable, colorless, nearly odorless, a high-boiling liquid with weak solvent action. In two-part urethane systems, the material acts as a viscosity reducer at room temperature. When the urethane mixture is heated to cure it, the diol reacts into the urethane matrix to eliminate solvent emissions.

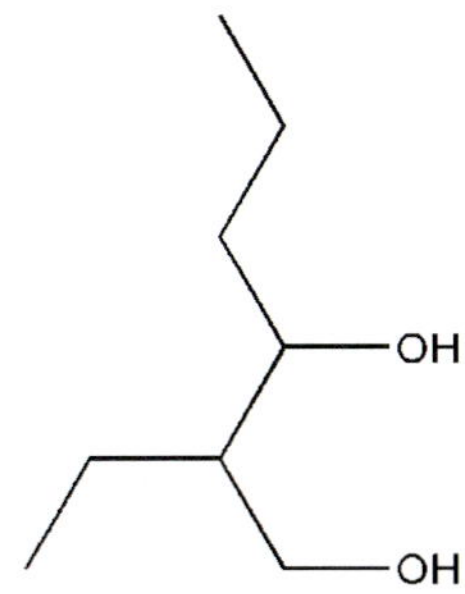

2-Ethylhexyl- *n.* An eight-carbon branched-chain radical of the formula $C_4H_9CH(C_2H_5)CH_2$–, often called *octyl* in the plastics industry. For example, the common plasticizer di-2-ethylhexyl phthalate is commonly referred to as dioctyl phthalate and by its abbreviation, DOP.

2-Ethylhexyl acetate (octyl acetate) *n.* A high-boiling retarder solvent with low evaporation rate and limited water solubility, used primarily in coating formulations based on cellulose nitrate. It is also used as a dispersant in vinyl organosols.

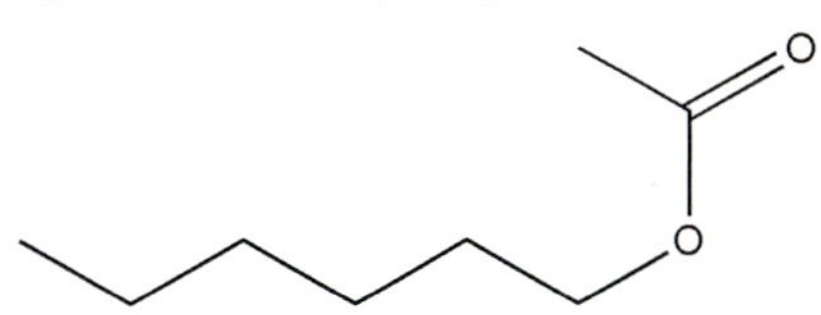

2-Ethylhexyl acrylate *n.* One of the monomers for acrylic resins, especially for those used in water-based paints.

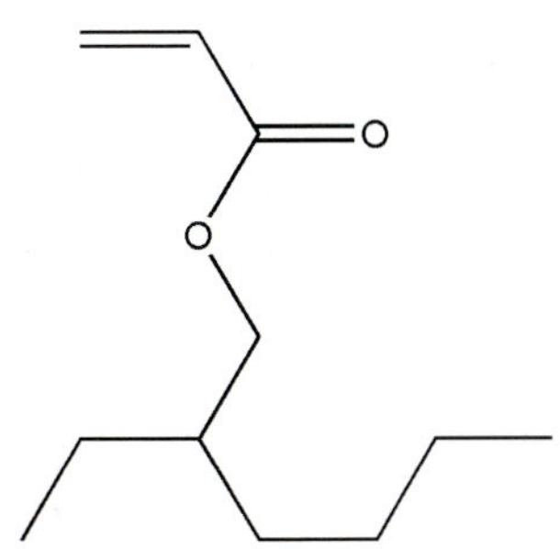

2-Ethylhexyl alcohol (2-ethylhexanol, octyl alcohol) *n.* $C_4H_9CH(C_2–H_5)CH_2OH$. An involatile solvent with many uses in the plastics industry. As a solvent, it is used in coatings for stenciling, silk screening, and dipping. As an intermediate, the alcohol is an important raw material for the production of the 2-ethylhexyl esters of dibasic acids used as plasticizers, such as dioctyl phthalate, adipate, and azelate.

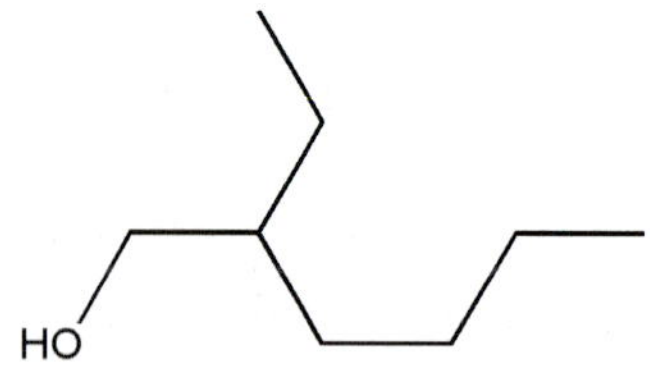

2-Ethylhexyl decyl phthalate *n.* A mixed diester plasticizer for cellulosics, polystyrene, PVC, and polyvinyl acetate.

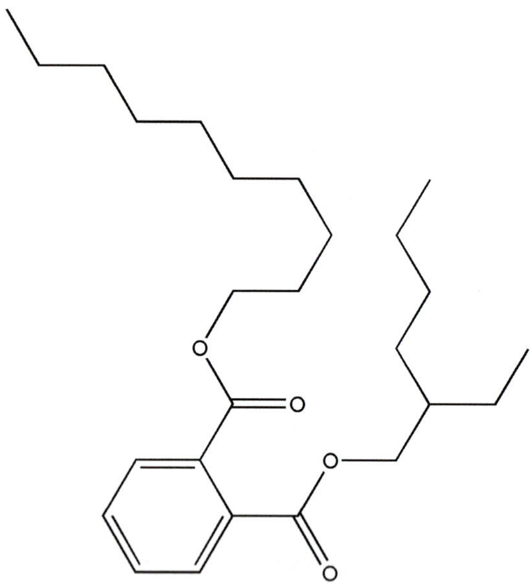

2-Ethylhexyl epoxytallate *n.* An epoxy ester used mainly as a combined plasticizer and stabilizer in vinyl compounds. At concentrations as low as 5 phr it reacts synergistically with many metallic stabilizers to provide stability comparable to similar combinations based on epoxidized soybean oils. As a partial replacement for other plasticizers, it imparts good low-temperature flexibility. It is also compatible with vinyl chloride–vinyl acetate copolymers, high-butyral cellulose acetate–butyrate resins, ethyl cellulose, polystyrene, and chlorinated rubbers.

2-Ethylhexyl isodecyl phthalate (octyl isodecyl phthalate) *n.* C_8H_{17}–$OOCC_6H_4COOC_{10}H_{21}$. A mixed ester compatible with PVC, vinyl chloride–acetate copolymers, cellulose acetate–butyrates with higher butyrate contents, cellulose nitrate, and, in lower concentrations, with polyvinyl butyral. In vinyls, it is somewhat less volatile than dioctyl phthalate and has equivalent low-temperature properties.

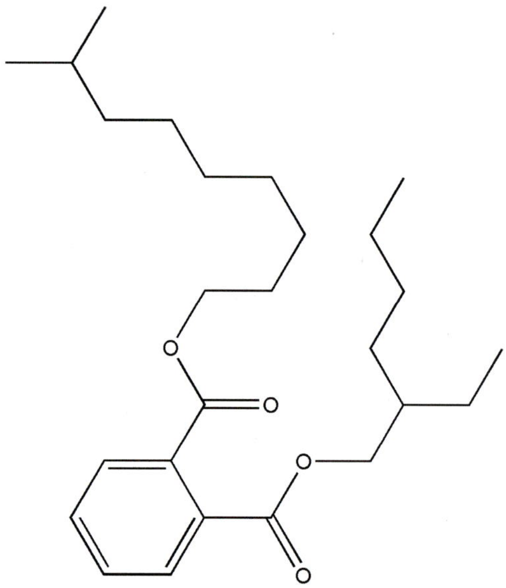

2-Ethylhexyl-*p*-hydroxybenzoate *n.* A plasticizer for polyamides.

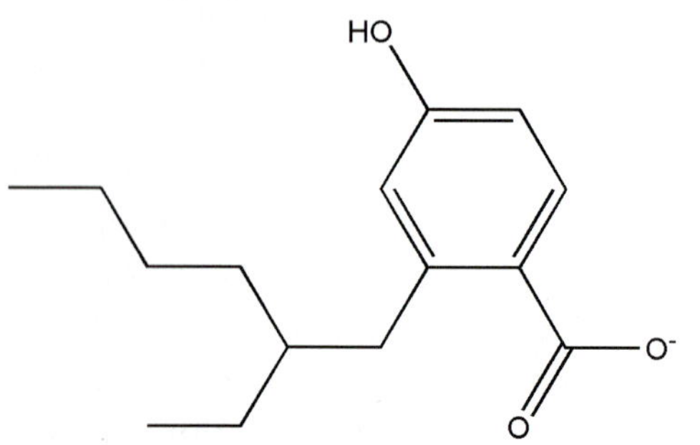

Ethyl hydroxide *See ethyl alcohol.*

Ethyl-α-hydroxyisobutyrate *n.* $C_2H_5OOC(OH)C(CH_3)_2$. A solvent for cellulose nitrate and cellulose acetate.

Ethylidene acetobenzoate (ethylidene benzoacetate) *n.* $C_6H_5COO(CH_3–CH)COCH_3$. A solvent for cellulosics and synthetic resins.

Ethyl lactate *n.* $CH_3CH(OH)COOC_2H_5$. Used as a solvent for nitrocellulose and cellulose acetate. Colorless liquid with a mold odor. Properties: mol wt. 118; bp, 154°C; Sp gr, 1.024/20°C; flp, 45°C (113°F); refractive index, 1.4111; vp. 5 minHg/233°C.

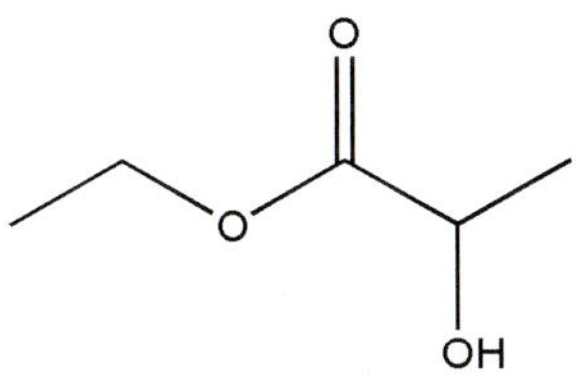

Ethyl levulinate *n.* $C_2H_5OOC(CH_2)_2COCH_3$. A solvent for cellulose acetate.

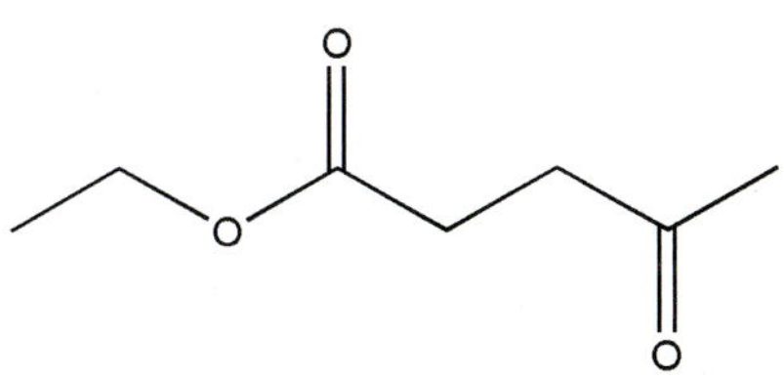

Ethyl methacrylate *n.* A readily polymerizable monomer used for certain types of acrylic resins.

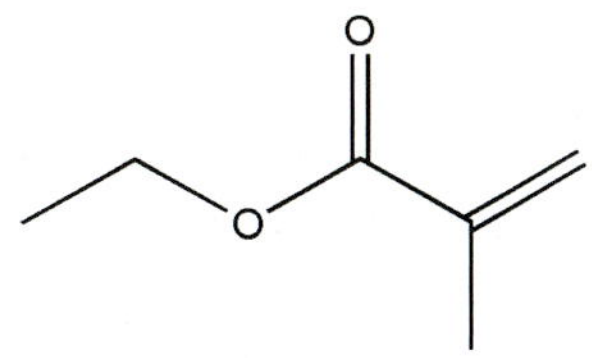

2-Ethyl-4-methylimidazole (EMI) *n.* An epoxy resin curing agent with the heterocyclic structure. EMI is used with epoxies formed from epichlorohydrin and bisphenol A or -F, and for novolac epoxy resins. It provides ease of compounding, long pot life, low viscosity, and non-staining characteristics, and yields castings with excellent mechanical and electrical properties.

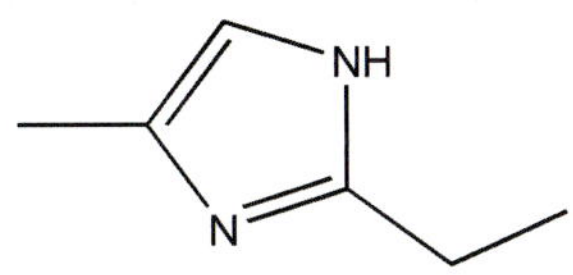

Ethyl-*n*-butyl ketone (2-heptanone) *n.* A stable, colorless liquid with medium volatility, used in solvent mixtures for cellulosic and vinyl resins. When used in vinyl organosols it imparts good lone-time viscosity stability.

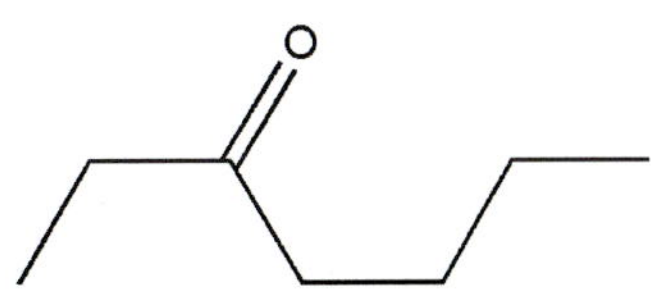

Ethyl oleate *n.* $C_2H_5OOCC_{17}H_{33}$. This monounsaturated fatty ester is a solvent, lubricant, and plasticizer.

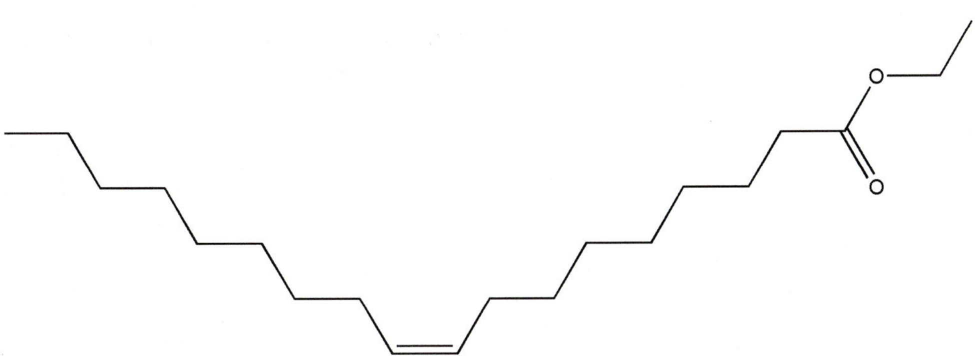

Ethyl oleate

E

Ethyl oxalate (oxalic acid diethyl ester) *n.* Colorless, unstable, oily, and aromatic liquid. Solvent for cellulose esters and ethers, many natural and synthetic resins. Sp gr., 1.0920/20°C; bp, 186°C; mp, −40.6°C; fp, 75.6°C (168°F).
Also known as diethyl oxalate.

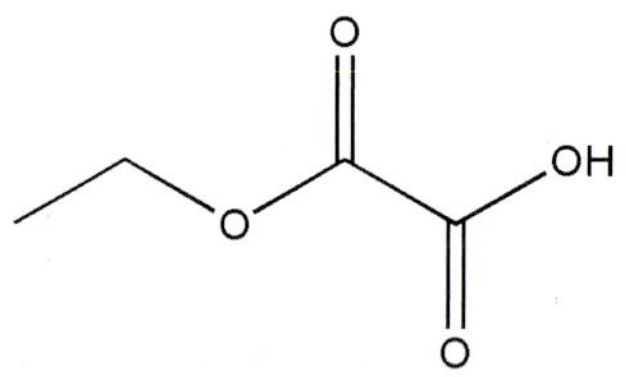

Ethyl phthalate *See diethyl phthalate.*

Ethyl phthalyl ethyl glycolate (2-ethoxy-1,3-dioxo-indan-2-carboxylic acid ester). A plasticizer compatible with PVC and most common thermoplastics. It has been approved by the FDA for use in food packaging.

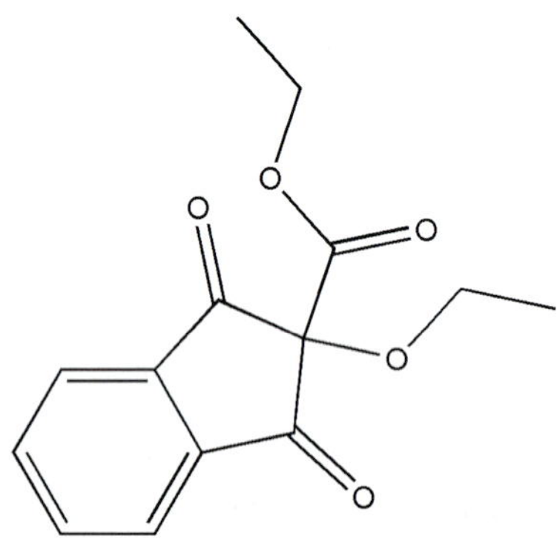

Ethyl propionate (propionic acid ethyl ester) *n.* Colorless liquid, fruity odor. Medium boiling solvent for nitrocellulose. Properties: bp, 98°C; Sp gr, 0.912/0°C; flp, 12°C (43.6°F); vp, 43 mmHg/30°C.

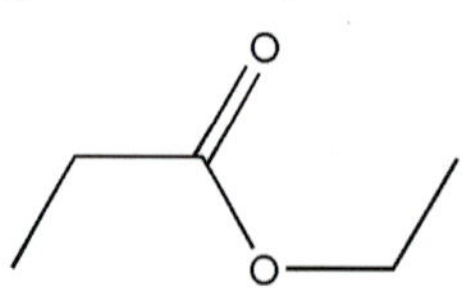

***N*-Ethyl-*p*-toluene sulfonamide** (2-ethyl-5-methyl-benzenesulfonamide) *n.* A solid plasticizer for rigid PVC.

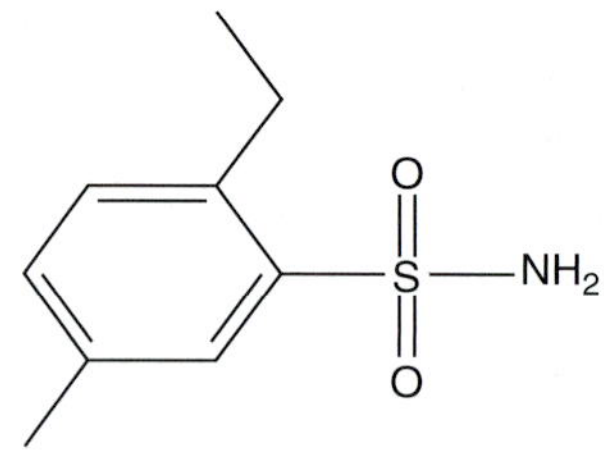

Ethyl succinate (succinic acid monoethyl ester) *n.* Saturated plasticizer. Properties: bp, 216°C; Sp gr, 1.049.
Also known as diethyl succinate.

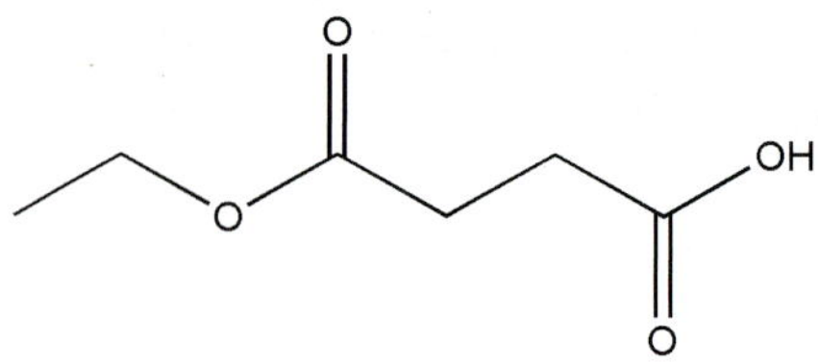

EtOH *See ethyl alcohol.*

Ettinghausen's effect [*Von Ettinghausen's*] *n.* When an electric current flows across the lines of force of a magnetic field an electromotive force is observed which is at right angles to both the primary current and the magnetic field: a temperature gradient is observed which has the opposite direction to the Hall electromotive force. Lide DR (ed) (2004) CRC handbook of chemistry and physics. CRC Press, Boca Raton, FL.

EU Abbreviation for polyether type of polyurethane rubber.

Euhedral *adj.* Euhedral crystals are those that are bounded by plane faces, cf., anhedral.

Eupolymer *n.* Polymer with a molecular weight over 10,000.

Eutectic \yú-ˈtek-tik\ [Gk *eutēktos* easily melted, fr. *eu-* +*tēktos* melted, fr. *tēkein* to melt] (1884) *adj.* A term applied to the specific mixture of two or more substances that has a lower melting point than that of any of its constituents alone or any other percentage composition of the constituents.

EVA (E/VAC) *n.* Copolymer from ethylene and vinyl acetate.
See ethylene–vinyl acetate copolymer.

Evaporation \i-ˈva-p(ə-)ˌrā-shən\ [ME, fr. L *evaporatus*, pp of *evaporare*, from *e-* + *vapor* steam, vapor] (15c) *v.* The changing from the liquid to the gaseous or vapor state, as when the solvent leaves the printed ink film.

Evaporation rate *n.* A measure of the length of time required for a given amount of a substance to evaporate, compared with the time required for an equal amount of ethyl ether or butyl acetate (rated at 100) to evaporate.

Evaporation rate, final *n.* Time interval for complete evaporation of all solvent.

Evaporation rate, initial *n.* Time internal during which low boiling solvent evaporates completely.

Evaporometer *n.* Instrument for measuring the evaporation rate of a liquid.

EVE *See vinylethyl ether.*

Evenness testing *n.* Determination of the variation in weight per unit length and thickness of yarns or fibers aggregates such as roving, sliver, or top.

EVOH *See ethylene-vinyl alcohol copolymer.*

Exa The SI prefix meaning $\times 10^{18}$.

Excessive clearer waste *n.* Higher that normal amount of short and regular fibers that become attached to the drafting rolls and are transferred to the clearer brushes to accumulate in abnormal amounts until they are removed manually.

Excitation \ˌek-ˌsī-ˈtā-shən\ (14c) *n.* In ultraviolet curing, the first state of the polymerization process, in which the photo-initiator, such as benzophenone amine, is stimulated by UV into a singlet or triplet state, with subsequent formation of free radicals.

Excitation purity *See purity, excitation.*

Excited state *n.* Any state higher in energy then the ground state.

Excluded volume *n.* The volume surrounding and including a given object, which is excluded to another object. This terminology comes from the statistical mechanics of gases, where this function arises in the leading order concentration expression (virial coefficient) for the pressure in the case of gas particles that repeal each other with a hard-core volume exclusion.

Exempt solvents *n.* Any solvent that has not been declared photochemically reactive by any of several regulatory agencies, most notably, the Los Angeles Air Pollution Control District. Many alcohols, many esters, some ketones, and mineral spirits are exempt under Rule 66.

Exfoliate \(ˌ)eks-ˈfō-lē-ˌāt\ [LL *exfoliatus*, pp of *exfoliare* to strip of leaves, fr. L *ex-* + *folium* leaf] (1612) *v.* Sealing from a surface in flakes or layers.

Exhaustion *n.* During wet processing, the ratio at any time between the amount of dye or substance taken up by the substrate and the amount originally available.

Exo A chemical-structure prefix meaning attachment to a side chain rather than to a ring. Compare Endo-.

Exotherm \ˌek-sō-ˈthərm\ *n.* (1) The temperature/time curve of a chemical reaction giving off heat, particularly the polymerization of a casting resin. ASTM test D 2471 delineates a procedure for measuring this curve, which in practice is strongly dependent on the amount of material present and the geometry of the casting. (2) The amount of heat given off per unit mass of the principal reactant. As yet (1992) no standard method for determining this heat in reacting plastics has been adopted.

Exothermic [ISV] (1884) *adj.* Denoting a chemical reaction that is accompanied by the evolution of heat. An example in the plastics industry is the curing reaction of an epoxy resin with an amine hardener. Opposite of endothermic.

Exothermic reaction *n.* A reaction, which liberates heat.

E

Expandable plastic *n.* A plastic formulated so as to be transformable into a cellular plastic by thermal, chemical, or mechanical means.
See cellular plastic.

Expanded polystyrene (XPS) *See polystyrene foam.*

Expanding agent *See blowing agent.*

Expansion coefficient The fractional change in length or volume of a material for a unit change in temperature.

Expansion of gases (Charles' law or Gay–Lussac's law) *n.* The volume of a gas at constant pressure increases proportionately to the absolute temperature. If V_1 and V_2 are volumes of the same mass of gas at absolute temperatures, T_1 and T_2,

$$\frac{V_1}{V_2} = \frac{T_1}{T_2}.$$

For an original volume V_o at 0°C the volume at t°C (at constant pressure) is

$$V_1 = V_o(1 + 0.00367t).$$

Expansion of gases (*General law for gases*) *n.*

$$p_t v_t = p_o v_o\left(1 + \frac{t}{273}\right),$$

where p_o, v_o, p_t, and v_t represent the pressure and value at 0° and t°C or

$$\frac{p_1 v_1}{T_1} = \frac{p_2 v}{T_2},$$

where p_1, v_1, and T_1 represent pressure, volume and absolute temperature in one case and p_2, v_2, and T_2 the same quantities for the same mass of gas in another. The law may also be expressed as $pv = RmT$, where m is the mass of gas at absolute temperature T. R is the *gas constant*, which depends on the units used. *Boltzmann's molecular gas constant* is obtained by expressing m in terms of the number of molecules. For volume in cm^3, pressure in $dyne/cm^2$ and temperature in Centigrade degrees on the absolute scale $R = 8.3136 \times 10^7$. Phillip R Watson (1997) Physical chemistry. John Wiley and Sons Inc., New York.

Expansivity *See coefficient of thermal expansion.*

Explosive limits *n.* When combustible vapor is mixed with air in the proper proportions, ignition will produce an explosion. This proper proportion is called the explosive range. The explosive range includes all concentrations of a mixture of flammable vapor or gas in air, in which a flash will occur or a flame will travel if the mixture is ignited. The lowest percentage at which this occurs is the lower explosive limit; and the highest percentage, the upper explosive limit. Explosive limits are express in percent by volume of vapor in air and, unless otherwise specified, under normal conditions of temperature and pressure.

Exposure meter (1891) *n.* A device for indicating correct photographic exposure under varying conditions of illumination.

Exposure rack *n.* Term given to a frame on which test panels are exposed for durability tests.

Exposure tests *n.* Tests, which are conducted to evaluate the durability of a coating or film. They include exposure to ultraviolet light, moisture, cold, heat, salt water, mildew, etc. They can be generated either naturally or artificially.

Expression *n.* Removal of a liquid from a solid by pressing, as in the manufacture of vegetable oils from meal cakes.

Extended length *n.* The length of a face pile yarn required to produce 1 in. of tufted carpet.

Extender \ik-ˈsten-dər\ (1611) *n.* (1) *See extender pigment.* (2) A transparent or semi-transparent white pigment or a varnish that is used to alter the color strength and working properties of an ink, without affecting its hue.

Extender pigment *n.* A specific group of achromatic pigments of low refractive index (between 1.45 and 1.70) incorporated into a vehicle system whose refractive index is in a range of 1.5–1.6. Consequently, they do not contribute significantly to the hiding power of paint. They are used in paint to: reduce cost, achieve durability, alter appearance (e.g., decrease in gloss), control rheology, and influence other desirable properties. If used at sufficiently high concentration, an extender may contribute dry hiding and increase reflectance.

Extensibility (1611) *n.* (1) The ability of a material to stretch or elongate upon application of sufficient tensile stress. It is expressed as a percentage of the original length. (2) The value of said ability just prior to rupture of the specimen; ultimate elongation.
See elongation.

Extensiometer *n.* A rheometer for measuring the extensional flow properties of molten polymers. In one early form, the *Cogswell rheometer,* useful at tensile viscosities over 10^5 Pa/s, unidirectional tensile force was exerted on a polymer rod by a dead-weight acting through a cam and pulley. As the cam rotated, the moment arm exerted by the weight on the rod decreased in proportion to the rod cross section so as to maintain constant stress.

Extensional strain rate *See elongational flow.*

Extensional viscosity *See elongational viscosity.*

Exterior finishes *n.* Coatings, which are expected to possess reasonable durability when exposed to natural weathering.
See exterior paints and varnishes.

Exterior paints and varnishes *n.* Material formulated for use in conditions exposed to the weather.

Exterior type plywood *n.* Plywood bonded with a fully waterproof glueline.

External mix Spray equipment in which fluid and air join outside of aircap.

External phase of an emulsion Another term for the continuous phase.

External plasticizer Post added plasticizer as opposed to plasticization by means of internally combined groups, such as copolymerization.
See internal plasticizers.

External undercut Any recess or projection on the outer surface of a molded part that prevents its direct removal from its mold cavity. Parts with such undercuts may be molded by splitting the mold vertically and opening the split to withdraw the part; or by providing *side draws,* i.e., mold parts that are withdrawn to the sides to relieve the undercuts.

Extinction \ik-ˈstiŋ(k)-shən\ (15c) *n.* If the orientation of the crystals that appear white or colored between crossed polars is changed by rotating the stage, all single crystals will be observed to disappear (become black) four times during complete rotation of the stage. The positions are 90° apart; they reveal the vibration directions of each crystal. These directions will parallel the vibration directions of the two polars when the crystal is extinct.

Extinction angle *n.* The angle between the nearer vibration direction and a prominent direction of the crystal. It never exceeds 45°.

Extinction coefficient (1902) *n.* An older term synonymous with absorption coefficient.
See absorption coefficient.

Extinction, oblique *n.* Extinction is oblique if the vibration directions are oblique to the long direction of the crystal.

Extinction, parallel *n.* Extinction is parallel if the vibration directions are parallel and perpendicular to the long direction of the crystal.

Extinction, symmetrical *n.* Extinction is symmetrical if the vibration directions bisect a prominent angle.

Extractable \ik-ˈstrak-tə-bəl\ [ME, fr. L *extractus*, pp of *extrahere*, fr. *ex-* + *trahere* to draw] (15c) *n.* The amount of soluble material extracted from a polymer specimen when it has been exposed to a solvent under specified conditions. In ASTM test D 4754, disks of the polymer and glass separator beads are alternately threaded onto a stainless-steel wire and slipped into a vial containing the test solvent. The tightly closed vial is placed in an oven, and the solvent is sampled and analyzed periodically.

Extraction *n.* The transfer of a constituent of a plastic mass to a liquid with which the mass is in contact. The process is generally performed with a solvent selected to dissolve one or more specific constituents; or it may occur as a result of environmental exposure to a solvent.

Extract printing *See printing, discharge printing.*

Extranuclear region All of an atom except the nucleus.

Extrudate (1) The product or material delivered from an extruder, for example film, pipe, profiles, and wire coatings. (2) The extruded melt just as it emerges from the die.

Extrudate roughness *See melt fracture.*

Extruded foam *n.* Cellular plastic produced by extrusion with the aid of a blowing agent – a decomposable, gas-generating chemical – or by injection into the extruder of a gas such as nitrogen or carbon dioxide, or a highly volatile liquid, such as a pentane.

Extruded shape (profile) *n.* Any of a huge variety of cross-sectional shapes produced in continuous or cut lengths by extrusion through profile dies. Some complex shapes reach their final form with the aid of post forming jigs that alter the shape before the extrudate has cooled and become firm.

Extruder \-ˈstrü-dər\ [L *extrudere*, fr. *es-* + *trudere* to thrust] (1566) *n.* (1) An extrusion machine. (2) A machine in which molten or semisoft materials are forced under pressure through a die to form continuous tubes, sheets, or fibers. It may consist of a barrel, heating elements, a screw, ram, or plunger, and a die through which the material is pushed to give it shape. (3) In fiber manufacture the machine that feeds molten polymer to an extrusion manifold or that first melts the polymer in a uniform manner then feeds it to a manifold and associated equipment for extrusion. (4) In plastics compounding, a material added to the mixture to reduce its cost per unit volume. The material may be a resin, plasticizer, or filler. (5) A substance, generally having some adhesive capacity, added to an adhesive formulation to reduce the amount of the primary (i.e., more costly) binder required per unit of bonding area. Strong AB (2000) Plastics materials and processing. Prentice-Hall, Columbus, OH.

See binder and filler. See also blending resin. Also see screw melter.

Extruder barrel (extruder cylinder) *n.* A thick-walled, cylindrical steel tube, lined with a special hard alloy to resist wear that forms the housing for the extruder screw and contains, between itself and the screw, the plastic material as it is conveyed from feed hopper to die. Barrels are usually surrounded by heating and cooling media, such as electrical heater bands, cast-in-aluminum calrods and tubing for coolant, induction heaters, or, rarely, by a compartmented jacket for the circulation of hot oil or steam. Electrically heated barrels are usually furnished with some means of cooling, such as air blowers or coils through which fluid may be circulated. Small holes drilled radically into (but not through) the barrel accommodate temperature sensors whose signals are used to control the means of heating and cooling.

Extruder burn *n.* The local decomposition of the plastic resin inside an extruder caused by the excessive mechanical heating. Strong AB (2000) Plastics materials and processing. Prentice-Hall, Columbus, OH.

Extruder breaker plate *See breaker plate.*

Extruder drive *n.* The system comprising an AC or DC motor, speed reducer (gearbox), screw-shaft bearings, coupling, and controls that supply power to the screw and regulate its speed. To facilitate startup and accommodate various operating conditions, modern extruders are always equipped with some means of varying the screw speed, with close control, over a wide range.

Extruder, dual-ram *n.* A modification of the original ram extruder employing two identical units, stroking alternately, and delivering to the same die. Aided by values, the result is the conversion of a batch operation into one that makes continuous extrusions from sinterable resins such as polytetrafluoroethylene and ultra-high-molecular-weight polyethylene.

Extruder, planetary screw *n.* A multi-screw device in which a number of satellite screws, generally six, are arranged around one longer central screw. The portion of the central screw extending beyond the satellite screws serves as the final pumping screw as in a single-screw extruder, while the planetary screws aid in plastication and permit the discharge of volatiles toward the hopper. A few of these have served in processing fine powders such as PVC dry-blend.

Extruder, ram *n.* An extruder in which the material is advanced through the barrel and die by means of a ram or plunger rather than by a screw. Melting is accomplished either by preheating the feedstock close to the fusing temperature and by conductive heating from the barrel wall, or both. The ram extruder was the earliest type to be used in the plastics industry, dating back to 1870 when cellulose nitrate was extruded into rods. Among plastics today, polytetrafluoroethylene, which softens to a gel but does not achieve a true melt state, is the one mainly processed with ram extruders.

Extruder screw *n.* A solid or cored shaft with a continuous helical channel (sometimes two channels) cut into it, usually extending from the feed throat of the extruder barrel to the die end of the barrel. In most screws, the channel varies in its volume per turn of the helix, being larger at the end receiving the feed with its low bulk density, to much less – roughly one-third – in the pumping section at the delivery end. The reduction in volume is usually accomplished by reducing the channel depth but can be done by reducing the helical lead; sometimes a combination of both has been used. The reduction in volume serves several purposes: feeding, compressing the particles and forcing the interstitial air back out the feed hopper, melting the polymer, and developing pressure to overcome resistance at the die. Extruder screws are made of tough steel – SAE 4140 is common – are usually chrome plated, and have flight tips hardened by one of several techniques (*See nitriding and hard-facing alloy*). Many designs have been, and continue to be, developed and marketed for extruder screws. Some terms used in describing extruder screws are defined below.

- **Barrier screw** *See solids-draining screw.*
- **Constant-lead screw** (uniform-pitch screw) *n.* A screw with a flight of constant helix angle over its whole length.
- **Constant-taper screw** *n.* A screw of constant lead and uniformly. **Cored screw.** A screw with a hole bored along its axis for circulation of heat-transfer medium or insertion of a heater. The core may extend only through the feed section or further, even to the screw tip.
- **Decreasing-lead screw** *n.* A screw in which the helix angle decreases steadily over the length of the screw. Channel depth is usually constant.
- **Metering-type screw** *n.* A screw whose final section is of constant lead and relatively shallow depth.
- **Multiple-flighted screw** (multi-flight screw) *n.* A screw having more than one flight, thus having two or more parallel channels.
- **Single-flighted screw** *n.* A screw having just one flight – the usual case, presumed if not otherwise stated.
- **Two-stage screw** *n.* Essentially two metering-type screws in series, typically used for vented operation. The first stage consists of a feed section, compression zone, and metering zone. The second stage consists of a deep, constant-depth section (decompression zone), usually running only fractionally full to permit expansion of bubbles and release of volatiles, followed by a short, steep compression zone and a metering zone of about one-third greater capacity than that of the first stage. There may be a restriction at the end of the first stage.
- **Vented screw** *n.* A two-stage screw with a screw vent in the decompression zone, permitting volatiles to escape through the screw core.
- **Water-cooled screw** *n.* A screw cored in its feed section to permit circulation of water there.
- **Extruder, single-screw** *n.* An extruder with one tubular barrel within which a solid or cored screw rotates. Strong AB (2000) Plastics: materials and processing. Prentice-Hall, New York. Pittance JC (ed) (1990) Engineering plastics and composites. SAM International, Materials Park, OH. James F (ed) (1993) Whittington's dictionary of plastics. Technomic Publishing Co. Inc., Carley.

Extruder size *n.* Traditionally, the nominal inside diameter of the extruder barrel, usually stated in inches or millimeters. However, the output power rating of the drive is more directly related to output capability than is the diameter.
See also L/D ratio.

Extruder, twin-screw (double-screw extruder) *n.* An extruder with a barrel consisting of two side-by-side intersecting cylinders internally open to each other along their intersection. There are two basic types. If the two internal cylinders are tangentially joined, the two screws are also nearly tangential and are normally of opposite "hands" and counter-rotating, turning downward at their juncture (Welding Engineers). This design permits the use of long vented sections and that feature, together with the milling action of the screws, makes it possible to remove large percentages of volatiles from the feedstock. One of the two screws is extended to become the metering section that provides die pressure. In the second basic type, the two cylinders of the barrel intersect more deeply and the screws intermesh with each other. Rotation may be co- or counter-, depending on whether the screws have the same or opposite "hands". In these machines the intermeshing of the screws traps the plastic and moves it much more positively than in the tangential type, with virtually no back flow, and shear working is less severe. The screws of both types of assembled from segments, each segment designed to perform a particular function. An important application for the second type has been extrusion of rigid PVC compounds from powder, heating gently by conduction and fluxing near the end of the screws and into the die entry and extruding at die pressures up to 100 MPa. Other applications are compounding fluffy polyolefin powders, volatiles extraction, and compounding. Twin-screw machines have also proved useful in reactive extrusion, in which chemical reactions, including polymerization, are performed with the extruder.

Extrusion [ML *extrusion-*, *extrusio*, fr. L *extrudere*] (1540) *n.* (1) A shaping/molding process use in plastics processing in which the material is melted and then pushed out the end of the machine, usually through a forming die. (2) Any process by which lengths of constant cross section are formed by forcing a material, e.g., a molten plastic, through a die. Typical shapes extruded are hose, tubing, flat films and sheets, wire and cable coatings, parisons for blow molding, filaments and fibers, strands cut hot or cold to make pellets for further processing, webs for coating and laminating, and many of the above in multiple layers by coextrusion. Strong AB (2000) Plastics materials and processing. Prentice-Hall, Columbus, OH. James F (ed) (1993) Whittington's dictionary of plastics. Technomic Publishing Co. Inc., Carley.

See compacting and strainer.

Extrusion, autothermal (autogenous extrusion, "adiabatic" extrusion) *n.* An extrusion operation in which the entire increase in enthalpy of the plastic, from feed throat to die, or very nearly all of it, is generated by the frictional action of the screw. In such an operation, which most commercial single-screw extrusions approach closely, the functions of the barrel heaters are to preheat the machine at startup and, during steady operation, to prevent heat loss from the plastic through the barrel to the surroundings.

Extrusion blow molding *n.* The most common process of blow molding in which the parison is formed by extrusion.

Extrusion casting A term sometimes employed in the industry for the process of extruding unsupported film, especially a composite of two or more integral resin layers formed by coextrusion. Such extrusion-cast composite films possess desired properties on each of the respective sides, e.g., heat-sealability on one side and stiffness on the other, or different levels of slip, or different colors.

Extrusion coating *n.* The process of coating a substrate by extruding a layer or molten resin onto the substrate with sufficient pressure to bond the two together without the use of an adhesive. A common application of the process is the coating of foil, paper, or fabric with polyethylene, by extruding a web directly into the nip of a pair of rolls through which the substrate is passing.

Extrusion coloring *n.* the method of adding colorants to a plastic compound by dry-blending the colorant with the solid granular resin, extruding the mixture into strands, and cutting these strands into pellets for use in subsequent processing operations.

Extrusion die *n.* The orifice-containing element, mounted at the delivery end of an extruder that shapes the extrudate. Elements of the die assembly are (1) the die block, (2) an adapter connecting the die to the extruder, (3) a manifold within the die that distributes the melt to the orifice, (4) in the case of dies for hollow sections, a mandrel inserted in the flow channel to form the interior surface of the extrudate, (5) a spider that holds the mandrel in position, and (6) the land section, i.e., the orifice that gives the extrudate its emergent shape. Extrusion dies are classified in four ways according to the relation between the screw axis and the direction of flow of the extrudate: straight (*in-line*), *offset*, *angle*, and *crosshead*. In an in-line die, the die-discharge channel is coaxial with the screw. In an offset die, those directions are parallel but not coaxial. In an angle die the axis of the die-discharge channel is at an angle, typically 45°, to that of the screw. A crosshead die is an angle die in which the two axes are perpendicular. Sheet and film dies are also classified as to the type of feed. In a *center-feed die*, the melt enters at the lateral center of the manifold and divides into two equal streams that move in opposite directions to the ends of the die. In an end-fed die, the melt enters one end of the manifold and flows toward the other. Theory tells that it is easier to maintain thickness uniformity over the width of the extruded film or sheet is the die is center-fed, but end-fed dies are sometimes used for logistical reasons, especially in extrusion coating.

Extrusion laminating *n.* A process in which a plastic layer is extruded between two layers of substrate(s) in a nip of a rubber roll and a chill roll.
See extrusion coating.

Extrusion moldings *n.* Moldings, which are made from plastic material by forcing the melted resin through a shaped orifice by means of pressure.

Extrusion plastometer (melt-indexer) *n.* A simple viscometer consisting of a heated vertical cylinder with two bores, a central one that contains a close-fitting piston and a recess for an orifice block, the other, nearby, for a thermoswitch. The orifice is 2.1 mm in diameter and 8 mm long. Plastic particles are loaded into the bore, allowed to heat for 6–8 min, then the weighted piston is released, and the extrudate is collected for a measured time internal. The melt-flow index (MFI) is stated as the rate of extrusion in grams/10 min. The instrument and its use are described in ASTM D 1238. Originally developed in 1953 for low-density polyethylene, the melt-indexer is now used with many other polymers with specific temperatures and piston weights. It is *essential* to state the condition (*A–X*) at which an MFI was measured. Most of the conditions result in shear rates far below those typical of commercial processing (MFI = 1 g/10 min corresponds to about 2 s^{-1}. Thus, while the measurement is useful for product identification and quality control, it is a poor indicator of processability. An estimate of a resin's pseudoplasticity may be obtained by running two or more tests with substantially different piston weights.

Extrusion pressure *n.* (1) Broadly, pressure indicated anywhere within an extruder. (2) The pressure at the delivery, or head end of the screw in a screw extruder, immediately upstream of the screen pack if one is present. (3) In a ram extruder, the pressure at the face of the ram in contact withy the plastic.

Exudation \ˌek-s(y)ú-ˈdā-shən\ (1612) *n.* The undesirable appearance on the surface of an article of one or more of its constituents that have migrated or exuded to the surface. In vinyls, such constituents may be residual emulsifier from the resin, stabilizer, lubricant, or plasticizer. Secondary plasticizers in particular have a tendency to exude when used in excessive percentages. Exudation may appear on a product shortly after it has been made, but more often it is delayed for periods ranging from several weeks to years. Products that do not exude for long periods under ideal storage conditions can be caused to exude by exposure to pressure, heat, high humidity, light, and other environmental agents.

E

Eyelet \ˈī-lət\ [alter. of ME *oilet*, fr. MF *oillet*, dim. of *oil* eye, fr. L *oculus*] (14c) *n.* (1) A series of small holes made to receive a string or tape. A buttonhole stitch is worked around the holes. (2) A type of yarn guide used on a creel. (3) A fabric style with areas of cut-outs surrounded by stitching. Complete textile glossary. Celanese Corporation, New York.

F

f \ˈef\ *n.* (1) SI abbreviation for femto. (2) Symbol for frequency of an oscillating system.

F (1) *n.* Chemical symbol for the element fluorine. (2) SI abbreviation for farad.

Fabric \ˈfa-brik\ [MF *fabrique*, fr. L *fabrica* workshop structure] (15c) (cloth) *n.* A flexible structure, usually thin relative to its width and length, made up o intermingled yarns, fibers, filaments, or wires. In woven fabrics, the elements are alternately crossed over and under one or more of those oriented in the other directions, typically two perpendicular directions. In non-woven fabrics, such as felts, fibers are randomly oriented. Complete textile glossary. Celanese Corporation, New York.

Fabricate \-ˈkāt\ [ME, fr. L *fabricatus*, pp of *fabricari*, fr. *fabrica*] (15c) *vt.* In the broadest sense, this term means to manufacture, devise, or to make an assembly of parts and sections. In the plastics industry it refers to the assembly or modification of preformed plastics articles by processes such as welding, heat sealing, adhesive joining, machining, and fastening. "Fabrication" is *not* generally used to mean basic manufacturing processes such as extrusion, calendering, molding and the like.

Fabric construction *n.* The details of structure of fabric. Includes such information as style, width, type of knit of weave, threads per inch in warp and fill, and weight of goods. Complete textile glossary. Celanese Acetate, New York.

Fabric crimp *n.* The angulation induced between a yarn and woven fabric via the weaving or braiding process. Vincenti R (ed) (1994) Elsevier's textile dictionary. Elsevier Science and Technology Books, New York.

Fabric crimp angle *n.* The maximum acute angle of a single weaving yarn's direction measured from a plane parallel to the surface of the fabric.

Fabric set The number of warp threads per inch, or other convenient unit.

Fabric stabilizer *n.* Resin or latex treatment for scrims used in coated fabric manufacture to stabilize the scrim for further processing. Tortora PG (ed) (1997) Fairchild's dictionary of textiles. Fairchild Books, New York.

Façade \fə-ˈsäd\ [F *façade*, fr. It *facciata*, fr. *faccia* face, fr. (ass.) VL *facia*] (ca. 1681) *n.* One of the faces of a building, usually that containing the main entrance.

Face \ˈfās\ [ME, fr. OF, fr. (assumed) VL *facia*, fr. L *facies*, fr. *facere*] (13c) *n.* The correct or better-looking side of a fabric.

Face color *n.* Term used to describe the color observed on paints (particularly, metallic paints) when they are viewed near to the normal (near to the perpendicular). It may also be referred to as the "top color". It is generally used in contrast to the term "flop color," which is the color observed at an angle near to the grazing angle (parallel to the surface).

Face putty *n.* Triangular fillet of glazier's putty on the exposed surface of glass.

Facial (fac-) isomer *n.* An isomer of an octahedral complex in which three adjacent octahedral positions are occupied by one kind of ligand.

Faciated yarn *n.* Yarns consisting of a core of discontinuous fibers with little or no twist and surface fibers wrapped around the core bundle.

Facing (15c) *v.* A lining or trim that protects the edges of a garment especially at collars, cuffs, and front closings.

Faconné *n.* A broad term for fabrics with a fancy-type weave made on a Jacquard or dobby loom.

Facsimile \fak-ˈsi-mə-lē\ [L *fac simile* make similar] (1691) *n.* The exact reproduction of a letter, document, or signature.

Factice *n.* Elastomeric products made by reaction of sulfur or sulfuric chloride with vegetable oils.

F

Factitious ultramarine Synthetic ultramarine blue.
See ultramarine blue.

Fadding *n.* Application of shellac lacquers by means of a pad known as a 'fad.'

Fadeometer *n.* (1) Laboratory device used to determine the fastness of a colored fabric to exposure to light. The test pieces are rotated around a light source simulating the sun's rays at 45°N latitude in July between the hours of 9 a.m. and 3 p.m. Fabrics are rated by visual comparison with a gray scale according to degree of fading. (2) An apparatus for determining the resistance of resins and other materials to fading. It accelerates the fading by subjecting the article to high-intensity ultraviolet wavelengths similar to those found in sunlight. *Also known as fugitometer.*

Fading [ME, fr. MF *fader,* fr. *fade* feeble, insipid, fr. (ass.) VL *fatidus,* alter. of L *fatuus* fatuous, insipid] (14c) *v.* Subjective term used to describe the lightening of the color of a pigmented paint following exposure to the effects of light, heat, time, temperature, chemicals, etc. The observed fading may result from deterioration of the pigment, from deterioration of the vehicle, or from a decrease in gloss. A separation of the vehicle from the pigment particle in the interior of the film, with the subsequent introduction of microvoids, which scatter light, may also be interpreted visually as fading. Koleske JV (ed) (1995) Paint and coating testing manual. American Society for Testing and Materials, Philadelphia, PA. Paint and coating testing manual (Gardner–Sward Handbook) MNL 17, 14th edn. ASTM, Conshohocken, PA, 1995.
See light resistance.

Fahrenheit scale \ˈfar-ən-ˌhīt, ˈfer-\ [Gabriel D. *Fahrenheit*] (1753) *adj.* Temperature scale on which the freezing point of water is represented by 32°F (0°C) and the bp by 212°F (100°C) under normal pressure at sea level. Whitten KW, Davis RE, Davis E, Peck LM, Stanley GG (2003) General chemistry. Brookes/Cole, New York.

Faille \ˈfī(ə)l\ [F, fr. OF] (1869) *n.* A soft, slightly glossy woven fabric made of silk, rayon, cotton, wool, or manufactured fibers or combinations of these fibers and having a light, flat cross-grain rib or cord made by using heavier yarns in the filling than in the warp.

Failure, adhesive *n.* The rupture of an adhesive bond, such that the separation appears to be at the adhesive adherend interface. *Note*—Sometimes termed failure in adhesion. Skeist I (ed) (1990) Handbook of adhesives. Van Nostrand Reinhold, New York.

Falling-ball viscometer (falling-sphere viscometer) *n.* An instrument well suited to determining polymer-melt viscosity at extremely low shear rates. That is, the limiting Newtonian viscosity. A sphere more dense than the melt is placed between two premolded slugs of the test polymer within a steel cylinder, which is then kept for a preset time in a temperature-controlled oven. From the initial and final positions of the sphere the viscosity can be calculated by Stokes' law (with corrections). By repeating the test with spheres of different densities, a range of low shear rates can be explored. Van Wazer JR, Lyons JW, Kim KY,

Colwell RE (1963) Viscosity and flow measurement. Interscience Publishers Inc., New York.

Falling bodies *n.* For bodies falling from rest conditions are as for uniformly accelerated motion except that $v_o = Q$ and g is the acceleration due to gravity. The formulate becomes – air resistance neglected,

$$v_t = gt, \quad s = \frac{1}{2}gt^2, \quad v_s = \sqrt{2gs}.$$

For bodies projected vertically upward, if v is the velocity of projection, the time to reach greatest height, neglecting the resistance of the air,

$$t = \frac{v}{g}.$$

Greatest height

$$h = \frac{v^2}{2g}.$$

Giambattista A, Richardson R, Richardson B (2003) College physics. McGraw-Hill Science/Engineering/Math, New York.

See projectiles.

Falling-dart impact test *n.* In addition to the ASTM tests mentioned at free-falling-dart test, several similar tests exist for products such as pipe and bottles as well as sheeting. One procedure is the *staircase method*, also known as the *up-and-down method*, which for a given quantity of testing, provides a good estimate of the impact energy at which 50% of such samples may be expected to break. In the *Probit method*, groups of samples are tested at pre-selected drop heights ranging from that at which most or all of the samples fail to that at which very few or none fail. This method also provides an estimate of the 50% point but, in addition, provides a better estimate of the standard deviation than does the staircase method. Combinations of the two methods have been used to optimize the amount of information per test specimen. Brown R (1999) Handbook of physical polymer testing, vol 50. Marcel Dekker, New York. Shah V (1998) Handbook of plastics testing technology. John Wiley and Sons, New York.

Falling sand abrasion test *n.* Abrasion resistance of coatings of paint, varnish, lacquer and related products is determined by the amount of abrasive (Ottawa) sand required to wear through a unit thickness of the coating, when the sand falls against it at a specified angle from a specified height through a guide tube. Koleske JV (ed) (1995) Paint and coating testing manual. American Society for Testing and Materials, Philadelphia, PA.

False Becke line *n.* A second bright line, which moves in the direction opposite to the Becke line. It is usually observed with thick particles or when the refractive index difference between particles and mountant is large. It appears as a concentration of reflected light at the low index side of the interface.

False body *n.* Thixotropic flow property of a suspension or dispersion. When a composition "thins down" on stirring and "builds up" on standing it is said to exhibit false body. The term "false body" is also used in practice for "buttery" materials, which are characterized by a relatively low viscosity and high yield value.

See thixotropy.

False neck *n.* In blow molding of containers, a neck construction that is additional to the neck finish of the container and that is only intended to facilitate the blow-molding operation. Afterwards the false neck is removed from the container.

False twisting *See texturing, false, twist method.*

Family mold (composite mold). A multicavity mold containing a variously shaped cavities, each of which produces a component of an item that is assembled from the components. For example, a family mold for a model-airplane kit would contain a cavity for each part, and components of a complete kit would be produced in one shot.

Fancy yarn *See novelty yarn.*

Fan gate A shallow gate becoming wider (and usually thinner) as it extends from the runner to the cavity.

Fantail die (fishtail die). An extrusion die, usually one making a wide strip or sheet, in which the flow passage diverges from the adapter to the die lip.

Farad \ˈfar-ˌad\ [Michael *Faraday*] (1873) (F). The SI unit of electrical capacitance. A capacitor with a 1 V potential between its plates and holding a charge of 1 C has a capacitance of 1 F. Thus 1 F = 1 C/V. Serway RA, Faugh JS, Bennett CV (2005) College physics. Thomas, New York.

Faraday effect \ˈfar-ə-ˌdā, -ə-dē-\ [Michael *Faraday*] (1904) *n.* The rotation of the plane of polarization produced when plane-polarized light is passed through a substance in a magnetic field, the light traveling in a direction parallel to the lines of force. For a given substance, the rotation is proportional to the thickness traversed by the light and to the magnetic field strength. Serway RA, Faugh JS, Bennett CV (2005) College physics. Thomas, New York.

Faraday (F) A unit of electrical charge: 1 F equals 9.65×10^4 C. Serway RA, Faugh JS, Bennett CV (2005) College physics. Thomas, New York.

Faraday's laws In the process of electrolytic changes equal quantities of electricity charge or discharge equivalent quantities of ions at each electrode. One gram equivalent weight of matter is chemically altered at each electrode for 96,501 int. Coulombs, or 1 F, of electricity passed through the electrolyte. Serway RA, Faugh JS, Bennett CV (2005) College physics. Thomas, New York.

Farnsworth–Munsell 100 hue test Test devised to check for both defective color vision and color discrimination ability. The test requires the subject to arrange four sets of randomized colors, contained in plastic caps, in order of hue change. The test can be administered in about fifteen minutes. It may be purchased from Munsell Color Co., Baltimore, MD. McDonald R (1997) Colour physics for industry, 2nd edn. Society of Dyers and Colourists, West Yorkshire, England.

Fashioning \ˈfash-niŋ\ (15c) *vt.* The process of shaping a fabric during knitting by increasing or decreasing the number of needles in action. Fashioning is used in manufacturing hosiery, underwear, and sweaters.

Fastness *See colorfastness.*

Fastness, fast The ability of a pigmented or dyed material to resist color change following exposure to the deleterious elements. *Also known as fast to light. See lightfastness and colorfastness.*

Fastopake inks Compositions for printing on waxed surfaces.

Fast ray The fast ray or fast component for a crystal corresponds to the lower refractive index.

Fast solvent Solvent that evaporates rapidly under atmospheric conditions.

Fast to light *See fastness, fast.*

Fat (14c) *n.* A solid or semisolid ester of the troll glycerol and fatty acids.

Fat edge Accumulation of paint in the form of a ridge at the edge of a painted surface.

Fatice Sometimes called "artificial rubber" or a "rubber substitute", fatice is made by vulcanizing with sulfur a vegetable oils such as soybean, rapeseed, or castor oil. It is used as a processing aid and extender in natural-rubber compounds and synthetic elastomers.

Fatigue \fə-ˈtēg\ [F, fr. MF, fr. *fatiguer* to fatigue, fr. L *fatigare*; akin to L af*fatim* sufficiently] (1669) *n.* (1) Fatigue refers to the resistance of a material to weakening or failure during alternate tension-compression cycles, i.e., in stretch yarns, the loss of ability to recover after having been stretched. (2) A plot of the maximum cyclic stress applied to a fatigue specimen versus the number of cycles to failure, the abscissa being a logarithmic scale (*S–N* curve). Typically the *S–N* curve is linear or slightly concave upward, sloping gently downward, sometimes flattening at the low-stress (right) end, suggesting that there may be an endurance limit. ASTM testing methods, www.astm.org. Shah V (1998) Handbook of plastics testing technology. John Wiley and Sons, New York.

Fatigue failure The cracking or rupture of a plastic article under repeated cyclic stress, at a stress well below the normal short-time breaking strength as measured in a static (0.5-cycle) test. Shah V (1998) Handbook of plastics testing technology. John Wiley and Sons, New York.

Fatigue life The number of cycles of specific alternating stress required to bring about the failure of a test specimen.

Fatigue limit Syn: endurance limit.

Fatigue notch factor The ratio of the fatigue strength of a specimen with no site of stress concentration (notch) to that of a duplicate specimen having a notch.

Fatigue ratio The ratio of fatigue strength at a given number of cycles of stated alternating tensile stress to the static tensile strength.

Fatigue strength The maximum-stress level at which a material subjected to cyclic alternating stress fails after a given number of cycles. This is a number read off a fatigue curve that is derived by measuring the cycles to failure of numerous specimens subjected to various known maximum alternating stresses. Shah V (1998) Handbook of plastics testing technology. John Wiley and Sons, New York.

Fatigue tests During testing, specimens are subjected to periodic varying of stresses by means of a mechanically operated device.

FATIPEC Acronym for Federation d'Associations de Techniciens des Industries des Peintures, Vernis, Emaux, et Encres d'Imprimerie de l'Europe Continentale (Federation of Associations of Technicians in the Paint, Varnish, Lacquer, and Printing Ink Industries of Continental Europe).

Fats Fats are triglycerides (*oils*) or glyceryl esters of higher fatty acids, such as stearic and palmitic. Oils (especially drying oils) are used for the production of alkyd resins, emollients, and cooking oils. Gooch JW (2002) Emulsification and polymerization of alkyd resins. Kluwer Academic/Plenum Publishers, New York. Paint: pigment, drying oils, polymers, resins, naval stores, cellulosics esters, and ink vehicles, vol 3. American Society for Testing and Material, Philadelphia, PA, 2001. Morrison RT, Boyd RN (1992) Organic chemistry, 6th edn. Prentice-Hall, Englewood Cliffs, NJ. Martens CR (1961) Alkyd resins. Reinhold Publishing Co., New York.

Fat turpentine Oxidized turpentine. Turpentine thickened by exposure to air at elevated temperature or for long periods at ambient temperature.

F

Fatty acid (ca. 1872) *n.* Organic acid of the general formula $C_nH_{2n}O_2$, e.g., butyric acid and stearic acid; organic acids of aliphatic or open chain structure. These are generally classified as saturated or unsaturated. The saturated fatty acids range from C_4 to C_{26}. Gooch JW (2002) Emulsification and polymerization of alkyd resins. Kluwer Academic/Plenum Publishers, New York. Morrison RT, Boyd RN (1992) Organic chemistry, 6th edn. Prentice-Hall, Englewood Cliffs, NJ.

Fatty acid pitches Pitch-like residues obtained from the distillation of fats and oils of animal or vegetable origin, or of their fatty acids. Fatty acid pitches include cottonseed pitch, palm oil pitch, wool grease pitch, etc. They have comparatively high acid and saponification values.

Fatty nitrogen products Nitrogen-containing compounds derived from vegetable and animal fatty acids.

Fatty paint Paint, which has thickened because of oxidation and polymerization of the drying oil vehicle during storage.

Fatty polyamide (Versamid®, oldest of many trade names). A polymer formed by the condensation of a dibasic acid having a bulky side group and from 13- to 21-carbon chains, or the dimmer acids, C-36, with di- or polyamines. The commercially important dimmer acids are addition products of unsaturated C-18 fatty acids and can take several forms, giving different structures to the polyamides. They are used in hot-melt adhesives, inks, as epoxy flexibilizers and, in amine form, as curing agents for epoxies. Skeist I (ed) (1990, 1977, 1962) Handbook of adhesives. Van Nostrand Reinhold, New York.

Faux Hois Painted decoration simulating marble, tortoise shell or wood grain.

Fay \ˈfā\ [ME *feien*, fr. OE *fēgan*; akin to OHGr *fuogen* to fit, L *pangere* to fasten] (before 12c) *v.* To smooth and fit together, as with two surfaces about to be lap-joined.

FDA Abbreviation for food and drug administration, the USA agency within the Department of Health, Education, and Welfare that is concerned with the safety of products marketed for consumer use, particularly those substances that might be ingested, applied to the skin, or used in therapy or prostheses.

See food and drug administration.

Fe Chemical symbol for the element iron (Latin: ferrum).

FE Fluorine-containing elastomers.

Feather edging Reducing the thickness of the edge of a dry paint film, e.g., the edge of a damaged area, prior to repainting.

Feathering (1) Operation of tapering off the edges of a coat of paint by laying off with a comparatively dry brush. (2) Printing with irregular edges to the design. (3) A ragged or feather edge which shows at the edge of type or cuts. It may be caused by poor ink distribution, bad impression, excessive ink or an ink not suitable for the paper. Printing ink handbook, 5th edn. National Association of Printing Ink Manufacturers Inc., Kluwer Academic Press, London, UK, 1999. Leach RH, Pierce RJ, Hickman EP, Mackenzie MJ, Smith HG (eds) (1993) Printing ink manual, 5th edn. Blueprint, New York.

Also called laying off.

Feculose A mixture of esters prepared by treating starch with glacial acetic acid and washing the product free of uncombined acid.

Federation of societies for coatings technology A not-for-profit organization comprised of affiliated members and members belonging to its constituent societies in USA, Canada, Mexico, and England, located at 492 Norristown Road, Blue Bell, PA 19422-2350 (www.coatingstech.org). Members are engaged in research, engineering, technical development, quality control, supervisory production, administrative management and sales for the manufacture, use or improvement of the finished products or raw materials of the coatings and printing ink industries. Prior to 1975, it was known as the Federation of Societies for Paint Technology.

Feedback control A system of controlling a machine or process in which the difference between a measured output variable and its target value is amplified and, through automation, causes an appropriate adjustment of an input machine variable or process condition that will move the output nearer to its target.

Feed block In coextrusion, a massive metal block in which the streams of the several extruders are brought together to form the layers of a single stream just before it enters the die.

Feed bushing Syn: sprue bushing.

Feedforward control Process control in which early process variables are monitored and their disturbances are fed to a process model that computes adjustments of the variables needed to provide the desired process outputs. Compare feedback control.

Feed hopper An inverted conical or pyramidal vessel mounted over the feed port of an extruder or injection-molding machine that contains a supply of pellets or powder being fed. Feed hoppers typically have a slender window from bottom to top along one side to permit observation of the feedstock level.

Feeding British term for livering.
See livering.

Feed plate In injection molds, the plate contacting the injection nozzle and containing the sprue and, usually, most of the runner system. Used withy a floating cavity plate, the system provides for separation of runners and sprue from the moldings and stripping of both into separate collectors or chutes.

Feed port An opening at the rear end of the barrel of an extruder of injection molder through which plastic powder or pellets fall into the rotating screw or, in older injection machines, in front of the withdrawn ram (raw now).

Feedscrew *See extruder screw.*

Feed zone The first (rear) zone of an extruder screw that is fed from the hopper, usually of constant lead and greater depth than other zones and officially terminating at the beginning of the compression zone.

Feel A journeyman's term for the working qualities of a paint.
See hand.

Feldspar \ˈfel(d)-ˌspär\ [mod. of obs. Gr *Feldspath* (now *Feldspat*), fr. *G Feld* field + obs. Gr *Spath* (now *Spat*) spar] (1772) *n.* $K_2O{\cdot}Al_2O_3{\cdot}6SiO_2$. Natural potassium aluminum silicate, which occurs in granite. The gradual decomposition of feldspar by natural agencies yields china clay and silica. Any of several anhydrous minerals containing aluminum silicates or alkali or alkaline-earth metals (Na, K, Ca, and Ba) which, when ground, make low-cost, modulus-raising, non-toxic fillers for plastics. McGraw-Hill dictionary of geology and mineralogy. McGraw-Hill, New York, 2002. Kirk–Othmer encyclopedia of chemical technology: pigments–powders. John Wiley and Sons, New York, 1996.

Fell (1) The end of a piece of fabric that is woven last. (2) In weaving, the last filling pick laid in the fabric at any time.

Felt [ME, fr. OE; akin to OHGr *filz* felt, L *pellere* to drive, beat] (before 12c) *n.* (1) Matted fibers of wood, cotton, fur, hair, etc., individually or in combination, compacted by rolling under pressure. (2) A non-woven, fibrous material made up of randomly oriented fibers held together by stitching, a chemical binder, or by action of heat or moisture.

Felt and gravel roofing *See built-up roofing.*

Felting (1686) *n.* (1) The process of exposing wool fibers alone or in combination with other fibers to mechanical and chemical action, pressure, moisture, and heat so that they tangle, shrink, and mat to form a compact material. Felting is generally carried out in a fulling mill. (2) *See needle-punching and needled fabric.*
Also see fulling.

Felting down Operation of flatting down a dry film of varnish or paint by means of a pad made of felt or similar material, charged with a very fine abrasive powder and lubricated with water or other suitable liquid.

Femto- \ˈfem(p)-tō\ [ISV, fr. Dan or N *femten* fifteen, fr. ON *fimmtān*; akin to OE *fīftēne* fifteen (f). The SI prefix meaning $\times 10^{-15}$.

Fenchol $C_{10}H_{17}OH$. High boiling alcohol present in terpene solvents. Bp of 201°C, Sp gr, 0.964/20°C, and mp of 38°C.
Known also as fenchyl alcohol.

Fenchone $C_9H_{16}C{=}O$. Ketonic constituent of terpene solvents, Bp, 193°C; Sp gr, 0.945, refractive index of 1.4625; mp, 5°C.

Fenchyl alcohol *See fenchol.*

FEP Abbreviation for fluorinated ethylene–propylene resin.

Fermat's principle The path followed by light (or other waves) passing through any collection of media from one specified point to another, is that path for which the time of travel is least.

Fermentation alcohol *See ethyl alcohol.*

Ferric \ˈfer-ik\ (1799) *adj.* (1) Of, relating to, or containing iron. (2) Being or containing iron usually with a valence of three.

Ferric ammonium citrate (ca. 1924) *n.* A complex salt containing varying amounts of iron and used for making blueprints.

Ferric hydroxide (1885) *n.* $Fe_2O_3{\cdot}nH_2O$. A hydrate of ferric oxide that is capable of acting both as a base and as a weak acid.

Ferric oxide (1882) *n.* Fe_2O_3. Iron (III) oxide. Stable anhydrous oxide of iron. This constitutes the major ingredient in the chemically prepared synthetic reds. Indian reds, iron oxides, etc., but in the natural red oxides the proportion of ferric oxide varies considerably and is frequently associated with hydrated forms of the oxide.
See iron oxides.

Ferriferro cyanide *See iron blue.*

Ferrite \ˈfer-ˌīt\ (1851) (hard ferrite) *n.* A compound having the general formula $MFe_{12}O_{19}$, in which M is usually a divalent ion such as barium or strontium. These materials are strongly magnetic and can be incorporated into plastics to make bonded permanent magnets, rigid or flexible, and in many forms, including strips.

Ferrite yellow $Fe_2O_3{\cdot}xH_2O$. Pigment Yellow 42 (77492) Synthetic type iron oxide with a color range from bright yellow to dark orange. Manufactured by oxidizing ferrous hydroxide to the desired shade of hydrated ferric oxide.
Also known as yellow iron oxide. See iron oxides, synthetic.

Ferrocene \ˈfer-ō-ˌsēn\ [*ferro-* + *cycl-* + *-ene*] (1952) (dicyclopentadienyl iron) *n.* $(C_5H_5)_2Fe$. A coordination compound of ferrous iron and cyclopentadiene, soluble in PVC and stable to 400°C. Its uses include smoke suppression, a curing agent for silicone resins, an intermediate for high-temperature polymers, and an ultraviolet absorber. Wickson EJ (ed) (1993) Handbook of polyvinyl chloride formulating. John Wiley and Sons Inc., New York.

Ferromagnetism \-ˈmag-nə-ˌti-zəm\ (1896) *adj.* A strong attraction into a magnetic field. Serway RA, Faugh JS, Bennett CV (2005) College physics. Thomas, New York.

Ferrous oxide (1873) *n.* FeO. A black easily oxidizable power that is the monoxide of iron.

Ferrous sulfate (1865) *n.* $FeSO_4{\cdot}7H_2O$.
Also known as copperas and green vitriol.

Ferrule \ˈfer-əl\ [alt. of ME *virole*, fr. MF, fr. L *viriola*, dim. of *viria* braclet, of Celtic origin; akin to OIr *fiar* oblique] (1611) *n.* Metal portion of the brush holding bristles, hair, feathers or other fibrous and flexible materials.

Festoon dryer A dryer in which cloth is suspended in loops over a series of supporting horizontal poles and carried through the heated chamber in this configuration.

Festooning Method of drying employed for heavy fabrics, impregnated with varnish or oils, which involves hanging the treated material over horizontal rods or poles in large drying rooms. While the impregnant is still wet, the fabric is moved gradually to avoid excessive accumulations of impregnant in the bottom of the folds. Process by which wallpaper is hung and dried after being printed. Sticks carry the paper, in long loops, through a drying chamber. Kadolph SJ, Langford AL (2001) Textiles. Pearson Education, New York.

Festooning oven An oven used to dry, cure, or fuse plastic-coated fabrics with uniform heating. The substrate is carried on a series of slowly rotating shafts with long loops or "festoons" between the shafts.

FF Abbreviation for furan-formaldehyde polymer.
See furan resin.

Fiber \ˈfī-bər\ [F *fibre*, fr. L *fibra*] (1540) (fibre) *n.* A single homogeneous strand of material having a length of at least 5 mm, that can be spun into a yarn or roving or made into a fabric by interlacing in a variety of methods. Fibers can be made by chopping filaments (converting). Staple fibers may be 1.2–8 cm in length with lineal density from 0.1 to 0.5 mg/m. The natural fibers used by mankind from the earliest times were first supplemented by rayon and acetate, both of which are derived from cellulose. The first commercially successful, wholly synthetic fiber was nylon, introduced in 1939. Then followed acrylic fibers in 1950, polyesters in 1951, and various other polymeric fibers in subsequent years. In 1967 the wholly synthetic, "man-made" fibers surpassed the natural fibers in volume produced. Complete textile glossary. Celanese Acetate LLC, New York, 2000.
See also manufactured fiber, natural fiber and synthetic fiber.

Fiber architecture The spatial arrangement of fibers in the preform. Each architecture has a definite repeating unit.

Fiberboard (1897) *n.* Building material composed of wood or other plant fibers bonded together and compressed into rigid sheets.

Fiber content The percent by volume of fiber in a fiber-filled molding compound, or a molding or a laminate. Fiber content is sometimes started as weight percent.

Fiber direction (fiber orientation). In a laminate, the direction(s) in which most of the fibers' lengths lie, relative to the length axis of the part.

Fiber distribution In a web, the orientation (random or parallel) of fibers and the uniformity of their arrangement.

Fiber fill Manufactured fibers that have been specially engineered for use as filling material for pillows, mattress pads, comforters, sleeping bags, quilted outerwear, etc. Polyester fibers are widely used.

F

Fiberfill molding (Fiberfil™). A term used for an injection-molding process employing as a molding material pellets containing short bundles of fiber surrounded by resin.

Fiberglass (Fiberglass™). *See glass-fiber reinforcement.*

Fiber migration *See migration* (2).

Fiber number The linear density of a fiber expressed in units such as denier or tex. *Also see fineness.*

F

Fiber optics (1956) *n.* A term employed for light-transmitting fibers of glass and some plastics, such as polymethyl methacrylate. Each fiber is coated with a material with a refractive index lower than that of the fiber itself, and many fibers may be gathered in a bundle that is jacketed with polyethylene or other flexible plastic. Such bundles transmit light from one end to the other even through curved. Applications are in aircraft and automobile instrument panels, telephone lines, electronics, displays, medical techniques, and packaging. Meeten GH (1986) Optical properties of polymers. Springer-Verlag, New York.

Fiber placement In general, refers to how the piles are laid into their orientation, i.e., by hand, by a textile process, by a tape layer, or by a filament winder. Tolerances and angles are specified. Microprocessor-controlled placement that gives precise control of each axis of motion permits more intricate winding patterns than are possible with conventional winding and is used to make composites that are more complex that usual filament-wound structures. Tortora PG (ed) (1997) Fairchild's dictionary of textiles. Fairchild Books, New York.

Fiber-reactive dyes *See dyes.*

Fiber-reinforced plastic (FRP) Any plastic material, part or structure that contains reinforcing fibers, such as glass, carbon, synthetic, or metal fibers generally having strength and stiffness much greater than that of the matrix resin, thereby improving those properties, Because glass fibers were used so early and widely, FRP is often used to mean *glass-fiber-reinforced plastics.* Harper CA (ed) (2002) Handbook of plastics, elastomers and composites, 4th edn. McGraw-Hill, New York.

See also advanced composites, composite laminates, and reinforced plastic.

Fiber–resin interface (fiber-matrix interface). The surfaces shared by the fibers and the resin in a fiber-reinforced plastic structure. This interface, and the effects of various *sizes* and chemical treatments on the interfacial bond, are subjects of many past and ongoing studies. Because of the pretreatment of fibers with sizes, the interface has a small but finite thickness.

Fiber–resin ratio An expression, as a ratio of fiber to resin, of the fiber content.

Fiber show (fiber prominence). In reinforced plastics, a condition in which ends of reinforcement strands, rovings, or bundle unwetted by resin appear on or above the surface. It is believed to be caused by a deficiency in the glass, and may not appear until the part is fully cured. Remedies include measures to improve wet-out, use of resins of optimum viscosity, and reducing exotherm rates, which cause stresses within the laminate, and gel coating after the main body of the part has partly cured. Harper CA (ed) (2002) Handbook of plastics, elastomers and composites, 4th edn. McGraw-Hill, New York.

Fiber spinning *See spinning.*

Fiber streak (fiber whitening). A group of fibers within a translucent laminate that were incompletely wetted by resin, appearing as a whitish defect.

Fiber stress The stress acting on the reinforcing fibers in a laminate under load (estimated). Harper CA (ed) (2002) Handbook of plastics, elastomers and composites, 4th edn. McGraw-Hill, New York.

Fibrets Very short, fine fibrillated fibers that are highly branched and irregular resulting in very high surface area. Filberts can be produced from a number of substances including acetate, polyester, nylon, and polyolefins. By selection of polymer type and incorporation of additives, they can be engineered to meet a range of specialized requirements. Harper CA (ed) (2002) Handbook of plastics, elastomers and composites, 4th edn. McGraw-Hill, New York.

Fibrids Short, irregular fibrous products, made by mixing a dilute polymer solution with a non-solvent with agitation. They can also be made by flash spinning and breaking up the resulting filaments. Used in felts, in papermaking, for filtration product, etc. *Also see fibrets.*

Fibril \ˈfī-brəl\ [NL *fibrilla*, dim. of L *fibra*] (1664) *n.* A short threadlike element of a synthetic or natural fiber. Merriam-Webster's collegiate dictionary, 11th edn. Merriam-Webster Inc., Springfield, MA, 2004.

Fibrillated-film yarn *See slit-film yarn.*

Fibrillation The phenomenon in which a filament or fiber shows evidence of smaller-scale fibrous structure by a longitudinal raveling of the filament under rapid, excessive tensile or shearing stress. Separate fibrils can then often be seen in the main filament trunk. The whitening of polyethylene when severely strained at room temperature is a manifestation of fibrillation. Vincenti R (ed) (1994) Elsevier's textile dictionary. Elsevier Science and Technology Books, New York.

Fibrils A single crystal in the form of a fiber.

Fibrous asbestos *See magnesium silicate, fibrous.*

Fibrous-glass reinforcement *See glass-fiber reinforcement.*

Fibrous magnesium silicate *See magnesium silicate, fibrous.*

Fick's law (Fick's first law). (1) The net diffusion rate of a gas across a membrane is proportional to the difference in partial pressure, proportional to the area of the membrane and inversely proportional to the thickness of the membrane. Fick's law is the basic law of diffusion of different molecular species into each other. (2) Combined with the diffusion rate from Graham's law, Fick's law provides the means for calculating exchange rates of gases across membranes. (3) Fick's law states that the flux of a given component will be in the direction in which the concentration of that component decreases most steeply (i.e., opposite the gradient), at a rate given by the product of the mutual diffusivity and the gradient. Fick's law, which has many equivalent forms, is the defining equation for diffusivity. Serway RA, Faugh JS, Bennett CV (2005) College physics. Thomas, New York. Perry RH, Green DW (1997) Perry's chemical engineer's handbook, 7th edn. McGraw-Hill, New York.

Field coat The coat(s) applied at the site of erection or fabrication.

Field diaphram The diaphram on the lamp housing which controls the size of the illuminated field of view in Köhler illumination.

Field painting Surface preparation and painting operations of structural steel or other materials conducted at the project site.

Fiery finish *See burned finish.*

Figure \ˈfi-gyər, *British and often USA* ˈfi-gər\ [ME, fr. OF, fr. L *figura*, fr. *fingere*] (13c) (Wood) *n.* The pattern produced in wood surface by irregular coloration and by annual growth rings, knots and such deviations from regular grain as interlocked and wavy grain.

Filament \ˈfi-lə-mənt\ [MF, fr. ML *filamentum*, fr. LL *filare* to spin] (1594) *n.* A variety of fiber characterized by extreme length, which permits its use in yarn with little or no twist and usually without the spinning operation required for fibers. Kadolph SJJ, Langford AL (2001) Textiles. Pearson Education, New York.

See also monofilament.

F

Filamentary composite A reinforced-plastic structure in which the reinforcement consists of filaments usually oriented to most efficiently withstand the stresses imposed on the structure. The filaments are not woven and in a single lamina they will all be parallel. Wallenberger FT, Weston NE (eds) (2003) Natural fibers, plastics and composites. Springer-Verlag, New York.

Filament count The number of individual filaments that make up a thread or yarn. Kadolph SJJ, Langford AL (2001) Textiles. Pearson Education, New York.

Filament number The linear density of a filament expressed in units such as denier or tex.

Also see fineness.

Filament winding A method of forming reinforced-plastic articles comprising winding continuous strands of resin-coated reinforcing material onto a mandrel. Reinforcements commonly used are single strands or rovings of glass, asbestos (rare today because of carcinogenicity fright), jute, sisal, cotton, and synthetic fibers. Polyester resins are most widely used, followed by epoxies, acrylics, nylon, and various others. To be effective, the reinforcing material must form a strong adhesive bond with the resin. The mandrels may be permanent structures remaining in the finished article, or of flexible or destructible material, or able to be disassembled, i.e., capable of being removed after curing. The process is performed by drawing the reinforcement from a spool or creel through a bath of resin, then winding it on the mandrel under controlled tension and in a predetermined pattern. The mandrel may be stationary, in which event the creel structure rotates about the mandrel; or it may be rotated on a lathe about one or more axes. By varying the relative amounts of resin and reinforcement, and the pattern of winding, the strength of filament-wound structures may be controlled to resist stresses in specific directions. After sufficient layers have been wound, the structure is cured at room temperature or with heat. Wallenberger FT, Weston NE (eds) (2003) Natural fibers, plastics and composites. Springer-Verlag, New York, 2003. Pittance JC (ed) (1990) Engineering plastics and composites. SAM International, Materials Park, OH.

Filament-wound Made by filament winding.

Filament yarn A yarn composed of continuous filaments assembled with or without twist.

Also see yarn.

Filiform \ˈfi-lə-ˌfórm\ (1757) *adj.* Slender as a thread.

Filiform corrosion A type of corrosion that occurs under coatings on metal substrates characterized by a definite thread-like structure and directional growth. Baboian R (2002) Corrosion engineer's handbook, 3rd edn. NACE International – The Corrosion Society, Houston, TX. Uhlig HH (2000) Corrosion and corrosion control. John Wiley and Sons Inc., New York.

Filiform corrosion resistance The ability of a coating to resist that type of corrosion of metal substrates characterized by a definite thread-like structure and directional growth that occurs under coatings. Baboian R (2002) Corrosion engineer's handbook, 3rd edn. NACE International – The Corrosion Society, Houston, TX.

See filiform corrosion.

Filing Manual filing is sometimes used to bevel, smooth, deburr, and fit the edges of plastic moldings and sheets. The process is limited to parts that cannot be tumbled easily, and to plastics with suitable hardness and heat resistance.

Fill *n, adj.* Syn: weft.

Fill-and-wipe A decorating process for articles molded with depressed designs, wherein the general area containing the designs is coated with paint by brushing, spraying, or rolling, then surplus paint is wiped from the undepressed areas surrounding the depressions.

Filled plastic Any plastic compound containing a significant percentage of a solid, usually not fibrous or resinous, material whose main purpose may be to dilute the resin, or to provide certain enhanced properties in the compound.

Filler \ˈfi-lər\ (15c) *n.* (1) A pigmented composition used for filling fine cracks and indentations to obtain a smooth, even surface preparatory to painting. (2) Synonymous with extender. (3) A relatively non-adhesive substance added to an adhesive to improve its working properties, permanence, strength, and other qualities. (4) Any compounding ingredient, usually in dry, powder form, added to rubber in substantial amount to improve quality of lower cost. Fillers have various effects; some are relatively inert, like calcium carbonate and silica, and provide loading for cost reduction; others, like carbon blacks, have a definite and desirable reinforcing effect with improvement in abrasion resistance and other properties. Fillers provide are added to a plastic compound to reduce its cost per unit volume and/or to improve such mechanical properties as hardness, modulus, and impact strength. (5) A filler differs from reinforcement in two respects. Filler particles are generally small and roughly equidimensional, and they do not markedly improve the tensile strength of a product. Reinforcements, on the other hand, are fibrous, having one dimension much longer than the others, and they do markedly improve tensile strength. Strong AB (2000) Plastics materials and processing. Prentice-Hall, Columbus, OH. Modern plastics encyclopedia. McGraw-Hill/Modern Plastics, New York, 1986-, 1990-, 1992-, 1993-editions. Kirk–Othmer encyclopedia of chemical technology: pigments–powders. John Wiley and Sons, New York, 1996.

Filler sheet *n.* A sheet of deformable or resilient material, that when placed between the assembly to be bonded and the pressure applicator, or when distributed within a stack of assemblies, aids in proving uniform application of pressure over the area to be bonded. Kirk–Othmer encyclopedia of chemical technology: pigments–powders. John Wiley and Sons, New York, 1996.

Filler coat A coat of paint, varnish, etc., used as a primer.

Filler rod (welding rod). A rod of plastic material used in hot-gas welding, made of the same material as the plastic to be welded.

Filler specks Visible particles of a filler, such as wood flour or asbestos, that stand out in color contrast against a background of plastic binder.

Fillet \ˈfi-lət\ [ME *filet*, fr. MF, dim. of *fil* thread] (14c) *n.* A concavely curved transition at the angle formed by the junction of two plane surfaces, i.e., a rounded inside corner. Also, the material making up the transition. Where the surfaces are likely to endure bending toward or away from each other, the fillet distributes and reduces the stress that would otherwise be magnified at the corner.

Filling In a woven fabric, the yarn running from selvage to selvage at right angles to the warp. Each crosswise length is called a pick. In the weaving process, the filling yarn is carried by the shuttle or other type of yarn carrier.

Filling band *See mixed end or filling.*

Filling Barré *See Barré.*

Filling bow *See skewness.*

Filling of coated abrasives Clogging of the abrasive coat by swarf. It can be reduced in many operations by using an open coat construction or a lubricant. Syn: loading.

Filling skewness *See skewness.*

Filling snarl *See snarl.*

Filling up (or filling in) A condition in the printing of halftones where the ink fills areas between the dots, and produces a solid rather than a sharp halftone print.

Filling up (1) Covering the non-printing areas (on a lithographic plate) with a partially dried ink film. Effect caused by an excessive drier in the ink. (2) A condition in the printing of halftones where the ink fills areas between the dots, and produces a solid, rather sharp halftone print. They may also occur in the printing of type matter. Printing ink handbook. National Association of Printing Ink Manufacturers Inc., 5th edn. Kluwer Academic Press, London, UK, 1999. Leach RH, Pierce RJ, Hickman EP, Mackenzie MJ, Smith HG (eds) (1993) Printing ink manual, 5th edn. Blueprint, New York.

Filling yarn *See split-film yarn and weft.*

Fill-type insulation Loose insulating material, which is applied by hand or blown into wall spaces mechanically.

Film \ˈfilm, *Southern also* ˈfi(ə)m\ {often attributive} [ME *filme*, fr. OE *filmen*; akin to Gk *pelma* sole of the foot, OE *fell* skin] (before 12c) *n.* (1) Customarily in the plastics industry, a web of plastic that is 0.25 mm or less in thickness. Thicker webs are called *sheet.* Films are made by extrusion, casting from solution, and calendaring. (2) In convective heat transfer, the thin, supposedly stagnant layer of fluid next to a heated or cooled surface (such as a pipe wall) that contributes part (or all) of the resistance to transfer of heat from the main body of the fluid to a medium on the opposite side of the wall (or to the wall itself. A closely related concept exists in mass transfer. (3) A layer of one or more coats of paint or varnish covering an object or surface. (4) Any supported or unsupported thin continuous covering or coating. (5) Unsupported, usually organic, non-fibrous, thin, flexible material of a thickness not exceeding 0.010 in. In excess of 0.010 in. thickness, such material is usually called sheet or sheeting. Pittance JC (ed) (1990) Engineering plastics and composites. SAM International, Materials Park, OH. Perry RH, Green DW (1997) Perry's chemical engineer's handbook, 7th edn. McGraw-Hill, New York. Wicks ZN, Jones FN, Pappas SP (1999) Organic coatings science and technology, 2nd edn. Wiley-Interscience, New York.

Film blowing (blown-film extrusion). The process of forming thermoplastic film wherein an extruded plastic tube is continuously inflated by internal air pressure, cooled, collapsed by rolls, and subsequently wound into rolls on thick cardboard cores. The tube is usually extruded vertically upward, and air is admitted through a passage in the center of the die as the molten tube emerges from the die. An air ring is always employed to speed and control the initial cooling close to the die. Air is contained within the blown bubble by a pair of pinch rolls that also serve to collapse and flatten the film. Thickness of the film is controlled not only by the die-lip opening but also by varying the internal air pressure and by the rates of extrusion and take-off. Extremely thin films (<0.01 mm) and films with considerable biaxial orientation can be produced by this method. Strong AB (2000) Plastics materials and processing. Prentice-Hall, Columbus, OH.

Film build The rheological property which coatings possess of providing thickness in applied films. Paint/coatings dictionary. Federation of Societies for Coatings Technology, Philadelphia, Blue Bell, PA, 1978.

Film casting The process of making an unsupported film or sheet by casting a fluid resin, a resin solution, or a plastic compound on a temporary carrier, usually an endless belt or circular drum, followed by solidification by cooling or drying, and removal of the film from the carrier. The term *film casting* is also used for the process of extruding a molten polymer through a slot die onto a chilled roll. Strong AB (2000) Plastics materials and processing. Prentice-Hall, Columbus, OH.

Film coefficient (1) In convective heat transfer, the rate of heat flow through a "stagnant" fluid film adjacent to a solid surface, per unit area of the film, divided by the temperature difference through the film. (2) A similarly structured definition applying to mass transfer through films at fluid interfaces. Perry RH, Green DW (1997) Perry's chemical engineer's handbook, 7th edn. McGraw-Hill, New York.

Film die A die for the extrusion of flat or blown film. Flat-film dies are usually of the crosshead and coathanger designs with one lip locally adjustable so as to achieve uniform thickness across the film (*see flexible-lip die*). Blown-film dies are cylindrical, end- or side-fed, with the concentricity/eccentricity of the core and body adjustable for circumferential uniformity of film thickness. That uniformity, which is critical in both types of films for winding even rolls, also depends on the performance of the air ring. Blown-film dies are often oscillated slowly about their axes to distribute the remaining non-uniformity evenly over the final roll width, in that way assuring that the unrolled film will lie flat in spite of slight thickness variations. Strong AB (2000) Plastics materials and processing. Prentice-Hall, Columbus, OH.

Film extrusion Making plastic films by extruding molten plastic through a film die by film blowing or film casting. Strong AB (2000) Plastics materials and processing. Prentice-Hall, Columbus, OH.

Film formation Term applied to an extruded sheet of a polymer.

Film former A type of resin with qualities of forming a tough, continuous dry film. Example: nitrocellulose.

Film forming Ability of a material to form a continuous dry film.

Film integrity Continuity of a coating free of defects.

Filmogen General term for film forming materials.

Film slitting *See slitting.*

Film thickness Thickness of any applied coating, wet or dry.

Film, thickness gauge Device for measuring film thickness; instruments for measuring either wet or dry films are available.

Filter aid Inert, insoluble material, more or less finely divided, used as a filter medium or to assist in filtration by maintaining adequate porosity of the filter cake.

Filter cake The solid mass remaining on a filter after the liquid that contained it has passed through.

Filter cloth Any cloth used for filtering purposed. Nylon, polyester, vinyon, PBI, and glass fibers are often used in such fabrics because they are not affected by most chemicals.

Filter fabrics *See geotextiles and filter cloth.*

Filter, optical Any uniform optical device, which transmits radiant energy of limited wavelengths and/or intensity. Filters may be made of glass, quartz, plastic, etc., containing absorbent material. They may consist of thin metallic coatings on glass or plastic; or they may consist of carefully made screens of uniform pore size. Optical filters may be used to modify the spectral distribution of radiant energy (selective filters) or the photometric intensity of radiant energy (neutral filters). Moller KD (2003) Optics. Springer-Verlag, New York. Giambattista A, Richardson R, Richardson RC, Richardson B (2003) College physics. McGraw-Hill Science/Engineering/Math, New York.

Filter press Apparatus for filtering, consisting of a number of flat chambers enclosed between sheets of filter cloth (paper or metal) through which the liquids is forced by a pump, leaving the solid matter in the chambers.

Filter spectrophotometer A spectrophotometer that uses filters of fixed, narrow band-pass transmissions of discrete wavelengths spaced across the spectrum, to measure the transmittance or reflectance of materials at these discrete wavelengths. The resulting special data arranged in order constitute an abridged spectrophotometric curve. Thus, the series of filters replaces the dispersion monochromator used in a continuous spectrophotometer. Willard HH, Merritt LL, Dean JA (1974) Instrumental methods of analysis. D. Van Nostrand, Company, New York.

Filtration \fil-ˈtrā-shən\ (1605) *n.* The operation of separating suspended solids from a liquid, or gas, by forcing the mixture through a porous barrier.
See straining.

Fin *n.* Overflow material protruding from surface of cured, molded articles, usually appearing at mold separation line or mold vent points.
See flash.

Findings *n.* (1) Miscellaneous items attached to garments and shoes during manufacture. Included are buttons, hooks, snaps, and ornaments. (2) Miscellaneous fabrics in garments such a zipper tapes, linings, pockets, waistbands, and facings.

Fine end *n.* (1) A warp yarn of smaller diameter than that normally used in the fabric. (2) A term for a defect in silk warp yarn consisting of thin places that occur when all the filaments required to make up the full ply are not present. This condition is generally caused by poor reeling.

Fine-etching *n.* A method of modifying a photomechanically prepared plate by controlled undercutting to change the size of the halftone dots.

Fine melt Process of running copals at high temperatures as distinct from low temperature running, which is described as a slack melt.

Fineness *n.* (1) A relative measure of fiber size expressed in denier or tex for manufactured fibers. For cotton, fineness is expressed as the mean fiber weight in micrograms per inch. For wool, fineness is the mean fiber width or mean fiber diameter expressed in microns (to the nearest 0.001 mm). (2) For yarn fineness. (3) For fineness of knit fabrics.
See yarn number.

Fineness of dispersion *n.* A measure of the size and prevalence of oversize particles in the coating or ink. The Hegman Gage is a widely used tool for measuring fineness of particles for coatings and inks supplied by Paul N. Gardner Company Inc., 316 N. E. Fifth Street, Pompano Beach, FL (www.gardco.com). Weismantal GF (1981) Paint handbook. McGraw-Hill Corporation Inc., New York. Printing ink handbook. National Association of Printing Ink Manufacturers Inc., 5th edn. Kluwer Academic Press, London, UK, 1999. Leach RH, Pierce RJ, Hickman EP, Mackenzie MJ, Smith HG (eds) (1993) Printing ink manual, 5th edn. Blueprint, New York.

Fineness of grind (deprecated) *n.* The degree of dispersion of a pigment in a printing ink vehicle, usually measured on a grindometer or grind gauge. Weismantal GF (1981) Paint handbook. McGraw-Hill Corporation Inc., New York.

Fineness of grind gauge *n.* A device to measure the fineness of dispersion of a pigment, based on drawing a paint down over a channel of tapered depth and observing the minimum depth at which pigment particles are observed to interfere with the smooth wet surface of the paint. ASTM provides standards for fineness of grind (www.astm.org). The Hegman Gage is a widely used tool for measuring fineness of grind for coatings and inks supplied by Paul N. Gardner Company Inc., 316 N. E. Fifth Street, Pompano Beach, FL (www.gardco.com). Tracton AA (ed) (2005) Coatings technology handbook. Taylor and Francis Inc., New York. Weismantal GF (1981) Paint handbook. McGraw-Hill Corporation Inc., New York.

Fines *n.* In the classification of powdered or granular materials according to particle size, fines are in portion of the material whose particles are smaller than a stated minimum size. When the particle-size distribution is determined by sieve analysis, the fines are those particles passing the finest sieve and found on the pan, usually designated as "minus 000 mesh", where 000 is the mesh number of that finest sieve. Provder T, Texter J (eds) (2004) Particle sizing and characterization. American Chemical Society, Washington, DC.

Fine structure *n.* Orientation, crystallinity, and molecular morphology of polymers, including fiber-forming polymers.

Finger mark *n.* A defect of woven fabrics that is seen as an irregular spot showing variation in picks per inch for a limited width. Causes are spreading of warp ends while the loom is in motion and pressure on the fabric between the reed and take-up drum. Kadolph SJJ, Langford AL (2001) Textiles. Pearson Education, New York. Vincenti R (ed) (1994) Elsevier's textile dictionary. Elsevier Science and Technology Books, New York.

Fingernail test *n.* Gouging a dried film with fingernail to form a subjective, qualitative estimate of the relative hardness and toughness.

Fingers \ˈfiŋ-gər\ [ME, fr. OD; akin to OHGR *fingar* finger] (before 12c) *n.* Classification of fossil copals, especially of the Congo type. The pieces of resin have long cylindrical shapes, similar to those of human fingers.

Finish \ˈfi-nish\ [ME *finisshen*, fr. MF *finiss-*, stem of *finir*, fr. L *finire*, fr. *finis*] *n.* (1) Refers to the degree of gloss or flatness of a print or any surface. (2) (size) In reinforced plastics, a compound containing a coupling agent and (optionally) a lubricant and/or binder, used to pretreat glass fibers prior to using them as reinforcements.

(3) The surface texture of a molding, a machined or polished surface, or other article. When measured, it is usually stated as the root-mean-square roughness in nanometers or microinches. (4) The final coat in a painting system. (5) Sometimes refers to the entire coating system: the texture, color, and smoothness of a surface, and other properties affecting appearance. Merriam-Webster's collegiate dictionary, 11th edn. Merriam-Webster Inc., Springfield, MA, USA, 2004. Lee SM (1989) Dictionary of composite materials technology. Technomic Publishing Co. Inc., Lancaster, PA. Ash M, Ash I (1982–1983) Encyclopedia of plastics polymers, and resins, vols 1–3. Chemical Publishing Co., New York. Weismantal GF (1981) Paint handbook. McGraw-Hill Corporation Inc., New York.

Finish coat *n.* (1) The final layer of plaster applied over a basecoat or other substrate. (2) *See top coat.*

Finish composition (yard) *n.* Physical and chemical analysis of the lubricant applied to yarns to reduce friction and improve processability.

Finished fabric Fabric that is ready for the market, having passed through the necessary finishing processes. Vigo TL (1994) Textile processing, dyeing, finishing and performance. Elsevier Science, New York.

Finishing *n.* (1) The removal of flash, gates, and defects from plastic articles. (2) The development of a desired texture and/or color on the surfaces of an article when such are not accomplished in compounding and forming the article.

See grind, polishing, sanding, and the processes listed under decorating.

Finishing bar *n.* A noticeable streak across the entire width of a fabric, usually caused by machine stoppage during processing.

Finishing spot *n.* A discolored area on a fabric caused by foreign material such as dirt, grease, or rust.

Finish insert (neck insert). In blow molding bottles, a removable part of the mold that aids in forming a specific neck finish of the bottle.

Finish turns The actual degree of twist in the final yarn product.

Finn oil *See tall oil.*

Fire clay *n.* An earthy or stony mineral aggregate, which has the essential constituent hydrous silicates of aluminum with or without free silica. It is plastic when sufficiently pulverized and wetted, rigid when subsequently dried, and of suitable refractoriness for use in commercial refractory products.

Fire point *n.* The temperature at which a material, when once ignited, continues to burn for a specified period of time. The fire point is several degrees of temperature higher than the flashpoint. It is the lowest temperature at which a liquid evolves vapors fast enough to support continuous combustion. Troitzsch J (2004) Plastics flammability handbook: principle, regulations, testing and approval. Hanser–Gardner Publications, New York. Babrauskas V (2003) Ignition handbook. Fire Science Publishers, New York. Tests for comparative flammability of liquids, UI 340. Laboratories Incorporated Underwriters, New York, 1997.

Fire resistance *n.* The property of a material or assembly to withstand fire or give protection from it. As applied to elements of buildings, it is characterized by the ability to confine a fire or to continue to perform a given structural function or both.

See flammability.

Fire-resisting finish *n.* The preferred term is "fire-retardant coating".

Fire resistive *n.* Refers to properties of materials or designs to resist the effects of any fire to which the material or structure may be expected to be subjected. Fire resistive materials or structures are non-combustible, but non-combustible materials are not necessarily fire resistive. Fire resistive implies a higher degree of fire resistance than non-combustible.

Fire retardant *n.* Descriptive term which implies that the described product, under accepted methods of test, will significantly: (a) Reduce the rate of flame spread on the surface of a material to which it has been applied. (b) Resist ignition when exposed to high temperatures. (c) Insulate a substrate to which it has been applied and prolong the time required to reach its ignition, melting, or structural-weakening temperature.

Fire-retardant chemical *n.* A chemical or chemical preparation used to reduce flammability or to retard the spread of flame.

Fire-retardant coating *n.* A coating, which will do one or more of the following: (1) Reduce the flame spread over which the coating is applied, sometimes at the sacrifice of the coating. (2) Resist ignition when exposed to high temperature. (3) Insulate substrate to which it is applied and thereby prolong time required to reach ignition, melting or structural-weakening temperature.

See intumescent coatings.

First coat First coating applied in any painting schedule; in some cases, it would be the sealing coat; in others, the priming coat.

First-down color *n.* In a multicolor printed material this is the first color printed on the substrate usually subsequently overprinted by other colors.

First-order kinetics *n.* A thermal transition that involves both a latent heat and a change in the heat capacity of the material.

First-order transition temperature *n.* The temperature at which a polymer freezes or melts.

First quality *See yarn quality.*

Fish eye *n.* (1) Paint defect which manifests itself by the crawling of wet paint into a recognized pattern resembling small "dimples" or "fish eyes." (2) A visible fault in transparent or translucent plastics, particularly films or thin sheets, appearing as a small globular mass (gel particle) and thought to be caused either by stray resin particles of much higher molecular weight than that of the polymer as a whole, or by inclusion of foreign particles. In rubber technology, such globules are known as "cat eyes". Koleske JV (ed) (1995) Paint and coating testing manual. American Society for Testing and Materials, Philadelphia, PA. Paint testing manual: physical and chemical examination of paints, varnishes, lacquers, and colors – STP 500. American Society for Testing and Materials, Philadelphia, PA, 1973. Hess M (1965) Paint film defects. John Wiley and Sons Inc., New York.

See pinhole.

Fish oil *n.* A natural oil extracted from fish and generally characterized by a rather large group of saturated fatty acids commonly associated with mixed triglycerides (Fish oil: a key to better coatings, manufacturing engineering magazine, 15 October 1987). The fatty acids derived from fish oils are three principal types: saturated, monounsaturated and polyunsaturated with carbon chain lengths ranging from C_{12} to C_{24}. Both types and relative amounts of the fatty chains vary widely among different species of fish and different geographical areas. The fish oil commonly used in the coatings industry is *menhaden oil*, produced from *menhaden* *fish* caught along the Atlantic Coast of USA. This oil has an iodine value of about 175. Other common types of fish oils and their iodine values are: herring – 134, anchovy – 199. In addition to glycerides of stearic and the lesser unsaturated fatty acids, fish oils, contain glycerides of clupanodonic acid, which appears to contain four double bonds. The iodine value varies over a wide range, approximately 130–1990. Fish oil is available in Asia, Europe, or anywhere a fishing industry is active. The tendency of fish oils films to yellow considerably is due to the presence of highly unsaturated groups in the molecule. Gooch JW (2002) Emulsification and Polymerization of Alkyd Resins. Kluwer Academic/Plenum Publishers, New York. Paint/coatings dictionary. Compiled by Definitions Committee of the Federation of Societies for Coatings Technology, 1978.

Fishtail die Syn: fantail die.

Fission \ˈfi-shᵊn, -zhᵊn\ [L *fission-*, *fission*, fr. *findere* to split] (ca. 1617) *n.* A nuclear reaction from which the atoms produced are each approximately half the mass of the parent nucleus. In other words, the atom is split into two approximately equal masses. There is also the emission of extremely great quantities of energy since the sum of the masses of the two new atoms is less than the mass of the parent heavy atom. The energy released is expressed by Einstein's equation. Serway RA, Faugh JS, Bennett CV (2005) College physics. Thomas, New York.

Fissure \ˈfi-shər\ [ME, fr. MF, fr. L *fissura*, fr. *fissus*] (14c) *n.* A term used in the cellular-plastics industry to denote a separation, crack, or split in a formed cellular article.

Fitch *n.* Long handled small brush bound with metal, with which nearly inaccessible areas are painted.

Five regions of viscoelasticity *n.* As an amorphous polymer is heated from an extremely low temperature it gains more dimensions of molecular motion and its mechanical behavior changes through five qualitative regions: glassy, slowed elastic, rubbery, rubbery flow, and viscous flow.

Fixation *n.* The process of setting a dye after dyeing of printing, usually by steaming or other heat treatment.

Fixative *n.* Solution which can be sprayed onto drawings rendered in pencil, chalk and other impermanent and easily removed materials so as to fix them and prevent smudging.

Fixed oil *See non-drying oil.*

Flag [ME *flagge* reed, rush] *n.* The end of a brush bristle, which divides into two or more branches. Flagging provides the brush with the ability to hold a greater amount of liquid coatings.

Flair A subjective term applied to the change in hue of a colored material when the light source is changed. Thus, a color may appear blue in daylight but change to a purple-blue in incandescent light; ultramarine blue is an example of such a blue, so it is said to "flair" red or to have a red "flair". Opposite of color consistency; not to be confused with metamerism, which applies to a pair of colors. McDonald R (1997) Colour physics for industry, 2nd edn. Society of Dyers and Colourists, West Yorkshire, England. Billmeyer FW, Saltzman M (1996) Principles of color technology. John Wiley and Sons Inc., New York.

See metamerism.

Flake [ME, of Scand origin; akin to Nor *flak* disk] *n.* (1) A term used to signify the dry, unplasticized, basic form of cellulosic plastics. (2) Glass flakes. (3) As used by Celanese, a term that refers to the granular form in which cellulose acetate and triacetate polymers exist prior to dissolving or feeding into the extrusion or molding unit. Paint: pigment, drying oils, polymers, resins, naval stores, cellulosics esters, and ink vehicles, vol 3. American Society for Testing and Material, Philadelphia, PA, 2001. Complete textile glossary. Celanese Acetate LLC, New York, 2000.

Flake board *n.* Same as particle board.

Flake shellac *n.* Orange shellac gum in flake form.

Flake white *n.* (1) Another name for white lead of fine textured prepared electrolytically. (2) Variety of white lead, frequently used by artists.

Flake yarn *n.* Yarn in which roving or short, soft staple fibers are inserted at intervals between long filament binder yarns.

Flaking *n.* That phenomenon manifested in paint films by the actual detachment of pieces of the film itself either from its substrate or from paint previously applied. Flaking (scaling) is generally preceded by cracking or checking or blistering and is the result of loss of adhesion, usually due to stress-strain factors coming into play. Syn: scaling.
See peeling.

Flaking resistance *n.* The ability of a coating to resist the actual detachment of film fragments, either from its undercoating or substrate. Flaking is generally preceded by cracking, checking or blistering and is the result of loss of adhesion.
Also known as scaling resistance. See flaking.

Flaky web *n.* A web at the card that shows thick and thin places, approximately 1–6 in.^{2} in size. This indicates that, instead of a free flow of fibers through the card, either an uneven amount has been fed into the card, or groups of fibers have hesitated in the card and then dropped back into production.

Flamboyant finish *n.* A glossy transparent coating with or without colorant, over a bright undercoat, metallic surface or metallic finish.

Flame \ˈflām\ [ME *flaume, flaumbe,* fr. MF *flamme* (fr. L *flamma*) & fr. OF, fr. *flamble,* fr. L *flammula,* dim. of *flamma* flame; akin to L *flagrare* to burn] (14c) *n.* The visible heat rays, which appears when the ignition of a material is reached. Hydrogen is one of the exceptions since the heat rays are not visible.

F

Flame cleaning *n.* Impingement of an intensely hot flame to the surface of structural steel resulting in the removal of mill scale and the dehydration of any remaining rust, leaving the surface in a condition suitable for wire brushing followed by the immediate application of paint.

Flame (flash) arrestor *n.* Devices utilized on vents for flammable liquid or gas tanks, storage containers, cans, gas lines or flammable liquid pipelines to prevent flashback (movement of flame) through the line or into the container, when a flammable of explosive mixture is ignited. Wire screen of 40 meshes is utilized on smaller openings. On larger openings, parallel metal plates or tubes are more effective.

Flame hardening *n.* A cheap method, obsolete today, of initially hardening flight tips of extruder screws in which the surface is rapidly chilled (tempered) after being heated with a flame (a process not easy to control). The hardness imparted is gradually lost over a few hundred hours of normal operation, requiring that the process be repeated frequently, defeating the initial saving.
Also called case hardening and surface hardening.

Flame polishing *n.* A method of finishing a plastic article, particularly a just formed extrudate, in which a carefully controlled flame or stream of hot gas is directed at the surface, melting a thin skin of resin that, when quenched, has a high gloss.

Flame pretreaters *n.* Equipment used in the flame pretreatment of polyolefins, by which the polyolefin molding is subjected to prior to the application of a decorative coating (usually in ink). Flame pretreatment is performed in order that the applied coating will have acceptable and permanent adhesion. Flaming consists of exposing the surface to be decorated to a suitable oxidizing flame. This treatment brings about a change to the polymer surface that makes it wet and permits a strong adhesive bond between the molding surface and the coating,

Flame proofing *See fire retardant.*

Flame resistance tests *See flammability tests.*

Flame resistant *adj.* A term used to describe a material that burns slowly or is self-extinguishing after removal of an external source of ignition. A fabric or yarn can be flame resistance because of the innate properties of the fiber, the twist level of the yarn, the fabric construction, or the presence of flame retardants, or because of a combination of these factors.

Also see flame-retardant and inherent flame resistance.

Flame retardant (1947) *n.* A material that reduces the tendency of plastics to burn. Flame retardants are usually incorporated as additives during compounding, but sometimes applied to surfaces of finished articles. Some plasticizers, particularly the phosphate esters and chlorinated paraffins, also serve as flame retardants. *Inorganic flame retardants* include antimony trioxide, hydrated alumina, monoammonium phosphate, dicyandiamide, zinc borate, boric acid, and ammonium sulfamate. Another group, called *reactive-type flame retardants*, includes bromine-containing polyols, chlorendic acid, and anhydride, tetrabromo- and tetrachlorophthalic anhydride, tetrabromo bisphenol A, diallyl chlorendate, and unsaturated phosphonated chlorophenols. A few neat resins, such as PVC and the fluoro- and chlorofluorocarbons, are flame-retardant. Elias HG (2003) An introduction to plastics. John Wiley and Sons, New York. Modern plastics encyclopedia. McGraw-Hill/Modern Plastics, New York, 1986-, 1990-, 1992-, 1993-editions.

See flammability.

Flame retardants *n.* A chemical or surface covering material that delays ignition and reduces flame spread.

Flame spray *n.* Any process whereby a material is brought to its melting point and sprayed onto a surface to produce a coating. The process includes: (1) Metallizing. (2) Thermospray. (3) Plasma flame.

Flame-spray coating *n.* A coating process utilizing powdered metals or plastics, in which the powdered materials are heated to the sintering temperature in a cone of flame enroute from a spray gun orifice to the article being coated.

Flame spraying *n.* Blowing a powder through a flame that partially melts the powder and fuses it as it hits the substrate.

Flame spread *n.* Flaming combustion along a surface; not to be confused with flame transfer by air currents.

Flame treating A method of rendering inert thermoplastics, particularly polyolefins, receptive to inks, lacquers, paints, and adhesive by briefly bathing the surface of the article in a highly oxidizing flame. This treatment oxidizes the surface slightly, creating carbonyl and possibly peroxide groups, thereby increasing its surface energy.

Flammability \ˌfla-mə-ˈbi-lə-tē\ (1646) *n.* With respect to plastics, flammability is a very broad term that has been the focus of a potpourri of tests and standards generated by many organizations, predominantly in USA by Underwriters Laboratory (UL) and ASTM. The behavior of various plastics when burning, and tests designed to evaluate flammability, encompass six categories: ignitability, burning rate, heat evolution, smoke production, products of combustion, and endurance of burning. Troitzsch J (2004) Plastics flammability handbook. Hanser–Gardner Publications, New York. Shah V (1998) Handbook of plastics testing technology. John Wiley and Sons, New York.
Also called flammability and burn tests.

Flammability and burn tests These tests are conducted for the purpose of evaluating polymeric materials for combustible properties. The American Society of Testing and Materials (www.astm.org) and National Institute for Standard and Technology (www.nist.gov) provide continuously updated method of evaluating flammability and combustible properties of polymeric materials (adhesives, coatings, plastics, etc.). Troitzsch J (2004) Plastics flammability handbook. Hanser–Gardner Publications, New York. Shah V (1998) Handbook of plastics testing technology. John Wiley and Sons, New York.

Flammable \ˈfla-mə-bəl\ [L *flammare* to flame, set on fire, fr. *flamma*] (1813) *adj.* A substance that is easily ignited, burns intensely, or has a rapid rate of flame spread. Flammable and inflammable are identical in meaning, however, the prefix "in" indicates negative in many words and can cause confusion. Flammable, therefore, is the preferred term. According to I.C.C. Regulations, liquids are flammable if their flash point is 100°F or lower. Troitzsch J (2004) Plastics flammability handbook. Hanser–Gardner Publications, New York.

Flammable limits *See explosive limits.*

Flammable liquid *n.* Any liquid having a flp below 37.8°C (100°F), except a mixture having components with flp of 37.8°C (100°F) or higher, the volume of which make up 99% or more of the total volume of the mixture. Babrauskas V (2003) Ignition handbook. Fie Science Publishers, New York. Tests for comparative flammability of liquids, UI 340. Laboratories Incorporated Underwriters, New York. Wray HA (ed) (1991) Manual for flash point standards and their use. American Society for Testing and Materials, Philadelphia, PA.
See combustible liquid.

Flange crimping *n.* Simultaneous crimping of two ends of yarn by using heated snubber pins, then combining both ends on a draw roll after they contact a rubber flange on the draw roll.

Flannel \ˈfla-nᵊl\ [ME *flaunneol*] (1503) *n.* Medium weight plain- or twill-weave, slightly napped fabric, usually of wool or cotton, but may be made of other fibers.

Flapper \ˈfla-pər\ (ca. 1570) *n.* The movable side of a fiber-crimping chamber that periodically opens or flaps to permit crimped fiber to be expelled from the chamber.

Flash (fin) *n.* The thin, surplus web of material that is forced into the parting line between mating mold surfaces during a molding operation and which remains attached to the molded article. For methods of removing flash.
see deflashing.

Flash aging *n.* A process for rapid reduction and fixation of vat dyes obtained when the printed fabric is padded with caustic soda and sodium hydrosulfite and immediately steamed in air-free steam.

Flash drying *n.* (1) Rapid method of drying brought about by the exposure of a coating surface to an elevated temperature or forced draft for a short period of time. (2) Drying at ambient temperature, preparatory for a period of forced drying at elevated temperatures.

Flash gate *n.* A long, shallow rectangular gate in an injection mold, extending from a runner that lies parallel to an edge of a molded part along the flash or parting line of the mold.

Flash groove (spew groove) *n.* A groove in a mold force that allows the escape of excess material during a compression-molding operation.

Flashing *n.* (1) A paint defect in a paint film in which patches glossier than the general finish develop, especially at joints or laps in the coating. (2) The non-uniform appearance, including spotty differences in color or gloss, usually due to improper or non-uniform sealing of a porous substrate. (3) Non-corrosive metal used around angles or junctions in roofs and exterior walls to prevent leaks.

Flash line *See parting line.*

Flash mold *n.* A mold in which the mating surfaces are perpendicular to the claming action of the press so that, as the clamping force increases, the distance between the mating surfaces decreases, thus permitting excess molding material to escape as flash as the mold closes.

Flash off *n.* Causing the greater part of the more volatile solvents in a sprayed coat of lacquer or enamel to evaporate before proceeding with the application of another coat.

Flash-off time *n.* Time allowed to elapse between the spray application of successive wet-on-wet coats or the time allowed for the evaporation of the bulk of the solvent before entering into a baking oven.
See flash drying.

Flashover (1892) *n.* (1) A flammability term. Flashover occurs when hot, combustible gases are generated in burning sections of a building, become mixed with sufficient oxygen upon spreading to non-burning areas, and ignite to cause total surface involvement, but without a progressive flame-spreading stage. (2) In the electrical industry, an electric discharge, around the edge or over the surface of insulation.

Flash point (1878). *n* The lowest temperature at which a combustible liquid will give off a flammable vapor that will momentarily burn when exposed to s small flame. The flash point can be determined by the open cup or the closed cup method. The flash point determined by the open cut method is usually somewhat higher than the closed cup method. Wray HA (ed) (1991) Manual for flash point standards and their use. American Society for Testing and Materials, Philadelphia, PA.
See flammability.

Flash points *n.* The lowest temperature of a liquid at which it gives off sufficient vapor to form an ignitable mixture with the air near the surface of the liquid or within the container used.

Flash ridge *n.* That part of a flash mold along which the excess material escapes until the mold is fully closed.

Flash spinning *n. See spinning* (5).

Flat *n.* In carding, one of the parts forming an endless chain that partially surrounds the upper portion of the cylinder and gives the name to a revolving flat card. Flats are made of cast iron, *T*-shaped in section, about 1 in. wide, and as long as the width of the cylinder. One side of the flat is nearly covered with fine card clothing, and the flats are set close to the teeth of the cylinder so as to work point against point. A chain of flats contains approximately 110 flats and operates at a surface speed of about 3 in./min.

See flatting down and rubbing.

Flat abrasion tester *See Stoll–Quartermaster universal wear tester.*

Flat card *n.* The type of card used for cotton fibers and for cotton-system processing. It is named for the flat wire brushes called flats that are assembled on an endless chain that partially surrounds the main cylinder. The staple is worked between the flats and cylinder, transferred to a doffer roll, and peeled off as a web that is condensed into a sliver. Vincenti R (ed) (1994) Elsevier's textile dictionary. Elsevier Science and Technology Books, New York.

Also see flat.

Flat coat *n.* Coat of filler (British) also incorrectly used when meaning prime coat under enamel. The preferred term is "enamel undercoater." An intermediate coat of paint used as a base for a topcoat.

Flat duck *See duck.*

Flat enamel *n.* Pigmented coating of low specular gloss, which has the leveling characteristics of a gloss enamel. Enamel has the connotation of both gloss and flow. *See enamel.*

Flat-entry die *n.* An extrusion die in which the approach to the land has no taper, i.e., one having a 180° entrance angle.

Flat film *n.* Film made by extrusion from a flat die onto a polishing roll. Not to be confused with lay-flat film.

Flat finish *See flat paint (finish).*

Flat grain *n.* Wood or veneer so sawed that the annual rings form an angle of less than 45° with the surface of the piece.

Flat knit fabric *n.* (1) A fabric made on a flat-knitting machine, as distinguished from tubular fabrics made on a circular-knitting machine. While tricot and milanese warp-knit fabrics (non-run) are knit in flat form, the trade uses the term flat-knit fabric to refer to weft-knits fabrics made on a flat machine, rather than warp-knit fabrics. (2) A term used in the underwear trade for plain stitch fabrics made on a circular-knitting machine. These fabrics have a flat surface and are often called flat-knit fabrics to differentiate them from ribbed-knit or Swiss rib fabrics. In this case, the term refers to the texture, not the type of machine on which the fabric was knit. Kadolph SJJ, Langford AL (2001) Textiles. Pearson Education, New York. Vincenti R (ed) (1994) Elsevier's textile dictionary. Elsevier Science and Technology Books, New York.

Flat knitting *See knitting.*

Flat knitting machine *n.* A weft-knitting machine with needles arranged in a straight line in a flat plate called the bed. The yarn travels alternately back and forth, and the fabric may be shaped or varied in width, as desired, during the knitting process. Lengthwise edges are selvages. Flat-knitting machines may be divided into two types: latch-needle machines for sweaters, scarves, and similar articles and fine spring-needle machines for full-fashioned hosiery.

Flat lacquer *n.* A lacquer having the appearance of having been rubbed after it has dried.

Flat paint (finish) *n.* Paint which dries to a surface which scatters the light falling on it, so as to be substantially free from gloss or sheen.

See matte finish. Also called flat finish.

Flat paint brush *See flat wall brush.*

Flat spot *n.* An imperfection on a glossy painted surface; a spot lacking gloss, usually caused by a porous spot on the undercoat.

Flat spotting *n.* A characteristic of certain tire cords. It occurs with all materials but

F

is more noticeable with nylon cord and is associated with nylon cord by users. Nylon exerts a shrinkage force as it becomes heated in tire operation. When the tire is stopped under load, the cord in the road-contact portion of the tire is under less tension than that in other portions of the tire, and it shrinks to conform to the flat surface of the road. When cooled in this position, the cord maintains the flat spot until it again reaches its glass transition temperature in use.

Flat stone mill *n.* Type of grinding mill in which the material to be disintegrated is fed between the grinding surfaces of two flat stones, one of which is caused to rotate. The grinding surfaces are specifically prepared with grooves or channels. These mills were used for either dry or wet grinding.

Flattening agent *See flatting agent.*

Flatting agent *n.* Material added to paints, varnishes and other coating materials to reduce the gloss of the dried film. Any material which, when added to a paint or plastic, lowers the gloss of the final surface. *See flatting pigment.*

Flatting down *vt.* Cutting or rubbing down the surface of a paint or varnish with fine abrasives to produce a smooth, dull surface.

Flatting oil A varnish-like composition made of heavy-bodied oil dissolved in a thinner, used to reduce paste paint to a flat paint.

Flatting or flattening *vt.* (1) Undesirable loss of gloss during drying. (2) Addition of a flatting pigment or agent to a paint or varnish. (3) Addition of a flatting oil to a paint or enamel.

Flatting pigment *n.* Any finely divided particle added to a paint formulation in order to decrease the gloss of the dried film. It may be an extruder-type (low refractive index) or a hiding pigment, and it is generally non-chromatic (neutral near-white), although not necessarily so. The reduction in gloss comes about from surface-light scattering, which occurs when light strikes the pigments protruding at the surface.
See extender (pigment) and dry-hiding.

Flatting (rubbing) varnish *n.* Varnish containing a high proportion of hard resin which can be rubbed down after application, to produce a smooth foundation for a finishing coat, or serve as the finish coat itself (not to be confused with a flat varnish). *Also known as rubbing varnish and polishing varnish.*

Flat top card *See flat card.*

Flat varnish *n.* Varnish formulated with a flatting agent so a to dry with a dull finish (not to be confused with a flatting or rubbing varnish). Syn: matte varnish.

Flat wall brush *n.* A paintbrush, usually 4–6 in. (10–15 cm) in width, with long, stiff bristles, usually made of synthetic fiber. Syn: flat paint brush.

Flax \ˈflaks\ [ME, fr. OD *fleax*; akin to OHGr *flahs* flax, L *plectere* to braid] (before 12c) *n.* The plant from which the cellulosic fiber linen is obtained.

Fleece fabric *n.* A fabric with a thick, heavy surface resembling sheep's wool. It may be a pile or napped fabric of either woven or knit construction.

Fleming's rule *n.* A simple rule for relating the directions of the flux, motion, and emf in an electric machine. The forefinger, second finger and thumb, placed at right-angles to each other, represent respectively the directions of flux, emf, and motion or torque. If the right hand is used the conditions are those obtaining in a generator and if the left hand is used the conditions are those obtaining in a motor. Weast RC (ed) (1978) CRC handbook of chemistry and physics, 59th edn. The Chemical Rubber Co., Boca Raton, FL.

Fleshing *See flushing.*

Flex abrasion tester *See Stoll–quartermaster universal wear tester.*

Flex-cracking *n.* Development of small cracks in flexible articles or coatings when these articles are subjected to repeated flexing or bending.

Flexibility *n.* (1) Degree to which a coating after drying is able to conform to movement or deformation of its supporting surface, without cracking or flaking. (2) A term relating to the hand of fabric, referring to ease of bending and ranging from pliable (high) to stiff (low).

Flexibility test *n.* Test applied to films to ascertain if they are able to accommodate elongations without fracture. Flexibility is usually determined by bending the film, applied to a suitable piece of thin metal, around one or a series of rods or mandrels or a special conical mandrel in a specified time. Syn: bend test.

Flexibilizer *n.* A term rarely used for an additive that makes a plastic more flexible. *See plasticizer.*

Flexible foam *See polyethylene foam, polyurethane foam, and vinyl foam.*

Flexible-lip die *n.* In film and sheet extrusion, a die in which a deep groove, reaching almost to the inside surface of the upper die body just behind the lip, has been machined. Adjusting bolts that can either push or pull on the lip pass through the gap from the upper body. The relative flexibility of the thin steel web from which the upper lip extends facilitates die adjustment to minimize sheet-thickness variations across the sheet.

See also auto-flex die.

Flexible mold *n.* A mold made of rubber, elastomer, or flexible thermoplastic, used for casting thermosetting plastics or other materials such as concrete and plaster. The mold can be stretched to permit removal of the cured casting, even one with undercuts.

Flexible wall coverings *n.* Those which are pliable, such as paper, manmade vinyls and fabrics, as opposed to "liquid" wall coatings, such as paint, or "rigid" like wood paneling or other solid wall products such as ceramic tile.

Flex life *n.* Informally, the number of bending-reversal cycles causing a part to fail in a particular service. Most specifically, the number of cycles to failure of a test specimen repeatedly bent in a prescribed manner. The ASTM test for plastics is D 671. The specimen, molded or cut from sheet, is subjected to load reversal at 30 Hz at a predetermined level of outer-fiber stress until it either fails or the test is discontinued. By setting up different stresses for successive specimens, one can develop a graph of stress at failure versus number of cycles to failure (usually plotted on semi-logarithmic coordinates), i.e., the flex-life curve of fatigue curve.

Flexographic decorating *n.* Printing technique by which images are transformed from a flexible raised plate directly to the material. Flexographic printing presses can process many different materials, including polyethylene, polypropylene, oriented polypropylene, polyester, and nylon. Film thicknesses range from 0.0004 to 0.008 in.

Flexographic ink *n.* Quick drying, low viscosity ink based on volatile solvents that are used in the flexographic printing process.

Flexographic printing *n.* A rotary process employing flexible rubber or elastomeric printing plates adhered to a roll, inked by a screen roll which in turn is coated from a feed roll immersed in ink.

Flexography \flek-ˈsä-grə-fē\ [*flexible* + *-o-* + *-graphy*] (1954) *n.* A typographic form of printing using rubber plates and relatively thin bodied resin-solvent inks. Formerly known as aniline printing.

Flexural fatigue *n.* A physical property expressed by the number of times a material can be bent on itself through a prescribed angle before it ruptures or loses its ability to recover.

Flexural modulus (flex modulus) *n.* The ratio, within the elastic limit, of the applied stress in the outermost fibers of a test specimen in three-point, static flexure, to the calculated strain in those outermost fibers, according to ASTM test D 790 or D 790M. For a given material and similar specimen dimensions and manufacture, the modulus values obtained will usually be a little higher than those found in a tensile test such as D 638, and may differ, too, from the moduli found in the cantilever-beam test, D 747.

Flexural rigidity *n.* This measure of a material's resistance to bending is calculated by multiplying the material's weight per unit area by the cube of its bending length.

Flexural strength (flexural modulus of rupture) *n* The maximum calculated stress in the outermost fibers of a test bar subjected to three-point loading at the moment of cracking or breaking. ASTM test D 790 and D 790M are widely used for measuring this property. For most plastics, flexural strength is usually substantially higher than the straight tensile strength. Also called modulus of rupture, bending strength.

Flight *n.* In an extruder screw, the helical ridge of metal remaining after machining the screw channel.

Flight clearance *n.* In screw extruders, half the difference between the inside diameter of the barrel and the diameter of the flight surface, usually about 0.1% of the nominal diameter in a new machine. Clearances in older machines may vary along the screw, intentionally or because of differential wear.

Flight depth (screw depth, channel depth) *n.* In screw extruders, the radial dimension, at any point along the screw, from the flight surface to the screw root. Some users of the term have taken it to mean half the difference between the internal diameter of the barrel and the screw-root diameter, but this difference is larger than the true flight depth by the amount of radial clearance between flight and barrel. In most (but not all) single-screw extruders, flight depth is much greater at the feed end, decreasing toward the delivery end of the extruder, where it is usually constant for at least several flights.

See channel–depth ratio and extruder screw.

Flint abrasive *n.* Not flint at all, but actually a natural quartz (silicon dioxide), which fractures into sharp-edged grains and is used on the common sandpaper for wood.

Flint paper *n.* An inexpensive paper using a natural quartz as the abrasive, for small sanding jobs on wood. It tends to clog quickly and cannot be used for wet sanding.

Flip-flop *n.* Decorative effect with automotive metallized finishes in which there is a difference in color depth and flash when the finish is viewed at different angles. Generally used in application to automotive metallic finishes.

See travel.

Flitter \ˈfil-tər\ *n, vt.* Syn: glitter.

Float *n.* (1) The portion of a warp or filling yarn that extends over two or more adjacent filling picks or warp ends in weaving for the purpose of forming certain designs. (2) In a knit fabric, a portion of yarn that extends for some length without being knitted in. (3) A fabric defect consisting of an end lying or floating on the cloth surface instead of being woven in properly. Floats are usually caused by slubs, knot-tails, knots, or fly waste, or sometimes by ends being drawn in heddle eyes incorrectly or being twisted around heddle wires.

Floating *vt.* Defect, which is sometimes apparent in colored paints containing one or more mixtures of different pigments. During drying or on storage, one or more of the pigments separates or floats apart from the others and concentrates in streaks or patches on the surface of the paint, producing a variegated effect. This effect also occurs with single pigment type consisting of multiple particle sizes.
See flooding.

Floating chase *n.* A mold member, free to move vertically, that fits over a lower plug or cavity, and into which an upper plug telescopes.

Floating coat *See brown coat.*

Floating ends *See float.*

Floating platen *n.* In compression molding, a platen located between the main head and the press table in a multi-daylight press and capable of being moved independently of them.

Floating punch *n.* A male mold member attached to the head of a press in such a manner that it is free to align itself in the female part of the mold when the mold is being closed.

Floats *n.* A term used in the past for asbestos filler in the form of very fine, short fibers with associated dust.

Float stitch *See miss-switch.*

Flocculate \ˈflä-kyə-ˌlāt\ (1877) *v.* Cluster of pigment particles formed in paint after the pigment has been wetted or dispersed. The spaces between the particles are filled with vehicle, as distinguished from agglomerates, where the spaces between the particles are filled with air. A group of two or more attached particles held together by physical force such as surface tension, adsorption, or similar forces. Flocculates are much affected by the sample preparation and technique.

Flocculation *n.* The formulation of clusters of pigment particles in a fluid medium, which may occur after dispersion has been effected. The condition is usually reversible and the particle clusters can be broken up by the application of relatively weak mechanical forces or by a change in the physical forces at the interface between the liquid and the solid dispersed particles. Flocculation is often visible, as a "Jack Frost" pattern in a flow-out of a dispersion; microscopically, it appears as a lacework or reticulum of loosely clustered particles. It results in more rapid settling although it is usually soft, shows loss of color strength and poor dispersion. A flocculated dispersion of sufficient pigment concentration shows yield value. Surface active agents are often useful in reducing the extent of flocculation and hence the yield value.
See agglomeration.

Flocculation of the vehicle *n.* In the printing ink industry, sometimes used to mean livering.

Flock \ˈfläk\ [ME *flok*, fr. OF *floc*, fr. L *floccus*] (13c) *n.* (1) Short fibers of cotton or synthetic fibers such as polyester, acrylic, or nylon. They are used as reinforcements in phenolic, allylic, and other thermosetting molding compounds, also for decorating plastics by the process of flocking. (2) Wall coverings imitating the surface of damask or cut velvet. Made by shaking finely chopped fibers over a pattern printed in varnish or some other sticky material.

Flock finish *n.* Finish produced by the application on a suitable adhesive coat of short fiber cotton, wood, or other fiber, giving a soft suede-like feel and appearance.

Flock gum Special spray gun for applying flock finishes.

Flocking (flock coating) *vt.* A method of finishing sometimes employed for plastics articles whereby the article is coated with a tacky, slow-drying adhesive, then is dusted with a fibrous material cut into very short lengths to give a finish resembling suede, plush, etc. Fibers for flocking are available in a wide range of materials including acrylic, nylon, polyester, polyolefins, and natural fibers. Machinery for flocking films and fabrics includes gravure printing stations for applying the adhesive in desired patterns and flock heads that distribute a precalculated layer of flock to the web, and retrieve and recirculate surplus flock.

Flood feeding *vt.* The usual way of feeding an extruder or screw-injection molder, in which the feed material flows from the feed hopper by gravity and completely fills the feed section of the screw. The actual throughput is thus controlled by screw design, die resistance and temperature conditions within the screw, in contrast to what occurs in starve feeding.

Flooding *vt.* (1) A concentration at the surface of a paint film of one of the ingredients of the pigment portion, giving rise to a change in color at the surface. (2) A differential separation of pigments in a dispersed pigment mixture. Floating is a differential separation of pigments (a) by gravity separation in the bulk dispersion; and (b) where the flooding results in a nonuniform or mottled surface coloration. While the terms "flooding" and "floating" are often used interchangeably, they also have been defined as separate, though related, phenomena. "Flooding" refers to uniform color changes and "floating" refers to local excess of one color. (3) Process of color change in which a paint or enamel undergoes homogeneously, from the freshly applied material to the finished dried film in which one or more pigments appear to be "flooding" the surface color. (4) An excess of ink on the printing plate caused by the ink fountain being open too much. In the case of lithographic or offset work, the use of too little water or the absence of an etching material in the water fountain.

See floating.

Floor and deck enamel *n.* An enamel designed for abrasion resistance.

Floor paint *See deck paint.*

Floor sealer *n.* Composition of resins, with or without oils, in a solvent; designed to penetrate the wood rather than produce a surface finish.

Floor varnishes *n.* Varnishes, which are formulated for the coating of flooring. Their main properties include rapid drying, tack-free surface, toughness and resistance to abrasion, washability, and receptiveness to wax and other floor polishes.

Flop Where two different painted panels appear to be a good match for color when viewed at a given angle, but appear different at all other angles.

Also known as geometric metamerism.

Flop color *n.* Used to describe the color observed on paints (particularly metallic paints) when they are viewed near to the grazing angle, i.e., nearly parallel to the surface. The term is generally used in contrast to the terms "face color" or "top color," the color observed when viewing the surface oat an angle near to the perpendicular.

See face color.

Florals (1897) *n.* Any wall covering design featuring recognizable flowers and foliage.

Florentine lake *See crimson lake.*

Flory–Fox theory (or Fox–Flory) *n.* This theory is relates viscosity to molecular dimensions by treating the polymer molecule as a hydrodynamic sphere. In a θ-solvent in which the molecular coil is compact, these authors write the intrinsic viscosity as,

$$[\eta] = KM^{0.5},$$

where K is a constant. When the polymer is dissolved in a solvent that not a θ-solvent, the equation is replaced with

$$[\eta] = KM^{0.5}\alpha_\eta,$$

where α_η is an expansions factor that takes into account the expansion of the polymer coil in going from a bad solvent to a good one. Also, the expansion in a good solvent is accompanied by an increase in viscosity (η) resulting expansion of the polymer molecule due to greater interactions between solvent and polymer.

Flory–Huggins parameter *n.* For polymers dissolved in solvents (may be monomers of the polymer), the Flory–Huggins parameter χ a measure of the interaction energy $\Delta\varepsilon$ (average energy gain per contact; $\Delta\varepsilon$ is a measure of the Gibbs energy and not of the enthalpy; and χ also contains an entropy contribution, which is often found to depend on the concentration of polymer or solute. Flory PJ (1953) Principles of polymer science. The Cornell University Press, Ithaca, New York.

Flory–Huggins theory *n.* (1) A thermodynamic theory of polymer solutions, first formulated independently by Flory and Huggins (in which the thermodynamic quantities of the solution are derived from a simple concept of combinational entropy of mixing and a reduced Gibbs energy parameter, the "χ parameter" (a dimensionless quantity). The Flory–Huggins theory lead to the equation,

$$\Delta F_1 = RT \ln a_1 = RT[\ln(1 - v_2) + (1 - 1/m)v_2 + \chi_2^2] \qquad \text{(a)}$$

for the partial free energy of the solvent, and to the corresponding equation for for the partial free energy of the polymer,

$$\Delta F_2 = RT[\ln v_2 - (m - 1)v_1 + m\chi v_1^2] \qquad \text{(b)}$$

When the logarithmic term in equation (a) is expressed as a series in v_2 and m is replaced by the degree of polymerization, P, the convenient equation results,

$$\Delta F_1 = RT[-(v_2/P) - (0.5 - \chi)v_2^2 - (v_2^3/3)\cdots]. \qquad \text{(c)}$$

A numerical parameter, χ, employed by the Flory–Huggins equation accounts for the contribution of the non-combinational entropy of mixing and for the enthalpy of mixing (Hans-Georg Elias, *Macromolecules*, Plenum Press, New York, 1977). The equation assumes that the $\Delta V_{mix} = 0$, and the enthalpy of mixing does not influence the value of ΔS_{mix}. The equation represents Flory–Huggins free energy of mixing per molecule. (2) Historical development of theory and equation: Polymer solutions show enormous deviations from Raoult's law. Polymer chemists early realized that the large deviations from the ideal behavior predicted by Raoult's law resulted from the flexibility of polymer molecules. A polymer molecule continuously changes shape, and, in the course of time, takes on a tremendous number of conformations. Therefore, when a polymer is dissolved, there is, in addition to entropy of solution, which results from simple interchange of position, conformational entropy. Flory and Huggins first worked out the statistical theory of polymer solutions, which took conformational entropy into account. They chose a lattice picture, widely used in statistical theories of simple liquids, as a model for the polymer solution. In the lattice picture, the space occupied by the solution is divided into boxes of equal size, and polymer segments and solvent molecules are fitted into these boxes. The number of ways, Ω that n_2 polymer molecules, which occupy n_2 sets of m contiguous boxes, can be arranged in a lattice can be computed by classical statistics. The number Ω is related to the entropy of mixing by Boltzmann's equation, $\Delta S = k \ln \Omega$, where k is Boltzmann's constant; if the entropy computed by this equation represents all of the entropy of mixing, and if ΔH^{m} is assumed to be positive and, following Van Laar, is set equal to $kt\chi n_1 v_2$, the free energy of mixing is given by,

$$\begin{aligned}\Delta F^{\mathrm{m}} &= -T\Delta S^{\mathrm{m}} + \Delta H^{\mathrm{m}} \\ &= kT(n_1 \ln v_1 + n_2 \ln v_2 + \chi n_1 v_2),\end{aligned}$$

where v_1 and v_2 are the volume fractions of solvent and polymer, respectively, n_1 and n_2 are the number of molecules of solvent and polymer, and χ is a dimensionless quantity called an interaction parameter that is related to the interaction energy characteristic of a particular polymer-solvent combination. Mark JE (ed) (1996) Physical properties of polymers handbook. Springer-Verlag, New York. Flory PJ (1969) Statistical mechanics of chain molecules. Interscience Publishers Inc., New York. Huggins ML (1958) Physical chemistry of high polymers. John Wiley and Sons Inc., New York. Flory PJ (1953) Principles of polymer science. The Cornell University Press, Ithaca, New York.

Flory–Krigbaum theory *n.* A theory of dilute polymer solutions which describes the distribution of polymer segments as clusters of segments separated by regions of pure solvent. A model for this structure is used which takes account of the excluded volume within the coils.

Flory–Rehner equation *n.* The correlation between thermodynamic parameters and the value of an equilibrium swelling of a polymer by a solvent is given by this equation:

$$\begin{aligned}&\ln(1-\varphi_2) + \varphi_2 + \chi_1\varphi_2^2 \\ &\qquad = -\frac{v_2}{V} Vs\left(\varphi_2^{1/3} - \frac{2\varphi_2}{f}\right),\end{aligned}$$

where φ_2 is the polymer volume fraction in a swollen sample, v_2/V the volume fraction of elasticity active chains, f the functionality, and χ the interaction parameter. The value of equilibrium swelling can be a practical criterion of solubility. The presence of even a small amount of cross-links hinders chain separation and polymer diffusion into solution. Solvent can penetrate into the polymer and cause swelling. Wypych G (ed) (2001) Handbook of solvents. Chemtec Publishing, New York.

Flory temperature *n.* The temperature at which, for a given polymer-solvent pair, the polymer exists in its unperturbed dimensions. Syn: theta temperature. Elias HG (1977) Macromolecules, vols 1–2, Plenum Press, New York. Flory PJ (1960) The statistical thermodynamics of solutions. John Wiley and Sons Inc., New York.

Flow \ˈflō\ [ME, fr. OE *flōwan*, akin to OHGr *flouwen* to rise, wash, L *pluere* to rain, Gk *plein* to sail, float] (before 12c) *v.* (1) Resistance to movement by a liquid material and divided rheologically into four categories: Newtonian (simple) flow, plastic flow, pseudoplastic flow, and dilatant flow.

(2) Movement of a coating during and after application and before the film is formed. (3) Property which a coating or ink possesses of leveling after application. Coatings may possess excessive flowing properties and may sag or run to an undesirable degree from vertical or inclined surfaces. (4) Inks of poor flow are classed as short or buttery in body, while inks of good flow are said to be long. Gooch JW (1997) Analysis and deformulation of polymeric materials. Plenum Press, New York. Patton TC (1965) Paint flow and pigment dispersion. Interscience Publishers Inc, New York. Eirich FR. Rheology – theory and applications, vol 1 (1956), vol 2 (1958), vol 3 (1967), vol 4 (1967), vol 5 (1969). Academic Press Inc., New York. Green H (1949) Industrial rheology and rheological structures. John Wiley and Sons Inc., New York. Mercurio A (1964) Rheology of acrylic paint resins. Canadian Paint and Varnish, ON.

See leveling.

Flow-behavior index *See power law.*

Flow birefrigence (streaming birefringence) *n.* The difference, Δn between the refractive indices of a flowing polymer solution or melt in the direction of flow and a direction perpendicular to the flow. Usually measured, using one of two techniques, by directing a light beam downward through the liquid in a rotational viscometer. The amount of birefringence is related to the degree of orientation of the polymer chains, in turn related to shear stress and first normal-stress difference in the flowing medium. The measurements are useful for testing molecular theories and rheological models, also for understanding processing problems such as extrudate roughness.

Flow coating *n.* A painting process in which the article to be painted is drenched with a paint, either by pouring or by spraying with a mist in a closed or semi-closed chamber. The parts are sometimes rotated during and after drenching to avoid sags and runs. The process is used for coating metallized parts and other irregularly shaped articles that are difficult to paint by ordinary spraying methods.

Flow curve *n.* A plot of applied force against the resulting rate of flow. A more in-depth definition is a graph, usually on bilogarithmic coordinates, of shear stress versus shear rate or, sometimes, of apparent or true (corrected) viscosity versus either shear rate or shear stress. For nearly all polymer melts, log–log plots of shear stress versus shear rate of sufficient range exhibit a Newtonian region of slope = 1 at extremely low shear rates, a brief transition region of decreasing slope ("knee"), and a higher-shear region in which the slope is nearly constant or very gradually decreasing and is in the range 0.7–0.25.

See also pseudoplastic fluid and viscosity.

Flower of antimony *See antimony oxide.*

Flowers of antimony *n.* Syn: antimony trioxide.

Flowing varnish *n.* A varnish designed to produce a smooth lustrous surface without rubbing or polishing.

Flow line Syn: weld line.

Flow mark *n.* A defect in a molded article characterized by a way surface appearance, caused by improper flow of the resin into the mold, which itself may have a number of causes.

Flow molding *n.* (1) A variation of injection molding used for thick-walled parts. A large gate is used and pressure is maintained on the injected melt so as to fore a little more melt into the mold as each part solidifies from the outside inward, thus minimizing shrinkage and improving consistency of the parts' final dimensions. (2) A process of heating material such as a cloth-backed, plasticized PVC with a high-frequency electric field while pressing the material into a mold made of silicone rubber. In 10–15 s, the PVC is formed to the contours and surface texture of the mold, which can give it the look of hand-tooled leather or similar effects.

Flow properties *See melt-flow index, viscosity, pseudo-plastic fluid, and rheology.*

Flue \ˈflü\ (1582) *n.* A passageway in a chimney for conveying smoke, gases or fumes to the outside air.

Fluffing (1875) *v.* A term describing the appearance of a carpet after loose fiber fragments left during manufacture have worked their way to the surface. Fluffing is not a defect; it is simply a characteristic of new carpets that disappears with vacuuming.

Fluid \ˈflü-əd\ (1661) *n.* A gas or a liquid, or, in the supercritical region, a hybrid.

Fluidity \flu-ˈi-də-tē\ (1603) *n.* (1) The ease with which a liquid flows under stress. (2) Specifically, the reciprocal of viscosity. The SI unit of fluidity is $(\mathrm{Pa\,s})^{-1}$ or 1/Pa s, replacing the deprecated cgs unit, the rhe. 1 rhe $= 10\ (\mathrm{Pa\,s})^{-1}$ (3) The reciprocal of viscosity. The cgs unit is the rhe, the reciprocal of the poise. Dimensions $[\mathrm{M}^{-1}\,\mathrm{L}\,\mathrm{T}]$.

Fluidization (ca. 1855) *vt.* A gas-solid or liquid–solid contacting process in which a stream of fluid is passed upwards through a bed of small solid particles, causing them to lift, expand, and behave as a boiling liquid. The process is widely used in the chemical industry for performing reactions in which the solid is either a reactant or a catalyst. In the plastics industry, the main application is in fluidized-bed coating.

Fluidized bed *n.* An expanded bed of solid particles, fluidized to create a system with similar properties of a liquid. Fluidized beds are used successfully in a multitude of processes both catalytic and non-catalytic. Among the catalytic uses are hydrocarbon cracking and reforming, oxidation of naphthalene to phthalic anhydride, and ammoxidation of propylene to acrylonitrile. A few of the non-catalytic uses are coating of sulfide ores, coking of petroleum residues, drying, and classification.

Fluidized-bed coating *n.* The process of applying plastics coatings to objects of other, higher-melting materials, often metals, wherein a powdered resin is placed in a container provided with a porous or perforated bottom through which a gas is directed upward to keep the resin particles in a state of agitated levitation. The part to be coated is preheated above the resin's softening temperature and lowered into the fluidized bed until a deposit of the desired thickness has formed, then the part is withdrawn and allowed to cool.

Fluorescein dye *See acid dyes.*

Fluorescence \flú-ˈre-sᵊn(t)s\ [*fluor*spar + opal*escence*] (1852) *n.* Fluorescence is a luminescence that is mostly found as an optical phenomenon in cold bodies, in which a molecule asbsorbs a high-energy photons (low wavelength such as ultraviolet) and emits photons at as a lower-energy photon (longer-wavelength) and usually visible light. The energy difference between the absorbed and emitted photons ($h\nu$) manifests itself as molecular vibrations or heat in the absorbing matter Usually the absorbed photon is ultraviolet and the emitted light (luminescence) is in the visible range so that humans can observe, but this depends on the absorbance curve and Stokes shift of the particular fluorosphore. Fluorescence is named after the mineral fluorite (calcium fluoride) that exhibits this phenomenon the expression below shows the fluorescence phenomenon,

$$S1 \rightarrow S2 + hv,$$

where, h is the Planck's constant, and v is the frequency of the fluorescing light. The Kasha–Vavilov rule describes the quantum yield of luminescence that is independent of the wavelength of exciting radiation, and the Jablonski diagram describes most the relaxation mechanism for excited state molecules. Serway RA, Faugh JS, Bennett CV (2005) College physics. Thomas, New York, www.Wikipedia.org.

Fluorescent brightening agent *See brightening agents.*

Fluorescent dyes *n.* luminescence, which is mostly found as an According to the Paint/Coatings Dictionary, dyestuffs which exhibit the phenomenon of fluorescence in the visible region of the spectrum. Dyes are characterized as fluorescent if: they absorb invisible radiation, such as ultraviolet and emit visible light; they also absorb primarily visible light and emit visible light; or they absorb visible light and emit invisible radiation, generally infrared. Some dyes fluoresce more strongly than others, i.e., some are only slight fluorescent and some are strongly fluorescent. The former may go undetected as fluorescent dyes. In order to measure the color materials containing fluorescent dyes, the material must be illuminated with light of defined spectral quality, and the light reflected or transmitted by the materials must be passed through the monochromating device after reflectance or transmission, using so-called "reversed optics." Strongly fluorescent pigments are frequently referred to as high visibility pigments; their combined emission and reflectance is generally much higher at certain wavelengths than the reflectance of the perfect diffuser. Paint/Coatings Dictionary. Compiled by Definitions Committee of the Federation of Societies for Coatings Technology, 1978. Morgans WM (1977) Pigments for paints and inks. Selection and Industrial Training Administration Ltd., London. Lubs HA (ed) (1965) The chemistry of synthetic dyes and pigments (ACS Monograph). Hafner Publishing Co., New York.

F

Fluorescent inks *n.* Inks which exhibit fluorescence, resulting in a very brilliant effect.

Fluorescent paint *n. See luminous paint and high visibility paints.*

Fluorescent pigment *n.* (1) Inorganic, usually coarse, crystalline, and materials, which emit light when activated by ultraviolet radiation, e.g., zinc sulfide, combined zinc, and cadmium sulfides. A type known as *daylight fluorescent pigments* (*dayglo pigments*) responds to radiation in both the ultraviolet and visible ranges, causing the effect of glowing in normal daylight. These pigments are comprised of fluorescent dyes incorporated in a clear-resin matrix, ground to powder form. Urea and melamine resins have been used as matrices, also a modified sulfonamide resin. (2) Organic dyes or pigments used to manufacture luminescent materials. These include spirit-soluble dyes, oil-soluble dyes, and metallic organic compounds, e.g., rhodamine, eosine, rhodamine tungstate, and zinc salt of 8-hydroxyquinoline. By absorbing unwanted wavelengths of light and converting them into light of desired wavelength, these pigments have the appearance of possessing an actual glow of their own. (3) Pigments, which exhibit the phenomenon of fluorescence in the visible region of the spectrum. Few commercial mineral pigments are fluorescent; most of the fluorescent pigments are made from fluorescent dyes incorporated into an insoluble matrix and ground to a small particle size suitable for use as a pigment.

See fluorescent dyes.

Fluorinated ethylene-propylene resin (FEP, PFEP) *n.* This member of the fluorocarbon family is a copolymer of tetrafluoroethylene and hexafluoropropylene, possessing most of the desirable properties of PTFE, yet truly meltable and, therefore, processable in conventional extrusion and injection-molding equipment. It is available in pellet form for those operations and as dispersions for spraying and dipping.

Fluorine-containing polymers *n.* Polymers in, which one or more hydrogen atoms have been replaced by fluorine. Often used in high-temperature wire and cable insulation, due to their resistance to chemicals and oxidation and broad useful temperature range.

Fluorite objective *n.* This objective is corrected both for spherical aberration and chromatic aberration at two wavelengths.

Fluorocarbon \ˌflúr-ō-ˈkär-bən\ (1937) *n.* Teflon tetrafluoroethylene (TFE) and Teflon fluorinated ethylene propylene (FEP). Both have remarkable chemical resistance.

Fluorocarbon blowing agent *n.* A family of inert, non-corrosive liquid compounds containing carbon, chlorine, and fluorine, originally developed as refrigerants. They are compatible with all resins and leave no residues in molds. For years they were widely used in structural-foam extrusion, in which they were incorporated with the polymer by direct injection through the barrel of the first of two tandem extruders. Fluorinated hydrocarbons are numbered by a three-digit system developed by DuPont for use with its trade name Freon®. The first digit is the number of carbon atoms in the molecule −1, omitted when it is 0 (the methane group), leaving just two digits for them. The second digit equals the number of hydrogens in the molecule +1; the third digit is the number of fluorine atoms. The remaining atoms in these saturated compounds are chlorine. Thus the once-common blowing agents were Freon 11 (trichlorofluoromethane), Freon 12 (dichlorodifluoro-methane), Freon 113 (trichlorotrifluoroethane), and Freon 114 (dichlorotetra-fluoroethane). Because there is strong evidence that these compounds when released to the atmosphere, migrate to the upper levels and catalyze the destruction of UV-blocking ozone by chain reactions, they are being phased out of use of international agreement and are being replaced, in the manufacture of plastics foams, by hydrocarbons, such as neopentane, and by hydrohalocarbons.

Fluorocarbon elastomer (fluoroelastomer) *n.* Any fluorocarbon polymer of low T_g and no crystallinity, therefore rubbery. As rubbers, these materials are resistant to high temperature and most chemicals and solvents. Most of the commercial materials are copolymers. In USA, DuPont's Viton® materials are most familiar.

Fluorocarbon resin *n.* Any of a family of thermoplastics chemically similar to the polyolefins, with all the hydrogen atoms replaced by fluorine. They are made by addition polymerization from olefinic monomers composed only of fluorine and carbon. The main members of the family are polytetrafluoroethylene, fluorinated ethylene-pro-pylene resin, and polyhexafluoropropylene.

Fluorocarbons *n.* Group of compounds containing fluorine atoms, fluoroplastics, or solvents.

Fluoroethylene Syn: vinyl fluoride.

Fluorohydrocarbon resin *n.* Any resin polymerized from an olefinic monomer composed of carbon, fluorine, and hydrogen only. Included are polyvinylidene fluoride, polyvinyl fluoride, poly-trifluorostyrene, and copolymers of halogenated and fluorinated ethylenes.

Fluoroplastic (fluoropolymer) *n.* A plastic based on polymers made from monomers containing one or more atoms of fluorine, or copolymers of such monomers with other monomers, the fluorine-containing monomer(s) being in the greatest amount by mass (ASTM D 883). This is a broad family including:

Chlorofluorocarbon resin
Chlorofluorohydrocarbon resin
Ethylene-chloro-tri-fluoroethylene copolymer
Ethylene-tetra-fluoroethylene copolymer
Fluorinated ethylene-propylene resin
Fluorocarbon resin
Fluorohydrocarbon resin
Perfluoroalkoxy resin
Polychlorotrifluoroethylene
Polytetrafluoroethylene
Polyvinylidene fluoride

Fluoroplastics *n.* Plastics, which contain the monomer of fluorine.
See fluorocarbons.

Fluoropolymers *n.* Polymers whose repeating units contain fluorine. Such polymers often have outstanding thermal, thermo-oxidative, and chemical resistance, both to chemical attack and to swelling by solvents.

Flushed color *n.* A color base in paste form prepared by flushing.

Flushed (flushing) colors *n.* Pigments which have been flushed.
See flushed pigment and flushing.

Flushed pigment *n.* A pigment obtained by the direct transfer of pigment particles that have been precipitated, from an aqueous phase to a non-aqueous stage (no intermediate drying or pulverizing stage being involved). In the process, the press cake (aqueous or water-wet pigment phase) is mixed and agitated with a non-aqueous vehicle (oil, solvent, and/or resin phase) in a heavy duty mixer. The pigment particles preferentially transfer (flush) to the non-aquerous phase, and the bulk of the essentially clear water is poured off. Any residual water is then vacuumed off at a temperature of about 50°C (122°F).

Flushing *v.* A method of transferring pigments from dispersions in water to dispersions in oil by displacement of the water by oil. The resulting dispersions are known as flushed colors. Syn: fleshing.
See flushed pigment.

Fluted core An integrally woven reinforcing material consisting of ribs between two skins, thus providing unitized sandwich construction.

Fluted mixing section In extrusion, a screw section in which several short barrier flights are placed so that the plastic must flow through the high-shear clearance between the flight tip and the barrel. Fluted sections may have the flights parallel to the screw axis or at an angle to the axis.

Fluting *n* (1611). Parallel concave grooves used as furniture decoration. Vertical and spiral flutings are common.

Flux \ˈfləks\ [ME, fr. MF & ML; MF, fr. ML *fluxus*, fr. L, flow, fr. *fluere* to flow] (14c) *n.* (1) In chemistry and metallurgy, a substance, e.g., borax or fluorspar, used to promote fusion of metals or minerals. (2) In plastics compounding, the term flux is sometimes used for an additive that improves flow properties, e.g., coumarone-indene resin in the milling of vinyl compounds. (3) In heat and mass transfer, the rate of transfer per unit of cross-sectional area perpendicular to the direction of transfer. (4) Luminous flux.

Flux (15c) *v.* To melt, fuse, or make liquid. In the early years of the vinyl-plastisol art this term was often used for "fuse" before the latter term came into general use.

Fluxing temperature Syn: fusion temperature.

Flux, luminous Radiant flux weighted for the relative luminous (visual) efficiency.

Flux, radiant *See radiant flux.*

F

Fly *n.* The short, waste fibers that are released into the air in textile processing operations such as picking, carding, spinning, and weaving.

Flyer *n.* (1) A device used to insert twist into subbing, roving, or yarn, and to serve as a guide for winding it onto a bobbin. The flyer is shaped like an inverted U that fits on the top of the spindle and revolves with it. One arm of the U is solid and the other is hollow. The yarn enters through the top of the hollow arm, travels downward, and emerges at the bottom where it is wound around a presser finger onto the take-up package. (2) *See loom fly.*

Flyer spinning *n.* A method of spinning by means of a driven flyer. It is used primarily for spinning worsted and coarser yarns. *Also see flyer* (1).

Flyer spinning frame *See spinning frame.*

Flyer waste *n.* During the roving operation, flyer waste refers to fibers that free themselves by centrifugal force from the regular bulk of roving and accumulate on the flyers and adjacent machinery.

Fly frame *See roving frame.*

Flying *n.* A condition wherein a fine mist or spray of ink is thrown off rapidly moving ink rollers. Syn: misting and spraying.

FMC color difference equations *n.* Letters "FMC" are the initials of Friele, MacAdam, and Chickering. There are two different equations, referred to as FMC-1 and FMC-2: FMC-1 is the same as FMC-2, except that the terms K_1 and K_2 are not included: FMC-2 is:

$$\Delta E = \left[(\Delta C)^2 + (\Delta L)^2\right]^{1/2},$$

where

$$\begin{aligned}
\Delta L &= K_2\,\Delta L_2,\\
\Delta C_1 &= \left[(\Delta C_{rg}/a)^2 + (\Delta C_{yb}/b)^2\right]^{1/2},\\
\Delta C_{rg} &= (Q\,\Delta P - P\,\Delta Q)/(P^2 + Q^2),\\
\Delta C_{yb} &= S\,\Delta L_1/(P^2 + Q^2)^{1/2} - \Delta S,\\
\Delta L_1 &= (P\,\Delta P + Q\,\Delta Q)/(P^2 + Q^2)^{1/2},\\
\Delta L_2 &= 0.279\,\Delta L_1/a,\\
K_1 &= 5.5669\times 10^{-1} + 4.9434\times 10^{-2}Y\\
&\quad - 8.2575\times 10^{-4}Y^2 + 7.9172\times 10^{-6}Y^3\\
&\quad - 3.0087\times 10^{-8}Y^4,\\
K_2 &= 1.7548\times 10^{-1} + 2.7556\times 10^{-2}Y\\
&\quad - 5.7262\times 10^{-4}Y^2 + 6.3893\times 10^{-6}Y^3\\
&\quad - 2.6731\times 10^{-8}Y^4,\\
a^2 &= 1.73\times 10^{-5}(P^2 + Q^2)/1\\
&\quad + 2.73P^2Q^2/(P^4 + Q^4),\\
b^2 &= 3.098\times 10^{-4}(S^2 + 2.015\times 10^{-1}Y^2),\\
P &= 0.724X + 0.382Y - 0.098Z,\\
Q &= -0.48X + 1.37Y + 0.1276Z,\\
S &= 0.686Z.
\end{aligned}$$

X, Y, Z are the tristimulus values for the standard, X_1, Y_1, Z_1 are the tristimulus values for the sample, ΔP, ΔQ, ΔS are the above formulate for P, Q, S substituting $\Delta X = X_1 - X$ for X, etc.

Foam \\ˈfōm\\ [ME *fome*, fr. OE *fām*; akin to OHGr *feim* foam, L *spuma* foam, *pumex* pumice] (before 12c) *n.* Dispersion of gas in a liquid or solid. The gas bubbles may be any size. The term covers a wide range of useful products such as insulating foam, cushions, etc. It also describes the undesirable froth in polymer melts, dyebaths, etc.

See cellular plastic.

Foam-backed *n.* A term describing a fabric laminated to or coated with a layer of rubber or plastic foam.

Foam casting (foam molding) *n.* A process withy many variations, depending on the polymers used. In general, a fluid resin or prepolymer containing catalyst is foamed before or during molding by mechanical frothing, or by gas dissolved in the mixture or vapor from a low-boiling liquid.

See also reaction injection molding.

Foamed plastic Plastics with an apparent density, which are significantly decreased by the presence of numerous cells disposed throughout its mass usually by an expanding gas foam blowing agent. Syn: cellular plastic.

Foam extrusion *See extruded foam.*

Foam fabrication The process of cutting large slabs, logs, or "buns" of foamed plastics into sections of desired dimensions. The raw slab is conveyed through an array of saws, knives, or hot wires that first remove the uneven top, bottom, and sides, then slice the remaining rectangular block into boards, finally cross-cutting the boards to standard lengths.

Foaming A dispersion of gas in a liquid or solid. Foams can be made by mechanically incorporating air, as in the food industry for whipped cream, egg whites, and ice cream.

Foaming agents (1) Materials that increase the stability of a suspension of gas bubbles in a liquid medium. (2) Blowing agents in rubber or plastics.

See blowing agents.

Foam-in-place Refers to deposition of foams at the site of the work, e.g., between two containing walls for insulation, as opposed to bringing the work to the foaming machine.

Fog \ˈfóg\ [prob. Scand origin, akin to Dn *fog* spray, shower] (1544) *n.* A colloidal dispersion of a liquid in a gas.

See bloom and blush.

Fogged coat Non-continuous spray coat.

See mist coat.

Fogged metal Metal, the luster of which has been sharply reduced by a film of corrosion products.

Fogging *See bloom and blushing.*

Foil [ME, leaf, fr. MF *foille* (fr. L *folia*, plural of *folium*) & *foil*, fr. L *folium*] (14c) *n.* (1) Refers to very thin membranes (less than 6 mils) of metal such as aluminum. Above 6 mils, the thin metal is called a sheet. (2) In wallcoverings, a very thin sheet of flexible metal on a paper or fabric back. Can be printed with transparent or opaque color, and mottled to resemble marble, tortoise shell, etc.

Foil decorating *See in-mold decorating.*

Folded selvage *n.* A curled selvage.

Folded yarn *See plied yarn.*

Folding machine *n.* A machine that folds sheet plastics such as cellulose acetate into shapes such as identification-card envelopes, sheets for ring binders, visible indexes, and the like. An electrically heated blade softens the plastic and folds it into a tight, 180° crease.

Fold testing *n.* Mechanical test to determine the ability of a polymer to retain its strength after being folded back and forth.

Food and drug administration (FDA) *n.* The FDA is the governmental body that is responsible for the approval of all food additives. All inks, coatings, and other packaging materials coming in direct contact with food or drugs must be shown to be non-migrating, or must be made only from raw materials which are known to be

harmless and which are listed in the Code of Federal Regulations, Title 21, in the paragraph covering the intended end use.
See fda.

Footcandle (1906) v_2. A deprecated unit of surface-lighting intensity, equal to 10.76391 lx. The lux (lx) is defined as 1 lumen per square meter (lm/m^2).
See also luminous flux.

F

Footing *n.* Concrete base on which a foundation sits.

Foot lambert *n.* Foot lambert is the unit of photometric brightness (luminance) equal to $1/\pi$ candle/ft^2.

Foots *n.* Sediment which settles from an oil on standing; chiefly albuminous matter. Originally used to describe those solid impurities, which precipitate from raw linseed oil during storage and then settle to the bottom or "foot" of a storage tank. Later used to denote material insoluble in a mixture of equal parts of acetone and linseed or other oil under test, and insoluble in calcium chloride solution under the specific conditions of test method. Also, the term is used to denote that material which is precipitated from the oil by phosphoric acid and which is insoluble in acetone under the specific conditions of test method.
See also tailings.

Force \\ˈfōrs, ˈfórs\\ [ME, fr. MF, fr. (assumed) VL *fortia*, fr. L *fortis* strong] (14c) *n.* (1) Either half of a compression mold (top force or bottom force), but usually the half that forms the concave surfaces of the molded part. (2) The male half of the mold, which enters the cavity and exerts pressure on the resin, causing it to flow. (3) A basic familiar quantity, familiar to everyone, customarily defined as that which changes a body's state of rest or motion, according to Newton's second law of motion:

$$F = \frac{\mathrm{d}}{\mathrm{d}t}(mv)/g_c = ma/g_c.$$

The second form applies when mass m is constant, as in all industrial applications. Here F is the force, v the velocity, t the time, a the body's acceleration is $\mathrm{d}v/\mathrm{d}t$, and $1/g_c$ is the proportionality constant relating the dimensions of the three primary quantities. Historically the units of the primary quantities, in whatever system, have been devised so that g_c is exactly 1 and it is often omitted from the equation. In the SI system, 1 N is defined as the force that will accelerate a 1-kg mass 1 s^{-1} (meter per second), making g_c exactly 1.0000 N s^2/(kg m). In the English system, however, where "pound" is used to mean either the pound-mass or the pound-force (lb or lb$_f$), the value of g_c = 32.174 lb$_f \cdot$ s^2/(lb ft)! *Weight* is the downward force exerted on bodies in the earth's gravitational field and results from the as yet unexplained mutual attraction of masses as set forth in Newton's law of universal gravitation. Because the earth is nearly spherical, the distances between bodies on and close to its surface and the earth's center are almost constant and so, therefore is the earth's attractive force on a unit mass at its surface. For convenience, the force is replaced by g, the standard "acceleration due to gravity", 9.806,650 m/s^2 (32.174 ft/s^2). Then, weight, $= mg/g_c$. For precise work, corrections are made for the variation in g with latitude and altitude, but g_c is always constant, even in outer space. Force has other aspects that are unconnected with rate of change of momentum. For example, the forces developed in a tightened bolt or the force exerted by a hydraulic ram on molding compound in a closed mold. One should keep in mind that force is *not identical* to mass times acceleration.

Force between two charges, Coulomb's law *n.* If two charges q and q' are at a distance r in a vacuum, the force between them is,

$$F = \frac{qq'}{r^2}.$$

The force will be given in dynes if q and q' are in electrostatic units and r in cm.

Force between two magnetic poles *n.* If two poles of strength m and m' are separated by a distance r in a medium whose permeability is μ (unity for a vacuum), the force between them is

$$F = \frac{mm'}{\mu r^2}.$$

Force will be given in dynes if r is in cm and m and m' are in cgs units of pole strength. The strength of a magnetic field at a point distance r from an isolated pole of strength m is

$$H = \frac{m}{\mu r^2}.$$

The field will be given in gauss if m and r are in cgs units.

Forced drying *n.* Drying coatings at a temperature between room temperature and 65.6°C (150°F) as opposed to air drying or baking.

Forced drying temperature *n.* Temperature between room temperature and 65.6°C (150°F).
See baking.

Force feeder *See Crammer–Feeder.*

Force plate *n.* The plate that carries the plunger or force plug of a mold and the guide pins or bushings. Since the force plate is usually drilled for steam or water lines, it is sometimes called the *steam plate.*

Force plug (plunger, piston) *n.* The portion of a mold that enters the cavity block and exerts pressure on the molding compound, designated as the *top force* or *bottom force* by position in the assembly.

Ford cup *n.* Efflux cup used for measuring the viscosity of paint or varnishes.
See efflux viscometer and viscometer.

Ford viscosity cups *n.* A series of three cylindrical cups with conical bottoms, differing only in the diameters of the orifices at the apexes of the cones, each cup having a capacity of about 100 ml. From the time or efflux, the sample volume and the orifice diameter, the kinematic viscosity of a liquid may be estimated. ASTM test D 1200 (Section 06.01) describes the Ford cup procedures to be used with paints and varnishes.

Foreign matter *n.* Anything visibly unrelated to the true nature of the substance under examination.

Foreign waste *n.* Thread waste or lint that is twisted in the yarn or woven in the fabric. If such foreign matter is of a different fiber, it may dye differently and thus show plainly.

Forgeability *n.* The ability of a solid material to flow without rupture under sudden intense compression. Some plastics have been forged into useful articles.

Forging *See cold forming and solid-phase forming.*

Form (sometimes Forme) \\ˈfóm, ˈfórm\\ (15c) *n.* (1) Type and other matter locked in a chase ready for printing. (2) In crystallography, a group of similar faces, e.g., cube, prism, dipyramid, etc. Also, a given polymorphic form of a substance, e.g., Form II of calcium carbonate is aragonite.

Formability *n.* The relative ease with which a plastic sheet or rod may be given another permanent shape.
See thermoformability.

Formal charge *n.* A somewhat arbitrary but useful way of indicating the approximate electrical characteristic or charge of an atom.

Formaldehyde \fór-ˈmal-də-ˌhīd, fər-\ [ISV *form-* + *aldehyde*] (1872) (formic aldehyde, methanal, oxymethylene) *n.* HCHO. A colorless gas with a pungent, suffocating odor, obtained most commonly by the oxidation of methanol or low-boiling petroleum gases such as methane, ethane, etc. The gas is difficult to handle, so it is sold commercially in the form of aqueous solutions (formalin), solvent solutions, as its oligomer, paraformaldehyde, and as the cyclic trimer, 1,3,5-trioxane (α-trioxym-ethylene). High-molecular-weight, commercial polymers of formaldehyde are called *polyoxymethylene* or acetal resin. Formaldehyde is also used in the production of other resins such as phenolic resin (phenol-formaldehyde) and amino resin (urea formaldehyde). Syn: methylene oxide, methanal.
See formalin.

Formalin \ˈfór-mə-lən, -ˌlēn\ [*Formalin*, a trademark] (1893) (formol) *n.* HCHO. Formaldehyde gas commercially available as a 37% wt. solution in water with a small amount of methanol for inhibiting polymerization.
See formaldehyde.

Formed fabric *See non-woven fabric.*

Form grinding *n.* A method of forming circularly symmetrical parts from plastic rod or tubing, employing a grinding wheel shaped to the inverse of the desired contour, a smaller hardened-steel regulating wheel that presses the plastic rod or tube against the grinding wheel, and work-rest blade that supports the work between the two wheels. Water is usually supplied as the coolant, from which the scrap powder can be recovered and reused.

Formica® \fór-ˈmī-kə, fər-\. A trademark of the Formica Corporation for high-pressure, decorative laminates of melamine-formaldehyde, phenolic and other thermosetting resins with paper, linen, canvas, glass cloth, etc., often misused by the public in a generic manner.

Formic acid (methanoic acid) *n.* HCOOH. The first of the aliphatic acids, with a pungent odor and probably existing mostly in the dimmer form, formic acid is produced by the reaction of carbon dioxide and dry sodium hydroxide followed by treatment with sulfuric acid. It is a solvent for phenol–formaldehyde resins, some polyesters, polyurethanes, and nylons.

Forming A general term encompassing processes in which the shapes of plastics pieces such as sheets, rods, or tubes are changed to a desired configuration, usually with the acid of heat. The term is not usually applied to operations such as molding, casting, or extrusion in which shapes or articles are made from molding materials and liquids.
See also fabricate and thermoforming.

Forming box *See vacuum sizing.*

Forming cake *n.* In filament winding, the collection (*package*) of glass-fiber strands on a mandrel during the operation.

Form lacquer *n.* Thin lacquer or varnish used to coat concrete forms to prevent concrete from adhering to the forms.

Formol Syn: formaldehyde.

Form roller *n.* The form roller is that roller in the ink distribution system of a printing press, which is in direct contact with the printing plate and transfers the ink to it.

Formula \ˈfór-myə-lə\ [L, dim. of *forma* form] (1618) (Chemical) *n.* A combination of symbols with their subscripts representing the constituents of a substance and their proportions by weight.

Formula unit The group of atoms indicated by the formula of a substance.

Formula weight (ca. 1920) (molecular weight) *n.* Of a chemical compound, the number obtained by summing the atomic weights of all the compound's atoms. The term is not used with polymers, which are always mixtures of compounds of different molecular weights.

Formula weight The sum of the masses of the atoms indicated in a formula, expressed in atomic mass units; the mass of one formula unit.

Fossil gum resin *n.* Fossil gums are the so-called hard gums or copals with were exuded from living plants in the form of liquids or semisolids, which have lain in the ground for centuries and thus hardened or fossilized with time. Some of the fossil gum resins are: Amber, Zanzibar, Kauri, Manila, Pontianak, and Congo. *See gym, natural and resin, natural.*

Fossil resin *n.* A natural resin obtained from fossilized remnants of plant or animal life. An example is amber, a fossilized resin derived from an extinct species of pine.

Foster charts *n.* Graphs for determining color differences in modified MacAdam units from tristimulus and ratio tristimulus measurements made on a tristimulus colorimeter. They are based on the Simon–Goodwin charts and give results, which are a close approximation to those which would be obtained with the S–G charts.

Foulard \fu-ˈlärd\ [F] (1830) *n.* A lightweight, lustrous 2/2 twill that is usually printed with small figures on a solid background, foulard is frequently used in men's ties. Foulards are made of silk, filament polyester, acetate, etc.

Fouling *n.* Sessile marine organisms on the hulls of ships.

Foundation Lower parts of walls on which the structure is built. Foundation walls of masonry or concrete are mainly below ground level.

Foundry resin *n.* A thermosetting resin used as a binder for sand in metals founding. The types most commonly used as water-soluble phenol–formaldehyde resins that become insoluble when cured, and cold-setting furfuryl alcohol resins that cure in the presence of an acid catalyst.

Fountain \ˈfaún-tᵊn\ [ME, fr. MF *fontaine*, fr. LL *fontana*, fr. L, feminine of *fontanus* of a spring, fr. *font-*, *fons*] (14c) *n.* Part of a printing press, which contains the ink to be fed to the distributing system. In lithography, it is also the part, which feeds the water or fountain solution to the dampening rollers.

Fountain roller *n.* The roller that revolves in the ink fountain. In lithography it is also the roller that revolves in the dampening solution.

Fountain solution *n.* In lithography, generally a mixture of water, acid, buffer and a gum to prevent the non-printing areas of the plate from receiving ink. To some fountain solutions, alcohols are added. *Also known as dampening solution.*

Four-dimensional braid *n.* A type of braided reinforcement used to achieve specially directed strength and resistance to interlaminar shear, often involving mixed fibers of different materials.

Fourier number *n* (N_F). A dimensionless group important in analysis of unsteady heat transfer in solids, such as sheets being heated or cooled in thermoforming, or cooling of extrudates and moldings, $N_F = \alpha \cdot t/x^2$, where α is the material's thermal diffusivity, t the heating or cooling time, and x is the a thickness or half-thickness in the direction of heat flow.

Fourier's law of heat conduction *n.* The fundamental equation for steady heat flow through solids. It is

$$q = -kA\frac{dT}{dx},$$

where q is the rate of heat flow, k the thermal conductivity of the material at temperature T, and A is the area through which heat flow is occurring, normal to coordinate x within the material in the direction of heat flow. This is the defining equation for thermal conductivity. The quotient q/A is known as the *heat flux* or *thermal flux*. If the thermal conductivity is constant over the range of temperatures involved or is linearly dependent on temperature, an integrated, simpler version of the law, with an average conductivity, can be used to find the heat flow through a layer of thickness Δx:

$$q = k_{avg}A(T_2 - T_1),$$

where T_2 and T_1 are the temperatures on the hot and cold sides of the layer.

Fourier-transform infrared spectroscopy (FTIR) *n.* Infrared (IR) spectroscopy is most commonly used for the identification of unknown pure organic compounds. In FTIR, infrared radiation of a broad range of wavelengths is passed through an interferometer and a pathlength difference is introduced into one part of the light beam. This IR beam is then passed through the sample, which absorbs light energies corresponding to various bond-vibration and rotation frequencies. The beam is then focused on a detector, and a computer calculates the absorption of the IR frequencies by the sample, identifying compounds present and their concentration in the sample. An example of a FTIR spectroscope is the Spectrum 100 FTIR Spectrophotometer by Perkin-Elmer Inc., image courtesy of Perkin-Elmer Inc., New York.

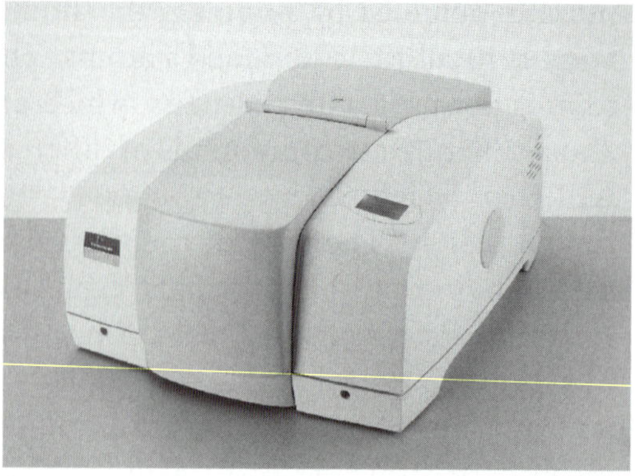

Foxy Pigment crystals are sometimes arranged in different order by method of application, causing different shades of color when viewed from different angles; this condition is often referred to as "foxy".

F.P.S. system *n.* The foot-pound-second system of units. The British system of physical units derived from the three fundamental units of length, mass, and time, i.e., the foot, pound, mass, and the second.

Fraction \ˈfrak-shən\ [ME *fraccioun*, fr. LL *fraction-*, *fractio* act of breaking, fr. L *frangere* to break] (14c) *n.* Solvent of definite boiling range obtained by fractional distillation.

Fractional distillation *n.* Separation of a liquid into its different components by collecting fractions or distillates of restricted boiling range.

Fractionation *n.* A method of determining the molecular-weight distribution of polymers based on the fact that polymers of higher molecular weight are less soluble than those of lower molecular weight. Two basic methods in use are (1) Precipitation fractionation, in which phases are separated from a solution of the polymer by incremental addition of non-solvents, stepwise lowering of the solution temperature, or volatilization of the solvent. (2) Extraction fractionation, in which fractions of increasing molecular weight are

preferentially extracted from a layer of polymer that has been deposited on a substrate. In either method, a series of fractions is obtained which must be recovered and characterized with respect to molecular weight. A third method that has had some success is ultracentrifugation. Fractionation is also used to prepare polymer specimens having narrow molecular-weight distributions.
See ultracentrifuge.

Fracture \ˈfrak-chər, -shər\ [ME, fr. L *fractura*, fr. *fractus*] (15c) *n.* The separation of a body, usually characterized as either brittle or ductile. In brittle fracture, the crack propagates rapidly with little accompanying plastic deformation. In ductile fracture, the crack propagates slowly, usually following a zigzag path along planes on which a maximum resolved shear stress occur, and their is substantial plastic deformation. Slower loading and higher temperature favor ductile behavior.

Fracture mechanics The study, both theoretical and experimental, of the behavior of cracks in stressed bodies. A basic principle is that fracture is driven by the energy released by the growth of the crack, which begins at a small imperfection such as is found in all bodies.

Fracture toughness (K_c) *n.* The critical value of the stress-intensity factor in a material beyond, which a crack will start to grow.

Frame *n.* (1) A general term for many machines used in yarn manufacturing such as the drawing frame, roving frame, and spinning frame. (2) *See tenter frame.*

Framework knitting *See knitting, weft knitting.*

Frankel–Acrivos equation *n.* An equation, derived wholly from theoretical considerations, giving the relative viscosity of suspensions of monodisperse spheres in Newtonian liquids. It is

$$\eta_r = \frac{\eta}{\eta_o} = \frac{9}{8}\left[\frac{(f/f_m)^{1/3}}{1-(f/f_m)^{1/3}}\right],$$

where η_o is the viscosity of the pure liquid, η the viscosity of the liquid containing a volume fraction f of spheres whose density is close to that of the liquid. f_m represents the maximum attainable volume fraction, which depends on the assumed geometry of packing, about 0.7–0.74. The equation well represents the best available data to loadings approaching $f/f_m = 1$. Compare Eilers equation and Mooney equation.

Frankincense \ˈfraŋ-kən-ˌsen(t)s\ [ME *fraunk encense*, fr. OF *franc encens*, fr. *franc* (perhaps in sense "of high quality") + *encens* incense] (14c) *n.* A fragrant gum resin from trees of a genus (*Boswellia* of the family *Burseracae*) of Somalia and coastal Arabia that is an important incense resin and was used in ancient times in religious rites and embalming.
See gum thus.

Fraunhofer's lines *n.* When sunlight is examined through a spectroscope it is found that the spectrum is traversed by an enormous number of dark lines parallel to the length of the slit. These dark lines are known as Fraunhofer's lines. Kirchoff conceived the idea that the sun is surrounded by layers of vapors which act as filters of the white light arising from incandescent solids within and which abstract those rays which correspond in their periods of vibration to those of the components of the vapors. Thus reversed or dark lines are obtained due to the absorption by the vapor envelope, in place of the bright lines found in the emission spectrum.

Fraying The slipping or raveling of yarns from unfinished edges of cloth.

Free energy *n.* Also Gibb's free energy or change in free energy of a reaction;

$$-\Delta G = T\,\Delta S - \Delta H,$$

where T is the temperature, S is entropy, and H is the enthalpy; if $-\Delta G$ the reaction can proceed, but if $+\Delta G$ then the reverse reaction can proceed; also referred to as the driving force of a reaction; and the decrease in free energy of a constant temperature, constant-pressure process is the measure of the tendency of the process to proceed spontaneously.

Free energy of formation, ΔG_f *n.* The free energy change for a reaction in which one mole of a compound is formed from its uncombined elements.

For the chemical reaction,

$$1/2N_2 + 2H_2 + 1/2Cl_2 \rightarrow NH_4Cl,$$

$$\Delta G = \Delta H - T\,\Delta S$$

$$\Delta G = (-314.4 \times 10^3\,\mathrm{J\,mol^{-1}}),$$

$$-\,(298.2)(-373.8\,\mathrm{J\,K^{-1}\,mol^{-1}}),$$

$$\Delta G = 2.030 \times 10^5\,\mathrm{J\,mol^{-1}},$$

where $-\Delta G$ indicates that the reaction occurs spontaneously.

Free energy, F (Gibbs free energy, *G*) *n.* A thermodynamic quantity, which measures the energy in a system which is available for doing work of expansion. The free energy of a system is defined as the difference between its enthalpy, *H*, and its temperature–entropy product: at constant temperature and pressure, $\Delta G = H - T\Delta S$. If $\Delta G < 0$, then a spontaneous reaction occurs, $\Delta G > 0$, non-spontaneous or a reversible reaction, $\Delta G = 0$, the system is at equilibrium. An example of $\Delta G < 0$ is the melting of ice above 0°C, and $\Delta G > 0$ for the freezing of water above 0°C. Consider a material undergoing a "change of state" at its equilibrium temperature. Gold metal undergoes a change of state (fusion) at its equilibrium temperature (melting temperature, T_m) and

$$\Delta S = \frac{\Delta H}{T_m},$$

where T_m can be calculated as,

$$T_m = \frac{12.36 \times 10^3\,\mathrm{J\,mol^{-1}}}{9.250\,\mathrm{K^{-1}}} = 1336\ \mathrm{K}$$

and 1336°K − 273 = 1063°C.

Free expansion *n.* Expansion of a substance, usually a gas, against no opposing pressure.

Free-falling-dart test *n.* A method of measuring the impact resistance of thermoplastic films by dropping a dart with a hemispherical head onto a film specimen held in a clamping frame. As described in ASTM D 1709, the dart is dropped from a fixed height onto each of ten specimens and the percent failure is noted. Another increment of weight is added and ten more specimens are tested. The process is repeated until 50% of the specimens fail. The weight of the dart at this point, times the drop height, is a measure of the film's impact strength. In ASTM D 4272, another falling-dart test for films, the dart is instrumented and the energy consumed in penetrating the film is computed. ASTM D 3029 describes a similar test for rigid sheets. *See also falling-dart impact test.*

Free fatty acid *n.* Unreacted fatty acid present in a coating vehicle.

Free forming *n.* A variant of sheet thermoforming in which a bubble is blown into the clamped, heat-softened sheet, either by applying a vacuum to the side that will be convex or pressure to the underside. The method has been used most with cast-acrylic sheet for applications where the best possible optical properties are foremost, such as airplane canopies.

Free phenol *n.* The uncombined phenol existing in a phenolic resin after curing, the amount of which is indicative of the degree of cure. The presence of such free phenol can be detected by the Gibbs indophenol test.

Free radical (1900) *n.* An atom or group of atoms having at least one unpaired electron. Most free radicals are short-lived intermediates with high reactivity and high energy, difficult to isolate. They are important agents in many polymerization processes and have been detected on corona-discharge-treated films by electron-spin resonance (ESR).

Free radical chain *n.* Polymerization of an atom or molecule, which has at least 1 e, which is not paired with another electron.

Free-radical polymerization *n.* A reaction initiated by a free radical derived from a polymerization catalyst. Polymerization proceeds by the chain-reaction addition of monomer molecules to the free-radical ends of growing chain molecules. Major polymerization methods such as bulk, suspension, emulsion, and solution polymerization involve free radicals. The free-radical mechanism is also useful in copolymerization, in which alternating monomeric units are promoted by the presence of free radicals. Lenz RW (1967) Organic chemistry of high polymers. Interscience Publishers, New York. Odian G (2004) Principles of polymerization, 4th edn. Wiley-Interscience, New York.

Free silica *n.* SiO_2. Silica generally present in small amounts in natural deposits of clay-like minerals and diatomaceous earth and usually considered to be a contaminant.

Free-thaw resistance *n.* Extent to which water-base paints, utilizing synthetic lattices or synthetic resin emulsions as vehicles: (1) Retain their original properties, free from detrimental changes in consistency. (2) Resist coagulation, or the formation of lumps and specks, when subjected to freezing and subsequent thawing.

Free volume *n.* In a liquid or amorphous solid, the specific volume minus the volume occupied by the molecules themselves. Free volume increases with rising temperature, causing viscosity to diminish. The f_{WFL} is calculated from the specific volume of the amorphous polymer (v^{o}_{am}) and the specific volume of the liquid polymer (v^{o}_{t}) in a polymer–monomer mixture during polymerization (Elias, 1977).

$$f_{WFL} = \frac{v^{o}_{am} - v^{o}_{t}}{v^{o}_{am}}.$$

The vacant sites of the solid polymer (amorphous) are called "free volume"; no long range order in extensive regions occurs in the amorphous state (e.g., amorphous material is not X-ray crystalline) although indications are that these regions possess a certain order; and theoretically a certain number of vacant sites must be present in the solid polymer. Free volume is mainly responsible for the compressibility of liquids and solids. Also called free volumes f: vacant free volume f_{vac}, Williams–Landell–Ferry free volume f_{WFL}, fluctuation free volume f_{fluc}, and free volume for thermal expansion f_{exp}. The effect of system variables on solubility is discussed in Handbook of Solvents, 2001. Miller (1968) noted that "the concept of free volume is easy to grasp, but, quantitatively, its definition runs into snares. Is free volume the specific volume of the liquid (solution) minus the volume of the molecules computed from Van der Waals radii, or minus the volume swept out by the segments as they rotate, or is it some other volume"? The free volume is generally, but necessarily, about 2.5% for all polymers.

Free-wheeling *n.* In reference to rolls, spinning without the application of either driving or braking force.

Freeze grinding *See cryogenic grinding.*

Freeze line Syn: frost line.

Freezing point (1747) *n.* Temperature at which a liquid material and solid are in equilibrium with one another, i.e., at a lower temperature the liquid will solidify. Amorphous materials, such as paints or paint vehicles, normally solidify over a temperature range referred to as freezing range.

Freezing-point depression *n.* $\Delta T = -K_{FP}m$, where T is the freezing point depression, K_{FP} the 1.86 for water, and m is the molality (mol solute/1000 g solvent), The lowering of the freezing point of a solvent brought about by the presence of a solute.

Frenchback *n.* A fabric with a corded twill backing of different weave than the face. The backing, which is frequently of inferior yarn, gives added weight, warmth, and stability to the cloth.

French blue *See ultramarine blue.*

French chalk *n.* $3MgO{\cdot}4SiO_2{\cdot}H_2O$. It is distinguished by its soapy or greasy feel. Chemically, it is substantially a hydrated magnesium silicate. It is supplied commercially as a white or gray powder with Sp gr of 2.4–2.6. Electromagnetic radiation to produce an exceedingly high gloss to wood surfaces. The use also of a little linseed oil on the polishing cloth is involved. *Known also as talc, soapstone, or steatite.*

French process zinc oxide *n.* Zinc oxide pigment made from zinc metal.
See zinc oxide.

French ultramarine *n.* Synthetic ultramarine blue.
See ultramarine blue.

French varnish *n* A hard-rubbed high-gloss finish, achieved by multiple applications of such varnish.
See French polish.

French veronese green *See hydrated chromium oxide.*

Freon® \ˈfrē-ˌän\ *See fluorocarbon blowing agent.*

Frequency \ˈfrē-kwən(t)-sē\ (1600) *n.* Number of vibrations per second, equal to the velocity of light (or any electromagnetic radiation) divided by the wavelength:

$$\nu = \frac{c}{\lambda},$$

where ν is the frequency, c the velocity 2.99793×10^8 m/s (for light and for all electromagnetic radiation), and λ is the wavelength which is specific for each type of radiation.

Frequency factor *n.* The quantity preceding the exponential term in the Arrhenius equation.

Frequency of vibrating strings *n.* The fundamental frequency of a stretched string is given by

$$n = \frac{1}{2l}\sqrt{\frac{T}{m}},$$

where l is the length, T the tension and m is the mass per unit length. For a string or wire of circular section of length l, tension T, density d, and radius r, the frequency of the fundamental is

$$n = \frac{1}{2rl}\sqrt{\frac{T}{\pi d}}.$$

The frequency in vibrations per second will be given if T is in dynes, r and l in cm and d in g/cm^3.

Fresco \ˈfres-(ˌ)kō\ [It, fr. *fresco* fresh, of Gmc origin; akin to OHGr *frisc* fresh] (1598) *n.*

Method of painting in which pigments mixed only with water are applied to a freshly laid surface of lime plaster. The painted surface is rendered durable by the pigment absorbing calcium hydroxide from the wet plaster, which is converted into insoluble calcium carbonate as the surface dries. The term is sometimes used incorrectly to denote any form of mural painting.

Fresco color *n.* Water paints used for wall decorations.

Fresnel reflection *n.* Phenomenon of reflection at the surface or interface where media or materials of different refractive indices join. Fresnel laws state that the angle of reflection is equal and opposite to the incident angle, and that the magnitude of the reflection depends on the angle of the incident light, on the relative refractive indices, and on the polarization of the incident light. When the surfaces are smooth and planar, specular (mirror like) reflection occurs, which is subjectively referred to as specular gloss. When the surfaces are rough, the reflectance at the air-material interface is diffuse, and the surface is observed to be matte-like.

Fret [ME, back-formation fr. *fret, fretted* adorned, interwoven, fr. MF *freté*, fr. OF, pp of *freter, ferter* to tie, bind, prob. fr. (assumed) VL *firmitare*, fr. L *firmus* firm] (14c) *vt.* A geometric band or border motif, consisting of interlacing or interlocking lines.

Also known as a key or meander pattern.

Friction \ˈfrik-shən\ [ME, fr. MF or L; MF fr. L *friction-, frictio*, fr. *fricare* to rub; akin to L to crumble] (1704) *n.* (1) The force resisting the sliding or rolling of one body relative to another with which it is in contact. The *coefficient of friction*, μ, is the quotient of the force required to cause or maintain motion divided by the normal force N exerted by the bodies upon each other, i.e., $\mu = F/N$. With most materials, the force required to initiate motion is somewhat greater than that required to maintain it, so a distinction is drawn between *static* and *dynamic* coefficients of friction. Coefficients of *rolling* friction are generally much lower than coefficients of sliding friction. Dynamic sliding friction is the most important one in plastics processing, but all three have roles in product design and operation. (2) In flowing fluids, particularly liquids, interlaminar friction is believed to be the basis of viscosity.

Frictional coefficient *n.* Resistance to sliding or rolling of surfaces of solid bodies in contact with each other.

Frictional heating *n.* (1) Heat evolved when two surfaces rub together. For a frictional force F and relative velocity V, the rate of heat evolution is FV (SI: J/s). (2) Viscous dissipation within liquids undergoing shear flow. At any point in a laminar-flowing liquid, the rate of viscous dissipation, per unit volume, is equal to the product of the shear rate times the shear stress, which is also equal to the product of the local viscosity and the square of the shear rate. Both these types of functional heating are important mechanisms of plastication in extruders and injection molders.

Friction calendering *n.* A process in which an elastomeric compound is forced into the interstices of woven or cord fabrics while passing through the rolls of a calender. The heated compound is fed into the top opening of three adjacent rolls, so that it will cling to the middle roll. The fabric to be impregnated is fed into the lower opening between the rolls. The distance between the rolls is regulated so as to squeeze the fabric without crushing it, and the rolls are operated at slightly different speeds so that the compound is wiped by friction into the meshes of the fabric.

Friction coefficient *n.* The coefficient of friction between two surfaces is the ratio of the force required to move one over the other to the total force pressing the two together. If F is the force required to move one surface over another, and W, is the force pressing the surfaces together, the coefficient of friction,

$$k = \frac{F}{W}.$$

Friction false-twist texturing *See texturing, false-twist method.*

Friction spinning *n.* A spinning system in which the yarn receives its twist by being rolled along the longitudinal axis in the nip between two revolving surfaces. The surfaces may rotate at the same or different speeds in the same or opposite directions depending on the particular machine design. Potential advantages include high production capacity, low stress on the fiber in processing, and the capacity to produce very fine counts.

Friction top can *n.* Can in which the lid is held in place by friction.

Friction welding (angular welding) *n.* A term encompassing spin welding and the newer process of applying rapid angular oscillations to heat the plastic parts to be joined. This variation of the spin-welding process is used for parts that are not symmetrical about an axis of rotation. The equipment must be programmed to stop when the parts of properly positioned for joining.

Friedel-catalysts *n.* Strongly acidic metal halides such as aluminum chloride, aluminum bromide, boron trifluoride, ferric chloride, and zinc chloride, used in the polymerization of unsaturated hydrocarbons, e.g., olefins (Friedel–Crafts reactions using such catalysis are named for Charles Friedel and James Crafts, who first used them in 1877). These acidic halides are *also known as Lewis acids.*

Friedel–Crafts catalyst *n.* Lewis Acid catalysts such as aluminum chloride or ferric chloride.

Friel–MacAdam-chickering color difference equations *See FMC color difference equations.*

Friezé \\ˈfrēz *or* frē-ˈzā\\ [ME *frise*, fr. MF, fr. MDu *vriese*] (15c) *n.* (1) Any decorative horizontal band, as along the upper part of a wall in a room. (2) A type of wallpaper popular in the early 1900s. Generally, a pictorial border, which ran above door height, or, in dining rooms, above the plate rail. (3) A term applied when the pile of a velvet, plush, velour, or other pile fabric is uncut. A friezé fabric is sometimes patterned by shearing the loops at different lengths. Friezé fabrics are widely used for upholstery. (4) A cut-pile carpet made of highly twisted yarns normally plied and heat-set. A kinked or curled yarn effect is achieved. Excellent durability results from the hard-twist pile yarns.

Fringed micelle *n.* A model for a partially crystalline polymer widely accepted as explaining many of the properties of such a polymer. Consists of crystallites (the "micelles") embedded in an amorphous matrix.

Frisket \\ˈfris-kət\\ [F *frisquette*, fr. MF] (ca. 1898) *n.* Device, usually of gummed paper placed on the working surface, to block out subsequent painting.

Frosting *n.* (1) Formation of a translucent finely wrinkled surface on a film of oil, ink, or paint during drying, particularly when exposed to gas fumes, etc. (2) Salt like deposit on the surface of a topcoat. (3) A clouding of the surface of some rubber and synthetic rubber goods, appearing within a few hours or days after vulcanization. The frosted appearance is different from bloom or blush and cannot be readily removed by washing with a solvent. It may disappear if the article is heated moderately but will generally reappear on cooling. It is thought to be caused by ozone in the air, which produces a maze of minute cracks. Some antioxidants have definite anti-frosting effects. *See color abrasion. See also chalking, bloom, and haze.*

Frosting marks *n.* A defect of woven fabric consisting of surface highlights that give a frosted appearance. Frost marks are caused by improper sizing or insufficient warp tension as a result of uneven bending of some warp ends over the picks.

Frost line (freeze line) *n.* In the extrusion of blown film, a ring-shaped transition zone of frosty appearance located at the level at which the film reaches its final diameter and is changing from melt to solid.

Frothing *n.* A technique for applying urethane foam in which blowing agents or small air bubbles are introduced under pressure into the liquid mixture of foam ingredients.

Frozen-in strain (residual strain) *n.* Strain that remains in an article after it has been shaped and cooled to its final form, due to a non-equilibrium configuration of the polymer molecules. Such strains occur when cooling its carried below a certain temperature before stresses of a molding or forming operation have been allowed to relax, Frozen-in strains/stresses can cause warping of large flat members and can lead to crazing at low applied stress levels. Aside from measures that can be taken during processing, post-annealing, sometimes done with the aid of shape-holding jigs, is used to complete the relaxation of stresses.

FRP *n.* Abbreviation for fiber-reinforced plastic.

FSCT *n.* Abbreviation for federation of societies for coatings technology. Paint technology prior to 1975.

Fuchsin \ˈfyük-sən, -ˌsēn\ [F *fuchsine*, prob. fr. NL *Fuchsia*; fr. its color] (1865) *n.* Synthetic rosaniline dyestuff, a mixture of rosaniline and pararosaniline hydrochlorides. Dark green powder of greenish crystals with a bronze luster; faint odor. Soluble in water and alcohol. Used in textile and leather industries; as a red dye; and as a pharmaceutical. Vigo TL (1994) Textile processing, dyeing, finishing, and performance. Elsevier Science, New York. *Also known as basic fuchsin magenta.*

Fugitive colors *n.* (1) Coloring matter that exhibits color change or color transfer while wet or during subsequent drying. (2) Inks made from pigments or dyes, which are not permanent, and change or lose color rapidly when exposed to light, heat, moisture, or other conditions. (3) Colorant, pigment or dyestuff, which changes color (fades) rapidly following exposure to light.

Fugitometer *See fadeometer.*

Full-bodied *See body.*

Full coat *n.* (1) Maximum film thickness of a particular coating, which can be properly applied. (2) Application of a coating at a specified film thickness designed to achieve a desired effect.

Fuller's earth *n.* $3MgO{\cdot}1.5Al_2O_3{\cdot}8SiO_2{\cdot}9H_2O$. Hydrated magnesium aluminum silicate (attapulgite) used for decolorizing solutions and oils, and as a substitute for absorbent charcoal and as a dusting powder. Density, 2.36 g/cm^3 (19.7 lb/gal). Refractive index, 1.50–1.55. Ash M, Ash I

(1998) Handbook of fillers, extenders and dilutents. Synapse Informtion Resources Inc., New York.

See hydrated magnesium aluminum silicate.

Full-fashioned *n.* A term applied to fabrics produced on a flat-knitting machine, such as hosiery, sweater, and underwear, that have been shaped by adding or reducing stitches.

F

Full-flighted screw *n.* An extruder screw in which the flights extend over the entire length of the screw.

Full gloss *See gloss.*

Fulling *n.* A finishing process used in the manufacture of woolen and worsted fabrics. The cloth is subjected to moisture, heat friction, chemicals, and pressure which cause it to mat and shrink appreciably in both the warp and filling directions, resulting in a denser, more compact fabric. Vigo TL (1994) Textile processing, dyeing, finishing, and performance. Elsevier Science, New York.

Fumaric resin A synthetic, hard resin formed by the reaction of fumaric acid and rosin. Ash M, Ash I (1982–1983) Encyclopedia of plastics polymers, and resins, vols 1–3. Chemical Publishing Co., New York.

Fumed silica (pyrogenic silica) *n.* An exceptionally pure form of silicon dioxide made by reacting silicon tetrachloride in an oxyhydrogen flame. Individual particles of fumed silica, ranging in size from 7 to 40 nm, tend to link together by a combination of fusion and secondary bonding to form chain-like aggregates with high surface areas that retard the flow of liquids in which they are dispersed. Thus fumed silica is a useful thickening agent, imparting thixotropy to liquid resins that are normally Newtonian, e.g., certain polyesters. Fumed silica is also used in dry molding powders to make them free-flowing and easier to disperse with colorants. Improved electrical properties, prevention of blocking, and reduction of plasticizer migration are other benefits attributed to fumed silica in vinyl compounds. Solomon DH, Hawthorne DG (1991) Chemistry of pigments and fillers. Krieger Publishing Co., New York. Ash M, Ash I (1998) Handbook of fillers, extenders and dilutents. Synapse Informtion Resources Inc., New York. Kirk–Othmer encyclopedia of chemical technology: pigments–powders. John Wiley and Sons, New York, 1996. Parfitt GD (1969) Dispersion of powders in liquids. Elsevier Publishing Co., New York.

Fume fading *See gas fading.*

Fumes *n.* A gas-like emanation containing minute solid particles arising from the heating of a solid body such as lead, in distinction to a gas or vapor. This physical change is often accompanied by a chemical reaction, such an oxidation. Fumes flocculate and sometimes coalesce. Odorous gases and vapors should not be called fumes. As distinguished from dusts, fumes are finely divided solids produced by other methods of subdividing, such as chemical processing, combustion, explosion, or distillation. Some solids, when heated to a liquid produce a vapor, which, while arising from the molten mass, immediately condenses to a solid without returning to its liquid state. Fumes are much finer than dusts, containing particles from 0.1 to 1.0 μm in size. Tests for comparative flammability of liquids, UI 340. Laboratories Incorporated Underwriters, New York, 1997.

Functional group *n.* A group of atoms in a molecule, which cause the molecule to undergo a set of characteristic reactions. Morrison RT, Boyd RN (1992) Organic chemistry, 6th edn. Prentice-Hall, Englewood Cliffs, NJ.

Functionality The ability of a molecule or group to form covalent bonds with another molecule or group in a chemical reaction. Compounds may be mono-, di-, tri-, or polyfunctional, depending on the number of functional groups capable of participating in a reaction. Morrison RT, Boyd RN (1992) Organic chemistry, 6th edn. Prentice-Hall, Englewood Cliffs, NJ.

Fundamental units *See mass, length, and time.*

Fungi \ˈfəŋ-gəs\ *n pl*, - *sing.* Fungus.
See fungus.

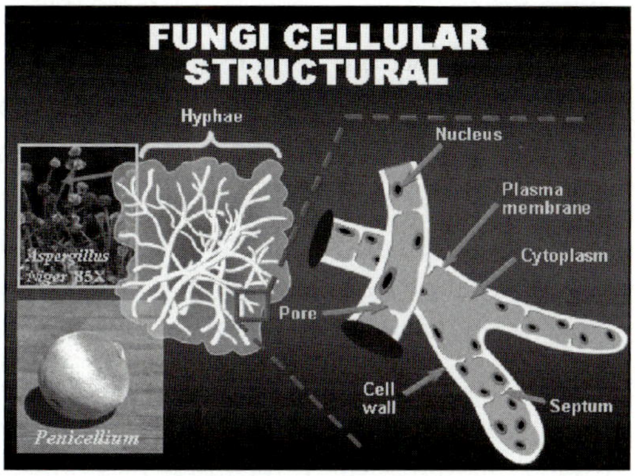

Fungicidal paint *n.* Some pigments are used as biocides. For example, zinc oxide I used as a fungicide. Paint, which discourages the growth of fungi on its dry applied film is a fungiciial paint or coating. The fungicidal properties are normally conferred by special additives, although certain pigments such zinc oxide, commonly used in paints, may themselves contribute to the fungicidal properties of the paint. Wicks ZN, Jones FN, Pappas SP (1999) Organic coatings science and technology, 2nd edn. Wiley-Interscience, New York.

Fungicidal wash Wash containing fungicides used before painting and designed to kill existing spores or germinations or to prevent their inception. Many of these substances are toxic (to human beings) used when in concentrated form and therefore need careful handling. Use of the term "antiseptic wash" is deprecated.

Fungicide \ˈfən-jə-ˌsīd\ fungicides [ISV] (1889) *n, pl.* An agent incorporated in a plastic compound to control fungus growth, usually by killing the organisms. Most plastics with a few exceptions, notably some of the cellulosics, are inherently resistant to fungus attack. However, many plasticizers are highly susceptible to attack. Examples of fungicides used in plastics are copper-8-quinolinate and tributyltin oxide. Agents that retard fungal growth without killing the organisms are called *fungistats.* Black JG (2002) Microbiology, 5th edn. John Wiley and Sons Inc., New York.
See also biocide.

Funginertness (fungus resistance). Not susceptible to the formation of fungus growth. James F (ed) (1993) Whittington's dictionary of plastics. Technomic Publishing Co. Inc., Carley.

Fungistat *n.* An agent incorporated in plastics compounds to control fungus growth without killing the fungi.
See also biocide.

Fungistatic \ˌfən-jə-ˈsta-tik\ (1922) *adj.* Preventing the growth of a fungus by the presence of some chemical or physical agent but not fungicidal.

Fungus \ˈfəŋ-gəs\ fungi [L] (1527) *n sing, -n pl.* Multicellular plant or plants of the phylum, *thallophytae* that contain no chlorophyll and reproduce by sexual or asexual spore. Fungi cells are eucarotic, each cell contains a nucleus. Fungi cause discoloration and growth in or on a coatings surface, i.e., mildew, mold. Black JG (2002) Microbiology, 5th edn. John Wiley and Sons Inc., New York.

Fungus resistance *See mildew (fungus) resistance.*

Furan *n.* Furfuran; oxole; tetrole; divinylene oxide, C_4H_4O, Molecular Weight: 68.07. Occurs in oils obtained by the distillation of rosin contg pine wood. Prepared by decarboxylation of 2-furancarboxylic acid. Furan has been prepared directly from furfural over hot soda-lime or by dropping furfural on a fused mixture of sodium and potassium hydroxides. Other physical properties: Liquid. $d_4^{19.4}$ 0.9371. bp_{760} 31.36°; bp_{758} 32°. $^{n}D^{20}$ 1.4216. Flash pt, closed cup: −32°F (−35°C). Absorption spectrum: Purvis J (1910) J Chem Soc 97:1648–1655. Insoluble in water. Freely soluble in alcohol and ether. Stable to alkalies; resinifies on evaporation or when in contact with mineral acids. Boiling point: bp_{760} 31.36°; bp_{758} 32°. Wilson J (1941) *Org. Syn.* coll. vol 1, 2nd edn. p274. Hurd R et al. (1932) J Am Chem Soc 54: 2532. Odian GC (2004) Principles of polymerization. John Wiley and Sons Inc., New York. Merck index (2001) 13th edn. Merck and Company Inc., Whitehouse Station, NJ.

Furan resin (furfuryl resin) *n.* A dark-colored, thermosetting resin obtained primarily by the condensation polymerization of furfuryl alcohol in the presence of a strong acid, sometimes in combination with formaldehyde or furfural (2-furaldehyde). The term also includes resins made by condensing phenol with furfuryl alcohol or furfural, and furfuryl–ketone polymers. The resins are available as liquids in a wide range of viscosities that cure to highly cross-linked, brittle substances. They are used for impregnating cured plaster structures, as binders for foundry-sand cores, for binding high explosives, and as wood adhesives. The cured resins exhibit good resistance to chemicals and solvents. Improved resin/catalyst systems have made the older furan systems obsolete and enabled the use of fire-retardant furans in hand layup, sprayup, and filament-winding, competing with polyesters. Odian GC (2004) Principles of polymerization. John Wiley and Sons Inc., New York. Harper CA (ed) (2002) Handbook of plastics, elastomers and composites, 4th edn. McGraw-Hill, New York.
See furfuryl.

Furan prepreg *n.* Latent catalysts announced in 1972 made it possible to produce furan-resin prepregs of acceptable shelf life, which avoided the difficulties experienced with the wet-layup process. These composites possess good heat and chemical resistance, excellent surface hardness and fire resistance. Harper CA (ed) (2002) Handbook of plastics, elastomers and composites, 4th edn. McGraw-Hill, New York.

Furfural \ˈfər-f(y)ə-ˌral\ [L *furfur* bran + ISV ^{3}*-al*] (1879) (2-furaldehyde, ant oil, furfuraldehyde) *n.* $C_4H_3O{\cdot}HO$. A liquid aldehyde with a bp of 161°C, mp of −36.5°C; Sp gr of 1.159/22°C, obtained by distilling acid-digested corn cobs or the hulls of oats, rice, or cottonseed, and having the structure shown below. Furfural is colorless liquid when first distilled, but darkens on exposure to air. It is used as a solvent, and in the production of furans and tetrahydrofurans.
See also furan resin.

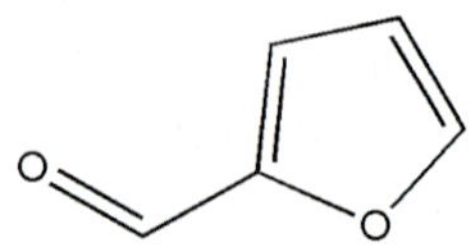

Furnace black *n.* A form of carbon black obtained by decomposing natural gas and/or petroleum oil under controlled

conditions in a furnace and precipitating the pigment in special chambers.
See carbon black.

Fuse Heating a coating component of dispersed resin(s) to a temperature at which it becomes homogeneous.

Fused acetate *n.* (1) A hard particle of acetate material of almost any shape or size other than recognizable fiber. Sometimes fused acetate particles resemble rock-like, hardened drops of acetate dope; in other cases fused acetate consists of particles covered with fiber clusters and completely hardened in the center. (2) Acetate yarns in which the individual filaments are coalesced. Tortora PG (ed) (1997) Fairchild's dictionary of textiles. Fairchild Books, New York.

Fused Congo *n.* Congo copal, which has been subjected to a fusing or running process, in order to induce oil solubility.

Fused driers *n.* Driers, which are prepared by the direct interaction of the main constituents, stimulated by heat, in the absence of added water. The term is used to distinguish them from driers, which are made by precipitation from aqueous solutions of metallic salts and alkali soaps. Weismantal GF (1981) Paint handbook. McGraw-Hill Corporation Inc., New York.

Fused ribbon *n.* Acetate fabrics in wide widths may be cut into narrow ones by the application of heat. A hot knife blade caused the edges to sear and bead, thereby doing away with selvages on the edges of the goods.

Fusel oil \ˈfyū-zəl\ *n.* An arid oily liquid occurring in insufficiently distilled alcoholic liquors, consisting mainly of amyl alcohol, and used especially as a source of alcohols and as a solvent. Alcohols such as normal, iso, and tertiary butyl alcohols, propyl, and isopropyl alcohols, and well as amyl and isoamyl alcohols may be present. Merriam-Webster's collegiate dictionary, 11th edn. Merriam-Webster Inc., Springfield, MA, 2004.

Fusing *n.* (1) Melting. (2) Uniting, as by melting together.

Fusion \ˈfyü-zhᵊn\ [L *fusion-*, *fusio*, fr. *fundere*] (1555) *n.* With respect to vinyl plastisols and organosols, fusion is the state attained by heating when the resin particles have completely dissolved in the plasticizers and solvents present, so that upon cooling a homogeneous solid solution results. Wypych G (ed) (2003) Plasticizer's data base. Noyes Publication, New York. Wickson EJ (ed) (1993) Handbook of polyvinyl chloride formulating. John Wiley and Sons Inc., New York.

Fusion (atomic) *n.* A nuclear reaction involving the combination of smaller atomic nuclei or particles into larger ones with the release of energy from mass transformation. This is also called a thermonuclear reaction by reason of the extremely high temperature required to initiate it.

Fusion bonding *n.* Any of several methods that create a thin layer of melt on the plastic surfaces to be joined, which are then pressed together. Fusion bonding is limited to identical or melt-compatible polymers. Strong AB (2000) Plastics materials and processing. Prentice-Hall, Columbus, OH.

Fusion point Temperature of fusion or melting.

Fusion temperature *n.* In vinyl dispersions, the temperature at which fusion occurs. The *optimum fusion temperature* is that at which thermal degradation has not occurred and maximum physical properties are obtained in the final product. Wickson EJ (ed) (1993) Handbook of polyvinyl chloride formulating. John Wiley and Sons Inc., New York.

Fustics \ˈfəs-tik\ [ME *fustyk* smoke tree, fr. MF *fustoc*, fr, Arabic *fustuq*, fr. Gk *pistakē* pistachio tree] (15c) *n.* Yellow coloring matters of natural origin. Old fustic is obtained from the species, *Morus tinctorial*, and consists of moric and morintannic acids. Young fustic is obtained from *Rhus cotinus*, and consists of fustic.

Fuzz *n.* An accumulation of short, broken filaments collected from passing glass strands, yarns, or rovings over a contact point. The fuzz may be collected, weighted and used as an inverse measure of abrasion resistance.

Fuzz ball *See balling up.*

Fuzziness *n.* (1) A term describing a woven fabric defect characterized by a hairy appearance due to broken fibers or filaments. Principle causes are underslashed warp; rough drop wires, heddles, or reed; fabric slippage on take-up drum; rough shuttles; cut glass, dents, or reeds in warped; and damage in slashing. (2) A term describing a fabric intentionally made with a hairy surface; such fabrics are usually produced from spun yarns.

G

g \ˈjē\ *n.* (1) SI abbreviation for gram. (2) Symbol for the acceleration due to gravity at the earth's surface, and, in particular, for the standard value, 9.806650 m/s^2. It is sometimes used, erroneously, for the proportionality constant (g_c) in Newton's law of momentum change. *See force* (3).

G *n.* (1) SI abbreviation for GIGA-. (2) Symbol for shear modulus. In dynamic testing, G′ symbolizes the "real" or in-phase part of G″ the "imaginary" or out-of-phase component.

Gabardine \ˈga-bər-ˌdēn\ [MF *gaverdine*] (1520) *n.* A firm, durable, warp-faced cloth, showing a decided twill line, usually a 45 or 63°C right-hand twill.

Gable \ˈgā-bəl\ [ME, fr. MF, of Gr origin; akin to ON *gafl* gable] (14c) *n.* Triangle formed by the edge of a ridged roof and the similarly shaped wall enclosed by the horizontal and raking cornices; called a pediment in classic architecture. Harris CM (2005) Dictionary of architecture and construction. McGraw-Hill Co., New York.

Gableboard *See bargeboard.*

Gadoleic acid *n.* $CH_3(CH_2)_9CH{=}CH(CH_2)_7COOH$. Fatty acid from marine oils.

Gage \ˈgāj\. Alternate spelling of gauge.

Gahn's ultramarine *See cobalt blue.*

Gaiting *n.* The spacing of the needles in the dial and cylinder in relation to each other on rib (double-knit) and interlock knitting machines. In rib gaiting, the dial needles are midway between the cylinder needles. For interlock gaiting the dial and cylinder needles are in direct alignment.

Gal [*Galileo* Galilei] (1914). *n.* 1 gal = cm/s/s. Therefore, where the value of gravity is 980 this is the same as 980 gal. The milligal is now quite commonly used since it is approximate one part in a thousand of the normal gravity of the earth.

Galatea \ˌga-lə-ˈtē-ə\ [L, fr. Gk *Galeteia*] *n.* A sturdy, serviceable, warp-effect, five-shaft, left-hand twill-weave fabric, frequently cotton or a cotton blend, used for children's play clothes. Kadolph J, Langford AL (2001) Textiles. Pearson Education, New York.

Galipot Another name for Bordeaux turpentine.

Galley \ˈga-lē\ [ME *galeie*, fr. OF *galie*, *galee*, ultimately fr. MGk *galea*] (13c) *n.* A shallow metal tray used for holding type.

Galley proof A proof taken of type standing in a galley, before it is made up into pages. Leach RH, Pierce RJ, Hickman EP, Mackenzie MJ, Smith HG (eds) (1993) Printing ink manual, 5th edn. Blueprint, New York.

Galliolino *n.* Old name for Naples yellow, which is substantially lead antimoniate.

Gallon, imperial *n.* English gallon, equivalent to 277.77 in.^{3} or 3.54 l. Equals 1.2 USA gallons. USA gallon = 0.833 imperial gallons.

Gallon, USA *n.* (1) Volume equal to 231 in.^{3} For paint, varnish, lacquer, and related products this is measured at 25°C (77°F). (2) Volume of 8.33 lb of water. (3) Equivalent to 3.785 l.

Galvanic anode \gal-ˈva-nik ˈa-ˌnōd\. Source of current for cathodic protection provided by a metal less noble than the one to be protected, e.g., magnesium, zinc or aluminum as used for cathodic protection of steel. Baboian R (2002) Corrosion engineer's handbook, 3rd edn. NACE International – The Corrosion Society, Houston, TX.
See cathodic protection.

Galvanic cell *n.* (1) Cell consisting of two or more dissimilar metals or alloys in contact with the same body of an electrolytic solution such as seawater. Upon electrically connecting the dissimilar metals, a current flows as the result of accelerated corrosion of the more active of the dissimilar metals or alloys. (2) Electrolytic cell capable of producing electrical energy by electrochemical action. Baboian R (2002) Corrosion engineer's handbook, 3rd edn. NACE International – The Corrosion Society, Houston, TX.

G

Galvanic corrosion *n.* Increased corrosion above normal corrosion of a metal that is associated with the flow of current to a less active metal in the same solution and in contact with the more active metal. Baboian R (2002) Corrosion engineer's handbook, 3rd edn. NACE International – The Corrosion Society, Houston, TX.

Galvanic protection *See cathodic protection.*

Galvanized steel Zinc plated steel.
See galvanizing.

Galvanize \ˈgal-və-ˌnīz\ (1802) *v.* To coat iron or steel with zinc by immersing in molten zinc to produce a coating of zinc–iron alloy.

Galvanizing *n.* Application of a coating of zinc to steel by a variety of methods: barrel-galvanized dipped, electrogalvanized, electro-zinc plated, flake-galvanized, hot-dipped, hot-dip galvanized, mechanically plated, peer-coated galvanized, tumbler-galvanized, and wean-galvanized.

Gama (γ) ray (1903) *n.* High-energy electromagnetic radiation emitted from a nucleus. Giambattista A Richardson R, Richardson RC, Richardson B (2003) College physics. McGraw-Hill Science/Engineering/Math, New York.

Gamboge \gam-ˈbōj, -ˈbüzh\ [NL *gambogium*, alteration of *cambugium*, fr. or akin to Portuguese *Camboja* Cambodia] (1712) *n.* $C_{30}H_{35}O_6$. Resinous exudation of intense yellow color obtained from the species, *Guttifera*, which grown in India, Ceylon, Siam, and East Indies. The product is a mixture consisting chiefly of a water insoluble resin (up to 80%), with a water soluble gum (up to 27%). The resin portion is chiefly gambogic acid.
Also known as cambogia.

Gamma- (1896) *adj.* A prefix usually abbreviated as the Greek letter γ- and usually ignored in alphabetizing compound names, signifying that the so-labeled substitution is on the third carbon away from the main functional group of the molecule, and generally synonymous with the label "4-". They may be emitted from radioactive substances. They are quanta of electromagnetic wave energy similar to but of much higher energy that ordinary X-rays. The energy of a quantum is equal to hv ergs, where h is Planck's constant (6.6254×10^{-27} erg s) and v is the frequency of the radiation. Gamma rays are highly penetrating, an appreciable fraction being able to traverse several centimeters of lead. Giambattista A, Richardson R, Richardson RC, Richardson B (2003) College physics. McGraw-Hill Science/Engineering/Math, New York.

Gamma cellulose *n.* One of the three forms of cellulose. With beta cellulose it is called hemicellulose. Paint: pigment, drying oils, polymers, resins, naval stores, cellulosics esters, and ink vehicles, vol 3. American Society for Testing and Material, Paul N. Gardner Co. Inc., FL, 2001.
Also see alpha cellulose and beta cellulose.

Gamma distribution *n.* A random variable x with density

$$f(x) = \frac{1}{\Gamma(\alpha)\beta^{\alpha}} x^{\alpha-1} e^{-x/\beta}$$

is called a Gamma random variable with parameter α and β, and its distribution function is called a gamma distribution with parameter α and β.

Gamma protein *n.* Protein derived from soya bean meal. The commercial product contains approximately 54% of actual protein. It disperses in water in the presence of a small amount of alkali and is used as an extender for casein in water paints and distempers.

Gamma ray (1903) *n.* A quantum of electromagnetic energy emitted from some radioactive materials as they decay. Gamma rays are similar to, but much shorter wavelengths (typically ≈ 0.03 nm) and much higher energy than ordinary X-rays. They are highly penetrating, capable of passing through several centimeters of lead. Serway RA, Faugh JS, Bennett CV (2005) College physics. Thomas, New York.

Gamma transition Syn: glass transition.

Gamut, color *n.* Total variety of colors that can be produced by any prescribed method, such as the mixtures of any particular sets of colorants. Gamuts are frequently described by plotting them on a chromaticity diagram.

Gap \ˈgap\ [ME, fr. ON, chasm, hole; akin to ON *gapa* to gape] (14c) *n.* In filament winding, an unintentional space between two windings that should lie next to each other.

Garcia nutans oil *n.* Vegetable drying oil obtained from the species, *Euphorbiaceae*, which grows in Mexico and Venezuela. It contains over 90% of eleostearic acid and gels in less than 8 min at 282°C. It possesses the following constants: Sp gr of 0.942/15°C, refractive index of 1.525, iodine value of 177.9, and saponification value of 189.

Garden rocket oil *n.* Oil with fair drying properties. It has an iodine value of 155, a Sp gr of 0.928, and a saponification value of 192.

Gardner color standards *n.* Liquid yellow to brown color standards for describing the color of drying oils, ASTM method D-1544.

Gardner–Holt bubble viscometer *See air-bubble viscometer.*

Gardner–Holt viscosity tubes *n.* Series of selected glass tubes of constant, standard diameter, which is filled with liquids of various viscosities, except for a small air space at the top. When these tubes are inverted, the air bubble travels through the liquid, and the rate of travel is a measure of the viscosity of the liquid. Empty tubes of similar standard dimensions are filled with liquids of which the viscosity is required, and the rates of travel of the bubbles compared with those of the tubes containing the liquids of known viscosities. Viscosities are compared under controlled temperature conditions (Paul N. Gardner Co. Inc., 316 N. E. Fifth Street, Pompano Beach, FL, www.gardco.com).

Gardner impact test (falling-weight test) *n.* A test for the impact resistance of rigid plastic sheets or parts. ASTM D 3029 describes two methods, F and G. In F, a weight with a hemispherical nose falls through a tube to strike the specimen, and G (for which the Gardner Impact Tester is approved), in which the falling weight hits a round-nose striker resting on the specimen. Either the weight or the height of drop may be varied, the former being recommended because of the sensitivity of most plastics to the velocity of impact (determined by height). Several tests on new specimens may be made at each impact-energy level (height × weight), and the fraction of breaks at each level is noted. This procedure efficiently provides information on both the mean energy to break and the standard deviation. Where the mean is of greatest interest, the up-and-down method provides a good estimate with less testing (Paul N. Gardner Co. Inc., 316 N. E. Fifth Street, Pompano Beach, FL, www.gardco.com).

Gardner mobilometer *n.* Viscometer used for determining the viscosity of all types of varnishes, oils, and pigmented compositions. The mobilometer consists of a cylinder into which is poured the paint product of which the viscosity is required. Into this is fitted a closely fitting disc attached to the end of a guide rod. Provision is made at the other end of the rod for weighting it, if required, when dealing with viscous products, by including a shallow tray into which weights may be placed. The time is determined for the disc and rod to travel a definite distance into the paint product, and is a measure of its viscosity.

G

Garnet abrasive \ˈgär-nət-\. Almandite, a type of garnet mineral occurring in New York State, is used in coated abrasives. Hardness and toughness are increased by heat treating. It fractures along the cleavage planes of the crude crystals, and the resulting grains have very sharp edges.

Garnet lac *n.* Form of shellac prepared from sticklac by a special refining process. The term "garnet" is merely an indication of the deep red color of the material when dissolved in alcohol.

Garnet paper (ca. 1902) *n.* The paper generally favored for hand sanding on wood. The abrasive is the same mineral as the semiprecious garnet gem, almandite.

Garnetting *n.* A process for reducing various textile waste materials to fiber by passing them through a machine called a garnett, which is similar to a card.

Gas \ˈgas\ [NL, alter. of L *chaos* space, chaos] (1779) *n.* A state of matter in which the material has a very low density and viscosity, can expand and contract greatly in response to changes in temperature and pressure, easily diffuses into other gases, and readily and uniformly distributes itself throughout any container. A gas can be changed to the liquid or solid state ink only by the combined effect of increased pressure and decreased temperature (below the critical temperature).

Gas black *n.* Another term for carbon black, so called from its manufacture from petroleum gas.
See carbon black.

Gas blue *n.* Variety of Prussian blue, so called from its origin in certain by-products of coal gas manufacture.

Gas checking *n.* Film phenomenon associated with compositions containing vegetable drying oils, or derivatives therefrom, and demonstrated by the formation of crowsfoot-like wrinkles, or wrinkles which have the appearance of crystalline patterns. The phenomenon occurs when susceptible films are allowed by dry in an atmosphere containing the products of combustion of coal gas or natural gas; hence the term, "gas checking". Syn: crystallizing, crowsfooting, webbing, and frosting.

Gas chromatography (GC) (1952) *n.* A method of chemical analysis in which the specimen is vaporized and introduced into a stream of carrier gas (usually helium), the stream then passed through a column packed with adsorbent particles that separate the stream into its constituent molecules. These fractions pass through the column – first adsorbed from, then desorbed into the ongoing stream of carrier gas – at characteristic rates. Their concentrations in the effluent are measured by a detector such as a thermal-conductivity cell. The recorded concentrations are

displayed on a strip chart with time as the abscissa. From the positions and areas of the peaks the identities and relative concentrations of the constituents in the sample may be determined. An example of a gas chromatograph combined with a mass spectrometer is the Clarus 500 GC Mass Spectrometer. Willard HH, Merritt LL, Dean JA (1974) Instrumental methods of analysis. D. Van Nostrand Co., New York.
See also chromatography.

Gas crazing *n.* The wrinkling of a tung oil film under certain drying conditions.
See gas checking.

Gas fading *n.* A change of shade of dyed fabric caused by chemical reaction between certain disperse dyes and acid gases from fuel combustion, particularly oxides of nitrogen.

Gas-injection molding *n.* A specialized technique for molding low-density structures in which a mixture containing cork particles, or glass or phenolic microspheres, glass fibers, and a thermosetting resin is injected into a mold by fluidizing it in a gas stream.

Gas–liquid chromatography (1952) *n.* A variation of gas chromatography in which the chromatographic column is packed with a finely divided solid impregnated with a non-volatile organic liquid. The sample to be analyzed is injected into the inlet of the column where it is quickly and completely vaporized. The carrier-gas stream carries it into the packed section, where the vapors contact the impregnated solids. They are absorbed by the non-volatile liquid phase, and then later desorbed into the carrier gas. The vapor of each compound spends a characteristic fraction of time in the condensed phase and the remainder in the mobile gas phase. Each chemical species will tend to migrate at its own rate and will be separated from other species in the time of emergence from the column. The detector senses their concentrations in the effluent and its signal strength is recorded and displayed versus time on a strip chart. Willard HH, Merritt LL, Dean JA (1974) Instrumental methods of analysis. D. Van Nostrand Co., New York.

Gas permeability *n.* The ease with which a gas or vapor passes through a membrane, e.g., a plastic sheet or film.
See permeability.

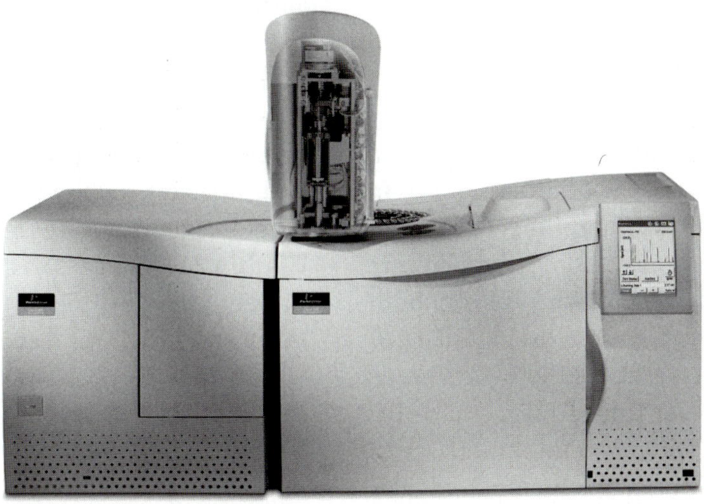

Gas chromatograph, courtesy of Perkin-Elmer Corporation

Gas-phase polymerization *n.* A polymerization process developed by Union Carbide for high-density polyethylene, particularly for a grade for making paper-like films. Purified ethylene and a highly active chromium-containing catalyst in dry-powder form are fed continuously into a fluidized-bed reactor. The resin forms as a powder, thereby avoiding the gel, discoloration, and contamination problems often associated with conventional polymerization processes.

Gasproof *n.* Coating composition, which does not gas check when exposed to an atmosphere containing the products of combustion of coal gas.
See gas checking.

Gassing (1852) *v. See singeing.*

Gas thermometer *n.* Where P_o, P_s, and P_z represent the total pressure with the bulb at 0°C, at the boiling-point of water and at the unknown temperature, respectively, t_s the temperature of steam and t_z the unknown temperature,

$$t_x = t_s \frac{P_x - P_o}{P - P_{os}}$$

(approximately). The total pressure on the gas in the bulb is the algebraic sum of barometric pressure at the time and that measured by the manometer.

Gas-transmission rate (GTR) *n.* The quantity of a given gas passing through a unit area of the parallel surfaces of a plastic film in unit time under the conditions of the test. These conditions, including temperature and partial pressure of the gas on both sides of the film, must be stated. The SI unit of GTR is mol/(m^2 s) but others, some of them involving mixed metric and English units, are still in common use. (www.astm.org) for standard method of gas-transmission rate. *See ASTM See also permeance and permeability.*

Gas welding *See hot-gas welding.*

Gate \ˈgāt\ [ME, fr. OE *geat*; akin to ON *gat* opening] (before 12c) *n.* In injection and transfer molding, the channel through which the molten resin flows from the runner into the cavity. It may be of the same cross-section as the runner, but most often is restricted to 3 mm or even much less. A gate whose diameter is less than 0.5 mm is known as a *pinpoint gate.* A *submarine gate* is shaped to conduct the melt below the parting line of the mold and into the cavity at a point just below its edge. Other types of gates are the *fan gate* (a diverging, thin gate), and the *tab gate* (one that extends the runner into the molded part). The term gate is also used for the portion of the plastic molding formed by the gate orifice. Strong AB (2000) Plastics materials and processing. Prentice-Hall, Columbus, OH.

Gate blush (gate splay) *n.* A blemish or disturbance in the gate area of an injection-molded article. It occurs when the melt fractures as it emerges from the gate due to sudden release of large elastic stresses.

Gauffrage *n.* Printing by pressure alone, without the use of a pigment, producing an embossed effect on the paper.

Gauge \ˈgāj\ [ME *gauge*, fr. ONF] (15c) (gage) *n.* (1) Any instrument that measures and indicates such quantities as thickness, pressure, temperature, or liquid level. (2) The thickness of a plastic sheet or film, usually given in mils or mm. (3) Any of the standard wire and sheet-metal scales in which the gauge numbers are inversely related to wire diameter or sheet thickness. (4) The number of wales per inch in a knit fabric. (5) On spinning or twisting frames, the distance from the center of one spindle to the center of the next spindle in the same row.

See strain gauge and wire gauge.

Gauge band *n.* A term used in the packaging-film industry for a thickness irregularity found in rolls of film. A thick area at some locality over the width of a flat film will produce a raised ring in a finished roll. Similarly, a thin area will cause a depressed ring. Such films, when unwound, tend not to lie perfectly flat. With shrinkable films, thin areas are troublesome because it is more likely that they will cause tearing or burn-through during the shrink cycle.

Gauge length *n.* On a dog bone-shaped, tensile-test specimen before stress is applied, the distance between two marks on the narrow part ("waist") of the specimen, perpendicular to the direction of pull that will be used to measure elongation. If the specimen is of uniform thickness and width, gauge length may be taken as the nip-to-nip distance between the clamping fixtures.

Gauge wire *n.* Used with an extra filling yarn during weaving, this type of standing wire controls the height of fabric pile.

Gauss \ˈgaüs\ [Karl French *Gauss*] (1882) *n.* The cgs of magnetic induction (flux density). It is equal to 1 Mx/cm^2. It has such a value that if a conductor 1 cm long moves through a magnetic field at a velocity of 1 cm, in an induction mutually perpendicular, the induced emf is one abvolt. Giambattista A, Richardson R, Richardson RC, Richardson B (2003) College physics. McGraw-Hill Science/Engineering/Math, New York.

Gauze \ˈgóz\ [MF *gaze*] (1561) *n.* A thin, sheer-woven fabric in which each filling yarn in encircled by two warp yarns twisted around each other, gauze is similar to cheesecloth. It may by made of silk, cotton, wool, or manufactured fibers. Cotton gauze is primarily for surgical dressings.

Gavan \ˈgä-vən\. Persian name for the plant, which yields gum tragacanth.

Gay–Lussac's law \ˈgā-lə-ˈsak-\ [Joseph-Louis *Gay-Lussac*, 1778–1850, F chemist and physicist]. Syn: Charles' law. Goldberg DE (2003) Fundamentals of chemistry. McGraw-Hill Science/Engineering/Math, New York.

Gay–Lussac's law of combining volumes *n.* If gases interact and form a gaseous product, the volumes of the reacting gases and the volumes of the gaseous products are to each other in simple proportions, which can be expressed by small whole numbers. Goldberg DE (2003) Fundamentals of chemistry. McGraw-Hill Science/Engineering/Math, New York.

Gear box (gear reducer) *n.* Syn: for speed reducer.

g_c The proportionality constant in Newton's law of momentum change, g_c is often needed to convert viscosities between force units and mass units: $\mu_f = g_c\mu_m$. In SI, $g_c = 1.000000\ \mathrm{N\,s^2/(kg\,m)}$, so a viscosity of, say 100 Pa s is also equal to 100 kg/(m s). *See force.*

Gear crimping *See texturing, gear crimping method.*

Gear pump A pump consisting of a sturdy housing within which two intermeshing toothed wheels rotate, and inlet and outlet ports. The gears typically have widths about equal to their diameters and may be spur, helical, or, rarely because of their cost, herringbone bears. The inlet port is attached at the waist of the 8-shaped casing where the gears are moving apart; the outlet is at the opposite side. Liquid trapped between the tight-fitting teeth and the casing is borne around to the discharge side where it is displaced and expelled by the meshing of the teeth. Small gear pumps have long been used in producing staple fiber, requiring extremely high melt pressures (up to 100 MPa). Since about the mid-1970s, larger ones have come into increasing use for general extrusion operations where die resistances or filtration requirements generate high pressures, or where close dimensional tolerances on extrudates are desired. Gear pumps have been successfully retrofitted to extruders, providing net higher output, better product quality, and quick return on the additional investment.

G

Geiger counter \ˈgī-gər-\ [Hans *Geiger* † 1945 Gr physicist] (1924) *n.* Detector for radioactivity depending upon ionized particles that affect its mechanism. As its indicates, it both detects and make a count of them possible. Goldberg DE (2003) Fundamentals of chemistry. McGraw-Hill Science/Engineering/Math, New York.

Gel \ˈjel\ [*gelatin*] (1899) *n.* A semisolid system consisting of a network (three-dimensional network polymer) of solid aggregates within which a liquid is trapped. The initial, jelly-like solid phase that develops during the formation of a resin from a liquid. With respect to vinyl plastisols, gel is a state between liquid and solid that occurs in the initial stages of heating, or upon prolonged storage. *Note*—All three types of gel have very low strength and do not flow like liquids. They are soft, flexible, and may rupture under their own weight unless supported externally (ASTM D 883). A defect in plastic film such as polyethylene or PVC characterized by tiny, hard, glassy particles that appear in an otherwise clear film. These gel particles are believed to be bits of resin of much higher-than-average molecular weight, perhaps cross-linked. Designed gels (three-dimensional network systems swell with solvents and change size predictably, and these properties can be of commercial value such as ion absorbers from aqueous. Bryson HC (1956) Paint faults and remedies. Scientific Surveys, London. Hess M (1965) Paint film defects. John Wiley and Sons Inc., New York. Paint/coatings dictionary. Compiled by Definitions Committee of the Federation of Societies for Coatings Technology, 1978. Addad JPC (ed) (1996) Physical properties of polymeric gels. John Wiley and Sons, New York.
See also gelation and three-deminsional network polymer.

Gelatin \ˈjel-lə-tᵊn\ [F *gélatine* edible jelly, gelatin, fr. It *gelatina*, fr. *gelato*, pp of *gelare* to freeze, fr. L] (1800) *n.* It is a protein product derived through partial hydrolysis of the collagen extracted from skin, bones, cartilage, ligaments, etc. The natural molecular bonds between individual collagen strands are broken down into a form that rearranges more easily. Gelatin melts when heated and solidifies when cooled again. Together with water it forms a semi-solid colloidal gel. Gelatin (also gelatine) is a translucent brittle solid substance, colorless or slightly yellow, nearly tasteless and odorless, which is created by prolonged boiling of animal skin and connective tissue. It has many uses in food, medicine, and manufacturing. Substances that contain or resemble gelatin are called gelatinous. Gelatin is protein type material extracted from animal skins, sinews, tendons, and from bones. Gelatin may be regarded as a pure glue. It dissolves in hot water to give a viscous solution with pronounced gelling characteristics.

Inedible gelatin has useful applications in the production of sizes for various purposes, and in emulsions. Morrison RT, Boyd RN (1992) Organic chemistry, 6th edn. Prentice-Hall, Englewood, Cliffs, NJ.

Gelatinous \jə-ˈlat-nəs, -ˈla-tᵊn-əs\ (1766) *adj.* Having the consistency of a very soft elastic solid, the nature and appearance of gelatin.

Gelation \ji-ˈlā-shən\ [L *gelation-*, *galatio*, fr. *gelare*] (1854) *n.* The formation of a GEL. With regard to vinyl plastisols and organosols, gelation is the change of state from the liquid suspension of the solid condition that occurs in the course of heating and/or aging, when the plasticizer has been mostly absorbed by the resin, resulting in a dry but weak and crumbly mass. Within normal proportions of resin and plasticizer, this state is attained when the resin particles have soaked up so much plasticizer that they touch each other. As heating progresses, the swollen particles begin to fuse together, resulting in some cohesive strength. Gelation is considered to continue until useful levels of mechanical properties are attained, such as have developed at the clear point. Much confusion has existed regarding the meaning of gelation because in the early years of the art, especially in Europe, *gelation* was used in place of the term fusion, now employed in USA. Wickson EJ (ed) (1993) Handbook of polyvinyl chloride formulating. John Wiley and Sons Inc., New York.

Gelation time *n.* With reference to thermosetting resins, the interval of time between the mixing in of a catalyst and the formation of a gel.

Gel cellophane *n.* Regenerated cellulose which has never been allowed to dry out, but has always been stored swollen with water or other solvent. Thin films are frequently used as the semi-permeable membrane in membrane Osmometry.

Gel coat *n.* (1) In reinforced plastics, a thin outer layer of resin, sometimes containing pigment, applied to give the structure its surface gloss and finish. It also serves as a barrier to liquids and ultraviolet radiation. The gel coat is the first to be applied to the mold in the lay-up process (after any mold-release agent), becoming permanently bonded to the succeeding layers of reinforcement and resins. In spray-up, the gel coat is applied last, sometimes after partly curing the mains structure. (2) High-build, chemical-resistant, thixotropic polyester coating. Starr TF (1993) Data book of thermoset resins for composites. Elsevier Science and Technology Books, New York.

Gel dyeing *See dyeing.*

Gel effect *n.* The increase in molecular weight as conversion progresses.
See autoacceleration and gel.

Gel filtration *n.* Method of polymer fractionation, both preparative and analytical, in which a sample of the polymer in dilute solution is injected into a stream of flowing solvent at the top of a packed column. The elution of the polymer is governed by the size of the polymer molecule. The polymer solute is frequently a biopolymer.

Gel, gel point, gelation *n.* The point at which a thermosetting system attains an infinite value of its average molecular weight; infinite viscosity and non-flowing. Allcock HR, Mark J, Lampe F (2003) Contemporary polymer chemistry. Prentice-Hall, New York. Alger MSM (1989) Polymer science dictionary. Elsevier Science Publishers, Barking, Essex, England. Elias HG (1977) Macromolecules, vols 1–2. Plenum Press, New York. Lenz RW (1967) Organic chemistry of synthetic high polymers. Interscience, Publishers Inc., New York. *See Trommsdorf effect.*

Gelling *n.* Any process whereby paint or varnish thickens to jelly-like consistency. *Also see livering.*

Gelling agent *See thickening agent.*

Gel particle *See gel* (4).

Gel-permeation chromatography (1966) *n.* This method is used to provide the molecular weigh and molecular weight distribution of a polymer by a fractionation technique. The polymer must be dissolved in a suitable solvent. It can also be defined as chromatography in which macromolecules such as polymers in a solution are separated by size on a column packed with a gel such as polystyrene. An example of a GPC chromatogram of a typical polymer is shown. Original name for, but now fading synonym of size-exclusion chromatography. An HPLC chromatograph equipped with a gel permeatioin column can measure the molecular weight of polymers. The highest molecular weight fraction exits the column first. Yau WW, Kirkland JJ, Bly DD (2001) Modern size-exclusion chromatography. John Wiley and Sons, New York. Collins EA, Bares J, Billmeyer FW Jr (1973) Experiments in polymer science. Wiley-Interscience, New York.

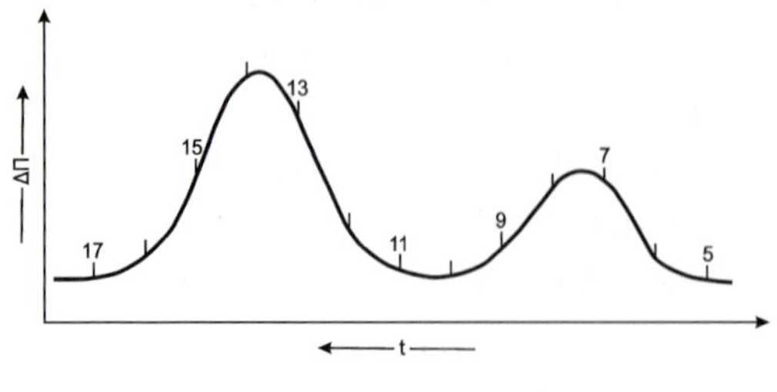

Gel point The stage at which a liquid begins to exhibit elastic properties and increased viscosity. This stage may be observed from the inflection point on a viscosity – time plot. *See gel* (2) (ASTM D 883).

Gel spinning *See spinning* (2).

Gel time (of a drying oil) *n.* Time required for the oil to form a solid gel under specified conditions of temperature.

Geminal *adj.* Description of substitution on adjacent atoms (1,2-chloro-) as opposed to same atom or vicinal (e.g., 1,1-chloro-) positions.

Genuine ultramarine *n.* Natural ultramarine blue. *See ultramarine blue.*

Geodesic \ˌjē-ə-ˈde-sik, -ˈdē-, -zik\ (1821) *n.* Pertaining to the shortest distance, lying on a surface, between two points on that surface, e.g., a straight line on a plane or great circle on a sphere.

Geodesic isotensoid A filamentary structure in which there exists a constant stress in any given filament at all points in its path. Iaac MD, Ishal O (2005) Engineering mechanics of composite materials. Oxford University Press, UK.

Geodesic-isotensoid contour *n.* In filament-wound, reinforced-plastic pressure vessels, a dome contour in which the filaments are placed on geodesic paths so that the filaments will experience uniform tensions throughout their lengths under pressure loading. Iaac MD, Ishal O (2005) Engineering mechanics of composite materials. Oxford University Press, UK.

Geodesic ovaloid *n.* A contour for end domes, the fibers forming geodesic lines on the surface of revolution. The forces exerted by the filaments are proportioned to meet hoop and meridional stresses at any point. Iaac MD, Ishal O (2005) Engineering mechanics of composite materials. Oxford University Press, UK.

Geometric metamerism \-mə-ˈta-mə-ˌri-zəm\ *n.* Phenomenon exhibited by a pair of colors, which appear to be a color match at one angle of illumination and viewing but which are no longer a match when the angle of illumination or viewing is changed. McDonald R (1997) Colour physics for industry, 2nd edn. Society of Dyers and Colourists, West Yorkshire, England.

See goniometachromatism, flop, and metamerism.

Geotextiles *n.* Manufactured fiber products made into fabrics of various constructions for use in a wide variety of civil engineering applications.

GEP *n.* Glass-fiber-reinforced epoxy resins.

Geranium lake \jə-ˈrā-nē-əm-\ *n.* Type of colorant produced by precipitation of certain acid dyestuffs with barium chloride followed by lead nitrate.

Germicidal-lamp test *n.* A quick screening method for evaluating the relative resistance of vinyl compounds to discoloration and degradation upon exposure to light and weather. The specimens are placed approximately three inches below a germicidal lamp with a principal UV wavelength of 253.7 nm. The materials are rated according to the degree of discoloration and plasticizer spewing after a specified interval of time, e.g., 24 or 48 h.

Gesso \ˈje-(ˌ)sō [It, literally, gypsum, fr. L *gypsum*] (1596) *n.* Traditional gesso is a material made from chalk and gelatin or casein glue, and painted on wooden panels as a surface for tempera work. Polymer gesso provides a flexible ground that can be applied to canvas and other materials.

GFK *n.* Glass–fiber-reinforced plastics.

Ghatti gum *n.* Gum obtained from India and Sri Lanka, from *Anogeissus latifolia*, from which it exudes.

Known also as Ghati gum and Indian gum.

Ghosting (ca. 1957) *n.* (1) A condition found in wall paints of low sheen and moderately strong colors where there is a slight but definitely perceptible bleaching in irregular patterns. (2) Presence of a faint image of a printed design appearing in areas which are not intended to receive that portion of the image. Paint/coatings dictionary. Federation of Societies for Coatings Technology, Philadelphia, Blue Bell, PA, 1978.

Gibbs' free energy *See free energy.*

Gibbsite *See aluminum oxide, hydrated.*

Gibbs' phase rule *n.* $F = C + 2 - P$, E (the number of degrees of freedom of a system) is the number of variable factors (temperature, pressure, and concentration) of the components, which must be arbitrarily fixed in order that the condition of the system may be perfectly defined. C, the number of the components of the system, is chosen equal to the smallest number of independently variable constituents by means of which the composition of each phase participating in the state of equilibrium can be expressed in the form of a chemical equation; the components must be chosen from among the constituents which are present when the system is in a state of true equilibrium, and which take part in that equilibrium; as components are chosen the smallest number of such constituents necessary to express the composition of each phase participating in the equilibrium, zero and negative quantities of components being permissible; in any system the number of components is definite, but may alter with changes in conditions of experiment; a qualitative but not quantitative freedom of selection of components is allowed, the choice being influenced by suitability and simplicity of application. P, the number of phases of the system, are the homogeneous, mechanically separate and physically distinct portions of a heterogeneous system; the number of phases capable of existence varies greatly in different systems; there can never be more than one gas or vapor phase since all gases are miscible in all proportions, a heterogeneous mixture of solid substances forms as many phases as there are substances present. Phillip R Watson (1997) Physical chemistry. John Wiley and Sons Inc., New York.

Giga- \\ˈji-gə, ˈgi-\\ {*combining form*} [ISV, fr. Gk *gigas* giant] (G). SI prefix meaning $\times 10^9$. Lide DR (ed) (2004) CRC handbook of chemistry and physics. CRC Press, Boca Raton, FL.

Gigging *See napping.*

Gilbert \\ˈgil-bərt\\ [William *Gilbert*] (1893) *n.* The cgs emu of magnetomotive force. 1 gilbert $= 10/4\pi$ A t.

Gilding \\ˈgild\\ [ME, fr. OE *gyldan*; akin to OE *gold* gold] (14c) *vt.* Art of covering substances, such as wood, with layers of gold leaf.

Gilling *See pin drafting.*

Gilsonite \\ˈgil-sə-ˌnīt {*trademark*}. A black asphaltic resinous material found in Utah and Colorado used in black printing ink and as an ingredient in the manufacture of cold molding compounds, paints, etc. It is hard, bituminous, and finds application in black japans, Brunswick blacks, stains, baking enamels, and electrical insulating varnishes of many types. Hibbard MJ (2001) Mineralogy. McGraw-Hill Co. Inc., New York. Ash M, Ash I (1998) Handbook of fillers, extenders and dilutents. Synapse Informtion Resources Inc., New York.
Also called vintahite.

g-Index *n.* A measure of polydispersity of molecular-weight distribution in polymers given by

$$g = (M_z/M_w - 1)^{0.5},$$

where M_z and M_w are the viscosity-average and weight-average molecular weights. Slade PE (2001) Polymer molecular weights, vol 4. Marcel Dekker, New York.

Gingelli oil *n.* Another name for sesame oil.

Gingham \\ˈgiŋ-əm\\ [modification of Malay *genggang* striped cloth] (1615) *n.* A woven fabric characterized by a block or check effect produced by weaving in dyed yarns at fixed intervals in both the warp and the filling.

Glacé \\gla-ˈsā\\ [F, fr. pp of *glacer* to freeze, ice, glaze, fr. L *glaciare*, fr. *glacies*] (1847) *adj.* A lustrous, glossy effect imparted to fabrics by finishing.

Glance pitch *n.* Another name for Manjak.

Glass \\ˈglas, ˈglás\\ {*often attributive*} [ME *glas*, fr. OE glœs; akin to OE *geolu* yellow] (before 12c) *n.* Any of a wide variety of rigid, amorphous solid materials, but in particular, any of the transparent vitreous glasses formed by cooling molten mixtures of silica, sodium oxide, other alkali-metal and alkaline-earth oxides, lead oxide, alumina, and, for special purposes, other oxides (many organic polymers experience "glassy states" at low temperatures but they are seldom referred to as "glasses"). Glass compositions used in fibrous form as plastics reinforcements include vitrified quartz (pure silica), C glass, E glass, and S glass. The glasses as a family are characterized by high moduli, low tensile strength (except as fibers), very low ultimate tensile elongation, brittleness, and good electrical properties and chemical resistance. Iaac MD, Ishal O (2005) Engineering mechanics of composite materials. Oxford University Press, UK. Lee SM (1989) Dictionary of composite materials technology. Technomic Publishing Co. Inc., Lancaster, PA.

Glass fiber (1882) *n.* A manufactured fiber in which the fiber-forming substance is glass (FTC definition). In the continuous filament process, glass marbles are melted in an electric furnace and the liquid flows in fine streams through small orifices at the bottom of the melting chamber. The resultant filaments are caught and drawn by a high-speed draw winding mechanism. In the staple fiber process, the streams of molten glass are attenuated into fibers by jets of high-pressure steam or air. These fibers are gathered on a revolving drum and then wound on tubes to form staple fiber sliver or bands that can be drafted, twisted, and plied. Glass fiber is incombustible and will tolerate heat up to 1000°F without material damage. Potential strength is not realized in woven fabrics or even in yarns, because the fiber is brittle and fracture points may develop, but nevertheless, very high tensile strength is obtained in woven fabrics, and is retained at elevated temperatures. The fiber originally was difficult to color but methods have been developed to accomplish this. Moisture absorption is low. Electrical and insulation resistance is high. Glass fiber is used for heat and electrical insulation, filter cloth in the chemical and dye industries, reinforcing belts in tires, novelty fabrics, tablecloths, and fireproof draperies. Because of its brittleness, it is not used in wearing apparel or in household fabrics that have to withstand frequent flexing. Lee SM (1989) Dictionary of composite materials technology. Technomic Publishing Co. Inc., Lancaster, PA.

Glass–fiber reinforcement *n.* Any of a family of reinforcing materials for plastics based on single, drawn filaments of glass ranging in diameter from 3 to 20 μm. Single filaments are produced by mechanically drawing molten-glass streams, then the filaments are usually coated with a size or coupling agent which protects them from abrasion (to which the uncoated filaments are very vulnerable) and improves bonding with resins. The filaments are then gathered into bundles called *strands* or larger bundles called *rovings*. The strands may be used in continuous form for filament winding or pultrusion; or chopped into short lengths for incorporation into molding compounds or during spray-up; or formed into fabrics and mats of various types for use in hand lay-up, matched-die molding, and other laminating processes. Glass–fiber reinforcements are classified according to their special properties. E glass is electrical-grade glass, the most common general-purpose type. C glass is the chemically resistant grade, and S glass denotes high tensile strength. Glass fibers coated with nickel by electron-beam deposition are used making molding compounds for electrically conductive articles. Another form of glass reinforcement is glass flakes. Lee SM (1989) Dictionary of composite materials technology. Technomic Publishing Co. Inc., Lancaster, PA.

Glass finish (size) *See coupling agent.*

Glass flakes *n.* A reinforcing filler produced by blowing molten Type E glass into very thin tubes, then smashing the tubes into small fragments. The flakes pack closely in thermosetting-resin systems, producing strong products with good moisture resistance. Lee SM (1989) Dictionary of composite materials technology. Technomic Publishing Co. Inc., Lancaster, PA.

Glassine \gla-ˈsēn\ (1916) *n.* Thin transparent paper treated with urea-formaldehyde resin, used for packaging.

Glass mat (fibrous-glass mat) *n.* A thin web of non-woven glass fibers that may or may not contain a small percentage of resin binder.

Glass microspheres *See microspheres.*

Glass-reinforced plastics (GRP) *See reinforced plastic.*

G

Glass–rubber transition Syn: glass transition.

Glass spheres *n.* Solid glass spheres of diameters ranging from 5 to 5000 μm are used as fillers and/or reinforcements in both thermosetting and thermoplastic compounds. The size used most frequently passes a 325-mesh sieve, with an average sphere diameter of 30 μm. The spheres are available coated with various silane coupling agents to improve bonding between polymer and glass. The spheres improve a physical properties and reduce costs of materials and end products. Some small glass spheres are considered to be microspheres. Lee SM (1989) Dictionary of composite materials technology. Technomic Publishing Co. Inc., Lancaster, PA.

Glass stress In a filament-would part, typically a pressure vessel, the stress calculated from the load and the cross-sectional area of the reinforcement only.

Glass transition (gamma transition, second-order transition, and glass–rubber transition) *n.* A reversible change that occurs in an amorphous polymer or in amorphous regions of a partly crystalline polymer when it is heated from a very low temperature into a certain range, peculiar to each polymer, characterized by a rather sudden change from a hard, glassy, or brittle condition to a flexible or elastomeric condition. Physical properties such as coefficient of thermal expansion, specific heat, and density, usually undergo changes in their temperature derivatives at the same time. During the transition, the molecular chains, normally coiled, tangled, and motionless at the lower temperatures, become free to rotate and slip past each other. This temperature varies widely among polymers. Groenewoud WM (2001) Characterization of polymers by thermal analysis. Elsevier Science and Technology Books, New York. Mark JE (ed) (1996) Physical properties of polymers handbook. Springer-Verlag, New York.

Glass-transition temperature (T_g) *n.* The approximate midpoint of the temperature range over which the glass transition occurs. T_g is not obvious (like a melting point), and is detected by changes, with rising temperature, in secondary properties such as the rate of change with temperature of specific volume or electrical or mechanical probe. Groenewoud WM (2001) Characterization of polymers by thermal analysis. Elsevier Science and Technology Books, New York.

Glass wood (1879) *n.* Glass fibers in a mass resembling wool and used for thermal insulation and air filters.

Glassy state *n.* A brittle solid state of a material, usually amorphous in structure, and prone to conchoidal fracture. Moreover, the observed T_g can vary significantly with the specific property chosen for observation and on experimental details such as the rate of heating or electrical frequency. A reported T_g should therefore be viewed as an estimate. The most reliable estimates are normally obtained from the loss peak in dynamic-mechanical tests or from dilatometric data. They change to the crystalline state, i.e., devitrify, in time or by the action of temperature, shock, scratching, etc. Glassy materials do not melt at a specific temperature but soften gradually and continuously on heating. Groenewoud WM (2001) Characterization of polymers by thermal analysis, Elsevier Science and Technology Books, New York. Mark JE (ed) (1996) Physical properties of polymers handbook. Springer-Verlag, New York.

Glaze \ˈglāz\ [ME *glasen*, fr. *glas* glass] (14c) *v.* Very thin coating of a paint product, usually a semitransparent tinted coating, applied on a previously painted surface to produce a decorative effect.

Glazing \ˈglā-ziŋ\ [ME *glasen*, fr. *glas* glass] (14c) *v.* (1) A finishing process that produces a smooth, highly polished, or lustrous surface on a fabric such as chintz. The fabric is treated with starch, glue, paraffin, or shellac, then friction calendered. Synthetic resins are used for a more permanent finish. (2) A shiny fabric appearance produced unintentionally, e.g., by pressing at excessive temperatures. Kadolph J, Langford AL (2001) Textiles. Pearson Education, New York.

Glazing compound *n.* A dough-like material consisting of pigment and vehicle, used for sealing window glass in frames. It differs from putty in that it retains its plasticity for an extended period. Weismantal GF (1981) Paint handbook. McGraw-Hill Inc., New York.

GLC *n.* Abbreviation of gas–liquid chromatography; a technique by which the separation and identification of volatile components of a mixture is possible.

Glitter \ˈgli-tər\ [ME *gliteren*, fr. ON *glitra*; akin to OE *geolu* yellow] (14c) (flitter, spangles) *vt.* A family of decorative pigments comprising light-reflective flakes of sizes large enough so that each flake produces an individually seen sparkle or reflection. They are incorporated into plastics during compounding.

Global radiation *n.* The wavelength distribution of sunlight under a given environment (e.g., under windowpane glass).

Gloss \ˈgläs, ˈglós\ [prob. of Scand origin; akin to Icelandic *glossa* to glow, akin to OE *geolu* yellow] (1538) *n.* Subjective term used to describe the relative amount and nature of mirror-like (specular) reflection. Different types of gloss are frequently arbitrarily differentiated, such as sheen, distinctness-of-image gloss, etc. Trade practice recognizes the following stages, in increasing order of gloss: *Flat* (or *matte*) – practically free from sheen, even when viewed from oblique angles (usually less than 15 on 85° meter); *Eggshell* – usually 20–35 on 60° meter; *Semigloss* – usually 35–70 on 60° meter; *Full gloss* – smooth and almost mirror-like surface when viewed from all angles, usually above 70 on 60° meter. Gloss of plastics is measured with a glossmeter described in ASTM test D 523. The same meter is used in measuring the resistance of shiny plastic surfaces to abrasion (D 673).
See also satin finish.

Gloss, distinctness-of-image *n.* The sharpness with which outlines are reflected by the surface of an object.

Glossing up *n.* Undesirable burnishing.
See burnish and burnish resistance.

Gloss ink *n.* An ink that dries with a minimum of penetration into the stock and yields a high luster.

Glossmeter *n.* An instrument for measuring the degree of gloss in relative terms. Such instruments measure the light reflected at a selected specular angle, such as 20° (from the perpendicular), 45°, 60°, 75°, or 85°. Results obtained are very dependent on the instrument design, calibration, technique used, types of samples, etc. In essence, a glossmeter is an abridged goniophotometer. Koleske JV (ed) (1995) Paint and coating testing manual. American Society for Testing and Materials, Paul N. Gardner Co. Inc., FL.

Gloss oil *n.* Solution of limed rosin or lined rosin acids in a volatile solvent, used chiefly in surface coatings (when made from tall oil, the source is usually indicated).

Gloss retention *n.* Degree to which the original sheen of a coating is retained. Paint and Coating Testing Manual, Koleske, J. V., ed., American Society for Testing and Materials, 1995. Paul N. Gardner Co. Inc., FL.

Gloss white *n.* A white mineral pigment used as an ink extender, made by co-precipitation of alumina hydrate and blanc fixe.

Glow discharge *See corona discharge.*

Glowing combustion *n.* Burning of a solid material without flame but with emission of light from the zone of burning.

Glue \ˈglü\ [ME *glu*, fr. MF, fr. LL *glut-*, *glus*; akin to L *gluten* glue] (14c) *n.* Originally, a hard gelatin obtained from hides, tendons, cartilage, bones, etc., of animals. Also, an adhesive prepared from this substance by heating with water. Through general use the term is now synonymous with the term "adhesive".

(*See also adhesive, mucilage, paste, and sizing*). (*v*) *See bond.*

Glue line (bond line) *n.* The layer of adhesive which attaches two adherents.

Glycerides \ˈgil-sə-ˌrīd\ (ca. 1864) *n.* The esters derived from glycerol. Glycerol is a trihydroxy alcohol and as such is capable of reacting with three monobasic acids. Such fully satisfied esters are described as the triglycerides, and the drying oils are typical members of this class. If only one hydroxy group is reacted with an acid, the resultant product is a monoglyceride, and if two are reacted a diglyceride is obtained. Salamone JC (ed) (1996) Polymeric materials encyclopedia. CRC Press, Boca Raton, FL.

Clycerin \ˈglis-rən, ˈgli-sə-\ [F *glycerine*, fr. Gk *glykeros* sweet; akin to Gk *glykys*] (1838) *n. See glycerol.*

Glycerol \ˈgli-sə-ˌról, -ˌrōl\ [*glycerin* + *-ol*] (1884) (1,2,3-propanetriol, glycerin, glycerine, glycyl alcohol) *n.* HO–CH_2CHOH-CH_2OH. The term *glycerol* applies to the pure product; *glycerin* (or *glycerine*) applies to commercial products containing at least 95% glycerol. Glycerol is a colorless, viscous liquid with a sweet taste, long produced as a by-product of soap manufacture (animal fats and vegetable oils are triglycerides of long-chain "fatty" acids). More recently, glycerol has been synthesized from propylene or sucrose (sugar). It uses in the plastics industry include the manufacture of alkyd resins (esters of glycerol and phthalic anhydride), the plasticizing of cellophane, and the production of urethane polymers. Several of its esters are plasticizers for various resins. It has a bp of 290°C, mp of 17°C, and Sp gr of 1.260/15°C. Salamone JC (ed) (1996) Polymeric materials encyclopedia. CRC Press, Boca Raton, FL. *Also known as glycerin.*

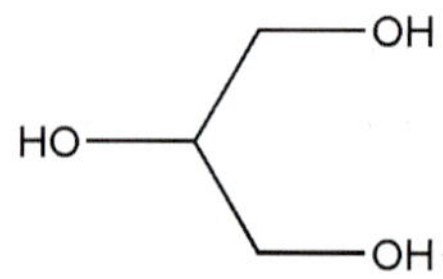

Glycerol diacetate *n.* Plasticizer; solvent for cellulose derivatives, alkyds, and shellac. It has a bp of 260°C, Sp gr of 1.176/15°C, flp of 146°C (295°F), refractive index of 1.44. *See diacetin.*

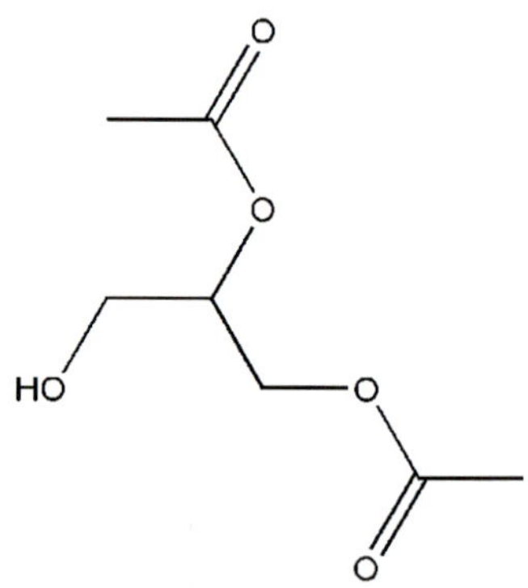

Glycerol ether acetate *n.* $C_3H_5(OCH_2\text{-}_3H_5(OCH_2CH_2OOCCH_3)_3$. A plasticizer for cellulosics and polyvinyl acetate.

Glycerol monoacetate *n.* CH_3COOCH_2-$CHOHCH_2OH$. A plasticizer for cellulose acetate, cellulose nitrate, and vinyl resins.

Glycerol monolactate diacetate *n.* CH_3CH $OHCOOC_3H_5(OOCCH_3)_2$. A plasticizer for cellulose acetate, imparting resistance to gasoline.

Glycerol monolaurate (glyceryl monolaurate, lauryl glycerin) *n.* C_{11}–$H_{23}COO CH_2CHOHCH_2OH$. A plasticizer for cellulosic and vinyl resins and polystyrene.

Glycerol monooleate *n.* $C_{17}H_{33}COOCH_2 CHOHCH_2OH$. A yellow oil, approved by the FDA as a plasticizer for food processing.

Glycerol monoricinoleate *n.* $C_6H1_3CHOH C_{10}H_{18}COOCH_2CHOHCH_2$–OH. A plasticizer for cellulose nitrate, ethyl cellulose, and polyvinyl butyral.

Glycerol-phthalic anhydride resin *n.* An alkyd resin made by modifying glycerol phthalate with an equal portion of oil, fatty acid, and natural or synthetic resin. It is used in varnishes, lacquers, and enamels.

Glycerol triacetate \-(ˌ)trī-ˈa-sə-ˌtāt\ (triacetin). A plasticizer for cellulosic resins, polymethyl methacrylate, and polyvinyl acetate. It has been approved by the FDA for food-contact use.

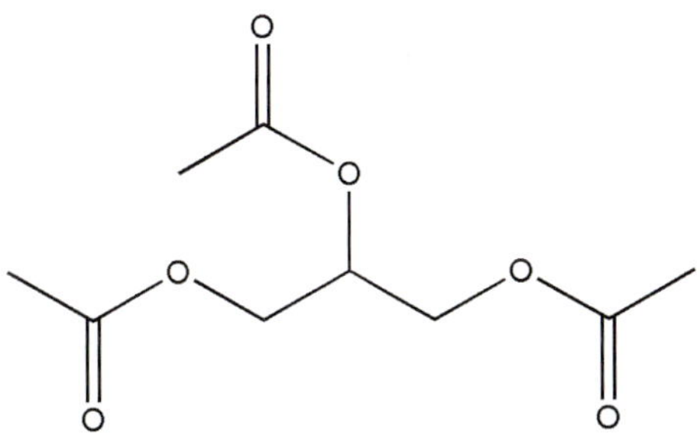

Glycerol tribenzoate (tribenzoin) *n.* A colorless, crystalline solid, a solid plasticizer for PVC.

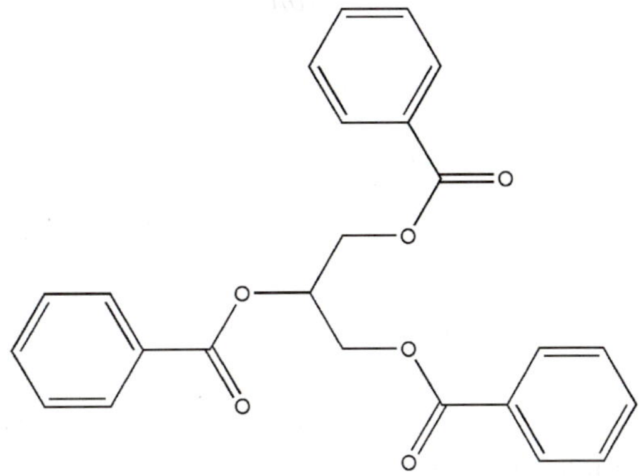

Glycerol tributyrate (tributyrin) *n.* A plasticizer for cellulose esters.

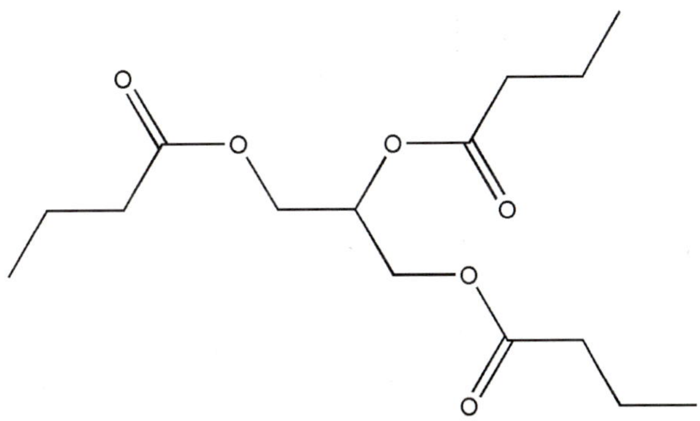

Glycerol tripropionate *n.* A plasticizer for cellulosic resins and polyvinyl acetate.

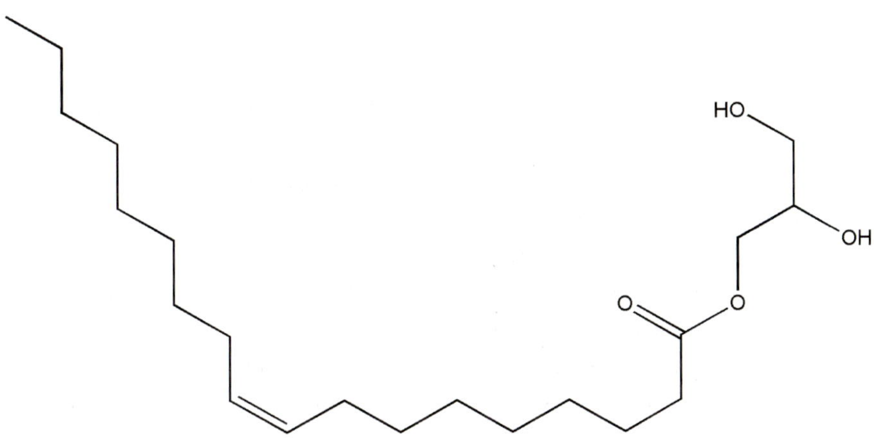

Glycerol monooleate

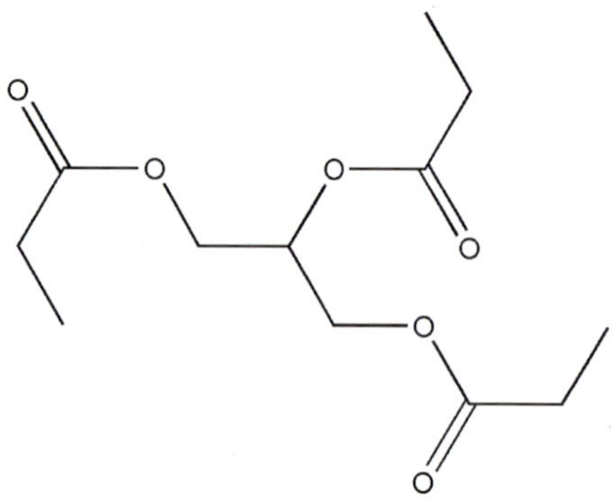

Glyceryl tri-(acetoxyricinoleate) *n.* $C_3H_5(OOCC_{10}H_{18}CHOHC_6H_{13}O–COCH_3)_3$. A plasticizer for cellulose nitrate, ethyl cellulose, and PVC.

Glyceryl tri-(12-acetoxystearate) *n.* $C_3H_5(OOCC_{17}H_{34}OCOCH_3)_3$. A plasticizer for cellulosic and vinyl resins.

Glycidol (2,3-epoxy-1-propanol) *n.* A stabilizer for vinyl resins.

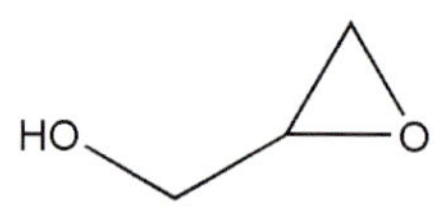

Glycidyl- *n.* The terminal epoxy group.

γ-Glycidoxypropyltrimethoxysilane *n.* A coupling agent for fibrous glass used in reinforced thermosetting and thermoplastic resins.

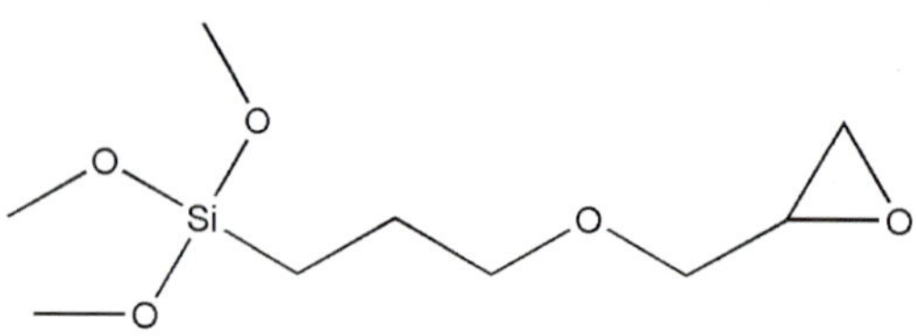

Glycidyl-ester resin *n.* Any of a family of epoxides resins derived from the condensation of epichlorohydrin with polycarboxylic acids, first available in commercial quantities in 1968. The preferred during agents for these resins are anhydrides, polycarboxylic acids, aromatic amines, and phenolics. They are resistant to electrical tracking and weather, have high strength and high modulus, yet are tough, have low viscosity, long pot life, and high reactivity at moderately elevated temperatures. Limitations are poorer properties above 100°C, higher shrinkage, poor alkali resistance, and inability to cure at room temperature. Salamone JC (ed) (1996) Polymeric materials encyclopedia. CRC Press, Boca Raton, FL.

Glycidyl-ether resin *See epoxyresin.*

Glycol \ˈglī-ˌkól, -ˌkōl\ [ISV *glyc-* + *-ol*] (1858) *n.* (1) An organic compound having hydroxyl groups on adjacent carbon atoms, for example, propylene glycol, $CH_3CHOHCH_2OH$. The name is also used for non-adjacent dihydric alcohols such as the 1,3-isomer of the above compound, $HOCH_2CH_2CH_2OH$, known as *trimethylene glycol.* (2) Specifically, ethylene glycol.

Glycol phthalate resin *n.* A type of thermoplastic polyester used mainly for fibers and oriented films.
See polyrester, saturated.

Glyptal \ˈglip-t^{ə}l\ (*trademark*) *n.* An alkyd resin, the reaction product of glycerol and phthalic anhydride (hence: gly p t al) An early product was the basis of a household and laboratory cement, trade named Glyptal. It has been used as a European generic term for alkyd resins and as an alternative to the term "alkyd".

Godet roll \gō-ˈdet-\ *n.* Roll used for transporting and controlling the movement of bundles of fibers and yarns in the processing of these materials.

Godet stand *n.* A device used in pairs downstream from the extruder when making monofilament and split-film fibers. One stand is placed before, and another following, a conditioning oven, each consisting of a vertical motor housing driving two or more horizontal drums. The filaments are wrapped repeatedly around the drums to prevent slipping. The post-oven drums are driven at speeds several times those of the first pre-oven set, causing the filaments to be stretched, oriented, and to become much stronger after cooling.

Gold bronzes *n.* Metallic powders of gold color produced from aluminum or copper and its alloys.

Golden chrome *See chrome yellow, light, and primrose.*

Gold ink *n.* A printing ink whose principal pigment consists of bronze powder.

Gold lacquer *n.* Vehicle of such color that it will give the effect of gold when applied over aluminum leaf.

Gold leaf (ca. 1741) *n.* Very thin leaves of metallic gold used for decorative purposes.

Gold leafing *v.* In antiquing, the art of applying a very thin sheet of gold over a "sized" area. It became fashionable in France in the 1700s.

Gold number *n.* A measure of the protective ability of hydrophilic colloids. Specifically, the number of milligrams of colloids which, when added to a standard gold solution, is just insufficient to prevent a change in color from red to blue.

Gold size *n.* Oleoresinous varnish which dries rapidly to a tacky condition, but which hardens slowly. It is used chiefly as an adhesive for fixing gold leaf to a surface.

Goniochromatic *n.* Adjective used to describe a colored material, which exhibits goniochromatism.

Goniochromatism *n.* The phenomenon where the color of a material changes as the angle of illumination and/or viewing is changed. McDonald R (1997) Colour Physics for Industry, 2nd edn. Society of Dyers and Colourists, West Yorkshire, England.

Goniometachromatic *n.* Adjective used to describe a pair of samples exhibiting goniometachromatism.

Goniometachromatism *n.* The phenomenon where the colors of *two* goniochromatic materials match exactly at one set of illumination-viewing angles but no longer appear to match if either the illumination or viewing angles, or both, are changed. Such pairs of samples are also called *geometric metamers.*

Goniophotometer *n.* An instrument used to measure the amount of light reflected as a function of the viewing angle when the incident angle is kept constant. Generally, the incident angle may be changed.

Goniophotometry *n.* A procedure for evaluating the manner in which materials geometrically redistribute light (*See www.astm.org for relevant standards*).

Goniospectrophotometer *n.* An instrument used to measure a spectrophotometric curve at varius angles of incidence and reflectance. The angles of incidence and reflectance may be changed. On goniochromatic samples, the results obtained are dependent on the angles of incidence and reflectance, and on samples exhibiting specular reflectance (glossy materials), the results are also dependent on the total subtended angle of viewing relative to the specular angle. Willard HH, Merritt LL, Dean JA (1974) Instrumental methods of analysis. D. Van Nostrand Co., New York.

Gouache \ˈgwäsh\ [F, fr. It *guazzo,* literally, puddle, prob. fr. L *aquatio* watering place, fr. *aquari* to fetch water, fr. *aqua* water] (1882) *n.* Method of painting with opaque colors which have been ground in water and tempered with a preparation of gum. Also, a term used to describe a method of water color painting in which white is employed as a pigment by contrast with the method of attaining whites by allowing the paper to show through. Merriam-Webster's collegiate dictionary, 11th edn.

Merriam-Webster Inc., Springfield, MA, 2004.

Gout \ˈgaüt\ [ME *goute*, fr. OF, gout, drop, fr. L *gutta* drop] (13c) *n.* Foreign matter that is accidentally woven into a fabric. It is usually fly or waste that drops into the loom during weaving or that catches in yarns during spinning.

GP *n.* (1) Abbreviation for general-purpose, sometimes used to denote types of resins and molding compounds suitable for a wide range of applications. (2) Abbreviation for Gutta–Percha.

GPC *n.* Abbreviation for gel-permeation chromatography.
See size-exclusion chromatography.

GR-1 *n.* Former symbol for butyl rubber.

Grab strength test *n.* A method for measuring the breaking strength of a fabric sample by mounting the sample in the tensile tester so that only a part of the width of the specimen is gripped in the clamps.

Gradient elution *n.* Elution in chromatography using a solvent whose composition is varied during a run.

Gradient-tube density method (density-gradient method) *n.* A convenient method, described in ASTM D 1505, for routinely measuring densities of small resin samples, e.g., a single pellet of molding compound. A graduated, vertical glass tube (the gradient tube) is carefully filled with graded solutions of two liquids, the densest at the bottom, the next densest above that, etc, to the least dense at the top. A specimen particle is introduced in the tube and falls to a position of equilibrium that indicates its density by comparison with the positions of calibrated hollow-glass floats. Over weeks or months, the gradient is gradually diminished by diffusion and the column must be reconstituted. *See www.astm.org for a description of this method.*

Graffito \grə-ˈfē-(ˌ)tō, gra-, grä-\ [It, incised inscription, fr. *graffiare* to scratch, prob. fr. *grafio* stylus, fr. L *graphium*] (1851) *n.* Last plaster coat of a fresco ground, which is colored before application. Merriam-Webster's collegiate dictionary, 11th edn. Merriam-Webster Inc., Springfield, MA, 2004.

Graft copolymer *n.* A copolymer having branches of varying length made up of different monomer units on a common "backbone" chain. Odian GC (2004) Principles of polymerization. John Wiley and Sons Inc., New York.

Grafting ratio *n.* In a graft copolymer, the weight of grafted side chains divided by the weight of the original polymer. Odian GC (2004) Principles of polymerization. John Wiley and Sons Inc., New York.

Graft polymer or copolymer *n.* A high polymer, the molecules of which consist of two or more polymeric parts of different composition, chemically united. Such a graft copolymer may be produced, for example, by polymerization of a given kind of monomer with subsequent polymerization of another kind of monomer onto the product of the first polymerization. The union of two different polymers by chemical reaction between their molecular end groups or by reaction producing cross-links between the different molecules would also produce a graft copolymer. Odian GC (2004) Principles of polymerization. John Wiley and Sons Inc., New York.

Grahamite *n.* Naturally occurring asphaltum which varies considerably in composition. Some types are completely soluble in carbon disulfide, whereas others may contain as much as 50% of mineral matter.

Graham's law [Thomas *Graham*, 1805–1869, Scottish chemist] (1833) *n.* The relative rates of diffusion of gases under the same conditions are inversely proportional to the square roots of the densities of those gases. Whitten KW, Davis RE, Davis E, Peck LM, Stanley GG (2003) General chemistry. Brookes/Cole, New York.

Grain *n.* The direction, size, arrangement, appearance or quality of the fibers in fibrous material such as paper, wood or veneer. To have a specific meaning the term must be qualified. Also, the direction of molecular orientation in a non-fibrous material.

See types of gain, i.e., cross grain, interlocked grain, coarse grain, close train, diagonal grain, and dip grain.

Grain alcohol (1889) *n.* Syn: ethyl alcohol.

Graining (1530) *v.* (1) A process for simulating a grain such as that of wood or marble on a painted surface by applying a stain, either translucent or pigmented, then working it into suitable patterns with tools such as graining combs, brushes, and rags. Weismantal GF (1981) Paint handbook. McGraw-Hill Inc., New York. Paint/coatings dictionary. Federation of Societies for Coatings Technology, Philadelphia, Blue Bell, PA, 1978.

Grain raising *n.* Swelling and standing up of the wood grain caused by absorbed water or solvent. Roughness of wood caused by the application of stains or other materials. Fibers of the wood raise due to swelling by water or certain solvents.

Gram \\ˈgram\\ [F *gramme*, fr. LL *gramma*, a small weight, fr. Gk *grammat-*, *gramma* letter, wirting, a small weight fr. *graphein* to write] (1810) *n.* Subunit of mass in the metric system of weights and measures, equal to 1/1000 of a standard kilogram (453.6 g = 1 pound). Lide DR (ed) (2004) CRC handbook of chemistry and physics. CRC Press, Boca Raton, FL. Merriam-Webster's collegiate dictionary, 11th edn. Merriam-Webster Inc., Springfield, MA, 2004.

Gram-atom (1927) (gram-atomic weight) *n.* The mass of an element in grams numerically equal to the elements atomic weight.

Gram break factor *See break factor.*

Gram equivalent (ca. 1897) *n.* The weight in grams of a substance displacing or reacting with 1.00797 g of hydrogen or combining with 15.9994 g of oxygen.

Gram molecular weight or gram molecule (ca. 1902) *n.* A mass in grams of a substance numerically equal to its molecular weight. Gram mole.

Gram mole, gram formula weight, gram equivalent *n.* Mass in grams numerically equal to the molecular weight, formula weight or chemical equivalent, respectively.

Granular (or powder) \\ˈgran-yə-lər\\ (1794) *adj.* Particles having equidimensional non-spherical shapes.

Granular polymerization *n.* A polymerization process in which the monomer, or mixture of monomers, is dispersed by mechanical agitation in a liquid phase, usually water, in which the monomer droplets are polymerized while they are dispersed by continuous agitation. Odian GC (2004) Principles of polymerization. John Wiley and Sons Inc., New York.

Also called suspension polymerization.

Granular structure *n.* Non-uniform appearance of finished plastic material due to retention of, or incomplete fusion of, particles of compound either within the mass or on the surface; or to the presence of coarse filler particles.

G

Granulate (1666) *v.* To reduce plastic sheet, chunks, or scrap to particles about 2–5 mm in size. (2, *n*) A molding compound in the form of small spheres or pellets.

Granulator (scrap grinder) *n.* A machine for cutting waste material such as sprues, runners, excess parison material, trim scarp from extrusion and thermoforming, and rejected parts into particles that can be reprocessed. The most common type of granulator is comprised of several thick knives bolted to a heavy cylindrical core, parallel to the core's axis, which is also its axis of rapid rotation. The rotating knives graze stationary knives mounted in the machine's housing, given an action that combines impact with shearing. A screen placed in the discharge opening controls the final particle size.

Granule \ˈgran-(ˌ)yü(ə)l\ [LL *granulum*, dim. of L *granum* grain] (1652) *n.* A small particle produced in various size and shapes by hot or cold cutting, of extruded strands or scrap, or by certain polymerization methods. Perry RH, Green DW (1997) Perry's chemical engineer's handbook, 7th edn. McGraw-Hill, New York.
Also see pellets.

Grapeseed oil *n.* Semidrying vegetable oil. Of the fatty acids present, up to 55% may be linoleic acid. Approximate constants for oil from various sources are: Sp gr of 0.925/15°C, refractive index of 1.4899/15°C, saponification value of 190, iodine value of 125–157.

Graphic arts coatings *n.* Coatings which are marketed solely for application to indoor or outdoor signs and include lettering enamels, poster colors and bulletin colors.

Graphite \ˈgra-ˌfīt\ [Gr *Graphit*, fr. Gk *graphein* to write] (1796) (black lead, plumbago) *n.* A crystalline form of carbon with atoms arranged hexagonally, characterized by a soft, greasy feel. It occurs naturally, but is produced by heating petroleum coke or other organic materials under controlled atmospheric conditions. In powder form, graphite is used as a lubricating filler for nylon and fluorocarbon resins. Too, it is added to compounds to make them electrically conductive. Pyrolytic graphite fibers, made by decomposing organic filaments at high temperatures in controlled atmospheres, are used as reinforcements for high-performance applications. The best graphite fibers are among the strongest and stiffest of all fibrous reinforcements, with strengths to 2.5 GPa and moduli to 500 GPa (360 kpsi and 72 Mpsi). Pierson HO (1994) Handbook of carbon, graphite, diamond and fullerenes. Noyes Data Corporation/Noyes Corporation, New York.

Graphite fiber *n.* Although the terms carbon and graphite are used interchangeably to describe these fibers, graphite fibers are more accurately defined as fibers that are 99+% carbonized (crystalline structure as opposed to non-crystalline carbon fiber) while the term carbon is used for any fiber carbonized to 93–95% or more. Pierson HO (1994) Handbook of carbon, graphite, diamond and fullerenes. Noyes Data Corporation/Noyes Corporation, New York.
See carbon fiber.

Graphite paint *n.* A painting compound consisting of powdered graphite and oil; used to coat metallic structures to inhibit corrosion. Pierson HO (1994) Handbook of carbon, graphite, diamond and fullerenes. Noyes Data Corporation/Noyes Corporation, New York.

Graphitic mica *See aluminum potassium silicate.*

Graphitization \ˌgra-fə-tə-ˈzā-shən\ (1899) *n.* Corrosion of gray iron in which metallic iron is converted into corrosion products, leaving a residue of intact graphite mixed with iron corrosion products and other insoluble constituents of cast iron and having approximately the original dimensions of the casting.

Also known as graphitic corrosion.

Graphitize (1899) *vt.* The pyrolysis of an organic material such as pitch or Polyacrylonitrile (PAN) in an inert atmosphere at temperatures between 2500 and 3000°C to produce graphite. Pyrolysis of PAN fibers is a principal source of graphite fibers.

Grass \ˈgras\ {*often attributive*} [ME *gras*, fr. OE *græs*; akin to OHGr *gras* grass, OE *grōwan* to grow] (before 12c) *n.* Sessile algae on the side of a ship.

Grass cloth (1857) *n.* Originally a handmade product imported from Japan, made by gluing woven native grasses onto paper backing. Printed wallpapers or dimensional wall coverings, especially vinyl, which simulates the same.

Grasshopper *n.* A stiff bunch of parallel strands in a fibrous mat.

Grass-tree gum *See accroides.*

Gravel \ˈgra-vəl\ [ME, fr. OF *gravele*, dim. of *grave*, *greve*, pebbly ground, beach] (13c) *n.* Rounded or semirounded particles of rock that will pass a 3-in. (76.2 mm) sieve and be retained on a No. 4 (4.75 mm) USA standard sieve.

Gravel roofing *See built-up roofing.*

Gravimetric feeder \ˌgra-və-ˈme-trik-\ (weigh feeder). A device for feeding extruders that continuously meters, by weighing, the rate at which feedstock enters the feed port of the extruder.

Gravitation \ˌgra-və-ˈtā-shən\ (ca. 1645) *n.* The universal attraction existing between all material bodies. The force of attraction between two masses m and m', separated by a distance r, k being the constant of gravitation,

$$F = k\frac{mm'}{r^2}$$

(if m and m' are given in gram and r in centimeter, F will be in dyne if $k = 6.670 \times 10^{-8}$).Giambattista A, Richardson R, Richardson RC, Richardson B (2003) College physics. McGraw-Hill Science/Engineering/Math, New York.

Gravure \grə-ˈvyür, grā-\ [F, fr. *graver* to grave, of Germanic origin; akin to OHGr *graban* to dig, engrave] (1893) *n.* One of the methods of printing using the intaglio process, i.e., the ink is placed in cells below the plate surface. Gravure inks are quick drying, low viscosity inks based on volatile solvents. Leach RH, Pierce RJ, Hickman EP, Mackenzie MJ, Smith HG (eds) (1993) Printing ink manual, 5th edn. Blueprint, New York.

Gravure coating (engraved-roll coating) *n.* A roller-coating process in which the amount of coating applied to the substrate web is metered by the depth of an all-over engraved pattern in the application roll. This process is frequently modified by interposing a resilient offset roll between the engraved roll and the web. Leach RH, Pierce RJ, Hickman EP, Mackenzie MJ, Smith HG (eds) (1993) Printing ink manual, 5th edn. Blueprint, New York.

Gravure printing *n.* Depressions in an engraved printing cylinder or plate are filled with ink, the excess on raised portions being wiped off by a doctor blade. Ink remaining in the depressions is deposited on the plastic film as it passes between the gravure roll and a resilient backup roll. Leach RH, Pierce RJ, Hickman EP, Mackenzie MJ, Smith HG (eds) (1993) Printing ink manual, 5th edn. Blueprint, New York.

G

Gravure tissue *n.* Sheet of paper coated with dichromated gelatin but containing no pigment. Leach RH, Pierce RJ, Hickman EP, Mackenzie MJ, Smith HG (eds) (1993) Printing ink manual, 5th edn. Blueprint, New York.

Gray (Gy) *n.* The SI unit of absorbed radiation dose, defined as the absorption of 1 J/kg of body mass, i.e., Gy = 1 J/kg. The former unit, the *rad*, now deprecated but still being used widely in the USA, equals 0.01 Gy.

Gray blast (commercial blast) *n. See NACE no. 3.*

Gray fabric *See greige fabric.*

Gray scale *n.* An archromatic scale ranging from black through a series of successively lighter grays to white. Such a series may be made up of steps, which appear to be equally distant from one another (such as the Munsell value scale) or may be arranged according to some other criteria such as a geometric progression based on lightness. Such scales may be used to describe the relative amount of difference between two similar colors. McDonald R (1997) Colour physics for industry, 2nd edn. Society of Dyers and Colourists, West Yorkshire, England.

Grazing angle *n.* Angle of light very nearly parallel to the surface of a material; used to describe incident and/or reflected angles.

Grease \ˈgrēs\ [ME *grese*, fr. OF *craisse*, *graisse*, fr. (assumed) VL *crassia*, fr. L *crassus* fat] (13c) *n.* A semisolid lubricant composed of emulsified petroleum oils and soluble hydrocarbon soaps.

Grease forming *See mechanical grease forming.*

Grease paint (1) Non-drying and non-oxidizing coating for void spaces in ships which can be easily removed for control of the steel sheets. (2) Theatrical make-up.

Grease proof inks and coatings *n.* Inks (and coatings) that are resistant to the action of fats, oils and greases.

Greasiness *See spewing.*

Greasing *n.* Sensitization of the non-printing areas of a lithographic plate resulting in undesirable adhesion of the ink to these areas.

Also called scumming.

Green B, naphthol *n.* Iron lake of 1-nitroso-2-naphthol-3-carboxylic acid.

Green B, pigment *n.* Pigment produced by treating 1-nitroso-2-naphthol with a ferrous salt.

Green, Brunswick *See chrome green.*

Green, chrome *See chrome green.*

Green cinnabar *See chromium oxide green.*

Green earth Naturally occurring earth pigment, which is primarily an iron silicate. Used to make lime green. Syn: terre verte.

See green, lime.

Green, emerald *See emerald green.*

Green gold *See nickel azo yellow.*

Green lake *n.* Mixture of Prussian blue and Yellow Lake.

Green, lime *n.* Term applied to those green pigments, which are employed for their fastness to alkali (lime). For a considerable time it consisted of the green produced from green earth but the term is now applied to other types, e.g., reduced Pigment Green B.

Green lumber *n.* Lumber which has not been dried or seasoned.

Green, Paris *n.* Pigment consisting essentially or copper acetoarsenite. It is sometimes used in anti-fouling compositions, but its main use is in insecticides.

Green rouge *See chromium oxide green.*

Green stock *n.* Raw or uncured plastic or rubber stock, which is ready for vulcanization. The term is not applied to crude rubber, synthetic rubber, or resin that has not been compounded.

Green strength *n.* Of a perform or incompletely cured molding, the ability to withstand handling without distorting.

Green tack *n.* A term used in fabric bonding for the preliminary bond created in the first stage of curing by the wet adhesive process. At this point, the bond is not fully cured and hence is "green".

Greentone ink blue *See iron blue.*

Green vitriol \-ˈvi-trē-əl\. $FeSO_4{\cdot}7H_2O$. Crystalline ferrous sulfate used in the manufacture of iron blue.

Green, zinc *n.* Range of greens made by mixing zinc chromate with iron blue. They are less bright and weaker than the Brunswick greens, but find application where lead-free greens are required.

Greenockite \ˈgrē-nə-ˌkīt\ [Charles M. Cathcart, Lord *Greenock* † 1859 English soldier] (1844) *n.* Naturally occurring cadmium sulfide mineral.

Greige fabric \ˈgrā(zh)-\ [F *grège* raw (of silk), fr. It *greggio*] (1926) *n.* An unfinished fabric just off the loom or knitting machine.

Grenadine \ˌgre-nə-ˈdēn, ˈgre-nə-ˌ\ [F, fr. *grenade* coarse silk fabric, pomegranate] (1852) *n.* (1) A fine, loosely woven fabric in leno weave made with dyed filling yarns and having a clipped dobby design. (2) A silk cord constructed by twisting together several twisted strands.

Grex (1) A unit of linear density equal to the weight in grams of 10 km of yarn, filament, fiber, or other textile strand. (2) The system of yarn numbering based on the use of grex units. Kadolph J, Langford AL (2001) Textiles. Pearson Education, New York.
Also see yarn number.

Grex number *n.* Like *cut* and *denier*, a deprecated measure of lineal density of yarns and fibers, the weight in grams of 10 km of the yarn. A grex number of 1 corresponds to an SI lineal density of 10^{-7} kg/m or 0.1 tex.
See also cut, denier, and tex.

Grey fabric *See greige fabric.*

Grid \ˈgrid\ [back-formation fr. *gridiron*] (1839) *n.* Channel-shaped, mold-supporting members.

Griffith theory *n.* The idea that, in brittle fracture, crack growth occurs, causing fracture, when the rate of decrease of stored elastic energy equals or exceeds the rate of creation of new fracture-surface energy. The Griffith equation states that, in thin sheets, tensile strength is given by: $\sigma = 2\gamma E/(\pi c)^{0.5}$, where γ is the fracture-surface energy per unit area, E the elastic modulus, and c is the half the crack length. Anderson TL (2005) Fracture mechanics. CRC Press, Boca Raton.

Grind \ˈgrīnd\ [ME, fr. OE *grindan*; akin to L *frendere* to crush, grind] (before 12c) *v.* (1) To reduce the size of breakable particles by impact or by squeezing them between relatively moving surfaces. (2) To remove material from a part by contact with an abrasive, to change either the part's shape or its finish.

Grinding *n.* (1) Process by which pigment particles are reduced in size, mechanically. (2) Its use to describe pigment dispersion in mill bases is deprecated.

Grinding media charge *See mill charge.*

Grinding medium *n.* Any medium which is used as a means of dispersing pigments or fillers by grinding or dispersing.

Grinding slab *n.* A flat piece, usually of glass or stone, on which color is ground from a coarse to a finely divided state, frequently with the medium that is to blend it as a paint.

Grinding-type resin *n.* A vinyl resin that must be ground to effect dispersion in plastisols or organosols.

G

G

Grinning *n.* (1) A flaw in fabric, especially a ribbed fabric, that occurs when warp threads show through the covering filling threads or when the threads have slipped leaving open spaces on either side. (2) A condition that occurs when the carpet backing shows through the pile. (3) A printing term referring to either poor cover where the background shade shows through the print, or to the "two-tone" appearance of a shade printed with incompatible dyes. Leach RH, Pierce RJ, Hickman EP, Mackenzie MJ, Smith HG (eds) (1993) Printing ink manual, 5th edn. Blueprint, New York.

Grinning through *n.* Showing through of the underlying surface due to the inadequate opacity of a paint film, which has been applied to it.

Gripper loams Shuttleless looms. These looms employ a projectile with a jaw that grips the end of the filling yarn during the insertion of the pick.

Grisaille \gri-ˈzī, -ˈzā(ə)l\ [F, fr. *gris* gray, fr. MF] (1848) *n.* Decorative painting in which objects are rendered in tones of one color, usually gray, often applied on glass, and intended to give the effect of sculptured relief. Merriam-Webster's collegiate dictionary, 11th edn. Merriam-Webster Inc., Springfield, MA, 2004.

Grit blasting *n.* A mold-finishing process in which abrasive particles are blown against mold surfaces in order to produce a controlled degree of roughness. The process is often used on molds for blow molding to assist air escape near the end of the blow, and on other types of molds to produce a desired texture in the product. Rosato DV (ed) (1992) Rosato's plastics encyclopedia and dictionary. Hanser–Gardner Publications, New York.

See blast cleaning.

Grit number *See mesh number.*

Gritty \ˈgri-tē\ (1598) *adj.* In a coating, roughness resembling grains of sand. Weismantal GF (1981) Paint handbook. McGraw-Hill Inc., New York.

Grooved barrel *n.* In an extruder or injection molder, a barrel that has shallow grooves over the first few screw diameters from the feed opening. The groves may be parallel to the machine axis or helical with direction opposite that of the screw. Grooves have been shown to improve feeding of difficult-to-feed materials and to reduce energy consumption per unit of throughput, but can cause overfeeding and attendant problems with screws and materials for which the grooves and screw were not designed. Weismantal GF (1981) Paint handbook. McGraw-Hill Inc., New York.

Grosgrain \ˈgrō-ˌgrān\ [F *gros grain* coarse texture] (1869) *n.* A heavy fabric with prominent ribs, grosgrain has a dressy appearance and is used in ribbons, vestments, and ceremonial cloths. Tortora PG (ed) (1997) Fairchild's dictionary of textiles. Fairchild Books, New York.

Ground *n.* (1) Surface applied to a canvas or other support, upon which a picture is painted. (2) Base on which organic agents are precipitated to form lakes. (3) Any surface, which is or will be painted. Gair A (1996) Artist's manual. Chronicle Books LLC, San Francisco. Mayer R, Sheehan S (1991) Artist's handbook of materials and techniques. Viking Adult, New York.

Ground color *n.* A term describing the plain background color against which a design is created. Mayer R, Sheehan S (1991) Artist's handbook of materials and techniques. Viking Adult, New York.

Groundnut oil Another name for arachis or peanut oil. *See peanut oil.*

Ground oyster shells *See calcium carbonate, natural.*

Ground perlite *See perlite, expanded.*

Ground silica *See silica, crystalline.*

Ground state (1926) *n.* The state of lowest energy of a particle.

Ground wire *n.* A wire attached between apparatus, containers and instruments and an electrical ground to dissipate an electric or electrostatic charge.

Group \ˈgrüp\ {*often attributive*} [F *groupe*, fr. It *gruppo*, of Germanic origin; akin to OHGr *kropf* craw] (1686) *n.* A vertical column of elements in the periodic table, sometimes called a family of elements. Whitten KW, Davis RE, Davis E, Peck LM, Stanley GG (2003) General chemistry. Brookes/Cole, New York.

Growth *See secondary creep.*

GRP *n.* Abbreviation for "glass-reinforced plastic".
See reinforced plastic.

Guaiacol *n.* 2-Methoxyphenol, is a natural organic compound with the molecular formula C7H8O2. It is the monomethyl ether of catectol. It is a yellowish oil drived from guaiacum or wood creosote. It is used medicinally as an expectorant, antiseptic, and local anesthetic. In coating technology, it is used for anti-skinning properties (www.wikipedia.org). Usmani AM (1997) Asphalt science and technology. Marcel Dekker, New York. Paint/coatings dictionary. Federation of Societies for Coatings Technology, Philadelphia, Blue Bell, PA, 1978.

Guanidine aldehyde resins *n.* Reaction products of guanidine aldehydes.

Guar gum \ˈgwär ˈgən\ (1950) *n.* The ground endosperms of *Cyanopsis tetra gonolobus.* The free-flowing powder is soluble in cold or hot water and can be gelled with borax. It is used as a thickening agent or protective colloid. Whistler JN, BeMiller JN (eds) (1992) Industrial gums: polysaccharides and their derivatives. Elsevier Science and Technology Books, New York.

Guide bar *n.* A mechanism on a warp-knitting machine that directs warp threads to the latch needles.

Guide coat *n.* Coat similar to the finish or color coat but of a different color to assure good coverage.

Guide eye *n.* In filament winding, a moving metal or ceramic loop (eye) through which the fiber passes as it flows from creel to mandrel.

Guide pin (dowel pin) *n.* In compression, transfer, and injection molding, usually one of two or more hardened steel pins that maintain proper alignment of the mold halves that open and close.

Guide-pin bushing (dowel-pin bushing) *n.* In molding, a hardened bushing that receives and guides the leader pin, controlling alignment of the mold halves as the mold closes. Strong AB (2000) Plastics materials and processing. Prentice-Hall, Columbus, OH.

Guides Fittings of various shapes for controlling the path of a thread line. Tortora PG (ed) (1997) Fairchild's dictionary of textiles. Fairchild Books, New York.

Guignet's green *See chrome-oxide green and hydrated chromium oxide.*

Guillotine \ˈgi-lə-ˌtēn\ [F, fr. Joseph *Guillotin* † 1814 F physician] (1793) *n.* Cutting device that consists of a single blade that descends between guides for chopping fibers, plastic strands, etc. Tortora PG (ed) (1997) Fairchild's dictionary of textiles. Fairchild Books, New York.

Guimet's blue *n.* Synthetic ultramarine blue.

Gum \ˈgəm\ [ME *gomme*, fr. MF, fr. L *cummi*, gummi, fr. Gk *kommi*, fr. Egyptian *qmyt*] (14c) (adhesives) *n.* Any of a class of colloidal substances, exuded by or prepared from plants, sticky when moist, composed of complex carbohydrates and organic acids, which are soluble or swell in water. The term gum is sometimes used loosely to denote various materials that exhibit gummy characteristics under certain conditions, for example. Gum balata, gum benzoin, and gum asphaltum. Gums are included by some in the category of natural resins. Skeist I (ed) (1990, 1977, 1962) Handbook of adhesives. Van Nostrand Reinhold, New York.

G

Gum (lithography) *n* . A water-soluble colloid such as gum arabic, cellulose gum, etc., used for coating a lithographic plate to make the non-image areas ink-repellent and to preserve the plate for future use. Leach RH, Pierce RJ, Hickman EP, Mackenzie MJ, Smith HG (eds) (1993) Printing ink manual, 5th edn. Blueprint, New York.

Gum acacia *See acacia gum.*

Gum accroides *See accroides.*

Gum animi *See animi.*

Gum benzoin *See benzoin.*

Gum bloom *n.* A defect in a painted surface, appearing as a lack of gloss or a haze resulting from the use of incorrect reducer.

Gum karaya *See karaya gum.*

Gum mastic *n.* A natural resin used for picture varnish to protect oil paintings, water colors and the like.

Gumming *n.* (1) Clogging of sandpaper by a coating. (2) Addition of a gum resin to a coating.

Gummy *n.* British term for brush drag.

Gum, natural *n.* A carbohydrate high polymer that is insoluble in alcohol and other organic solvents, but generally soluble or dispersible in water. Natural gums are hydrophilic polysaccharides composed of monosaccharide units joined by glycosidic bonds. They occur as exudations from various trees and shrubs in tropical areas or as phycocolloids (algae), and differ from natural resins in both chemical composition and solubility properties. Some contain acidic components and others are neutral. Their chief use is as protective colloids and emulsifying agents in food products and pharmaceuticals; as sizing for textiles; and in electrolytic deposition of metals. Examples are: Arabic, tragacanth, guar, and karaya. The word "gum", often used as an adjective, seems to acquire a different meaning from the noun. For example, the resinous products obtained from pine pitch (produced by the parenchyma cells of softwoods) are conventionally called "gum turpentine" and "gum rosin". There are also such "gum reins" as gum benzoin, gum camphor, and others. The so called ester gum is a semi synthetic reaction product of rosin and a polyhydric alcohol. All these are actually resinous products having properties quite different from hose of natural gums. Still further complicating the matter is the common application of the word "gum" to such plant lattices as chicle and natural rubber, which are different from both carbohydrate gums and resins. Langenheim JH (2003) Plant resins: chemistry, evolution ecology and ethnobotany. Timber Press, Portland, OR. Whistler JN, BeMiller JN (eds) (1992) Industrial gums: polysaccharides and their derivatives. Elsevier Science and Technology Books, New York.

Gum rosin (1712) *n.* Rosin which is produced from the gum turpentine obtained by tapping living pine trees, as distinguished from the wood rosin obtained from the stump and branches of dead trees.

Gum rubber (pure gum) *n.* Raw, unvulcanized rubber recovered from the latex of the *hevea* tree or from polymerization of synthetic rubber. Gum runner has almost no useful properties prior to being vulcanized. Langenheim JH (2003) Plant resins: chemistry, evolution ecology and ethnobotany. Timber Press, Portland, OR. Whistler JN, BeMiller JN (eds) (1992) Industrial gums: polysaccharides and their derivatives. Elsevier Science and Technology Books, New York.

Gums *n.* Although rarely used in the plastics industry, this term is used in kindred industries to include materials that can be dissolved in water to produce viscous or mucilaginous solutions. Thus, water-soluble polymers such as polyvinyl pyrrolidone, polyvinyl alcohol, ethylene oxide resins, polyacrylic acid, and polyacrylamide are regarded as gums. Langenheim JH (2003) Plant resins: chemistry, evolution ecology and ethnobotany. Timber Press, Portland, OR. Whistler JN, BeMiller JN (eds) (1992) Industrial gums: polysaccharides and their derivatives. Elsevier Science and Technology Books, New York.

Gum sandarac \-ˈsan-də-ˌrak\. Soft resin obtained from *Tetraclinis articulate*, used in the manufacture of paints, linoleum, and oil cloth. Langenheim JH (2003) Plant resins: chemistry, evolution ecology and ethnobotany. Timber Press, Portland, OR. Whistler JN, BeMiller JN (eds) (1992) Industrial gums: polysaccharides and their derivatives. Elsevier Science and Technology Books, New York.

Also known as juniper gum.

Gum spirits *See turpentine.*

Gum thus *n.* Botanically, the oleoresin from trees of *Boswellia* species native to Arabia and Somaliland. As applied to the naval stores industry, the term refers to the crystallized pine oleoresin or "scarpe" collected from scarified "faces" of trees being worked for turpentine. Langenheim JH (2003) Plant resins: chemistry, evolution ecology and ethnobotany. Timber Press, Portland, OR. Whistler JN, BeMiller JN (eds) (1992) Industrial gums: polysaccharides and their derivatives. Elsevier Science and Technology Books, New York.

Also known as olibanum and frankioncense.

Gum turpentine (1926) *n. See turpentine.*

Guncotton \-ˈkä-tᵊn\ (1846) *n.* A highly inflammable and explosive form of cellulose nitrate made by digesting clean cotton in a mixture of one part nitric acid and three parts sulfuric acid, the latter acting as a catalyst and scavenger of the water generated in the reaction. As it ages in contact with air, dry guncotton becomes gradually more unstable and dangerous to handle. Goldberg DE (2003) Fundamentals of chemistry. McGraw-Hill Science/Engineering/Math, New York.

Gunk \ˈgeŋk\ [fr. *Gunk*, trademark for a cleaning solvent] (1943) *n.* A slang term for premix.

Gunk molding *See premix molding.*

GUP *n.* Glass-fiber-reinforced polyester resins.

Gusset \ˈgə-sət\ [ME, piece of armor covering the joints in a suit of armor, fr. MF *gousset*] (ca. 1570) *n.* A tuck in the side of a bag, usually made in symmetrical pairs in both paper and plastic-film bags, that permit the bag to assume nearly rectangular-boxy form when opened. Gussets may be formed in the tubular blown film from which bags are made just before the film enters the pinch rolls.

Gutta–Percha \ˌgə-tə-ˈpər-chə\ [Malay *gètah-pèrcha*, fr. *gètah* sap, latex + *pèrcha* scrap, rag] (1845) (GP, PI, and *trans*-1,4-polyisoprene) *n.* A rubber-related, polymeric substance extracted from the milky sap of leaves and bark of certain trees belonging to the family *Sapotaceae*, genera *Palaquium*, and *Payena*, plants native to Malaysia. Its mer has the same empirical formula

as natural rubber, its *cis* isomer. Gutta–Percha is a tough, horny substance at room temperature, but becomes soft and tacky when warmed to 100°C. In the past it was used in compounds for golf-ball covers, electrical insulation, cutlery handles, and machinery belting, and as a stiffening agent in natural rubber, but has been replaced in many of these applications by synthetics. Langenheim JH (2003) Plant resins: chemistry, evolution ecology and ethnobotany. Timber Press, Portland, OR.
See also balata.

Gutta–Percha, synthetic *See polyisoprene.*

Gutter \ˈgə-tər\ [ME *goter*, fr. MF *goutiere* fr. *goute* drop, fr. L *gutta*] (14c) *n.* A channel at the eaves for conveying away rain water. Harris CM (2005) Dictionary of architecture and construction. McGraw-Hill Co., New York.

Gutzeit test *n.* Method for testing and estimating small quantities of arsenic and antimony in other materials. Details are given in the *British pharmacopoeia.*

Gym finish *n.* Composition of oils and resins in a solvent; designed to produce a high gloss, tough surface finish on wood floors. Weismantal GF (1981) Paint handbook. McGraw-Hill Inc., New York.

Gynocardia oil *n.* A drying oil, smelling like linseed oil, and obtained from *Gynocardia odorata*, which grows in India and Pakistan. Its component acids include isolinolenic, linolenic, lin oleic, oleic, and palmitric acids. It has a Sp gr of 0.925/25°C, iodine value of 153–160. Paint: pigment, drying oils, polymers, resins, naval stores, cellulosics esters, and ink vehicles, vol 3. American Society for Testing and Material, Paul N. Gardner Co. Inc., FL, 2001.

Gypsum \ˈjip-səm\ [L, fr. Gk *gypsos*] (14c) *n.* $CaSO_4{\cdot}2H_2O$. Natural crystalline calcium sulfate used as an extender in paint manufacture. Hibbard MJ (2001) Mineralogy. McGraw-Hill Co. Inc., New York. Solomon DH, Hawthorne DG (1991) Chemistry of pigments and fillers. Krieger Publishing Co., New York.
See calcium sulfate. Also known as mineral white.

Gypsum, calcined *n.* $CaSO_4{\cdot}½G_2O$. Gypsum partially dehydrated by heat.
See plaster of Paris.

Gypsum cement *See plaster of Paris.*

Gypsum plaster *See cement plaster.*

Gypsum wallboard *n.* A sheet or slab having an incombustible core, essentially gypsum, surfaced with paper suitable to receive decoration. Harris CM (2005) Dictionary of architecture and construction. McGraw-Hill Co., New York.
Also known as plasterboard.

H

h \ˈāch\ *n.* (1) SI abbreviation for hour and for the prefix hecto-. (2) Also H, the symbol for channel depth of an extruder screw.

H (1) The chemical symbol for the element hydrogen. (2) Abbreviation for the SI unit of magnetic inductance, the henry. (3) Symbol for enthalpy. Lide DR (ed) (2004) CRC handbook of chemistry and physics. CRC Press, Boca Raton, FL.

Habit \ˈha-bət\ [ME, fr. OF, fr. L *habitus* condition, character, from *habēre* to have, hold] (13c) *n.* The general shape of a crystal. The same descriptive terms are used in each of the six crystal systems, e.g., rods, plates, tablets, needles, flakes, etc. Hibbard MJ (2001) Mineralogy. McGraw-Hill Co. Inc., New York.

Hackberry tree seed oil \ˈhak-ˌber-ē-\. Obtained from the kernels of the tree, *Celtis occidentalis.* Its main constituent fatty acid is linoleic acid, present to the extent of over 70%. It has a Sp gr of 0.9204/25°C, refractive index of 1.4794/25°C, saponification value of 191, and an iodine value of 150.

Hackles \ˈha-kəl\ [ME *hakell*; akin to OHGr *hāko* hook] (15c) *n.* Thin, needle-like or sliver-like protrusions (ranging from 3 to 6 μm) found on steel plates, which have been blasted with steel or grit. Anderson TL (2005) Fracture mechanics: fundamentals and applications. CRC Press, Boca Raton, FL.

Haco oil *n.* Specifically processed whale oil, of German origin, with good drying properties.

Hagen–Poiseuille equation (Poiseuille equation) *n.* The equation of steady, laminar, Newtonian flow through circular tubes:

$$Q = \pi R^4 \Delta P / 8\eta L,$$

where Q is the volumetric flow rate, R and L are the tube radius and length, ΔP the pressure drop (including any gravity head) in the direction of low, and η is the fluid viscosity. With the roles of Q and η interchanged, this is the basic equation of capillary viscometry. Any *consistent* system of units may be used. This important equation was first derived theoretically in 1839 by G. Hagen and, a year later, inferred from experimental measurements by J. L. Poisuille. In a laminar flow through a circular tube, a simple force balance shows that the shear stress at the wall, $\tau_w = \Delta PR/2L$. By Newton's law of viscosity (*see viscosity*), the shear rate at the wall, γ_w, must equal the shear stress divided by the viscosity. Solving the above equation for $\Delta PR/2\eta L$, one obtains $\gamma_w = 4Q/\pi R^3$. By applying the Rabinowitsch correction to this expression for Newtonian shear rate, one can get the true shear rate at the wall for a non-Newtonian liquid. An often seen, equivalent, but slightly different form for the Newtonian shear rate at a tube wall is $8V/D$, where V is the average fluid velocity $= Q/(\pi D^2/4)$ and $D = 2R$. Goodwin JW, Goodwin J, Hughes RW (2000) Rheology for chemists. Royal Society of Chemistry, Cambridge, UK. Patton TC (1979) Paint flow and pigment dispersion: a rheological approach to coating and ink technology. John Wiley and Sons, New York. Parfitt GD (1969) Dispersion of powders in liquids. Elsevier Publishing Co., New York. Anderson TL (2005) Fracture mechanics: fundamentals and applications. CRC Press, Boca Raton, FL. Van Wazer JR, Lyons JW, Kim KY, Colwell RE (1963) Viscosity and flow measurement. Interscience Publishers Inc., New York.

Hair-cracking *See cracking.*

Hairy *See fuzziness.*

Half-life (1907) *n.* The period of time necessary for one-half of a reactant to be consumed in reaction or process. Used to measure the rate of radioactive decay of disintegration. The time lapse during which a radioactive mass losses one half of its radioactivity.

Half-reaction An equation for an oxidation or a reduction which specifically indicates the electrons lost or gained.

H

Halftone \ˈhaf-ˌtōn\ (1651) *n.* A printed image composed of dots of varying frequency (number per in.^{2}), size or density, producing tonal gradations. The term is also applied to the process, "plates" and ink used to produce this image.

Halftone screens *n.* Transparent plates ruled diagonally with opaque lines at right angles to one another, the distance between them and their thickness being approximately equal.

Halide \ˈha-ˌlīd\ (1876) *n.* Any compound including one (or more) of the halogen elements (F, Cl, Br, and I) in its –1 valence state. The term halide is often used to indicate that any of the four principal halogens may be interchangeably present.

Hall effect When a steady current is flowing in a steady magnetic field, electromotive forces are developed which are all right angles both to the magnetic force and to the current and are proportional to the product of the intensity of the current, the magnetic force and the sine of the angle between the directions of these quantities.

Halocarbon plastic \ˈha-lə-ˌkär-bən-\ (1950) *n.* A term listed by ASTM (D 883) to mean a polymer containing only carbon and one or more halogens. The primary members of the family are the chlorofluorocarbon and fluorocarbon resins.

Halo effect (ca. 1928) *n.* Piling up of ink at the edges of printed letters and dots. Also colored or sometimes uncolored areas adjacent to them, caused by the spread of colored or uncolored vehicles. Leach RH, Pierce RJ, Hickman EP, Mackenzie MJ, Smith HG (eds) (1993) Printing ink manual, 5th edn. Blueprint, New York.

Halogen \ˈha-lə-jən\ [Sw, fr. *hal-* + *-gen*] (1842) *n.* The elements of group 7a of the periodic table; fluorine, chlorine, bromine, and iodine (F, Cl, Br, and I). The fifth member, astatine, is rare, radioactive, and unstable, with a half-life less than 9 h, so it is never seen in commerce.

Halogenated solvents *n.* The solvents containing halogen (usually chlorine) have improved solvency compared with the hydrocarbons from which they are derived and in addition flammability is reduced. Some of these are highly toxic, and precautions must be taken to avoid inhalation of their vapors.

Halogenation (1882) *n.* Process or reaction wherein any members of the halogen group are introduced into an organic compound, either by simple addition or substitution.

Halogen/phosphorus flame retardant *n.* Chlorine- and bromine-substituted esters of phosphoric acid or phosphonic acid, used mostly in polyurethane foams to impart fire resistance.

Hammer finish \ˈha-mər-\. A paint finish which appears to have been applied over hammered metal; produced by the use of non-leafing metallic pigment, plus tinting pigments, which are mixed in a special binder.

Hammer mill (1610) *n.* Type of mill used for pulverizing dry materials in which the disintegration is caused by the action of a number of small hammers.

Hammett equation \ˈha-met-\. An equation used in organic chemistry to predict reactions for aromatic compounds.

$$\log\frac{K}{K_o} = \rho\sigma \text{ or } \log\frac{k}{k_o} = \rho\sigma,$$

where *K* or *k* refers to the reaction of a *m*- or *p*-substituted phenyl compound (e.g., ionization of a substituted benzoic acid) and K_o or k_o refers to the same reaction of the unsubstituted compound (e.g., ionization of benzoic acid), the substituent group constant ρ is a number (+ or −) indicating the relative electron-withdrawing or electron-releasing effect of a particular substituent, the reaction constant σ is a number (+ or −) indicating the relative need of a particular reaction for electron withdrawal or electron release; a linear free energy relationship since it is based on the fact that a linear relationship exists between free energy change and the effect exerted by a substituent. Smith MB, March J (2001) Advanced organic chemistry, 5th edn. John Wiley and Sons, New York. Morrison RT, Boyd RN (1992) Organic chemistry, 6th edn. Prentice-Hall, Englewood Cliffs, NJ.

Hand \ˈhand\ {*often attributive*} ME, fr. OE, akin to OHGr *hant* hand] (before 12c) *n.* The feel of film, fabric, or coated fabric – its flexibility, smoothness, and softness – as judged by the touch of a person.

Hand-blocked print *n.* A fabric that has been printed by hand with wooden or linoleum blocks.

Also see printing.

Hand-blocking *See block printing.*

Hand lay-up *n.* Method of forming reinforced plastics articles comprising the steps of placing a web of the reinforcement, which may or may not be preimpregnated with a resin, in a mold or over a form and applying fluid resin to impregnate and/or coat the reinforcement, followed by curing of the resin and extraction of the cured article from the mold. *See lay-up molding.*

Handle *See hand.*

Hand mold *n.* A mold that is removed from the press after each shot for extraction of the molded article; generally used only for short runs and experimental moldings.

Hand-screening *See silk-screening.*

Hanger wire *n.* The wire employed to suspend the acoustical ceiling from the existing structure (wood joists, steel bar joists, steel beams, concrete slabs, etc.). Harris CM (2005) Dictionary of architecture and construction. McGraw-Hill Co., New York.

Hang pick *n.* A pick that is caught on a warp yarn knot for a short distance, which produces a triangular hole in the fabric. Hang picks usually result from knots that are tied incorrectly, shuttle tension that is too loose, or harness that is timed too early.

Hang shot *See hang pick.*

Hangup *n.* (1) Stray bits of extrudate that cling to the face of the die and eventually have to be cleaned off. (2) Failure of feed material in a hopper to fall out of the hopper, caused sometimes by arching when the feed opening is only a few times larger than the pellet, but more often by clumping together of sticky particles.

Hank \ˈhaŋk\ [ME, of Scand origin; akin to ON *hǫnk* hank; akin to OE *hangian* to hand] (14) *n.* (1) A skein of yarn. (2) A standard length of slubbing, roving, or yarn. The length is specified by the yarn numbering system in use; e.g., cotton hanks have a length of 840 yards. (3) A term applied to slubbing or roving that indicates the yarn number (count); e.g., a 1.5 hank roving.

Hank shellac Form of bleached shellac, which, instead of being ground and dried, is pulled into hanks. Contains up to 25% moisture.

Hansa yellow \ˈhan(t)sə-\ *n.* Range of yellow pigment dyestuffs which are characterized by clean tones and good fastness to light in mass tones but poor fastness to light in pale tints. Heat resistance is also poor and therefore these pigments should not be used in baking finishes. They are produced by coupling diazotized bases with acetoacetanilide or its substitution products.

Hanus iodine number *n.* Method claimed to measure the complete unsaturation of both conjugated and non-conjugated substances. The method employs as a reagent a solution of iodine and bromine in glacial acetic acid.

H

Hypalon® Chlorosulfonated polyethylene, manufactured by DuPont Performance Elastomers. A versatile, high-performance elastomer used extensively in the automotive and general rubber industries.

Hardboard \ˈhärd-ˌbōrd\ (1925) *n.* A generic term for a smooth, grainless panel manufactured primarily from interfelted lignocellulosic fibers (usually wood), consolidated under heat and pressure in a hot-press to a density of 31 lb/ft.^{3} (specific gravity 0.50) or greater, and to which other materials may have been added during manufacture to improve certain properties. In its most common form, it is smooth on one side and has a light waffle-like texture on the reverse side. Used for interior panels or durable siding. It is also available with a smooth surface on both sides, and with a variety of embossed or textured finishes, as well as a perforated form that accepts hooks and fixtures, for converting any wall into a storage area.

Hard dry *See dry-hard.*

Hardener \ˈhärd-nər\ (1611) *n.* Additive (cross-linking agent, resin, or other modifier) used to promote or control the hardening or curing reaction of a coating, plastic, adhesive or resin system. {G Härter m, F durcisseur m, S endurecedor m, I induritore m}. Odian GC (2004) Principles of polymerization. John Wiley and Sons Inc., New York.
See curing agent, catalyst, and cross-linking.

Hard-facing alloy *n.* Any of several metals applied by welding to the contact surfaces (screw-flight tips) to reduce the rate of wear. Colmonoys® are nickel-based, containing boron, chromium, iron, and silicon. Stellites® are cobalt-based, with chromium, and tungsten. They are usually applied into a machined helical groove prior to cutting the main channel, then finish-ground afterward.

Hard ferrite *See ferrite.*

Hard fiber *n.* Stiff, elongated fibers obtained from leaves or stems of plants. Coarse and stiff, they are used in matting and industrial products. Vincenti R (ed) (1994) Elsevier's textile dictionary. Elsevier Science and Technology Books, New York.

Hard gloss paint *n.* A high-gloss enamel, formulated with a hard-drying resin vehicle.

Hard gums *n.* Resins exuded from living plants centuries ago, Some of the hard gums are Amber, Zanzibar, Kauri, Pontianak, Manila, and Congo.

Hardinge mill *n.* Conical type of ball mill used for grinding rock and similar materials.

Hard lac resin *n.* Resin resulting from shellac when treated by fractional solvent extraction to remove the soft resin component. Hard lac resin is claimed to give films, which are harder, and more durable, than those of ordinary shellac. In addition, hard lac resin polymerizes more easily when heated and has better water resistance. Weismantal GF (1981) Paint handbook. McGraw-Hill Inc., New York.

Hardness \-nəs\ (before 12c) *n.* (1) The resistance to local deformation. (2) Ability of a coating film, as distinct from its substrate, to resist cutting, indentation, or penetration by a hard object. An arbitrary scale of hardness is based upon ten selected minerals. For metals the diameter of the indentation made by a hardened steel sphere (Brinnell) or the height of rebound of a small drop hammer (Shore Scleroscope) serve to measure hardness. {G Härte f, F dureté f, S dureza f, I durezza f}. *See barcol hardness, brinell hardness, durometer, indentation hardness, knoop hardness number, knoop microhard-ness, Mohn hardness, pendulum-rocker hardness, modulus of elasticity, rockwell hardness, scratch hardness, and vickers hardness.*

Hardness scale *See Mohs, scale, knoop hardness number, pencil hardness, and rocker hardness tester.*

Hard putty *n.* (1) Linseed oil putty hardened with litharge or basic lead carbonate. (2) A quick setting putty containing water and plaster of Paris.

Hard rubber *n.* The material obtained by heating a highly unsaturated diene rubber with a high percentage of sulfur. The first hard rubber was Ebonite, made from natural rubber, and black, as its name implies. Similar products have been made from several of the synthetics. The sulfur is mostly contained in 3- and 4-carbon rings with a few –C–S–C– cross-links.

Hard size *n.* A condition found in areas of fabric where the warp contains an excessive quantity of sizing. Vincenti R (ed) (1994) Elsevier's textile dictionary. Elsevier Science and Technology Books, New York.

Hard stopping *See hard putty.*

Hard-surfacing *See hard-facing alloy and case hardening.*

Hardware (ca. 1515) *n.* The physical parts of mechanical devices and electronic control systems.

Hardwoods (1568) *n.* Generally, the botanical group of trees that have broad leaves, in contrast to the conifers or softwoods. The term has no reference to the actual hardness of the wood. Angiosperms is the botanical name for hardwoods.

Harmonic motion (1867) *n.* Simple harmonic motion and angular harmonic motion.

Harness \ˈhär-nəs\ [ME *herneis* baggage, gear, fr. OF] (14c) *n.* A frame holding the heddles in position in the loom during weaving. Vincenti R (ed) (1994) Elsevier's textile dictionary. Elsevier Science and Technology Books, New York.

Harness chain *n.* A mechanism used to control the vertical movements of the harness, or shaft, on a loom.

Harsh fiber *n.* Fiber that is rough or coarse to the touch, but not fused or bonded filaments. Vincenti R (ed) (1994) Elsevier's textile dictionary. Elsevier Science and Technology Books, New York.

Hashab *n.* Name applied to a fine pale grade of gum acacia.

Haul-off *n.* In sheet extrusion, the three-roll stand and cooling-conveyor assembly that polishes and cools the molten sheet emerging from the die, often extended in scope to include the rubber pull rolls and winder. The term has analogous meaning for other extruded products, such as pipe. *See, too, sheet line.*

Hazardous substance *n.* A substance which, by reason of being explosive, flammable, poisonous, corrosive, oxidizing, or otherwise harmful, is likely to cause death or injury. Element or compound which, when discharged in any quantity, presents an imminent and substantial danger to the public health or welfare.

Haze *n.* Haze and luminous transmittance are tests to measure the light transmittancy of different types and grades of plastic materials; Haze is defined in ASTM D 1003 as the percentage of transmitted light passing through the specimen which deviates more than 2.5° from the incident beam by

H

forward scattering; luminous transmittance is defined as the ratio of transmitted to incident light. The cloudy or turbid appearance of an otherwise transparent specimen caused by light scattered from within the specimen or from its surfaces (www.astm.org). Koleske JV (ed) (1995) Paint and coating testing manual. American Society for Testing and Materials, West Conshohocken, PA. Paint testing manual: physical and chemical examination of paints, varnishes, lacquers, and colors – STP 500. American Society for Testing and Materials, West Conshohocken, PA, 1973.

Hazen color scale *n. See apha color scale.*

Hazy (1) A relatively small amount of non-settling, finely dispersed matter which is not visibly homogeneous with the mass of the liquid specified, even through the liquid is transparent and transmits most of the light incident upon it. (2) Displaying haze.

HDPE *n.* Abbreviation for high-density polyethylene.
See polyethylene.

Head *n.* (1) In any extrusion operation, the delivery end of the extruder, usually fitted with a hinged gate that may contain a breaker plate and screen pack, to which the adapter and die are attached. (2) In blow molding, the entire apparatus by which the molten plastic is shaped into a tubular parison. This may include an adapter, a parison die, and a melt accumulator.

Head end *n.* (1) The beginning of a new piece of fabric in the loom that bears appropriate identification. (2) A small sample of fabric that may be submitted to a customer for approval.

Headers *n.* Double wood pieces supporting joists in a floor, or double wood members placed on edge over windows and doors to transfer the roof and floor weight to the studs.

Head pressure *n.* In extrusion, the pressure of the melt at the delivery, or head end of the screw, typically as signaled by a pressure transducer mounted in the barrel opposite the tip of the screw. Head pressure is one of the most important indicators of the state of any extrusion operation.

Head-to-head polymer *n.* A configuration whereas the functional groups are on adjacent carbon atoms on a polymer chain. A polymer in which the monomeric units are alternately reversed produced from the monomer, $CH_2 = CHR$ Odian GC (2004) Principles of polymerization. John Wiley and Sons Inc., New York.
(*See image*).

Head-to-tail polymer *n.* A structure of the type

$$\sim\sim CH_2CHX–CHXCH_2\sim\sim$$

in a vinyl or related type of polymer. During polymerization, it results from a growing chain $\sim\sim CH_2C^*HX$ adding on a monomer ($CH_2{=}CHX$) to give $–CH_2CHX–CHXCH_2^*$, the substituted carbon atom being designated the head (*See image*).

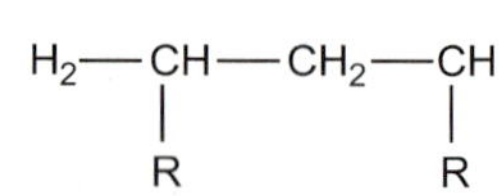

{G Kopf-Schwanz-Polymerisation f, F polymérisation tête-bêche, polymérisation f, S polimerización cabeza a cola, polimerización f, I polimerizzazione testa-coda,

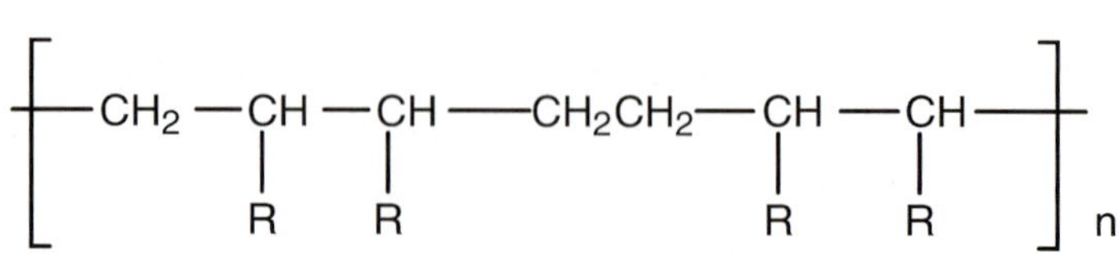

polimerizzazione f}. Odian GC (2004) Principles of polymerization. John Wiley and Sons Inc., New York.

Heartwood \-ˈwu̇d\ (1810) *n.* The wood extending from the pith to the sapwood, the cells of which no longer participate in the life processes of the tree. Heartwood may be infiltrated with gums, resins and other materials, which usually make it darker and more decay resistant than sapwood.

Heat \ˈhēt\ [ME *hete*, fr. OE *hîte*, *hîtu*; akin to OE *hāt* hot] (before 12c) *n.* Energy which is transit from a hot to a cold object because of the temperature difference, to become warm or hot. Heat is a form of energy associated with the motion of atoms, molecules and other particles matter is composed of. It can be created by chemical reactions (such as burning), nuclear reactions (such as fusion reactions taking place inside the Sun), electromagnetic dissipation (as in electric stoves), or mechanical dissipation (such as friction). Heat can be transferred between objects by radiation, conduction and convection. Temperature is used to indicate the level of elementary movement associated with heat. Heat can only be transferred between objects, or areas within an object, that have different temperatures. By common knowledge, the term heat has been used in connection with the warmth, or hotness, of surrounding objects. The concept that warm objects "contain heat" is not uncommon. During its 350 year development, the science of thermodynamics had established a physical quantity named temperature to quantify the level of "warmth", whereas heat (also improperly called heat change) was defined as a transient form of energy that quantifies the spontaneous transfer of thermal energy due to a temperature difference or gradient. The SI unit for heat is the joule; an alternative unit still in use in the United States and other countries is the British thermal unit. The amount of heat exchanged by an object when its temperature varies by 1° is called heat capacity. Heat capacity is specific to each and every object. When referred to a quantity unit (such as mass or moles), the heat exchanged per degree is termed specific heat, and depends primarily on the composition and physical state (phase) of objects. Fuels generate predictable amounts of heat when burned; this heat is known as heating value and is expressed per unit of quantity. Upon transitioning from one phase to another, pure substances can exchange heat without their temperature suffering any change. The amount of heat exchanged during a phase change is known as latent heat and depends primarily on the substance and the initial and final phase. Heat is a process quantity, as opposed to being a state quantity, and is to thermal energy as work is to mechanical energy. Heat flows between regions that are not in thermal equilibrium with each other; it spontaneously flows from areas of high temperature to areas of low temperature. All objects (matter) have a certain amount of internal energy, a state quantity that is related to the random motion of their atoms or molecules. When two bodies of different temperature come into thermal contact, they will exchange internal energy until the temperature is equalized (that is, until they reach thermal equilibrium). The amount of energy transferred is the amount of heat exchanged. It is a common misconception to confuse heat with internal energy: heat is related to the change in internal energy and the work performed by the system. The term heat is used to describe the flow of energy, while the term internal energy is used to describe the energy itself. Understanding this difference is a necessary part of understanding the first law of thermodynamics. Infrared radiation is often linked to heat, since objects at room temperature or above will emit radiation mostly concentrated in the mid-infrared band (*see black body*). Smith JM, Abbott MM, Abbot M, Van Ness HC (2004) Introduction to chemical engineering thermodynamics. McGraw-Hill Co., New York. Lide DR (ed) (2004) CRC Handbook of chemistry and physics. CRC Press, Boca Raton, FL. Hudson RD Jr (1969) Infrared system engineering. John Wiley and Sons, New York.

(*See also www.wikipedia.org*).

Heat bodied *n.* Thickening or polymerizing a drying oil by heat processing.

Heat-bodied oil *See oil, heat-bodied.*

Heat capacity (specific heat) *n.* The amount of heat required to raise the temperature of a unit mass of a substance one degree. In the SI system, the unit of heat capacity is J/kg K, but kJ/kg K, or J/g K are often more convenient. Conversions from older units are: 1 cal/g°C = 1 Btu/lb°F = 4.186 J/g K. Most neat resins have heat capacities (averaged from room temperature to about 100°C) between 0.92 J/g K for polychlorotrifluoroethylene and 2.9 for polyolefins (The heat capacity of water, one of the highest of all materials, is 4.18 J/g K at room temperature.) A term loosely used as a synonymous with heat capacity but not truly so is specific heat.

Heat cleaning *n.* (1) A batch or continuous process in which sizing on glass fibers is vaporized off. (2) Cleaning residual polymer from small extruder screws, dies or other small parts by immersing them in fused salts such as sodium nitrate.

Heat content *n.* Syn: enthalpy.

Heat-convertible resins *n.* Thermosetting resins which on controlled heating become infusible and insoluble. This phenomenon is associated with cross-linking of the resin molecules.

Heat distortion *n.* The degree of deformation of a plastic material under load for a short time at elevated temperature.

Heat-distortion point *n.* The former name, now deprecated and fading from use, of deflection temperature.

Heated-tool welding *See hot-plate welding.*

Heat effect *n.* The heat in calories developed in a circuit by an electric current of I amperes flowing through a resistance of R ohms, with a difference of potential E volts for a time t seconds,

$$H = \frac{RI^2 t}{4.18} = \frac{EIt}{4.18}.$$

Weast RC (ed) (1971) Handbook of chemistry and physics, 52nd edn. CRC Press, Boca Raton, FL.

Heat equivalent, or latent heat, of fusion *n.* The quantity of heat necessary to change one gram of solid to a liquid with no temperature change. Dimensions $[L^2 T^{-2}]$.

Heater band *n.* An electrical heating unit shaped to fit extruder barrels, adapters, die surfaces, etc, for maintaining high temperatures of the banded items and furnishing heat to the plastics within them. Nichrome wires within the bands are embedded in insulation such as magnesium oxide and their ends are welded to screw-type terminals at the end of the band. The ends of cylindrical bands are brought together by Monel bolts to tighten them snugly against the surface to be heated. Service life is greatly increased by operating them at 80% or less of rated wattage, the usual practice.

See also cast-in heater.

Heat forming *See thermoforming.*

Heather yarn \ˈhe-Thər-\ *n.* A term describing mottled or mélange-type yarns.

Heating cylinder (heating chamber) *n.* In elderly injection molders, that part of the machine in which the cold plastic pellets were heated to the molten condition before injection. Until the late 1950s, heating cylinders for injection machines were static devices equipped internally with *torpedoes* or *spreaders*, sometimes heated separately that caused the charge to be distributed in a thin annulus with the cylinder. This hastened the heating but was still slow and could degrade the plastic. Today, all new commercial machines are screw-injection types, accomplishing the heating mainly by the frictional action of the screw. This heating and melting process is not only faster but produces a melt of more uniform temperature, less likely to have been scorched. Strong AB (2000) Plastics materials and processing. Prentice-Hall, Columbus, OH.

Heat mark *n.* An extremely shallow depression or groove in the surface of a plastic article, visible because of a sharply defined rim or a roughened surface.
See also sink mark.

Heat of combustion *n.* The amount of heat evolved by the combustion of a unit mass, usually one gram-molecular weight, of the substance (kJ/mol). The former unit of heat of combustion, deprecated now but still widely used, was the kcal/mol, equal to 4.186 kJ/mol. Heats of combustion, which are tabulated for many pure compounds, are an important segment of the database of physical chemistry and chemical engineering, for it is from sums and differences in heats of combustion that most heats of reaction are estimated. Ready RG (1996) Thermodynamics. Plenum Publishing Co., New York.

Heat of formation, ΔH_f *n.* The change in enthalpy for a reaction in which 1 mol of a compound in formed from its uncombined elements (*enthalpy of formation*). Ready RG (1996) Thermodynamics. Plenum Publishing Co., New York.

Heat of fusion (enthalpy of fusion) *n.* The quantity of heat needed to melt a unit mass of a solid at constant temperature (J/kg or J/mol). Because all meltable plastics consist of broad mixtures of homologous molecules having different molecular weights, and because melting points of homologs increase with molecular weight, it is not possible to melt a plastic at a constant temperature. Instead, melting occurs over a range of temperature. With crystalline resins, such as polyethylene, the range may be relatively small, about 20°C, and it is really only for such resins that "heat of fusion" is meaningful and measurable. To estimate the heat of fusion, enthalpy (J/g) is measured beginning at a temperature well below that at which melting begins and carried on to temperatures well above that at which the plastic has completely melted. Plots (or equations) of enthalpy versus temperature for both ranges are extrapolated to the center of the melting range, the "melting point". The difference between the higher and lower enthalpies at that point is taken as the heat of fusion of the plastic. Ready RG (1996) Thermodynamics. Plenum Publishing Co., New York.

Heat of polymerization (enthalpy of polymerization). The difference between the enthalpy of 1 mol of monomer and the enthalpy of the products of the polymerization reaction. Addition polymerizations are exothermic, values ranging from about 35 to 100 kJ/mol, and removal of this heat is an important aspect of reactor design. The reported value of an enthalpy of polymerization may be referred either to the temperature at which the polymerization is usually carried out or to a standard temperature, such as 25 or 18°C. Ready RG (1996) Thermodynamics. Plenum Publishing Co, New York.

Heat quantity *n.* The cgs unit of heat is the *calorie*, the quantity of heat necessary to change the temperature of 1 g of water from 3.5 to 4.5°C (called a small calorie). If the temperature change involved is from 14.5 to 15.5°C, the unit is the normal calorie. The mean calorie is $\frac{1}{100}$ the quantity of heat necessary to raise one gram of water from 0 to 100°C. The large calorie is equal to 1000 small calories. The British thermal unit is the heat required to raise the temperature of one pound of water at its maximum density, 1°F. It is equal to about 252 cal. Dimensions of energy $[M\,L^2T^{-2}]$. Lide DR (ed) (2004) CRC handbook of chemistry and physics. CRC Press, Boca Raton, FL. Ready RG (1996) Thermodynamics. Plenum Publishing Co., New York.

Heat-reactive phenolic resin *n.* Phenol-formaldehyde resin which has been manufactured under alkaline conditions and which reacts with oils, rosin, or other unsaturated components during varnish making, through the free methylol groups in its structure along with further condensation of the methylol groups in the resin.

Heat resistance *n.* A property of certain fibers or yarns whereby they resist degradation at high temperature. Heat resistance may be an inherent property of the fiber-forming polymer or it may be imparted by additives or treatment during manufacture. *Also see heat stabilized.*

Heat-resistant finishes Finishes designed to resist deterioration on continuous or intermittent exposure to a predetermined elevated temperature.

Heat seal *n.* A method of uniting two or more surfaces by fusion, either of the coatings or the base materials, under controlled conditions of temperature, pressure, and dwell time.

Heat sealing *n.* The process of joining two or more thermoplastic films or sheets by heating areas in contact with each other to the temperature at which fusion occurs, usually aided by pressure. When the heat is applied by dies or rotating wheels maintained at a constant temperature, the process is called *thermal sealing*. In *melt-bead sealing*, a narrow strand of molten polymer is extruded along one surface, trailed by a wheel that presses the two surfaces together. In *impulse sealing*, heat is applied by resistance elements that are applied to the work when relatively cool, then are rapidly heated. Simultaneous sealing and cutting can be performed in this way. *Dielectric sealing* is accomplished with polar materials by inducing heat within the films by means of radio-frequency waves. When heating is performed with ultrasonic vibrations, the process is called *ultrasonic sealing*.
See also welding.

Heat-seal strength *n.* With heat-sealed flexible films, the force required to pull apart a heat-sealed joint divided by the joined area tested. The strength of a heat seal is sometimes expressed as a percentage of the film's tear strength or tensile strength.

Heat-sensitive paints *n.* Coatings, which change color at a specific temperature.

Heat sensitivity *n.* The tendency of a plastic to undergo changes in properties, color, or even to degrade at elevated temperatures. Severity of change is always a matter of both temperature and time. ASTM D 794 describes the procedures to be used in determining the permanent effects on plastics of elevated-temperature exposure.

Heat set inks *n.* Letterpress and lithographic inks which dry under the action of heat by evaporation of their high boiling solvent.

Heat-shrinkable film *n.* A film that is stretched and oriented while it is being cooled so that later, when used in packaging, it will, upon being rewarmed, shrink tightly around the package contents. Blown film made from plasticized PVC is the largest-volume shrink film. Heat-shrinkable tubing of several polymers if widely used in the electronics industry to protect bundles of wiring. ASTM D 2671 (section 10.01) describes a method for testing such tubing.

Heat sink (1936) *n.* A device for the absorption or transfer of heat away from a critical part or assembly.

Heat stability *n.* The resistance to change in color or other properties as a result of heat encountered by a plastic compound or article either during processing or in service. Such resistance may be enhanced by the incorporation of a stabilizer.

Heat stability test *n.* An accelerated test used to predict viscosity stability of coatings with time. The test generally involves measuring viscosity change after the paint has been heat-aged about a week to 10 days at 50–60°C. An excessive increase in viscosity is unacceptable.

Heat stabilizers *n.* A stabilizer additive to reduce or eliminate thermal degradation or its effects. Most commonly used with chlorinated polymers, which are particularly thermally sensitive due to their tendency to dehydrochlorinate.

Heat test *n.* Test employed for assessing the purity of tung and similar rapidly polymerizing oils. The test involves heating the oil at a specified temperature and noting the time required for gelation to occur. Syn: browne heat test.
See gel time.

Heat transfer *n.* The movement of energy as heat from a hotter body to a cooler body. The three basic mechanisms of heat transfer are radiation, conduction, and convection. Radiation heating occurs when heat passes from the emitting body to the receiving body through a medium, such as air, that is not warmed. Conduction heating is the flow of heat from a hot region to a cooler one in either single homogeneous substances or two substances in close contact with each other. Convection is the transfer of heat by flow of a fluid, either a gas or a liquid, and either by natural currents caused by differences in density or by forced movement caused by a fan, pump, or stirrer. All three modes of transfer are important in plastics processing {*heat transfer coefficient* G Härte f, F dureté f, S dureza f, I durezza f}.

Heat-transfer medium *See thermal fluid.*

Heat transfer printing *n.* A method, whereby a printed image is transferred from a carrier to a receiving substrate by the use of heat. In the process, as currently performed, the ink is made up of sublimable dyes in conventional ink vehicles; the carrier is paper, and the receiving substrate is a synthetic fabric.
Also known as thermal printing.

Heavy-bodied *n.* Having a thick consistency or high viscosity.

Heavy bodied inks *n.* Inks of a high viscosity or stiff consistency.

Heavy-centered pattern *n.* Spray pattern having most paint in center, less at edges.

Heavy end *n.* (1) The higher boiling fraction in distillation. (2) *See coarse thread.*

Heavy filling *See coarse thread.*

Heavy metals (1974) *n.* Metallic elements with high molecular weights generally toxic to plant and animal life. Examples: mercury, chromium, arsenic, lead, etc.

Heavy pick *See coarse thread.*

Heavy solvent naphtha *n.* High-boiling coal-tar naphtha generally distilling below 190°C (90/190°C is a usual grade).

Heavy spar *See barium sulfate, natural.*

Hecto- {*combining form*} [F, irreg. fr. Gk *hekaton*] (h) *n.* The SI prefix meaning $\times 100$.

Heddle \ˈhe-dəl\ [prob. alt. of ME *helde*, fr. OE *hefeld*; akin to ON *hafald* heddle, OE *hebban* to lift] (1513) *n.* A cord, round steel wire, or thin flat steel strip with a loop or eye near the center through which one or more warp threads pass on the loom so that the thread movement may be controlled in weaving. The heddles are held at both ends by the harness frame. They control the weave pattern and shed as the harnesses are raised and lowered during weaving.

Hegman gauge *See fineness of grind gauge.*

Hehner number (value). A number expressing the percentage (i.e., grams per hundred grams) of water-insoluble fatty acids in an oil or fat.

Heisenberg theory of atomic structure \ˈhī-zən-bərg-\ *n.* The currently accepted view of the structure of atoms, formulated by Heisenberg in 1934, according to which the atomic nuclei are built of nucleons, which may be protons or neutrons, while the extranuclear shells consist of electrons only. If m and m' are given in grams, and r in centimeters, F will be in dynes if $k = 6.670 \times 10^{-8}$.

$$F = k\frac{mm'}{r^2}.$$

Goldberg DE (2003) Fundamentals of chemistry. McGraw-Hill Science/Engineering/Math, New York.

Heisenberg uncertainty principle [Gr. Werner *Heisenberg*] (1939) *n.* The product of the uncertainty in the position of a particle times that in its momentum is a constant. It is impossible to determine simultaneously both the position and the momentum of a particle exactly. Goldberg DE (2003) Fundamentals of chemistry. McGraw-Hill Science/Engineering/Math, New York.

Helical screw feeder \ˈhe-li-kəl-\ *n. See crammer-feeder, screw conveyor.*

Helical transition *n.* In the transition zone (compression zone) of an extruder screw, the root surface of the screw describing an advancing spiral surface of increasing radius.

See also conical transition.

Helical winding *n.* A winding in which the filament or band advances along a spiral path, not necessarily at a constant angle except in the case of a cylindrical article.

Helio fast rubine 4BL lake (58055) *n.* This pigment is marketed as an alumina hydrate lake; the parent toner is hard and horny when dried and, consequently, too difficult to disperse. It is black in masstone but yields a purpose tint on reduction with white. Its good lightfastness in dark shades, high color intensity, good chemical and bleed resistance make Helio Fast Rubine 4BL lake particularly suitable for shading full shade durable maroon finishes. On the other hand, its poor lightfastness in very weak pastel shades plus its poor bake resistance at high temperatures make it unsuitable as a white toning agent for high bake exterior white finishes.

Helio-klischograph *n.* A method of engraving gravure cylinders by the use of an electronic scanning system which transmits a signal modulated by the density of a positive copy to a diamond cutting head effecting the engraving mechanically rather than by chemical etching. Leach RH, Pierce RJ,

Hickman EP, Mackenzie MJ, Smith HG (eds) (1993) Printing ink manual, 5th edn. Blueprint, New York.

Helix angle \ˈhē-liks-\ *n.* (1) Of an extruder screw, the lead angle of the flight with respect to the screw axis, the angle whose tangent h = $t/\pi D$, where t is the axial distance the flight advances per turn (lead) and D is the major (or nominal) screw diameter. (2) In filament winding, the angle between the axis of rotation and the filament at its point of contact with the winding.

Hellige system or comparator *n.* Standard instrument for color determination of liquids such as varnishes, oils, etc., based on comparisons which are made against a series of discs of colored glass, and which are carefully graded.

Hematite \ˈhē-mə-ˌtīt\ (1540) *n.* Fe_2O_3. Natural oxide of iron.
See ferric oxide.

Hemiacetal \ˌhe-mē-ˈa-sə-ˌtal\ (1893) *n.* Any class of compounds characterized by the grouping C(OH)(OR) where R is an alkyl group and usually formed as intermediates in the preparation of acetals from aldehydes or ketones. Morrison RT, Boyd RN (1992) Organic chemistry, 6th edn. Prentice-Hall, Englewood Cliffs, NJ.

Hemiacetal group *n.* Functional groups derived from carbonyl groups by addition of one molecule of an alcohol, of the general structure shown: –C–OH–OR–. Morrison RT, Boyd RN (1992) Organic chemistry, 6th edn. Prentice-Hall, Englewood Cliffs, NJ.

Hemicelluloses \ˌhe-mi-ˈsel-yə-ˌlōs\ [ISV] (1891) *n.* The principal non-cellulosic polysaccharides in wood. Wood contains 28–35% hemicelluloses, the balance being cellulose and lignin. Morrison RT, Boyd RN (1992) Organic chemistry, 6th edn. Prentice-Hall, Englewood Cliffs, NJ.

Hemicolloid \-ˈkä-ˌlóid\ *n.* Colloidal particle of specified size, both as regards number of molecules in association, and length of chain. This includes substances involving between 20 and 100 molecules, and possessing an overall chain length not exceeding 25 nm.

Hemihydrate \-ˈhī-ˌdrāt\ (ca. 1901) *n.* A hydrate which contains one-half molecule of water to one molecule of the compound; the most common such material is partially dehydrated gypsum (plaster of Paris).

Hemimorphic \-ˈmór-fik\ [ISV] (ca. 1959) *adj.* A crystal with unlike faces at opposite ends of the same axis.

Hemipolymer *n.* A readily soluble polymer with molecular weights between 1000 and 10,000.

Hemisymmetric \-sə-ˈme-trik\ *n.* A crystal on which one-half the faces of at least one form are missing.

Hemp \ˈhemp\ [ME, fr. OE *hȋnep*; akin to OHGr *hanaf* hemp, Gk *kannabis*] (before 12c) *n.* A coarse, durable bast fiber of *Cannabis sativa* found all over the world. Used primarily for twines, cordage, halyards, and tarred riggings.

Hempseed oil *n.* A semidrying oil obtained from the seeds of *Cannabis sativa,* which grows in India, Manchuria, China, Japan, and some parts of Europe. It is usually classed with soybean, poppyseed, sunflower, and walnut oils, and has an iodine value of approximately 160, arising in the main from the presence of practically 50% of linoleic acid and 25% of linolenic acid. It has some application in manufacture of paints and varnishes.

Henderson–Hasselbalch equation *n.* A formula relating the pH value of a solution to the pK_a value of the acid in the solution and the ratio of the acid and the conjugate base concentrations: $pH = pK_a +$

H

log([A⁻]/[HA]), where [A⁻] is the concentration of the conjugate base and [HA] is the concentration of the protonated acid. For the bicarbonate buffer system in blood,

$$pH = pK + \log([HCO_3^-]/[CO_2]).$$

Whitten KW, Davis RE, Davis E, Peck LM, Stanley GG (2003) General chemistry. Brookes/Cole, New York.

Henry \ˈhen-rē\ [for Joseph *Henry*] (H) *n.* The SI unit of electric inductance, the inductance of a closed circuit in which an electromotive force of one volt is produced when the electric current in the circuit varies uniformly at a rate of 1 A/s. The magnetic flux in this situation will be just 1 weber (Wb). Thus, 1 H = 1 V s/A = 1 Wb/A. Weast RC (ed) (1971) Handbook of chemistry and physics, 52nd edn. CRC Press, Boca Raton, FL.

Henry's law [for W. *Henry*] *n.* The mass of a slightly soluble gas that dissolves in a definite mass of a liquid at a given temperature is directly proportional to the partial pressure of that gas. This law holds only for gases that do not chemically react with the solvent. Phillip R Watson (1997) Physical chemistry. John Wiley and Sons Inc., New York. Weast RC (ed) (1971) Handbook of chemistry and physics, 52nd edn. CRC Press, Boca Raton, FL.

Heptanol-2 *n.* Another name for methyl amyl carbinol. It has a bp of 160.4°C, vp of 0.9 mmHg/20°C, and a Sp gr of 0.8187/20°C.

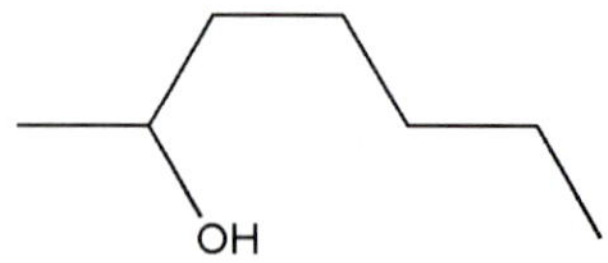

Heptyl acetate *n.* Bp, 191°C; Sp gr, 0.874.

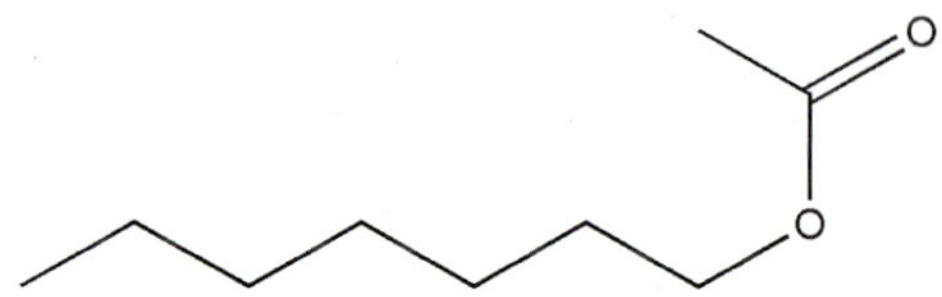

***n*-Heptyl *n*-decyl phthalate** *n.* $C_7H_{15}OOC CF_6H_4COOC_{10}H_{21}$. A general purpose plasticizer for PVC and several other thermoplastics, with lower volatility and better low-temperature performance than dioctyl phthalate. It is excellent for use in vinyl plastisols.

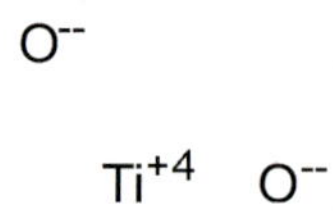

***n*-Heptyl *n*-dinonyl trimellitate** *n.* An aromatic plasticizer similar to trioctyl trimellitate with better low-temperature performance and lower volatility.

***n*-Heptyl *n*-nonyl adipate** *n.* A plasticizer for PVC similar to dioctyl adipate but with better low-temperature performance and lower volatility. It is also approved for use in contact with foods.

Herbicide \ˈ(h)ər-bə-ˌsīd\ [L *herba* + ISV – *cide*] (1899) *n.* An agent used to destroy or inhibit plant growth.

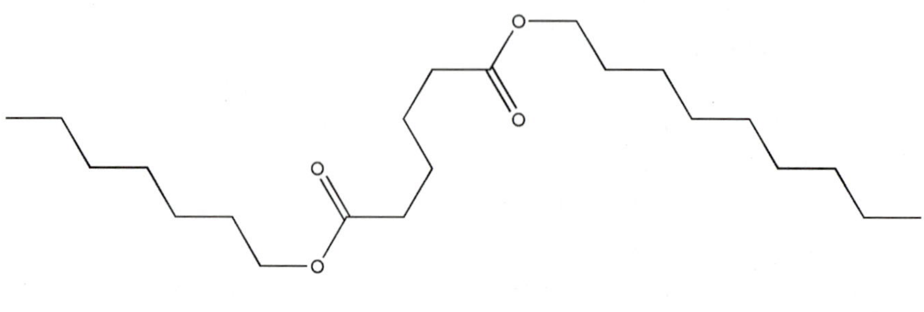

Herringbone \ˈher-iŋ-ˌbōn\ {*often attributive*} (1659) *n.* A broken twill weave characterized by a balanced zigzag effect produced by having the rib run first to the right and then to the left for an equal number of threads.

Hertz \ˈhərts\ [Heinrich R. *Hertz*] (ca. 1928) (Hz) *n.* Cycles per second of any periodic phenomenon. The term and its multiple, megahertz (MHz), were in the past used most frequently for waves in the radio-frequency range, but the hertz is now the SI unit of frequency.

Hessian \ˈhe-shən\ {*chiefly British*} (1729) *n.* A name for burlap used in the United Kingdom, India, and parts of Europe.
Also see burlap.

Hess's law of constant heat summation *n.* The amount of heat generated by a chemical reaction is the same whether the reaction takes place in one step or in several steps, or all chemical reactions, which start with the same original substances and end with the same final substances liberate the same amounts of heat, irrespective of the process by which the final state is reached. Ready RG (1996) Thermodynamics. Plenum Publishing Co., New York.

Heterochain polymers *n.* Polymers which contain more than one type of atom in the polymer chain.

Heterocyclic compound *n.* A compound whose molecule includes at least one ring that contains one or more elements other than carbon and hydrogen. A simple example is *pyridine*, C_5H_5N, a benzene molecule in which one carbon is replaced by nitrogen.

Heterocyclic monomers *n.* Monomer whose molecule includes at least one ring that contains one or more elements other than carbon and hydrogen. A simple example is pyridine, C_5H_5N, a benzene molecule in which one carbon is replaced by nitrogen.

Heterofillament *Also called heterofil.*
See bicomponent fibers.

Heterogeneous \ˌhe-tə-rə-ˈjē-nē-əs\ [ML *heterogeneus*, fr. Gk *heterogenēs*, fr. *heter-* + *genos* kind] (1630) *adj.* Composed of two or more phases.

Heteropolymerization *n.* A special case of addition polymerization involving the combination of two dissimilar unsaturated monomers, the product being a *heteropolymer.*

Hevea rubber *See rubber, natural.*

Hexabromide value \-ˈbrō-ˌmīd ˈval(ˌ)yü\ *n.* Hexabromide value is a measure of the amount of acids of the linolenic types which are present in a mixture of natural fatty acids. Bromine, under suitable conditions, forms addition compounds through addition at the unsaturated linkages. The hexabromides are separated from the dibromides and tetrabromides by reason of the insolubility of the former in cold ether. The hexabromide value is the percentage of hexabromide obtainable from a given material.

Hexabromobiphenyl \-ˈbrō-(ˌ)mō-(ˌ)bī-ˈfe-nᵊl\ *n.* $[2,4,6\text{-}(Br)_3C_6H_2\text{–}]_2$. A flame retardant suitable for use in thermosetting resins and thermoplastics such as acrylonitrile–butadiene–styrene resin, nylons, polycarbonate, polyolefins, PVC, polyphenylene oxide, and polystyrene–acrylonitrile. It is insoluble in water, heat-stable, and furnishes a high bromine content in the end product.

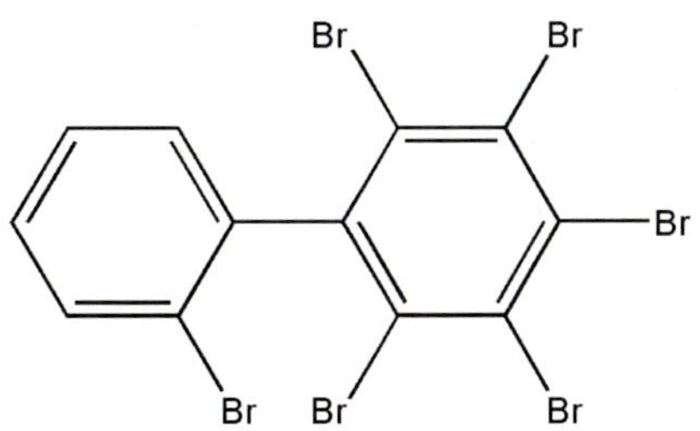

Hexachloroethane (carbon hexachloride, perchloroethane) *n.* A substitute for camphor in celluloid manufacture.

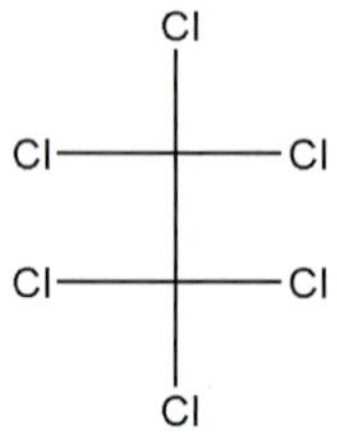

Hexachlorophene *n.* $(C_6HCl_3OH)_2CH_2$. A white, essentially odorless, free-flowing powder widely used as a bacteriostat in many thermoplastics including vinyls, polyolefins, acrylics, and polystyrene.

Hexafluoroacetone trihydrate *n.* A solvent cement, active at room temperatures, for bonding acetal resin articles to themselves and to other polymers such as nylon, acrylonitrile–butadiene–styrene, styrene–acrylonitrile, polyester, cellulosics, and natural or synthetic rubber. It is also a toxic irritant, so it must be handled with care.

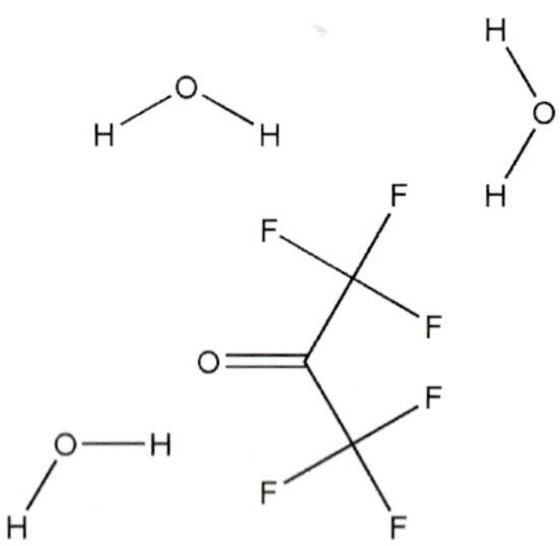

Hexagonal \hek-ˈsa-gə-nᵊl\ (1571) *adj.* (1) Of a closed plane figure, having six sides. (2) Having the properties of a *regular* hexagon, i.e., all six sides and all six angles equal. (3) One of the six crystal systems, in which there are four principal axes, three in one plane at 120° to each other, the fourth perpendicular to the others. Atomic spacing is equal along the planar axes, different along the fourth. Hibbard MJ (2001) Mineralogy. McGraw-Hill Co. Inc., New York.

Hexahydrophenol *n.* Syn: cyclohexanol.

Hexahydrophthalic anhyride (HHPPA) $C_6H_{10}(CO)_2O$. A curing agent for epoxy resins and an intermediate for the production of alkyd resins.

Hexalin *See cyclohexanol.*

Hexalin acetate *See cyclohexyl acetate.*

Hexamethylene \ˌhek-sə-ˈme-thə-ˌlēn\. Syn: cyclohexane.

Hexamethylene adipamide (nylon 6/6). A nylon made by condensing hexamethylene diamine with adipic acid.
See nylon 6/6.

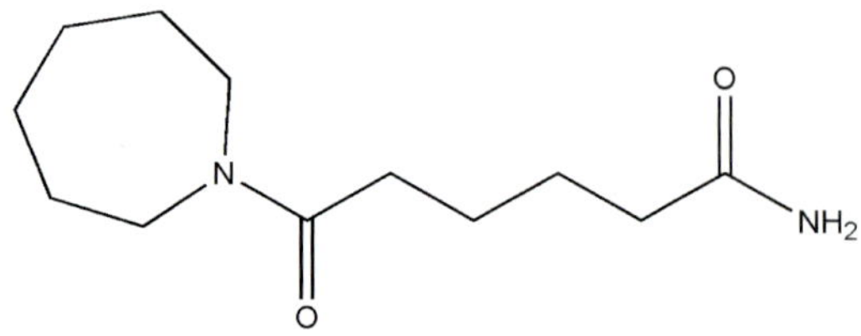

Hexamethylene diamine \-ˈdī-ə-ˌmēn\ (1,6-diaminohexane). A colorless solid in leaflet form, which, when condensed with adipic acid, forms nylon 6/6, It has also been used to cure epoxy resins, especially for coatings and usually in a modified form, for example as an epoxy resin adduct.

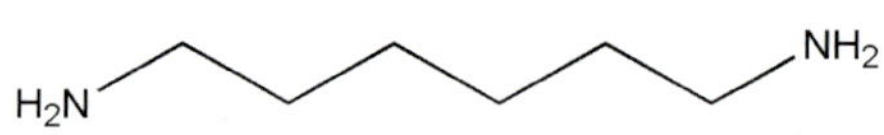

Hexamethylene-1,6-diisocyanate (HDI) A colorless liquid, the first aliphatic diisocyanates to be used commercially in the production of urethanes. When used with certain metal catalysts, it produces urethane polymers with good resistance to discoloration, hydrolysis, and thermal degradation.

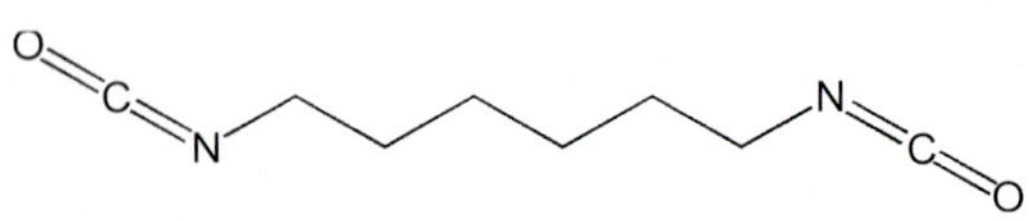

Hexamethylenetetramine (Hexamine) \-ˈte-trə-ˌmēn\ [ISV *hexa-* + *methylene* + *tetra-* + *amine*] (1888) (HMT, methenamine, and urotropine) *n.* $(CH_2)_6$–N_4. A bicyclic compound, the reaction product of ammonia and formaldehyde. It is used as a basic catalyst and accelerator for phenolic and urea resins, and a solid, catalytic-type curing agent (hardener) for epoxies (*See image*).

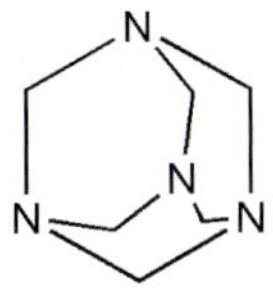

Hexamethylphosphoric triamide $[N(CH_3)_2]_3PO$. A pale, water-soluble liquid used as an ultraviolet absorber in PVC compounds.

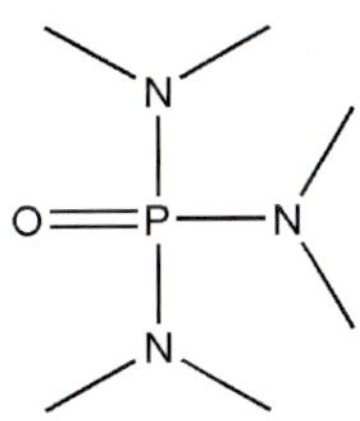

Hexane \ˈhek-ˌsān\ [ISV] (1877) *n.* C_6H_{14}. A straight-chain hydrocarbon, extracted from petroleum or natural gas. Commercial grades contain other hydrocarbons such as cyclohexane, methyl cyclopentane, and benzene. Hexanes are used as catalyst-carrying solvents in the polymerization of olefins and elastomers. Sp gr, 0.660; bp, 69°C; solidification range −95 to −100°C; flp, −22°C (−70°F).

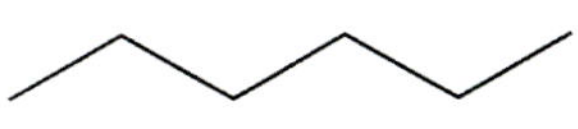

Hexanedioic acid Syn: adipic acid.

Hexanol $CH_3(CH_2)_4CH_2OH$. High-boiling solvent (144–156°C) with a vp of 4 mmHg at 30°C; Sp gr, 0.8186.

Hexone *See methyl isobutyl ketone.*

Hexyl- \ˈhek-səl\ [ISV] (1869) *n.* The straight-chain radical of hexane, C_6H_{13}–.

Hexyl acetate $CH_3COO(CH_2)_5{\cdot}CH_3$. Cellulose nitrate, cellulose acetate-butyrate, polyvinyl acetate, polystyrene, phenolics, alkyds, and coumarone-indene resins. It has a bp of 169°C, Sp gr, of 0.890, and vp of 5 mmHg/20°C.

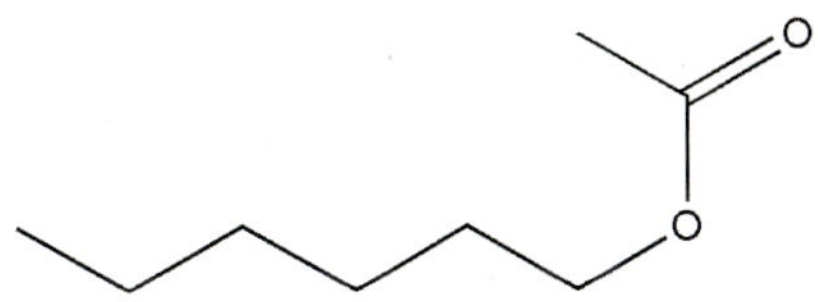

Hexyl ether High boiling solvent. It has a bp of 226°C, Sp gr of 0.794/20°C, vp of 0.07 mmHg/20°C, and flp of 77°C (170°F).

***n*-Hexylethyl *n*-decyl phthalate (NODP)** *See n-octyl n-decyl phthalate.*

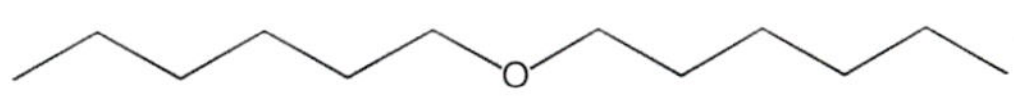

Hexyl methacrylate \-ˌme-ˈtha-krə-ˌlāt\ *n.* A monomer used in producing acrylic resins.

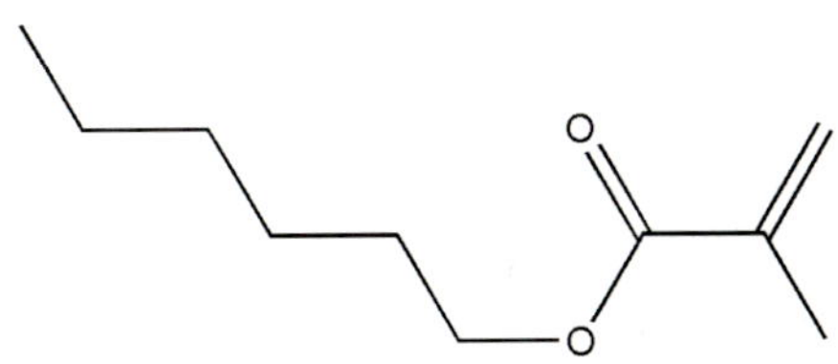

HF *n.* Abbreviation for high frequency.

HF preheating *See dielectric heating.*

Hickies \ˈhi-kēs\ (ca. 1915) *n.* Defects in a print appearing as specks surrounded by an unprinted "halo".

Hide finishes *n.* Designed for use on leather. They are based on pigmented nitrocellulose, and completely cover the surface.

Hiding, complete *n.* The state of uniform application of paint at such a thickness that, when dry, the substrate is no longer affecting the color, i.e., thickness such that increasing the thickness even slightly does change the *perceived* color. Complete hiding is frequently judged by application of a uniform film over a white and black substrate such as a Morest Chart. It is sometimes defined as the state when a contrast ratio of 0.98 is obtained (although this is seldom complete hiding as judged visually). Visual complete hiding is not necessarily spectrophotometric complete hiding. Koleske JV (ed) (1995) Paint and coating testing manual. American Society for Testing and Materials, West Conshohocken, PA.

See contrast ratio.

Hiding, incomplete *n.* The state of application of a coating at such a thickness that the color of the substrate has an effect on the color, which the coating would exhibit if it were applied at complete hiding. Increasing the thickness of the coating thus changes the perceived color. The relative degree of incomplete hiding is frequently described by the contrast ratio. Koleske JV (ed) (1995) Paint and coating testing manual. American Society for Testing and Materials, West Conshohocken, PA.

See contrast ratio and hiding, complete.

Hiding power *n.* The ability of a paint, ink, etc., to hide or obscure a surface over which it has been applied uniformly. When expressed numerically, it is generally in terms of the number of square feet over which a gallon of paint, or pound of pigment, as used, can be uniformly spread to produce a specified contrast ratio. Hiding power is usually expressed in square feet per gallon or square meters per liter. Koleske JV (ed) (1995) Paint and coating testing manual. American Society for Testing and Materials, West Conshohocken, PA.

See contrast ratio.

Hiding power chart *n.* (1) A paper chart, partially black and partially white. A paint film applied over such a chart gives a visual evaluation of the degree of hiding. If a measurement of the reflectance of the film over the black portion is divided by the measurement of the reflectance over the white portion, the contrast ratio (opacity) is obtained. (2) A chart prepared calculation of the Kubelka–Munk equation for a white substrate of specified reflectance for various reflectances at complete hiding (R_∞) and for various reflectances over black, various contrast ratios, scattering-thickness values (SX). From such a chart, using measurements of reflectances over black and white, or measurements over black and of R_∞, changes in thickness, pigment volume concentration, R_∞ for obtaining a desired contrast ratio (degree of hiding) or R_∞, can be determined. Such charts are frequently called Judd Charts. Koleske JV (ed) (1995) Paint and coating testing manual. American Society for Testing and Materials, West Conshohocken, PA.

Hiding, spectrophotometric, complete *n.* The state of uniform application of a paint film which, when dry, exhibits identical spectrophotometric curves over a perfect white of 100% reflectance and a perfect black of 0% reflectance. In the Kulblka–Munk equations, spectrophotometric hiding is defined as R_∞ at all wavelengths. Koleske JV (ed) (1995) Paint and coating testing manual. American Society for Testing and Materials, West Conshohocken, PA.

Hiding, spectrophotometric, incomplete *n.* The state of a uniform paint film in which the spectrophotometric curves measured over white and black areas are not identical; addition of more paint is required to achieve complete spectrophotometric hiding or identity of reflectance curves of the film over white and black substrates. Films, which appear visually to be at complete hiding are not necessarily at complete spectrophotometric hiding.

Hi-flash naphtha *See naphthas.*

High-boiling solvent *n.* A solvent with an initial boiling point above 150°C (302°F).

High build *n.* Proceeding thick, dry films per coat.

See build.

High-build coating *n.* A coating composed of a series of uniform, tile-like films which are applied in thicknesses (minimum 5 mils) greater than those normally associated with paint films and thinner than those normally applied with a trowel.

See also tile-like coating.

High-bulk yarn *See textured yarns.*

High density (1960) *n.* A term to describe a material with heavier than normal weight per unit volume.

Also see density.

High-density overlay *n.* An overlay consisting of paper that is impregnated with a thermosetting resin and then applied to plywood; provides a smooth, hard, wear-resistant surface for high-quality concrete formwork and decking.

High-density polyethylene (HDPE) *n.* This term is generally considered to include polyethylenes ranging in density from about 0.94 to 0.96 g/cm^3 and higher. Whereas the molecules in low-density polyethylenes are branched and linked in random fashion, those in the higher-density polyethylenes are linked in longer chains with fewer side branches, resulting in higher-modulus materials with greater strength, hardness, and chemical resistance, and higher softening temperatures.

See also polyethylene.

Higher fatty acid *See fats.*

High-frequency (1892) *n.* Pertaining to the part of the electromagnetic spectrum between 3 and 200 MHz, employed in plastics welding, sealing and preheating operations. Frequencies of 30 MHz and below are the most used.

High-frequency heating *n.* Syn: dielectric heating.

High-frequency welding *n.* A method of welding thermoplastic articles in which the surfaces to be joined are heated by contact with electrodes of a high-frequency electrical generator.

High gloss *See gloss. Also called full gloss.*

High-impact polystyrene (HIPS) *n.* Polystyrene whose impact strength has been elevated by the incorporation of rubber particles. The best grades, produced by polymerization of styrene containing dissolved rubber, have better impact resistance than those made by mixing rubber and polystyrene together. Whereas crystal PS has notched-Izod impact between 0.13 and 0.24 J/cm, HIPS resins range between 0.37 and 2.1 J/cm.

High-intensity mixer *n.* A type of mixer consisting of a large bowl with a high-speed rotor in its bottom, used for producing dry blends of PVC with liquid and powder additives, also for producing other powder blends. Though there is considerable heating of the charge during mixing, the main constituent is not melted and the mix is discharged in powder form. Compare with internal mixer.

Highlight \ˈhī-ˌlīt\ (ca. 1889) *n.* The light or open areas of a halftone print.

Highlighting *n.* In antiquing, accenting raised areas on furniture with a lighter tone than the crevices. Produced by rubbing harder. Emphasizing the impression of relief by making certain parts of a finished surface lighter than the general color of that surface.

High-load melt index *n.* (1) The rate of flow of a molten resin through an orifice 2.096 mm in diameter and 8.000 mm long at 190°C when subjected to a pressure difference of 2.982 MPa. Shah V (1998) Handbook of plastics testing technology. John Wiley and Sons, New York.
Also known as the flow-behavior index. See power law.

Highloft *n.* General term for a fiber structure containing more air than fiber. Specifically, a lofty, low-density non-woven structure that is used for applications such as fiberfill, insulation, health care, personal protection, and cleaning material.

High-low profile *n.* A pile construction characterized by the presence of two or more pile heights. High-low pile carpets sometimes combine looped and cut surface yarns.

High modulus *n.* A term that refers to a material with a higher than normal resistance to deformation.
Also see modulus.

High-molecular-weight high-density polyethylene (HMWHDPE) *n. See polyethylene* {G Niederdruck-Polyethylen n; Polyethylen hoher Dichte, Polyethylen n, F polyéthylène haute densité (PEHD), polyéthylène m S polietileno de alta densidad, polietileno m, I polietilene alta densità (PE a.d.), polietilene m}.

High-performance composite *See advanced composite.*

High-performance plastic (advanced plastic) *n.* Any neat, filled, or reinforced resin, thermoplastic or thermoset, that maintains stable dimensions and mechanical properties above 100°C. Some writers consider many of the engineering plastics to be high-performance plastics and the distinction is blurred. Others have limited the term to specialty thermoplastics (generally high-priced) such as polyimide, polytetrafluoroethylene, polyphenylene oxide, and sulfide, liquid-crystal polymers, ultra-high molecular-weight polyethylene, etc.
See also advanced resin.

High-performance polyethylene fiber *n.* A strong, stiff fiber made from ultra-high-molecular-weight polyethylene by gel-spinning and drawing to a highly oriented state. Produced in the USA by Allied-Signal as Spectra® fiber.

High pH finish *n.* A finish, basic in nature rather than acid or neutral, that is applied to yarn or fiber.

High polymer (1942) *n.* A polymer with molecules of high molecular weight, sometimes arbitrarily designated as greater than 10,000. All materials commonly regarded as plastics are high polymers, but not all high polymers are plastics.
See polymer.

High-pressure laminate *n.* A laminate molded and cured at pressures not lower than 6.895 MPa (1000 psi), and more commonly in the range of 8–14 MPa.
See also decorative board and laminate.

High-pressure molding *n.* A method of molding or laminating in which the pressure used is greater than 1.4 MPa (200 psi) (ASTM D 883).

High-pressure powder molding *n.* Some polymers in powder form can be molded by high-pressure compaction at room temperature followed by heating to complete sintering, curing or polymerization reactions. The process is limited to fairly simple shapes, and to polymers that do not release vapors when heated. It has been most successful with semi-crystalline polymers that can be post-heated for a sufficient time at a temperature within the crystalline endotherm of the polymer. Examples of such polymers are polyphenylene oxide, polytetrafluoroethylene, and ultra-high-molecular-weight polyethylene. Strong AB (2000) Plastics materials and processing. Prentice-Hall, Columbus, OH.

High-pressure spot *n.* A defect in reinforced plastics: an area containing very little resin, usually due to an excess of reinforcing material.

High-shrink staple *n.* Staple with a higher degree of potential shrinkage than regular staple of the same generic fiber. When blended with regular staple and treated (in yarn or fabric form) to induce shrinkage, it produces a high degree of bulk in the product.

High-temperature dyeing *See dyeing.*

High-temperature plasticizer *n.* Any plasticizer that imparts higher than the normal resistance to high temperatures to plastics compounds in which it is incorporated. An example is di-tridecyl phthalate, which permits vinyl compounds to be used at temperatures up to 136°C.

High tenacity *n.* A term to describe a material with a higher than normal tensile strength.
Also see tenacity.

High-visibility paints *n.* Generally applied to paints containing fluorescent pigments or to paints having high reflectivity because of the use of retroreflective pigments in them. *See fluorescent pigments and retroreflective.*

High-visibility pigments *See fluorescent dyes.*

Hindered isocyanate \-ˌī-sō-ˈsī-ə-ˈnāt\ *n. See isocyanate generator.*

HIPS *n.* Abbreviation for high-impact polystyrene.

Hiroe Grade of pale East India resin exported from Macassar.

Hitch-back *See sticker* (1).

HLB value *See hydrophile–lipophile balance value.*

HMT *n.* Abbreviation for hexamethylenetetramine.

HMWPE *n.* Abbreviation for high-molecular-weight polyethy-lene.
See polyethylene.

Hob \ˈhäb\ [ME *hobbe*, fr. *Hobbe*, nickname for Robert] *n.* A master model of hardened steel that is pressed into a block of softer metal to form a number of identical mold cavities. The hobbed cavities are inserted into recesses in a steel mold base and are connected by a runner system. (2, *v*) To form a mold cavity by forcing a hardened steel *hob*, having the inverse shape of the cavity into a soft metal block (that may subsequently be hardened).

Hold-down groove *n.* A small groove cut into the side wall of a mold to assist in retaining the molding in the mold as it opens.

Hold out (1585) *v.* Ability (or property) to prevent soaking into the substrate.

Hold paint *n.* Marine paint for holds of ships.

Hole \ˈhōl\ [ME, fr. OE *hol* (fr. neuter of *hol*, adj, hollow) & *holh*; akin to OHGr *hol*, adj, hollow; perhaps akin to OE *helan* to conceal] (before 12c) *n.* The absence of an electron in a semiconductor.

Holes (Tow) *n.* In tow opening processes, partial or complete filament breakage within a confined spread of tow, usually circular or oval in shape. Not to be confused with splitting or partial crimp deregistration, which are linear.

Holiday detector *n.* Device for detection of pinholes or voids.

Holidays Application defect whereby small areas are left uncoated. Syn: misses, skips, voids, discontinuities, and vacations.

Holland \ˈhä-lənd\ {*often capitalized*} [ME *holand*, fr. *Holand*, county in the

H

Netherlands, fr. MD *Holland*] (14 c) *n. See shade cloth.*

Holland finish *n.* A glazed or unglazed finish containing oil and a filling material. The finish is applied to cotton fabrics to make them opaque or semi-opaque. The resultant fabric resembles a beetled linen fabric called Holland fabric.

Hollow filament fibers *n.* Manufactured, continuous filament fibers, having voids created by introduction of air or other gas in the polymer solution or by melt spinning through specially designed spinnerets.

Holomicroscopy \-mī-ˈkräs-kə-pē\ *n.* A three-dimensional photomicrographic process utilizing laser beams and time-differential interferometry to study miscroscopic particles in three dimension. This technology has potential for studying crystalline structures in polymeric materials. Loveland RP (1981) Photomicrograsphy. Krieger Publishing Co., New York. Rhodes G (1999) Crystallography made crystal clear: a guide for users of macromolecular models. Elsevier Science and Technology Books, New York.

Homespun \-ˌspən\ (1591) *adj.* Coarse plain-weave fabric of uneven yarns that have a handspun appearance.

Homochain polymers *n.* A polymer containing only one type of atom in the polymer chain.

Homogeneous \-ˌjē-nē-əs\ [ML *homogeneus, homogenus,* fr. Gk *homogenēs,* fr. *hom-* + *genos* kind] (1641) *adj.* Composed of a single phase.

Homogeneous polymerization *n.* Polymerization in which a homopolymer is fanned.

Homogenizer \hō-ˈmä-jə-ˌnīz-ər\ (1886) *n.* A machine used to break up agglomerates and disperse elemental particles in fluids, consisting of a positive-displacement pump capable of attaining very high pressure, an orifice through which the material is forced at high velocity, and an impact ring on which the stream impinges, Homogenizers are used in the preparation of monomer emulsions and coating compounds, and for dispersing pigments and plasticizers into resin. The term is not restricted to one type of machine, and is used to describe agitators and masticators as well as high-speed rotor-stator type precision machines with close clearances. Perry RH, Green DW (1997) Perry's chemical engineer's handbook, 7th edn. McGraw-Hill, New York.
See also colloid mill.

Homologous series \hō-ˈmä-lə-gəs\ *n.* A family of organic compounds that have identical functionality with each succeeding member having one more $-CH_2-$ group in its molecule than the preceding member. An example is the series *methanol* (CH_3OH), *ethanol* (CH_3CH_2OH), *propanol* ($C_3H_7{-}OH$), *butanol* (C_4H_9OH), etc. Morrison RT, Boyd RN (1992) Organic chemistry, 6th edn. Prentice-Hall, Englewood Cliffs, NJ.

Homologous temperature *n.* The ratio of the absolute temperature of a material sample to the absolute temperature at which the material melts.

Homonuclear \ˌhō-mə-ˈnü-klē-ər\ (1930) *adj.* Possessing nuclei of atoms of the same element.

Homopolymer (1946) *n.* High polymers consisting of molecules that contain (neglecting the ends, branch junctions, and other minor irregularities) either a single type of unit or two or more chemically different types in regular sequence. Odian GC (2004) Principles of polymerization. John Wiley and Sons Inc., New York.
See copolymer and polymer.

Homopolymerization *n.* A polymer consisting of a single species of monomer, as polyadenylic acid or polyglutamic acid. Odian GC (2004) Principles of polymerization. John Wiley and Sons Inc., New York.

Honan \ˈhə-ˌnän\ *n.* A pongee-type fabric of the very best Chinese wild silk. Honan is sometimes woven with blue edges.

Honeycomb \ˈhə-nē-ˌkōm\ (before 12c) *n.* Manufactured product consisting of sheet metal or resin-impregnated sheet material (paper, fibrous glass, etc.) that has been formed into a network of open-ended, hexagonal cells, each cell's walls being shared with its immediate neighbors. Honeycombs are used as cores for sandwich constructions.

Honeycomb cores *n.* Plastic impregnated woven or non-woven fiber fabric material that serves as a core in sandwich construction.

Honeycombing *n.* (1) Checks often visible at the surface that occurs in the interior of a piece of wood, usually along the wood drays. (2) Lack of vertical film integrity; formation of cell structure; voids.

Hookean elasticity (ideal elasticity) *n.* Stress–strain behavior in which stress and strain are directly proportional, in accordance with Hooke's law. Serway RA, Faugh JS, Bennett CV (2005) College physics. Thomas, New York.

Hookean spring *n.* A concept visualized as a coil spring whose extension is proportional to the applied load, useful by analogy in modeling the viscoelastic behavior of polymers. Serway RA, Faugh JS, Bennett CV (2005) College physics. Thomas, New York.

Hooke model *See Hookean spring.*

Hooke's law [Robert *Hooke*] (1853) *n.* The observation, by Robert Hooke, that, in a body placed in tension, the fractional increase in length is proportional to the applied load divided by the body's cross-sectional area perpendicular to the load. It was later extended to shear and compressive loading. Thomas Young later contributed the idea of the modulus and formulated Hooke's law as $e = \sigma/E$, where σ is the applied stress, E the modulus of elasticity of the material, and e is the relative elongation, i.e., the change in length divided by the original length. Few plastics conform to Hooke's law beyond deformations greater than 1 or 2%. Serway RA, Faugh JS, Bennett CV (2005) College physics. Thomas, New York.

Hook reed *See reed.*

Hoop stress *c.* The circumferential stress in a cylindrical body, such as a pipe, that is subjected to internal (or external) pressure. Hoop stress (σ) is given as $\sigma = PD/2t$, where P is the pressure, D the mean diameter of the cylinder $(D_o + D_i)/2$, and t is the wall thickness $= (D_o - D_i)/2$. In pipes under internal pressure, the hoop stress is twice the lengthwise stress, so dominates the processes of creep and failure. Hoop stress is an often used term when designing and testing plastic pipe and vessels. Pittance JC (ed) (1990) Engineering plastics and composites. SAM International, Materials Park, OH. Perry RH, Green DW (1997) Perry's chemical engineer's handbook, 7th edn. McGraw-Hill, New York. Chung DD (1994) Carbon fiber compositers. Elsevier Science and Technology Books, New York.

Hopper \ˈhä-pər\ (13c) *n.* In extrusion injection molding, the bin mounted over the feed opening that holds a supply of molding material. The hopper may be intermittently filled or continuously fed (*See hopper loader*). Feeding from the hopper is ordinarily by gravity, but it may be aided by vibrators, stirrers, or screw feeders. In some setups the hopper feeds into a metering device such as a gravimetric feeder or vibratory feeder that meters the rate at which the feedstock enters the extruder or molder.

Hopper dryer *n.* A hopper through which hot air flows upward, drying and heating the feedstock. To improve extraction of moisture from the plastic the air may be passed through a desiccant prior to entering the hopper. The attendant preheating also reduces the amount of power the extruder must furnish per unit mass of material fed and can significantly increase the extruder output. Strong AB (2000) Plastics materials and processing. Prentice-Hall, Columbus, OH.

Hopper loader (hopper filler) *n.* A device for automatically feeding molding powders to hoppers of extruders and injection molders, and maintaining the level of feedstock in the hopper. The functions of drying and blending color concentrates with the feedstock are also sometimes accomplished by loaders. There are two general types: mechanical and pneumatic. Strong AB (2000) Plastics materials and processing. Prentice-Hall, Columbus, OH.

Hopsacking \ˈhäp-ˌsak-\ [ME *hopsak* sack for hops, fr. *hoppe hop* + *sak* sack] (1888) *n.* A coarse, open, basket-weave fabric that gets its name from the plain-weave fabric of jute or hemp used for sacking in which hops are gathered.

Horizontal line *See ring.*

Hot air shrinkage *n.* Generally, the reduction in the dimensions of a fabric, yarn, or fiber induced by exposure to dry heat. Specifically, a fundamental property of fibers.

Hotbench test *n.* A method of determining gelation properties of plastisols, employing a bar or plate whose temperature rises from one side to the other ("temperature-gradient plate"). The sample is spread on the bar and from the positions at which various changes occur in its state, the temperatures of those changes may be estimated.

Hot-gas welding *n.* A welding process for plastics analogous to that used for metals, except that a stream of hot gas is used instead of a flame. Welding guns for plastics consist of a blower behind a heating element, much like a hair drier, and a nozzle that focuses the stream at the weld zone. Either air, or, better, dry nitrogen is used. The heated gas is directed at the joint that has been prepared for welding, while a rod of the plastic being welded is applied to the heated zone and melted into the groove.

Hot head press *n.* A pressing machine capable of generating high temperatures and pressures. Used for pressing and processing permanent-press fabrics.

Hot-leaf stamping *See hot stamping.*

Hot-manifold mold *n.* An injection mold equipped with an internal heater located in the center of the melt stream in the manifold and nozzle system. This type of mold was developed for thermally sensitive resins to provide gentler heating and avoid the decomposition problems experienced with external heating techniques because of excessive temperature differences.

Hot-melt \ˈhät-ˌmelt\ (1939) *n.* A fast-drying non-volatile adhesive applied hot in the molten state. Skeist I (ed) (1990, 1977, 1962) Handbook of adhesives. Van Nostrand Reinhold, New York.

Hot-melt adhesive *n.* A thermoplastic adhesive applied to the surfaces to be joined in the molten state, then allowed to cool, usually under pressure. They are convenient, require no drying, and leave no voids in the joint. But they can creep and generally have lower strength than standard adhesives. ASTM has developed a number of tests for hot-melt adhesives, most of them in section 15.06 of the Annual Book of Standards. Skeist I (ed) (1990, 1977, 1962) Handbook of adhesives. Van Nostrand Reinhold, New York.

Hotmelt coatings *n.* Compositions, which liquify readily on heating and are applied to various surfaces in molten condition.

Hot oil expression *n.* Method of obtaining vegetable oils from seeds by subjecting the oil-bearing seeds to pressure in a suitable press. To facilitate expression of the oil and/or to increase the yield, the seeds are carefully and uniformly preheated before expression. Oils obtained from this process are known as hot-expressed oils.

Hot-plate welding (hot-tool welding) *n.* Two plastic surfaces to be joined are first held lightly against a heated metal surface, which may be coated with polytetrafluoroethylene to prevent sticking, until the surface layers have melted. The surfaces are then quickly joined and held under light pressure until the joint has cooled.
See also thermoband welding and welding.

Hot-pressing *n.* The pressure-forming, between heated platens, of plywood, laminates, particleboard, fiberboard, etc.; usually requires thermosetting resins and heat for curing.

Hot-runner mold (insulated-runner mold) *n.* An injection mold for thermoplastics in which the runners and sometimes the secondary sprues are insulated from the chilled cavity plate so that they remain hot during the entire cycle. The plastic in the runners remains molten and is not ejected with the molded part, thus avoiding the normal handling, grinding, and reprocessing of sprues and runners.

Hot-setting adhesive *See adhesive, hot-setting.*

Hot-short *adj.* Inelastic, not stretchable, and easily broken in tension when hot.

Hot spray *n.* Process in which paint is heated prior to spraying to reduce viscosity so that higher solids may be applied.

Hot spraying *n.* Spraying of hot lacquers or paints, the viscosities of which have been reduced to spraying consistency by means of heat instead of by the addition of volatile solvents. By such a process it is possible to apply materials with higher solid contents and therefore better build.

Hot stage *n.* A microscope stage equipped with heating and cooling at controllable rates, enabling one to observe changes in morphology with temperature, such as spherulite growth in polymers as they are cooled from the melt.

Hot stamping (roll-leaf stamping, gold-leaf stamping) *n.* A method of marking plastics in which a special pigmented, dyed, or metallized foil is pressed against the plastic article by a heated die, welding selected areas of the foil to the article. The term also includes a process of impressing inked type into the material when the type is heated.

Hot surface *n.* (1) An alkaline surface such as plaster, concrete, etc. (2) An abnormally absorbent surface (British).

Hot-tack strength *n.* In heat sealing, the strength of the seal just at the end of the dwell when the die halves part (clearly more a concept than a measurable property).

Hot-tip-gate molding *n.* An injection-molding technique used for thin, large-area articles molded in a single cavity. In conventional molding, a sprue connects the machine nozzle with runners leading to each gated cavity. In hot-tip-gate molding, the sprue is eliminated and material is injected directly from the nozzle through a heated bushing that serves as the sprue and gate for the individual cavity. Advantages are faster molding cycles, less material waste and reprocessing, fewer post-molding operations, and reduced sink marks and flow lines. Strong AB (2000) Plastics materials and processing. Prentice-Hall, Columbus, OH.

Hot wire cutter *n.* A process for splitting some types of plastic foam blocks into smaller pieces whereas heated wires slowly melt through the block as they are fed into the wires by gravity or conveyor belt.

H

Houndstooth \ˈhaún(d)z-ˌtüth-\ (1937) *n.* A term describing a medium-sized broken-check effect; the check is actually a four pointed star.

Housekeeping \ˈhaús-ˌkēp-piŋ\ (1550) *n.* Cleanliness, neatness, and orderliness of an area, with the designation of a proper place for everything and everything in its proper place.

House paint *n.* Coating designed for use on large exterior surfaces of a building; generally of lower gloss than the coating used on trim areas. Tracton AA (ed) (2005) Coatings technology handbook. Taylor and Francis Inc. New York. Wicks ZN, Jones FN, Pappas SP (1999) Organic coatings science and technology, 2nd edn. Wiley-Interscience, New York.

HT-1 A type of nylon made from phenylenediamine and iosphthalic or terephthalic acid, with good high-temperature properties. Carley JF (ed) (1993) Whittington's dictionary of plastics. Technomic Publishing Co. Inc., Lancaster, PA.

Hubl solution *n.* Reagent used for the determination of Hubl iodine values, involving iodine and mercuric chloride.

Huckaback \ˈhə-kə-ˌbak\ (1690) *n.* A heavy, serviceable toweling made with slackly twisted filling yarns to aid absorption. The cloth has a honeycomb effect. Tortora PG (ed) (1997) Fairchild's dictionary of textiles. Fairchild Books, New York.

Hue \ˈhyü\ [ME *hewe*, fr. OE *hīw*; akin to ON *hȳ* plant down, Gothic *hiwi* form] (before 12c) *n.* The specific quality distinguishing one color from other, such as yellow, red, or blue (color, character, and dominant wavelength) Blue, green, red, etc. White, black, and grays possess no hue.

Hue, Munsell *See Munsell hue.*

Huemann's blue *n.* Synthetic ultramarine blue.

See ultramarine blue.

Huggins constant (k') *n.* The slope coefficient of the Huggins equation found to be constant for a series of homologous polymers of different molecular weights dissolved in a particular solvent. Huggins ML (1958) Physical chemistry of high polymers. John Wiley and Sons Inc., New York.

Huggins equation *n.* In dilute polymer solutions $\eta_{sp}/c = [\eta] + k'[\eta]^2 c$, where η_{sp} and $[\eta]$ are the specific and intrinsic viscosities, c the concentration in g/dL, and k' the Huggins constant; one of several equations (Schulz, Blasche, Huggins, and Kraemer) to extracted values $c \to 0$ or infinite dilution of solute. By making viscosity measurements at several concentrations, plotting η_{sp}/c versus c, and extrapolating the line obtained to $c = 0$, the intrinsic viscosity can be evaluated. Then k' = the slope of the line divided by $[\eta]^2$. Kamide K, Dobashi T (2000) Physical chemistry of polymer solutions. Elsevier, New York. Huggins ML (1958) Physical chemistry of high polymers. John Wiley and Sons Inc., New York. *See also dilute-solution viscosity, Staudinger index, and Mark–Houwink equation.*

Hull paint *n.* Ships' hull paint, bottom paint.

Humectant \hyü-ˈmek-tənt\ [L *humectant-*, *humectans*, pp of *humectare* to moisten, fr. *humectus* moist, fr. *humēre* to be moist] (ca. 1857) *n.* A substance, e.g., sorbitol that promotes retention of moisture. Humectants have been used in antistatic coatings for plastics. Merriam-Webster's collegiate dictionary, 11th edn. Merriam-Webster Inc., Springfield, MA, 2004. Bart J (2005) Additives in polymers: industrial analysis and applications. John Wiley and Sons, New York.

Humidity \hyü-ˈmi-də-tē\ (15c) *n.* (1) The amount of a volatile compound, normally a liquid at the prevailing ambient conditions, present as its vapor in a gas. (2) Specifically, the amount of water vapor present in air, with absolute humidity expressed as the mass of water per unit mass of dry air.
See relative humidity.

Humidity, absolute (1867) *n.* Mass of water vapor present in unit volume of the atmosphere, usually measured as grams per cubic meter. It may also be expressed in terms of the actual pressure of the water vapor present.

Humidity blush *See blushing.*

Humidity ratio *n.* In a mixture of water vapor and air, the weight of water vapor per unit weight of dry air.
Also called specific humidity.

Humidity, relative (1820) *n.* The ratio of the pressure of water vapor present to the pressure of saturated water vapor at the same temperature. The ratio is generally expressed as a percentage or as a decimal fraction.

Humidity, specific *See humidity ratio.*

Hund's rule The rule which states that two electrons tend to remain unpaired and in separate orbitals of the same energy, rather than paired in the same orbital.

Hungary blue \ˈhəŋ-g(ə-)rē\ *n. See cobalt blue.*

Hungry surface *n.* Surface, the absorptive powers of which have not been fully satisfied by the coats of paint applied to it, usually resulting in a patchy film.
Also known as starved surface.

Hunter color difference equation *n.* Generally refers to the color difference equation used for calculating color differences on certain color difference meters (especially those made by Hunter Labs and Gardner Instruments, VA, USA). It is a simplified equation to approximate the NBS equation:

$$\Delta E = [(\Delta a_{\mathrm{L}})^2 + (\Delta b_{\mathrm{L}})^2 + (\Delta L)^2]^{1/2}, \quad (1)$$

where $L = 10Y^{1/2}$, $a_{\mathrm{L}} = 17.5\,(1.02X - Y)/Y^{1/2}$, $b_{\mathrm{L}} = 7.0\,(Y - 0.847Z)/Y^{1/2}$, and the deltas are obtained as the difference of the sample minus the standard. Positive a or Δa indicates redness, negative indicates greenness; plus b or Δ indicates yellowness, minus indicates blueness. Other instruments may use the following version:

$$\Delta E = [(\Delta a_{\mathrm{R}})^2 + (\Delta b_{\mathrm{R}})^2 + (\Delta R)^2]^{1/2}, \quad (2)$$

where $R = Y$, $a_{\mathrm{R}} = 2.75f_y\ (1.02X - Y)$, $b_{\mathrm{R}} = 0.70f_y\ (Y - 0.847Z)$, and $f_y = 0.50\ (21 + 0.2Y)/(1 + 0.2Y)$. The particular equation used must be specified. McDonald R (1997) Colour physics for industry, 2nd edn. Society of Dyers and Colourists, West Yorkshire, England. Billmeyer FW, Saltzman M (1966) Principles of color technology. John Wiley and Sons Inc., New York.

Huygens' theory of light \ˈhī-gənz-\ *n* This theory states that light is a disturbance traveling through some medium such as the ether. Thus light is due to wave motion in ether. Every vibrating point on the wave-front is regarded as the center of a new disturbance. These secondary disturbances traveling with equal velocity, are enveloped by a surface identical in its properties with the surface from which the secondary disturbances start and this surface forms the new wave-front. Serway RA, Faugh JS, Bennett CV (2005) College physics. Thomas, New York. Saleh BEA, Teich MC (1991) Fundamentals of photonics. John Wiley and Sons, New York.

H

Hybrid composite \ˈhī-brəd\ *n.* Advanced composite with a combination of different high-strength continuous filaments in the matrix. Also, composite in which continuous and staple fibers are used in the same matrix.

Hybrid fabric *n.* Fabric for composite manufacture in which two or more different yarns are used in the fabric construction. This provides design flexibility to meet performance requirements and controls cost by permitting some lower priced fibers to be used.

Hybrid orbital *n.* An atomic orbital formed by combining or mixing two or more ground-state atomic orbitals.

Hybrid yarn *n.* In aerospace textiles, a yarn having more than one component.
Also see commingled yarn.

Hydantoin epoxy resin *See epoxy resin.*

Hydrate \ˈhī-ˌdrāt\ (1802) *n.* Compound in which molecules of water, as such, are present. Many crystalline salts contain what is described as water of crystallization. Each molecule of salt is associated with one or more molecules of water. These salts are known as hydrated salts.

Hydrated alumina \-əˈlü-mə-nə\ *n.* Syn: alumina trihydrate.

Hydrated aluminum silicate *See aluminum silicate (clay).*

Hydrated chromium oxide *n.* $Cr_2O_3 \cdot 2H_2O$. Pigment Green 18 (77289). Permanent green pigment with excellent lightfastness but lower temperature resistance than chromium oxide and less tinting strength. Density, 2.9–3.7 g/cm^3 (24.2–30.8 lb/gal); O.A., 110. Syn: Guignet's green, permanent green, viridian, vert emeraude (transparent), French veronese green, emerald oxide of chromium, Mittler's green, native green, Pelletier's green, emerald green, and smeraldino.

Hydrated lime *See calcium hydroxide.*

Hydrated magnesium aluminum silicate *n.* $3MgO \cdot 1.5Al_2O_3 \cdot 8SiO_2 \cdot 9H_2O$. Natural adsorptive clay with a primary constituent of attapulgite clay, which has rodlike particle shape. Density, 2.36 g/cm^3 (19.7 lb/gal); O.A., 80–190, particle size, 0.13 μm. Syn: fuller's earth, attapulgite, and palygorskite.

Hydration \hī-ˈdrā-shən\ (1850) *n.* The addition of water to another substance. The water may react with the other material, as in the hydration of lime; may be taken up in a mixed chemical/physical affinity (absorption), as in 6/6 nylon and numerous other plastics, in time reaching an equilibrium value for the prevailing conditions; or may simply be drawn into pores of the material by capillary action (wetting).
See, too, solvation.

Hydraulic entanglement *See hydroentangling.*

Hydraulic spraying *n.* Spraying by hydraulic pressure.
See airless spraying.

Hydric \ˈhī-drik\ (1926) *adj.* Containing or relating to, hydrogen in combination.

Hydride \ˈhī-ˌdrīd\ (1869) *n.* A compound containing hydrogen in the –1 oxidation state.

Hydroabietyl alcohol Obtained by hydrogenation of Abietic acid (a triple-ring, aromatic acid extracted from rosin) and used as a plasticizer for PVC and some cellulosic resins.

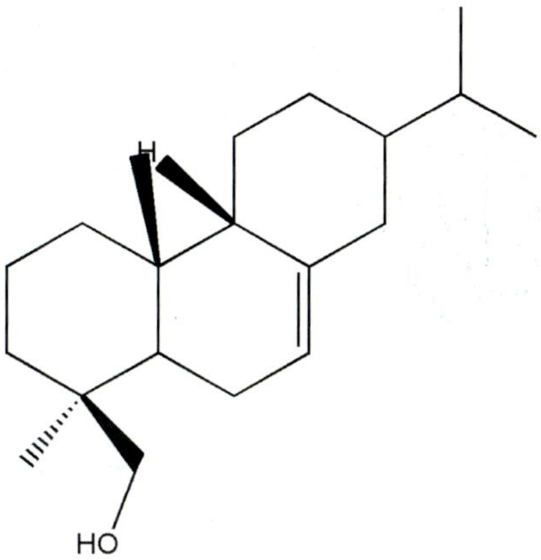

Hydroblasting *n.* Cleaning with high-pressure water jet.

Hydrocarbon \ˈhī-drō-ˌkär-bən\ (1826) *n.* An organic compound (as acetylene or butane) containing only carbon and hydrogen and often occurring in petroleum, natural gas, coal, and bitumens.

Hydrocarbon plastics *n.* Plastics based on resins composed of hydrogen and carbon alone. *Note*—Oxygen and other elements may be present as impurities in the resins, less than 1% by wt.

Hydrocarbon resins *n.* Polymers consisting solely of hydrogen and carbon, specifically denoting, within the trade, those resins which are hard and friable, which give good adhesive strength, and lack the cohesive strength typical of elastomeric polymers.

Hydroentangling *n.* Process for forming a fabric by mechanically wrapping and knotting fibers in a web through the use of high-velocity jets or curtains of water. *Also see spunlaces fabric.*

Hydroextractor *See centrifuge.*

Hydrogel *n.* A three-dimensional network of a hydrophilic polymer, generally covalently or ionically cross-linked. The most widely used example is poly(hydroxyethyl methacrylate), especially in medical applications such as implants, blood bags, and syringes.

Hydrogenated methyl abietate *n.* $C_{19}H_{31}COOCH_3$. A derivative of Abietic acid, which is extracted from pine rosin, used as a plasticizer for cellulose nitrate, ethyl cellulose, acrylic and vinyl resins, and polystyrene.

Hydrogenated naphthalenes *n.* Two commercial solvents are included in this description, namely tetrahydronaphthalene and decahydronaphthalene. The former is a partly hydrogenated naphthalene, whereas the latter is completely hydrogenated.

Hydrogenated naphthas *n.* Series of solvents obtained by subjecting the sulfur dioxide extracts of petroleum to hydrogenation.

Hydrogenated rosin *n.* Modified rosin obtained by the hydrogenation of rosin at high pressures in the presence of a catalyst.

Hydrogenated rubber *n.* Rubber reacted in solution under pressure, in the presence of a suitable catalyst, to yield hydrogenated rubber, in which the unsaturation of natural rubber has been substantially satisfied.

Hydrogenation \hī-ˌdrä-jə-ˈnə-shən\ (1809) *n.* The process of passing hydrogen into an unsaturated chemical in the presence of a catalyst to convert the material to a more saturated state (i.e., containing more combined hydrogen).

Hydrogenation of oils *n.* Method of treatment employed to convert unsaturated fatty acids or oils into saturated types.

Hydrogen bonds (1923) *n.* A very strong attraction between a hydrogen atom, which is attached to an electronegative atom, and an electronegative atom which is usually on another molecule. Jeffrey GA (1997) An introduction to hydrogen bonds. Oxford University Press, New York. Pauk A (2001) Orbital interaction theory of organic chemistry, 2nd edn. Wiley Interscience, New York.

Hydrogen bromide (1885) *n.* HBr. A colorless irritating gas that fumes in moist air and yields hydrobromic acid when dissolved in water.

Hydrogen chloride (1869) *n.* HCl. A colorless pungent poisonous gas that fumes in moist air and yields hydrochloric acid when dissolved in water.

Hydrogen cyanide (1882) *n.* A poisonous usually gaseous compound HCN that has the odor of bitter almonds.

Hydrogen equivalent *n.* The number of replaceable hydrogen atoms in one molecule of a substance, or the number of atoms of hydrogen with which one molecule could react.

Hydrogen fluoride (1885) *n.* A colorless corrosive gaseous compound HF that yields hydrofluoric acid when dissolved in water.

Hydrogen iodide (1885) *n.* An acrid colorless gas HI that fumes in moist air and yields hydriodic acid when dissolved in water.

Hydrogen ion (1896) *n.* The cation H^+ of acids consisting of a hydrogen atom whose electron has been transferred to the anion of the acid.

Hydrogen ion concentration (pH) *n.* The concentration of hydrogen ions in solution when the concentration is expressed as gram-ionic weights per liter. A convenient form of expressing hydrogen ion concentration is in terms of the negative logarithm of this concentration. The negative logarithm of the hydrogen ion concentration is called pH. The significance of pH is still in dispute. Water at 25°C has a concentration of H ion of 10^{-7} and of OH ion of 10^{-7} mol/l. Thus, the pH of water is 7 at 24°C. A greater accuracy is obtained if one substitutes the thermodynamic activity of the union for its concentration.
See pH.

Hydrogen plastics *n.* (1) Plastics based on polymers made from monomers containing only carbon and hydrogen. (2) In the plastics industry, hydrocarbon resins are considered to be those thermoplastic resins of low molecular weight made from relatively impure monomers that are derived from coal-tar fractions, cracked-petroleum distillates, and turpentine. The family includes coumarone–indene resins, cyclopentadiene resins, petroleum resins, terpene resins, and many others of little commercial importance. Having little strength, most hydrocarbon resins are rarely used alone. Their primary applications are as binders in asphalt flooring, processing aids in elastomers and polyolefins, and coating additives. Ash M, Ash I (1982–1988) Encyclopedia of plastics polymers, and resins, vols 1–3. Chemical Publishing Co., New York.

Hydrogen peroxide (1872) *n.* H_2O_2. An unstable compound used as an oxidizing and bleaching agent.

Hydrogen value *n.* Method of determining hydrogen absorption for unsaturated oils or similar compounds. Hydrogen is reacted with unsaturated oils, and iodine values are calculated from the amount of hydrogen absorbed.

Hydrolysis \hī-ˈdrä-lə-səs\ [NL] (1880) *n.* Disruptive reaction consisting of splitting a compound into two parts, one of which combines with the H ion of water and the other combines with the OH ion of water. For example, when vegetable oils or fats are split, the esters are broken down into an avid component and an alcohol component, by taking unto themselves the H and OH ions of water. When alkaline reagents are present, the acids do not appear as such at the end of the reaction, but as the corresponding alkali metal salts. Whitten KW, Davis RE, Davis E, Peck LM, Stanley GG (2003) General chemistry. Brookes/Cole, New York.

Hydrolysis (hydrolytic) constant, K_h *n.* The equilibrium constant for a hydrolysis equilibrium.

Hydrolytic degradation *n.* Any breakdown of a plastic in which reaction with water or water vapor (steam) plays a role.

Hydrometer \hī-ˈdrä-mə-tər\ (1675) *n.* An instrument that senses and indicates the density of a liquid in order to measure the specific gravity of a liquid or solution. A simple type consists of a glass tube with a bulb at its bottom, fine lead shot in the bulb, and a graduated scale within the tube. The amount of shot is adjusted before the top of the tube is sealed so that, when the hydrometer is floated in a liquid of density within its calibrated range, the scale reads the density of the liquids at the meniscus.

Hydronium ion \hī-ˈdrō-nē-əm ˈī-ən\ *n.* The hydrated hydrogen ion; represented by H_3O^+. Whitten KW, Davis RE, Davis E, Peck LM, Stanley GG (2003) General chemistry. Brookes/Cole, New York.
Also known as the oxonium ion.

Hydroperoxides \hī-ˈdrō-pə-ˈräk-ˌsīd\ (1937) *n.* Compounds of the type R–O–OH, where R is an alkyl group. Useful as free radial initiators for free radical polymerization. Morrison RT, Boyd RN (1992) Organic chemistry, 6th edn. Prentice-Hall, Englewood Cliffs, NJ.

Hydrophile *See emulsoid.*

Hydrophile–lipophile balance (HLB) value *n.* An empirical value (arbitrary units of 0–20) which is based on the premise that all surfactants combine hydrophilic and lipophilic groups in one molecule and that the proportion between the weight percentages of these two groups for non-ionic surfactants is an indication of the behavior that may be expected from the product. McCutcheon's Emulsifiers and Detergenst (2000) North American Edition, vol 1. McCutcheon Division of McCutcheon Publishing Co., NJ. Gooch JW (2002) Emulsification and polymerization of alkyd resins. Kluwer Academic/Plenum Publishers, New York.

Hydrophilic \ˌhī-drə-ˈfi-lik\ [NL *hydrophilus*, fr. Gk *hydr-* + *-philos* –philous] (1916) *adj.* (1) A "water-loving", i.e., material having an affinity for, and readily wetted by water. (2) In lithography, referring to that property of a substance that makes it more receptive to water and fountain solutions than to oils and inks. Goldberg DE (2003) Fundamentals of chemistry. McGraw-Hill Science/Engineering/Math, New York.

Hydrophobic (Chemistry) \-ˈfō-bik\ (1807) *adj.* A substance, which does not absorb or exhibit affinity for water; non-wettable. Goldberg DE (2003) Fundamentals of chemistry. McGraw-Hill Science/Engineering/Math, New York.

Hydrophobic (Lithography) *n.* Hydrophobic refers to that property of a substance that makes it more receptive to oils and inks than to water and fountain solutions.
Also called lipophilic.

Hydroquinone \-kwi-ˈnōn\ [ISV] (ca. 1872) (*p*-dihydroxybenzene, hydroquinol, *p*-hydroxyphenol, qui-nol) *n.* $C_6H_4(OH)_2$. A white crystalline material derived from aniline, used, as are many of its derivatives, as an inhibitor of free-radical polymerization in unsaturated polyester resins and in monomers such as vinyl acetate.

Hydroquinone is almost colorless and can retain its inhibitory action even in the presence of oxygen. Odian GC (2004) Principles of polymerization. John Wiley and Sons Inc., New York.
(*See image*).

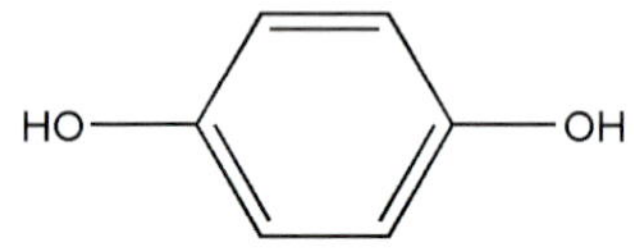

Hydroquinone di(β-Hydroxyethyl) ether (HQEE) *n.* A white solid material used as a reactant in the preparation of polyesters, polyolefins, and polyurethanes. As a chain extender in urethane prepolymers, HQEE increases to 150°C the high-temperature resistance of parts molded from the prepolymer.

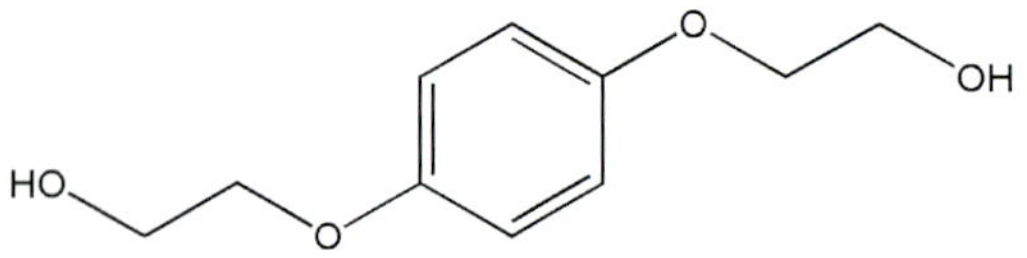

Hydroscopic *adj.* Having the ability to absorb moisture from the atmosphere. All fibers have this property in varying degrees. Complete textile glossary. Celanese Acetate LLC, New York, 2000. Goldberg DE (2003) Fundamentals of chemistry. McGraw-Hill Science/Engineering/Math, New York.

Hydrosol \ˈhī-drə-ˌsäl\ [*hydr-* + *sol* (fr. *solution*] (1864) *n.* (1) In physical chemistry, a colloidal suspension in water. (2) In the plastics industry, a suspension of resin such as PVC or nylon in water, not necessarily of colloidal nature.
See also latex.

Hydrostatic \-ˈsta-tik\ [prob. fr. NL *hydrostaticus*, fr. *hydra-* + *staticus* static] (1666) *adj.* Of or relating to fluids at rest or to the pressures they exert or transmit.

Hydrostatic design basis (HDB) *n.* One of a series of 10^5-h or 50-year strength values and forming the basis for choosing the wall thickness of pipe for various diameters and service pressures and temperatures (American Society for Testing and Materials, West Conshohocken, PA, www.astm.org).

Hydrostatic design stress *n.* The estimated sustained hoop stress that can exist in a pipe at expected service conditions and over the life of the pipe with a high degree of certainty that failure will not occur.
See preceding entry.

Hydrostatic pressure *n.* At a distance *h* from the surface of a liquid of density *d*,

$$P = hdg.$$

The total force on an area *A* due to hydrostatic pressure,

$$F = PA = Ahdg.$$

Force in dynes and pressure in dynes/cm^2 will be given if *h* is in cm, *d* in g/cm^3 and *g* in cm/s^2. Munson BR, Young DF, Okiishi TH (2005) Fundamentals of fluid mechanics. John Wiley and Sons, New York.

Hydrostatic strength *n.* The hoop, stress calculated from the following equation, at which a pipe fails because of rising internal pressure, usually in about one minute. The equation is: $\sigma = PD_m/2t$, in which *P* is the internal (gauge) pressure at rupture, D_m the initial mean diameter (=outer diameter - *t*), and *t* is the initial wall thickness. This strength is usually less than the tensile strength of the pipe material as determined in a standard tensile test because of the presence in the pipe of longitudinal stress equal to half the hoop stress. The pressure at which the failure occurs is known as the *quickburst pressure.* Munson BR, Young DF, Okiishi TH (2005) Fundamentals of fluid mechanics. John Wiley and Sons, New York.

Hydroterpins *n.* Hydrogenation products derived from turpentine. They differ from the parent material in that they possess greater stability, i.e., less tendency to oxidation, higher boiling ranges, and slower evaporation rates.

Hydrous \ˈhī-drəs\ (1826) *adj.* Containing water usually in chemical association (as in hydrates).

Hydrous alumina *n.* Syn: alumina trihydrate.

Hydrous oxides *n.* Poorly characterized compounds formed from the combination of certain oxides with water or by precipitation from the addition of base to certain ions in aqueous solution.

Hydrox-, hydroxy- A chemical prefix denoting the presence of the –OH group in a compound.

Hydroxide \hī-ˈdräk-ˌsīd\ [ISV] (1851) *n.* Hydrated metallic oxide; a base; a compound which will give hydroxyl (OH) ions in solution.

(2-Hydroxpropyl) methyl cellulose *n.* A mixed ether of cellulose containing both hydropropyl and methyl groups.

2(2′-Hydroxy-3,5-ditertiarybutylphenyl)-7-chlorobenzotriazol *n.* An off-white, nontoxic, crystalline powder with high thermal stability, used as an ultraviolet absorber for polyolefins, PVC, polyurethanes, polyamides, and polyesters.

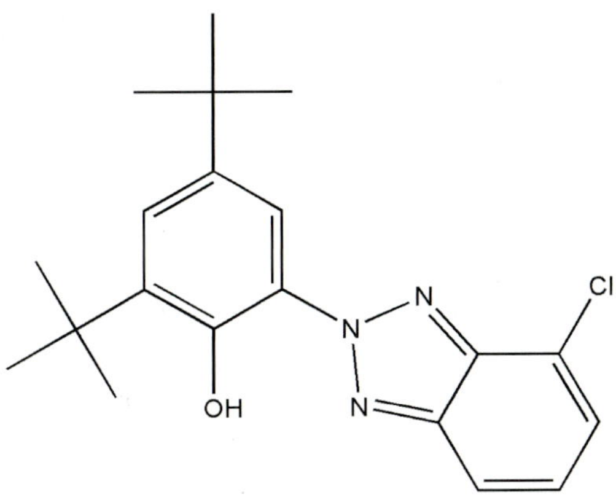

Hydroxyethyl acetamide *See N-acetyl ethanolamine.*

Hydroxyethyl cellulose *n.* Any of a family of polymeric ethers formed by reacting alkali cellulose with ethylene oxide. Water solubility and applications depend on the degree of substitution of hydroxyenthyl groups for –OH groups, and include textile sizes, adhesives, thickeners, and stabilizers for vinyl polymers.

Hydroxyethyl cellulose resins *n.* Prepared by reaction of alkali cellulose with ethylene oxide, which then forms long polyoxyethylene chains by continued reaction with the hydroxyethyl groups.

Hydroxyethylmethyl methacrylate *n.* A monomer that polymerizes to a hydrophilic polymer that is rigid when dry but when saturated with water becomes a soft, clear material (Hydron®). Applications include masonry coatings, soft contact lens, and other biomedical devices.

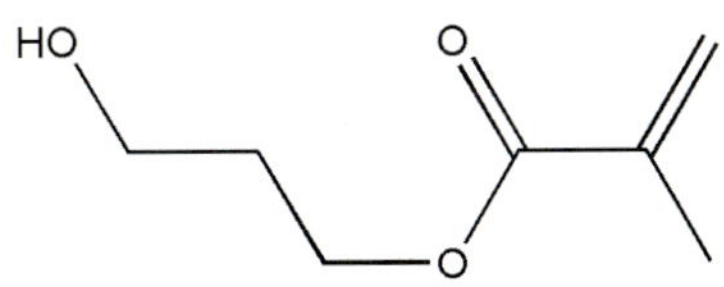

Hydroxyl \hī-ˈdräk-səl\ [*hydr-* + *ox-* + *-yl*] (1869) *n.* The chemical group or ion OH that consists of one atom of hydrogen and one of oxygen and is neutral or positively charged.

Hydroxyl end group *n.* A polymer chain-terminating (–OH) group.

Hydroxyl group *n.* –OH. The monovalent group characteristic of hydroxides and alcohols.

2-Hydroxyl-4-methoxy-5-sulfonbenzophenone *n.* An ultraviolet absorber for thermoplastics.

Hydroxyl value *n.* A measurement of hydroxyl groups (–OH) in an organic material (American Society for Testing and

Materials, West Conshohocken, PA, www.astm.org). In plasticizers, the hydroxyl value includes –OH groups present in any free unesterified alcohol as well as those of the plasticizer molecule itself. In some plasticizers, large hydroxy values signal that the plasticizer may become incompatible on aging. In urethane technology, hydroxyl number is an important factor in the selection of polyols to achieve desired characteristics in elastomers and foams. Wypych G (ed) (2003) Plasticizer's data base. Noyes Publication, New York.

H

2-Hydroxy-4-methoxybenzophenone *n.* An ultraviolet absorber for numerous thermoplastics.

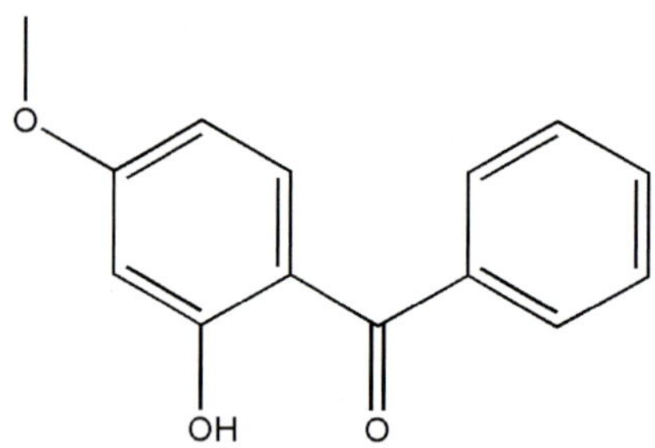

2(2′-Hydroxy-4′-methylphenyl) benzotriazole ("Tinuvin P") *n.* A non-toxic crystalline powder with high thermal stability, an ultraviolet absorber for polystyrene, acrylics, PVC, polyesters, and polycarbonates.

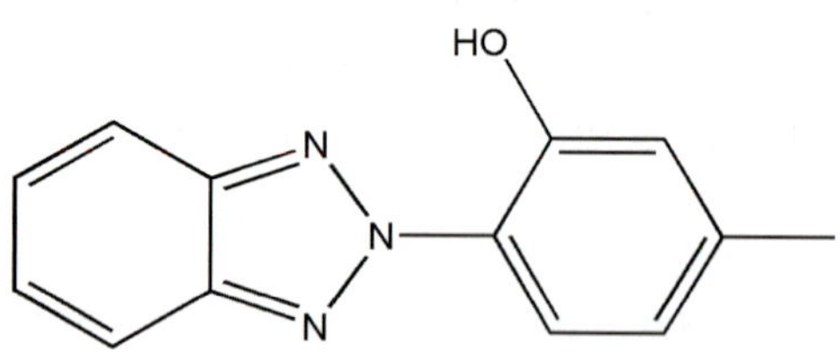

2-Hydroxy-4-*n*-octoxybenzophenone *n.* $C_{21}H_{26}O_3$. A pale yellow powder, an ultraviolet absorber for PVC, and several other plastics. It is compatible with highly plasticized vinyls and has a very low order of toxicity.

Hydroxypropylglycerin *n.* A pale-straw-colored liquid used as a plasticizer for cellulosic resins and as an intermediate for alkyd and polyester resins.

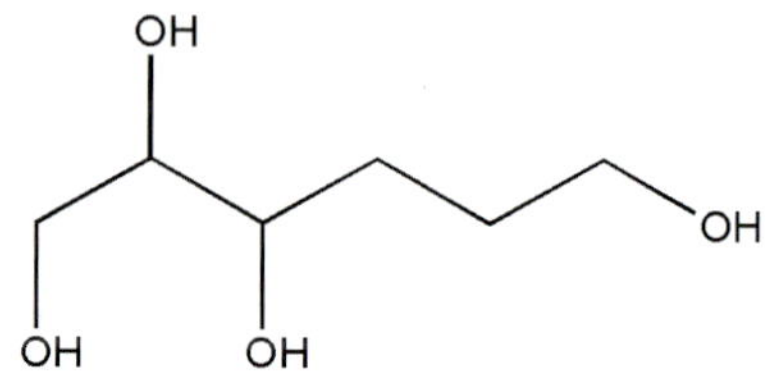

Hydroxypropyl methacrylate (HPMA) *n.* A reactive monomer copolymerizable with a wide variety of acrylic and vinyl monomers, used for thermosetting resins and surface coatings.

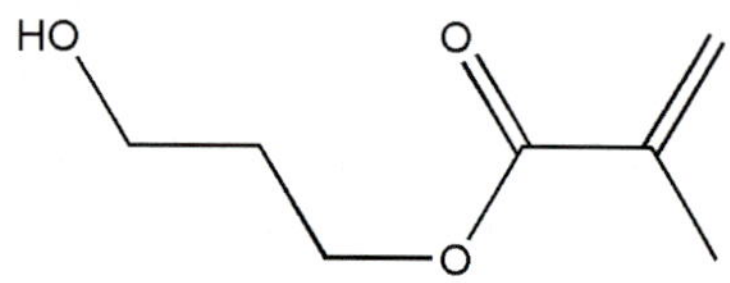

Hygrometer \hī-ˈgrä-mə-tər\ [prob. fr. F *hygromètre*, fr. *hygr-* + *-mètre* –meter] (1670) *n.* An instrument that senses and indicates the relative humidity (of moisture) in air.

Hygroscopic \ˌhī-grə-ˈskä-pik\ [*hygroscope*, an instrument showing changes in humidity + [1]*-ic*; fr.] (1790) *adj.* Having a strong tendency to absorb moisture from the air. Some resins are hygroscopic, and therefore usually require drying before being extruded or molded.
See hydroscopic.

Hyperon \ˈhī-pə-ˌrän\ [prob. fr. *hyper-* + [2]*-on*] (1953) *n.* Any article with mass intermediate between that of the neutron and the deuteron. Giambattista A, Richardson R, Richardson RC, Richardson B (2003) College physics. McGraw-Hill Science/Engineering/Math, New York.
See meson.

Hypogoeic acid *n.* $C_{15}H_{29}COOH$. Unsaturated fatty acid found in certain vegetable oils. It has a mp of 33°C, and a bp of 236°C/15 mmHg.

Hypohalous Compound in which a hydroxyl group is combined with a halogen atom.

Hysteresis \ˌhis-tə-ˈrē-səs\ [NL, fr. Gk *hysterēsis* shortcoming, fr. *hysterein* to be late, fall short, fr. *hysteros* later] (1881) *n.* (1) The lagging of the physical effect on a body behind its cause, particularly during repeated cycling. Hysteresis is accompanied by dissipation of energy as heat. If a fatigue specimen is cycled at too high a frequency/ or stress, the resulting heat buildup can seriously bias the test and even soften the specimen. (2) The influence of prior history that causes the repetition of a process or its reversal to take a different path from the original. In tensile loading of plastics past the proportional limit but short of failure, the path (stress versus strain) followed upon unloading will generally be at lower stresses than were recorded during loading at the same strains. Open or closed loops may be observed. Shah V (1998) Handbook of plastics testing technology. John Wiley and Sons, New York. Brown R (1999) Handbook of physical polymer testing, vol 50. Marcel Dekker, New York.

Hysteresis loop *n.* Flow curve for a dilatent (shear thining) material obtained by measurements on a rotational viscometer showing for each value of rate of shear, two values of shearing stress, one for an increasing rate of shear and the other for a decreasing rate of shear. A hysteresis loop characterizes a dilatant material. If no hysteresis loop is obtained, the material is non-dilatant. The area within the loop is a measure of dilatancy (increasing shear thining with increasing shear stress). Dilatancy should not be confused with thixoptropy (shear thining at constant shear stress over time). Munson BR, Young DF, Okiishi TH (2005) Fundamentals of fluid mechanics. John Wiley and Sons, New York. Patton TC (1979) Paint flow and pigment dispersion: a rheological approach to coating and ink technology. John Wiley and Sons, New York. Van Wazer JR, Lyons JW, Kim KY, Colwell RE (1963) Viscosity and flow measurement. Interscience Publishers Inc., New York.

Hytrel® *n.* DuPont trade name for a family of copolyester elastomers. Typical reactants from which the elastomers are derived are terephthalic acid, polytetramethylene glycol, and 1,4-butanediol. Powders and pellets are available for extrusion and molding. The products are highly resilient, have good flex-fatigue life at low and high temperatures, and are resistant to oils and chemicals. Some grades termed *segmented copolyesters* are excellent modifiers for PVC, improving processability and imparting resistance to abrasion, impact, and fungi.

Hz *n.* SI abbreviation for Hertz. Lide DR (ed) (2004) CRC handbook of chemistry and physics. CRC Press, Boca Raton, FL.

H

I

i \ˈī\ *n.* (1) (also I) Symbol for electric current. (2) Symbol for $\sqrt{-1}$, the coefficient of "imaginary", i.e., out-of-phase, components in complex quantities.

I *n.* (1) Chemical symbol for the element iodine. (2) The symbol for electric current. (3) The symbol for moment of inertia. Whitten KW, Davis RE, Davis E, Peck LM, Stanley GG (2003) General chemistry. Brookes/Cole, New York. Giambattista A, Richardson R, Richardson RC, Richardson B (2003) College physics. McGraw-Hill Science/Engineering/Math, New York. Giambattista A, Richardson R, Richardson RC, Richardson B (2003) College physics. McGraw-Hill Science/Engineering/Math, New York.

Iatrochemistry [Greek *iatros*, "doctor" + chemistry] *n.* Eighteenth-century chemistry was derived from and remained involved with questions of mechanics, light and heat as well as iatrochemistry and notions of medical therapy and the interaction between substances and the formations of new substances. This period was also the time of alchemy (approximately 1550–1750 AD). Iatrochemistry attempted to explain life and physiological processes using the knowledge of physics at that time. From the root word was derived: iatrogenic [Greek *latros "doctor"* + -genic] meaning "caused by a doctor – a symptom or illus brought on untentionally by something that a doctor does or says (www.britannica.com and Microsoft Encarta 2004).

Iceland moss \ˈi-slən(d)-, -ˌslan(d)-\ (1805) *n.* Lichen grown in Iceland, Norway and Sweden, which yields a mucilaginous extract, used in sizes and the like. Syn: Cetraria.

Iceland spar (1771) *n.* Natural, transparent, double refracting, crystalline form of calcium carbonate. Hibbard MJ (2001) Mineralogy. McGraw-Hill Co. Inc., New York.

Ichnography *n.* (1) Art of making decorative drawings by the use of compass and rule. (2) The term also refers to the art of tracing plans and illustrations.

ICI *n.* (1) Older abbreviation for international commission on illumination (Use international abbreviation CIE).

Ideal co-polymerization *n.* A co-polymerization in which, in the simple binary case, the product of the monomer reactivity ratios r_A and r_B for monomers A and B (r_A, r_B) equals unity. Thus each type of growing center $\sim\sim A^*$ and $\sim\sim B^*$ shows the same preference for adding on or the other monomer during propagation. Odian GC (2004) Principles of polymerization. John Wiley and Sons Inc., New York.

Ideal elastomer (ideal rubber) *n.* An elastomer, which on deformation at constant, e.g., by stretching by application of a force *f*, satisfies the conditions

$$(\partial H/\partial l)_{T,P} = 0 \quad \text{and} \quad f = -T(\partial S/\partial l)_{T,P},$$

where l is the length, T the temperature, P the pressure, S the entropy, and H is the enthalpy. Mark JE (ed) (1996) Physical properties of polymers handbook. Springer-Verlag, New York.

Ideal gas *n.* A gas whose behavior is described by the ideal-gas law, $PV = nRT$. Goldberg DE (2003) Fundamentals of chemistry. McGraw-Hill Science/Engineering/Math, New York.

Ideal-gas law *n.* The combination of Charles' and Boyle's laws, usually stated in the form $PV = nRT$, where P is the absolute pressure, V the gas volume, n the number of moles of gas present, T the absolute temperature, and R is the universal molar gas (or energy) constant. R has many numerical values, to be consistent with the units chosen for the variables. In SI, $R = 8.31439\,\text{K/mol K}$. Goldberg DE (2003) Fundamentals of chemistry. McGraw-Hill Science/ Engineering/Math, New York.

Ideal liquid *n.* This term has several meanings in physical chemistry and hydrodynamics. The one relevant to plastics is Newtonian liquid. Munson BR, Young DF, Okiishi TH (2005) Fundamentals of fluid mechanics. John Wiley and Sons, New York. Goldberg DE (2003) Fundamentals of chemistry. McGraw-Hill Science/Engineering/Math, New York.

Ideal solid *n.* A material that obeys Hooke's law. Serway RA, Faugh JS, Bennett CV (2005) College physics. Thomas, New York.

Ideal solution *n.* Solutions that exhibit no change of internal energy on mixing and complete uniformity of cohesive forces. The behavior of ideal solutions is described by Raoult's law over all ranges of temperature and concentration.

Ignition \ig-ˈni-shən\ (1612) *n.* The beginning of burning. Babrauskas V (2003) Ignition handbook. Fie Science Publishers, New York. Goldberg DE (2003) Fundamentals of chemistry. McGraw-Hill Science/ Engineering/Math, New York.

Ignition loss *See ASTM D2584 at flammability.*

Ignition time Babrauskas V (2003) Ignition handbook. Fie Science Publishers, New York.
See flammability.

IIR *n.* Butyl rubber.

Illuminance \i-ˈlü-mə-nən(t)s\ (ca. 1938) *n.* The luminous flux incident per unit area of a surface. Lide DR (ed) (2004) CRC handbook of chemistry and physics. CRC Press, Boca Raton, FL.

Illuminant \-nənt\ (1644) *n.* Mathematical description of the relative spectral power distribution of a real or imaginary light source, that is, the relative energy emitted by a source at each wavelength in its emission spectrum. Popularly used synonymously with the term "light source" or "lamp". Such usage is not recommended. Lide DR (ed) (2004) CRC handbook of chemistry and physics. CRC Press, Boca Raton, FL.

Illuminant A (CIE) *n.* Incandescent illumination, yellow orange in color, with a correlated color temperature of 2856 K. It is defined in the wavelength range of 380–770 nm.
See correlated color temperature.

Illuminant B (CIE) *n.* An approximation of mean noon sunlight, approximately neutral in hue, with a correlated color temperature of 4874 K. It is defined in the wavelength range of 380–770 nm.
See correlated color temperature.

Illuminant C (CIE) *n.* An approximation of average daylight, bluish, with a correlated color temperature of 6774 K. It is defined in the wavelength range of 380–770 nm.
See correlated color temperature.

Illuminant D (CIE) *n.* Daylight illuminants, defined from 300 to 830 nm, the UV portion 300–380 nm being necessary to describe correctly colors which contain fluorescent dyes or pigments. They are designated as D with a subscript to describe the correlated color temperature: D_{65} having a correlated color temperature of 6504 K, close to that of Illuminant C, is the most commonly used. They are based on actual measurements of the spectral distribution of daylight.

Illuminating \-ˌnā-tiŋ\ [ME, fr. L *illuminatus*, pp of *illuminare*, fr. *in* + *luminare* to light up, fr. *lumen-*, *lumen* light] (15c) *vt.* Hand decoration of books, as done in medieval times, with drawings and miniature paintings. Also ornamentation and embellishments, usually in red, blue and gold, which were sometimes added to initial letters and borders. Giambattista A, Richardson R, Richardson RC, Richardson B (2003) College physics. McGraw-Hill Science/Engineering/Math, New York.

Illumination \i-ˌlü-mə-ˈnā-shən\ (14c) *n.* The illumination on any surface is measured by the luminous flux incident on unit area. The units in use are: the *lux* (abbreviation lx) 1 lm/m^2; the *photo* (abbreviation ph) 1 lm/cm^2; the *Foot-candle* (abbreviation fc) 1lm/ft^2. Giambattista A, Richardson R, Richardson RC, Richardson B (2003) College physics. McGraw-Hill Science/Engineering/Math, New York.

Illumination, diffuse *n.* Non-specular illumination or non-direct incident light source projected onto the object of observation. Indirect or reflected rays of light are often used as diffuse sources. Also, a scattering lens over a light source will diffuse the rays to reduce specular light.

Ilmenite \ˈil-nə-ˌnīt\ [Gr *Ilmenit*, fr. *Ilmen* range, Ural Mts., Russia] (ca. 1827) *n.* $FeO{\cdot}TiO_2$. Natural iron titanate. It is an iron-black mineral and is one of the main sources of titanium dioxide.

Ilmenite black *n.* Opaque black pigment, prepared from the mineral of the same name. Hibbard MJ (2001) Mineralogy. McGraw-Hill Co. Inc., New York.

Image, real *n.* The image, which would register on a ground glass screen or photographic film placed in that plane. Moller KD (2003) Optics. Springer-Verlag, New York.

Image, virtual *n.* This image would appear by construction to be in a given plane, but a ground glass screen or photographic film placed in that plane would show no image. Moller KD (2003) Optics. Springer-Verlag, New York.

Imbibition \ˌim-bə-ˈbi-shən\ (15c) *n.* Swelling of a solid, semisolid or gel by absorption of a liquid.

Imitation gold ink *n.* A simulated gold ink that uses aluminum powder to produce the metallic luster.

Immediate elastic deformation *n.* Recoverable deformation that is essentially independent of time, i.e., occurring in (a time approaching) zero time and recoverable in (a time approaching) zero time after removal of the applied load.

Immediate set *n.* The deformation found by measurement immediately after the removal of the load causing the deformation in a short-time test.

Immersion objective *n.* A microscope objective that is used with a liquid of refractive index greater than 1.00 between the specimen and objective, and usually between the specimen and substage condenser. Immersion objectives are used when a numerical aperture greater than 1.00 is desired, since this cannot be achieved with a dry objective. Moller KD (2003) Optics. Springer-Verlag, New York.

Immiscible \(ˌ)i(m)-ˈmi-sə-bəl\ (1671) *adj.* Not miscible. Any liquid, which will not mix with another liquid, in which case it forms two separate layers or exhibits cloudiness or turbidity. *Compare incompatible.*

Impact adhesive *n. See contact adhesive.*

Impact modifiers A general term for any additive, usually an elastomer or plastic of different type, incorporated in a plastic compound to improve the impact resistance of finished articles. The improvement is customarily assessed by performance of test specimens in standard tests.

Impact polystyrene *n.* An alternative name for high impact polystyrene; usually polystyrene modified with thermoplastic elastomer to provide a tougher material. *See high-impact polystyrene.*

Impact resistance *n.* (1) Ability of a coating to resist a sudden blow. Ability to resist deformation from impact. (2) The relative durability of plastics article to fracture under stresses applied at high speeds. A widely used ASTM impact test, www.astm.org, employs the Izod pendulum striker swung from a fixed height to strike a specimen in the form of a notched bar mounted vertically as a cantilever beam. The Charpy tester, an alternative in D 256, uses a specimen in the form of a horizontal beam supported at both ends. ASTM lists some several different impact tests for plastics and plastics products.

See also brittleness temperature, drop-weight test, free-falling-dart test, and tensile-impact test.

Impact strength *n.* The quantitative measure of the ability of a material to withstand shock loading in a standard test. For plastics the test is usually either the Izod or Charpy test described in ASTM (www.astm.org), and the result is calculated as the energy expended (work done) in breaking a specimen, divided by its width or thickness. Specimens of both tests are usually notched on the wide opposite that where they are struck, though the notch position may be reversed and unnotched specimens may be tested. In SI the convenient reporting unit is J/cm of notch width, or, for unnotched specimens, J/cm. To provide a much simpler stress distribution free of notch effects, researchers, developed the tensile-impact test. Shah V (1998) Handbook of plastics testing technology. John Wiley and Sons, New York.

Impact test *n.* Used to assess the adhesion and flexibility of applied coatings. A weighted plunger is allowed to fall from a specified height onto the front (direct) or back (reverse) of a coated panel. The extent of the damage caused by the impact is used as a basis for assessing the adhesion and flexibility. Paint and coating testing manual (Gardner–Sward handbook) MNL 17, 14th edn. ASTM, Conshohocken, PA, 1995.

Impasto \im-ˈpas-(ˌ)tō, -ˈpäs-\ [It, fr. *impastare*] (1784) *n.* This term implied a thick application (paste, impasted) of paint or coloring material. Particularly thick or heavy application of paint. The thinly painted and impasted areas of an oil painting provide texture. Gair A (1996) Artist's manual. Chronicle Books LLC, San Francisco.

Imperial gallon *See gallon, imperial.*

Impingement mixing *n.* Very intense, rapid and thorough mixing accomplished by causing two fast-moving liquid streams or sprays to meet within a confined space. Typically the streams are resin and curing agent or resin streams containing catalyst in one and accelerator in the other. The technique is used in reaction–injection molding, in spray-up of reinforced plastics, and in foam-in-place molding.

Impregnated cloth *n.* A cloth impregnated with resin, varnish shellac, etc.

Impregnated fabric *n.* A fabric in which the interstices between the yarns are completely filled, as compared with sized or coated material where the interstices are not completely filled. Not included in the definition is a woven fabric constructed from impregnated yarns, rather than one impregnated after weaving.

Impregnating insulating varnish *n.* First coating insulating varnish applied to electrical equipment by means of vacuum–pressure impregnation, or by dipping. Its main purpose is to fill up the voids between the windings and elsewhere. The name distinguishes it from finishing insulating varnishes, which are used primarily to give an effective seal to the outside of windings, to prevent ingress of moisture.

Impregnation \im-ˈpreg-ˌnāt, ˈim-ˌ\ [LL *impraegnatus*, pp of *impraegnare*, fr. L *in-* + *praegnas* pregnant] (1605) *vt.* The process of thoroughly soaking a material of an open or porous nature with a resin. When webs or shapes of reinforcing fibers are impregnated with a thermosetting resin in the B stage or with a thermoplastic, and such webs are intended for subsequent shaping or laminating, the masses are called sheet-molding compounds or prepreg. The main difference between impregnation and encapsulation is that in encapsulation an outer protective coating is formed with little or no penetration of the resin into the article, whereas in impregnation there is little or no protective coating.

Impression \im-ˈpre-shən\ (14c) *n.* (1) The printing pressure necessary for ink transfer. (2) Also refers to a single print.

Impression cylinder *n.* The cylinder on a printing press that holds the material being printed against the printing plate, cylinder, or blanket.

Impsonite Naturally occurring asphaltum, found in Oklahoma.

Impulse sealing (thermal-impulse sealing) *n.* The process of joining thermoplastic sheets or films by pressing them between elements equipped to provide a pulse of intense heat to the sealing area for a very short time, followed immediately by cooling. The heating element may be a length of thin resistance wire such as nichrome, or a bar heated by a high-frequency electric field and cored for water cooling.
See also heat sealing.

Incandescent \ˌin-kən-ˈdes-sᵊnt\ [prob fr. F, fr. L *incandescent-*. *incandescens*, pre. part. of *incandescere* to become hot, fr. *in-* + *candescere* to become hot, fr. *candēre* to glow] (1794) *adj.* Luminous in the yellow-to-white range because of being at a high temperature (1250–1550°C).

Inching (1599) *v.* A very low rate of mold closing used in many molding operations and often automatically included in the cycle, for the last millimeter or so before the mold faces meet.

Incident angle (1628) *n.* The angle between the incident ray and a normal to the surface at the point of incidence. It is the same as "entrance angle" in SAE automotive nomenclature.

Incident ray *n.* The center line of the light beam incident upon a surface.

Inclusion \in-ˈklü-zhən\ [L *inclusion-*, *inclusio*, fr. *includere*] (1600) *n.* (1) A foreign body or impurity phase in a solid. (2) Presence of foreign material in the finished material.

Inclusion complex *See adduct.*

Incompatibility \ˌin-kəm-ˌpa-tə-ˈbi-lə-tē\ (1611) *n.* The effect when two or more components of a composition do not blend properly to produce a uniform or homogeneous mixture. It can be characterized by gelatin, curding, precipitation, cloudiness, seediness, and land loss in gloss of blooming oin resulting film.

Incompatible \ˌin-kəm-ˈpa-tə-ˈbəl\ [ME, fr. MF & ML; MF. fr. ML incompatibilis, fr. L *in-* + ML *compatibilis* compatible] (15c) *adj.* A term that describes two materials, such as two resins, or a resin and plasticizer, that are incapable of forming a solution of a two-phase blend and that tend to separate after being mixed.

Incompatible polyblends *n.* Polymers that do not alloy to form stable compositions.

Incomplete hiding *See hiding, incomplete.*

Indanthrone blue *n.* A member of a group of vat dye pigments. Relative to phthalocyanine blue, it is considerably redder in hue and superior in bronze resistance; lightfastness is excellent. It shows no tendency toward crystal growth in conventional paint thinners or solvents.

Indelible ink \in-ˈde-lə-bəl-\. Compositions, which are made water- and alkali-fast. Used on cloth to withstand laundering.

Indene *n.* C_9H_8. Colorless liquid with a Sp gr of 1.006 (20/4°C), mp of 3.5°C, bp of 182°C, refractive index of 1.5726, and a flp of 78.3°C. Insoluble in water, soluble in most organic solvents, rapidly absorbs oxygen from the air, forms polymers by exposure to air and sunlight. Derivation: Contained in the fraction of crude coal tar distillates, which boils from 176 to 182°C. Uses: Preparation of synthetic resins.
See coumarone–indene resins.

Indene resins *n.* A colorless, liquid hydrocarbon, C_9H_8, obtained from coal tar by fractional distillation: used in synthesizing resins.

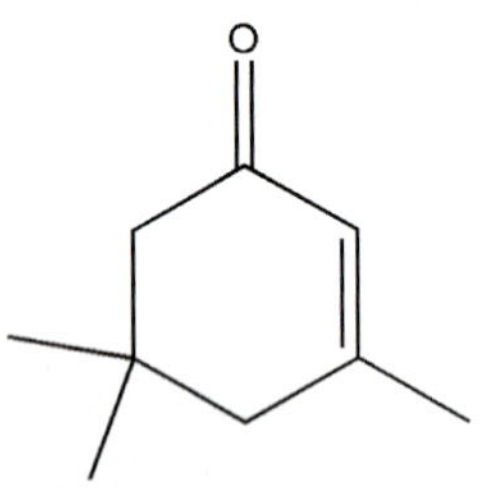

See coumarone–indene resins.

Indentation hardness *n.* Resistance to penetration by an indenter. The hardness of a material as determined by either the size of an indentation made by an indenting tool under a fixed load, or the load needed to produce penetration of an indenter to a predetermined depth. The instruments commonly used with plastics are the Shore Durometer (indenter A for soft resins and elastomers, D for hard materials) described in ASTM D 2240, and the Barcol Impressor, ASTM D 2583. In D 2240, the authors say, "No simple relationship exists between indentation hardness determined by this method and any fundamental property of the material tested. For specification purposes it is recommended that Test Method D 1415 (Section 09.01) be used for soft materials and method A of D 530 or test D 785 be used for hard materials". D 530 and D 785 use the Rockwell test in which a ball of suitable diameter is pressed into the sample with known force, and the indentation depth is noted. A similar test format is used in D 1415.
See knoop hardness number and pfund hardness number.

Index of metamerism \-mə-ˈta-mə-ˌri-zəm\. An index to describe the degree of metamerism existing between two samples. There is no standard accepted index. However, a frequently computed index is the color difference, using a specified equation, which exists under a second illuminant when a pair of colors is an exact match under a primary or first illuminant.

Index of refraction (1829) (refractive index) *n.* The ratio of the velocity of light in a vacuum to its velocity in the transparent material of interest. It generally varies with the wavelength of the light, being higher at the short wavelengths, also with temperature. When a light beam passes from a less dense medium to a denser medium at an acute angle θ_1 with the interfacial normal, it will be bent closer to the normal in the more dense medium, defining the angle of refraction, θ_2. The index of refraction is given by sin θ_1/sinθ_2. This relationship is the basis for measurement of refractive index with an instrument such as the Abbé refractometer (www.astm.org) for transparent plastics). Index of refraction is useful for identification of plastics, as well as minerals and liquids, and is involved in many quantitative analytical methods. Serway RA, Faugh JS, Bennett CV (2005) College physics. Thomas, New York. Moller KD (2003) Optics. Springer-Verlag, New York. Weast RC (ed) (1978) CRC handbook of chemistry and physics, 59th edn. The Chemical Rubber Co., Boca Raton, FL.

India ink {*often cap 1st* I} (1665) *n.* A liquid ink made of lampblack mixed with a binding material and water.
Also called Chinese ink.

Indian ochre \-ˈō-kər\. *See iron oxides, natural.*

Indian red (ca. 1753) *n.* Natural, Pigment Red 101 (77491). Synthetic Pigment Red 101 (77491). Crystalline copperas (ferrous sulfate in crystal form) is converted by calcinations to ferric oxide, producing a range of light to dark red oxide pigments. The dark shades are known as Indian red. Original Indian reds were a variety of natural red iron oxides obtained from India.

Indian red ochre *See iron oxides, natural.*

Indian yellow *n.* Purree. A yellow pigment containing the magnesium salt of euxanthic acid.

India rubber {*often cap I*} (1790) *n. See rubber, natural.*

Indicator \ˈin-də-ˌkā-tər\ (1666) *n.* A conjugate acid-base pair in which at least one of the pair is highly colored.

Indicators *n.* Substances, which change from one color to another when the hydrogen ion concentration reaches a certain value, different for each indicator.

Indicatrix *n.* A three-dimensional construction of the optical relationships in the crystal. Radii proportional in length to refractive index values or velocities may be used to represent principal index directions of the crystal.

Indicia \in-ˈdi-sh(ē)ə\ [L, plural of *indicium* sign, fr. *indicare*] (ca. 1626) *n, pl.* Any markings such as symbols, lettering, small pictures, etc., applied to a plastic article to identify it.

Indigo \ˈin-di-ˌgō\ [It dialect, fr. L *indicum*, fr. Gk. *indikon*, fr. neuter of *indikos* Indic, fr. *Indos* India] (1555) *n.* $C_{16}H_{10}N_2O_2$ (molecular weight, 262.26). Dark blue, transparent dye produced either by fermentation from the genus, *Indigofera*, indigenous to the Far East, or synthetically.

Indophenol test \ˌin-dō-ˈfē-ˌnōl-\ [ISV] *n. See Gibbs indophenol test.*

Induced decomposition reaction *n.* Chain transfer in a free radical polymerization by reaction of a growing active center with an undissociated initiator molecule, especially with a peroxide.

Induced electromotive force *n.* In a circuit is proportional to the rate of change of magnetic flux through the circuit

$$E = \frac{d\phi}{dt},$$

where $d\phi$ is the change of magnetic flux in a time dt. The induced current will be given by

$$I = \frac{d\phi}{R\,dt},$$

where R is the resistance of the circuit. Weast RC (ed) (1971) Handbook of chemistry and physics, 52nd edn. The Chemical Rubber Co., Boca Raton, FL.

Inductance \in-ˈdək-tən(t)s\ (1886) *n.* The change in magnetic field due to the variation of a current in a conducting circuit causing an induced counter electromotive force in the circuit itself. The phenomenon is known as *self-induction.* If an electromotive force is induced in a neighboring circuit the term mutual injection is used.

Inductance may thus be distinguished as self- or mutual and is measured by the electromotive force produced in a conductor by unit rate of variation of the current. Units of inductance are the centimeter (absolute electromagnetic) and the Henry, which is equal to 10^9cm of inductance. The *Henry* is that inductance in which an induced electromotive force of 1 V is produced when the inducing current is changed at the rate of 1 A/s. Thus, 1 H = 1 V/A/s = 1 V s/A = 1 Wb/A. Dimensions: $[E^{-1}L^{-1}T^2]$, $[\mu L]$. Weast RC (ed) (1971) Handbook of chemistry and physics, 52nd edn. The Chemical Rubber Co., Boca Raton, FL.

I

Induction \-shən\ (14c) *n.* Any change in the intensity or direction of a magnetic field causes an electromotive force in any conductor in the field. The induced electromotive force generates an induced current if the conductor forms a closed circuit. Weast RC (ed) (1971) Handbook of chemistry and physics, 52nd edn. The Chemical Rubber Co., Boca Raton, FL.

Induction heating (1919) *n.* A method of heating electrically conductive materials, usually metallic parts, by placing the part or material is an alternating magnetic field generated by passing an alternating current through a primary coil. The alternating magnetic field induces eddy currents in the piece that generate hysteretic heat. Plastics, being poor conductors, cannot be heated directly by induction, but the process is used indirectly in welding of plastics and has been used to heat extruders and dies. Weast RC (ed) (1971) Handbook of chemistry and physics, 52nd edn. The Chemical Rubber Co., Boca Raton, FL.

Induction period *n.* The time period before a chemical reaction (e.g., polymerization) proceeds that usually includes the production of free radicals or ions; rise in temperature and other reaction conditions before an initiation of polymerization occurs; other necessary periods of time preceding a chemical reaction between one or more chemical species prior to production of a different species. Odian GC (2004) Principles of polymerization. John Wiley and Sons Inc., New York.

Induction period (of drying) *n.* That period prior to set-to-touch in the drying process where there is but a small change in consistency of the binder and the natural inhibitors of the oil are removed by oxidation.

Induction welding *n.* A method of welding thermoplastic materials by placing a conductive metal insert between two plastic surfaces to be joined, applying pressure to hold the surfaces together, heating the metallic insert in an alternating magnetic field until the adjacent plastic is softened and welded, then killing the field and cooling the joint.

Induline *n.* Range of blue or black dyestuffs belonging to the safranine group. They can be prepared in oil, spirit, and water-soluble forms. They are used as dyes for leather and in lake manufacture.

Induline bases *n.* Substances prepared from amidoazobenzene aniline and aniline hydrochloride.

Industrial fabric *n.* A broad term for fabrics used for non-apparel and non-decorative uses. They fall into several classes: (1) A broad group including fabrics employed in industrial processes (e.g., filtering, polishing, and absorption). (2) Fabrics combined with other materials to produce a different type of product (e.g., rubberized fabric for hose, belting, and tires; fabric combined with synthetic resins to be used for timing gears and electrical machinery parts; coated or enameled fabrics for automobile tops and book bindings; and fabrics impregnated with adhesive and dielectric compounds for application in the electrical industry). (3) Fabrics incorporated directly in a finished product (e.g., sails, tarpaulins, tents, awnings, and specialty belts for agricultural machinery, airplanes, and conveyors). Fabrics developed for industrial uses cover a wide variety of widths, weights, and constructions and are attained, in many cases, only after painstaking research and experiment. Cotton and manufactured fibers are important fibers in this group, but virtually all textile fibers have industrial uses. The names mechanical fabrics or technical fabrics sometimes have been applied to certain industrial fabrics.

Industrial finishes or coatings *n.* Coatings which are applied to factory-made articles (before or after fabrication), usually with the help of special techniques for applying and drying – as opposed to trade sales paints.
See architectural coatings.

Industrial maintenance paints *n.* High performance coatings, which are formulated for the purpose of heavy abrasion, water immersion, chemical, corrosion, temperature, electrical, or solvent resistance.
See also maintenance paints.

Industrial methylated spirits *See methylated spirit.*

Industrial talc *n.* $Mg_3Si_4O_{10}(OH)_2$. A mineral product varying in composition from that approaching the theoretical formula of talc (which is non-fibrous), to mixtures of talc and other naturally associated minerals, some of which may be fibrous as defined in ASTM D 2946 and/or asbestos.
See asbestos, magnesium silicate, and fibrous.

Industrial talc, asbestos-free *n.* Industrial talc of which less than 2/100 particles are asbestos, where "asbestos fiber" is defined as being both by a fiber by ASTM D 2946 and one of the asbestiform varieties of serpentine, riebeckite, and cummingtonite (which are chrysotile, crocidolite, and amosite, respectively), anthopyllite, tremolite, or actinolite. The non-asbestiform varieties of these same materials are not asbestos.

Industrial waste water *n.* Water discharged from an industrial process as a result of formation or utilization in that process.

Industrial water *n.* Water (including its impurities) used directly or indirectly in industrial processes.

Inert \i-ˈnərt\ [L *inert-*, *iners* unskilled, idle, fr. *in-* + *art-*, *ars* skill] (1647) *adj.* The term applied to various extended pigments such as asbestine, barites, silica, calcium sulfate, mica, talc, etc. In general, they have poor hiding power but they are inert from a chemical and physical standpoint. While they contribute some desirable properties to paint, they are primarily used to lower the cost.

Inert additive *n.* A material added to a plastic compound to improve properties, but which does not react chemically with any other constituents of the composition.

Inert base *n.* A paint base, which does not provide hiding, color, or drying properties. Its main function is to provide solid, usually at low cost.

Inert gas (1902) *n.* Any of a group of rare gases that include helium, neon, argon, krypton, xenon, and sometimes radon and that exhibit heat stability and extremely low reaction rates.
Also known as Nobel gas.

Inertia \i-ˈnər-shə, -shē-ə\ [NL, fr. L, lack of skill, fr. *inert-*, *iners*] (1713) *n.* The resistance shown by all matter to any attempt to change its state of motion.

Inert pair *n.* The pair of s electrons in an atom with a valence shell containing at least one p electron.

Inert pigment *n.* (1) A pigment which remains relatively inactive or chemically unchanged in paints under stated conditions. The term has little significance unless the conditions are stated. (2) Frequently used for extender.

Inflammable \in-ˈfla-mə-bəl\ [F, fr. ML *inflammabilis*, fr. L *inflammare*] (1605) (deprecated) *adj.* A general term once used to describe combustible gases, liquids or solids. Now obsolete.
See flammable.

Inflatable structures Structures opened or enlarged by input of air and, once enlarged, able to retain the air to maintain the distended position.

Inflow quench *n.* Cooling air for extruded polymer filaments that is directed radically inward across the path of the filaments. The threadline is completely enclosed in a quench cabinet in inflow quenching.

Information retrieval *n.* Recovery of data from a collection for the purpose of obtaining information. Retrieval includes all the procedures used to identify, search, find, and remove specific information or data stored.

Infrared \ˌin-frə-ˈred, -(ˌ)frä-, -fə-\ (1881) *adj.* (1) The region of the electromagnetic spectrum including wavelengths from 0.78 to about 300 μm. Infrared radiation serves as a source of heat. (2) Rays which are longer than visible red light rays and are felt as heat. Infrared radiant energy is used to cure coatings.

Infrared drying (or IR drying) *n.* Method of heating for drying, curing, and/or fusing a coating. The radiations involved in this form of heating have wavelengths greater than 780 nm, and they may be derived either from specially designed filament bulbs or special types of gas ovens.

Infrared heating *n.* A heating process used mostly in sheet thermoforming and for drying coatings, employing lamps, or heating elements that emit invisible radiation at wavelengths of about 2 μm.

Infrared (or I.) spectrophotometry \-ˌspek-trō-fə-ˈtä-mə-tər\ *n.* An analytical instrumental technique based on selective absorption of infrared radiation by organic and inorganic materials, which helps identify them. An example of an infrared spectrum of toluene is shown. Stuart BH (2004) Infrared spectroscopy. John Wiley and Sons, New York.
See FTIR.

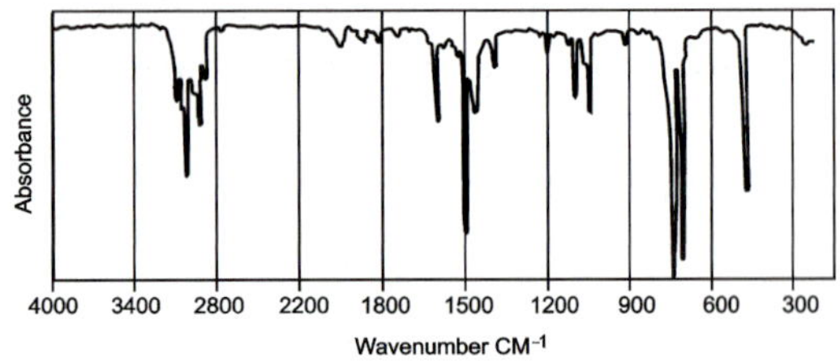

Infrared polymerization index (IRPI) *n.* A number representing the degree of cure of phenolic resins, defined as the ratio of absorbances by the sample, corrected for background absorbance, at wavelengths of 12.2 and 9.8 μm (far infrared). The test is often used in conjunction with the Marquardt index.

Infrared pyrometer \-pī-ˈrä-mə-tər\ *n.* A narrow- or broad-band instrument that senses the peak wavelength, λ (μm) of IR radiation emanating from a warm surface. By Wien's displacement law, the absolute temperature (K) is given by $2884/\lambda$.

Infrared spectroscopy \-spek-ˈträs-kə-pē\ *n.* A technique to identify and quantitatively determine many organic substances such as plastics. All chemical compounds have characteristic intramolecular vibratory motions and can absorb incident radiant energy if such energy is sufficient to increase the vibrational motions of the atoms. With most organic molecules, vibrational motions of the substituent groups within the molecules coincide with the electromagnetic frequencies of the infrared region. An infrared spectrophotometer directs IR radiation through a film or layer or solution of the sample and measures and records the relative amount of energy absorbed by the sample as a function of the wavelength or frequency of the radiation. The chart produced is compared with charts of known substances to identify the sample. Peak area is a measure of the percentage of the various compounds present as generated by the Spectrum 100 FTIR (courtesy of Perkin-Elmer Inc., MA, USA).

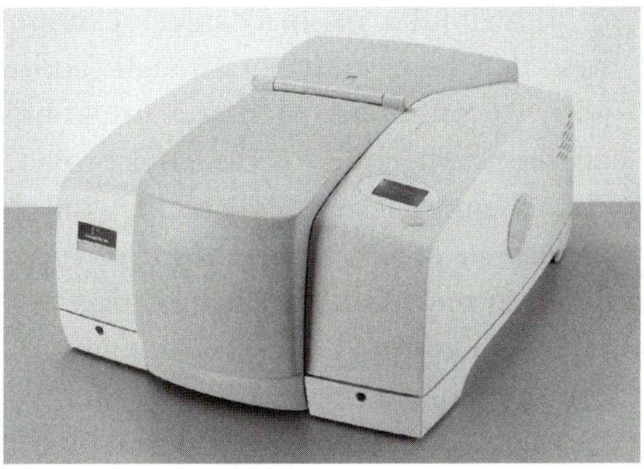

Infrared thermography \-(ˌ)thər-ˈmä-grə-fē\ (thermographic NDT). A type of nondestructive testing in which damage development during fatigue testing of composites may be detected without interrupting the test by measuring locally generated infrared radiation. The damage is associated with heat developed through two mechanisms: hysteresis and frictional heating from rubbing at the surface of developing cracks.

Infusible \(ˌ)in-ˈfyü-zə-bəl\ (1555) *adj.* Not capable of melting when heated, as are all cured thermosetting resins and a few special thermoplastics such as ultrahigh-molecular-weight polyethylene, polybenzimidazole, and aramid resins.

Infusorial earth \(ˌ)in-fyü-ˈzōr-ē-əl-\ (1882) *n. See diatomaceous silica.*

Ingrain \ˈin-ˌgrān\ (1766) *adj. See dyeing.*

Inherent flame resistance *n.* As applied to textiles, flame resistance that derives from an essential characteristic of the fiber from which the textile is made.

Inherent viscosity (logarithmic viscosity number) *n.* In measurement of dilute-solution viscosities, the inherent viscosity is the ratio of the natural logarithm of the relative viscosity to the concentration of the polymer in g/dL of solvent: $\ln(\eta_{\text{ref}})/c$.

See dilute-solution viscosity.

Inhibition, inhibitors *n.* A substance capable of retarding or stopping an undesired chemical reaction. Inhibitors are used in certain monomers and resins to prolong storage life. When used to retard degradation of plastics by heat and/or light, an inhibitor functions as a stabilizer.

Inhibitive pigment *n.* Pigment which assists in the prevention of cedar staining or Redwood staining, mildew, corrosion, yellowing, etc.

See corrosion-inhibitive pigments.

Inhibitor \in-ˈhi-bə-tər\ (ca. 1611) *n.* (1) General term for compounds or materials that have the effect of slowing down or stopping an undesired chemical change such as corrosion, oxidation or polymerization, drying, skinning, mildew growth, etc. (2) A negative catalyst, which prevents or retards vulcanization or oxidation. A good inhibitor will retard vulcanization at room temperature to prevent precure but will not retard at normal curing temperatures.

Also see corrosion inhibitor and inhibitive pigment.

Initial boiling point *n.* Temperature indicated by the distillation thermometer at the instant the first drop of condensate leaves the condenser tube.

Initial modulus The modulus of elasticity extrapolated to zero strain from measurements in the low-strain region.

Initial viscosity *n.* A term used in the vinyl plastisol industry to denote the viscosity measured immediately after the plastisol has been mixed. The viscosity normally rises at a declining rate after mixing.

Initiation \i-ˌni-shē-ˈā-shən\ (1583) *n.* The initial stage of polymerization usually comprising a compound (e.g., peroxide or hydroperoxide), heat or irradiation or combination that produced a free radial or ion that causes monomers to successfully attach (polymerization phase) in a chain reaction; not condensation polymerization which utilizes a catalyst without an initiator.

Initiator \i-ˈni-shē-ˌāt\ [LL *initiatus*, pp of *initiare*, fr. L, to induct, fr. *initium*] (ca. 1573) *vt.* An agent that causes a chemical reaction to commence and that enters into the reaction to become part of the resultant compound. Initiators differ from catalysts in that catalysts do not combine chemically with the reactants. Initiators are used in many polymerization reactions, especially in emulsion polymerizations. Initiators most commonly used in polymerizing monomers and resins having ethenic unsaturation (–C=C–) are the organic peroxides. Odian GC (2004) Principles of polymerization. John Wiley and Sons Inc., New York.

Initiator activator *n.* It is usually necessary to add an activator (e.g., $FeSO_4.7H_2O$) to attain reasonable rates of polymerization when redox initiators (lauroÿl peroxide and fructose) are used at zero degree temperature or below. Odian GC (2004) Principles of polymerization. John Wiley and Sons Inc., New York.

Injection blow molding *n.* A blow-molding process in which parisons are formed over mandrel by injection molding, after which the mandrels and parisons are transferred to blow molds where the final shape is blown. While the parts are being blown, cooled, and ejected, another set of parisons is being injection-molded. Advantages of the process are that it delivers a completely finished part requiring no neck trimming, etc., closer tolerances are possible, and parison-wall thicknesses can be locally varied as desired.

Injection mold *n.* A mold used in the process of injection molding. The mold usually comprises two main sections held together by a hydraulic or mechanical clamping press that have sufficient strength and rigidity to contain the high pressure of molten plastic when it is injected. It is provided with channels for mold venting and for circulation of temperature-control media, and with means for product ejection.

Injection molding (1932) *n.* The method of forming objects from granular or powdered plastics, most often thermoplastics, in which the material is fed from a hopper into a screw-type plasticator (or heating cylinder in elderly machines), after which the screw or a ram forces the molten compound into a chilled mold. Pressure is maintained until the mass has hardened sufficiently for removal from the mold (in a variation called *flow molding*, a small additional amount of melt is forced into the mold during cooling of the initial charge to offset shrinkage). Machines employing screws for plastication (as must do now days) are either single- or double-stage. In single-stage machines, plastication and injection are done in the same cylinder, injection pressure being generated by forward motion of the screw while rotation is stopped. This process is called *reciprocating-screw injection molding.* In double-stage machines, the material is plasticated by a constantly rotating screw that delivers the melt through a check valve to an accumulator cylinder, from which it is injected into the mold by a piston. This process is called *screw-and-piston injection molding.*

See also reaction injection molding, hot-tip-gate molding, and two-shot injection molding.

Injection molding, molds *n.* Comprises two main sections held together by a hydraulic or mechanical clamping press, that have sufficient strength and rigidity to contain the high pressure of molten plastic when it is injected. It is provided with channels for mold venting and for circulation of temperature-control media, and with means for product ejection.

Injection molding nozzles *n.* A short, thick-walled, hardened-steel tube that conducts the molten plastic emerging from an injection cylinder into the mold. It usually terminates in a hemispherical tip that closely mates a recess in the mold and is enclosed by band heaters that are controlled by a temperature sensor.

Injection-molding pressure *n.* (1) The pressure applied to the plastic by the injection screw (or ram) during the injection stroke. (2) The pressure of the melt within the mold cavity just prior to the freezing of the gate.

Injection nozzle *n.* A short, thick-walled, hardened-steel tube that conducts the molten plastic emerging from an injection cylinder into the mold. It usually terminates in a hemispherical tip that closely mates a recess in the mold called a *sprue bushing* and is enclosed by band heaters that are controlled by a temperature sensor.

Injection ram *n.* In two-stage injection molders and elderly machines having no screws but using contact heating, the piston that applies pressure to the plastic during injection. In single-stage screw-injection machines, the screw doubles as the ram during a brief period of stopped rotation.

Injection stamping *n.* A little used modification of injection molding wherein first the plastic melt is injected under relatively low pressure into a mold that is vented during this stage, then, after the cavity is filled, additional clamping pressure is applied to completely close the mold and compress or "stamp" the molded shape. Molds are designed so that even in the venting position no material exudes onto the land areas. Advantages claimed for the process are lower injection time and pressure, and shorter cycle because the injection screw can resume plastication as soon as the mold has been charged.

Injector molder *n.* (1) A person or company that operates injection-molding machines. (2) An injection-molding machine.

Ink \ˈiŋk\ [ME *enke*, fr. OF, fr. LL *encaustum*, fr. neuter of L *encaustus* burned in, fr. Gk *enkaustos*, verbal of *enkaiein* to burn in] (13c) *n.* A fluid or viscous substance used for writing or printing. Printing ink handbook. National Association of Printing Ink Manufacturers Inc., NJ, 1976.

Ink adhesion *n.* The tendency of printing ink to stick to the printed surface and not smear or rub off. An informal test of ink adhesion consists of pressing pressure-sensitive tape over the printed area, waiting a few minutes, then stripping it off and observing the a real fraction of ink coming away on the tape. Leach RH, Pierce RJ, Hickman EP, Mackenzie MJ, Smith HG (eds) (1993) Printing ink manual, 5th edn. Blueprint, New York.

Ink compounds *n.* Additives to remedy deficiencies in the ink vehicle or pigment. Apps E (1963) Ink technology for printers and students, vol 3. L. Hill, London.

Ink-embossed wallpaper *n.* Differs from regular embossed wallpaper in that the ink colors are applied at the time the paper is being embossed.

Ink fountain *See fountain.*

Inkometer *n.* An instrument indicating the tack of an ink in terms of the torque developed by a system of rotating ink-wet rollers. Useful in determining tack differences in series of inks for multicolor wet printing and in the standardization of ink consistency.

Ink receptivity *n.* That property of a sheet of paper (or other material) which causes it to accept and/or absorb ink. Wolfe HJ (1957) Printing and litho inks. MacNair-Dorland Co., New York.

Inks, cover *n.* Generally opaque, heavy bodied and have a high pigment concentration.

Inks, dull *n.* Inks that dry with a dull or matte finish.

Inks, halftone *n.* Inks formulated for good reproduction of fine detail such as halftone dots on coated stock. They generally have high tinctorial strength and are finely dispersed.

Inks, job *n.* Heavy bodied inks formulated to print on uncoated stock using small sheet-fed presses.

Inks, quick-setting *n.* Inks for letterpress and offset which dry by either filtration, coagulation, selective absorption or often a combination of these with some of the other drying methods. The vehicles are generally special resin–oil combinations, which, after the inks have been printed, separate into a solid material, which remains on the surface as a dry film and an oily material which penetrates rapidly into the stock. This rapid separation gives the effect of very quick setting or drying.

Inlay \(ˌ)in-ˈlā, ˈin-ˌ\ (1596) *n.* Pattern or design formed by inserting cut shapes of contrasting materials into a surface, usually at the same level.

Inlay printing *See valley printing.*

In-mold decorating *n.* The process of applying labels or decorations to plastic articles during the molding operation by which they are formed. Two basic methods, each with many variations, are in use. The first employs a preprinted label of plastic film, paper, or cloth that is positioned in the mold prior to molding. During the molding cycle, the label or its printed image fuses to and becomes an integral part of the article. In the second method, the image is printed directly onto the mold surface with wet or dry ink, or applied to the mold by an offset process. In-mold decorating is done with blow molding, injection molding, and casting.

Inoculum \i-ˈnä-kyə-ləm\ [NL, fr. L *inoculare*] (1902) *n.* That portion of the pathogen which is transferred to the plant or plant part for inoculation purposes.

Inorganic \ˌi-(ˌ)nór-ˈga-nik\ (1794) *adj.* Designation of compounds that generally do not contain carbon. Source: matter other than vegetable or animal. Examples: sulfuric acid and salt. Exceptions are carbon monoxide and carbon dioxide and their derivatives.

Inorganic coatings *n.* Coatings based on silicates or phosphates and usually used pigmented with metallic zinc.

Inorganic fibers *n.* Fibers that are composed of matter other than hydrocarbons and their derivatives and are not of plant or animal origin.

Inorganic pigments *n.* (1) Any pigment derived from naturally occurring minerals or synthesized from inorganic substances. They are always opaque, unlike organic pigments and dyes, and are usually resistant to heat and light. Examples of those used in plastics are titanium dioxide, iron oxides, ultramarines, lead chromates, and cadmium compounds. (2) A class of pigments used in printing ink manufacture consisting of compounds of the various metals. These pigments do not contain carbon such as iron oxide. Example: Chrome yellow.

Inorganic polymer *n.* Within the scope of the plastics industry, an inorganic polymer may be defined as a polymer without carbon in its main chain and of a degree of polymerization sufficient for the polymer to exhibit considerable mechanical strength, plastic, or elastomeric properties, and the ability to be formed by processes used with plastics. Organic-group side chains may be present. The inorganic polymers of greatest commercial importance are the silicones. Leach RH, Pierce RJ, Hickman EP, Mackenzie MJ, Smith HG (eds) (1993) Printing ink manual, 5th edn. Blueprint, New York.

Inpaint *n.* To renew damaged areas on paintings or painted surfaces by repainting.

Insect wax *See Chinese wax.*

Insert \ˈin-ˌsərt\ (ca, 1889) *n.* An article of metal or other material that is incorporated in a plastic molding either by pressing the insert into a recess in the finished molding or by placing the insert into the cavity so that it becomes a part of the molding. A common example is a metal bushing, knurled on the outside, threaded on the inside, for attaching the molded article to another article with a screw or bolt.

Insert bonding *n.* Method of incorporating an article of metal or other material in a plastic molding either by pressing the insert into a recess in a finished molding or by placing the insert into the cavity so that it becomes a part of the molding.

Insert molding *n.* The process by which components may be molded into a part.
See double-shot molding.

Inserts An important part of a plastic molding consisting of metal, plastic, or other material that may be molded into position or may be pressed into the molding after the molding is fabricated.

In-situ foaming \(ˌ)in-ˈsī-(ˌ)tü\ [L, in position] (1740) *adv, adj.* The technique of depositing a foamable plastic (prior to foaming) into the volume where it is intended that foaming shall occur. An example is the placing of foamable plastics into cavity brickwork to provide insulation. Shortly after being poured, the liquid mix foams and fills the cavity.
See also cellular plastic.

Insoluble \(ˌ)in-ˈsäl-yə-bəl\ [ME *insolible*, fr. L *insolubilis*, fr. *in-* + *solvere* to free, dissolve] (14c) *adj.* Of low solubility; slightly soluble; and sparingly soluble.

Inspection \in-ˌspek-shən\ (14c) *n.* The process of examining textiles for defects at any stage of manufacturing and finishing.

Instron *n.* A tensile or other testing machine made by the Instron Corp of Canton, MA. Their machines have been so omnipresent in plastics testing that "Instron" has taken on a near-generic character.

Instron tensile tester *n.* A high precision electronic test instrument designed for testing a variety of material under a broad range of test conditions. It is used to measure and chart the load-elongation properties of fibers, yarns, fabrics, webbings, plastics, films, rubber, leather, paper, etc. It may also be used to measure such properties as tear resistance and resistance to compression.

Insulated-runner mold *See hot-runner mold.*

Insulating varnish *n.* Varnish often used in combination with materials such as mica, fabric, paper, etc., to provide electrical insulation for electrical equipment. These are formulated to have a high resistance to electrical passage.

Insulation resistance *n.* (1) The electrical resistance between two conductors or systems of conductors separated only by an insulating material. The resistance of a particular insulation may be measured by dividing the voltage difference applied to two electrodes in contact with, or embedded in a unit area of the specimen by the current flowing between the electrodes. Tests for thermoplastics include ASTM D 2633, section 10.02. (2) *See thermal resistance.* *See also resistivity.*

I

Intaglio \in-ˈtal-(ˌ)yō\ [It, fr. *intagliare* to engrave, cut fr. ML *intaliare*, fr. L *in-* + LL *taliare* to cut] (1644) *n.* (1) Printing style in which the design is cut into the surface of the cylinder and is thus below the surface. (2) A lustrous, brocade pattern knitted in a tricot fabric.

Integral-skin molding *n.* A method of producing urethane-foam articles with substantially non-porous integral skins in one operation. Whereas normal urethane foams are inflated by carbon dioxide generated by reaction of isocyanate with excess water in the mixture (*see urethane foam*), integral-skin foam is expanded by the vapor of a volatilized solvent such as hexane. The mold must be heat conductive so that a layer next to the mold surfaces can be chilled to prevent foaming. The reaction mixture, typically consisting of a polyol, an isocyanate, and the blowing solvent, is introduced rapidly into a closed mold located near the geometric center of the mold. The mold is usually preheated to between 40 and 65°C, but the main heat needed to foam and cure the mass is provided by the reaction exotherm. After the mass has gelled, the skin should be pierced to equalize pressure and prevent shrinkage. The solvent blowing agent can be removed from the finished article by allowing it to stand at ambient conditions for about a day, or by oven drying for half an hour at 140°C. In some instances, integral-skin moldings have replaced composites of vinyl-covered urethane foam articles and cast vinyl skins filled with urethane foam, such as automotive arm rests, crash pads, and instrument-panel covers.

Integrating sphere *n.* A sphere coated inside with a highly reflective, diffuse material and used for the measurement of luminous flux. If the internal surface is perfectly diffuse, the intensity of any part of the sphere is the same. Many instruments used for reflectance measurements utilize such a device for measuring the diffuse or total reflectance from a sample material relative to a reference material.

Integration (colorimetry) *n.* The process of summing the products of the three values: the illuminant, the observer response characteristics and the reflectance or transmittance of an object at specified wavelengths in order to obtain the tristimulus values.

See tristimulus values.

Intensity \in-ˈten(t)-sə-tē\ (1665) *n.* (1) The amount of energy per unit (space, charge, and time). (2) The brilliance of a color. (3) The brightness of light.

Intensity of illumination (properly called illumination) *n.* Illumination in *lux* of a screen by a source of illuminating power P at a distance r meter, for normal incidence,

$$I = \frac{P}{r^2}.$$

If two sources of illuminating powers P_1 and P_2 produce equal illumination on a screen when at distances r_1 and r_2, respectively,

$$\frac{P_1}{r_1^2} = \frac{P_2}{r_2^2} \quad \text{or} \quad \frac{P_1}{P_2} = \frac{r_1^2}{r_2^2}.$$

If I_0 is the illumination when the screen is normal to the incident light, then I is the illumination when the screen is at an angle θ. Thus,

$$I = I_0 \cos \theta.$$

Intensity of magnetization *n.* The rate of transfer of energy across unit areas by the radiation. In all forms of energy transfer by waves (radiation) the intensity I is given by $I = 0$; Uv, where U is the energy density of the wave in the medium, and v is the velocity of propagation of the wave. The energy density U is always proportional to the square of the wave amplitude.

Intensity of segregation *n.* In mixing, the average measure of the deviation in concentration of a component at any point in a mixture from the mean concentration. The standard deviation as determined by sampling is usually used.

Intensity of sound *n.* Depends upon the energy of the wave motion. The intensity is measured by the energy in ergs transmitted per second through $1\,\text{cm}^2$ of surface. The energy in ergs/cm^3 in a sound wave is given by

$$E = 2\pi^2 dn^2 a^2,$$

where d is the density in g/cm^3, n is frequency in vibration per sec and a is the amplitude in cm. The energy reaching the ear in unit time will also be proportional to the velocity of propagation.

Intensive mixer *See internal mixer.*

Interaction parameter *n.* Interaction parameter χ in solution theory; Flory–Huggins parameter χ; a measure of the interaction energy ΔE, which is a measure of the Gibbs energy, but not of the enthalpy.

Intercoat adhesion *n.* The adhesion between two coats of paint.

Intercoat contamination *n.* Presence of foreign matter between successive coats.

Interdeterminancy principle (uncertainty principle) *n.* The postulate that it is impossible to determine simultaneously both the exact position and the exact momentum of an electron. So this aspect of electronics can only be expressed as a probability.

I

Interface \ˈin-tər-ˌfās\ (1882) *n.* The common surface separating two different phases, e.g., the resin-glass interface in glass–fiber-reinforced plastics.

Interfacial angle *n.* That angle between two adjoining faces of a crystal.

Interfacial angle (polar) *n.* That angle between the normals to two adjoining faces of a crystal.

Interfacial polycondensation *n* Involves polymer formation at or near the interface between two immiscible monomer solutions under very mild reaction conditions.

Interfacial polymerization *n.* A polymerization reaction that occurs at or near the interfacial surfaces of two immiscible solutions. A simple example is the often-performed demonstration of making nylon thread from a beaker containing a lower layer of a solution of sebacyl chloride in carbon tetrachloride and an upper layer of hexamethylene diamine solution in water. A pair of tweezers is gently lowered through the upper layer, closed on the interfacial layer of polymer, and then drawn upward to pull with it a continuous strand of nylon 6/10.

Interfacial tension (interfacial surface energy) *n.* (1) The tension in, or energy of, the interfacial surface between two immiscible liquids. Measurable, like surface tension by capillary rise or with a du Noüy tensiometer. (2) The molecular attractive force between unlike molecules at an interface. It may be expressed in dynes per centimeter.

Interfacing *See interlining.*

Interference color *n.* Color which appears in thin films, illuminated with composite light as the result of the reinforcement of some colors (wavelengths) and weakening or elimination of others, depending on the phase differences and amplitudes of the light waves involved. The color produced depends on the angle of incidence, thickness of the film, and the refractive index. Interference color also appears when light strikes narrow slits, the principle of the diffraction grating.

Interference, constructive *n.* The retardation of two light beams by exactly one wavelength or even multiples of one wavelength. The two waves reinforce each other, resulting in brightness.

Interference, destructive *n.* The retardation of two light beams exactly an odd number of half wavelengths, resulting in darkness as the two waves are perfectly out of phase.

Interference figure *n.* The conoscopic pattern of extinction positions of a crystal superimposed on the pattern of interference colors corresponding to the full cone of directions by which the crystal is illuminated, each direction showing its own interference color.

Interference filter *n.* Optical filter made with extremely thin alternate layers of metals and dielectrics so that it passes only very narrow wavelength bands. Such filters are frequently used in abridged spectrophotometers where they serve to isolate discrete wavelengths for measuring, thus effectively replacing a dispersion device in a spectrophotometer.

See spectrophotometer, abridged.

Interference microscopy *n.* Optical microscopy in the most common arrangement of which the incident beam is split into two non-parallel oppositely polarized beams. The beams are reflected back from the specimen and its substrate, respectively, and recombined. The sample beam follows a shorter path length and the beams recombine destructively producing a dark image.

Interference pigments *n.* Materials used as pigments, which depend on the interference of light for their color. They are generally made from platelets coated with a thin film of higher refractive index. For example, thin films of titanium dioxide coated on mica flakes are one type manufactured and used today.

See pearlescent pigments.

Interferometry \ˌin-tə(r)-fə-ˈrä-mə-tər\ [ISV] (1897) *n.* Any system of measurement based on wave interference between split rays of a light beam, one of which takes a longer or more optically dense path than the other, thus delaying it and generating a phase difference between the two rays. Interferometry has been used to measure a great variety of quantities, e.g., the speed of light, film thicknesses, and chemical concentrations in solutions.

Interfusion \ˌin-tər-ˈfyüz\ [L *interfusus*, pp of *interfundere* to pour between, fr. *inter-* + *fundere* to pour] (1593) (infusion) *v.* A novel method for emplacing hard-surfacing alloys within extruder barrels and, in future, on other parts. Few details have been released (as of late 1992) but it appears the hard-facing material in powder form is distributed around the barrel surface by centrifugal action while induction heating,

and perhaps high pressure, are applied to speed the diffusion of that material into the barrel near its inner surface. The process marketers, Inductamentals of Chicago, claim much slower barrel wear than is experienced by cast-in liners.
See also bimetallic cylinder.

Interhalogen *n.* A compound of two different halogens.

Interlaminar shear stress *n.* Shear stress between layers of a laminate, an important cause of laminate failure and delamination.

Interlayer *n.* An intermediate sheet in a laminate.

Interlining *n.* A padding or stiffening fabric used in garment manufacture to provide shape retention. Interlining is sandwiched between layers of fabric.

Interlocked grain *n.* Wood in which the fibers incline in one direction in a number of rings of annual growth, then gradually reverse and incline in an opposite direction in succeeding rings and then reverse again.

Interlock knit *n.* To produce an interlock knit, long and short needles are arranged alternately in both the dial and cylinder; the needles in the dial and cylinder are also positioned in direct alignment. When the long and short needles knit in alternate feeds in both needle housings, a fabric with a type of cross 1×1 rib effect is produced.

Interlock twiner *n.* A machine for making three-dimensional braided performs in which yarn bundles are entwined by braiders to form the desired shape.

Intermediate (1650) *n.* A compound produced from raw materials that is to be used to synthesize end products. For example, benzene, originally distilled from coal tar and now made from petroleum constituents, and its derivatives, cyclohexane, cyclohexanol, and adipic acid are all intermediates in the manufacture of nylon 6/6.

Intermediate temperature setting adhesive *n. See adhesive, intermediate temperature setting.*

Intermingling *n.* (1) Use of air jets to create turbulence to entangle the filaments of continuous filaments yarns, without forming loops, after extrusion. Provides dimensional stability and cohesion for further processing but is not of itself a texturing process. It is compatible with high-speed spin-drawing and high-speed take-up. When compared with twisting processes, it also permits increased take-up package size. (2) Combining two or more yarns via an intermingling jet. This process can be used to get special effect yarns, i.e., mixing dye variants to get heather effects upon subsequent dyeing.

Intermittent pattern *n.* A pattern occurring in interrupted sequence.

Intermolecular polymerization *n.* Polymerization which occurs as the result of association between molecules.

Internal bubble coating *n.* In blown-film production, the circulation of chilled air or carbon dioxide inside the film bubble to substantially reduce the cooling time and thereby increase the production rate. The same practices have been used in blow molding.

Internal dye variability *n.* The change from point to point in dye uniformity across the diameter and along the length of the individual filaments. Affects appearance of the dyed product and is a function of fiber, dye, dyeing process, and dye bath characteristics.

Internal lubricant *n.* A lubricant that is incorporated into the compound or resin prior to processing, as opposed to one that is applied to the mold or die. Examples of internal lubricant are waxes, fatty acids and their amines, and metallic stearates such as

I

calcium, lead, lithium, magnesium, and zinc stearates. The lubricants reduce friction and adhesion between polymers and metal surfaces, improve flow characteristics, and enhance knitting at weld surfaces and wetting properties of compounds. They are used primarily in rigid and flexible PRV, high-molecular-weight polyethylene, polystyrene, and acrylonitrile–butadiene–styrene, melamine, and phenolic resins.

Internally plasticized *n.* When a product is synthesized from a reaction involving two or more raw materials, it may be said to be internally plasticized if one of the raw materials is able to confer plasticity or flexibility to it. In other words, the product is plasticized because it is build up from a component, which is naturally plastic. For example, an oil-modified alkyd, in which phthalic anhydride and glycerol are combined with drying oil fatty acids, is internally plasticized by reason of the presence of the fatty acid component. If a Congo varnish were made by interaction of run Congo and linseed oil monoglyceride, it could be described as internally plasticized, in contrast to a varnish made by simple dissolution of run Congo in linseed oil, which would be externally plasticized. Also a polyvinyl acetate–acrylic co-polymer is internally plasticized while a polyvinyl acetate homopolymer, into which a plasticizer has been stirred, is externally plasticized.
See internal plasticizer.

Internal mandrel cooling *n.* In extrusion of tubing and pipe, the mandrel, which is an extension of the die core, is usually cooled internally so that the hot extrudate, cooled from both surfaces, cools faster than if cooled only on the outside, and with better control of dimensions.

Internal mix spray gun *n.* Spray gun in which the fluid and air are combined before they leave the gun.

Internal mixer *n.* A heavy-duty machine in which the materials to be mixed are strenuously worked and fused by one or more rotors designed so that all parts of the charge pass repeatedly through zones of high shear. The shell and rotors may be cored to permit circulation of heat-transfer liquids to control batch temperature. One type, the banbury mixer has long been used in the compounding plastics and rubbers, doing a good job of both *dispersive* and *distributive* mixing. Internal mixers have the inherent advantage of minimizing dust and fume hazards. The disadvantage of most types is that processing is batch wise rather than continuous.
See banbury mixer.

Internal phase *n.* In an emulsion, the discontinuous phase. For example, in an oil-in-water emulsion, the oil is the internal phase.

Internal plasticization *n.* Plasticization by means of internally combined groups, such as by co-polymerization.

Internal plasticizer *n.* An agent incorporated in, or co-polymerized with, a resin during its polymerization to make it softer and more flexible, as opposed to a plasticizer added to the resin during compounding. Common examples occur in oil-modified alkyds, and such materials have considerable practical advantages over externally plasticized systems, where the plasticizer may be lost by evaporation or leached out by water or other agents.

Internal reflection spectroscopy *n.* A technique for exposing a sample to an infrared beam in spectroscopy, where a thin sample film is placed in direct contact with the reflecting surface of a prism of high refractive index. The output generated is a spectrum similar to the transmission spectrum of the surface layers of the sample.

Internal stabilizer *n.* An agent incorporated in a resin during its polymerization to make it more resistant to high temperatures and other processing and environmental conditions, as opposed to a stabilizer added to the resin during compounding.

Internal undercut *n.* Any restriction that prevents a molded part from being directly removed from its core (force).

International gray scale *n.* A scale distributed through AATCC that is used as a comparison standard to rate degrees of fading from 5 (negligible or no change) to 1 (severe change). The term is sometimes applied to any scale of quality in which 5 is excellent and 1 is poor.

Interpenetrating polymer *n.* A material with two co-continuous phases or a material with two parts intertwined throughout.

Interpenetrating polymer network (IPN) *n.* A kind of blend formed by swelling a cross-linked polymer with a monomer, than inducing polymerization of the monomer. Another route is to infuse a cross-linking monomer into a cross-linkable polymer. Zaccaria VK, Utracki L (2003) Polymer blends. Springer-Verlag, New York. Galina H, Spiegel S, Meisel I, Kniep CS, Grieve K (2001) Polymer networks. John Wiley and Sons, New York.

Interpolymer A type of co-polymer in which the two monomer units are so intimately distributed in the polymer that the substance is essentially homogeneous in chemical composition. An interpolymer is sometimes called a *true co-polymer.*

Inter-society color council (ISCC) *n.* A USA council made up of delegates from about 30 "member-bodies", i.e., national organizations and professional societies, and of individual members, all of whom have an interest in information about color and in basic concepts of color. Member-body organizations and individual members represent: (1) Creators of color effects in art and design. (2) Manufacturers of colored material. (3) Producers of color reproduction processes. (4) Fields of science and technology. The work of the Council is carried out in study groups, called "problem subcommittees", and is directed by specialists in the color problems under investigation. News of Council activities and newsworthy items in the world of color are published in the ISCC Newsletter, published bimonthly. Annual Council meetings provide a forum for discussion and resolution of color problems. Specialized interest meetings are held periodically. The ISCC is the official USA member of the International Colour Association – Association Internationale de la Couleur.

Interstice \in-ˈtər-stəs\ [ME, fr. L *interstitium,* fr. *inter-* + *-stit-*, *-stes* standing (as in *superstes* standing over)] (15c) *n.* A space between objects; such as between atoms in a crystal.

Intimate blend *n.* A technique of mixing two or more dissimilar fibers in a very uniform mixture. Usually the stock is mixed before or at the picker.

Intramolecular polymerization *n.* Polymerization which occurs within an individual molecule. For example, drying oils consist of triglyceride esters, involving up to three unsaturated fatty acids attached to a single glycerol molecule. Under certain conditions, polymerization is believed to occur between neighboring fatty acid chains in a single molecule.

Intrinsic viscosity \in-ˈtrin-zik-\ (limiting viscosity number) *n.* In measurements of dilute-solution viscosity, intrinsic viscosity is the limit of the reduced and inherent viscosities as the concentration of polymer solute approaches zero. It represents the capacity of the polymer to increase viscosity. Interactions between solvent and polymer molecules give rise to different intrinsic viscosities for a given polymer in different solvents. Intrinsic viscosity is related to polymer molecular weight by the equation $[\eta] = K'M^a$, where the exponent a lies between 0.5 and 1.0, and, for many systems, between 0.6 and 0.8.

Also known as limiting viscosity number. See also dilute-solution viscosity, Huggins equation, and viscosity-average molecular weight.

I

Introfaction *n.* The change in fluidity and wetting capability of an impregnating material, produced by addition of an introfier.

Introfier *n.* A chemical that will convert a colloidal solution into a molecular one, by improving the solubility of the colloidal material.

Intumesce *n.* To form a voluminous char on ignition. Foaming or swelling when exposed to heat.

Intumescence \ˌin-tu-ˈme-sən(t)s\ [F, fr. L *intumescere* to swell up, fr. *in-* + *tumescere* to swell] (ca. 1656) *n.* The foaming and swelling of a plastic when exposed to high surface temperatures or flames. It has particular reference to ablative urethanes used on rocket nose cones, and to intumescent coating. Kidder RC (1994) Handbook of fire retardant coatings and fire testing services. CRC Press, Boca Raton, FL.

Intumescent coatings *n.* A coating that, when exposed to flame or intense heat, decomposes and bubbles into a foam that protects the substrate and prevents the flame from spreading. Such coatings are used, for example, on reinforced-plastics building panels. Examples of such coating materials are magnesium oxychloride cement used on urethane foams, and certain epoxy coatings used on polyester panels. Kidder RC (1994) Handbook of fire retardant coatings and fire testing services. CRC Press, Boca Raton, FL.

Inventory \ˈin-vən-ˌtōr-ē\. In injection molding or extrusion, the amount of plastic contained in the heating cylinder or barrel.

Inverse emulsion phase polymerization *n.* The reversal of emulsion polymerization phases; feeding the aqueous phase into a monomer phase (hydrophobic) where the water particles are dispersed in a monomer phase while agitating (and polymerizing monomers) until the phase shifts to an aqueous phase, and the monomer particles becomes dispersed in an aqueous phase; the phases shift when the aqueous component is more than 50% by volume than the monomer phase. Becher P (2001) Emulsions: theory and practice. American Chemical Society, Washington, DC. Lovell PA, El-Aasser (eds) (1997) Emulsion polymerization and emulsion polymers. John-Wiley and Sons Inc., New York.

Inversion temperature (1921) *n.* A temperature at which free expansion of a real gas produces neither heating nor cooling of the gas.

Investment casting (lost-wax process) *n.* A metals-casting method in which patterns made from wax or other expendable material are mounted on sprues, then "invested", i.e., covered with a ceramic slurry that sets at room temperature. The set slurry is then heated to melt away the pattern, leaving a mold into which metal is poured.

Iodine number (value) *n.* A number expressing the percentage (i.e., gram per 100 g) of iodine absorbed by a substance. It is a measure of the proportion of unsaturated linkages present and is usually determined in the analysis of oils and fats.

See iodine value.

Iodine test *n.* Test for the detection of starches and dextrins. It involves the addition of a small amount of a solution of iodine in potassium iodide to an aqueous solution of the suspected substance. Starches give purpose or blue colors according to the degree of hydrolyzation, whereas erythrodextrin gives an orange or reddish-brown coloration.

Iodine value (iodine number) *n.* The number of grams of iodine that 100g of an unsaturated compound will absorb in a given time under arbitrary conditions. It is used to indicate the residual unsaturation in epoxy hardeners (and many oils); a high value implies a high degree of unsaturation.

Ion \ˈī-ən, ˈī-ˌän\ [Gk, neuter of *iōn*, pp of *ienai* to go] (ca. 1834) *n.* An atom, molecule or radical that has become electrically charged by having either gained or lost an electron. When an electron is gained the negatively charged ion is called an *anion.* A positively charged ion is called a *cation.*

Ion current *n.* The migration of ions between the electrodes of an electrochemical cell.

Ion exchange (1923) *n.* A reversible interchange of ions between a solid phase and a liquid phase in which there is no permanent change in the structure of the solid phase. In the leading application, water softening, an ion-exchange resin extracts "hard", soap-precipitating calcium, magnesium, and iron ions from the water, replacing them with equivalent amounts of soluble sodium ions. Subsequently, the resin loaded with hard ions may be treated with salt solution (regenerated) to bring it back to the original sodium form, ready for reuse.

Ion-exchange resins (1943) *n.* Any of several small granular or bead-like resins consisting of two principal parts: a resinous matrix serving as a structural portion, and an ion-active group serving as the functional portion. The functional group may be acidic or basic. Complete deionization of water is accomplished by use of both acidic and basic resins, in sequence or in mixed beds. Ion-exchange resins are also used for other chemical processes such as electrodialysis.

Ionic \ ī -ˈä –nik\ [ISV] (1890) *adj.* Pertaining to an atom, radical, or molecule that is capable of being electrically charged, either negatively (anionic) or positively (cationic) or both (amphoteric), or to a material whose atoms already exist in charged state. Whitten KW, Davis RE, Davis E, Peck LM, Stanley GG (2003) General chemistry. Brookes/Cole, New York.

Ionic atmosphere The space around a given ion in solution, occupied largely by counterions.

Ionic bond (1939) *n.* The electrostatic attraction between ions of opposite electrical charge. Whitten KW, Davis RE, Davis E, Peck LM, Stanley GG (2003) General chemistry. Brookes/Cole, New York.

Ionic bonding *n.* Ionic bonding contributes strength and adhesive to end-use properties by providing a moderate form of cross-linking.

Ionic initiator *n.* A substance providing either carbonium ions (cationic) or carbanions (anionic) that attack the reactive double bonds of vinyl monomers and add on, regenerating the ion species on the propagating chain. Odian GC (2004) Principles of polymerization. John Wiley and Sons Inc., New York.

Ionic polymerization (cationic polymerization, anionic polymerization) *n.* A polymerization conducted in the presence of electrically charged ions that become attached to carboxylic groups on carbon atoms in the polymer chain. The carboxylic groups are produced along the polymer chain by co-polymerization, providing the anionic portion of the ionic cross-links. The cationic portion is provided by metallic ions added to the polymerization mixture. The electrostatic forces binding the chains together are much stronger than the covalent bonds between the molecules in conventional polymers. Some polymers produced by cationic polymerization are polyisobutylene, butyl rubber, polyvinyl ethers, and coumarone–indene resins. A typical product of anionic polymerization is polybutadiene, prepared with an alkali-metal catalyst. Odian GC (2004) Principles of polymerization. John Wiley and Sons Inc., New York.

Ionic polyurethane *n.* A urethane resin containing electrical charges in its backbone or side chain. Cationic urethanes (those containing positive changes) can be formed by reacting diisocyanates with diols containing tertiary nitrogen to yield urethanes that are then treated by a quaternization reaction that forms positive charges in the macromolecular backbone or in side chains.

Ionic solid *n.* A solid composed of anions and cations in its lattice.

Ionization [ISV] (1898) *v.* (1) Loss of an electron by an atom, molecule, or ion. (2) Dissociation of an electrolyte. It also can be written as a chemical change by which ions are formed from a neutral molecule of an inorganic solid, liquid, or gas.

Ionization chamber (1904) *n.* A partially evacuated tube provided with electrodes so that its conductivity due to the ionization of the residual gas reveals the presence of ionizing radiation.

Ionization energy *n.* The energy used to remove an electron from a gaseous, isolated, ground-state atom (or, sometimes, ion).

Also known as ionization potential.

Ionization foaming *n.* The process of foaming polyethylene by exposing it to ionizing radiation, which evolves hydrogen from the molten polymer, causing it to foam.

Ionization potential *n.* The work (expressed in electron volts) required to remove a given electron from its atomic orbit and place it at rest at an infinite distance. It is customary to list values in electron volts (eV) $1\,\text{eV} = 23{,}053\,\text{cal/mol}$.

Also known as ionization energy.

Ionomer *n.* Polymers with repeating ionic groups, which tend to form ionic domains that act as physical cross-links. The domains disassociate on heating allowing the material to be processed as a thermoplastic. Odian GC (2004) Principles of polymerization. John Wiley and Sons Inc., New York.

Ionomer resin *n.* A polymer containing interchain ionic bonding. In particular, commercial thermoplastics based on metal salts of co-polymers of ethylene and methacrylic acid, produced in the USA by the DuPont Co under the trade name Surlyn®. Ionomers are tough and flexible, the many grades ranging in modulus from 14 to 590 MPa. They have been approved for food-contact applications by the FDA, have outstanding resistance to puncture and impact. Applications include sporting goods (most golf-ball covers), footwear, packaging, automotive, and foams. Mark JE (ed) (1996) Physical properties of polymers handbook. Springer-Verlag, New York. Modern plastics encyclopedia. McGraw-Hill/Modern Plastics, New York.

Ion pair *n.* Pairs of oppositely charged ions held together by columbic attraction without formation of a covalent bond. Experimentally, an ion pair behaves as one unit in determining conductivity, kinetic behavior, osmotic properties, etc. Whitten KW, Davis RE, Davis E, Peck LM, Stanley GG (2003) General chemistry. Brookes/Cole, New York.

Ion plating *n.* A process for deposition of metals for dielectric films onto plastic substrates with a highly adherent bond. The process is performed in a tank similar to a vacuum-metallizing tank. A negative charge is developed on a metal bias plate located behind the plastic substrate. Next, the plating material is converted to a plasma of positive ions by filament or radio-frequency heating. At this point a phenomenon known as Crooke's dark space appears, enveloping the entire surface of the substrate, and establishing a large potential difference between the ions and the charged plastic surface. This causes the ions of the plating material to strike the plastic with high kinetic energy and to form strong bonds. James F (ed) (1993) Whittington's dictionary of plastics. Technomic Publishing Co. Inc., Carley.

Ion product *n.* (1) The product of the concentrations of hydrogen (hydronium) ions and hydroxide ions in water. (2) The mass-action expression for a solubility equilibrium.

IPA Abbreviation for isophthalic acid.

IPN Abbreviation for interpenetrating polymer network.

IR *n.* (1) Abbreviation for infrared. (2) Abbreviation for isoprene rubber (British Standards Insitution), the *cis*-1,4-type of polyisoprene.

Iridescence \ˌir-ə-ˈde-sᵊn(t)s\ (1804) *n.* Color produced from light interference phenomenon. Serway RA, Faugh JS, Bennett CV (2005) College physics. Thomas, New York. McDonald R (1997) Colour physics for industry, 2nd edn. Society of Dyers and Colourists, West Yorkshire, England. *See interference color.*

Irish gum *See carrageen.*

Iron blue *n.* $FeNH_4F_3(CN)_6$. Any of various pigments prepared by precipitating ferrous ferrocyanide from a soluble ferrocyanide and ferrous sulfate. Kirk–Othmer encyclopedia of chemical technology: pigments–powders. John Wiley and Sons, New York, 1996. Solomon DH, Hawthorne DG (1991) Chemistry of pigments and fillers. Krieger Publishing Co., New York.

Iron driers *n.* Iron salts of naphthenic acid, 2-ethyl hexoic acids or other acids, which actively accelerate polymerization at elevated temperatures, but are only feebly active at room temperature. The presence of the very dark ferric ion makes them applicable in dark or tinted finishes. The British call then iron soaps. Wicks ZN, Jones FN, Pappas SP (1999) Organic coatings science and technology, 2nd edn. Wiley-Interscience, New York.

Iron mill *n.* Paint mill consisting of a corrugated steel disc revolving tightly against a stationary steel shell. As the pigment particles pass between these moving steel parts, the pigments are dispersed into the vehicle.

Iron oxides (1885) *n.* Fe_2O_3. Pigments, which are substantially oxides of iron. (1) *Natural*: Ochres, raw and burnt umbers; raw and burnt siennas; red oxides; metallic browns; maroon oxides; and black oxides. Range in Fe_2O_3 content: 20–99%. (2) *Synthetic*: Yellow, red, brown, and black oxides. *Red oxide* – Fe_2O_3; *Yellow oxide* – $Fe_2O_3 \cdot H_2O$; *Brown oxide* – $FeO_3 \cdot x FeO$; *Black oxide* –Fe_2O_4. Paint/coatings dictionary. Federation of Societies for Coatings Technology, Philadelphia, Blue Bell, PA, 1978.

Iron oxides, synthetic (e.g., $Fe_2O_3 \cdot H_2O$) *n.* Iron oxides manufactured by a synthetic process. Kirk–Othmer encyclopedia of chemical technology: pigments–powders.

John Wiley and Sons, New York, 1996. Solomon DH, Hawthorne DG (1991) Chemistry of pigments and fillers. Krieger Publishing Co., New York.

Iron phosphate coating *See chemical conversion coating.*

Iron soaps (British) *n. See iron driers.*

Iron titanate *See ilmenite.*

Irradiance \i-ˈrā-dē-ən(t)s\ (1667) *n.* The quotient of the radiant flux incident on an infinitesimal surface element containing the point in question, by the area of that surface element. Serway RA, Faugh JS, Bennett CV (2005) College physics. Thomas, New York. Moller KD (2003) Optics. Springer-Verlag, New York.

I

Irradiation of polymers (1901) *n.* The subjection of a material to high-energy particle radiation for the purpose of producing a desired change in properties or of determining the effects of radiation on the material. Thermosetting resins such as unsaturated polyesters, acrylic-modified polyesters, and acrylic-modified epoxies can be cured rapidly at room temperature and without catalysts by exposure to ionizing radiation. However, if the polymer is overdosed, serious degradation can occur. Radiation sources most widely used in the plastics industry are electron accelerators and radioisotopes such as cobalt-60. Zaiko GE (ed) (1995) Degradation and stabilization of polymers. Nova Science Publishers Inc., New York.

Irregular block *n.* A block (in a polymer structure) that cannot be described by only one species of constitutional repeating unit in a single sequential arrangement. Odian GC (2004) Principles of polymerization. John Wiley and Sons Inc., New York.

Irregular polymer *n.* A polymer whose molecules cannot be described by only one species of constitutional unit in a single sequential arrangement. Odian GC (2004) Principles of polymerization. John Wiley and Sons Inc., New York.

Isanolic acid *n.* Eighteen-carbon fatty acid with acetylene triple bonds at the 9 and 11 positions, and having also an ethylene bond but no hydroxyl substituent. Occurs in isano oil, along with closely related bolekic acid, which is a hydroxy acid.
Also known as erythrogenic acid.

Isano oil *n.* A fatty oil extracted from an African tree of the same name, used as a flame retardant for acrylic resins. When heated to 200°C it polymerizes and may explode.
See isanolic acid.

ISCC Abbreviation for inter-society color council.

Iso- (1) The strict meaning of this prefix according to chemical nomenclature is "one methyl group on the next-to-last carbon atom, and no other branches". In the plasticizer field, the prefix is used to denote an isomer of a compound, specifically an isomer having a single, simple branching, not limited to methyl, at the end of a straight chain. (2) Equal, same, or constant, as in *isothermal*, at constant temperature. Morrison RT, Boyd RN (1992) Organic chemistry, 6th edn. Prentice-Hall, Englewood Cliffs, NJ.

ISO *n.* Abbreviation for international organization for standardization, headquartered in Geneva, Switzerland. ISO publishes standards in many fields, including hundreds on plastics in Field 170. ISO standards are available from American National Standards Institute.

Isoamyl acetate Rectified amyl acetate.

Isoamyl butyrate *n.* $C_5H_{11}OOCC_3H_7$. A colorless liquid derived by treating isoamyl alcohol with butyric acid, used as a solvent and as a plasticizer for cellulose acetate.

Isoamyl salicylate (amyl salicylate, orchidae) *n.* $C_5H_{11}OOCC_6H_4OH$. A colorless liquid used as a plasticizer.

Isobar \ˈī-sə-ˌbär\ [ISV *is-* + *-bar* (fr. Gk *baros* weight); akin to Gk *barys* heavy] (ca. 1864) *n.* For chemistry, elements of the same atomic mass but of different atomic numbers. The sum of their nucleons is the same but there are more protons in one than in the other.

Isobaric \ˌī-sə-ˈbär-ik, -ˈbar-\ [ISV] (1878) *adj.* Taking place without change in pressure.

Isobutane \ˌī-sō-ˈbyü-ˌtān\ [ISV] (1876) (isobutylene, 2-methylpropene) *n.* $(CH_3)_2C{=}CH_2$. A colorless, highly volatile liquid or liquidified gas derived from petroleum, easily polymerized to form polybutene. It has a bp of −6.9°C; a fp of −139°C, a flp of −105°C, and Sp gr of 0.6 (20°C). Uses: Production of isooctane, butyl rubber, polyisobutene resins, *tert*-butanol, methacrylates, co-polymer resins with butadiene, and acrylonitrile.

Isobutyl acetate $(CH_3)_2CHCH_2OOCCH_3$. A colorless liquid with a fruity odor, used as a solvent for cellulosic plastics and lacquers. Its properties are similar to those of butyl acetate, except that its evaporation rate is higher.

Isobutyl alcohol *n.* $(CH_3)_2CHCH_2OH$. Colorless flammable liquid used as a solvent. Sp gr, 0.806; bp, 108.39.

Also known as 2-methyl-1-propanol and isopropyl carbinol.

Isobutylene \ˈbyü-tᵊl-ˌēn\ [ISV] (1872) *n.* A colorless, very volatile liquid or flammable gas, C_4H_8, used chiefly in the manufacture of butyl rubber.

Isobutyl isobutyrate *n.* $(CH_3)_2CHCOOCH_2CH(CH_3)_2$. A colorless liquid with a fruity odor and slow evaporation rate, giving resin solutions with good flow and leveling characteristics. It is used as a solvent for nitrocellulose and vinyl resins.

Isocyanate \ˌī-sō-ˈsī-ə-nāt, -nət\ [ISV] (1872) *n.* A compound containing the isocyanate group, –N=C=O, attached to an organic radical or hydrogen. Isocyanates containing just one –N=C=O group (monoisocyanates) have limited uses in the plastics industry. The term is often used to mean a compound containing two –N=C=O groups (diisocyanate) or several such groups (polyisocyanate). However, in the case of a trimer compound containing three –N=C=O groups in a six-membered ring, the term *isocyanurate* is used. Morrison RT, Boyd RN (1992) Organic chemistry, 6th edn. Prentice-Hall, Englewood Cliffs, NJ.

See also diisocyanate.

Isocyanate-alcohol reaction *n.* Reaction involving a very reactive chemical group, –N=C=O with an alcohol to form a highly cross-linked polymer, importance in polyurethane formation. Odian GC (2004) Principles of polymerization. John Wiley and Sons Inc., New York.

Isocyanate foam *See urethane foam.*

Isocyanate generator (hindered isocyanate) *n.* A mixture of an isocyanate, a phenol, and a polyester that remains stable at room temperature. When heated to 70°C, the phenol and isocyanate components dissociate and react with the polyester to form a polyurethane.

Isocyanate plastic *n.* A plastic based on polymers made by the polycondensation of organic isocyanates with other compounds. Reaction of isocyanates with hydroxyl-containing compounds produces polyurethanes having the urethane group –NHC(=O)O–. Reaction of isocyanates with amine-containing compounds produces polyurea having the urea group –NHCONH– (ISO). Odian GC (2004) Principles of polymerization. John Wiley and Sons Inc., New York.

See also urethane and polyurethane foam.

Isocyanate resins *n.* Resins synthesized from isocyanates (–N=C=O) and alcohols (–OH). The reactants are joined through the formation of the urethane linkage and hence this field of technology is generally known as urethane chemistry. Odian GC (2004) Principles of polymerization. John Wiley and Sons Inc., New York.

See polyurethanes.

I

Isocyanates *n.* Compounds containing the isocyanate group –N=C=O, that are very reactive, especially with compounds containing active hydrogen atoms, such as in amines, alcohols, carboxylic acids and water. James F (ed) (1993) Whittington's dictionary of plastics. Technomic Publishing Co. Inc., Carley.

Isocyanurate foam *n.* A foam prepared from an isocyanurate. The unmodified foams have excellent flame resistance but are brittle and of little commercial value. However, isocyanurate foams modified withepoxides, polyimides, or (most commonly) urethane groups and polyols possess flame resistance far superior to that of conventional urethane foams and can be processed into a variety of foam products suitable for insulation. James F (ed) (1993) Whittington's dictionary of plastics. Technomic Publishing Co. Inc., Carley.

Isocyanurate plastic *n.* A plastic based on isocyanate polymers in which trimerization of the isocyanates incorporates six-membered isocyanurate-ring groups in a chain. Elias HG (1977) Macromolecules, vols 1–2, Plenum Press, New York.

Isocyaurate *n.* A trimer of an isocyanate, formed by the catalytic cyclization of three isocyanate molecular groups into a six-membered ring. Elias HG (1977) Macromolecules, vols 1–2, Plenum Press, New York.

Isodecyl octyl adipate *n.* $C_{10}H_{21}OOCC_4H_8COOC_8H_{17}$. A light-colored, oily liquid used as a plasticizer for vinyls.

Isodecyl octyl phthalate *n.* $C_{10}H_{21}OOCC_6H_4COOC_8H_{17}$. A primary plasticizer for vinyls, cellulose nitrate, polystyrene, and ethyl cellulose. In vinyls, it performs better than diethylhexyl phthalate (DOP).

Isoelectric heating Form of heating by means of electrical energy, in which a special form of element is employed, and which is submerged in liquid raw materials. In practice a series of such elements is employed.

Isoelectric point *n.* The pH value at which a substance or system (e.g., a protein solution) is electrically neutral; at this value, electrophoresis does not occur when a direct electric current is applied. Goldberg DE (2003) Fundamentals of chemistry. McGraw-Hill Science/Engineering/Math, New York.

Isogyres In a uniaxial interference figure, the two black bars that form a cross and represent the pattern of extinction positions of the crystal. In a biaxial interference figure, the two black, curved bars (brushes) that intersect the isochromatic curves (lemniscates) and represent the pattern of extinction positions of the crystal. Hibbard MJ (2001) Mineralogy. McGraw-Hill Co. Inc., New York. Rhodes G (1999) Crystallography made crystal clear: a guide for users of macromolecular models. Elsevier Science and Technology Books, New York.

Isoindolinone *n.* Any of a small family of organic pigments, available in bright yellows and reds. They have good lightfastness, heat stability and bleed resistance.

Isolated system *n.* A system, which can exchange neither matter nor energy in any form with its surroundings. Whitten KW, Davis RE, Davis E, Peck LM, Stanley GG (2003) General chemistry. Brookes/Cole, New York.

Isomer \ˈī-sə-mər\ [ISV, back-formation fr. *isomeric*] (1866) *n.* From the Greek *isos* (the same, equal, alike and *meros* (part or portion), isomers are substances comprising molecules that contain the same number and kinds of atoms, and have the same chemical formula, but that differ in structure, so that they form materials whose properties can differ widely. For example, the gas dimethyl ether, CH_3OCH_3, and the liquid ethyl alcohol. CH_3CH_2OH differ in both their chemical and physical properties, yet both have the empirical formula C_2H_6O. Isomeric polymers are formed by polymerizing isomonomers that link together in different ways. Morrison RT, Boyd RN (1992) Organic chemistry, 6th edn. Prentice-Hall, Englewood Cliffs, NJ.

Isomeric polymers *n.* Polymers which have essentially the same percentage composition, but differ with regard to their molecular architecture. Odian GC (2004) Principles of polymerization. John Wiley and Sons Inc., New York. Elias HG (1977) Macromolecules, vols 1–2. Plenum Press, New York.

Isomerism (1839) *n.* Existence of molecules having the same number and kinds of atoms but in different configurations. Smith MB, March J (2001) Advanced organic chemistry, 5th edn. John Wiley and Sons, New York. Morrison RT, Boyd RN (1992) Organic chemistry, 6th edn. Prentice-Hall, Englewood Cliffs, NJ.

Isomerized oils *n.* Oils which have been isomerized from a molecular rearrangement by several means, such as the catalytic action of alkalis, sulfur dioxide, nickel, etc. Isomerized oils exhibit greater reactivity, more rapid polymerization and greater drying rate than their non-isomerized parent oils. Paint: pigment, drying oils, polymers, resins, naval stores, cellulosics esters, and ink vehicles, vol 3. American Society for Testing and Material, Conshohocken, PA, 2001. Shahidi F, Bailey AE (eds) (2005) Bailey's industrial oil and fat products. John Wiley and Sons.

Isomerized rubber *n.* Rubber which has undergone cyclic rearrangement by heating in solution in the presence of a suitable catalyst. James F (ed) (1993) Whittington's dictionary of plastics. Technomic Publishing Co. Inc., Carley.

Also referred to as cyclized rubber.

Isomorphism \ˌī-sə-ˈmór-ˌfi-zəm\ [ISV] (ca. 1828) *n.* The existence of two or more chemical compounds with the same molecular formula but having different properties, owing to different arrangement of atoms within the molecule, e.g., ammonium cyanate NH_4CNO, and urea, $CO(NH_2)_2$, are isomers.

Isooctyl adipate *See diisooctyl adipate.*

Isooctyl palmitate *n.* $C_8H_{17}OOCC_{15}H_{31}$. A plasticizer for polystyrene and cellulosic plastics.

Isophorone *n.* A cyclic, unsaturated ketone with the structure.

It is a powerful solvent for vinyl and cellulosic resins, with moderate power to dissolve nearly all common thermoplastic and (uncured) thermosetting resins.

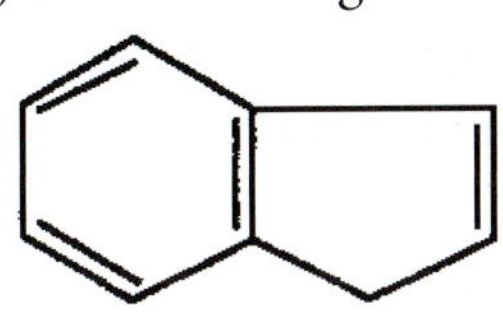

Isophorone diisocyanate (IPDI) *n.* An isocyanate used in the production of urethane elastomers and foams. It is less volatile than toluene diisocyanate, therefore easier to maintain at low levels in workers' airspace and safer to work with. Its structure is modified from that of isophorone, above, in that the oxygen has been replaced by an –N=C=O group, the double bond is gone, and the top carbon atom in the ring has an additional $-CH_2N=C=O$ linked to it.

Isophthalic acid (benzene-1,3-dicarboxylic acid, IPA) *n.* $C_6H_4-(COOH)_2$. Used instead of phthalic anhydride in making unsaturated polyester resins that, when cured, have good stiffness and resistance to heat and chemicals. Molecular weight, 166.13. Crystalline powder, which melts at 345–348°C; sublimes without decomposition. Merck index, 13th edn. Merck and Co. Inc., Whitehouse Station, NJ, 2001.

I

Isoprene \ˈī-sə-ˌprēn\ [prob. fr. *is-* + *propyl* + *-ene*] (1860) (3-methyl-1,3-butadiene, 2-methyl-1,3-butadiene) *n.* $CH_2=C-(CH_3)CH=CH_2$. A colorless, volatile liquid derived from propylene or from coal gases or tars, chemically similar to the mer unit of natural rubber. Its polymer of the *cis*-1, 4-type of polyisoprene is chemistry's nearest approach to synthesizing the natural product and it has sometimes been called "synthetic natural rubber". It has a molecular weight of 68.06, a mp of −120°C, and a bp of 34°C. Ash M, Ash I (1982–1983) Encyclopedia of plastics polymers, and resins, vols 1–3. Chemical Publishing Co., New York.

Also called 2-methyl-1,3-butadiene, β-methylbivinyl; hemiterpene.

Isoprene rubber *n.* The *cis*-1,4-type of polyisoprene.

Isopropyl acetate *n.* $(CH_3)_2CHOOCCH_3$. A colorless, fragrant liquid used as a solvent for cellulose nitrate, ethyl cellulose, polyvinyl acetate, polymethyl methacrylate, polystyrene, and certain phenolic and alkyd resins. It has bp of 90°C, Sp gr of 0.930, flp of 33°F, and a vp of 45 mmHg/20°C.

Isopropyl alcohol (1872) (2-propanol, dimethyl carbinol) *n.* $(CH_3)_2CHOH$. A colorless solvent boiling at 82.4°C, moderately polar, which, because of its low toxicity, is enjoying greater use today than formerly in the plastics industry. It has a bp of 82°C, Sp gr of 0.791/15°C, a flp of 62°F, and a vp of 34 mmHg/20°C.

Also known as propanol-2.

Isopropylbenzene (isopropylbenzol) *n.* Syn: cumene.

Isopropyl butyrate *n.* $(CH_3)_2CHCOOC_3H_7$. It has a bp of 128°C and Sp gr of 0.879/0°C.

Isopropyl cellosolve *n.* $(CH_3)_2CHOCH_2$-CH_2OH. It has a bp of 141°C, Sp gr of 0.906/20°C, and a vp of 8 mm Hg per 30°C.

***p,p′*-Isopropylidenediphenol** *See bisphenol A.*

Isopropyl myristate $(CH_3)_2CHOOCC_{13}H_{27}$. A plasticizer for cellulosic resins.

Isopropyl oleate *n.* $(CH_3)_2CHOOCC_{17}H_{33}$. A plasticizer for cellulose nitrate, ethyl cellulose, and polystyrene, and, with partial compatibility, vinyl resins.

Isopropyl palmitate *n.* $(CH_3)_2CHOOCC_{15}H_{31}$. A plasticizer for cellulose nitrate and ethyl cellulose.

Isostatic pressing *n.* The process of molding solid articles from fine powders by enclosing the powder in a closed rubber mold, then immersing the mold in a hydraulic fluid and holding it at a controlled temperature and high pressure for 5–30 min. The pressure is equal in all directions and therefore a consistent-molded object is formed. The method though more costly than force-in-mold pressing and much more limited in the shapes producible, provides pieces of more uniform density and strength. Strong AB (2000) Plastics materials and processing. Prentice-Hall, Columbus, OH. James F (ed) (1993) Whittington's dictionary of plastics. Technomic Publishing Co. Inc., Carley. Rosato DV (ed) (1992) Rosato's plastics encyclopedia and dictionary. Hanser–Gardner Publications, New York.

Isotactic \ˌī-sə-ˈtak-tik\ (1955) *adj.* (1) Denoting a polymer structure in which monomer units attached to a polymer backbone are identical on one side and/or the other side of the backbone. (2) Pertaining to a type of polymeric molecular structure containing a sequence of regularly spaced asymmetric atoms arranged in like configuration in a polymer chain (ASTM D 883). Materials containing isotactic molecules may exist in highly crystalline form because of the high degree of order that may be imparted to such structures. Odian GC (2004) Principles of polymerization. John Wiley and Sons Inc., New York.

See also stereospecific, syndiotactic.

Isotactic polymer *n.* A tactic polymer, the base unit of which possesses, as a component of the main chain, a carbon (or similar) atom with two different lateral substituents. These atoms being so arranged that a hypothetical observer advancing along the bonds constituting the main chain finds each of its substituents in the same steric order. *Note*—Here, hydrogen is counted a substituent (IUPAC). Odian GC (2004) Principles of polymerization. John Wiley and Sons Inc., New York.

Also see atactic polymer, syndiotactic polymer, and tactic polymer.

Isotactic polypropylene *n.* Polypropylene in which each mer unit has the pendant $-CH_3$ group on the same side of the chain backbone. Commercial PPs are about 90% isotactic, conferring high crystallinity and softening range. In contrast, atactic PP is rubbery and weak. Odian GC (2004) Principles of polymerization. John Wiley and Sons Inc., New York.

Isotherm \ˈī-sə-ˌthərm\ [F *isotherme, adj*] (1859) *n.* Constant temperature line used on graphs of climatic conditions or thermodynamic relations, such as pressure–volume relations at constant temperature. Ready RG (1996) Thermodynamics. Plenum Publishing Co., New York.

Isothermal \ˌī-sə-ˈthər-məl\ [F *isotherme,* fr. *is-* + *-therme* fr. Gk *thermos* hot] (1826) *adj.* When a gas passes through a series of pressure and volume variations without change of temperature, the changes are called isothermal. Ready RG (1996) Thermodynamics. Plenum Publishing Co., New York.

Isotope \ˈī-sə-ˌtōp\ [*is-* + Gk *topos* place] (1913) *n.* In chemistry, one of two or more forms of an element ("nuclides") having the same number of protons in the nucleus but differing in mass number because of different numbers of neutrons. Natural elements are usually mixtures of isotopes; thus the observed atomic weights are average values weighted by isotopic relative abundance. Serway RA, Faugh JS, Bennett CV (2005) College physics. Thomas, New York.

Isotope dating A method of determining a period in time using the rate of deterioration of an unstable isotope. Serway RA, Faugh JS, Bennett CV (2005) College physics. Thomas, New York.

See carbon dating.

Isotope effect The dependence of a property such as reaction rate on the mass number of an element. Serway RA, Faugh JS, Bennett CV (2005) College physics. Thomas, New York.

Isotopes *n.* Atoms of an element having different numbers of neutrons in their nuclei, and hence, different mass numbers. Isotopes of a given element have the same number of nuclear protons but differing numbers of neutrons. Naturally occurring chemical elements are usually mixtures of isotopes so that observed (non-integer) atomic weights are average values for the mixture. Serway RA, Faugh JS, Bennett CV (2005) College physics. Thomas, New York. *See atomic number.*

Isotropic [ISV] (ca. 1860) *adj.* Substances showing a single refractive index at a given temperature and wavelength no matter what the direction of light through the particle. They show no interference colors between crossed polars. Examples are unstrained glasses, unoriented polymers, and compounds in the cubic system. Meeten GH (1986) Optical properties of polymers. Springer-Verlag, New York.

I

Isotropic laminate *n.* A laminate in which the strength properties are equal, or approximately equal, in all directions. Iaac MD, Ishal O (2005) Engineering mechanics of composite materials. Oxford University Press, UK.

Isotropy [ISV] (ca. 1860) *adj.* A materials state in which the material properties are the same in all directions. Meeten GH (1986) Optical properties of polymers. Springer-Verlag, New York.

Itaconic acid \ˌi-tə-ˈkä-nik-\ [ISV, anagram of *aconitic acid*, $C_3H_3(COOH)_3$, fr. *aconite*] (ca. 1872) (methylenesuccinic acid HOO CC $(=CH_2)CH_2COOH$ *n.* A white crystalline powder usually obtained by the oxidative fermentation of sucrose or glucose with *Aspergillyus terreus.* It is capable of polymerization alone, or as a comonomer with acrylic acid, acrylonitrile, styrene, methyl methacrylate, and vinylidene chloride. It is used as an additive in acrylic resins to increase their adhesion to cellulose. By polycondensation of itaconic acid with diols, polyesters are obtained that contain methylene side groups. Merck index, 13th edn. Merck and Co. Inc., Whitehouse Station, NJ, 2001.
Also called methylenebutanedioic acid and methylene-succinic acid.

IUPAC *n.* Abbreviation for international union of pure and applied chemistry, headquartered in Oxford, England. Progress in polymer science and technology: 2002 IUPAC World Polymer Congress, Beijing, China, July 7–12, 2002, John Wiley and Sons, New York. IUPAC handbook, 2000–2001. International Union of Pure and Applied Chemistry, Oxford, England, 2000.

IVE *Abbreviation for vinylisobutyl ether.*

Izod impact strength *n.* A widely used measure of impact strength (www.astm.org) determined by the difference in energy of a swinging pendulum before and after it breaks a notched specimen clamped vertically as a cantilever beam. The pendulum is released from a vertical height of 0.61 m, and the vertical height to which it rises after breaking the specimen is used to calculate the energy expended in that breakage. The notch across the width of the specimen is usually on the side opposite the side impacted, but the reverse setup is provided for and specimens may also be tested unnotched. www.astm.org. Shah V (1998) Handbook of plastics testing technology. John Wiley and Sons, New York. Brown R (1999) Handbook of physical polymer testing, vol 50. Marcel Dekker, New York.

J

J *n.* (1) The SI abbreviation for joule. (2) In most references to technical publications whose title contains the word "Journal", the abbreviation of that word. Weast RC (ed) (1978) CRC handbook of chemistry and physics, 59th edn. CRC Press, Boca Raton, FL.

J-acid *n.* Fusion product of beta-naphthylamine-3,6-bisulfonic acid with caustic. Used in the manufacture of dyes. *Also known as 2-amino-5-naphthol-7-sulfonic acid and 6-amino-1-naphthol-3-sulfonic acid.*

Jack \ˈjak\ [ME *Jacke*, familiar term of address to a social inferiorm nickname for *Johan* John] (1548) *n.* (1) A blade having high and/or low butts used to actuate the movement of latch knitting needles. (2) Part of a dobby head designed to serve as a lever in the operation of the harness of a loom. Vincenti R (ed) (1994) Elsevier's textile dictionary. Elsevier Science and Technology Books, New York.

Jacket \ˈja-kət\ [ME *jaket*, fr. MF *jacquet*, dimin. of *jaque* short jacket, fr. *jacque* peasant, fr. the name *Jacques* James] (15c) *n.* (1) A woven or felted tubular sleeve for covering and shrinking on a machine roll. (2) A short coat. (3) In polymer manufacture, an external shell around a reaction vessel. For example, jacketed vessels are used when heat-transfer medium is circulated around the vessel. Vincenti R (ed) (1994) Elsevier's textile dictionary. Elsevier Science and Technology Books, New York.

Jacquard \ˈja-ˌkärd\ (1841) *n.* A system of weaving that utilizes a highly versatile pattern mechanism to permit the production of large, intricate designs. The weave pattern is achieved by a series of punched cards. Each card perforation controls the action of one warp thread for the passage of one pick. The machine may carry a large number of cards, depending upon the design, because there is a separate card for each pick in the pattern. Jacquard weaving is used for tapestry, brocade, damask, brocatelle, figured necktie and dress fabrics, and some floor coverings. A similar device is used for the production of figured patterns on some knit goods. Vincenti R (ed) (1994) Elsevier's textile dictionary. Elsevier Science and Technology Books, New York.

Jalousie \ˈja-lə-sē\ [F] (1766) *n.* A window with movable, horizontal glass slats angled to admit ventilation and keep out rain. The term is also used for outside shutters of wood constructed in this way. Harris CM (2005) Dictionary of architecture and construction. McGraw-Hill Co., New York.

Jamb \ˈjam\ [ME *jambe*, fr. MF, LL *gamba*] (14c) *n.* Vertical face inside an opening to the full thickness of a wall; vertical slide members of a window frame, door frame, or lining. Harris CM (2005) Dictionary of architecture and construction. McGraw-Hill Co., New York.

Jamba seed oil *n.* Non-drying oil, similar to rape oil, obtained from the seeds of *Eruca sativa*. Its main constituent acids are erucic, oleic, and linoleic. Approximate constants: Sp gr, 0.916/15°C; iodine value, 100; saponification value, 173.

Japan \jə-ˈpan\ (1688) *n.* Glossy black enamel, either air drying or baking, based on asphaltum and drying oil. Japan should not be confused with black enamels which are produced by pigmentation and do not have an asphaltum base. Ash M, Ash I (1982–1983) Encyclopedia of plastics, polymers, and resins, vols 1–3. Chemical Publishing Co., New York.
Also known as black Japan.

J

Japan color *n.* Paste containing pigment and a grinding Japan vehicle, used for lettering and decorating.

Japan drier Resinate-base liquid drier. *See resinates, metallic.*

Japanese lacquer *n.* A glossy coating obtained by tapping the sap from the Japanese varnish tree (*Rhus vernicifera*) or sumac. Langenheim JH (2003) Plant resins: chemistry, evolution ecology and ethnobotany. Timber Press, Portland, OR. Weismantal GF (1981) Paint handbook. McGraw-Hill Corp. Inc., New York.

Japan, grinding *n.* Rapid, hard-drying varnish suitable for use as a vehicle for Japan colors; frequently contains shellac. Weismantal GF (1981) Paint handbook. McGraw-Hill Corp. Inc., New York.

Japanners' brown *n.* A brown pigment made by high temperature oxidation of ferrous hydroxide precipitated from a solution of an iron salt. Kirk–Othmer encyclopedia of chemical technology: pigments–powders. John Wiley and Sons, New York, 1996.

Japanning Process of finishing with a baking black Japan.

Japan wax (1859) *n.* Product obtained from the berries of trees indigenous to Japan and China. Although described as a wax, it differs from waxes in composition, in that it is a mixture of triglycerides. Mp, 52°C; Sp gr, 0.987/15°C; saponification value, 215; iodine value, 11. Langenheim JH (2003) Plant resins: chemistry, evolution ecology and ethnobotany. Timber Press, Portland, OR. Ash M, Ash I (1982–1983) Encyclopedia of plastics polymers, and resins, vols I–III. Chemical Publishing Co., New York.

Jar mill *n.* A small ball mill utilizing a portable jar of porcelain or metal rather than a fixed cylinder for containing the material to be ground and the grinding media. After being charged and tightly closed, the jar is placed on a pair of rubber rollers – one driven, the other idling – and rotated for the desired time, typically overnight. Perry RH, Green DW (1997) Perry's chemical engineer's handbook, 7th edn. McGraw-Hill, New York. Weismantal GF (1981) Paint handbook. McGraw-Hill Corp. Inc., New York. *See ball mill.*

Jasmine oil \ˈjaz-mən-\ [Arabic yāsamīn, from Persian] An essential oil in perfumery, fragranced, and flavoring, dextrorotatory *n.* Origin: any of numerous often climbing scrubs (genus *Jasminum*) of the olive family that usually have extremely fragrant flowers. Langenheim JH (2003) Plant resins: chemistry, evolution ecology and ethnobotany. Timber Press, Portland, OR. Shahidi F, Bailey AE (eds) (2005) Bailey's industrial oil and fat products. John Wiley and Sons, New York. Merriam-Webster's collegiate dictionary, 11th edn. Merriam-Webster Inc., Springfield, MA, 2004.

Jaspé *n.* (1) A fabric used for suiting, draperies, or upholstery characterized by a series of faint stripes formed by dark, medium, and light yarns of the same color. (2) A term describing carpets having a faint striped effect. Tortora PG, Merkel RS (2000) Fairchild's dictionary of textiles, 7th edn. Fairchild Publications, New York.

Javalle water \zha-ˈvel\ *n.* Sodium hyprochlorite (NaHClO) dissolved in water, a disinfectant and bleaching solution. Merriam-Webster's collegiate dictionary, 11th edn. Merriam-Webster Inc., Springfield, MA, 2004.

J-box *n.* A J-shaped holding device used in continuous operations to provide varying amounts of intermediate material storage such as in wet processing of fabrics and in tow production. The material is fed to the top and pleated to fill the long arm before being withdrawn from the short arm. Vigo TL (1994) Textile processing, dyeing, finishing and performance. Elsevier Science, New York.

JCT *n.* Abbreviation for the journal of coatings technology.

J-cut *n.* In tufting cut-pile carpet constructions, uneven cutting of the loops caused by poor adjustment of knives and hooks or excessive tension.

Jean \ˈjēn\ [short for *jean fustian*, fr. ME *Gene* Genoa, Italy + *fustian*] (1577) *n.* Cotton twill fabric, similar to denim, but lighter and finer, in a 2/1 weave for sportswear and linings. Tortora PG (ed) (1997) Fairchild's dictionary of textiles. Fairchild Books, New York.

Jeffamine® *n.* Trade mark (Texaco, TX, USA) for polyoxypropyleneamines, grades are 3-diamine, 1-triamaine, liquid form, and corrosive. Used as curing agent for epoxy resin systems in adhesives, elastomers, and foam formulations, and as an intermediate for textile and paper-treating chemicals. Ash M, Ash I (1996) Handbook of paint and coating raw materials: trade name products – chemical products dictionary with trade name cross-references. Ashgate Publishing Ltd., New York.

Jersey \ˈjər-zē\ (1587) *n.* (1) A circular-knit or flat-knit fabric made with a plain stitch in which the loops intermesh in only one direction. As a result, the appearance of the face and the back of a jersey fabric is wholly different. (2) A tricot fabric made with a simple stitch, characterized by excellent drape and wrinkle recovery properties. Vincenti R (ed) (1994) Elsevier's textile dictionary. Elsevier Science and Technology Books, New York.

Jet \ˈjet\ [ME, fr. MF *jaiet*, fr. L *gagates*, fr. Gk *gagatēs*, fr. *Gagas*, town and river in Asia Minor] (14c) *n.* Term used to describe the blackness or intensity of the mass tone of black or near black surfaces. It is frequently used to describe dark blues, blue-black, or black pigments. Billmeyer FW, Saltzman M (1966) Principles of color technology. John Wiley and Sons Inc., New York.

Jet A device used to bulk yarns by introducing curls, coils, and loops that are formed by the action of a high velocity stream, usually of air or steam. Vigo TL (1994) Textile processing, dyeing, finishing and performance. Elsevier Science, New York.

Jet abrader *n.* Device used for measuring the abrasion resistance of organic coating materials, in terms of the time required for a controlled jet of fine abrasive particles to abrade through the coating to the substrate.

Jet-abrasion test *n.* A test for the abrasion resistance of coatings in which the time required for an air blast of fine abrasive particles to wear through the coating is measured (ASTM D 658, Section 06.01).

Jet dyeing machine *n.* A high temperature piece dyeing machine that circulates the dye liquor through a Venturi jet, thus imparting a driving force to move the fabric. The fabric, in rope form, is sewn together to form a loop.

Jet loom *n.* A shuttleless loom that employs a jet of water or air to carry the filling yarn through the shed.
Also weft insertion.

Jet molding (offset molding) *n.* A modification of injection molding designed for molding thermosets. An elongated nozzle or "jet" is attached to the front of the molding cylinder and is provided with a high-watt-density heating element and means for rapid cooling. It is also necessary to control cylinder temperatures carefully to prevent premature hardening of the resin. Strong AB (2000) Plastics materials and processing. Prentice-Hall, Columbus, OH.

Jet printing *n.* A process, wherein charged ink droplets are emitted from a nozzle and deflected vertically and horizontally by positively and negatively charged electrodes. The operation is analogous to electron beam tracing as in television tubes and, consequently, is extremely fast with speeds of over 1200 words per min easily attained. Leach RH, Pierce RJ, Hickman EP, Mackenzie MJ, Smith HG (eds) (1993) Printing ink manual, 5th edn. Blueprint, New York.

Jet spinning *n.* For most purposes, similar to melt spinning of staple fiber. Hot-gas-jet spinning uses a directed jet of hot gas to "pull" molten polymer from a die lip and instantly draw it into fine fibers. Kadolph SJJ, Langford AL (2001) Textiles. Pearson Education, New York.

Jetting *n.* In injection molding, a wriggly flow of resin from a small gate into the mold cavity, mistakenly referred to as turbulence and probably related to melt fracture. Jetting is the antithesis of the desired laminar flow forming a smooth flow front across the mold. It can cause strength problems in molded parts because of incomplete welding of the wormlike surfaces of the jet. Strong AB (2000) Plastics materials and processing. Prentice-Hall, Columbus, OH.

Jewelers' rouge \ˈjü-ə-lərs ˈrüzh *esp. Southern* ˈrüj\ *n.* Very fine and pure ferric oxide (Fe_2O_3) powder used for polishing metals and plastics. *Rouge paper* contains the same abrasive glued to paper. It also comes in cloth form, called *crocus cloth.* James F (ed) (1993) Whittington's dictionary of plastics. Technomic Publishing Co. Inc., Carley.

Jig \ˈjig\ [perhaps fr. MF *giguer* to frolic, fr. *gigue* fiddle, of Gr origin; akin to OHGr *gīga* fiddle; akin to ON *geiga* to turn aside] (ca. 1560) *n.* (1) A device for positioning component parts while they are being assembled or otherwise worked on, or for holding tools. (2) A clamping device used to secure a bonded assembly until the adhesive has set. (3) A restraining frame into which freshly molded parts are placed to prevent their warping during annealing or final cooling. Merriam-Webster's collegiate dictionary, 11th edn. Merriam-Webster Inc., Springfield, MA, 2004. James F (ed) (1993) Whittington's dictionary of plastics. Technomic Publishing Co. Inc., Carley.

Joggles \ˈjä-gəl\ [frequentative of [1]*jog*] (1513) (keys) *v.* A term sometimes employed for matching inserts that exactly position the parts of a multipiece mold. James F (ed) (1993) Whittington's dictionary of plastics. Technomic Publishing Co. Inc., Carley.

Joining \ˈjói-niŋ\ (14c) *n.* The process of assembling plastic parts by means of mechanical fastening devices such as rivets, screws, clamps, etc.
See also fabricate.

Joint \ˈjóint\ [ME *jointe,* fr. OF, fr. *joindre*] (13c) *n.* The location where two separately made parts are joined with each other by adhesive bonding, welding, or fastening. Skeist I (ed) (1990) Handbook of adhesives. Van Nostrand Reinhold, New York.
See also butt joint, lap joint, and scarf joint.

Joint filler *n.* Sealant inserted between abutting ends of wallboard.

Joint, scarf *n.* A joint made by cutting away similar angular segments of two adherents and bonding the adherents with the cut areas fitted together. Skeist I (ed) (1990) Handbook of adhesives. Van Nostrand Reinhold, New York.
See also joint, lap.

Joint, starved *n.* A joint that has an insufficient amount of adhesive to produce a satisfactory bond. *Note* – This condition may result from too thin a spread to fill the gap between the adherents, excessive penetration of the adhesive into the adherent, too short an assembly time, or the use of excessive pressure.

See starved joint.

Joint tape *n.* Paper or paper-faced cotton tape, metal, fabric, glass mesh, or other material, sometimes embossed or perforated, which is fixed over the joints between wallboards, to conceal the joints and provide a smooth surface for painting.

Joist \ˈjóist\ (ME *joiste*, fr. MF *giste*, fr. (assumed) VL *jacitum*, fr. L *jacēre*] (15c) *n.* A small rectangular sectional member arranged parallel from wall to wall in a building, or resting on beams or girders. They support a floor or the laths or furring strips of a ceiling.

Jolly balance *n.* Spring balance for determining the specific gravity of a solid by weighing it alternatively in air and water.

Joule \ˈjü(ə)l *also* ÷ˈjaü(ə)l\ [James P. *Joule*] (1882) (J) *n.* The SI unit of work and energy, equal to 1 Nm, that replaces a variety of not-quite-equal older joules as well as numerous calories, all of which equal 4.18–4.19 J. Six different British thermal units are all about equal to 1055 J. Serway RA, Faugh JS, Bennett CV (2005) College physics. Thomas, New York.

Joule–Thomson effect *n.* The cooling which occurs when a highly compressed gas is allowed to expand in such a way that no external work is done is known as the Joule–Thomson effect. This cooling is inversely proportional to the square of the absolute temperature. Serway RA, Faugh JS, Bennett CV (2005) College physics. Thomas, New York.

Journal of coatings technology (JCT) *n.* The official journal of the Federation of Societies for Coatings Technology.

Journeyman painter *n.* One who has had at least 3 years experience and schooling as an apprentice.

Juniper gum \ˈjü-nə-pər-\. Whistler JN, BeMiller JN (eds) (1992) Industrial gums: polysaccharides and their derivatives. Elsevier Science and Technology Books, New York.

See gum sandarac.

Juta hycica resin *n.* Brazilian resin occasionally used in certain types of varnishes. It is a very hard and pale-colored material, insoluble in most of the common solvents. Langenheim JH (2003) Plant resins: chemistry, evolution ecology and ethnobotany. Timber Press, Portland, OR. Paint: pigment, drying oils, polymers, resins, naval stores, cellulosics esters, and ink vehicles, vol 3. American Society for Testing and Material, Conshohocken, PA, 2001.

J

Jute (Burlap) \ˈjüt\ [Bengali *jhuto*] (1746) *n.* A fiber obtained from the stems of several species of the plant *Corchorus* grown mainly in India and Pakistan. It is used in the form of fiber, yarn, and fabric for reinforcing phenolic and polyester resins. Kadolph SJJ, Langford AL (2001) Textiles. Pearson Education, New York.

Jute count *n.* The weight in pounds of a spindle of 14,400 yards of yarn.

Jute seed oil *n.* Obtained from the seed of *Corchorus capsularis*, grown chiefly in Indian, Egypt, and China. It contains linoleic and oleic acids. Iodine value, 103; Sp gr, 0.921/15°C; saponification value, 185; acetyl value 27. Langenheim JH (2003) Plant resins: chemistry, evolution ecology and ethnobotany. Timber Press, Portland, OR. Paint: pigment, drying oils, polymers, resins, naval stores, cellulosics esters, and ink vehicles, vol 3. American Society for Testing and Material, Conshohocken, PA, 2001.

K

k \ˈkā\ *n.* (1) Abbreviation for SI prefix, kilo-. (2) Symbol for thermal conductivity. Ready RG (1996) Thermodynamics. Plenum Publishing Co., New York.

k′ Symbol for Huggins constant. Huggins ML (1958) Physical chemistry of high polymers. John Wiley and Sons Inc., New York.

K (1) Abbreviation for kelvin. (2) Chemical symbol for potassium (Latin: kalium). (3) Symbol for bulk modulus.

Kakemono \ˌkä-ki-ˈmō-(ˌ)nō\ [Japanese] (1890) *n.* A painting mounted on a margin of brocade; hung by its top when in use, and rolled up when not in use.

Kalrez® Dupont's trade name for fluoroelastomers made from tetrafluoroethylene, perfluorovinylmethyl ether, and a small percentage of cross-linkable monomer. These elastomers combine the rubbery properties of viton with the thermal stability, chemical resistance, and electrical characteristics of tetrafluoroethylene resin. Harper CA (ed) (2002) Handbook of plastics, elastomers and composites, 4th edn. McGraw-Hill, New York. Ash M, Ash I (1982–1983) Encyclopedia of plastics polymers, and resins, vols 1–3. Chemical Publishing Co., New York.

Kalsomine *See calcimine.*

Kaolin \ˈkā-ə-lən\ [F *kaolin*, fr. *Gaoling* hill in China] (ca. 1741) (China clay, bolus alba) *n.* A variety of clay consisting essentially of the minerals *kaolinite*, *dickite*, and *nacrite* (all are $Al_2O_3{\cdot}2SiO_2{\cdot}2H_2O$). The name kaolin comes from the Chinese *kaoling*, meaning high hill, the name of the mountain in China, which yielded the first kaolin sent to Europe. Hibbard MJ (2001) Mineralogy. McGraw-Hill Co. Inc., New York.

Kaolinite \-lə-ˌnīt\ (1867) *n.* $Al_2O_3{\cdot}2SiO_2{\cdot}2H_2O$. A finely divided crystalline form of hydrated aluminum silicate that occurs as monoclinic crystals with a basal cleavage, resulting chiefly from the alteration of feldspars under conditions of hydrothermal or pneumatolytic metamorphism. It is an important clay mineral. Density, 2.58 g/cm^3 (21.5 lb/gal); refractive index, 156; O.A., 32–55; Mohs hardness, 2.5. Hibbard MJ (2001) Mineralogy. McGraw-Hill Co. Inc., New York. Solomon DH, Hawthorne DG (1991) Chemistry of pigments and fillers. Krieger Publishing Co., New York.

Kapok \ˈkā-ˌpäk\ [Malay] (ca. 1750) *n.* Short, lightweight cotton-like fibers from the seed pod of trees of the family *Bombacabeae.* A very brittle fiber, it is generally not spun. It is used for stuffing cushions, mattresses, etc., and for life jackets because of its buoyancy and moisture resistance. Kadolph SJJ, Langford AL (2001) Textiles. Pearson Education, New York. Vincenti R (1994) Elsevier's textile dictionary. Elsevier Science and Technology Books, New York.

Karaya gum \kə-ˈrī-ə-\ [Hindi *karāyal* resin] (1916) *n.* Dry exudation from *Sterculia urens*, which grows in India. It swells in water, and has some resemblance to gum tragacanth.

Also known as gum karaya. Whistler JN, BeMiller JN (1992) Industrial gums: polysaccharides and their derivatives. Elsevier Science and Technology Books, New York.

Karl Fischer reagent A colored solution of iodine, sulfur dioxide, and pyridine in methanol. It reacts quantitatively with water, becoming colorless. It is used to determine small amounts of water in a wide range of materials, including many polymerics. Goldberg DE (2003) Fundamentals of chemistry. McGraw-Hill Science/Engineering/Math, New York.

Kauri \ˈkau(-ə)r-ē\ [Maori *kawri*] (1823) *n.* A fossil copal resin used in oleo-resinous varnishes found in New Zealand. Langenheim JH (2003) Plant resins: chemistry, evolution ecology and ethnobotany. Timber Press, Portland, OR. Paint: pigment, drying oils, polymers, resins, naval stores, cellulosics esters, and ink vehicles, vol 3. American Society for Testing and Material, Conshohocken, PA, 2001.

Kauri-butanol value Volume in ml at 25°C (77°F) of a solvent, corrected to a defined standard, required to produce a defined degree of turbidity when added to 20g of a standard solution of kauri resin in normal butyl alcohol. For kauri-butanol values of 69 and above, the standard is toluene and has an assigned value of 105. For kauri-butanol values below 60, the standard is a blend of 75% *n*-heptane and 25% toluene and has an assigned value of 40. Abbreviation is KB value. Paint: pigment, drying oils, polymers, resins, naval stores, cellulosics esters, and ink vehicles, vol 3. American Society for Testing and Material, Conshohocken, PA, 2001. Flick EW (1991) Industrial synthetic resins handbook. Williams Andrews Publishing/Noyes, New York.

Kauri reduction test Test for measuring the flexibility of a varnish. Paint: pigment, drying oils, polymers, resins, naval stores, cellulosics esters, and ink vehicles, vol 3. American Society for Testing and Material, Conshohocken, PA, 2001.

Kauri glue Urea-formaldehyde resin, manufactured by BASF, Germany.

Kautex Polyvinyl chloride, manufactured by Kautex Werke, Germany.

Kautschin *See dipentine.*

KB value A numerical measure of the solvent power of hydrocarbon solvents and oils using a kauri-butanol reagent. The values range from 20, which is a poor solvent, to a high of 105, which is an excellent solvent. Paint: pigment, drying oils, polymers, resins, naval stores, cellulosics esters, and ink vehicles, vol 3. American Society for Testing and Material, Conshohocken, PA, 2001.

See kauri-butanol value.

Kel-F elastomer Co-polymer from vinylidene fluoride and trifluorochloroethylene. Manufactured by M. W. Kellog, USA.

Kelvin \ˈkel-vən\ [William Thomson, Lord *Kelvin* (1824–1907)] (1968) (K) *n.* The SI unit of both temperature and difference between temperatures, equal to 1/273.16 of the thermodynamic triple point of water, i.e., the temperature and pressure at which all three phases of water – ice, liquid, and vapor – are in equilibrium. A change or difference of 1K is exactly equal to 1° difference on the Celsius (formerly centigrade) scale, and the temperature 0°C corresponds to 273.15K. Symbol K; the name "degree of Kelvin" (symbol K) was discontinued by international agreement in 1967.

Kelvin temperature scale An absolute temperature scale, in which the unit is the Kelvin (K), defined as 1/273.16 of the temperature difference between absolute zero (0K) and the triple point of water. $K = °C + 273$. This scale is used to describe the correlated color temperature of light sources and illuminants in color designations and color rendition. Symbol K. *See correlated color temperature and Kelvin.* Kelvin is used extensively in gas laws and calculation for expansion of gases. Theoretically, all molecular motions cease at 0K or absolute zero temperature. Paint: pigment, drying oils, polymers, resins, naval stores, cellulosics esters, and ink vehicles, vol 3. American Society for Testing and Material, Conshohocken, PA, 2001.

Keratin \ˈker-ə-tᵊn\ [ISV] (ca. 1849) *n.* (1) The protein derived from feathers, hair, hoofs, horns, etc. of animals by calcinations. It is sometimes used as filler in plastics, particularly urea-formaldehyde molding compounds, in which it reduces brittleness and permits drilling and tapping. (2) A class of natural fibrous proteins occurring in vertebrate animals and man, characterized by their high content of several amino acids, especially crystine, arginine, and serine. They are generally harder than the fibrous collagen group of proteins. Keratins are insoluble in organic solvents but do absorb and hold water. The molecules contain both acidic and basic groups and are thus amphoteric. Morrison RT, Boyd RN (1992) Organic chemistry, 6th edn. Prentice-Hall, Englewood Cliffs, NJ.

Kernel \ˈkər-nᵊl\ [ME, fr. OE *cyrnel*, dim. of *corn*] (before 12c) *n.* All of an atom except for its valence shell of electrons; also called the *core.*

Kerosene \ˈker-ə-ˌsēn\ [Gk *kēros* + E *–ene* (as in *camphene*] (1854) (*or* kerosene) *n.* A low viscosity oil distilled from petroleum or shale oil, used as a fuel, paint thinner, and alcohol denaturant. Wypych G (2001) Handbook of solvents. Chemtec Publishing, New York.

Kerr effect *n.* When plane polarized light is incident on the pole of an electromagnet, polished so as to act like a mirror, the plane of polarization of the reflected light is not the same when the magnet is "on" as when it is "off". It was found that the direction of rotation was opposite to that of the currents exciting the pole from which the light was reflected. Weast RC (1979) CRC handbook of chemistry and physics, 59th edn. CRC Press, Boca Raton, FL.

Kerrolic acid *n.* $C_{15}H_{27}(OH)_4COOH$. Monobasic tetrahydroxy acid constituent of shellac.

Kersey \ˈkər-zē\ [ME, fr. *Kersey*, England] (14c) *n.* A heavily milled woolen fabric having a high lustrous nap and a "grainy" face, kersey is frequently used in overcoats.

Ketohexamethylene Syn: cyclohexanone.

Ketone \ˈkē-ˌtōn\ [Gr *Keton*, alter. of *Aceton* acetone] (1851) *n.* An organic compound containing a carbonyl group (C=O) bound to two carbon atoms. The simplest one is acetone $(CH_3)_2C{=}O$. It and the other lower ketones are widely used as solvents for vinyl and cellulosic resins, and as intermediates in the production of resins. Odian GC (2004) Principles of polymerization. John Wiley and Sons Inc., New York.

Ketone-based resins *n.* Consist of ether and ketone groups combined with phenyl rings in different sequences. The rigid ketone and phenyl groups produce high thermal stability. Odian GC (2004) Principles of polymerization. John Wiley and Sons Inc., New York.

Ketone condensation resins *n.* Resins produced as the result of condensation of ketones and aldehydes, e.g., the resin obtained from methyl ethyl ketone and formaldehyde. Odian GC (2004) Principles of polymerization. John Wiley and Sons Inc., New York.

Ketones *n.* A class of strong organic solvents used in gravure inks; for example, acetone, methyl ethyl ketone (MEK). Wypych G (ed) (2001) Handbook of solvents. Chemtec Publishing, New York.

Kettle \ˈke-tᵊl\ [ME ketel, fr. ON ketill (akin to OE *cietel* kettle), both from a prehistoric Germanic word borrowed fr. L *catillus*, dimin. of *catinus* bowl] (13c) *n.* Reaction vessel for varnish or resin manufacture.

Kevlar® *n.* DuPont's trade name for poly-(*p*-phenylene terephthalamide) fibers. *See aramid.*

K-factor *n.* A term sometimes used (incorrectly) for insulation value or used for thermal conductivity.

K film *n.* A chemical wood pulp made by the sulfate process, or paper or paperboard made from such pulp.

kg SI abbreviation for kilogram. Lide DR (ed) (2004) CRC handbook of chemistry and physics. CRC Press, Boca Raton, FL.

Khaki \ˈka-kē, ˈkä-, *Canad often* ˈkär-\ [Hindi *khākī* dust-colored, fr. *khāk* dust, fr Persian] (1857) *n.* (1) A light yellowish brown. (2) A khaki-colored cloth of cotton, wool, or combinations of these fibers with manufactured fibers used primarily in military uniforms and work clothes. Vincenti R (ed) (1994) Elsevier's textile dictionary. Elsevier Science and Technology Books, New York.

K

KHN *See knoop hardness number.*

Kibbled \ˈki-bəl\ (ca. 1790) *vt.* Broken into small lumps, about 1/4in. diameter, e.g., kibbled glue.

Kick board *See baseboard.*

Kidney oil *n.* Fraction boiling between 250 and 270°C, obtained by the destructive distillation of rosin.

Kienle's functionality theory *n.* Fundamental postulate which covers, among other things, the likelihood of reactions between compounds to form products of high molecular weight; and the relation between size and shape of reacting molecules and physical properties of the reaction products. Fundamentals of polymer science: an introductory text. CRC Press, Boca Raton, FL, 1998. Kroschwitz JI (ed) (1990) Concise dictionary of polymer science and engineering. John Wiley and Sons, New York.

Kier *n.* A large metal tank, capable of being heated uniformly, used for wet processing.

Kier boiling Process of boiling cellulosic materials in alkaline liquors in a kier at or above atmospheric pressure.

Kieselguhr \ˈkē-zəl-ˌgur\ [Gr *Kieselgur*] (1875) *n.* Alternate name for diatomite. *See diatomaceous silica.*

Kiln \ˈkiln, ˈkil\ [ME *kilne*, fr. OE *cyln*, fr. L *culina* kitchen, fr. *coquere* to cook] (before 12c) *n.* An oven, furnace, or heated enclosure used for processing a substance by burning, firing, or drying.

Kiln dried *n.* Lumber dried in chambers. Heat is controlled to prevent cracking and warping. Syn: kiln seasoned.

Kiln seasoned *See kiln dried.*

kilo- \ˈkē-(ˌ)lō *also* ˈki-\ *combining form* [F, mod. of Gk *chilioi*] (k). The SI prefix meaning $\times 10^3$.

Kilogram \-ˌgram\ [F *kilogramme*, fr. *kilo* + *gramme* gram] (1797) (kg) *n.* One of the basic units of SI, the mass of a particular platinum cylinder kept at the International Bureau of Weights and Measures in Paris. One avoirdupois pound = 0.4535924kg. Lide DR (ed) (2004) CRC handbook of chemistry and physics. CRC Press, Boca Raton, FL.

Kinel Polyimide, manufactured by Rhone Poulenc, France.

Kinematics \ˌki-nə-ˈma-tiks\ [F. *cinématique*, fr. Gk *kinēmat-*, *kinēma* motion, fr. *kinein* to move] (1840) *n. plural but singular in construction.* A branch of dynamics that deals with aspects of motion apart from considerations of mass and force.

Kinematic viscosity (kinetic viscosity) *n.* The absolute (dynamic) viscosity of a fluid divided by the density of the fluid. The SI unit is m^2/s, but the cgs unit, the *stoke*, which equals $10^{-4} m^2/s$, is still in wide use, as is its submultiple, the centistoke. Goodwin JW, Goodwin J, Hughes RW (2000) Rheology for chemists. Royal Society of Chemistry, UK.

Kinetic energy (1870) *n.* Energy associated with the motion of an object. An object of mass *m* moving at velocity *v* has kinetic energy $1/2\,mv^2$. Serway RA, Faugh JS, Bennett CV (2005) College physics. Thomas, New York.

Kinetic-energy correction *n.* (1) In the Izod, Charpy, and tensile-impact tests, a subtraction of the kinetic energy imparted to the broken-off part of the specimen. (2) In measurement of dilute-solution viscosity a correction for the energy required to accelerate the liquid in the reservoir to its higher velocity in the capillary. A similar correction is theoretically needed in melt rheometry, but has so far been found to be much smaller than the errors of measurement. Kamide K, Dobashi T (2000) Physical chemistry of polymer solutions. Elsevier, New York.

Kinetics \kə-ˈne-tiks *also* kī-\ (ca. 1859) *n.* A branch of dynamics concerned with the relations between the movement of bodies and the forces acting upon them. Connors KA (1990) Chemical kinetics. John Wiley and Sons, New York.

Kinetic theory (1864) *n.* Either of two theories in physics based on the fact that the minute particles of a substance are in vigorous motion. The first theory is that the particles of a gas move in straight lines with high average velocity, continually encounter one another and thus change their individual velocities and directions, and cause pressure by their impact against the walls of a container.

Also known as the kinetic theory of gases. The second theory is that the temperature of a substance increases with an increase in either the average kinetic energy of the particles or the average potential energy of separation (as in fusion) of the particles or in both when heat is added.

Also known as the kinetic theory of heat. Connors KA (1990) Chemical kinetics. John Wiley and Sons, New York.

Kink \ˈkiŋk\ [D; akin to MLGr *kinke* kink] (1678) *n.* (1) In fabrics, a place where a short length of yarn has spontaneously doubled back on itself. (2) In yarn, *see snarl.*

Kinking *n.* The doubling back of yarn on itself to relieve torque imparted by twisting or texturing.

See kink.

Kira-ye [Japanese]. A print with a mica background.

Kirchhoff's law (emissivity). $1 = \varepsilon_R + \varepsilon_T + \varepsilon_A$, where ε_R is the energy reflected, ε_T the energy transmitted through the material, and ε_A is the energy absorbed by the material, which is reemitted. The ability to reemit the energy is the emissivity. The common emissivity (ε) of a material is equal to $1 - \varepsilon_R$. This is an indicator of the property of a material to absorb radiation such as white light or infrared radiation and a perfect absorber of energy is a black body, where $\varepsilon = 1$. Driggers RC, Cox P, Edwards T (1998) Introduction to infrared and electro-optical systems. Artech House Inc., Aylesford, UK. Klocek P (ed) (1991) Handbook of infrared materials. Marcel Dekker, New York.

Kirksite An alloy of aluminum and zinc, easily castable at relatively low temperatures, often used for molds for blow molding. Its high thermal conductivity hastens cooling.

Kling test A method for determining the degree of fusion between flexible vinyl sheets, coated fabrics, and thin sections of cast or molded parts, by immersing the folded specimen in a solvent and observing the elapsed time at which disintegration commences. Useful solvent systems comprise methyl ethyl ketone, tetrahydrofuran, ethyl acetate, and carbon tetrachloride. The preferred solvent system is one that will initiate degradation within 5–10 min in a fully fused specimen. Wickson EJ (ed)

(1993) Handbook of polyvinyl chloride formulating. John Wiley and Sons Inc., New York.

Kneader \ˈnēd-\ [ME *kneden*, fr. OE *cnedan*; akin to OHGr *knetan* to knead] (before 12c) *n.* A mixer with a pair of intermeshing blades, often S-shaped, used for working plastic masses of semidry or rubbery consistency. Perry RH, Green DW (1997) Perry's chemical engineer's handbook, 7th edn. McGraw-Hill, New York.

Knee break-out test *n.* A method to evaluate the performance of fabrics, especially boys' wear, when subjected to abrasion, stretch, and impact forces under conditions which simulate ordinary wear at the knee. Tortora PG (ed) (1997) Fairchild's dictionary of textiles. Fairchild Books, New York.

K

Kneeing *n.* Abnormal behavior of a spinning threadline (especially in melt spinning) in which one or more filaments form an angle (knee).

Knife coating *n.* A method of coating a substrate in which the substrate, in the form of a continuous moving web, is coated with a plastic whose thickness is controlled by an adjustable knife or bar set at a suitable angle to the substrate.
See spread coating and air-knife coating.

Knife mark *See doctor mark.*

Knife test *n.* Test for brittleness, toughness, and tendency to ribbon, by cutting a narrow strip of the coating from the test panel, with a knife. It is not recommended as a test for adhesion, as other methods are more accurate. Weldon DG (2001) Failure analysis of paints and coatings. John Wiley and Sons, New York. Paint and coating testing manual (Gardner–Sward handbook) MNL 17, 14th edn. ASTM, Conshohocken, PA, 1995.

Knifing filler *n.* Filling composition suitable for application with a filling knife as distinct from one made for brush application.

Knit-de-knit *See texturing.*

Knit fabric *n.* A structure produced by interlooping one or more ends of yarn or comparable material.
Also see knitting.

Knit-miss *n.* A form of tricot knitting in which yarns on each bar of a two-bar machine are knit at alternate courses only. This type of knitting permits the use of heavy-denier yarns without creating undesirable bulkiness in the fabric. Solomon DH, Hawthorne DG (1991) Chemistry of pigments and fillers. Krieger Publishing Co., New York.

Knitting *n.* A type of knitting in which the yarns run lengthwise in the fabric. The yarns are prepared as warps on beams with one or more yarns for each needle. Examples of this type of knitting are tricot, milanese, and raschel knitting. (a) *Milanese Knitting*: A type of run-resistant warp knitting with a diagonal rib effect using several sets of yarns. (b) *Raschel knitting*: A versatile type of warp knitting made in plain and Jacquard patterns; the latter can be made with intricate eyelet and lacy patterns and is often used for underwear fabrics. Raschel fabrics are coarser than other warp-knit fabrics, but a wide range of fabrics can be made. Raschel knitting machines have one or two sets of latch needles and up to 30 sets of guides. (c) *Tricot knitting*: A run-resistant type of warp knitting in which either single or double sets of yarn are used. *Also see tricot.* (d) *Weft knitting*: A common type of knitting, in which one continuous thread runs crosswise in the fabric making all of the loops in one course. Weft knitting types are circular and flat knitting. (e) *Circular knitting*: The fabric is produced on the knitting machine in the form of a tube, the threads running continuously around the fabric. (f) *Flat knitting*: The fabric is produced on the knitting machine in flat form, the threads alternating back and forth a cross the fabric. The fabric can be given shape in the knitting process by increasing or decreasing loops. Full-fashioned garments are made on a flat knitting machine. *Also see flat-knit fabric* {*knitted fabric* G Gewirke n, F tissu tricoté, tissu m, S género de punto, género m, I tessuto a maglia, tessuto m}. Kadolph SJJ, Langford AL (2001) Textiles. Pearson Education. New York. Tortora PG (ed) (1997) Fairchild's dictionary of textiles. Fairchild Books, New York. Vigo TL (1994) Textile processing, dyeing, finishing and performance. Elsevier Science, New York. Vincenti R (ed) (1994) Elsevier's textile dictionary. Elsevier Science and Technology Books, New York. Tortora PG (ed) (1997) Fairchild's dictionary of textiles. Fairchild Books, New York.

Knockout \ˈnäk-ˌau̇t\ (1887) *n.* Any part or mechanism of a mold whose function is to eject the molded article. Strong AB (2000) Plastics materials and processing. Prentice-Hall, Columbus, OH.

Knoop hardness number, KHN *n.* The indentation hardness determined with a Knoop indenter, and calculated as follows:

$$\mathrm{KHN} = \frac{L}{A_p} = \frac{L}{l^2 C_p},$$

where L is the load in kilogram applied to the indenter, l^2 the measured length of long diagonal of the indention in millimeters, C_p the indenter constant relating l^2 to A_p, and A_p is the projected area of indention in mm^2. www.astm.org. Brown R (1999) Handbook of physical polymer testing, vol 50. Marcel Dekker, New York.

Knoop hardness test *n.* An indentation hardness test using calibrated machines to force a rhombic-based pyramidal diamond indenter having specific edge angles, under specified conditions, into the surface of the material under test and to measure the long diagonal after removal of the load. www.astm.org. Brown R (1999) Handbook of physical polymer testing, vol 50. Marcel Dekker, New York.

Knoop indenter *n.* Pyramidal diamond of prescribed dimensions used for testing the indentation hardness of organic coatings. In a more restricted sense, a type of diamond hardness indenter having edge angles of 172°, 30°, and 130°. www.astm.org. Brown R (1999) Handbook of physical polymer testing, vol 50. Marcel Dekker, New York.

Knoop microhardness *n.* A test employing a diamond indenter whose point is an obtuse pyramid that makes an indentation of length seven times its width. This tester makes smaller, shallower indentations than Brinell or Vickers, suiting it to testing hardnesses of surfaces, as in case-hardened steel, or coatings. The Knoop hardness number (KHN) is equal to $14.2P/L^2$, where P is the applied load in grams and L is the length in millimeters of the long axis of the indentation. www.astm.org. Brown R (1999) Handbook of physical polymer testing, vol 50. Marcel Dekker, New York.

Knot sealer *n.* Solutions of various resins in alcohol, used to seal knots in new wood.

Knot tenacity (knot strength) *n.* The strength of a yarn specimen containing an overhand knot to measure, by comparison with the strength of the unknotted yarn, its sensitivity to compression or shearing. Kadolph SJJ, Langford AL (2001) Textiles. Pearson Education, New York.

Knotting Fairly concentrated solution of shellac in alcohol, used for sealing knots in new wood. Paint/coatings dictionary. Federation of Societies for Coatings

Technology, Philadelphia, Blue Bell, PA, 1978.

Knuckle area *n.* In reinforced plastics, the area of transition between sections of different geometry in a filament-wound part. Iaac MD, Ishal O (2005) Engineering mechanics of composite materials. Oxford University Press, UK.

Kodar Polyester, manufactured by Eastman, USA.

Kodel-2 Polyester from terephthalic acid and 1,4-dimethylol cychlohexane, manufactured by Eastman, USA.

Kodel-10 Poly(ethylene terephthalate), manufactured by Eastman, USA.

Koettstorfer number *See saponification value.*

Kohinoor test *n.* A test for scratch hardness employing a series of pencils of different hardnesses.

K

Kohlrausch–Williams–Watts equation *n.* This empirical equation aids designers of load-bearing products made of plastics and reinforced plastics.

$$G(t) = G_0 e^{-(t/t_0)^m},$$

where $G(t)$ is the stress-relaxed modulus of the test piece at time t from application of the load, G_0 the modulus at the reference time t_0, usually in the range of a set of measurements on the logarithmic time scale, and m is the material-specific constant between 0.33 and 0.5 for many polymers and composites. Brown R (1999) Handbook of physical polymer testing, vol 50. Marcel Dekker, New York. Shah V (1998) Handbook of plastics testing technology. John Wiley and Sons, New York.

Kordofan gum \ˌkór-də-ˈfan-\ *n.* Another name for gum acacia, derived from the Sudanese providence of Kordofan. Whistler JN, BeMiller JN (ed) (1992) Industrial gums: polysaccharides and their derivatives. Elsevier Science and Technology Books, New York.

Kraemer equation For dilute polymer solutions, an equation relating the inherent viscosity to intrinsic viscosity and concentration. It is

$$\eta_{inh} = (\ln \eta_r)/c = [\eta] + k''[\eta]^2 c,$$

where η_r is the reduced viscosity, $[\eta]$ the intrinsic viscosity, and c the concentration in g/dl.

See also Huggins equation and dilute-solution viscosity. Huggins' constant, k', and Kraemer's, k, are related by: $k' - k'' = 0.5$. Thus, since k' is often between 0.6 and 0.8, k'' will often lie between 0.1 and 0.3. Kamide K, Dobashi T (2000) Physical chemistry of polymer solutions. Elsevier, New York. Huggins ML (1958) Physical chemistry of high polymers. John Wiley and Sons Inc., New York.

Kraemer–Sarnow method (K&S) *n.* Softening point. Method of determining the softening points of resins and aliphatic materials. www.astm.org. Paint: pigment, drying oils, polymers, resins, naval stores, cellulosics esters, and ink vehicles, vol 3. American Society for Testing and Material, Conshohocken, PA, 2001.

Kraftcord This yarn produced by tightly twisting plant fiber is sometimes used in carpet backings. Kadolph SJJ, Langford AL (2001) Textiles. Pearson Education, New York. Vincenti R (ed) (1994) Elsevier's textile dictionary. Elsevier Science and Technology Books, New York.

Kraft pulps *n.* Pulps prepared in the alkaline liquor consisting of sodium hydroxide, sodium carbonate, and sodium sulfide. *Also called sulfate pulp.*

Kraft yarn *n.* A yarn made by twisting a strip of paper manufactured from kraft pulp.

Kralastic *n.* ABS, manufactured by USA Rubber, USA.

Krebs–Stormer viscometer *n.* This is the most commonly used viscometer in the paint industry. Paint consistency is measured by its resistance to stirring by a two-vane paddle. The two vanes are offset to avoid channeling in viscous paints. Approximate weights are added to a platform on a string attached to a pulley, which is connected to the paddle until a speed of 100 rpm is reached, generally measured by a stroboscopic timer. The grams required to reach this speed are then converted, by using a conversion table, to Krebs units. Paint and coating testing manual (Gardner–Sward handbook) MNL 17, 14th edn. ASTM, Conshohocken, PA, 1995.
See viscometer.

Krebs unit *n.* A unit used in reporting viscosity measurements made with the weight-driven stormer viscometer used primarily for evaluating the viscosity of formulated paint and coating materials. A Krebs unit is the weight in grams that will turn a paddle-type rotor, submerged in the sample, 100 rev in 30 s.

Kroy shrinkproofing process® *n.* Continuous process for shrinkproofing wool tops in which there is a direct chlorination step with no intervening chemical reaction followed by anti-chlorination and neutralization. Provides better hand and strength than conventional shrinkproofing. Vigo TL (1994) Textile processing, dyeing, finishing and performance. Elsevier Science, New York.

Kryston *n.* Polyester, manufactured by Goodrich, USA.

Krytox *n.* Perfluorinated polyether, manufactured by DuPont, USA.

KU *See Krebs units.*

Kubelka–Munk equation (apply to a single wavelength of light) *n.*

$$R = \frac{1 - R_g(a - b\coth bSX)}{A - R_g + b\coth bSX}$$

coth = hyperboliccontangent; abbreviation $(e^u + 1/e^u)/(e^u - 1/e^u)$ (1)

$$SX = \frac{1}{b}\left(\operatorname{arc\,coth}\frac{a-R}{b} - \operatorname{arc\,coth}\frac{a-R_g}{b}\right),$$

arc coth = inverse hyperboliccontangent, (2)

where R is the reflectance over substrate of reflectance R_g, R_g the reflectance of substrate,

$$a = \frac{S+K}{S} = \frac{1}{2}\left(\frac{1}{R_\infty} + R_\infty\right),$$

$$b = (a^2 - 1)^{1/2} = \frac{1}{2}\left(\frac{1}{R_\infty} + R_\infty\right),$$

S the unit scattering coefficient expressed in $X - 1$, X the thickness (or product of thickness times concentration), R_∞ the reflectance at infinite thickness (complete hiding), and K is the unit absorption coefficient expressed in $X-1$.
When hiding in complete (substrate is completely obscured), the first equation reduces to

$$\frac{(1-R_\infty)^2}{2R_\infty} = \frac{K}{S}$$

and for a mixture of pigments

$$\left(\frac{K}{S}\right)_{\text{MIX}} = \frac{C_1K_1 + C_2K_2 + C_3K_3 + \cdots}{C_1S_1 + C_2S_2 + C_3S_3 + \cdots}, \quad (3)$$

where the C's refer to relative concentrations of pigments in the mixture (Ec's = 100%) and subscripts identify the pigments. When all of the scatterings come from a single component (such as white pigment in a pasted color or the fiber in a textile), the denominator of the above equation is determined by the S of this component, and the equation becomes simply

$$\left(\frac{K}{S}\right)_{\text{MIX}} = C_1\left(\frac{K}{S}\right)_1 + C_2\left(\frac{K}{S}\right)_2 + C_3\left(\frac{K}{S}\right)_3 + \cdots \quad (4)$$

The latter is referred to as the single constant K–M equation compared to the third one using separate K's and S's, which is called the two-constant equation. The above equations are those most widely used for computer color matching. McDonald R (1997) Colour physics for industry, 2nd edn. Society of Dyers and Colourists, West Yorkshire, England. Paint/coatings dictionary. Federation of Societies for Coatings Technology, Philadelphia, Blue Bell, PA, 1978.

K

Kubelka–Munk theory *n.* A theory describing the optical behavior of materials containing small particles, which scatter and absorb radiant energy. It is widely used for color matching calculations. The mathematical equation describes the reflectance or transmittance in terms of an absorption coefficient, K, and a scattering coefficient, S. The K–M theory is based on the assumption of multiple scattering, that is, reflectances from one particle to other particles before the reflected radiant energy is observed. This assumed behavior is in contrast to the Mie theory, which is based on an assumption of single isolated scattering of individual particles. McDonald R (1997) Colour physics for industry, 2nd edn. Society of Dyers and Colourists, West Yorkshire, England.

Kusters dyeing range *n.* Continuous dye range for carpets. The unit wets the carpet, applies dyes and auxiliary chemicals by means of a doctor blade, fixes the dyes in a festoon steamer, and washes and dries the carpet in one pass through the range. An optional auxiliary unit may be installed to randomly drip selected dyes onto the background shade for special styling effects. This process is called TAK dyeing. Vigo TL (1994) Textile processing, dyeing, finishing and performance. Elsevier Science, New York.

***K*-value** *n.* An alternate name for thermal conductivity.

Kynol *n.* Phenol-formaldehyde fiber, manufactured by Carborundium, USA.

L

l \ˈel\ {*often capitalized, often attributive*} (before 12c) (also L). Symbol for length.

L (1) Abbreviation for SI-permitted (but discouraged) volume unit, the liter, redefined in 1964 to be exactly 1 dm^3. (2) Symbol for length or magnetic inductance. (3) Abbreviation for lightness, generally used in color order systems to describe the perception of the amount of light reflected or transmitted by materials.
See lightness and luminance.

Labeling \ˈlā-b(ə-)liŋ\ (1601) *vt.* To affix a precut, printed, flexible material to the surface of a product.

Labradorite \ˈla-brə-ˌdór-ˌīt\ [*Labrador* Peninsula, Canada] (1814) *n.* Calcium sodium aluminum silicate, hardness 6 Mohs scale, used as a wear resistant filler in deck paints. Herbst W, Hunger K (2004) Industrial organic pigments. John Wiley and Sons, New York.

Lac \ˈlak\ [Persian *lak* & Hindi *lākh*] (1598) *n.* Resin secreted by the female of the insect, *Lacccifer iacca*, and deposited on the twigs of various species of trees in Indian and Indochina. After refining, the resin is known as shellac. Langenheim JH (2003) Plant resins: chemistry, evolution ecology and ethnobotany. Timber Press, Portland, OR.

Lace \ˈlās\ [ME, fr. MF *laz*, fr. L *laqueus* snare] (14c) *n.* Ornamental openwork fabric, made in a variety of designs by intricate manipulation of the fiber by machine or by hand. Tortora PG (ed) (1997) Fairchild's dictionary of textiles. Fairchild Books, New York.

Lace stitch In this knitting stitch structure, loops are transferred from the needles on which they are made to adjacent needles to create a fabric with an open or a raised effect. Tortora PG (ed) (1997) Fairchild's dictionary of textiles. Fairchild Books, New York.

L acid *See cleve's acid.*

Lac dye Red dye obtained by the maceration of lac.

Lacquer \ˈlac-kər\ [Portuguese lacré sealing wax, fr. *laca* lac, fr. Arabic *lakk*, fr. Persian *lak*] (1592) *n.* (1) A solution of a film-forming natural or synthetic resin in a volatile solvent, with or without color pigment, which when applied to a surface forms an adherent film that hardens solely by evaporation of the solvent. The dried film has the properties of the resin used in making the lacquer. The word derives from the *lac* insect, which secreted the resinous substance from which *shellac* solutions were (and stilla are) made. Today most lacquers are made with cellulosic, alkyd, acrylic, and vinyl resins. (2) Finish on Chinese and Japanese lacquer ware. Wicks ZN, Jones FN, Pappas SP (1999) Organic coatings science and technology, 2nd edn. Wiley-Interscience, New York.
See spirit varnish.

Lactam [ISV *lact-* + *am*ide] *n.* A cyclic amide obtained by removing one molecule of water from an amino acid. An example is caprolactam.

Lactic acid (1790) (milk acid, α-hydroxypropionic acid) *n.* CH_3CHOH COOH. A colorless or yellowish liquid with several applications in plastics. Reacted with glycerine, it forms an alkyd resin. It is a catalyst for vinyl polymerizations, and an additive for phenolic casting resins. It has a bp of 122°C/15 mmHg, mp of 18°C, and Sp gr of 1.2
Also known as alpha-hydroxypropionic acid and milk acid.

Lac wax Wax obtained from lac consisting of myricyl and ceryl alcohols, free and combined with various fatty acids.

Ladder polymers (double-stranded polymer). A polymer comprising chains made up of fused rings. Examples are cyclized (acid-treated) rubber and polyimidazopyrrolone.

Laid-in fabric A knit fabric in which an effect yarn is tucked in, not knitted into, the fabric structure. The laid-in yarns are held in position by the knitted yarns.
See axial yarn.

Laitance \ˈlā-tᵊn(t)s\ [F, fr. *lait* milk, fr. L *lact-*, *lac*] (ca. 1902) *n.* Milky white deposit on new concrete; efflorescence.

Lake A type of organic pigment prepared from water-soluble acid dyes, precipitated on an inert substrate by means of a metallic salt, tannin, or other reagents. Lakes were used in plastics at one time, but have been replaced by more permanent pigments.

Lake \ˈlāk\ [F laque lac, fr. OP *laca*, fr. Arabic *lakk*] (1598) *n.* Special type of pigment consisting essentially of an organic soluble coloring mater combined more or less definitely with an inorganic base or carrier. It is characterized generally by a bright color and more or less pronounced translucency when made into an oil paint. Under this term are included two (or perhaps three) types of pigments: (a) The older original type composed of hydrate of alumina dyed with a solution of the natural organic color. (b) The more modern and far more extensive type made by precipitating from solution various coal-tar colors by means of a metallic salt, tannin, or other suitable reagents, upon a base or carrier either previously prepared or coincidently formed. (c) A number combining both types in varying degree might be regarded as a third class.

Lake Bordeaux B Red dyestuff produced from 2,3-hydroxy naphthoic acid and 2-naphthylamine-1-sulfonic acid. The calcium and manganese toners are known as Bordeaux R Toner and Maroon Toner B.B., respectively. The latter has good lightfastness.

Lake dyes Dyes used for the making of lakes by combination with, or adsorption on, salts of calcium, barium, chromium, aluminum, phosphotungstic acid, or phosphomolbybdic acid.

Lake of acid yellow 1 Acid Yellow 1, Lake (10316). A nitro dye prepared by precipitation from an aqueous solution by a metal salt, i.e., aluminum or barium chloride.

Lake red C Pigment Red 53 (15585). Red dyestuff derived from 6-chlor-3-toluidine-4-sulfonic acid, and β-naphthol. When laked with barium chloride, bronze scarlet is obtained.

Lake red P Pigment produced by coupling diazotized *p*-nitroaniline-*o*-sulfonic acid with betanaphthol.

Laketine A transparent ink used for extending letterpress or lithographic inks.

Lallemantia oil Drying oil obtained from the seeds of *Lallemantia iberica*, found in parts of Asia and Europe. It has Sp gr of 0.934/20°C, iodine value of 190, and saponification value of 190. Its main constituents are linolenic and linolenic glycerides.

LALLS *See low-angle laser-light scattering.*

Lambda zero (λ_0) In dispersion staining, the wavelength at which both particle and liquid have the same refractive index.

Lambert \ˈlam-bərt\ [Johann H. *Lambert* † 1777 German physicist & philosopher] (1915) *n.* A deprecated unit of illumination equal to 1lm/cm^2. The SI unit is the *lux* (lx), equal to 1lm/m^2, i.e., 10^{-4} lambert.
See also luminous flux.

Lambert's law of absorption *See Bouguer's law.*

Lambert's law of reflection The flux reflected per unit solid angle is proportional to the cosine of the angle measured from the normal (perpendicular) to the surface. If the reflected flux is isotropic, the surface is said to be a perfect Lambertian reflector or a perfect diffuser.

Lamé \lä-ˈmā, la-\ [F] (1922) *n.* A fabric woven with flat metal threads, usually silver or gold, that form either the background or the pattern.

Lamellae \lə-ˈme-lə\ [NL, fr. L, dim. of *lamina* thin plate] (1678) *n.* Thin, flat scales or plates.

Lamellae structure Plate-like single crystals that exist in most crystalline polymers. A thin, flat scale or part.

Lamina \ˈla-mə-nə\ [L] (ca. 1656) *n.* A single layer or ply within a laminate.

Laminar flow Flow without turbulence, i.e., the movement of one layer of fluid past another layer with no eddying between them. Most melt flow, even at high velocities, is laminar. With sudden changes in melt velocity, such as may occur at some extrusion-die entries and at injection-mold gates, laminar flow may be disrupted by melt fracture.

See Reynolds number. Also see jetting.

Laminar scale Rust formation in heavy layers.

Laminate \ˈla-mə-ˌnāt\ (1665) (1, *n*). A product made by bonding together two or more layers of material or materials. The term most usually applies to preformed layers joined by adhesives or by heat and pressure. However, some authors apply the term to composites of plastic films, with other films, foils, and papers, even though they have been made by spread coating or by extrusion coating. In the reinforced-plastics industry, the term refers mainly to superimposed layers of resin-impregnated or resin-coated fabrics or fibrous reinforcements that have been bonded together, usually by heat and pressure, to form a single piece. Wallenberger FT, Weston NE (eds)(2003) Natural fibers, plastics and composites. Springer-Verlag, New York. Pittance JC (ed) (1990) Engineering plastics and composites. SAM International, Materials Park, OH. Skeist I (ed) (1990) Handbook of adhesives. Van Nostrand Reinhold, New York.

Laminate, high pressure Laminates molded and cured at pressures not lower than 1000 psi, and commonly in the range of 1200–200 psi.

Laminated fabric Fabric composed of a high-strength reinforcing scrim or base fabric between two plies of flexible thermoplastic film. Usually open scrims are used to permit the polymer to flow through the interstices and bond during calendering.

Laminated glass A structure consisting of two or more parallel sheets or shells of glass interleaved with, and bonded to, layers of tough, sticky plastic, typically polyvinyl butyral or polycarbonate. The former resin is used in the "safety-glass" windshields of USA-made cars. The glass will splinter under a heavy blow, but it is very resistant to penetration and the shards stick to the interlayer.

Laminated, cross A laminate in which some of the layers of materials are oriented at right angles to the remaining layers with respect to the grain or strongest direction in tension. *Note* – Balanced construction of the laminations about the center line of the thickness of the laminate is normally assumed.

See also laminated, parallel.

Laminated, parallel A laminate in which all the layers of materials are oriented approximately parallel with respect to the grain or strongest direction in tension.

See also laminated, cross.

Laminates Products made by bonding together two or more layers of material and materials.

Lamination (ca. 1676) *n.* The process of preparing a laminate. Also, any layer in a laminate.

Lampblack \-ˈblak\ (1598) *n.* Pigment Black 6 (77266). A bulky black soot obtained from the incomplete combustion of creosote or fuel oils, of duller and less intense blackness than channel black and other carbon blacks, and having a blue undertone and a small oil content. Syn: Paris black.

Land (1) The horizontal bearing surface of a semipositive or flash mold by which excess material escapes. (2) The bearing surface along the top of the flights of an extruder screw. (3) The final shaping surface of an extrusion die, usually parallel to the direction of melt flow. (4, *pl*) The mating surfaces of any mold, adjacent to the cavity depressions that, when in contact, prevent the escape of material.

Land area The area of those surfaces of a mold that contact each other when the mold is closed, measured in a plane perpendicular to the direction of application of the closing pressure.

Land width (flight width). Of an extruder screw, the distance across the tip of the flight, perpendicular to the flight faces.

Lane length In an extrusion die, the distance across the land in the direction of melt flow between the lands.

Lanital Fiber from milk albumin. Manufactured by Snia Viscosa, Italy.

Lanolin \ˈla-nᵊl-ən\ [L *lana* wood + ISV ³*-ol* + ¹*-in*] (1885) *n.* Purified wool grease. *See adepts lanae and anhydrous.*

Landed force A force with a shoulder that seats on the land in a landed positive mold.

Lap (1800) *n.* (1) Region where a coat extends over an adjacent fresh coat. The object of the painter is usually to effect a joint between the two coats without showing the lap. (2, *v*) To place one coat of finishing material alongside another, partly extending over it, causing increased thickness where the two coats are present. (3) To overlap or partly cover one surface with another, as in shingling. (4) The length of the overlap, as the distance one tile extends over another.

Lap joint (1823) *n.* A joint made by placing one surface to be joined partly over another surface and bonding or fastening the overlapping portions. Compare butt joint and scarf joint.
See joint, lap.

Lap siding *See clapboard.*

Lap winding A variant of filament winding consisting of convolutedly winding a resin-impregnated tape onto a mandrel of the desired configuration. The process has been used for making large chemical- and heat-resistant, conical or hemispherical parts such as heat shields for atmospheric-reentry vehicles.

Lapis lazuli \ˌlap-əs-ˈla-zə-lē, -ˈla-zhə-\ [ME, fr. ML, fr. L *lapis* + ML *lazuli*, general of *lazuhum* lapis lazuli, fr. Arabic *lāzaward*] (15c) *n.* A rich blue semiprecious stone; either used decoratively or ground and powdered for use as an ultramarine pigment. Kirk–Othmer encyclopedia of chemical technology: pigments–powders. John Wiley and Sons, New York, 1996.

Lapis lazuli blue Natural ultramarine blue. Kirk–Othmer encyclopedia of chemical technology: pigments–powders. John Wiley and Sons, New York, 1996.

Lapping A term describing the movement of yarn guides between needles, at right angles to the needle bar, or laterally in relation to the needle bar, or laterally in relation to the needle bar during warp knitting.

Lapping time *See wet-edge time.*

Lase \ˈlāz\. An acronym for load at specified elongation: the load required to produce a given elongation of a yarn or cord.

Laser \ˈlā-zər\ [*l*ight *a*mplification by *s*timulated *e*mission of *r*adiation] (1960) *n.* Laser is an acronym coined from the bold-face letters in light application by stimulated emission of radiation. Early lasers were made of synthetic-ruby rod, silvered at one end face, semisilvered at the other, and surrounded by a toroidal flash lamp. Today, the (neodymium–yttrium–aluminum–garnet (NdYAG) laser, with higher productivity than the ruby laser, has taken over many of its jobs. Lasers have been useful in drilling, perforating, cutting, and welding operations with plastics and other materials. Medium-power CO_2 lasers are preferred for machining plastics because they produce light at a wavelength of 10.6 mm, which is completely absorbed by plastics. Lasers have been useful in the analysis of polymers by laser mass spectrometry, in initiating polymerizations, and in curing polymers. Giambattista A, Richardson R, Richardson RC, Richardson B (2003) College physics. McGraw-Hill Science/Engineering/Math, New York. James F (ed) (1993) Whittington's dictionary of plastics. Technomic Publishing Co. Inc., Carley. Saleh BEA, Teich MC (1991) Fundamentals of photonics. John Wiley and Sons, New York.

Laser-ionization mass spectrometry A technique for chemical analysis of transparent and opaque plastics, capable of detecting all the elements but hydrogen and helium.

Lashed-in filling *See pulled-in filling.*

Lastrile Fiber from co-polymers with 10–15% acrylonitrile and an aliphatic diene. Lastrile is a generic name.

Lastril fiber A manufactured fiber in which the fiber-forming substance is a co-polymer of acrylonitrile and a diene composed of at least 10% by weight, but not more than 50% by weight, of acrylonitrile [–CH_2–CH (CN)–] units (FTC definition).

Latch needle One of the two types of knitting machine needles. The latch needle has a small terminal hook with a latch that pivots automatically in knitting to close the hook. The fabric loop is cast off. The latch then opens, allowing a new loop to be formed by the hook, and loop-forming and casting-off proceed simultaneously. *Also see spring needle.*

Latch plate A plate used for retaining a removable mold core of relatively large diameter, or for holding insert-carrying pins on the upper part of a mold. Release of the pins or core is effected by moving the latch plate.

Latent crimp Crimp in fibers that can be developed by a specific treatment. Fibers are prepared specially to crimp when subjected to specific conditions, e.g., tumbling in a heated chamber or wet processing.

Latent heat (ca. 1757) *n.* The quantity of heat necessary to change 1 g of liquid to vapor (latent heat of vaporization) without change of temperature; heat necessary for change of state being negative or positive heat (e.g., fusion, evaporation, and melting). For example, the latent heat of fusion of ice water is 80 cal/g. Ready RG (1996) Thermodynamics. Plenum Publishing Co., New York.

Latent heat of fusion *See heat of fusion.* Ready RG (1996) Thermodynamics. Plenum Publishing Co., New York.

Latent heat of vaporization The quantity of heat required to change a unit mass (or sometimes a mole) of a liquid to its vapor, the two phases remaining in equilibrium at constant temperature, most commonly the *normal boiling point*, i.e., the temperature of boiling at a pressure of 101.325 kPa. The convenient SI units are J/g, kJ/kg, or kJ/mol. Because high polymers decompose without boiling, their heats of vaporization cannot be directly measured.

Latent solvent An organic liquid that has little or no solvent effect on a particular resin until it is activated by either heat or a mixture with a true solvent.

Latex \ˈlā-ˌteks\ [NS *latic-*. *latex*, from L. fluid] (1835) (*pl* lattices or latexes) *n.* (1) An emulsion of a polymeric substance in an aqueous medium. (2) The sap of the *hevea* (rubber) tree and other plants, or emulsions prepared from the same. Latices of interest to the coatings and plastics industry are based mainly on styrene-butadiene co-polymers, polystyrene, acrylics, and vinyl polymers and co-polymers. (3) Fine dispersion of rubber or resin, natural or synthetic, in water; the synthetic is made by emulsion polymerization. Latex and emulsion are often used synonymously in the paint industry. Emulsified monomers once polymerized become solids or plasticized gel particles and not emulsions but aqueous suspensions. Lovell PA, El-Aasser MS (eds) (1997) Emulsion polymerization and emulsion polymers. John Wiley and Sons, New York. Martens CR (1964) Emulsion and water-soluble paints and coatings. Reinhold Publishing Co., New York. Vanderhoff JW, Gurnee EF (1956) Motion picture investigation of polymer latex phenomena. TAPPI 39 (2):71-77. Vanderhoff JW, Tarkowski HL, Jenkins MC, Bradford EG (1966) Theoretical considerations of the interfacial forces involved in the coalescence of latex particles. J Macromol Chem 1 (2):361-397.

See also hydrosol

Latex mechanical stability The ability of latex to resist coagulation under influence of mechanical agitation.

Latex paint A paint containing a stable aqueous dispersion of synthetic resin, produced by emulsion polymerization, as the principal constituent of the binder. Modifying resins may also be present. Lovell PA, El-Aasser MS (eds)(1997) Emulsion polymerization and emulsion polymers. John Wiley and Sons, New York. Martens CR (1964) Emulsion and water-soluble paints and coatings. Reinhold Publishing Co., New York.

Lath \ˈlath *also* ˈlath\ [ME, fr. (ass.) OE *lœthth-*; akin to OHGr *latta* lath, Welsh *llath* yard] (13c) *n.* One of a number of thin narrow strips of wood nailed to rafters, ceiling joists, wall studs, etc., to make a groundwork or key for slates, tiles or plastering. Harris CM (2005) Dictionary of architecture and construction. McGraw-Hill, New York.

Lattice energy The energy necessary to separate 1 mol of a crystalline solid into a gaseous collection of the units at the lattice points (*exception*: in the case of a metallic solid a gaseous collection of *atoms* is formed). Rhodes G (1999) Crystallography made crystal clear: a guide for users of macromolecular models. Elsevier Science and Technology Books, New York.

Lattice pattern In filament winding, a pattern with a fixed arrangement of open voids producing a basket-weave effect. Rhodes G (1999) Crystallography made crystal clear: a guide for users of macromolecular models. Elsevier Science and Technology Books, New York.

Laundry blue Synthetic ultramarine blue.

See ultramarine blue.

Lauric acid \ˈlór-ik-, ˈlär-\ [ISV, fr. L *laurus*] (1873) *n.* $CH_3(CH_2)_{10}COOH$. Fatty acid occurring in many vegetable fats as the glyceride, especially in cocoanut oil and laurel oil. It has a Sp gr of 0.833, mp of 44°C, and refractive index of 1.4323.

Also called dodecanoic acid.

Lauroyl peroxide $[CH_3(CH_2)_{10}C(O){-}]_2$. A peroxide used as an initiator in free-radical polymerizations of styrene, vinyl chloride, and acrylic monomers.

Lauryl alcohol \ ˈlór-əl-, ˈlär-\ (1922) *n.* $C_{12}H_{26}O$ A liquid mixture of this and other alcohols used in making detergents.

Lauryl methacrylate $H_2C{=}C(CH_3)COO(CH_2)_{11}CH_3$. A monomer used in the production of acrylic resins.

Lawn \ˈlón\ [ME, fr. *Laon*, France] (15c) *n.* A light, thin cloth made of carded or combed yarns, this fabric is given a crease resistant, crisp finish. Lawn is crisper than voile but not as crisp as organdy.

Law of constancy of interfacial angles In all crystals of the same substance (in the absence of polymorphism), angles between corresponding faces are identical. Rhodes G (1999) Crystallography made crystal clear: a guide for users of macromolecular models. Elsevier Science and Technology Books, New York.

Law of mixtures (rule of mixtures). Properties of binary mixtures lie between the corresponding properties of the pure components and are proportional to the volume fractions v (sometimes weight or mole fractions) of the components. For example, the density ρ of a blend is given by:

$$\rho_m = \rho_1 v_1 + \rho_2 v_2 = \rho_2 + (\rho_1 - \rho_2).$$

The "law" works well for properties of unidirectional composites, such as modulus μ, but often fails for melt viscosities of blends, where maxima and minima above or below the viscosities of the neat resins are common. Another rule suggested by Arrhenius for mixture viscosities has the form:

$$\log \mu m = x_i \log \mu_1 + x_2 \log \mu_2,$$

where the x_i are the mole fractions. While this rule has a different form than the preceding one and fits the mixture data of some systems, it, too, cannot provide maxima or minima.Pittance JC (ed) (1990) Engineering plastics and composites. SAM International, Materials Park, OH.

Law of rational indices The lengths of intercepts of different crystal faces on any crystallographic axis are in ratios of small integers. Rhodes G (1999) Crystallography made crystal clear: a guide for users of macromolecular models. Elsevier Science and Technology Books, New York.

Lay \ˈlā\ (1590) *n.* (1) The length of twist produced by stranding singly or in groups, such as fibers or rovings; or the angle that such filaments make with the axis of the strand during a stranding operation. The length of twist of a filament is usually measured as the distance parallel to the axis of the strand between corresponding points on successive turns of the filament. (2) The term is also used in the packaging of glass fibers for the spacing of the roving bands in the package expressed as the number of bands per inch.

Lay-flat film Film that has been extruded as a wide, thin-walled, circular tube, usually blown, cooled, then gathered by converging sets of rollers and wound up in flattened form.

Lay-flat width In blown-film manufacture, half the circumference of the inflated film tube.

Laying off Final light strokes of the brush during a painting operation.
See feathering.

Lay-up The tailoring and placing of reinforcing material – mat or cloth – in a mold prior to impregnating it with resin. The reinforcement is usually cut and fitted to the mold contours.
See also sheet-molding compound.

Lay-up molding (hand-lay-up molding). A method of forming reinforced plastics articles comprising the steps of placing a web of the reinforcement, which may or may not be preimpregnated with a resin, in a mold or over a form and applying fluid resin to impregnate and/or coat the reinforcement, followed by curing of the resin and extraction of the cured article from the mold. When little or no pressure is used in the curing process, it is sometimes called *contact-pressure molding*. When pressure is applied during curing, the process is often named for the method of applying pressure, e.g., *vacuum-bag molding* or *autoclave molding*. A related process is spray-up molding. Strong AB (2000) Plastics materials and processing. Prentice-Hall, Columbus, OH. James F (ed) (1993) Whittington's dictionary of plastics. Technomic Publishing Co. Inc., Carley.

L

LCP *See liquid-crystal polymer.*

LC polymer Abbreviation for liquid-crystal polymer.

LC_{50} The concentration, in parts per million, of a substance in air that is lethal to 50% of the laboratory animals exposed to it.

LC_{50} test (lethal dose 50%). A toxicity test based on the results of animal experiments, the results of which mean that half the animals die from a given dosage of the substance tested.

See oral dose (LD_{50}).

LDPE Abbreviation for low-density polyethylene.

Also see polyethylene.

L/D ratio In an extruder, the ratio of the flighted length of the screw to its nominal diameter.

Lea (1) One-seventh of an 840-yard cotton hank, i.e., 120 yards. (2) A standard skein with 80 rev of 1.5 yards each (total length of 120 yards). It is used for strength tests. (3) A unit of measure, 300 yards, used to determine the yarn number of linen yarn. The number of leas in 1 lb is the yarn number. Kadolph SJJ, Langford AL (2001) Textiles. Pearson Education, New York.

Leaching (1796) *v.* The process of the extraction of a component from a solid material by treating the material with a solvent that dissolves the component of interest but not the remaining principle material. Natural leaching occurs when rain water dissolves soluble salts from soil. Some components of a paint coating leach from the dried film after application to a substrate when exposed to rain or when immersed in water.

Leaching rate Milligrams of water-soluble material released per cm^2/day from an immersed anti-fouling paint in seawater.

Leacril Poly(acrylonitrile), manufactured by ACSA, Italy.

Lead \\ˈled\\ [ME *leed*, fr. OE *lēad*; akin to MHGr *lōt* lead] (before 12c) *n.* Element (Pb) found mostly in combination and used in pipes, cable sheaths, batteries, solder, and shields against radioactivity. Powders of this metal have also been employed as fillers in plastics. Hibbard MJ (2001) Mineralogy. McGraw-Hill, New York.

Lead \\ˈlēd\\. Of an extruder screw, the distance parallel to the screw axis from any point on the screw thread to the corresponding point on the next turn of that thread, or pitch.

Lead acetate (1885) *n.* $Pb(C_2H_3O_2)_2{\cdot}3H_2O$. Crystalline salt, soluble in water and used in the manufacture of lead pigments; also used in production of varnishes. Kirk–Othmer encyclopedia of chemical technology: pigments–powders. John Wiley and Sons, New York, 1996. Gooch JW (1993) Lead based paint handbook. Plenum Press, New York.

Lead azide (1918) *n.* $Pb(N_3)_2$ A crystalline explosive compound used as a detonating agent.

Lead carbonate (1873) *n. See basic lead carbonate.*

Lead carbonate, black *See carbonate white lead.*

Lead chrome green *See chrome green.*

Lead chrome pigment Any of a series of inorganic pigments including yellows, oranges, and greens, used in PVC, polyolefins, cellulosics, acrylics, and polyesters.

Lead dioxide (1885) *n.* PbO_2. A poisonous compound used as an oxidizing agent and as an electrode in batteries.

Lead drier One of many organic lead salts, which are soluble in paints and varnishes; used to speed the drying and hardening of the oil vehicle.

Leaded zinc oxide Mixed pigment comprising zinc oxide and basic lead sulfate. Several grades are possible and vary in the proportions of the individual ingredients. Leaded zinc oxides are either mechanical mixtures of the two ingredients, or are made by heating together roasted zinc ore, lead sulfide ore and coal. Kirk–Othmer encyclopedia of chemical technology: pigments–powders. John Wiley and Sons, New York, 1996. Gooch JW (1993) Lead based paint handbook. Plenum Press, New York.

Leader-pin bushing *See guide-pin bushing.*

Leading flight face (leader flight). The forward or front side of the flight of an extruder screw, the rear side being the *trailing flight face.*

Lead in oil Basic carbonate white lead ground in linseed oil; formerly wide use, now replaced largely by titanium dioxide pigments. Gooch JW (1993) Lead based paint handbook. Plenum Press, New York.

Lead monoxide (ca. 1909) *n.* PbO A yellow to brownish red poisonous compound used in rubber manufacture and glassmaking. *See litharge.*

Lead naphthenate Lead soap of naphthenic acids. Used as a drier.

Lead nitrate $Pb(NO_3)_2{\cdot}2H_2O$. Crystalline salt soluble in water and used in the manufacture of lead chromes.

Lead octoate Lead soap of 2-ethylhexanoic acid. Used as a drier.

Lead oxide (ca. 1926) *n.* Either yellow lead oxide, PbO (litharge) or red lead oxide, Pb_3O_4. Both are used as pigments, though much less today in USA than formerly because of concerns about lead's toxicity and the need to keep it out of the environment. The oxides are sometimes used as fillers in radiation-shielding applications. *See litharge.*

Lead phosphite, dibasic $2PbO{\cdot}PbHPO_3{\cdot}1/2H_2O$. Fine white acicular crystals. Pigment grade is the dibasic lead salt of phosphorous acid. Used as an anti-corrosive white pigment, and for its ability to eliminate bleed-through of cedar and redwood stains in latex paints. Kirk–Othmer encyclopedia of chemical technology: pigments–powders. John Wiley and Sons, New York, 1996.

Lead-restricted paint Normally, a paint having a lead content below a given limit, as used for applications where the presence of lead may be harmful. Gooch JW (1993) Lead based paint handbook. Plenum Press, New York.

Lead salicylate (dibasic lead salicylate) $Pb(C_6H_4OHCOO)_2{\cdot}H_2O$. A white crystalline material formerly used as a heat stabilizer.

Lead stabilizer Any of a large family of highly effective heat stabilizers that are limited to use in applications where toxicity, sulfur staining, and lack of clarity are not objectionable. Examples are: basic lead carbonate, basic lead sulfate complexes, basic silicate, or white lead, co-precipitated lead silicate, and silica gel, dibasic lead maleate.

Lead stearate $Pb(C_{17}H_{35}COO)_2$. A white powder used as a vinyl-resin stabilizer and lubricant in extrusion compounds, and, earlier, in phonograph-record compounds. Wickson EJ (ed) (1993) Handbook of polyvinyl chloride formulating. John Wiley and Sons, New York.

Lead sulfide (ca. 1898) *n.* PbS. Mineral which occurs as galena and is one of the chief sources of lead. It can be prepared by passing hydrogen sulfide gas into an acid solution of lead nitrate. It is the compound responsible for darkening of paints when a lead-pigmented paint is exposed to sulfur-containing gases. Hibbard MJ (2001) Mineralogy. McGraw-Hill, New York.

Lead titanate $PbTiO_3$. A yellow buff-colored pigment, which possesses good resistance to weather and chalking. Pale yellow solid; insoluble in water. Its Sp gr is 7.52. Derivation: interaction of oxides of lead and titanium at a high temperature. Contains lead sulfate and lead oxide as impurities.

L

Lead white *See carbonate white lead.*

Leaf green *See chromium oxide green.*

Leafing A phenomenon produced by metallic pigments forming a layer parallel to the surface of the print and providing a high metallic luster.

Leafing The ability of an aluminum or gold bronze to provide a brilliant or silvery appearance. The flat pigment particles align themselves parallel with the coated surface providing a high reflection of light.

Leather cloth A term sometimes used, especially in Europe, for plastic-coated fabric with a leather-like texture.

Le Châtelier's principle When a system at equilibrium experiences a stress, it will adjust, if possible, so as to minimize the effect of the stress.

Lecithin \ˈle-sə-thən\ [ISV, fr. Gk *lekithos* yolk of an egg] (1861) *n.* A liquid, obtained in refinement of soya beans or cottonseed; used in paints to promote pigment wetting and to control pigment settling and flow properties. Weismantal GF (1981) Paint handbook. McGraw-Hill, New York.

Lecyar model A model for viscosities for polymer blends, which do not obey simple mixture laws. The Lecyar model can accommodate a viscosity minimum, maximum, or both in the composition range from 0 to 100% of one polymer intimately blended with another. For a given shear rate and temperature it has the form:

$$\ln \eta_B = Am_1^3 + Bm_2^3 + Cm_1^2 m_2 + Dm_1 m_2^2,$$

which A and B are the natural logarithms of the neat resins 1 and 2, C and D are the natural logarithms of interaction viscosities that must be evaluated from measurements on blends, and m_1 and m_2 are the mass fractions of the two components ($m_1 + m_2 = 1$). Shenoy AV (1996) Thermoplastics melt rheology and processing. Marcel Dekker, New York.

LEFM Abbreviation for linear elastic fracture mechanics.

Legging *n.* The drawing of filaments or strings when adhesive-bonded substrates are separated.

Legs The stringy effect that is apparent when cemented surfaces are separated shortly after the bond is made. Long legs or strings are indicative often of a weak bond, whereas short legs indicate a strong bond. Skeist I (ed) (1990) Handbook of adhesives. Van Nostrand Reinhold, New York.

Leguval Unsaturated polyester, manufactured by Bayer, Germany.

Leipsic yellow *See chrome yellow, light and primrose.*

Leithner's blue *See cobalt blue.*

Lemon yellow (1807) *n.* Any of two families of yellow pigments, one being mixtures of barium chromate with zinc carbonate, the other mixtures of lead chromate and lead carbonate.

Length The property of an ink whereby it can be stretched out into a long thread without breaking. Long inks have good flow in the fountain. Leach RH, Pierce RJ, Hickman EP, Mackenzie MJ, Smith HG (eds)(1993) Printing ink manual, 5th edn. Blueprint, New York.

Leno weave \ˈlē-(ˌ)nō-\ [per. fr. F *linon* linen fabric, lawn, fr. MF *lin* flax, linen, fr. L *linum* flax] (1821) *n.* A weave in which the warp yarns are arranged in pairs with one twisted around the other between picks of filling yarn as in marquisette. This type of weave gives firmness and strength to an open weave fabric and prevents slippage and displacement of warp and filling yarns. Kadolph SJJ, Langford AL (2001) Textiles. Pearson Education, New York.

Let down (1) The process of paint manufacturing in which the pigment paste (mill base) is reduced (let down) by the addition of the remaining ingredients of the formula. (2) Reducing the intensity or depth of a colored pigment through the addition of a white, or sometimes colorless, pigment.

Let-go An area in laminated glass over which an initial adhesion between interlayer and glass has been lost. Wicks ZN, Jones FN, Pappas SP (1999) Organic coatings science and technology, 2nd edn. Wiley-Interscience, New York. Weismantal GF (1981) Paint handbook. McGraw-Hill, New York.

Let-off A device used in coating by calendering or extrusion to suspend a coil or reel from which the material to be coated is fed to the coating machine.

Letterpress \ˈle-tər-ˌpres\ (ca. 1765) *n.* The process of printing from an inked raised surface when the paper is impressed directly upon the surface. Wijnekur FJM (1967) Elsevier's dictionary of the printing and allied industries in four languages. Research Institute for the Graphic and Allied Industries, TNO, Amsterdam, Elsevier Publishing Co., New York.

Letterpress printing The process used for paper is adapted to plastics by the use of special inks and transfer rolls, and possibly a modification of press speed. Flexible printing plates are usually employed, made of vinyl or rubber. It is the oldest method of printing flat plastic substrates and uses a press for relief printing from metal type or raised surfaces formed from wood, metal, or linoleum. Leach RH, Pierce RJ, Hickman EP, Mackenzie MJ, Smith HG (eds) (1993) Printing ink manual, 5th edn. Blueprint, New York.

Letterset *See dry offset.*

Levapren Ethylene/vinyl acetate co-polymer. Manufactured by Bayer, Germany.

Leveler Group of steel rollers that bend any edges that are not flat prior to coating. Also, a process in the aluminum industry that flattens strip so that the thickness is the same from one side to the other.

Leveling (1) The measure of the ability of a coating to flow out after application so as to obliterate any surface irregularities such as brush marks, orange peel, peaks, or craters which have been produced by the mechanical processes of applying or coating. (2) The property of a freshly spread aqueous polish to dry to a uniform and streak-free appearance. Paint/coatings dictionary. Federation of Societies for Coatings Technology, Philadelphia, Blue Bell, PA, 1978. Martens CR (1968) Technology of paints, varnishes and lacquers. Reinhold Publishing Co., New York.

Levigation \ˌle-və-ˈgā-shən\ [L *levigatus*, pp of *levigare* to make smooth, fr. *levis* smooth (akin to Gk *leios* smooth and perhaps to L *linere* to smear) + *-igare* (akin to *agere* to drive)] (1612) *n.* Process of mixing a pigment with water and washing. The percentage of salts and heavy particles is reduced or eliminated by controlled sedimentation. Paint/coatings dictionary. Federation of Societies for Coatings Technology, Philadelphia, Blue Bell, PA, 1978.

Levorotatory \ˌle-və-ˈrō-tə-ˌtōr- ē\ (1873) *adj.* Turning toward the left or counterclockwise: esp: rotating the plane of polarization toward the left – compare *dextrorotatory.* Morrison RT, Boyd RN (1992) Organic chemistry, 6th edn. Prentice-Hall, Englewood Cliffs, NJ.

Lewis acid \ˈlü-əs-\ [Gilbert N. *Lewis* † (1946) American chemist] (1944) *n.* A substance that is capable of accepting an unshared pair of electrons from a base to form a covalent bond.

Lewis-acid catalyst *See Friedel–Crafts catalysts.*

Lewis–Nielsen equation An equation derived from the theory of mixtures that provides an estimate of the modulus of a short-fiber, thermoplastic composite. It is given below.

$$E_c = \frac{E_r(1 + 2AB\phi_f)}{1 - B\psi\phi_f},$$

where

$$B = \frac{(E_f/E_r) - 1}{(E_f/E_r) + 2A}$$

and

$$\psi = 1 + \frac{(1 - V)\phi_{fr}}{2V_r}.$$

In these expressions, E_c, E_f, and E_r are the moduli of the composite, fiber, and resin, ϕ_f is the volume fraction of fiber, A is the average aspect ratio of the fiber, and ψ is the maximum packing fraction. Application of the equation is limited to small strains. Wallenberger F, Weston NE (eds) (2003) Natural fibers, plastics and composites. Springer-Verlag, New York. Murphy J (1998) Reinforced plastics handbook. Elsevier Science and Technology Books, New York. Pittance JC (ed) (1990) Engineering plastics and composites. SAM International, Materials Park, OH.

Lewis structure A method of indicating the assignment of valence electrons in an atom, molecule, or ion by representing them as dots placed around symbols, which represent the cores, or kernels, of the atoms. Whitten KW, Davis RE, Davis E, Peck LM, Stanley GG (2003) General chemistry. Brookes/Cole, New York.

Lexan® General electric's trade name for their polycarbonate resins produced by reacting bisphenol A and phosgene, the first commercial PC resins.

See also polycarbonate resin.

Leyden blue *See cobalt blue.*

Li Chemical symbol for the element lithium.

Licanic acid $CH_3(CH_2)_3(CH{=}{=}CH)_3(CH_2)_4COCH_2COOH$. Constitutes 75–80% of the fatty acid of oiticica oil; mp, 75°C. Shahidi F, Bailey AE (eds) (2005) Bailey's industrial oil and fat products. John Wiley and Sons, New York. Paint: pigment, drying oils, polymers, resins, naval stores, cellulosics esters, and ink vehicles, vol 3. American Society for Testing and Material, Conshohocken, PA, 2001.

Lieberman–Storch method Procedure for the qualitative detection of rosin in vehicles. The rosin may be present as either free rosin (abietic acid), esterified rosin, or metal salts. In Europe, the Lieberman–Storch–Morawski test is more commonly used. Paint: pigment, drying oils, polymers, resins, naval stores, cellulosics esters, and ink vehicles, vol 3. American Society for Testing and Material, Conshohocken, PA, 2001.

Lift (1) The complete set of moldings produced in one cycle of a molding press.

Lifting Softening and raising or wrinkling of a previous coat by the application of an additional coating; often caused by the solvents.

Ligand \ˈli-gənd, ˈlī-\ [L *ligandus*, gerundive of *ligare*] (1949) *n.* An atom, molecule, or ion bonded to the central atom in a complex.

Light \ˈlīt\ [ME, fr. OE lēoht; akin to OHGr *lioht* light, L *luc-*, *lux* light, *lucēre* to shine, Gk *leukos* white] (before 12c) *n.* (1) Electromagnetic radiation of which a human observer is aware through the visual sensations that arise from the stimulation of the retina of the eye. This portion of the spectrum includes wavelengths from about 380 to 770nm. Thus, it is incorrect to speak of ultraviolet "light" because the human observer cannot see radiant energy in the ultraviolet region. (2) *Adj.* meaning high reflectance, transmittance, or level of illumination as contrasted to dark, or low level of intensity. Moller KD (2003) Optics. Springer-Verlag, New York. Johnson SF (2001) History of light and colour measurement: a science in the shadows. Taylor and Francis, London, UK. McDonald R (ed) (1997) Colour physics for industry, 2nd edn. Society of Dyers and Colourists, West Yorkshire, England. Saleh BEA, Teich MC (1991) Fundamentals of photonics. John Wiley and Sons, New York.

Light absorbance (absorbance, absorptivity). The percentage of the total luminous flux incident upon a test specimen that is neither reflected from, nor transmitted through the specimen. Compare light reflectance and light transmittance.

Light end (1) The low boiling fraction in distillation. (2) *See fine end* (1).

Light reflectance (reflectance, reflectivity). The fraction of the total luminous flux incident upon a surface that is reflected, generally a function of the color (wavelength) of the light. Johnson SF (2001) History of light and colour measurement: a science in the shadows. Taylor and Francis, London, UK. Saleh BEA, Teich MC (1991) Fundamentals of photonics. John Wiley and Sons, New York.

See also light absorbance and light transmittance.

Light resistance (light fastness, color fastness). The ability of a plastic material to resist fading, darkening, or degradation upon exposure to sunlight or ultraviolet light. Nearly all plastics tend to change color under outdoor conditions, due to characteristics of the polymeric material and/or pigments incorporated therein. Tests for light resistance are made by exposing specimens to natural sunlight or to artificial light sources such as the carbon arc, mercury lamp, germicidal lamp, or xenon arc lamp. ASTM D 4459 and D 4674 describe procedures for testing light fastness of plastics for indoor applications. Zaiko GE (ed) (1995) Degradation and stabilization of polymers. Nova Science Publishers Inc., New York.

See also artificial weathering.

Light scattering (Rayleigh, Mie, and Frankhoffer) In a dilute polymer solution, light rays are scattered and diminished in intensity by a number of factors including fluctuations in molecular orientation of the polymer solute. Observations of the intensity of light scattered at various angles provide the basis for an important method of measuring weight-average molecular weights of high polymers. Kokhanovsky

AA (2004) Light scattering media optics. Springer-Verlag, New York. Kamide K, Dobashi T (2000) Physical chemistry of polymer solutions. Elsevier, New York. Berne BJ (2000) Dynamic light scattering: applications to chemistry, biology and physics. Dover Publications, New York. Brown W (1996) Light scattering: principles and development. Oxford University Press, UK. Pethrick RA (ed) (1999) Modern techniques for polymer characterization. John Wiley and Sons, New York. Elias (1977) Macromolecules, vols 1–2. Plenum Press, New York. Miller ML (1966) The structure of polymers. Reinhold Publishing Co., New York.

See also low-angle laser-light scattering.

Light source An object, which emits light or radiant energy to which the human eye is sensitive. The emission of a light source can be described by the relative amount of energy emitted at each wavelength in the visible spectrum, thus defining the source as an illuminant, or the emission may be described in terms of its correlated color temperature. Johnson SF (2001) History of light and colour measurement: a science in the shadows. Taylor and Francis, London, UK. Saleh BEA, Teich MC (1991) Fundamentals of photonics. John Wiley and Sons, New York.

See illuminant and correlated color temperature.

Light stabilizer An agent added to a plastic compound to improve its resistance to light-induced changes.

See also stabilizer and ultraviolet stabilizer.

Light transmittance (luminous transmittance, light transmissivity). The ability of a material to pass incident light through it, whether specular or diffuse. ASTM (www.astm.org) prescribes several tests of this property in plastics. *Transmissivity* is the ratio of the intensity of the transmitted light to that of the unreflected incident light. Fox AM (2001) Optical properties of solids. Oxford University Press, UK.

See also opacity.

Light-fastness ad (1950). The resistance of coated, printed or colored material to the action of sunlight or artificial light. Syn: colorfastness.

Lightness (before 12c) *n.* (1) Achromatic dimension necessary to describe the three-dimensionality of color, the others being hue and saturation. Sometimes the lightness dimension is called "brightness". In the Munsell color order system, the lightness dimension is called "value". (2) Perception by which white objects are distinguished from gray and light objects from dark ones.

Lignin \ˈlig-nən\ (1822) *n.* The major non-carbohydrate constituent of wood and woody plants, functioning in nature as a binder to hold the matrix of cellulose fibers together. Lignins are obtained commercially from by-products of coniferous woods, for example, by treating wood flour with a derivative of lignosulfonic acid. They are used as extenders in phenolic resins, and sometimes as reactants in the production of phenol-formaldehyde resins. Hoadley RB (2000) Understanding wood. The Taunton Press, Newtown, CT.

Lignin plastic A plastic based on lignin resins.

Lignin resin A resin made by heating lignin or by reaction of lignin with chemicals or resins, the lignin being in greatest amount of mass.

Lignostone Compressed wood. Manufactured by Röchling, Germany.

Ligroin (ligroine, benzine). Any of several saturated petroleum–naphtha fractions boiling in the range 60–135°C (68–275°F), used as solvents. The term *benzine* is depreciated due to confusion with *benzene*, and should not be used.

Lime \ˈlīm\ [ME, fr. OE *līm*; akin to OHGr *līm* birdlime, L *limus* mud, slime, and perhaps to L *linere* to smear] (before 12c) *n.* A caustic highly infusible solid that consists of calcium oxide often together with magnesium oxide, that is obtained by calcining forms of calcium carbonate (as shells or limestone). Syn: calcium oxide.

Lime blue Mixture of ultramarine and terra alba. Another type of lime blue is made from methylene blue by adsorption on natural earth.

Lime green *See green, lime.*

Lime putty *See putty.*

Lime red Lake produced by adsorbing magenta on a natural earth.

Lime yellow Lake produced by adsorbing auramine or other yellow dyestuff on a natural earth.

Limed rosin Commercial calcium resinate made by the direct interaction of lime and rosin.

Limestone *See calcium carbonate and calcium carbonate, natural.*

Lime washing Coating with limewash made from hydrated lime or by slaking quick lime, to which tallow is sometimes added. Syn: whitewash, whitening.

Limiting oxygen index A relative measure of flammability that is determined as follows. A sample is ignited in an oxygen/nitrogen atmosphere. The oxygen content is adjusted until the minimum required to sustain steady burning is found. The higher the value, the lower the flammability. Troitzsch J (2004) Plastics flammability handbook: principle, regulations, testing and approval. Hanser–Gardner Publications, New York. Tests for comparative flammability of liquids, UI 340. Laboratories Incorporated Underwriters, New York, 1997.

Limiting viscosity number The IUPAC term for intrinsic viscosity.

Limonite \ˈlī-mə-ˌnīt\ [Gr *Limonit*, fr. Gk *leimōn* wet meadow; akin to Gk *limnē* pool] (1823) *n.* A native hydrous ferric oxide of variable composition that is a major ore of iron.

See iron oxides, natural.

Line etching A print made up of lines or pigmented areas and lighter spaces free from shading.

Lineal density The meaning of Denier. Mass per unit length expressed as grams per centimeter, pounds per foot, or equivalent units. It is the quotient obtained by dividing the mass of a fiber or yarn by its length.

Linear elastic fracture mechanics (LEFM) A theory of fracture, applicable to brittle plastics (and other brittle materials), based on the assumption that the material is Hookean up to the point of fracture, with yielding restricted to a small volume near crack tips in the stressed material.

Linear expansion *See coefficient of thermal expansion.*

Linear low-density polyethylene (LLDPE) The original low-density polyethylene (LDPE), produced at high pressure, has a highly branched structure. Using Ziegler–Natta catalysts and low pressure, with a small percentage of 1-butene or other comonomer, one can produce a more linear PE with density between 0.919 and 0.925 g/cm^3. LLDPE films have the gloss and clarity of LDPE films, but are stronger, so can be blown thinner to carry design loads. Because of higher melt viscosity, screw modifications are usually necessary for processing LLDPE in extruders designed for LDPE. Odian GC (2004) Principles of polymerization. John Wiley and Sons, New

York. Mark JE (ed) (1996) Physical properties of polymers handbook. Springer-Verlag, New York.

Linear polymer A polymer in which the molecules form long chains without branches or cross-linking. The molecular chains of a linear polymer may be intertwined, but the forces tending to hold the molecules together are physical rather than chemical and therefore can be weakened by heating. Linear polymers are thermoplastic. Odian GC (2004) Principles of polymerization. John Wiley and Sons, New York. Mark JE (ed) (1996) Physical properties of polymers handbook. Springer-Verlag, New York.

Linear unsaturated polyesters *See polyester, unsaturated.*

L

Linear viscoelasticity Viscoelasticity is characterized by a linear relationship between stress, strain, and strain rate.

Linen \ˈli-nən\ [ME, fr. OE *līnen*, fr. *līn* flax, fr. L *linum* flax; akin to Gk *linon* flax, thread] (before 12c) *adj.* Cellulosic fibers derived from the stem of the flax plant or a fabric made from these fibers. Linen fibers are much stronger and more lustrous that cotton; they yield cool, absorbent fabrics that wrinkle easily. Fabrics with linen-like texture and coolness but with good wrinkle resistance can be produced from manufactured fibers and blends.

Linen lea The number of 300-yard hanks contained in 1 lb.

Liner (1) A continuous, usually flexible coating on the inside surface of a filament-wound pressure vessel, used to protect the laminate from chemical attack or to prevent leakage under stress. (2) In extruders and injection molders, the hard-alloy interior surface of the cylinder. Decades ago, some of these were separately fabricated and pressed into the steel cylinders. Today they are centrifugally cast into cylinders.

Linet A French-made lining fabric of unbleached linen.

Lining fabrics (1) Muslin or canvas used underneath fine wallpapers to avoid small cracks possibly opening up in a plaster wall and showing through. (2) Fabric that is used to cover inner surfaces, especially when the inner surface is of a different material than the outer. May refer to garment lining, lining for boxes, coffins, etc. Generally of smooth, lustrous appearing fabrics, but also of felt and velvet. Both manufactured fibers and natural fibers are used.

Lining paper Plain paper applied before the wallpaper. Assures a smoother surface and better adhesion.

Lining tool (British). A small flat fitch with a slanting edge, used for painting lines with the help of a run. Syn: liner.

Linoleates \lə-ˈnō-lē-ˌāt\ (ca. 1865) *n.* Generally the salts or soaps of linseed fatty acid. Cobalt, lead, and manganese linoleates are widely used as driers in printing ink.

Linoleic acid \ˌli-nə-ˈlē-ik-\ [Gk *linon* flax + ISV *oleic (acid)*] (1857) *n.* $C_{18}H_{32}O_2$. *cis, cis*-9,12 Octadecadienoic acid. An 18-carbon, straight-chain fatty acid with two double bonds that may be in the 9 and 12 or 9 and 11 positions. It is found in nature as its glyceryl trimester in many vegetable oils and is a starting material for some plasticizers for plastics. It has a mol wt of 280.44, bp of 230°C, iodine value of 181.1.

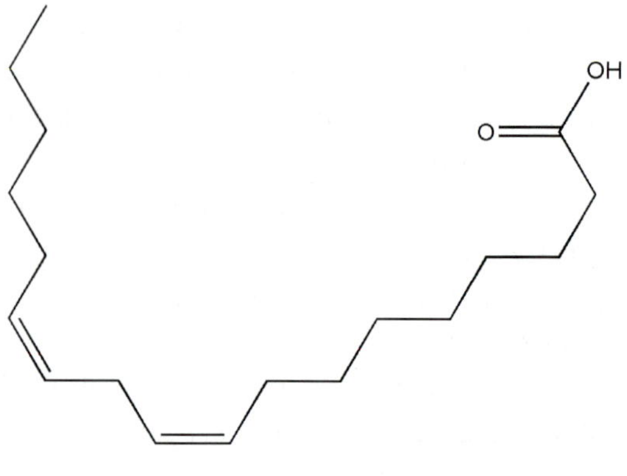

Linolein Glyceride of linoleic acid. It is one of the constituents of linseed oil, which induces the drying properties.

Linolenate driers Certain metal salts of, and soaps of, linseed fatty acids.

Linolenic acid \ˌli-nə-ˈlē-nik-\ [ISV, irreg. fr. *linoleic*] (1887) *n.* $CH_3CH_2CH{=}CHCH_2CH{=}CHCH_2CH{=}CH(CH_2)_7CO\text{–}OH$. *cis, cis, cis*-9,12,15-Octadecatrienoic acid. Tripoly-unsaturated fatty acid component of linseed and other drying oils, It has a bp of 230°C/17 mmHg, an acid value of 201.6, and iodine value of 273.7. Paint: pigment, drying oils, polymers, resins, naval stores, cellulosics esters, and ink vehicles, vol 3. American Society for Testing and Material, Conshohocken, PA, 2001.

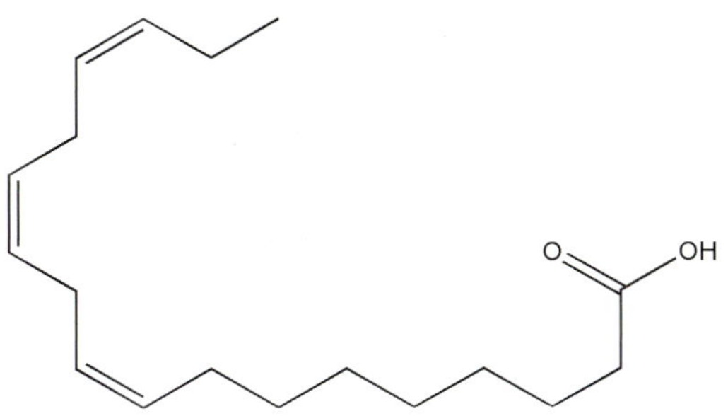

Linoleum and oilcloth varnishes Special highly flexible and elastic varnishes.

Linoleum, floor and wall covering Made from oxidized linseed oil or combinations of drying oils, wood flour and/or ground cork, resins and pigment, rolled out and compressed onto an asphalt saturated felt, burlap, or other backing. Heat, which fuses and sets the oils and resins to form strong binding agents, is applied to the mixture during compression.

Linon A jour A gauze-like linen fabric used as dress goods.

Linoxyn Semisolid, highly oxidized linseed oil; used in the manufacture of linoleum.

Linseed oil (15c) *n.* Drying oil from seeds of the flax plant *Linum usitatissimum*, a mixture of glyceryl esters of linolenic (25%), oleic (5%), linoleic (62%), stearic (3%), and palmitic (5%) acids. The oil is refined by treatments, which remove water and mucilaginous material and is then described as refined oil, according to the method of treatment. Further processing produces boiled oil, blown oil or bodied oil. This best known and most widely used oil in the paint industry if characterized by its relatively short drying time. Its high degree of unsaturation, to which its good drying characteristics can be partially ascribed, is due to the presence of large percentages of linolenic and linoleic triglycerides. Many years ago the oil was obtained from seed by mechanical pressure, including both hydraulic presses and later expellers. In recent years, the more modern solvent extraction is used. Oils thus obtained show lower percentages of impurities and better overall quality. Linseed oil responds very readily to a variety of refining techniques and is used in the paint industry both as a drying oil and as an ingredient in a wide array of modified resins of many varieties. Paint: pigment, drying oils, polymers, resins, naval stores, cellulosics esters, and ink vehicles, vol 3. American Society for Testing and Material, Conshohocken, PA, 2001.

Lint \ˈlint\ [ME] (14c) *n.* Particles and short fibers that fall from a textile product during the stresses of use.

Lint ball *See balling up.*

Lintel \ˈlin-təl\ (ME, fr. MF, fr. LL *limitaris* threshold, fr. L, constituting a boundary mfr. *limit-*, limes boundary] (14c) *n.* Horizontal beam supported at each end and spanning an opening. Usually supports the structure above it.

Linters (ca. 1889) *n.* Short fibers that adhere to cotton seeds after ginning. Used in rayon manufacture, as fillers for plastics, and as a base for the manufacture of cellulosic plastics.

Lipophilic livering \ˌlī-pə-ˈfi-lik-\. An increase in the consistency of a paint resulting in a rubbery or coagulated mass.

Liquid chromatography (LC or HPLC) Any chromatographic process in which the moving phase is a liquid, in contrast to the moving gas phase of gas chromatography. Materials, which can be run through a liquid chromatograph are liquids or solutions. An example of a liquid chromatograph is the Series 200 LC Plus Diode Array System by Perkin-Elmer. Kroschwitz JI (ed) (1990) Polymers: polymer characterization and analysis. John Wiley and Sons, New York.

Liquid crystal (1891) *n.* A liquid in which the molecules are oriented parallel to each other resulting in birefringence and interference patterns visible in polarizing light. Collins PJ (1997) Introduction to liquid crystals: chemistry and physics, vol 1. Taylor and Francis, New York.

L

Liquid-crystal polymer (LC polymer, liquid-crystalline polymer, and mesomorphic polymer). A polymer capable of forming regions of highly ordered structure (*mesophase*) while in the liquid (melt or solution) phase. The degree of order is somewhat less than that of a regular solid crystal. Four types have been identified: rod-like, including aromatic polyamides, esters, azomethines, and benzobisoxazoles; helical, mostly natural materials such as polypeptides; side-chain (*comb polymers*); and block copolymers with alternating rigid and flexible units. These polymers are described as *nematic*, in which the mesogens (ordered regions) show no positional order, only long-range order; *Cholesteric* or *chiral*, a modified nematic phase in which the orientation direction changes from layer to layer in a helical pattern; and *smectic*, in which the mesogens have both long-range order and 1- or 2-dimensional positional order. Liquid-crystal polymers are difficult to get into the molten condition because

TCLC plus system

the solid crystals generally decompose before melting. The most commercially successful ones to date are those processed in solution, e.g., poly(*p*-phenylene terephthalamide) (Kevlar). LC polymers are also classified as *lyotropic* and *thermotropic.* Lyotropic ones show their liquid-crystalline character only in solution, while thermotropic ones can show it in the melt without the presence of a solvent. Collins PJ (1997) Introduction to liquid crystals: chemistry and physics, vol 1. Taylor and Francis, New York.

Liquid crystal polymer resins Self-reinforcing plastics because of their densely packed fibrous polymer chains. Odian GC (2004) Principles of polymerization. John Wiley and Sons, New York.

Liquid driers Solutions of soluble driers in volatile organic solvents, usually hydrocarbons.

Liquid emulsion polymer coating Synthetic resin emulsion (latex) which produce hard, tough coatings having a high luster. Usually non-buffable or semibuffable, and referred to and sold as self-polishing floor wax.

Liquid injection molding (LIM) A process of injection-molding thermosetting resins in which the uncured resin components are metered, mixed, and injected at relatively low pressures through nozzles into mold cavities, the curing or polymerization taking place in the mold cavities. The process is most widely used with resins that cure by addition polymerization such as polyesters, epoxies, silicones, alkyds, diallyl phthalate, and (occasionally) urethanes.

Liquid-junction potential A voltage produced across the junction between two dissimilar liquids.

Liquid reaction molding (LRM) Older Syn: reaction injection molding.

Liquid resin (deprecated). *See tall oil.*

Liquid rosin Syn: tall oil.

Liquid solvent wax Mixture of waxes and other ingredients in a solvent base. Must be polished for luster and are occasionally colored. *Sometimes called, simply, solvent wax.*

Liquid water emulsion wax Dispersion of wax and other modifying materials in water.

Liquifying stress The ease with which concentrated inks can be liquefied by remilling after storage.

Liquor ratio In wet processing the ratio of the weight of liquid used to the weight of goods treated.

Lisle yarn [*Lisle* Lille, F] (1851) *n.* A high-quality cotton yarn made by plying yarns spun from long combed staple. Lisle is singed to hive it a smooth finish.

Lithium \\ˈli-thē-əm\\ [NL, fr. *lithia* oxide of lithium, fr. Gk *lithos*] (1818) (Li) *n.* Element number 3, the least dense of all the metals (density $= 0.534 g/cm^3$), with valence of +1, and highly reactive. Lithium aluminum hydride ($LiAlH_4$) is an important catalyst in organic reductions and lithium is a component of many greases, e.g., the high-temperature lubricant, lithium stearate.

Lithium carbonate (1873) *n.* Li_2CO_3. A crystalline salt used in the glass and ceramic industries.

Lithium fluoride (1944) *n.* LiF. A crystalline salt used in making prisms and ceramics and as a flux.

Lithium niobate [*niob*ium + *¹-ate*] (1966) *n.* $LiNbO_3$. A crystalline material whose physical properties change in response to pressure or the presence of an electric field and which is used in fiber optics and as a synthetic gemstone.

Lithium stearate $LiOOCC_{17}H_{35}$. A white crystalline material used as a lubricant in plastics.

L

Lithographic inks Inks used in the lithographic process. The principal characteristic of a good lithographic ink is its ability to resist excessive emulsification by the fountain solution.

Lithographic varnish Heat-bodied refined linseed oils used as vehicles for printing and lithographic inks.

Lithography \li-ˈthä-grə-fē\ [Gr *Lithographie*, fr. *lith-* + *-graphie* –graphy] (1813) *n.* A process of planographic printing involving two different areas on the plate, one receptive to ink, the other receptive to fountain solution. Leach RH, Pierce RJ, Hickman EP, Mackenzie MJ, Smith HG (eds) (1993) Printing ink manual, 5th edn. Blueprint, New York. Printing ink handbook. National Association of Printing Ink Manufacturers, Inc., NJ, 1976.

Also see stereolithography for three-dimensional formation of objects.

Lithographed paper Wallpaper made by the printing process used for billboards and posters. Lithography exploits the affinity of color for the greasy material in which the design is put on the roller. a transfer process.

Lithographic chalk A greasy crayon, composed of soap, wax, oil and lampblack, for drawing designs on a lithographic plate prior to etching.

Lithographic inks Inks used in the lithographic process. The principal characteristic of a lithographic ink is its ability to resist excessive emulsification by the fountain solution.

Lithographic stones Slabs of limestone, three or four inches thick, the surfaces of which are smoothed or grained, for use in lithographic printing.

Lithographic varnishes Heat-bodied refined linseed oils used as vehicles for printing and lithographic inks.

Lithol red $C_{20}H_{13}N_2O_4SM^*$ (M^* = Na, Ba, or Ca). Pigment Red 49 (15630). Pigment made by combining the intermediates, tobias acid and beta-naphthol. This type of red is available as sodium, barium, and calcium toners, also lakes; the sodium is the lightest shade, the barium is what may be termed a medium shade, and the calcium lithols are deep reds and maroons.

Lithol rubine Pigment Red 57 (15850). Azo pigment made by diazotizing paratoluidine-*meta*-sulfonic acid, and coupling with 3-hydroxy-2-naphthoic acid. Pigment is used as the calcium salt. Most frequently encountered in the resinated form, it is characterized by its distinctly brilliant deep masstone and its blue intense tint.

Lithopone \ˈli-thə-ˌpōn\ [ISV *lith-* + Gk *ponos* work] (ca. 1884) *n.* $BaSO_4{\cdot}ZnS$. Pigment White 5 (77115) (Charlton white, Orr's white, and zinc baryta). A mixed pigment obtained by the interaction (*metathesis*) of equimolar solutions of barium sulfide and zinc sulfate, from which precipitate barium sulfate and zinc sulfide, both white.

Lithrage PbO. Oxide of lead made by controlled heating of metallic lead. Pure litharge has Sp gr of 9.53, mp of 888°C, and mol wt of 223.21. Used as a raw material in the manufacture of pigments and driers and infrequently as a catalyst in pains.

Litre \ˈlē-tər\ [*variant* of liter]. The volume of a kilogram of water at 4°C, equal to 1.06 quarts or 61.02 $in.^3$.

Little Joe A one- or two-station dry offset proof press.

Live centers Point at the very center at each end of the roll. This is usually a center point of the journal at which the roll turns. The roll is driven from this point and is considered concentric from the same point.

Live edge *See wet edge.*

Live-feed molding *See multi-live-feed molding.*

Livering The progressive, irreversible increase in consistency of a pigment-vehicle combination. Livering in the majority of cases arises from a chemical reaction of the vehicle and the solid dispersed material, but it may also result from polymerization of the vehicle.

Living polymers A polymerization reaction in which there is no termination, and the polymer chains continue to grow as long as there are monomer molecules to add to the growing chain.

Living ring *See revolving spinning ring.*

LLDPE Abbreviation for linear low-density polyethylene.

Ln (1) Abbreviation for Lumen. (2) Abbreviation for natural logarithm, i.e., logarithm to the base e (=2.71828. . .).

Load cell An instrument, most often part of a machine for testing mechanical properties and some rheometers, that senses the force applied to the specimen (or piston).

Locking pressure The pressure applied to an injection or transfer mold to keep it closed during molding.

Locking ring A slotted plate in an injection or transfer mold that locks the parts of the mold together and prevents the mold from opening while the plastic is being injected.

Loft The properties of firmness, resilience, and bulk of a fiber batting, yarn, fabric, or other textile material.

Logarithm \\ˈló- gə-ˌri-thəm\\ *n.* The exponent that indicates the power to which a number is raised to produce a given number. A common logarithm (log) has a base of 10 and a natural (ln) or Napierian logarithm has a base of e (irrational number 2.71828. . .). Conversion of log to ln: $\ln N = 2.303 \log N$.

Logarithmic decrement \\ˌló-gə-ˈrith-mik ˈde-krə-mənt\\ (Δ) In a damped, vibrating system, the natural logarithm of the ratio of the amplitude of any oscillation to the amplitude of the succeeding oscillation. Where damping is mild, the ratio of the amplitude of any oscillation to the amplitude of the succeeding oscillation. Where damping is mild, the ratio of amplitudes several vibrations apart will usually give a more accurate estimate. The equations are:

$$\Delta = \ln(A_i/A_i + 1)$$

and

$$\Delta = (l/n)\ln(A_i/A_i + n).$$

Logarithmic viscosity number The IUPAC term for inherent viscosity.

Log normal distribution (logarithmic normal distribution). A statistical probability-density function, characterized by two parameters, that can sometimes provide a faithful representation of a polymer's molecular-weight distribution or the distribution of particle sizes in ground, brittle materials. It is a variant of the familiar normal or Gaussian distribution in which the logarithm of the measured quantity replaces the quantity itself. Its mathematical for is

$$f(x) = \frac{1}{\sqrt{2\pi}\beta} x^{-1} e^{-(\ln x - \alpha)/2\beta^2} dx$$

or

$$f(\ln x) = \frac{1}{\sqrt{2\pi}\beta} e^{-(\ln x - \alpha)/2\beta^2} d \ln x,$$

α and β are the mean and standard deviation of $\ln x$. Anti-$\ln \alpha$ is called the *log mean* of x. To test the suitability of this distribution, one plots the cumulative percent of members having weights or sizes below x, versus x, on lognormal probability paper and looks for linearity in the plot.

Logotype (or logo) \\ˈló-gə-ˌtīp, ˈlä-\\ (ca. 1816) *n.* Name of a company or product in a unique design used as a trademark in advertising.

Logu Sluggish, low snap or recovery. A condition formed in poorly cured or overloaded vulcanizers.

London dispersion forces (London forces) [fr. Fritz London (1930) *n.* Identifies weak intermolecular forces based on transient dipole interactions. One of *van der Waals forces*, also called dispersion forces, but distinct from dipole–dipole forces. These forces arise from momentary fluctuations in the electron charge cloud density in a atom or molecule. Changes in symmetry of the electron cloud that causes a momentary dipole moment and attractive/repulsive charges. The larger a molecule is and the more electrons it has, the more polarizable it will be, and thus the larger the London forces can be. Molecular shape and other factors are also important. Whitten KW, Davis RE, Davis E, Peck LM, Stanley GG (2003) General chemistry. Brookes/Cole, New York.

L

Lone pair A pair of electrons, which belongs to only one atom and hence is not shared (VB model).

Long oil High ratio of oil to resin in a medium. (1) *Long oil alkyd* – an alkyd resin containing more than 60% of oil as a modifying agent. (2) *Long oil varnish* – an oleoresinous varnish, other than alkyd, containing more than 25gal of oil/100lb of resin. A long oil varnish is usually slower drying, tougher and more elastic than a short oil varnish.

See also short oil varnish and medium oil varnish.

Long-chain branching In a polymer's structure, the presence of arms (branches) off the main chain that are about as long as the main chain. In making low-density polyethylene, a typical molecule may contain 50 short branches and only one or zero long branch, yet the presence of long branches greatly broadens the molecular-weight distribution. Polymers containing long branches tend to be less crystalline than the corresponding polymers without long branches.

Long-fiber-reinforced thermoplastic A palletized thermoplastic resin for injection molding, usually nylon 6/6 or polypropylene, produced by pultrusion from continuous-filament glass yarn, and cut to lengths of 9–13mm (about three times the length of short-fiber pellets). Nylon containing 50wt.% longer-fiber glass is about 15% stronger and stiffer than its short-fiber mate, with double the notched-Izod impact strength.

Long staple A long fiber. In reference to cotton, long staple indicates a fiber length of not less that 1–1/8in. In reference to wool, the term indicates fiber 3–4in. long suitable for combing.

Longo A colloquialism used in the filament-winding industry, designating an article that is wound longitudinally or with a low-angle helix.

Loom \ˈlüm\ [ME *lome* tool, loom, fr. OE *gelōma* tool; akin to MD al*lame* tool] (15c) *n.* A machine for weaving fabric by interlacing a series of vertical, parallel threads (the warp) with a series of horizontal, parallel threads (the filling). The warp yarns from a beam pass through the heddles and reed, and the filling is shot through the "shed" of warp threads by means of a shuttle or other device and is settled in place by the reed and lay. The woven fabric is then wound on a cloth beam. The primary distinction between different types of looms is the manner of filling insertion. The principal elements of any type of loom are the shedding, picking, and beating-up devices.

In shedding, a path is formed for the filling by raising some warp threads while others are left down. Picking consists essentially of projecting the filling yarn from one side of the loom to the other. Beating-up forces the pick that has just been left in the shed, up to the fell of the fabric. This is accomplished by the reed, which is brought forward with some force by the lay. Kadolph SJJ, Langford AL (2001) Textiles. Pearson Education, New York.

Loom Barré A repeated unevenness in the fabric, usually running from selvage to selvage, and caused by uneven let-off or take-up or by a loose crank arm.

Loom-finished A term describing fabric that is sold in the condition in which it comes from the loom.

Loom fly Waste fibers that are inadvertently woven into a fabric.

Loop elongation The maximum extension of a looped yarn at maximum load, expressed as a percentage of the original gauge length. Kadolph SJJ, Langford AL (2001) Textiles. Pearson Education, New York.

Loop pile Carpet construction in which the tufts are formed into loops from the supply yarn.

Loop selvage A weaving defect at the selvage of excessive thickness or irregular filling loops that extend beyond the outside selvages.

Loop tenacity The strength of a compound strand formed when one strand of yarn is looped through another strand, then broken. It is the breaking load in grams divided by twice the measured yarn denier or decitex. Loop tenacity, when compared with standard tenacity measurements, is an indication of the brittleness of a fiber.

Loop test A simple test (ASTM 3291) for evaluating the compatibility of vinyl resin plasticizers based on the fact that a material under compressive stress will exude plasticizer more rapidly. A specimen in sheet form is folded double, forming a loop with internal radius equal to the sheet thickness. At intervals, the bend of the loop is reversed 360° and the former inside surface of the loop is examined for evidence of plasticizer spewing.

Looped filling A woven-in loop caused by the filling sloughing off the quill or by the shuttle rebounding in the box.

Looped pile A pile surface made of uncut looped yarns.

Looped year *See kink.*

Looping (1832) *v.* Generally, a method of uniting knit fabrics by joining two courses of loops on a machine called a looper.

Looping bar A bar inserted in the bottom of an extrusion métier around which the dried filaments pass as they leave the spinning cabinet.

Loopy yarn *See textured yarns.*

Loose edge *See slack selvage.*

Loose end *See tight or loose end.*

Loose filling A fabric defect that is usually seen as short, loose places in the filling caused by too little tension on the yarn in the shuttle or by the shuttle rebounding in the box. Loose filling can often be felt by an examiner when passing a hand over the surface of the fabric.

Loose pick *See slack pick.*

Loose punch A male portion of a mold constructed so that it remains attached to the molding when the press is opened, to be removed from the part after demolding.

Lopac Co-polymer from methacrylonitrile and styrene or α-methyl styrene (9:1). Manufactured by Monsanto, USA.

Loss angle The inverse tangent of the electrical dissipation factor.
See dielectric loss angle.

L

Loss compliance The "imaginary" part of the complex compliance.
See compliance and complex modulus.

Loss dielectric A loss of energy evidenced by the rise in heat of a dielectric placed in an alternating electric field. It is usually observed as a frequency-dependent conductivity.

Loss factor The product of a power factor and dielectric constant of a dielectric material.

Loss modulus "*G*" – the component of applied shear stress, which is 90° out of phase with the shear strain, divided by the strain. Syn: viscous modulus, hysteretic modulus, imaginary modulus, and out-of-phase modulus.

Loss of drier *See drier dissipation.*

Loss of gloss A paint defect in which a dried film of paint loses gloss, usually over a period of several weeks.

L

Lost end An end on a section or tricot beam that has been broken at some stage in warping and has not been repaired by a knot.

Lost-wax process *See investment casting.*

Lot \ˈlät\ [ME, fr. OE *hlot*; akin to OHGr *hlōz*] (before 12c) *n.* A unit of production or a group of other units or packages that is taken for sampling or statistical examination, having one or more common properties and being readily separable from other similar units.

Lot number The number used by the manufacturer to identify an entity of production.

Louvers \ˈlü-vər-\ [ME *lover*, fr. MF *lovier*] (14c) *n.* Slats placed at an angle, as in shutters.

Lovibond color system A system of color specification by means of numbers proportional to the optical densities of three glass filters (yellow, red, and blue) required to modify a standard light source (such as daylight or incandescent lamp light) to produce a color match. The carefully calibrated glasses should be used as described by the manufacturer, Tintometer Ltd., UK, in order that results in different laboratories will agree. Lovibond specifications have long been used as the basis for describing the colors of edible oils and vegetable oils used for paint vehicles.

Lovibond tintometer Optical comparison instrument manufactured by Tintometer Ltd., for use of Lovibond glasses.
See lovibond color system.

Low-angle laser-light scattering A technique for determining weight-average molecular weights of polymers in solution. The low angle – 2° to 10° – reduces the number of measurements needed and simplifies their interpretation, as compared with conventional, wide-angle light scattering. Kamide K, Dobashi T (2000) Physical chemistry of polymer solutions. Elsevier, New York. Berne BJ (2000) Dynamic light scattering: applications to chemistry, biology and physics. Dover Publications, New York.

Low-density polyethylene (LDPE) This term is generally considered to include polyethylenes having densities between 0.915 and 0.925 g/cm^3. In LDPE, the ethylene molecules are linked in random fashion, with many side branches, mostly short ones. This branching prevents the formation of a closely knit pattern, resulting in material that is relatively soft, flexible, and tough, and which will withstand moderate heat.
See also high-density polyethylene and polyethylene.

Lower explosive limit Lower limit of flammability or explosibility of a gas or vapor at ordinary ambient temperatures expressed in percent of the gas vapor in air by volume. Tests for comparative flammability of liquids, UI 340. Laboratories Incorporated Underwriters, New York, 1997.
See explosive limits.

Low-pressure injection molding A term sometimes used for the process of injecting a fluid material such as a vinyl plastisol into a closed mold, using a grease gun or similar low-pressure equipment.

Low-pressure laminate Various definitions place the upper limit of pressure for this term at from 6.9MPa down to pressures obtained by mere contact of the piles. According to ASTM D 883, the upper limit is 1.4MPa (200psi). The Decorative Board Section of the National Electrical Manufacturers' Association (NEMA) has recommended abandonment of the term "low-pressure laminate" in favor of *decorative board* in the case of "...a product resulting from the impregnation or coating of a decorative web of cloth, paper, or other carrying media with a thermosetting resin and consolidation of one or more of these webs with a cellulosic substrate under heat and pressure of less than 500psi". This includes all boards that were formerly called low-pressure melamine and polyester laminates, but not vinyls.

See also contact-pressure molding and laminate.

Low-pressure molding Molding or laminating in which the pressure is 1.4MPa (200psi) or less (ASTM D 883).

Low-pressure resin *See contact-pressure resin.*

Low rows A carpet defect characterized by rows of unusually low pile height across the width of the goods.

Low-temperature flexibility All plastics that are flexible at room temperature become less so as they are chilled, finally becoming brittle at some low temperature. This property is often measured by torsional tests over wide ranges of temperature, from which apparent moduli of elasticity are calculated. Some relevant ASTM tests are D 1043, D 3295, D 3296, D 3374 (Section 07.02), and D 1055 (Section 09.01).

See also brittleness temperature and clashberg point.

***L*-sealer** A heat-sealing device, used in packaging, that seals a length of flat, folded film on the edge opposite the fold and simultaneously seals a strip across the width at 90° from the edge seals. The article to be packaged may be inserted between the two layers of folded film prior to sealing. When it is desired to sever the continuous length of sealed compartments into individual packages, a heated wire or knife is incorporated between two sealing bars that form the bottom of the *L*. These bars then make the top seal of the filled bag and the bottom seal of the next bag to be filled.

Luana A fabric characterized by a crosswise rib effect, usually made with a filament yarn warp and a spun yarn filling.

Lumband oil This oil is obtained from the nuts of the tree, *Aleurites moluccana*. Although a product of an Aleurites tree, it contains no elaeostearin. It dries somewhat better than soybean oil.

Also called candlenut oil.

Lubricant \ˈlü-bri-kənt\ (ca. 1828) *n.* A substance that tends to make surfaces slippery, reduce friction, and prevent adhesion.

Lubricant bloom *See bloom.* The term *lubricant bloom* should only be used when the exudation is known to be caused by a lubricant contained in the plastic compound or applied to it during processing.

Lucite® \ˈlü-ˌsīt\ Poly(methyl methacrylate). DuPont's trade name for methacrylate-ester monomers and polymers (*acrylic*), including PMMA and several other resins, and for certain products made from such resins.

Lumen \ˈlü-mən\ [NL *lumin-*, *lumen*, fr. L, light, air shaft, opening] (1873) (lm) *n.* The SI unit of luminous flux.

Luminance \ˈlü-mə-nən(t)s\ (1880) *n.* The luminous intensity of light reflected or transmitted by a material in a given direction per unit of projected area of the material, as viewed from that direction.

Luminescent (1889) *n.* Emitting light not due to high temperature, usually caused by excitation by rays of a shorter wavelength. Saleh BEA, Teich MC (1991) Fundamentals of photonics. John Wiley and Sons, New York.

Luminescent pigment A pigment that produces striking effects in darkness or light. Dainth J (2004) Dictionary of chemistry. Oxford University Press, UK. Solomon DH, Hawthorne DG (1991) Chemistry of pigments and fillers. Krieger Publishing Co., New York.

See fluorescent pigment and phosphorescent pigment.

Luminosity curve, CIE Graph representing the luminous intensity of wavelengths in the visible spectrum relative to the maximum intensity at the same wavelength. The relative luminosity curve is described by the *Y*-color matching function in the CIE System. The shape of the curve depends on the area of the retina being stimulated and on the intensity of the incident light. Thus, the CIE luminosity curves for the 2° and 10° observers are slightly different, and those for daylight vision, i.e., photopic vision, and for dark-adapted vision, i.e., scotopic vision, are extremely different.

Luminous \ˈlü-mə-nəs\ [ME, fr. L *luninosus*, fr. *lumin-*, *lumen*] (15c) *adj.* (1) Adjective used to imply dependence on the spectral response characteristic of the Standard Observer defined in the CIE System. Thus, the luminous reflectance or the luminous transmittance is described by the *Y*-tristimulus value in the CIE System. The adjective is applied to many measures of light, such as intensity, density, etc., and always indicates the measures are weighted for the relative luminous sensitivity of the human observer. (2) Material emitting or spearing to be emitting, visible radiant energy.

Luminous directional reflectance Reflectance of a surface for specified directions of illumination and view is the ratio of the brightness of the surface to the brightness that an ideally diffusing, perfectly white, surface would have if illuminated and viewed in the same manner. Saleh BEA, Teich MC (1991) Fundamentals of photonics. John Wiley and Sons, New York.

Luminous energy (ca. 1931) *n.* Energy transferred in the form of visible radiation. Saleh BEA, Teich MC (1991) Fundamentals of photonics. John Wiley and Sons, New York.

Luminous flux (1925) *n.* The total visible energy emitted by a source per unit time. The SI unit is the *lumen* (lm), defined as the luminous flux emitted in a solid angle of 1 steradian (sr, the solid central angle that cuts out of a spherical surface a square whose side is equal to the radius) by a point source having a uniform intensity of 1 cd. Therefore, $1\,\text{lm} = 1\,\text{cd}\,\text{sr}$.

See flux, luminous.

Luminous paint (ca. 1889) *n.* Paint which exhibits fluorescence. (1) *Fluorescent paint* – contains pigments, which are capable of absorbing energy from the blue or ultraviolet end of the spectrum and reemitting it in the form of light in the visible wavelengths. A fluorescent paint ceases to "glow" if the activating source is removed. (2) *Phosphorescent paint* – contains pigments (phosphors) which absorb energy at one wavelength and emit it over a period of time, in the form of light at a longer wavelength in the visible spectrum. It differs from a fluorescent paint in that it continues to glow after the stimulating source has been removed. (3) *Radioactive* or *self-luminous paint* – normally, this is a phosphorescent paint containing a portion of radioactive compounds, and in such a paint the phosphor is permanently activated by absorbing energy from the bombardment by the radioactive rays and emits light in the visible spectrum. Paint/coatings dictionary. Federation of Societies for Coatings Technology, Philadelphia, Blue Bell, PA, 1978.

Luminous transmittance Syn: light transmittance.

Luparen Poly(propylene), manufactured by BASF, Germany.

Luphen Phenoplast, manufactured by BASF, Germany.

Lupolen Poly(ethylene) (high pressure), manufactured by BASF, Germany.

Luran Co-polymer from styrene/acrylonitrile, manufactured by BASF, Germany.

Luster \ˈləs-tər\ [MF *lustre*, from OI *lustro*, fr. *lustrare* to brighten, fr. L. to purify ceremonially, fr. *lustrum*] (ca. 1522) *n.* (1) Type of surface reflectance, or gloss, where the ratio of specular reflectance to diffuse reflectance is relatively high, but not so high as that from a perfect specular reflector (mirror). (2) Another term for gloss.

Lustering (1582) *v.* The finishing of yarn or fabric by means of heat, pressure, steam, friction, calendering, etc., to produce luster.

Lusterless (ca. 1522) *adj.* An adjective describing a non-glossy or non-reflecting surface with respect to illumination of the surface, dull in appearance.

Lux \ˈləks\ [L, light] (1889) (lx) *n.* The SI unit of illuminance, defined as the illuminance produced by a luminous flux of 1 lm uniformly distributed over a surface of $1\,m^2$. That is, $1\,lx = 1\,lm/m^2$.

Lycra \ˈlī-krə\. Elastomer from segments of polyether and polyurethane, manufactured by DuPont.

Lyocell fiber A manufacturing cellulose fiber made by direct dissolution of wood pulp in an amine oxide solvent, *N*-methylmorpholine-*N*-oxide. The clear solution is extruded into a dilute aqueous solution of amine oxide, which precipitates the cellulose in the form of filaments. The fiber is then washed before it is dried and finished. The solvent spinning process for making lyocell fiber is considered to be environmentally friendly because the non-toxic spinning solvent is recovered, purified, and recycled as an integral part of the manufacturing process. No chemical intermediates are formed, the minimal waste in not hazardous, and energy consumption is low. Wood pulp is a renewable resource, and the fiber is biodegradable. Characteristics: lyocell fiber is stronger than other cellulosic fibers. It is inherently absorbent, having a water imbibition of 65–75%. Lyocell retains 85% of its dry tenacity when wet, making it stronger when wet than cotton. The fiber has a density of $1.15\,g/cm^3$. End uses: lyocell fiber is suitable for blending with cotton or other manufactured fibers. Because of its molecular structure, lyocell has the tendency to develop surface fibrils that can be beneficial in the manufacture of hydroentabled and other non-wovens, and in specialty papers. For apparel uses, the fiber's unique fibrillation characteristic has enabled the development of fabrics with a soft luxurious hand. The degree of fibrillation is controlled by cellulose enzyme treatment.

Lyophilic \ˌlī-ə-ˈfi-lik\ (1911) *adj.* Describing a substance that easily forms colloidal suspensions. Such ability when the suspending medium is water is called *hydrophilic*. A PVC plastisol is an example of a lyophilic suspension.

Lyophobic \ˌlī-ə-ˈfō-bik\ (1911) *adj.* Characterizing a material, which exists in the colloidal state without any significant affinity for the medium.

Lyotropic *See liquid-crystal polymer.*

Lyotropin polymer Polymers that decompose before melting but that form liquid crystals in solution under appropriate condition. They can be extruded from high concentration dopes to give fibers of high modulus and orientation for use in advanced composites, tire cord, ballistic protective devices, etc.

M

m \ˈem\ *n.* (1) Abbreviation for meter. (2) Abbreviation for the SI prefix milli-. (3) (usually italicized) Abbreviation for chemical positional prefix meta-.

M *n.* (1) Abbreviation for prefix mega-. (2) Symbol for molecular weight. (3) Symbol for being moment.

mA *n.* Abbreviation for milliampere.

MAC *n.* Maximum allowable concentrations of solvent vapors, also known as *threshold limit values.* These values refer to air-borne concentrations of substances and represent conditions to which it is believed that nearly all workers may be repeatedly exposed, day after day, without adverse effect. *See maximum allowable concentration.*

MacAdam color difference equation *n.* A color difference equation developed by David MacAdam, which is now used as modified by Hugh Davidson and Fred Simon to incorporate the effect of lightness on the chromaticity differences:

$$\Delta E = \left[1/K(g_{11}\,\overline{\Delta x^2} + 2g_{12}\,\Delta x\,\Delta y + g_{22}\,\overline{\Delta y^2} + G\overline{\Delta Y^2}\right]^{1/7},$$

where g_{11}, $2G_{11}$, and g_{22} are the constants depending on the chromaticity coordinates, x and y, and K and G are the constants depending on the luminous reflectance or transmittance, Y. This color difference is frequently calculated from charts prepared by Simon and Goodwin, which have the required constants built-in.

MacAdam limits *n.* The theoretical limit or gamut of colors, which can be obtained at various limits of luminance (Y). Thus, the gamut of colors, which can be obtained theoretically, decreases steadily as the luminance (Y) increases.

Machinability \mə-ˌshē-nə-ˈbi-lə-tē\ (ca. 1864) *vt.* (1) In fabricating materials by such operations as drilling, lathe-turning, and milling, the ease with which the material is removed.

Machine-printing *n.* The method by which the bulk of modern wallpapers are produced. Machine-printing employs a rotary press and a series of cylinders or rollers to turn out wallpaper at high speeds. Raw paper stock is first given a coating of the ground color by a special machine, after which the paper proceeds in a continuous web to the rotary press where the top colors are applied, and it is then festooned on specially heated drying racks.

Machine shot capacity *See shot capacity.*

Machine twist *n.* A hard-twist sewing thread, usually of three-ply construction spun with S-twists and plied with Z-twist, especially made for use in sewing machines.

M

Machining of plastics *n.* Many of the machining operations used for metals are applicable to rigid plastics, with appropriate variations in tooling and speeds [*see machinability* (1), above]. Among such operations are blanking, boring, drilling, grinding, milling, planning, punching, routing, sanding, sawing, shaping, tapping, threading, and turning.

Mach number (N_{Ma}) *n.* The ratio of a fluid velocity or the relative velocity of an object moving through a fluid to the velocity of sound in the fluid. All fluids (liquids and gases) have Mach numbers.

Macrolattice *n.* A repeating structure in very small microfibrils of alternating crystalline and amorphous regions. Yarn properties are thought to be governed by morphology at the macrolattice scale.

Macromolecule \ˌma-krō-ˈmä-li-ˌkyü(ə)l\ [ISV] (ca. 1929) *n.* The large ("giant") molecules that make up high polymers, both natural and synthetic. Each macromolecule may contain hundreds of thousands of atoms.

See polymer.

Macromonomers *n.* High molecular weight monomers.

Also called macromens.

Macroscopic \ˌma-krə-ˈskä-pik\ [ISV *macr-* + *scopic* (as in *microscopic*)] (1872) *adj.* Visible to the naked eye, as opposed to *microscopic.*

Madder lake *n.* Lightfast, non-bleeding, red-colored pigment prepared from the coloring matter of madder root.

Madder lakes *n.* A class of solvent-resistant and lightfast pigments; generally dirty in appearance.

Madras \ˈma-drəs; mə-ˈdras, -ˈdräs\ [*Madras*, India] (ca. 1830) *n.* A lightweight, plain weave fabric with a striped, checked, or plaid pattern. True madras is "guaranteed to bleed".

Magdala red *n.* $C_{30}H_{21}N_4Cl$. Red dyestuff.

Known also as naphthalene red.

Magnesia \mag-ˈnē-shə, -zhə\ [NL, fr. *magnes carneus*, a white earth, literally, flesh magnet] (1755) *n.* MgO. (1) Magnesium oxide. (2) Sometimes used incorrectly in the printing ink industry to mean magnesium carbonate. Syn: magnesium oxide.

Magnesite \ˈmag-nə-ˌsīt\ (1815) *n.* $MgCO_3$. Mineral, magnesium carbonate, principally used as a filler or extender.

Magnesite floor *n.* Hard composition floors in which magnesium oxychloride is the binder. This binder is formed in laying the floor, when magnesium oxide is combined with a strong solution of magnesium chloride. Fillers which may be added to this binder are: asbestos, cork, sand, wood flour, marble dust, talc, leather, etc. This great variety of fillers produces magnesite floors having variable porosity, resiliency, appearance, and durability.

Magnesium carbonate (1903) (magnesia alba, precipitated magnesium carbonate) *n.* $MgCO_3$. A white powder of low density, prepared by metathesis, used as a filler or modifier in phenolic resins. This carbonate also occurs naturally as *magnesite.*

Magnesium carbonate, precipitated *n.* Chemically, this is the same as magnesite, but physically it has a much better color, in bulk being a very intense white. It is usually a very fine light powder of rather high oil absorption.

Magnesium chloride (ca. 1910) *n.* A bitter deliquescent salt $MgCl_2$ used especially as a source of magnesium metal.

Magnesium glycerophosphate *n.* $MgPO_4$ $C_3H_5(OH)_2$. A colorless powder, derived by the action of glycerophosphoric acid on magnesium hydroxide, used as a stabilizer for plastics.

Magnesium hydrogen phosphate trihydrate *n.* Dibasic magnesium phosphate, magnesium monohydrogen *ortho*phosphoric acid with magnesium oxide, used as a non-toxic stabilizer for plastics.

Magnesium hydroxide (ca. 1909) *n.* $Mg(OH)_2$. Used as a thickening agent for polyester resins. Its action is slower than that of magnesium oxide.

Magnesium hydroxychloride cement (Sorel cement, magnesium oxychloride cement) *n.* A mixture of magnesium chloride and magnesium oxide that reacts with water to form a solid mass, presumed to be magnesium hydroxychloride, Mg(OH)Cl. It has been useful as an intumescent coating for urethane foams and other materials such as polystyrenes, nylons, acetals, polyesters, and silicones.

Magnesium oxide (ca. 1909) (magnesia, periclase) *n.* A white powder used as filler and as a thickening agent in polyester resins. It occurs naturally as the mineral *periclase*, but it is usually made in purer form by calcining magnesium hydroxide or carbonate.

Magnesium phosphate, dibase *n.* See *magnesium hydrogen phosphate trihydrate.*

Magnesium phosphate, monobasic (magnesium dihydrogen phosphate) *n.* $Mg(H_2PO_4)_2{\cdot}2H_2O$. A white, hygroscopic, crystalline powder derived by reacting phosphoric acid with magnesium hydroxide. It is used as a flame retardant and stabilizer for plastics.

Magnesium phosphate, tribasic *n.* $Mg_3(PO_4)_2{\cdot}8H_2O$ or $\cdot 4H_2O$. A fine, soft white powder derived by reacting magnesium oxide and phosphoric acid at a high temperature, used as a non-toxic stabilizer.

Magnesium silicate, fibrous *n.* $3MgO{\cdot}2SiO_2{\cdot}2H_2O$. A fibrous chrysotile mineral white to gray powder, chemically inert used as extender and/or filler in paints and caulks. Pigment grades are used for their high temperature resistance, high oil absorption and water demand. Density, 2.48–2.56 g/cm^3 (20.7–21.3 lb/gal); O.A., 50–180. Syn: asbestos, chrysotile, and fibrous asbestos.

Magnesium silicate, non-fibrous *n.* $3MgO{\cdot}4SiO_2{\cdot}H_2O$. Pigment White 26 (77718). A hydrated magnesium silicate extender of filler of wide range of composition. Soft white, gray or yellow shade. Natural product (talc) used in paint, rubber, ceramics, paper and roofing compounds. Density, 2.7–2.8 g/cm^3 (22.5–23.3 lb/gal); O.A., 30–50; particle size, 0.5–2.5 μm. Syn: talc and asbestine.

Magnesium soap *n.* A magnesium salt of a fatty acid, e.g., magnesium stearate, precipitated by an inorganic magnesium salt from a solution of sodium or potassium soaps. *See also soap, metallic.*

Magnesium soaps Saponification products of magnesium and various fatty acids.

Magnesium stearate *n.* $Mg(OOCC_{17}H_{35})_2$. A white, soft powder used as a lubricant and stabilizer.

Magnetic field due to a current *n.* The intensity of the magnetic field in oersted at the center of a circular conductor of radius r in which a current I in absolute electromagnetic units is flowing,

$$H = \frac{2\pi I}{r}.$$

If the circular coil has n turns the magnetic intensity at the center is

$$H = \frac{2\pi n I}{r}.$$

The magnetic field in a long solenoid of n turns per centimeter carrying a current I in absolute electromagnetic units

$$H = 4\pi n I.$$

If I is given in amperes the above formulae becomes

$$H = \frac{2\pi I}{10r}, \quad H = \frac{2\pi n I}{10r}, \quad H = \frac{4\pi I}{10}.$$

Magnetic field due to a magnet *n.* At a point on the magnetic axis prolonged, at a distance r cm from the center of the magnet of length $2l$ whose poles are $+m$ and $-m$ and magnetic moment M, the field strength is oersted is

$$H = \frac{4mlr}{(r^2 - l^2)^2}.$$

If r is large compared with l,

$$H = \frac{2M}{r^3}.$$

At a point on a line bisecting the magnet at right angles, with corresponding symbols,

$$H = \frac{2ml}{(r^2 + l^2)^{3/2}}.$$

For large value of *r*,

$$H = \frac{M}{r^3}.$$

Magnetic field intensity or magnetizing force *n.* It is measured by the force acting on unit pole. Unit field intensity, the oersted is that field which exerts a force of 1 dyne on unit magnetic pole. The field intensity is also specified by the number of lines of force intersecting unit area normal to the field, equal numerically to the field strength in oersted. Magnetizing force is measured by the space rate of variation of magnetic potential and as such its unit may be the **Gilbert per centimeter**. The gamma (γ) is equivalent to 0.00001 oersted. Dimensions

$$\left[\varepsilon^{1/2}\,\mathrm{M}^{1/2}\,\mathrm{L}^{1/2}\,\mathrm{T}^{-2}\right], \quad \left[\mu^{-1/2}\,\mathrm{M}^{1/2}\,\mathrm{L}^{-1/2}\,\mathrm{T}^{-1}\right].$$

Magnetic filler *n.* Any permanently magnetizable material in powder form that may be incorporated into plastics to produce molded or extruded-strip magnets. Major ones in use are Alnico, rare earths, and, most used in plastics, hard ferrite.

Magnetic flux (1896) *n.* Through any area perpendicular to a magnetic field is measured as the product of the area by the field strength. The unit of magnetic flux, the **Maxwell,** is the flux through a square centimeter normal to a field of 1 G. The line is also a unit of flux. It is equivalent to the Maxwell. Dimensions

$$\left[\varepsilon^{-1/2}\,\mathrm{M}^{1/2}\,\mathrm{L}^{1/2}\right], \quad \left[\mu^{1/2}\,\mathrm{M}^{1/2}\,\mathrm{L}^{1/2}\,\mathrm{T}^{-1}\right].$$

Magnetic induction *n.* Results when any substance is subjected to a magnetic field is measured as the magnetic flux per unit area taken perpendicular to the direction of the flux. The unit is the Maxwell per square centimeter or its equivalent, the gauss. Dimensions

$$\left[\varepsilon^{-1/2}\,\mathrm{M}^{1/2}\,\mathrm{L}^{3/2}\right], \quad \left[\mu^{1/2}\,\mathrm{M}^{1/2}\,\mathrm{L}^{-1/2}\,\mathrm{T}^{-1}\right].$$

If a substance of permeability of μ is placed in a magnetic field *H*, then the magnetic induction in the substance is

$$M = \mu H.$$

If *I* is the magnetic moment for unit volume, or intensity of magnetization

$$M = H + 4\pi I.$$

The susceptibility,

$$k = \frac{I}{H}, \quad \mu = 1 + 4\pi\kappa.$$

Magnetic inks *n.* Inks made with pigments, which can be magnetized after printing. The printed characters can be recognized later by electronic reading equipment.

Magnetic moment (1865) *n.* The magnetic moment of a magnet is measured by the torque experienced when it is at right angles to a uniform field of unit intensity. The value of the magnetic moment is given by the product of the magnetic pole strength by the distance between the poles. Unit magnetic moment is that possessed by a magnet formed by two poles of opposite signs and of unit strength, 1 cm apart. Dimensions

$$\left[\mu^{1/2}\,\mathrm{M}^{1/2}\,\mathrm{L}^{-1/2}\mathrm{T}^{-1}\right], \quad \left[\varepsilon^{-1/2}\,\mathrm{M}^{1/2}\,\mathrm{L}^{3/2}\right].$$

If the poles are separated by a distance, which is great compared with the dimensions of the magnet, then the magnetic moment of a magnet of length *1* whose poles have values of $+m$ and $-m$ is $m = ml$.

Magnetic permeability *n.* A property of materials modifying the action of magnetic poles placed therein and modifying the magnetic induction resulting when the material is subjected to a magnetic field or magnetizing force. The permeability of a substance may be defined as the ratio of the magnetic induction in the substance to the magnetizing field to which it is subjected. The permeability of a vacuum is unity. Dimensions

$$[\varepsilon^{-1}\,L^{-2}\,T^{2}], \quad [\mu].$$

Magnetic pole or quantity of magnetism *n.* Two unit quantities of magnetism concentrated at points unit distance apart in a vacuum repeal each other with unit force. If the distance involved in 1 cm and the force 1 dyne, the quantity of magnetism at each point is one cgs unit of magnetism. Dimensions

$$\left[\varepsilon^{-1/2}\,M^{1/2}L^{1/2}\right], \quad \left[\mu^{1/2}\,M^{1/2}\,L^{3/2}\,T^{-1}\right].$$

Magnetic potential or magnetomotive force *n.* At a point is measured by the work required to bring until positive pole from an infinite distance (zero potential) to the point. The unit is the *Gilbert*, that magnetic potential against which an erg of work is done when unit magnetic pole is transferred. Dimensions

$$\left[\varepsilon^{1/2}\,M^{1/2}\,L^{3/2}\,T^{-2}\right], \quad \left[\mu^{-1/2}\,M^{1/2}\,L^{1/2}\,T^{-1}\right].$$

Magnetic quantum number, m_l (1923) *n.* A quantum number, which indicates the orbital occupied by an electron.

Magnetic separator *n.* A device that removes tramp iron and steel from a stream of mainly non-magnetic material, such as reground plastic or mixed wastes, by passing the stream close to strong magnets. Some design parameters for magnetic separators are given in Section 21 of Perry and Green (1997). Perry RH, Green DW (eds) (1997) Chemical engineers' handbook, 6th edn. (and the two preceding editions). McGraw-Hill, New York

Magnetite black \\ˈmag-nə-ˌtīt ˈblak\\ (1851) *n.* Fe_3O_4. Magnetic iron oxide.
See black iron oxide.

Magnification, empty *n.* A higher magnification than necessary to resolve detail.

Magnification, maximum useful (MUM) *n.* The maximum magnification necessary to resolve detail. Magnification in excess of MUM gives no additional resolving power. It can usually be estimated as being 1000 times the NA of the objective.

Magnifying power *n.* The magnifying power of an optical instrument is the ratio of the angle subtended by the image of the object seen through the instrument to the angle subtended by the object when seen by the unaided eye. In the case of the microscope or simple magnifier the object as viewed by the unaided eye is supposed to be a distance of 25 cm (10 in.).

M

Mahlstick \\ˈmól-\\ *variant of maulstick.* Long stick, padded at one end, on which a painter can rest his hand to steady if when working.

Mahogany sulfonates *n.* Soaps, the sodium salts of sulfonic acids from petroleum refining sludge; used in synthetic resin production, as are sorbitan oleates and laurates, polyoxyethylene esters.

Maintenance paints *n.* Coatings used to maintain manufacturing plants, offices, stores and other commercial structures, hospitals and nursing homes, schools and universities, government and public buildings, and both building and non-building requirements in such areas as public utilities, railroads, roads, and highways, and including industrial paint, other than the original coatings, the primary function of which is protection. Residential maintenance is excluded.
See also industrial maintenance paints.

Makeready *n.* The preparation and correction of the printing plates, before starting the printing run, to insure uniformly clean impressions of optimum quality. All preparatory operations preceding a production run.

Makrolon Polycarbonate from bisphenol A and phosgene base units. Manufactured by Bayer, Germany.

Malachite \ˈma-lə-ˌkīt\ *n.* [alt of ME *melochites*, fr. L *molochites*, fr. Gk *molchitēs*, fr. *molochē*, *malchē* mallow] (1656) $CuCO_3 \cdot Cu(OH)_2$. Basic carbonate of copper, which occurs naturally. The color varies from a bright emerald to a dark green.

Malachite green *n.* (1) Bluish-green dyestuff made from dimethyl aniline and benzaldehyde. It is often sold in the form of its oxalate. (2) A green lake pigment produced for a basic dye, used in the manufacture of printing inks.

M

Maleic Acid \mə-ˈlē-ik-, -ˈlā-\ *n.* [F, *acide maléique*, alter. of *acide malique* malic acid, fr. its formation by dehydration of malic acid] (1857). $COOH(CH)_2COOH$. Dibasic acid used in the manufacture of synthetic resins.

Maleic anhydride (1857) (2,5-furandione) *n.* A compound crystallizing as colorless needles, obtained by passing a mixture of benzene and air over a heated vanadium pentoxide catalyst, and having the structure shown below. It has many applications in plastics, including the production of alkyd, polyester, and vinyl-copolymer resins, and as a curing agent for thermosetting resins such as phenolics and ureas. About half the maleic anhydride produced in the USA is used in the manufacture of unsaturated polyester resins, to which it imparts fast curing and high strength. Used in manufacturing synthetic resins and maleinized oils, mp, 56°C; bp, 202°C; acid value, 1.143.

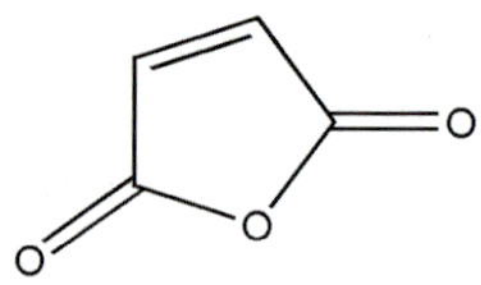

Maleic anhydride value *See diene value or number.*

Maleic ester resin *n.* A synthetic resin made from maleic acid or maleic anhydride and a polyhydric alcohol.

Maleic resin *n.* A resin made from a natural resin and maleic anhydride or maleic acid.

Maleic resins *n.* A class of resins obtained from the condensation of maleic anhydride with rosin, terpenes, etc.

Maleic value Another name for diene value.

Maleinized oil *n.* Oil which has been reacted, through its double bonds, with maleic anhydride.

Mallory fatigue test *n.* A test to measure the endurance properties of tire cord.

Maltese cross (1877) *n.* A dark shadow, having the shape of a maltese cross, seen in polymer (e.g., polyethylene) spherulities when viewed under a polarizing microscope.

MAN Abbreviation for methacrylonitrile.

Mandrel \ˈman-drəl\ [prob. mod. of F *mandrin*] (1665) *n.* (1) The core around which paper, fabric, or resin-impregnated fibrous glass is wound to form pipes or tubes. (2) In extrusion, an extension of the core of a pipe or tubing die, internally cooled by circulating water or other fluid, that guides and cools the internal surface of the tube as it emerges from the die proper. The mandrel is an important determiner of the final internal diameter of the tube.

Mandrel test *n.* Test for determining the flexibility and adhesion of surface coatings, so named because it involves the bending of coated metal panels around mandrels.

Manganese black \ˈmaŋ-gə-ˌnēz \. MnO_2. Manganese dioxide. A black pigment.

Principal uses are as a drier and as a colorant for ceramics.

Manganese brown *n.* There are two types of manganese browns: (1) Burnt turkey umber. (2) A brown oxide pigment, made artificially as a by-product from chlorine manufacture.

Manganese dioxide (1882) *n.* MnO_2. A dark insoluble compound used especially as an oxidizing agent, as a depolarizer of dry cells, and in making glass and ceramics.

Manganese driers *n.* (1) Material containing chemically combined manganese used to accelerate the oxidation and polymerization of an ink film. (2) These include manganese dioxide, the hydrated oxide, manganese acetate, sulfate and borate. The organic driers are salts of various organic acids such as naphthenic or 2-ethyl hexoic. Manganese driers are characterized by their reddish-brown colors and their surface drying activity.

Manganese green *n.* Strong green pigment, with good alkali resistance, prepared by roasting manganese dioxide and barium hydroxide together under oxidizing conditions.

Manganese violet *See mineral violet.*

Manifold \ˈma-nə-ˌfōld\ (1855) *n.* A pipe or channel with several inlets or outlets. With reference to blow molding, extrusion, and injection molding, a manifold is a piping or distribution system that receives the outflow of the extruder or molder and divides or distributes it to feed several blow-molding heads or injection nozzles.

Manila \mə-ˈni-lə\ (1834) *adj.* Fiber obtained from the leaf stalks of the abaca plant. It is generally used for cordage.

Manila copal *n.* Natural resins, two types of which are used in varnish manufacture, in which they are described as hard and soft manilas. The hard type requires running and is used to some extent in oil varnishes. The soft type is readily soluble in industrial alcohol and forms the basis of spirit paper and other air-drying varnishes; the native name for the soft type, obtained by tapping, is Melengket.

Manjak *n.* Intense black, naturally occurring asphaltum, obtained from Barbados. It differs from other asphaltums used in the trade by reason of its unusual staining power and difficult solubility. Prolonged high-temperature treatment is necessary in order to effect a reasonable solution in drying oils, and a substantial amount of mineral matter always remains undissolved. It is used alone, or with gilsonite, in black bituminous finishes of many types. Syn: glance pitch.

Man-made fiber Syn: synthetic fiber.

Mannich reaction *n.* The condensation of ammonia or a primary or secondary amine with formaldehyde and a compound containing at least one hydrogen atom of pronounced activity. The active hydrogen replaced by an aminomethyl or substituted aminomethyl group. This reaction has been employed in producing "mannich polyols" for use in making urethane foams.

Mannite \ˈma-ˌnīt\ [F, fr. *manna*, fr. LL] (1830) *n.*

See mannitol.

Mannitol \ˈma-nə-ˌtól\ [ISV] (1879) *n.* $C_6H_8(OH)_6$. A hexahydric alcohol which has been used in the production of synthetic oils and alkyd resins. Bp, 278°C/1 mmHg; mp, 166°C.

Known also as mannite.

Manufactured fiber *n.* A class name for various genera of fibers (including filaments) produced from fiber-forming substances which may be: (1) Polymers synthesized from chemical compounds, e.g., acrylic, nylon, polyester, polyethylene, polyurethane, and polyvinyl fibers. (2) Modified or transformed natural polymers, e.g., alginic and cellulose-based fibers such as acetates and rayons. (3) Minerals, e.g., glasses. The term manufactured usually refers to all chemically produced fibers to distinguish them from the truly natural fibers such as cotton, wool, silk, flax, etc.

Manufactured unit *n.* A quantity of finished adhesive or finished adhesive component, processed at one time. *Note* — The manufactured unit may be a batch or a part thereof.

Marble \ˈmär-bəl\ [ME, fr. OF *marbre*, fr. L *marmor*, fr. Gk *marmaros*] (12c) *n.* (1) Limestone that has crystallized to varying extent, often with veined inclusions, and occurring in many colors. Its preponderant constituent is calcium carbonate. (2) A smooth round sphere of any hard non-metal in the size range from about 0.7 to 2.5 cm.

Marble flour *See calcium carbonate, natural.*

Marbling, marbleizing *n.* Imitating with finishing materials, as in antiquing, the figure and texture of polished marble or other decorative stones, usually by stippling or mottling in conjunction with graining, scratching and spattering.

March, non-conditional *See non-conditional match.*

Margaric acid *See daturic acid.*

Marine borers *n.* Mollusks and crustaceans that attack submerged wood in salt and brackish water.

Marine coatings *n.* Paints and varnishes specifically formulated to withstand water immersion and exposure to marine atmosphere. *See also spar varnish.*

Marine varnishes *See marine coatings and spar varnish.*

Marker *n.* In the floor coverings industry, a distinctive threadline in the back of a carpet that enables the installer to assemble breadths of carpet so that the pile lays in one direction or so that patterns match.

Mark–Houwink equation *n.* Also referred to as Kuhn–Mark–Houwink–Sakurada equation; allows prediction of the viscosity average molecular weight M_v for a specific polymer in a dilute solution of solvent by $[\eta] = KM_v^a$, where K is a constant for the respective material and a is a branching coefficient; K and a (sometimes a^η) can be determined by a plot of log $[\eta]$ versus $\log M_v^a$ and the slope is a and intercept on the Y-axis is K. Kamide K, Dobashi T (2000) Physical chemistry of polymer solutions. Elsevier, New York. Mark JE (ed) (1996) Physical properties of polymers handbook. Springer-Verlag, New York. Elias HG (1977) Macromolecules, vols 1–2. Plenum Press, New York.

Marking nut oil *n.* Oil that resembles cashew nut shell liquid in that it is phenolic and quite unlike the glyceride vegetable oils. *Known also as dhobi marking nut oil or bhilawan oil.*

Marl A yarn made from two rovings of contrasting colors drafted together, then spun. Provides a mottled effect.

Marlex Poly(ethylene), manufactured by Phillips, USA.

Marouflage *v.* To glue a canvas to a wall which is to be covered by a mural painting.

Marquardt index *n.* In an infrared-absorption study of the cure advancement of a phenolic resin, the Marquardt index is the numerical difference in percent transmission between the absorption peaks at 12.2 and 13.3 μm. As the resin cure progresses, the intensity of the 13.3-μm absorption increases more rapidly than that of the initially stronger 12.2-μm peak.

Marquetry \ˈmär-kə-trē\ [MF *marqueterie*, fr. *marqueter* to checker, inlay, fr. *marque* mark] (1563) *n.* Decorative inlay.

Marquisette \ˌmär-kwə-ˈzet\ [*marquise* + *-ette*] (1908) *n.* A lightweight, open-mesh fabric made of cotton, silk, or manufactured fibers in a leno, doup, or gauze weave. Marquisettes are used for curtains, dresses, mosquito nets, and similar end uses.

Mar resistance *n.* The resistance of a glossy plastic surface to abrasive action. It is measured (ASTM D 673) by abrading a specimen to a series of degrees, then measuring the gloss of the abraded spots with a glossmeter and comparing the results to that of the unbraided area of the specimen. *See also gloss.*

Married fiber clump *n.* A defect that occurs in converter top. It consists of a group of unopened, almost coterminous fibers with the crimp in register.

Martens heat-deflection temperature *n.* The temperature at which, under four-point loading, a bar of polymer deflects by a specified amount. For amorphour polymers, the Martens temperature is about 20°C below the glass-transition temperature. *Compare deflection temperature.*

Martius yellow *n.* Calcium derivative of naphthalene yellow.

Mask \ˈmask\ [MF *masque*, fr. OIt *maschera*] (1534) *n.* A stencil used for spray-painting plastics, consisting of a relatively thin sheet shaped to fit the part to be painted with openings for areas to be painted.

Masking *n.* Temporarily covering that part of a surface to which it is not desired to apply a coating.

Masking tape *n.* Adhesive backed paper tape used to mask or protect parts of a surface not to be finished.

Masonry \ˈmā-sən-rē\ (13c) *n.* The art of the mason in shaping, arranging and uniting stone, brick, building blocks, tile and similar materials, to form walls and other parts of a building.

Masonry conditioner *n.* A solvent-based, pigmented primer coating formulated to have great penetrating power so as to prepare masonry (especially chalky stucco) to receive finish coats. Particularly important under latex paints.

Masonry paint *n.* An alkali-resistant coating, usually a latex paint, used for masonry substrates.

Mass \ˈmas\ [ME *masse*, fr. MF, fr. L *massa*, fr. Gk *maza*; akin to Gk *massein* to kneed] (15c) *n.* (1) Quantity of matter, whose unit, the kilogram, is one of seven base units of the SI system. The term is often confused with *weight* in everyday use, probably because, when weighed on an equal-arm balance, the mass being determined is compared with standard masses, ordinarily referred to as "weights". Although the kilogram-force (*kilopond*) has long been used and is still being used, it has no place in the SI system. (2) Units of mass – the gram is 1/1000 the quantity of matter in the International Prototype kilogram; one of the three fundamental units of the cgs system. The British standard of mass is the pound, of which a standard is preserved by the government. The USA standard mass is the avoirdupois pound defined as 0.45359 Kg. Giambattista A, Richardson R, Richardson B (2003) College physics. McGraw-Hill Science, New York. Kricheldorf HR, Swift G, Nuyken O, Huang SJ (2004) Handbook of polymer synthesis. CRC Press, Boca Raton, FL.

See also weight and force.

Mass-action expression, *Q* The product of the concentrations or partial pressures (or, better, activities) of the products in a reaction, divided by those of the reactants. Each term is raised to an exponential power corresponding to the coefficient written before the corresponding substance or species in the balanced equation. Pure solids and liquids are omitted, as are substances present in large excess, and therefore almost constant concentration.

Mass–action law *n.* For a homogeneous reacting system, the rate of chemical reaction is proportional to the active masses of the reacting substances, the molecular concentration of a substance in a gas or liquid being taken as its active mass.

Mass action, law of *n.* At a constant temperature the product of the active masses on one side of a chemical equation when divided by the product of the active masses on the other side of the chemical equation is a constant, regardless of the amounts of each substance present, at the beginning of the action. At constant temperature the rate of the reaction is proportional to the concentration of each kind of substance taking part in the reaction.

M

Mass by weighing on a balance with unequal arms *n.* If W_1 is the value for one side, W_2 the value for the other, the true mass,

$$W = \sqrt{W_1 W_2}.$$

Mass color *n.* The color, when viewed by reflected light, of a pigment-vehicle mixture of such thickness as to obscure completely the background.

Sometimes called over-tone or mass-tone.

Mass defect (ca. 1923) *n.* Difference between atomic mass and mass number of a nuclide.

See packing fraction.

Mass dyeing *See spin drying.*

Mass–energy equivalence *n.* The equivalence of a quantity of mass and a quantity of energy when the two quantities are related by the equation $E = mc^2$. The conversion factor c^2 is the square of the velocity of light. The relationship was developed from **relativity theory**, but has been experimentally confirmed.

Mass (fiber) strength *n.* The force per unit of lineal density required to break a fiber. The SI measure is newton per (kilogram/meter), or Nm/kg. Long used in the staple-fiber industry has been the unit gram-force per denier. 1 g_f/denier = 88,259 Nm/kg.

Massicot (massocot) *n.* Another name for lead monoxide.

See litharge.

Mass number (1923) *n.* The total number of nucleons (protons and neutrons) in an atom.

Mass polymerization *See bulk polymerization.*

Mass spectrometry (spectroscopy) (1943) (MS) *n.* Mass spectrometry is an analytical technique in which a material (e.g., a polymer) is pyrolyzed, the fragment molecules are injected into a vacuum chamber where they are ionized with an electron gun, accelerated in an electric field, and forced through a magnetic field, the paths of the more massive molecules deflecting (curving) less than the lighter ones. A detector registers the mass number and ion count at each mass number and from this information develops a spectrum. An analyst can determine the composition of the original polymer from his interpretation of the spectrum fragments. The MS method may be supplemented by gas chromatography, which can identify the types of chemical structures in the fragments. An example of a MS spectrum of toluene is shown. Kroschwitz JI (ed) (1990) Polymers: polymer characterization and analysis. John Wiley and Sons, New York. Willard HH, Merritt LL, Dean JA (1974) Instrumental methods of analysis. D. Van Nostrand Co., New York.

Mass tone *n.* The color produced by a single color dispersed full strength in a suitable vehicle.

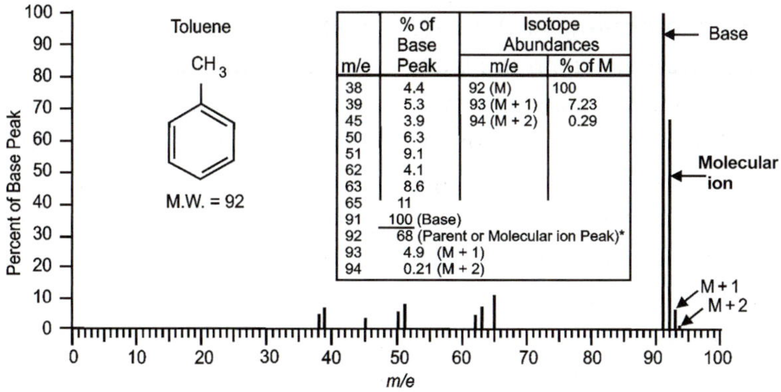

m/e	% of Base Peak	Isotope Abundances m/e	% of M
38	4.4	92 (M)	100
39	5.3	93 (M + 1)	7.23
45	3.9	94 (M + 2)	0.29
50	6.3		
51	9.1		
62	4.1		
63	8.6		
65	11		
91	100 (Base)		
92	68 (Parent or Molecular ion Peak)*		
93	4.9 (M + 1)		
94	0.21 (M + 2)		

Mass spectrometer spectrum of toluene (Courtesy of John Wiley & Sons)

Masstone *n.* (1) A pigment-vehicle mixture which contains a single pigment only. (2) Occasionally, this term is used more loosely to describe a pigment-vehicle mixture, which contains no white pigment. *See mass color.*

Masstone color *n.* The color of a masstone paint applied at complete hiding.

Masterbatch A term used in the rubber industry for rubber compounds containing high percentages of pigments and/or other additives, to be added in small amounts to batches during compounding. The term is often used in the plastics industry for color concentrate.

Master curve *n.* The curve one gets by applying the principle of time–temperature equivalence to viscoelastic data on, say, relaxation modulus or creep.

Mastic \\ˈmas-tik\\ [ME *mastik*, fr. L *mastiche*, fr. Gk *mastichē*, prob. back-formation fr. *mastichan*] (14c) *n.* (1) A solid resinous material obtained from the mastic tree (*Pistacia lentiscus*) and used in adhesives and lacquers. (2) *Asphalt mastic*, a composition of mineral matter with resin and solvent. (3) Any pasty material used as a waterproof coating or as cement for setting tile.

Masticate \\ˈmas-tə-ˌkāt\\ [LL *masticatus*, pp of *masticare*, fr. Gk *mastichan* to gnash the teeth; akin to Gk *masasthai* to chew] (1649) *v.* To work rubber on a mixing mill or in an internal mixer until it becomes soft and plastic. Synonymous with breakdown.

Mastication *n.* Intense shearing of unvulcanized rubber by working in a roll mill or internal mixer to reduce its molecular weight preparatory to compounding and molding.

Mastication of resins *n.* Process of hot working of resins, which is believed to reduce molecular complexity and to confer solubility. It has been used in natural copals, and advantages of this treatment are that thermal cracking is avoided and pale colors maintained. Patented methods include treatment between rotating rollers, sometimes in the presence of solvents.

Mat *n.* A fabric or felt of glass or other reinforcing fibrous material cut to the contour of a mold, for use in reinforced-plastics processes such as matched-die molding, hand lay-up, or contact-pressure molding. The mat is usually impregnated with resin just before or during the molding process.

Matched-die molding *n.* A reinforced plastic manufacturing process in which close-fitting metal matching male and female molds are used to form the part using pressure, temperature, and time cycle.

Matched-mold thermoforming *n.* A sheet-thermoforming process in which the heated plastic sheet is shaped between male and female halves of a matched mold. The molds may be of metal or inexpensive materials such as plaster, wood, epoxy resin, etc., and must be vented to permit the escape of air as the mold closes.
See sheet thermoforming.

Matching, color *n.* Act of making one material appear to match another color. If the achieved match is dependent on the conditions of illumination and viewing, the match is termed conditional or metameric. If the achieved is independent of the quality of the illuminant viewer, or viewing conditions, the match is termed non-conditional or non-metameric.

Matelassé *n.* A soft, double or compound fancy-woven fabric with a quilted appearance. Heavier types are used as draperies and upholsteries. Crepe matelassé is used for dresses, wraps, and other apparel. Matelassé is usually woven on a Jacquard loom.

M

Maximum permissible stress *See allowable stress and factor of safety.*

Maxwell \ˈmaks-ˌwel\ [James Clerk *Maxwell*] (1900) *n.* The cgs emu magnetic flux is the flux through a cm^2 normal to a field at 1 cm from a unit magnetic pole.

Maxwell model (Maxwell element) *n.* A concept useful in modeling the deformation behavior of viscoelastic materials. It consists of an elastic spring in series with a viscous dashpot. When the ends are pulled apart with a definite force, the spring deflects instantaneously to its stretched position then motion is steady as the dashpot opens. A simple combination of these two types provides a fair analogic representation of real viscoelastic behavior under stress.
See also Voigt model.

Maxwell's rule *n.* A law stating that every part of an electric circuit is acted upon by a force tending to move it in such a direction as to enclose the maximum amount of magnetic flux.

MBK *n.* Abbreviation for methyl butyl ketone.

MBS *n.* Abbreviation for methacrylate-butadiene-styrene resin. These are mixtures of PMMA and butadiene-styrene co-polymers, formulated in a variety of types with markedly different characteristics according to their composition and molecular weight. MBS resins can be processed by all the usual thermoplastics processes.

Mc *n.* Abbreviation for megacycle, one million cycles, loosely used to mean 1 MHz, one million cycles *per second.*

MC *n.* Methyl cellulose.

MD *n.* (1) Abbreviation for machine direction. (2) Abbreviation for methylene dianiline, little used because of its carcinogenicity.

MDI *n.* Abbreviation for diphenylmethane-4,4′-diisocyanate.
See diisocyanate.

MDPE *n.* Abbreviation for medium-density polyethylene.
See polyethylene.

Measling *n.* The appearance of spots or stars under the surface of the resin portion of an epoxy/glass–fiber laminate (from *measles*).

Mechanical adhesion *n. See adhesion, mechanical and adhesion specific.*

Mechanical equivalent of heat *n.* A conversion factor that transforms work or kinetic energy into heat. Probably the best known one is 788 ft-lb per British thermal unit; others are 2545 Btu per horsepower-hour, 4.186×10^7 ergs/cal, and 3413 Btu/kW h. In SI there is no need for such factors because work, heat, and electrical energy are all measured in joules (1 J = 1 m N = 1 W s).

Mechanical finishing *n.* Changing the appearance or physical properties of a fabric by a mechanical process such as calendering, embossing, bulking, compacting, or creping.

Mechanical grease forming *n.* A method of sheet thermoforming used with acrylic sheet when excellent opticals are imperative and the shape desired cannot be produced by free forming. The mold surface is covered with a 1- to 2-mm-thick layer of felt soaked with melted grease that must be cleaned off the sheet after forming.

Mechanically foamed plastic *n.* A cellular plastic in which the cells have been produced by gases introduced by physical means.

See also cellular plastic.

Mechanical properties *n.* Those properties of a material that are associated with elastic and inelastic reaction when force is applied, or that involve the relationship between stress and strain.

Mechanical property *n.* Any property of a material that defines its response to a particular mode of stress or strain. Such properties include elastic moduli, strength, and ultimate strain in several modes, impact strength, abrasion resistance, creep, ductility, coefficient of friction, hardness, cyclic fatigue strength, tear strength, and machinability. Many ASTM tests in Section 08 are devoted to the mechanical properties of plastics.

Mechanical spectrometer *n.* An instrument (Rheometrics Inc., NJ, USA) capable of applying an alternating tensile/compressive (or flexural or torsional) deformation of constant amplitude to a plastic specimen in the frequency range from 0.002 to 80 Hz and measuring the variation of force so caused and the phase angle between the deformation and the force. For this information one can calculate the "real" and "imaginary" parts of the various moduli (*See ASTM, www.astm.org*).

Mechanism *n.* The set of steps (elementary processes) which together comprise an overall reaction.

Mechanisms *n.* Step-by-step pathway from reactants to products showing which bonds break and which bonds form in what order.

Media *n.* Aggregate used to effect dispersion in certain types of production equipment, such as ball, pebble, and sand mills. The media vary in size and composition. Some examples are: steel balls, natural stones or pebbles, synthetic ceramic balls, glass beads, and sand.

Media mill *n.* Any mill using any one of the various types of grinding media, e.g., sand, steel ball, pebble, etc.

Median \ˈmē-dē-ən\ *n.* The value in an arrayed set of repeated measurements that divides the set into two equal-numbered groups. If the sample size is odd, the medium is the middle value. The median is a useful measure of the center when the distribution is strongly skewed toward low or high values. *Compare arithmetic mean.*

Medium \ˈmē-dē-əm\ [L, fr. neuter of *medius* middle] (1593) *n.* In paints or enamels, the continuous phase in which the pigment is dispersed; thus, in the liquid paint in the can, it is synonymous with vehicle, and in the dry film it is synonymous with binder.

Medium (art) *n.* In a general sense, the particular material with which a work of art is executed: oils, water color, chalks, lithographic stone, pen and ink, etc. It may also refer to the liquid with which powdering pigments are ground to make artist's paint, and in a more restricted sense, to the liquid used to render such paint more fluid and workable.

M

Medium oil varnish *n.* Varnish of medium oil content usually containing from 18 to 25 gal of oil per 100 lb of resin.
See long oil and short oil.

Medium yellow *n.* A pigment based on pure, monoclinic lead chromate.

Mega- [Gk, fr. *megas* large] (M) *adj combining form.* The SI prefix meaning $\times 10^6$.

Megahertz \ˈme-gə-ˌhərts, -ˌherts\ [ISV] (1941) *n.* A unit of vibrational frequency equal to 10^6 cycles/s, i.e., 10^6 Hz.

Megapoise *n.* One million poises. This unit is used for materials of very high viscosity, e.g., asphalts.
See viscosity and poise.

MEK *n.* Abbreviation for methyl ethyl ketone.

MEKP *n.* Abbreviation for methyl ethyl ketone peroxide.

Melamine \ˈme-lə-ˌmēn\ [Gr *Melamin*] (ca. 1835) (2,4,6-triamino-1,3,5-triazine) *n.* $C_3N_3(NH_2)_3$. A cyclic unsaturated compound, derived from cyanuric acid, with the structure shown below. It reacts with formaldehyde to give a series of heat reactive resins. Melamine's main use is for melamine-formaldehyde resins.

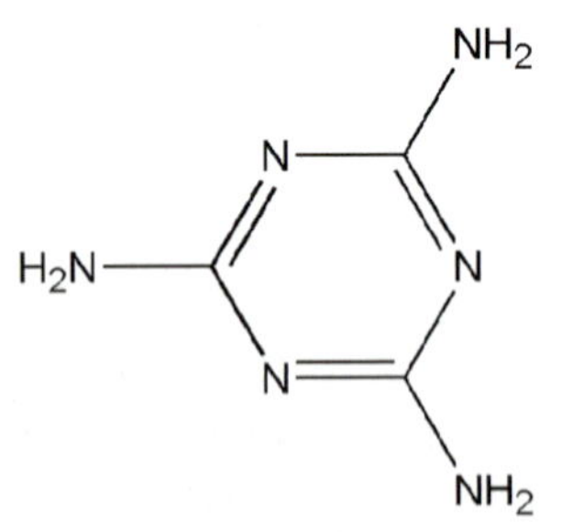

Melamine–formaldehyde resins (melamine resin) *n.* Any of a group of thermosetting resin of the amino-resin family, made by reacting melamine with formaldehyde. The lower-molecular-weight uncured melamine resins are water-soluble syrups, used for impregnating paper, laminating, etc. High-molecular-weight resins, usually cellulose-filled, are powders widely used from 1950 to 1970 for plastic tableware.

Melamine/phenolic resin *n.* A mixture of melamine- and phenol-formaldehyde resins that combines the dimensional stability and ease of molding of phenolics with the wider range of colorability of the melamine resins.

Melamine resins (1939) *n.* Any of the class of thermosetting resins formed by the interaction of melamine and formaldehyde.

Melan, melamin *n.* Melamine/formaldehyde pre-polymers. Manufactured by Henkel, Germany.

Melatrope *n.* The center of rotation of the isogyres in biaxial interference figures representing the point of emergence of rays that, in the crystal, travel along the optic axes.

Melbrite *n.* Melamine/formaldehyde resin. Manufactured by Montedison, Italy.

Melded fabric *n.* A non-woven fabric of a base fiber and a thermoplastic fiber. The web is hot-calendered or embossed at the softening point of the thermoplastic fiber to form the bond.

Meldola blue *n.* Methylene blue type of dye.
Known also as new blue or naphthol blue.

Melengket *n.* Native name for soft Manila copal obtained by tapping.

Melissic acid *n.* $CH_3(CH_2)_{28}COOH$. Monobasic fatty acid constituent of beeswax. Mp, 90°C.

Melt \ˈmelt\ (1854) *n.* A material, solid at room temperature, that has been heated to a molten condition.

Melt-bead sealing *See extruder-bead sealing.*

Melt blend *See biconstituent fiber.*

Melt blowing *n.* The formation of a non-woven by extruding molten polymer through a die then attenuating and breaking the resulting filaments with hot, high-velocity air or steam. This results in short fiber lengths. Short fibers are then collected on a moving screen where they bond during cooling.

Melt coating *See extrusion coating.*

Melt-draining screw *See solids-draining screw.*

Melt-dyed *See dyeing, mass colored.*

Melt extruder *n.* A short extruder, typically of constant channel depth and lead throughout, designed to receive a molten feed and raise its pressure for extrusion through a die, such as a pelletizing die.

Melt flow *n.* The rate of extrusion of molten resin through a die of specified length and diameter. The conditions of the test (e.g. temperature and load) should be given. Frequently, however, the manufacturer's data lists only the value, not the condition as well.

Melt-flow index (MFI, melt index) *n.* The rate of flow, in gram per 10 min, of a molten resin through an orifice 2.096 mm in diameter and 8.000 mm long at a specified temperature and weight of piston pressing on the melt. Numerous combinations of temperatures and weights are listed in ASTM (www.astm.org) various thermoplastics. This single-point flow measurement is useful in controlling production quality and resin purchasing, but most of the MFI conditions are at much lower shear than those prevailing in commercial processing, so MFI is not a reliable guide to processing behavior. MFI is inversely related to viscosity and decreases rapidly as the molecular weight in a resin family increases.

Melt fracture *n.* In extrusion, the distortion of the extrudate as it emerges from a die. The effect ranges from minor, regular ridges and valleys at 45° or 90° to the axis of the extrudate to violent wriggling and curling and, at its most extreme, breaking up of the extrudate into fragments.

Melting point (melting range) *n.* In pure compounds, the temperature at which the transition from solid to liquid occurs, requiring heat input. Polymers, being broad mixtures of homologs, melt over a substantial range of temperature, the shorter chains melting first with rising temperature, the longer ones later. Crystalline polymers have narrower, more distinct melting ranges than amorphous polymers.
See also heat of fusion.

Melting zone *n.* In a well-designed extruder screw, the section, to be coincident with the transition section, in which most, if not all, of the melting of the feedstock occurs. The pumping section, in which the plastic is presumed to be fully melted, is sometimes called the *melt zone.*

Melt instability (melt-flow instability) *n.* A term applied to the early manifestations of melt fracture.

Melton \ˈmel-tᵊn\ [*Melton* Mowbray, town in England] (1823) *n.* A heavily full, hard, plain coating fabric that was originally all wool but is now also seen in wool blends.

Melt pressure *n.* The gauge pressure exerted at any point in a processing apparatus that develops pressure. In extruders, melt pressure in the head is usually monitored. In injection machines the location is analogous but melt pressures have also been measured in mold cavities. Not to be confused with (though related to) injection-molding pressure.

Melt spinning *See spinning.*

Melt spinning process *n.* Molten polymer is pumped first through sand-bed filters, then through one to thousands of tiny orifices, called jets or spinnerets by small gear pumps operating at extremely high pressures. The fibers are then oriented to realize their optimal strength and modulus, four times or more that of the unoriented fibers. *See spinning.*

Melt strength *n.* The strength of a plastic while in the molten state. This property is pertinent to extrusion of a parisons for blow molding, to drawing extrudates from dies, as in making monofilaments and cast film, and to sheet thermoforming. It is also important when a plastic film is reheated for shrink-packaging. This property is very difficult to measure because of the ease with which a filament stretches in elongational flow at the temperatures of interest.

Melt temperature *n.* The temperature of molten or softened plastic at any point within the material being processed. In extrusion and injection molding, melt temperature is an important indicator of the state of the material and the process. Many types of instruments, most of them based on thermocouples or resistance thermometers, have been employed in extruders, where melt temperature is usually measured in the head and sometimes in the die. In thermoforming, temperatures of softened sheets are measured with infrared pyrometers.

M

Melt viscosity *n.* The resistance to shear in a molten resin, quantified as the quotient of shear stress divided by shear rate at any point in the flowing material. Elongational viscosity, which comes into plan in the drawing of extrudates, is analogously defined. In polymers, the viscosity depends not only on temperature and, less strongly, on pressure, but also on the level of shear stress (or shear rate).

See viscosity, power law, and pseudoplastic fluid.

Membrane osmometry *n.* The pressure difference between a solution and the pure solvent is measured for the case where the solvent is separated from the solution by a semipermeable membrane, isothermally; the measurement yields Δp (change in pressure) which corresponds to M_n number average molecular weight – a colligative property of polymer solutions:

$$dG = V\,dp - S\,dT,$$

where G is the Gibbs energy ($H-TS$), p the pressure, S the entropy, H the enthalpy, and T is the temperature.

Memory zone (elastic memory, plastic memory) *n.* The tendency of a plastic article to revert in dimensions to a size previously existing at some stage in its manufacture. For example, a film that has been oriented by hot stretching and chilled while under tension, will, upon reheating, tend to revert to its original pre-stretched size due to its "memory".

See also orientation.

Menaccanite Old name for the titanium mineral, ilmenite.

Mending *n.* A process in woven fabric manufacture in which weaving imperfections, tears, broken yarns, and similar defects are repaired after weaving; especially on woolen and worsted fabrics to prepare them for dyeing, finishing, or other processing.

Mer \mər\ [ISV, fr. Gk *meros* part] *n.* Derived from the Greek *meros*, meaning a part or unit, the mer is the smallest repeating structural unit (mono + mer) of a polymer (poly + mer). In addition polymers such as polyethylene the mer weight is the same as the monomer's molecular weight. Saving a small correction for end groups, the molecular weight of a polymer chain equals the mer weight times the degree of polymerization. Dimers, trimers, tetramers, oligomers, and polymers contain two, three, four, several, and many mer units, respectively.

See monomeric unit.

Meraklon *n.* Poly(propylene). Manufactured by Montecatini, Italy.

Mercadium orange *See cadmium–mercury sulfides.*

Mercerization \ˈmər-sə-ˌrīz\ [John *Mercer* † 2866 English calico printer] (1859) *vt.* A treatment of cotton yarn or fabric to increase its luster and affinity for dyes. The material is immersed under tension in a cold sodium hydroxide (caustic soda) solution in warp or skein form or in the piece, and is later neutralized in acid. The process causes a permanent swelling of the fiber and thus increases its luster.

Mercurials \(ı)mər-ˈkyur-ēl-əls\ (1676) *n.* Fungicides and bactericides containing mercury.

Mercuric chloride (1874) (corrosive sublimate, mercury bichloride) *n.* $HgCl_2$. White crystals, used as a polymerization catalyst for PVC. Mercuric chloride is highly toxic, so must be handled with care and requires special disposal procedures.

Mercuric sulfide *n.* HgS. Pigment Red 106 (77766). (1) A naturally occurring mineral, cinnabar. (2) Synthetically produced by addition of Hg to alkali sulfides depending on temperature, the colors vary from red orange to bluish red. Has good alkali resistance. Density, 8.0 g/cm^3 (66.6 lb/gal); O.A., 11–15. Syn: are vermilion, carmine vermilion, Chinese vermilion, English vermilion, patent vermilion, cinnabar, cenobrium, cinaper, cinoper, cynoper, vermiculus, zunsober, and red cinnabar.

Mercury–cadmium lithopones *n. See cadmium–mercury lithopones.*

Merge *n.* A group to which fiber production is assigned based on properties and dyeability. All fibers within a merge can be expected to behave uniformly, and for this reason, can be mixed or used interchangeably.

Meridional (mer-) isomer *n.* An isomer of an octahedral complex in which a plane contains three identical ligands and the central ion.

Merino \mə-ˈrē(ı)nō\ [Spanish] (1810) *n.* (1) Wool from purebred Merino sheep. Merino wool usually has a mean fiber diameter of 24 μm or less. (2) A yarn of blended wool and cotton fibers.

Merinova *n.* Casein fiber, manufactured by Snia Viscosa, Italy.

Mesh \ˈmesh\ [ME, prob. fr. MD *maesche*; akin to OH Gr *masca* mesh, Lithuanian *mazgos* knot] (14c) *n.* (1) The square opening of a sieve. (2) The number of apertures per linear inch in a woven or electroformed metal screen or sieve, made especially for laboratory testing of high gravity dry powders or pigments for fineness and impurity content. Such screens are available to a mesh size of about 400.

Mesh analysis *See sieve analysis.*

Mesh fabrics *n.* A broad term for fabric characterized by open spaces between the yarns. Mesh fabrics may be woven, knit, lace, net, crochet, etc.

Mesh number *n.* (1) The designation of size of an abrasive grain. Its name is derived from the openings per linear inch in the control sieving screen. Syn: grit number. (2) The deprecated, but still widely used (in USA) nomenclature for screen sizes, meaning the number of wires per inch of screen width. In standard square-mesh screens used in sieve analysis, the count and wire diameter are the same in both directions. Thus, the widths of the standard-screen openings (inches) are in approximate inverse proportion to the mesh numbers, ≈ 0.6 (mesh number)$^{-1}$. Modern nomenclature, in accordance with SI, designates open-mesh screens by the minimum width of the openings in millimeters.

Mesitylene *n.* $C_6H_3(CH_3)_3$. Powerful high-boiling hydrocarbon solvent. Bp, 165°C.

Mesityl oxide (4-methyl-3-pentene-2-one) *n.* $CH_3COCH{=}C(CH_3)_2$. An oily, colorless liquid used as a powerful solvent for cellulosic and vinyl resins, and as an intermediate

M

in the production of plasticizers. Bp, 130°C; flp 25°C (78°F); vp, <10 mmHg per 30°C.

Mesocolloid *n.* Between hemicolloids and encolloids in size. Colloid particles limited to aggregates of from 100 to 1,000 molecules and from 25 to 250 nm long.

Mesomerism *n.* Essentially synonymous with resonance. The term is particularly associated with the picture of pi electrons as less localized in an actual molecule than in a Lewis formula. The term is intended to imply that the correct representation of a structure is intermediate between two or more Lewis formulae.

Meson \ˈme-ˌzän, ˈmā-, ˈmē-, -ˌsän\ [ISV *mes-* + *²-on*] (1939) *n.* Two types of particles of mass intermediate between that of the electron and proton have been discovered in cosmic radiation and in the laboratory. The one particle with mass about $215m_e$ is called μ-meson, the other with about $280m_e$ π-meson. Mesons of both positive and negative charge have been found and there is now reasonably good evidence for neutral mesons. Both types of mesons decay spontaneously. Some evidence exists for a meson of mass about $100m_e$.

Mesopic vision *n.* Vision at luminosities intermediate between luminosities required for completely photopic or completely scotopic vision; sometimes called twilight vision.

Meta- prefix [NL & ML, fr. L or Gk; L, fr. Gk, among, with, after, fr. *meta* among, with, after; akin to OE *mid, mith* with, OH Gr *mit*] (*m-*). A prefix used in naming aromatic organic compounds, ignored in alphabetization that designates the 3- and 5-positions relative to the substituted 1-position in a benzene ring. *Compare ortho- and para-* (3).

Metafiltration *n.* Edge filtration through superimposed metallic strips with beveled edges, involving a change from coarse filtration (due to the strips) to fine filtration (due to the filter bed formed in their interstices).

Metal \ˈme-tᵊl\ [ME, fr. OF, fr. L *metallum* mine, metal, fr. Gk. *metallon*] (14c) *n.* An element, which has high electrical and thermal conductivities, a characteristic luster, and a low ionization energy, electron affinity, and electronegativity.

Metal alloying *n.* Combining two or more metals into an alloy, materials with different advantages exist. The number of available alloys increases factorially, each with its specific set of properties.

Metal chelate polymers *n.* A polymer which contains metal atoms bonded to organic functional groups by coordinate bonds.

Metal decorating *n.* The process of lithographic printing on metal. The term most often applies to the can coating industries and includes the coating of can liners, outside basecoats and overprint varnishes as well as lithographic printing. The coatings are generally applied by direct roller coating or by spray onto tinplate or aluminum, on individual metal sheets, or by continuous coil coating.

Metallic bond *n.* Bonding present in metals. Metallic bonding can be described as the movement of valence electrons throughout the metal lattice.

Metallic brown *See brown iron oxide pigment.*

Metallic elements *n.* Are distinguished from the non-metallic elements by their luster, malleability, conductivity and usual ability to form positive ions. Non-metallic elements are not malleable, have low conductivity and never form positive ions.

Metallic fiber *n.* A manufactured fiber composed of metal, plastic-coated metal, metal-coated plastic, or a core completely covered by metal (FTC definition). They are available in "yarn" form as well as in staple form for spinning with other fibers. A core yarn with a metal surface is produced by twisting a strip of metal around yarn of natural or manufactured fibers. The most important characteristic of metallic fiber and the chief reason for its use in textiles is glitter. Metallic fibers are used as a decorative accent in fabrics for apparel, bedspreads, towels, draperies, and upholstery. A relatively new application for metallic fibers is in carpet pile, where they are being used in small percentages for control of static electricity.

Metallic fillers *n.* Powdered nickel, etc., to impart special properties, usually conductivity.

Metallic-flake pigment *n.* Flat, thin particles of either aluminum, copper or copper alloy that reflect light specularly when incorporated into a plastic substance or coating vehicle with their reflecting surfaces approximately parallel. The aluminum pigments reflect very strongly throughout the visible spectrum, producing brilliant blue-white highlights. The copper-based pigments, called gold bronzes but actually brasses, range from the characteristic red copper to progressively more yellow with rising zinc content.

Metallic inks *n.* Inks composed of aluminum or bronze powders in varnish to produce gold or silver color effects.

Metallic paint *n.* Paint which, on application, gives a film with a metallic appearance. This effect is normally produced by the incorporation of fine flakes of such metals as copper, bronze or aluminum. The aluminum used may be leafing or non-leafing, the former giving a far more brilliant metallic effect. These metals can be used in tinted or colored media to give polychromatic finishes.

Also known as metallic pigmented paint.

Metallic pigment *n.* Particles or flakes of non-oxidized metals or alloys used as pigments to modify the optical characteristics of a paint, to hide the substrate, modify the color or adjust other properties. The metals most commonly used as aluminum, bronze, and zinc.

See also metallic paint.

Metallic pigmented paint *See metallic paint.*

Metallic pigments *n.* A class of pigments consisting of thin opaque aluminum flakes or copper alloy flakes. Added to plastics, they produce unusual silvery and other metal-like attractive effects.

Metallic soap *n.* Salts derived from metals and organic acids, usually fatty acids. They include not only the sodium and potassium salts, which are popularly known as soaps, but compounds such as lead linoleate, calcium resinate, aluminum stearate, etc.

See soap, metallic.

Metallic solid *n.* A solid in which positive ions are bonded together by delocalized electrons.

Metallized glass *n.* Glass spheres, flaks, or fibers that have been coated with silver or aluminum and, as fillers, provide increased electrical conductivity and light-reflecting pigmentation.

Metallizing \ˈme-tᵊl-ˌīz\ (1594) *vt.* (1) Applying a thin coating of metal to a nonmetallic surface. May be done by chemical deposition or by exposing the surface to vaporized metal in a vacuum chamber. (2) *See flame spray.* (3) Adding metallic pigments, such as aluminum, to a paint. (4) A term covering all processes by which plastics (and some other base materials) are coated with metal. The most commonly used processes are described under *electroless plating, silver-spray process, and*

M

vacuum metallizing. Other methods include spraying with metallic pigments, chemical reduction, gas plating and vapor pyrolysis.

Metallizing agents *See electroplating chemicals.*

Metalloid *n.* An element which has properties that are intermediate between those of a typical metal and those of a typical non-metal.

Also called a semimetal.

Metal marking *n.* In the coil coating industry, the degree or amount of a defect left on the surface of a film when the edge of a piece of metal is pulled across its surface. Pencil line marks on a painted surface caused by scratching with metal.

Also known as coin marking or marring.

Metal oxide catalysts *n.* CrO_3, CrO_2Cl_2, etc.

Metal primer *n.* The first coat of paint on metal; a primer. One coat.

M

Metal spraying *n.* Application of a spray coat of metal (usually zinc or aluminum) onto a prepared surface (usually shot blasted mild steel). The metal to be sprayed is rendered molten by passing it, in wire or powder form, through a flame pistol which projects the semi-molten metal onto the surface by means of a jet of compressed air.

Metamer *n.* (1) One of a pair of colors which matches the other color when viewed in a described way but which does not match under all viewing conditions. For example, if the light source, observer or angle of viewing is changed, the color will no longer match the other color of the pair. (2) From the Greek *meta* (change, transposition, and transfer) and *meros* (part or portion), the term metamer was formerly used in chemistry for a specific kind of isomer having to do with group-positional differences in molecules of the same composition and functionality. The term isomer is now used in this limited sense (as well as in broader ones).

See metameric pair.

Metameric color match *n.* A color match between two materials in which the colors are identical under some lighting conditions but not under others. Metameric color matches are common when different pigments or dyestuffs are used to color the two materials.

Metameric match *n.* A conditional identity of color exhibited by a pair of colors, each with different spectral distribution curves.

Also known as conditional match. See metamerism.

Metameric pair *n.* A pair of colors, which match when viewed in a described way but which do not match if the viewing conditions are changed. thus, a metameric pair of samples exhibit the same tristimulus values for a described set of viewing conditions (observer, light source, geometry of the illumination, and viewing arrangement) but have different spectral distributions. Hence, they exhibit a match, which is conditional only.

See conditional match and metamerism.

Metamerism *n.* A term sometimes used in the color industry for the phenomenon exhibited by two surfaces that appear to be of the same color when viewed under one light source (e.g., sunlight), but that appear different when viewed under a different light source (e.g., incandescent lamp). The term *geometric metamerism* refers to a change in perceived color of a surface with a change in viewing angle. Metamerism should not be confused with "flair" or color constancy, which terms apply to the apparent color change exhibited by a single color when the spectral distribution of the light source is changed

or when the angle of illumination or viewing is changed.
See metameric pair.

Metamerism, degree of *n.* Metamerism exists in varying amounts, depending on the magnitude of the differences in the spectral distribution curves of the two colors. Thus, the degree of metamerism may be slight to moderate to severe and is generally described by one of a number of types of metamerism (or metameric) indices.

Metamerism index *See index of metamerism.*

Metap weave-knit process *n.* A technique combining weaving and knitting in one operation with two independent yarn systems wound on warp beams. In the fabrics produced, woven strips are linked together with wales of stitches. Generally the fabrics have 75–85% woven and 25–15% knitted structure.

Metastable \-ˈstā-bəl\ [ISV] (1897) *adj.* A temporary state of structure in a plastic, such as a crystalline plastic in which the final crystallinity is attained after passage of hours or days following molding. No physical or mechanical tests should be made while the test material is in a metastable condition (unless data regarding that condition are desired).

Meter \ˈmē-tər\ *n* [F *mètre*, fr. *Gk metron* measure] (1797). (1) (m, meter) The SI unit of length, one of the seven basic units of the system, defined as 1,680,763.73 wavelengths of the radiation in vacuum corresponding to the transition between the levels $2p_{10}$ and $5d_5$ of the krypton-86 atom (an orange spectral line). One foot equals (exactly) 0.3048 m. (2) Any device for measuring a physical or chemical quantity in which the measurement is indicated digitally, or analogically on a scale. In this sense, -meter is often used as a suffix, as in *thermometer.*

Metering pump *n.* A positive displacement device that pumps a measured amount of polymer solution to the spinnerets.

Metering screw *n.* An extruder screw whose final section, from four to ten flights, has a shallow channel of constant depth and lead. As its name suggests, the metering section of such a screw is intended to regulate the amount delivered per rotation of the screw. It also provides time for the equalization of melt temperature and helps to control the steadiness of the extrusion rate.

Metering zone (metering section) *n.* The final portion of a metering screw that builds pressure to force the melt through the screens and die. The metering section usually has a constant lead and a shallower channel than the preceding sections of the screw.

Methacrylate ester \ˌme-ˈtha-krə-ˌlāt\ [ISV] (1865) *n.* Any of the esters of methacrylic acid having the general formula $CH_2{=}C(CH_3)COOR$, wherein R is usually methyl, ethyl, isobutyl, or *n*-butyl to *n*-octyl. These esters are polymerizable to acrylic resins.

Methacrylate plastic *See acrylic resin.*

Methacrylate resins *n.* A class of resins produced by the polymerization of methacrylate esters.

Methacrylic acid \ˌme-thə-ˈkri-lik-\ [ISV] (1865) (α-methacrylic acid, 2-methyl-2-propenoic acid) *n.* $CH_2{=}C(CH_3)COOH$. A colorless liquid prepared by the acid hydrolysis of acetone, from which are derived all of the methacrylate compounds. Most important of these are the esters, especially methyl methacrylate.

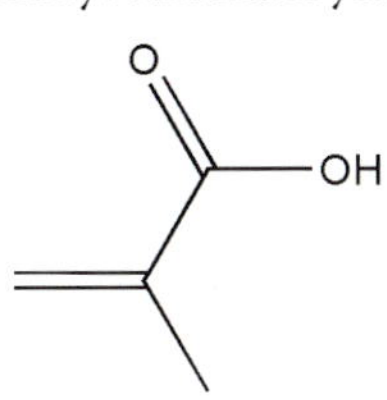

Methacrylonitrile (MAN, α-methyl acrylonitrile) *n.* A vinyl monomer containing the nitrile group whose homopolymers are true thermoplastics with good mechanical strength and high resistance to solvents, acids, and alkalis. Modified properties can be obtained through blending, grafting, or co-polymerization with other monomers such as styrene and methyl methacrylate. **MAN** is also used as a replacement for acrylonitrile in preparing nitrile elastomers.

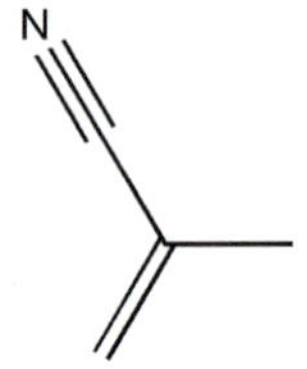

γ-Methacryloxypropyltrimethoxy silane *n.* $CH_2{=}CHCOO(CH_2)_3SI(O{-}CH_3)_3$. A silane coupling agent used in reinforced polyesters, epoxies, and many thermoplastics to achieve improved adhesion between resin and glass fibers.

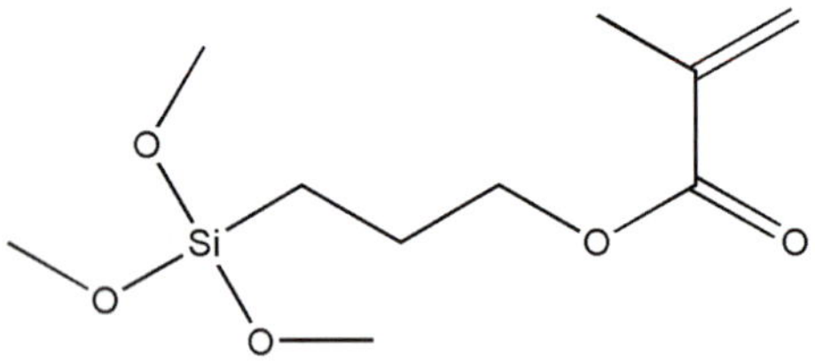

Methanol \ˈme-thə-ˌnól, -ˌnōl\ [ISV] (1894) (carbinol, methyl alcohol, and wood alcohol) *n.* CH_3OH. A colorless, toxic liquid usually obtained by synthesis from hydrogen and carbon monoxide. It is sometimes called *wood alcohol*, but the methanol obtained from the destructive distillation of wood also contains additional, contaminating compounds. Methanol is used as an intermediate in producing formaldehyde, phenolic, urea, melamine, and acetal resins, and as a solvent for cellulose nitrate, ethyl cellulose, polyvinyl acetate, and polyvinyl butyral.
Also known as methyl alcohol, carbinol, wood alcohol, colonial spirits, and MeOH. Syn: formaldehyde.

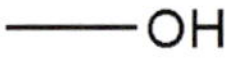

Methenamine pill test *See flammability tests.*

Method of least squares *See regression analysis.*

Methoxybutyl acetate *n.* $CH_3OCH(CH_3)CH_2CH_2COOCH_3$. High-boiling solvent. Bp, 169°C; flp, 60°C (140°F).

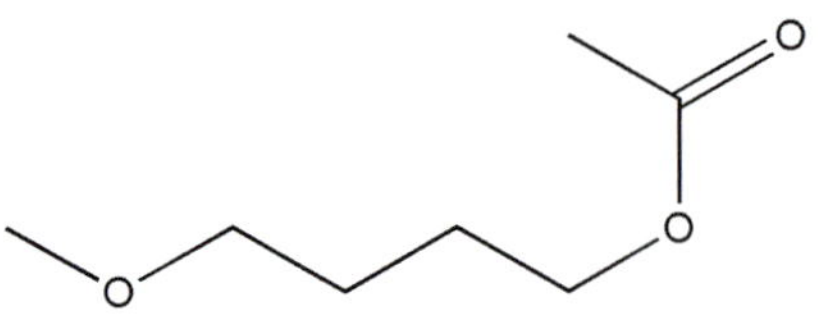

Methoxyethylacetoxy stearate *n.* $C_{17}H_{34}(OCOCH_3)COOCH_2CH_2OCH_3$. A plasticizer for vinyl and cellulosic resins.

Methoxyethylacetyl ricinoleate *n.* A plasticizer for cellulosic and vinyl resins. *n* (1,2-propylene glycol monostearate) $C_{17}H_{35}{-}COOCH_2CH_2OCH_3$. A solvent and plasticizer for cellulosic plastics.

Methoxyl group *n.* The monovalent group, $-OCH_3$, characteristic of methyl alcohol and its esters or ethers.

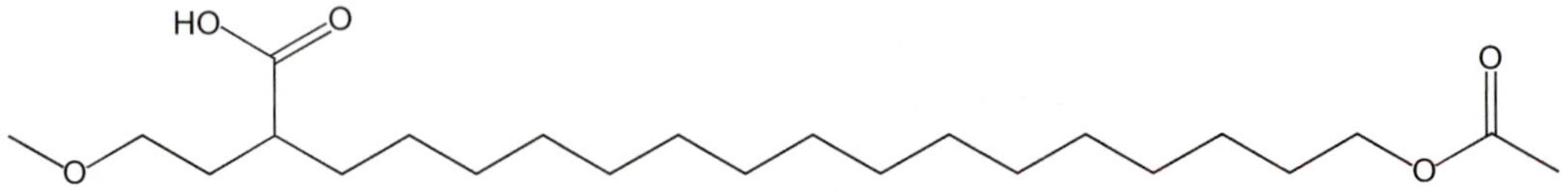

Methoxyethylacetoxy stearate

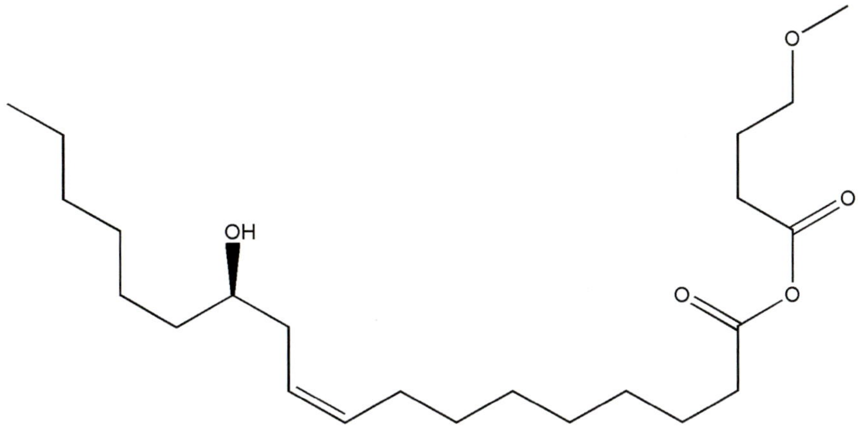

Methoxyethylacetyl ricinoleate

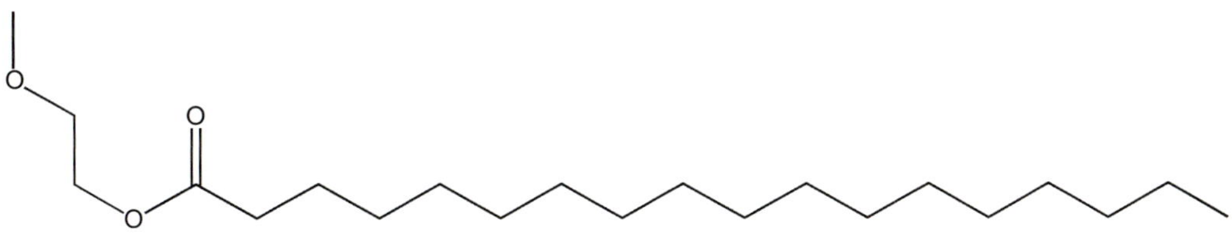

Methyl abietate *n.* $C_{19}H_{29}COOCH_3$. A derivative of abietic acid (from rosin) used as a plasticizer for cellulosic, acrylic, and vinyl resins, polystyrene, and urea–formaldehyde resins.

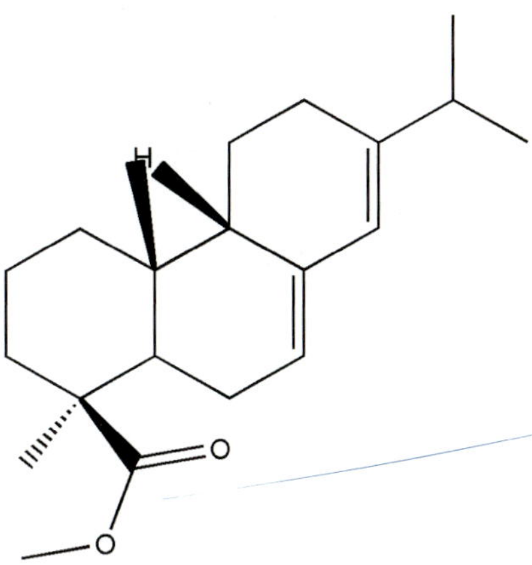

N-Methyl acetamide (NMA) $CH_3CONHCH_3$ *n.* A solvent useful in making aromatic-mer polymers, such as polyimides.

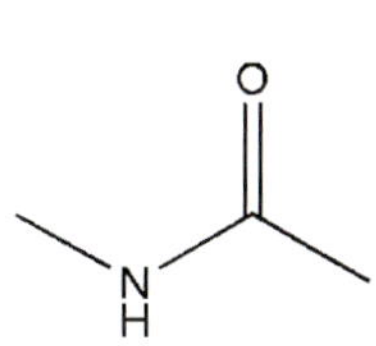

Methyl acetate (1885) *n.* CH_3COOCH_3. A colorless, volatile liquid with a fragrant odor, a solvent for acetyl cellulose and cellulose esters. A low-boiling ester solvent, exhibiting rapid evaporation; bp, 58°C, vp, 173 mmHg per 20°C.

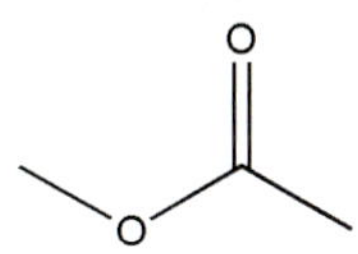

Methylacetyl ricinoleate *n.* $C_{17}H_{32}(OCOCH_3)COOCH_3$. A plasticizer for some vinyl resins and polystyrene.

Methyl acrylate *n.* $CH_2{=}CHCOOCH_3$. A colorless, volatile liquid, a monomer for

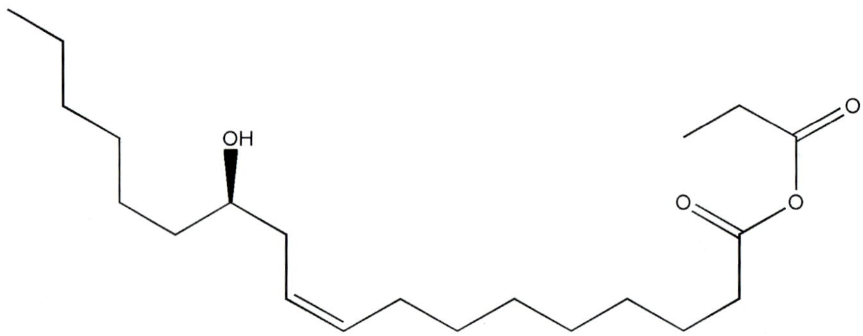

Methylacetyl ricinoleate

acrylic resins used in the manufacture of synthetic resins. Sp gr, 0.953; bp, 80.5°C.

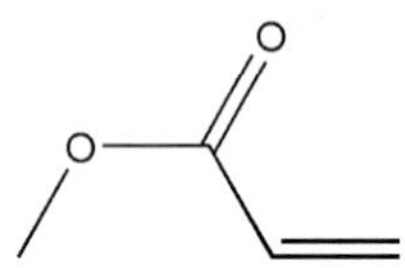

Methyl alcohol (ca. 1847) *n.* Syn: methanol.

Methyl amyl carbinol *See heptanol-2.*

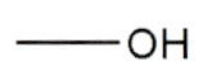

Methyl amyl ketone *n.* Boiling range, 147–153°C; flp, 106°F; vp, 4 mmHg per 20°C. *Also known as MAK and 2-heptanone.*

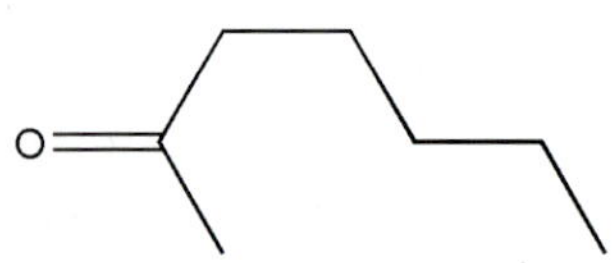

Methylated spirit *n.* A mixture of ethyl alcohol and a small amount of methyl alcohol; used industrially as a solvent for paints, lacquers, and varnishes.

Methyl benzene (methyl benzol) *n.* Syn: toluene.

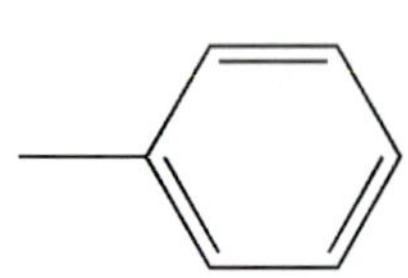

Methyl butadiene *n.* Syn: isoprene.

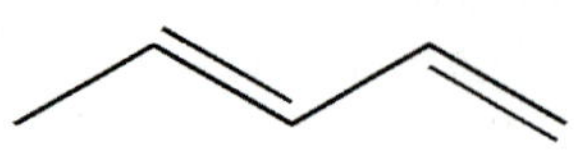

Methyl butyl ketone (MBK, propylacetone) *n.* $CH_3COC_4H_9$. A solvent for vinyl and many other resins, often used in conjunction with methyl ethyl ketone to control the drying rate of lacquers. a higher content of MBK slows the rate. Boiling range, 114–137°C; flp. 73°F; vp, 10 mmHg per 20°C. Also known as *MBK and 2-hexanone.*

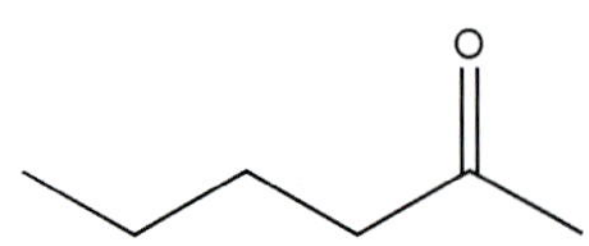

Methyl butynol *n.* $HC{=}CCOH(CH_3)_2$. A viscosity stabilizer and solvent for some nylons.

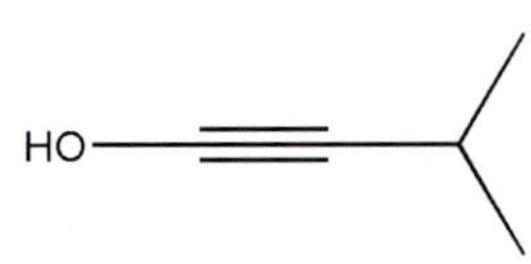

Methyl butyrate *n.* $CH_3(CH_2)_2COOCH_3$. A medium-boiling solvent for ethyl cellulose and cellulose nitrate. Bp, 102°C.

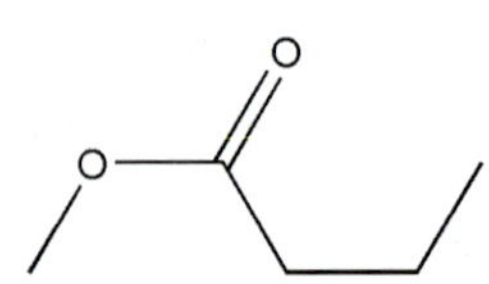

Methyl cellosolve *See ethylene glycol monoethyl ether.*

Methyl cellosolve®, acetate Ether-ester solvent. Bp, 143°C; vp, 6 mmHg per 30°C.

Methyl cellosolve®, methoxyethanol *n.* $CH_3OC_2H_4OH$. Ether-alcohol solvent, Bp, 124°C.

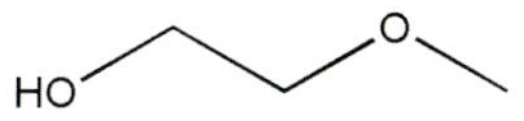

Methyl cellulose *n.* A cellulose ether in which some of the cellulosic –OH groups have been replaced by $-OCH_3$. The degree of substitution determines properties and uses as thickeners and emulsifiers. Commercially, a granular, white, flakey material, which acts as a water-soluble thickener and stabilizer; used in water-based paints.

Methyl chavicol *n.* Ether constituent found in some terpene solvents. Boiling range, 215–216°C.

Methyl citrate (tri) *n.* Plasticizer with solvent properties, Bp, 176°C per 16 mmHg.

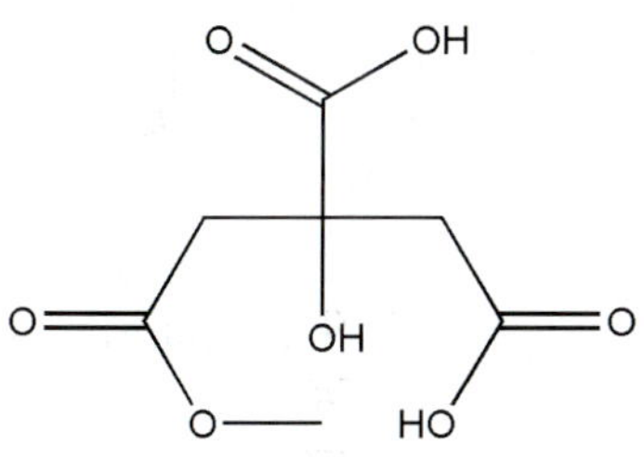

Methyl-2-cyanoacrylate *n.* A fast setting adhesive used for bonding cellulosics, nylon, polyesters, acrylics, polystyrene, and polyurethanes to each other and to other materials such as woods, metals, and glass. Catalyzed by atmospheric moisture or lightly applied methanol, the adhesive polymerizes without loss of solvent. For best results, the surfaces to be bonded should mate closely.

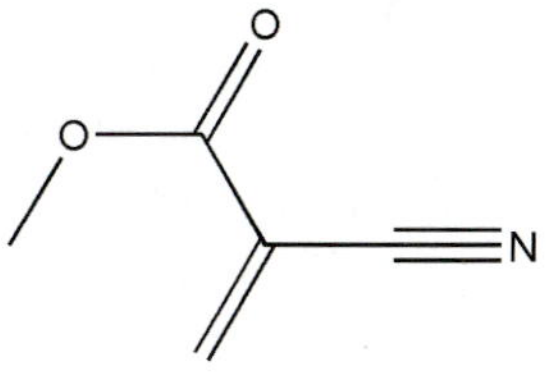

Methyl cyclohexane *n.* $CH_3C_6H_{11}$. Hydrocarbon solvent. Bp, 101°C.

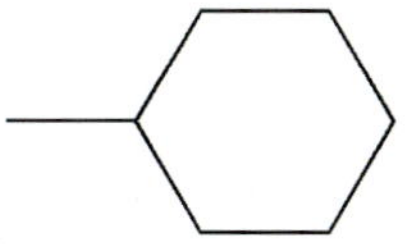

Methyl cyclohexanol *n.* $CH_3C_6H_{10}OH$. Alcoholic solvent. Because of its high boiling range, 160–180°C, it is used to improve flowing properties. Small quantities added to weaker solvents often exert a very pronounced effect on the viscosity of a given solution. Flp, 60°C (140°F).

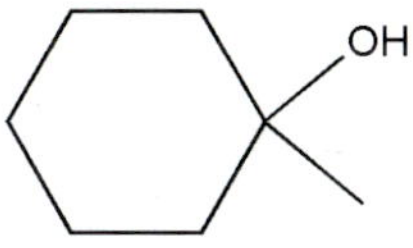

Methyl cyclohexanone *n.* High-boiling solvent. Boiling range, 160–170°C; flp, 49°C (120°F); vp, 4 mmHg per 30°C. It is often used in small amounts to improve flow or brushing properties.

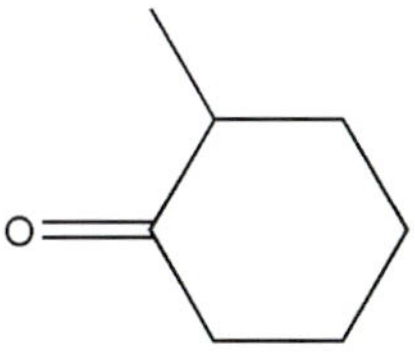

Methyl cyclohexyl acetate *n.* Boiling range, 175–190°C; flp, 68°C (155°F).

Known also as methyl hexaline acetate and sextate.

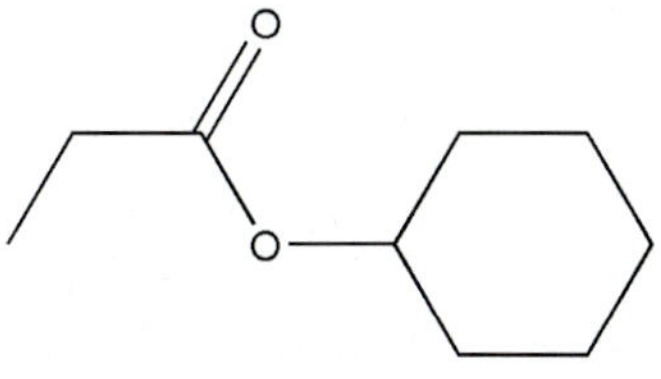

Methyl cyclohexyl stearate *n.* $CH_3(CH_2)_{16}COOC_6H_{10}CH_3$. Boiling range for the commercial product, 220–240°C per 4 mmHg; flp, 170°C (338°F).

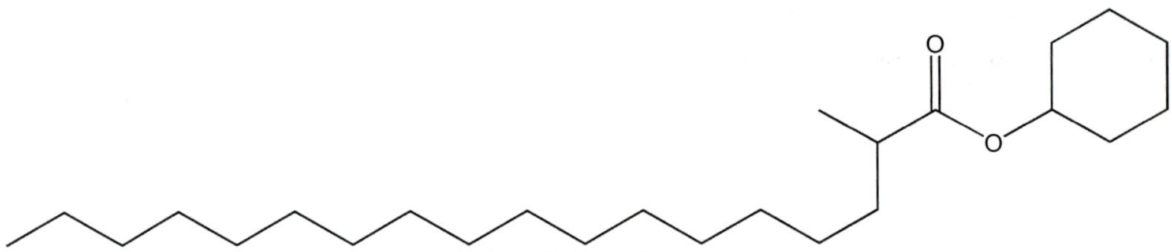

Methyl cyclohexyl stearate

2,2′-Methylene-bis-(cyclohexyl isocyanate) ($H_{12}MDI$) *n.* A diisocyanate used in making urethane elastomers and foams.

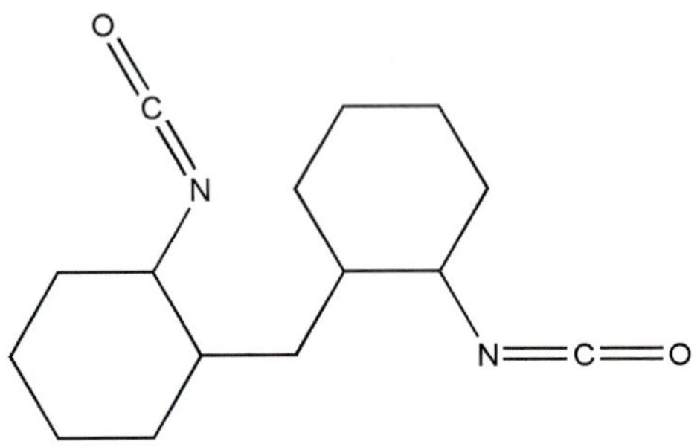

2,2′-Methylene-bis-(6-*tert*-butyl-4-ether phenol) *n.* An antioxidant for acrylonitrile-butadiene-styrene packaging, appliances, pipe, and automotive items.

2,2′-Methylene-bis-(6-*tert*-butyl-4-methyl phenol) *n.* A phenolic-type antioxidant for polyolefins and acrylonitrile-butadiene-styrene resins.

Methylene blue *n.* Blue 9 (52015). A thiazine dye used to make a type of lime blue.

Methylene chloride (dichloromethane, methylene dichloride) *n.* $CH_2–Cl_2$. A low-boiling chlorinated hydrocarbon which is a colorless, fairly dense, non-flammable liquid used as a solvent for cellulose triacetate and vinyl resins, a solvent in the polymerization of polycarbonate resins, and as a reactant for certain phenolic resins. It was widely used as a paint stripper and solvent for cured epoxy resins, but is less used now in the effort to keep chlorinated solvents out of the atmosphere. Bp, 40°C; vp. 230 mmHg per 10°C; Sp gr. 1.34/4°C.

Also known as dichloromethane.

Methylene group *n.* The radical $–CH_2–$ or $=CH_2$, existing only in combination.

Methyl ethyl ketone (1876) (MEK, 2-butanone) *n.* $CH_3COC_2H_5$. A low-boiling colorless, flammable solvent, possessing all the properties of acetone without its extremely high volatility. One of the most widely used solvents for several thermoplastics including cellulosics, acrylics, polystyrene, and vinyl co-polymers. Bp, 80°C; flp, 3°C (35°F); vp, 77 mmHg per 20°C.

Also known as MEK and butanone.

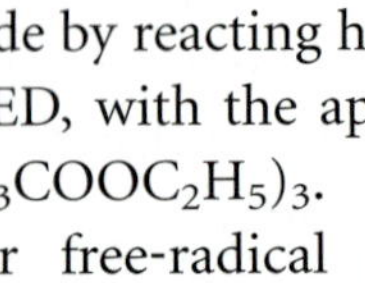

Methyl ethyl ketone peroxide (MEKP, MED, and peroxide) *n.* A complex peroxide mixture made by reacting hydrogen peroxide with MED, with the approximate formula $(CH_3COOC_2H_5)_3$. MEKP is an initiator for free-radical polymerization and a curing agent for polyester resins. In combination with an accelerator such as cobalt naphthenate, MEKP can bring about cure at room temperature. Because it is unstable, it is often handled in solution. MEKP should be kept only in small quantities and stored in a freezer when not in use.

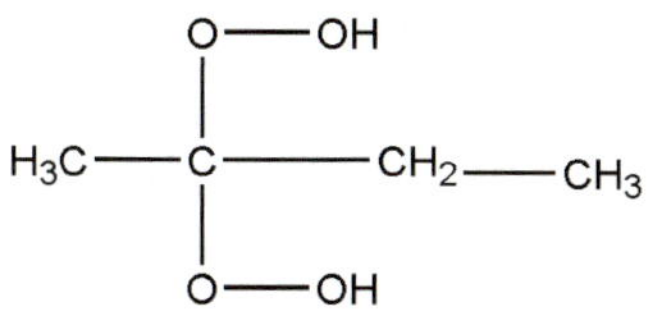

Methyl glucoside *n.* $CH_2OHCH(CHOH)_3CHOOCH_3$. A plasticizer for alkyd, amino and phenolic resins. It is also used as a polyol for urethane-foam production.

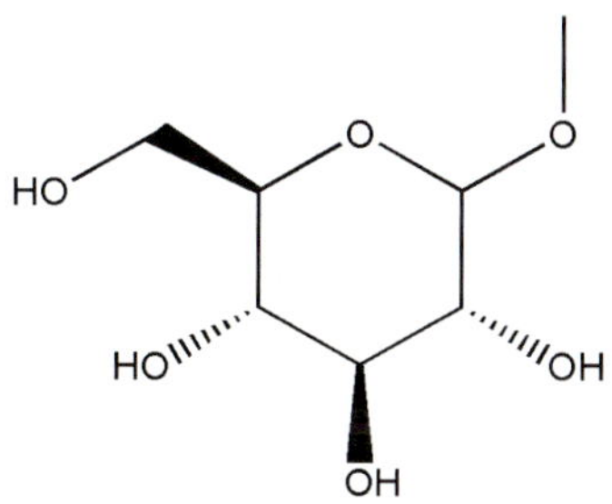

Methyl glycol phthalate (di) *n.* Plasticizer. Boiling range, 210–260°C per 20 mmHg; flp, 174°C (345°F).

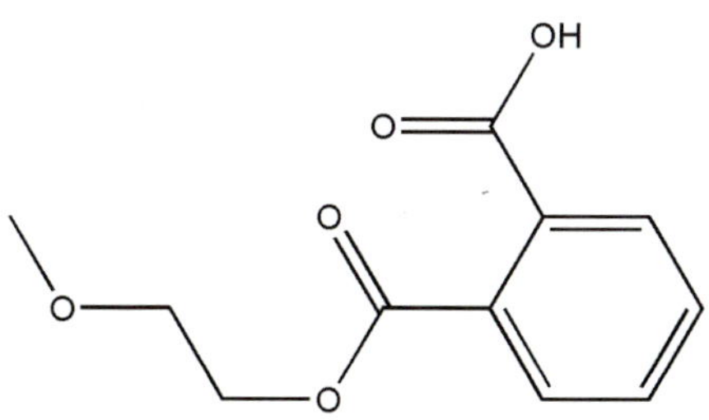

Methyl group *n.* The radical $-CH_3$, existing only in combination.

Methyl hexyl ketone (2-octanone) *n.* $CH_3COC_6H_{13}$. A colorless, high-boiling liquid with a pleasant odor, used as a solvent for epoxy coatings. Boiling range, 169–173°C; vp. 3 mmHg per 30°C.

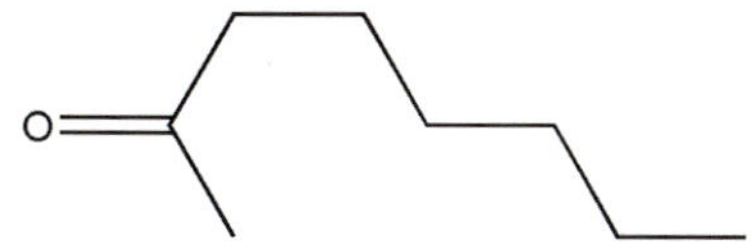

Methylimidazole (EMI) *n.* An epoxy-resin curing agent with a heterocyclic structure. EMI is used with epoxies formed from epichlorohydrin and bisphenol A or -F, and for novolac epoxy resins. It provides ease of compounding, long pot life, low viscosity, and non-staining characteristics.

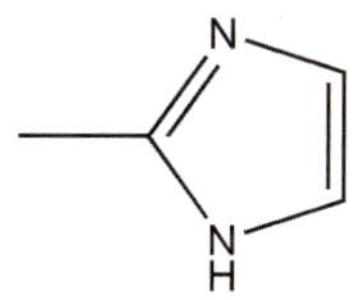

Methyl isoamyl ketone (5-methyl-2-hexanone) *n.* $CH_3COC_2H_4CH(CH_3)_2$. A colorless liquid with a pleasant odor, used as a solvent for cellulose esters, acrylic resins, and certain vinyl polymers. It has a high solvent power and low evaporation rate, making it useful as a retarder solvent that promotes flow-out of coatings and reduces blushing.

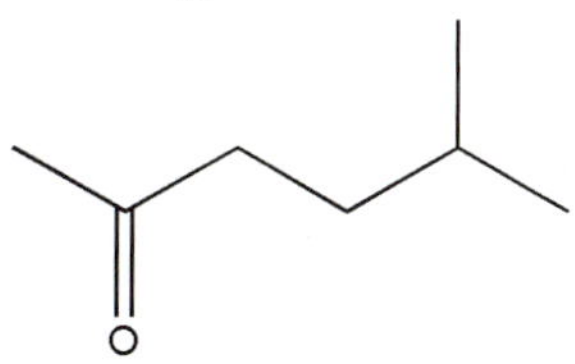

Methyl isobutyl ketone (MIBK, hexanone, 4-methyl-2-pentanone) *n.* $(CH_3)CHCH_2COCH_3$. A solvent with a moderate evaporation rate, used with cellulosic, vinyl, alkyd, acrylic, phenolic, and coumarone-indene resins, and polystyrene. Bp, 118°C; flp, 74°F; vp. 13 mmHg per 20°C. *Known also as hexone and MIBK.*

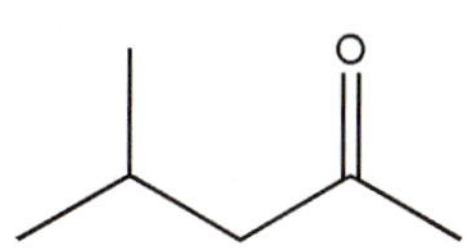

M

Methyl isopropenyl ketone *n.* $CH_2{=}C(CH_3)COCH_3$. A flammable liquid used as a co-polymerizable monomer.

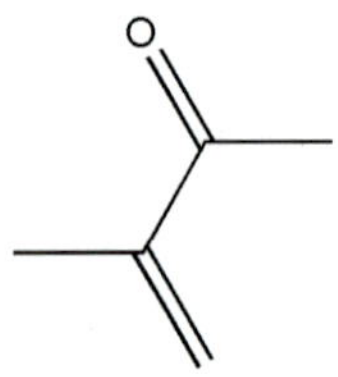

Methyl methacrylate (1933) *n.* $C_5H_8O_2$. A volatile flammable liquid that polymerizes readily and is used as a monomer for resins.

Methylol phenol \ˈme-thəl-ōl ˈfē-ˌnōl\ *n.* A phenol having one or more $–CH_2OH$ groups in its ring, a first stage in the formation of phenolic resin by reaction of phenol with formaldehyde.

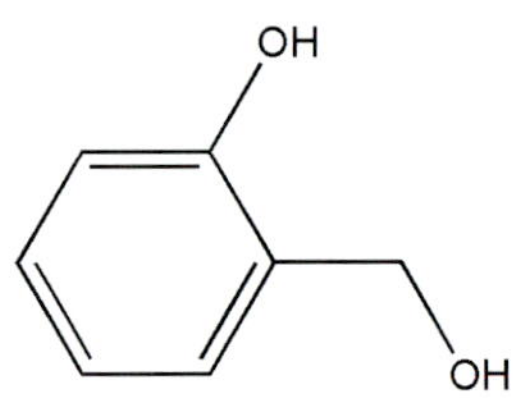

Methylol urea *n.* $H_2NCONHCH_2OH$. Colorless crystals derived from combination of urea with formaldehyde, the first stage in the production of urea–formaldehyde resins.

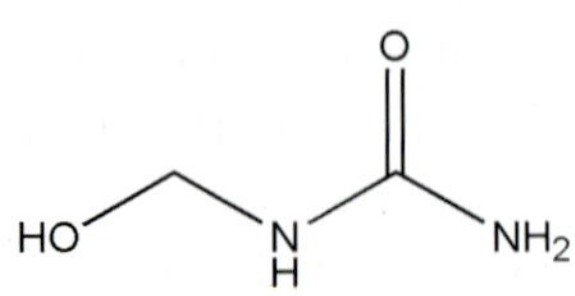

Methylolurea *n.* $H_2NCONHCH_2OH$. First stage in the formation of urea formaldehyde resins. Reaction products of urea and formaldehyde. Monomethylol urea is obtained by reacting one molecule of formaldehyde and one molecule of urea.

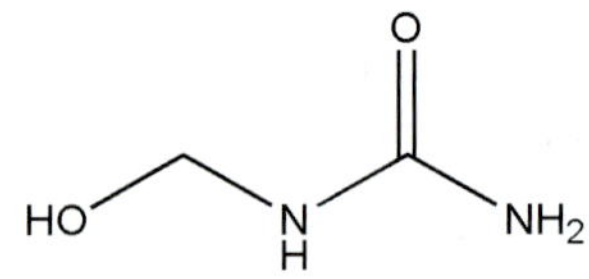

Methylpentene resin Syn: poly(4-methylpentene-1).

1-Methyl-2-pyrrolidinone (NMP) *n.* $(CH_3)\overline{NCH_2CH_2CH_2C} = O$. A solvent with a low order of inhalation toxicity, good thermal and chemical stability, and a high flash point. It is capable of dissolving resistant resins such as polyamide-imides, epoxies, urethanes, nylon, and PVC. It is a solvent of choice for spinning PVC fibers from solution. Previously known an *N*-methyl-2-pyrrolidone.

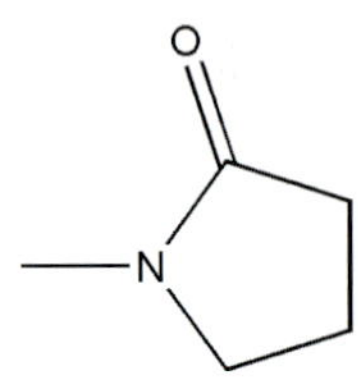

Methyl ricinoleate *n.* $CH_3(CH_2)_5CH(OH)CH_2CH{=}CH(CH_2)_7COOCH_3$. A plasticizer for cellulosic resins, polyvinyl acetate, and polystyrene with the composition of methyl ester of ricinoleic acid. Bp. 245°C per 10 mmHg; Sp gr. 0.9236.

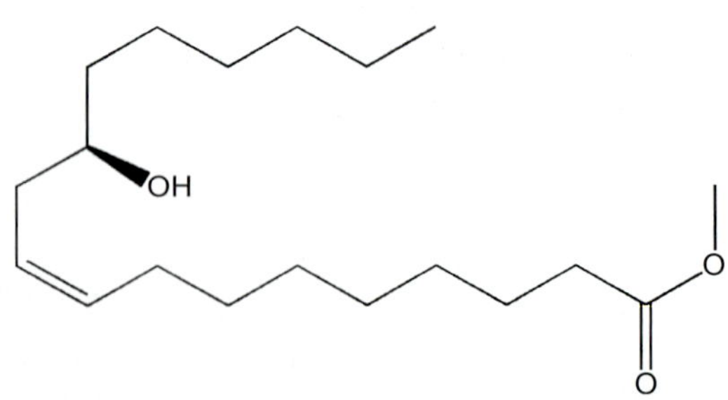

Methyl rubber *n.* Poly(2,3-dimethylbutadiene). Manufactured by Bayer, Germany.

α-Methylstyrene *n.* $C_6H_5C(CH_3){=}CH_2$. A colorless liquid, easily polymerizable by heat or with catalysts, and typically co-polymerized with methyl methacrylate or styrene.

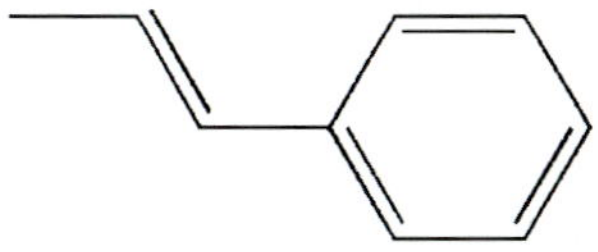

Methyl tartrate (di) *n.* A plasticizer with a bp of 280°C.

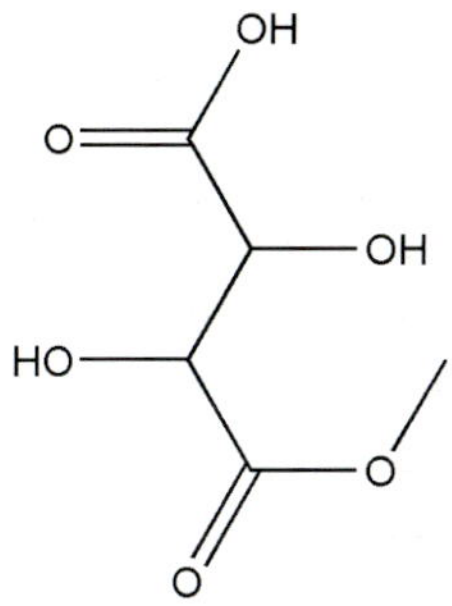

Methyl violet *n.* A class of strong purple dyes and pigments.

Metier \ˈme-ˈtyā, me-ˈ\ [F, fr. (assumed) VL *misterium*, alter. of L *ministerium* work, ministry] (1792) *n.* A spinning machine for producing manufactured fibers. The bank of cells or compartments and associated equipment used in the dry spinning of fibers, such as cellulose acetate and cellulose triacetate.

Metier twist *n.* The amount of twist present in yarn wound at the métier.

Metre \ˈmē-tər\ *chiefly British variant of meter.* The basic unit of length in the International System of Units (SI), equal to 39.37 in. SI spelling of meter.

Metrication \ˌme-tri-ˈkā-shən\ (1965) *n.* Any act tending to increase the use of the International System of Units (SI).

Metric count *n.* The number of kilometers per kilogram of yarn.

Metrology , *n.* (from greek 'metron' (measure), and -logy) is the science of measurement. Metrology includes all theoretical and practical aspects of measurement. It is a field of study which has been highly politicized and nationalized with sharp even severe rhetoric.

Mev Abbreviation for million electron volts, a measure of kinetic energy for subatomic particles (1 MeV $= 1.60219 \times 10^{-13}$ J).

Meyer bar *n. See wire-wound rod and equalizer rod.*

Mezzotints \ˈmet-sō-ˌtint, ˈmed-zō-\ [mod. of It *mezzatinta*, fr. *mezzo* (feminine of *mezzo*) + *tinta* tint] (1738) *n.* Prints made from ground (e.g., crossed in several directions with knife-edge cuts) copper plates, the burred surface of which is scraped away to create the design.

MF *See melamine–formaldehyde resin.*

MFC *n.* Abbreviation for multifunctional concentrate.

See color concentrate.

MFI *n.* Abbreviation for melt-flow index.

Mg *n.* Chemical symbol for the element magnesium.

M-glass *n.* A high-modulus glass whose fibers are sometimes used for reinforcing plastics when high modulus at moderate cost is desired. Major constituents are SiO_2 54%, CaO 13%, MgO 9%, BeO, and TiO_2 8% each, Li_2O and CeO_2 3% each, and ZrO_2 2%. Fiber density is 2.89 g/cm^3, modulus (E) is 110 GPa, and tensile strength is 3.5 GPa.

MHz *n.* Abbreviation for megahertz.

MI *n.* Abbreviation for melt index, a term replaced by melt-flow index.

MIBK *n.* Abbreviation for methyl isobutyl ketone.

Mica \ˈmī-kə\ [NL, fr. L, grain, crumb; perhaps akin to Gk. *mikros* small] (1777) *n.* Any of a family of crystalline silicate minerals characterized physically by a perfect basal cleavage, consisting essentially of orthosilicates of aluminum and potassium. They occur naturally, mainly as the minerals *muscovite* (white mica), *phlogopite* (amber mica), and *biotite*; and are also synthesized from potassium fluorilicate and alumina. Micas are used as fillers in thermosetting resins, imparting good electrical properties and heat resistance. A grade having high aspect ratios (HAR) with flakes 3–5 μm thick and aspect ratios as high as 200 can be processed, although the optimum aspect ratio appears to be about 70. The larger flakes increase flexural modulus and strength, have lower moisture content, and raise the deflection temperature of compounds containing them. *See aluminum potassium silicate.*

Micaceous iron oxide *n.* Naturally occurring iron ore which resembles mica only in appearance and not chemically. It is converted to a grayish pigment, which has value as a constituent of anti-corrosive paints and protective coatings. Density, 4.90 g/cm^3 (40.8 lb/gal); O.A., 11/200 lb; hardness (Mohs), 6.0; ph, 7.7.

Mica fillers *n.* High-aspect-ratio minerals used traditionally in thermosets as phenolics and epoxies for mechanical improvements, as well as for electrical and heat insulation benefits.

Micelle \mīi-ˈsel\ *n* [NL *micella*, fr. L *mica*] (1881) A colloidal particle formed by the reversible aggregation of dissolved molecules. Micelles may be in the shape of spheres, cylinders, or platelets. Soaps, detergents, and other emulsifying agents used in emulsion polymerization contain micelles generally composed of from 50 to 100 molecules of emulsifier, within which the polymerization reaction may be initiated.

Michel–Levy chart *n.* A chart relating thickness, birefringence, and retardation so that any one of these variables can be determined when the other two are known.

Micro- (μ) *adj.* The SI prefix meaning $\times 10^{-6}$.

Microballoons *n.* (1) Tiny, hollow plastic spheres used to reduce evaporation of liquids such as oils by floating a layer of spheres on the surfaces of stored liquids. (2) Syn: microspheres.

Microbial degradation *n. See biodegradation and pink staining.*

Microcrystalline \ˈmī-krō-ˌkris-tᵊl\ (1886) *n.* Pertaining to crystallinity that is visible only under a microscope, sometimes taken to mean that the crystals referred to are no larger than 1 μm.

Microcrystalline silica *See silica, microcrystalline.*

Microcrystalline silicate *n.* A derivative of chrysotile asbestos, consisting of tiny rod-shaped particles of hydrated magnesium silicate. The particles have hydroxyl groups on their surfaces that bond with hydrogen-bonding sites on the molecules of a fluid in which they are incorporated. The material has also been used as a viscosity-building agent in unsaturated polyester and other resins.

Microcrystalline wax *n.* Any of a group of petroleum-derived waxes that differ from paraffin waxes in having finer crystal structure, higher melting points – between 60 and 93°C, higher liquid viscosities, and greater ductility. They are used in fiberboard coatings, paper-container linings, and polishes.

Microdenier Refers to fibers having less than 1 denier per filament or 0.1 tex per filament.

Microemulsions *n.* Transparent solutions of water and oil, which are thermodynamically stable and which spontaneously form when the components are brought in contact.

Microencapsulation \-in-ˈkap-sə-ˌlāt\ (1963) *vt.* The process of encasing a small solid particle or a discrete amount of liquid or gas in a capsule. The term applies to capsules ranging in diameter from a few micrometers to about 500 μm. The capsule is usually made of a synthetic plastic, although waxes, glass, and metals are also used. Methods used for forming polymeric microcapsules fall into three broad classes: phase separation, interfacial reaction, and physical methods. Phase separation methods include coacervation, applying meltable dispersions, and spray-drying of a suspension of the material in a vaporizable solvent, in-situ polymerization, and chemical-vapor deposition. The physical methods include fluidized-bed coating processes, spray coating, electrostatic coating methods, and extrusion. Typical examples of microencapsulation are "carbonless" carbon paper, timed-release drugs and fertilizers, and battery separators.

Microgel *n.* A small particle of cross-linked polymer of very high molecular weight and containing closed loops. Microgels may be present in trace amounts due to impurities in monomers, and can influence polymer properties and molecular-weight studies.

Micrometer (μm) \mī-ˈkrä-mə-tər\ [F *micromètre*, fr. *micr-* + *-mètre* –meter] (1670) *n.* (1) Instrument for measuring small lengths under the microscope. (2) Micrometer caliper; instrument used in measuring the dry film thickness of a coating or the thickness of metal or other materials. (3) The usual unit of length for light microscopical measurements ($1\ \mu m = 10^{-3}\ mm$): it is still often referred to by its former name, "micron".

Micrometre *n.* A unit of length measure used to describe the wavelength of radiant energy equal to one-thousandth part of a millimeter or one-millionth of a meter, 1000 nm. This term is replacing the term micron. Micrometer is designated as μm.

Micron \ˈmī-ˌkrän\ [NL, fr. Gk *mikron*, neuter of *mikros* small] (1885) *n.* This long deprecated but still used length unit and its abbreviation, the Greek letter μ, were dropped by action of the General Conference on Weights and Measures on October 13, 1967. The symbol "μ" is to be used solely as the abbreviation for the prefix micro-. The old micron should now be spoken as micrometer (μm) or 10^{-6} m.

Micronaire method *n.* A means of measuring fiber fineness by determining the resistance of a sample to a flow of air forced through it.

Micronized pigment Pigment with a narrow, particle-size distribution.

Micronizer mill Used for the dry grinding of pigments. Pre-crushed crudes or solids are fed into a relatively flat, cylindrical grinding chamber, the opening being in the bottom or top plate. The fluid medium, superheated compressed air or steam, enters the grinding chamber through nozzles located in the peripheral wall. The grinding action is done by the pigment particles hitting each other. The fines are carried out through an opening in the center of the mill, the coarser, larger particles return for further grinding.

Microorganisms \-ˈór-gə-ˌni-zəm\ [ISV] (1880) *n.* In paint technology, bacteria and fungi, which are harmful to liquid paint and dry paint films. Bactericides and fungicides are added to paints to inhibit the growth of these organisms.

Microphotograph \-ˈfō-tə-ˌgraf\ [ISV] (1858) *n.* A small, microscopic photograph, in which the image is minified; it requires enlarging or the use of a lens system in order to view it (cf., photomicrograph).

Microporous \ˈmī-krə-ˌpōr, -ˌpór\ [ISV] (1884) *n.* Having pores of microscopic dimensions. Some plastic films and fabric coatings are rendered microporous in order to permit the passage of water vapor ("breathing") while preventing the penetration of raindrops.

Microscopy \mī-ˈkräs-kə-pē\ (ca. 1665) *n.* The application of any tool or technique helpful in characterizing microscopic objects.

Microspheres \ˈmī-krə-ˌsfirs\ (1894) *n.* Tiny, hollow spheres of glass or plastic used as fillers to impart low density to plastics, such plastics being known as *syntactic foams.* Plastics used to make microspheres include phenolic, epoxy and a co-polymer of vinylidene chloride and acrylonitrile. The last contains a heat-activated blowing agent that expands the spheres either before their incorporation into a matrix polymer or afterward. The co-polymer spheres impart better mechanical properties to the matrix than do the glass or epoxy microspheres.

See also glass spheres.

Microstructure \ˈmī-krō-ˌstrək-chər\ [ISV] (1885) *n.* The detailed structure of plastics as seen through light and electron microscopes, approximately the magnification range of 100–100,000×, including such features as crystalline form, spherulites, voids, distribution of filler and pigment particles, discontinuous-phase particles in blends, and, in reinforced plastics, configuration, length distribution, and cross-section distribution of yarns and, within the yarns, the filament ends, etc.

Microvoids (1) A region in a polymer of lower electron density than its surroundings, of about 100 Å in size and amounting to about 1% of the volume. (2) Small voids or holes in a paint medium of such size that when filled with air (or some other material of different refractive index) they scatter light much as a pigment does because of the difference in refractive index between the material in the void and the vehicle. Some air-filled microvoids account for the white color of snow, for example. *See bubble coating.*

Microwaveable (1973) *vt.* Said of plastics for kitchen use, and of the utensils made from them, that are heated little or not at all by the direct action of the high-frequency waves generated by microwave ovens, and that withstand many repeated heating by the foods contained in them without warping, shrinking, or staining.

Microwave drying *n.* Drying of printing inks by the use of microwave radiation. The presence of polar materials is a prerequisite.

Microwave heating *n.* A heating process similar to dielectric heating, but using frequencies in the 10^9–10^{10} Hz (radar) range. The Federal Communications commission has allocated the specific frequencies 915, 2450, and 5850 MHz for industrial use. Microwave ovens similar to those used in restaurants and households for rapidly cooking foods have been used experimentally for preheating molding powders, vacuum-bag curing, autoclave molding, and curing of nylon over wraps. Plastic films coated with water-containing materials such as polyvinylidene chloride can be dried rapidly and economically by microwave energy. Line speeds about 5 m/s have been attained with polyethylene film, by means of a microwave cabinet only 2.4 m long.

Middle chrome *See chrome yellow, light, and primrose.*

Mie theory *n.* Theory which relates the scattering of a single spherical particle in a medium to the diameter of the particle, the difference in refractive index between the particle and the medium, and the wavelength of radiant energy in the medium which is incident on the particle.

This theory relates to the direct observation of the scattering of a single particle as compared to the Kubelka–Munk theory, which relates to multiple scattering between particles. Mie theory also takes into account the absorption, which the particle may also exhibit.

Migration *n.* (1) Movement of certain materials in a coating or plastic (e.g., plasticizers in vinyl) to the surface or into an adjacent material. (2) Movement of dye from one area of dyed fabric to another. Includes movement of color from the dyed area to the undyed area of cloth. (3) Movement of fibers which go from the center to the outside surface of yarn and back again periodically.

Migration of plasticizer *n.* In plasticized thermoplastics or elastomers, the movement of molecules of plasticizer from their interior locations when the article was originally formed to the surface layer of the article, where the plasticizer appears as a greasy or oily layer and may be rubbed off or dissolved away. The phenomenon occurs most often in vinyl compounds containing incompatible plasticizers.

Mil \ˈmil\ [L *mille* thousand] (1721) *n.* A unit of thickness equal to 0.001 in., often used for specifying diameters of wires and glass fibers, and thicknesses of films. It is gradually being replaced by the SI units, the millimeter and micrometer. 1 mil = 0.0254 mm = 25.4 μm.

Milanese knitting *See knitting.*

Mildew \ˈmil-ˌdü, -ˌdyü\ [ME, fr. OE *meladēaw* honeydew; akin to OH Gr *militou* honeydew] (14c) *n.* Superficial growth produced by fungi on various surfaces forms of organic matter and on living plants which are exposed to moisture: it results in discoloration and decomposition of the surface. Two types are common: (1) Spore type, which resembles caviar in appearance and (2) Mycelium or filament type.

Mildewcide *n.* Chemical agent which destroys retards or prevents the growth of mildew.

Mildew (fungus) resistance The ability of a coating to resist fungus growth that can cause discoloration and ultimate decomposition of a coating's binding medium.
See mildew.

Mileage *n.* The surface area covered by a given quantity of ink or coating material.

Milk acid *See lactic acid.*

Milkiness *n.* Whitish or translucent appearance in an unpigmented liquid coating or film which could normally be transparent.

Mill [ME *mille*, fr. OE *mylen*, fr. LL *molina*, molinum, fr. feminine and neuter of *molinus* of a mill, of a millstone, fr. L *mola* mill, millstone; akin to L *molere* to grind] (before 12c) (1, *n*). In the plastics industry, the term mill is generally taken to refer to a roll mill such as a two-roll mill used in compounding. More broadly, it includes all mechanical devices for converting raw materials into a conduction ready for use, as well as machine tools that cut materials with rotating bits and many types of size-reduction machines (1552) (2, *v*) To process components of a plastic mixture in a two-roll mill.

Mill base *n.* The portion of the coating formulation which is charged in the dispersion mill.

Milled fibers Small lengths of glass filaments produced by hammer-milling continuous glass strands. They are useful as anticrazing and reinforcing fillers for adhesives.

Mill end *n.* A remnant or short length of finished fabric.

Miller indices The notations usually used for naming crystal faces; they have the form *hkl.* These notations are based on the assignment of crystallographic axes and on an expression of the intercepts of the face on the three axes (hexagonal has four.)

Milli- *combining form* [F, fr. Latin *milli-* thousand, fr. *mille*] (m). The SI prefix meaning $\times 10^{-3}$.

Millimeter of mercury (mmHg) *n.* A unit of pressure: 1 mmHg $= \frac{1}{760}$ atm.
Also known as Torr.

Millimicron \ˌmi-lə-ˈmī-ˌkrän\ [ISV] (1904) (deprecated) *n.* A unit of length used to describe the wavelength of electromagnetic radiation, particularly that in the visible region, equal to 10^{-9} m. The use of this term is largely being replaced by the term nanometer.

Milling *n.* (1) The process of treating fabric in a fulling mill, i.e., fulling. (2) In silk manufacturing the twisting of the filaments into yarn. (3) A grinding process, i.e., ball-milling of dyes and pigments.

M

Millipoise *n.* One-thousandth part of a poise (measure of viscosity) or 1/10 of a centipoise.

Millitre *n.* One-thousandth of a liter.

Mill run *n.* A yarn, fabric, or other textile product that has not been inspected or that does not come up to the standard quality.

Mills, ball, and pebble *n.* Roll mills, such as two-roll mills used in compounding.

Mill scale *n.* The heavy oxide layer formed during hot fabrication or heat treatment of metals.

Milori blue A green shade iron blue pigment. A pigment of the ferriferrocyanide family.
Also called Chinese blue. See iron blue.

Milori green *See chrome green.*

Mineral \ˈmin-rəl, ˈmi-nə-\ [ME, fr. ML *minerale*, fr. neuter of *mineralis*] (15c) *n.* Any naturally occurring, homogeneous inorganic substance having a definite chemical composition and characteristic crystalline structure, color and hardness.

Mineral acids *n.* Strong inorganic acids, e.g., nitric, sulfuric, hydrochloric, etc.

Mineral black *n.* Black pigment made by grinding and/or heating black slate, shale, or salty coal coke, or coal. It usually contains a high percentage of carbon, mixed with substantial amounts of mineral matter, which may include aluminum silicate, and oxides of silicon, iron, calcium, and magnesium. By comparison with true carbon blacks, its staining strength is inferior.
See also biddiblack.

Mineral brown *See brown iron oxide pigment.*

Mineral fibers A generic term for all non-metallic, inorganic fibers, which may be natural, such as asbestos, or manufactured from such sources as rock, ore, alloys, slag, or glass.

Mineral fillers *n.* Minerals compounded into plastics that provide one or more mechanical or thermal property improvements to the polymer matrix. They can be either extenders or reinforcing fillers.

Mineral oil (1805) *n.* Any liquid product of petroleum within the viscosity range of products commonly called oils and consisting of high molecular weight hydrocarbons.

Mineral orange *See orange mineral.*

Mineral pigments *See earth pigment.*

Mineral spirits (1927) (naphtha) *n plural but singular or plural in construction.* An aliphatic-hydrocarbon fraction of petroleum evolved in the distillation range of about 150–200°C. An example is "VM&P naphtha", used as a diluent in organosols. It is a petroleum fraction with boiling range between 300 and 400°F. Due to having a low aromatic hydrocarbon content, with volatility, flash point, and other properties this makes it suitable as a thinner and solvent in paints, varnishes, and similar products. Syn: is petroleum spirits.

See also odorless solvent.

Mineral surfaced roofing *n.* Felt or fabric saturated with bitumen, coated on one or both sides with a bituminous coating and surfaced on its weather side with mineral granules.

Mineral violet (77742) *n.* A complex manganese ammonium phosphate. Its very low tinting strength, low color intensity and low hiding power largely restrict its use to toning white finishes, in which service it offers low cost and excellent bake and bleed resistance; however, tints fade severely in steam sterilizers. Because of its poor alkali resistance, it is not suitable for some household appliance finishes; in such cases, quinacridone violet can be used to advantage. Syn: manganese violet, permanent, fast violet, and nurnberg violet.

Mineral white *See gypsum.*

Miniemulsion *n.* A form of emulsion in which an organic liquid is dispersed into a continuous aqueous phase. Unlike a conventional emulsion, where the droplets may be of the order of 10 μm or greater, the droplets sizes are submicron. This is accomplished through a combination of high shear, and surfactant/costabilizer combination. The surfactant prevents emulsion degradation via calescence (as in a conventional emulsion), while the costabilizer prevents diffusional degradation caused by Ostwald ripening.

Minimized spangle *n.* Galvanized sheet obtained by treating the regular galvanized sheet during the solidification of the zinc to restrict the normal spangle formation. This product usually has a dull appearance not characterized by a high degree of uniformity, and dissimilarity from coil to coil is not unusual. This minimizes the crystalline pattern from photographing through the applied coating resulting in a smoother appearing finish.

Minimum care *n.* A term describing home laundering methods. Minimum care fabrics, garments, and household textile articles can be washed satisfactorily by normal home laundering methods and can be used or worn after light ironing. Light ironing denotes ironing without starching or dampening and with a relatively small expenditure of physical effort.

Minimum detectable amount (MDA) *n.* In chemical analysis, the least amount of a substance being sought that balances two risks, type I, the risk of falsely finding the substance to be present when in fact it is not, and type II, the risk of not detecting that least amount. Typically, the two risks are made equal and, if both are 5%, the MDA is very nearly four times the standard deviation of the method. Lowering the risk increases MDA.

Minimum deviation *n.* The deviation or change of direction of light passing through a prism in a minimum when the angle of incidence is equal to the angle of emergence. If D is the angle of minimum deviation and A the angle of the prism, the index of refraction of the prism for the wavelength used is

$$n = \frac{\sin \frac{1}{2}(A + D)}{\sin \frac{1}{2}A}.$$

Minimum perceptible difference *n.* The minimum color difference, which can be observed between two colors. The magnitude is dependent on the conditions of viewing, i.e., size, illumination level and character, surrounding area, adaptation, etc.

See just noticeable difference.

Minium *n.* Naturally occurring and synthetically prepared red lead oxide; used as a pigment.

Minute value *n.* Voltage with a unit thickness of insulator (e.g., varnish) will withstand for 1 min without breakdown.

Mipolam *n.* Poly(vinyl chloride). Manufactured by Dynamit Nobel, Germany.

Mirbane oil *n.* Nitrobenzene. Liquid with an almond-like smell, used as a deodorant, chiefly in polishes.

See nitrobenzene and oil of mirbane.

Mired *n.* A unit used to measure the reciprocal of color temperature, equal to the reciprocal of a color temperature of 10 K. Derived from micro-reciprocal-degree.

Mirlon *n.* Polyamide, manufactured by Viscose-Suisse, Switzerland.

M

Miscibility \ˈmi-sə-bəl\ [ML *miscibilis*, fr. L *miscēre* to mix] (1570) (solubility) *adj.* The greatest percentage of one liquid or polymer that forms a true, homogeneous solution, i.e., a single phase, in another liquid of polymer. Few binary polymer systems are miscible over the entire range of composition, but many have limited miscibility at either end of the range. Miscibility usually increases with rising temperature.

See also compatibility.

Misclip *See scalloped selvage.*

Mispick *n.* A weaving defect in which a pick is improperly interlaced, resulting in a break in the weave pattern. Mispicks can result from starting the loom on the incorrect pick after a pick-out.

Misses *See holidays.*

Missing end *See end out.*

Miss-stitch *n.* A knitting construction formed when the needle holds the old loop and does not receive new yarn. It connects two loops of the same course that are not in adjacent wales.

Also known as float-stitch.

Mist coat *n.* (1) Very thin sprayed coat. (2) A thin coat of volatile thinners, with or without a small amount of lacquer, which is sprayed over a dry lacquer film to improve the smoothness and luster.

Misting *See flying.*

Mitt *See painter's mitt.*

Mittler's green *See hydrated chromium oxide.*

Mixed aniline point *n.* Minimum equilibrium solution temperature of a mixture of two volumes of aniline, one volume of sample, and one volume of normal heptane of specified purity. Refer to ASTM D 1012.

Mixed end or filling *n.* Warp or filling yarn differing from that normally used in the fabric, e.g., yarn with the incorrect twist or number of plies, yarn of the wrong color, or yarn from the wrong lot.

Mixer \ˈmik-sər\ (ca. 1611) *n.* Any of a wide variety of devices used to intermingle two or more materials to some defined state of uniformity. Some equipment intended mainly to provide size reduction may also accomplish mixing. Types used in the plastics industry are:

Ball mill	Internal mixer
Banbury mixer	Kneader
Centrifugal impact mixer	Mill
Change-can mixer	Propeller mixer
Colloid mill	Ribbon blender
Conical dry-blender	Rod mill
Disk-and-cone agitator	Roll mill
Drum tumbler	Sand mill
High-intensity mixer	Static mixer
Homogenizer	Tumbling agitator
Intensive mixer	Vibratory mill

Mixing, entropy, free energy *n.* The change in free energy ΔG_{mix}, for mixing two substances at certain temperature *T*, is given by the Gibbs function:

$$\Delta G_{mix} = \Delta H_{mix} - T\,\Delta S_{mix}.$$

Because the solution process is an endothermic process for most solutions, the heat of mixing, ΔH_{mix}, is a positive quantity. The entropy of mixing, ΔS_{mix}, is normally positive because of the more random nature of solutions compared with that of the unmixed components. A necessary, although not sufficient, condition for a polymer to dissolve is that $\Delta G_{mix} \leq 0$, and if $\Delta G_{mix} \approx 0$ then mixing is marginal, and if $\Delta G_{mix} \geq 0$ then components will not mix.

Mixing screw *n.* Any extruder screw that incorporates some modification (from standard designs) intended to improve mixing, mainly distributive mixing but sometimes improving dispersion, too. One simple method is to insert one or more rings of closely spaced pegs arranged circumferentially in the screw channel and having nearly the same height as the flight. The pegs divide and re-divide the melt streaming in a complex but regular path down the channel, accomplishing a kind of braiding of substreams.
See also Dulmadge mixing section, Maddock mixing section, and cavity-transfer mixer.

Mixing varnish *n.* General term for a variety of varnishes, or paint vehicles, that are used by mixing with: (1) Aluminum pigment. (2) Pigments paste in oil or colors in oil. (3) Other vehicle types, such as gloss oil (if the mixing varnish is long in oil), or a spar varnish (if the mixing varnish is short in oil) to produce a desired oil length for the complete vehicle. (4) The mill base discharged from pigment-dispersing equipment, as contrasted to the "grinding" varnish for vehicle in which the pigment has been dispersed. (5) Paint, to alter its properties.
See aluminum mixing varnish.

Mixing white *n.* A white ink, either transparent or opaque, used in making tints.

Mixture \ˈmiks-chər\ [ME, fr. MF, fr. OF *misture*, fr. L *mixtura*, fr. *mixtus*] (15c) *n.* A combination of two or more different substances intermingled with varying percentage composition (unlike a true solution), in which each component retains its chemical identity.

MKS system *n.* A system of units derived from the meter, kilogram, and second. Now superceded for scientific purposes by the SI units, which are based on the MKS system.

MMA *n.* Abbreviation for methyl methacrylate.

M_n *n.* (1) Abbreviation for number-average molecular weight. (2) Chemical symbol for the element manganese.

Mo *n.* Chemical symbol for the element molybdenum.

Mobility *n.* The property of a material which allows it to flow when a shearing force larger than the yield value has been applied. The coefficient of mobility is the rate of shear induced by a shearing force per square cm of 1 dyne in excess of the yield value. Mobility pertains to plastic materials and is the analogue of fluidity. It is calculated from the slope of the straight-line portion of the flow curve. The coefficient of mobility is the reciprocal of the coefficient of plastic viscosity.

Mobilometer *n.* Rheological instrument for measuring the consistency of paints and similar products. The "mobility" is determined by the time in seconds required for a loaded perforated disk to pass through a specified depth of the sample contained in the cylinder.

See Gardner mobilometer.

MOCA® *n.* DuPont's trade name for methylene-bis-*o*-chloroaniline, much used until about 1980 as a curing agent for urethane rubbers and epoxy resins, prior to its being declared to be a carcinogen by OSHA.

Mock dyeing *n.* A heat stabilization process for yarns. The yarns are wound onto packages and subjected to package dyeing conditions (water, pressure, and temperature) but without dye and chemicals in the bath.

Mock leno *n.* A combination of weaves having interlacings that tend to form the warp ends into groups (with empty spaces intervening) in the cloth, thereby giving an imitation of the open structure that is characteristic of leno fabrics. Mock leno fabrics are used for summer shirts, dresses, and other apparel, and as a shading medium in Jacquard designs.

MOD *n.* Modacrylic fibers (EEC abbreviation).

Modacrylic fibers \ˌmä-də-ˈkri-lik-\ [*modi*fied *acrylic*] (1960) *n.* A manufactured fiber in which the fiber-forming substance is any long-chain synthetic polymer composed of less than 85% but at least 35% by weight of acrylonitrile units (Federal Trade Commission). Characteristics: although modacrylics are similar to acrylics in properties and application, certain important differences exist. Modacrylics have superior resistance to chemicals and combustion, but they are more heat sensitive (lower safe ironing temperature) and have a higher specific gravity (less cover). End uses: the principal applications of modacrylic fibers are in pile fabrics, flame-retardant garments, draperies, and carpets.

Modacrylics *n.* Generic name for a fiber containing between 35 and 85% acrylonitrile repeating units in the polymer chain [poly (acrylonitrile)], excluding rubbers, manufactured by Viscose-Suisse, Switzerland.

Modal Generic name for fibers from regenerated cellulose of modified structure.

Moderator \ˈmä-də-ˌrā-tər\ (ca. 1560) *n.* A material used for slowing down neutrons in an atomic pile or reactor. Usually graphite or "heavy water" (deuterium oxide).

Modes of appearance *n.* Various manners in which colors can be perceived, depending on spatial distributions and temporal variations of the light causing the sensation. Five modes are generally recognized: surface mode, volume mode, film (or aperture) mode, illumination mode and illuminant mode. The first two of these, surface mode and volume mode, together make up the object mode. Color perceived as falling on an object, thus filling space around the object, describes the illumination mode. The illuminant mode is the appearance of the glow from a light source. The film or aperture mode describes color perceived in space, where no object is discernible.

Modified phenolic resins *n.* Resins in which the basic phenolaldehyde product has been modified by the introduction of rosin or other natural resin. The products are often esterified.

Modified resin *n.* Any synthetic resin into which has been incorporated a natural resin, an elastomer, or an oil that alters the processing characteristics or physical properties of the material.

Modifier \ˈmä-də-ˌfī(-ə)r\ (1583) *n.* Any chemically inert ingredient added to an adhesive formulation that changes its properties.

See also additive, filler, extender, and plasticizer.

Modulus \ˈmä-jə-ləs\–li [NL, fr. L, small measure] (1753) *n, pl.* (1) A modulus is a measure of a mechanical property of a material, most frequently a stiffness property. (2) The absolute value of a complex number or quantity, equal to the square root of the sum of the squares of the "real" and "imaginary" parts. (3) Modulus at 300% *n*: The tensile stress required to elongate a specimen to three times its original length (200% elongation) divided by 2. Although other elongations are used, 300% is the one most often employed for rubbers and flexible plastics.

See compressive modulus, flexural modulus, shear modulus, modulus of elasticity and modulus of resilience.

Modulus in compression *See compressive modulus.*

Modulus in flexure *See flexural modulus.*

Modulus in shear *See shear modulus.*

Modulus of elasticity *n.* (1) (elastic modulus, tensile modulus, and Young's modulus) The ratio of nominal tensile stress to the corresponding elongation below the proportional limit of a material. Since elongation is dimensionless, modulus has the units of stress. The relevant ASTM test is D 638. In contrast to structural metals such as mild steel, the stress–strain graphs for many plastics exhibit some curvature, even at very low strains. Since there is then no significant linear region whose slope would give the modulus, a secant modulus at 1–3% elongation may be reported for stiff materials. (2) More generally, any of the several elastic moduli characterizing behavior in shear (torsion), flexure, or change in volume under pressure. In SI, all types of elastic moduli are reported in pascals, usually megapascals (MPa) 1000 psi = 6.894,757 MPa.

Also known as elastic modulus or Young's modulus. See hardness. See bulk modulus.

Modulus of resilience *n.* The energy that can be absorbed per unit volume of a stressed specimen without creating a permanent deformation. It is equal to the area under the stress–strain graph from zero to the elastic limit divided by the volume of specimen undergoing deformation.

Modulus of rigidity *See shear modulus.*

Modulus of rupture (MOR) *n.* During a stress versus strain test, tensile pull test, of a material a stress (force or load) is approached where the material ruptures, breaks and separates which is sometimes referred to as the modulus (stress/strain) at rupture. Also referred to as the ultimate tensile strength and force at break.

Mohair \ˈmō-ˌhar, -ˌher\ [mod. of obs. It *mocaiarro*, fr. Arabic *mukhayyar*, literally, choice] (1619) *n. See angora* (1).

M

Mohr balance Balance used to measure the specific gravity of fluids.

Mohs hardness \ˈmōz-, ˈmōs-, ˈmō-səz-\ [Friedrich *Mohs* † 1839 German mineralogist]. A system of ranking materials according to their ability to scratch, and resist being scratched by, lower-ranking materials, diamond being the hardest material known and having the highest rank. Mohs' original scale ranked diamonds as 10, corundum as 9, etc., and talc as 1. The scale has been modified to recognize some newer hard materials ranking in the large gap between corundum and diamond. The modified scale is listed below, in order of decreasing a scratch hardness. There is a strong positive correlation between rank on the Mohs scale and *Knoop microhardness*.

See also scratch hardness.

Mohs scale [Friedrich *Mohs* † 1839 German mineralogist] (1879) *n.* Scale for

Modified Mohs number	Material
15	Diamond
14	Boron carbide
13	Silicon carbide
12	Fused alumina
11	Fused zirconia
10	Garnet
9	Topaz
8	Quartz or stellite®
7	Vitreous silica
6	Orthoclase
5	Apatite
4	Fluorite
3	Calcite
2	Gypsum
1	Talc

determining the relative hardness of a mineral, according to is resistance to scratching by one of the following minerals: (1) Talc. (2) Gypsum. (3) Calcite. (4) Fluorite. (5) Apatite. (6) Feldspar. (7) Vitreous silica (8) Quartz. (9) Topaz. (10) Garnet. (11) Fused zirconia. (12) Fused alumina. (13) Silicone carbide. (14) Boron carbide. (15) Diamond. Other useful hardnesses are: fingernail, slightly more than (2); penny, about (3); pocket knife, slightly more than (5); window glass (5.5) and a steel file (6.5).
See hardness.

Moiety \ˈmói-ə-tē\ [ME *moite*, fr. OF *moité*, fr. LL *medietat-*, *medietas*, fr. L *medius* middle] (15c) *n.* An indefinite amount of a constituent present in a material or compound.

Moil A rarely used Syn: molding flash.

Moiré \mó-ˈra, mwä\ [F *moiré*, fr. *moiré* like moire, fr. *moire*] (1818) *n.* A wavy or watered effect on a textile fabric, especially a corded fabric of silk, rayon, or one of the manufactured fibers. Moiré is produced by passing the fabric between engraved cylinders, which press the design into the material, causing the crushed and uncrushed parts to reflect light differently.

Moire effect *n.* An optical effect, which results from light interference, exhibiting a pattern of light and dark areas. The effect of superimposing a repetitive design, such as a grid, on the same or a different design to produce a pattern distinct from its components.

Moire papers *n.* Wallpapers having a watered silk sheen effect.
See Moire effect.

Moisture absorption *n.* The pickup of water vapor by a material upon exposure for a definite time internal to an atmosphere of specified humidity and temperature. No ASTM test exists for this property. Moisture absorption should not be confused with water absorption, for which there *is* an ASTM test.

Moisture barrier *n.* Treated paper or metal that retards or bars water vapor, used to keep moisture from passing into walls or floors.

Moisture content The amount of moisture in a material under prescribed conditions and expressed as a percent of the mass of the moist specimen that is, the mass of the dry substance plus the moisture. Also, it can be described as the water in solid waste. Expressed as the percentage of weight lost when a sample is dried at more than 100°C until it reaches a constant weight.

Moisture equilibrium *n.* The condition reached by a sample when the net difference between the amount of moisture absorbed and the amount desorbed, as shown by a change in weight, shows no trend and becomes insignificant.

Moisture-free weight *n.* (1) The constant weight of a specimen obtained by drying at a temperature of 105°C in a current of desiccated air. (2) The weight of a dry substance calculated from an independent determination of moisture content (e.g., by distillation with an immiscible solvent or by titration with Fischer reagent.

Moisture properties *n.* All fibers when exposed to the atmosphere pick up some moisture; the quantity varies with the fiber type, temperature, and relative humidity. Measurements are generally made at standard conditions, which are fixed at 65% RH and 70°F. Moisture content of a fiber or yarn is usually expressed in terms of percentage regain after partial drying.

Moisture regain *n.* The loss of weight on drying, expressed as percent of dry weight, of a pre-dried material exposed for a specified time to a specified humidity and temperature, then over-dried at a temperature above 100°C. ASTM D 885 (Section 07.01) describes a procedure recommended for rayon yarns and tire cords.

Moisture sensitivity *n.* (1) The degree to which the performance of a plastic part or product is affected by changes in its moisture content or, for some persons, by changes in the relative humidity of the environment in which the product is situated. (2) The degree to which processing performance is affected by moisture pickup prior to processing.

Moisture-set inks *n.* Inks that dry or set principally by precipitation. The vehicle consists of a water insoluble resin dissolved in a hygroscopic solvent. Drying occurs when the hygroscopic solvent has absorbed sufficient moisture either from the atmosphere, substrate or external application to precipitate the binder. An important characteristic of these inks is their low odor.

See steam-set ink.

Moisture vapor permeability *See specific permeability (of a film to moisture).*

Moisture–vapor transmission Rate of movement of moisture vapor through a membrane.

See water–vapor transmission rate.

Moisture–vapor transmission rate (MVTR) *See water–vapor transmission rate.*

Molality (molal concentration; *m*) *n.* A concentration unit: the number of moles of solute per kilogram of solvent.

Molal solution *n.* A solution that contains 1 mol of the solute per kilogram of the solvent.

Molar heat of fusion, ΔH_{fus} *n.* The heat necessary to melt 1 mol of a substance.

Molar heat of solution, ΔH_{sol} *n.* The heat liberated (if negative) or absorbed (if positive) when 1 mol of solute dissolves in a solvent.

Molar heat of vaporization, ΔH_{vap} *n.* The heat necessary to vaporize 1 mol of a substance.

Molarity (molar concentration, *M*) *n.* A concentration unit: number of moles of solute per liter of solution.

Molar solution (1 M) *n.* A solution that contains 1 mol of solute per liter of solution.

Molar volume, V_m *n.* The volume occupied by 1 mol of a substance. For an ideal gas, 22.4 l mol^{-1}.

Mold \ˈmōld\ [ME, mod. of OF *modle*, fr. L *modulus*, dim. of *modus* measure] (13c) *n.* A hollow form or matrix into which a liquid or molten plastic material is placed and which imparts to the material, upon cooling or curing, its final shape as a finished article (14c) *v.* To impart shape to a plastic mass by means of a confining cavity or matrix, by a process usually involving high pressure and changes in temperature. The term *molding* is usually employed for processes using dry thermoplastic or thermosetting compounds, and in injection or transfer molding. The term *casting* is

M

preferred for processes employing liquids – solutions or suspensions – that are sufficiently fluid to be poured into a mold and to fill it by gravity flow. Glenz W (ed) (2001) A glossary of plastics terminology in 5 languages, 5th edn. Hanser–Gardner Publications Inc., Cinicinnati. Carley JF (ed) (1993) Whittington's dictionary of plastics. Technomic Publishing Co. Inc., Lancaster, PA.

Mold base *n.* An assembly of ground-flat steel plates, usually containing dowel pins, bushings and other components of injection or compression molds excepting the cavities and cores.

Mold-clamping force *See clamping force.*

Mold efficiency *n.* In a multimold blow-molding system, the percentage of the total turn-around time of the mold actually required for forming, cooling, and ejecting the part.

M

Molding *n.* (1) A projecting or depressed surface, generally used at or around edges of walls, furniture, etc., either plain or decorated, employed to ornament a wall surface, cornice or capital. (2) In plastics processing, any process at some stage of which the plastic is softened or melted, usually by heating, and forced to flow into a shaped cavity or mold that essentially determines all the final dimensions of the product. Carley JF (ed) (1993) Whittington's dictionary of plastics. Technomic Publishing Co. Inc., Lancaster, PA. Glenz W (ed) (2001) A glossary of plastics terminology in 5 languages, 5th edn. Hanser–Gardner Publications Inc., Cinicinnati.

Molding compound *n.* Granules or pellets of a resin containing all desired additives such as plasticizes, stabilizers, colorants, processing aids, and fillers, prepared by blending these ingredients with the neat resin, then reducing the hot mix to pellets by extrusion or milling, cutting, and chilling, ready for further processing into finished products.
See also molding powder.

Molding cycle *n.* (1) The sequence of operations necessary on a molding press to produce a set of moldings. (2) The period of time occupied by the complete sequence of operations required for the production of one set of moldings.

Molding index *n.* A practical measure of the difficulty of molding of thermosetting compound. A calculated weight of the candidate molding powder, is placed into a flash-type cup mold that has been preheated to the temperature prescribed for the material. The mold is closed and the total minimum force required to close it is reported as the molding index of the compound.

Molding powder *n.* This term usually denotes pellets or granules of a neat resin or a molding compound.
Also see dry blend.

Molding pressure *n.* The pressure applied to the ram of an injection machine or press to force the softened plastic to completely fill the mold cavities. It is expressed in force per unit of cross-sectional area of the ram surface acting upon the material.

Molding shrinkage (mold shrinkage, shrinkage, and contraction) *n.* The fractional difference in corresponding dimensions between a mold cavity and the molding made in the cavity, both the cavity and the molding being at normal room temperature when measured. Shrinkage is often found to be different in different directions. It may be expressed as a percent, in mils/in., or mm/m.

Mold pressure *n.* The pressure measured inside a mold cavity, usually by a flush-mounted pressure transducer, at any time during a molding cycle, but in particular the highest pressure recorded during the cycle. *Compare with melt pressure.*

Mold release *See parting agent.*

Mold seam *n.* A visible line on a molded or laminated piece, often very slightly raised above the general surface, impressed by the parting line of the mold.

Mole \\ˈmōl\\ [Gr *Mol*, short for *Molekulargewicht* molecular weight, fr. *molekular* molecular + *Gewicht* weight] (1902) (mol) *n.* (1) In SI, the mole is defined as the amount of a substance of a system that contains as many elementary entities as there are atoms in 0.012 kg of carbon-12. The elementary entities must be specified and may be atoms, molecules, ions, electrons, other particles, or specified groups of such particles. (2) In practical usage, a mass of a substance equal to its molecular weight. To agree with the SI definition, the mass must be in grams. However, engineers often find it convenient to work with pound-mole, kilogram-mole, and even ton-mole of materials. If the mass unit is not prefixed, the mole is always the gram-mole. Whitten KW, Davis RE, Davis E, Peck LM, Stanley GG (2003) General chemistry. Brookes/Cole, New York. Perry RH, Green DW (1997) Perry's chemical engineer's handbook, 7th edn. McGraw-Hill, New York.

Molecular formula (ca. 1903) *n.* A formula expressing the actual number of atoms of each element in one molecule.

Molecularity *n.* The number of molecules which collide to form the activated complex in an elementary process.

Molecular orbital (MO) (1932) *n.* An electronic energy level in a molecule and the corresponding charge-cloud distribution in space.

Molecular orientation *See orientation.*

Molecular sieve (1926) *n.* A porous mineral or synthetic inorganic material, such as a zeolite (hydrous silicate), usually in the form of porous pellets or fine granules, having the ability to strongly absorb molecules of other (fluid) materials.

Molecular solid *n.* A solid in which molecules are held together by dipole–dipole or London forces.

Molecular volume (molar volume) *n.* The volume occupied by 1 mol, numerically equal to the gram-molecular weight divided by the density at the prevailing pressure and temperature.

Molecular weight (1880) (formula weight, molecular mass) *n.* The sum of the atomic weights of all atoms in a molecule. In most non-polymeric compounds the molecular weight is a known constant value. In high polymers, the individual molecules range widely in the number of atoms they contain, and therefore, in molecular weight. Hence an average must be used to characterize a particular sample of polymer. The two averages most commonly used are *number-average molecular* weight (M_n) and weight-average molecular weight (M_w). Methods for determining these averages include measurements of light scattering and osmotic pressure in solutions, sedimentation in an ultracentrifuge, and depression of freezing points and vapor pressures of solutions, dilute-solution viscosity, end-group titration, and spectroscopy. Goldberg DE (2003) Fundamentals of chemistry. McGraw-Hill Science/Engineering/Math, New York. Slade PE (2001) Polymer molecular weights, vol 4. Marcel Dekker, New York.

Molecular-weight distribution *n.* The percentages by number (or weight) of molecules of various molecular weights that comprise a given specimen of a polymer. Two samples of a given polymer having the same number-average molecular weight

M

may perform quite differently in processing because one has a broader distribution of molecular weights than the other. Two basic groups of methods are used for measuring molecular-weight distribution. Fractionation methods, which actually divide the specimen into portions having relatively narrow ranges of molecular weight, include: fraction precipitation and fraction solution (the two most widely used), chromatography, liquid-liquid partition, ultra-centrifugation, zone refining, and thermo gravimetric diffusion. After fractionation by any of these methods, the weight percent of each fraction is plotted versus the average molecular weight for that fraction to obtain a histogram of the distribution, which may be smoothed into a curve. Non-fractionation methods include light-scattering studies, electron microscopy, dilute-solution viscosity, size-exclusion chromatography, ultra-centrifugation, and diffusion. A popular measure of the breadth of a distribution is the ratio of weight-average to number-average molecular weight, M_w/M_n. Odian G (2004) Principles of polymerization. Wiley Interscience, New York. Lenz RW (1967) Organic chemistry of synthetic high polymers. Interscience Publishers, New York.

See also polydispersity.

M

Molecular-weight distribution, poisson *n.* The ratio of the weight-average (M_w) to the number-average (M_n); gives the range of the distribution. MWD gives a general picture of the ratio of the large, medium, and small molecular chains in the plastic. Odian G (2004) Principles of polymerization. Wiley Interscience, New York. Lenz RW (1967) Organic chemistry of synthetic high polymers. Interscience Publishers, New York.

Molecular weights of polymers *n.* The sum of the atomic weights of all atoms in a molecule. In most non-polymeric materials, the molecular weight is a fixed constant value. In high polymers, the molecular weight weights of individual molecules vary widely so that must be expressed as averages, shown by a bar above the symbol, e.g., $\overline{M}n$. Number-average molecular weight,

$$M_n = \sum n_i M_i / \sum n_i = \sum w_i / \sum (w_i/M_i),$$

where n_i is the number of molecules with molecular weight M_i, w_i the weigh fraction of material having molecular weight, M, Σn is the total number of molecules and weight-average M_w

$$M_w = \sum w_i M_i / \sum w_i.$$

Z-average M_z

$$M_z = \sum w_i M_i^2 / \sum w_i M_i.$$

Methods of determining molecular weight include osmotic pressure, light scattering, gel permeation chromatography, dilute solution viscosity, vapor pressure, freezing temperatures, and others. Odian G (2004) Principles of polymerization. Wiley Interscience, New York. Lenz RW (1967) Organic chemistry of synthetic high polymers. Interscience Publishers, New York.

Molecule \ˈmä-li-ˌkyü(ə)l\ [F *molécule*, fr. NL *molecula*, dim. of L *moles* mass] (1794) *n.* The smallest unit quantity of a compound that can exist by itself and still retain the chemical identity of the substances as a whole. Goldberg DE (2003) Fundamentals of chemistry. McGraw-Hill Science/Engineering/Math, New York.

Mole fraction (*X*) *n.* A concentration unit: the number of moles of one component in a solution divided by the total number of moles of all components. Goldberg DE (2003) Fundamentals of chemistry. McGraw-Hill Science/Engineering/Math, New York.

Mole percent (mol%) *n.* Mole fraction multiplied by 100. Goldberg DE (2003) Fundamentals of chemistry. McGraw-Hill Science/Engineering/Math, New York.

Moleskin \ˌskin\ (1668) *n.* A heavy sateen-weave fabric made with heavy, soft-spun filling yarns. The fabric is sheared and napped to produce a suede effect.

Moltopren *n.* Polyester or polyether + diisocyanate + water (foam). Manufactured by Bayer, Germany.

Mol volume *n.* The volume occupied by a mol or a gram molecular weight of any gas measured at standard conditions is 22.414 l. A molal solution contains 1 mol/1000 g of solvent. A molar solution contains 1 mol or gram molecular weight of the solute in 1 l of solution.

Molybdate orange \mə-ˈlib-ˌdāt\ *n.* 25Pb $CrO_4 \cdot 4PbMoO_4 \cdot 1PbSO_4$. Pigment Red 104 (77605). Bright orange, inorganic pigment, which is a solid solution of lead chromate, lead molybdate, and lead sulfate. Exhibits excellent hiding and tinting strength, brilliant hue, good permanency, but only fair chemical resistance.

Molybdate orange pigment *n.* Any of a range of solid solutions of lead chromate, lead molybdate, and lead sulfate, used as dark-orange to light-red pigments for plastics. Their advantages are high opacity, bright color, light-fastness, good heat stability, and freedom from bleeding.

Molybdenum disulfide (ca. 1931) (molybdic sulfide, molybdenum sulfide) *n.* MoS_2. A black, shiny, flaky-crystalline material used as a filler in nylons, fluorocarbons, and polystyrene to improve stiffness and strength, and, principally, to provide lubricity. The compound occurs naturally as the ore *molybdenite.*

Molybdenum FR Any of several molybdenum compounds, such as the oxide or ammonium dimolybdate $(NH_2)_2Mo_2O_7$, added to plastics to improve their fire retardancy (FR) and smoke suppression.

Mombassa gum *See animi.*

Moment of force or torque *n.* The effectiveness of a force to produce rotation about an axis, measured by the product of the force and the perpendicular distance from the line of action of the force to the axis. cgs unit – the dyne-centimeter. If a force F acts to produce rotation about a center at a distance d from the line in which the force acts, and force has a torque,

$$L = Fd.$$

Moment of inertia *n.* A measure of the effectiveness of mass in rotation. In the rotation of a rigid body not only the body's mass, but also the distribution of the mass about the axis of rotation determines the change in the angular velocity resulting from the action of a given torque for a given time. Moment of inertia in rotation is analogous to mass (inertia) in simple translation. The cgs unit is $g\,cm^2$. Dimensions – $[M\,L^2]$. If m_1, m_2, m_3, etc., represent the masses of infinitely small particles of a body; r_1, r_2, r_3, etc., their respective distances from an axis of rotation, the moment of inertia about this axis will be

$$I = m_1 r_1^2 + m_2 r_2^2 + m_3 r_3^2 + \cdots$$

or

$$I = \sum mr^2.$$

Giambattista A, Richardson R, Richardson RC, Richardson B (2003) College physics. McGraw-Hill Science, New York.

M

Momentum \mō-ˈmen-təm, mə-\ [NL, fr. L, movement] (1610) *n.* Quality of motion measured by the product of mass and velocity. cgs unit –1 g cm/s. A mass m moving with velocity v has a momentum, $M = mv$. If a mass m has its velocity changed from v_1 to v_2 by the action of a force F for a time t_1,

$$mv_2 - mv_1 = Ft.$$

Momentum flux *n.* In hydrodynamics, an interpretation of shear stress τ in which each of the six shear components of the stress tensor is viewed as the rate of flow, per unit of shear area, and perpendicular to that area, of momentum directed along a principal axis in the surface of shear. In laminar flow through a circular orifice, with radial coordinate r and axial coordinate z, the only non-zero shear component is t_{rz}, the flux of z-directed momentum in the r-direction. Newton's law of viscosity for this situation becomes

$$t_{rz} = -\mu(\mathrm{d}v_z/\mathrm{d}r),$$

where v_z is the fluid velocity at any radius r and the viscosity μ has the dimensions: (momentum/area-time) shear rate, and for which the SI unit is $1\ [\mathrm{kg\,m/s/m^2\,s}]/\mathrm{s}^{-1} = 1\ \mathrm{kg/m\,s} = 1\ \mathrm{Pa\,s}$. Goodwin JW, Goodwin J, Hughes RW (2000) Rheology for chemists. Royal Society of Chemistry, UK. Kamide K, Dobashi T (2000) Physical chemistry of polymer solutions. Elsevier, New York. Weast RC (ed) (1978) CRC handbook of chemistry and physics, 59th edn. The CRC Press, Boca Raton, FL.

Monkey *n.* Trade term for a batch of resin which has become thermoset and unusable during processing. Syn: giff.

Monk's cloth *n.* Wallcoverings that simulate a basket-weave material of cotton, jute, or flax.

Mono- A prefix designating the entity that follows it as the only one or as containing only one of that kind, e.g., monomer, or monohydric alcohol.

Monobasic \ˌmä-nə-ˈbā-sik\ [ISV] (1842) *adj.* (1) Pertaining to acids having one active hydrogen per molecule, e.g., hydrochloric acid, HCl. (2) Designating an acid salt in which one hydrogen (of two or three) has been replaced by a metal, e.g., potassium monophosphate, KH_2PO_4.

Monocarboxylic acid \-ˌkär-(ˌ)bäk-ˈsi-lik\ (ca. 1909) *adj.* Any organic acid containing a single –COOH group in the molecule. Many of the larger acids of this type are derived from natural fats and oils and are used in the production of alkyd resins and polyesters. Esters of oleic, stearic, pelargonic and richinoleic acid, all monocarboxylic, are widely used as plasticizers.

Monochloroethylene *n.* Syn: vinyl chloride.

Monochromatic color scheme *n.* Combining of colors in a room based on one color used in various values and saturation or chromas.

Monochromatic emissive power *n.* The ratio of the energy of certain defined wavelengths radiated at definite temperatures to the energy of the same wavelengths radiated by a black body at the same temperature and under the same conditions.

Monochromatic light *n.* Light of a single wavelength. It may be obtained by the use of a laser or by gaseous discharge tubes in combination with proper filters. An approximation is obtained by interference filters or monochromators. The degree of monochromaticness depends upon the dispersion device and accompanying optical scheme used to separate the different wavelengths from a continuous or "white" light source. Emission lines from elements which, when excited, emit light, such as mercury, sodium, etc., are most nearly true monochromatic light. Moller KD (2003) Optics. Springer-Verlag, New York.

Monochromator \ˌmä-nə-ˈkrō-ˌmā-tər\ [*monochromatic* + illumin*ator*] (1909) *n.* A device for dispersing white light into individual wavelengths, generally a prism or diffraction grating. Moller KD (2003) Optics. Springer-Verlag, New York.

Monochrome \ˈmä-nə-ˌkrōm \ [ML *monochroma*, fr. L, feminine of *monochromos* of one color, fr. Gk *monochrōmos*, fr. *mono-* + *-chrōmos* –chrome] (1662) *n.* Painting in different values and chromas of one hue. Moller KD (2003) Optics. Springer-Verlag, New York.

Monoclinic \ˌmä-nə-ˈkli-nik\ [ISV] (ca. 1864) *adj.* Of a crystal (or crystal system) having two axes that are mutually perpendicular to the third one, but not to each other. An example is the β-form of elemental sulfur, which is stable between 112 and 119°C, but slowly converts to the rhombic form below 112°C. Rhodes G (1999) Crystallography made crystal clear: a guide for users of macromolecular models. Elsevier Science and Technology Books, New York. Hibbard MJ (2001) Mineralogy. McGraw-Hill Co. Inc., New York.

Monocyclic terpenes *n.* Heterogeneous mixture of monocyclic, bicyclic and other related terpene hydrocarbons recovered or removed in the fractionation of certain terpenes or other essential oils, or as a by-product in the chemical conversion of pinenes; generally sold under trade names.

Monodisperse \ˌmä-nō-dis-ˈpərs\ [*mon-* + *disperse*, adj, fr. *disperse*, v.] (1925) *adj.* Of a polymer, all the molecules having the same molecular weight. It has long been possible to make (nearly) monodisperse polystyrene of various molecular weights and a few other monodisperse polymers are now available as laboratory chemicals. Slade PE (2001) Polymer molecular weights, vol 4. Marcel Dekker, New York.
Compare polydisperse.

Monodispersity *n.* Refers to a polymer system, which is homogeneous in molecular weight, i.e., lacks molecular weight distribution. Slade PE (2001) Polymer molecular weights, vol 4. Marcel Dekker, New York.

Monofil *See monofilament.*

Monofilament \ˌmä-nə-ˈfi-lə-mənt\ (1940) *n.* A single filament of indefinite length, strong enough to function as a yarn in textile operations or as an entity in other applications. Monofilaments are generally produced by extrusion. Their outstanding uses are in the fabrication of brush bristles, surgical sutures, fishing lines, racquet strings, screen materials, ropes, and nets. The finer monofilaments are woven and knitted on textile machinery. Kadolph SJJ, Langford AL (2001) Textiles. Pearson Education, New York.

Monoglyceride \ˌmä-nə-ˈgli-sə-ˌrīd\ (1860) *n.* Partial ester of glycerol in which one of the three hydroxyl groups is reacted with an organic acid.

Monolithic flooring *See seamless flooring.*

Monomer \ˈmä-nə-mər\ [ISV] (1914) *n.* A relatively simple compound that can react with itself or other compounds to form long-chain compounds by either. (1) Utilizing its C=C bonds for addition. (2) By having two or more functional groups that can react with receptive groups in other molecules. Monomers are the basic building blocks of polymers. Monomer is sometimes referred to as *mer.* Odian GC (2004) Principles of polymerization. John Wiley and Sons Inc., New York.
See monomeric unit (mer) and polymer.

Monomeric Pertaining to a monomer.

Monomeric cement *See adhesive.*

Monotropic *n.* Of a material or element having two forms, one of which is metastable toward the other. The metastable form tends to change spontaneously to the stable form but the reverse change does not occur.

Monotype printing *n.* Process of printing from mechanically assembled pieces of type, each of which bears one character.

Montan wax \ˈmän-tᵊn-\ [L *montanus* of a mountain] (1908) (lignite wax) *n.* A hard, white wax derived from *lignite* (a lower-grade hydrocarbon fossil mineral between peat and bituminous coal). It is sometimes described as a bitumen wax. The crude product is very dark in color, almost black, but after refining it becomes pale yellow. Its mp varies from 72 to 82°C; its acid value, from 25 to 99; and its saponification value, from 58 to 104. The wax is used as a mold lubricant.

M

Monthier's blue *See iron blue.*

Mooney equation (M. Mooney) *n.* An empirical modification of the Einstein equation applicable to higher solids concentrations, and relating the viscosity of a suspension of monodisperse spheres η_f to that of the pure liquid η_o.

$$\ln(\eta_f/\eta_o) = 2.5f/(1 - Sf),$$

where f is the volume fraction of solids and $S \cong 1.4$ for spheres (ASTM, www.astm.org). Goodwin JW, Goodwin J, Hughes RW (2000) Rheology for chemists. Royal Society of Chemistry, UK.

See also Eilers equation.

Mooney scorch time *n.* For a rubber specimen tested in a Mooney viscometer, the time elapsing after the minimum torque has been reached for torque to increase by five "Mooney units".

Mooney viscosimeter *n.* An instrument invented by M. Mooney in 1924, used to measure the effects of time of shearing and temperature on the comparative viscosities of rubber compounds. It consists of a motor driven disk, tooth-surfaced on the sides, enclosed within a die cavity formed by two halves maintained at controlled temperature and closing force. The specimen is a double disk, joined at the edges and trapped between the die halves and the rotor (ASTM, www.astm.org).

Mop board *See baseboard.*

Moplen *n.* Poly(ethylene) and poly(propylene), manufactured by Montedison, Italy.

Mop polishing *n.* (1) Application of a modified French polish with a camel hair "mop" to carved work. (2) Polishing by friction with a rotary mop.

Mordant dyes *n.* Dyes, which develop their characteristic colors when precipitated on suitable bases to form lakes. It is often possible to obtain considerable variation in the color of the final lake by altering the metallic radical of the base. Mordant dyes in their original state are colorless.

Mordants \ˈmór-dᵊnt\ [ME, fr. MF, pp of *mordre* to bite, fr. L *mordēre*; perhaps akin to Sanskrit *mrdnāti* he presses, rubs] (1791) *n.* Substances capable of uniting with both dyes and textile fibers so as to improve the bond between dye and textile and give improved textures and alter the colors.

Moresque *n.* A multicolored yarn formed by twisting or plying single strands of different colors.

Morphological analysis *n.* The identification of particles based on easily observed microscopical characteristics.

Morphology \mór-ˈfä-lə-jē\ [Gr *Morphologie*, fr. *morph-* + *-logie* –logy] (1830) *n.* The study of the physical form and structure of a material. This includes a wide range of characteristics, extending from the external size and shape of large articles to dimensions of crystal lattices but, with polymers, it most often refers to microstructure. Kamide K, Dobashi T (2000) Physical chemistry of polymer solutions. Elsevier, New York. Mark JE (ed) (1996) Physical properties of polymers handbook. Springer-Verlag, New York.

Mortar \ˈmór-tər\ [ME *morter*, fr. OF *mortier*, fr. L *mortarium*] (14c) *n.* (1) Material used in a plastic state which can be troweled, and becomes hard in place, to bond units of masonry structures (*Note* — The word "mortar" is used without regard to the composition of the material, and is defined only with reference to its use as a bonding material as contrasted with the words "stucco" and "plaster"). (2) A mixture of gypsum plaster with aggregate or hydrated lime, or both, and water is to produce a trowelable fluidity.

Mosaic \mō-ˈzā-ik\ [ME *musycke*, fr. MF *mosaique*, fr. OIt *mosaico*, fr. ML *musaicum*, alter. of LL *musivum*, fr. L *museum*, *musaeum*] (15c) *n.* (1) Picture or surface decoration made by embedding small pieces of colored stone, marble, pottery, or glass in mortar.

Mosley's law *n.* The frequencies of the characteristic X-rays of the elements show a strict linear relationship with the square of the atomic number.

Mote *n.* A small piece of seed or vegetable matter in cotton. Motes are removed by boiling the fiber or fabric in sodium hydroxide, then bleaching. When not removed, they can leave a dark spot in the fabric.

Motion, laws of *See Newton's law of motion.*

Motionless mixer *See static mixer.*

Mottle \ˈmä-tᵊl\ [prob. back-formation fr. *motley*] (1676) *n.* (1) An irregular distribution or mixture of colorants or colored materials giving a more or less distinct appearance of specks, spots, or streaks of color. Mottling is often deliberate although it may occur accidentally due to inadequate mixing. (2) Presence of irregularly shaped and randomly distributed areas of nonuniform appearance in color, gloss or sheen. Syn: blotching.

Mouldrite *n.* Urea, phenol, and melamine/formaldehyde resins manufactured by ICI, Great Britain.

Mountain blue *See azurite.*

Mounting plate *n.* In blow molding, the plate to which the mold is attached. *See also clamping plate.*

Movable platen *n.* The large back plate of an injection molding machine to which the back half of the mold is secured during operation. This platen is moved either by a hydraulic ram or a toggle mechanism.

M

Movil, Mowil *n.* Poly(vinyl chloride), manufactured by Polymer Ind., Italy.

Moviol *n.* Poly(vinyl alcohol), manufactured by Hoechst, Germany.

Mowilith *n.* Poly(vinyl acetate), manufactured by Hoechst, Germany.

MS Abbreviation for mass spectrometry.

Mucilage \ˈmyü-s(ə)lij\ [ME *muscilage*, fr. LL *mucilago* mucus, musty juice, fr. L *mucus*] (15c) *n.* (1) Flocculant or slimy deposit which separates from unrefined vegetable oils on heating. (2) An adhesive prepared from a gum and water. Also in a more general sense, a liquid adhesive which has a low order of bonding strength.

Mudcracking *n.* Paint film defect characterized by a broken network of cracks in the film.

Muff *n.* A loose skein of textured yarn prepared for dyeing or bulking. In the bulking operation, the yarn contracts and the resulting skein resembles a muff.

Muff dyeing *See dyeing.*

Mullen bursting strength *n.* An instrumental test method that measures the ability of a fabric to resist rupture by pressure exerted by an inflated diaphragm.

Muller \ˈmə-lər\ [alter. of ME *molour,* prob. fr. *mullen* to grind] (1612) *n.* An instrument usually of glass used for dispersing pigments in varnish for test purposes. The two glass grinding surfaces are generally ground glass. Pigment and vehicle are placed between them and the two rubbed together, either by hand or by making power. Power driven automatic mullers have two circular glass grinding surfaces.

Mullion \ˈməl-yən\ [prob. alter. of *monial* mullion] (1567) *n.* Narrow dividers between windows or glass panels.

Multicavity mold (multiple-cavity mold, multiple-impression mold) *n.* A mold having several to hundreds of cavities so that many parts may be molded with each shot. In many cases, the parts are identical, but that need not be so. In a type of multicavity mold known as a family mold, some of the cavities may be identical while others are different, or they may all be different.

M

Multicolor finish *n.* A speckled coating containing flecks of small, individual colored particles different from the base color. *Also called speckled finish.*

Multicomponent polymerization *n.* A polymerization which in general involves the use of more than one co-monomer. The term is usually reserved for co-polymerization of three or more co-monomers. Odian GC (2004) Principles of polymerization. John Wiley and Sons Inc., New York.

Multifilament *n.* A yarn consisting of many continuous filaments or strands, as opposed to monofilament which is one strand. Most textile filament yarns are multifilament.

Multifilament yarn *n.* A manufactured yarn composed of many fine continuous filaments or strands.

Multifunctional concentrate *n.* A plastic compound that contains high percentages of at least two of such additives as colorants, stabilizers, flame retardants, lubricants, antistatic agents, anti-blocking agents, blowing agents, fillers, etc, that will be diluted in base resin to provide a tailored compound with the desired final concentrations of the additives in the extruded or molded product.

Multigated *n.* Of an injection-mold cavity, having two or more gates. Multigating is common in molding large or complex parts that would be difficult to fill through one gate. Also, shrewd placement of the several gates can direct the weld lines to areas of the part where the stresses expected in service will be low.

Multilayer fabric *n.* A fabric for reinforced-plastic structures formed by braiding to and fro or overlapping in one direction. Layers may be biaxial or triaxial, fibers mixed, and braid angles varied.

Multilayer film *n.* A film comprised of layers of two or more different materials, all polymeric. The goal is to make a film that has a combination of properties not achievable with a single polymer. Most multilayer films are made by co-extrusion. The principal layers are usually joined by an adhesive resin compatible with both the adjacent principals and co-extruded with them.

Multiple proportions, law of *n.* If two elements form more than one compound, the weights of the first element which combine with a fixed weight of the second element are in the ratio of integers to each other.

Multiple-regression analysis *See regression analysis.*

Multiple scattering *See scattering multiple.*

Munsell book notation *n.* The Munsell color notation of a specimen obtained by visual or computational comparison with the Munsell hue, value, and chroma scales of the Munsell Book of Color.

See Munsell renotation.

Munsell book of color *n.* A collection of color chips, which illustrate the Munsell Color System, manufactured by the Munsell Color Co. (MD, USA) in both matte and glossy finishes. It consists of pages of constant hue at 2.5 hue step intervals, with darkest colors (of lowest value at the bottom) increasing in value upwards to the top and colors of lowest chroma in the center fold increasing to highest chroma at the outside edge of the page.

Munsell chroma *n.* The color term applied to the psychological sensation dimension of saturation. In Munsell notation, neutrals, white, graphs, etc., have a chroma of zero. Increasing deviation from gray to purest color in equally perceptible steps are given increasing numerical notations, with intermediate steps designated by decimals proportional to the perceptibility differences.

Munsell value *n.* (1) That portion of Munsell notation that corresponds to the sensation of lightness. It is described by a number from zero for a perfect black to ten for a perfect white and is calculated from the equation

$$Y_0 = 1.2219V - 0.23111V^2 + 0.23951V^3 - 0.021009V^4 + 0.0008404V^5.$$

(2) The daylight reflectance of a specimen expressed on a scale extending from 0 for ideal black to 10 for ideal white by steps of approximately equal visual importance. Achromatic or neutral colors are designated *N* followed by the value notation, thus: $N = 5.0$. McDonald R (ed) (1997) Colour physics for industry, 2nd edn. Society of Dyers and Colourists, West Yorkshire, England.

Mu oil *n.* Name sometimes given to tung oil obtained from *Aleurites Montana.*

Mural painting *n.* Usually, large picture which is painted on a wall, or fastened to a wall surface. Gair A (1996) Artist's manual. Chronicle Books LLC, San Francisco.

Muscovite \ˈməs-kə-ˌvīt\ [ML or NL *Muscovia, Moscovia* Moscow] (1535) *n.*

See aluminum potassium silicate.

Muslin \ˈməz-lən\ [F *mousseline*, fr. It *mussolina*, fr. Arabic *mawsilīy* of Mosul, fr. al-*Mawsil* Mosul, Iraq] (1609) *n.* A broad term describing a wide variety of plain-weave cotton or polyester/cotton fabrics ranging from lightweight seers to heavier shirting and sheeting.

Muted colors Colors whose chroma or saturation has been lessened or moderated by use of their complementary color or a neutral. *Also called grayed colors or tones,* which, on losing most of their character as a color, approach *Neutrals.*

MVP *n.* Abbreviation for moisture vapor permeability.

MVT *n.* Abbreviation for moisture vapor transmission.

See specific permeability (of a film to moisture).

MVTR *n.* Abbreviation for moisture–vapor-transmission rate.

See water–vapor-transmission rate.

M_w *n.* Symbol for weight-average molecular weight.

MW Abbreviation for molecular weight.

M

Mycology \mī-ˈkä-lə-jē\ [NL *mycologia*, fr. *myc-* + Latin *-logia* -logy] (1836) *n.* The science dealing with fungi.

Mylar \ˈmī-ˌlär\ *tradename* Polyester (film). DuPont's registered trade name for biaxially oriented film composed of polyethylene glycol terephthalate.

Myristoyl peroxide $(C_{13}H_{27}CO)O_2$. A soft, granular powder, used as a polymerization catalyst for vinyl monomers.

M_v *n.* Symbol for viscosity-average molecular weight.

Printing: Krips bv, Meppel
Binding: Stürtz, Würzburg